Living in the Environment

Principles, Connections, and Solutions

TWELFTH EDITION

G. TYLER MILLER, JR.

President, Earth Education and Research

Adjunct Professor of Human Ecology
St. Andrews Presbyterian College

BROOKS/COLE

THOMSON LEARNING

Australia • Canada • Mexico • Singapore • Spain • United Kingdom • United States

BROOKS/COLE

THOMSON LEARNING

Publisher: *Jack Carey*
Assistant Editor: *Mark Andrews*
Editorial Assistant: *Karen Hansten*
Production Management: *Electronic Publishing Services Inc., NYC*
Marketing Manager: *Rachel Alvelais*
Marketing Communications Project Manager: *Carla Martin-Falcone*
Print Buyer: *Mary Noel*
Photo Researcher and Permissions Editor: *Linda Rill*
Copy Editor: *Electronic Publishing Services Inc., NYC*
Cover Design: *Vernon T. Boes*
Cover Photo: *© Leland Howard. Fall Creek Falls, Snake River, Idaho.*
Interior Illustration: *Electronic Publishing Services Inc., NYC; Precision Graphics; Sarah Woodward; Darwin and Vally Hennings; Tasa Graphic Arts, Inc.; Alexander Teshin Associates; John and Judith Waller; Raychel Ciemma; and Victor Royer*
Typesetting: *Electronic Publishing Services Inc., NYC*
Printing and Binding: *R. R. Donnelley and Sons, Willard*

Title Page Photographs: *Left: Monarch butterfly (Frans Lanting/Bruce Coleman Collection). Right: Gray Wolf (Tom J. Ulrich/Visuals Unlimited). Bottom: Crater Lake, Oregon (Jack Carey).*
Part Opening Photographs:
Part I: *Composite satellite view of Earth. © Tom Van Sant/The GeoSphere Project*
Part II: *Endangered green sea turtle. © David B. Fleetham*
Part III: *Satellite view of Nile River and delta near Egypt. Stone/Landstat Satellite/Nigel Press*
Part IV: *Area of forest in Czechoslovakia killed by acid deposition and other air pollutants. Silvestris Fotoservice/NHPA*
Part V: *Highly endangered Florida panther. John Cancalosi/Peter Arnold, Inc.*
Part VI: *Antipollution demonstration by Louisiana residents against Monsanto Chemical Co. Sam Kitner/SIPA Press*

Library of Congress Cataloging-in-Publication Data

Miller, G. Tyler (George Tyler),
   Living in the environment : principles, connections, and solutions / G. Tyler Miller. -- 12th ed.
      p. cm.
   Includes bibliographical references and index.
   ISBN 0-534-37697-5
   1. Environmental sciences   2. Human ecology.
   3. Environmental policy.   I. Title
GE105.M547   2002
363.7—dc21   00-031187

Wadsworth/Thomson Learning
10 Davis Drive
Belmont, CA 94002-3098
USA

For more information about our products, contact us:
Thomson Learning Academic Resource Center
1-800-423-0563
http://www.brookscole.com

International Headquarters
Thomson Learning
International Division
290 Harbor Drive, 2nd Floor
Stamford, CT 06902-7477
USA

UK/Europe/Middle East/South Africa
Thomson Learning
Berkshire House
168-173 High Holborn
London WC1V 7AA
United Kingdom

Asia
Thomson Learning
60 Albert Street, #15-01
Albert Complex
Singapore 189969

Canada
Nelson Thomson Learning
1120 Birchmount Road
Toronto, Ontario M1K 5G4
Canada

For Instructors and Students

How Did I Become Involved with Environmental Problems? In 1966, I heard a scientist give a lecture on the problems of overpopulation and pollution. Afterward I went to him and said, "If even a fraction of what you have said is true, I will feel ethically obligated to give up my research on the corrosion of metals and devote the rest of my life to research and education on environmental problems and solutions. Frankly, I don't want to believe a word you have said, and I'm going into the literature to try to prove that your statements are either untrue or grossly distorted."

After 6 months of study I was convinced of the seriousness of these problems. Since then, I have been studying, teaching, and writing about them. This book summarizes what I have learned in more than three decades of trying to understand environmental principles, problems, connections, and solutions.

What Is My Philosophy of Education? In our lifelong pursuit of knowledge, I believe we should do three things:

- *Question everything and everybody, as any good scientist does.*

- *Develop a list of principles, concepts, and rules to serve as guidelines in making decisions,* and continually evaluate and modify this list on the basis of experience. This is based on my belief that the key goal of education is to learn how to sift through mountains of facts and ideas to find the few that are most useful and worth knowing. We need to be *wisdom seekers,* not information vessels. This takes a firm commitment to learning how to think logically and critically. This book is full of facts and numbers, but they are useful only to the extent that they lead to an understanding of key ideas, scientific laws, concepts, principles, and connections.

- *Interact with what you read as a way to sharpen your critical thinking skills.* I do this by marking key sentences and paragraphs with a highlighter or pen. I put an asterisk in the margin next to something I think is important and double asterisks next to something that I think is especially important. I write comments in the margins, such as *Beautiful, Confusing, Misleading,* or *Wrong.* I fold down the top corner of

pages with highlighted passages and the top and bottom corners of especially important pages. This way, I can flip through a book and quickly review the key passages. I urge you to interact in such ways with this book.

Redefining Environmental Science for the 21st Century This is a *science based* book designed for introductory courses on environmental science. It treats environmental science as an *interdisciplinary* study, combining ideas and information from *natural sciences* (such as biology, chemistry, and geology) and *social sciences* (such as economics, politics, and ethics) to present a general idea of how nature works and how things are interconnected. It is a study of *connections in nature.*

This new edition is the most significant revision of this book since the first edition appeared in 1975. This new edition redefines this course by emphasizing the following major shifts in environmental education and environmental policy that have taken place over the past 25 years and that will accelerate in this century.

- *Increased emphasis on science-based approaches to understanding and solving environmental problems.* Since its first edition in 1975, this book has led the way in using scientific laws, principles, models, and concepts to help us **(1)** understand environmental and resource problems and possible solutions and **(2)** see how these concepts, problems, and solutions are connected. The first edition had five chapters on basic scientific concepts when other books had a single chapter. In this 12th Edition, eight chapters and 245 pages are devoted to the treatment of basic scientific principles and concepts—far more than any other text. This emphasis on basic science will become increasingly important throughout this century. I have introduced only the concepts and principles necessary for understanding the material in the book, and I have tried to present them simply but accurately.

- *Increased emphasis on solutions.* The emphasis in this century is on finding and implementing solutions to environmental problems. This text has stressed solutions as a major theme for many years. In this new edition, 254 pages are devoted to presenting and evaluating solutions to environmental problems—far more than any other textbook.

- *Increased emphasis on prevention and precaution.* Since its first edition this book has consistently categorized proposed solutions to environmental problems as either **(1)** *input* (prevention) solutions such as pollution prevention and waste reduction or **(2)** *output* (cleanup) solutions such as pollution control and waste management. Both approaches are needed, but so far most emphasis has been on output or management solutions. There is a growing awareness of the need to put more emphasis on input or prevention approaches. This edition increases this emphasis based on using the *precautionary principle* as a guideline for solutions to urgent environmental problems when there is insufficient scientific information. This shift in environmental thinking will accelerate during this century.

- *Increased emphasis on decentralized micropower.* I highlight the shift from large centralized sources of electricity (mostly coal and nuclear plants) to a dispersed array of smaller micropower plants, including gas turbines, solar cell arrays, and fuel cells. This shift, discussed in Section 15-9, p. 389, is underway and will accelerate during this century.

- *Increased recognition of the connections between poverty, environmental harm, poor human health, and loss of life quality.* Increasingly, there is recognition that solving most environmental problems requires that we attempt to sharply reduce poverty (Section 26-6, p. 706).

- *Greater integration of economics and environment.* I emphasize the increased use of emissions trading, environmental accounting, full-cost pricing, and evolving eras of environmental management in the greening of businesses. This trend, discussed in Chapter 26 (p. 688), is underway and will increase rapidly as businesses are learning that improving environmental quality is one of the greatest investment opportunities of this century.

To help ensure that the material is accurate and up to date, I have consulted more than 10,000 research sources in the professional literature and about the same number of internet sites. I have also benefited from the more than 200 experts and teachers (see list on pp. xi–xiii) who have provided detailed reviews of this and my other two books in this field.

How Have I Attempted to Eliminate Bias? There are at least two sides to all controversial issues. The challenge for an author is to give a fair and balanced view of opposing viewpoints without injecting personal bias. This allows students to make up their own minds about important issues. Studying a subject as important as environmental science and ending up with no conclusions, opinions, and beliefs means that both the teacher and student have failed. However, such conclusions should be based on using critical thinking to evaluate opposing ideas.

In this new edition, the publisher and I enlisted the aid of 25 new reviewers, charged with trying to improve this book and to focus especially on detecting and eliminating any hint of bias.

A few examples of my efforts to give a balanced presentation of opposing viewpoints are **(1)** the pros and cons of reducing birth rates (pp. 250–252), **(2)** the Pro/Con boxes on genetically modified food (pp. 275–276) and oil development in the Arctic National Wildlife Refuge (p. 343), **(3)** Section 18-3 (pp. 453–454), on global warming, **(4)** discussion of acid deposition (pp. 432–433), **(5)** the asbestos problem (pp. 436–437), **(6)** problems with the Superfund law (pp. 543–545), **(7)** pros and cons of pesticides (Sections 20-2, p. 504, and 20-3, p. 505), and **(8)** 20 Pro/Con summary diagrams.

However, bias can be subtle, and I invite instructors and students to write me and point out any remaining bias.

What Are Some Key Features of This Book? This book is *science based, solutions oriented,* and *flexible.* The book is divided into six major parts (see Brief Contents, p. x). After the introductory chapters in Part I and the scientific principles and concepts chapters in Part II have been covered, the rest of the book can be used in almost any order. In addition, most chapters and many sections within these chapters can be moved around or omitted to accommodate courses with different lengths and emphases.

Each chapter begins with a brief *Earth Story,* a case study designed to capture interest and set the stage for the material that follows. In addition to these 28 case studies, 66 other case studies are found throughout the book (some in special boxes and others within the text); they provide a more in-depth look at specific environmental problems and their possible solutions. Eighteen *Guest Essays* present an individual researcher's or activist's point of view, which is then evaluated through Critical Thinking questions.

Other special boxes found in the text include **(1)** *Pro/Con boxes* that present both sides of controversial environmental issues, **(2)** *Connections boxes* that show connections in nature and between environmental concepts, problems, and solutions, **(3)** *Solutions boxes* that summarize a variety of solutions to environmental problems proposed by various analysts, **(4)** *Spotlight boxes* that highlight and give insights into key environmental problems and concepts, and **(5)** *Individuals Matter boxes* that describe what people have done to help solve environmental problems. To encourage critical thinking and integrate it throughout the book, all boxes (except Individuals Matter) end with Critical Thinking questions.

This book is an integrated study of environmental problems, connections, and solutions. The eight integrated themes in this book are **(1)** *biodiversity and natural resources (ecosystem services)*, **(2)** *sustainability*, **(3)** *connections in nature*, **(4)** *pollution prevention*, **(5)** *population and exponential growth*, **(6)** *energy and energy efficiency*, **(7)** *solutions to environmental problems*, and **(8)** *the importance of individuals working together to bring about environmental change*.

I hope you will start by looking at the brief table of contents (p. x) to get an overview of this book. Then I suggest that you look at the Concepts and Connections diagram inside the back cover, which shows the major components and relationships found in environmental science. In effect, it is a map of the book.

The book's 599 illustrations are designed to present complex ideas in understandable ways and to relate learning to the real world. They include 490 full-color diagrams (220 of them new to this edition and 50 of them maps) and 109 carefully selected color photographs (15 of them new to this edition).

Students and teachers also have access to *InfoTrac*® College Edition, a fully searchable online database with access to complete environmental articles from more than 700 periodicals. I have put two practice exercises at the end of each chapter to help users learn how to navigate this valuable source of information.

I have not cited specific sources of information. This is rarely done for an introductory-level text in any field, and it would interrupt the flow of the material. Instead, readings providing backup for almost all the information in this book and serving as springboards to further information and ideas are provided for each chapter on the website for this book. Placing these references on the website allows me to update them between editions of this book. This edition also has a greatly expanded and improved interactive World Wide website at **(http:// www.brookscole.com/product/0534376975s)** that can be used as a source of further information and ideas.

Instructors wanting a shorter book covering this material with a different emphasis and organization can use one of my two other books written for various types of environmental science courses: **(1)** *Environmental Science*, 8th edition (549 pages, Brooks/Cole, 2001) or **(2)** *Sustaining the Earth: An Integrated Approach*, 5th edition (385 pages, Brooks/Cole, 2002).

What Are the Major Changes in the Twelfth Edition? *This edition represents the most significant revision since the first edition came out in 1975.* Detailed changes by chapter are listed in the annotated material in the front of the instructor's version of this book and on this book's website. Major changes include the following:

Content

- Updated and revised material throughout the book.

- 220 new figures.

- Three new chapters on *Environmental History: An Overview* (Chapter 2), *Geology: Processes, Hazards, and Soils* (Chapter 10), and *Sustaining Aquatic Biodiversity* (Chapter 24).

- Expanded coverage of biodiversity, with a new chapter (Chapter 24), *Sustaining Aquatic Biodiversity*. This is the only introductory text to devote an entire chapter to this important and neglected environmental problem.

- Addition of 250 new topics. See the front insert in the instructor's version of this book and this book's website for a detailed list of these topics listed by chapter. Examples include the following:

  - Effects of globalization (p. 10).

  - Ecological footprints of countries (Figure 1-10, p. 11).

  - Ecological and economic services provided by terrestrial and aquatic systems. Examples include Figures 4-36 (p. 99), 13-11 (p. 304), 24-4 (p. 633), and 24-5 (p. 633).

  - Precautionary principle (pp. 193–195).

  - Genetically modified food (pp. 274–276).

  - Drip irrigation (p. 312).

  - Hybrid gas-electric and fuel-cell car engines (p. 365), Figure 15-9 (p. 365), and Figure 15-10 (p. 366).

  - Using the internet to save energy and reduce global warming (p. 369).

  - Micropower electricity production (pp. 389–391).

  - Selling environmental services instead of goods (p. 527).

  - Brownfields (pp. 544–546).

  - Ways in which humans value nature (pp. 560–562).

  - Sustainable timber certification (p. 598).

  - Biodiversity hot spots (p. 617).

  - Adaptive ecological management (pp. 621–623).

  - Integrated coastal management (p. 643).

  - Smart urban growth (pp. 678–679).

  - Failure of Biosphere 2 (p. 740).

- Increased emphasis on quantitative and research data by adding 45 new figures with graphs or other data presentations.

- 25 new reviewers to improve content and balance. See the front insert in the instructor's version of this book and this book's website for a detailed list by chapter of how balance on controversial issues has been achieved.

- 20 Pro/Con summary diagrams summarize a large amount of complex information in an easy-to-understand manner. Examples include Figures 12-30 (p. 289), 14-25 (p. 342), 14-26 (p. 342), 14-28 (p. 344), 14-31 (p. 346), 14-35 (p. 349), 15-22 (p. 375), 15-25 (p. 378), 15-27 (p. 379), 15-31 (p. 382), 15-33 (p. 384),15-35 (p. 387), 21-11 (p. 536), 21-13 (p. 538), 21-14 (p. 538), and 21-15 (p. 539).

- Four new Guest Essays (see pp. 28, 117, 377, and 743).

Learning Aids

- Greater use of numbered and bulleted lists to make the book simpler and help students comprehend and review key material.

- Use of questions as titles for all subsections to spark interest and serve as a built-in list of learning objectives for readers.

- Cross-references by page number to link concepts and material throughout the book. This **(1)** emphasizes the basic ecological concept that everything is connected and **(2)** serves as a textbook version of interconnected Web links.

- Addition of comprehensive Review Questions at the end of each chapter.

Welcome to Controversy and Challenge Despite much research, we still know little about how nature works at a time when we are altering nature at an accelerating pace. This uncertainty, and the complexity and importance of these issues to current and future generations of humans and other species, make many environmental issues highly controversial.

Another source of disagreement is that science advances through controversy and careful scrutiny of its results until there is a general consensus about their validity (p. 48). What is important is not what the experts disagree on (the frontiers of scientific knowledge that are still being developed, tested, and argued about) but what they generally agree on—the *scientific consensus*— about concepts, problems, and possible solutions.

Controversy also arises because environmental science is an interdisciplinary blend of natural and social sciences that sometimes questions the ways we view and act in the world around us.

Study Aids Each chapter begins with a few general questions to reveal how it is organized and what students will be learning. When a new term is introduced and defined, it is printed in boldface type. A glossary of all key terms is located at the end of the book.

Questions are used as titles for all sub-sections so that readers know the focus of the material that follows. In effect, this is a built-in set of learning objectives.

Each chapter ends with **(1)** a set of Review Questions covering all of the material in the chapter as a study guide for students and **(2)** a set of questions to encourage students to think critically and apply what they have learned to their lives. The Critical Thinking questions are followed by several projects that individuals or groups can carry out. Many additional projects are given in the *Instructor's Manual* and in the *Critical Thinking and the Environment* supplement available with this book. See front insert in the instructor's version of this book and the website for a detailed list of supplementary materials.

Readers who become especially interested in a particular topic can consult the *further readings* for each chapter, found on the book's website. The website also contains a list of important publications and some key environmental organizations and government and international agencies.

Students can also access World Wide Web for material in the book marked with the icon

www.

- "Flash Cards," which allow you to test your mastery of the Terms and Concepts to Remember for each chapter.

- "Tutorial Quizzes," which provide a multiple-choice practice quiz.

- "Student Guide to InfoTrac," which will lead you to Critical Thinking Projects that use InfoTrac College Edition as a research tool.

- "References," which lists the major books and articles consulted in writing this chapter.

- "Hypercontents," which takes you to an extensive list of websites with news, research, and images related to individual sections of the chapter.

Help Me Improve This Book Let me know how you think this book can be improved; if you find any errors, bias, or confusing explanations please send them to Jack Carey, Biology Publisher, Brooks/Cole, 10

Davis Drive, Belmont, CA 94002 (E-mail: jack.carey@ brookscole.com). He will forward them to me. Most errors can be corrected in subsequent printings of this edition rather than waiting for a new edition.

Annenberg/CPB Television Course This textbook is being offered as part of the Annenberg/CPB Project television series *Race to Save the Planet*, a 10-part public broadcasting series and a college-level telecourse examining the major environmental questions facing the world today. The series takes into account the wide spectrum of opinion about what constitutes an environmental problem and discusses the controversies about appropriate remedial measures. It analyzes problems and emphasizes the successful search for solutions. The course develops a number of key themes that cut across a broad range of environmental issues, including sustainability, the interconnections of the economy and the ecosystem, short-term versus long-term gains, and the trade-offs involved in balancing problems and solutions. A study guide and a faculty guide, both available from Brooks/Cole, integrate the telecourse and this text.

For further information about available television course licenses and duplication licenses, contact PBS Adult Learning Service, 1320 Braddock Place, Alexandria, VA 22314-1698 (1-800-ALS-ALS-8).

For information about purchasing videocassettes and print material, contact the Annenberg/CPB Collection, P.O. Box 2284, South Burlington, VT 05407-2284 (1-800-LEARNER).

Acknowledgments I wish to thank the many students and teachers who responded so favorably to the 11 previous editions of *Living in the Environment*, the 8 editions of *Environmental Science*, and the 5 editions of *Sustaining the Earth* and who corrected errors and offered many helpful suggestions for improvement. I am also deeply indebted to the reviewers, who pointed out errors and suggested many important improvements in this book. Any errors and deficiencies left are mine.

The members of the talented production team, listed on the copyright page, have made vital contributions as well. My thanks also go to **(1)** copyeditor Carol Anne Peschke, **(2)** production editors Brooks Ellis and Hal Humphrey, **(3)** Eileen Mitchell, Michael Gutch, and other members of the staff of Electronic Publishing Services who have improved the design of this edition and made my life much easier, **(4)** photo researcher Linda L. Rill, **(5)** Brooks/Cole's hard-working sales staff, and **(6)** Pat Waldo and her talented colleagues who develop multimedia associated with this book.

I also thank **(1)** Paul M. Rich for serving as coauthor of the basic science chapters in the 11th edition, **(2)** C. Lee Rockett and Kenneth J. Van Dellen for developing the *Laboratory Manual* to accompany this book, **(3)** Jane Heinze-Fry for her outstanding work on concept mapping, *Environmental Articles*, *Critical Thinking and the Environment: A Beginner's Guide*, and the *Internet Booklet*, **(4)** Richard K. Clements for his excellent work on the *Instructor's Manual*, and **(5)** the people who have translated this book into five different languages for use throughout much of the world.

My deepest thanks go to Jack Carey, biology publisher at Brooks/Cole, for his encouragement, help, 34 years of friendship, and superb reviewing system. It helps immensely to work with the best and most experienced editor in college textbook publishing.

I dedicate this book to the earth and to Kathleen Paul, my research assistant and fiancée.

G. Tyler Miller, Jr.

Guest Essayists and Reviewers

Guest Essayists The following are authors of Guest Essays: **(1) M. Kat Anderson**, ethnoecologist with the National Plant Center of the USDA's Natural Resource Conservation Service; **(2) Lester R. Brown**, chair of the board, Worldwatch Institute; **(3) Alberto Ruz Buenfil**, environmental activist, writer, and performer; **(4) Robert D. Bullard**, professor of sociology and director of the Environmental Justice Resource Center at Clark Atlanta University; **(5) Herman E. Daly**, senior research scholar at the School of Public Affairs, University of Maryland; **(6) Lois Marie Gibbs**, director, Center for Health, Environment, and Justice; **(7) Garrett Hardin**, professor emeritus of human ecology, University of California, Santa Barbara; **(8) Paul Hawken**, environmental author and business leader; **(9) Jane Heinze-Fry**, author, teacher, and consultant in environmental education; **(10) Amory B. Lovins**, energy policy consultant and director of research, Rocky Mountain Institute; **(11) Lester W. Milbrath**, director of the research program in environment and society, State University of New York, Buffalo; **(12) Peter Montague**, director, Environmental Research Foundation; **(13) Norman Myers**, tropical ecologist and consultant in environment and development; **(14) David W. Orr**, professor of environmental studies, Oberlin College; **(15) David Pimentel**, professor of entomology, Cornell University; **(16) Andrew C. Revkin**, environmental author and environmental reporter for *The New York Times*; **(17) Nancy Wicks**, ecopioneer and director of Round Mountain Organics; and **(18) Donald Worster**, environmental historian and professor of American history, University of Kansas.

Cumulative Reviewers Barbara J. Abraham, Hampton College; Donald D. Adams, State University of New York at Plattsburgh; Larry G. Allen, California State University, Northridge; Susan Allen-Gil, Ithaca College; James R. Anderson, U.S. Geological Survey; Mark W. Anderson, University of Maine; Kenneth B. Armitage, University of Kansas; Samuel Arthur, Bowling Green State University; Gary J. Atchison, Iowa State University; Marvin W. Baker, Jr., University of Oklahoma; Virgil R. Baker, Arizona State University; Ian G. Barbour, Carleton College; Albert J. Beck, California State University, Chico; W. Behan, Northern Arizona University; Keith L. Bildstein, Winthrop College; Jeff Bland, University of Puget Sound; Roger G. Bland, Central Michigan University; Grady Blount II, Texas A&M University, Corpus Christi; Georg Borgstrom, Michigan State University; Arthur C. Borror, University of New Hampshire; John H. Bounds, Sam Houston State University; Leon F. Bouvier, Population Reference Bureau; Daniel J. Bovin, Universitè Laval; Michael F. Brewer, Resources for the Future, Inc.; Mark M. Brinson, East Carolina University; Dale Brown, University of Hartford; Patrick E. Brunelle, Contra Costa College; Terrence J. Burgess, Saddleback College North; David Byman, Pennsylvania State University, Worthington-Scranton; Lynton K. Caldwell, Indiana University; Faith Thompson Campbell, Natural Resources Defense Council, Inc.; Ray Canterbery, Florida State University;

Ted J. Case, University of San Diego; Ann Causey, Auburn University; Richard A. Cellarius, Evergreen State University; William U. Chandler, Worldwatch Institute; F. Christman, University of North Carolina, Chapel Hill; Preston Cloud, University of California, Santa Barbara; Bernard C. Cohen, University of Pittsburgh; Richard A. Cooley, University of California, Santa Cruz; Dennis J. Corrigan; George Cox, San Diego State University; John D. Cunningham, Keene State College; Herman E. Daly, The World Bank; Raymond F. Dasmann, University of California, Santa Cruz; Kingsley Davis, Hoover Institution; Edward E. DeMartini, University of California, Santa Barbara; Charles E. DePoe, Northeast Louisiana University; Thomas R. Detwyler, University of Wisconsin; Peter H. Diage, University of California, Riverside; Lon D. Drake, University of Iowa; David DuBose, Shasta College; Dietrich Earnhart, University of Kansas; T. Edmonson, University of Washington; Thomas Eisner, Cornell University; Michael Esler, Southern Illinois University; David E. Fairbrothers, Rutgers University; Paul P. Feeny, Cornell University; Richard S. Feldman, Marist College; Nancy Field, Bellevue Community College; Allan Fitzsimmons, University of Kentucky; Andrew J. Friedland, Dartmouth College; Kenneth O. Fulgham, Humboldt State University; Lowell L. Getz, University of Illinois at Urbana-Champaign; Frederick F. Gilbert, Washington State University; Jay Glassman, Los Angeles Valley College; Harold Goetz, North Dakota State University; Jeffery J. Gordon, Bowling Green State University; Eville Gorham, University of Minnesota; Michael Gough, Resources for the Future; Ernest M. Gould, Jr., Harvard University; Peter Green, Golden West College; Katharine B. Gregg, West Virginia Wesleyan College; Paul K. Grogger, University of Colorado at Colorado Springs; L. Guernsey, Indiana State University; Ralph Guzman, University of California, Santa Cruz; Raymond Hames, University of Nebraska, Lincoln; Raymond E. Hampton, Central Michigan University; Ted L. Hanes, California State University, Fullerton; William S. Hardenbergh, Southern Illinois University at Carbondale; John P. Harley, Eastern Kentucky University; Neil A. Harriman, University of Wisconsin, Oshkosh; Grant A. Harris, Washington State University; Harry S. Hass, San Jose City College; Arthur N. Haupt, Population Reference Bureau; Denis A. Hayes, environmental consultant; Stephen Heard, University of Iowa; Gene Heinze-Fry, Department of Utilities, State of Massachusetts; Jane Heinze-Fry, environmental educator; John G. Hewston, Humboldt State University; David L. Hicks, Whitworth College; Kenneth M. Hinkel, University of Cincinnati; Eric Hirst, Oak Ridge National Laboratory; Doug Hix, University of Hartford; S. Holling, University of British Columbia; Donald Holtgrieve, California State University, Hayward; Michael H. Horn, California State University, Fullerton; Mark A. Hornberger, Bloomsberg University; Marilyn Houck, Pennsylvania State University; Richard D. Houk, Winthrop College; Robert J. Huggett, College of William and Mary; Donald Huisingh, North Carolina State

University; Marlene K. Hutt, IBM; David R. Inglis, University of Massachusetts; Robert Janiskee, University of South Carolina; Hugo H. John, University of Connecticut; Brian A. Johnson, University of Pennsylvania, Bloomsburg; David I. Johnson, Michigan State University; Mark Jonasson, Crafton Hills College; Agnes Kadar, Nassau Community College; Thomas L. Keefe, Eastern Kentucky University; Nathan Keyfitz, Harvard University; David Kidd, University of New Mexico; Pamela S. Kimbrough; Jesse Klingebiel, Kent School; Edward J. Kormondy, University of Hawaii-Hilo/West Oahu College; John V. Krutilla, Resources for the Future, Inc.; Judith Kunofsky, Sierra Club; E. Kurtz; Theodore Kury, State University of New York, Buffalo; Steve Ladochy, University of Winnipeg; Mark B. Lapping, Kansas State University; Tom Leege, Idaho Department of Fish and Game; William S. Lindsay, Monterey Peninsula College; E. S. Lindstrom, Pennsylvania State University; M. Lippiman, New York University Medical Center; Valerie A. Liston, University of Minnesota; Dennis Livingston, Rensselaer Polytechnic Institute; James P. Lodge, air pollution consultant; Raymond C. Loehr, University of Texas at Austin; Ruth Logan, Santa Monica City College; Robert D. Loring, DePauw University; Paul F. Love, Angelo State University; Thomas Lovering, University of California, Santa Barbara; Amory B. Lovins, Rocky Mountain Institute; Hunter Lovins, Rocky Mountain Institute; Gene A. Lucas, Drake University; Claudia Luke; David Lynn; Timothy F. Lyon, Ball State University; Stephen Malcolm, Western Michigan University; Melvin G. Marcus, Arizona State University; Gordon E. Matzke, Oregon State University; Parker Mauldin, Rockefeller Foundation; Marie McClune, The Agnes Irwin School (Rosemont, Pennsylvania); Theodore R. McDowell, California State University; Vincent E. McKelvey, U.S. Geological Survey; Robert T. McMaster, Smith College; John G. Merriam, Bowling Green State University; A. Steven Messenger, Northern Illinois University; John Meyers, Middlesex Community College; Raymond W. Miller, Utah State University; Arthur B. Millman, University of Massachusetts, Boston; Fred Montague, University of Utah; Rolf Monteen, California Polytechnic State University; Ralph Morris, Brock University, St. Catherine's, Ontario, Canada; Angela Morrow, Auburn University; William W. Murdoch, University of California, Santa Barbara; Norman Myers, environmental consultant; Brian C. Myres, Cypress College; A. Neale, Illinois State University; Duane Nellis, Kansas State University; Jan Newhouse, University of Hawaii, Manoa; Jim Norwine, Texas A&M University, Kingsville; John E. Oliver, Indiana State University; Eric Pallant, Allegheny College; Charles F. Park, Stanford University; Richard J. Pedersen, U.S. Depart-ment of Agriculture, Forest Service; David Pelliam, Bureau of Land Management, U.S. Department of Interior; Rodney Peterson, Colorado State University; Julie Phillips, De Anza College; William S. Pierce, Case Western Reserve University; David Pimentel, Cornell University; Peter Pizor, Northwest Community College; Mark D. Plunkett, Bellevue Community College; Grace L. Powell, University of Akron; James H. Price, Oklahoma College; Marian E. Reeve, Merritt College; Carl H. Reidel, University of Vermont; Charles C. Reith, Tulane University; Roger Revelle, California State University, San Diego; L. Reynolds, University of Central Arkansas; Ronald R. Rhein, Kutztown University of Pennsylvania; Charles Rhyne, Jackson State University; Robert A. Richardson, University of Wisconsin; Benjamin F. Richason III, St. Cloud State University; Ronald Robberecht, University of Idaho; William Van B. Robertson, School of Medicine, Stanford University; C. Lee Rockett, Bowling Green State University; Terry D. Roelofs, Humboldt State University; Christopher Rose, California Polytechnic State University; Richard G. Rose, West Valley College; Stephen T. Ross, University of Southern Mississippi; Robert E. Roth, The Ohio State University; Arthur N. Samel, Bowling Green State University; Floyd Sanford, Coe College; David Satterthwaite, I.E.E.D., London; Stephen W. Sawyer, University of Maryland; Arnold Schecter, State University of New York, Syracuse; Frank Schiavo, San Jose State University; William H. Schlesinger, Ecological Society of America; Stephen H. Schneider, National Center for Atmospheric Research; Clarence A. Schoenfeld, Univ-ersity of Wisconsin, Madison; Henry A. Schroeder, Dartmouth Medical School; Lauren A. Schroeder, Youngstown State University; Norman B. Schwartz, University of Delaware; George Sessions, Sierra College; David J. Severn, Clement Associates; Paul Shepard, Pitzer College and Claremont Graduate School; Michael P. Shields, Southern Illinois University at Carbondale; Kenneth Shiovitz; F. Siewert, Ball State University; E. K. Silbergold, Environmental Defense Fund; Joseph L. Simon, University of South Florida; William E. Sloey, University of Wisconsin, Oshkosh; Robert L. Smith, West Virginia University; Val Smith, University of Kansas; Howard M. Smolkin, U.S. Environmental Protection Agency; Patricia M. Sparks, Glassboro State College; John E. Stanley, University of Virginia; Mel Stanley, California State Polytechnic University, Pomona; Norman R. Stewart, University of Wisconsin, Milwaukee; Frank E. Studnicka, University of Wisconsin, Platteville; Chris Tarp, Contra Costa College; Roger E. Thibault, Bowling Green State University; William L. Thomas, California State University, Hayward; John D. Usis, Youngstown State University; Tinco E. A. van Hylckama, Texas Tech University; Robert R. Van Kirk, Humboldt State University; Donald E. Van Meter, Ball State University; Gary Varner, Texas A&M University; John D. Vitek, Oklahoma State University; Harry A. Wagner, Victoria College; Lee B. Waian, Saddleback College; Warren C. Walker, Stephen F. Austin State University; Thomas D. Warner, South Dakota State University; Kenneth E. F. Watt, University of California, Davis; Alvin M. Weinberg, Institute of Energy Analysis, Oak Ridge Associated Universities; Brian Weiss; Margery Weitkamp, James Monroe High School (Granada Hills, California); Anthony Weston, SUNY at Stony Brook; Raymond White, San Francisco City College; Douglas Wickum, University of Wisconsin, Stout; Charles G. Wilber, Colorado State University; Nancy Lee Wilkinson, San Francisco State University; John C. Williams, College of San Mateo; Ray Williams, Rio Hondo College; Roberta Williams, University of Nevada, Las Vegas; Samuel J. Williamson, New York University; Ted L. Willrich, Oregon State University; James Winsor, Pennsylvania State University; Fred Witzig, University of Minnesota at Duluth; George M. Woodwell, Woods Hole Research Center; Robert Yoerg, Belmont Hills Hospital; Hideo Yonenaka, San Francisco State University; Malcolm J. Zwolinski, University of Arizona.

Brief Contents

Detailed Contents

Paul W. Johnson/Biological Photo Service

Temperate deciduous forest, Fall, Rhode Island

Paul W. Johnson/Biological Photo Service

Temperate deciduous forest, Winter, Rhode Island

Hawaiian monk seal's mouth caught in plastic

Don Alcorn/National Maritime Fisheries

Rocky Mountain Institute, Colorado

Robert Millman/Rocky Mountain Institute

Pat Armstrong/Visuals Unlimited

An earth-sheltered house, Will County, IL.

Gary Milburn/Tom Stack & Associates

Cotton-top tamarin

Mutualism between oxpeckers and a rhinoceros

Tree farm, North Carolina

U.S. Department of Agriculture

Fast growing Kenaf, Texas

U.S. Department of Agriculture

Boll weevil

U.S. Windpower

Wind farm, California

National Archives/EPA Documerica

Monoculture cropland, Blythe, California

Mt. St. Helens before volcano eruption

Mt. St. Helens after volcano eruption

Ocean Arcs International

John Todd at solar sewage plant, Rhode Island

Dan Kline/Visuals Unlimited

Wolf spider

Living in the Environment

PART I

HUMANS AND SUSTAINABILITY: AN OVERVIEW

The environmental crisis is an outward manifestation of a crisis of mind and spirit. There could be no greater misconception of its meaning than to believe it is concerned only with endangered wildlife, human-made ugliness, and pollution. These are part of it, but more

importantly, the crisis is concerned with the kind of creatures we are and what we must become in order to survive.

LYNTON K. CALDWELL

1 ENVIRONMENTAL ISSUES, THEIR CAUSES, AND SUSTAINABILITY

Living in an Exponential Age

Once there were two kings from Babylon who enjoyed playing chess, with the winner claiming a prize from the loser. After one match, the winning king asked the loser to pay him by placing one grain of wheat on the first square of the chessboard, two on the second, four on the third, and so on. The number of grains was to double each time until all 64 squares were filled.

The losing king, thinking he was getting off easy, agreed with delight. It was the biggest mistake he ever made. He bankrupted his kingdom and still could not produce the incredibly large number of grains of wheat he had promised. In fact, it's probably more than all the wheat that has ever been harvested!

This is an example of **exponential growth**, in which a quantity increases by a fixed percentage of the whole in a given time. As the losing king learned, exponential growth is deceptive. It starts off slowly, but after only a few doublings it grows to enormous numbers because each doubling is more than the total of all earlier growth.

Here is another example. Fold a piece of paper in half to double its thickness. If you could do this 42 times, the stack would reach from the earth to the moon, 386,400 kilometers (240,000 miles) away. If you could double it 50 times, the folded paper would almost reach the sun, 149 million kilometers (93 million miles) away!

The environmental issues we face—**(1)** *population growth,* **(2)** *increasing resource use,* **(3)** *destruction and degradation of wildlife habitats,* **(4)** *premature extinction of plants and animals,* **(5)** *poverty, and* **(6)** *pollution*—are interconnected and are growing exponentially. For example, world population has more than doubled in 50 years, from 2.5 billion in 1950 to 6.1 billion in 2000. Unless death rates rise sharply, it may reach 8 billion by 2028, 9 billion by 2054, and 10–14 billion by 2100 (Figure 1-1). Global economic output, much of it environmentally damaging, increased 17-fold between 1900 and 2000 and almost eightfold between 1950 and 2000.

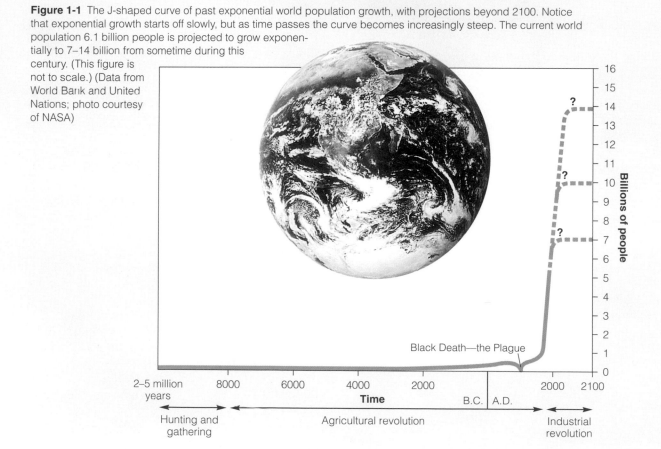

Figure 1-1 The J-shaped curve of past exponential world population growth, with projections beyond 2100. Notice that exponential growth starts off slowly, but as time passes the curve becomes increasingly steep. The current world population 6.1 billion people is projected to grow exponentially to 7–14 billion from sometime during this century. (This figure is not to scale.) (Data from World Bank and United Nations; photo courtesy of NASA)

Alone in space, alone in its life-supporting systems, powered by inconceivable energies, mediating them to us through the most delicate adjustments, wayward, unlikely, unpredictable, but nourishing, enlivening, and enriching in the largest degree—is this not a precious home for all of us? Is it not worth our love?

BARBARA WARD AND RENÉ DUBOS

This chapter is an overview of environmental issues, their root causes, the controversy over their seriousness, and ways we can live more sustainably. It addresses these questions:

- What are natural resources and why are they important? What is an environmentally sustainable society?
- How fast is the human population increasing?
- What is the difference between economic growth, economic development, and environmentally sustainable economic development?
- What are the earth's main types of resources? How can they be depleted or degraded?
- What are the principal types of pollution? How can pollution be reduced and prevented?
- What are the root causes of today's environmental problems and how are these causes connected?
- Is our current course sustainable? How can we live more sustainably?

1-1 LIVING MORE SUSTAINABLY

What Is the Difference Between Environment, Ecology, and Environmental Science? **Environment** is everything that affects a living organism (any unique form of life). **Ecology** is a biological science that studies the relationships between living organisms and their environment.

This textbook is an introduction to **environmental science***. It is an interdisciplinary science that uses concepts and information from *natural sciences* such as ecology, biology, chemistry, and geology and *social sciences* such as economics, politics, and ethics to **(1)** help us understand how the earth works, **(2)** learn how we are affecting the earth's life-support systems (environment) for us and other forms of life, and **(3)** propose and evaluate solutions to the environmental problems we face. Many different groups of people are concerned about environmental issues (Spotlight, right).

What Keeps Us Alive? Our existence, lifestyles, and economies depend completely on the sun and the earth,

---

*Some people prefer to call this *environmental studies* and reserve the term *environmental science* only for the physical science aspects of studying the environment. However, most people use the broader definition of *environmental science* given here.

Cast of Players in the Environmental Drama

SPOTLIGHT

The cast of major characters you will encounter in this book include the following:

- **Ecologists**, who are biological scientists studying relationships between living organisms and their environment.

- **Environmental scientists**, who use information from the physical sciences and social sciences to **(1)** understand how the earth works, **(2)** learn how humans interact with the earth, and **(3)** develop solutions to environmental problems.

- **Conservation biologists**, who in the 1970s created a multidisciplinary science to **(1)** investigate human impacts on the diversity of life found on the earth (biodiversity) and **(2)** develop practical plans for preserving such biodiversity.

- **Environmentalists**, who are concerned about the impact of people on environmental quality and believe that some human actions are degrading parts of the earth's life-support systems for humans and many other forms of life. Some of their beliefs and proposals for dealing with environmental problems are based on scientific information and concepts and some are based on their social and ethical environmental beliefs (environmental worldviews). Environmentalists are a broad group of people from different economic groups (rich, middle-class, poor) and with different political persuasions (conservative and liberal).

- **Preservationists**, concerned primarily with setting aside or protecting undisturbed natural areas from harmful human activities.

- **Conservationists**, concerned with using natural areas and wildlife in ways that sustain them for current and future generations of humans and other forms of life.

- **Restorationists**, devoted to the partial or complete restoration of natural areas that have been degraded by human activities.

Many people consider themselves members of several of these groups.

Critical Thinking

Which, if any, of these groups do you most identify with? Why?

---

a blue and white island in the black void of space (Figure 1-1). To economists *capital* is wealth used to sustain a business and to generate more wealth. By analogy, we can think of energy from the sun as **solar capital** and the planet's air, water, soil, wildlife, minerals, and natural purification, recycling, and pest control processes as **natural resources** or **natural capital**

(Guest Essay, p. 6). **Solar energy** is defined broadly to include direct sunlight and indirect forms of solar energy such as **(1)** wind power, **(2)** hydropower (energy from flowing water), and **(3)** biomass (direct solar energy converted to chemical energy stored in biological sources of energy such as wood).

What Is an Environmentally Sustainable Society?

To survive and maintain good health, all forms of life must have enough food, clean air, clean water, and shelter to meet their *basic needs.* Additional needs for humans include enough income to meet basic needs, respectable and safe work, health care, recreation, cultural opportunities, education, and freedom from physical danger.

An **environmentally sustainable society** satisfies the basic needs of its people without depleting or degrading its natural resources and thereby preventing current and future generations of humans and other species from meeting their basic needs. *Living sustainably* means living off the natural income replenished by soils, plants, air, and water and not depleting the natural capital that supplies this income (Guest Essay, p. 6).

For example, imagine that you inherit $1 million. Invest this capital at 10% interest per year and you will have a sustainable annual income of $100,000 without depleting your capital. If you spend $200,000 a year, your $1 million will be gone early in the 7th year; even if you spend only $110,000 a year, you will be bankrupt early in the 18th year.

The lesson here is a very old one: *Don't kill the goose that lays the golden egg,* or *protect your capital.* Deplete your capital, and you move from a sustainable to an unsustainable lifestyle.

The same lesson applies to the earth's natural capital. Environmentalists and many leading scientists believe that we are living unsustainably by depleting and degrading the earth's natural capital at an accelerating rate as our population (Figure 1-1) and demands on the earth's resources and life-sustaining processes increase exponentially.

In 1999, the World Wildlife Fund, the New Economics Foundation, and the World Conservation Monitoring Centre issued *The Living Planet Report.* It was designed to be an "environmental Dow Jones index" measuring the environmental health of the world's forests, rivers, lakes, and oceans. According to this study, the world lost about one-third of its natural capital between 1970 and 1999 because of a combination of exponential growth in population and in the use of the earth's natural resources.

Other analysts do not believe that we are living unsustainably. They contend **(1)** that environmentalists have exaggerated the seriousness of population and environmental problems and **(2)** that any population, resource, and environmental problems we face can be overcome by human ingenuity and technological advances.

1-2 POPULATION GROWTH, ECONOMIC GROWTH, ENVIRONMENTALLY SUSTAINABLE DEVELOPMENT, AND GLOBALIZATION

What Is the Difference Between Linear Growth and Exponential Growth? Suppose you hop on a train that accelerates by 1 kilometer (0.6 mile) per hour every second. After 30 seconds, your speed would be 30 kilometers (19 miles) per hour. This is an example of **linear growth**, in which a quantity increases by a constant amount per unit of time, as in the sequence 1, 2, 3, 4, 5— or 1, 3, 5, 7, 9—and so on. If plotted on a graph, such growth in speed or growth of money in a savings account yields a straight line that slopes upward (Figure 1-2).

However, suppose the train has a motor strong enough to double its speed every second. After 30 seconds, you would be traveling a billion kilometers (620 million miles) per hour!

This is an example of the astounding power of exponential growth. Any quantity growing exponentially by a fixed percentage, even as small as 0.001% or 0.1%, will experience extraordinary growth as its base of growth doubles again and again. If plotted on a graph, continuing exponential growth eventually yields a graph shaped somewhat like the letter J (Figure 1-2).

How long does it take to double the world's resource use or population size or money in a savings account that is growing exponentially? A quick way to calculate this **doubling time** in years is to use the **rule of 70**: 70/percentage growth rate = doubling time in years (a form-

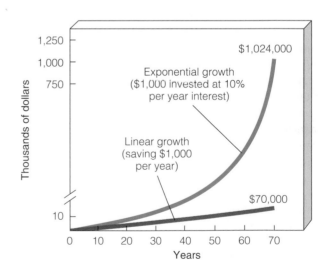

Figure 1-2 Linear and exponential growth. If you save $1,000 a year for a lifetime of 70 years, the resulting linear growth will allow you to save $70,000 (lower curve). If you invest $1,000 each year at 10% interest for 70 years and reinvest the interest, your money will grow exponentially to $1,024,000 (upper curve). If resource use, economic growth, or money in a savings account grows exponentially for 70 years (a typical human lifetime) at a rate of 10% a year, it will increase 1,024-fold.

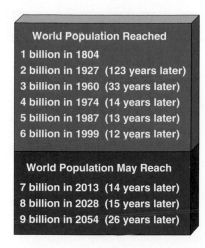

World Population Reached

1 billion in 1804
2 billion in 1927 (123 years later)
3 billion in 1960 (33 years later)
4 billion in 1974 (14 years later)
5 billion in 1987 (13 years later)
6 billion in 1999 (12 years later)

World Population May Reach

7 billion in 2013 (14 years later)
8 billion in 2028 (15 years later)
9 billion in 2054 (26 years later)

Figure 1-3 World population milestones. (Data from United Nations Population Division, *World Population Prospects*, 1998)

ula derived from the basic mathematics of exponential growth). For example, in 2000 the world's population grew by 1.35%. If that rate continues, the earth's population will double in about 52 years (70/1.35 = about 52 years). In the example of exponential growth of savings by 10% a year (Figure 1-2), your money would double roughly every 7 years (70/10 = 7).

How Rapidly Is the Human Population Growing?
The increasing size of the human population is an example of exponential growth (Figures 1-1 and 1-3 and Spotlight, right). The main reason for the rapid growth of the earth's human population over the past 100 years has been a much greater drop in death rates (mostly because of increases in food supply and better health and nutrition) than in birth rates.

Recent studies by researchers at Conservation International suggest that roughly *73% of the earth's habitable land surface (that which is not bare rock, ice, or drifting sand) has been partially or heavily disturbed by human activities* (Figure 1-4, p. 8). What will happen to the earth's remaining wildlife habitat and wildlife species if the human population increases from 6.1 billion to 8 billion between 2000 and 2028 and perhaps to 9 billion by 2054?

What Is Economic Growth? Almost all countries seek **economic growth**: an increase in their capacity to provide goods and services for people's final use. This increase is accomplished by population growth (more consumers and producers), more consumption per person, or both.

Economic growth usually is measured by an increase in several indicators:

- **Gross national product (GNP):** the market value in current dollars of all goods and services produced *within* and *outside* a country by the country's businesses during a year

- **Gross domestic product (GDP):** the market value in current dollars of all goods and services produced *within* a country during a year

- **Gross world product (GWP):** the market value in current dollars of all goods and services produced in the world each year

SPOTLIGHT

Current Exponential Growth of the Human Population

The world's population is growing exponentially at a rate of about 1.35% per year. The relentless ticking of this population clock means that in 2000 the world's population of 6.1 billion grew by 82 million people (6.1 billion × 0.0135 = 82 million), an average increase of 226,000 people a day, or 9,400 an hour.

At this 1.35% annual rate of exponential growth, it takes only about

- 5 days to add the number of Americans killed in all U.S. wars

- 2 months to add as many people as live in the Los Angeles basin

- 1.6 years to add the 129 million people killed in all wars fought in the past 200 years

- 3.4 years to add 276 million people (the population of the United States in 2000)

- 15 years to add 1.26 billion people (the population of China, the world's most populous country, in 2000)

How much is a billion? If you could live for a billion minutes, you would be 1,902 years old. To travel 1 billion kilometers (0.6 billion miles), you would have to circle the earth about 25,000 times.

Critical Thinking

Some economists argue that population growth is good because it provides more workers, consumers, and problem solvers to keep the global economy growing. Environmentalists argue that population growth threatens economies and the earth's life-support systems through increased pollution and environmental degradation. What is your position? Why?

To show one person's slice of the economic pie, economists calculate the **per capita GNP**: the GNP divided by the total population.

Economic development is the improvement of living standards by economic growth. The United Nations classifies the world's countries as economically developed or developing based primarily on their degree of industrialization and their per capita GNP (Figure 1-5, p. 9).

The **developed countries** include the United States, Canada, Japan, Australia, New Zealand, and all the countries of Europe. Most are highly industrialized and have high average per capita GNPs (above $10,000 per year, except for industrialized countries in eastern Europe and some in northern and southern Europe). These countries,

www.brookscole.com/product/0534376975s   5

Paul G. Hawken

GUEST ESSAY

Paul G. Hawken understands both business and ecology. In addition to founding Smith & Hawken, a retail company known for its environmental initiatives, he has written seven widely acclaimed books, including Growing a Business *(1987),* The Ecology of Commerce *(1993),* Factor 10, The Next Industrial Revolution *(1998, with Amory and Hunter Lovins), and* Natural Capitalism *(1999, with Amory and Hunter Lovins). He produced and hosted* Growing a Business, *a series for public television shown nationwide on 210 stations and now shown in 115 countries.* The Ecology of Commerce *was hailed as the best business book of 1993 and one of the most important books of the 20th century. In 1987,* Inc. *magazine named him one of the 12 best entrepreneurs of the 1980s, and in 1995 he was named by the* Utne Reader *as one of the 100 visionaries who could change our lives.*

Great ideas, in hindsight, seem obvious. The concept of natural capital is such an idea. *Natural capital* is the myriad necessary and valuable resources and ecological processes that we rely on to produce our food, products, and services.

The concept of natural capital is not a new one. Economists have long noted that natural capital is a factor in industrial production, but a marginal factor.

A new view is emerging: Our economic systems cannot long endure without taking the flow of renewable and nonrenewable resources through economies into account. The value of natural capital is becoming paramount to the success of all business.

This revision of neoclassical economics, yet to be accepted by most mainstream academicians, provides business and public policy with a powerful new tool for the continued prosperity of business and the preservation and restoration of the earth's living natural systems.

Most Americans are filled with cornucopian fantasies of technological prowess, where human ingenuity bypasses natural limits and creates unimagined abundance. Optimism easily intertwines with the belief that nanotechnology, biotechnology, computers, and technologies yet to developed will eliminate hunger, disease, and want.

Dreams of alleviating human suffering are worthy. However, they usually overlook the absolute necessity for fertile soil, ocean fisheries, a stable climate, biological diversity, and pure water, all of which we are degrading and none of which can be created by any human-made technology known or imagined.

In our pursuit of dominance over the natural world, we have not taken into account the basic principle that industrialism, for all its sophistication, is enormously inefficient with respect to resources, energy, and waste. It is difficult for neoclassical economists, whose hypotheses and theories originated in a time of resource abundance, to understand that the very success of linear industrial systems based on increasing economic growth by increasing the rate of flow of materials and energy through economic systems has laid the groundwork for the next stage in economic evolution.

This shift is profoundly biological. It involves incorporating the cycling of material resources that supports natural systems into our ways of making things and our ways of dealing with the waste matter produced by our current linear industrial systems. This shift is going to happen because cyclical industrial systems work better than linear ones. They close the loop and reincorporate wastes as part of the production cycle. There are no landfills in a cyclical society.

with 1.2 billion people (20% of the world's population in 2000), **(1)** have about 85% of the world's wealth and income, **(2)** use about 88% of its natural resources, and **(3)** generate about 75% of its pollution and waste.

All other nations are classified as **developing countries**, most of them in Africa, Asia, and Latin America. Their 4.9 billion people (80% of the world's population in 2000) **(1)** have only about 15% of the wealth and income and **(2)** use only about 12% of the world's natural resources. Some are *middle-income, moderately developed countries* with average per capita GNPs of $1,000 to $10,000 per year [Figure 1-5]. Examples are South Africa, Mexico, Argentina, Brazil, Saudi Arabia, Malaysia, and Thailand. Others are *low-income countries* with per capita GNPs less than $1,000 per year. Examples include India, China, Bangladesh, Pakistan, Vietnam, Nicaragua, Haiti, Bolivia, and most of the countries in western, eastern, and central Africa.

More than 95% of the projected increase in the world's population is expected to take place in developing countries (Figure 1-6), *where 1 million people are added every 5 days.* The primary reason for such rapid population growth in developing countries (1.7% compared with 0.1% in developed countries) is the *large percentage of people who are under age 15* (34% compared with 19% in developed countries in 2000). As these young people move into their prime reproductive years over the next several decades, they will fuel rapid population growth.

What Is Environmentally Sustainable Economic Development? Between 1900 and 2000, the global output of goods and services as measured by the global

If there is so much inefficiency in our current system, why isn't it more apparent? The inefficiencies are masked by a financial system in which money, prices, and markets give us inaccurate information. Markets are not giving us correct information about how much our suburbs, cars, and plastic drinking water bottles truly cost.

Instead, we are getting warning signals from the beleaguered airsheds and watersheds, the overworked and eroded soils, the life-degrading inner cities and rural counties, the breakdown of stability worldwide, and the conflicts based on resource shortages. These feedbacks from nature are providing the information that our prices should give us but don't.

Prices don't give us good information for a simple reason: improper accounting. Natural capital has never been placed on the balance sheets of companies or the countries of the world. To paraphrase G. K. Chesterton, it could fairly be said that capitalism might be a good idea except that we have not tried it yet. Capitalism cannot be fully attained or practiced until we have an accurate balance sheet, as any accounting student will tell us.

As it stands, our economic system is based on accounting principles that would bankrupt a company. When natural capital is placed on the balance sheet, not as a free resource of infinite supply but as an integral and valuable part of the production process, everything changes. The nearly obsessive pursuit of improvement in human productivity becomes balanced by the need for improved resource productivity. Using more and more resources to make fewer people more productive flies in the face of what we need to improve our society and the environment. After all, it is people we have more of, not natural resources, so it is people we must use to reduce the flow of matter and energy resources through economies and the resulting pollution and loss

of natural capital. Moving from linear industrial systems to cyclical ones that mimic nature accomplishes this.

Many people sincerely believe that an economic system based on the integrity of natural systems is unworkable. To answer that concern, we may want to reverse the question and ask, "How have we created an economic system that tells us it is cheaper to destroy natural capital than to maintain it?" We know this is not the way to take care of our cars, houses, and bridges, but somehow we have managed to overlook a pricing system that discounts the future and sells off the past. Or to put it another way, "How did we create an economic system that confuses capital with income?"

Can we devise and implement a more rational economic system? I think so. It is right before us. It requires no new theories, only common sense. It is based on the simple but powerful proposition that *all capital must be valued.*

There may be no *right* way to value a forest or a river, but there is a *wrong* way, which is to give it no value at all. If we have doubts about how to value a 500-year-old tree, we need only ask how much it would cost to make a new one from scratch. Or a new river. Or a new atmosphere.

The work of the future is absorbing and integrating the worth of living systems into every aspect of our culture and commerce so that human systems mimic natural systems. Only if we do this can our cultures reflect growth and harmony rather than damage and discord.

Critical Thinking

If you were in charge of the world's economy, what are the three most important things you would do? Compare your answers with those of other members of your class.

world product (GWP) increased from $2.3 trillion to $42 trillion and is projected to triple by 2050. Most of this growth occurred after 1950 (Figure 1-7). During this same period the global income per person rose from $1,500 to $6,700.

This economic growth has allowed billions of people to (1) live longer (with average life expectancy increasing from 35 to 66 years), (2) lead healthier lives, and (3) enjoy many comforts that were unimaginable in 1900. This economic growth, coupled with population growth, has (1) also led to the environmental problems we face today and (2) not wiped out poverty for more than half the world's people, who must live on an income equal to $3 per day ($1,095 per year) or less.

Some analysts have called for shift from emphasis on traditional economic development fueled by

economic growth of essentially any type to emphasis on **environmentally sustainable economic development***. This type of development (1) *encourages* environmentally sustainable forms of economic growth that meet the basic needs of the current generations of humans and other species without preventing future generations of humans and other species from meeting their basic needs and (2) *discourages* environmentally harmful and unsustainable forms of economic growth.

---

*This definition is based on the one proposed in 1987 in *Our Common Future*, a study made by the World Commission on Environment and Development to examine the relationships between economic development and the environment and to suggest ways to make the two more compatible.

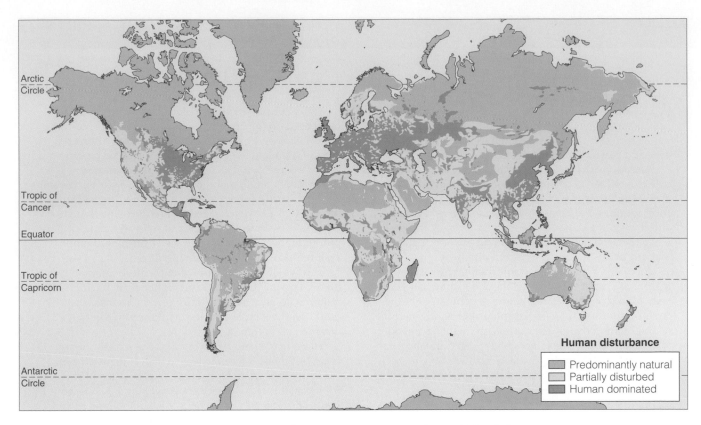

Figure 1-4 Human disturbance of the earth's land area. Green indicates undisturbed areas. Yellow indicates partially disturbed areas. Orange indicates seriously disturbed land areas, including deforestation, farmland, overgrazed grasslands, and urban areas. Excluding uninhabitable areas of rock, ice, desert, and steep mountain terrain, *only about 27% of the planet's land area remains undisturbed by human activities.* (Data from Lee Hannah and David Lohse, *1993 Annual Report*, Conservation International, Washington, DC)

In other words, environmentally sustainable economic development is the economic component of *environmentally sustainable societies* (p. 4). Using environmentally sustainable economic development as the basis for developing more environmentally sustainable societies requires that governments, businesses, and individuals integrate social, economic, and environmental goals and policies in their decision making (Figure 1-8).

What Is the Wealth Gap? Since 1960, and especially since 1980, the gap between the per capita GNP of the rich, middle income, poor, and acutely poor has widened. According to the World Bank, about

- 20% (1.2 billion) of the world's 6.1 billion people have a high per capita income.

- 25% (1.5 billion) have a moderate per capita income.

- 30% (1.8 billion) have a low per capita income of around $2–3 per day.

- 25% (1.5 billion) have a very low per capita income of no more than $1 per day.

According to the United Nations, about 1.2 billion people—one person in five—are hungry or malnourished (Figure 1-9) and lack access to clean water, decent housing, and adequate health care. About two-thirds of humanity lacks sanitary toilets and one of every four adults (1.3 billion people) is illiterate.

Daily life is a harsh struggle for the world's poor and acutely poor people in developing countries who try to survive on an income of $1–3 per day. Many poor parents have many children as a form of economic security to **(1)** help them grow food, **(2)** gather fuel (mostly wood and dung), **(3)** haul drinking water, **(4)** tend livestock, **(5)** work, **(6)** beg in the streets, and **(7)** help them survive in their old age (typically their 50s or 60s. Poor people

- may deplete and degrade local forests, soil, grasslands, wildlife, and water supplies for short-term survival even though they know it may lead to disaster in the long run.

- often have to live in areas with the highest levels of air and water pollution and with the greatest risk of natural disasters such as floods, earthquakes, hurricanes, and volcanic eruptions.

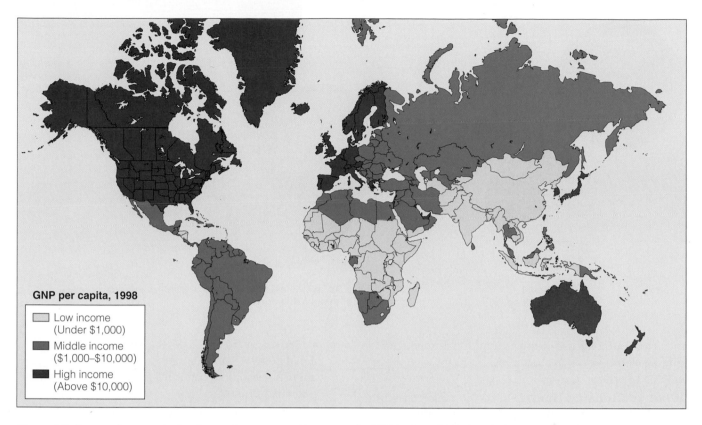

Figure 1-5 Degree of economic development as measured by per capita GNP in 1998. (Data from United Nations and the World Bank)

■ spend an average of **(1)** 4–6 hours per day searching for and carrying fuelwood and **(2)** 4–6 hours per week drawing and carrying water.

■ must take jobs (if they can find them) that subject them to unhealthy and unsafe working conditions at very low pay.

According to the World Health Organization (WHO), each year, at least 10 million of the desperately poor die prematurely of **(1)** malnutrition (lack of protein and other nutrients needed for good health), **(2)** increased susceptibility to infectious diseases because of their weakened condition from malnutrition, and **(3)** infectious diseases from drinking contaminated water. *This premature death of at least 27,400 human beings per day is equivalent to 69 jumbo jet planes, each carrying 400 passengers, crashing every day with no*

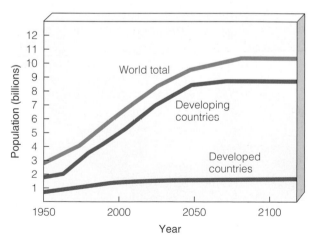

Figure 1-6 Past and projected population size for developed countries, developing countries, and the world, 1950–2120. More than 95% of the addition of 3.6 billion people between 1990 and 2030 is projected to occur in developing countries. (Data from United Nations)

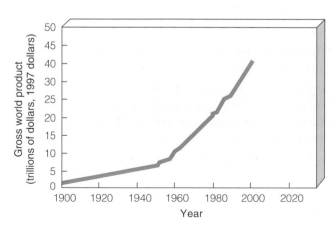

Figure 1-7 Growth of the gross world product, 1900–2000. (Data from United Nations and the World Bank)

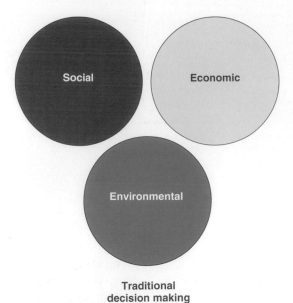

Figure 1-8 Types of decision making in traditional and sustainable societies. The traditional decision making in most societies involves treating social, economic, and environmental issues separately (left). Environmentally sustainable economic development calls for integrating social, economic, and environmental issues and concepts to find *sustainable solutions* to problems (right).

Traditional
decision making

Decision making in a
sustainable society

survivors. Half of those dying are children under age 5 (Figure 1-9).

What is Globalization? One of the major trends since 1950 and especially since 1970 is globalization. **Globalization** is the broad process of global social, economic, and environmental change that leads to an increasingly integrated world.

Here are a few indicators of globalization:

- Between 1950 and 2000, the global economy grew from $6.7 trillion to $42 trillion.

- Between 1950 and 2000, international trade of goods and increased from $311 million to $5.5 trillion and accounted for 13% of the gross world product in 2000 (up from 5% in 1950).

- Between 1970 and 2000, the number of transnational corporations operating in three or more countries grew from 7,000 to an estimated 53,600.

- Between 1960 and 2000, the number of lines linking noncellular phones grew from 89 million to 850 million, and the number of cellular phone connections has soared since 1990.

- Since 1995, the internet has grown by about 50% per year. By 2000, roughly 1 in every 36 people in the world had internet access, and this figure is growing rapidly.

- Between 1950 and 2000, the number of passenger-kilometers flown internationally grew from 28 million to 2.6 trillion, and the amount of international air freight increased 135-fold.

- Between 1961 and 2000, the number of international refugees receiving United Nations assistance grew from 1.4 million to 22.4 million.

- Between 1956 and 2000, the number of international nongovernment organizations (NGOs) operating in at least three countries grew from 985 to an estimated 23,000.

- Since 1950, there has been a massive increase in the number of species and infectious disease organisms (microbes) being transported across international borders by trade and travel.

- Since 1950, long-lived pollutants such as DDT, PCBs, radioactive particles, and acidic chemicals have been transferred throughout parts of the globe by wind, rainfall patterns, ocean currents, and rivers. On an even larger scale, nations now face the global threats of **(1)** widespread ocean pollution, **(2)** deple-

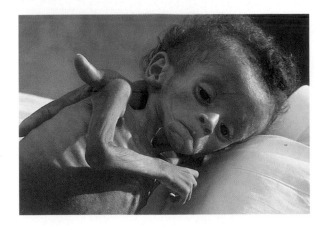

Figure 1-9 One in every three children under age 5, such as this Brazilian child, suffers from malnutrition. According to the World Health Organization, each day at least 13,700 children die prematurely from malnutrition and infectious diseases, most from drinking contaminated water and a weakened condition from malnutrition—an average of 10 preventable deaths each minute. Some analysts put the estimated death toll at almost twice this number. (John Bryson/Photo Researchers, Inc.)

Figure 1-10 Relative ecological footprints of the United States, the Netherlands, and India. An *ecological footprint* is the amount of land needed to produce the resources needed by an average person in a country. The total ecological footprint for the 16 million people living in the Netherlands is 15 times the country's area. Indeed, it would take the land area of about three earths if all the world's 6.1 billion people consumed the same amount of resources as is consumed by the 276 million people in the United States.

tion of ozone in the upper atmosphere (stratosphere) that keeps much of the sun's harmful ultraviolet radiation from reaching the earth's surface, and **(3)** global and regional climate change caused by chemicals released into the environment by human activities.

These numbers are hard to take in. Taken together, they indicate the enormous and increasingly rapid changes that are taking place on a global scale. As a result, the environmental impact or *ecological footprint* of each person on the planet, especially those in developed countries, is large and is growing rapidly (Figure 1-10).

1-3   RESOURCES

What Are Ecological and Economic Resources?
An **ecological resource** is anything an organism needs for normal maintenance, growth, and reproduction. Examples include habitat, food, water, and shelter.

An **economic resource** is anything obtained from the environment to meet human needs and wants. Examples include food, water, shelter, manufactured goods, transportation, communication, and recreation.

Some resources, such as solar energy, fresh air, wind, fresh surface water, fertile soil, and wild edible plants, are directly available for use. Other resources, such as petroleum (oil), iron, groundwater (water found underground), and modern crops, are not directly available. They become useful to us only with some effort and technological ingenuity. For example, petroleum was a mysterious fluid until we learned how to find, extract, and convert (refine) it into gasoline, heating oil, and other products that could be sold at affordable prices.

On our short human time scale, we classify the material resources we get from the environment as **(1)** *perpetual*, **(2)** *renewable*, or **(3)** *nonrenewable* (Figure 1-11; also see the orange boxes in the Concepts and Connections diagram inside the back cover).

What Are Perpetual and Renewable Resources?
Solar energy is called a **perpetual resource** because on a human time scale it is continually renewed. It is

expected to last at least 6 billion years as the sun completes its life cycle.

On a human time scale, a **renewable resource** can be replenished fairly rapidly (hours to several decades) through natural processes as long as it is not used up faster than it is replaced. These resources are *flow resources* that pass through plants, economies, and other systems. They are endlessly renewable but only at a rate at which nature provides them. Examples are **(1)** forests, **(2)** grasslands, **(3)** wild animals, **(4)** fresh water, **(5)** fresh air, and **(6)** fertile soil.

However, renewable resources can be depleted or degraded. The highest rate at which a renewable

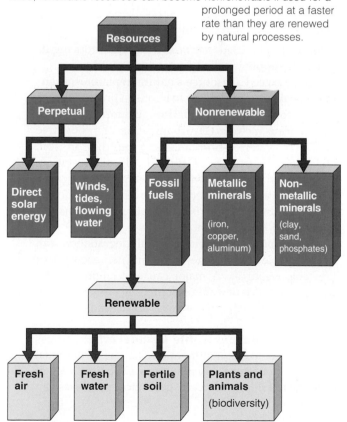

Figure 1-11 Major types of material resources. This scheme is not fixed; renewable resources can become nonrenewable if used for a prolonged period at a faster rate than they are renewed by natural processes.

Free-Access Resources and the Tragedy of the Commons

CONNECTIONS

One cause of environmental degradation is the overuse of **common-property** or **free-access resources.** Such resources are owned by no one (or jointly by everyone in a country or area) but are available to all users at little or no charge.

Examples include **(1)** clean air, **(2)** the open ocean and its fish, **(3)** migratory birds, **(4)** wildlife species, **(5)** publicly owned lands (such as national forests, national parks, and wildlife refuges), **(6)** gases of the lower atmosphere, and **(7)** space.

In 1968, biologist Garrett Hardin (Guest Essay, p. 252) called the degradation of renewable free-access resources the **tragedy of the commons**. It happens because each user reasons, "If I do not use this resource, someone else will. The little bit I use or pollute is not enough to matter and such resources are renewable."

With only a few users, this logic works. However, the cumulative effect of many people trying to exploit a free-access resource eventually exhausts or ruins it. Then no one can benefit from it, and therein lies the tragedy.

Two solutions to this problem are to

■ *Use free-access resources at rates well below their estimated sustainable yields or overload limits by reducing population, regulating access, or both.* This prevention approach is rarely used because **(1)** it entails establishing and enforcing regulations that restrict resource use or population growth, and **(2)** it is difficult and expensive to determine the sustainable yield of a forest, grassland, or animal population because yields vary with weather, climate, and unpredictable biological factors.

■ *Convert free-access resources to private ownership.* The reasoning is that owners of land or some other resource have a strong incentive to protect their investment. However, this approach is not practical for global common resources (such as the atmosphere, the open ocean, most wildlife species, and migratory birds) that cannot be divided up and converted to private property.

Experience shows that there is also another possibility. Just because a resource is easily available to a community does not always mean that people have free and unregulated access to that resource. There are many examples in which communities have established a set of rules and traditions to regulate and share their access to a common-property resource such as fisheries, grazing lands, and forests.

Critical Thinking

Give three examples of how you cause environmental degradation as a result of the tragedy of the commons. How should we deal with this problem? Explain.

---

resource can be used *indefinitely* without reducing its available supply is called **sustainable yield**.

If we exceed a resource's natural replacement rate, the available supply begins to shrink, a process known as **environmental degradation**. Examples of such degradation include **(1)** urbanization of productive land, **(2)** waterlogging and salt buildup in soil, **(3)** excessive erosion of topsoil, **(4)** deforestation, **(5)** depletion of groundwater, **(6)** overgrazing of grasslands by livestock, **(7)** reduction in the earth's forms of wildlife (biodiversity) by elimination of habitats and species, and **(8)** pollution.

Such forms of environmental degradation can change usable renewable resources into nonrenewable or unusable resources. A major cause of environmental degradation of renewable resources is a phenomenon known as the *tragedy of the commons* (Connections, above).

What Are Nonrenewable Resources? Resources that exist in a fixed quantity or stock in the earth's crust are called **nonrenewable resources**. On a time scale of millions to billions of years, geological processes can renew such resources. However, on the much shorter human time scale of hundreds to thousands of years, these resources can be depleted much faster than they are formed.

These exhaustible resources include **(1)** *energy resources* (such as coal, oil, and natural gas, which cannot be recycled), **(2)** *metallic mineral resources* (such as iron, copper, and aluminum, which can be recycled), and **(3)** *nonmetallic mineral resources* (such as salt, clay, sand, and phosphates, which usually are difficult or too costly to recycle).

A **mineral** is any hard, usually crystalline material that is formed naturally. We know how to find and extract more than 100 nonrenewable minerals from the earth's crust. We convert these raw materials into many everyday items and then we discard, reuse, or recycle them.

Figure 1-12 shows the production and depletion cycle of a nonrenewable energy or mineral resource. We never completely exhaust a nonrenewable mineral resource, but such a resource becomes *economically depleted* when the costs of extracting and using what is

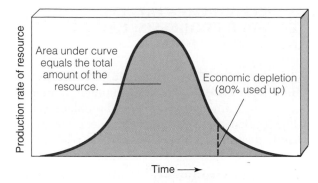

Figure 1-12 Full production and exhaustion cycle of a nonrenewable resource such as copper, iron, oil, or coal. Usually, a nonrenewable resource is considered *economically depleted* when 80% of its total supply has been extracted and used. Normally, it costs too much to extract and process the remaining 20%.

left exceed its economic value. At that point, we have six choices: **(1)** try to find more, **(2)** recycle or reuse existing supplies (except for nonrenewable energy resources, which cannot be recycled or reused), **(3)** waste less, **(4)** use less, **(5)** try to develop a substitute, or **(6)** wait millions of years for more to be produced.

Some nonrenewable material resources, such as copper and aluminum, can be recycled or reused to extend supplies. **Recycling** involves collecting and reprocessing a resource into new products. For example, glass bottles can be crushed and melted to make new bottles or other glass items. **Reuse** involves using a resource over and over in the same form. For example, glass bottles can be collected, washed, and refilled many times.

Recycling nonrenewable metallic resources takes much less energy, water, and other resources and produces much less pollution and environmental degradation than exploiting virgin metallic resources. Reusing such resources takes even less energy and other resources and produces less pollution and environmental degradation than recycling.

Nonrenewable energy resources, such as coal, oil, and natural gas, cannot be recycled or reused. Once burned, the useful energy in these fossil fuels is gone, leaving behind waste heat and polluting exhaust gases.

1-4 POLLUTION

What Is Pollution? Any addition to air, water, soil, or food that threatens the health, survival, or activities of humans or other living organisms is called **pollution**. Pollutants can enter the environment **(1)** naturally (for example, from volcanic eruptions) or **(2)** through human (anthropogenic) activities (for example, from burning coal). Most pollution from human activities occurs in or near urban and industrial areas, where pollutants are concentrated. Industrialized agriculture also is a major source of pollution.

Some pollutants contaminate the areas where they are produced; others are carried by wind or flowing water to other areas. Pollution does not respect local, state, or national boundaries.

Where Do Pollutants Come from and What Types of Harm Do They Cause? There are two types of pollutant sources:

- **Point sources**, where pollutants come from single, identifiable sources. Examples are the **(1)** smokestack of a coal-burning power plant, **(2)** drainpipe of a factory, or **(3)** exhaust pipe of an automobile.

- **Nonpoint sources**, where pollutants come from dispersed (and often difficult to identify) sources. Examples are **(1)** runoff of fertilizers and pesticides (from farmlands, golf courses, and suburban lawns and gardens) into streams and lakes and **(2)** pesticides sprayed into the air or blown by the wind into the atmosphere.

It is much easier and cheaper to identify and control pollution from point sources than from widely dispersed nonpoint sources.

Unwanted effects of pollutants include the following:

- Disruption of life-support systems for humans and other species

- Damage to wildlife, human health, and property

- Nuisances such as noise and unpleasant smells, tastes, and sights

Solutions: What Can We Do About Pollution? There are two basic approaches to dealing with pollution: **(1)** prevent it from reaching the environment or **(2)** clean it up if it does. **Pollution prevention** or **input pollution control** reduces or eliminates the production of pollutants, often by using less harmful chemicals or processes. We can prevent (or at least reduce) pollution by following the four *R*s of resource use: *refuse (do not use)*, *reduce*, *reuse*, and *recycle*.

Pollution cleanup or **output pollution control** involves cleaning up pollutants after they have been produced. Environmentalists have identified three problems with relying primarily on pollution cleanup:

- *It is only a temporary bandage as long as population and consumption levels grow without corresponding improvements in pollution control technology.* For example, adding catalytic converters to car exhaust systems has reduced air pollution. However, increases in the number of cars and in the total distance each travels have reduced the effectiveness of this cleanup approach.

- *It often removes a pollutant from one part of the environment only to cause pollution in another.* For example, we can collect garbage, but the garbage is then (1) burned (perhaps causing air pollution and leaving a toxic ash that must be put somewhere), (2) dumped into streams, lakes, and oceans (perhaps causing water pollution), or (3) buried (perhaps causing soil and groundwater pollution).

- *Once pollutants have entered and become dispersed into the environment at harmful levels, it usually costs too much to reduce them to acceptable levels.*

Both pollution prevention and pollution cleanup are needed. However, environmentalists and some economists urge us to emphasize prevention because it works better and is cheaper than cleanup. As Benjamin Franklin reminded us long ago, "An ounce of prevention is worth a pound of cure." An increasing number of businesses have found that *pollution prevention pays.*

Governments can encourage both pollution prevention and pollution cleanup by

- Using incentives such as various subsidies and tax write-offs

- Using regulations and taxes

Most analysts believe that a mix of both approaches is best because excessive regulation and too much taxation can cause a political backlash. Achieving the right balance is difficult.

1-5 ENVIRONMENTAL AND RESOURCE PROBLEMS: CAUSES AND CONNECTIONS

What Are Key Environmental Problems and Their Root Causes? We face a number of interconnected environmental and resource problems (Figure 1-13). The first step in dealing with these problems is to identify their underlying causes (Figure 1-14).

How Are Environmental Problems and Their Causes Connected? Once we have identified environmental problems and their root causes, the next step is to understand how they are connected to one another. The three-factor model in Figure 1-15 is a starting point.

According to this simple model, the environmental impact (I) of population on a given area depends on three key factors: (1) the number of people (P), (2) average resource use per person (affluence, A), and (3) the environmental effects of the technologies used to provide and consume each unit of resource (T).

In developing countries, population size and the resulting degradation of potentially renewable resources (as the poor struggle to stay alive) tend to be the key factors in total environmental impact (Figure 1-15, top). In such countries per capita resource use is low.

In developed countries, high rates of per capita resource use and the resulting high levels of pollution and environmental degradation per person usually are the key factors determining overall environmental impact (Figure 1-15, bottom). For example, it is estimated that the average U.S. citizen consumes 35 times as much as the average citizen of India and 100 times as much as the average person in the world's poorest countries. *Thus, poor parents in a developing country would need 70–200 children to have the same lifetime resource consumption as 2 children in a typical U.S. family.*

Some forms of technology, such as polluting factories and motor vehicles and energy-wasting devices, increase environmental impact by raising the T factor in the equation. Other technologies, such as pollution control, solar cells, and energy-saving devices, lower environmental impact by decreasing the T factor in the equation. In other words, some forms of technology

Air Pollution
- Global climate change
- Stratospheric ozone depletion
- Urban air pollution
- Acid deposition
- Outdoor pollutants
- Indoor pollutants
- Noise

Biodiversity Depletion
- Habitat destruction
- Habitat degradation
- Extinction

Major Environmental Problems

Water Pollution
- Sediment
- Nutrient overload
- Toxic chemicals
- Infectious agents
- Oxygen depletion
- Pesticides
- Oil spills
- Excess heat

Waste Production
- Solid waste
- Hazardous waste

Food Supply Problems
- Overgrazing
- Farmland loss and degradation
- Wetlands loss and degradation
- Overfishing
- Coastal pollution
- Soil erosion
- Soil salinization
- Soil waterlogging
- Water shortages
- Groundwater depletion
- Loss of biodiversity
- Poor nutrition

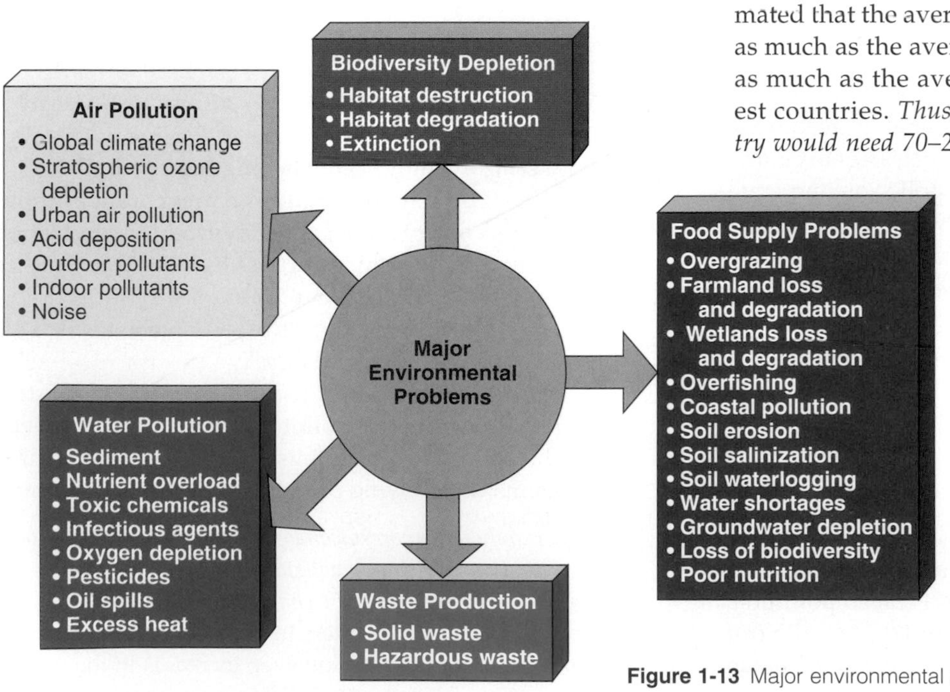

Figure 1-13 Major environmental and resource problems.

Figure 1-14 Environmentalists have identified five root causes of the environmental problems we face.

- Rapid population growth
- Unsustainable resource use
- Poverty
- Not including the environmental costs of economic goods and services in their market prices
- Trying to manage and simplify nature with too little knowledge about how it works

are *environmentally harmful* and some are *environmentally beneficial*.

For example, the greater affluence and development of environmentally beneficial technologies have made it possible for people in developed countries to devote more financial and creative resources to improving environmental quality. In addition, these environmentally beneficial technologies can be transferred to developing countries to help humanity lighten its ecological footprint on the world (Figure 1-10).

The three-factor model in Figure 1-15 can help us understand how key environmental problems and some of their causes are connected. However, these problems involve a number of poorly understood interactions among many more factors than those in this simplified model, as outlined in Figure 1-16. A more detailed model is given inside the back cover.

1-6 IS OUR PRESENT COURSE SUSTAINABLE?

Are Things Getting Better or Worse? There is good news and bad news about the environmental problems we face, as summarized in more detail in Appendix 3. Experts have conflicting views about how serious our

Developing Countries

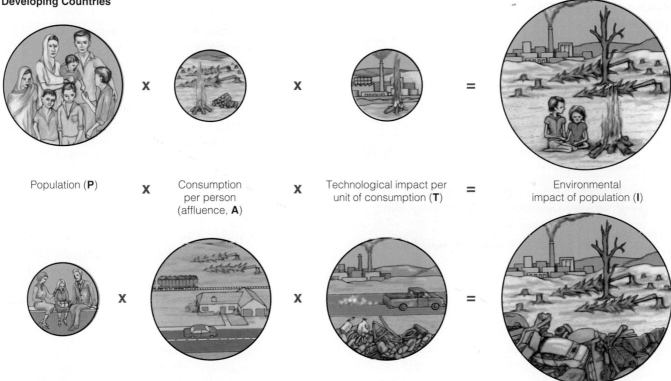

Population (**P**) X Consumption per person (affluence, **A**) X Technological impact per unit of consumption (**T**) = Environmental impact of population (**I**)

Developed Countries

Figure 1-15 Simplified model of how three factors—population, affluence, and technology—affect the environmental impact of population in developing countries (top) and developed countries (bottom). According to this model, the damage we do to the earth's life-support systems (environmental impact, **I**) is equal to **(1)** the number of people (population, **P**), **(2)** multiplied by the amount of resources each person uses (affluence, **A**), **(3)** multiplied by the environmental effects of the technologies used to provide and consume each unit of resource (technology, **T**). Circle size shows the relative importance of each factor. We can reduce the size of the *T* factor by improving technology for controlling and preventing pollution, resource waste, and environmental degradation.

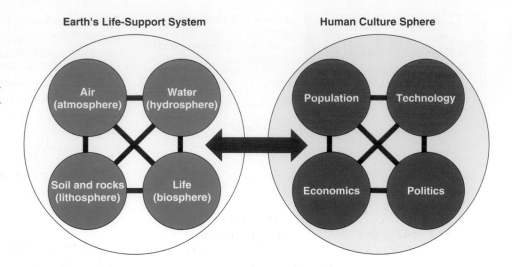

Figure 1-16 Major components and interactions within and between the earth's life-support system and the human sociocultural system (culture sphere). The goal of environmental science is to learn as much as possible about these complex interactions.

Earth's Life-Support System

Air (atmosphere)

Water (hydrosphere)

Soil and rocks (lithosphere)

Life (biosphere)

Human Culture Sphere

Population

Technology

Economics

Politics

population and environmental problems are and what we should do about them.

Some analysts believe that human ingenuity and technological advances will allow us **(1)** to clean up pollution to acceptable levels, **(2)** find substitutes for any resources that become scarce, and **(3)** keep expanding the earth's ability to support more humans, as we have done in the past. They accuse environmentalists of exaggerating the seriousness of the problems we face and failing to appreciate the progress we have made in improving quality of life and protecting the environment (see Appendix 3).

According to the late Julian L. Simon, a professor of economics and business administration and critic of environmentalists,

In almost every respect important to humanity, the trends have been improving not deteriorating. Here are a few encouraging examples:

- *People are living longer and living more healthy than ever before.*
- *A resource shortage usually leaves us better off than if the shortage had never arisen. For example, if firewood had not become scarce in 17th-century England, coal would not have been developed.*

- *The prices of food, metals, and other raw materials have been declining since the beginning of the 19th century.*

- *Many people are still hungry, but the food supply has been improving since World War II.*

- *There is no statistical evidence for predictions of rapid losses of plant and animal wildlife species in the near future.*

- *Threats of air and water pollution have been vastly overblown.*

I do not say that all is well for everywhere and that all will be rosy in the future. However, I am confident

that the nature of the physical world permits continued improvement in humankind's economic lot. We now have in our hands the technology to feed, clothe, and supply energy to an ever-growing population for the next 7 billion years.*

On the other hand, environmentalists and many leading scientists contend that we are disrupting the earth's life-support system for us and other forms of life at an accelerating rate (Guest Essay, p. 18, and Appendix 3).

In 1991, the Ecological Society of America, made up of many of the world's leading ecologists, issued the following warning to humanity:

Environmental problems resulting from human activities have begun to threaten the sustainability of Earth's life-support systems.

On November 18, 1992, some 1,680 of the world's senior scientists from 70 countries, including 102 of the 196 living scientists who are Nobel laureates, signed and sent an urgent "World Scientists' Warning to Humanity" to government leaders of all nations. According to this warning,

Our massive tampering with the world's interdependent web of life—coupled with the environmental damage inflicted by deforestation, species loss, and climate change—could trigger widespread adverse effects,

---

*Simon later said that he meant to say that there would be enough resources for the world's population to keep growing at its current rate for 7 million years. However, Albert A. Bartlett, a physicist at the University of Colorado, illustrated the absurdity of this estimate. He calculated that if the current world population of 6.1 billion people continued to grow at a rate of 1% per year (the current growth rate is 1.5%), in only 17,000 years the number of people on the earth would be equal to all the atoms estimated to be in the universe.

including unpredictable collapses of critical biological systems whose interactions and dynamics we only imperfectly understand. . . . No more than one or a few decades remain before the chance to avert the threats we now confront will be lost and the prospects for humanity immeasurably diminished.

Also in 1992, the U.S. National Academy of Sciences and the Royal Society of London, two of the world's leading scientific organizations, issued a joint statement titled "Population Growth, Resource Consumption and a Sustainable World." According to this statement,

If current predictions of population growth prove accurate and patterns of human activity on the planet remain unchanged, science and technology may not be able to prevent either irreversible degradation of the environment or continued poverty for much of the world. . . . Sustainable development can be achieved, but only if irreversible degradation of the environment can be halted in time.

These three major warnings are not the views of a small number of scientists but the consensus of the mainstream scientific community, consisting of most of the world's key researchers on environmental problems.

Whom Should We Believe? A Clash of Environmental Worldviews There is no easy answer to this question. Conflicts over how serious our environmental problems are and what we should do about them arise mostly out of differing **environmental worldviews**: how people think the world works, what they think their role in the world should be, and what they believe is right and wrong environmental behavior (**environmental ethics**).

People with widely differing environmental worldviews or beliefs can take the same data, be logically consistent, and arrive at quite different conclusions (see Appendix 3) because they start with different assumptions.

There are many different environmental worldviews, as discussed in more detail in Chapter 28. However, most are variations of two major opposing environmental worldviews.

Most people in today's industrial consumer societies have a **planetary management worldview**, which has become increasingly common in the past 50 years. According to this environmental worldview, human beings, as the planet's most important and dominant species, can and should manage the planet mostly for their own benefit.

The basic environmental beliefs of this worldview include the following:

- *We are the planet's most important species, and we are apart from and in charge of the rest of nature.*

- *The earth has an unlimited supply of resources for use by us through science and technology.* If we deplete a resource, we will find substitutes. To deal with pollutants, we can invent technology to clean them up, dump them into space, or move into space ourselves. If we extinguish other species, we can use genetic engineering to create new and better ones.

- *Economic growth increases human well-being, and the potential for economic growth is essentially limitless.*

- *Our success depends on how well we can understand, control, and manage the earth's life-support systems for our benefit.*

This worldview is widely supported because it is considered to be the primary driving force behind the major improvements in the human condition during the past 300 years and especially during the last 50 years. Several variations of this environmental worldview are discussed in Section 28-1, p. 741.

Another environmental worldview, known as the **environmental wisdom worldview,** is based on the following major beliefs, which are the opposite of those on which the planetary management worldview is based:

- *We are part of nature, and nature does not exist just for us.* We need the earth, but the earth does not need us.

- *The earth's resources are limited, should not be wasted, and should be used efficiently and sustainably for us and other species.*

- *Some forms of technology and economic growth are environmentally beneficial and should be encouraged, but some are environmentally harmful and should be discouraged.*

- *Our success depends on (1) learning how the earth sustains itself and adapts to ever-changing environmental conditions and (2) integrating such scientific lessons from nature (environmental wisdom) into the ways we think and act.*

Many of the ideas in this environmental worldview are incorporated in the concept of environmentally sustainable economic development (p. 7). Several variations of this worldview are discussed in Section 28-2, p 742.

People with this and related environmental worldviews call for us to launch an *environmental or sustainability revolution* to take place over the next 50 years (Guest Essay, p. 18). This new cultural change would involve shifting our efforts from

- Pollution cleanup to pollution prevention (cleaner production)

- Waste disposal (mostly burial and burning) to waste prevention and reduction

- Protecting species to protecting the places (habitats) where they live

- Environmental degradation to environmental restoration

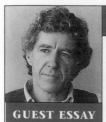

Lester R. Brown

Lester R. Brown is chairman of the board of the Worldwatch Institute, a private non-profit research institute he founded in 1974 that is devoted to analyzing global environmental issues. Under his leadership, since 1984 the institute has published the annual State of the World Report, *considered by many environmentalists and world leaders to be the best source of information about key environmental issues. It also publishes monographs on specific topics,* World Watch *magazine, and a series of* Environmental Alert *books. He has been awarded 20 honorary degrees and is author of 18 books. In addition, he has received more than 30 national and international prizes and awards, including the McArthur Foundation Genius Award, the 1987 UN Environment Prize, and the 1994 Blue Planet Prize. He has been described by the* Washington Post *as "one of the world's most influential thinkers."*

Two challenges facing us in this new century are that the human population is 4 times as large as it was a century ago [Figure 1-1] and the world economy is 17 times as large [Figure 1-7]. This growth has allowed advances in living standards that our ancestors could not have imagined, but it has also undermined natural systems in ways they could not have feared. The policy decisions we make in the years immediately ahead will determine whether our children live in a world of development or decline.

There is no precedent for the rapid and substantial change we need to make. Building an environmentally sustainable future depends on **(1)** a restructuring of the global economy so that it does not destroy or degrade its natural support systems, **(2)** major shifts in human

reproductive behavior, and **(3)** dramatic changes in values and lifestyles all within a few decades. If this *environmental* or *sustainability revolution* succeeds, it will rank as one of the great economic and social transformations in human history.

The two overriding challenges facing our global civilization as the new century begins are to stabilize population and stabilize climate. The exciting thing about the population and climate challenges is that we already have the knowledge and technologies to succeed at both.

The key to stabilizing world population is for national governments to formulate strategies for stabilizing population humanely rather than waiting for nature to intervene with its inhumane methods based on rising death rates, as in Africa. Once these strategies are developed, it is in the interest of the international community to support the stabilization effort.

Stabilizing climate means shifting away from a fossil fuel- or carbon-based economy to a solar-hydrogen economy. This new energy system taps a mix of renewable sources of energy from the sun, such as direct sunlight, hydropower, wind power, and wood. Electricity produced by wind turbines and photovoltaic (solar) cells can be used to produce clean-burning hydrogen gas from water to meet most of our energy needs. Making this shift is the greatest investment opportunity in history. Its success depends on stabilizing human population size to reestablish a balance between people and the natural systems on which they depend.

Caught up in the excitement of unprecedented economic growth and the information economy, many people have lost sight of the deterioration of the earth's environmental systems and resources. Whereas eco-

- Increased resource use to more efficient (less wasteful) resource use

- Population growth to population stabilization

The Solutions box (p. 20) gives some guidelines various analysts have suggested for living more sustainably by using environmental wisdom to work with the earth.

What Can We Do When Faced with Fundamental Disagreement and Uncertainty? Suppose a doctor tells a smoker that if she continues to smoke, she risks getting lung cancer. No one can be sure that this will happen until it does. If she assumes "it will never happen to me" and continues the risk of smoking because of the pleasure it provides, she could be right (a good result) or wrong (a catastrophic result). *This is a win big or lose big strategy.*

However, she could take a *precautionary approach* and stop smoking. If right, she has probably saved her life. If wrong, she has given up some pleasure and

saved a lot of money to avoid taking an unnecessary risk. This is a *win big or win pretty good strategy*.

This choice is similar to the one we face over conflicting environmental worldviews, except on a global scale. The problem is that there is *fundamental uncertainty* about which of the two general types of opposing worldviews is right because each group starts from fundamentally different assumptions.

Technological optimists supporting the *planetary management worldview* are convinced they are right. They point to how far the human species has come and assume that such technological progress can be sustained indefinitely. If they are right, things will get better and better (*we win big*). If they are wrong, we are headed for ecological and economic disaster (*we lose big*).

Technological skeptics with environmental wisdom worldviews also are convinced that they are right. They assume that **(1)** we cannot use technology to find adequate or affordable substitutes for the natural capital that supports the earth's life and economies, **(2)** there are

nomic indicators such as investment, production, and trade are consistently positive, the key environmental indicators are increasingly negative. Forests are shrinking, water tables are falling, soils are eroding, plant and animal species are disappearing, wetlands are disappearing, fisheries are collapsing, rangelands are deteriorating, rivers are running dry, coral reefs are dying, temperatures are rising, glaciers are melting, and there are more destructive storms and more hungry people in the world today than ever before. The global economy as now structured cannot continue to expand much longer if the natural systems on which it depends continue to deteriorate at the current rate.

On the economic front, the signs are equally ominous: Soil erosion, deforestation, and overgrazing are adversely affecting farming, forestry, and livestock productivity and slowing overall economic growth in agriculturally based economies. The decline in living standards that was once predicted by some ecologists from the combination of continuing rapid population growth and spreading environmental degradation has become a reality for one-sixth of humanity.

Suppose we used a more comprehensive system of national economic accounting that incorporated losses of natural capital [Guest Essay, p. 6], such as topsoil and forests, the destruction of productive grasslands, the extinction of plant and animal species, and the health costs of air and water pollution and increased ultraviolet radiation. Such an analysis might well show that most of humanity suffered a decline in living conditions since the 1980s.

The key environmental limits we face are fresh water, forests, rangelands, ocean fisheries, biological diversity, and the global atmosphere. Our numbers expand, but earth's natural systems do not. Will we recognize the world's natural limits and adjust our economies accordingly, or will we expand our ecological footprint [Figure 1-10] until it is too late? Nature has no reset button.

Some good news is that there is a growing worldwide recognition outside the environmental community that the economy we now have cannot take us where we want to go. Three decades ago, only environmental activists were speaking out on the need for change. Now the ranks of activists have broadened to include CEOs of major corporations, government ministers, prominent scientists, and intelligence agencies.

The goal is to develop a new type of economy. It is a solar-powered, bicycle- and rail-centered, reuse and recycle economy that uses energy, water, land, and materials much more efficiently and wisely than we do today. In addition to helping sustain the earth's life-support systems, such an economy can lead to greater economic security, healthier lifestyles, and a worldwide improvement in the human condition. This challenge to humanity rivals any in our history.

Critical Thinking

1. Do you agree with the author that we need to bring about an environmental or sustainability revolution within a few decades? Explain.

2. Do you believe that this can be done by making minor adjustments in the global economy or by restructuring the global economy to put less strain on the earth's natural systems? Explain.

limits to the growth of the human population and resource use on a finite planet, **(3)** past technological advances can't automatically be extrapolated as a straight line of advancement into the future, and **(4)** the future will be full of mostly unpredictable discontinuities and ecological and economic surprises as we continue stressing the earth's life-support systems.

These skeptics are not opposed to technology. However, they call for us to use our economic and political systems **(1)** to encourage environmentally beneficial forms of technology and economic growth and **(2)** to discourage environmentally harmful forms of technology and economic growth.

Suppose the technological skeptics are right. If we take their advice and they are right, we will be much better off than if ecological and economic disaster occurred because we took the advice of the technological optimists and they were wrong (*we win big*).

However, suppose we take the advice of the technological skeptics and they are wrong. Then we would still be in good shape because most of the things that they call for will improve human health and the health of the earth's life-support systems (*we win pretty good*).

Thus, in the face of the *fundamental uncertainty* about which view is correct, we have the best chance of avoiding possible environmental and economic catastrophe by going with the technological skeptics. This is a *precautionary* position. It assumes that when gambling with the earth's life-support system for us and other species a *win big or win pretty good strategy* is better than a *win big or lose big strategy*. In gambling at this planetary level, *we get only one time up at bat*, and *nature always bats last and owns the stadium*.

This chapter has presented an overview of the problems most environmentalists and many of the world's most prominent scientists believe we face and their root causes. It has also summarized the controversy over how serious environmental problems are and presented two opposing environmental worldviews. The rest of this book presents a more detailed

Some Guidelines for Working with the Earth

- ■ Leave the earth as good as or better than we found it.
- ■ Take no more than we need.
- ■ Try not to harm life, air, water, or soil.
- ■ Sustain the variety of the earth's life forms (biodiversity).
- ■ Help maintain the earth's capacity for self-repair and adaptation.
- ■ Do not use potentially renewable resources (soil, water, forests, grasslands, and wildlife) faster than they are replenished.
- ■ Do not waste resources.
- ■ Do not release pollutants into the environment faster than the earth's natural processes can dilute or recycle them.
- ■ Emphasize pollution prevention and waste reduction.
- ■ Slow the rate of population growth.
- ■ Have the market prices of all goods and services include all of their harmful environmental costs.
- ■ Reduce poverty.

Specific things you can do to work with the earth by trying to implement such guidelines are listed in Appendix 6.

Critical Thinking

Which of these guidelines do you agree with and which do you disagree with? Why? Can you add any other guidelines?

---

analysis of these problems, the controversies they have created, and solutions analysts propose.

Try not to be overwhelmed or immobilized by the *bad environmental news* because there is also some *great environmental news*. We are learning a great deal about how nature works and sustains itself, and we have numerous scientific, technological, and economic solutions available to deal with the environmental problems we face, as you will learn in this book.

The challenge is to make creative use of our economic and political systems to implement such solutions. One key is to recognize that most economic and political change comes about as a result of individual actions and individuals acting together to bring about change. Anthropologist Margaret Mead summarized our potential for change: "Never doubt that a small group of thoughtful, committed citizens can change the world. Indeed, it is the only thing that ever has."

We live in exciting times during what might be called a *hinge of cultural history*. Indeed, if I had to pick a time to live, it would be the next 50 years as we face the challenge of developing more environmentally sustainable societies.

What's the use of a house if you don't have a decent planet to put it on?

HENRY DAVID THOREAU

REVIEW QUESTIONS

1. Define the boldfaced terms found in this chapter.

2. What is *exponential growth*? What is the connection between exponential growth and environmental problems?

3. Distinguish between *environment, ecology,* and *environmental science.* Distinguish between *ecologists, environmental scientists, conservation biologists, environmentalists, preservationists, conservationists,* and *restorationists.*

4. Distinguish between *solar capital* and *natural capital (natural resources).*

5. What are your *basic needs*? What is an *environmentally sustainable society*? Distinguish between living on principal and living on interest and relate this to the sustainability of **(a)** the earth's life-support system and **(b)** your lifestyle.

6. Distinguish between *exponential growth* and *linear growth,* and give an example of each type.

7. What are *doubling time* and the *rule of 70*? Use the rule of 70 to calculate how many years it would take for the population of a country to double if it was growing at 2% per year.

8. Define *economic growth, gross national product, gross domestic product, per capita GNP,* and *economic development.* Distinguish between *developed countries* and *developing countries.*

9. What is *environmentally sustainable economic development*? How does it differ from traditional economic growth and economic development?

10. What is the *wealth gap*? Describe changes in the wealth gap since 1960 and connect them to environmental degradation and premature deaths of poor people.

11. Why does it make sense for a poor family to have a large number of children?

12. Distinguish between *ecological resources* and *economic resources.* Distinguish between *perpetual resources, renewable resources,* and *nonrenewable resources,* and give an example of each.

13. What are *sustainable yield* and *environmental degradation*? Give four examples of environmental degradation.

14. Define and give three examples of *common-property resources.* What is the *tragedy of the commons*? Give three examples of this tragedy on a global scale and explain how your lifestyle contributes to these examples. List three ways to deal with the tragedy of the commons.

15. What is a *mineral*? Distinguish between a *physically depleted nonrenewable resource* and an *economically depleted nonrenewable resource*. Distinguish between *reuse* and *recycling*. Draw a depletion curve for a nonrenewable resource.

16. What is *pollution*? Distinguish between *point sources* and *nonpoint sources* of pollution. Distinguish between *pollution prevention* (*input pollution control*) and *pollution cleanup* (*output pollution control*). What are three problems with relying primarily on pollution cleanup? Why is pollution prevention better than pollution control?

17. According to environmentalists, what are five root causes of the environmental problems we face?

18. Describe a simple model of relationships between population, resource use per person, resource use technology, and overall environmental impact. How do these factors differ in developed and developing countries?

19. What is an *environmental worldview*? Distinguish between the *planetary management* and *environmental wisdom environmental worldviews*. Analyze these world views in terms of *win-big, lose-big*, and *win pretty good strategies*.

20. List six major changes that environmentalists believe should take place over the next 50 years as part of an *environmental* or *sustainability revolution*.

21. List major guidelines that environmentalists have suggested for working with the earth.

CRITICAL THINKING

1. Do you believe that the society you live in is on an unsustainable path? Explain. Do you believe that it is possible for the society you live in to become a sustainable society within the next 50 years? Explain.

2. Do you favor instituting policies designed to reduce population growth and stabilize **(a)** the size of the world's population as soon as possible and **(b)** the size of the U.S. population (or the population of the country where you live) as soon as possible? Explain. If you agree that population stabilization is desirable, what three major policies do you believe should be implemented to accomplish this goal?

3. Explain why you agree or disagree with the following propositions:

a. High levels of resource use by the United States and other developed countries are more beneficial than harmful.
b. The economic growth from high levels of resource use in developed countries provides money for more financial aid to developing countries for reducing pollution, environmental degradation, and poverty.
c. Stabilizing population is not desirable because without more consumers, economic growth would stop.
d. The world will never run out of potentially renewable resources and most currently used nonrenewable resources because technological innovations will produce substitutes, reduce resource waste, or allow use of lower grades of scarce nonrenewable resources.

4. List three forms of economic growth that you believe are environmentally unsustainable and three forms that you believe are environmentally sustainable.

5. Is the wealth gap good or bad for **(a)** you, **(b)** a poor person, and **(c)** the environment? If you believe that the wealth gap is bad, what three things should be done to narrow this gap? What obligation, if any, should a wealthy person, corporation, or country have to **(a)** reduce poverty and **(b)** improve environmental quality? Explain.

6. When you read that at least 27,400 human beings die prematurely each day (19 per minute) from preventable malnutrition and infectious disease, do you **(a)** doubt whether it is true, **(b)** not want to think about it, **(c)** feel hopeless, **(d)** feel sad, **(e)** feel guilty, or **(f)** want to do something about this problem?

7. How do you feel when you read that **(1)** the average American consumes about 50 times more resources than the average Chinese citizen, **(2)** human activities lead to the premature extinction of at least 10 species per day, **(3)** humans have disturbed about 73% of the earth's habitable land (Figure 1-4, p. 8), and **(4)** human activities are projected to make the earth's climate warmer: **(a)** skeptical about their accuracy, **(b)** indifferent, **(c)** sad, **(d)** helpless, **(e)** guilty, **(f)** concerned, or **(g)** outraged? Which of these feelings help perpetuate such problems and which can help alleviate these problems?

8. Do you agree or disagree with the five root causes of environmental problems listed in Figure 1-14, p. 15? Explain. List any other root causes you believe should be added.

9. Explain how during his or her lifetime a single child born in the United States can have a greater environmental impact than about 35 children born in India and 100 children born in one of Africa's poorest countries.

10. One of the tragic characters in Greek mythology is Cassandra. The god Apollo gave her the gift of being able to foretell the future but then added the curse that no one would believe her. Have the environmental problems we face been **(a)** overblown by prophets of doom (Cassandras) or **(b)** reduced in severity because enough people have listened to their warnings and acted to prevent the prophecies from coming true? Explain.

11. Explain why you agree or disagree with **(a)** the position of Julian Simon on p. 16 and **(b)** the 1992 warning to humanity from the world's senior scientists (p. 16).

12. Explain why you agree or disagree with each of the following statements: **(a)** humans are superior to other forms of life; **(b)** humans are in charge of the earth; **(c)** all economic growth is good; **(d)** the value of other species depends only on whether they are useful to us; **(e)** because all species eventually become extinct we should not worry about whether our activities cause the premature extinction of a species; **(f)** all species have an inherent right to exist; **(g)** nature has an almost unlimited storehouse of resources for human use; **(h)** technology can solve our environmental problems; **(i)** I do not believe I have any obligation to future generations; and **(j)** I do not believe I have any obligation to other species.

13. Explain why you agree or disagree with each of the beliefs of **(a)** the planetary management worldview on p. 17 and **(b)** the environmental wisdom worldview on p. 17.

14. What are the basic beliefs of your environmental worldview? Are the beliefs of your environmental worldview consistent with your answers to question 12? Are your environmental actions consistent with your environmental worldview?

15. Do you believe that your current lifestyle is sustainable? If the answer is yes, explain why and include the impact of the world's other 6.1 billion people on your ability to sustain your current lifestyle. If your answer is no, explain why and list five things you could do now to make your lifestyle more sustainable (see Appendix 6). Which of these things do you actually plan to do?

PROJECTS*

1. What are the major resource and environmental problems in **(a)** the city, town, or rural area where you live and **(b)** the state where you live? Which of these problems affect you directly? Have these problems gotten better or worse during the last 10 years?

2. Roughly what percentages of key resources such as water, food, and energy used by your local community come from the following places: nearby, another state, or another country? How do these resource inputs affect the long-term environmental and economic sustainability of your community?

3. Make a list of the resources you truly need. Then make another list of the resources you use each day only because you want them. Finally, make a third list of resources you want and hope to use in the future. Compare your lists with those compiled by other members of your class and relate the overall result to the tragedy of the commons.

4. Most environmentalists favor the concept of sustainable development. However, some environmentalists contend that it is being used by some governments and businesses as a verbal smokescreen to continue various forms of unsustainable development. Use the library and the internet to evaluate the validity of this charge.

5. When the "World Scientists' Warning to Humanity" was released to the press it was almost ignored by the television networks and major papers in the United States and Canada, with the *Washington Post* and the *New York Times* in the United States rejecting the story as not newsworthy. Use the library or internet to see what the major news stories were in the *Washington Post* and the *New York Times* on November 18, 1992, and evaluate their importance relative to "World Scientists' Warning to Humanity."

6. Use the library or the internet to find out bibliographic information about *Lynton B. Caldwell, Barbara*

Ward, René Dubos, and *Henry David Thoreau,* whose quotes appear at the beginning and end of this chapter.

7. Write two-page scenarios describing what your life and that of any children you choose to have might be like 50 years from now if **(a)** we continue on our present path or **(b)** we shift to more environmentally sustainable societies throughout most of the world.

8. Make a concept map of this chapter's major ideas, using the section heads and subheads and the key terms (in boldface). Look at the inside back cover and on the website for this book for information about making concept maps.

INTERNET STUDY RESOURCES AND RESOURCES FOR FURTHER READING AND RESEARCH

The website for this book contains helpful study aids and many ideas for further reading and research. Log on to:

http://www.brookscole.com/product/0534376975s

and click on the Chapter-by-Chapter area. Choose Chapter 1 and select a resource:

- "Flash Cards" allows you to test your mastery of the Terms and Concepts to Remember for this chapter.

- "Tutorial Quizzes" provides a multiple-choice practice quiz.

- "Student Guide to InfoTrac" will lead you to Critical Thinking Projects that use InfoTrac College Edition as a research tool.

- "References" lists the major books and articles consulted in writing this chapter.

- "Hypercontents" takes you to an extensive list of sites with news, research, and images related to individual sections of the chapter.

INFOTRAC COLLEGE EDITION

Improve your skills with InfoTrac College Edition, a searchable online database of articles from more than 700 periodicals. Log on to:

http://www.infotrac-college.com

or access InfoTrac through the website for this book.

Try the following articles:

Kates, R.W. 2000. Population and consumption: what we know, what we need to know. *Environment* vol. 42, no. 3, pp. 10–19. (subject guide: environmental problems)

Maniates, M.F. and J.C. Whissel. 2000. Environmental studies: the sky is not falling. *BioScience* vol. 50, no. 6, pp. 509–517. (subject guide: environmental sciences)

---

*These are either laboratory exercises or individual or class projects.

2 ENVIRONMENTAL HISTORY: AN OVERVIEW

Near Extinction of the American Bison

In 1500, before Europeans settled North America, 30–60 million North American bison grazed the plains, prairies, and woodlands over much of the continent.

These animals were once to numerous that in 1932 a traveler wrote, "As far as my eye could reach the country seemed absolutely blackened by innumerable herds." A single herd on the move might thunder past for hours.

For centuries, several Native American tribes depended heavily on bison, and typically they killed only the animals they needed for food, clothing, and shelter. The dried feces of these animals, known as "buffalo chips," were used as fuel. This state of ecological balance was maintained because these Native Americans hunted only with lances and bows and arrows, and occasionally drove the bison over cliffs.

By 1906, however, the once-vast range of the bison had shrunk to a tiny area, and the species had been driven nearly to extinction (Figure 2-1). How did this happen? First, settlers moving west after the Civil War upset the sustainable balance between Native Americans and bison. Several plains tribes traded bison skins to settlers for steel knives and firearms, so they began killing more bison.

However, the new settlers caused the most relentless slaughter. As railroads spread westward in the late 1860s, railroad companies hired professional bison hunters—including Buffalo Bill Cody—to supply construction crews with meat. Passengers also gunned down bison from train windows for sport, leaving the carcasses to rot.

Commercial hunters shot millions of bison for their hides and tongues (considered a delicacy), leaving most of the meat to rot. "Bone pickers" collected the bleached bones that whitened the prairies and shipped them east to be ground up as fertilizer.

Farmers shot bison because they damaged crops, fences, telegraph poles, and sod houses. Ranchers killed them because they competed with cattle and sheep for pasture. The U.S. Army killed at least 12 million bison as part of its campaign to subdue the plains tribes by killing off their primary source of food.

Between 1870 and 1875 at least 2.5 million bison were slaughtered each year. Only 85 bison were left by 1892. They were given refuge in Yellowstone National Park and protected by an 1893 law against the killing of wild animals in national parks.

In 1905, 16 people formed the American Bison Society to protect and rebuild the captive population. Soon thereafter, the federal government established the National Bison Range near Missoula, Montana. Today there are an estimated 200,000 bison, about 97% of them on privately owned ranches.

Some wildlife conservationists have suggested restoring large herds of bison on public lands in the North American plains. This idea has been strongly opposed by ranchers with permits to graze cattle and sheep on federally managed lands. The history of humanity's relationship to the environment provides many important lessons that can help us understand and deal with today's environmental problems and not repeat past mistakes.

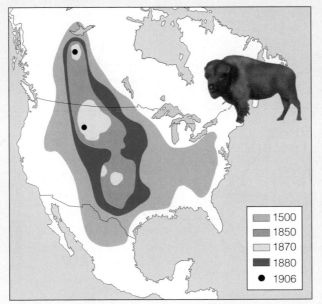

1500
1850
1870
1880
● 1906

Figure 2-1 The historical range of the bison shrank severely between 1500 and 1906, mostly because of unregulated and deliberate over hunting.

A continent ages quickly once we come.
ERNEST HEMINGWAY

This chapter addresses the following questions:

- What major effects have hunter-gatherer societies, agricultural societies, and industrialized societies had on the environment? What might be the environmental impact of the current information and globalization revolution?

- What are the major phases in the history of land and wildlife conservation, public health, and environmental protection in the United States?

- What is Aldo Leopold's land ethic?

2-1 CULTURAL CHANGES AND THE ENVIRONMENT

What Major Human Cultural Changes Have Taken Place? Evidence from fossils and studies of ancient cultures suggests that the current form of our species, *Homo sapiens sapiens*, has walked the earth for only about 60,000 years (some recent evidence suggests 90,000 to 176,000 years), an instant in the planet's estimated 4.6-billion-year existence.

Until about 12,000 years ago, we were mostly hunter-gatherers who typically moved as needed to find enough food for survival. Since then, there have been three major cultural changes: **(1)** the *agricultural revolution* (which began 10,000–12,000 years ago), **(2)** the *industrial revolution* (which began about 275 years ago), and **(3)** the *information and globalization revolution* (which began about 50 years ago).

These major cultural changes have

- Given us much more energy and new technologies with which to alter and control more of the planet to meet our basic needs and increasing wants

- Allowed expansion of the human population, mostly because of increased food supplies and longer life spans (Figure 2-2)

- Increased our environmental impact because of increased resource use, pollution, and environmental degradation

How Did Ancient Hunting-and-Gathering Societies Affect the Environment? During most of our 60,000-year existence, we were **hunter-gatherers** who survived by collecting edible wild plant parts, hunting, fishing, and scavenging meat from animals killed by other predators. Our hunter-gatherer ancestors typically lived in small bands (of fewer than 50 people) who worked together to get enough food to survive. Many groups were *nomadic*, picking up their few possessions and moving seasonally from place to place to find enough food.

The earliest hunter-gatherers (and those still living this way today) survived through expert knowledge and understanding of their natural surroundings. They discovered **(1)** a variety of plants and animals that could be eaten and used as medicines, **(2)** where to find water, **(3)** how plant availability changed throughout the year, and **(4)** how some game animals migrated to get enough food. Because of high infant mortality and an estimated average life span of 30–40 years, hunter-gatherer populations grew very slowly (Figure 2-2).

Advanced hunter-gatherers had a greater impact on their environment than did early hunter-gatherers. They **(1)** used more advanced tools and fire to convert forests into grasslands, **(2)** contributed to

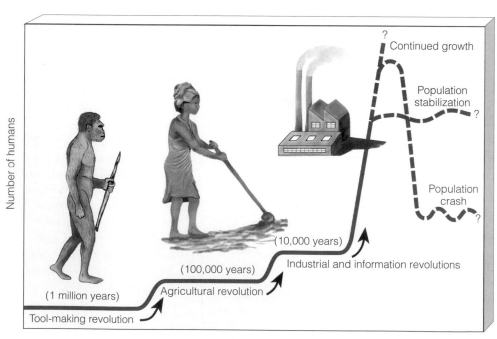

Continued growth

Population stabilization

Population crash

(10,000 years)

Industrial and information revolutions

(100,000 years)

Agricultural revolution

(1 million years)

Tool-making revolution

Number of humans

Time

Figure 2-2 Techological innovations have led to greater human control over the rest of nature and an expanding human population. Dashed lines represent three alternative population futures: **(1)** continued growth (top), **(2)** stabilization (middle), and **(3)** a crash and stabilization at a much lower level.

the extinction of some large animals (including the mastodon, saber-toothed tiger, giant sloth, cave bear, mammoth, and giant bison), and **(3)** altered the distribution of plants (and animals feeding on such plants) as they carried seeds and plants to new areas.

Early and advanced hunter-gatherers exploited their environment to survive. However, their environmental impact usually was limited and local because of **(1)** their small population sizes, **(2)** low resource use per person, **(3)** migration, which allowed natural processes to repair most of the damage they caused, and **(4)** lack of technology that could have expanded their impact.

How Has the Agricultural Revolution Affected the Environment? Some 10,000-12,000 years ago, a cultural shift known as the **agricultural revolution** began in several regions of the world. It involved a gradual move from usually nomadic hunting-and-gathering groups to settled agricultural communities in which people domesticated wild animals and cultivated wild plants.

Plant cultivation probably developed in many areas, especially in the tropical forests of Southeast Asia, northeast Africa, and Mexico. People discovered how to grow various wild food plants from roots or tubers (fleshy underground stems). To prepare the land for planting, they cleared small patches of tropical forests by cutting down trees and other vegetation and then burning the underbrush (Figure 2-3). The ashes fertilized the nutrient-poor soils in this **slash-and-burn cultivation**.

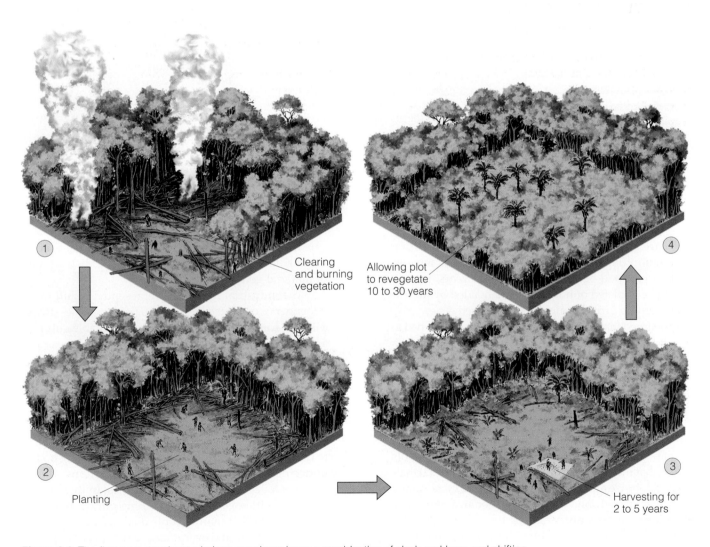

Figure 2-3 The first crop-growing technique may have been a combination of slash-and-burn and shifting cultivation in tropical forests. This method is sustainable only if small plots of the forest are cleared, cultivated for no more than 5 years, and then allowed to regenerate for 10-30 years to renew soil fertility. Indigenous cultures have developed many variations of this technique and have found ways to make some nondestructive uses of former plots while they are being regenerated.

Consequences of the Agricultural Revolution

Here are some of the beneficial and harmful effects of the agricultural revolution:

- *Using domesticated animals to plow fields, haul loads, and perform other tasks increased the ability to expand agriculture and support more people.*

- *People (1) cut down vast forests to supply wood for fuel and building materials, (2) plowed up large expanses of grassland to grow crops, and (3) built irrigation systems to transfer water from one place to another.* Such extensive land clearing degraded or destroyed the habitats of many wild plants and animals, causing or hastening their extinction.

- *Soil erosion, salt buildup in irrigated soils, and overgrazing of grasslands by huge herds of livestock helped turn fertile land into desert; topsoil washed into streams, lakes, and irrigation canals.* This environmental degradation was a factor in the downfall of many great civilizations in the Middle East, North Africa, and the Mediterranean.

- *People began accumulating material goods.* Nomadic hunter-gatherers could not carry many possessions in their travels, but farmers living in one place could acquire as much as they could afford.

- *Farmers could grow more than enough food for their families.* They could store the excess for a "rainy day" or use it to barter for other goods and services.

- *Urbanization—the formation of villages, towns, and cities—became practical.* Some villages grew into towns and cities, which served as centers for trade, government, and religion. Towns and cities concentrated sewage and other wastes, polluted the air and water, and greatly increased the spread of diseases.

- *Increased production and use of material goods created growing volumes of wastes.*

- *Conflict between societies became more common as ownership of land and water rights became crucial economic issues.* Armies and their leaders rose to power and conquered large areas of land and water supplies. These rulers forced powerless people (slaves and landless peasants) to do the hard, disagreeable work of producing food and constructing irrigation systems, temples, and walled fortresses.

- *The survival of wild plants and animals, once vital to humanity, became less important.* Wild animals, which competed with livestock for grass and fed on crops, became enemies to be killed or driven from their habitats. Wild plants invading cropfields became weeds to be eliminated.

Critical Thinking

Would we be better off if agricultural practices had never been developed and we were still hunters and gatherers? Explain.

Early growers also used various forms of **shifting cultivation** (Figure 2-3) primarily in tropical regions. After a plot had been used for several years, the soil would be depleted of nutrients or reinvaded by the forest. Then the growers moved and cleared a new plot. They learned that each abandoned patch normally had to be left fallow (unplanted) for 10–30 years before the soil became fertile enough to grow crops again. While patches were regenerating, growers used them for tree crops, medicines, fuelwood, and other purposes. In this manner, most early growers practiced *sustainable cultivation*.

These early farmers had fairly little impact on the environment because (1) their dependence mostly on human muscle power and crude stone or stick tools meant that they could cultivate only small plots, (2) their population size and density were low, and (3) normally there was enough land to move to other areas and leave abandoned plots unplanted for the several decades needed to restore soil fertility. The gradual shift from hunting and gathering to farming had several significant effects (Connections, above).

How Has the Industrial Revolution Affected the Environment? The next cultural shift, the **industrial revolution**, began in England in the mid-1700s and spread to the United States in the 1800s. It led to a rapid expansion in the production, trade, and distribution of material goods.

The industrial revolution represented a shift from dependence on (1) *renewable* wood (with supplies dwindling in some areas because of unsustainable cutting) and flowing water to (2) dependence on machines running on *nonrenewable* fossil fuels (first coal and later oil and natural gas). This led to a switch from small-scale, localized production of handmade goods to large-scale production of machine-made goods in centralized factories in rapidly growing industrial cities.

Factory towns grew into cities as rural people came to the factories for work. There they worked long hours under noisy, dirty, and hazardous conditions. Other workers toiled in dangerous coal mines. In these early industrial cities, coal smoke belching out of chimneys was so heavy that many people died of lung ailments. Ash and soot covered everything, and on some days the smoke was so thick that it blotted out the sun.

Fossil fuel–powered farm machinery, commercial fertilizers, and new plant-breeding techniques increased per acre crop yields. This helped protect biodiversity by reducing the need to expand the area of cropland to grow food. Because fewer farmers were needed, more people migrated to cities. With a larger and more reliable food supply and longer life spans, the size of the human population began the sharp increase that continues today (Figure 1-1, p. 2, and Figure 1-3, p. 5).

After World War I (1914–18), more efficient machines and mass production techniques were developed. These technologies became the basis of today's advanced industrial societies in places such as the United States, Canada, Japan, Australia, and western Europe. Advanced industrial societies have provided numerous benefits along with environmental problems (Connections, right).

How Might the Information and Globalization Revolution Affect the Environment? We are in the midst of a new cultural shift, the **information and globalization revolution**, in which new technologies such as the telephone, radio, television, computers, the internet, automated databases, and remote sensing satellites are enabling people to have increasingly rapid access to much more information on a global scale. It's estimated that scientific information now doubles about every 12 years and general information doubles about every 2.5 years. The World Wide Web contains hundreds of millions of electronic pages and grows by roughly a million electronic pages per day.

What the information and globalization revolution means for the environment is not yet clear. On the *positive side*, today's information technologies

- Help us understand more about how the earth, economies, and other complex systems work and how such systems might be affected by our actions.

- Allow us to respond to environmental problems more effectively and rapidly.

- Allow us to use remote sensing satellites to survey resources and monitor changes in the world's forests, grasslands, oceans, rivers, polar regions, cities, and other systems.

- Allow us to develop sophisticated computer models and computer-generated maps of the earth's environmental systems.

- Can reduce pollution and environmental degradation by substituting data for materials and energy and communication for transportation.

- Allow environmental researchers and activists to exchange data and information rapidly.

On the *negative side*, information technologies

- Provide an overload of information.

- Cause confusion, distraction, and a sense of hopelessness as we try to identify useful environmental information and ideas in a rapidly growing sea of information.

CONNECTIONS

Consequences of Advanced Industrial Societies

The *good news* is that advanced industrial societies provide a variety of benefits to most people living in them, including

- Mass production of many useful and economically affordable products

- A sharp increase in agricultural productivity

- Lower infant mortality and longer life expectancy because of better sanitation, hygiene, nutrition, and medical care

- A decrease in the rate of population growth

- Better health, birth control methods, and education

- Methods for controlling pollution

- Greater average income and old-age security

However, the *bad news* is the resource and environmental problems we face today (Figure 1-13, p. 14), mostly because of the rise of advanced industrial societies.

Critical Thinking

1. On balance, do you believe that the advantages of the industrial revolution have outweighed its disadvantages? Explain.

2. What three major things would you do to reduce the harmful environmental impacts of advanced industrial societies?

- Increase environmental degradation and decrease cultural diversity as a globalized economy spreads over most of the earth and homogenizes the world's cultures.

- According to Christian de Duve, a Nobel Prize winner in medicine, "I think the greatest threat to humanity is the rapid growth of knowledge. We are creating and introducing new technologies much faster than we can evaluate their impacts."

2-2 ENVIRONMENTAL HISTORY OF THE UNITED STATES: THE TRIBAL AND FRONTIER ERAS

What Happened During the Tribal Era? The environmental history of the United States can be divided into four eras: **(1)** tribal, **(2)** frontier, **(3)** conservation, and **(4)** environmental.

During the *tribal era* North America was occupied by 5–10 million tribal people for at least 10,000 years before European settlers began arriving in 1607. These

Lessons in Native American Plant Gathering

M. Kat Anderson

GUEST ESSAY

Kat Anderson is an ethnoecologist with the National Plant Center of the USDA's Natural Resource Conservation Service. She holds a research and lecturer position in the Department of Environmental Horticulture at University of California, Davis. She is co-editor of the book Before the Wilderness *, an anthology that documents the environmental management practices of California Indians.*

Some of the instructions of how to live with nature lie in the rich cultural heritage of the more than 800 Indian nations living in the United States when Columbus set foot on this continent. These Native American cultures acquired vast experience in how to use, manage, respect, and coexist with other forms of life.

Many kinds of native plants were managed with a variety of horticultural techniques including tilling, pruning, burning, sowing, weeding, irrigating, and selective harvesting. Today Native Americans from various tribes continue to gather, use, conserve, and tend plants and animals within their surroundings, reenacting age-old traditions.

Indigenous interactions with nature are an important part of the land's ecological history. This storehouse of knowledge about the natural world is called *traditional ecological knowledge*, and it has helped sustain tremendous biological diversity for more than a hundred centuries. In-depth evaluation of ancient indigenous traditional ecological knowledge and land management systems may provide a set of ecological principles and applications useful to Western culture.

Lesson No. 1: *Respect and understanding of nature comes through repeated, judicious use.* According to contemporary Native Americans, it is only through interaction and relationships with native plants that mutual respect is established. Through trial and error, keen observation, and repeated, tempered harvesting and tending, one gains knowledge of a plant's reproductive biology and how the plant responds to harvesting and management. For example, in addition to providing useful objects, the weaving of a watertight basket or the carving of an elkhorn spoon is a way of paying homage to nature and celebrating one's own humanness. If the same shrub or tree is harvested for many generations, it is also a way of connecting humans with their past. A healthy 600-year-old tree, 400-year-old shrub, or century-old fern is living proof that harvesting with restraint and temperance is respectful of the natural world.

Lesson No. 2: *The rate of harvesting should not exceed the biological capacity of plant populations to regenerate.* Many Native American cultures gathered renewable plant parts in the form of fruits, branches, leaves, flowers, tubers, and stems year after year, while leaving individual plants and plant parts behind to ensure plant replacement. For example, in what is present-day New York, the Seneca, after gathering herbaceous plants, would break off the seed stalks and drop the pods into dug holes along with a handful of leaf mold. After the Potawatomi gathered the roots of plants for medicines, they placed the seed heads in the holes from which they removed the roots and covered them back up. When digging, the Dena'ina of Alaska purposefully left fragments of underground swollen stems in the soil to ensure the growth of new plants.

---

indigenous people, called Indians by the Europeans and now often called Native Americans, practiced hunting and gathering, burned and cleared fields, and planted crops (Guest Essay, above). Because of their small populations and simple technology, they had a fairly low environmental impact.

Although there were exceptions, many Native American cultures had a deep respect for the land and its animals and did not believe in land ownership, as indicated by these quotes:

My people, the Blackfeet Indians, have always had a sense of reverence for nature that made us want to move through the world carefully, leaving as little mark behind as possible. My mother once told me: "A person should never walk so fast that the wind cannot blow away his footprints." (Jamake Highwater, Blackfoot)

From our childhood we are taught that the animals and even the trees and other plants that we share a

place with are our brothers and sisters. So when we speak of land, we are not speaking of property, territory, or even a piece of property upon which our houses sit and our crops are grown. We are speaking of something truly sacred. (Jimmie Durham, Cherokee)

What Happened During the Frontier Era (1607–1890)? The frontier era began in the early 1600s frontier when European colonists began settling North America. The colonists found a vast continent with abundant forests and wildlife and rich soils.

American settlers responded to this seemingly inexhaustible resource abundance with a **frontier environmental worldview**. They viewed most of the continent as a wilderness to be conquered by clearing and planting and with vast resources to be used. Forests were cleared not only for timber and cropland but also because they were seen as a hostile wilderness full of dangerous savages and wild beasts. Populations of animals, such as the American bison, were decimated

Lesson No. 3: *Help ensure the future abundance of native plants by gathering, pruning, or burning them at certain seasons of the year.* The time of year a plant is harvested affects its long-term productivity. Tribes of the Rocky Mountain region harvested large quantities of the prairie turnip (*Psoralea esculenta*) when the tops were browning in July or August, only after the plants had reseeded a site. California Indian basketweavers still prune or burn native shrubs such as sourberry (*Rhus trilobata*), buckbrush (*Ceanothus cuneatus*), and redbud (*Cercis occidentalis*) in the fall or winter, after the leaves have dropped, a time that is least detrimental to the vital processes of such plants.

Lesson No. 4: *Regulate the frequency of harvest to enable plant and animal regeneration.* The Cree rotate their animal traplines every year, allowing 4-year rest periods to reduce the possibility of overhunting. A 3-year rest period was specified for gathering western red cedar (*Thuja plicata*) roots among the Klickitat basketmakers of southern Washington. Navajo medicine men still refrain from harvesting plants from the same stand 2 years running.

Lesson No. 5: *Set fires to help maintain specific vegetation types and increase ecological biodiversity.* Many habitats were deliberately maintained by, and essentially dependent on, ongoing Indian-set fires. In the Pacific Northwest, prairies were burned to keep them open and increase the yields of camas and bracken fern rhizomes eaten by various tribes. In the Sierra Nevada of California, Native Americans used repeated burning to influence the size of black oak–ponderosa pine forests and montane meadows. Within the boreal forest of Alberta, Canada, Slavey, Cree, and Beaver Indians regularly burned meadows to keep them from being taken over by brush and forest. In the Southwest, fires were set by the Shoshone to favor native grass seeds and prevent invasion by surrounding woody vegetation.

Native American cultures also used fire to create or enhance ecological effects such as **(1)** recycling nutrients, **(2)** decreasing plant competition, **(3)** increasing the abundance, density, and diversity of plant species, **(4)** augmenting seed, fruit, or bulb production, **(5)** increasing the quality and quantity of forage for wildlife, **(6)** reducing insect infestations and diseases, **(7)** reducing catastrophic fires, and **(8)** decreasing detritus (dead and rotting material). For example, Western Mono, Sierra Miwok, and Foothill Yokuts tribes managed black oak–ponderosa pine forests in the Sierra Nevada of California to **(1)** increase mushroom production, **(2)** facilitate acorn collection, **(3)** promote rapid elongation of certain types of oak branches used to make various items, **(4)** reduce the incidence of insect pests that inhabit acorns, **(5)** promote useful understory grasses and forbs, **(6)** promote a vegetative structure that increases acorn production, and **(7)** inhibit catastrophic fires by eliminating brush.

Lesson No. 6: *Profound understanding and reverance are essential for responsible management of vegetation.* These valuable ecological lessons and more detailed experimental research can help Western society use nature in ways that maintain the diversity of wild plant and animal genetic resources and preserve the renewal capacity of the land. Renewed respect for Native American ecological wisdom may also promote efforts to restore traditional land and resource rights to indigenous peoples.

Critical Thinking

Explain how these traditional land management techniques discussed in this essay could be used in modern management of public lands in the United States (Spotlight, p. 32).

(p. 23), forests were cut, and soils eroded from land cleared for agriculture.

In the 1600s, the British colonies in New England traded beaver skins, lumber, and other natural resources with Europe in exchange for manufactured goods. As coastal areas were depleted of animal populations and forests, the settlers moved the frontier further inland.

This world view contrasted sharply with that of some native American cultures (Guest essay, above). According to Luther Standing Bear, a Sioux, "Only to the white man was nature a 'wilderness' and only to him was the land infested with 'wild animals' and 'savage' people. To us it was tame and bountiful."

Another factor accelerating settling of the continent and use of its resources was the transfer of vast areas of public land to private interests between 1850 and 1900. In 1850, the U.S. government owned about 80% of the total land area of the territorial United States, with tribal cultures occupying about 4% of the land, mostly in reservations designated by the government.

By 1900, more than half of the country's public land had been given away or sold cheaply to railroad, timber, and mining companies, land developers, states, schools, universities, and homesteaders to encourage settlement across the country. Under the Homestead Act of 1862, each qualified settler in the Great Plains was given 65 hectares (160 acres) of land free of charge.

This frontier view prevailed for more than 280 years, until the government declared the frontier officially closed in 1890. However, this environmental worldview remains part of the American culture.

2-3 ENVIRONMENTAL HISTORY OF THE UNITED STATES: THE EARLY CONSERVATION ERA (1832–1960)

Who Were Some Early Conservationists (1832–1870)? Between 1832 and 1870, some people became alarmed at the scope of resource depletion and

Figure 2-4 Henry David Thoreau (1817–1862) was an American writer and naturalist who kept journals about his excursions into wild nature throughout parts of the northeastern United States and Canada and at Walden Pond in Massachusetts. He sought self-sufficiency, a simple lifestyle, and a harmonious coexistence with nature.

degradation in the United States. They urged that part of the unspoiled wilderness on public lands owned jointly by all people (but managed by the government) be protected as a legacy to future generations.

Two of these early conservationists were *Henry David Thoreau* (1817–62) and *George Perkins Marsh* (1812-1939). Thoreau (Figure 2-4) was alarmed at the loss of numerous wild species from his native eastern Massachusetts. To gain a better understanding of nature, he **(1)** built a cabin in the woods on Walden Pond near Concord, Massachusetts, **(2)** lived there alone for 2 years

and **(3)** wrote *Life in the Woods*, an environmental classic.

In 1864, George Perkins Marsh, a scientist and congressperson from Vermont, published a book, *Man and Nature*, that helped legislators and influential citizens see the need for resource conservation. Marsh questioned the idea that the country's resources were inexhaustible and used scientific studies and case studies to show how the rise and fall of past civilizations were linked to the use and misuse of their resource base. He also formulated basic resource conservation principles still used today.

What Happened Between 1870 and 1930?

Between 1870 and 1930, a number of actions increased the role of the federal government and private citizens in resource conservation and public health (Figure 2-5). The *Forest Reserve Act of 1891* was a turning point in estab-

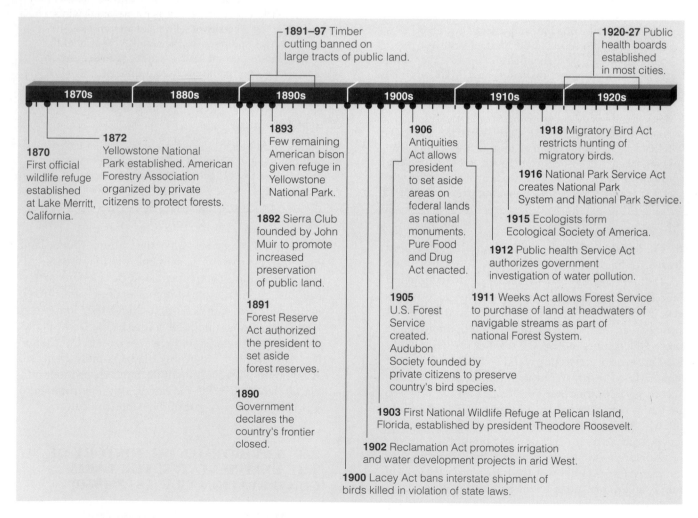

Figure 2-5 Examples of the increased role of the federal government in resource conservation and public health and establishment of key private environmental groups, 1870–1930.

Figure 2-6 John Muir (1838–1914) was a geologist, explorer, and naturalist. He spent 6 years studying, writing journals, and making sketches in the wilderness of California's Yosemite Valley and then went on to explore wilderness areas in Utah, Nevada, the Northwest, and Alaska. He was largely responsible for the establishment of Yosemite National Park in 1890. He also founded the Sierra Club and spent 22 years lobbying actively for conservation laws.

lishing the responsibility of the federal government for protecting public lands from resource exploitation.

In 1892, nature preservationist and activist *John Muir* (Figure 2-6) founded the Sierra Club. He became the leader of the *preservationist movement* advocating that large areas of wilderness on public lands should be protected from human intervention, except for low-impact recreational activities such as hiking and camping. This idea was not enacted into law until 1964. He also proposed and lobbied for creation of a national park system on public lands, an idea that became law in 1912 (two years before his death).

Mostly because of political opposition, effective protection of forests and wildlife did not begin until *Theodore Roosevelt* (Figure 2-7), an ardent conservationist, became president. His term of office, 1901–9, has been called the country's Golden Age of Conservation. Some of his major contributions to conservation include

- Persuading Congress to give the president power to designate public land as federal wildlife refuges

- Establishing the first federal refuge at Pelican Island off the east coast of Florida for preservation of the endangered brown pelican in 1903 and adding 35 more reserves by 1904

- Designating the Grand Canyon as one of the first 16 national parks

- Establishing the U.S. Bureau of Reclamation

- More than tripling the size of the national forest reserves and transferring their administration from the Department of the Interior, known for lax enforcement, to the Department of Agriculture

In 1905 Congress created the *U.S. Forest Service* to manage and protect the forest reserves. Roosevelt appointed *Gifford Pinchot* (1865–1946) as its first chief. Pinchot pioneered scientific management of forest resources on public lands, using the principles of **(1)** *sus-tainable yield* (cutting trees no faster than they could regenerate) and **(2)** *multiple use* (using the lands for a variety of purposes, including resource extraction, recreation, and wildlife protection).

In 1906 Congress passed the *Antiquities Act*, which allows the president to protect areas of scientific or historical interest on federal lands as national monuments. Roosevelt then used this act to protect the Grand Canyon and other areas that would later become national parks.

In 1907 Congress, upset because Roosevelt had added vast tracts to the forest reserves, banned further executive withdrawals of public forests. On the day before the bill became law, Roosevelt defiantly reserved another 6.5 million hectares (16 million acres).

Early in the 20th century the U.S. conservation movement split over how the beautiful Hetch Hetchy Valley (in what is now Yosemite National Park) was to be used (Spotlight, p. 32). The *wise-use* or *conservationist* school, led by Roosevelt (Figure 2-7) and Pinchot, believed that all public lands should be managed wisely and scientifically to provide needed resources. The *preservationist* school, led by Muir (Figure 2-6), believed that wilderness areas on public lands should be left untouched. This controversy over how public lands should be used continues today.

In 1916, while Woodrow Wilson (1856–1924) was president, Congress passed the *National Park Service Act*, which **(1)** declared that the parks were to be maintained in a manner that leaves them unimpaired for future generations and **(2)** established the National Park Service (within the Department of the Interior) to manage the system. Under its first head, Stephen T. Mather (1867–1930), the dominant park policy was to encourage tourist visits by allowing private concessionaires to operate facilities within the parks.

During the early 1900s improvements in public health (Figure 2-5) were led by women such as *Jane Addams* (1860-1935) and *Alice Hamilton* (Individuals Matter, p. 33).

After World War I (1914-18) the country entered a new era of economic growth and

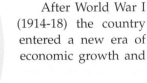

Figure 2-7 Theodore ("Teddy") Roosevelt (1858–1919) was a writer, explorer, naturalist, avid bird-watcher, and 26th president of the United States. He was the first national political figure to bring the issues of conservation to the attention of the American public. According to many historians, he has contributed more than any other president to the natural resource conservation in the United States.

How Should Public Land Resources Be Used? Preservationists vs. Conservationists

SPOTLIGHT

In 1901, conservationists, led by Gifford Pinchot and San Francisco mayor James D. Phelan, proposed to dam the Tuolumne River running through Hetch Hetchy Valley to supply drinking water for San Francisco. Preservationists, led by John Muir (Figure 2-6), were opposed.

After a bitter 12-year battle, Pinchot's views won and in 1913 the dam was built and the valley was flooded. Today the controversy continues, with preservationists pressing to have the dam removed.

Preservationists would keep large areas of public lands untouched so they can be enjoyed today and passed on unspoiled to future generations. After Muir's death in 1914, preservationists were led by forester *Aldo Leopold* (see Figure 2-15, p. 40), who said that the role of the human species should be to protect nature, not conquer it (Section 2-5, p. 39).

Another effective supporter of wilderness preservation was *Robert Marshall* (1901–39) of the U.S. Forest Service. In 1935, he and Leopold founded the Wilderness Society. More recent preservationist leaders

include **(1)** *David Brower* (1912–2000), former head of the Sierra Club and founder of both Friends of the Earth and Earth Island Institute, and **(2)** *Howard Zahniser* (1906–64), who as head of the Wilderness Society helped draft the Wilderness Act of 1964 and lobby Congress for its passage.

In contrast, *conservationists* see wilderness and other public lands as resources to be used to enhance the nation's economic growth and to provide the greatest benefit to the greatest number of people. In their view the government should protect these lands from harm by managing them efficiently and scientifically, using the principles of sustainable yield and multiple use.

Roosevelt and Pinchot thought conservation experts should form an elite corps of resource managers within the federal bureaucracy. Shielded from political pressure, they could develop scientific management strategies. Pinchot angered Muir and other preservationists when he stated his "wise use" principle:

> *The first great fact about conservation is that it stands for development. There has been a*

fundamental misconception that conservation means nothing but the husbanding of resources for future generations. There could be no more serious mistake. . . . The first principle of conservation is the use of the natural resources now existing on this continent for the benefit of the people who live here now.

Despite their basic differences, both groups opposed delivering public resources into the hands of a few for private profit. Both groups have been disappointed. Since 1910 rights to extract water and minerals, graze livestock, and harvest trees from public lands routinely have been given away or sold by Congress at below-market prices to large corporate farms, ranches, mining companies, and timber companies.

Critical Thinking

1. Why do the conservationist and preservationist philosophies lead to different management practices for public lands?

2. Which philosophy do you favor? Explain.

---

expansion. During the administrations headed by presidents Warren Harding, Calvin Coolidge, and Herbert Hoover, the federal government promoted increased resource removal from public lands at low prices to stimulate economic growth.

President Hoover went even further and proposed that the federal government return all remaining federal lands to the states or sell them to private interests for economic development. However, the Great Depression (1929–41) made owning such lands unattractive to state governments and private investors.

What Happened Between 1930 and 1960? A second wave of national resource conservation and improvements in public health began in the early 1930s (Figure 2-9), as President *Franklin D. Roosevelt* (1882–1945) strove to bring the country out of the

Great Depression. Massive federal government programs designed to provide jobs and restore the environment included

- Low-cost purchase of large tracts of public land from cash-poor landowners.

- The *Civilian Conservation Corps* (CCC), established in 1933 to put 2 million unemployed people to work **(1)** planting trees, **(2)** developing and maintaining parks and recreation areas, **(3)** restoring silted waterways, **(4)** building levees and dams for flood control, **(5)** controlling soil erosion, and **(6)** protecting wildlife.

- Establishing the *Tennessee Valley Authority* (TVA) to provide jobs, replant forests, and build dams for flood control and hydroelectric power in the economically depressed Tennessee Valley.

Alice Hamilton (1869–1970) was the country's first influential expert in industrial medicine (Figure 2-8). After graduating from medical school she became professor of pathology at the Woman's Medical School of Northwestern University. She became interested in the neglected and poorly understood field of industrial medicine after hearing numerous stories about health hazards in stockyards and factories.

Hamilton began investigating various hazardous industries. Despite little information, company resistance, and workers' failure to report health problems from fear of losing their jobs, Hamilton's persistence and resourcefulness as a researcher paid off. During the next several decades she became the country's leading investigator of occupational hazards.

In 1919, Hamilton was appointed assistant professor of industrial medicine at Harvard University, the first teaching appointment of a woman at this institution. In the 1920s she published her classic text *Industrial Poisons in the United States* and became the country's most effective advocate for investigating and dealing with the environmental consequences of industrial activity.

Four decades before the widespread concern about pesticides and other industrial chemicals in the 1960s and 1970s, Hamilton was warning workers that they were being exposed to a variety of new chemicals whose effects on human health were unknown.

She unsuccessfully opposed the use of tetraethyl lead in gasoline in the 1920s, arguing that there is no safe exposure to lead. Her position was vindicated in the late 1980s, when tetraethyl lead was phased out of gasoline in the United States. Her efforts were an important factor in the introduction of workers' compensation laws.

Alice Hamilton was a strong advocate of *pollution prevention*. In a 1925 article she expressed the hope "that the day is not far off when we shall take the next step and investigate a new danger in industry before it is put to use, before any fatal harm has been done."

Figure 2-8 Alice Hamilton (1869–1970) was the first and foremost expert on industrial disease in the United States. (Schliesinger Library, Radcliffe College)

- Building and operating many large dams in the arid western states, including Hoover Dam on the Colorado River, to provide jobs, flood control, cheap irrigation water, and cheap electricity for industry.

- Enacting the Soil Conservation Act of 1935, which established the *Soil Erosion Service* as part of the Department of Agriculture to correct the enormous erosion problems that had ruined many farms in the Great Plains states. Its name was later changed to the *Soil Conservation Service*, and it is now called the *Natural Resources Conservation Service*.

There were few new developments in federal resource conservation and public health policy during the 1940s and 1950s, mostly because of preoccupation with World War II (1941–45) and economic recovery after the war.

Between 1933 and 1960 improvements in public health included **(1)** establishment of public health boards and agencies at the municipal, state, and federal levels, **(2)** increased public education about health issues, **(3)** introduction of vaccination programs, and **(4)** a sharp reduction in waterborne infectious disease, mostly because of improved sanitation and garbage collection.

2-4 ENVIRONMENTAL HISTORY OF THE UNITED STATES: THE ENVIRONMENTAL ERA (1960–2000)

What Happened During the 1960s? There were a number of important milestones in American environmental history during the 1960s (Figure 2-10). In 1962, biologist *Rachel Carson* (1907–64) published *Silent Spring*, which documented the pollution of air, water, and wildlife from pesticides such as DDT (Individuals Matter, p. 36). This influential book helped broaden the concept of resource conservation to include preservation of the *quality* of the air, water, soil, and wildlife.

Many historians mark this wake-up call as the beginning of the modern **environmental movement**, in which a growing number of citizens at the grassroots level organized to demand that political leaders enact laws and develop policies to **(1)** curtail pollution, **(2)** clean up polluted environments, and **(3)** protect pristine areas from environmental degradation.

In 1964, Congress passed the *Wilderness Act*, inspired by the vision of John Muir more than 80 years earlier. The act authorized the government to protect undeveloped tracts of public land as part of the National Wilderness System unless Congress later decides they

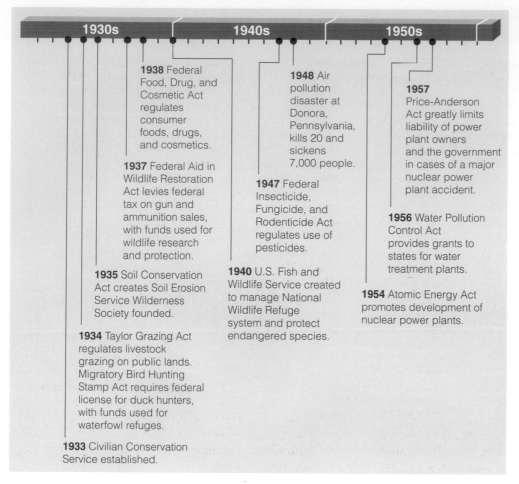

Media attention, public concern about environmental problems, scientific research, and action to address these concerns grew rapidly during *the 1970s, sometimes called the first decade of the environment.* Figure 2-12 summarizes some important events during this decade.

1938 Federal Food, Drug, and Cosmetic Act regulates consumer foods, drugs, and cosmetics.

1937 Federal Aid in Wildlife Restoration Act levies federal tax on gun and ammunition sales, with funds used for wildlife research and protection.

1935 Soil Conservation Act creates Soil Erosion Service Wilderness Society founded.

1934 Taylor Grazing Act regulates livestock grazing on public lands. Migratory Bird Hunting Stamp Act requires federal license for duck hunters, with funds used for waterfowl refuges.

1933 Civilian Conservation Service established.

1948 Air pollution disaster at Donora, Pennsylvania, kills 20 and sickens 7,000 people.

1947 Federal Insecticide, Fungicide, and Rodenticide Act regulates use of pesticides.

1940 U.S. Fish and Wildlife Service created to manage National Wildlife Refuge system and protect endangered species.

1957 Price-Anderson Act greatly limits liability of power plant owners and the government in cases of a major nuclear power plant accident.

1956 Water Pollution Control Act provides grants to states for water treatment plants.

1954 Atomic Energy Act promotes development of nuclear power plants.

Figure 2-9 Some important conservation and environmental events between 1930 and 1960.

The first annual *Earth Day* was held on April 20, 1970. During this event, proposed by Senator *Gaylord Nelson* (born 1916), some 20 million people in more than 2,000 communities took to the streets to heighten awareness and to demand improvements in environmental quality.

President *Richard Nixon* (1913–94) responded to the rapidly growing environmental movement by **(1)** establishing the *Environmental Protection Agency* (EPA) in 1970 and **(2)** supporting passage of the *Endangered Species Act of 1973*, which greatly strengthened the role of the federal government in protecting endangered species.

An eye-opening event occurred in 1973 when the Arab members of the Organization of Petroleum Exporting Countries (OPEC)* reduced oil exports to the West and banned oil shipments to the United States because of its support for Israel in the 18-day Yom Kippur War with Egypt and Syria. This *OPEC oil embargo,* lasting until March 1974, sharply raised the price of crude oil. The result was **(1)** double-digit inflation in the United States and many other countries, **(2)** high interest rates, **(3)** soaring international debt, and **(4)** a global economic recession. In 1979, a second reduction in oil supplies and a sharp price increase occurred when Iran's Islamic Revolution shut down most of Iran's oil production.

are needed for the national good. Land in this system is to be used only for nondestructive forms of recreation such as hiking and camping.

Between 1965 and 1970, the emerging science of *ecology* received widespread media attention. At the same time, the popular writings of biologists such as *Paul Ehrlich, Barry Commoner,* and *Garrett Hardin* (Guest Essay, p. 252) awakened people to the interlocking relationships between population growth, resource use, and pollution (Figure 1-15, p. 15).

During that same period, a number of events increased public awareness of pollution (Figure 2-10). The public also became aware that pollution and loss of habitat were endangering well-known wildlife species such as the North American bald eagle, grizzly bear, whooping crane, and peregrine falcon.

During a 1969 U.S. *Apollo* mission to the moon, a photograph of the earth was taken from space showing the earth as a tiny blue and white planet in the black void of space (Figure 1-1, p. 2). This widely publicized photo led to the development of the *spaceship-earth environmental worldview* that we had better take care of the earth because it is all that we have.

*OPEC was formed in 1960 so that developing countries with much of the world's known and projected oil supplies could get a higher price for this resource. Its 11 members are Algeria, Indonesia, Iran, Iraq, Kuwait, Libya, Nigeria, Qatar, Saudi Arabia, the United Arab Emirates, and Venezuela. In 1973, OPEC produced 56% of the world's oil and supplied about 84% of all oil imported by other countries.

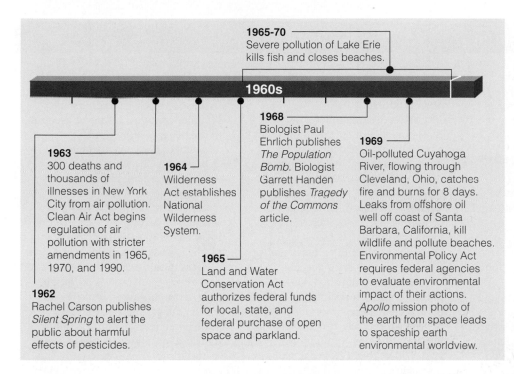

The following is the content within the timeline image (Figure 2-10):

1965-70
Severe pollution of Lake Erie kills fish and closes beaches.

1960s

1963
300 deaths and thousands of illnesses in New York City from air pollution. Clean Air Act begins regulation of air pollution with stricter amendments in 1965, 1970, and 1990.

1962
Rachel Carson publishes *Silent Spring* to alert the public about harmful effects of pesticides.

1964
Wilderness Act establishes National Wilderness System.

1965
Land and Water Conservation Act authorizes federal funds for local, state, and federal purchase of open space and parkland.

1968
Biologist Paul Ehrlich publishes *The Population Bomb*. Biologist Garrett Handen publishes *Tragedy of the Commons* article.

1969
Oil-polluted Cuyahoga River, flowing through Cleveland, Ohio, catches fire and burns for 8 days. Leaks from offshore oil well off coast of Santa Barbara, California, kill wildlife and pollute beaches. Environmental Policy Act requires federal agencies to evaluate environmental impact of their actions. *Apollo* mission photo of the earth from space leads to spaceship earth environmental worldview.

Figure 2-10 Some important environmental events during the 1960s.

In 1978 the *Federal Land Policy and Management Act* gave the *Bureau of Land Management (BLM)* its first real authority to manage the public land under its control, 85% of which is in 12 western states. This law angered a number of western interests whose use of these lands was being restricted for the first time. In the late 1970s a coalition of ranchers, miners, loggers, developers, farmers, some elected officials, and others launched a political campaign known as the *sagebrush rebellion* against government regulation of the use of public lands. Its primary goal was to remove most public lands in the Western United States from federal ownership and management and turn them over to the states. Then the plan was to persuade state legislatures to sell or lease the resource-rich lands at low prices to ranching, mining, timber, land development, and other private interests.

When *Jimmy Carter* (born 1924) was president between 1977 and 1981, he

- Persuaded Congress to create the *Department of Energy* to develop a long-range energy strategy to reduce the country's heavy dependence on imported oil.

- Appointed a number of competent and experienced administrators, drawn heavily from environmental and conservation organizations, to key posts in the EPA, the Department of the Interior, and the Department of Energy.

- Consulted with environmental leaders on environmental and resource policy matters.

- Helped create a *Superfund* as part of the *Comprehensive Environment Response, Compensation, and Liability Act of 1980* to clean up abandoned hazardous waste sites, including the Love Canal near Niagara Falls, New York.

- Used the Antiquities Act of 1906 to triple the amount of land in the National Wilderness System and double the area in the National Park System (primarily by adding vast tracts in Alaska). He used the Antiquities Act to protect more public land in all 50 states from development than any other president.

What Happened During the 1980s? Figure 2-13, p. 38 summarizes some key environmental events during the 1980s. During this decade a strong *anti-environmental movement* developed. It was funded by farmers and ranchers and leaders of the oil, automobile, mining, and timber industries who opposed many of the environmental laws and regulations developed in the 1960s and 1970s.

In 1981, *Ronald Reagan* (born 1911), a self-declared *sagebrush rebel* and advocate of less federal control, became president. During his 8 years in office he

- Appointed to key federal positions people who opposed most existing environmental and public and land use laws and policies

- Greatly increased private energy and mineral development and timber cutting on public lands

- Drastically cut federal funding for research on energy conservation and renewable energy resources and eliminated tax incentives for residential solar energy and energy conservation enacted during the Carter administration

- Lowered automobile gas mileage standards and relaxed federal air and water quality pollution standards

Although Reagan was an immensely popular president, many people strongly opposed his environmental and resource policies, which prompted **(1)** strong opposition in Congress, **(2)** public outrage, and **(3)** legal challenges by environmental and conservation organizations, whose memberships soared during this period.

INDIVIDUALS MATTER

Rachel Carson

Rachel Carson (Figure 2-11) began her professional career as a biologist for the Bureau of U.S. Fisheries (later to become the U.S. Fish and Wildlife Service). In that capacity, she **(1)** carried out research on oceanography and marine biology, **(2)** wrote articles about the oceans and topics related to the environment, and **(3)** became editor-in-chief of the bureau's publications in 1949.

In 1951, she wrote *The Sea Around Us*, which described in easily understandable terms the natural history of oceans and the harm that humans were doing them. The book was on the best-seller list for 86 weeks, sold more than 2 million copies, was translated into 32 languages, and won a National Book Award.

During the late 1940s and throughout the 1950s, the use of DDT and related compounds—to kill insects that ate food crops, attacked trees, bothered people, and transmitted diseases such as malaria—expanded rapidly.

In 1958, DDT was sprayed to control mosquitoes near the home and private bird sanctuary of Olga Huckins, a good friend of Carson. After the spraying, Huckins witnessed the agonizing deaths of several of her birds, and in distress she asked Carson whether she could find someone to investigate the effects of pesticides on birds and other wildlife.

Carson decided to look into the issue herself and quickly found that almost no independent research on the environmental effects of pesticides existed. As a well-trained scientist, Carson **(1)** surveyed the scientific literature, **(2)** became convinced that pesticides could harm wildlife and humans, and **(3)** methodically built a case against the widespread use of pesticides.

In 1962 she published her findings in popular form in *Silent Spring*, an allusion to the silencing of "robins, catbirds, doves, jays, wrens, and scores of other bird voices" because of their exposure to pesticides. She pointed out that "for the first time in the history of the world, every human being is now subjected to dangerous chemicals, from the moment of conception until death."

Carson's book was read by many scientists, politicians, and policy makers and was embraced by the public. However, the chemical industry viewed the book as a serious threat to booming pesticide sales and mounted a $250,000 campaign to discredit Carson. A parade of critical reviewers and industry scientists claimed that her book **(1)** was full of inaccuracies, **(2)** made selective use of research findings, and **(3)** failed to give a balanced account of the benefits of pesticides.

Some critics even claimed that, as a woman, she was incapable of understanding the highly scientific and technical subject of pesticides. Others charged that she was a hysterical woman and a radical nature lover trying to scare the American public in order to sell books.

During this period of intense controversy Carson was suffering from terminal cancer, but she was able to defend her research and strongly counter her critics. She died in 1964, about 18 months after the publication of *Silent Spring*, without knowing that many historians considered her work a key element in the birth the modern environmental movement in the United States.

Figure 2-11 Biologist Rachel Carson (1907–64) was a pioneer in increasing public awareness of the importance of nature and the threat of pollution. She died without knowing that her efforts were a key in beginning the modern era of environmentalism in the United States. (©1962 Eric Hartmann/Magnum Photos)

In 1988, an industry-backed anti-environmental coalition called the *Wise-Use movement* was formed with the major goals of **(1)** weakening or repealing most of the country's environmental laws and **(2)** destroying the effectiveness of the environmental movement in the United States. Major tactics of the U.S. anti-environmental movement are given in Appendix 4.

Upon his election in 1989, *George Bush* (born 1924) promised to be "the environmental president." However, he received criticism from environmentalists for

- Failing to provide leadership on such key environmental issues as population growth, global warming, and loss of biodiversity

- Continued support of exploitation of valuable resources on public lands at giveaway prices

- Allowing some environmental laws to be undercut by the powerful influence of industry, mining, ranching, and real estate development officials

What Happened During the 1990s? Figure 2-14 lists some other significant environmental events of the 1990s. In 1993, *Bill Clinton* (born 1946) became president and promised to provide national and global environmental leadership. During his 8 years in office he

- Implemented various environmental policies on the advice of Vice President Al Gore, who has a comprehensive understanding of environmental issues.

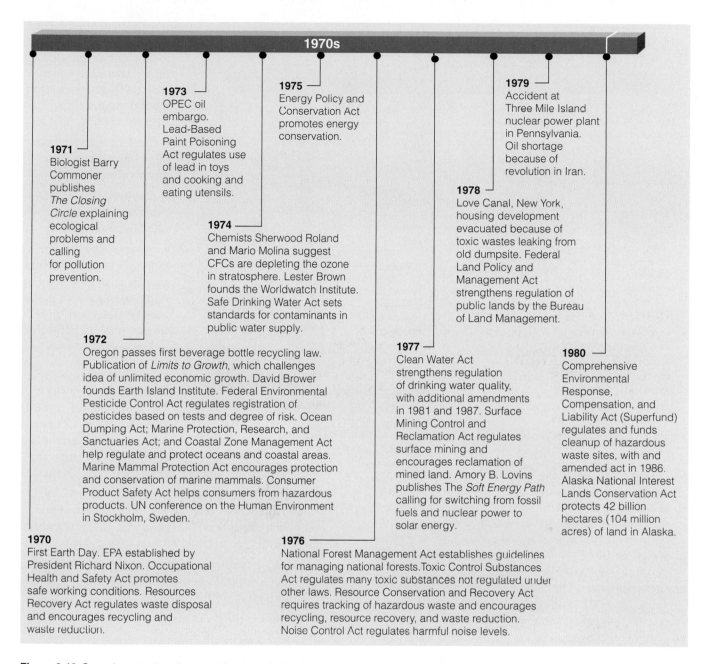

1970s

1971
Biologist Barry Commoner publishes *The Closing Circle* explaining ecological problems and calling for pollution prevention.

1973
OPEC oil embargo. Lead-Based Paint Poisoning Act regulates use of lead in toys and cooking and eating utensils.

1975
Energy Policy and Conservation Act promotes energy conservation.

1979
Accident at Three Mile Island nuclear power plant in Pennsylvania. Oil shortage because of revolution in Iran.

1974
Chemists Sherwood Roland and Mario Molina suggest CFCs are depleting the ozone in stratosphere. Lester Brown founds the Worldwatch Institute. Safe Drinking Water Act sets standards for contaminants in public water supply.

1978
Love Canal, New York, housing development evacuated because of toxic wastes leaking from old dumpsite. Federal Land Policy and Management Act strengthens regulation of public lands by the Bureau of Land Management.

1972
Oregon passes first beverage bottle recycling law. Publication of *Limits to Growth*, which challenges idea of unlimited economic growth. David Brower founds Earth Island Institute. Federal Environmental Pesticide Control Act regulates registration of pesticides based on tests and degree of risk. Ocean Dumping Act; Marine Protection, Research, and Sanctuaries Act; and Coastal Zone Management Act help regulate and protect oceans and coastal areas. Marine Mammal Protection Act encourages protection and conservation of marine mammals. Consumer Product Safety Act helps consumers from hazardous products. UN conference on the Human Environment in Stockholm, Sweden.

1977
Clean Water Act strengthens regulation of drinking water quality, with additional amendments in 1981 and 1987. Surface Mining Control and Reclamation Act regulates surface mining and encourages reclamation of mined land. Amory B. Lovins publishes The *Soft Energy Path* calling for switching from fossil fuels and nuclear power to solar energy.

1980
Comprehensive Environmental Response, Compensation, and Liability Act (Superfund) regulates and funds cleanup of hazardous waste sites, with and amended act in 1986. Alaska National Interest Lands Conservation Act protects 42 billion hectares (104 million acres) of land in Alaska.

1970
First Earth Day. EPA established by President Richard Nixon. Occupational Health and Safety Act promotes safe working conditions. Resources Recovery Act regulates waste disposal and encourages recycling and waste reduction.

1976
National Forest Management Act establishes guidelines for managing national forests. Toxic Control Substances Act regulates many toxic substances not regulated under other laws. Resource Conservation and Recovery Act requires tracking of hazardous waste and encourages recycling, resource recovery, and waste reduction. Noise Control Act regulates harmful noise levels.

Figure 2-12 Some important environmental events during the 1970s, sometimes called the *environmental decade*.

- Appointed respected environmentalists to key positions in environmental and resource agencies.

- Consulted with environmentalists about environmental policy.

- Vetoed most of the anti-environmental bills (or other bills passed with anti-environmental riders attached) passed by a Republican-dominated Congress between 1995 and 2000.

- Announced regulations requiring sport utility vehicles (SUVs) to meet the same air pollution emission standards as cars.

- Used an executive order to make forest health the primary priority in managing national forests.

- Used an executive order to declare many roadless areas in national forests off limits to roads and logging.

- Used the Antiquities Act of 1906 to protect various parcels of public land in the West from development and resource exploitation as national monuments. He has protected more public land as national monuments in the lower 48 states than any other president, including Teddy Roosevelt and Jimmy Carter.

However, environmentalists criticized Clinton for failing to push hard enough on key environmental issues such as global warming and global biodiversity protection.

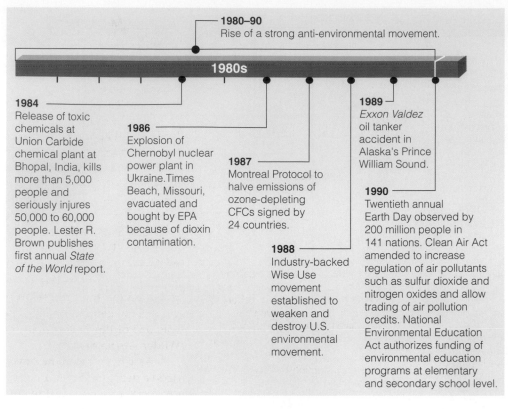

1980–90
Rise of a strong anti-environmental movement.

1980s

1984
Release of toxic chemicals at Union Carbide chemical plant at Bhopal, India, kills more than 5,000 people and seriously injures 50,000 to 60,000 people. Lester R. Brown publishes first annual *State of the World* report.

1986
Explosion of Chernobyl nuclear power plant in Ukraine. Times Beach, Missouri, evacuated and bought by EPA because of dioxin contamination.

1987
Montreal Protocol to halve emissions of ozone-depleting CFCs signed by 24 countries.

1988
Industry-backed Wise Use movement established to weaken and destroy U.S. environmental movement.

1989
Exxon Valdez oil tanker accident in Alaska's Prince William Sound.

1990
Twentieth annual Earth Day observed by 200 million people in 141 nations. Clean Air Act amended to increase regulation of air pollutants such as sulfur dioxide and nitrogen oxides and allow trading of air pollution credits. National Environmental Education Act authorizes funding of environmental education programs at elementary and secondary school level.

Figure 2-13 Some important environmental events during the 1980s.

During the 1990s the anti-environmental movement strengthened because of **(1)** continuing support from its backers and **(2)** the 1994 federal election, which gave Republicans (many of whom were generally unsympathetic to environmental concerns) a majority in Congress.

For the most part, the 1990s were disappointing to environmentalists. They had to spend much of their time and funds **(1)** fighting efforts to discredit the environmental movement and to weaken or eliminate most environmental laws passed during the 1960s and 1970s and **(2)** countering claims by anti-environmental groups that major environmental problems such as global warming and ozone depletion are hoaxes or not very serious.

During this decade membership in many of the older and established major environmental groups such as Greenpeace, the Wilderness Society, and the Sierra Club declined because of **(1)** the anti-environmental backlash and **(2)** failure of some of these large organizations to keep in touch with and support grassroots environmental actions. With a decline in members and contributions, some of these organizations have had to close offices and lay off employees.

On the other hand, during the 1990s

▪ Many newer, smaller, and mostly local grassroots environmental organizations sprung up, mostly to deal with environmental threats in their local communities. Currently, there are more than 6,000 active environmental groups in the United States.

▪ Interest in environmental issues increased on many college campuses.

▪ Environmental studies programs at colleges and universities expanded.

▪ Awareness of important environmental issues such as sustainability, population growth, biodiversity protection, and threats from global warming increased.

What Are the Major Components of the U.S. Environmental Agenda for the 21st Century? Environmental leaders believe that the four most important environmental issues to be faced in the 21st century are

▪ The threat of climate change and ecosystem and economic disruption from enhanced global warming (Section 18-2, p. 450, and Section 18-4, p. 458)

▪ Growing water shortages and political conflicts over water supplies in many local and regional areas (Section 13-3, p. 299)

▪ Continuing population growth

▪ Continuing biodiversity loss (Chapters 22–24)

Major goals of U.S. environmental organizations for the early part of the 21st century are to

▪ Focus on the four major problems just listed

▪ Protect an additional 40 million hectares (100 million acres) of land in the United States

▪ End commercial logging in U.S. national forests and use these forests primarily for recreation and conservation (Solutions, p. 603)

▪ Halt urban sprawl and build more livable and sustainable cities (Section 25-5, p. 681)

▪ Build enough public support for these and other environmental issues to counter opposition by the anti-environmental movement

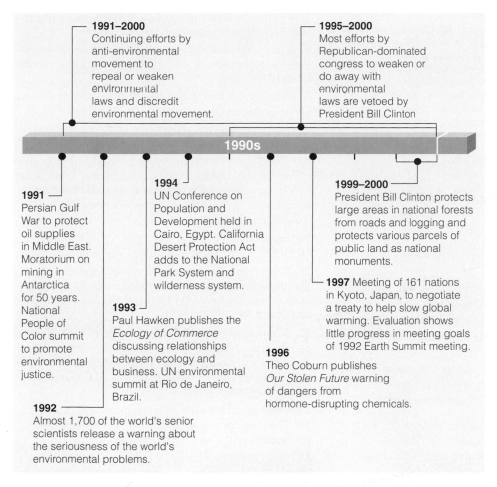

1991–2000 Continuing efforts by anti-environmental movement to repeal or weaken environmental laws and discredit environmental movement.

1995–2000 Most efforts by Republican-dominated congress to weaken or do away with environmental laws are vetoed by President Bill Clinton

1990s

1991 Persian Gulf War to protect oil supplies in Middle East. Moratorium on mining in Antarctica for 50 years. National People of Color summit to promote environmental justice.

1994 UN Conference on Population and Development held in Cairo, Egypt. California Desert Protection Act adds to the National Park System and wilderness system.

1993 Paul Hawken publishes the *Ecology of Commerce* discussing relationships between ecology and business. UN environmental summit at Rio de Janeiro, Brazil.

1992 Almost 1,700 of the world's senior scientists release a warning about the seriousness of the world's environmental problems.

1996 Theo Coburn publishes *Our Stolen Future* warning of dangers from hormone-disrupting chemicals.

1997 Meeting of 161 nations in Kyoto, Japan, to negotiate a treaty to help slow global warming. Evaluation shows little progress in meeting goals of 1992 Earth Summit meeting.

1999–2000 President Bill Clinton protects large areas in national forests from roads and logging and protects various parcels of public land as national monuments.

Figure 2-14 Some important environmental events during the 1990s.

- Build a pro-environmental coalition in Congress by electing pro-environmental Democrats and Republicans to the U.S. Congress

- Use the political and economic system to improve environmental quality by **(1)** phasing out harmful government environmental subsidies, **(2)** replacing taxes on wealth and income with taxes on environmental pollution, and **(3)** having the market prices of all goods and services include their harmful environmental costs

2-5 CASE STUDY: ALDO LEOPOLD AND HIS LAND ETHIC

Who Was Aldo Leopold? *Aldo Leopold* (Figure 2-15) is best known for being a strong proponent of *land ethics*, a philosophy in which humans as part of nature have an ethical responsibility to preserve wild nature. Leopold was born on a farm in Iowa in 1886. In 1909, after earning a master's degree in forestry from Yale University, he joined the U.S. Forest Service. During this period he had extensive field experience in protecting wilderness areas and managing wildlife in forests in New Mexico and Arizona.

He was alarmed by overgrazing and land deterioration on public lands where he worked and became convinced that the United State was losing too much of its mostly untouched wilderness lands. In 1924 Leopold convinced the Forest Service to protect 202,000 hectares (500,000 acres) of New Mexico's Gila National Forest as wilderness. It was the National Forest System's first officially designated wilderness area.

In 1933, Leopold became a professor of game management at the University of Wisconsin and founded the profession of game management. In 1935, he was one of the founders of the Wilderness Society.

As years passed, he developed a deep understanding and appreciation for wildlife and urged that nature should be included in our ethical concerns. Through his writings and teachings he became one of the founders of the *conservation* and *environmental movements* of the 20th century.

In 1935, Leopold purchased a run-down sand farm in central Wisconsin. For years thereafter he and his family spent the weekends on the farm at a rebuilt cabin (once a chicken coop) called the Shack. They planted thousands of trees to help restore the health of the land.

Leopold died in 1948 while fighting a brush fire at a neighbor's farm. His weekends of planting, hiking, and observing nature at the farm provided material he used to write his most famous book, *A Sand County Almanac*, published posthumously in 1949. Since then more than 2 million copies of this important book have been sold.

Today his farm and the surrounding preserve are a picture of environmental health, with maturing forests and restored grasslands. His land ethic still serves as guiding light for wilderness protection and conservation.

What Is Leopold's Concept of Land Ethics? The following quotes from his writings reflect Leopold's land ethic and form the basis of many of the beliefs of the modern *environmental wisdom worldview* (p. 17):

All ethics so far evolved rest upon a single premise: that the individual is a member of a community of interdependent parts.

Figure 2-15 Aldo Leopold (1887–1948) was a forester, writer, and conservationist. His book, *A Sand County Almanac* (published after his death) is considered an environmental classic that has inspired the modern environmental movement. His *land ethic* expanded the role of humans as protectors of nature. (Robert McCabe, University of Wisconsin-Madison Archives)

That land is a community is the basic concept of ecology, but that land is to be loved and respected is an extension of ethics.

The land ethic simply enlarges the boundaries of the community to include soils, waters, plants, and animals, or collectively the land. . . . In short, a land ethic changes the role of Homo sapiens *from conqueror of the land-community to plain member and citizen of it.*

A land ethic, then, reflects the existence of an ecological conscience, and this in turn reflects a conviction of individual responsibility for the health of the land. Health is the capacity of the land for self-renewal. Conservation is our effort to understand and preserve this capacity.

We abuse land because we regard it as a commodity belonging to us. When we see land as a community to which we belong, we may begin to use it with love and respect.

Anything is right when it tends to preserve the integrity, stability, and beauty of the biotic community. It is wrong when it tends otherwise.

Thank God, they cannot cut down the clouds!

HENRY DAVID THOREAU

REVIEW QUESTIONS

1. What were the key factors in the near extinction of the American bison from the Great Plains of the United States?

2. Who are *hunter-gatherers*, and what were their major environmental impacts?

3. What is the *agricultural revolution*? What are its major benefits and environmental drawbacks?

4. What are *slash-and-burn cultivation* and *shifting cultivation*? Under what conditions can these practices be a sustainable form of agriculture?

5. What is the *industrial revolution*? What are its major benefits and environmental drawbacks?

6. What is the *information and globalization revolution*? What are its potential major benefits and environmental drawbacks?

7. What are the four major eras of environmental history in the United States?

8. What major events happened during the *tribal era* of environmental history in North America?

9. What major events happened during the *frontier era* of environmental history in the United States? What is the *frontier environmental worldview*? How did it contrast with the environmental worldview of some Native American cultures?

10. Summarize the contributions of *early conservationists* **(a)** Henry David Thoreau and **(b)** George Perkins Marsh.

11. What major environmental events happened during the *conservation era* of the environmental history of the United States between **(a)** 1870 and 1930 and **(b)** 1930 and 1960?

12. Summarize the major contributions of the following people during the *conservation era* of the environmental history of the United States: **(a)** John Muir, **(b)** Theodore Roosevelt, **(c)** Gifford Pinchot, **(d)** Alice Hamilton, and **(e)** Franklin D. Roosevelt.

13. Distinguish between *preservationists* and *conservationists*. Describe the ongoing controversy between these two groups in the environmental community.

14. What is the *environmental movement* in the United States? What major environmental events happened during the *environmental era* of the environmental history of the United States during the **(a)** 1960s, **(b)** 1970s, **(c)** 1980s, and **(d)** 1990s?

15. Summarize the major contributions of the following people during the *environmental era* of the environmental history of the United States: **(a)** Rachel Carson, **(b)** Jimmy Carter, and **(c)** Bill Clinton.

16. What is the *spaceship-earth environmental worldview*? What is the *sagebrush rebellion*? Describe what Ronald Reagan did during his presidency to further this movement and weaken environmental laws.

17. Describe why the 1990s were largely disappointing for the environmental movement and encouraging for the anti-environmental movement in the United States.

18. What do environmental leaders believe are the four most important environmental issues to be faced during the 21st century?

19. What are seven major goals of the U.S. environmental movement during the early 21st century?

20. What major contributions did *Aldo Leopold* make to the environmental history of the United States? What is his *land ethic*?

CRITICAL THINKING

1. Explain how in one sense the roots of our present environmental and resource problems began with the invention of agriculture about 10,000 years ago. Would we be better off if agricultural practices had never been developed and we were still hunters and gatherers? Explain.

2. List the benefits and drawbacks of the *frontier environmental worldview*. List three major ways in which U.S. history might have been different without this worldview. Is this environmental worldview still useful today? Explain.

3. List the most important benefits and drawbacks of an advanced industrial society such as the United States. Do the benefits outweigh the drawbacks? Explain. What are the alternatives?

4. On balance do you believe that the potential environmental benefits of the current *information and globalization revolution* will outweigh its potentially harmful environmental effects? Explain.

5. Summarize the major contributions of the follow-ing people to the conservation and environmental movements: **(a)** John Muir, **(b)** Theodore Roosevelt, **(c)** Franklin D. Roosevelt, **(d)** Rachel Carson, and **(e)** Aldo Leopold.

6. What one person do you believe has made the greatest and longest-lasting contribution to the conservation and environmental movements in the United States? Explain.

7. What are the major benefits and drawbacks of the *spaceship-earth environmental worldview*?

8. Public forests, grasslands, wildlife reserves, parks, and wilderness areas are owned by all citizens and managed for them by federal and state governments in the United States. In terms of the management policies for most of these lands, would you classify yourself as a **(a)** preservationist, **(b)** conservationist, or **(c)** advocate of transferring most public lands to private enterprise? Explain.

9. Do you favor or oppose efforts to greatly weaken or repeal most U.S. environmental laws? Explain.

10. Explain why you agree or disagree with the seven major goals of U.S. environmental organizations during the early 21st century that were listed on pp. 38–39.

11. Some analysts believe that the world's remaining hunter-gatherer societies should be given title to the land on which they and their ancestors have lived for centuries and should be left alone by modern civilization. They contend that we have created protected reserves for endangered wild species, so why not create reserves for these endangered human cultures? What do you think? Explain.

PROJECTS

1. Use the library or internet to analyze speeches and writings that indicate continuation of the frontier environmental worldview in American culture.

2. What major changes (such as a change from agricultural to industrial, from rural to urban, or changes in population size, pollution, and environmental degradation) have taken place in your locale during the past 50 years? On balance, have these changes improved or decreased **(a)** the quality of your life and **(b)** the quality of life for members of your community as a whole?

3. Use the library or internet to summarize the major goals and accomplishments of the anti-environmental movement in the United States during the 1980s and 1990s.

4. Use the library or the internet to find bibliographic information about *Ernest Hemingway* and *Henry David Thoreau*, whose quotes appear at the beginning and end of this chapter.

5. Make a concept map of this chapter's major ideas, using the section heads and subheads and the key terms (in boldface). Look at the inside back cover and on the website for this book for information about making concept maps.

INTERNET STUDY RESOURCES AND RESOURCES FOR FURTHER READING AND RESEARCH

The website for this book contains helpful study aids and many ideas for further reading and research. Log on to:

http://www.brookscole.com/product/0534376975s

and click on the Chapter-by-Chapter area. Choose Chapter 2 and select a resource:

■ "Flash Cards" allows you to test your mastery of the Terms and Concepts to Remember for this chapter.

■ "Tutorial Quizzes" provides a multiple-choice practice quiz.

■ "Student Guide to InfoTrac" will lead you to Critical Thinking Projects that use InfoTrac College Edition as a research tool.

■ "References" lists the major books and articles consulted in writing this chapter.

■ "Hypercontents" takes you to an extensive list of sites with news, research, and images related to individual sections of the chapter.

INFOTRAC COLLEGE EDITION

Improve your skills with InfoTrac College Edition, a searchable online database of articles from more than 700 periodicals. Log on to:

http://www.infotrac-college.com

or access InfoTrac through the website for this book.

Try the following articles:

Golden, F. 2000. A century of heroes. *Time* vol. 155, no. 17, pp. 54+. (subject guide: Rachel Carson)

Vaughn, G.F. 1999. The land economics of Aldo Leopold. *Land Economics* vol. 75, no. 1, pp. 156–159. (subject guide: Aldo Leopold)

Animal and vegetable life is too complicated a problem for human intelligence to solve, and we can never know how wide a circle of disturbance we produce in the harmonies of nature when we throw the smallest pebble into the ocean of organic life.

GEORGE PERKINS MARSH

3 SCIENCE, SYSTEMS, MATTER, AND ENERGY

Two Islands: Can We Treat This One Better?

Easter Island (Rapa Nui) is a small, isolated island in the great expanse of the South Pacific. It was first colonized by Polynesians about 2,500 years ago.

The civilization they developed was based on the island's towering palm trees, which were used for shelter, tools, boats to catch fish, fuel, food, rope, and clothing. Using these resources, they developed an impressive civilization and a technology capable of making and moving large stone structures, including their famous statues (Figure 3-1).

The people flourished, with the population peaking at about 10,000 (with estimates ranging from 7,000 to 20,000) by 1400. However, they used up the island's precious trees faster than they were regenerated—an example of the tragedy of the commons (p. 12). Each person who cut a tree reaped immediate personal benefits while helping to doom the civilization as a whole.

Once the trees were gone the islanders could not build canoes for hunting porpoises and catching fish. Without the forest to absorb and slowly release water, springs and streams dried up, exposed soils eroded, crop yields plummeted, and famine struck.

The starving people turned to warfare and, possibly, cannibalism. Both the population and the civilization collapsed. When Dutch explorers first reached the island on Easter Day, 1722, they found only about 2,000 inhabitants, struggling under primitive conditions on a mostly barren island.

Like Easter Island at its peak, the earth is an isolated island (in the vastness of space) with no other suitable planet to migrate to. As on Easter Island, our population is growing, and we are consuming exhaustible and potentially renewable resources at a rapid pace.

Will the humans on Earth Island re-create the tragedy of Easter Island on a grander scale, or will we learn how to live sustainably on this planet that is our only home?

Scientific knowledge is a key in learning how to live more sustainably. Thus, it is important to **(1)** know what science is, **(2)** understand the behavior of complex systems studied by scientists, and **(3)** have a basic knowledge of the nature of the matter and energy that make up the earth's living and nonliving resources.

Figure 3-1 These massive stone figures on Easter Island are the remains of the technology created by an ancient civilization of Polynesians. This civilization collapsed because the people used up the trees (especially large palm trees) that were the basis of their livelihood. More than 200 of these stone statues once stood on huge stone platforms lining the coast. At least 700 additional statues were abandoned in rock quarries or on ancient roads between the quarries and the coast. No one knows how the early islanders (with no wheels, no draft animals, and no sources of energy except their own muscles) transported these gigantic structures for miles before erecting them. However, it is presumed that they did this by felling large trees and using them to roll and erect the statues. (George Holton/Photo Researchers, Inc.)

Science is an adventure of the human spirit. It is essentially an artistic enterprise, stimulated largely by curiosity, served largely by disciplined imagination, and based largely on faith in the reasonableness, order, and beauty of the universe.

WARREN WEAVER

This chapter addresses the following questions:

- What are science and critical thinking? What are some limitations of environmental science?

- What are major components and behaviors of complex systems?

- What are the basic forms of matter? What is matter made of? What makes matter useful to us as a resource?

- What are the major forms of energy? What makes energy useful to us as a resource?

- What are physical and chemical changes? What scientific law governs changes of matter from one physical or chemical form to another?

- What three main types of nuclear changes can matter undergo?

- How can exposure to radioactivity affect human health?

- What two scientific laws govern changes of energy from one form to another?

- How are the scientific laws governing changes of matter and energy from one form to another related to resource use and environmental disruption?

3-1 SCIENCE, ENVIRONMENTAL SCIENCE, AND CRITICAL THINKING

What Is Science and What Do Scientists Do? *Science* is a pursuit of knowledge about how the world works. It is an attempt to discover order in nature and use that knowledge to make predictions about what is likely to happen in nature.

As physicist Albert Einstein once said, "The whole of science is nothing more than a refinement of everyday thinking." Figure 3-2 and the Guest Essay on p. 48 summarize the more systematic version of the everyday critical thinking process used by scientists.

The first thing scientists must do is ask a question or identify a problem to be investigated. Then scientists working on this problem collect **scientific data**, or facts, by making observations and measurements. The resulting scientific data or facts must be confirmed by repeated observations and measurements, ideally by several different investigators.

This concept of *reproducibility* is important in science to detect errors in measurement and conscious and unconscious bias by investigators. This is done by repeating observations and measurements and publishing the results of scientific research for others to examine, repeat, verify, and criticize.

The primary goal of science is not facts themselves but a new idea, principle, or model that **(1)** connects and explains certain facts and **(2)** leads to useful predictions about what is likely to happen in nature. Scientists working on a particular problem try to come up with a variety of possible or tentative explanations, or **scientific hypotheses**, of what they (or other scientists) observe in nature.

To be accepted, a scientific hypothesis not only must explain scientific data and phenomena but also should make predictions that can be used to test the validity of the hypothesis. Once a scientific hypothesis is invented, experiments are conducted (and repeated to be sure they are reproducible) to test its validity and predictions. One method scientists use to test a hypothesis is to develop a **model**, an approximate representation or simulation of a system being studied, as discussed on page 50.

Scientists want lots of evidence before they are willing to accept the accuracy of any data or the usefulness of a particular hypothesis or model. Usually a number of scientists working on the same problem subject each other's data, hypotheses, and models

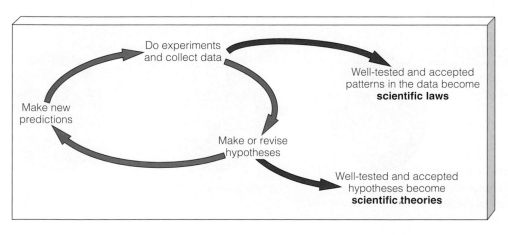

Figure 3-2 What scientists do. In the scientific process, a form of critical thinking, **(1)** facts (data) are gathered and verified by repeated experiments, **(2)** data are analyzed to see whether there is a consistent pattern of behavior that can be summarized as a *scientific law*, **(3)** hypotheses are proposed to explain the data, and **(4)** deductions or predictions are made and tested to evaluate each hypothesis. A hypothesis that is supported by a great deal of evidence and is widely accepted by the scientific community becomes a *scientific theory*.

Do experiments and collect data

Well-tested and accepted patterns in the data become **scientific laws**

Make new predictions

Make or revise hypotheses

Well-tested and accepted hypotheses become **scientific theories**

to careful scrutiny. The process takes this form: observe → hypothesize → argue → test → hypothesize → argue → test. This continues until the scientists involved reach some general consensus.

If many experiments by different scientists support a particular hypothesis, it becomes a **scientific theory**: an idea, principle, or model that **(1)** usually ties together and explains many facts that previously appeared to be unrelated and **(2)** is supported by a great deal of evidence. To scientists, theories are not to be taken lightly. They are ideas or principles stated with a high degree of certainty because they are supported by extensive evidence.

Nonscientists often use the word *theory* incorrectly when they mean to refer to a *scientific hypothesis*, a tentative explanation that needs further evaluation. The statement, "Oh, that's just a theory," made in everyday conversation, implies a lack of knowledge and careful testing—the opposite of the scientific meaning of the word.

Another important result of science is a **scientific** or **natural law**: a description of what we find happening in nature over and over in the same way, without known exception. For example, after making thousands of observations and measurements over many decades, scientists discovered what is called the *second law of energy* or *thermodynamics*. One simple way of stating this law is that heat always flows spontaneously from hot to cold—something you learned the first time you touched a hot object. *Scientific laws* describe what we find happening in nature in the same way, whereas *scientific theories* are widely accepted explanations of data and laws.

What Is the Difference Between Accuracy and Precision in Scientific Measurements? All scientific observations and measurements have some degree of uncertainty because people and measuring devices are not perfect. In determining the uncertainty involved in a measurement it is important to distinguish between

- **Accuracy:** how well a measurement agrees with the accepted or correct value for that quantity, and

- **Precision:** a measure of *reproducibility,* or how closely a series of measurements of the same quantity agree with one another

The dartboard analogy shown in Figure 3-3 illustrates the difference between precision and accuracy. *Accuracy* depends on how close the darts are to the bull's-eye. *Precision* depends on how close the darts are to each other.

How Do Scientists Learn About Nature? We often hear about *the* scientific method. In reality, there are many **scientific**

methods: ways scientists gather data and formulate and test scientific hypotheses, models, theories, and laws (Figure 3-2). Instead of being a recipe, each scientific method involves trying to answer a set of questions, with no particular guidelines for answering them:

- What is the question about nature to be answered?

- What relevant facts and data are already known?

- What new data (observations and measurements) should be collected, and how should this be done?

- After the data are collected, can they be used to formulate a scientific law or model?

- How can a hypothesis be invented that explains the data and predicts new facts? Is this the simplest and only reasonable hypothesis?

- What new experiments can be done to test the hypothesis (and modify it if necessary) so it can become a scientific theory?

New discoveries happen in many ways. Some follow this sequence: data → law → hypothesis → theory. At other times, scientists simply follow a hunch, bias, or belief and then do experiments to test the idea or hypothesis. Some discoveries occur when an experiment gives totally unexpected results and the scientist investigates to find out what happened.

According to physicist Albert Einstein, "There is no completely logical way to a new scientific idea." Intuition, imagination, and creativity are as important in science as they are in poetry, art, music, and other great adventures of the human spirit.

Most processes or parts of nature that scientists seek to understand are influenced by a number of *variables* or *factors*. One way scientists test a hypothesis about the effects of a particular variable is to conduct a *controlled experiment* to isolate and study the effect of a single vari-

Figure 3-3 The distinction between accuracy and precision. Good precision is necessary for accuracy but does not guarantee it. The closely spaced darts may be far away from the bull's-eye (right). In scientific measurements, a measuring device that has not been tested (calibrated) to determine its accuracy may give precise or reproducible results that are not accurate.

Good accuracy
and good precision

Poor accuracy
and poor precision

Poor accuracy
and good precision

able. This *single-variable analysis* is done by setting up two groups: **(1)** an *experimental group*, in which the chosen variable is changed in a known way, and **(2)** a *control group*, in which the chosen variable is not changed.

The experiment is designed so that all components of each group are as identical as possible and experience the same conditions, except for the single factor being varied in the experimental group. If the experiment is designed properly, any difference between the two groups should result from a variable that was changed in the experimental group (Connections, right).

Another example of a controlled experiment, called a *double-blind* experiment, is used to test most new drugs. One group of patients is given the new drug (the experimental group). A second, similar set of patients (the control group) is given a *placebo*, a harmless starch pill similar in shape, size, and color to the pill being tested. To avoid bias, neither the patients nor the physicians involved know who is getting the experimental drug and who is getting the placebo, which is why this is called a double-blind approach. The secret code used to identify the experimental and control patients is revealed only after the experiment is over and the results are in.

A basic problem is that many of the questions investigated by environmental scientists involve a huge number of interacting variables. In such cases, it is often difficult or impossible to carry out meaningful controlled experiments.

However, this limitation is being overcome by the use of *multivariable analysis*, made possible by the development of computing and communication technologies. This new and exciting way to do science uses mathematical models run on high-speed computers to analyze the interactions of many variables without having carry out traditional controlled experiments.

What Types of Reasoning Do Scientists Use? Scientists arrive at certain conclusions with varying degrees of certainty by using inductive and deductive reasoning. **Inductive reasoning** involves using observations and measurements to arrive at generalizations or hypotheses. For example, suppose we observe that a variety of different types of birds fly. We might then use inductive reasoning to conclude that *all birds fly*. Depending on the number of observations made there may be a high degree of certainty in this conclusion. However, what we are really saying is that "All of the birds that we or other observers have seen fly." Although it's highly unlikely, we cannot be absolutely sure that someone will not discover a type of bird that cannot fly. When scientists say that something had been proved or established by inductive reasoning they do not mean that it is absolutely true but that there is a very high probability or degree of certainty that it is true.

What's Harming the Robins?

CONNECTIONS

Suppose a scientist observes an abnormality in the growth of robin embryos in a certain area. She knows that the area has been sprayed with a pesticide and suspects that the chemical may be causing the abnormalities she has observed.

To test this hypothesis, the scientist carries out a *controlled experiment*. She maintains two groups of robin embryos of the same age in the laboratory. Each group is exposed to exactly the same conditions of light, temperature, food supply, and so on, except that the embryos in the experimental group are exposed to a known amount of the pesticide in question.

The embryos in both groups are then examined over an identical period of time for the abnormality. If there is a significantly larger number of the abnormalities in the experimental group than in the control group, then the results support the idea that the pesticide is the culprit.

To be sure there were no errors in the procedure, the experiment should be repeated several times by the original researcher, and ideally by one or more other scientists.

Critical Thinking

Can you find flaws in this experiment that might lead you to question the scientist's conclusions? (*Hint* : What other factors in nature—not the laboratory—and in the embryos themselves could possibly explain the results?)

Deductive reasoning involves using logic to arrive at a specific conclusion based on a generalization or premise. It goes from the general to the specific. For example:

Generalization or premise: All birds have feathers.

Example: Eagles are birds.

Deductive conclusion: All eagles have feathers.

This conclusion of this *syllogism* (a series of logically connected statements) is valid as long as the premise is correct and we do not use faulty logic to arrive at the conclusion.

However, inductive reasoning can be used to arrive at a logically correct but untrue statement. For example,

Generalization or premise: All animals with feathers are mammals.

Example: Eagles have feathers.

Deductive conclusion: Eagles are mammals.

Critical Thinking and Environmental Studies

Jane Heinze-Fry

Jane Heinze-Fry has a Ph.D. in environmental education and teaches environmental science and biology at Emerson College in Boston, Massachusetts. She is author of

GUEST ESSAY

Critical Thinking and Environmental Studies: A Beginner's Guide *(with G. Tyler Miller as coauthor). Previously, she taught and directed environmental studies at Sweet Briar College. She also taught biology to students at the junior high, high school, and college levels. Her interdisciplinary orientation is reflected in her concept maps, including the ones inside the back cover of this textbook.*

Learning how to think critically is essential in helping you evaluate the validity and usefulness of what you read in newspapers, magazines, and books (such as this textbook), what you hear in lectures and speeches, and what you see and hear on the news and in advertisements.

Learners engaged in *critical thinking* try to

- Connect new knowledge to prior knowledge and experience

- Evaluate the validity of claims made by people

- Relate what they have learned to their own life experiences

- Understand and evaluate their environmental worldviews

- Take positions on issues

- Develop and implement strategies for dealing with problems

Whenever we are faced with new information, we need to evaluate it by using critical thinking. Do we believe the information or not, and why? Do the claims seem reasonable or exaggerated? Here are some rules for evaluating scientific evidence and claims:

1. Gather all the information you can.

2. Understand the definitions of all key terms and concepts.

3. Question how the information (data) was obtained.

   - Were the studies well designed and carried out?

   - Was there an experimental group and a control group? Were the control and experimental groups treated identically except for the variable changed in the experimental group?

   - Did the investigators repeat their experiments several times and get essentially the same results?

   - Were the results verified by one or more other investigators?

4. Question the conclusions derived from the data.

   - Do the data support the claims, conclusions, and predictions?

   - Are there other possible interpretations? Are there more reasonable interpretations?

In this case the conclusion must be valid if the first two statements are valid. However, experience reveals that the first statement is not true. Thus, the conclusion is false.

How Valid Are the Results of Science? Scientists can do two major things: (1) disprove things and (2) establish that a particular model, theory, or law has a very high degree of validity or probability of being true. However, like scholars in any field, scientists cannot prove that their theories, models, and laws are *absolutely* true.

When people say that something has or has not been "scientifically proven," they can mislead us by falsely implying that science yields absolute proof or certainty. Although it may be extremely low, there is always some degree of uncertainty involved in any scientific theory, model, or law.

In addition, although scientists strive for objectivity independent of their personal beliefs or biases, they (like all people) have personal biases that can con-

sciously or unconsciously affect their objectivity. The goal of the rigorous scientific process is to reduce the degree of uncertainty and lack of objectivity as much as possible. The high standards of evidence required for reaching scientific conclusions accepted by most scientists in a field mean that science is the best way we have to get reliable knowledge about how nature works.

How Does Frontier Science Differ from Consensus Science? News reports often focus on new scientific "breakthroughs" and on disputes among scientists over the validity of preliminary (untested) data, hypotheses, and models (which are by definition tentative). These preliminary results, called **frontier science**, are controversial because they have not been widely tested and accepted.

By contrast, **consensus science** consists of data, theories, and laws that are widely accepted by scientists considered experts in the field involved. This aspect of science is very reliable but is rarely considered newsworthy. One way to find out what scientists generally agree on is to seek out reports by scientific bodies such

- Are the conclusions based on the results of original research by experts in the field involved, or are they drawn by reporters or scientists in other fields?

- Are the conclusions based on stories or reports of isolated events (*anecdotal information*) or on careful analysis of a large number of related observations?

5. Try to determine the assumptions and biases of the investigators and then question them.

- Do the investigators have a monetary or political advantage in the outcome of the investigation or issue involved?

- Would investigators with different basic assumptions or worldviews take the same data and come to different conclusions?

6. Are the data, claims, and conclusions based on the tentative results of *frontier science* or the more reliable and widely accepted results of *consensus science?*

7. Based on these steps, take a position by either rejecting or conditionally accepting the claims.

Other ways to improve your critical thinking skills involve using

- *Thinking strategies* such as constructing models, brainstorming, creating alternative solutions, and visualizing future possibilities

- *Attitude and value strategies* such as reflecting on the effects of your lifestyle on the environment

and understanding and evaluating your environmental worldview

- *Action strategies* such as evaluating alternative solutions, creating plans of action, and developing strategies for implementing action plans

In the environmental course you are taking, there are many opportunities to develop your critical thinking skills. Your textbook offers Critical Thinking questions at the ends of chapters and in most boxes. If your course uses the supplement *Critical Thinking and Environmental Studies: A Beginner's Guide,* you will learn the critical thinking strategies mentioned here.

Critical Thinking

1. Can you come up with an example in which critical thinking has helped you make a major change in one or more of your beliefs or helped you make an important personal decision? Can you think of a decision that may have come out better if you had used critical thinking skills such as those discussed in this essay?

2. Rote learning often involves the "memorize and spit back" strategy. Meaningful learning (including critical thinking) goes far beyond memorization and requires us to evaluate the validity of what we learn. Currently, about what percentage of your learning involves rote learning and what percentage involves critical thinking as discussed in this essay?

as the U.S. National Academy of Sciences and the British Royal Society that attempt to summarize consensus among experts in key areas of science.

To give a sense of balance and fairness, the media often quote one or more of a handful of scientists who criticize the consensus view of the vast majority of scientists in a particular field. Sometimes these critics are not experts in the field they are criticizing. Instead of balance, this can lead to a biased presentation because a minority view is given about the same weight as the consensus view held by most scientists in a particular field.

The history of science shows that occasionally the scientific consensus about an idea can be modified or overturned by new information or better ideas. However, until such an event occurs, the current scientific consensus is our most useful guideline.

What Are Some Limitations of Environmental Science? As mentioned earlier (p. 3), *environmental science* integrates knowledge from a number of disciplines to help us understand how natural systems and human

societies operate and interact. In other words, it is an interdisciplinary study of *connections* and *interactions* between humans and the rest of nature (see inside back cover of this book and Figure 1-16, p.16).

There is controversy over some of the knowledge provided by environmental science, for much of it falls into the realm of frontier science. At the preliminary frontier stage, it is normal and healthy for reputable scientists in a field to state contradictory opinions about the meaning and accuracy of scientific data and the validity of various hypotheses.

One problem involves *arguments over the validity of data.* There is no way to measure accurately how many metric tons of soil are eroded worldwide, how many hectares of tropical forest are cut, how many species become extinct, or how many metric tons of certain pollutants are emitted into the atmosphere or bodies of water each year.

The point environmental scientists want to make is that the trends in these phenomena are significant enough to be evaluated and addressed. Such environmental data should not be dismissed because they are "only estimates" (which are all we can ever have). However, this does not relieve investigators of the responsibility to get the best estimates possible and point out that they *are* estimates.

Another limitation is that *most environmental problems involve so many variables and such complex interactions that we often do not have enough information or sufficiently sophisticated mathematical models to aid in understanding them very well.* However, at some point we must use critical thinking skills (Guest Essay, p. 48) to

- evaluate the available (but always inadequate) scientific information about a particular environmental problem.

- list the possible solutions (including doing nothing).

- predict positive and negative consequences of each alternative solution and the probability that each consequence will occur.

- evaluate the alternatives and choose the best solution.

3-2 MODELS AND BEHAVIOR OF SYSTEMS

What Is a System and What Are Its Major Components? A **system** is a set of components that (1) function and interact in some regular and theoretically predictable manner and (2) can be isolated for the purposes of observation and study. The environment consists of a vast number of interacting systems involving living and nonliving things.

Most *systems* have the following key components:

- **Inputs** of things such as matter, energy, or information into the system.

- **Flows** or **throughputs** of matter, energy, or information within the system at certain rates.

- **Stores** or **storage areas** within a system where energy, matter, or information can accumulate for various lengths of time before being released. For example, your body stores various chemicals with different residence times, and water vapor typically remains in the lower atmosphere for about 10 days before it is replaced.

- **Outputs** of certain forms of matter, energy, or information that flow out of the system into *sinks* in the environment (such as the atmosphere, bodies of water, underground water, soil, and land surfaces).

Why Are Models of Complex Systems Useful? Over time, people have learned the value of using models as approximate representations or simulations of real systems to find out how systems work and to evaluate which ideas or hypotheses work.

Some of the most powerful and useful technologies invented by humans are mathematical models, which are used to supplement our mental models. *Mathematical models* consist of one or more equations used to describe the behavior of a system and to make predictions about the behavior of a system.

Making a mathematical model usually requires going many times through three steps: (1) Make a guess and write down some equations, (2) compute the predictions implied by the equations, and (3) compare the predictions with observations, the predictions of mental models, existing experimental data, and scientific hypotheses, laws, and theories.

Mathematical models are important because they can give us improved perceptions and predictions, especially in situations where our mental models are weak. Research has shown that mental models tend to be especially unreliable when (1) there are many interacting variables, (2) we attempt to extrapolate from too few experiences to a general case, (3) consequences follow actions only after long delays, (4) the consequences of actions lead to other consequences, (5) responses vary from one time to the next, and (6) controlled experiments (Connections, p. 47) are impossible, too slow, or too expensive to conduct.

Most of our effects on the environment involve these characteristics that limit the usefulness of most mental models. By using well-designed mathematical models of some aspect of the environment, we can analyze the interactions of multiple variables and gain the equivalent of hundreds or thousands of years of experience in a few weeks or months.

After building and testing a mathematical model, scientists use it to predict what is *likely* to happen under a variety of conditions. In effect, they use mathematical models to answer *if-then* questions: "*If* we do such and such, *then* what is likely to happen now and in the future?"

Despite its usefulness, a mathematical model is nothing more than a set of hypotheses or assumptions about how we think a certain system works. Such models (like all other models) are no better than (1) the assumptions built into them and (2) the data fed into them to make projections about the behavior of complex systems.

How Do Feedback Loops Affect Systems? In making and using mathematical and other models, it is important to know how systems operate and change. Systems undergo change as a result of feedback loops. A **feedback loop** occurs when one change leads to some other change, which eventually either reinforces or slows the original change.

Feedback loops occur when an output of matter, energy, or information is fed back into the system as an input. For example, recycling aluminum cans involves melting aluminum and feeding it back into an economic system to make new aluminum products. This feedback loop of matter reduces (1) the need to find, extract, and process virgin aluminum ore and (2) the flow of waste matter (discarded aluminum cans) into the environment.

Feedback loops can be either positive or negative. A **positive feedback loop** is a runaway cycle in which a change in a certain direction provides information that causes a system to change further in the same direction. Often such loops, called *vicious circles*, destabilize a system.

An example involves projected global warming of the atmosphere. Such warming appears to be shrinking the reflective white Arctic ice pack and exposing more open ocean. Because the open ocean absorbs much more heat than ice pack, as the ice pack diminishes the oceans will warm more rapidly. This can in turn melt more ice and lead to even faster warming of the ocean's surface in a positive feedback loop.

Another example involves depositing money in a bank at compound interest. In this case, the interest increases the balance, which through a positive feedback loop leads to more interest (Figure 1-2, p. 4).

In a **negative feedback loop**, one change leads to a lessening of that change. For example, to survive, you must maintain your body temperature within a certain range, regardless of whether the temperature outside is steamy or freezing. This phenomenon is called **home-** **ostasis**: the maintenance of favorable internal conditions despite fluctuations in external conditions. Homeostatic systems consist of one or more negative feedback loops that help maintain constant internal conditions when changes occur. Normally negative feedback is desirable because it helps stabilize a system.

Most systems contain one or a series of *coupled positive and negative feedback loops*. A negative (or corrective) feedback loop coupled to a positive feedback loop can dampen or even halt a positive feedback loop of runaway growth. Generally, at any given time, one of the two loops dominates the other depending on the state of the system.

For example, the temperature-regulating system of your body involves coupled negative and positive feedback loops (Figure 3-4). Normally a negative feedback regulates your body temperature. However, if your body temperature exceeds 42°C (108°F), your built-in negative feedback temperature control system breaks down as your body produces more heat than your sweat-dampened skin can get rid of. Then a positive feedback loop caused by overloading the system (Figure 3-4) overwhelms the negative or corrective feedback loop. These conditions produce a net gain in body heat, which produces even more body heat, and so on, until you die from heatstroke.

The tragedy on Easter Island (p. 44) also involved the coupling of positive and negative feedback loops. As the abundance of trees turned to a shortage of trees, the positive feedback loop (more births than deaths) became weaker as death rates rose, and the negative feedback loop (more deaths than births) eventually dominated and caused a dieback of the human population.

Figure 3-4 Coupled negative and positive feedback loops involved in temperature control of the human body. Homeostasis works in a limited range only: Above a certain body temperature, body metabolic rates get out of control and generate large amounts of heat. This positive (runaway) feedback loop generates more heat than the negative feedback loop can get rid of, and body temperature increases out of control, resulting in death.

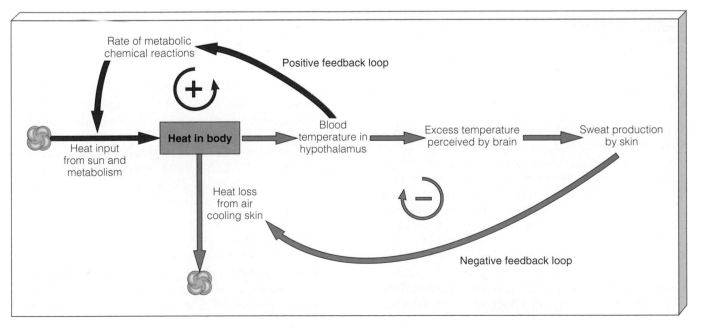

How Do Time Delays Affect Complex Systems?
Complex systems often show **time delays** between the input of a stimulus and the response to it. A long time delay can mean that corrective action comes too late. For example, a smoker exposed to cancer-causing chemicals in cigarette smoke may not get lung cancer for 20–30 years.

Time delays allow a problem to build up slowly until it reaches a *threshold level* and causes a fundamental shift in the behavior of a system, in what system analysts call a *discontinuity*. Examples in which prolonged delays dampen the negative feedback mechanisms that might slow, prevent, or halt environmental problems are **(1)** population growth, **(2)** leaks from toxic waste dumps, **(3)** depletion of ozone in the stratosphere (the second layer of the earth's atmosphere), **(4)** global warming and climate change from carbon dioxide and other chemicals we add to the atmosphere, and **(5)** degradation of forests from prolonged exposure to air pollutants.

How Can We Overcome Resistance to Change in Complex Systems?
Because of negative feedback, systems have a built-in resistance to change that helps maintain their stability. For example, complex systems, such as economic and political systems, are resistant to change because of multiple feedback loops. However, a careful analysis of such systems usually reveals a few components for which small changes can yield a large improvement. Such *leverage* often comes from making it easy and rewarding for people to do the right thing (Connections, right).

Two major goals of environmental science are to discover the positive feedback loops in complex natural, economic, and political systems that amplify

- *beneficial actions* that help sustain the earth's life-support systems and thus our own species and economies, and

- *environmentally harmful effects* and to find and promote negative feedback loops to correct or diminish such effects.

What Is Synergy and How Can It Affect Complex Systems?
In arithmetic, 1 plus 1 always equals 2. However, in some of the complex systems found in nature, 1 plus 1 may add up to more than 2 because of synergistic interactions. A **synergistic interaction** occurs when two or more processes interact so that the combined effect is greater than the sum of their separate effects.

Synergy can result when two people work together to accomplish a task. For example, suppose you and I need to move a 140-kilogram (300-pound) tree that has fallen across the road. By ourselves, each of us can lift only, say, 45 kilograms (100 pounds). However, if we cooperate and use our muscles properly, together we can move the tree out of the way. That is using synergy to solve a problem. Research in the social sciences suggests that most political changes or changes in cultural beliefs are brought about by only about 5% (and rarely more than 10%) of a population working together (synergizing) and expanding their efforts to other people.

Synergy amplifies the action of positive feedback loops and thus can promote change we believe is favorable. However, synergy also can amplify harmful changes. For example, in 1998 flooding of China's Yangtze River displaced about 223 million people and caused $30 million in damages. These damages were higher than expected because of the synergistic interaction of dense populations of people living on the river's floodplains and by deforestation in which the Yangtze River basin had lost 85% of its forest cover. As a result, water ran off of the denuded land at a faster rate, causing more extensive flooding. The deforestation also interacted synergistically to speed up soil erosion and increase water pollution from the resulting silt.

By identifying potentially harmful synergistic interactions and the leverage points that can activate them, we may be able to **(1)** counter harmful synergisms, **(2)** promote beneficial ones, and **(3)** improve the quality of life on the earth.

How Can We Anticipate Environmental Surprises?
It is comforting to think of environmental decline as gradual and predictable. However, this is not always the case. One of the basic principles of environmental science is that *we can never do one thing*. Any action in a complex system has multiple and often unpredictable effects.

In recent years, we have experienced unforeseen *environmental surprises* such as the **(1)** sudden dying of large areas of forest after years of exposure to air and soil pollutants, **(2)** a rapid decline in the health of coral reefs, and **(3)** the appearance of severe thinning of the ozone layer over Antarctica several months each year.

Environmental surprises are the result of

- *Discontinuities*, or abrupt shifts in a previously stable system when some *environmental threshold* is crossed. By analogy, you may be able to lean back in a chair and balance yourself on two of its legs for a long time with only minor adjustments. However, if you pass a certain threshold of movement, your balanced system suffers a discontinuity or sudden shift and you may find yourself on the floor.

- *Synergistic interactions*, in which two or more factors interact to produce effects greater than the sum of their effects acting separately.

- *Unpredictable or chaotic events* such as hurricanes, earthquakes, invasions of ecosystems by nonnative species, sudden shifts in climate, or slowly building but so far unknown environmental problems.

Using Taxes to Reduce Pollution

Levying taxes on pollution emissions gives business leaders incentives to find ways to reduce emissions and thus their taxes. They can do this by (1) inventing new technologies for removing pollutants (*pollution control*) or (2) redesigning the process to reduce or eliminate production of pollutants (*pollution prevention*).

Pollution taxes eventually pay for themselves because the cleaner environment leads to healthier people, animals, and plants. This eventually reduces the cost of health care, insurance, and cleaning.

Taxing emissions of pollutants is a *polluter-pays* approach. Understandably, polluters resist this approach because it means that they would have to bear the costs of pollution.

A *taxpayer- or consumer-pays* approach passes the costs of pollution on to taxpayers and people who buy products and to future generations. This means that they eventually pay for the costs of pollution in the form of poorer health, higher medical bills, and higher health insurance costs. However, these costs are hidden because they are not included in the market prices of things people buy.

Taxing pollutant emissions changes this system by incorporating most or all of the harmful environmental costs of products into their market prices—something called *full-cost pricing*. Consumers would pay more up front for most things, but their overall costs in the form of higher health and insurance costs should decrease.

Critical Thinking

Why do you think that the polluter-pays system based on full-cost pricing is not widely used? What lever points in the political system could you use to institute full-cost pricing?

Environmental scientists warn that we are undoubtedly in for some nasty environmental surprises and should expect the unexpected. Strategies to help deal with and reduce such surprises include the following:

- Greatly increasing research on environmental thresholds and synergistic interactions

- Developing models to understand the components and behavior of complex living systems and our economic and political systems

- Formulating scenarios of possible environmental surprises and developing a range of strategies for dealing with them (as defense departments and emergency management agencies do)

- Acting to prevent or lessen the effects of possible surprises through (1) pollution prevention, (2) more efficient and environmentally benign use of resources, and (3) reduced population growth

3-3 MATTER: FORMS, STRUCTURE, AND QUALITY

What Are Nature's Building Blocks? Matter is anything that has mass (the amount of material in an object) and takes up space. Matter includes the solids, liquids, and gases around us and within us. Matter is found in two *chemical forms*:

- **Elements:** the distinctive building blocks of matter that make up every material substance

- **Compounds:** two or more different elements held together in fixed proportions by attractive forces called *chemical bonds*

Various elements, compounds, or both can be found together in **mixtures**.

All matter is built from the 115 known chemical elements (92 of them occur naturally and the other 23 have been synthesized in laboratories). To simplify things, chemists represent each element by a one- or two-letter symbol; hydrogen (H), carbon (C), oxygen (O), nitrogen (N), phosphorus (P), sulfur (S), chlorine (Cl), fluorine (F), bromine (Br), sodium (Na), calcium (Ca), lead (Pb), mercury (Hg), and uranium (U) are but a few. These are the elements discussed in this book. Chemists have developed a way to classify elements in terms of their chemical behavior by arranging them in a *periodic table of elements*, as discussed in Appendix 3.

If you had a supermicroscope capable of looking at individual elements and compounds, you could see that they are made up of three types of building blocks

- **Atoms:** the smallest unit of matter that is unique to a particular element

- **Ions:** electrically charged atoms or combinations of atoms

- **Molecules:** combinations of two or more atoms of the same or different elements held together by chemical bonds

Some elements are found in nature as molecules. Examples are nitrogen and oxygen, which together make up about 99% of the volume of air you breathe. Two atoms of nitrogen (N) combine to form a gaseous molecule, with the shorthand formula N_2 (read as "N-two"). The subscript after the element's symbol indicates the number of atoms of that element in a molecule. Similarly, most of the oxygen gas in the atmosphere exists as O_2 (read as "O-two") molecules. A small amount of oxygen, found mostly in the second layer of the atmosphere (stratosphere), exists as O_3 (read as "O-three") molecules, a gaseous form of oxygen called *ozone*.

Matter is also found in three *physical states*: solid, liquid, and gas. For example, water exists as ice, liquid water, or water vapor depending on its temperature and pressure. The three physical states of matter differ in the spacing and orderliness of its atoms, ions, or molecules, with solids having the most compact and orderly arrangement and gases the least compact and orderly arrangement (Figure 3-5).

What Are Atoms Made Of? If you increased the magnification of your supermicroscope, you would find that each different type of atom contains a certain number of *subatomic particles*. The main building blocks of an atom are **(1)** positively charged **protons** (*p*), **(2)** uncharged **neutrons** (*n*), and **(3)** negatively charged **electrons** (*e*).

Each atom consists of an extremely small center, or **nucleus**, containing protons and neutrons, and one or more electrons in rapid motion somewhere outside the nucleus. The location of an electron in an atom can be described only in terms of its probability of being at any given place outside the nucleus. This is analogous to saying that a certain number of tiny gnats are found somewhere in a cloud, without being able to identify their exact positions.

Atoms are incredibly small. For example, more than 3 million hydrogen atoms could sit side by side on the period at the end of this sentence.

Each atom has an equal number of positively charged protons (inside its nucleus) and negatively charged electrons (outside its nucleus). Because these electrical charges cancel one another, *the atom as a whole has no net electrical charge.*

Each element has its own specific **atomic number**, equal to the number of protons in the nucleus of each of its atoms. The simplest element, hydrogen (H), has only 1 proton in its nucleus, so its atomic number is 1. Carbon (C), with 6 protons, has an atomic number of 6, whereas uranium (U), a much larger atom, has 92 protons and an atomic number of 92.

Because atoms are electrically neutral, the atomic number of an atom tells us the number of positively charged protons in its nucleus and the equal number of negatively charged electrons outside its nucleus. For example, an uncharged hydrogen atom with an atomic number of 1 has 1 positively charged proton in its nucleus and 1 negatively charged electron outside its

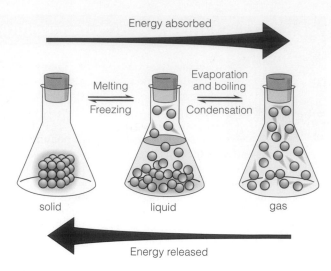

Figure 3-5 Comparison of the solid, liquid, and gaseous physical states of matter.

nucleus and thus no overall electrical charge. Similarly, each atom of uranium with an atomic number of 92 has 92 protons in its nucleus and 92 electrons outside, and thus no net electrical charge.

Because electrons have so little mass compared with the mass of a proton or a neutron, *most of an atom's mass is concentrated in its nucleus.* The mass of an atom is described in terms of its **mass number**: the total number of neutrons and protons in its nucleus. For example, a hydrogen atom with 1 proton and no neutrons in its nucleus has a mass number of 1, and an atom of uranium with 92 protons and 143 neutrons in its nucleus has a mass number of 235.

Although all atoms of an element have the same number of protons in their nuclei, they may have different numbers of uncharged neutrons in their nuclei, and thus may have different mass numbers. Various forms of an element having the same atomic number but a different mass number are called **isotopes** of that element. Isotopes are identified by attaching their mass numbers to the name or symbol of the element. For example, hydrogen has three isotopes: hydrogen-1 (H-1), hydrogen-2 (H-2, common name *deuterium*), and hydrogen-3 (H-3, common name *tritium*). A natural sample of an element contains a mixture of its isotopes in a fixed proportion or percentage abundance by weight (Figure 3-6).

What Are Ions? Atoms of some elements can lose or gain one or more electrons to form **ions**: atoms or groups of atoms with one or more net positive (+) or negative (−) electrical charges. For example, an atom of sodium (Na, atomic number 11) with 11 positively charged protons and 11 negatively charged electrons can lose one of its electrons. It then becomes a sodium ion with a positive charge of 1 (Na^+) because it now has

Hydrogen (H)

 1e

Mass number = 0 + 1 = 1
Hydrogen-1
(99.98%)

 1e

Mass number = 1 + 1 = 2
Hydrogen-2
or deuterium (D)
(0.015%)

 1e

Mass number = 2 + 1 = 3
Hydrogen-3
or tritium (T)
(trace)

Uranium (U)

 92 e

Mass number = 143 + 92 = 235
Uranium-235
(0.7%)

 92 e

Mass number = 146 + 92 = 238
Uranium-238
(99.3%)

Figure 3-6 Isotopes of hydrogen and uranium. All isotopes of hydrogen have an atomic number of 1 because each has one proton in its nucleus; similarly, all uranium isotopes have an atomic number of 92. However, each isotope of these elements has a different mass number because its nucleus contains a different number of neutrons. Figures in parentheses indicate the percentage abundance by weight of each isotope in a natural sample of the element.

11 positive charges (protons) but only 10 negative charges (electrons). An atom of chlorine (Cl, with an atomic number of 17) can gain an electron and become a chlorine ion with a negative charge of 1 (Cl^-) because it then has 17 positively charged protons and 18 negatively charged electrons.

The number of positive or negative charges on an ion is shown as a superscript after the symbol for an atom or a group of atoms. Examples of other positive ions are hydrogen ions (H^+), calcium ions (Ca^{2+}) and ammonium ions (NH_4^+); other common negative ions are nitrate ions (NO_3^-), sulfate ions (SO_4^{2-}), and phosphate ions (PO_4^{3-}). These are the ions discussed in this book.

The amount of a substance in a unit volume of air, water, or other medium is called its **concentration**. The concentration of hydrogen ions (H^+) in a water solution is a measure of its acidity or alkalinity. **pH** is a measure of the concentration of H^+ in a water solution. On a *pH scale* of 0 to 14, *acids* have a pH less than 7, *bases* have a pH greater than 7, and a *neutral solution* has a pH of 7 (Figure 3-7).

What Holds the Atoms and Ions in Compounds Together? Most matter exists as compounds. Chemists use a shorthand **chemical formula** to show the number of atoms (or ions) of each type in a compound. The formula contains the symbols for each of the elements present and uses subscripts to represent the number of atoms or ions of each element in the compound's basic structural unit. Compounds made up of oppositely charged ions are called *ionic compounds*, and those made up of molecules of uncharged atoms are called *covalent* or *molecular compounds*.

Sodium chloride (table salt), an *ionic compound*, is represented by NaCl. It consists of a three-dimensional array of oppositely charged *ions* (Na^+ and Cl^-). The forces of attraction between these oppositely charged ions are called *ionic bonds*, as discussed in more detail in Appendix 3.

Water, a *covalent* or *molecular compound*, consists of molecules made up of uncharged atoms of hydrogen (H) and oxygen (O). Each water molecule consists of two hydrogen atoms chemically bonded to an oxygen atom, yielding H_2O (read as "H-two-O") molecules. The bonds between the atoms in such molecules are called *covalent bonds*, as discussed in Appendix 3. There are also weaker forces of attraction, called *hydrogen bonds*, between the molecules of covalent compounds (such as water), as discussed in Appendix 3.

What Are Organic and Inorganic Compounds? Table sugar, vitamins, plastics, aspirin, penicillin, and many other important materials have one thing in common: They are **organic compounds**, containing carbon atoms combined with each other and with atoms of one or more other elements such as hydrogen, oxygen, nitrogen, sulfur, phosphorus, chlorine, and fluorine. Almost all organic compounds are molecular compounds held together by covalent bonds. Organic compounds can be either natural or synthetic (such as plastics and many drugs made by humans).

The millions of known organic (carbon-based) compounds include the following:

- *Hydrocarbons*: compounds of carbon and hydrogen atoms. An example is methane (CH_4), the main component of natural gas.

- *Chlorinated hydrocarbons*: compounds of carbon, hydrogen, and chlorine atoms. Examples are DDT ($C_{14}H_9Cl_5$, an insecticide) and toxic polychlorinated biphenyls (such as $C_{12}H_5Cl_5$), oily compounds used as insulating materials in electric transformers.

- *Chlorofluorocarbons* (CFCs): compounds of carbon, chlorine, and fluorine atoms. An example is Freon-12 (CCl_2F_2), until recently widely used as a coolant in refrigerators and air conditioners, as an aerosol propellant, and as a foaming agent for making some plastics.

- *Simple carbohydrates* (simple sugars): certain types of compounds of carbon, hydrogen, and oxygen atoms. An example is glucose ($C_6H_{12}O_6$), which most plants and animals break down in their cells to obtain energy.

Larger and more complex organic compounds, called *polymers,* consist of a number of basic structural or molecular units (*monomers*) linked by chemical bonds, somewhat like cars linked in a freight train. The three major types of organic polymers are complex carbohydrates, proteins, and nucleic acids.

Complex carbohydrates are made by linking a number of simple carbohydrate molecules such as glucose ($C_6H_{12}O_6$). Examples are the complex starches in rice and potatoes and cellulose found in the walls around plant cells. Simple and complex carbohydrates are broken down in cells to supply energy.

Proteins are produced in cells by the linking of different sequences of about 20 different monomers known as *alpha-amino acids*, whose number and sequence in each protein are specified by the genetic code found in DNA molecules in an organism's cells. From only about 20 alpha-amino acid molecules, the earth's life-forms can make tens of millions of different protein molecules.

Nucleic acids, such as DNA and RNA, are made by linking hundreds to thousands of five different types of monomers, called *nucleotides,* as discussed in Appendix 3. DNA molecules contain the hereditary instructions for assembling new cells and assembling the proteins each cell needs to survive and reproduce.

Genes consist of specific sequences of nucleotides in a DNA molecule. Each gene carries codes (each consisting of three nucleotides) needed to make various proteins. These coded units of genetic information about specific traits are passed on from parents to offspring during reproduction. Collectively, the complete set of genetic information for an organism is called a **genome**. You have about 100,000 genes in your body. Occasionally one or more of the nucleotide bases in a gene sequence are deleted, added, or replaced. Such changes, called **gene mutations**, can be helpful or harmful to an organism and its offspring or not affect them at all.

Chromosomes are combinations of genes that make up a single DNA molecule, together with a number of proteins. Each chromosome typically contains thousands of genes. Genetic information coded in your chromosomal DNA is what makes you different from an oak leaf, an alligator, or a flea and from your parents. The relationships of genetic material to cells are depicted in Figure 3-8.

All other compounds are called **inorganic compounds**. Such compounds do not have carbon-carbon or carbon-hydrogen covalent bonds. Some of the inorganic compounds discussed in this book are sodium chloride (NaCl), water (H_2O), nitrous oxide (N_2O), nitric oxide (NO), carbon monoxide (CO), carbon dioxide (CO_2), nitrogen dioxide (NO_2), sulfur dioxide (SO_2), ammonia (NH_3),

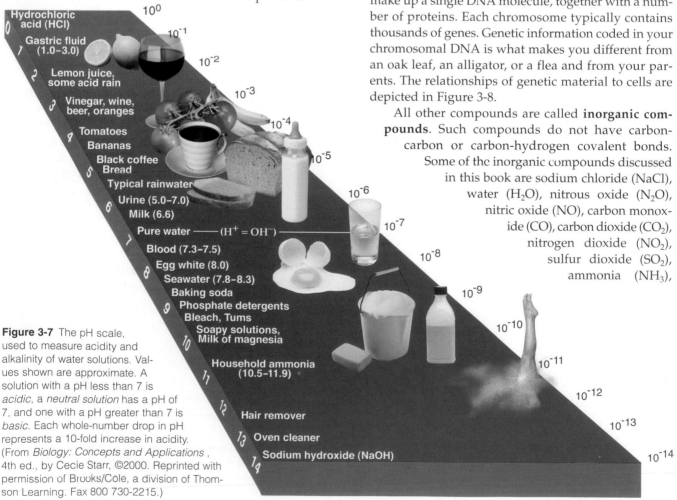

Figure 3-7 The pH scale, used to measure acidity and alkalinity of water solutions. Values shown are approximate. A solution with a pH less than 7 is *acidic,* a *neutral solution* has a pH of 7, and one with a pH greater than 7 is *basic.* Each whole-number drop in pH represents a 10-fold increase in acidity. (From *Biology: Concepts and Applications*, 4th ed., by Cecie Starr, ©2000. Reprinted with permission of Brooks/Cole, a division of Thomson Learning. Fax 800 730-2215.)

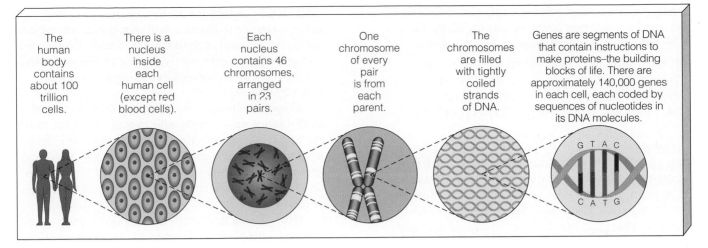

The human body contains about 100 trillion cells.

There is a nucleus inside each human cell (except red blood cells).

Each nucleus contains 46 chromosomes, arranged in 23 pairs.

One chromosome of every pair is from each parent.

The chromosomes are filled with tightly coiled strands of DNA.

Genes are segments of DNA that contain instructions to make proteins–the building blocks of life. There are approximately 140,000 genes in each cell, each coded by sequences of nucleotides in its DNA molecules.

Figure 3-8 Relationships between cells, nuclei, chromosomes, DNA, and genes.

hydrogen sulfide (H_2S), sulfuric acid (H_2SO_4), and nitric acid (HNO_3). These are the major compounds discussed in this book.

What Are Matter Quality and Material Efficiency?
Matter quality is a measure of how useful a form of matter is to us as a resource, based on its availability and concentration. **High-quality matter** is (1) organized, (2) concentrated, (3) usually found near the earth's surface and (4) has great potential for use as a matter resource. **Low-quality matter** is (1) disorganized, (2) dilute, (3) often deep underground or dispersed in the ocean or the atmosphere and (4) usually has little potential for use as a matter resource (Figure 3-9).

An aluminum can is a more concentrated, higher-quality form of aluminum than aluminum ore containing the same amount of aluminum. That's why it takes less energy, water, and money to recycle an aluminum can than to make a new can from aluminum ore.

Entropy is a measure of the disorder or randomness of a system or its environment. The greater the disorder of a sample of matter, the higher its entropy, and the greater its order, the lower its entropy. Thus, an aluminum can has a lower entropy (more order) than aluminum ore with the same amount of aluminum mixed with other materials. Similarly, a piece of ice in which the water molecules are held in an ordered solid structure has a lower entropy (more order) than the highly dispersed and rapidly moving water molecules in water vapor (Figure 3-5).

Material efficiency or **resource productivity** is the total amount of material needed to produce each unit of goods or services. Although resource productivity has been improving, only about 2–6% of the matter resources flowing through the economies of developed countries ends up providing useful goods and services. Because of such waste, business expert Paul Hawken (Guest Essay, p. 6) and physicist Amory Lovins (Guest

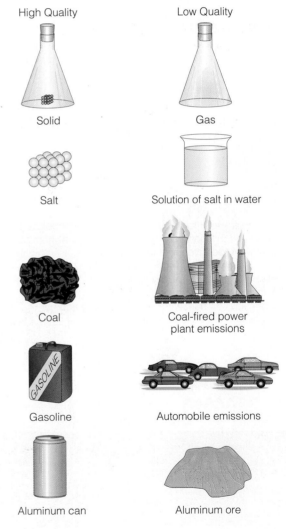

Figure 3-9 Examples of differences in matter quality. High-quality matter (left-hand column) is fairly easy to extract and is concentrated; low-quality matter (right-hand column) is more difficult to extract and is more dispersed than high-quality matter.

Essay, p. 361) contend that resource productivity in developed countries could be improved by 75–90% within two decades using existing technologies, as discussed in more detail in the solutions box on p. 525.

3-4 ENERGY: FORMS AND QUALITY

What Are Different Forms of Energy? **Energy** is the capacity to do work and transfer heat. Work is performed when an object—be it a grain of sand, this book, or a giant boulder—is moved over some distance. Work, or matter movement, also is needed to **(1)** boil water (to change it into the more dispersed and faster-moving water molecules in steam) or **(2)** to burn natural gas to heat a house or cook food. Energy is also the heat that flows automatically from a hot object to a cold object when they come in contact.

Scientists classify energy as either kinetic or potential. **Kinetic energy** is the energy that matter has because of its mass and its speed or velocity. It is energy in action or motion. Examples are **(1)** wind (a moving mass of air), **(2)** flowing streams, **(3)** heat flowing from a body at a high temperature to one at a lower temperature, **(4)** electricity (flowing electrons), **(5)** electromagnetic radiation (Figure 3-10),

(6) heat (the total kinetic energy of all the moving atoms, ions, or molecules within a given substance, excluding the overall motion of the whole object), and **(7)** temperature (the average speed of motion of the atoms, ions, or molecules in a sample of matter at a given moment).

Potential energy is stored energy that is potentially available for use. A rock held in your hand, an unlit stick of dynamite, still water behind a dam, the chemical energy stored in gasoline molecules, and the nuclear energy stored in the nuclei of atoms all have potential energy because of their position or the position of their parts.

Potential energy can be changed to kinetic energy. When you drop a rock, its potential energy changes into kinetic energy. When you burn gasoline in a car engine, the potential energy stored in the chemical bonds of its molecules changes into heat, light, and mechanical (kinetic) energy that propels the car.

What Is Energy Quality? **Energy quality** is a measure of an energy source's ability to do useful work (Figure 3-11). **High-quality energy** is organized or concentrated and can perform much useful work. Examples are **(1)** electricity, **(2)** the chemical energy stored in coal and gasoline, **(3)** concentrated sunlight, **(4)** nuclei of uranium-235 used as fuel in nuclear power plants, and **(5)** heat concentrated in small amounts of matter so that its temperature is high.

By contrast, **low-quality energy** is disorganized or dispersed and has little ability to do useful work. An example is heat dispersed in the moving molecules of a large amount of matter (such as the atmosphere or a large body of water) so that its temperature is low. For example, the total amount of heat stored in the Atlantic Ocean is greater than the amount of high-quality chemical energy stored in all the oil deposits of Saudi Arabia. Yet the ocean's heat is so widely dispersed that it

Figure 3-10 The *electromagnetic spectrum*: the range of electromagnetic waves, which differ in wavelength (distance between successive peaks or troughs) and energy content. Cosmic rays, gamma rays, X rays, and ultraviolet radiation are called *ionizing radiation* because they have enough kinetic energy to knock electrons from atoms and change them to positively charged ions. The resulting highly reactive electrons and ions can **(1)** disrupt living cells, **(2)** interfere with body processes, and **(3)** cause many types of sickness, including various cancers. The other forms of electromagnetic radiation (right side) do not contain enough kinetic energy to form ions and are called *nonionizing radiation*.

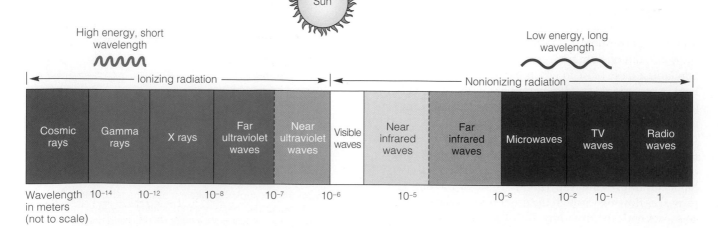

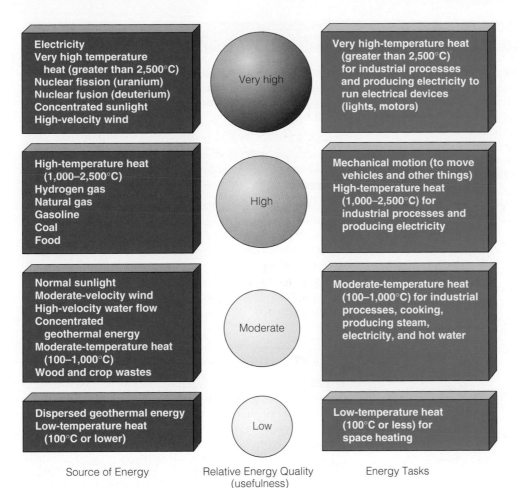

Source of Energy	Relative Energy Quality (usefulness)	Energy Tasks
Electricity Very high temperature heat (greater than 2,500°C) Nuclear fission (uranium) Nuclear fusion (deuterium) Concentrated sunlight High-velocity wind	Very high	Very high-temperature heat (greater than 2,500°C) for industrial processes and producing electricity to run electrical devices (lights, motors)
High-temperature heat (1,000–2,500°C) Hydrogen gas Natural gas Gasoline Coal Food	High	Mechanical motion (to move vehicles and other things) High-temperature heat (1,000–2,500°C) for industrial processes and producing electricity
Normal sunlight Moderate-velocity wind High-velocity water flow Concentrated geothermal energy Moderate-temperature heat (100–1,000°C) Wood and crop wastes	Moderate	Moderate-temperature heat (100–1,000°C) for industrial processes, cooking, producing steam, electricity, and hot water
Dispersed geothermal energy Low-temperature heat (100°C or lower)	Low	Low-temperature heat (100°C or less) for space heating

Figure 3-11 Categories of the quality (usefulness for performing various energy tasks) of different sources of energy. *High-quality energy* is concentrated and has great ability to perform useful work; *low-quality energy* is dispersed and has little ability to do useful work. To avoid unnecessary energy waste, it is best to match the quality of an energy source with the quality of energy needed to perform a task.

cannot be used to move things or to heat things to high temperatures. It makes sense to match the quality of an energy source with the quality of energy needed to perform a particular task (Figure 3-11) because doing so saves energy and usually money unless goverment subsidies or taxes have distorted the energy market place.

3-5 PHYSICAL AND CHEMICAL CHANGES AND THE LAW OF CONSERVATION OF MATTER

What Is the Difference Between a Physical and a Chemical Change? A **physical change** involves no change in chemical composition. Cutting a piece of aluminum foil into small pieces is one example. Changing a substance from one physical state to another is a second example. When solid water (ice) is melted or liquid water is boiled, none of the H_2O molecules involved are altered; instead, the molecules are organized in different spatial (physical) patterns (Figure 3-5).

In a **chemical change** or **chemical reaction**, the chemical compositions of the elements or compounds are altered. Chemists use shorthand chemical equations to represent what happens in a chemical reaction. For example, when coal burns completely, the solid carbon (C) it contains combines with oxygen gas (O_2) from the atmosphere to form the gaseous compound carbon dioxide (CO_2):

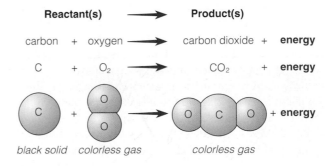

Reactant(s)	⟶	Product(s)
carbon + oxygen ⟶		carbon dioxide + **energy**
C + O_2 ⟶		CO_2 + **energy**

black solid   *colorless gas*   *colorless gas*

Energy is given off in this reaction, making coal a useful fuel. The reaction also shows how the complete

In keeping with the law of conservation of matter, each side of a chemical equation must have the same number of atoms of each element involved. When this is the case, the equation is said to be *balanced*. The equation for the burning of carbon ($C + O_2 \longrightarrow CO_2$) is balanced because there is one atom of carbon and two atoms of oxygen on both sides of the equation.

Consider the following chemical reaction: When electricity is passed through water (H_2O), the latter can be broken down into hydrogen (H_2) and oxygen (O_2), as represented by the following equation:

$$H_2O \longrightarrow H_2 + O_2$$

2 H atoms   2 H atoms   2 O atoms
1 O atom

This equation is unbalanced because there is one atom of oxygen on the left but two atoms on the right.

We cannot change the subscripts of any of the formulas to balance this equation because then we would be changing the arrangements of the atoms involved. Instead, we could use different numbers of the *molecules* involved to balance the equation. For example, we could use two water molecules:

$$2\,H_2O \longrightarrow H_2 + O_2$$

4 H atoms   2 H atoms   2 O atoms
2 O atom

This equation is still unbalanced because even though the numbers of oxygen atoms on both sides are now equal, the numbers of hydrogen atoms are not.

We can correct this by having the reaction produce two hydrogen molecules.

$$2\,H_2O \longrightarrow 2\,H_2 + O_2$$

4 H atoms   4 H atoms   2 O atoms
2 O atoms

Now the equation is balanced, and the law of conservation of matter has not been violated. We see that for every two molecules of water through which we pass electricity, two hydrogen molecules and one oxygen molecule are produced.

Try to balance the chemical equation for the reaction of nitrogen gas (N_2) with hydrogen gas (H_2) to form ammonia gas (NH_3).

Critical Thinking

1. Balancing equations is based on the law of conservation of matter. Do you believe that this is an ironclad law of nature or one that through new scientific discoveries could be overthrown? Explain.

2. Imagine that you have the power to revoke the law of conservation of matter. List the three major ways this would affect your life.

---

burning of coal (or any of the carbon-containing compounds in wood, natural gas, oil, and gasoline) gives off carbon dioxide gas, which is a key gas that can lead to warming of the lower atmosphere (troposphere).

The Law of Conservation of Matter: Why Is There No "Away"? People commonly talk about consuming or using up material resources, but the truth is that *we do not consume matter*; we only use some of the earth's resources for a while. We take materials from the earth, carry them to another part of the globe, and process them into products that are used and then discarded, burned, buried, reused, or recycled.

We may change various elements and compounds from one physical or chemical form to another, but in no physical and chemical change can we create or destroy any of the atoms involved. All we can do is rearrange them into different spatial patterns (physical changes) or different combinations (chemical changes). The italicized statement, based on many thousands of measurements, is known as the **law of conservation of matter**. In describing chemical reactions, chemists use a shorthand system to make sure atoms are neither created nor destroyed, as required by the law of conservation of matter (Spotlight, above).

The law of conservation of matter means that there really is no "away" in "to throw away." *Everything we think we have thrown away is still here with us in one form or another.* We can collect dust and soot from the smokestacks of industrial plants, but these solid wastes must then be put somewhere. We can remove substances from polluted water at a sewage treatment plant, but the gooey sludge must be **(1)** burned (producing some air pollution), **(2)** buried (possibly contaminating underground water supplies), or **(3)** cleaned up and applied to the land as fertilizer (dangerous if the sludge contains nondegradable toxic metals such as lead and mercury). Banning use of the pesticide DDT in the United States but still selling it abroad means that it can return to the United States as DDT residues in imported coffee, fruit, and other foods, or as fallout from air masses moved long distances by winds.

How Harmful Are Pollutants? We can make the environment cleaner and convert some potentially harmful chemicals into less harmful physical or chemical forms. However, the *law of conservation of matter means that we will always face the problem of what to do*

Table 3-1 Equivalents of Some Trace Concentration Units

Unit	1 part per million	1 part per billion	1 part per trillion
Time	1 minute in two years	1 second in 32 years	1 second in 320,000 centuries
Money	1¢ in $10,000	1¢ in $10,000,000	1¢ in $10,000,000,000
Weight	1 pinch of salt in 10 kilograms (22lbs.) of potato chips	1 pinch of salt in 10 tons of potato chips	1 pinch of salt in 10,000 tons of potato chips
Volume	1 drop in 1,000 liters (265 gallons) of water	1 drop in 1,000,000 liters (265,000 gallons) of water	1 drop in 1,000,000,000 liters (265,000,000 gallons) of water

with some quantity of wastes and pollutants. By placing much greater emphasis on pollution prevention, waste reduction, and more efficient resource use, we can greatly reduce the amount of wastes and pollution we add to the environment.

Thus, regardless of what we do we will always have certain pollutants that can cause potential harm to humans or other forms of life. Three factors that determine how severe the harmful effects of a pollutant are its

- *Chemical nature.*

- *Concentration*, which is sometimes expressed in *parts per million (ppm)*, with 1 ppm corresponding to 1 part pollutant per 1 million parts of the gas, liquid, or solid mixture in which the pollutant is found. Smaller concentration units are parts per billion (ppb) and parts per trillion (ppt) (Table 3-1). The concentration of a pollutant can be reduced by dumping it into the air or a large volume of water, but there are limits to the effectiveness of this dilution approach.

- **Persistence**, or how long it stays in the air, water, soil, or body.

Pollutants can be classified into three categories based on their persistence as

- **Degradable**, or **nonpersistent, pollutants** that are broken down completely or reduced to acceptable levels by natural physical, chemical, and biological processes. Complex chemical pollutants broken down (metabolized) into simpler chemicals by living organisms (usually specialized bacteria) are called **biodegradable pollutants**. Human sewage in a river, for example, is biodegraded fairly quickly by bacteria if the sewage is not added faster than it can be broken down.

- **Slowly degradable**, or **persistent, pollutants** that take decades or longer to degrade. Examples include the insecticide DDT and most plastics.

- **Nondegradable pollutants** that cannot be broken down by natural processes. Examples include the toxic elements lead and mercury.

3-6 NUCLEAR CHANGES

What Is Natural Radioactivity? In addition to physical and chemical changes, matter can undergo a third type of change known as a **nuclear change**. This occurs when nuclei of certain isotopes spontaneously change or are made to change into one or more different isotopes. Three types of nuclear change are (1) natural radioactive decay, (2) nuclear fission, and (3) nuclear fusion.

Natural radioactive decay is a nuclear change in which unstable isotopes spontaneously emit fast-moving chunks of matter (called particles), high-energy radiation, or both at a fixed rate. The unstable isotopes are called **radioactive isotopes** or **radioisotopes**. Radioactive decay into various isotopes continues until the original isotope is changed into a stable isotope that is not radioactive.

Radiation emitted by radioisotopes is damaging ionizing radiation (Figure 3-10). The most common form of ionizing energy released from radioisotopes is **gamma rays**, a form of high-energy electromagnetic radiation. High-speed ionizing particles emitted from the nuclei of radioactive isotopes are most commonly of two types: (1) **alpha particles** (fast-moving, positively charged chunks of matter that consist of two protons and two neutrons) and (2) **beta particles** (high-speed electrons).

Figure 3-12 depicts the relative penetrating power of alpha, beta, and gamma ionizing radiation. All of us are exposed to small amounts of harmful ionizing radiation from both natural and human sources.

Each type of radioisotope spontaneously decays at a characteristic rate into a different isotope. This rate of decay can be expressed in terms of **half-life**: the time needed for *one-half* of the nuclei in a radioisotope to decay and emit their radiation to form a different isotope (Figure 3-13). The decay continues, often producing a series of different radioisotopes, until a nonradioactive isotope is formed. Each radioisotope has a characteristic half-life, which may range from a few millionths of a second to several billion years (Table 3-2).

An isotope's half-life cannot be changed by temperature, pressure, chemical reactions, or any other known factor. Half-life can be used to estimate how long

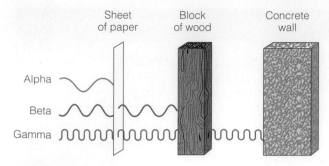

Figure 3-12 The three principal types of ionizing radiation emitted by radioactive isotopes differ greatly in their penetrating power.

Table 3-2 Half-Lives of Selected Radioisotopes		
Isotope	Radiation Half-Life	Emitted
Potassium-42	12.4 hours	Alpha, beta
Iodine-131	8 days	Beta, gamma
Cobalt-60	5.27 years	Beta, gamma
Hydrogen-3 (tritium)	12.5 years	Beta
Strontium-90	28 years	Beta
Carbon-14	5,370 years	Beta
Plutonium-239	24,000 years	Alpha, gamma
Uranium-235	710 million years	Alpha, gamma
Uranium-238	4.5 billion years	Alpha, gamma

a sample of a radioisotope must be stored in a safe container before it decays to what is considered a safe level. A general rule is that such decay takes about 10 half-lives. Thus, people must be protected from radioactive waste containing iodine-131 (which concentrates in the thyroid gland and has a half-life of 8 days) for 80 days (10 × 8 days). Plutonium-239 (which has a half-life of 24,000 years, is produced in nuclear reactors, and is used as the explosive in some nuclear weapons) can cause lung cancer when its particles are inhaled in minute amounts. Thus, it must be stored safely for 240,000 years (10 × 24,000 years)—about four times longer than the latest version of our species has existed.

How Much Ionizing Radiation Are We Exposed To? Each year people are exposed to some ionizing radiation (Figure 3-10) from natural or background sources and from human activities (Figure 3-14). Sources of natural ionizing radiation include cosmic rays from outer space, soil, rocks, air, water, and food.

Nuclear power plants provide very low exposure if they are operating properly. However, a serious nuclear accident, such as the one that occurred in 1986 at the Chernobyl nuclear power plant in Ukraine (p. 350), can release large quantities of radioactive materials, killing and harming people and rendering large areas uninhabitable.

Most human-caused exposure to ionizing radiation comes from medical X rays and from diagnostic tests and treatments using radioactive isotopes. The federal government estimates that one-third of the X rays taken each year in the United States are unnecessary.

What Are the Effects of Ionizing Radiation? Ionizing radiation can cause harm by **(1)** penetrating a human cell, **(2)** knocking loose (ionizing) one or more electrons from a cellular chemical, and **(3)** altering molecules needed for normal cellular functioning. Exposure to ionizing radiation can damage cells in two ways:

- *Genetic damage* from mutations in DNA molecules that alter genes and chromosomes (Figure 3-8). If the mutation is harmful, genetic defects can become apparent in the next generation of offspring or several generations later.

- *Somatic damage* to tissues, which causes harm during the victim's lifetime. Examples include burns, miscarriages, eye cataracts, and certain cancers.

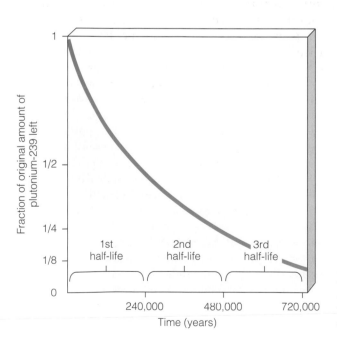

Figure 3-13 The radioactive decay of plutonium-239, which is produced in nuclear reactors and used as the explosive in some nuclear weapons, has a half-life of 240,000 years. The amount of radioactivity emitted by a radioactive isotope decreases by one-half for each half-life that passes. Thus, after three half-lives, amounting to 720,000 years, one-eighth of a sample of plutonium-239 would still be radioactive.

The effects of ionizing radiation vary with the **(1)** type and penetrating power (Figure 3-12), **(2)** source (outside or inside the body), and **(3)** half-life of the radioisotope (Figure 3-13 and Table 3-2). Alpha particles lack the penetrating power of beta particles but have more energy; thus, alpha-emitting isotopes are particularly dangerous when breathed in or ingested with food or water. Alpha particles outside the body can cause skin cancer but cannot penetrate the skin and reach vital organs.

Because beta particles can penetrate the skin, a beta emitter outside the body can damage internal organs. In general, radioisotopes with intermediate half-lives pose the greatest threat to human health; they stay around long enough to reach and enter the human body but decay fast enough to cause damage while they are around.

According to the National Academy of Sciences, exposure over an average lifetime to average levels of ionizing radiation from natural and human sources (Figure 3-14) causes about 1% of all fatal cancers and 5–6% of all normally encountered genetic defects in the U.S. population.

Is Nonionizing Electromagnetic Radiation Harmful? The simple answer is that we don't know. *Electromagnetic fields (EMFs)* are low-energy, nonionizing forms of electromagnetic radiation (Figure 3-10) given off when an electric current passes through a wire or a motor.

Sources of these weak electrical and magnetic fields include **(1)** overhead power lines and **(2)** household electrical appliances (such as microwave ovens, hair dryers, electric blankets, waterbed heaters, electric razors, computer monitors, and TV sets).

Since the late 1960s there has been growing public concern and controversy over the possibility that EMFs could have harmful health effects on humans. Numerous epidemiological studies have suggested that prolonged exposure to EMFs could lead to increased risk from **(1)** some cancers (including childhood leukemia, brain tumors, and breast cancer), **(2)** miscarriages, **(3)** birth defects, and **(4)** Alzheimer's disease. More recent studies show an insignificant correlation between EMFs and such health effects. It may be decades before these contradictory results and evaluations are resolved.

What Are Some Useful Applications of Radioisotopes? Scientists can use radioisotopes to estimate the age of ancient rocks, bones, and fossils. One common method, called *radiocarbon dating*, uses radioactive carbon-14 to estimate the age of plants, wood, teeth, bone, fossils, and other carbon-containing substances from dead plants and animals.

Radioisotopes are also used as *tracers* in pollution detection, agriculture, and industry. Suppose that a leak occurs somewhere in an underground pipeline carrying water or some other chemical. One way to locate the leak (without digging up the pipeline until the leak is found) is to mix a harmless and fairly rapidly decaying radioisotope with the material being transported in the pipeline. Then a radiation detector can be used to scan the ground above the pipeline until a high level of radioactivity is measured (where the pipe is leaking).

The U.S. Department of Agriculture exposed millions of screwworm flies (an insect pest that can kill cattle) to radioisotopes in laboratories, resulting in sterile male flies. Once released in large numbers into the natural population, these flies can outnumber the fertile males, reducing the pest population.

A special branch of medicine, called *nuclear medicine*, uses radioisotopes for diagnosis and treatment of various diseases. For example, radioactive sodium-24 can be injected in a salt solution into the bloodstream. Then its radioactivity can be measured to detect constrictions or blockages in the blood vessels.

Excessive exposure to radioactivity can cause cancer, but it is also useful in treating cancer by using high doses of radioactivity from radioisotopes (such as cobalt-60 or cesium-137) to kill rapidly growing cancer cells. Because the radioactivity also kills noncancerous cells, people exposed to radioactivity to kill cancer cells often have nausea, diarrhea, hair loss, and a low white blood cell count (which reduces their ability to fight infections).

What Is Nuclear Fission? Splitting Nuclei Nuclear fission is a nuclear change in which nuclei of certain isotopes with large mass numbers (such as uranium-235) are split apart into lighter nuclei when struck by neutrons; each fission releases two or three more neutrons and energy (Figure 3-15). Each of these neutrons, in turn,

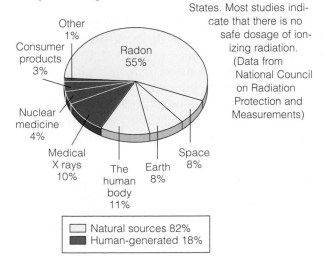

Figure 3-14 Natural and human sources of the average annual dosage of ionizing radiation received by people in the United States. Most studies indicate that there is no safe dosage of ionizing radiation. (Data from National Council on Radiation Protection and Measurements)

Other 1%
Consumer products 3%
Radon 55%
Nuclear medicine 4%
Medical X rays 10%
The human body 11%
Earth 8%
Space 8%

Natural sources 82%
Human-generated 18%

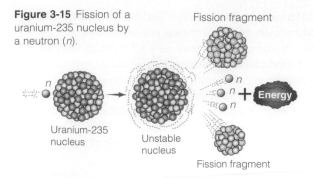

Figure 3-15 Fission of a uranium-235 nucleus by a neutron (*n*).

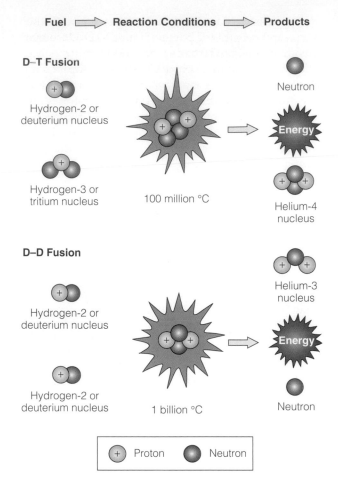

Figure 3-17 The deuterium–tritium (D–T) and deuterium–deuterium (D–D) nuclear fusion reactions, which take place at extremely high temperatures.

can cause an additional fission. For these multiple fissions to take place, enough fissionable nuclei must be present to provide the **critical mass** needed for efficient capture of these neutrons.

Multiple fissions within a critical mass form a **chain reaction**, which releases an enormous amount of energy (Figure 3-16). Living cells can be damaged by the ionizing radiation released by the radioactive lighter nuclei and by high-speed neutrons produced by nuclear fission.

In an atomic bomb, an enormous amount of energy is released in a fraction of a second in an uncontrolled nuclear fission chain reaction. This reaction is initiated by an explosive charge, which suddenly pushes two masses of fissionable fuel together, causing the fuel to reach the critical mass needed for a chain reaction.

In the reactor of a nuclear power plant, the rate at which the nuclear fission chain reaction takes place is controlled so that under normal operation only one of every two or three neutrons released is used to split another nucleus. In conventional nuclear fission reactors, the splitting of uranium-235 nuclei releases heat, which produces high-pressure steam to spin turbines and thus generate electricity.

Figure 3-16 A nuclear chain reaction initiated by one neutron triggering fission in a single uranium-235 nucleus. This figure illustrates only a few of the trillions of fissions caused when a single uranium-235 nucleus is split within a critical mass of uranium-235 nuclei. The elements krypton (Kr) and barium (Ba), shown here as fission fragments, are only two of many possibilities.

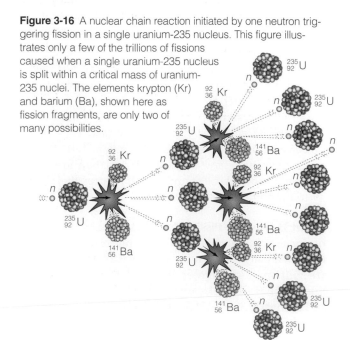

What Is Nuclear Fusion? Forcing Nuclei to Combine Nuclear fusion is a nuclear change in which two isotopes of light elements, such as hydrogen, are forced together at extremely high temperatures until they fuse to form a heavier nucleus, releasing energy in the process. Temperatures of at least 100 million °C are needed to force the positively charged nuclei (which strongly repel one another) to fuse.

Nuclear fusion is much more difficult to initiate than nuclear fission, but once started it releases far more energy per unit of fuel than does fission. Fusion of hydrogen nuclei to form helium nuclei is the source of energy in the sun and other stars.

After World War II, the principle of *uncontrolled nuclear fusion* was used to develop extremely powerful hydrogen, or thermonuclear, weapons. These weapons use the D–T fusion reaction, in which a hydrogen-2, or deuterium (D) nucleus and a hydrogen-3 (tritium, T) nucleus are fused to form a larger, helium-4 nucleus, a neutron, and energy (Figure 3-17).

Scientists have also tried to develop *controlled nuclear fusion*, in which the D–T reaction is used to produce heat that can be converted into electricity. Despite more than 50 years of research, this process is still in the laboratory stage. Even if it becomes technologically and economically feasible, many energy experts don't expect it to be a practical source of energy until perhaps 2100, if then.

3-7 THE TWO IRONCLAD LAWS OF ENERGY

What Is the First Law of Energy? You Cannot Get Something for Nothing Scientists have observed energy being changed from one form to another in millions of physical and chemical changes, but they have never been able to detect the creation or destruction of any energy (except in nuclear changes). The results of their experiments have been summarized in the **law of conservation of energy**, also known as the **first law of energy** or the **first law of thermodynamics**: *In all physical and chemical changes, energy is neither created nor destroyed, but it may be converted from one form to another.*

This scientific law tells us that when one form of energy is converted to another form in any physical or chemical change *energy input always equals energy output.* No matter how hard we try or how clever we are, we cannot get more energy out of a system than we put in; in other words, *we cannot get something for nothing in terms of energy quantity.*

What Is the Second Law of Energy? You Cannot Even Break Even Because the first law of energy states that energy can be neither created nor destroyed, it's tempting to think that there will always be enough energy. Yet if we fill a car's tank with gasoline and drive around, or use a flashlight battery until it is dead, something has been lost. If it is not energy, what is it? The answer is *energy quality* (Figure 3-11), the amount of energy available that can perform useful work.

Countless experiments have shown that when energy is changed from one form to another, a decrease in energy quality always occurs. The results of these experiments have been summarized in what is called the **second law of energy** or the **second law of thermodynamics**: *When energy is changed from one form to another, some of the useful energy is always degraded to lower-quality, more dispersed, less useful energy.* This degraded energy usually takes the form of heat given off at a low temperature to the surroundings (environment). There it is dispersed by the random motion of air or water molecules and becomes even more disorderly and less useful. Another way to state the second law of energy is that *heat always flows spontaneously from hot (high-quality energy) to cold (lower-quality energy).*

Basically, this law says that in any energy conversion, we always end up with *less* usable energy than we started with. So not only can we not get something for nothing in terms of energy quantity, *we cannot even break even in terms of energy quality because energy always goes from a more useful to a less useful form.* The more energy we use, the more low-grade energy (heat)—or entropy (disorder)—we add to the environment. No one has ever found a violation of this fundamental scientific law (see quote at the end of this chapter).

Here are three examples of the second energy law in action.

- When a car is driven, only about 10% of the high-quality chemical energy available in its gasoline fuel is converted into mechanical energy (to propel the vehicle) and electrical energy (to run its electrical systems). The remaining 90% is degraded to low-quality heat that is released into the environment and eventually lost into space.

- When electrical energy flows through filament wires in an incandescent light bulb, it is changed into about 5% useful light and 95% low-quality heat that flows into the environment. In other words, this so-called *light bulb* is really a *heat bulb*.

- In living systems, solar energy is converted into chemical energy (food molecules) and then into mechanical energy (moving, thinking, and living). During each of these conversions high-quality energy is degraded and flows into the environment as low-quality heat (Figure 3-18).

The second law of energy also means that *we can never recycle or reuse high-quality energy to perform useful work.* Once the concentrated energy in a serving of food, a liter of gasoline, a lump of coal, or a chunk of uranium is released, it is degraded to low-quality heat that is dispersed into the environment. We can heat air or water at a low temperature and upgrade it to high-quality energy, but the second law of energy tells us that it will take more high-quality energy to do this than we get in return.

Energy efficiency or **energy productivity** is a measure of how much useful work is accomplished by a particular input of energy into a system. As with material efficiency (p. 57), there is plenty of room for improvement. For example, scientists estimate that only about 16% of the energy used in the United States ends up performing useful work. The remaining 84% is either unavoidably wasted because of the second law of energy (41%) or unnecessarily wasted (43%).

Connections: How Does the Second Energy Law Affect Life? To form and maintain the highly ordered arrangement of molecules and the organized biochemical processes in your body, you must continually get

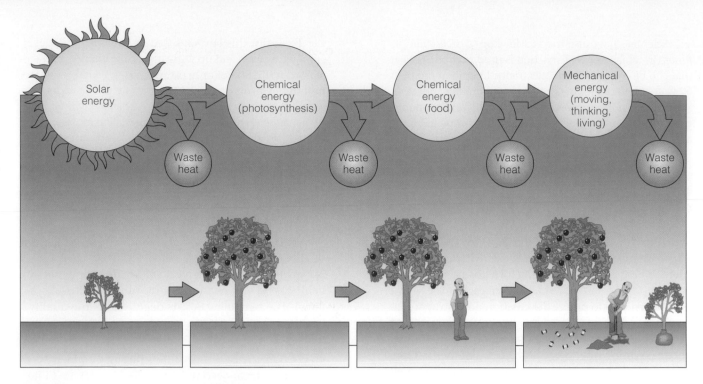

Figure 3-18 The second energy law in action in living systems. Each time energy is changed from one form to another, some of the initial input of high-quality energy is degraded, usually to low-quality heat that disperses into the environment.

and use high-quality matter and energy resources from your surroundings. As you use these resources, you add low-quality heat and low-quality waste matter to your surroundings. Your body continuously gives off heat roughly equal to that of a 100-watt incandescent light bulb; this is why a closed room full of people gets warm. You also continuously break down solid, large molecules (such as glucose) into smaller molecules of carbon dioxide gas and water vapor, which are dispersed in the atmosphere.

Planting, growing, processing, and cooking food all require high-quality energy and matter resources that add low-quality (high-entropy) heat and waste materials to the environment. In addition, enormous amounts of low-quality heat and waste matter are added to the environment when concentrated deposits of minerals and fuels are extracted from the earth's crust, processed, and used. Thus, *all forms of life are tiny pockets of order (low entropy) maintained by creating a sea of disorder (high entropy) in their environment.*

3-8 CONNECTIONS: MATTER AND ENERGY LAWS AND ENVIRONMENTAL PROBLEMS

What Are High-Throughput Economies? As a result of the law of conservation of matter and the second law of energy, individual resource use automatically adds some waste heat and waste matter to the environment.

Most of today's advanced industrialized countries have **high-throughput (high-waste) economies** that attempt to sustain ever-increasing economic growth by increasing the flow of matter and energy resources through their economic systems (Figure 3-19). These resources flow through their economies, to planetary *sinks* (air, water, soil, organisms), where pollutants and wastes end up and can accumulate to harmful levels.

The scientific laws of matter and energy discussed in this chapter tell us that if more and more people continue to use and waste more and more energy and matter resources at an increasing rate, eventually the

Figure 3-19 The high-throughput (high-waste) economies of most developed countries are based on maximizing the rates of energy and matter flow. This process rapidly converts high-quality matter and energy resources into waste, pollution, and low-quality heat.

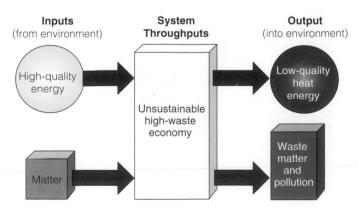

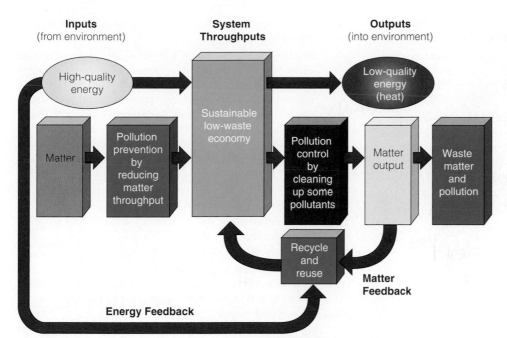

Inputs
(from environment)

System Throughputs

Outputs
(into environment)

High-quality energy

Matter

Pollution prevention by reducing matter throughput

Sustainable low-waste economy

Pollution control by cleaning up some pollutants

Matter output

Low-quality energy (heat)

Waste matter and pollution

Recycle and reuse

Matter Feedback

Energy Feedback

Figure 3-20 Lessons from nature. A low-throughput (low-waste) economy, based on energy flow and matter recycling, works with nature to reduce throughput. This is done by **(1)** reusing and recycling most nonrenewable matter resources, **(2)** using renewable resources no faster than they are replenished, **(3)** using matter and energy resources efficiently, **(4)** reducing unnecessary consumption, **(5)** emphasizing pollution prevention and waste reduction, and **(6)** controlling population growth.

capacity of the environment to dilute and degrade waste matter and absorb waste heat will be exceeded.

What Are Matter-Recycling Economies? A stopgap solution to this problem is to convert a high-throughput economy to a **matter-recycling economy**. The goal of such a conversion is to allow economic growth to continue without depleting matter resources or producing excessive pollution and environmental degradation.

Even though recycling matter saves energy, the two laws of energy tell us that *recycling matter resources always requires expenditure of high-quality energy (which cannot be recycled) and adds waste heat (entropy) to the environment.*

There is also a limit to the number of times some materials, such as paper fiber, can be recycled before they become unusable. Changing to a matter-recycling economy is an important way to buy some time, but it does not allow more and more people to use more and more resources indefinitely, even if all of them were somehow perfectly recycled.

What Are Low-Throughput Economies? Learning from Nature The three scientific laws governing matter and energy changes suggest that the best long-term solution to our environmental and resource problems is to shift from an economy based on maximizing matter and energy flow (throughput) to a more sustainable **low-throughput (low-waste) economy**, as summarized in Figure 3-20.

The next five chapters are devoted to applying the three basic scientific laws of matter and energy to living systems and looking at some *biological principles* that can teach us how to work with nature.

The second law of thermodynamics holds, I think, the supreme position among laws of nature. . . . If your theory is found to be against the second law of thermodynamics, I can give you no hope.

ARTHUR S. EDDINGTON

REVIEW QUESTIONS

1. Define the boldfaced terms found in this chapter.

2. Describe what happened to the people on Easter Island and how this may relate to the current situation on the earth.

3. Define *science* and explain how it works. Distinguish between *scientific data*, *scientific hypothesis*, *scientific model*, *scientific theory*, and *scientific law*. What is *reproducibility* in science, and why is it important? Distinguish between *accuracy* and *precision* in scientific measurements. Explain why a scientific theory should be taken seriously.

4. What are *scientific methods*, and what five questions do scientists try to answer in studying some aspect of nature? What is a *controlled experiment*? What is *multivariable analysis*?

5. Distinguish between *inductive reasoning* and *deductive reasoning* and give an example of each.

6. What does it mean to say that something has been scientifically proven? If scientists cannot establish absolute proof, what do they establish?

7. Distinguish between *frontier science* and *consensus science*.

8. What are two major limitations of environmental science?

9. What is a *mathematical model*, and how is such a model made? Why are mathematical models important?

10. What is a *system*? Distinguish between the *inputs*, *flows* or *throughputs*, *stores*, and *outputs* of a system.

11. What is a *feedback loop*? Distinguish between a *positive feedback loop* and a *negative feedback loop* and give an example of each. What is *homeostasis* and how is it maintained in a system?

12. Define and give an example of a *time delay* in a system. Explain how *leverage* can be used to change a complex system, and give an example of its use in controlling pollution.

13. Define *synergy* and give an example of how it can change a system. List three environmental surprises that we have encountered and three causes of environmental surprises.

14. Distinguish between *matter, elements, compounds*, and *mixtures*.

15. Distinguish between *atoms, ions*, and *molecules* and give an example of each. What three major types of subatomic particles are found in atoms? Which two of these particles are found in the nucleus and which is found outside the nucleus?

16. Distinguish between *atomic number* and *mass number*. What is an *isotope* of an atom?

17. What is the *concentration* of a chemical? What is *pH*?

18. What is a *chemical formula*? Distinguish between *ionic compounds* and *covalent compounds* and give the names and chemical formulas for an example of each of these types of compounds.

19. Distinguish between *organic compounds* and *inorganic compounds* and give an example of each type. Distinguish between *hydrocarbons, chlorinated hydrocarbons, chlorofluorocarbons, simple carbohydrates, polymers, complex carbohydrates, proteins, nucleic acids*, and *nucleotides*.

20. Distinguish between *genes, gene mutations*, and *chromosomes*.

21. Distinguish between *high-quality matter* and *low-quality matter* and give an example of each. What is *entropy*, and how does it differ for high-quality and low-quality matter?

22. What is *energy*? Distinguish between *kinetic energy* and *potential energy* and give an example of each. List three types of electromagnetic radiation. Distinguish between *ionizing radiation* and *nonionizing radiation*. Distinguish between *heat* and *temperature*.

23. Distinguish between *high-quality energy* and *low-quality energy* and give an example of each.

24. Distinguish between a *physical change* and a *chemical change* and give an example of each.

25. What is the *law of conservation of matter*? What is its environmental significance? Explain why there is no "away" as a repository for pollution. What is a *balanced chemical equation*, and how is it related to the law of conservation of matter?

26. What three factors determine the harm caused by a pollutant? Distinguish between concentrations of *parts per million, parts per billion*, and *parts per trillion*. What is the *persistence* of a pollutant? Distinguish between *degradable (nonpersistent), biodegradable, slowly degradable (persistent)*, and *nondegradable pollutants* and give an example of each type.

27. Distinguish between a *chemical change* and a *nuclear change*. What is the *law of conservation of matter and energy*? Distinguish between *natural radioactive decay, radioisotopes, gamma rays, alpha particles*, and *beta particles*. What is the *half-life* of a radioactive isotope? For how many half-lives should radioactive material be stored safely before it decays to an acceptable level of radioactivity? What are the major sources of human exposure to ionizing radiation? What are three useful applications of radioisotopes?

28. Distinguish between *nuclear fission* and *nuclear fusion*. Distinguish between *critical mass* and a *nuclear chain reaction*.

29. Distinguish between the *first law of energy (thermodynamics)* and the *second law of energy (thermodynamics)* and give an example of each law in action. Use the second law of thermodynamics to explain why energy cannot be recycled.

30. Distinguish between a *high-throughput (high-waste) economy*, a *low-throughput (low-waste) economy*, and a *matter-recycling economy*. Use the law of conservation of matter and the first and second laws of thermodynamics to explain the need to shift from a high-throughput economy to a matter-recycling economy and eventually to a low-throughput economy.

CRITICAL THINKING

1. Respond to the following statements:
   a. It has never been scientifically proven that anyone has ever died from smoking cigarettes.
   b. The greenhouse theory—that certain gases (such as water vapor and carbon dioxide) warm the atmosphere—is not a reliable idea because it is only a scientific theory.

2. See whether you can find an advertisement or an article describing or using some aspect of science in which (a) the concept of scientific proof is misused, (b) the term *theory* is used when it should have been *hypothesis*, and (c) a consensus scientific finding is dismissed or downplayed because it is "only a theory."

3. (a) List two examples of negative feedback loops not discussed in this chapter, one that is beneficial and one that is detrimental. Compare your examples with those of other members of your class. (b) Give two examples of positive feedback loops not discussed in this chapter. Include one that is beneficial and one that is detrimental. Compare your examples with those of other members of your class.

4. Describe two examples of actions that help at first but have undesirable consequences later on. Describe two actions that hurt at first but end up as net beneficial changes.

5. How does a scientific law (such as the law of conservation of matter) differ from a societal law (such as one imposing maximum speed limits for vehicles)? Can each be broken?

6. Explain why we do not really consume anything and why we can never really throw matter away.

7. A tree grows and increases its mass. Explain why this is not a violation of the law of conservation of matter.

8. If there is no "away," why isn't the world filled with waste matter?

9. Methane (CH_4) gas is the major component of natural gas. Write and balance the chemical equation for the burning of methane when it combines with oxygen gas in the atmosphere to form carbon dioxide and water. Use this equation to explain why burning natural gas can contribute to projected warming of the atmosphere (global warming).

10. Suppose you have 100 grams of radioactive plutonium-239 with a half-life of 24,000 years. How many grams of plutonium-239 will remain after **(a)** 12,000 years, **(b)** 24,000 years, and **(c)** 96,000 years?

11. Someone wants you to invest money in an automobile engine that will produce more energy than the energy in the fuel (such as gasoline or electricity) you use to run the motor. What is your response? Explain.

12. Use the second energy law to explain why a barrel of oil can be used only once as a fuel.

13. **(a)** Sometimes the first law of energy (thermodynamics) is summarized as "you cannot get something for nothing in terms of energy quantity." What does this phrase mean? **(b)** Sometimes the second law of energy (thermodynamics) is summarized as "you cannot break even in terms of energy quality." What does this phrase mean?

14. **(a)** Use the law of conservation of matter to explain why a matter-recycling economy sooner or later will be necessary. **(b)** Use the first and second laws of energy to explain why, in the long run, we will need a low-throughput economy, not just a matter-recycling economy.

15. **(a)** Imagine that you have the power to violate the law of conservation of energy (the first energy law) for 1 day. What are the three most important things you would do with this power? **(b)** Repeat this process, imagining that you have the power to violate the second law of energy for 1 day.

PROJECTS

1. As a class, identify a change that would improve the quality of the environment on your campus. Work together to come up with a plan for bringing about this change. Cooperate (synergize) to bring about the proposed change, and use feedback to modify your plan as needed to accomplish your goal.

2. If you have the use of a sensitive balance, try to demonstrate the law of conservation of mass in a physical change. Weigh a container with a lid (a glass jar will do), add an ice cube and weigh it again, and then allow the ice to melt and weigh it again.

3. Use the library or internet to find examples of various perpetual motion machines and inventions that allegedly

violate the two laws of energy (thermodynamics) by producing more high-quality energy than the high-quality energy needed to make them run. What has happened to these schemes and machines (many of them developed by scam artists to attract money from investors)?

4. Use the library or the internet to find bibliographic information about *George Perkins Marsh*, *Warren Weaver*, and *Arthur S. Eddington*, whose quotes appear at the beginning and end of this chapter.

5. Make a concept map of this chapter's major ideas, using the section heads and subheads and the key terms (in bold-face). Look at the inside back cover and on the website for this book for information about making concept maps.

INTERNET STUDY RESOURCES AND RESOURCES FOR FURTHER READING AND RESEARCH

The website for this book contains helpful study aids and many ideas for further reading and research. Log on to:

http://www.brookscole.com/product/0534376975s

and click on the Chapter-by-Chapter area. Choose Chapter 3 and select a resource:

■ "Flash Cards" allows you to test your mastery of the Terms and Concepts to Remember for this chapter.

■ "Tutorial Quizzes" provides a multiple-choice practice quiz.

■ "Student Guide to InfoTrac" will lead you to Critical Thinking Projects that use InfoTrac College Edition as a research tool.

■ "References" lists the major books and articles consulted in writing this chapter.

■ "Hypercontents" takes you to an extensive list of sites with news, research, and images related to individual sections of the chapter.

INFOTRAC COLLEGE EDITION

Improve your skills with InfoTrac College Edition, a searchable online database of articles from more than 700 periodicals. Log on to:

http://www.infotrac-college.com

or access InfoTrac through the website for this book.

Try the following articles:

Riley, M.T. 1992. Dimitri Ivanovich Mendeleev. *Great Thinkers of the Western World* pp. 395–397. (subject guide: periodic system)

Asimov, I. 1994. Second law of thermodynamics. *Asimov's Chronology of Science & Discovery, Edition 1*, p. 362. (subject guide: thermodynamics)

4 ECOSYSTEMS: COMPONENTS, ENERGY FLOW, AND MATTER CYCLING*

Have You Thanked the Insects Today?

Insects have a bad reputation. We classify many insect species as *pests* because they compete with us for food, spread human diseases (such as malaria), and invade our lawns, gardens, and houses. Some people have "bugitis," fear all insects, and think that the only good bug is a dead bug. However, this view fails to recognize the vital roles insects play in helping sustain life on earth.

A large proportion of the earth's plant species (including many trees) depend on insects to pollinate their flowers (Figure 4-1, right). In turn, we and other land-dwelling animals depend on plants for food, either by eating them or by consuming animals that eat them. If there were no pollinating insects, there would be very few fruits and vegetables for us and plant-eating animals to eat.

Insects, such as the praying mantis (Figure 4-1, left), that eat other insects help control the populations of at least half the species of insects we call pests. This free pest control service is an important part of the natural resources or ecological services that help sustain us.

Suppose all insects disappeared today. Within a year most of the earth's animals would become extinct because of the disappearance of so much plant life. The earth would be covered with rotting vegetation and animal carcasses being decomposed by unimaginably huge hordes of bacteria and fungi.

Fortunately, this is not a realistic scenario because insects, which have been around for at least 400 million years, are phenomenally successful forms of life. They were the first animals to invade the land and, later, the air. Today they are by far the planet's most diverse, abundant, and successful animals.

Insects can rapidly evolve new genetic traits, such as resistance to pesticides. They also have an exceptional ability to evolve into new species when faced with new environmental conditions, and they are extremely resistant to extinction. Because of their ability to quickly develop genetic resistance to pesticides, using such chemical warfare to control insect pests will fail eventually. Moreover, many of the pesticides we use reduce the populations of pest insect species that kill large numbers of beneficial insects that help keep pest insects under control. Although insects can thrive without newcomers such as us, we and most other land organisms would perish quickly without them.

Learning about the roles insects play in nature requires us to understand how insects and other organisms living in a biological *community* (such as a forest or pond) interact with one another and with the nonliving environment. *Ecology* is the science that studies such relationships and interactions in nature, as discussed in this chapter and the six chapters that follow.

---

*Paul M. Rich, associate professor of ecology and evolutionary biology and environmental studies at the University of Kansas, is coauthor of this chapter.

Figure 4-1 Insects play important roles in helping sustain life on earth. The bright green caterpillar moth feeding on pollen in a crocus (right) and other insects pollinate flowering plants that serve as food for many plant eaters. The praying mantis eating a monarch butterfly (left) and many other insect species help control the populations of at least half of the insect species we classify as pests. (Left: Pat Andersen, Visuals Unlimited; right: Stephen Hopkin/Planet Earth Pictures, Ltd.)

The earth's thin film of living matter is sustained by grand-scale cycles of energy and chemical elements.

G. EVELYN HUTCHINSON

This chapter addresses the following questions:

- What is ecology?

- What basic processes keep us and other organisms alive?

- What are the major components of an ecosystem?

- What happens to energy in an ecosystem?

- What happens to matter in an ecosystem?

- How do scientists study ecosystems?

- What are ecosystem services and how do they affect the sustainability of the earth's life-support systems?

4-1 THE NATURE OF ECOLOGY

What Is Ecology? **Ecology** (from the Greek words *oikos*, "house" or "place to live," and *logos*, "study of") is the study of how organisms interact with one another and with their nonliving environment. In effect it is a study of *connections in nature*.

What Are Organisms and Species? Ecologists focus on trying to understand the interactions between organisms, populations, communities, ecosystems, and the biosphere (Figure 4-2).

An **organism** is any form of life. The **cell** is the basic unit of life in organisms. Organisms may consist of a single cell (bacteria, for instance) or many cells.

On the basis of their cell structure, organisms can be classified as either *eukaryotic* or *prokaryotic*. Each cell of a **eukaryotic** organism **(1)** is surrounded by a membrane, **(2)** has a distinct *nucleus* (a membrane-bounded structure containing genetic material in the form of DNA), and **(3)** has several other internal parts called *organelles* (Figure 4-3a). All organisms except bacteria are eukaryotic.

The cell of a **prokaryotic** organism is surrounded by a membrane, but inside the cell there is no distinct nucleus or other internal parts enclosed by membranes (Figure 4-3b). All bacteria are single-celled prokaryotic organisms. Although most familiar organisms are eukaryotic, they could not exist without hordes of prokaryotic organisms (bacteria; Connections, p. 74). These bacteria are examples of *microorganisms*, so small that they can be seen only with the aid of a microscope.

Organisms can be classified into **species**, or groups of organisms that resemble one another in appearance, behavior, chemistry, and genetic makeup. Species differ in how they produce offspring. **Asexual reproduc-**tion is common in species such as bacteria with only one cell, which divides to produce two identical cells that are clones or replicas of the original cell.

Sexual reproduction occurs in organisms that produce offspring by combining sex cells or gametes (such as ovum and sperm) from both parents. This produces offspring that have combinations of genetic traits from each parent. Sexual reproduction usually gives the offspring a greater chance of survival under changing environmental conditions than the genetic clones produced by asexual reproduction.

Organisms that reproduce sexually are classified in the same species if, under natural conditions, they can actually or potentially breed with one another and produce live, fertile offspring. Scientists use a special system to name each species, as discussed in Appendix 4.

We do not know how many species exist on the earth. Estimates range from 5 million to 100 million. Most are insects (left) and microorganisms too small to be seen with the naked eye (Connections, p. 74). Excluding hordes of bacterial species, a best guess of the number of species is about 10-14 million.

So far biologists have identified and named about 1.8 million species, not including bacteria. About 42% (751,000) of these known species are insects, 15% (270,000) are plants, 0.5% (9,000) are birds, and only 0.25% (4,500) are mammals. Thousands of newly identified species are added each year by *taxonomists*, biologists who specialize in identifying and cataloging the earth's species (Appendix 4). Biologists know a fair amount about roughly one-third of the known species but understand the detailed roles and interactions of only a few.

What Is a Population? A **population** consists of a group of interacting individuals of the same species that occupy a specific area at the same time (Figure 4-4). Examples are all sunfish in a pond, all white oak trees in a forest, and all people in a country. In most natural populations, individuals vary slightly in their genetic makeup, which is why they do not all look or behave exactly alike—a phenomenon called **genetic diversity** (Figure 4-5). Populations are dynamic groups that change in **(1)** size, **(2)** age distribution (number of individuals in each age group), **(3)** density (number of individuals per unit of space), and **(4)** genetic composition as a result of changes in environmental conditions.

The place where a population (or an individual organism) normally lives is its **habitat**. It may be as large as an ocean or prairie or as small as the underside of a rotting log or the intestine of a termite.

What Are Communities, Ecosystems, and the Biosphere? Populations of all the different species occupying a particular place make up a **community**, or **biological community**. It is a complex interacting network of plants, animals, and microorganisms.

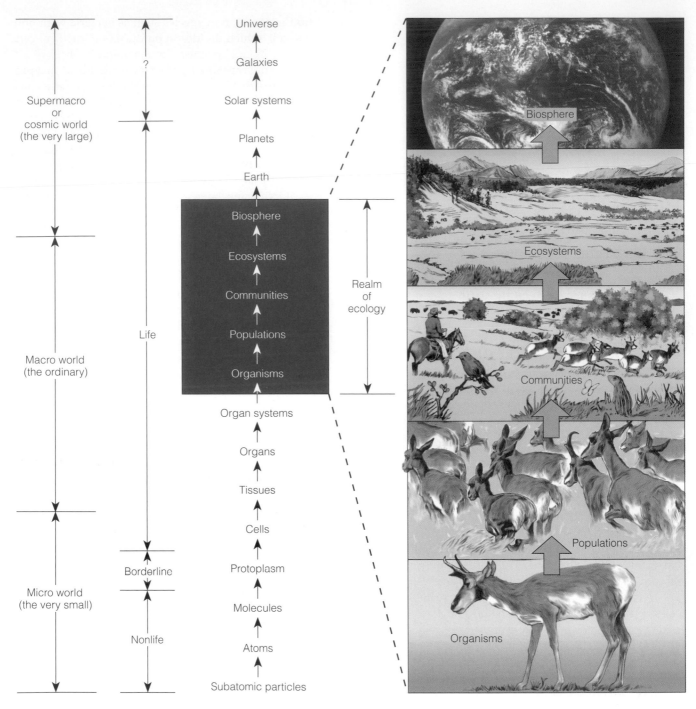

Figure 4-2 Model of levels of organization of matter in nature. Note that ecology focuses on five levels of this hierarchical model.

An **ecosystem** is a community of different species interacting with one another and with their nonliving environment of matter and energy. Ecosystems can range in size from a puddle of water to a stream, a patch of woods, an entire forest, or a desert. All of the earth's ecosystems together make up what we call the **biosphere**. Ecosystems can be natural or artificial (human-created). Examples of human-created ecosystems are cropfields, farm ponds, and reservoirs.

4-2 THE EARTH'S LIFE-SUPPORT SYSTEMS

What Are the Major Parts of the Earth's Life-Support Systems? We can think of the earth as being made up of several spherical layers (Figure 4-6). The **atmosphere** is a thin envelope of air around the planet. Its inner layer, the **troposphere**, extends only about 17 kilometers (11 miles) above sea level but contains most

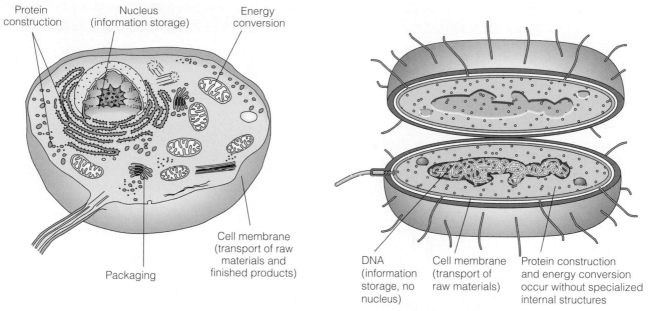

Protein construction   Nucleus (information storage)   Energy conversion

Cell membrane (transport of raw materials and finished products)

Packaging

DNA (information storage, no nucleus)   Cell membrane (transport of raw materials)   Protein construction and energy conversion occur without specialized internal structures

(a) Eukaryotic Cell   **(b) Prokaryotic Cell**

Figure 4-3 (a) Generalized structure of a eukaryotic cell. The parts and internal structure of cells in various types of organisms such as plants and animals differ somewhat from this generalized model. **(b)** Generalized structure of a prokaryotic cell. Note that prokaryotic cells lack a distinct nucleus. (Adapted from *Biology: The Unity and Diversity of Life*, 8/e by Cecie Starr and Ralph Taggart ©1998. Reprinted with permission from Wadsworth, a division of Thomson Learning. Fax 800 730-2215)

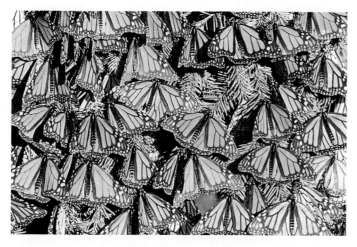

Figure 4-4 A population of monarch butterflies wintering in Michoacán, Mexico. The geographic distribution of this butterfly coincides with that of the milkweed plant, on which monarch larvae and caterpillars feed. (Frans Lanting/Bruce Coleman Collection.)

of the planet's air, mostly nitrogen (78%) and oxygen (21%). The next layer, stretching 17–48 kilometers (11–30 miles) above the earth's surface, is the **stratosphere**. Its lower portion contains enough ozone (O_3) to filter out most of the sun's harmful ultraviolet radiation, thus allowing life to exist on land and in the surface layers of bodies of water.

The **hydrosphere** consists of the earth's **(1)** liquid water (both surface and underground), **(2)** ice (polar ice,

icebergs, and ice in frozen soil layers, or permafrost), and **(3)** water vapor in the atmosphere. The **lithosphere** is the earth's crust and upper mantle; the crust contains nonrenewable fossil fuels and minerals we use as well as potentially renewable soil chemicals (nutrients) needed for plant life, as discussed in more detail in Section 10-5, p. 219, and Section 10-6, p. 228.

The **biosphere** is the portion of the earth in which living (biotic) organisms exist and interact with one another and with their nonliving (abiotic) environment. The biosphere includes most of the hydrosphere and parts of the lower atmosphere and upper lithosphere, reaching from the deepest ocean floor, 20 kilometers (12 miles) below sea level, to the tops of the highest mountains. If the earth were an apple, the biosphere would be no thicker than the apple's skin. *The goal of ecology*

Figure 4-5 The genetic diversity among individuals of one species of Caribbean snail is reflected in the variations in shell color and banding patterns. (Alan Solem)

Microbes: The Invisible Rulers of the Earth

CONNECTIONS

They are everywhere and there are trillions of them. Billions are found inside your body, on your body, in a handful of soil, and in a cup of river water.

These mostly invisible rulers of the earth are *microbes*, a catchall term for many thousands of species of bacteria, protozoa, fungi, and yeasts, most of which are too small to be seen with the naked eye.

Most microbes do not get the respect they deserve. Most of us think of them as threats to our health in the form of **(1)** infectious bacteria or "germs," **(2)** fungi that cause athlete's foot and other skin diseases, and **(3)** protozoa that cause killer diseases such as malaria. However, these potentially harmful microbes are in the minority.

In truth, most of the earth's hordes of microbes not only are harmless but make the rest of life possible. Some of them play a vital role in producing foods such as bread, cheese, yogurt, vinegar, tofu, soy sauce, beer, and wine. Others provide us with food by converting nitrogen gas in the atmosphere into forms that plants can take up from the soil as nutrients.

Bacteria and fungi in the soil decompose organic wastes into nutrients that can be taken up by plants. Bacteria in your intestinal tract break down the food you eat. Some microbes in your nose prevent harmful bacteria from reaching your lungs.

Some scientists are collecting microbes from extreme environments such as deep-sea hydrothermal vents to look for chemicals that could kill cancer cells and viruses. Other microbes have been the source of disease-fighting antibiotics, including penicillin, erythromycin, and streptomycin.

Another vital ecological service provided by some microbes is the control of some plant diseases and populations of insect species that attack food crops. Enlisting some of these microbes for pest control can reduce the use of potentially harmful chemical pesticides (Section 20-3, p. 505).

In addition, bioengineering is being used to develop microbes than can **(1)** extract metals from ores, **(2)** break down various pollutants, and **(3)** help clean up toxic waste sites.

Harvard biologist Edward O. Wilson, who has developed many important ecological theories and is one of the world's experts on ants, says that if he were starting over he would study microbes.

Critical Thinking

1. A bumper sticker reads, "Have You Thanked Microbes Today?" Give reasons for doing so and explain why microbes are the real rulers of the earth.

2. What are some potentially harmful effects of **(a)** using genetic engineering to design microbes to break down oil and toxic chemicals and **(b)** overusing antibacterial soaps, sprays, and antibiotics.

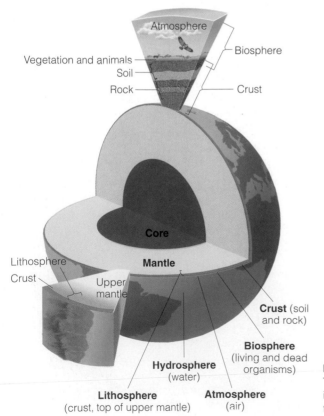

Vegetation and animals
Soil
Rock
Atmosphere
Biosphere
Crust

Lithosphere
Crust
Upper mantle
Core
Mantle

Crust (soil and rock)
Biosphere (living and dead organisms)
Hydrosphere (water)
Atmosphere (air)
Lithosphere (crust, top of upper mantle)

Figure 4-6 The general structure of the earth. The atmosphere consists of several layers, including the tropo-sphere (innermost layer) and the strato-sphere (second layer).

is to understand the interactions in this thin, life-supporting global skin of air, water, soil, and organisms.

What Sustains Life on Earth? Life on the earth depends on three interconnected factors (Figure 4-7):

- The *one-way flow of high-quality (low-entropy) energy* from the sun, **(1)** through materials and living things in their feeding interactions, **(2)** then into the environment as low-quality (high-entropy) energy (mostly heat dispersed into air or water molecules at a low temperature), and **(3)** eventually back into space as heat.

- The *cycling of matter* (the atoms, ions, or molecules needed for survival by living organisms) through parts of the biosphere. The earth is closed to significant inputs of matter from space. Thus, essentially all the nutrients used by organisms are already present on earth and must be recycled again and again for life to continue.

Figure 4-7 Life on the earth depends on **(1)** the *one-way flow of energy* (dashed lines) from the sun through the biosphere, **(2)** the *cycling of crucial elements* (solid lines around circles), and **(3)** *gravity*, which keeps atmospheric gases from escaping into space and draws chemicals downward in the matter cycles. This simplified model depicts only a few of the many cycling elements.

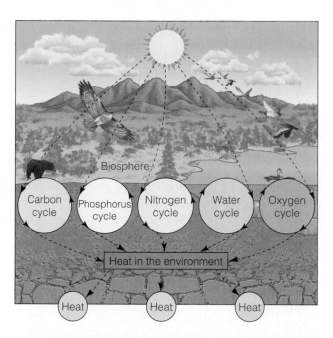

- *Gravity*, which allows the planet to hold onto its atmosphere and causes the downward movement of chemicals in the matter cycles.

How Does the Sun Help Sustain Life on Earth?

The sun is a middle-aged star whose energy

- Lights and warms the planet

- Supports *photosynthesis*, the process used by green plants and some bacteria to make compounds such as carbohydrates that keep them alive and that feed most other organisms

- Powers the cycling of matter

- Drives the climate and weather systems that distribute heat and fresh water over the earth's surface

The sun is a gigantic fireball of hydrogen (72%) and helium (28%) gases. Temperatures and pressures in its inner core are so high that hydrogen nuclei fuse to form helium nuclei (Figure 3-17, p. 64), releasing enormous amounts of energy.

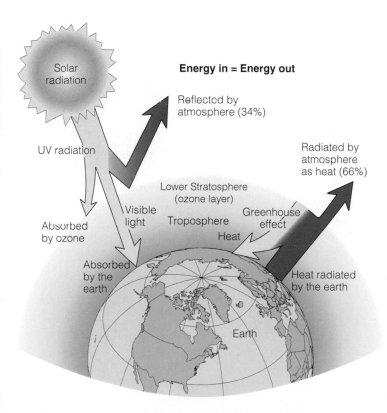

Figure 4-8 The flow of energy to and from the earth. The ultimate source of energy in most ecosystems is sunlight.

The sun, then, is really a gigantic nuclear fusion (thermonuclear) reactor running on hydrogen fuel. This enormous reactor radiates energy in all directions as electromagnetic radiation (Figure 3-10, p. 58). Moving at the speed of light, this radiation makes the 150-million-kilometer (93-million-mile) trip between the sun and the earth in slightly more than 8 minutes.

Because the earth is a tiny sphere in the vastness of space, it receives only about one-billionth of this output of energy. Much of this energy is either reflected away or absorbed by chemicals in its atmosphere. Most of what reaches the troposphere is visible light, infrared radiation (heat), and the small amount of ultraviolet radiation that is not absorbed by ozone in the stratosphere. About 34% of the solar energy reaching the troposphere is reflected back into space by clouds, chemicals, dust, and the earth's surface land and water (Figure 4-8).

Most of the remaining 66% of solar energy **(1)** warms the troposphere and land, **(2)** evaporates water and cycles it through the biosphere, and **(3)** generates winds. A tiny fraction (about 0.023%) is captured by green plants, algae, and bacteria to fuel photosynthesis and make the organic compounds most forms of life need to survive.

Most of this unreflected solar radiation is degraded into infrared radiation (which we experience as heat) as it interacts with the earth. Greenhouse gases (such as water vapor, carbon dioxide, methane, nitrous oxide, and ozone) in the atmosphere reduce this flow of heat back into space. This helps warm the earth by acting somewhat like the glass in a greenhouse or the windows in a closed car, which allow a buildup of heat. Without this **natural greenhouse effect**, the earth would be nearly as cold as Mars, and life as we know it could not exist.

4-3 ECOSYSTEM CONCEPTS AND COMPONENTS

What Are Biomes and Aquatic Life Systems?

Viewed from outer space, the earth resembles an enormous jigsaw puzzle consisting of large masses of land and vast expanses of ocean (p. 1 and Figure 1-1, p. 2). Biologists have classified the terrestrial (land) portion of the biosphere into **biomes** ("BY-ohms"). They are large regions such as forests, deserts, and grasslands characterized by a distinct climate and specific life-forms, especially vegetation, adapted to it (Figure 4-9).

Climate—long-term patterns of weather-is the main factor determining what type of life, especially what plants, will thrive in a given land area. Each biome consists of a patchwork of many different ecosystems whose communities have adapted to differences in climate, soil, and other factors throughout the biome.

Marine and freshwater portions of the biosphere can be divided into **aquatic life zones**, each containing numerous ecosystems. Aquatic life zones are the aquatic equivalent of biomes. Examples include **(1)** *freshwater life zones* (such as lakes and streams) and **(2)** *ocean or marine*

life zones (such as estuaries, coastlines, coral reefs, and the deep ocean). The earth's major land biomes and aquatic life zones are discussed in more detail in Chapters 6 and 7.

Do Ecosystems Have Distinct Boundaries? For convenience, scientists often consider an ecosystem under study as an isolated unit. However, natural ecosystems rarely have distinct boundaries and are not truly self-contained, self-sustaining systems.

Instead, one ecosystem tends to merge with the next in a transitional zone called an **ecotone**, a region containing a mixture of species from adjacent ecosystems and often species not found in either of the bordering ecosystems. For example, a marsh or wetland found between dry land and the open water of a lake or ocean is an ecotone (Figure 4-10). Another example is the zone of grasses, small shrubs, and scattered small trees found between a grassland and a forest.

What Are the Major Components of Ecosystems?
The biosphere and its ecosystems can be separated into two parts: **(1) abiotic**, or nonliving, components (water,

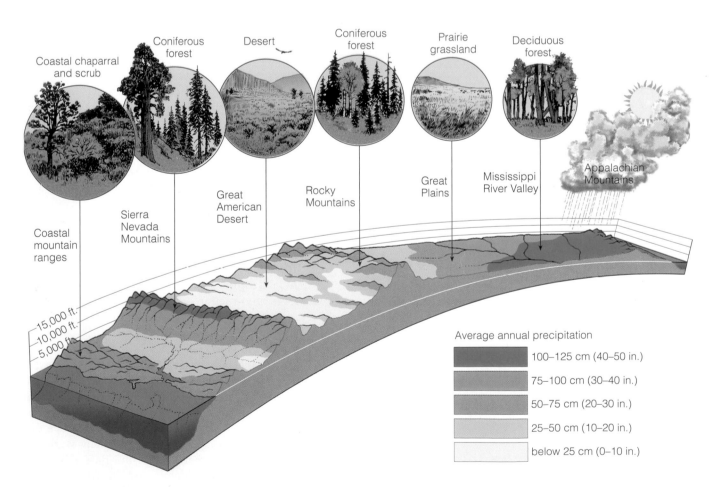

Figure 4-9 Major biomes found along the 39th parallel across the United States. The differences reflect changes in climate, mainly differences in average annual precipitation and temperature (not shown).

air, nutrients, and solar energy) and **(2) biotic**, or living, components (plants, animals, and microorganisms, sometimes called *biota*). Figures 4-11 and 4-12 are greatly simplified diagrams of some of the biotic and abiotic components in a freshwater aquatic ecosystem and a terrestrial ecosystem.

What Are the Major Nonliving Components of Ecosystems? The nonliving, or abiotic, components of an ecosystem are the physical and chemical factors that influence living organisms in land (terrestrial) ecosystems and aquatic life zones (Figure 4-13).

Different species thrive under different physical conditions. Some need bright sunlight, and others thrive better in shade. Some need a hot environment and others a cool or cold one. Some do best under wet conditions and others under dry conditions.

Each population in an ecosystem has a **range of tolerance** to variations in its physical and chemical environment (Figure 4-14). Individuals within a population may also have slightly different tolerance ranges for temperature or other factors because of small differences in genetic makeup, health, and age. Thus, although a trout population may do best within a narrow band of temperatures (*optimum level or range*), a few individuals can survive above and below that band. As Figure 4-14 shows, tolerance has its limits, beyond which none of the trout can survive.

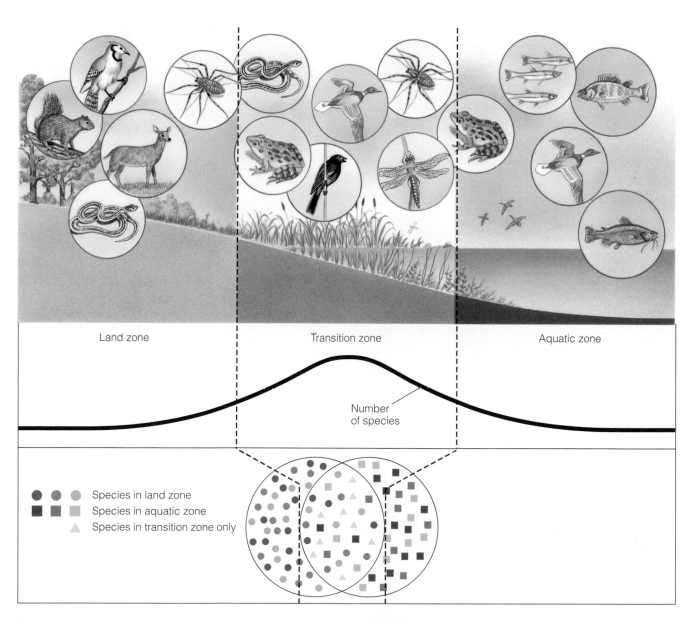

Land zone

Transition zone

Aquatic zone

Number of species

- ● ● ● Species in land zone
- ■ ■ ■ Species in aquatic zone
- ▲ Species in transition zone only

Figure 4-10 Ecosystems rarely have sharp boundaries. Two adjacent ecosystems such as dry land and an open lake often contain a marsh—an ecotone or transitional zone—between them. This zone contains a mixture of species found in each ecosystem and contains some species not found in either ecosystem.

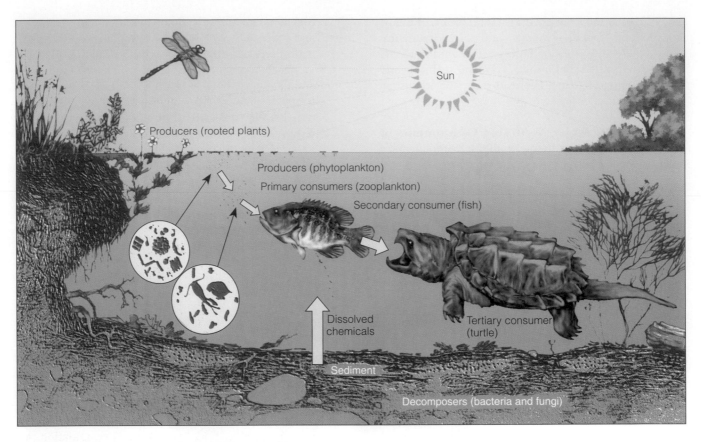

Figure 4-11 Major components of a freshwater ecosystem.

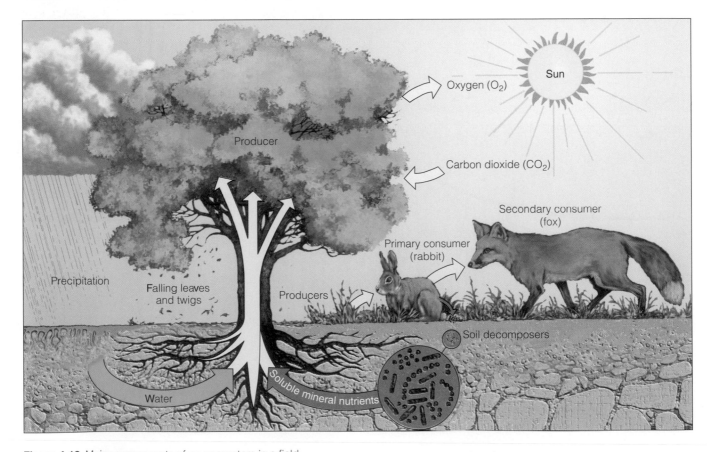

Figure 4-12 Major components of an ecosystem in a field.

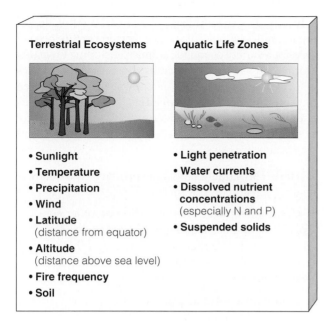

Terrestrial Ecosystems

- **Sunlight**
- **Temperature**
- **Precipitation**
- **Wind**
- **Latitude**
  (distance from equator)
- **Altitude**
  (distance above sea level)
- **Fire frequency**
- **Soil**

Aquatic Life Zones

- **Light penetration**
- **Water currents**
- **Dissolved nutrient concentrations**
  (especially N and P)
- **Suspended solids**

Figure 4-13 Key physical and chemical or abiotic factors affecting terrestrial ecosystems (left) and aquatic life zones (right).

ers. Most organisms are least tolerant during juvenile or reproductive stages of their life cycles. *Highly tolerant species can live in u variety of habitats with widely different conditions.*

A variety of factors can affect the number of organisms in a population. However, sometimes one factor, known as a **limiting factor**, is more important in regulating population growth than others factors. This ecological principle, related to the law of tolerance, is called the **limiting factor principle**: *Too much or too little of any abiotic factor can limit or prevent growth of a population, even if all other factors are at or near the optimum range of tolerance.*

On land, precipitation often is the limiting factor. Lack of water in a desert limits plant growth. Soil nutrients also can act as a limiting factor on land. Suppose a farmer plants corn in phosphorus-poor soil. Even if water, nitrogen, potassium, and other nutrients are at optimum levels, the corn will stop growing when it uses up the available phosphorus.

Too much of an abiotic factor can also be limiting. For example, too much water or too much fertilizer can kill plants, a common mistake of many beginning gardeners.

The limiting factor for a particular population can change. For example, at the beginning of a plant's growing season, temperature may be the limiting factor; later on, the supply of a particular nutrient may limit growth; if a drought occurs, water may be the limiting factor.

These observations are summarized in the **law of tolerance**: *The existence, abundance, and distribution of a species in an ecosystem are determined by whether the levels of one or more physical or chemical factors fall within the range tolerated by that species.* In other words, there are minimum and maximum limits for physical conditions (such as temperature) and concentrations of chemical substances, called **tolerance limits**, beyond which no members of a particular species can survive.

A species may have a wide range of tolerance to some factors and a narrow range of tolerance to oth-

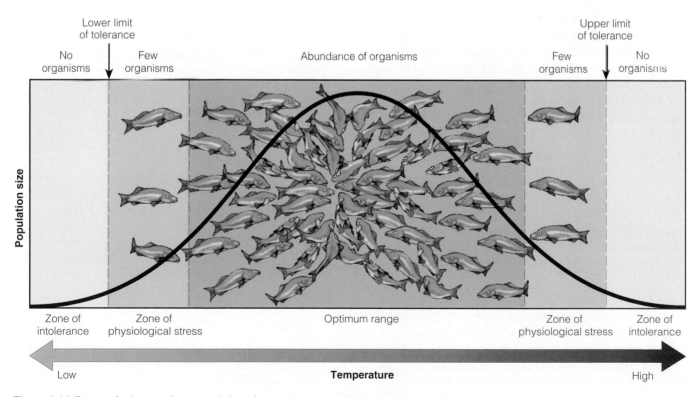

Figure 4-14 Range of tolerance for a population of organisms to an abiotic environmental factor-in this case, temperature.

Important limiting factors for aquatic ecosystems include **(1)** temperature, **(2)** sunlight, **(3) dissolved oxygen content** (the amount of oxygen gas dissolved in a given volume of water at a particular temperature and pressure), and **(4)** nutrient availability. Another limiting factor in aquatic ecosystems is **salinity** (the amounts of various inorganic minerals or salts dissolved in a given volume of water).

Seawater is about 3.4% salt by weight, and it has probably been less than 4% as far back as we have data. If it ever rose to more than 6%, all life in the sea (except possibly some archaebacteria) would be killed.

What Are the Major Living Components of Ecosystems? Living organisms capture and transform matter and energy from their environment to supply their needs for survival, growth, and reproduction. The complete set of chemical reactions that carries out this role in cells and organisms is called **metabolism**.

Living organisms in ecosystems usually are classified as either *producers* or *consumers*, based on how they get food. **Producers**, sometimes called **autotrophs** (self-feeders), make their own food from compounds obtained from their environment. All other organisms are consumers, which depend directly or indirectly on food provided by producers.

On land, most producers are green plants. In freshwater and marine ecosystems, algae and plants are the major producers near shorelines; in open water the dominant producers are floating and drifting *phytoplankton*, most of them microscopic.

Most producers capture sunlight to make carbohydrates (such as glucose, $C_6H_{12}O_6$) and other complex organic compounds from inorganic (abiotic) nutrients in the environment. This process for using sunlight to make carbohydrates is called **photosynthesis**.

Although hundreds of chemical changes take place during photosynthesis, the overall reaction can be summarized as follows:

carbon dioxide + water + **solar energy** $\longrightarrow$ glucose + oxygen

$6\,CO_2$ + $6\,H_2O$ + **solar energy** $\longrightarrow$ $C_6H_{12}O_6$ + $6\,O_2$

A few producers, mostly specialized bacteria, can convert simple compounds from their environment into more complex nutrient compounds without sunlight, a process called **chemosynthesis**.

In one such case, the source of energy is heat generated by the decay of radioactive elements deep in the earth's core. This heat is released at hot-water (hydrothermal) vents in the ocean's depths, where new crust is constantly being formed and re-formed. In the pitch darkness around such vents, large populations of specialized producer bacteria use this geothermal energy to convert dissolved hydrogen sulfide (H_2S) and carbon dioxide into organic nutrient molecules. These bacteria in turn become food for a variety of aquatic animals, including huge tube worms and various clams, crabs, mussels, and barnacles.

All other organisms in an ecosystem are **consumers** or **heterotrophs** ("other-feeders"), which get their energy and nutrients by feeding on other organisms or their remains. Based on their primary source of food, consumers are classified as

- **Herbivores** (plant eaters) or **primary consumers** feeding directly on producers.

- **Carnivores** (meat eaters) feeding on other consumers, with those feeding only on primary consumers called **secondary consumers** and those feeding on other carnivores called **tertiary (higher-level) consumers**.

- **Omnivores** (such as pigs, rats, foxes, bears, cockroaches, and humans) that eat plants and animals.

- **Scavengers** (such as vultures, flies, hyenas, and some species of sharks and ants) feeding on dead organisms.

- **Detritivores** (detritus feeders and decomposers) feeding on **detritus** ("di-TRI-tus"), or parts of dead organisms and cast-off fragments and wastes of living organisms (Figure 4-15).

- **Detritus feeders** (such as crabs, carpenter ants, termites, and earthworms) that extract nutrients from partly decomposed organic matter in leaf litter, plant debris, and animal dung.

- **Decomposers** (mostly certain types of bacteria and fungi) that recycle organic matter in ecosystems. They do this by breaking down (*biodegrading*) dead organic material (detritus) to get nutrients and releasing the resulting simpler inorganic compounds into the soil and water, where they can be taken up as nutrients by producers.

Figures 4-11 and 4-12 show various types of producers and consumers.

Both producers and consumers use the chemical energy stored in glucose and other organic compounds to fuel their life processes. In most cells, this energy is released by **aerobic respiration**, which uses oxygen to convert organic nutrients back into carbon dioxide and water. The net effect of the hundreds of steps in this complex process is represented by the following reaction:

glucose + oxygen $\longrightarrow$ carbon dioxide + water + **energy**

$C_6H_{12}O_6$ + $6\,O_2$ $\longrightarrow$ $6\,CO_2$ + $6\,H_2O$ + **energy**

Although the detailed steps differ, the net chemical change for aerobic respiration is the opposite of that for photosynthesis.

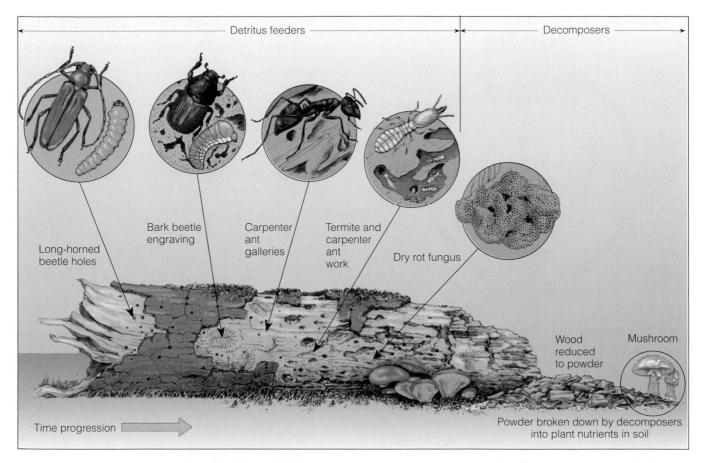

Figure 4-15 Some detritivores, called *detritus feeders*, directly consume tiny fragments of this log. Other detritivores, called *decomposers* (mostly fungi and bacteria), digest complex organic chemicals in fragments of the log into simpler inorganic nutrients. These nutrients can be used again by producers if they are not washed away or otherwise removed from the system.

Some decomposers get the energy they need by breaking down glucose (or other organic compounds) in the absence of oxygen. This form of cellular respiration is called **anaerobic respiration** or **fermentation**. Instead of carbon dioxide and water, the end products of this process are compounds such as **(1)** methane gas (CH_4, the main component of natural gas), **(2)** ethyl alcohol (C_2H_6O), **(3)** acetic acid ($C_2H_4O_2$, the key component of vinegar), and **(4)** hydrogen sulfide (H_2S, when sulfur compounds are broken down).

The survival of any individual organism depends on the *flow of matter and energy* through its body. However, an ecosystem as a whole survives primarily through a combination of *matter recycling* (rather than one-way flow) and *one-way energy flow* (Figure 4-16).

Decomposers complete the cycle of matter by breaking down detritus into inorganic nutrients that are used by producers. Without decomposers, the entire world would soon be knee-deep in plant litter, dead animal bodies, animal wastes, and garbage. Most life as we know it would no longer exist.

What Is Biodiversity and Why Is It Important?
One important renewable resource is **biological diversity** or **biodiversity**: the different life-forms (species) and life-sustaining processes that can best survive the variety of conditions currently found on the earth. Kinds of biodiversity include

- **Genetic diversity** (variety in the genetic makeup among individuals within a species)

- **Species diversity** (variety among the species or distinct types of living organisms found in different habitats of the planet; Figure 4-17)

- **Ecological diversity** (variety of forests, deserts, grasslands, streams, lakes, oceans, wetlands, and other biological communities)

- **Functional diversity** (biological and chemical processes or functions such as energy flow and matter

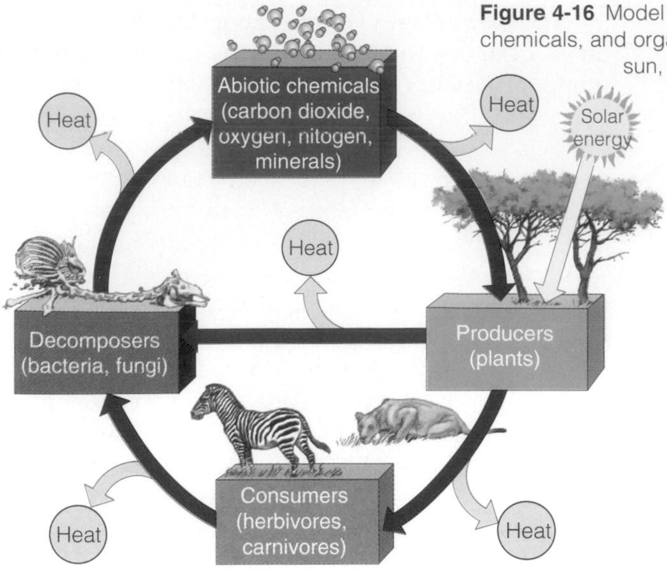

Figure 4-16 Model showing how an ecosystem's main structural components (energy, chemicals, and organisms) are linked by matter recycling and the flow of energy from the sun, through organisms, and back to the environment as low-quality heat. Each type of organism in an ecosystem plays a unique role in the processes of energy flow and matter cycling.

the availability of ecosystem services and decreases the ability of species, communities, and ecosystems to adapt to changing environmental conditions. Biodiversity is nature's insurance policy against disasters.

Some people also include *human cultural diversity* as part of the earth's biodiversity. The variety of human cultures represents numerous social and technological solutions to changing environmental conditions.

4-4 CONNECTIONS: FOOD WEBS AND ENERGY FLOW IN ECOSYSTEMS

What Are Food Chains and Food Webs? All organisms, whether dead or alive, are potential sources of food for other organisms. A caterpillar eats a leaf, a robin eats the caterpillar, and a hawk eats the robin. Decomposers consume the leaf, caterpillar, robin, and hawk after they die. As a result, *there is little matter waste in natural ecosystems.*

The sequence of organisms, each of which is a source of food for the next, is called a **food chain**. It determines

cycling needed for the survival of species and biological communities, figure 4-16)

This rich variety of genes, species, biological communities, and life-sustaining biological and chemical processes

- gives us food, wood, fibers, energy, raw materials, industrial chemicals, and medicines, all of which pour hundreds of billions of dollars into the world economy each year end.

- provides us with free recycling, purification, and natural pest control services.

Every species here today **(1)** contains genetic information that represents thousands to millions of years of adaptation to the earth's changing environmental conditions and **(2)** is the raw material for future adaptations. Loss of biodiversity reduces

Figure 4-17 Two species found in tropical forests are part of the earth's biodiversity. On the right is the world's largest flower, the flesh flower (*Rafflesia arnoldi*), growing in a tropical rain forest in Sumatra. The flower of this leafless plant can be as large as 1 meter (4.3 feet) in diameter and weigh 7 kilograms (15 pounds). The plant gives off a smell like rotting meat, presumably to attract flies and beetles that pollinate its flower. After blossoming once a year for a few weeks, the flower dissolves into a slimy black mass. On the left is a cotton top tamarin. (Right, Mitschuhiko Imanori/Nature Production; left, Gary Milburn/Tom Stack & Associates)

how energy and nutrients move from one organism to another through the ecosystem (Figure 4-18).

Ecologists assign each organism in an ecosystem to a *feeding level*, or **trophic level** (from the Greek word *trophos*, "nourishment"), depending on whether it is a producer or a consumer and on what it eats or decomposes. Producers belong to the first trophic level, primary consumers to the second trophic level, secondary consumers to the third, and so on. Detritivores and decomposers process detritus from all trophic levels.

Real ecosystems are more complex than this. Most consumers feed on more than one type of organism, and most organisms are eaten by more than one type of consumer. Because most species participate in several different food chains, the organisms in most ecosystems form a complex network of interconnected food chains called a **food web** (Figure 4-19). Trophic levels can be assigned in food webs just as in food chains.

How Can We Represent the Energy Flow in an Ecosystem? Pyramids of Energy Flow Each trophic level in a food chain or web contains a certain amount of **biomass**, the dry weight of all organic matter contained in its organisms. In a food chain or web, chemical energy stored in biomass is transferred from one trophic level to another, with some usable energy degraded and lost to the environment as low-quality heat in each transfer. Thus, **(1)** only a small portion of what is eaten and digested is actually converted into an organism's bodily material or biomass, and **(2)** the amount of usable energy available to each successive trophic level declines.

The percentage of usable energy transferred as biomass from one trophic level to the next is called **ecological efficiency**. It ranges from 5% to 20% (that is, a loss of 80–95%) depending on the types of species and the ecosystem involved, but 10% is typical.

Assuming 10% ecological efficiency (90% loss) at each trophic transfer, if green plants in an area manage to capture 10,000 units of energy from the sun, then only about 1,000 units of energy will be available to support herbivores and only about 100 units to support carnivores.

The more trophic levels or steps in a food chain or web, the greater the cumulative loss of usable energy as energy flows through the various trophic levels. The **pyramid of energy flow** in Figure 4-20 illustrates this energy loss for a simple food chain, assuming a 90% energy loss with each transfer. Figure 4-21 shows the pyramid of energy flow during 1 year for an aquatic ecosystem in Silver Springs, Florida.* Pyramids of energy flow *always* have an upright pyramidal shape because of the automatic degradation of energy quality required by the second law of energy.

Energy flow pyramids explain why the earth can support more people if they eat at lower trophic levels by consuming grains, vegetables, and fruits directly (for example, grain ⟶ human) rather than passing such crops through another trophic level and eating grain eaters (grain ⟶ steer ⟶ human).

---

*Because such pyramids represent energy flows, not energy storage, they should not be called pyramids of energy (a common error in many biology and environmental science textbooks).

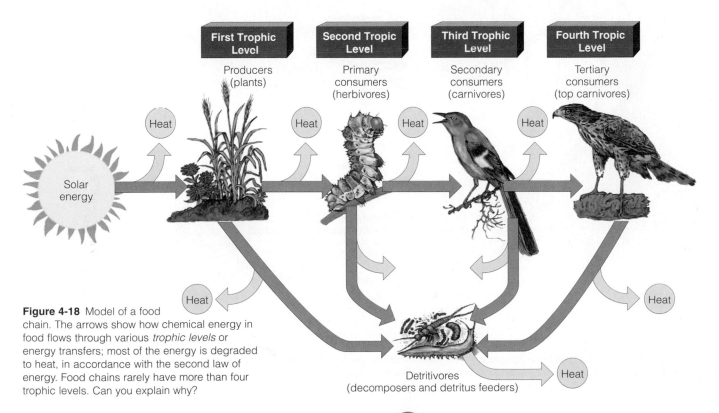

Figure 4-18 Model of a food chain. The arrows show how chemical energy in food flows through various *trophic levels* or energy transfers; most of the energy is degraded to heat, in accordance with the second law of energy. Food chains rarely have more than four trophic levels. Can you explain why?

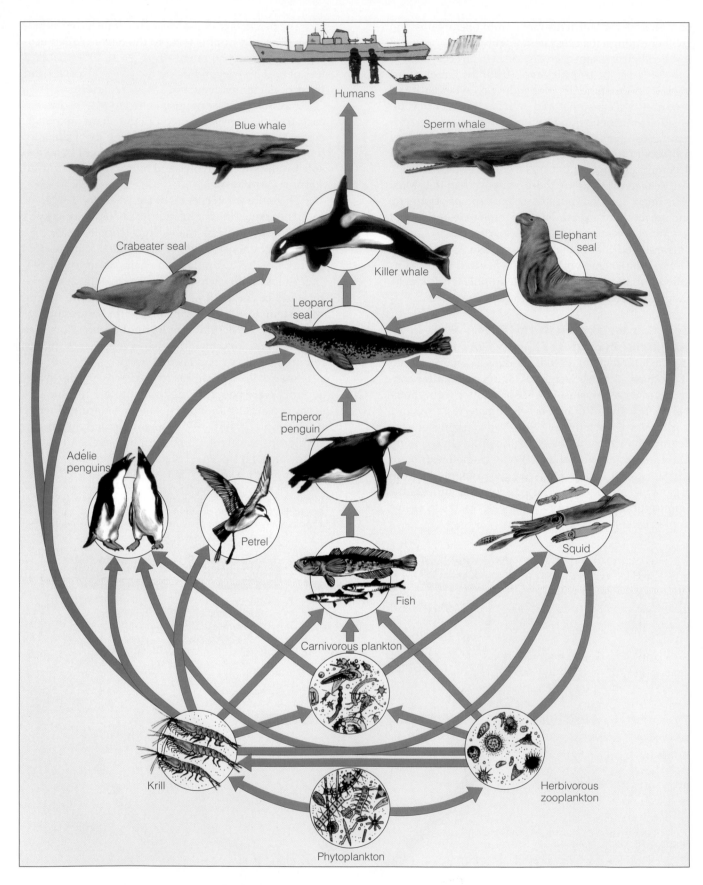

Figure 4-19 Model of a greatly simplified food web in the Antarctic. Many more participants in the web, including an array of decomposer organisms, are not depicted here.

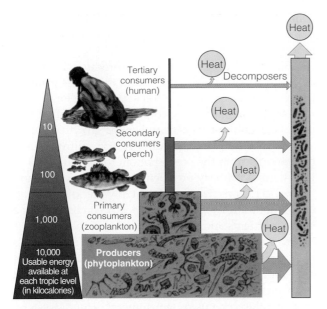

Figure 4-20 Generalized *pyramid of energy flow* showing the decrease in usable energy available at each succeeding trophic level in a food chain or web. In nature, ecological efficiency varies from 5% to 20%, with 10% efficiency being common. This model assumes a 10% ecological efficiency (90% loss in usable energy to the environment, in the form of low-quality heat) with each transfer from one trophic level to another. Because of the degradation of energy quality required by the second law of energy, these models always have a pyramidal shape.

The large loss in energy between successive trophic levels also explains why food chains and webs rarely have more than four or five trophic levels. In most cases, too little energy is left after four or five transfers to support organisms feeding at these high trophic levels. This explains **(1)** why there are so few top carnivores such as eagles, hawks, tigers, and white sharks, **(2)** why such species usually are the first to suffer when the ecosystems that support them are disrupted, and **(3)** why these species are so vulnerable to extinction.

How Can We Represent Biomass Storage in an Ecosystem? Pyramids of Biomass and Numbers

The storage of biomass at various trophic levels in an ecosystem can be represented by a **pyramid of biomass** (Figure 4-22). Ecologists estimate biomass by harvesting organisms from random patches or narrow strips in an ecosystem. The sample organisms are then sorted according to trophic levels, dried, and weighed. These data are used to plot a pyramid of biomass.

For most land ecosystems, the total biomass at each successive trophic level decreases, yielding a pyramid of biomass with a large base of producers, topped by a series of increasingly smaller biomasses at higher trophic levels (Figure 4-22, left). In the open waters of aquatic ecosystems, however, the

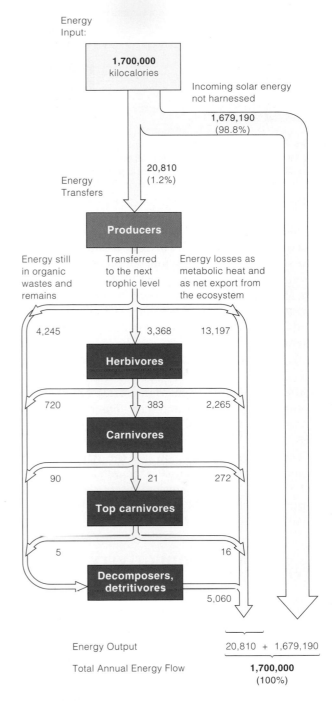

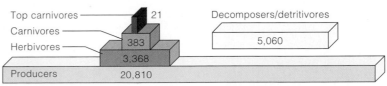

Figure 4-21 Annual pyramid of energy flow (in kilocalories per square meter per year) for an aquatic ecosystem in Silver Springs, Florida. The pyramid is constructed by using the data on energy flow through this ecosystem shown in the top drawing. (From *Biology: Concepts and Applications*, 4/e by Cecie Starr ©2000)

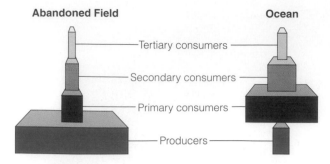

Abandoned Field **Ocean**

Tertiary consumers

Secondary consumers

Primary consumers

Producers

Figure 4-22 Generalized graphs of *biomass of organisms* in the various trophic levels for two ecosystems. The size of each tier in this conceptual model represents the dry weight per square meter of all organisms at that trophic level.

biomass of primary consumers (zooplankton) can exceed that of producers. The reason is that the producers are microscopic phytoplankton that grow and reproduce rapidly, not large plants that grow and reproduce slowly. The zooplankton eat the phytoplankton almost as fast as they are produced so that the producer population is never very large, and the graph is not an upright pyramid (Figure 4-22, right).

By estimating the number of organisms at each trophic level, ecologists can also create a **pyramid of numbers** for an ecosystem (Figure 4-23). Numbers of organisms for grasslands and many other ecosystems taper off from the producer level to the higher trophic levels, forming an upright pyramid (Figure 4-23, left).

For other ecosystems, however, the graph can take a different shape. For example, a temperate forest (Figure 4-23, right) has a few large producers (the trees) that support a much larger number of small primary consumers (insects) that feed on the trees.

4-5 PRIMARY PRODUCTIVITY OF ECOSYSTEMS

How Rapidly Do Producers in Different Ecosystems Produce Biomass? The *rate* at which an ecosystem's producers convert solar energy into chemical energy as biomass is the ecosystem's **gross primary productivity (GPP)**. In effect, it is the rate at which plants or other producers use photosynthesis to make more plant material (biomass).

Figure 4-24 shows how this productivity varies across the earth. This figure shows that gross primary productivity generally is greatest in (1) the shallow waters near continents, (2) along coral reefs where abundant light, heat, and nutrients stimulate the growth of algae, and (3) where upwelling currents bring nitrogen and phosphorus from the ocean bottom to the surface. The lowest gross primary productivity is in (1) deserts and other arid regions because of their low precipitation and high temperatures and (2) the open ocean because of a lack of nutrients and sunlight except near the surface.

To stay alive, grow, and reproduce, an ecosystem's producers must use some of the total biomass they produce for their own respiration. Only what is left, called **net primary productivity (NPP)**, is available for use as food by other organisms (consumers) in an ecosystem:

Net primary productivity = Rate at which producers store chemical energy as biomass (produced by photosynthesis) − Rate at which producers use chemical energy stored as biomass (through aerobic respiration)

Net primary productivity is the *rate* at which energy for use by consumers is stored in new biomass (cells, leaves, roots, and stems). It is measured in units of the energy or biomass available to consumers in a specified area over a given time. It is typically measured in (1) kilocalories per square meter per year ($kcal/m^2/yr$) or (2) grams of biomass created per square meter per year ($g/m^2/yr$).

Various ecosystems and life zones differ in their net primary productivity (Figure 4-25). The most productive are (1) estuaries, (2) swamps and marshes, and (3) tropical rain forests. The least productive are (1) open ocean, (2) tundra (arctic and alpine grasslands), and (3) desert. Despite its low net primary productivity, there is so much open ocean that it produces more of the earth's net primary productivity per year than any of the other ecosystems and life zones shown in Figure 4-25.

Agricultural land is a highly modified and managed ecosystem in which we try to increase the net primary productivity and biomass of selected crop plants by adding water (irrigation) and nutrients (fertilizers). Nitrogen as nitrate (NO_3^-) and phosphorus as phosphate (PO_4^{3-}) are the most common nutrients in fertilizers because they are most often the nutrients limiting crop growth. Despite such inputs, the net primary productivity of

Grassland (summer) **Temperate Forest** (summer)

Tertiary consumers

Secondary consumers

Primary consumers

Producers

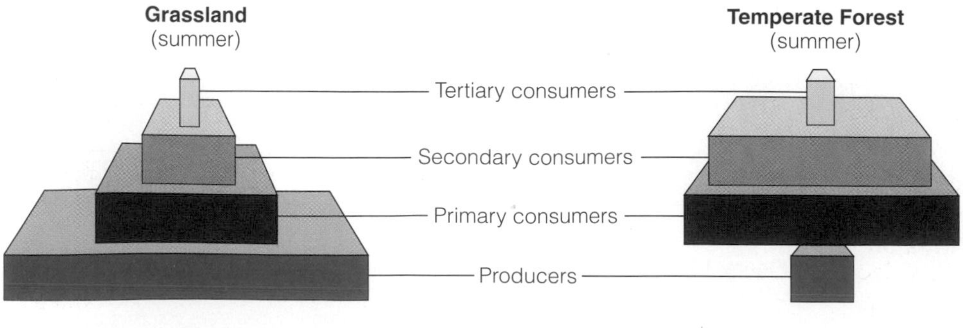

Figure 4-23 Generalized graphs of *numbers of organisms* in the various trophic levels for two ecosystems.

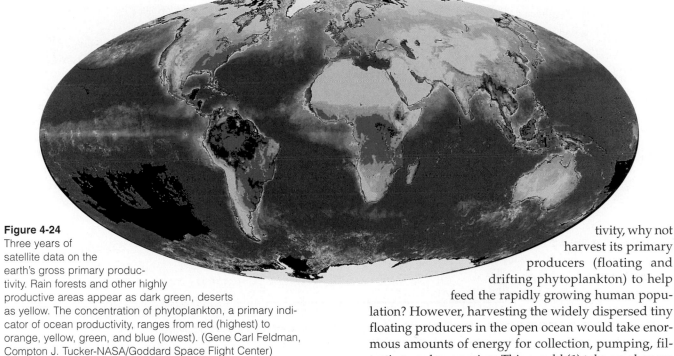

Figure 4-24
Three years of satellite data on the earth's gross primary productivity. Rain forests and other highly productive areas appear as dark green, deserts as yellow. The concentration of phytoplankton, a primary indicator of ocean productivity, ranges from red (highest) to orange, yellow, green, and blue (lowest). (Gene Carl Feldman, Compton J. Tucker-NASA/Goddard Space Flight Center)

agricultural land is not particularly high compared with that of other ecosystems (Figure 4-25).

How Does the World's Net Rate of Biomass Production Limit the Populations of Consumer Species? Ultimately, the planet's net primary productivity limits the number of consumers (including humans) that can survive on the earth. In other words, *the earth's net primary productivity is the upper limit determining the planet's carrying capacity for all consumer species.*

It is tempting to conclude from Figure 4-25 that a good way to feed the world's hungry millions would be to harvest plants in estuaries, swamps, and marshes. This is not a good idea because **(1)** most plants in estuaries, swamps, and marshes cannot be eaten by people and **(2)** these plants are vital food sources (and spawning areas) for fish, shrimp, and other aquatic life-forms that provide us and other consumers with protein.

We might also conclude from Figure 4-25 that we could grow more food for human consumption by clearing tropical forests and planting food crops. According to most ecologists, this is also a bad idea. The basic problem is that in tropical forests most of the nutrients needed to grow food crops are stored in the vegetation rather than in the soil. When the trees are removed, the nutrient-poor soils are rapidly depleted of their nutrients by frequent rains and growing crops. Crops can be grown only for a short time without massive and expensive applications of commercial fertilizers.

Because the earth's vast open oceans provide the largest percentage of the earth's net primary productivity, why not harvest its primary producers (floating and drifting phytoplankton) to help feed the rapidly growing human population? However, harvesting the widely dispersed tiny floating producers in the open ocean would take enormous amounts of energy for collection, pumping, filtration, and processing. This would **(1)** take much more fossil fuel and other types of energy than the food energy we would get and **(2)** disrupt the food webs of the open ocean (Figure 4-19) that provide us and other consumer organisms with important sources of energy and protein from fish and shellfish.

How Much of the World's Net Rate of Biomass Production Do We Use? Excluding uninhabitable areas of rock, ice, desert, and steep mountain terrain, humans have taken over, disturbed, or degraded about 73% of the earth's land surface (Figure 1-4, p. 8). Peter Vitousek and other ecologists estimate that humans now use, waste, or destroy about **(1)** 27% of the earth's total potential net primary productivity and **(2)** 40% of the net primary productivity (Figure 4-26) of the planet's terrestrial ecosystems.

This is the main reason why we are crowding out or eliminating the habitats and food supplies of a growing number of other species. What might happen to us and to other consumer species if the human population doubles over the next 40-50 years and per capita consumption of resources such as food, timber, grassland rises sharply?

4-6 CONNECTIONS: MATTER CYCLING IN ECOSYSTEMS

What Are Biogeochemical Cycles? A nutrient is any atom, ion, or molecule an organism needs to live, grow, or reproduce. Some elements (such as carbon, oxygen, hydrogen, nitrogen, phosphorus, sulfur, and

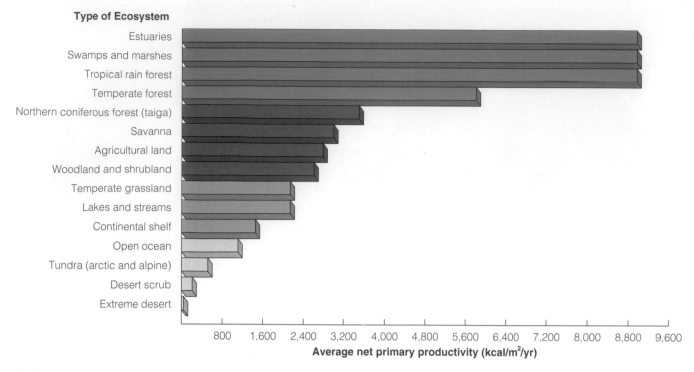

Type of Ecosystem

Average net primary productivity (kcal/m²/yr)

Figure 4-25 Estimated annual average *net primary productivity* (NPP) per unit of area in major life zones and ecosystems, expressed as kilocalories of energy produced per square meter per year (kcal/m²/yr). (Data from *Communities and Ecosystems*, 2/E by R. H. Whittaker, 1975, New York: Macmillan,)

calcium) are needed in fairly large amounts, whereas others (such as sodium, zinc, copper, and iodine) are needed in small or even trace amounts.

These nutrient atoms, ions, and molecules are cycled continuously from the nonliving environment (air, water, soil, and rock) to living organisms (biota) and then back again in what are called **biogeochemical cycles** (literally, life-earth-chemical cycles). These cycles, driven directly or indirectly by incoming solar energy and gravity, include the carbon, oxygen, nitrogen, phosphorus, and hydrologic (water) cycles (Figure 4-7, p. 75).

The earth's chemical cycles also connect past, present, and future forms of life. Some of the carbon atoms in your skin may once have been part of a leaf, a dinosaur's skin, or a layer of limestone rock. Some of the oxygen molecules you just inhaled may have been inhaled by your grandmother, Plato, or a hunter-gatherer who lived 25,000 years ago.

What Are the Major Types of Nutrient Cycles?
There are three general types of nutrient cycles: hydrologic, atmospheric, and sedimentary.

In the *hydrologic*, or *water*, *cycle*, water in the form of ice, liquid water, and water vapor cycles through the biosphere. In this case, the hydrosphere is the main storehouse. This cycle operates at the local, regional, and global levels.

In an *atmospheric cycle*, a large portion of a given element exists in gaseous form in the atmosphere. Examples are nitrogen gas (N_2) and carbon dioxide gas

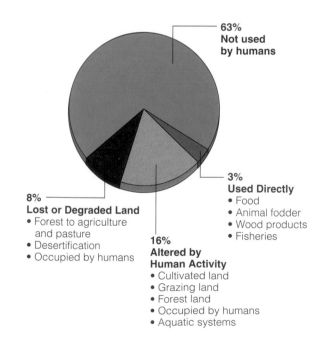

63% Not used by humans

3% Used Directly
• Food
• Animal fodder
• Wood products
• Fisheries

8% Lost or Degraded Land
• Forest to agriculture and pasture
• Desertification
• Occupied by humans

16% Altered by Human Activity
• Cultivated land
• Grazing land
• Forest land
• Occupied by humans
• Aquatic systems

Figure 4-26 Human use of the biomass produced by photosynthesis. Humans destroy, alter, and directly use about 27% of the earth's total net primary productivity and about 40% of the net primary productivity of the earth's terrestrial ecosystems. (Data from Peter Vitousek)

Element	Main nonliving reservoir	Main forms in living organisms	Other nonliving reservoir
Carbon (C)	*Atmospheric:* carbon dioxide (CO_2)	Carbohydrates $(CH_2O)_n$ and all other organic molecules	*Hydrologic:* dissolved carbonate (CO_3^{2-}) and bicarbonate (HCO_3^-) *Sedimentary:* carbon-containing minerals in rocks
Nitrogen (N)	*Atmospheric:* nitrogen gas (N_2)	Proteins and other nitrogen-containing organic molecules	*Hydrologic:* dissolved ammonium (NH_4^+), nitrate (NO_3^-), and nitrite (NO_2^-) in water and soils
Phosphorus (P)	*Sedimentary:* phosphate (PO_4^{3-}) containing minerals in rocks	DNA, other nucleic acids (e.g., ATP), and phospholipids	*Hydrologic:* dissolved phosphate (PO_4^{3-})
Sulfur (S)	*Sedimentary:* rocks (e.g., iron disulfide and pyrite) and minerals (e.g., sulfate [SO_4^{2-}])	Sulfur-containing amino acids in most proteins, and some vitamins	*Atmospheric:* hydrogen sulfide (H_2S), sulfur dioxide (SO_2), sulfur trioxide (SO_3), and sulfuric acid (H_2SO_4) *Hydrologic:* sulfate (SO_4^{2-}) and sulfuric acid (H_2SO_4)

Figure 4-27 Major nonliving and living storehouses of elemental nutrients.

(CO_2), which cycle fairly rapidly from the atmosphere, through soil and organisms, and back into the atmosphere. Because they involve the atmosphere, such cycles operate at local, regional, and global levels.

In a *sedimentary cycle*, an element does not have a gaseous phase, or its gaseous compounds do not make up a significant portion of its supply. In this case, the earth's crust is the main storehouse. Such elements cycle quite slowly, moving mostly from the land to sediments in the seas and then back to the land through long-term geological uplifting of the earth's crust over millions to hundreds of million of years.

Phosphorus and most nonrenewable solid minerals are circulated in such cycles. The slow rate of cycling of such nutrients explains why availability of phosphorus in soil often limits plant growth. Because they have no (or little) circulation in the atmosphere, such cycles tend to operate only on a local and regional basis. Figure 4-27 shows the major nonliving and living storehouses for the carbon, nitrogen, phosphorus, and sulfur cycles.

On a long-term geological scale of thousands to millions of years, the earth's biogeochemical cycles do not have a fixed balance or homeostasis. These cycles have undergone various changes in response to changes in environmental conditions such as major climate changes and collisions of large asteroids with the earth. Most of these changes have occurred over thousands to millions of years.

However, there is growing evidence that these major cycles are now being altered by human activities on a much shorter time scale of decades to several hundred years. Types of changes include (1) depletion of certain chemicals currently stored in cycle biogeochemical reservoirs (Figure 4-27), (2) buildup of certain chemicals in biogeochemical cycle reservoirs, and (3) changes in the rate at which chemicals are cycled. Ecologists are working to find (1) what human activities are affecting these cycles, (2) where and how such changes are taking place within these cycles, and (3) how such changes might affect our species, other species, and national and global economies.

How Is Water Cycled in the Biosphere? In the words of Leonardo da Vinci, "Water is the driver of nature." Without water, the other nutrient cycles would not exist in their present forms, and current forms of life on the earth—consisting mostly of water-containing cells and tissues—could not exist. The **hydrologic cycle**, or **water cycle**, which collects, purifies, and

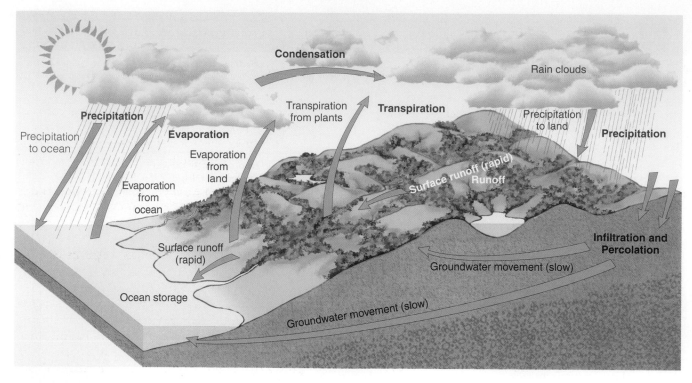

Figure 4-28 Simplified model of the hydrologic cycle.

distributes the earth's fixed supply of water, is shown in simplified form in Figure 4-28.

The main processes in this water recycling and purifying cycle are **(1)** *evaporation* (conversion of water into water vapor), **(2)** *transpiration* (evaporation from leaves of water extracted from soil by roots and transported throughout the plant), **(3)** *condensation* (conversion of water vapor into droplets of liquid water, **(4)** *precipitation* (rain, sleet, hail, and snow), **(5)** *infiltration* (movement of water into soil), **(6)** *percolation* (downward flow of water through soil and permeable rock formations to groundwater storage areas called aquifers), and **(7)** *runoff* (downslope surface movement back to the sea to resume the cycle).

The water cycle is powered by energy from the sun and by gravity. Incoming solar energy evaporates water from oceans, streams, lakes, soil, and vegetation. About 84% of water vapor in the atmosphere comes from the oceans and the rest comes from land.

The amount of water vapor air can hold depends on its temperature, with warm air holding more water vapor than cold air. **Absolute humidity** is the amount of water vapor found in a certain mass of air and is usually expressed as grams of water per kilogram of air. **Relative humidity** is the amount of water vapor in a certain mass of air, expressed as a percentage of the maximum amount it could hold at that temperature. For example, a relative humidity of 60% at 27°C (80°F) means that each kilogram (or other unit of mass) of air

contains 60% of the maximum amount of water vapor it could hold at that temperature.

Winds and air masses transport water vapor over various parts of the earth's surface, often over long distances. Falling temperatures cause the water vapor to condense into tiny droplets that form clouds or fog. For precipitation to occur, air must contain **condensation nuclei**: tiny particles on which droplets of water vapor can collect. Volcanic ash, soil dust, smoke, sea salts, and particulate matter emitted by factories, coal-burning power plants, and vehicles are sources of such particles. The temperature at which condensation occurs is called the **dew point**.

Some of the fresh water returning to the earth's surface as precipitation becomes locked in glaciers. Most of the precipitation falling on terrestrial ecosystems becomes *surface runoff* flowing into streams and lakes, which eventually carry water back to the oceans, where it can be evaporated to cycle again.

Besides replenishing streams and lakes, surface runoff also causes soil erosion, which moves soil and weathered rock fragments from one place to another. Water is thus the primary sculptor of the earth's landscape. Because water dissolves many nutrient compounds, it is a major medium for transporting nutrients within and between ecosystems.

Some of the water returning to the land soaks into (infiltrates) the soil and porous rock and then percolates downward, dissolving minerals from porous rocks on the way. This water is stored as *groundwater* in the pores and cracks of rocks. Where the pores are joined, a network of water channels allows water to flow through the porous rock. Such water-laden rock is called an *aquifer*, and the level of the earth's land crust to which it is filled is called the *water table*. This underground water flows slowly downhill through rock pores and seeps out into streams and lakes or comes out in springs. Eventually, this water evaporates or reaches the sea to continue the cycle.

Throughout the hydrologic cycle, many natural processes act to purify water. Evaporation and subsequent precipitation act as a natural distillation process that removes impurities dissolved in water. As water flows above ground through streams and lakes, and below ground in aquifers, it is naturally filtered and purified by chemical and biological processes. Thus, the hydrologic cycle also can be viewed a cycle of natural renewal of water quality.

How Are Human Activities Affecting the Water Cycle? During the past 100 years, we have been intervening in the earth's current water cycle in several ways:

■ Withdrawing large quantities of fresh water from streams, lakes, and underground sources. In heavily populated or heavily irrigated areas, withdrawals have led to groundwater depletion or intrusion of ocean salt water into underground water supplies.

■ Clearing vegetation from land for agriculture, mining, road and building construction, and other activities. This **(1)** increases runoff, **(2)** reduces infiltration that recharges groundwater supplies, **(3)** increases the risk of flooding, and **(4)** accelerates soil erosion and landslides.

■ Modifying water quality by adding nutrients (such as phosphates) and other pollutants and by changing ecological processes that purify water naturally. Chapter 19 (p.476) examines issues of water pollution and water quality.

How Is Carbon Cycled in the Biosphere? Carbon is essential to life as we know it. It is the basic building block of the organic compounds necessary for life, including carbohydrates, fats, proteins, and nucleic acids such as DNA and RNA.

The **carbon cycle** (Figure 4-29), a *global gaseous cycle*, is based on carbon dioxide gas, which makes up 0.036% of the volume of the troposphere and is also dissolved in water. Carbon dioxide is a key component of nature's thermostat. If the carbon cycle removes too much CO_2 from the atmosphere, the atmosphere will cool; if the cycle generates too much, the atmosphere will get warmer. Thus, even slight changes in the carbon cycle can affect climate and ultimately the types of life that can exist on various parts of the planet.

Terrestrial producers remove CO_2 from the atmosphere, and aquatic producers remove it from the water. They then use photosynthesis to convert CO_2 into complex carbohydrates such as glucose ($C_6H_{12}O_6$).

The cells in oxygen-consuming producers, consumers, and decomposers then carry out aerobic respiration, which breaks down glucose and other complex organic compounds and converts the carbon back to CO_2 in the atmosphere or water for reuse by producers. This linkage between photosynthesis in producers and aerobic respiration in producers, consumers, and decomposers circulates carbon in the biosphere and is a major part of the global carbon cycle. Oxygen and hydrogen, the other elements in carbohydrates, cycle almost in step with carbon.

Over millions of years, buried deposits of dead plant matter and bacteria are compressed between layers of sediment, where they form carbon-containing *fossil fuels* such as coal and oil (Figure 4-29). This carbon is not released to the atmosphere as CO_2 for recycling until these fuels are extracted and burned or until long-term geological processes expose these deposits to air. In only a few hundred years, we have extracted and burned fossil fuels that took millions of years to form. This explains why fossil fuels are nonrenewable resources on a human time scale.

The largest storage reservoir for the earth's carbon is sedimentary rocks such as limestone ($CaCO_3$) deposited as sediments on the ocean floor and on continents. This carbon reenters the cycle very slowly, when some of the sediments dissolve and form dissolved CO_2 gas that can enter the atmosphere. Geologic processes can also bring bottom sediments to the surface, exposing the carbonate rock to chemical attack by oxygen and converting it to carbon dioxide gas. Carbon dioxide also is released into the atmosphere when acidic rain falls on and dissolves exposed limestone rock.

The oceans are the second largest storage reservoir in the carbon cycle. Oceans also play a major role in regulating the level of carbon dioxide in the atmosphere. Some carbon dioxide gas, which is readily soluble in water, stays dissolved in the sea, some is removed by photosynthesizing producers, and some reacts with seawater to form carbonate ions (CO_3^{2-}) and bicarbonate ions (HCO_3^-). As water warms, more dissolved CO_2 returns to the atmosphere, just as more carbon dioxide fizzes out of a carbonated beverage when it warms.

In marine ecosystems, some organisms take up dissolved CO_2 molecules, carbonate ions, or bicarbonate ions from ocean water. These ions can then react with calcium ions (Ca^{2+}) in seawater to form slightly soluble carbonate compounds such as calcium carbonate ($CaCO_3$) to build the shells and skeletons of marine organisms. When these organisms die, tiny particles of their shells and bone drift slowly to the ocean depths and are buried for eons (as long as 400 million years) in deep bottom

Figure 4-29 Simplified model of the global carbon cycle. The left portion shows the movement of carbon through marine systems, and the right portion shows its movement through terrestrial ecosystems. Carbon reservoirs are shown as boxes and processes that change one form of carbon to another are shown in unboxed print. (From *Biology: Concepts and Applications*, 4/e by Cecie Starr ©2000)

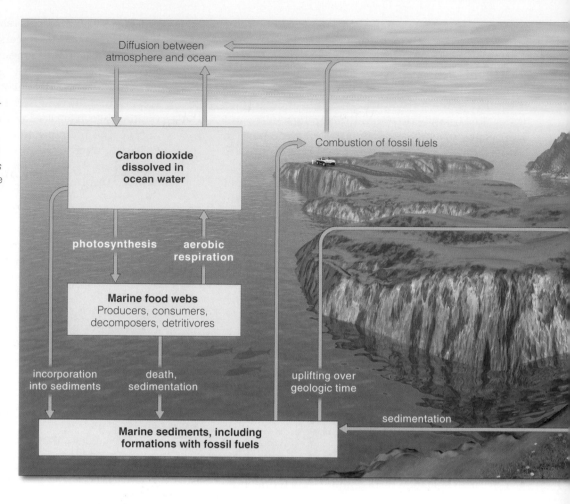

Diffusion between atmosphere and ocean

Carbon dioxide dissolved in ocean water

Combustion of fossil fuels

photosynthesis   aerobic respiration

Marine food webs
Producers, consumers, decomposers, detritivores

incorporation into sediments

death, sedimentation

uplifting over geologic time

sedimentation

Marine sediments, including formations with fossil fuels

sediments (Figure 4-29), where under immense pressure they are converted into limestone rock.

How Are Human Activities Affecting the Carbon Cycle? Since 1800 and especially since 1950, as world population and resource use have soared, we have been intervening in the earth's current carbon cycle in two ways that add carbon dioxide to the atmosphere:

- Clearing trees and other plants that absorb CO_2 through photosynthesis

- Adding large amounts of CO_2 by burning fossil fuels and wood

Computer models of the earth's climate systems suggest that increased concentrations of atmospheric CO_2 and other gases we're adding to the atmosphere could enhance the planet's natural greenhouse effect that helps warm the lower atmosphere (troposphere) and the earth's surface (Figure 4-8). The resulting *global warming* could disrupt global food production and wildlife habitats and raise the average sea level in various parts of the world, as discussed in more detail in Section 18-4 (p. 458).

How Is Nitrogen Cycled in the Biosphere? Bacteria in Action Organisms use nitrogen to make vital organic compounds such as amino acids, proteins, DNA, and RNA. Nitrogen is an essential plant nutrient. However, in both terrestrial and aquatic ecosystems, it is typically in short supply and limits the rate of primary production. This is why most commercial inorganic fertilizers contain biologically useful compounds of nitrogen such as ammonium nitrate (NH_4NO_3).

Nitrogen is the atmosphere's most abundant element, with chemically unreactive nitrogen gas (N_2) making up 78% of the volume of the troposphere. However, N_2 cannot be absorbed and used (metabolized) directly as a nutrient by multicellular plants or animals.

Thus, nitrogen must be "fixed" or combined with hydrogen or oxygen to provide compounds that plants can use. Fortunately, atmospheric electrical discharges in the form of lightning (that causes nitrogen and oxygen in the atmosphere to react and produce oxides of nitrogen, such as $N_2 + O_2 \longrightarrow 2NO$) and certain bacteria in the soil and aquatic systems convert

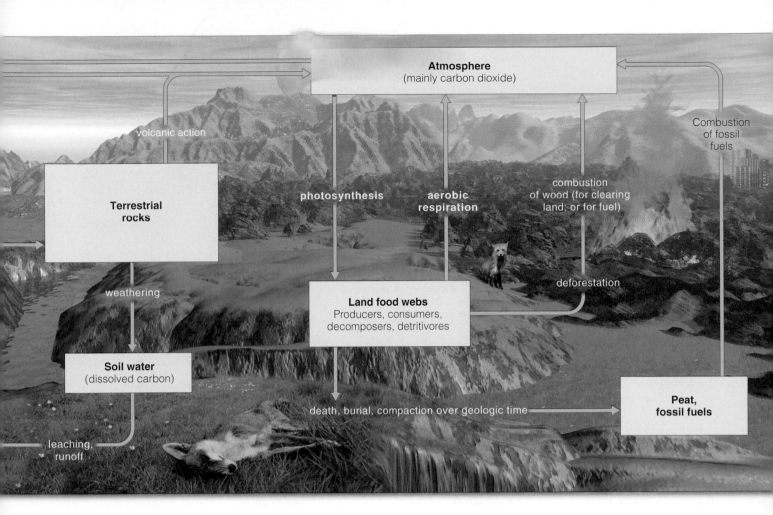

Atmosphere
(mainly carbon dioxide)

volcanic action

Terrestrial rocks

photosynthesis

aerobic respiration

combustion of wood (for clearing land; or for fuel)

Combustion of fossil fuels

weathering

Land food webs
Producers, consumers, decomposers, detritivores

deforestation

Soil water
(dissolved carbon)

death, burial, compaction over geologic time

Peat, fossil fuels

leaching, runoff

nitrogen gas into compounds that can enter food webs as part of the **nitrogen cycle** (Figure 4-30). This *global gaseous cycle* is the most complex of the earth's biogeochemical cycles.

In the first step in the nitrogen cycle, called *nitrogen fixation*, specialized bacteria convert gaseous nitrogen (N_2) to ammonia (NH_3) that can be used by plants by the reaction $N_2 + 3H_2 \longrightarrow 2NH_3$. This is done mostly by (1) cyanobacteria in soil and water and (2) *Rhizobium* bacteria living in small nodules (swellings) on the root systems of a wide variety of plant species, including soybeans and alfalfa.

In a two-step process called *nitrification*, most of the ammonia in soil is converted by specialized aerobic bacteria to nitrite ions (NO_2^-), which are toxic to plants, and then to nitrate ions (NO_3^-), which are easily taken up by plants as a nutrient.

In a process called *assimilation*, plant roots then absorb inorganic ammonia, ammonium ions, and nitrate ions formed by nitrogen fixation and nitrification in soil water. They use these ions to make nitrogen-containing organic molecules such as DNA, amino

acids, and proteins. Animals in turn get their nitrogen by eating plants or plant-eating animals.

After nitrogen has served its purpose in living organisms, vast armies of specialized decomposer bacteria convert the nitrogen-rich organic compounds, wastes, cast-off particles, and dead bodies of organisms into (1) simpler nitrogen-containing inorganic compounds such as ammonia (NH_3) and (2) water-soluble salts containing ammonium ions (NH_4^+). This process is known as *ammonification*.

In a process called *denitrification*, other specialized bacteria (mostly anaerobic bacteria in waterlogged soil or in the bottom sediments of lakes, oceans, swamps, and bogs) then convert NH_3 and NH_4^+ back into nitrite (NO_2^-) and nitrate (NO_3^-) ions and then into nitrogen gas (N_2) and nitrous oxide gas (N_2O). These are then released to the atmosphere to begin the cycle again.

How Are Human Activities Affecting the Nitrogen Cycle? Over the last 100 years, human activities have more than doubled the amount of fixed nitrogen entering the nitrogen cycle. Almost 60% of this

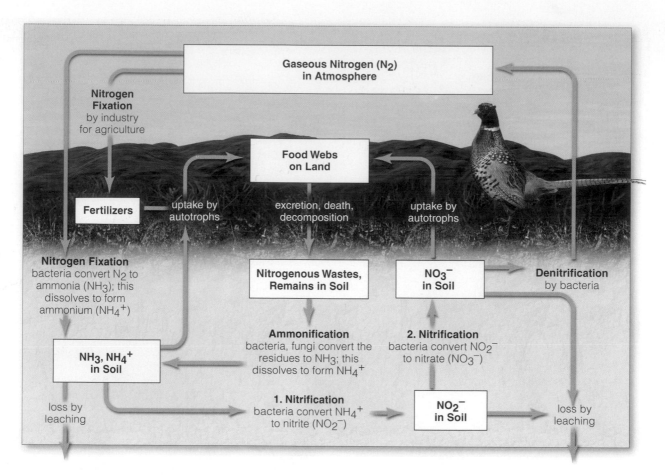

Figure 4-30 Greatly simplified model of the nitrogen cycle in a terrestrial ecosystem. Nitrogen reservoirs are shown as boxes and processes changing one form of nitrogen to another are shown in unboxed print. (From *Biology: The Unity and Diversity of Life*, 9/e by Starr & Taggart ©2001)

human input of nitrogen comes from commercial inorganic fertilizer.

According to a 1997 study by a team of ecologists headed by Peter M. Vitousek of Stanford University, this planetary nitrogen overload is having a number of harmful ecological effects in every place on the earth.

Major human interventions in the earth's current nitrogen cycle over the past 100 years include

- Adding large amounts of nitric oxide (NO) into the atmosphere when we burn any fuel ($N_2 + O_2 \longrightarrow 2NO$). In the atmosphere, this nitric oxide combines with oxygen to form nitrogen dioxide gas (NO_2), which can react with water vapor to form nitric acid (HNO_3). Droplets of HNO_3 dissolved in rain or snow are components of *acid deposition*, commonly called *acid rain*. Nitric acid, along with other air pollutants, can (1) damage and weaken trees, (2) upset aquatic ecosystems, (3) corrode metals, and (4) damage marble, stone, and other building materials, as discussed in Section 17-4 (p. 427).

- Adding nitrous oxide (N_2O) to the atmosphere through the action of anaerobic bacteria on livestock wastes and commercial inorganic fertilizers applied to the soil. When N_2O reaches the stratosphere, it can (1) help warm the atmosphere by enhancing the natural greenhouse effect and (2) contributes to depletion of the earth's ozone shield, which filters out harmful ultraviolet radiation from the sun.

- Removing nitrogen from the earth's crust and soil when we mine nitrogen-containing mineral deposits.

- Removing nitrogen from topsoil when we (1) harvest nitrogen-rich crops, (2) irrigate crops, and (3) burn or clear grasslands and forests before planting crops (Case Study, right).

- Adding nitrogen compounds to aquatic ecosystems in agricultural runoff and discharge of municipal sewage. This excess of plant nutrients stimulates rapid growth of photosynthesizing algae and other aquatic plants. The subsequent breakdown of dead

Effects of Deforestation on Nutrient Cycling

CASE STUDY

In the 1960s, F. H. Bormann of Yale University, Gene Likens of Cornell University, and their colleagues began carrying out a controlled experiment to compare the loss of water and nutrients from an uncut forest ecosystem (the control system) with one that was stripped of its trees (the experimental system).

To do this, V-shaped concrete catchment dams were built across the creeks at the bottom of several valleys in the Hubbard Brook Experimental Forest in New Hampshire. The dams were anchored on impenetrable bedrock so that all surface water leaving each forested valley ecosystem had to flow across the dams, where its volume and dissolved nutrient content could be measured.

The first project was to measure the amounts of water that entered and left an undisturbed (control) forest and the amount of dissolved nutrients in this inflow and outflow. These baseline data showed that an undisturbed mature forest ecosystem is very efficient at retaining chemical nutrients.

The next experiment was to disturb the system and observe any changes that occurred. One winter the investigators cut down all trees and shrubs in one valley. They left them where they fell and were careful not to disturb the soil. The cut area was sprayed with herbicides to prevent regrowth. The inflow and outflow of water and nutrients in this modified experimental valley were then compared with those in the control valley for 3 years.

With no plants to absorb and transpire water from the soil, water runoff in the deforested valley increased by 30–40%. As this excess water ran rapidly over the surface of the ground, it eroded soil and carried nutrients out of the ecosystem. Overall, the loss of minerals from the cut forest was six to eight times that in a nearby undisturbed forest.

For example, chemical analysis of the water flowing through the dams showed a 60-fold rise in the concentration of nitrate ions (NO_3^-; Figure 4-31). So much nitrogen as nitrate (NO_3^-) was lost from the experimental valley that (1) the water flowing out of it was unsafe to drink, and (2) the overfertilized stream below this valley became covered with populations of cyanobacteria and algae. After a few years, however, vegetation grew back and nitrate levels returned to normal (Figure 4-31).

Critical Thinking

What do you think would be the effect on phosphate levels in runoff water from a Hubbard-Brook valley whose trees were destroyed by a major forest fire?

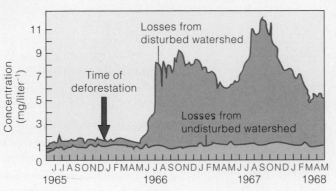

Figure 4-31 Loss of nitrate ions (NO_3^-) from a deforested watershed in the Hubbard Brook Experimental Forest in New Hampshire. The concentration of nitrate ions in runoff from the deforested experimental watershed was many times greater than in a nearby unlogged watershed used as a control. (Data from F. H. Bormann and Gene Likens: Presentation from *Biology: Concepts and Applications* 4/e by Cecie Starr ©2000)

algae by aerobic decomposers can (1) deplete the water of dissolved oxygen and (2) disrupt aquatic ecosystems by killing some types of fish and other oxygen-using (aerobic) organisms, as discussed in Section 19-2 (p. 478). Excess nitrogen in coastal waters can trigger explosions of algal blooms, with some types consisting of toxic dinoflagellates and diatoms that have killed billions of fish and caused some human health problems.

- Accelerating the deposition of acidic nitrogen compounds (such as NO_2 and HNO_3) from the atmosphere onto terrestrial ecosystems. This excessive input of nitrogen can stimulate the growth of

weedy plant species, which can outgrow and perhaps eliminate other plant species that cannot take up nitrogen as efficiently.

How Is Phosphorus Cycled in the Biosphere?

Phosphorus circulates through water, the earth's crust, and living organisms in the **phosphorus cycle** (Figure 4-32). In this *sedimentary cycle*, phosphorus moves slowly from phosphate deposits on land and in shallow ocean sediments to living organisms, and then much more slowly back to the land and ocean.

Bacteria are less important here than in the nitrogen cycle. Very little phosphorus circulates in the

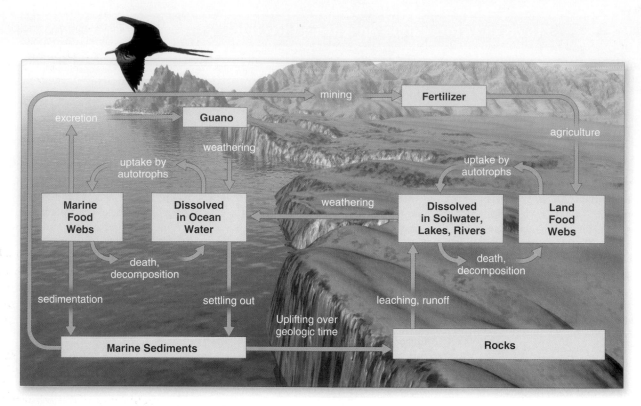

Figure 4-32 Simplified model of the phosphorus cycle. Phosphorus reservoirs are shown as boxes and processes that change one form of phosphorus to another are shown in unboxed print. (From *Biology: The Unity and Diversity of Life,* 9/e by Starr & Taggart ©2001)

atmosphere because at the earth's normal temperatures and pressures, phosphorus and its compounds are not gases. Phosphorus is found in the atmosphere only as small particles of dust. In contrast to the carbon cycle, the phosphorus cycle is slow, and on a short human time scale much phosphorus flows one way from the land to the oceans.

Phosphorous typically is found as phosphate salts containing phosphate ions (PO_4^{3-}) in terrestrial rock formations and ocean bottom sediments. Because most soils contain little phosphate, it is often the *limiting factor* for plant growth on land unless phosphorus (as phosphate salts mined from the earth) is applied to the soil as a fertilizer. Phosphorus also limits the growth of producer populations in many freshwater streams and lakes because phosphate salts are only slightly soluble in water. This explains why adding phosphate compounds to lakes greatly increases their biological productivity.

How Are Human Activities Affecting the Phosphorus Cycle? During the past 100 years humans have been intervening in the earth's current phosphorus cycle by

- Mining large quantities of phosphate rock for use in commercial inorganic fertilizers and detergents.

- Reducing the available phosphate in tropical forests by removing trees. When such forests are cut and burned, most remaining phosphorus and other soil nutrients are washed away by heavy rains, and the land becomes unproductive.

- Adding excess phosphate to aquatic ecosystems in **(1)** runoff of animal wastes from livestock feedlots, **(2)** runoff of commercial phosphate fertilizers from cropland, and **(3)** discharge of municipal sewage. Too much of this nutrient causes explosive growth of cyanobacteria, algae, and aquatic plants. When these plants die and are decomposed they use up dissolved oxygen and disrupt aquatic ecosystems.

How Is Sulfur Cycled in the Biosphere? Sulfur circulates through the biosphere in the **sulfur cycle**, which is a *gaseous cycle* (Figure 4-33). Much of the earth's sulfur is stored underground in rocks and minerals, including sulfate (SO_4^{2-}) salts buried deep under ocean sediments.

Sulfur also enters the atmosphere from several natural sources. Hydrogen sulfide (H_2S) is a colorless, highly poisonous gas with a rotten-egg smell. It is released from active volcanoes and by the breakdown

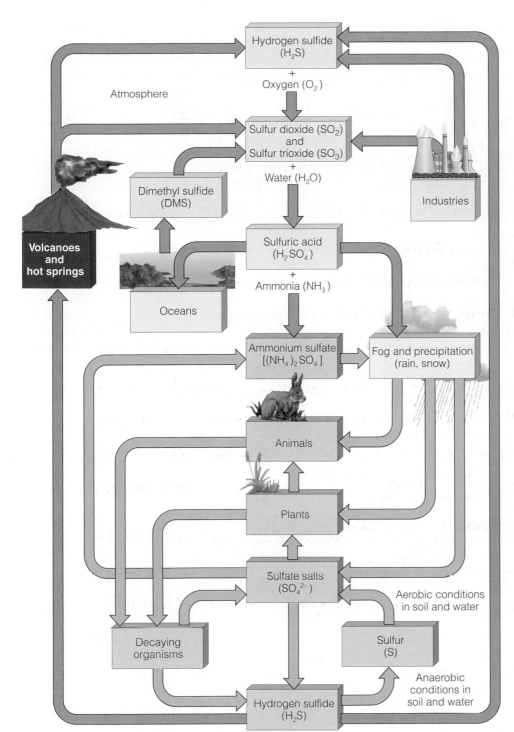

Figure 4-33 Simplified model of the sulfur cycle. Green shows the movement of sulfur compounds in living organisms, blue in aquatic systems, and orange in the atmosphere.

droplets of DMS serve as nuclei for the condensation of water into droplets found in clouds. Thus, changes in DMS emissions can affect cloud cover and thus climate.

In the atmosphere, sulfur dioxide reacts with oxygen to produce sulfur trioxide gas (SO₃). Some of the sulfur trioxide then reacts with water droplets in the atmosphere to produce tiny droplets of sulfuric acid (H₂SO₄). Sulfur dioxide also reacts with other chemicals in the atmosphere such as ammonia to produce tiny particles of sulfate salts. These droplets and particles fall to the earth as components of *acid deposition*, which along with other air pollutants can harm trees and aquatic life, as discussed in Section 17-4, p. 427).

How Are Human Activities Affecting the Sulfur Cycle? Over the past 100 years we have been intervening in the atmospheric phase of the earth's current sulfur cycle by

of organic matter in swamps, bogs, and tidal flats caused by decomposers that do not use oxygen (anaerobic decomposers). Sulfur dioxide (SO₂), a colorless, suffocating gas, also comes from volcanoes. Particles of sulfate (SO₄²⁻) salts, such as ammonium sulfate, enter the atmosphere from sea spray.

Certain marine algae produce large amounts of volatile dimethyl sulfide, or DMS (CH₃SCH₃). Tiny

- Burning sulfur-containing coal and oil to produce electric power, producing about two-thirds of the human inputs of sulfur dioxide

- Refining petroleum

- Using smelting to convert sulfur compounds of metallic minerals into free metals such as copper, lead, and zinc

4-7 HOW DO ECOLOGISTS LEARN ABOUT ECOSYSTEMS?

What Is Field Research? *Field research*, sometimes called muddy-boots biology, involves going into nature and observing and measuring the structure of ecosystems and what happens in them. Most of what we know about the structure and functioning of ecosystems described in this chapter has come from such research.

Increasingly, ecologists are using new technologies to collect field data. These include *remote sensing* from aircraft and satellites and *geographic information systems* (*GISs*), in which information gathered from broad geographic regions is stored in a spatial databases (Figure 4-34). Then computers and GIS software can analyze and manipulate the data and combine them with ground and other data to produce computerized maps of **(1)** forest cover and health, **(2)** water resources, **(3)** air pollution emissions, **(4)** coastal changes, **(5)** relationships between cancer and other health effects and sources of pollution, and **(6)** changes in global sea temperatures. Researchers can also can use computers with such GIS information to look at environmental changes with time. The World Wildlife Fund has used GIS to develop a global map of the world's terrestrial, marine, and freshwater ecosystems.

Most satellite sensors use either reflected light or reflected infrared radiation to gather data. However, some new satellites have radar sensors that measure the reflection of microwave energy from the earth. These microwaves can "see" in the dark and penetrate smoke, clouds, haze, and water. This method is also being used to map the topography of the ocean floor and provide information about ocean currents and upward flows of nutrients from the ocean bottom (upwellings) that sustain fisheries.

What Is Laboratory Research? In the past 50 years, ecologists have increasingly supplemented field research by using *laboratory research* to set up, observe, and make measurements of model ecosystems and populations under laboratory conditions. Such simplified systems

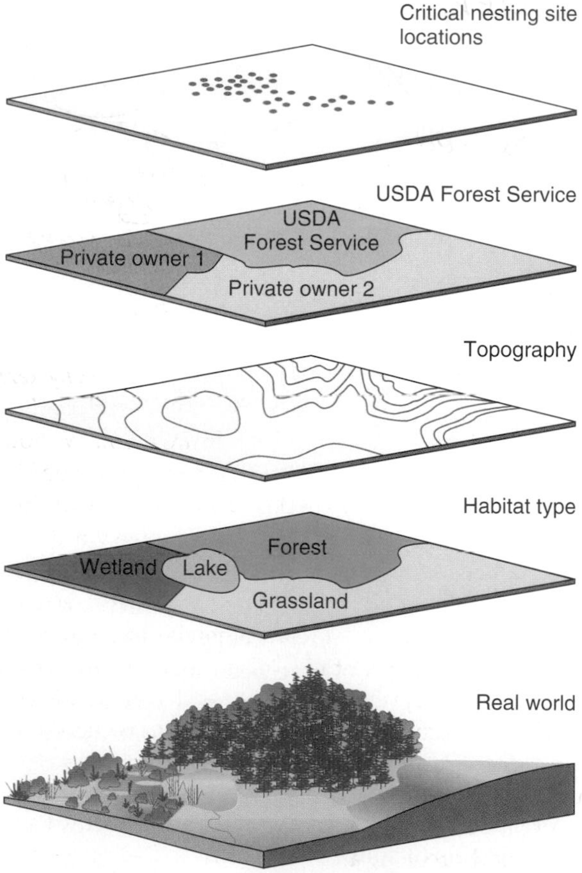

Figure 4-34 Geographic information systems (GISs) provide the computer technology for organizing, storing, and analyzing complex data collected over broad geographical areas. GISs enable scientist to overlay many layers of data (such as soils, topography, distribution of endangered populations, and land protection status).

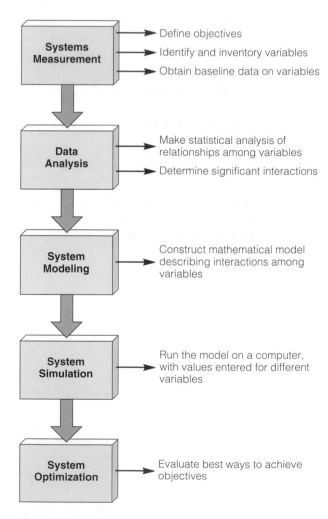

Figure 4-35 Major stages of systems analysis. (Modified data from Charles Southwick)

have been set up in containers such as culture tubes, bottles, aquarium tanks, and greenhouses and in indoor and outdoor chambers where temperature, light, CO_2, humidity, and other variables can be controlled carefully.

In such systems, it is easier for scientists to carry out controlled experiments. In addition, such laboratory experiments often are quicker and cheaper than similar experiments in the field.

However, it is important to consider whether what scientists observe and measure in a simplified, controlled system under laboratory conditions takes place in the same way in the more complex and dynamic conditions found in nature. Thus, the results of laboratory research must be coupled with and supported by field research.

What Is Systems Analysis? Since the late 1960s ecologists have made increasing use of *systems analysis* to develop mathematical and other models that simulate ecosystems. Computer simulation of such models can help us understand large and very complex systems (such as rivers, oceans, forests, grasslands, cities, and climate) that cannot be adequately studied and modeled in field and laboratory research. Figure 4-35 outlines the major stages of systems analysis.

Researchers can change values of the variables in their computer models to **(1)** project possible changes in environmental conditions, **(2)** help anticipate environmental surprises, and **(3)** analyze the effects of various alternative solutions to environmental problems.

However, simulations and predictions made using ecosystem models are no better than the data and assumptions used to develop the models. Thus, careful field and laboratory ecological research must be used to provide the baseline data and determine the causal relationships between key variables needed to develop and test ecosystem models.

4-8 ECOSYSTEM SERVICES AND SUSTAINABILITY

What Are Ecosystem Services? We depend on nature for food, air, water, and almost everything else we use. Ecosystems provide us and other species with a number

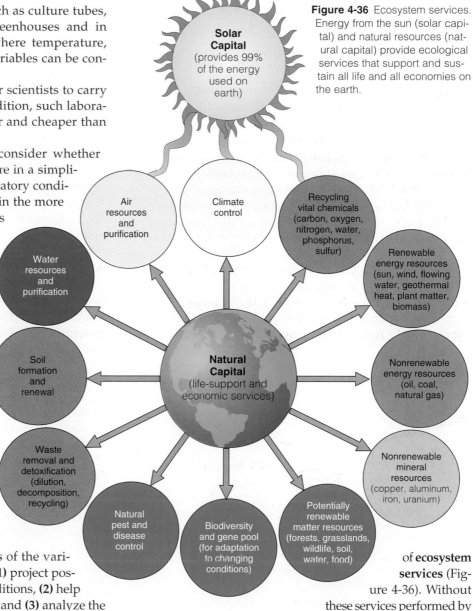

Figure 4-36 Ecosystem services. Energy from the sun (solar capital) and natural resources (natural capital) provide ecological services that support and sustain all life and all economies on the earth.

of **ecosystem services** (Figure 4-36). Without these services performed by diverse communities of species, we would be starving, gasping for breath, and drowning in our own wastes.

It is very expensive to **(1)** build sewage treatment plants to replace free water purification services provided by wetlands, **(2)** add commercial inorganic fertilizers to replace plant nutrients lost by erosion and excessive crop growing, **(3)** rely mostly on pesticides rather than natural biological controls to control crop and forest pests, **(4)** try and save species whose premature extinction could have been prevented, **(5)** replant forested areas to improve flood control and reduce soil erosion, and **(6)** try to restore degraded ecosystems. In addition, these costly replacements are rarely as effective as the free ecological services provided by nature.

What Are the Two Basic Principles of Ecosystem Sustainability? In this chapter we have seen that almost all natural ecosystems and the biosphere itself achieve *sustainability* by

- Using renewable solar energy as their energy source

- Recycling the chemical nutrients its organisms need for survival, growth, and reproduction

These two principles for sustainability arise from the **(1)** structure and function of natural ecosystems (Figures 4-7 and 4-16), **(2)** law of conservation of matter (p. 60), and **(3)** the two laws of energy (p. 65). Thus, the results of basic research in both the physical and biological sciences provide us with the same guidelines or lessons from nature on how we can live more sustainably on the earth, as summarized in Figure 3-20, p. 67.

In this chapter we have learned about **(1)** how the earth's life-support systems work, **(2)** the living and nonliving components of ecosystems, **(3)** how energy flows through ecosystems, **(4)** how rapidly biomass is produced in different ecosystems, **(5)** how matter cycles in ecosystems, **(6)** how ecologists learn about ecosystems, and **(7)** how ecosystem services sustain life. In Chapter 5 we will learn about life developed on the earth and reached its current level of biodiversity.

All things come from earth, and to earth they all return.
MENANDER (342–290 B.C.)

REVIEW QUESTIONS

1. Define all boldfaced terms in this chapter.

2. Why are insects important for many forms of life and for you and your lifestyle?

3. What is *ecology*, and what five levels of organization are the main focus of ecology?

4. Distinguish between *organism, eukaryotic organism, prokaryotic organism, species, population, genetic diversity, habitat, community, ecosystem*, and *biosphere*.

5. Explain why microbes (microorganisms) are so important.

6. Distinguish between the *atmosphere, troposphere, stratosphere, hydrosphere, lithosphere*, and *biosphere*.

7. What three processes sustain life on earth?

8. How does the sun help sustain life on the earth? How is this related to the earth's natural greenhouse effect?

9. What are *biomes* and how are they related to climate? What are *aquatic life zones*? What is an *ecotone*?

10. Distinguish between the *abiotic* and *biotic* components of ecosystems and give three examples of each.

11. Distinguish between *range of tolerance* for a population in an ecosystem, the *law of tolerance*, and *tolerance limits*. How does each of these factors affect the composition (structure) of ecosystems? What is a *limiting factor*, and how do such factors affect the composition of

ecosystems? What are two important limiting factors for terrestrial ecosystems and for aquatic ecosystems?

12. Distinguish between *producers* and *consumers* in ecosystems and give three examples of each type. What is *photosynthesis*, and why is it important to both producers and consumers? What is *chemosynthesis*?

13. Distinguish between *primary consumers (herbivores), secondary consumers (carnivores), tertiary consumers, omnivores, scavengers, detritivores, detritus feeders*, and *decomposers*. Why are decomposers important, and what would happen without them?

14. Distinguish between *aerobic respiration* and *anaerobic respiration*.

15. What are the four components of biodiversity? Why is biodiversity important to **(a)** the earth's life-support systems and **(b)** the economy?

16. Distinguish between a *food chain* and a *food web*.

17. What is *biomass*? What is the *pyramid of energy flow* for an ecosystem? What is the effect of the second law of energy (thermodynamics) on the flow of energy through an ecosystem and on the amount of food energy available to top carnivores and humans?

18. Distinguish between the *pyramid of biomass* and the *pyramid of numbers*.

19. Distinguish between *gross primary productivity* and *net primary productivity*. Explain how net primary productivity affects the number of consumers in an ecosystem and on the earth. List two of the most productive ecosystems or aquatic life zones and two of the least productive ecosystems or aquatic life zones. Use the concept of net primary productivity to explain why harvesting plants from estuaries, clearing tropical forests to grow crops, and harvesting the primary producers in oceans to feed the human population are not good ideas.

20. About what percentages of total potential net primary productivity of **(a)** the entire earth and **(b)** the earth's terrestrial ecosystems are used, wasted, or destroyed by humans?

21. What is a *biogeochemical cycle*? How do such cycles connect past, present, and future forms of life? Distinguish between *hydrologic, atmospheric*, and *sedimentary biogeochemical cycles* and give an example of each type.

22. Describe the *water cycle*. Distinguish between *absolute humidity* and *relative humidity* and between *condensation nuclei* and *dew point*. What is *groundwater*? What is an *aquifer*?

23. List three human activities that alter the water cycle.

24. Describe the *carbon cycle* and list two human activities that alter this cycle.

25. Describe the *nitrogen cycle*. Distinguish between *nitrogen fixation, nitrification, assimilation, ammonification*, and *denitrification*. Explain why the level of nitrogen in soil often limits plant growth. List six ways in which humans alter this cycle.

26. Describe the controlled ecological experiment carried out at the Hubbard-Brook Experimental Forest in New Hampshire and summarize the results.

27. Describe the *phosphorus cycle*. Explain why the level of phosphorus in soil often limits plant growth on land and why phosphorus also limits the growth of producers in many freshwater streams and lakes. List three ways in which humans alter this cycle.

28. Describe the *sulfur cycle,* and list three ways in which humans alter this cycle.

29. Distinguish between *field research, laboratory research,* and *systems analysis* as methods for learning about ecosystems. What are *geographic information systems* and how are they used to learn about ecosystems?

30. Define *ecosystem services* and list nine examples of such services.

31. What are two basic principles of ecosystem sustainability?

CRITICAL THINKING

1. (a) A bumper sticker asks, "Have you thanked a green plant today?" Give two reasons for appreciating a green plant. **(b)** Trace the sources of the materials that make up the bumper sticker and decide whether the sticker itself is a sound application of the slogan. **(c)** Explain how decomposers help keep you alive.

2. (a) How would you set up a self-sustaining aquarium for tropical fish? **(b)** Suppose you have a balanced aquarium sealed with a clear glass top. Can life continue in the aquarium indefinitely as long as the sun shines regularly on it? **(c)** A friend cleans out your aquarium and removes all the soil and plants, leaving only the fish and water. What will happen?

3. Using the second law of energy, explain why there is such a sharp decrease in usable energy as energy flows through a food chain or web. Doesn't an energy loss at each step violate the first law of energy? Explain.

4. Using the second law of energy, explain why many poor people in developing countries live on a mostly vegetarian diet.

5. Why could the total amount of animal flesh on the earth never exceed the total amount of plant flesh, even if all animals were vegetarians?

6. Which causes a larger loss of energy from an ecosystem: a herbivore eating a plant or a carnivore eating an animal? Explain.

7. Why are there more mice than lions in an African ecosystem supporting both types of animals?

8. What would happen to an ecosystem if **(a)** all its decomposers and detritus feeders were eliminated or **(b)** all its producers were eliminated?

PROJECTS

1. Visit several types of nearby aquatic life zones and terrestrial ecosystems. For each site, try to determine **(a)** the major producers, consumers, detritivores, and decomposers and **(b)** the shapes of the pyramids of energy flow, biomass, and numbers.

2. Write a brief scenario describing the sequence of consequences to us and to other forms of life if each of the following nutrient cycles stopped functioning: **(a)** carbon, **(b)** nitrogen, **(c)** phosphorus, and **(d)** water.

3. Use the library or the internet to find out bibliographic information about *G. Evelyn Hutchinson* and *Menander*, whose quotes are found at the beginning and end of this chapter.

4. Make a concept map of this chapter's major ideas, using the section heads and subheads and the key terms (in boldface). Look at the inside back cover and on the website for this book for information about making concept maps.

INTERNET STUDY RESOURCES AND RESOURCES FOR FURTHER READING AND RESEARCH

 www.

The website for this book contains helpful study aids and many ideas for further reading and research. Log on to:

http://www.brookscole.com/product/0534376975s

and click on the Chapter-by-Chapter area. Choose Chapter 4 and select a resource:

- "Flash Cards" allows you to test your mastery of the Terms and Concepts to Remember for this chapter.

- "Tutorial Quizzes" provides a multiple-choice practice quiz.

- "Student Guide to InfoTrac" will lead you to Critical Thinking Projects that use InfoTrac College Edition as a research tool.

- "References" lists the major books and articles consulted in writing this chapter.

- "Hypercontents" takes you to an extensive list of sites with news, research, and images related to individual sections of the chapter.

INFOTRAC COLLEGE EDITION

Improve your skills with InfoTrac College Edition, a searchable online database of articles from more than 700 periodicals. Log on to:

http://www.infotrac-college.com

or access InfoTrac through the website for this book.

Try the following articles:

M. Gatto, G.A. De Leo. 2000. Pricing biodiversity and ecosystem services: the never-ending story. *BioScience* vol. 50, no. 4, pp. 347–356. (subject guide: ecosystem services)

Chaffin, T. 1998. Whole-earth mentor: a conversation with Eugene P. Odum. *Natural History* vol. 107 no. 8, pp. 8–11. (subject guide: ecology, fundamentals)

5 EVOLUTION AND BIODIVERSITY: ORIGINS, NICHES, AND ADAPTATION*

Earth: The Just-Right, Resilient Planet

Like Goldilocks tasting porridge at the Three Bears' house, life on the earth as we know it (Figure 5-1) needs a certain temperature range: Venus is much too hot and Mars is much too cold, but the earth is *just right*. (Otherwise, you would not be reading these words.)

Life as we know it depends on liquid water. Again, temperature is crucial; life on the earth needs average temperatures between the freezing and boiling points of water, between 0°C and 100°C (32°F and 212°F) at the earth's range of atmospheric pressures.

The earth's orbit is the right distance from the sun to provide these conditions. If the earth were much closer, it would be too hot—like Venus—for water vapor to condense to form rain. If it were much farther away, its surface would be so cold—like Mars—that its water would exist only as ice. The earth also spins; if it did not, the side facing the sun would be too hot and the other side too cold for

*Paul M. Rich, associate professor of ecology and evolutionary biology and environmental studies at the University of Kansas, is coauthor of this chapter.

Figure 5-1 The earth is a blue and white planet in the black void of space. Currently, it has the right physical and chemical conditions to allow the development of life as we know it today. (NASA)

water-based life to exist. So far, like Baby Bear's porridge, the temperature has been just right.

The earth is also the right size; that is, it has enough gravitational mass to keep its iron–nickel core molten and to keep the gaseous molecules in its atmosphere from flying off into space. (A much smaller earth would be unable to hold onto an atmosphere consisting of such light molecules as N_2, O_2, CO_2, and H_2O.)

The slow transfer of its internal heat (geothermal energy) to the surface also helps keep the planet at the right temperature for life. And thanks to the development of photosynthesizing bacteria more than 2 billion years ago, an ozone sunscreen protects us and many other forms of life from an overdose of ultraviolet radiation.

On a time scale of millions of years, the earth is enormously resilient and adaptive. During the 3.7 billion years since life arose, the average surface temperature of the earth has remained within the narrow range of 10–20°C (50–68°F), even with a 30–40% increase in the sun's energy output. During this period the temperature has not varied from the mean by more than 5°C (9°F). In short, the earth is just right for life as we know it.

Over the past 3.7 billion years species have come and gone in response to changing environmental conditions. The net result of all of these interactions and adaptations is the incredible biodiversity found on the earth today.

We can summarize the 3.7-billion-year biological history of the earth in one sentence. *Organisms convert solar energy to food, chemicals cycle, and a variety of species with different biological roles (niches) have evolved in response to changing environmental conditions.*

Each species here today represents a long chain of evolution, and each of these species plays a unique ecological role in the earth's communities and ecosystems.

This chapter is devoted to helping us understand how the earth's species evolved and the nature of their niches or biological roles. This information is important for helping us **(1)** predict the effects of human actions on wild species and **(2)** protect species—including the human species—from premature extinction.

There is a grandeur to this view of life . . . that, whilst this planet has gone cycling on . . . endless forms most beautiful and most wonderful have been, and are being, evolved.

CHARLES DARWIN

This chapter addresses the following questions:

- How do scientists account for the emergence of life on the earth?

- What is evolution, and how has it led to the current diversity of organisms on the earth?

- How does evolution affect the way organisms fit into their environment?

- What is an ecological niche, and how does it relate to changing environmental conditions?

- How do extinction of species and formation of new species affect biodiversity?

5-1 ORIGINS OF LIFE

How Did Life Emerge on the Earth? How did a barren planet become a living jewel in the vastness of space? How did life on the earth evolve to its present incredible diversity species, living in an interlocking network of matter cycles, energy flows, and species interactions? We do not know the full answer to these questions, but a growing body of evidence suggests what might have happened.

Evidence about the earth's early history comes from chemical analysis and measurements of radioactive elements in primitive rocks and fossils. Chemists have conducted laboratory experiments showing how simple inorganic compounds in the earth's early atmosphere might have reacted to produce organic molecules such as amino acids, simple sugars, and other building-block molecules for large biopolymer molecules (such as proteins, complex carbohydrates, RNA, and DNA) needed for life.

When this diverse and continually accumulating evidence is pieced together, an important idea emerges: *The evolution of life is linked to the physical and chemical evolution of the earth.*

A second conclusion from this evidence is that life on the earth developed in two phases over the past 4.7–4.8 billion years (Figure 5-2):

- *Chemical evolution* of the organic molecules, biopolymers, and systems of chemical reactions needed to form the first protocells (taking about 1 billion years)

- *Biological evolution* from single-celled prokaryotic bacteria (Figure 4-3b, p. 73), to single-celled eukaryotic creatures (Figure 4-3a, p. 73), and then to multicellular organisms (taking about 3.7–3.8 billion years)

The earth's environment has also changed and evolved. The earth has cooled since its formation, the temperature has stabilized within a fairly narrow range (left), and the composition of the atmosphere has changed from one with no free oxygen to today's atmosphere with 21% oxygen by volume.

How Did Chemical Evolution Take Place? Here is an overview of how scientists believe life may have formed and evolved on the earth, based on the current evidence. Some 4.6–4.7 billion years ago a cloud of cosmic dust condensed into the earth, which soon turned molten from meteorite impacts and from the heat produced by the radioactive decay of chemical elements in its interior. The earth slowly cooled, and about 3.8 billion years ago the outermost portion of the molten sphere solidified to form a thin, hardened crust of rock, devoid of atmosphere and oceans.

Comets hitting the lifeless earth pierced its thin crust, and energy from the earth's highly radioactive contents released water vapor, carbon dioxide, and other gases from the molten interior. Eventually the crust cooled enough for the water vapor to condense and fall to the surface as rain. This rain eroded minerals

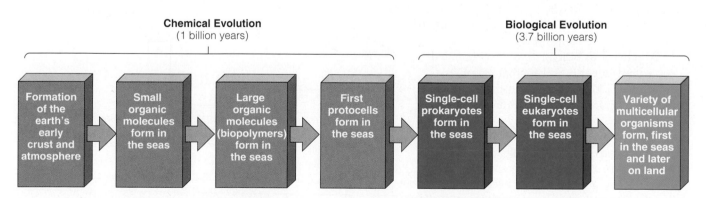

Figure 5-2 Summary of the chemical and biological evolution of the earth. This drawing is not to scale. Note that the time span for biological evolution is almost four times longer than that for chemical evolution.

from rocks, and solutions of these minerals collected in depressions to form the early oceans that covered most of the globe.

Research suggests that most of the planet's first atmosphere had formed by 4.4 billion years ago. The exact composition of this primitive atmosphere is unknown, but atmospheric scientists believe it was dominated by carbon dioxide (CO_2), nitrogen (N_2), and water vapor (H_2O). Trace amounts of methane (CH_4), ammonia (NH_3), hydrogen sulfide (H_2S), and hydrogen chloride (HCl) probably were present (although scientists disagree about the relative amounts of these gases).

Whatever the composition of this primitive atmosphere, scientists agree that it had no oxygen gas (O_2) because this element is so chemically reactive that it would have combined into compounds. The only reason today's atmosphere has so much O_2 is that plants and some aerobic bacteria produce it in vast quantities through photosynthesis. However, this is getting ahead of the story.

Energy from electrical discharges (lightning), heat from volcanoes, and intense ultraviolet (UV) light and other forms of solar radiation was readily available for the synthesis of biologically important organic molecules from the inorganic chemicals found in the earth's primitive atmosphere.

In a number of experiments conducted since 1953, various mixtures of gases believed to have been in the earth's early atmosphere have been put in closed, sterilized glass containers. Then they were subjected to spark discharges to simulate lightning and heat (Figure 5-3). In these experiments, compounds necessary for life—various amino acids (the building-block molecules of proteins), simple carbohydrates, nucleic acids (the building-block molecules of DNA and RNA), and other small organic compounds—formed from the inorganic gaseous molecules.

Another possibility is that simple organic molecules necessary for life formed on dust particles in space and reached the earth on meteorites or comets (or on countless interplanetary dust particles floating around in space when the earth was formed). In 1997, researchers found amino acids on a meteorite that struck Australia in 1969.

Another possibility is that life originated deep within the earth. Still another hypothesis is that life could have arisen in the form of heat-loving prokaryotic microbes around mineral-rich and very hot *hydrothermal vents*, which sit atop cracks in the ocean floor leading to subterranean chambers of molten rock. We do not know which of these processes might have produced the organic molecules necessary for life, but all of these hypotheses are reasonable explanations.

However these building-block organic molecules formed, they accumulated and underwent countless chemical reactions in the earth's warm, shallow waters. After several hundred million years of different chemical combinations in this hot organic soup, conglomerates of proteins, RNA, and other biopolymers may have combined to form membrane-bound *protocells*: small globules that could take up materials from their environment and grow and divide (much like living cells).

Suppose the building-block molecules of life formed around thermal hydrothermal vents on the ocean bottom instead of in the atmosphere. Then scientists hypothesize that the biopolymer molecules necessary for life and the first protocells formed after hundreds of millions of years of different chemical reactions in the *deep ocean* rather than in *shallow pools* on the earth's surface.

Again, there are several hypotheses explaining how these first protocells formed. With these forerunners of living cells, the stage was set for the drama of biological evolution (Figure 5-4 and Spotlight, p. 106).

Figure 5-3 Synthesis of organic compounds necessary for life from the gases believed to be in the earth's primitive atmosphere. Stanley Miller and Harold Urey used a similar apparatus to show that under the influence of electrical sparks simulating lightning, simple inorganic compounds believed to be in the earth's primitive atmosphere could be converted into amino acids (the building blocks of proteins) and a variety of other simple organic chemicals necessary for life.

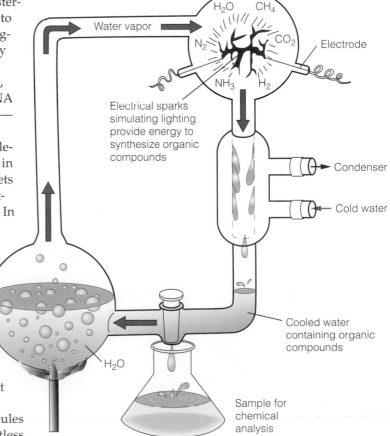

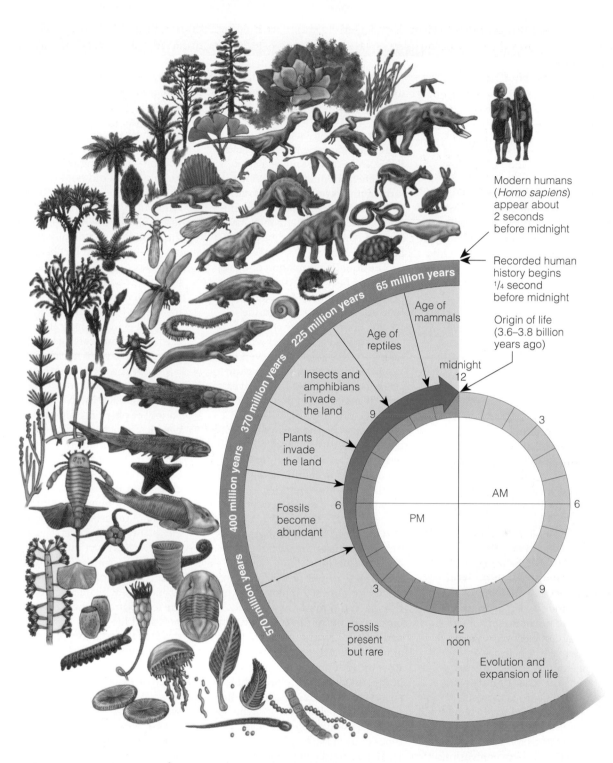

Figure 5-4 Greatly simplified overview of the biological evolution of life on the earth, which was preceded by about 0.5–1 billion years of chemical evolution. The early span of biological evolution on the earth, between about 3.7 billion and 570 million years ago, was dominated by microorganisms (mostly bacteria and, later, protists) that lived in water. Plants and animals evolved first in the seas and moved onto land about 400 million years ago. Humans arrived on the scene only a very short time ago. If we compress the earth's roughly 3.7-billion-year history of biological evolution to a 24-hour time scale, the first human species (*Homo habilis*) appeared about 47–94 seconds before midnight, and our species (*Homo sapiens sapiens*) appeared about 2 seconds before midnight. Agriculture began only 0.25 second before midnight, and the industrial revolution has been around for only 0.007 second. (Adapted from *Life: An Introduction to Biology*, 2d ed. by George Gaylord Simpson and William S. Beck. New York: Harcourt Brace Jovanovich, 1965)

How Did Life First Evolve?

SPOTLIGHT

About 3.2 billion years ago the earth was a very hostile environment for life. Hot lava spewed from the earth's surface and beneath the sea. Much of the land was dotted with boiling hot springs, and the atmosphere was thick with steam and carbon dioxide. Despite such conditions, simple single-celled life-forms began to multiply in the earth's seas.

Over time, it is believed that the protocells evolved into single-celled, bacterialike prokaryotes having the properties we describe as life. However, the details of how this might have happened are hotly debated by scientists.

These anaerobic cells probably evolved **(1)** in the muddy sediments of tidal flats, **(2)** at least 10 meters (30 feet) below the ocean's surface, or **(3)** in the deep ocean near hydrothermal vents. All these environments would protect these single-celled forms of life from the intense UV radiation that bathed the earth at that time.

The early single-celled organisms probably obtained their energy directly by absorbing hydrogen sulfide molecules. Soon thereafter, they developed the process of *fermentation* or anaerobic respiration to break down sugars and other organic compounds to release energy for their use and to store some of this energy in adenosine triphosphate (ATP) molecules.

As the earth's first chemists, these bacteria learned to make new kinds of molecules, including proteins and enzymes to speed up chemical reactions between various molecules. They also developed the ability to exchange bits of DNA by touching and dissolving a portion of their cell walls.

Scientists believe that these single-celled anaerobic bacteria multiplied and underwent genetic changes (mutated) for about a billion years in the earth's warm, shallow seas or in the deep ocean. The result was a variety of new types of prokaryotic cells that began forming about 2.8-3.5 billion years ago.

Scientists contend that during this long early period, life could not have survived on land because there was no ozone layer to shield the DNA and other molecules of early life from intense ultraviolet radiation. However, some recent evidence suggests that some primitive prokaryotic bacteria may have survived on land as early as 2.6 billion years ago.

The aquatic population of these fermenter bacteria expanded to the point at which they exceeded their food supplies. This created the world's first *population crisis*.

This crisis was relieved when different types of bacteria evolved that produced their food through *photosynthesis*. About 2.3–2.7 billion years ago, the evolution of these photosynthetic prokaryotes, called *cyanobacteria*, in the ocean drastically changed the earth. These cells could remove carbon dioxide from the water and (powered by sunlight) combine it with water to make the carbohydrates they needed. In the process, they released oxygen (O_2) into the ocean and the atmosphere. This was a major breakthrough for life on the earth.

As these photosynthesizers multiplied for almost 2 billion years in the water and mud, they created large amounts of dissolved oxygen and oxygen that escaped into the atmosphere. This oxygen reacted with the food molecules used by both fermenting and photosynthesizing bacteria. This resulted in the world's first *pollution crisis*, which caused a huge dieback of both types of bacteria.

This spurred a new round of biological innovation. Fermenters survived by living in deep water and mud flats to avoid contact with the oxygen-rich air. However, the most significant event was the development of bacteria that used *aerobic respiration*, in which oxygen molecules are used to break up food molecules to obtain energy. At this point, the earth's single-celled organisms had developed three ways to feed themselves: *fermentation*, *photosynthesis*, and *aerobic respiration*.

Over the next half a billion years, the oxygen released by photosynthesizers built up in the atmosphere. However, because of oxygen-consuming respiration, oxygen levels did not become high enough for the entire atmosphere to burst into flames. The resulting *oxygen revolution* opened the way for the evolution of a great variety of oxygen-using (aerobic) bacteria. Complex organisms came later, first in the seas and then increasingly on land (after a protective ozone layer formed in the stratosphere).

Fossil evidence indicates that at least 1.2 billion years ago the first *eukaryotic cells* (with nuclei, Figure 4-3, left p. 73) emerged in earth's shallow seas. Because eukaryotes could reproduce sexually, they produced a variety of offspring with different genetic characteristics. These first eukaryotes were the building blocks for the more complex multicelled life-forms that followed. Genetic changes in these eukaryotic cells eventually spawned an amazing variety of protists, fungi, and eventually plants and animals (Figure 5-4).

As oxygen accumulated in the atmosphere, some was converted by incoming solar energy into ozone (O_3), which began forming in the lower stratosphere. This shield protected life-forms from the sun's deadly UV radiation, allowing green plants to live closer to the ocean's surface.

Fossil evidence suggests that about 400–500 million years ago UV levels were low enough for the first plants (probably waxy-coated algae) to emerge from the underwater world and exist on the land. Over the next several hundred million years a variety of land plants and animals arose, followed by mammals and eventually the first humans (Figure 5-4).

Critical Thinking

If photosynthesizing bacteria had not developed, what kinds of life would you expect to find on the earth today? Where would most of it be found?

How Do We Know What Organisms Lived in the Past? Most of what we know of the earth's life history comes from **fossils**: mineralized or petrified replicas of skeletons, bones, teeth, shells, leaves, and seeds, or impressions of such items. Such fossils give us physical evidence of organisms that lived long ago and show us what their internal structures looked like.

Despite its importance, the fossil record is uneven and incomplete. Some life-forms left no fossils, some fossils have decomposed, and others are yet to be found. So far we have found fossils representing only about 1% of the species believed to have ever lived.

Other sources of information include chemical and radioactive dating (p. 63) of fossils, nearby ancient rocks, and material in cores drilled out of buried ice. Another record of evolutionary history is written in the DNA of organisms alive today. Unfortunately, ancient mineralized fossil specimens rarely include any of the original tissues, cells, or molecules found in the organism's body.

5-2 EVOLUTION AND ADAPTATION

What Is Evolution? According to scientific evidence, the major driving force of adaptation to changes in environmental conditions is **biological evolution**, or **evolution**: the change in a population's genetic makeup (gene pool) through successive generations. Note that *populations*, not individuals, evolve by becoming genetically different.

According to the **theory of evolution**, all species descended from earlier, ancestral species. This widely accepted scientific theory explains how life has changed over the past 3.7 billion years and why life is so diverse today.

Biologists use the term *microevolution* to describe the small genetic changes that occur in a population. The term *macroevolution* is used to describe long-term, large-scale evolutionary changes in groups of species, through which new species are formed from ancestral species and other species are lost through extinction.

How Does Microevolution Work? The first step in evolution is the development of *genetic variability* in a population. Recall that **(1)** genetic information in *chromosomes* is contained in various sequences of chemical units (called *nucleotides*) in DNA molecules, and **(2)** genes found in chromosomes are segments of DNA that are coded for certain traits that can be passed on to offspring (Figure 3-8, p. 57).

A population's **gene pool** is the set of all genes in the individuals of the population of a species. *Microevolution* is a change in a population's gene pool over time.

Although members of a population generally have the same number and kinds of genes, a particular gene may have two or more different molecular forms, called **alleles**. Sexual reproduction leads to a random shuffling or recombination of alleles. As a result, each individual in a population has a different combination of alleles. Without such genetic variability, evolution as we know it could not occur.

Microevolution works through a combination of four processes that change the genetic composition of a population:

- **Mutation**, which involves random changes in the structure or number of DNA molecules in a cell and is the ultimate source of genetic variability in a population

- **Natural selection**, which occurs when some individuals of a population have genetically based traits that cause them to survive and produce more offspring than other individuals

- **Gene flow**, which involves movement of genes between populations and can lead to changes in the genetic composition of local populations

- **Genetic drift**, which involves change in the genetic composition of a population by chance and is especially important for small populations

What Is the Role of Mutation in Microevolution? Genetic variability in a population originates through **mutations**: random changes in the structure or number of DNA molecules in a cell. Mutations can occur in two ways:

- Exposure of DNA to external agents such as radioactivity, X rays, and natural and human-made chemicals (called *mutagens*).

- Random mistakes sometimes occur in coded genetic instructions when DNA molecules are copied (each time a cell divides and whenever an organism reproduces).

Mutations can occur in any cells, but only those in reproductive cells are passed on to offspring.

Some mutations are harmless, but most are harmful and alter traits so that an individual cannot survive (lethal mutations). Every so often, a mutation is beneficial. The result is new genetic traits that give their bearer and its offspring better chances for survival and reproduction, either under existing environmental conditions or when such conditions change.

It is important to understand that mutations are **(1)** random and unpredictable, **(2)** the only source of totally new genetic raw material (alleles), and **(3)** rare events. Once created by mutation, new alleles can be shuffled together or recombined *randomly* to create new combinations of genes in populations of sexually reproducing species.

What Role Does Natural Selection Play in Microevolution? The process of **natural selection** occurs when some individuals of a population have genetically based traits that increase their chances of survival and their ability to produce offspring. This idea was developed in 1846 by Charles Darwin and published in 1859 in his now-famous book *On the Origin of Species by Means of Natural Selection*. Darwin recognized that three conditions are necessary for evolution of a population by natural selection to occur:

- There must be natural *variability* for a trait in a population.

- The trait must be *heritable*, meaning that it must have a genetic basis such that it can be passed from one generation to another.

- The trait must somehow lead to **differential reproduction**, meaning that it must enable individuals with the trait to leave more offspring than other members of the population.

Natural selection causes any allele or set of alleles that result in a beneficial trait to become more common in succeeding generations and other alleles to become less common. A heritable trait that enables organisms to better survive and reproduce under a given set of environmental conditions is called an **adaptation**, or **adaptive trait**.

When faced with a change in environmental conditions, a population of a species has three alternatives: **(1)** adapt to the new conditions through natural selection, **(2)** migrate (if possible) to an area with more favorable conditions, or **(3)** become extinct.

It is important to understand that environmental conditions do not create favorable heritable characteristics. Instead, natural selection favors some individuals over others by acting on inherited genetic variations (alleles) already present in the gene pool of a population.

The process of microevolution can be summarized as follows: *Genes mutate, individuals are selected, and populations evolve.* The genetic characteristics of populations of a species also can be changed through *artificial selection* (Spotlight, right).

What Is an Example of Microevolution by Natural Selection? One of the best-documented examples of evolution by natural selection involves camouflage coloration in the peppered moth, which is found in England (Figure 5-5).

This example illustrates the important points of evolution by natural selection. Natural selection occurred because **(1)** there were two color forms (*variability*), **(2)** color form was genetically based (*heritability*), and **(3)** there was greater survival and reproduction by one of the color forms (*differential reproduction*). In this case, first an environmental change in the form of soot caused a change in the background color of tree trunks. This environmental change then allowed bird predators to find and eat the moths with the coloration that no longer blended in with the background.

What Is Artificial Selection?

SPOTLIGHT

Artificial selection is the process by which humans select one or more desirable genetic traits in the population of a plant or animal and then use *selective breeding* to end up with populations of the species containing large numbers of individuals with the desired traits. This process has been used to develop essentially all of the domesticated breeds of plants and animals from wild populations.

Artificial selection involves several steps. First, breeders of a particular food crop species (such as wheat), livestock species (such as cattle), or pet species (such as a dog) decide on genetic traits they would like to have in the species. Examples might be **(1)** wheat that grows short so it will not topple over, **(2)** cattle that need less water, or **(3)** dogs with short legs.

Next they examine existing populations of species (such as wheat, cattle, or dogs) to identify individuals that exhibit more of the desired genetic trait than other members of the population.

Then the selected individuals are bred. Offspring that exhibit more of the desired trait than their parents are *selected* to be breeders for the next generation, and other offspring are prevented from breeding.

When this process of selection and breeding is carried out over many generations, more and more of the offspring have the selected or desired traits.

Artificial selection results in many different domesticated breeds or hybrids of the same species, all originally developed from a particular wild species. For example, despite their widely different genetic traits, all of the hundreds of different breeds of dogs are members of the same species because they can potentially interbreed and produce fertile offspring.

Critical Thinking

How does artificial selection differ from natural selection?

What Are Three Types of Natural Selection? Biologists recognize three types of natural selection that take place under different environmental conditions and lead to a different genetic outcome for the populations involved (Figure 5-6).

In *directional natural selection* (Figure 5-6, left) changing environmental conditions cause allele frequencies to shift so that individuals with traits at one end of the normal range become more common than

Figure 5-5 Two varieties of peppered moths found in England illustrate one kind of adaptation: camouflage. Before the industrial revolution in the mid-1800s, the speckled light-gray form of this moth was prevalent. When these night-flying moths rested on light-gray lichens on tree trunks during the day, their color camouflaged them from their predators. A dark-gray form also existed but was quite rare. During the industrial revolution, soot and other pollutants from factory smokestacks began killing lichens and darkening tree trunks. As a result, the dark form of moth became the common one, especially near industrial cities. In this new environment, the dark form of moth blended in with the blackened trees, whereas the light form of moth was highly visible to predators. Through natural selection, the dark form began to survive and reproduce at a greater rate than its light-colored kin. (Both varieties appear in each photo. Can you spot them?) (Right, Michael Tweedie/Natural History Photographic Agency; left, Kim Taylor/Bruce Coleman Collection)

midrange forms. Examples of this "it pays to be different" type of natural selection are **(1)** the changes in the varieties of peppered moths (Figure 5-5) and **(2)** the evolution of genetic resistance to pesticides among insects and to antibiotics among disease-carrying bacteria. This type of natural selection is most common during periods of environmental change or when members of a population migrate to a new habitat with different environmental conditions.

Stabilizing natural selection tends to eliminate individuals on both ends of the genetic spectrum and favor individuals with an average genetic makeup (Figure 5-6, center). This "it pays to be average" type of natural selection occurs when an environment changes little and when most members of the population are well adapted to that environment. Individuals with unusual alleles have no advantage and tend to be eliminated.

Diversifying natural selection (also called disruptive natural selection) occurs when environmental conditions favor individuals at both extremes of the genetic spectrum and eliminate or sharply reduce numbers of individuals with normal or intermediate genetic traits (Figure 5-6, right). In this "it does not pay to be normal" type of natural selection, a population is split into two groups.

This type of natural selection can occur when a shift in the food supply selects against the average individuals. For example, most members of a population of finches have a certain beak length and width that allows them to eat certain fruits, seeds, and insects. Suppose that an environmental disturbance eliminates most foods except those that can be eaten only by the few birds with wider or stronger beaks. Then more birds with these extreme variations survive than birds with intermediate and weaker beaks, and diversifying natural selection occurs.

What Is Coevolution? Some biologists have proposed that interactions between *species* also can result in microevolution in each of their populations. According to this hypothesis, when populations of two different species interact over a long time, changes in the gene pool of one species can lead to changes in the gene pool of the other species. This process is called **coevolution**.

Suppose that certain individuals in a population of carnivores (such as owls) become better at hunting prey (such as mice). Because of genetic variation, certain individuals of the prey have traits that allow them to escape or hide from their predators, and they pass these adaptive traits on to some of their offspring. However, a few individuals in the predator population also may have traits (such as better eyesight or quicker reflexes) that allow them to hunt the better-adapted prey successfully. They would then pass these traits on to some of their offspring.

Similarly, individual plants in a population may evolve defenses, such as camouflage, thorns, or poisons, against efficient herbivores. In turn, some herbivores

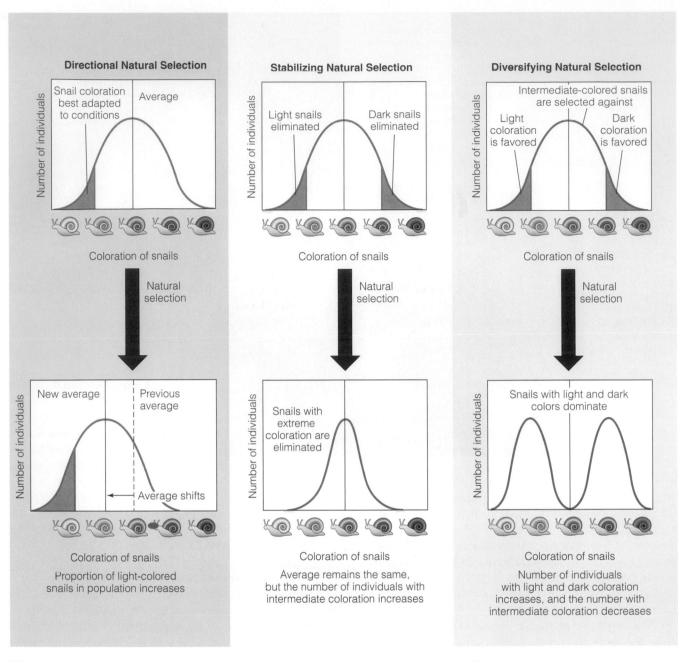

Figure 5-6 Three ways in which natural selection can occur, using the trait of coloration in a population of snails. In *directional natural selection*, changing environmental conditions select organisms with alleles that deviate from the norm so that their offspring (lighter-colored snails) make up a larger proportion of the population. In *stabilizing selection*, environmental factors eliminate fringe individuals (light- and dark-colored snails) and increase the number of individuals with average genetic makeup (intermediate-colored snails). In *diversifying natural selection*, environmental factors favor individuals with uncommon traits (light- and dark-colored snails) and greatly reduce those with average traits (intermediate colored snails).

in the population may have genetic characteristics that enable them to overcome these defenses and produce more offspring than those without such traits.

In coevolution, adaptation follows adaptation in something like an ongoing, long-term arms race between individuals in interacting populations of different species.

5-3 ECOLOGICAL NICHES AND ADAPTATION

What Is an Ecological Niche? If asked what role a certain species such as an alligator plays in an ecosystem, an ecologist would describe its **ecological niche**, or simply **niche** (pronounced "nitch"), the species' way

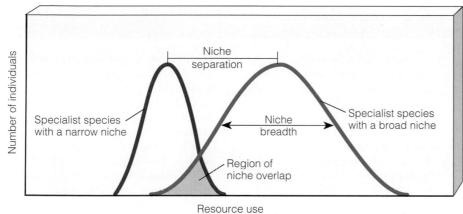

Figure 5-7 Overlap of the niches of two different species: a specialist and a generalist. In the overlap area the two species compete for one or more of the same resources. As a result, each species can occupy only a part of its fundamental niche and thus occupy its realized niche. Generalist species have a broad niche (right), and specialist species have a narrow niche (left).

of life or functional role in an ecosystem. A species' niche involves everything that affects its survival and reproduction. This includes **(1)** the range of tolerance for various physical and chemical conditions, such as temperature or water availability (Figure 4-14, p. 79), **(2)** the types and amounts of resources it uses, such as food or nutrients and space, **(3)** how it interacts with other living and nonliving components of the ecosystems in which it is found, such as what it eats or is eaten by, and **(4)** the role it plays in the flow of energy and cycling of matter in an ecosystem (Figure 4-16, p. 82).

The ecological niche of a species is different from the **habitat** of the species. Habitat is the actual physical location where organisms making up a species live. Ecologists often say that a niche is like a species' occupation, whereas habitat is like its address.

What Is the Difference Between a Species' Fundamental Niche and Its Realized Niche? A species' **fundamental niche** is the full potential range of physical, chemical, and biological conditions and resources it could theoretically use if there were no direct competition from other species. However, the niche of a species in a particular ecosystem tends to overlap with niches of other species. Niche overlap leads to competition for one or more of the same resources.

To survive and avoid competition for the same resources, a species usually occupies only part of its fundamental niche in a particular community or ecosystem— what ecologists call its **realized niche**. By analogy, you may be capable of being president of a particular company (your *fundamental professional niche*), but competition from others may mean that you may become only a vice president (your *realized professional niche*).

Why Is Understanding the Niches of Species Important? Understanding a species' niche is important because it **(1)** can help us prevent it from becoming prematurely extinct and **(2)** can be useful in helping us

assess the environmental changes we make in terrestrial and aquatic systems. For example, how will the niches of various species be changed by clearing a forest, plowing up a grassland, filling in a wetland, or dumping pollutants into a lake or stream?

Unfortunately, we have little knowledge about the ecological roles of most of the earth's species because determining the niche of a species is difficult, expensive, and time-consuming.

Is It Better to Be a Generalist or a Specialist Species? Broad and Narrow Niches The niches of species can be used to broadly classify them as *generalists* or *specialists*. **Generalist species** have broad niches (Figure 5-7, right curve): They can live in many different places, eat a variety of foods, and tolerate a wide range of environmental conditions. Flies, cockroaches (Spotlight, p. 112), mice, rats, white-tailed deer, raccoons, coyotes, copperheads, channel catfish, and humans are generalist species.

Specialist species have narrow niches (Figure 5-7, left curve): They may be able to live in only one type of habitat, tolerate only a narrow range of climatic and other environmental conditions, or use only one or a few types of food. This makes them more prone to becoming endangered when environmental conditions change. Examples of specialists are **(1)** *tiger salamanders*, which can breed only in fishless ponds so their larvae will not be eaten, **(2)** *red-cockaded woodpeckers*, which carve nest-holes almost exclusively in old (at least 75 years) longleaf pines, **(3)** *spotted owls*, which need old-growth forests in the Pacific Northwest for food and shelter, and **(4)** China's highly endangered giant pandas, which feed almost exclusively on various types of bamboo.

Is it better to be a generalist than a specialist? It depends. When environmental conditions are fairly constant, as in a tropical rain forest, specialists have an advantage because they have fewer competitors. However, under rapidly changing environmental conditions, the generalist usually is better off than the specialist.

How Does Ecological Niche Relate to Adaptation? Evolution by natural selection leads to a remarkable fit between organisms and their environment. In terms of the ecological niche of a particular species, this fit involves having a set of traits that enables individuals to survive and reproduce in a particular environment.

Species that have similar niches tend to evolve similar sets of traits, even if they are unrelated species growing in different parts of the world. Such *convergent evolution* has led to similar organisms in different parts of the world with similar environmental conditions. For example, in desert biomes, shrubs tend to converge on a set of traits that enable them to grow in hot, dry conditions. Examples of such traits or adaptations include (1) deep roots to access water, (2) small leaves to reduce heat loads, (3) hairs or waxy coatings that protect against intense sunlight, (4) tolerance to high daytime temperatures and low night temperatures, and (5) flowering during wet periods, when insect pollinators are most active. Chapter 6 discusses the biomes of the world, where similar environmental conditions have led to similar sets of ecological niches in different parts of the world.

What Limits Adaptation? Shouldn't evolution lead to perfectly adapted organisms? Shouldn't adaptations to new environmental conditions allow our skin to become more resistant to the harmful effects of ultraviolet radiation, our lungs to cope with air pollutants, and our livers to become better at detoxifying pollutants? The answer to these questions is *no* because of the following limits to adaptations in nature:

- *A change in environmental conditions can lead to adaptation only for traits already present in the gene pool of a population.*

- *Even if a beneficial heritable trait is present in a population, that population's ability to adapt can be limited by its reproductive capacity.* Populations of genetically diverse species that reproduce quickly—such as weeds, mosquitoes, rats, bacteria, or cockroaches— often can adapt to a change in environmental conditions in a short time. In contrast, populations of species such as elephants, tigers, sharks, and humans, which cannot produce large numbers of offspring rapidly, take a long time (typically thousands or even millions of years) to adapt through natural selection.

- *Even if a favorable genetic trait is present in a population, most of the population would have to die or become sterile so that individuals with the trait could predominate and pass the trait on.* This is hardly a desirable solution to the environmental problems the human species faces.

5-4 SPECIATION, EXTINCTION, AND BIODIVERSITY

How Do New Species Evolve? Under certain circumstances natural selection can lead to an entirely new species. In this process, called **speciation**, two species arise from one.

The most common mechanism of speciation (especially among animals) takes place in two phases: geographic isolation and reproductive isolation. **Geographic isolation** occurs when two populations of a species or two groups of the same population become physically separated for fairly long periods into areas with different environmental conditions. For example, part of a population may migrate in search of food and then begin living in another area with different environmental conditions (Figure 5-8). Populations also may become separated **(1)** by a physical barrier (such as a mountain range, stream, lake, or road), **(2)** by a change such as a volcanic eruption or earthquake, or **(3)** when a few individuals are carried to a new area by wind or water.

The second phase of speciation is **reproductive isolation**. It occurs as mutation and natural selection operate independently in two geographically isolated populations and change the allele frequencies in different ways, a process called *divergence*. If divergence continues long enough, members of the geographically and reproductively isolated populations may become so different in genetic makeup that they cannot interbreed or, if they do, they cannot produce live, fertile offspring. Then one species has become two, and *speciation* has occurred through *divergent evolution*.

In a few rapidly reproducing organisms this type of speciation may occur within hundreds of years. However, with most species speciation takes from tens of thousands to millions of years. Given this time scale, it is difficult to observe and document the appearance of a new species. As a result, there are many controversial hypotheses about the details of speciation.

How Do Species Become Extinct? After evolution, the second process affecting the number and types of species on the earth is **extinction**. When environmental conditions change, a species may either evolve (become better adapted) or cease to exist (become extinct).

The earth's long-term patterns of speciation and extinction have been affected by several major factors: **(1)** large-scale movements of the continents (continental drift) over millions of years (Figure 5-9), **(2)** gradual climate changes caused by continental drift and slight shifts in the earth's orbit around the sun, and **(3)** rapid climate change caused by catastrophic events (such as large volcanic eruptions, huge meteorites and asteroids crashing into the earth, and release of large amounts of methane trapped beneath the ocean floor). Some of these events create dust clouds that shut down or sharply reduce photosynthesis long enough to eliminate huge numbers of producers and, soon thereafter, the consumers that fed on them.

Biologists estimate that 99.9% of all the species that have ever existed are now extinct. As local conditions change, a certain number of species disappear at a low rate, called **background extinction**. In contrast, **mass extinction** is an abrupt rise in extinction rates above the background level. It is a catastrophic, widespread (often global) event in which large groups of existing species (perhaps 30–90%) are wiped out.

Most mass extinctions are believed to result from global climate changes that kill many species and leave behind those able to adapt to the new conditions. Fossil and geological evidence indicates that the earth's species have experienced five great mass extinctions (20–60 million years apart) during the past 500 million years (Figure 5-10).

A crisis for one species is an opportunity for another. The existence of millions of species today

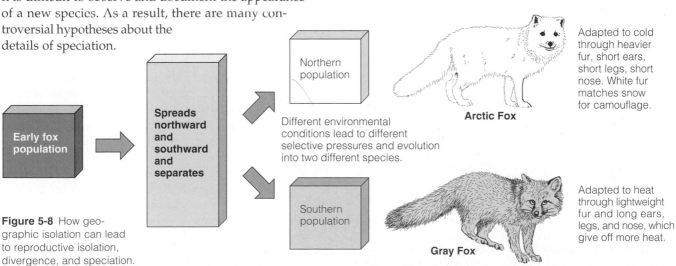

Figure 5-8 How geographic isolation can lead to reproductive isolation, divergence, and speciation.

Northern population

Different environmental conditions lead to different selective pressures and evolution into two different species.

Arctic Fox

Adapted to cold through heavier fur, short ears, short legs, short nose. White fur matches snow for camouflage.

Early fox population

Spreads northward and southward and separates

Southern population

Gray Fox

Adapted to heat through lightweight fur and long ears, legs, and nose, which give off more heat.

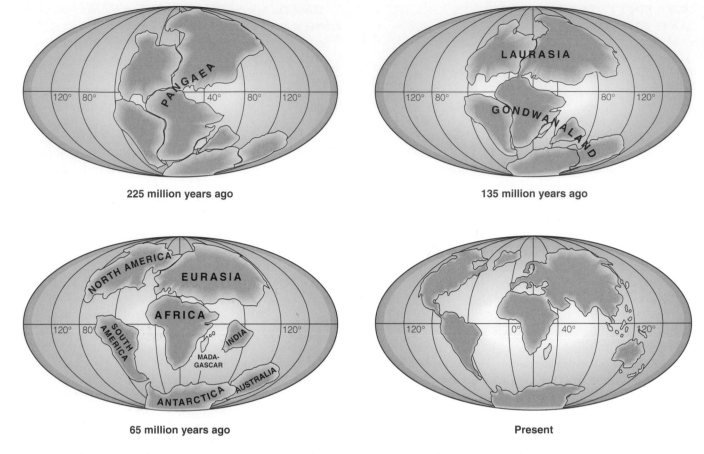

225 million years ago

135 million years ago

65 million years ago

Present

Figure 5-9 Continental drift, the extremely slow movement of continents over millions of years on several gigantic plates (discussed in more detail in Section 10-2, p. 212). This process plays a role in the extinction of species and the rise of new species. Populations are geographically and eventually reproductively isolated as land masses float apart and new coastal regions are created. Rock and fossil evidence indicates that about 200–250 million years ago all of the earth's present-day continents were locked together in a supercontinent called Pangaea (top, left). About 180 million years ago, Pangaea began splitting apart as the earth's huge plates separated and eventually resulted in today's locations of the continents (bottom, right).

means that speciation, on average, has kept ahead of extinction. Evidence shows that the earth's mass extinctions have been followed by periods of recovery called **adaptive radiations**, in which numerous new species have evolved over several million years to fill new or vacated ecological roles or niches in changed environments (Figure 5-11).

Fossil records suggest that it takes 10 million years or more for adaptive radiations to rebuild biological diversity after a mass extinction. According to biologists, our species owes its existence to the mass extinction 65 million years ago, when an incredible variety of dinosaur species died and a small, primitive mammal that eventually led to us happened to survive.

Does Macroevolution Take Place Gradually or in Bursts? Until recently it was widely accepted that macroevolutionary change occurred gradually over many millions of years as a result of steady and small microevolutionary changes. This is called the *gradualist model of evolution*. However, fossils rarely document such gradual changes in lineage. Instead, most species appear suddenly, persist for a fairly long time with little change, and then become extinct.

In the early 1970s, Stephen Jay Gould of Harvard University and Niles Eldredge of the American Museum of Natural History offered an alternative explanation of how macroevolution takes place called the *punctuated equilibrium hypothesis*. According to this model, evolution consists of long periods of little change in species (equilibrium) punctuated by brief periods of rapid change (thousands to tens of thousands of years). During periods of rapid change, more species become extinct and more new species arise.

There is intense debate over which of these models of evolution best explains the data. However, most evolutionary biologists believe that the rate and mode of evolution vary and fall on a spectrum

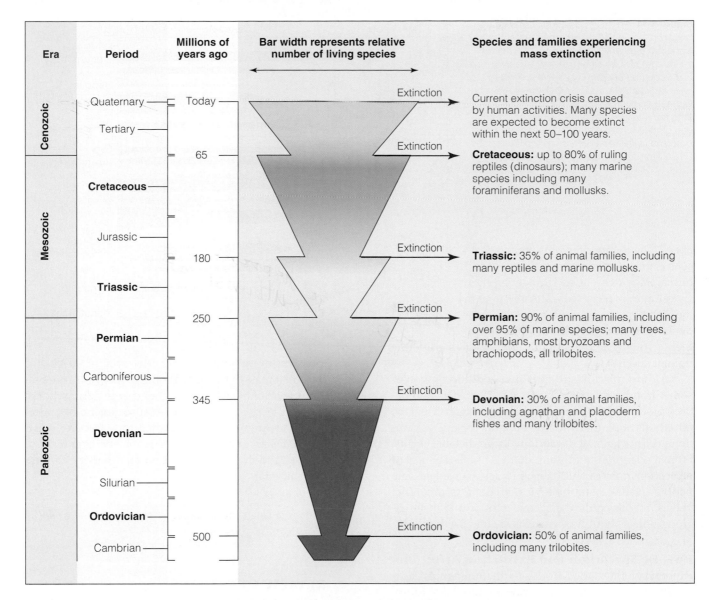

Era	Period	Millions of years ago	Bar width represents relative number of living species	Species and families experiencing mass extinction

Cenozoic — Quaternary, Tertiary
Mesozoic — Cretaceous, Jurassic, Triassic
Paleozoic — Permian, Carboniferous, Devonian, Silurian, Ordovician, Cambrian

Today
65
180
250
345
500

Extinction — Current extinction crisis caused by human activities. Many species are expected to become extinct within the next 50–100 years.

Extinction — **Cretaceous:** up to 80% of ruling reptiles (dinosaurs); many marine species including many foraminiferans and mollusks.

Extinction — **Triassic:** 35% of animal families, including many reptiles and marine mollusks.

Extinction — **Permian:** 90% of animal families, including over 95% of marine species; many trees, amphibians, most bryozoans and brachiopods, all trilobites.

Extinction — **Devonian:** 30% of animal families, including agnathan and placoderm fishes and many trilobites.

Extinction — **Ordovician:** 50% of animal families, including many trilobites.

Figure 5-10 Over millions to hundreds of millions of years, macroevolution has consisted of dramatic exits (extinctions) and entrances (speciation and radiations) of large groups of species. Fossils and radioactive dating indicate that five major *mass extinctions* (indicated by arrows) have taken place over the past 500 million years. Mass extinctions leave large numbers of organism roles (niches) unoccupied and create new ones. As a result, each mass extinction has been followed by periods of recovery (represented by the wedge shapes) called *adaptive radiations* in which, over 10 million years or more, new species evolve to fill new or vacated ecological roles (niches). Many scientists say that we are now in the midst of a sixth mass extinction, caused primarily by overhunting and the increasing elimination, degradation, and fragmentation of wildlife habitats as a result of human activities as discussed on pp. 558-559.

between the extremes of the gradualist and punctuated equilibrium models.

What Are Three Common Misconceptions About Evolution? One misconception about evolution arises from the interpretation of the expression "survival of the fittest" (which biologists almost never use), sometimes used to describe how natural selection works. This often is misinterpreted as "survival of the strongest." To biol-

ogists, however, *fitness* is a measure of reproductive success, so that the fittest individuals are those that leave the most descendants. Instead of "tooth and claw" competition, natural selection favors populations of species that *avoid* direct competition by producing offspring that can occupy ecological roles (niches) different from those of other species.

Some people have misinterpreted the theory of evolution as asserting that "humans evolved from

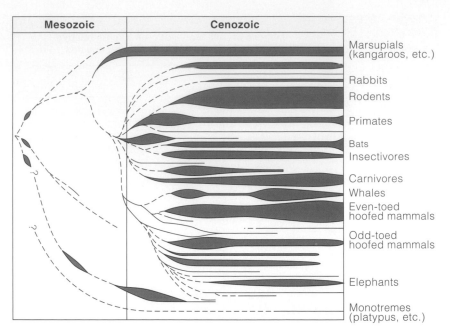

Figure 5-11 Adaptive radiation of mammals began in the first 10-12 million years of the Cenozoic era (which began about 65 million years ago) and continues today. This evolution of a large number of new species is thought to have resulted when huge numbers of new and vacated ecological niches became available after the mass extinction of dinosaurs near the end of the Mesozoic era. (Adapted from *Biology: The Unity and Diversity of Life*, 8th ed. by Cecie Starr and Ralph Taggart © 1998)

apes." The theory of evolution makes no such claim. Instead, it states that apes and humans are descended from a common ancestor. In other words, at some time in the distant past a particular population of organisms had descendants, with some of them evolving into apes and others evolving into the human species.

A third misconception is that evolution involves some grand plan of nature in which species become progressively more perfect. This ignores the fact that the mutations and other processes that drive microevolution occur as a result of random, unpredictable events. From a scientific standpoint, there is no plan or goal of perfection in the evolutionary process. However, some people (creationists) believe that there is a conflict between the scientific theory of evolution and their religious beliefs about how life was created on the earth.

How Do Speciation and Extinction Affect Biodiversity? Speciation minus extinction equals *biodiversity*, the planet's genetic raw material for future evolution in response to changing environmental conditions. In this long-term give-and-take between extinction and speciation, mass extinctions temporarily reduce biodiversity. However, they also create evolutionary opportunities for surviving species to undergo adaptive radiations to fill unoccupied and new biological roles or niches (Figure 5-11).

Although extinction is a natural process, there is much evidence that humans have become a major force in the premature extinction of species. As the human population and resource consumption increase over the next 50 years and we take over more and more of the planet's surface (Figure 1-4, p. 8) and more of the earth's net primary productivity (Figure 4-25, p. 88), we may cause the extinction of up to a quarter of the earth's current species. If this happens it will constitute a *sixth mass extinction*, caused by us (Guest Essay, right).

On our short time scale, such catastrophic losses cannot be recouped by formation of new species; it took tens of millions of years after each of the earth's five great mass extinctions for life to recover to the previous level of biodiversity. Genetic engineering cannot stop this loss of biodiversity because genetic engineers do not create new genes. Rather, they transfer existing genes or gene fragments from one organism to another and thus rely on natural biodiversity for their raw material.

Nothing in biology makes sense except in the light of evolution.
THEODOSIUS DOBZHANSKY

REVIEW QUESTIONS

1. Define all of the boldfaced terms in this chapter.

2. Describe the conditions that make life on the earth just right for life as we know it.

3. Distinguish between *chemical evolution* and *biological evolution*.

4. Summarize how scientists think chemical evolution took place on the earth.

5. Summarize how scientists think biological evolution took place on the earth. What was the *oxygen revolution*, and what is its importance to life as we know it?

6. What are *fossils*, and how do they help us formulate ideas about how life developed on the earth?

7. Distinguish between *biological evolution*, *the theory of evolution*, *microevolution*, and *macroevolution*.

8. Use the concepts of *genes*, *gene pool*, *alleles*, *mutations*, *gene flow*, and *genetic drift* to explain how microevolution takes place.

9. Define *natural selection*, *differential reproduction*, and *adaptation* and explain their roles in microevolution.

The Future of Evolution*

GUEST ESSAY

Norman Myers

Norman Myers is a tropical ecologist and international consultant in environment and development, with emphasis on conservation of wildlife species and tropical forests. He is one of the world's leading environmental experts. His research and consulting have taken him to 80 countries. He has served as a consultant for many development agencies and research organizations, including the U.S. National Academy of Sciences, the World Bank, the Organization for Economic Cooperation and Development, various UN agencies, and the World Resources Institute. Among his many publications (see Further Readings) are The Primary Source: Tropical Forests and Our Future *(1992),* The Gaia Atlas of Planet Management *(1993),* Scarcity or Abundance: A Debate on Environment *(1994),* Ultimate Security: The Environmental Basis of Political Security *(1996), and* Perverse Subsidies *(2001).*

Human activities have brought the earth to a biotic crisis. Many biologists have commented that this crisis will result in the loss of large numbers of species, possibly half within the lifetimes of students reading this book. However, surprisingly few biologists have recognized that in the longer term these extinctions will alter the future of evolution. The environmental changes of the next few decades will impoverish evolution's course for several million years.

So the future of evolution should be regarded as one of the most challenging issues humankind has ever encountered. Many biologists seem to give scant attention to this unique challenge, or indeed to the unprecedented opportunity for research into evolution during a state of extreme turmoil. After all, we are effectively conducting a planet-scale experiment, with little clue as to how it might turn out, except that it will prove irreversible and will severely reduce human well-being. We could get by without half of all mammals and other vertebrates, but if we lost half of all insects with their pollinating functions [p. 70], let alone their many other services, we would be in trouble in the first crop-growing season.

In addition, the mass extinction under way is the biggest of our environmental problems in terms of the duration of its impact and the numbers of people to be affected. All our other problems are potentially reversible. If we wanted to clean up acid rain, we could do it within a few decades. We could push back the deserts, restore topsoil, and allow the ozone layer to be repaired within a century or so. We could probably restore climate stability in the wake of global warming within a thousand years. But once a species is gone, it is gone for good.

Of course, in the long run evolution will generate replacement species with numbers and variety to match

*For scientific details, see Norman Myers, "The Biodiversity Crisis and the Future of Evolution," *The Environmentalist*, 16 (1996): 37-47.

today's [Figures 5-10 and 5-11]. But that is likely to take millions of years. As Michael Soulé put it, "Death is one thing, an end to birth is something else." We are witnessing the gross reduction if not the elimination of entire sectors of biomes, notably tropical forests, coral reefs, and wetlands, all of which may have served as powerhouses of evolution—centers of new speciation—in the prehistoric past.

Suppose, as has happened after the five mass extinctions of the prehistoric past [Figure 5-10], that the bounce-back period lasts at least 5 million years. This would be 20 times longer than humans have been a species. Suppose that the average number of people on the earth during that period is 2.5 billion people, as opposed to the 6.1 billion today. Then the total number of people affected by what we do (or don't do) to protect the biosphere during the next few decades will be about 500 trillion, in contrast with the 50 billion people who have ever existed. Even 1 trillion is a big number: Figure out how long a period of time in years is made up of 1 trillion seconds.

In short, we are engaged in by far the biggest decision ever made by one human community on behalf of future human communities. In certain respects, it is bigger than all such decisions put together since we came out of our caves. Yet the issue is almost entirely disregarded, whether scientifically, ethically, or otherwise. How often do you hear leading scientists even mention the issue?

Despite our gross ignorance of what lies ahead, we can venture a few hypotheses:

- *A temporary outburst of speciation.* As large numbers of niches are vacated, there could be an outburst of speciation, although not nearly enough to match the extinction spasm.

- *A proliferation of opportunistic species* such as cockroaches [Spotlight, p. 112], rats, flies, and others that prosper when new niches open up. This proliferation will be enhanced by the likely elimination of species that naturally control opportunistic species. In turn, this could lead to a "pest and weed" ecology.

- *An end to large vertebrates.* After the crash of 65 million years ago, when the dinosaurs and associated species were eliminated, the only mammals and other vertebrates to survive the catastrophe were creatures weighing no more than 10 kilograms (22 pounds)—about the size of a small dog.

- *An end to speciation of large vertebrates.* Even if larger vertebrates were to survive the extinction spasm ahead, our largest protected areas will prove far too small for further speciation of elephants, rhinoceroses, apes, bears, and the bigger cats, among other large vertebrates. We are bidding goodbye to whatever new and possibly more spectacular forms of these species might eventually have emerged.

(continued)

What does all this imply for our conservation efforts? By far the predominant strategy of conservationists is to save as many species as possible. This has an obvious and entirely sensible rationale. But we now need to safeguard evolutionary processes as well. A prime goal is to look out especially for endemic species (found only in a particular place) or species confined to small habitats. Examples include the California condor, the black-footed ferret, the giant panda, and the gorilla.

However, the paleoecological background shows that endemic species often turn out to be evolutionary dead ends: Generally they do not throw off new species. So should we shift our conservation priority from endemic species to broader-ranging species in the hope that they have more genetic variability and thus more of a diversified resource stock on which natural selection can work its creative impact?

Similarly, should we devote more attention to the evolutionary powerhouses such as the forests, coral reefs, and wetlands of the tropics? All these are in dire trouble and may be all but eliminated within just a few decades. Do they deserve preferential treatment ahead of, say, temperate-zone woodlands and grasslands and boreal forests with their lack of species, ecological complexity, and evolutionary potential?

These tough questions have scarcely been addressed by scientific leaders. But the times are changing. The author organized an international conference in March 2000 to tackle these challenges, under the auspices of the U.S. National Academy of Sciences.

If within your lifetime we allow the current biotic crisis to proceed unchecked (which is what the recent record suggests), it is possible that your children will ask you a key question: "When the evolutionary debacle was becoming all too plain at the start of the new millennium, what did you do about it that was sufficient to ward off disaster?" I hope you will engage yourself in dealing with this crucial issue.

Critical Thinking

1. Do you agree or disagree with the thesis of this essay? Explain.

2. If you agree, list three things you could do to help prevent the outcomes described in this essay.

---

10. Give an example of microevolution by natural selection. Describe three types of natural selection.

11. What is *coevolution*, and what is its importance?

12. What is the *ecological niche* of a species, and why is it important to understand the niches of species? What is the difference between a species' *habitat* and its *niche*? What is the difference between a species' *fundamental niche* and its *realized niche*?

13. What is *convergence* or *convergent evolution*, and how does it relate to adaptation? List three factors that limit adaptation.

14. What is *speciation*? Distinguish between *geographic isolation* and *reproductive isolation* and explain how they can lead to speciation through divergent evolution.

15. What is *extinction*? List three factors that have affected the earth's long-term patterns of speciation and extinction. Distinguish between *background extinction* and *mass extinction*. What is believed to have caused the five great mass extinctions in the past?

16. What is an *adaptive radiation*? How can such a radiation lead to recovery after a mass extinction? What evidence suggests that human activities may be bringing about a sixth mass extinction? What are some consequences of such an extinction for the potential for future evolution on the earth?

17. Explain how speciation and extinction result in the planet's biodiversity.

18. Describe the controversy over whether macroevolution takes place gradually or in bursts.

19. What are three common misconceptions about evolution?

20. How do speciation and extinction affect biodiversity?

CRITICAL THINKING

1. How would you respond to someone who tells you **(a)** that they do not believe in evolution because it is "just a theory" and **(b)** not to worry about air pollution because through natural selection the human species will develop lungs that can detoxify pollutants.

2. What would happen if extinction had never occurred during evolutionary history? Do you think we would have more species than we do now? Why? What do you think would happen to biodiversity if extinction rates occurred throughout evolutionary history at the same rate they have been occurring during the past 50 years?

3. The peppered moth (*Biston betularia*) responded to industrial pollution in England by short-term evolution (Figure 5-5, p. 109). Consider species such as northern spotted owls, California condors, or grizzly bears. Would you expect such species that are being heavily affected by human activity to be able to respond by short-term evolution? Why or why not? Consider coyotes, deer, and cockroaches. Would you expect these species to respond to human impacts by fairly short-term evolution? Explain.

4. How would you respond to someone who says that extinction is a natural process and that we should not worry about the loss of biodiversity?

5. Why is it important for somebody studying environmental science to understand the basics of evolution?

6. Why is the realized niche of a species narrower, or more specialized, than its fundamental niche?

7. As well as you can, describe the major differences between the ecological niches of humans and cockroaches. Are these two species in competition? If so, how do they manage to coexist?

8. In what ways do humans occupy generalist niches and in what ways do they occupy specialist niches? Do you think humans should become more dependent on a greater variety of foods and energy resources to promote long-term survival and sustainability? Explain and discuss the short- and long-term economic and political consequences of such a shift.

9. By analogy, use the concepts of generalist and specialist to evaluate the roles of humans in today's societies. In general, the role of college and graduate education is to create specialists in a particular field or a narrow portion of a field. What are the pros and cons of relying mainly on this approach? Is there a need for more generalists? Or are they people who may know a lot about many things (and about connections between things) but not enough about anything in particular? Can they serve a useful role and make a satisfactory living in today's increasingly specialized societies? Explain your answers.

PROJECTS

1. Go to the library and locate a replica of the first edition of Charles Darwin's *On the Origin of Species by Means of Natural Selection*. Why did Darwin feel compelled to publish his book more quickly than he had originally planned, and why does he consider the book only an "abstract"? Find a later edition of the book. What did Charles Darwin change in later editions of his book?

2. An important adaptation of humans is a strong opposable thumb, which allows us to grip and manipulate things with our hands. As a demonstration of the importance of this trait, fold each of your thumbs into the palm of its hand and then tape them securely in that position for an entire day. After the demonstration, make a list of the things you could not do without the use of your thumbs.

3. The Texas horned lizard of North America (*Phrynosoma cornutum*) and the thorny devil lizard (*Moloch horridus*) of Australia are distantly related lizard species that live in desert environments and eat ants as their main source of food. Do a Web search for these species and compare photographs and information about their natural history. Describe major characteristics of the ecological niche and the appearance of each of these species.

4. Visit a local forest, pond, or lake, choose a particular organism, and then use the library or the internet to describe as much as you can about its niche, including what species it depends on and what species help sup-

port its survival. Predict what might happen if your selected species disappeared from the local environment.

5. Use the library or the internet to find out bibliographic information about *Charles Darwin* and *Theodosius Dobzhansky*, whose quotes appear at the beginning and end of this chapter.

6. Make a concept map of this chapter's major ideas, using the section heads and subheads and the key terms (in boldface). Look at the inside back cover and on the website for this book for information about making concept maps.

INTERNET STUDY RESOURCES AND RESOURCES FOR FURTHER READING AND RESEARCH

The website for this book contains helpful study aids and many ideas for further reading and research. Log on to:

http://www.brookscole.com/product/0534376975s

and click on the Chapter-by-Chapter area. Choose Chapter 5 and select a resource:

- "Flash Cards" allows you to test your mastery of the Terms and Concepts to Remember for this chapter.

- "Tutorial Quizzes" provides a multiple-choice practice quiz.

- "Student Guide to InfoTrac" will lead you to Critical Thinking Projects that use InfoTrac College Edition as a research tool.

- "References" lists the major books and articles consulted in writing this chapter.

- "Hypercontents" takes you to an extensive list of sites with news, research, and images related to individual sections of the chapter.

INFOTRAC COLLEGE EDITION

Improve your skills with InfoTrac College Edition, a searchable online database of articles from more than 700 periodicals. Log on to:

http://www.infotrac-college.com

or access InfoTrac through the website for this book.

Try the following articles:

Cowen, R. 2000. Meteoric wallop may have diversified life. *Science News* vol. 157, no. 11, p. 165. (subject guide: biological diversity, causes)

Boake, C.R.B. 2000. Flying apart: mating behavior and speciation. *BioScience* vol. 50, no. 6, p. 501–508. (keyword: speciation)

6 BIOGEOGRAPHY: CLIMATE, BIOMES, AND TERRESTRIAL BIODIVERSITY*

Connections: Blowing in the Wind

One of the things that connects all life on the earth is *wind*, a vital part of the planet's circulatory system. Without wind, most of the earth would be uninhabitable: The tropics would be unbearably hot, and most of the rest of the planet would freeze.

Winds also transport nutrients from one place to another. Dust rich in phosphates blows across the Atlantic from the Sahara Desert in Africa (Figure 6-1), helping to replenish rain forest soils in Brazil and build up agricultural soils in the Bahamas. Iron-rich dust blowing from China's Gobi Desert falls into the Pacific Ocean between Hawaii and Alaska and stimulates the growth of phytoplankton, the minute producers that support ocean food webs. That is the *good news*.

---

*Paul M. Rich, associate professor of ecology and evolutionary biology and environmental studies at the University of Kansas, is coauthor of this chapter.

The *bad news* is that wind also transports harmful substances. Particles of reddish-brown soil and pesticides banned in the United States are blown from Africa's deserts and eroding farmlands into the sky over Florida. This makes it difficult for the state to meet federal air pollution standards and is suspected to be a factor in degrading or killing coral reefs in the Florida Keys and the Caribbean.

Industrial pollution and dust from rapidly industrializing China and central Asia blow across the Pacific Ocean and degrade air quality over the western United States, especially Washington and Oregon. Studies show that Asian pollution contributes as much as 10% to West Coast smog.

There's *mixed news* as well. Particles from volcanic eruptions ride the winds, circle the globe, and change the earth's climate for a while. Emissions from the 1991 eruption of Mount Pinatubo in the Philippines, the largest eruption this century, cooled the earth slightly for 3 years, temporarily masking signs of global warming. On the other hand, volcanic ash, like the blowing desert dust, adds valuable trace minerals to the soil where it settles.

The lesson, once again, is that *there is no "away,"* and wind—acting as part of the planet's circulatory system for heat, moisture, and plant nutrients—is one reason. Movement of soil particles from one place to another by wind and water is a natural phenomenon, but when we disturb the soil and leave it unprotected we hasten the process.

Wind is also an important factor in climate through its influence on global air circulation patterns. Climate, in turn, is crucial for determining what kinds of plant and animal life are found in the major biomes of the different geographic regions of the biosphere, as we shall see in this chapter.

Figure 6-1 Some of the dust shown here blowing from Africa's Sahara Desert can end up as soil nutrients in Amazonian rain forests. (NOAA.USGS/MND EROS Data Center)

To do science is to search for repeated patterns, not simply to accumulate facts, and to do the science of geographical ecology is to search for patterns of plant and animal life that can be put on a map.

ROBERT H. MACARTHUR

This chapter addresses the following broad questions about geographic patterns of ecology:

- What key factors determine the earth's weather and climate?

- How does climate determine the major geographic patterns of ecology (biomes) on the earth?

- What are the major types of desert biomes, and how are they being affected by human activities?

- What are the major types of grassland biomes, and how are they being affected by human activities?

- What are the major types of forest biomes, and how are they being affected by human activities?

- Why are mountain and arctic biomes important, and how are they being affected by human activities?

- Why lessons can we learn from a geographic perspective of ecology?

6-1 WEATHER AND CLIMATE: A BRIEF INTRODUCTION

What Is Weather? At every moment at any spot on the earth, the *troposphere* (the inner layer of the atmosphere containing most of the earth's air) has a particular set of physical properties. Examples are **(1)** temperature, **(2)** pressure, **(3)** humidity, **(4)** precipitation, **(5)** sunshine, **(6)** cloud cover, and **(7)** wind direction and speed. These short-term properties of the troposphere at a particular place and time are **weather**.

Meteorologists use devices such as weather balloons, aircraft, ships, radar, and satellites to obtain data on variables such as atmospheric pressures, precipitation, temperatures, wind speeds, and locations of air masses and fronts. These data are fed into computer models to draw weather maps for each of seven levels of the troposphere, ranging from the ground to 19 kilometers (12 miles) up. Computer models use the map data to forecast the weather in each box of the seven-layer grid for the next 12 hours. Other computer models project the weather for the next several days by calculating the probabilities that air masses, winds, and other factors will move and change in certain ways.

Why Is Weather So Changeable? Masses of air that are **(1)** warm or cold, **(2)** wet or dry, and **(3)** contain air at high or low pressure are constantly moving across the land and sea. Weather changes as one air mass replaces or meets another. Cold air is more dense (weighs more per volume) than warm air. Thus, cold air tends to sink through warmer, less dense air.

The most dramatic changes in weather occur along a **front**, the boundary between two air masses with different temperatures and densities. A **warm front** is the boundary between an advancing warm air mass and the cooler one it is replacing. Because warm air is less dense than cool air, an advancing warm front will rise up over a mass of cool air.

As the warm front rises, its moisture starts to condense into droplets to form layers of clouds at different altitudes. High, wispy clouds are the first signs of an advancing warm front. Gradually the clouds thicken, descend to a lower altitude, and often release their moisture as rainfall. A moist warm front can bring days of cloudy skies and drizzle.

A **cold front** is the leading edge of an advancing mass of cold air. Because cold air is more dense than warm air, an advancing cold front stays close to the ground and wedges underneath less dense warmer air. An approaching cold front produces rapidly moving, towering clouds called *thunderheads*.

As the overlying mass of warm air is pushed upward it cools, and its water vapor condenses to form large and heavy droplets that fall to the earth's surface as precipitation. As a cold front passes through, we often experience high surface winds and thunderstorms. After the front passes through, we usually experience cooler temperatures and a clear sky.

Weather is also affected by changes in atmospheric pressure. An air mass with high pressure, called a *high*, contains cool, dense air that descends toward the earth's surface and becomes warmer. Fair weather follows as long as the high-pressure air mass remains over an area.

In contrast, a low-pressure air mass, called a *low*, produces cloudy and sometimes stormy weather. This happens because less dense warm air spirals inward toward the center of a low-pressure air mass. Because of its low pressure and low density, the center of the low rises, and its warm air expands and cools. When the temperature drops below the dew point, the moisture in the air condenses and forms clouds. If the droplets in the clouds coalesce into large and heavy drops, then precipitation follows.

In addition to normal weather, we sometimes experience *weather extremes* such as violent storms called **(1)** *tornadoes* (which form over land) and **(2)** *tropical cyclones* (which form over warm ocean waters and sometimes pass over coastal land). Tropical cyclones are called *hurricanes* in the Atlantic and *typhoons* in the Pacific. Figure 6-2 shows the

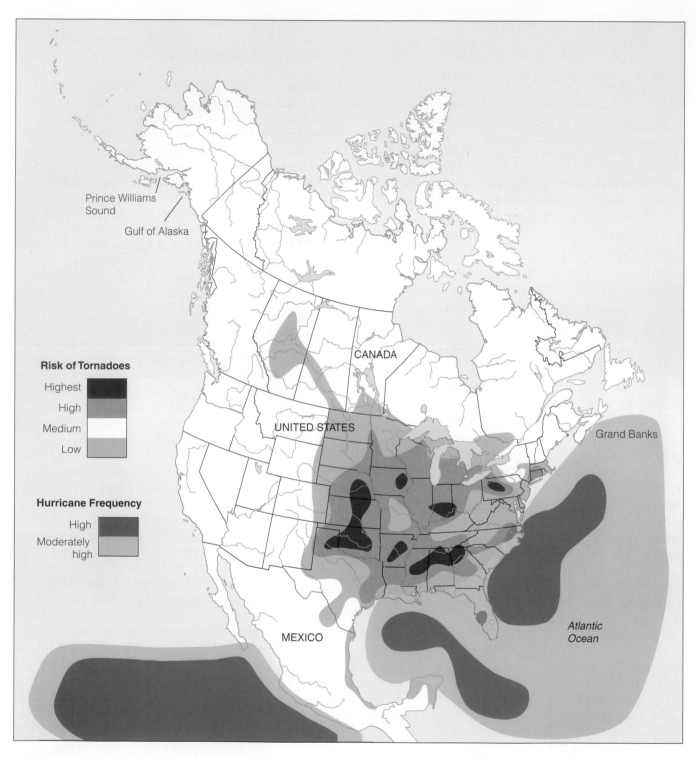

Figure 6-2 Areas of North America most susceptible to tornadoes and hurricanes. (Data from NOAA and U.S. Geological Survey)

areas of North America most susceptible to tornadoes and hurricanes.

Hurricanes and typhoons can kill and injure people and damage property and agricultural production. However, in some cases a hurricane can have long-term ecological and economic benefits that can exceed its short-term, negative effects.

For example, in parts of Texas along the Gulf of Mexico, coastal bays and marshes normally are closed off from freshwater and saltwater inflows. In August

1999, this coastal area was struck by Hurricane Brett. According to marine biologists, this hurricane **(1)** flushed excess nutrients from land runoff and dead sea grasses and rotting vegetation from coastal bays and marshes and **(2)** carved 12 channels through the barrier islands along the coast that allowed huge quantities of fresh seawater to flood the bays and marshes. This flushing out **(1)** reduced brown tides, consisting of explosive growth of algae feeding on excess nutrients, **(2)** increased growth of sea grasses, which serve as nurseries for shrimp, crabs, and fish and food for millions of ducks wintering in Texas bays, and **(3)** increased production of commercially important species of shellfish and fish.

What Is Climate? Climate is a region's general pattern of atmospheric or weather conditions, including seasonal variations and weather extremes (such as hurricanes or prolonged drought or rain) over a long period. Figure 6-3 shows the major effects of climate, and Figure 6-4 is a generalized map of the earth's major climate zones.

Average temperature and *average precipitation* are the two main factors determining a region's climate. The temperature and precipitation patterns that lead to different climates (Figure 6-4) are caused primarily by the way air and water circulate on a round planet spinning on a tilted axis and heated by the sun most strongly at the equator.

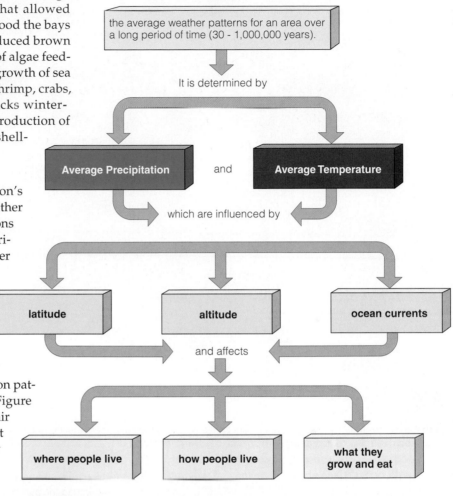

Figure 6-3 Climate and its effects. (Data from National Oceanic and Atmospheric Administration)

How Does Global Air Circulation Affect Regional Climates? The following factors determine global air circulation patterns:

- *Uneven heating of the earth's surface* because air is heated much more at the equator (where the sun's rays strike directly throughout the year) than at the poles (where sunlight strikes at an angle and thus is spread out over a much greater area). These differences in incoming solar energy help explain why **(1)** tropical regions near the equator are hot, **(2)** polar regions are cold, and **(3)** temperate regions in between generally have intermediate average temperatures (Figure 6-4).

- *Seasonal changes in temperature and precipitation* because the earth's axis (an imaginary line connecting the north and south poles) is tilted. As a result, various regions are tipped toward or away from the sun

as the earth makes its annual revolution around the sun (Figure 6-5). This creates opposite seasons in the northern and southern hemispheres.

- *Rotation of the earth on its axis*, which prevents air currents from moving due north and south from the equator. Forces in the atmosphere created by this rotation deflect winds (moving air masses) to the right in the northern hemisphere and to the left in the southern hemisphere in what is called the *Coriolis effect*. The result is six huge convection cells of swirling air masses, three north and three south of the equator, that transfer heat and water from one area to another (Figure 6-6).

- *Long-term variations in the amount of solar energy striking the earth*. These are caused by occasional

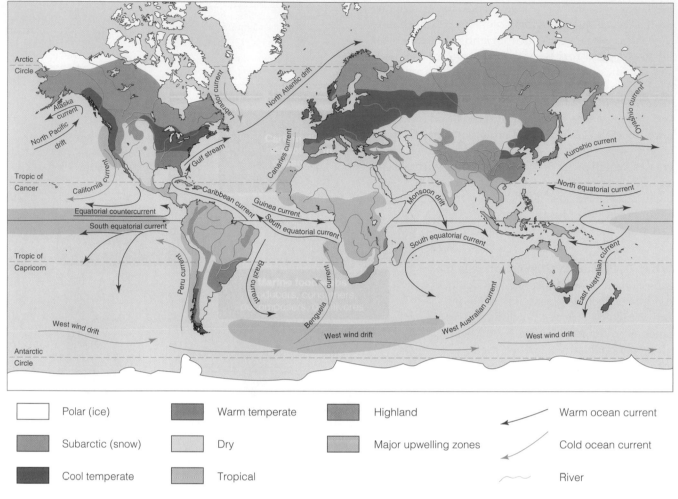

Polar (ice)	Warm temperate	Highland	Warm ocean current
Subarctic (snow)	Dry	Major upwelling zones	Cold ocean current
Cool temperate	Tropical		River

Figure 6-4 Generalized map of global climate zones, showing the major contributing ocean currents and drifts. Large variations in climate are dictated mainly by temperature (with its seasonal variations) and by the quantity and distribution of precipitation.

Figure 6-5 The effects of the earth's tilted axis on climate. As the planet makes its annual revolution around the sun on an axis tilted about 23.5°, various regions are tipped toward or away from the sun. The resulting variations in the amount of solar energy reaching the earth create the seasons.

changes in solar output and slight planetary shifts in which the earth's axis wobbles (22,000-year cycle) and tilts (44,000-year cycle) as it revolves around the sun.

■ *Properties of air and water.* Heat from the sun evaporates ocean water and transfers heat from the oceans to the atmosphere, especially near the hot equator. This creates cyclical convection cells that transport heat and water from one area to another (Figure 6-7). The resulting convection cells circulate air, heat, and moisture both vertically and from place to place in the troposphere, leading to different climates and patterns of vegetation (Figure 6-8).

How Do Ocean Currents Affect Regional Climates? The factors just listed, plus differences in water density, create warm and cold ocean currents

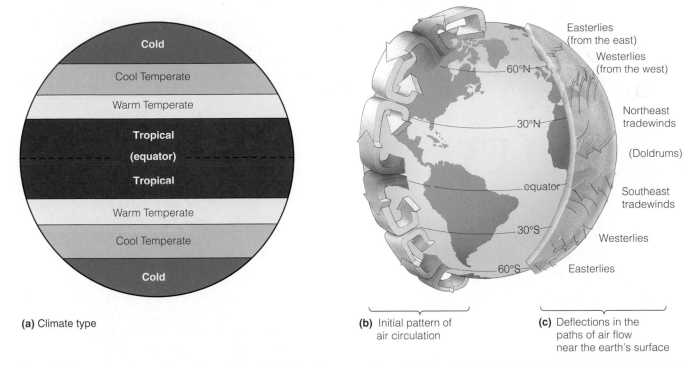

(a) Climate type

(b) Initial pattern of air circulation

(c) Deflections in the paths of air flow near the earth's surface

Figure 6-6 Formation of prevailing surface winds, which disrupt the general flow of air from the equator to the poles and back to the equator. As the earth rotates, its surface turns faster beneath air masses at the equator and slower beneath those at the poles. This deflects air masses moving north and south to the west or east, creating six huge convection cells in which air swirls upward and then descends toward the earth's surface at different latitudes. The direction of air movement in these cells sets up belts of prevailing winds that distribute air and moisture over the earth's surface. These winds affect the general types of climate found in different areas and drive the circulation of ocean currents. (From *Biology: Concepts and Applications*, 4th ed. by Cecie Starr © 2000)

(Figure 6-4). These currents, driven by winds and the earth's rotation (Figure 6-6) redistribute heat received from the sun and thus influence climate and vegetation, especially near coastal areas.

For example, without the warm Gulf Stream, which transports 25 times more water than all the world's rivers, the climate of northwestern Europe would be subarctic. Warm currents moving up from the equator make Alaska and northern Japan warmer than one would predict based on their location. Currents also help mix ocean waters and distribute nutrients and dissolved oxygen needed by aquatic organisms.

Along some steep western coasts of continents, almost constant tradewinds blow offshore, pushing surface water away from the land. This outgoing surface water is replaced by an **upwelling** of cold, nutrient-rich bottom water (Figure 6-9). Upwellings, whether far

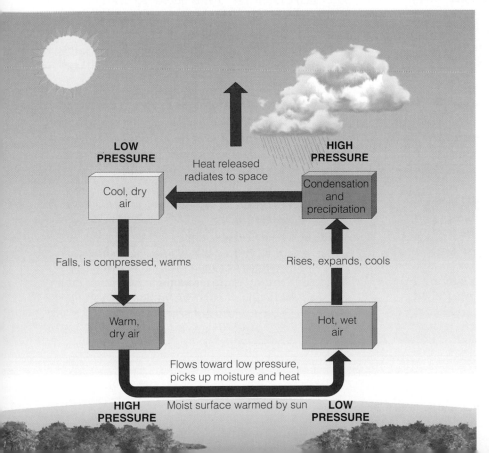

Figure 6-7 Transfer of energy by convection in the atmosphere. *Convection* occurs when matter warms, becomes less dense, and rises within its surroundings. This very efficient means of heat transfer occurs in the planet's interior, oceans, and atmosphere (as shown here). Distribution of heat and water occurs in the atmosphere because vertical convection currents stir up air in the troposphere and transport heat and water from one area to another in circular convection cells. The relative humidity increases as the air rises (right side) and decreases as it falls (left side).

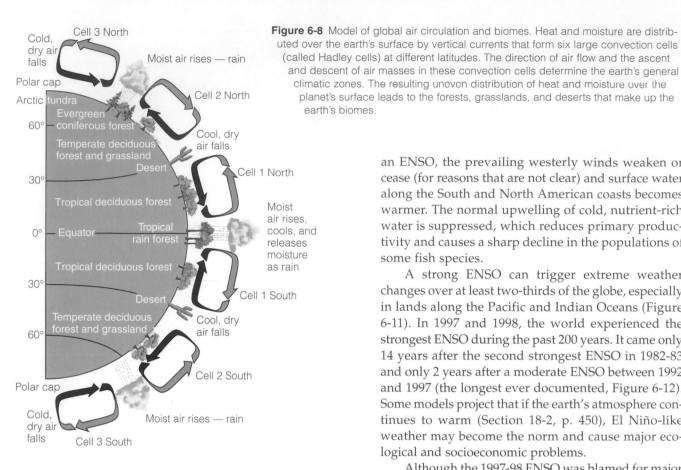

Figure 6-8 Model of global air circulation and biomes. Heat and moisture are distributed over the earth's surface by vertical currents that form six large convection cells (called Hadley cells) at different latitudes. The direction of air flow and the ascent and descent of air masses in these convection cells determine the earth's general climatic zones. The resulting uneven distribution of heat and moisture over the planet's surface leads to the forests, grasslands, and deserts that make up the earth's biomes.

from shore or near shore (Figure 6-4), **(1)** bring plant nutrients from the deeper parts of the ocean to the surface and **(2)** support large populations of phytoplankton, zooplankton, fish, and fish-eating seabirds.

What Is the El Niño-Southern Oscillation? Every few years in the Pacific Ocean, normal coastal upwelling (Figure 6-10, left) is affected by changes in climate patterns called the *El Niño-Southern Oscillation* or *ENSO* (Figure 6-10, right), often called *El Niño*. In

an ENSO, the prevailing westerly winds weaken or cease (for reasons that are not clear) and surface water along the South and North American coasts becomes warmer. The normal upwelling of cold, nutrient-rich water is suppressed, which reduces primary productivity and causes a sharp decline in the populations of some fish species.

A strong ENSO can trigger extreme weather changes over at least two-thirds of the globe, especially in lands along the Pacific and Indian Oceans (Figure 6-11). In 1997 and 1998, the world experienced the strongest ENSO during the past 200 years. It came only 14 years after the second strongest ENSO in 1982-83 and only 2 years after a moderate ENSO between 1992 and 1997 (the longest ever documented, Figure 6-12). Some models project that if the earth's atmosphere continues to warm (Section 18-2, p. 450), El Niño-like weather may become the norm and cause major ecological and socioeconomic problems.

Although the 1997-98 ENSO was blamed for major storms in California and deadly tornadoes in Florida, it produced an unusually mild winter in the northern and midwestern states and blocked hurricanes along the Atlantic coast. According to climatologist Stanley A. Changon, these weather changes meant that the 1997-98 ENSO **(1)** caused 189 deaths but saved nearly 850 lives that would otherwise have been lost and **(2)** resulted in damages of $4.2-4.5 billion but provided nearly $20 billion in benefits for home and business owners, insurers, and governments.

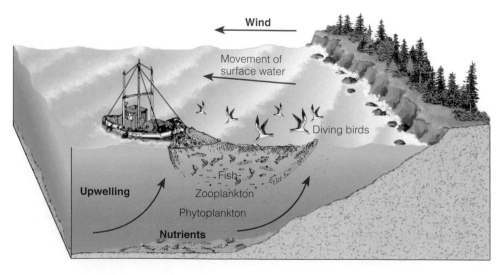

Figure 6-9 A *shore upwelling* (shown here) occurs when deep, cool, nutrient-rich waters are drawn up to replace surface water moved away from a steep coast by wind-driven currents. Such areas support large populations of phytoplankton, zooplankton, fish, and fish-eating birds. *Equatorial upwellings* occur in the open sea near the equator when northward and southward currents interact to push deep waters and their nutrients to the surface, thus greatly increasing primary productivity in such areas (Figure 4-24, p. 87, and Figure 4-25, p. 88).

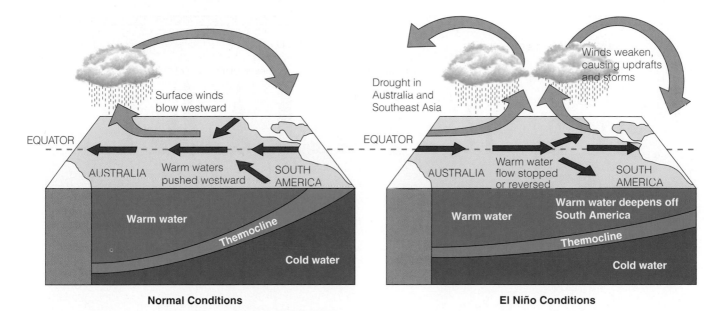

Normal Conditions

El Niño Conditions

Figure 6-10 Normal surface winds blowing westward cause shore upwellings of cold, nutrient-rich bottom water in the tropical Pacific Ocean near the coast of Peru (left). The warm and cold water are separated by a zone of gradual temperature change called the *thermocline*. Every few years a climate shift known as the El Niño-Southern Oscillation (ENSO) disrupts this pattern. Westward surface winds weaken, which depresses the coastal upwellings and warms the surface waters off South America (right). When an ENSO lasts 12 months or longer, it severely disrupts populations of plankton, fish, and seabirds in upwelling areas and can trigger extreme weather changes over much of the globe (Figure 6-11, below).

What Is La Niña Sometimes an El Niño is followed by its cooling counterpart, *La Niña*, as occurred from July 1998 into 1999 (Figure 6-12). Typically a La Niña means (1) more Atlantic Ocean hurricanes, (2) colder winters in Canada and the Northeast, (3) warmer and drier winters in the southeastern and southwestern United States, (4) wetter winters in the Pacific Northwest, (5) torrential rains in Southeast Asia, (6) lower wheat yields in Argentina, and (7) more wildfires in Florida.

Typically a La Niña is worse than an El Niño for the United States because (1) there are more Atlantic coast hurricanes and (2) there is an increase in tornadoes spread out over a wider area. According to some climate experts, there are 20- to 30-year climate cycles during which there are more La Niñas and fewer El Niños. There is some preliminary evidence that the United States may be moving into such a pattern for the next 2-3 decades.

How Does the Chemical Makeup of the Atmosphere Lead to the Greenhouse Effect? Small amounts of gases such as water vapor (H_2O), carbon dioxide (CO_2), ozone (O_3), methane (CH_4), nitrous oxide (N_2O), and chlorofluorocarbons (CFCs) play a key role in determining the earth's average temperatures and thus its climates.

Together, these gases, known as **greenhouse gases**, act somewhat like the glass panes of a greenhouse: They allow light, infrared radiation, and some ultraviolet

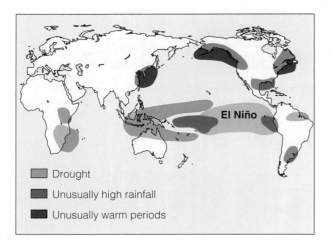

Drought

Unusually high rainfall

Unusually warm periods

El Niño

Figure 6-11 Typical global climatic effects of an El Niño-Southern Oscillation. During the 1996-98 ENSO, huge waves battered the California coast, and torrential rains caused widespread flooding and mudslides. In Peru, hundreds of people were killed by floods and mudslides, about 250,000 people were left homeless, and harvests were ruined. Drought in Brazil, Indonesia, and Australia led to massive wildfires in tinder-dry forests. India and parts of Africa also experienced severe drought. A catastrophic ice storm hit Canada and the northeastern United States, but the southeastern United States had fewer hurricanes. (Data from United Nations Food and Agriculture Organization)

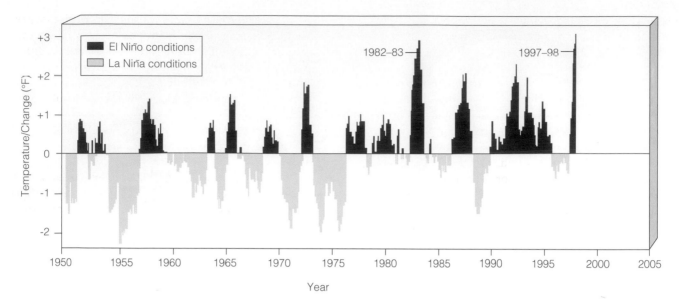

Figure 6-12 El Niño and La Niña conditions between 1950 and 1999. (Data from U.S. National Weather Service)

(UV) radiation from the sun (Figure 3-10, p. 58) to pass through the troposphere. The earth's surface absorbs much of this solar energy and degrades it to longer-wavelength, infrared radiation (heat), which then rises into the troposphere (Figure 4-8, p. 75).

Some of this heat escapes into space, and some is absorbed by molecules of greenhouse gases and emitted into the troposphere as infrared radiation, which warms the air. This natural warming effect of the troposphere is called the **greenhouse effect** (Figure 6-13).

The amount of heat in the troposphere depends primarily on the concentrations of greenhouse gases and the length of time they stay in the atmosphere. The primary greenhouse gas in the atmosphere is *water vapor*. However, because its concentration in the atmosphere is high (1–4%), inputs of water vapor from human activities have little effect on this chemical's greenhouse effects. By contrast, the concentration of carbon dioxide in the atmosphere is so small (0.036%) that the fairly large input of CO_2 from human activities can significantly affect the amount of heat in the lower atmosphere.

The basic principle behind the greenhouse effect is well established. Indeed, without its current

(a) Rays of sunlight penetrate the lower atmosphere and warm the earth's surface.

(b) The earth's surface absorbs much of the incoming solar radiation and degrades it to longer-wavelength infrared radiation (heat), which rises into the lower atmosphere. Some of this heat escapes into space and some is absorbed by molecules of greenhouse gases and emitted as infrared radiation, which warms the lower atmosphere.

(c) As concentrations of greenhouse gases rise, their molecules absorb and emit more infrared radiation, which adds more heat to the lower atmosphere.

Figure 6-13 The *greenhouse effect*. Without the atmospheric warming provided by this natural effect, the earth would be a cold and mostly lifeless planet. According to the widely accepted greenhouse theory, when concentrations of greenhouse gases in the atmosphere rise, the average temperature of the troposphere also rises. (From *Biology: Concepts and Principles*, 4th ed. by Cecie Starr © 2000)

greenhouse gases (especially water vapor), the earth would be a cold and mostly lifeless planet with an average surface temperature of $-18°C$ ($0°F$) instead of its current $15°C$ ($59°F$).

We and other species currently benefit from a comfortable level of greenhouse gases that typically undergo only minor, slow fluctuations over hundreds to thousands of years. However, mathematical models of the earth's climate indicate that natural or human-induced global warming taking place over a few decades could be disastrous for human societies and many forms of life, as discussed in Section 18-4, p. 458.

How Does the Chemical Makeup of the Atmosphere Create the Ozone Layer? In a band of the stratosphere 16–26 kilometers (11–16 miles) above the earth's surface, oxygen (O_2) is continuously converted to ozone (O_3) and back to oxygen by a sequence of reactions initiated by UV radiation from the sun ($3O_2 + UV \rightleftharpoons 2O_3$). The result is a thin veil of protective ozone at very low concentrations (up to 12 parts per million). Normally, the average levels of ozone in this life-saving layer do not change much because the rate of ozone destruction is equal to its rate of formation.

UV radiation reaching the stratosphere is composed of three bands: A, B, and C. The ozone layer blocks out **(1)** nearly all the highest-energy, shortest-wavelength radiation (UV-C), **(2)** approximately half of the next highest band (UV-B), and **(3)** a small part of the lowest-energy radiation (UV-A). Stratospheric ozone prevents at least 95% of the sun's harmful UV radiation from reaching the earth's surface (Figure 4-8, p. 75).

This ozone also creates warm layers of air that prevent churning gases in the troposphere from entering the stratosphere. This *thermal cap* is important in determining the average temperature of the troposphere and thus the earth's current climates. There is much evidence that chemicals added to the atmosphere by human activities are decreasing levels of protective ozone in the stratosphere, as discussed in more detail in Section 18-6 (p. 465).

How Do Topography and Other Features of the Earth's Surface Create Microclimates? Various topographic features of the earth's surface create local climatic conditions, or **microclimates**, that differ from the general climate of a region. For example, mountains interrupt the flow of prevailing surface winds and the movement of storms. When moist air blowing inland from an ocean reaches a mountain range, it cools as it is forced to rise and expand. This causes the air to lose most of its moisture as rain and snow on the windward (wind-facing) slopes. As the drier air mass flows down the leeward (away from the wind) slopes, it draws moisture out of the plants and soil over which it passes. The lower precipitation and the resulting semiarid or arid conditions on the leeward side of high mountains are called the **rain shadow effect** (Figure 6-14).

Vegetation also creates microclimates because it takes up and releases water, changes the movement of wind near the ground, and casts shadows. For example, compared with nearby areas of open land, forests are warmer in winter and cooler in summer, and they have lower wind speeds and higher humidity.

Figure 6-14 The rain shadow effect is a reduction of rainfall on the side of high mountains facing away from prevailing surface winds. It occurs when warm, moist air in prevailing onshore winds loses most of its moisture as rain and snow on the windward (wind-facing) slopes of a mountain range. This leads to semiarid and arid conditions on the leeward side of the mountain range and the land beyond. The Mojave Desert, east of the Sierra Nevada in California, is produced by this effect. Blue numbers represent average annual precipitation (in centimeters) in California's Sierra Nevada, and white numbers signify elevation (in meters). (From *Biology: Concepts and Principles*, 4th ed. by Cecie Starr ©2000)

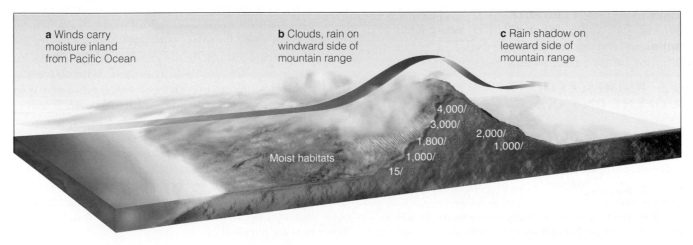

a Winds carry moisture inland from Pacific Ocean

b Clouds, rain on windward side of mountain range

c Rain shadow on leeward side of mountain range

Moist habitats

4,000/
3,000/
1,800/
2,000/
1,000/
1,000/
15/

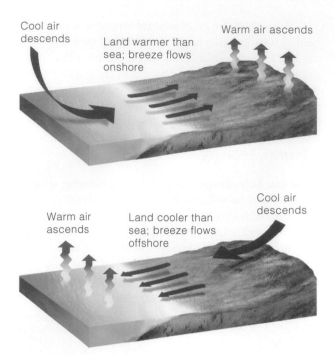

Cool air descends

Land warmer than sea; breeze flows onshore

Warm air ascends

Warm air ascends

Land cooler than sea; breeze flows offshore

Cool air descends

Figure 6-15 Sea and land breezes. Land heated by the sun during the day heats ground-level air, which rises and is replaced by cooler air drawn from the ocean. As a result, surface wind blows from the ocean to the land, creating a *sea breeze* (top). At night the land loses its heat quickly through radiation while the ocean water remains at about the same temperature. As the land surface cools below the sea surface temperature, warmer air rises over the ocean and is replaced by air from the land. Thus, surface wind blows from the land to the ocean, creating a *land breeze* (bottom). (From *Biology: Concepts and Principles*, 4th ed. by Cecie Starr © 2000)

Cities also create distinct microclimates. Bricks, concrete, asphalt, and other building materials absorb and hold heat, and buildings block wind flow. Motor vehicles and the climate control systems of buildings release large quantities of heat and pollutants. As a result, cities tend to have more haze and smog, higher temperatures, and lower wind speeds than the surrounding countryside. This is known as the *heat island effect.*

Land-ocean interactions affect the local climates of coastal by creating ocean-to-land breezes (called *sea breezes*) during the day and land-to-ocean breezes (called *land breezes*) at night (Figure 6-15).

6-2 BIOMES: CLIMATE AND LIFE ON LAND

Why Are There Different Organisms in Different Places? Why is one area of the earth's land surface a desert, another a grassland, and another a forest? Why are there different types of deserts, grasslands, and forests?

The general answer to these questions is differences in *climate* (Figure 6-4), caused mostly by differences in average temperature and precipitation caused by global air circulation (Figure 6-8). Different climates promote different communities of organisms.

Figure 6-16 and the photo on p. 1 show global distributions of *biomes*: terrestrial regions with characteristic types of natural, undisturbed ecological communities adapted to the climate of the region. By comparing Figure 6-16 with Figure 6-4, you can see how the world's major biomes vary with climate. Figure 4-9 (p. 76) shows major biomes in the United States as one moves through different climates along the 39th parallel.

For plants, *precipitation generally is the limiting factor that determines whether a land area is desert, grassland, or forest.* Taken together, average annual precipitation and temperature (along with soil type) are the most important factors in producing tropical, temperate, or polar deserts, grasslands, and forests (Figure 6-17).

On maps such as the one in Figure 6-16, biomes are presented as having sharp boundaries and as being covered with the same general type of vegetation. In reality, most biomes do not have sharp boundaries and blend into one another in transitional zones or *ecotones* (Figure 4-10, p. 77). Also, the types and numbers of plants in a biome vary from one location to another because of small variations in climate (microclimates), soil types, and natural and human-caused disturbances (Figure 1-4, p. 8).

As a result, *biomes are not uniform.* They consist of a *mosaic of patches*, all with somewhat different biological communities but with similarities unique to the biome.

Climate and vegetation vary with **latitude** (distance from the equator) and **altitude** (elevation above sea level). If you travel from the equator toward either pole, you will generally encounter colder climates and zones of vegetation adapted to those climates. Similarly, as elevation above sea level increases, climate becomes colder. Thus, if you climb a tall mountain from its base to its summit, you can observe changes in plant life similar to those you would encounter in traveling from equator to poles (Figure 6-18, p. 133).

Why Do Plant Sizes, Shapes, and Survival Strategies Differ? Arctic soils are wet and nutrient rich. So why are there no trees in the Arctic, and why are the plants there so close to the ground? Why are there no leaves on desert plants such as cacti? Why do trees in most forests found in both the warm tropics and in cold areas such as Canada and Sweden keep their leaves year-round, whereas most trees in temperate forests lose their leaves in winter?

Plants exposed to cold air year-round or during winter have traits that keep them from losing too much heat and water. For example, trees or tall plants cannot survive in the cold, windy arctic grasslands (tundra) because they would lose too much of their heat for survival.

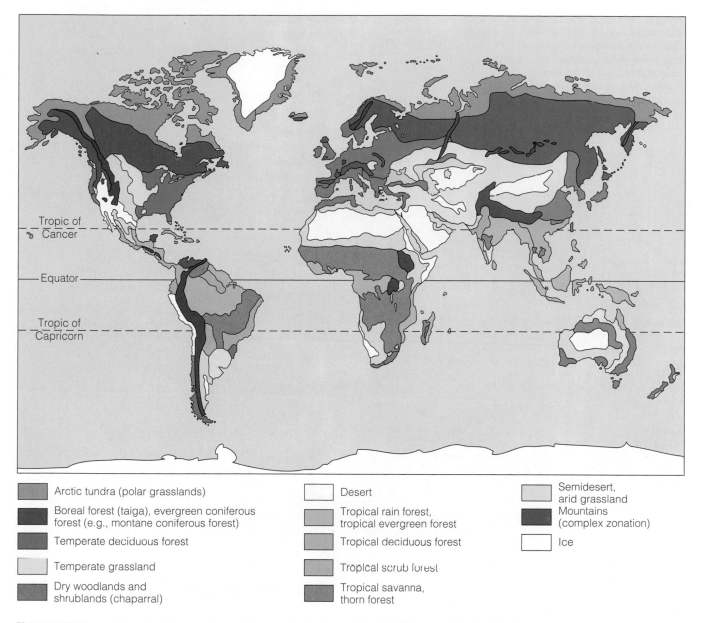

▨ Arctic tundra (polar grasslands)	▨ Desert	▨ Semidesert, arid grassland
▨ Boreal forest (taiga), evergreen coniferous forest (e.g., montane coniferous forest)	▨ Tropical rain forest, tropical evergreen forest	▨ Mountains (complex zonation)
▨ Temperate deciduous forest	▨ Tropical deciduous forest	▨ Ice
▨ Temperate grassland	▨ Tropical scrub forest	
▨ Dry woodlands and shrublands (chaparral)	▨ Tropical savanna, thorn forest	

Figure 6-16 The earth's major biomes—the main types of natural vegetation in different undisturbed land areas-result primarily from differences in climate. Each biome contains many ecosystems whose communities have adapted to differences in climate, soil, and other environmental factors. In reality, people have removed or altered much of this natural vegetation in some areas for farming, livestock grazing, lumber and fuelwood, mining, and construction, thereby altering the biomes (Figure 1-4, p. 8).

Desert plants exposed to the sun all day long must be able to lose enough heat so that they do not overheat and die. They must also conserve enough water for survival. **Succulent** (fleshy) **plants**, such as the saguaro ("sah-WAH-row") cactus, survive in dry climates by having a vertical orientation of most surfaces, no leaves, and the ability to store water and synthesize food in their expandable, fleshy tissue. The plant's shape and lack of leaves give it a small surface area exposed to sunlight and a large area away from the sun for radiating heat out. Succulent plants also reduce water loss by opening their pores (stomata) to take up carbon dioxide (CO_2) only at night.

Trees of wet tropical rain forests tend to be **broadleaf evergreen plants**, which keep most of their broad leaves year-round. The large surface area of the leaves allows them to collect ample sunlight for photosynthesis and radiate heat during hot weather.

In a climate with a cold (and sometimes dry) winter, keeping such leaves would cause plants to lose too much heat and water for survival. In such climates, **broadleaf deciduous plants**, such as oak and maple

Figure 6-17 Average precipitation and average temperature, acting together as limiting factors over a period of 30 or more years, determine the type of desert, grassland, or forest biome in a particular area. Although the actual situation is much more complex, this simplified diagram explains how climate determines the types and amounts of natural vegetation found in an area left undisturbed by human activities. (From *Earth* by Derek Elsom, 1992. Copyright © 1992 by Marshall Editions Developments Limited. New York: Macmillan. Used by permission.)

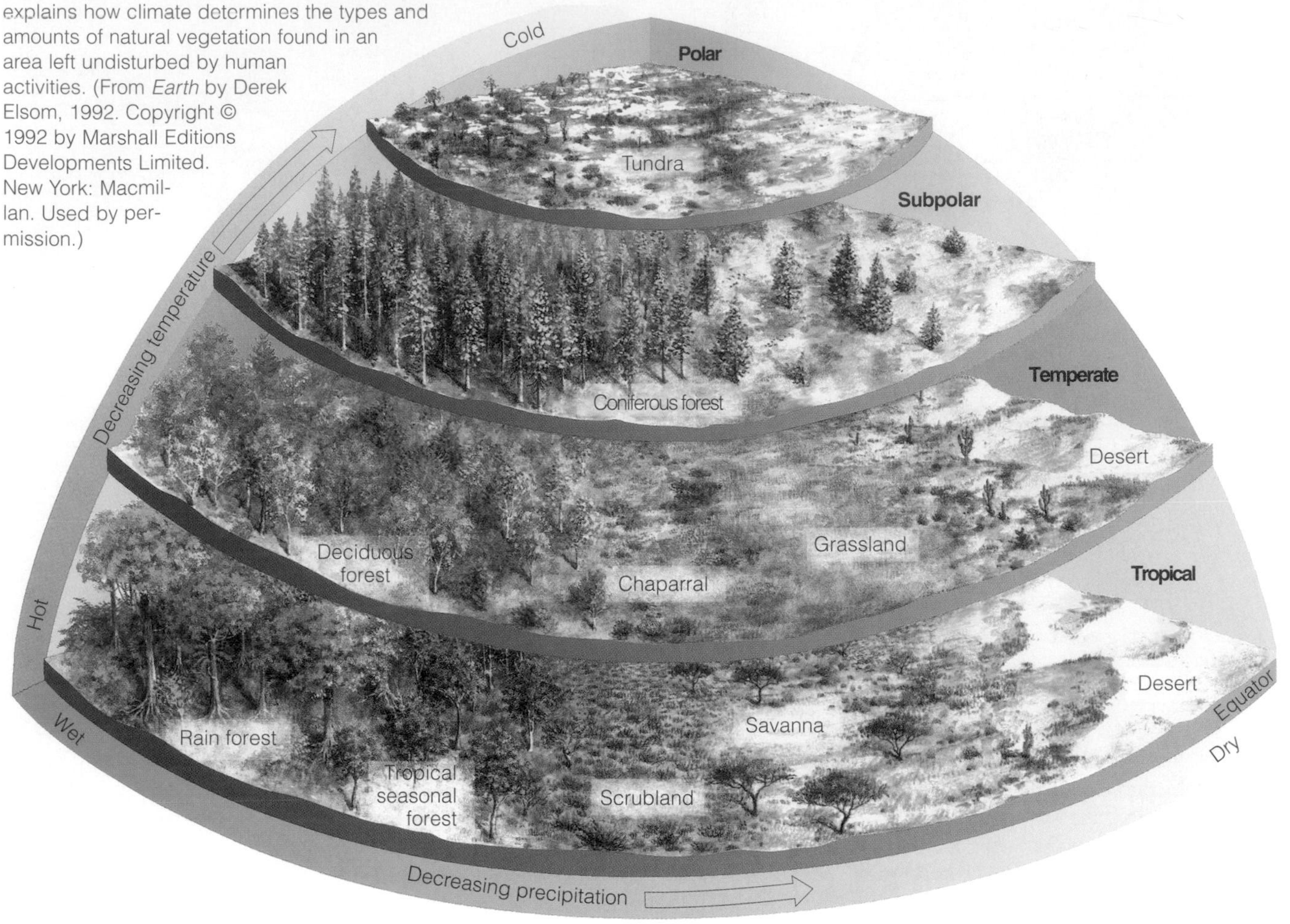

trees, survive drought and cold by shedding their leaves and becoming dormant during such periods.

If we move further north to areas such as Canada and Sweden, where summers are cool and short, this strategy is less successful. Instead evolution has favored **coniferous** (cone-bearing) **evergreen plants** (such as spruces, pines, and firs). These plants keep some of their narrow, pointed leaves (needles) all year. The waxy coating, shape, and clustering of conifer needles slow down heat loss and evaporation during the long, cold winter. Additionally, by keeping their leaves all winter, such trees are ready to take advantage of the brief summer without having to take time to grow new needles.

Thus, communities with similar climates contain plants (and other organisms) with similar evolutionary adaptations, even though the individual species may be unrelated. For example, plants in both American and African deserts tend to have thick body parts, protective spines, and small leaves even though plants in these two areas evolved from different ancestors.

6-3 DESERT BIOMES

What Are the Major Types of Deserts? A **desert** is an area where evaporation exceeds precipitation. Precipitation typically is less than 25 centimeters (10 inches) a year and often is scattered unevenly throughout the year. Deserts have sparse, widely spaced, mostly low vegetation.

Deserts cover about 30% of the earth's land, and are situated mainly between tropical and subtropical regions north and south of the equator, at about 30° north and 30° south latitude (Figure 6-16).

In these areas, air that has lost its moisture over the tropics falls back to the earth (Figure 6-8). The largest deserts are in the interiors of continents, far from moist sea air and moisture-bearing winds. Other, more local deserts form on the downwind sides of mountain ranges because of the rain shadow effect (Figure 6-14).

The baking sun warms the ground in the desert during the day. At night, however, most of the heat stored in the ground radiates quickly into the atmosphere because

Alpine
Tundra

Montane
Coniferous
Forest

Deciduous
Forest

Tropical
Forest

Tropical Forest   Temperate Deciduous Forest   Northern Coniferous Forest   Arctic Tundra

High ◄──────────────────── Moisture Availability ────────────────────► Low

Low ▲ Elevation ▼ High

Figure 6-18 Generalized effects of latitude and altitude on moisture and biomes in North America. Parallel changes in vegetation type occur when we travel from the equator to the poles or from lowlands to mountaintops. Plant types in these areas also vary with mean annual air temperature and soil types. (From *The Unity and Diversity of Life*, 4th ed. by Cecie Starr and Ralph Taggart © 2001)

desert soils have little vegetation and moisture and the skies usually are clear. This explains why in a desert you may roast during the day but shiver at night.

A combination of low rainfall and different average temperatures creates tropical, temperate, and cold deserts (Figures 6-17 and 6-19). In *tropical deserts*, such as the southern Sahara in Africa, temperatures usually are high year-round and there is little rain, which typically falls during only 1 or 2 months of the year (Figure 6-19, left). These driest places on earth typically have few plants and a hard, windblown surface strewn with rocks and some sand.

In *temperate deserts*, such as the Mojave in southern California (Figure 6-14), daytime temperatures are high in summer and low in winter, and there is more precipitation than in tropical deserts (Figure 6-19, center). The vegetation is sparse, consisting mostly of widely dispersed, drought-resistant shrubs and cacti or other succulents, and animals are adapted to the lack of water and temperature variations (Figure 6-20).

In *cold deserts*, such as the Gobi Desert in China, winters are cold, summers are warm or hot, and precipitation is low (Figure 6-19, right). In the semi-arid zones between deserts and grasslands, we find

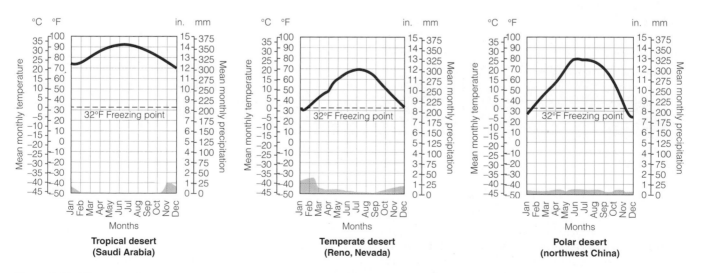

Figure 6-19 Climate graphs showing typical variations in annual temperature and precipitation in tropical, temperate, and polar (cold) deserts.

Red-tailed hawk

Gambel's quail

Yucca

Jack rabbit

Collared lizard

Agave

Roadrunner

Prickly pear cactus

Diamondback rattlesnake

Darkling beetle

Bacteria

Fungi

Kangaroo rat

Producer to primary consumer	Primary to secondary consumer	Secondary to higher-level consumer	All producers and consumers to decomposers

Figure 6-20 Some components and interactions in a temperate desert biome. When these organisms die, decomposers break down their organic matter into minerals used by plants. Transfers of matter and energy between producers, primary consumers (herbivores), and secondary (or higher-level) consumers (carnivores) are indicated by colored arrows. Organisms are not drawn to scale.

semidesert. This biome is dominated by thorn trees and shrubs adapted to a long dry spells followed by brief, sometimes heavy rains.

How Do Desert Plants and Animals Survive?

Adaptations for survival in the desert have two themes: "Beat the heat" and "Every drop of water counts." Some desert plants are evergreens with wax-coated leaves (creosote bush) that minimize transpiration. Most desert perennials tend to have small leaves (coachman's whip) or no leaves (cacti), which helps them conserve water. Perennial shrubs such as mesquite and creosote plants grow deep roots to tap into groundwater, and they drop their leaves to survive in a dormant state during long dry spells.

Other perennials such as short (prickly pear, Figure 6-20) and tall (saguaro) cacti use widely spread, shallow roots to collect water after brief showers and store it in their spongy tissues. Most of these succulents are armed with sharp spines to discourage herbivores from feeding on their water-storing, fleshy tissue. The spines also reduce overheating by reflecting some sunlight and by providing shade and insulation. Some desert plants, such as the creosote bush and sagebrush, also secrete toxins in the soil. This reduces competition for water and soil nutrients from nearby plants of other species.

Many desert plants are annual wildflowers and grasses that store much of their biomass in seeds during dry periods and remain inactive (sometimes for years) until they receive enough water to germinate. Shortly after a rain the seeds **(1)** germinate, **(2)** grow, **(3)** carpet such deserts with a dazzling array of colorful flowers, **(4)** produce new seed, and **(5)** die, all in only a few weeks.

Other, less visible desert plants are mosses and lichens, which can tolerate extremely high temperatures, dry out completely, and become dormant until the next rain falls. In museums, some desert moss specimens have been known to recover and grow after 250 years without water.

Most desert animals are small. They beat the heat and reduce water loss by evaporative cooling. They hide in cool burrows or rocky crevices by day and come out at night or in the early morning.

Birds, ants, rodents, and other seed-eating herbivores are common, feeding on the multitudes of seeds produced by the desert's annual plants. Some deserts have a few large grazing animals such as gazelle and the endangered Arabian oryx.

Major carnivores in temperate North American deserts are coyotes, kit and gray foxes, and various species of snakes and owls that come out mainly at night to prey on the desert's many rodent species. The few daytime animals, such as fast-moving lizards

The Kangaroo Rat: Water Miser and Keystone Species

The kangaroo rat (Figure 6-20) is a remarkable mammal superbly adapted for conserving water in its desert environment.

As the desert's chief seed eater, it is also a keystone species that helps support other desert species and helps keep desert shrubland from becoming grassland.

This rodent comes out of its burrow only at night, when the air is cool and water evaporation has slowed. It seeks dry seeds that it quickly stuffs into its cheek pouches.

After a night of foraging it empties its cache of seeds into its cool burrow, where they soak up water exhaled in the rodent's breath. When the rodent eats these seeds, it gets this water back.

The kangaroo rat does not drink water; its water comes from the recycled moisture in the seeds and from water produced when sugars in the seeds undergo aerobic respiration during digestion.

Some of the water vapor in the rat's breath also condenses on the cool inside surface of its nose. This condensed water then diffuses back to its body.

Kangaroo rats have no sweat glands, so they do not lose water by perspiration. In addition, they save water by excreting hard, dry feces and thick, nearly solid urine produced by their extremely efficient kidneys.

Critical Thinking

Water is scarce in much of the southwestern United States, where the kangaroo rat lives. However, this area has one of the highest rates of human population growth. As this happens, what ecological lesson can we learn from the kangaroo rat about how to survive in this area (and other water-short areas throughout the world)?

and some snake species, are preyed upon mostly in the early morning and late afternoon by hawks and roadrunners.

Some desert animals have physical adaptations for conserving water (Spotlight, above). Insects and reptiles have thick outer coverings to minimize water loss through evaporation. They also reduce water loss by having dry feces and excreting a dried concentrate of urine.

Some of the smallest desert animals, such as spiders and insects, get their water only from dew or from the food they eat. Some desert animals become dormant during periods of extreme heat or drought and are active only during the cooler months of the year. Arabian oryxes survive by licking the dew that accumulates at night on rocks and on one another's hair.

What Impacts Do Human Activities Have on Desert Ecosystems? Deserts take a long time to recover from disturbances because of their **(1)** slow plant growth, **(2)** low species diversity, **(3)** slow nutrient cycling (because of little bacterial activity in their soils), and **(4)** water shortages. Desert vegetation destroyed by livestock overgrazing and off-road vehicles may take decades to grow back. For example, tracks left by tanks practicing in the California desert during World War II are still visible. Vehicles can also collapse underground burrows where many desert animals live.

Some major human impacts on deserts are as follows:

- Rapid growth of large desert cities. Examples are in Saudi Arabia and Egypt and in the southwestern United States (such as Palm Springs, California; Las Vegas, Nevada; and Phoenix, Arizona). Increasingly, people in such cities are destroying fragile desert soil, plants, and animal burrows with four-wheel-drive vehicles, motorcycles, and urban development.

- Irrigation of some desert areas to grow crops. As the water evaporates, salts may accumulate in the soil (salinization) and limit crop productivity.

- Depletion of underground water (aquifers) as desert cities and irrigation expand.

- Disruption and pollution by extraction of oil, minerals, and building materials (such as road stone and sand).

- Use of remote desert areas as sites for **(1)** storage of toxic and radioactive wastes, **(2)** underground testing of nuclear weapons, and **(3)** maneuvers by heavy tanks and other military vehicles.

Abundant sunlight is the largest and most untapped resource of deserts. Over the next 40-50 years many analysts expect industrialized societies to make a transition from reliance on nonrenewable fossil and nuclear fuels to greatly increased use of renewable solar energy. If such a shift takes place, large areas of many deserts near urban areas will be covered with arrays of solar collectors and solar cells (p. 375) to produce electricity and hydrogen gas (p. 384). Great care must be taken to do this in ways that do not seriously disrupt desert ecosystems.

6-4 GRASSLAND, TUNDRA, AND CHAPARRAL BIOMES

What Are the Major Types of Grasslands? Grasslands are regions with enough average annual precipitation to allow grass (and in some areas, a few trees) to prosper but with precipitation so erratic that drought and fire prevent large stands of trees from growing. Most grasslands are found in the interiors of continents (Figure 6-16).

Grasslands persist because of a combination of **(1)** seasonal drought, **(2)** grazing by large herbivores, and **(3)** periodic fires, all of which keep large numbers of shrubs and trees from invading and becoming established. If not overgrazed by large herbivores, grasses (many of them perennials) in these biomes are renewable resources because these plants grow out from the bottom. This allows their stems to grow again after being nibbled off by grazing animals. The three main types of grasslands—tropical, temperate, and polar (tundra)—result from combinations of low average precipitation and various average temperatures (Figures 6-17 and 6-21).

What Are Tropical Grasslands and Savannas? *Tropical grasslands* are found in areas with high average temperatures, low to moderate precipitation, and a prolonged dry season. They occur in a wide belt on either side of the equator beyond the borders of tropical rain forests (Figure 6-16).

One type of tropical grassland, called a *savanna*, usually has warm temperatures year-round, two prolonged dry seasons, and abundant rain the rest of the year (Figure 6-21, left). The largest savannas are in central and southern Africa, but they are also found in central South America, Australia, and Southeast Asia (Figure 6-16).

Most savannas consist of grasslands punctuated by stands of deciduous shrubs and trees, which shed their leaves during the dry season and thus avoid excessive water loss. African savanna often is dotted with flat-topped deciduous trees called acacias. In Australian savanna (bush), the predominant tree is the eucalyptus (whose leaves are eaten by herbivores such as the koala), and the predominant grazers are native kangaroos and nonnative rabbits introduced by settlers in 1859.

African tropical savannas contain enormous herds of *grazing* (grass- and herb-eating) and *browsing* (twig- and leaf-nibbling) hoofed animals, including wildebeests, gazelles, zebras, giraffes, antelopes, and other animals that graze in the four types of grassland habitat found in African savanna (Figure 6-22). These and other large herbivores have evolved specialized eating habits that minimize competition for resources between species for vegetation. For example, **(1)** giraffes eat leaves and shoots from the tops of trees, **(2)** elephants eat leaves and branches further down, **(3)** Thompson's gazelles and wildebeests prefer short grass, and **(4)** zebras graze on longer grass and stems.

During the dry season wildebeest and other large grazing animals migrate to find enough water and

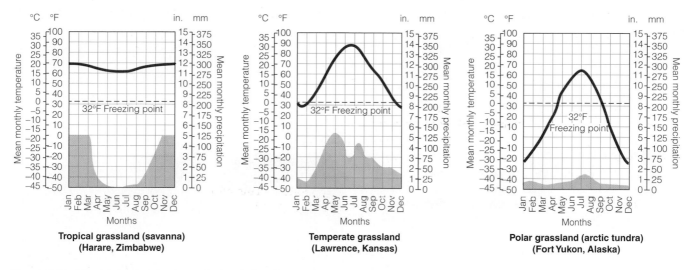

Figure 6-21 Climate graphs showing typical variations in annual temperature and precipitation in tropical, temperate, and polar (arctic tundra) grasslands.

high-quality grasses, and smaller animals become dormant or survive by eating plant seeds. The large grazing animals (and many small ones) are preyed upon by predators such as cheetahs, lions, hyenas, eagles, and hawks. Many large savanna animal species are killed for their economically valuable coats and parts (tigers), tusks (rhinoceroses), and ivory tusks (elephants).

What Are Temperate Grasslands? *Temperate grasslands* cover vast expanses of plains and gently rolling hills in the interiors of North and South America, Europe, and Asia (Figure 6-16). Unlike savannas, such grasslands lack trees and have seasonal extremes of hot and cold instead of wet and dry. Winters there are bitterly cold and summers are hot and dry (Figure 6-21, center). Annual precipitation averages only 25–100 centimeters (10–39 inches) and falls unevenly throughout the year.

Windswept temperate grasslands tend to dry out in summer and fall and have recurrent fires. As a result, grassland plants have evolved the ability to grow back rapidly after a fire.

Drought, occasional fires, and intense grazing inhibit the growth of trees and bushes, except along rivers. Because the aboveground parts of most of the grasses die and decompose each year, organic matter accumulates to produce a deep, fertile soil. This soil is held in place by a thick network of intertwined roots of drought-tolerant perennial grasses unless the topsoil is plowed up and allowed to blow away by prolonged exposure to high winds found in these biomes.

Types of temperate grasslands are the **(1)** *tallgrass prairies* (Figure 6-23) and *short-grass prairies* of the midwestern and western United States and Canada, **(2)** South American *pampas*, **(3)** African *veldt*, and **(4)** *steppes* of central Europe and Asia. Here winds blow almost continuously, and evaporation is rapid, often leading to fires in the summer and fall.

Because of their thick and fertile soils, temperate grasslands are widely used to grow crops. However, plowing breaks up the complex soil structure and leaves it vulnerable to erosion by wind and water.

Before the plow and the cow arrived, the tall-grass prairie predominated in the humid eastern portion of the Great North American Prairie. The short-grass prairie predominated in the arid West, with mixed-grass prairie between.

Huge numbers and varieties of beetles, spiders, grasshoppers, and other insects and invertebrate animals live among the plants in North American temperate grasslands. Ants and earthworms are abundant in the soil. Primary consumers include a variety of small animals such as prairie dogs, deer mice, jackrabbits, squirrels, and meadowlarks. Most of these animals live in burrows or escape predators running or hopping swiftly. These small animals are preyed upon by carnivores such as coyotes, bobcats, kit foxes, ferrets, snakes, and hawks. Wolves and pumas once preyed upon these larger animals, but most have been killed or driven out by farmers, ranchers, and hunters.

What Are Polar Grasslands? *Polar grasslands*, or *arctic tundra*, covering about 10% of the earth's land area, occur just south of the arctic polar ice cap (Figure 6-16). During most of the year these treeless plains are bitterly cold, swept by frigid winds, and covered with ice and snow (Figure 6-21, right). Winters are long and dark, and the scant precipitation falls mostly as snow.

Beisa oryx

Cape buffalo

Wildebeest

Topi

Warthog

Thompson's gazelle

Waterbuck

Grant's zebra

Dry Grassland

Moist Grassland

Giraffe

African elephant

Gerenuk

Black rhino

Dik-dik

East African eland

Blue duiker

Greater kudu

Bushbuck

Dry Thorn Scrub

Riverine Forest

Figure 6-22 Some of the grazing animals found in different parts of the African savanna. Each species has a different feeding niche that allows them to share vegetation resources.

	Producer to primary consumer		Primary to secondary consumer		Secondary to higher-level consumer		All producers and consumers to decomposers

Figure 6-23 Some components and interactions in a temperate tall-grass prairie ecosystem in North America. When these organisms die, decomposers break down their organic matter into minerals used by plants. Transfers of matter and energy between producers, primary consumers (herbivores), and secondary (or higher-level) consumers (carnivores) are indicated by colored arrows. Organisms are not drawn to scale.

Long-tailed jaeger

Grizzly bear

Caribou

Mosquito

Snowy owl

Horned lark

Arctic fox

Willow ptarmigan

Dwarf willow

Lemming

Mountain cranberry

Moss campion

Producer to primary consumer

Primary to secondary consumer

Secondary to higher-level consumer

All consumers and producers to decomposers

Figure 6-24 Some components and interactions in an arctic tundra (polar grassland) ecosystem. When these organisms die, decomposers break down their organic matter into minerals used by plants. Transfers of matter and energy between producers, primary consumers (herbivores), and secondary (or higher-level) consumers (carnivores) are indicated by colored arrows. Organisms are not drawn to scale.

This biome is carpeted with a thick, spongy mat of low-growing plants, primarily grasses, mosses, and dwarf woody shrubs (Figure 6-24). These hardy plants are adapted to (1) lack of sunlight and water, (2) freezing temperatures, and (3) constant high winds. Most of the annual growth of these plants occurs during the 6- to 8-week summer, when sunlight shines almost around the clock.

To retain water and survive the winter cold, most tundra plants grow close to the ground, and some have leathery evergreen leaves coated by waxes that reduce heat loss. Other plants survive the long, cold winter underground as roots, stems, bulbs, and tubers; some, such as lichens, dehydrate during winter to avoid frost damage.

One effect of the extreme cold is **permafrost**, a perennially frozen layer of the soil that forms when the water there freezes. In summer, water near the surface thaws, but the permafrost soil layer below stays frozen and prevents liquid water at the surface from seeping into the ground. Thus, during the brief summer the soil above the permafrost layer remains waterlogged, forming a large number of shallow lakes, marshes, bogs, ponds, and other seasonal wetlands. Hordes of mosquitoes, blackflies, and other insects thrive in these shallow surface pools. They feed large colonies of migratory birds (especially waterfowl) that return from the south to nest and breed in the bogs and ponds.

In North American arctic tundra, caribou herds arrive to feed on the summer vegetation, bringing with them their wolf predators. In arctic tundra in Europe and Asia, reindeer occupy the grazing niche of caribou.

The arctic tundra's permanent animal residents are mostly small herbivores such as lemmings, hares, voles, and ground squirrels, which burrow underground to escape the cold. They are eaten by predators such as the lynx, weasel, snowy owl, and arctic fox (Figure 5-8, p. 113). Most tundra animals do not hibernate because the summer is too short for them to accumulate adequate fat reserves. Animals in this biome survive the intense winter cold through adaptations such as (1) thick coats of fur (arctic wolf, arctic fox, and musk oxen), (2) feathers (snowy owl), (3) compact bodies to expose as little surface as possible to the air, and (4) living underground (arctic lemming).

Because of the cold, decomposition is slow. Partially decomposed organic matter forms soggy peat bogs (the source of gardener's peat moss), which contain about 95% of this biome's carbon. Because of low decomposer populations, the soil is poor in organic matter and in nitrates, phosphates, and other minerals.

Figure 6-25 Replacement of a temperate grassland with a monoculture crop near Blythe, California. When the tangled root network of natural grasses is removed, the fertile topsoil is subject to severe wind erosion unless it is covered with some type of vegetation. If global warming accelerates over the next 50 years, many of these grasslands may become too hot and dry for farming, thus threatening the world's food supply. (National Archives/EPA Documerica)

What Is Alpine Tundra? Another type of tundra, called *alpine tundra*, occurs above the limit of tree growth but below the permanent snow line on high mountains (Figure 6-18). The vegetation there is similar to that found in arctic tundra, but it gets more sunlight than arctic vegetation and has no permafrost layer.

For a few weeks each summer the land blazes with color as wildflowers burst into bloom. The small plants that survive in this biome are grazed by herbivores such as elk, mountain goats, and sheep, while golden eagles soar above looking for marmots and ground squirrels.

What Impacts Do Human Activities Have on Grassland Ecosystems? Some major human impacts on grasslands are as follows:

- Burning, plowing up, and converting some areas of savanna into cropland. By releasing large quantities of carbon dioxide into the atmosphere, this may contribute to the greenhouse effect as much as (if not more than) the more publicized clearing and burning of tropical rain forests.

- Overgrazing by livestock in tropical and temperate grasslands as governments and aid agencies encourage the drilling of water wells. With more water, the livestock population increases, and pastures around wells are overgrazed and trampled by thousands of hooves. This can convert grassland into less productive desert and semidesert.

- Plowing large areas of fertile temperate grasslands in North America, western Europe, and Ukraine and converting them to highly productive cropland (Figure 6-25). As long as temperate grasslands keep their fertile soil and the climate does not change, they can continue producing much of the world's cereal grains. However, poor farming practices, overgrazing, mismanagement, and occasional prolonged droughts lead to severe wind erosion and loss of topsoil, which can convert temperate grasslands into desert or semidesert shrubland.

- Damaging the fragile arctic tundra in Alaska and Siberia. This is caused by oil exploration and drilling, air pollution, spills or leaks of oil and toxic wastes, and disruption of soil and vegetation by vehicles.

What Is Chaparral? Some temperate areas have a biome known as *temperate shrubland* or *chaparral*. This biome occurs along coastal areas with what is called a *Mediterranean climate*: winters are mild and moderately rainy and summers are long, hot, and dry. It is found mainly along **(1)** parts of the Pacific coast of North America, **(2)** in southern Texas and northeastern Mexico, and **(3)** in the coastal hills of Chile, the Mediterranean, southwestern Africa, and southwestern Australia (Figure 6-16).

This biome usually is dominated by an almost impenetrably dense growth of low-growing evergreen shrubs with leathery leaves that resist water loss and large underground root systems. Many of these low-growing shrubs produce compounds that leach into the soil and inhibit the germination and growth of competing plant species. Some areas also have a sprinkling of small, drought-resistant trees such as pines and scrub oak. This biome supports large populations of small rodents, most of which store seeds in underground burrows.

During the long, hot dry season chaparral vegetation is dormant and becomes very dry and brittle. In the fall, fires started by lightning or human activities spread with incredible swiftness through the dry brush and litter of leaves and fallen branches.

Research reveals that chaparral is adapted to and maintained by periodic fires. Many of the shrubs store food reserves in their fire-resistant root crown and have seeds that sprout only after a hot fire. With the first rain, annual grasses and wildflowers spring up and use nutrients released by the fire. New shrubs grow quickly and crowd out the grasses. Within a decade or two after a fire, the natural chaparral community is restored.

People like living in this biome because of its favorable climate. However, those living in chaparral assume the high risk of losing their homes (and possibly their lives) to the frequent fires associated with it. After fires often comes the hazard of flooding; when heavy rains come, great torrents of water pour off the unprotected burned hillsides to flood lowland areas.

6-5 FOREST BIOMES

What Are the Major Types of Forests? Undisturbed areas with moderate to high average annual precipitation tend to be covered with **forest**, which contains various species of trees and smaller forms of vegetation. The three main types of forest—*tropical, temperate,* and *boreal* (polar)—result from combinations of this precipitation level and various average temperatures (Figures 6-17 and 6-26).

What Are Tropical Rain Forests? *Tropical rain forests* are a type of broadleaf evergreen forest (Figure 6-27) found near the equator (Figure 6-16), where hot, moisture-laden air rises and dumps its moisture (Figure 6-8). The world's largest tropical rain forest is in the Amazon River basin in South America. These forests

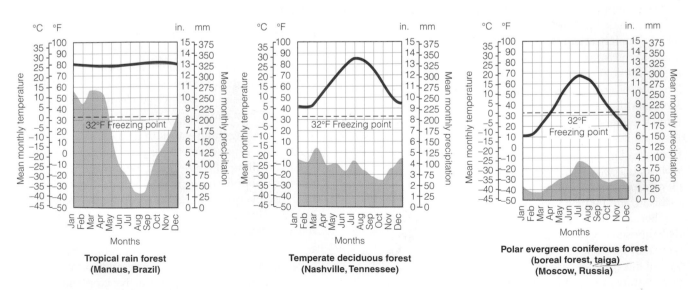

Figure 6-26 Climate graphs showing typical variations in annual temperature and precipitation in tropical, temperate, and polar forests.

| | Producer to primary consumer | | Primary to secondary consumer | | Secondary to higher-level consumer | | All producers and consumers to decomposers |

Figure 6-27 Some components and interactions in a tropical rain forest ecosystem. When these organisms die, decomposers break down their organic matter into minerals used by plants. Transfers of matter and energy between producers, primary consumers (herbivores), and secondary (or higher-level) consumers (carnivores) are indicated by colored arrows. Organisms are not drawn to scale.

have a warm annual mean temperature (which varies little, daily or seasonally), high humidity, and heavy rainfall almost daily (Figure 6-26, left).

Tropical rain forests have incredible biological diversity. These diverse forms of life occupy a variety of specialized niches in distinct layers, based mostly on their need for sunlight (Figure 6-28). Much of the animal life, particularly insects, bats, and birds, lives in the sunny *canopy* layer, with its abundant shelter and supplies of leaves, flowers, and fruits (Figure 6-27). To study life in the canopy, ecologists climb trees and build platforms and boardwalks.

Only dim light penetrates to the *understory* layer of small trees adapted to low light. Beneath this is a *shrub layer* of shrubs and short plants adapted to even less light. The dark and wet *forest floor* is open and free of vegetation. The popular image of a rain forest floor as a tangled, almost impenetrable jungle is accurate only along the banks of rivers, near clearings, or where a large tree has fallen and sunlight reaches the ground.

Many plants have evolved specialized ways to grow in tropical rain forests. Climbing vines, called *lianas*, most rooted in the soil, wind upward around the trunks of larger trees until their leaves reach the sunlit canopy. Orchids, bromeliads, and other *epiphytes* attach themselves to the trunks and branches of canopy trees and obtain nutrients from bits of organic matter falling from the canopy. Many plants dwelling in the pale light of the understory and shrub layer (such as philodendrons) survive by using huge, dark green leaves to capture enough sunlight. This ability to thrive under low light levels makes them good house plants.

The roots of even the largest trees tend to be shallow and spread out in the nutrient-poor, moist, and thin layer of soil. Many of the large trees are supported by large bulges at their bases called *buttresses*.

Figure 6-28 Stratification of specialized plant and animal niches in various layers of a tropical rain forest. The presence of these specialized niches enables species to avoid or minimize competition for resources and results in the coexistence of a great variety of species (biodiversity).

The stratification of specialized plant and animal niches in various layers of a tropical rain forest enables coexistence of a great variety of species (biodiversity). Although tropical rain forests cover only about 2% of the earth's land surface, they are habitats for 50-80% of the earth's terrestrial species.

Dropped leaves, fallen trees, and dead animals decompose quickly because of the warm, moist conditions and hordes of decomposers. This rapid recycling of scarce soil nutrients is why there is usually little litter on the ground. Instead of being stored in the soil, most minerals released by decomposition are taken up quickly by plants. Thus, most of a tropical rain forest's nutrients are stored in the biomass of its living organisms.

Because of the dense vegetation, little wind blows in tropical rain forests, eliminating the possibility of wind pollination. Many of the plants have evolved elaborate flowers (Figure 4-17, right, p. 82) that attract particular insects, birds, or bats as pollinators.

What Are Tropical Deciduous Forests? Moving a little farther from the equator (Figure 6-16), we find *tropical deciduous forests* (sometimes called *tropical monsoon forests* or *tropical seasonal forests*). These forests are warm year-round and get most of their plentiful rainfall during a wet (monsoon) season that is followed by a long dry season.

Tropical deciduous forests have a lower canopy than tropical rain forests. They contain a mixture of deciduous trees (which lose their leaves to survive the dry season) and drought-tolerant evergreen trees (which retain most of their leaves year-round). Where the dry season is especially long, we find *tropical scrub forests* (Figure 6-16) containing mostly small deciduous trees and shrubs.

What Are Temperate Deciduous Forests? *Temperate deciduous forests* (Figure 6-29) grow in areas with moderate average temperatures that change significantly with the season (Figure 6-26, center). These areas have (1) long, warm summers, (2) cold but not too severe winters, and (3) abundant precipitation, often spread fairly evenly throughout the year.

This biome is dominated by a few species of broadleaf deciduous trees such as oak, hickory, maple, poplar, and sycamore. They survive cold winters by dropping their leaves in the fall and becoming dormant. Each spring they grow new leaves that change in the fall into an array of reds and golds. Because of the fairly low rate of decomposition, these forests accumulate a thick layer of slowly decaying leaf litter that is a storehouse of nutrients.

Compared with tropical rain forests, temperate deciduous forests have a simpler structure and contain fewer tree species. However, the penetration of more sunlight supports a richer diversity of plant life at ground level. Because of the fairly low rate of decomposition, these forests accumulate a thick layer of slowly decaying leaf litter that is a storehouse of nutrients.

The temperate deciduous forests of the eastern United States were once home for such large predators as bears, wolves, foxes, wildcats, and mountain lions (pumas). Today most of the predators have been killed or displaced, and the dominant mammal species often is the white-tailed deer, along with smaller mammals such as squirrels, rabbits, opossums, raccoons, and mice.

Warblers, robins, and other bird species migrate to these forests during the summer to feed and breed. However, many of these species are declining in numbers because of loss or fragmentation of their summer and winter habitats, which also makes them more vulnerable to predators and parasitic cowbirds. Small mammals and birds are preyed upon by owls, hawks, and, in remote areas, bobcats and foxes.

What Are Evergreen Coniferous Forests? *Evergreen coniferous forests*, also called *boreal forests* and *taigas* (pronounced "TIE-guhs"), are found just south of the arctic tundra in northern regions across North America, Asia, and Europe (Figure 6-16). In this subarctic climate, winters are long, dry, and extremely cold; in the northernmost taiga, sunlight is available only 6–8 hours a day. Summers are short, with mild to warm temperatures (Figure 6-26, right), and the sun typically shines 19 hours a day.

Most boreal forests are dominated by a few species of evergreen conifer trees such as spruce, fir, cedar, hemlock, and pine. The small, needle-shaped, waxy-coated leaves of these trees can withstand the intense cold and drought of winter when snow blankets the ground. Plant diversity is low in these forests because few species can survive the winters when soil moisture is frozen.

Beneath the stands of trees, there is a deep layer of partially decomposed conifer needles and leaf litter. Decomposition is slow because of the low temperatures, the waxy coating of conifer needles, and the high acidity. As the conifer needles decompose, they make the thin, nutrient-poor soil acidic and prevent most other plants (except certain shrubs) from growing on the forest floor.

These biomes contain a variety of wildlife (Figure 6-30). The animal populations of these forests consist mostly of seed eaters (such as squirrels and nutcrackers), insect herbivores, and larger browsers (such as elk and moose). Predators include wolves, wolverines, grizzly bears, and black bears. In North America, caribou migrate from the tundra to this biome in winter.

During the brief summer the soil becomes waterlogged, forming acidic bogs, or *muskegs*, in low-lying areas of these forests. Warblers and other insect-eating birds feed on hordes of flies, mosquitoes, and caterpillars.

What Are Temperate Rain Forests? *Coastal coniferous forests* or *temperate rain forests* are found in scattered

Figure 6-29 Some components and interactions in a temperate deciduous forest ecosystem. When these organisms die, decomposers break down their organic matter into minerals used by plants. Transfers of matter and energy between producers, primary consumers (herbivores), and secondary (or higher-level) consumers (carnivores) are indicated by colored arrows. Organisms are not drawn to scale.

Producer to primary consumer

Primary to secondary consumer

Secondary to higher-level consumer

All producers and consumers to decomposers

Producer to primary consumer	Primary to secondary consumer	Secondary to higher-level consumer	All producers and consumers to decomposers

Figure 6-30 Some components and interactions in an evergreen coniferous (boreal or taiga) forest ecosystem. When these organisms die, decomposers break down their organic matter into minerals used by plants. Transfers of matter and energy between producers, primary consumers (herbivores), and secondary (or higher-level) consumers (carnivores) are indicated by colored arrows. Organisms are not drawn to scale.

Figure 6-31 Tree farm in North Carolina. Some diverse virgin (old-growth) or second-growth forests are cleared and replanted with a single tree species (monoculture), often for harvest as Christmas trees, timber, or wood converted to pulp to make paper. (Gene Alexander/USDA)

coastal temperate areas with ample rainfall or moisture from dense ocean fogs. Along the coast of North America, from Canada to northern California, undisturbed areas of biomes are dominated by dense stands of large conifers such as Sitka spruce, Douglas fir, and redwoods.

The ocean moderates the temperature so that winters are mild and summers are cool. The trees in these moist forests depend on frequent rains and moisture from summer fog that rolls in off the Pacific.

What Impacts Do Human Activities Have on Forest Ecosystems? Some major human impacts on forests are as follows:

- Clearing and degradation of tropical rain forests and tropical seasonal forests for timber, grazing land, and agriculture, leading to severe erosion of their already nutrient-poor soils. Within 50 years, only scattered fragments of these diverse forests might remain.

- Clearing of *temperate deciduous forests* in Europe, Asia, and North America, mostly for timber, cropland, and urban development. In North America, about 99.9% of the original stands of temperate deciduous forests have been cleared for such purposes. Many have grown back as less diverse secondary forests. Some have been converted to managed *tree farms* or *tree plantations*, where a single species is grown for timber, pulpwood, or Christmas trees (Figure 6-31).

- Clearing of large areas of *evergreen coniferous forests* by loggers in North America, Finland, Sweden, and Canada. Within a decade most of the vast boreal forests of Siberia and Russia may disappear because of logging, mining, and flooding by the reservoirs created for hydroelectric projects. Because trees grow slowly in the cold northern climate, these forests can take a long time to recover from disturbance.

6-6 MOUNTAIN BIOMES

Why Are Mountains Ecologically Important? Some of the world's most spectacular and important environments are mountains, which make up about 20% of the earth's land surface. Mountains are places where dramatic changes in altitude, climate, soil, and vegetation take place over a very short distance (Figure 6-18). It is estimated that each 100-meter (300-foot) gain in elevation on a mountainside is roughly equivalent to a 100-kilometer (62-mile) change in latitude. Above a certain altitude, known as the *snow line*, temperatures are so cold that the mountain is almost permanently covered by snow and ice (except in places too steep for snow to cling to).

Because of the steep slopes, mountain soils are especially prone to erosion when the vegetation holding them in place is removed by natural disturbances (such as landslides and avalanches) or human activities.

Many freestanding mountains are *islands of biodiversity* surrounded by a sea of lower-elevation landscapes transformed by human activities. As a result, many mountain areas contain endemic species found nowhere else on earth; they are also sanctuaries for animal species driven from lowland areas.

The ice and snow of mountaintops help regulate the earth's climate by reflecting solar radiation back into space. Sea levels depend on the melting of glacial ice, most of which is locked up in Antarctica, the most mountainous of all continents.

Mountain regions also contain the majority of the world's forests, which contain much of the world's biodiversity. Mountains also play a critical role in the hydrologic cycle (Figure 4-28, p. 90) by gradually releasing melting ice, snow, and water stored in the soils and vegetation of mountainsides to small streams. These streams flow into larger streams (rivers) that in turn empty into the ocean to begin the cycle again.

What Impacts Do Human Activities Have on Mountain Ecosystems? Despite their ecological, economic, and cultural importance, the fate of mountain ecosystems has not been a high priority of governments or many environmental organizations. Mountain ecosystems are coming under increasing pressure from several major trends:

- Rapidly increasing population, especially in developing countries. This is forcing landless poor people, refugees, and minority populations to migrate uphill and try to survive on less stable soils. These newcomers often use mountain soils and forests unsustainably in a desperate struggle to survive or because of a lack of knowledge about how to grow food, raise livestock, and harvest wood in these habitats.

- Increased commercial extraction of timber and mineral resources. People often do this with little regard for the resulting environmental consequences such as **(1)** excessive soil erosion, **(2)** flooding in the valleys below, **(3)** loss of wildlife habitat, and **(4)** air and water pollution from mining activities.

- A growing number of hydroelectric dams and reservoirs. Rivers in mountains are attractive sites for dams and reservoirs because the elevation and slope of mountains increase the force of flowing water. These reservoirs flood mountain slopes, and the dams alter the types and abundance of species in rivers.

- Environmental degradation from the global boom in skiing, trekking, and other forms of recreation and tourism in mountainous areas.

- Increased air pollution from growing urban and industrial centers and increased use of automobiles. Trees and other vegetation at high elevations (especially conifers such as spruce) are bathed year-round in air pollutants such as ozone and acidic compounds, carried there by prevailing winds from cities, factories, and coal-burning power plants.

- Changes in climate and levels of UV radiation brought about by human activities. If global warming occurs as projected during the next century, many species of mountain plants could be displaced by lowland species that have higher rates of growth and reproduction. Similarly, any increase in UV radiation brought about by depletion of the ozone layer may have a pronounced effect on mountain life, which is already exposed to high levels of UV radiation because of altitude.

6-7 LESSONS FROM GEOGRAPHIC ECOLOGY

Ecology involves understanding connections in time and space. In this chapter we examined the connections between weather, climate, and the distribution of the earth's biomes. Three general ecological lessons emerge from this study:

- Different climates occur as a result of currents of air and water flowing over an unevenly heated planet spinning on a tilted axis.

- Different climates result in different communities of organisms, or biomes.

- Everything is connected.

Several more particular lessons are as follows:

- A general climate map for the earth (Figure 6-4) can be drawn based on broad patterns of temperature and precipitation. The distribution of temperature and precipitation results from the interplay between incoming solar energy and the earth's orbit (Figure 6-5) and spin (Figure 6-6), which in turn lead to large-scale patterns of air circulation (Figure 6-8) and ocean currents (Figure 6-4).

- A general biome map for the earth (Figure 6-16) can be drawn that largely follows the general climate map (Figure 6-4). This is because temperature and precipitation tolerances are major factors in determining the ecological niches of biological organisms. Because of similar climate, species within biomes in very different parts of the world have similar ecological niches and display similar adaptations.

- Understanding the general characteristics of each biome leads to a general understanding of **(1)** the range of biodiversity on the earth, **(2)** how this biodiversity is distributed, and **(3)** how all biomes are connected through global climate patterns, energy flow and chemical cycling (Figure 4-7, p. 75, and Figure 4-16, p. 82).

Scientists have made a good start in understanding important aspects of geographic ecology. However, to further our understanding of ecological connections in time and space we need more studies and accurate maps of **(1)** species distributions, **(2)** changes in climate, **(3)** changes in species distribution, and **(4)** how and where human activities are affecting climate, biomes, and species distribution. Obtaining and analyzing such information is vital in helping us learn how to live more sustainably.

Threats to the global climate system and possible solutions are discussed in Chapter 18. Threats to biodiversity and the sustainability of terrestrial and aquatic ecosystems and possible solutions to these problems are discussed in Chapters 22, 23, and 24.

When we try to pick out anything by itself, we find it hitched to everything else in the universe.

JOHN MUIR

REVIEW QUESTIONS

1. Define the boldfaced terms in this chapter.

2. How does wind affect climate, global distribution of nutrients, and pollution?

3. What is *weather*?

4. Distinguish between **(a)** a *warm front* and a *cold front*, **(b)** a *high-pressure* and a *low-pressure* air mass, **(c)** a *tornado* and a *tropical cyclone*, and **(d)** a *hurricane* and a *typhoon*.

5. What is *climate*? What two main factors determine a region's climate?

6. What five factors affect global air circulation? How does each factor affect global air circulation? What causes opposite seasons in the northern and southern hemispheres?

7. How do oceans affect regional climates? What would happen to the climate of northwestern Europe if the Gulf Stream did not exist? What is an *upwelling,* and why are upwellings important to life?

8. What is the *El Niño-Southern Oscillation (ENSO)?* How does it affect ocean life and weather in various parts of the world? What is *La Niña,* and how does it affect weather in various parts of the world?

9. What is the *greenhouse effect,* and how does it affect the earth's climate? What would life on the earth be like without the natural greenhouse effect? Why is there concern over human activities enhancing the natural greenhouse effect?

10. What is the *ozone layer,* and how does it help protect life on the earth and affect the earth's climate?

11. What are *microclimates?* What is the *rain shadow effect,* and how can it affect the microclimate on each side of high mountains? How can vegetation affect microclimates? Why do urban areas have different microclimates than surrounding areas, and how do these microclimates differ from those of surrounding areas? How do *land breezes* and *sea breezes* affect the microclimates of coastal areas?

12. What is a *biome?* What limiting factor usually determines whether a land area is a desert, grassland, or forest? What two factors determine whether an area of the earth's surface is a tropical, temperate, or polar desert, grassland, or forest? What is an *ecotone?* How do climate and vegetation vary with latitude and altitude?

13. Why are there no trees in the Arctic, and why are the plants there so close to the ground? Why are there no leaves on desert plants such as cacti? Why do trees in most forests found in both the warm tropics and in cold areas such as Canada and Sweden keep their leaves year-round, whereas most trees in temperate forests lose their leaves in winter? What are *succulent plants,* where are they found, and how do they survive? What are *broadleaf evergreen plants,* where are they found, and how do they survive? What are *broadleaf deciduous plants,* where are they found, and how do they survive? What are *coniferous evergreen plants,* where are they found, and how do they survive?

14. What is a *desert?* What are the three major types of desert, and how do they differ in climate and biological makeup? How do desert plants and animals survive heat and a lack of water? Why are desert ecosystems vulnerable to disruption, and what five types of human activities have harmful impacts on deserts?

15. What is *grassland?* What are the three major types of grassland, and how do they differ in climate and biological makeup? Distinguish between *arctic tundra* and *alpine tundra.* Why are grasslands vulnerable to disruption, and what four types of human activities have harmful impacts on grasslands? What is *chaparral,* and what is the importance of fire in this biome?

16. What is a *forest?* What are the three major types of forest, and how do they differ in climate and biological makeup? Distinguish between *tropical rain forests, tropical deciduous forests,* and *temperate rain forests.* What three types of human activities have harmful impacts on forests?

17. Why are mountains ecologically important, and what factors make them vulnerable to ecological disruption? What six types of human activities have harmful impacts on mountains?

18. List three general lessons and three specific lessons that we can learn from geographic ecology.

CRITICAL THINKING

1. Why might **(a)** the microclimate of a north-facing slope differ from that of a south-facing slope, **(b)** a ridgetop differ from that of a valley bottom, and **(c)** an isolated mountaintop differ from a mountaintop surrounded by other mountains?

2. List a limiting factor for each of the following ecosystems: **(a)** a desert, **(b)** arctic tundra, **(c)** alpine tundra, **(d)** the floor of a tropical rain forest, and **(e)** a temperate deciduous forest.

3. Why do deserts and arctic tundra support a much smaller biomass of animals than do tropical forests?

4. Well drilling for water in desert areas has allowed many traditional nomadic tribes to raise more livestock by not having to migrate. Is this desirable or undesirable from **(a)** an ecological standpoint, **(b)** an economic standpoint, and **(c)** a cultural standpoint? Explain.

5. Why do you think there are no amphibians and reptiles in arctic tundra?

6. Suppose that global warming shifts the climate of the temperate grasslands (which now provide most of the world's food) northward to temperate evergreen forest biomes. If we were to clear these forests and plant wheat, explain why wheat might not grow very well despite the favorable climate.

7. Some biologists have suggested restoring large herds of bison (p. 23) on public lands in the North American plains as a way of restoring remaining tracts of tallgrass prairie. This idea has been strongly opposed by ranchers with permits to graze cattle and sheep on federally managed lands. What do you think about the idea of restoring large numbers of bison to the plains of North America?

8. Why do most animals in a tropical rain forest live in its trees?

9. What factors in your lifestyle contribute to the destruction and degradation of tropical forests?

10. What biomes are best suited for **(a)** raising crops and **(b)** grazing livestock?

11. Compare the general types of vegetation and species found in **(a)** temperate deserts (Figure 6-20, p. 134) and arctic tundra (Figure 6-24, p. 140), **(b)** temperate grasslands (Figure 6-23, p. 139) and temperate deciduous forests (Figure 6-29, p. 146), **(c)** arctic tundra (Figure 6-24, p. 140) and evergreen coniferous forests (Figure 6-30, p. 147), and **(d)** tropical rain forests (Figure 6-27, p. 143) and evergreen coniferous forests (Figure 6-30, p. 147).

12. How might technologies such as remote sensing and geographic information systems (GIS; Figure 4-34, p. 98) help our understanding of ecology and our environment? Both remote sensing and GIS involve expensive technology. Do you think it is worth investing in such technologies to help solve environmental problems, or are there better ways we could spend our money? Explain.

PROJECTS

1. How has the climate changed in the area where you live during the past 50 years? Investigate the beneficial and harmful effects of these changes. How have these changes benefited or harmed you personally?

2. What type of biome do you live in or near? What effects have human activities over the past 50 years had on the characteristic vegetation and animal life normally found in the biome you live in? How is your own lifestyle affecting this biome?

3. Visit the ecotone between the biome in which you live and one of the adjacent biomes. What does the ecotone look like and how broad is it?

4. Go to the library and peruse the section devoted to geography and cartography. Locate vegetation and climate maps. How were these maps produced? How do the ways in which they were produced differ?

5. Use the library or the internet to find bibliographic information about *Robert H. MacArthur* and *John Muir*, whose quotes appear at the beginning and end of this chapter.

6. Make a concept map of this chapter's major ideas, using the section heads and subheads and the key terms (in boldface). Look at the inside back cover and on the website for this book for information about making concept maps.

INTERNET STUDY RESOURCES AND RESOURCES FOR FURTHER READING AND RESEARCH

The website for this book contains helpful study aids and many ideas for further reading and research. Log on to:

http://www.brookscole.com/product/0534376975s

and click on the Chapter-by-Chapter area. Choose Chapter 6 and select a resource:

- "Flash Cards" allows you to test your mastery of the Terms and Concepts to Remember for this chapter.

- "Tutorial Quizzes" provides a multiple-choice practice quiz.

- "Student Guide to InfoTrac" will lead you to Critical Thinking Projects that use InfoTrac College Edition as a research tool.

- "References" lists the major books and articles consulted in writing this chapter.

- "Hypercontents" takes you to an extensive list of sites with news, research, and images related to individual sections of the chapter.

INFOTRAC COLLEGE EDITION

Improve your skills with InfoTrac College Edition, a searchable online database of articles from more than 700 periodicals. Log on to:

http://www.infotrac-college.com

or access InfoTrac through the website for this book.

Try the following articles:

Galtie, A. 1999. Is El Niño now a man-made phenomenon? *The Ecologist* vol. 29, no. 2, pp. 64–67. (subject guide: el niño, environmental aspects)

Middleton, N. 2000. Shifting sands. *Geographical* vol. 72, no. 4, pp. 24–30. (subject guide: desert ecology)

7 AQUATIC ECOLOGY: BIODIVERSITY IN AQUATIC SYSTEMS*

Why Should We Care About Coral Reefs?

In the shallow coastal zones of warm tropical and subtropical oceans we often find **coral reefs** (Figure 7-1, left). These beautiful natural wonders are among the world's oldest, most diverse, and most productive ecosystems.

Coral reefs are formed by massive colonies of tiny animals called *polyps* that are close relatives of jellyfish. They slowly build reefs by secreting a protective crust of limestone (calcium carbonate) around their soft bodies. When they die their empty crusts or outer skeletons remain as a platform for more reef growth. The result is an elaborate network of crevices, ledges, and holes that serve as calcium carbonate "condominiums" for a variety of marine animals.

Coral reefs involve a mutually beneficial relationship between the polyps and tiny single-celled algae called *zooxanthellae* ("zoh-ZAN-thel-ee") that live in the tissues of the polyps. The algae provide the polyp with color, food, and oxygen through photosynthesis. The polyps in turn provide a well-protected home for the algae.

Coral reefs provide a number of important ecological and economic services, including the following:

- Removing some of carbon dioxide from the atmosphere as part of the carbon cycle (when coral polyps form limestone shells).

- Acting as natural barriers that **(a)** help protect 15% of the world's coastlines from erosion by battering waves and storms and **(b)** allow the ocean to replenish beaches with sand.

- Supporting at least one-fourth of all identified marine species and 65% of marine fish species even though such reefs occupy less than 1% of the ocean floor.
- Providing fish and shellfish, jobs, and building materials for some of the world's poorest countries.
- Supporting fishing and tourism industries worth billions of dollars each year. For example, tourism generates about $1 billion per year at Australia's Great Barrier Reef and about $1.7 billion a year at Florida's reefs.
- Giving us an underwater world to study and enjoy.

According to a 1999 study by the World Resources Institute, nearly 60% of the world's coral reefs are threatened by human activities such as coastal development, overfishing, pollution, and warmer ocean temperatures.

One problem is *coral bleaching* (Figure 7-1, top), which occurs when a coral becomes stressed and expels most of its colorful algae. This occurs because of stresses such as increased water temperature and runoff of silt that covers the coral and prevents photosynthesis. This loss of algae exposes the colorless coral animals and the underlying ghostly white skeleton of calcium carbonate. Unable to grow or repair themselves, the corals eventually die unless the stress is removed and they are recolonized by algae.

Coral reefs sometimes are called the aquatic equivalent of tropical rain forests because they harbor such a high species biodiversity with myriad ecological interrelationships. These oceanic sentinels are warning us about the health of their habitats.

*Paul M. Rich, associate professor of ecology and evolutionary biology and environmental studies at the University of Kansas, is coauthor of this chapter.

Figure 7-1 A healthy coral reef in the Philippines covered by colorful algae (bottom) and a bleached coral reef in the Bahamas that has lost most of its algae (top) because of changes in the environment (such as cloudy water or too warm temperatures). With the algae gone, the white limestone of the coral skeleton becomes visible. If the environmental stress is not removed, the corals die, and recovery may take 100 to 1,000 years. These diverse and productive ecosystems are being damaged and destroyed at an alarming rate. (Top, ©Robert Wicklund; bottom, Karl & Jill Wallin/FPG International)

If there is magic on this planet, it is contained in water.

LOREN EISLEY

This chapter addresses the following questions:

- What are the basic types of aquatic life zones, and what factors influence the kinds of life they contain?
- What are the major types of saltwater life zones, and how do human activities affect them?
- What are the major types of freshwater life zones, and how do human activities affect them?
- How can we help sustain aquatic life zones?

7-1 AQUATIC ENVIRONMENTS: TYPES, COMPONENTS, AND LIMITING FACTORS

What Are the Two Major Types of Aquatic Life Zones? The aquatic equivalents of biomes are called *aquatic life zones*. The major types of organisms found in aquatic environments are determined by the water's *salinity* (the amounts of various salts such as sodium chloride [NaCl] dissolved in a given volume of water). As a result, aquatic life zones are divided into two major types: **(1)** *saltwater* or *marine* (particularly estuaries, coastlines, coral reefs, coastal marshes, mangrove swamps, the ocean above the continental shelf, and the deep ocean) and **(2)** *freshwater* (particularly lakes and ponds, streams and rivers, and inland wetlands). Figure 7-2 shows the distribution of the world's major oceans, lakes, coral reefs, rivers, and mangroves.

What Are the Main Kinds of Organisms in Aquatic Life Zones? Saltwater and freshwater life zones contain several major types of organisms: **(1)** weakly swimming, free-floating **plankton**, **(2)** strongly swimming consumers (**nekton**) such as fish, turtles, and whales, **(3)** bottom-dwellers (**benthos**) such as barnacles and oysters that anchor themselves to one spot, worms that burrow into the sand or mud, and lobsters and crabs that walk about on the bottom, and **(4) decomposers** (mostly bacteria) that break down the organic compounds in the dead bodies and wastes of aquatic organisms into simple nutrient compounds for use by producers. Plankton are divided into three categories:

- *Phytoplankton* ("FIE-toe-plank-ton"), or *plant plankton*: free-floating, microscopic cyanobacteria and many types of algae that are the producers supporting most aquatic food chains and food webs (Figure 4-19, p. 84).

- *Nanoplankton* ("NAN-oh-plank-ton"): smaller recently discovered and poorly understood producers.

- *Zooplankton* ("ZOE-oh-plank-ton"), or *animal plankton*: a mixture of nonphotosynthetic primary consumers (herbivores) that feed on phytoplankton and secondary consumers that feed on other zooplankton. They range from single-celled protozoa to large invertebrates such as jellyfish.

What Are Some Major Differences Between Life on Land and Life in Water? Research has revealed a number of differences in the nature and diversity of aquatic and terrestrial life.

Physical and Chemical Characteristics

- Water provides buoyancy. This **(1)** provides physical support, **(2)** reduces the need for large supporting structures such as legs and trunks, and **(3)** helps aquatic animals move verically.

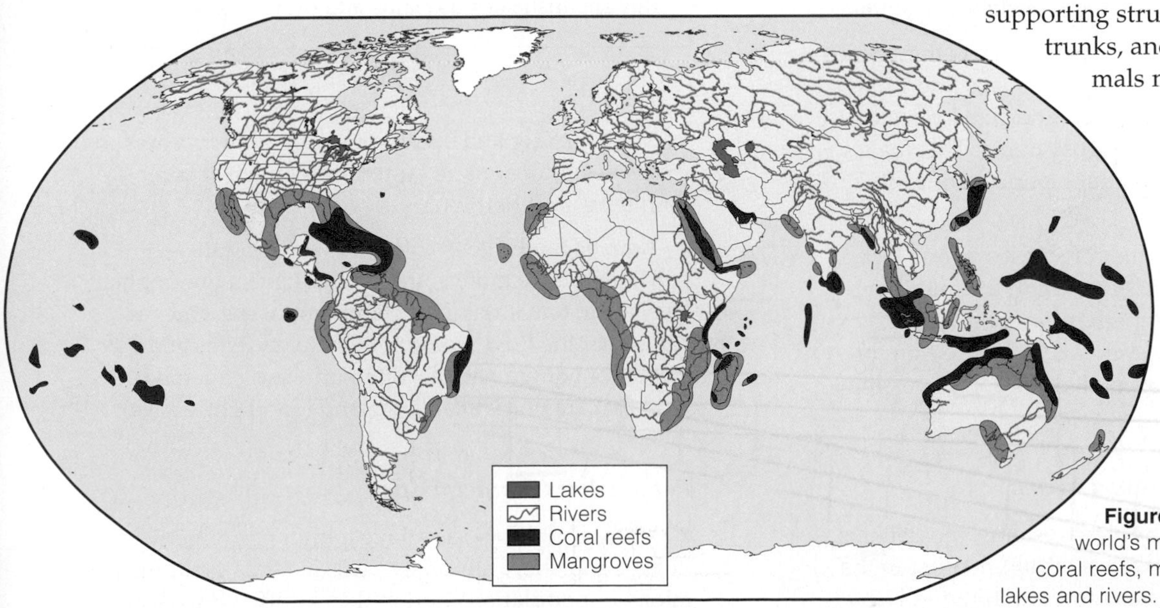

Figure 7-2 Distribution of the world's major saltwater oceans, coral reefs, mangroves, and freshwater lakes and rivers.

Lakes
Rivers
Coral reefs
Mangroves

- Water acts as a thermostat that keeps aquatic life from drying out or becoming too hot or too cold. However, because of this property most aquatic organisms have evolved the ability to tolerate only a narrow range of temperatures (Figure 4-14, p. 79).

- The great dissolving power of water allows aquatic plants to get their nourishment directly from the water surrounding them instead of from a combination of air and soil.

- Water provides a connected and circulating medium that disperses organisms and their larvae and eggs to new habitats more readily than occurs for most terrestrial organisms (except those dispersed by wind).

- Because of the ready availability of water, aquatic species have not had to evolve ways of getting and conserving water.

- Because water filters out ultraviolet radiation (except near the surface), aquatic organisms are less susceptible to harm from UV radiation than most surface-dwelling terrestrial organisms.

- Water dilutes and disperses most potentially toxic pollutants and metabolic wastes secreted by aquatic organisms or added by human activities.

- Dissolved pollutants tend to concentrate in the microlayer on the surface and in bottom sediments of oceans and lakes instead of in air.

Species and Habitat Diversity

- The expanded third dimension (depth) of water supports a variety of organisms not found on land.

- Most aquatic systems (except perhaps coral reefs and the ocean bottom) have a smaller number of distinctly different habitats than terrestrial ecosystems.

- Because of the fluid nature of water, communities in aquatic systems tend to shift in space and time and thus have less pronounced and fixed physical boundaries than terrestrial ecosystems. This makes it difficult to count and manage populations of aquatic organisms.

- Endemism, the restriction of species to one location or habitat, is much less common in water habitats than on land, except perhaps among some bottom-dwelling aquatic animals. A key reason for this is that water provides a medium for dispersing eggs and larvae.

Trophic Structure and Food Webs

- In the open water of aquatic systems, most plants are microscopic plants (*phytoplankton*) floating in the water instead of larger plants rooted in the soil.

Biofiltration

SPOTLIGHT

Some aquatic organisms get food by filtering it out of water. Examples of such *filter feeders* are barnacles, clams, oysters, sponges, and baleen whales.

Some species do this by passively sieving the water and others by actively pumping it through their bodies. One of the most efficient biofilter species is the sponge, which is capable of pumping an amount of water equal to its own body volume every 10-20 seconds and filtering out 99% of the particulate matter.

While feeding themselves, sponges also play an important role in keeping water over coral reefs clean and clear. It is estimated that most of the water overlying a coral reef could be filtered through its existing sponges in 2-3 days.

Some shellfish such as clams use their muscular foot to burrow down into the sand or mud and then extend input and output tubes, called siphons, up into the water. Water is "inhaled" through an incoming siphon, filtered for food particles and dissolved oxygen, and then "exhaled" through an outgoing siphon.

As huge baleen whales (such as the blue whale) move through the water, zooplankton are trapped in large comblike filters, called *baleen*, found in their mouths. The whales then use their tongues to slurp down the trapped zooplankton.

There are no equivalent types of filter-feeding organisms in terrestrial ecosystems, except that humans have copied nature by developing fishnets to strain food from the water.

Critical Thinking

Why do some health scientists warn us not to eat raw shellfish such as clams and oysters?

- Most floating and swimming animal herbivores (*zooplankton*) in aquatic systems are much smaller than terrestrial herbivores.

- Aquatic food chains and food webs (Figure 4-19, p. 84) tend to be more complex and have more trophic levels than terrestrial food chains and webs. One reason is that the fluid medium of water systems and the variety of bottom habitats open up ways of getting food that are not available on land (Spotlight, above).

Population Characteristics

- Many aquatic species have a high reproductive output and short life cycles. This leads to great fluctuations in populations and makes it difficult to distin-

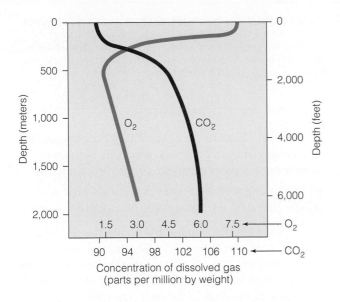

Figure 7-3 Variations in concentrations of dissolved oxygen (O_2) and carbon dioxide (CO_2) in parts per million (ppm) with water depth. Dissolved O_2 is high near the surface because oxygen-producing photosynthesis takes place there. Because photosynthesis cannot take place below the sunlit layer, O_2 levels fall because of aerobic respiration by aquatic animals and decomposers. In contrast, levels of dissolved CO_2 are **(1)** low in surface layers because producers use CO_2 during photosynthesis and **(2)** high in deeper, dark layers where aquatic animals and decomposers produce CO_2 through aerobic respiration.

guish between natural and human-related causes of population declines.

- Because they are dispersed by water, many aquatic species are separated from their parents earlier in life than some terrestrial animals that raise and tend their young.

Monitoring and Protection

- It is more difficult to monitor and study aquatic ecosystems (especially marine systems) than most terrestrial systems because of their size and because they are largely hidden from view.

- Greater uncertainty about the structure and behavior of aquatic systems than about many terrestrial systems means that preventing pollution and environmental degradation is the most effective way to protect aquatic biodiversity.

What Factors Limit Life at Different Depths in Aquatic Life Zones? Most aquatic life zones can be divided into three layers: **(1)** surface, **(2)** middle, and **(3)** bottom. Important environmental factors determining the types and numbers of organisms found in these layers are **(1)** *temperature,* **(2)** *access to sunlight for photosynthesis,* **(3)** *dissolved oxygen content,* and **(4)** *availability of nutrients* such as carbon (as dissolved CO_2 gas), nitrogen (as NO_3^-), and phosphorus (mostly as PO_4^{3-}) for producers. Photosynthesis is confined mostly to the upper layer, or **euphotic zone,** of deep aquatic systems because it is filtered out by suspended algal cells and suspended particles of clay and silt in nearshore waters. The depth of the euphotic zone in oceans and deep lakes can be reduced by excessive algal growth (algal blooms) that make water cloudy.

O_2 enters an aquatic system from the atmosphere and through photosynthesis by aquatic producers and

is removed by aerobic respiration of producers, consumers, and decomposers. CO_2 enters an aquatic system from the atmosphere and through aerobic respiration by producers, consumers, and decomposers and is removed by photosynthesizing producers. This removal of the greenhouse gas CO_2 from the atmosphere by aquatic producers helps keep the earth's average atmospheric temperature from rising as a result of the natural greenhouse effect (Figure 6-13, p. 128).

Some dissolved CO_2 forms carbonate ions (CO_3^{2-}). They are stored mostly as calcium carbonate ($CaCO_3$) for long periods in sediments, minerals, and the shells and skeletons of living aquatic animals as part of the carbon cycle (Figure 4-29, p. 92). The concentration of oxygen in the atmosphere varies little. However, the amount of oxygen dissolved in water can vary widely, depending on factors such as **(1)** temperature, **(2)** number of producers (which add O_2), and **(3)** number of consumers and aerobic decomposers (which remove O_2).

Many aquatic organisms, especially fish, die when dissolved oxygen levels fall below 5 ppm. The concentrations of dissolved O_2 and CO_2 in water vary in different ways with depth (Figure 7-3) because of differences in the rates of **(1)** photosynthesis (which produces O_2 and consumes CO_2) and **(2)** aerobic respiration (which consumes oxygen and produces CO_2).

In shallow waters in streams, ponds, and oceans, there are usually ample supplies of nutrients for primary producers. By contrast, in the open ocean nitrates, phosphates, iron, and other nutrients often are in short supply and limit net primary productivity (Figure 4-25, p. 88). However, net primary productivity is much higher in parts of the open ocean where upwellings (Figure 6-4, p. 124, and Figure 6-9, p. 126) bring such nutrients from the ocean bottom to the surface for use by producers.

7-2 SALTWATER LIFE ZONES

Why Are the Oceans Important? A more accurate name for Earth would be Ocean because saltwater oceans cover about 71% of the planet's surface (Figure 7-4).

Ocean hemisphere   Land–ocean hemisphere

Figure 7-4 The ocean planet. The salty oceans cover about 71% of the earth's surface. About 97% of the earth's water is in the interconnected oceans, which cover 90% of the planet's mostly ocean southern hemisphere (left) and 50% of its land-ocean northern hemisphere (right). The average depth of the world's oceans is 3.8 kilometers (2.4 miles).

Three key roles that oceans play in the survival of almost all life on the earth are as follows:

- Regulating the earth's climate by **(1)** distributing solar heat through ocean currents (Figure 6-4, p. 124), **(2)** evaporating ocean water as part of the global hydrologic cycle (Figure 4-28, p. 90), and **(3)** serving as a giant reservoir for removing carbon dioxide (a greenhouse gas) from the atmosphere.

- Providing habitats for about 250,000 known species of marine plants and animals, which are sources of food for many other organisms (including humans).

- Dispersing and diluting many human-produced wastes that flow into or are dumped into the ocean.

Oceans have two major life zones: the *coastal zone* and the *open sea* (Figure 7-5). The **coastal zone** is the warm, nutrient-rich, shallow water that extends from the high-tide mark on land to the gently sloping, shallow edge of the *continental shelf* (the submerged part of the continents). This zone has numerous interactions with the land and thus is easily affected by human activities.

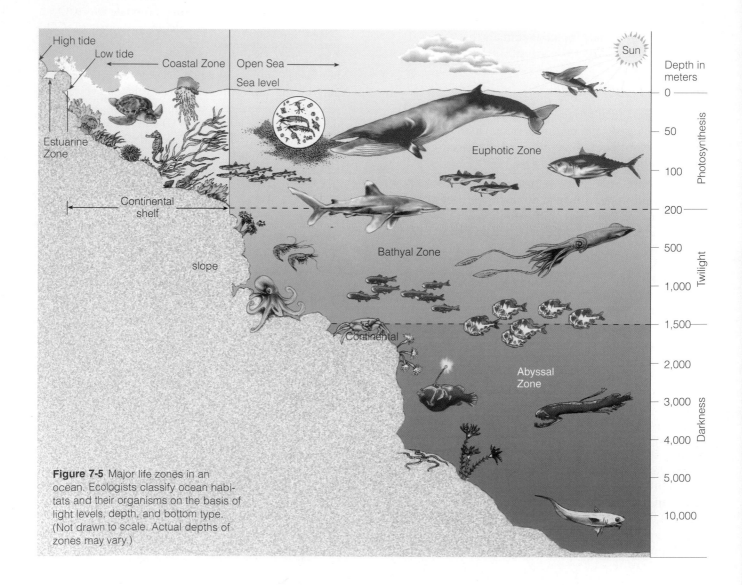

Figure 7-5 Major life zones in an ocean. Ecologists classify ocean habitats and their organisms on the basis of light levels, depth, and bottom type. (Not drawn to scale. Actual depths of zones may vary.)

Although it makes up less than 10% of the ocean's area, the coastal zone contains 90% of all marine species and is the site of most large commercial marine fisheries. Most ecosystems found in the coastal zone have a very high primary productivity (Figure 4-25, p. 88). This occurs because of the zone's ample supplies of sunlight and plant nutrients (flowing from land and distributed by wind and ocean currents).

The coastal zone provides a variety of habitats for marine life because **(1)** its bottom can be sandy, muddy, or rocky, **(2)** shorelines can vary from smooth and sandy beaches to jagged and rocky cliffs, **(3)** coastal wetlands covered with water all or part of the time can vary from dense stands of mangrove trees (Figure 7-6) along shores in tropical areas to grass-dominated saltwater marshes in temperate areas, and **(4)** water depth can vary from a few centimeters to kilometers.

In their desire to live near the coast, people are destroying or degrading the resources that make coastal areas so enjoyable and economically and ecologically valuable. Currently, about 40% of the world's population live along coasts or within 160 kilometers (100 miles) of a coast. By 2030, more than 6.3 billion people—more than the current global population—are expected to live in or near coastal areas.

What Are Estuaries and Coastal Wetlands? One highly productive area in the coastal zone is an **estuary**, a partially enclosed area of coastal water where seawater mixes with fresh water and nutrients from rivers, streams, and runoff from land (Figure 7-7). It is an eco-

Figure 7-6 Mangroves in Columbia, South America. Mangroves (Figure 7-2) are important coastal systems that protect coastlines and provide important habitats for aquatic species. About half of the world's mangroves have been cleared, mostly for growing crops and raising shrimp in aquaculture ponds. (Alan Watson/Forest Light)

tone (Figure 4-10, p. 77) between the marine environment and the land where large volumes of fresh water from land and salty ocean water mix. Estuaries and their associated **coastal wetlands** (land areas covered with water all or part of the year) include **(1)** river mouths, **(2)** inlets, **(3)** bays, **(4)** sounds, **(5)** mangrove forest swamps in tropical waters (Figures 7-2 and 7-6), and **(6)** salt marshes in temperate zones (Figures 7-8 and 7-9). Temperature and salinity levels vary widely in estuaries and coastal wetlands because of **(1)** the daily rhythms of the tides, **(2)** seasonal variations in the flow of fresh water into the estuary, and **(3)** unpredictable flows of fresh water from coastal land and rivers after heavy rains and of saltwater from the ocean as a result of storms, hurricanes, and typhoons.

The constant water movement stirs up the nutrient-rich silt, making it available to producers. This explains why estuaries and their associated coastal wetlands are some of the earth's most productive ecosystems (Figure 4-25, p. 88).

Figure 7-7 View of an estuary taken from space. The photo shows the sediment plume at the mouth of Madagascar's Betsiboka River as it flows through the estuary and into the Mozambique Channel. Because of its topography, heavy rainfall, and the clearing of forests for agriculture, Madagascar is the world's most eroded country. (NASA)

Figure 7-8 Salt marsh of an estuary in a temperate area consists of several connected coastal life zones. (From *Biology: Concepts and Applications*, 4th ed. by Cecie Starr © 2000)

In addition to their high net primary productivity, estuaries and their associated coastal wetlands provide many other ecological and economic services by:

- Acting as breeding grounds and habitats for a variety of waterfowl and other wildlife

- Helping maintain the quality of coastal waters by diluting, filtering, and settling out sediments, excess nutrients, and pollutants

- Serving as popular areas for recreational activities such as birdwatching, nature photography, boating, fishing, and hunting

- Protecting lives and property during floods by absorbing and slowing the flow of water and buffering shores against damage and erosion

What Impacts Do Human Activities Have on Estuaries and Coastal Wetlands? Since 1780, about 53% of the area of estuaries and coastal wetlands in the contiguous United States has been destroyed or damaged. California alone has lost 91% of its original coastal wetlands, but Florida has lost the largest area of such wetlands in the United States. Major causes of such destruction and degradation are (1) dredging, (2) filling in for coastal development, and (3) contamination from industrial and sewage discharges, runoff from land, and airborne pollutants falling on coastal waters. Dam construction and diversion of river water for human consumption and irrigation have also altered these areas. Coastal wetlands and estuaries are particularly vulnerable to toxic contamination because they trap pesticides, heavy metals, and other pollutants and concentrate them to high levels.

Since the mid-1960s, some tropical coastal countries have lost half or more of their mangrove forests (Figures 7-2 and 7-6) because of (1) industrial logging for timber and fuelwood, (2) conversion to ponds for raising fish and shellfish (aquaculture), (3) conversion to rice fields and other agricultural land, and (4) urban development. Since 1960, Southeast Asia has lost half of its mangroves and the Philippines has lost 75%.

What Are Rocky and Sandy Shores? The area of shoreline between low and high tides is called the **inter-tidal zone**. The organisms that live in this stressful zone must be able to avoid being (1) swept away or crushed by waves, (2) immersed during high tides, and (3) left high and dry (and much hotter) at low tides. They must also cope with changing levels of salinity when heavy rains dilute salt water. To deal with such stresses, most intertidal organisms hold onto something, dig in, or hide in protective shells. The fluctuating tides expose different parts of the intertidal zone to different levels of water, sunlight, and air. This leads to a variety of ecological niches found in fairly clear zones or different-colored bands reflecting the colors of dominant species.

Some coasts have steep *rocky shores* pounded by waves. The numerous pools and other niches in the rocks in the intertidal zone of rocky shores contain a great variety of species (Figure 7-10, top).

Other coasts have gently sloping *barrier beaches*, or *sandy shores*, with niches for different marine organisms, including crabs, lugworms, clams, ghost shrimp, sand dollars, and flounder (Figure 7-10, bottom). Most of them are hidden from view and survive by burrowing, digging, and tunneling in the sand. Microalgae that wash in with the tides are the major producers in sandy shore food webs. These sandy beaches and their adjoining coastal wetlands are also home to a variety of shorebirds that feed in specialized niches on crustaceans, insects, and other organisms.

Biodiversity in rocky shore and sandy beaches is reduced by (1) violent storms, hurricanes, and typhoons, (2) inflows of polluted water from the land and the ocean, (3) overharvesting of fish, shellfish, and seaweed, and (4) coastal development. These beaches usually can recover from storm damage because storms do not occur regularly. However, recovery from pollution, overfishing, and development is difficult because such activities are almost continuous and are increasing in many areas.

One or more rows of natural sand dunes on undisturbed barrier beaches (with the sand held in place by the roots of grasses) serve as the first line of defense against the ravages of the sea (Figure 7-11). However, when coastal developers remove the protective dunes or build behind the first set of dunes, storms can flood and even sweep away seaside buildings and severely erode the sandy beaches.

Herring gulls

Peregrine falcon

Snowy egret

Cordgrass

Short-billed dowitcher

Marsh periwinkle

Phytoplankton

Smelt

Zooplankton and small crustaceans

Soft-shelled clam

Bacteria

Clamworm

Producer to primary consumer

Primary to secondary consumer

Secondary to higher-level consumer

All producers and consumers to decomposers

Figure 7-9 Some components and interactions in a salt marsh ecosystem in a temperate areas such as the United States. When these organisms die, decomposers break down their organic matter into minerals used by plants. Transfers of matter and energy between consumers (herbivores) and secondary (or higher-level) consumers (carnivores) are indicated by colored arrows. Organisms are not drawn to scale.

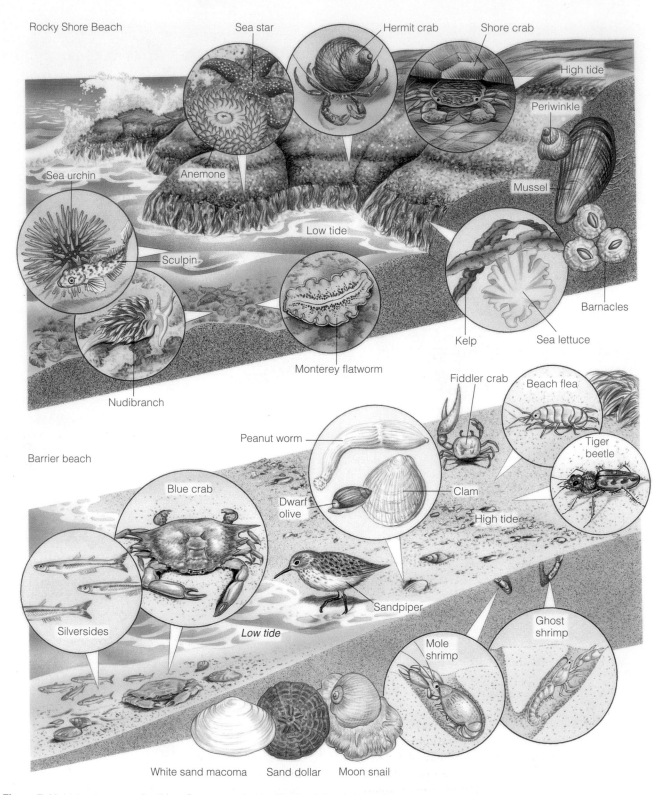

Figure 7-10 Living between the tides. Some organisms with specialized niches found in various zones on rocky shore beaches (top) and barrier or sandy beaches (bottom). Organisms are not drawn to scale.

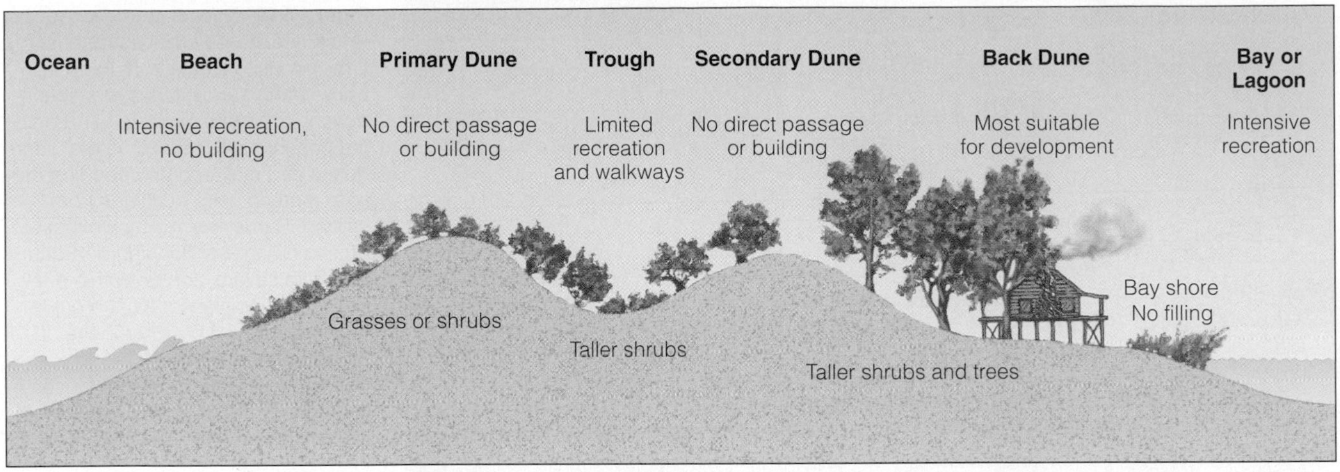

Ocean
Beach
Primary Dune
Trough
Secondary Dune
Back Dune
Bay or Lagoon
Intensive recreation, no building
No direct passage or building
Limited recreation and walkways
No direct passage or building
Most suitable for development
Intensive recreation
Grasses or shrubs
Taller shrubs
Taller shrubs and trees
Bay shore No filling

Figure 7-11 Primary and secondary dunes on gently sloping sandy beaches play an important role in protecting the land from erosion by the sea. The roots of various grasses that colonize the dunes help hold the sand in place. Ideally, construction and development should be allowed only behind the second strip of dunes; walkways to the beach should be built over the dunes to keep them intact. This not only helps preserve barrier beaches but also protects structures from being damaged and washed away by wind, high tides, beach erosion, and flooding from storm surges. This type of protection is rare, however, because the short-term economic value of oceanfront land is considered to be much higher than its long-term ecological and economic values.

What Are Barrier Islands? **Barrier islands** are long, thin, low offshore islands of sediment that generally run parallel to the shore. They are found along some coasts such as most of North America's Atlantic and Gulf coasts. These islands help protect the mainland, estuaries, and coastal wetlands by dispersing the energy of approaching storm waves. Their low-lying beaches are constantly shifting, with gentle waves building them up and storms flattening and eroding them. Currents running parallel to the beaches constantly take sand from one area and deposit it in another. Sooner or later, many of the structures humans build on low-lying barrier islands, such as Atlantic City, Miami Beach, and Ocean City, Maryland (Figure 7-12), are damaged or destroyed by flooding, severe beach erosion, or major storms (including hurricanes).

What Are Coral Reefs? Coral reefs (Figure 7-1) form in clear, warm coastal waters of the tropics and subtropics (Figure 7-2). After tropical rain forests, coral reefs are world's most biologically diverse life zones. Coral reefs are also ecologically complex in terms of the many interactions among the diverse organisms that live there (Figure 7-13). These organisms fall into three main groups: **(1)** attached organisms (such as corals, algae, and sponges) that give the reef its structure, **(2)** fishes, and **(3)** small organisms that bore into, attach to, or hide within a reef's many nooks and crannies.

Coral reefs are vulnerable to damage because they **(1)** grow slowly, **(2)** are disrupted easily, and **(3)** thrive only in clear, warm, and fairly shallow water of constant high salinity. Corals can live only in water with a temperature of 18–30°C (64–86°F), and coral bleaching (Figure 7-1, top) can be triggered by an increase of just 1°C (1.8°F) above the maximum.

Thus, the health and survival of coral reefs are closely connected to projected global warming caused by increases in atmospheric concentrations of gases such as CO_2. Rising temperatures in tropical oceans also can reduce the levels of calcium needed for reef growth. According to a 1999 study by Joan Kleypas, the doubling of current atmospheric levels of CO_2 expected by 2065 could lead to a 30% drop in the amount of calcium that tropical oceans can hold.

The biodiversity of coral reefs can be reduced by natural disturbances such as severe storms, freshwater floods, and invasions of predatory fish. Throughout their very long geologic history, coral reefs have been able to adapt to such natural environmental changes. However, scientists are concerned that these and other changes such as increases in the surface temperature of ocean waters are occurring so rapidly (over decades) and over such a wide area that many of the world's coral systems may not have enough time to adapt.

Today the biggest threats to the biodiversity of many of the world's coral reefs come from human activities. These include the following:

- Deposition of eroded soil produced by land development, deforestation, agriculture, mining, and

Figure 7-12 A developed barrier island: Ocean City, Maryland, host to 8 million visitors a year. To keep up with shifting sands, taxpayers spend millions of dollars to pump sand onto the beaches and rebuild natural sand dunes and may end up spending millions more to keep buildings from sinking. Barrier islands lack effective protection against flooding and damage from severe storms. Within a few hours, barrier islands may be cut in two or destroyed by a hurricane. If global warming raises average sea levels, as projected, most of these valuable pieces of real estate will be under water. (G. H. Demetrakas/O. C. Camera)

dredging along increasingly populated coastlines. Suspended soil sediment washing downriver to the sea or eroding from coastal areas smothers coral polyps or blocks their sunlight.

- Runoff of fertilizers that promote algae growth in ocean water, which reduces sunlight the polyps need. Reefs starved for sunlight can then be colonized by seaweeds. There is also evidence that nutrient-polluted water can enhance the destruction of reefs by diseases. An example is black band, cyanobacteria that secrete highly toxic sulfides, which kill the coral.

- Destruction of mangroves (Figures 7-2 and 7-6), which filter silt and pollution and foster coral growth.

- Coral reef bleaching (Figure 7-1, top), caused mainly by silting and increased ocean temperatures from a combination of strong El Niños (Figure 6-12, p. 128) and enhanced global warming (Chapter 18, p. 446). In 1998, higher than normal sea temperatures led to some bleaching all of the world's coral reef systems. According to a 1999 report by the Coral Reef Research Institute, unless global warming is slowed, widespread coral bleaching will occur annually by 2030.

- Potential flooding of some reefs to levels too deep for survival if global warming causes sea levels to rise faster than coral growth.

- Increased ultraviolet radiation resulting from depletion of stratospheric ozone (Section 18-6, p. 465).

- Destruction or weakening of coral through use of cyanide or dynamite to stun and harvest coral reef fish for food and aquariums.

- Removal of coral for building material, aquariums, and jewelry.

- Oil spills.

- Physical damage from tourist divers, anchors, and ships colliding with reefs.

There is much evidence that these human-caused stresses are occurring more rapidly than the reefs can respond. Marine biologists estimate that human activities have seriously degraded about 10% of the world's coral reefs (especially in Southeast Asia and the Caribbean). Another 30% of the remaining reefs are in critical condition and 30% more are threatened; only 30% are stable.

According to the United Nations Environment Programme, at least two-thirds of these oases of biodiversity could collapse ecologically within 20–40 years, and few might remain by 2100. In addition to the biodiversity losses, such a collapse would sharply reduce fish harvests and cause a substantial increase in storm damage on the coasts of tropical and warm temperate areas.

Some 300 coral reefs in 65 countries are protected as reserves or parks, and another 600 have been recommended for protection. The *good news* is that protected coral reefs often can recover (Connections, p. 164). However, protecting reefs is difficult and expensive, and only half of the countries with coral reefs have set aside reserves that receive some protection from human activities.

What Biological Zones Are Found in the Open Sea? The sharp increase in water depth at the edge of the continental shelf separates the coastal zone from the vast volume of the ocean called the **open sea**. Based primarily on the penetration of sunlight, it is divided into three vertical zones (Figure 7-5):

- *Euphotic zone*: the lighted upper zone where photosynthesis occurs mostly by phytoplankton, nutrient levels are low (except around upwellings, Figure 6-9,

Producer to primary consumer

Primary to secondary consumer

Secondary to higher-level consumer

All consumer and producers to decomposers

Figure 7-13 Some components and interactions in a coral reef ecosystem. When these organisms die, decomposers break down their organic matter into minerals used by plants. Transfers of matter and energy between producers, primary consumers (herbivores), and secondary (or higher-level) consumers (carnivores) are indicated by colored arrows. Organisms are not drawn to scale.

All of the news about coral reefs is not bad. In 1998, researchers found that in some cases coral bleaching (Figure 7-1, top) may not be as fatal to reefs as once thought.

Researchers have been puzzled because corals in shallow water in some areas might be healthy, whereas at a slightly greater depth there is extensive bleaching. Marine biologist Rob Rowan studied DNA extracted from coral at different depths and locations and identified up to three different algae species (which he labeled A, B, and C) in the same coral.

He also noticed a pattern in the distribution of these algae species.

Two species, which he called A and B, preferred areas on coral with lots of light, and species C was adapted to lower light at greater depths. He hypothesized that coral with species A and B in shallow water may have some resistance to bleaching. In deeper and darker water, species C can grow on top of the corals.

Recent research also suggests that bleaching may help one type of coral to switch to new algae species more adapted to the changed water conditions.

Additional research indicates that some types of coral that appear dead from bleaching may have a hidden reserve of algae that may eventually allow the coral to come back to life. However,

these and other researchers caution that we still have much to learn about the complex interactions between various types of algae and coral.

More *good news* is the growing evidence that coral reefs can recover when given a chance. When localities or nations have imposed restrictions on reef fishing or reduced inputs of nutrients and other pollutants, reefs have rebounded.

Critical Thinking

Does the research described here mean that we do not need to worry so much about protecting coral reefs? Explain.

---

p. 126), and levels of dissolved oxygen are high (Figure 7-3). It is populated by large, fast-swimming predatory fish such as swordfish, sharks, and bluefin tuna.

- *Bathyal zone*: dimly lit middle zone that does not contain photosynthesizing producers because of a lack of sunlight but is populated by various types of zooplankton and smaller fish, many of which migrate to feed on the surface at night.

- *Abyssal zone*: dark lower zone that is very cold, has little dissolved oxygen (Figure 7-3), and has enough nutrients on the ocean floor to support about 98% of the 250,000 identified species living in the ocean.

Dead and decaying organisms fall to the ocean floor to feed microscopic decomposers and scavengers such as crabs and sea urchins. Some of these organisms (such as many worms) are *deposit feeders*, which take mud into their guts and extract nutrients from it. Others (such as oysters and mussels) are *filter feeders*, which pass water through or over their bodies and extract nutrients from it (Spotlight p. 154). On portions of the dark, deep ocean floor near hydrothermal vents, scientists have found communities of organisms where specialized bacteria use chemosynthesis to produce their own food and food for other organisms feeding on them (Figure 14-3, p. 322)

Average primary productivity and net primary productivity per unit of area are quite low in the open sea (Figure 4-25, p. 88) except at an occasional equatorial upwelling, where currents bring up nutrients from the ocean bottom. However, because the open sea covers so much of the earth's surface (Figure 7-4), it makes the largest contribution to the earth's overall net primary productivity.

7-3 FRESHWATER LIFE ZONES

What Are Freshwater Life Zones? Freshwater life zones occur where water with a dissolved salt concentration of less than 1% by volume accumulates on or flows through the surfaces of terrestrial biomes. Examples are **(1)** *standing* (lentic) bodies of fresh water such as lakes, ponds, and inland wetlands and **(2)** *flowing* (lotic) systems such as streams and rivers. These bodies of water cover only a small part of the earth's surface (Figure 7-2), and their locations are largely unrelated to climate. Runoff from nearby land provides freshwater life zones with an almost constant input of organic material, inorganic nutrients, and pollutants. Thus, these life zones are closely connected to nearby terrestrial biomes.

What Life Zones Are Found in Freshwater Lakes? **Lakes** are large natural bodies of standing fresh water formed when precipitation, runoff, or groundwater seepage fills depressions in the earth's surface. Causes of such depressions include **(1)** glaciation (the Great Lakes of North America), **(2)** crustal displacement (Lake Nyasa in East Africa), and **(3)** volcanic activity (Crater Lake in Oregon, Figure 19-6, left, p. 483). Lakes are fed by rainfall, melting snow, and streams that drain the surrounding

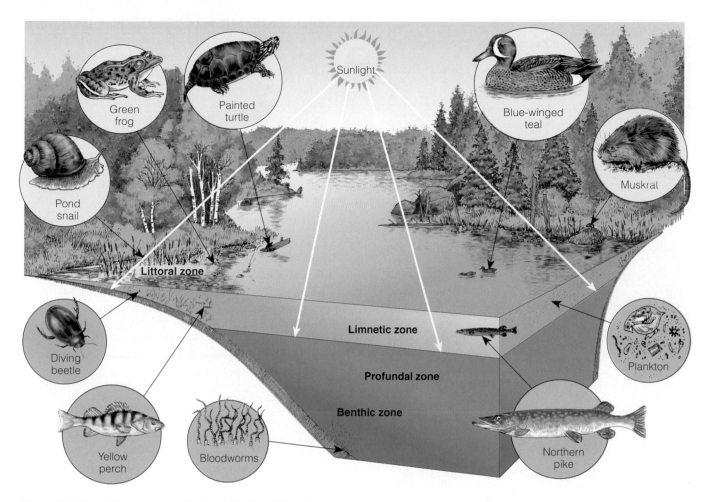

Figure 7-14 The distinct zones of life in a fairly deep temperate zone lake.

watershed. Because ponds (Figure 4-11, p. 78) are shallow, sunlight often penetrates to the bottom so that ponds usually have only one zone. In contrast, lakes normally consist of four distinct zones that are defined by their depth and distance from shore (Figure 7-14).

The *littoral zone* ("LIT-torc-el") consists of the shallow sunlit waters near the shore to the depth at which rooted plants stop growing. It has a high biological diversity, containing a variety of **(1)** phytoplankton, **(2)** rooted plants that extend above the water's surface (such as cattails and water lilies), **(3)** totally submerged rooted plants (such as muskgrass), **(4)** various species of floating plants (such as duckweed), **(5)** large numbers of decomposers, and **(6)** frogs, snails, insects, fish, and other consumers.

The *limnetic zone* ("limb-NET-ic") is the open, sunlit water surface layer away from the shore that extends to the depth penetrated by sunlight. It is the main photosynthetic body of a lake and thus produces food and oxygen that support most of the lake's consumers.

The *profundal zone* ("pro-FUN-dahl") is the deep, open water where it is too dark for photosynthesis. Without sunlight and plants, oxygen levels are low. It

is inhabited by fish adapted to its cooler, darker water, and most of their food is produced in the limnetic and littoral zones.

The *benthic zone* ("BEN-thick") at the bottom of a lake is inhabited mostly by organisms that tolerate cool temperatures and low oxygen levels. This includes decomposers (bacteria and fungi), snails, clams, crayfish, catfish, and wormlike insect larvae (such as those of the mayfly and dragonfly) that emerge in the spring and become flying insects.

How Do Plant Nutrients Affect Lakes? Ecologists classify lakes according to their nutrient content and primary productivity. A newly formed lake generally has a small supply of plant nutrients and is called an **oligotrophic** (poorly nourished) **lake** (Figure 7-15, top). This type of lake is often deep, with steep banks. Because of its low net primary productivity, such a lake usually has crystal-clear blue or green water and small populations of phytoplankton and fish (such as small-mouth bass and trout). Over time, sediment washes into an oligotrophic lake and plants grow and decompose to form bottom sediments. A lake with a large or

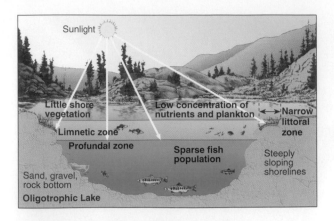

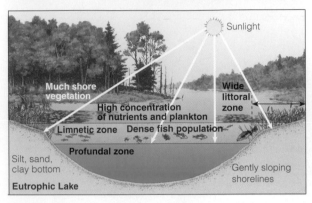

Figure 7-15 An oligotrophic, or nutrient-poor, lake (top) and a eutrophic, or nutrient-rich, lake (bottom). Mesotrophic lakes fall between these two extremes of nutrient enrichment. Nutrient inputs from human activities can accelerate eutrophication.

excessive supply of nutrients (mostly nitrates and phosphates) needed by producers is called a **eutrophic (well-nourished) lake** (Figure 7-15, bottom). Such lakes typically are shallow and have murky brown or green water with very poor visibility. Because of their high levels of nutrients, these lakes **(1)** have a high net primary productivity and **(2)** contain large populations of phytoplankton (especially cyanobacteria), many zooplankton, and diverse populations of fish (such as bass, sunfish, and yellow perch). In warm months the bottom layer of a eutrophic lake often is depleted of dissolved oxygen. Many lakes fall somewhere between the two extremes of nutrient enrichment and are called **mesotrophic lakes**.

Human inputs of nutrients from the atmosphere and from nearby urban and agricultural areas can accelerate the eutrophication of lakes, a process called *cultural eutrophication*. Water quality experts consider excessive eutrophication, mostly from human activities, to be one of the world's major water quality problems (as discussed on p. 481).

What Seasonal Changes Occur in Temperate Lakes? Most substances become denser as they go

from gaseous to liquid to solid physical states. Water does not follow this typical behavior; it is densest as a liquid at 4°C (39°F). In other words, solid ice at 0°C (32°F) is less dense than liquid water at 4°C (39°F), which is why ice floats on water. This unusual property of water causes *thermal stratification* of deep lakes in temperate areas (Figure 7-16). During the summer these lakes have the following three distinct layers characterized by different temperatures:

- *Epilimnion* ("ep-eh-LIM-knee-on"): an upper layer of warm water with high levels of dissolved oxygen.

- *Thermocline* ("THUR-moe-cline"): where the water temperature changes rapidly with depth and with moderate levels of dissolved oxygen.

- *Hypolimnion* ("high-poe-LIM-knee-on"): a lower layer of colder, denser water, usually with a lower concentration of dissolved oxygen because it is not exposed to the atmosphere. During summer the thermocline acts as a barrier preventing the transfer of nutrients and dissolved oxygen between the epilimnion and hypolimnion.

In fall, the surface water gradually cools, becomes denser, and sinks to the bottom when it cools to 4°C (39°F), and the thermocline disappears (Figure 7-16, top right). This mixing, or *fall overturn*, **(1)** brings nutrients from bottom sediments to the surface, **(2)** brings dissolved oxygen from the surface to the bottom, and **(3)** allows fish to live at various depths.

During winter the colder temperatures cause the lake to separate into layers of different density (Figure 7-16, bottom left). In spring, the lake's surface water reaches maximum density when it warms to 4°C (39°F) and sinks through and below the cooler, less dense water. Winds blowing across the lake's surface cause strong vertical currents that mix the surface and bottom water. This brings dissolved oxygen from the surface to the bottom and nutrients from the bottom to the surface. During this brief *spring overturn*, the temperature of the lake and dissolved oxygen levels are roughly the same at all depths (Figure 7-16, bottom right).

What Are the Major Characteristics of Freshwater Streams and Rivers? Precipitation that does not sink into the ground or evaporate is **surface water**. It becomes **runoff** when it flows into streams. The land area that delivers runoff, sediment, and dissolved substances to a stream is called a **watershed**, or **drainage basin**. Small streams join to form rivers, and rivers flow downhill to the ocean (Figures 7-2 and 7-17) as part of the hydrologic cycle (Figure 4-28, p. 90). In many areas, streams begin in mountainous or hilly areas that collect and release water falling to the earth's surface as rain or snow.

The downward flow of surface water and groundwater from mountain highlands to the sea takes place in three different aquatic life zones with different environmental conditions: the *source zone*, *transition zone*, and *floodplain zone* (Figure 7-17).

In the first, narrow *source zone*, headwater or mountain highland streams of cold, clear water rush over waterfalls and rapids. As this turbulent water flows and tumbles downward, it dissolves large amounts of oxygen from the air so that photosynthesis is a less important source of oxygen than it is in ponds and lakes. Here plants such as algae and mosses are attached to rocks and it is populated by cold-water fish (such as trout in some areas), which need lots of dissolved oxygen. Many fish and other animals in headwater streams have compact and flattened bodies that allow them to live under stones.

In the *transition zone* (Figure 7-17), the headwater streams merge to form wider, deeper streams that flow down gentler slopes with fewer obstacles. The warmer water and other conditions in this zone support more producers (phytoplankton) and a variety of cool-water and warm-water fish species (such as black bass) with slightly lower oxygen requirements. Figure 7-18 show some species and interactions in the transition zone of a river in a tropical forest.

In the *floodplain zone* (Figure 7-17), streams join into wider and deeper rivers that meander across broad, flat valleys. Water in this zone usually has higher temperatures and less dissolved oxygen than water in the first two zones. These slow-moving rivers sometimes support fairly large populations of producers such as algae and cyanobacteria and rooted aquatic plants along the shores. Because of increased erosion runoff over a larger area, water in this zone often is muddy and contains high concentrations of suspended particulate matter (silt). The main channels of these slow-moving, wide, and murky rivers support distinctive varieties of fish (carp and catfish), whereas their backwaters support species similar to those present in lakes. At its mouth, a river may divide into many channels as it flows

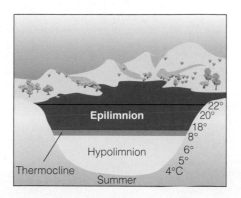
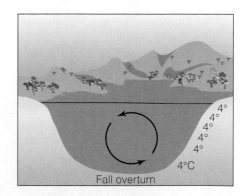
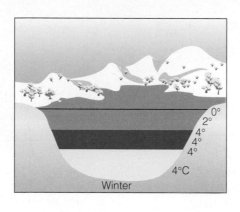
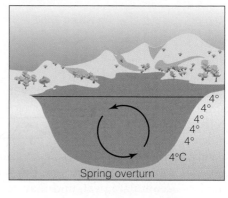

Dissolved O₂ concentration   ■ High   ■ Medium   □ Low

Figure 7-16 During the summer and winter, the water in deep temperate zone lakes becomes stratified into different temperature layers, which do not mix. Twice a year, in the fall and spring, the waters at all layers of these lakes mix in overturns that equalize the temperature at all depths. These overturns bring **(1)** oxygen from the surface water to the lake bottom and **(2)** nutrients from the lake bottom to the surface waters.

through coastal wetlands and estuaries, where the river water mixes with ocean water (Figure 7-7).

As streams flow downhill, they become powerful shapers of land. Over millions of years the friction of moving water levels mountains and cuts deep canyons, and the rock and soil the water removes are deposited as sediment in low-lying areas.

Streams are fairly open ecosystems that receive many of their nutrients from bordering land ecosystems. Such nutrient inputs come from falling leaves, animal feces, insects, and other forms of biomass washed into streams during heavy rainstorms or by melting snow. To protect a stream or river system from excessive inputs of nutrients and pollutants, one must protect its watershed—the land around it. Human impacts on streams are discussed on p. 478.

Why Are Freshwater Inland Wetlands Important? Inland wetlands are lands covered with fresh water all or part of the time (excluding lakes, reservoirs, and streams) and located away from coastal areas (Figure 7-19). They include **(1)** *marshes* with few

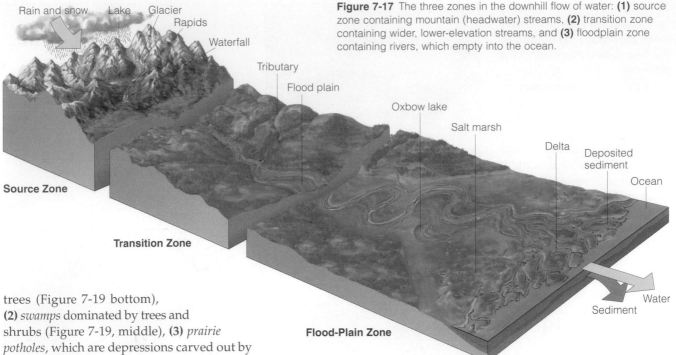

Figure 7-17 The three zones in the downhill flow of water: **(1)** source zone containing mountain (headwater) streams, **(2)** transition zone containing wider, lower-elevation streams, and **(3)** floodplain zone containing rivers, which empty into the ocean.

trees (Figure 7-19 bottom), **(2)** *swamps* dominated by trees and shrubs (Figure 7-19, middle), **(3)** *prairie potholes*, which are depressions carved out by glaciers (Figure 7-19, top), **(4)** *floodplains*, which receive excess water during heavy rains and floods, **(5)** *bogs* and *fens* that have waterlogged soils, which tend to accumulate peat, and may or may not have trees, and **(6)** *wet arctic tundra* in summer.

Some wetlands are huge; others are small. Swamps and marshes generally have shallow flowing water and are among the world's most productive ecosystems (Figure 4-25, p. 88), which is why they are sometimes called "biological supermarkets" for a variety of species. Bogs and fens usually have standing water and low productivity. Bogs are fed solely by precipitation and fens by surface runoff and groundwater.

Some of these wetlands are covered with water year-round. Others, called *seasonal wetlands*, usually are underwater or soggy for only a short time each year. They include prairie potholes (Figure 7-19, top), flood-plain wetlands, and bottomland hardwood swamps. Some stay dry for years before being covered with water again. In such cases, scientists must use the composition of the soil or the presence of certain plants (such as cattails, bulrushes, or red maples) to determine that a particular area is really a wetland.

Inland wetlands play the following important ecological and economic roles:

- Provide food and habitats for fish, migratory waterfowl, shorebirds, and a variety of other wildlife, including one-third of the endangered and threatened species in the United States.

- Improve water quality by filtering, diluting, and degrading toxic wastes, excess nutrients, sediments,

and other pollutants. The Audubon Society conservatively estimates that inland wetlands in the United States provide water quality protection worth at least $1.6 billion per year.

- Reduce flooding and erosion by absorbing stormwater and releasing it slowly and by absorbing overflows from streams and lakes. According to Audubon Society estimates, if the remaining wetlands in the United States were destroyed, additional flood control costs such as dredging and levees would be $7.7–31 billion per year.

- Help replenish stream flows during dry periods.

- Play significant roles in the global carbon, nitrogen, sulfur, and water biogeochemical cycles.

- Keep down atmospheric levels of carbon dioxide by storing carbon within their plant communities and soils instead of releasing it to the atmosphere as CO_2.

- Provide valuable natural products such as fish and shellfish, blueberries, cranberries, timber, wild rice, and medicines derived from wetland soils and plants.

- Provide recreation for birdwatchers, nature photographers, boaters, anglers, and waterfowl hunters.

What Impacts Do Human Activities Have on Inland Wetlands? Despite the ecological importance of year-round and seasonal inland wetlands, many are drained, dredged, filled in, or covered over. In the United States, 53% of the inland wetlands estimated to have existed in the lower 48 states during the 1600s have

Capybara

Cayman alligator

Water lily

Plant plankton

Terecay turtle

Amazonian dolphin

Animal plankton

Neon tetra

Piranha

Leporinus fish

Arapaima

Producer to primary consumer	Producer to secondary consumer	Secondary to higher-level consumer	All producers and consumers to decomposers

Figure 7-18 Some components and interactions in a river in a tropical forest. When these organisms die, decomposers break down their organic matter into minerals used by plants. Transfers of matter and energy between producers, primary consumers (herbivores), and secondary (or higher-level) consumers are indicated by colored arrows. Organisms are not drawn to scale.

Prairie Pothole

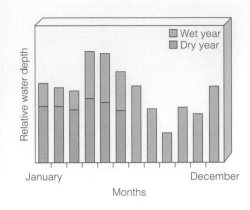

Cypress swamp

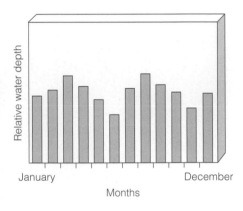

Freshwater marsh

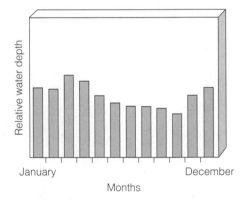

Figure 7-19 Inland wetlands are not always wet. The graphs show the fluctuating water levels of three types of inland wetlands. (Data from Jon A. Kusler, William J. Mitsch, and Joseph S. Larson, 1994)

been drained and converted to other uses. The largest losses of wetlands occurred between the mid-1950s and the mid-1970s and have been an important factor in increased flood and drought damage in the United States. Although the rate of loss has decreased, each year an estimated 400 square kilometers (150 square miles) of inland wetland in the United States are lost, about 80% to agriculture and the rest to mining, forestry, oil and gas extraction, highways, and urban development. Other countries have suffered similar losses.

7-4 SUSTAINABILITY OF AQUATIC LIFE ZONES

Why Is an Understanding of Aquatic Life Zones Important? The survey of the various saltwater and freshwater environments in this chapter gives us the opportunity to grapple with both the general principles of ecology and the unique characteristics and biodiversity of each of these life zones. The lessons from such a study are many, but as in terrestrial biomes

(Chapter 6), the grand lesson is that *everything is connected*. Most aquatic life zones are connected by flows of water and nutrients from one type to another. In addition, nutrients and pollutants flowing from the land and deposited from the atmosphere affect the ecological processes of aquatic systems.

How Sustainable Are Aquatic Ecosystems?

From an ecological perspective, protecting and enhancing the sustainability of aquatic life zones requires an understanding of how human activities affect **(1)** energy and nutrient flow, **(2)** trophic relationships, and **(3)** biodiversity. The *bad news* in terms of human impacts is that each stream, river, and lake reflects the sum of all that occurs in the watersheds above. Also, many of the nutrients, wastes, and pollutants produced by human activities end up in the ocean. Recent research also shows that many of the chemicals reaching aquatic systems come from the atmosphere. Human-induced degradation of marine and freshwater life zones is much greater and more widespread than previously thought.

The *good news* is that aquatic life zones are constantly renewed because **(1)** water is purified by natural hydrologic processes (Figure 4-28, p. 90), **(2)** nutrients cycle in and out, and **(3)** populations of biological organisms can be replenished, given sufficient opportunity and time. However, these life-sustaining processes work only if they are not overloaded with pollutants and excessive nutrients and are not overfished. These threats to the biodiversity and sustainability of aquatic life zones and possible solutions to these problems are discussed in more detail in Chapter 24, p. 629.

All at last returns to the sea—to Oceanus, the ocean river, like the ever-flowing stream of time, the beginning and the end.

RACHEL CARSON

REVIEW QUESTIONS

1. Define the boldfaced terms in this chapter.

2. What are *coral reefs* and *coral polyps*? How are coral reefs formed? List six ecological and economic services provided by coral reefs. What is *coral bleaching*, and what are its main causes?

3. What are the two major types of aquatic life zones? What two factors determine the major types of organisms found in aquatic systems?

4. Distinguish between *plankton*, *nekton*, *benthos*, and *decomposers* in aquatic life zones. Distinguish between *phytoplankton*, *nanoplankton*, and *zooplankton*.

5. Describe major differences between life on land and life in water in terms of **(a)** physical and chemical differences, **(b)** species and habitat diversity, **(c)** trophic structure and food webs, **(d)** population characteristics, and **(e)** monitoring and protection.

6. List four major factors determining the types and numbers of organisms found in the surface, middle, and bottom layers of aquatic systems. How do the concentrations of carbon dioxide and dissolved oxygen vary in these three layers?

7. List three reasons why oceans are important to life on the earth.

8. What is the *coastal zone*? What are *estuaries* and *coastal wetlands*? List three reasons why temperature and salinity vary widely in estuaries and coastal wetlands and explain why they have such high net primary productivity. List four major ecological and economic services provided by estuaries and coastal wetlands. What are three major causes of the destruction and degradation of estuaries and coastal wetlands?

9. What is the *intertidal zone*? Distinguish between *rocky shores* and *sandy beaches* and describe the major types of aquatic life found in each. What four factors reduce the biodiversity found in rocky shores and sandy beaches? Why is it important to preserve the dunes on barrier beaches? What are *barrier islands*, why are they so attractive for human development, and why are human structures built there so vulnerable to destruction?

10. What are the three main groups of organisms found in coral reefs? List three reasons why coral reefs are vulnerable to damage. List ten harmful impacts of human activities on coral reefs.

11. What is the *open sea*, and what are its three major zones? Why is the net primary productivity per unit of area so low in the open sea?

12. What is a *freshwater life zone*, and what are the two major types of such zones?

13. What is a *lake*? Distinguish between the *littoral*, *limnetic*, *profundal*, and *benthic zones* of a lake and describe the major forms of life found in each zone.

14. What are the three types of lakes, based on their nutrient content and primary productivity? Describe the general properties and types of life found in each type.

15. Describe the seasonal changes that can take place in deep lakes in northern temperate areas. What three layers are found in such lakes during the summer? What happens to these layers and to levels of dissolved and nutrients in such lakes during fall, winter, and spring?

16. Distinguish between *surface water*, *runoff*, and a *watershed*. Describe the properties and general forms of life found in the three zones of a river as it flows from mountain highlands to the sea.

17. What are *inland wetlands*? Describe three types of such wetlands. List three examples of *seasonal wetlands*. List eight important ecological and economic services provided by inland wetlands.

18. What are the major human impacts on inland wetlands?

19. Explain why a study of aquatic life zones reinforces the basic ecological principle that everything is connected.

CRITICAL THINKING

1. List a limiting factor for each of the following: **(a)** the surface layer of a tropical lake, **(b)** the surface layer of the open sea, **(c)** an alpine stream, **(d)** a large, muddy river, **(e)** the bottom of a deep lake.

2. Consider the differences in selective pressures in aquatic and temperate environments. Why do terrestrial organisms evolve tolerances to broader temperature ranges than aquatic organisms? Why do aquatic plants tend to be very small (e.g., phytoplankton), whereas most terrestrial plants tend to be larger (e.g., trees) and have more specialized structures for growth (e.g., stems and leaves)? Why are some aquatic animals (especially marine mammals) extremely large compared with terrestrial animals?

3. Describe the changes in physical environment for organisms that live in the intertidal zone. Do you expect the extreme fluctuations in physical and chemical conditions in this zone to lead to a higher or a lower diversity of species than in other aquatic zones?

4. How would you respond to someone who proposes that we use the deep portions of the world's oceans to deposit our radioactive and other hazardous wastes because the deep oceans are vast and are located far away from human habitats? Give reasons for your response.

5. What factors in your lifestyle contribute to the destruction and degradation of coastal and inland wetlands?

6. Someone tries to sell you several brightly colored pieces of dry coral. Explain in biological terms why this transaction is probably a rip-off.

7. A dam is being built across a river with a fast current. Which of the current types of organisms in the river above and below the dam are likely to be affected? How might they be affected?

8. Developers want to drain a large area of inland wetlands in your community and build a large housing development. List **(a)** the main arguments the developers would use to support this project and **(b)** the main arguments ecologists would use in opposing this project. If you were an elected city official, would you vote for or against this project? Can you come up with a compromise plan?

9. You are a defense attorney arguing in court for sparing an undeveloped old-growth tropical rain forest and a coral reef from severe degradation or destruction by development. Write your closing statement for the defense of each of these ecosystems. If the judge decides you can save only one of the ecosystems, which one would you choose, and why?

PROJECTS

1. Search for information about mangrove trees, using the internet and your library. Are the different species of mangrove trees closely related? What characteristics do they have in common? What characteristics do they have that make them a good place for fish to breed?

2. If possible, visit a nearby lake, pond, or reservoir. Would you classify it as oligotrophic, mesotrophic, or eutrophic? What are the primary factors contributing to

its nutrient enrichment? Which of these factors are related to human activities?

3. Examine a topographic map for the area around a stream or lake near where you live to define the watershed for the stream. What human activities occur in the watershed? What influence, if any, do you expect these activities to have on the ecology of the stream or lake?

4. Use the library or the internet to find bibliographic information about *Loren Eisley* and *Rachel Carson*, whose quotes appear at the beginning and end of this chapter.

5. Make a concept map of this chapter's major ideas, using the section heads and subheads and the key terms (in boldface). Look at the inside back cover and on the website for this book for information about concept making maps.

INTERNET STUDY RESOURCES AND RESOURCES FOR FURTHER READING AND RESEARCH

The website for this book contains helpful study aids and many ideas for further reading and research. Log on to:

http://www.brookscole.com/product/0534376975s

and click on the Chapter-by-Chapter area. Choose Chapter 7 and select a resource:

- "Flash Cards" allows you to test your mastery of the Terms and Concepts to Remember for this chapter.

- "Tutorial Quizzes" provides a multiple-choice practice quiz.

- "Student Guide to InfoTrac" will lead you to Critical Thinking Projects that use InfoTrac College Edition as a research tool.

- "References" lists the major books and articles consulted in writing this chapter.

- "Hypercontents" takes you to an extensive list of sites with news, research, and images related to individual sections of the chapter.

INFOTRAC COLLEGE EDITION

Improve your skills with InfoTrac College Edition, a searchable online database of articles from more than 700 periodicals. Log on to:

http://www.infotrac-college.com

or access InfoTrac through the website for this book.

Try the following articles:

Hayden, T. 2000. A growing coral crisis: overfishing and global warming are killing reefs around the world. Is it too late to save them? *Newsweek*, Oct. 30, 2000, p. 79. (subject guide: coral reefs)

Becker, K. 2000. Seeding the oceans with observatories. *Oceanus* vol. 42, no. 1, p. 2–5. (subject guide: marine sciences)

8 COMMUNITY ECOLOGY: STRUCTURE, SPECIES INTERACTIONS, SUCCESSION, AND SUSTAINABILITY*

Flying Foxes: Keystone Species in Tropical Forests

The durian (Figure 8-1, bottom right) is one of the most prized fruits growing in Southeast Asian tropical forests. The odor of this football-sized fruit is so strong that it is illegal to have them on trains and in many hotel rooms in Southeast Asia. However, its custard-like flesh has been described as "exquisite," "sensual," "intoxicating," and "the world's finest fruit."

Durian fruits come from a wild tree that grows in the tropical rain forest. The tree depends on nectar- and pollen-feeding flying foxes (Figure 8-1, top right) to pollinate the flowers that hang high in the durian trees (Figure 8-1, bottom left). Pollination by flying foxes is an example of *mutualism*: an interaction between two species in which both species benefit. Hundreds of tropical plant species are entirely dependent on various flying fox species for pollination and seed dispersal.

Many species of flying foxes are now listed as *endangered*, and most populations are much smaller than historic numbers. One reason for these population declines is *deforestation*. Another is *hunting* of the bats for their meat, which is sold in China and other parts of Asia. The bats also are viewed as pests and are killed to keep them from eating commercially grown fruits (even though these fruits are picked green). Flying foxes are easy to hunt because they tend to congregate in large numbers when they feed or sleep.

According to ecologists, flying foxes are *keystone species* in tropical forest ecosystems. They are important not only for the plant species they pollinate and the plant seeds they disperse in their droppings but also for the many other species that depend on them. Flying foxes are the key to maintaining the plant diversity of many areas. A study in Samoa found that 80–100% of the seeds landing on the ground during the dry season were deposited by flying foxes. They are also critical in regenerating deforested areas because other tropical forest animals are reluctant to venture into the open.

Rain forest ecologists are concerned that the decline of flying fox populations could lead to a cascade of linked extinctions. Their decline also has important economic effects. Studies have also shown that flying foxes are economically important for many products including fruits (such as the durian and wild bananas), other foods, medicine, timber (such as ebony and mahogany), fibers, dyes, medicines, animal fodder, and fuel.

The story of flying foxes and durians illustrates the unique role (niche) of each species in a community or ecosystem and shows how interactions between species can affect ecosystem structure and function. This chapter deals with these topics and the topics of *community structure, ecosystem sustainability*, and *ecological succession*.

Figure 8-1 Flying foxes (top right) are bats that play key ecological roles in tropical rain forests in Southeast Asia by pollinating (bottom left) and spreading the seeds of durian trees that produce durians, a highly prized tropical fruit (bottom right). (Top and bottom right, Dr. Merlin Tuttle/Photo Researchers, Inc.; Bottom left, Christer Friedriksson/Bruce Coleman Collection)

*Paul M. Rich, associate professor of ecology and evolutionary biology and environmental studies at the University of Kansas, is coauthor of this chapter.

What is this balance of nature that ecologists talk about?
Stuart L. Pimm

This chapter addresses the following questions:

- What determines the number of species in a community?
- How can we classify species according to their roles?
- How do species interact with one another?
- How do communities and ecosystems change as environmental conditions change?
- Does high species diversity increase the stability of ecosystems?

8-1 COMMUNITY STRUCTURE: APPEARANCE AND SPECIES DIVERSITY

What Is Community Structure? One property of a community or ecosystem is the *structure* or *spatial distribution* of its individuals and populations. Ecologists usually describe the structure of a community or ecosystem in terms of four characteristics:

- *Physical appearance*: relative sizes, stratification, and distribution of its populations and species

- *Species diversity or richness*: the number of different species

- *Species abundance*: the number of individuals of each species

- *Niche structure*: the number of ecological niches (Section 5-3, p. 110), how they resemble or differ from each other, and how they interact (species interactions)

How Do Communities Differ in Physical Appearance and Population Distribution? The types, relative sizes, and stratification of plants and animals vary in different terrestrial communities and biomes (Figure 8-2). There are also marked differences in the physical structures of different types of aquatic life zones such as (1) oceans (Figure 7-5, p. 156), (2) rocky shore and sandy beaches (Figure 7-10, p. 160), (3) lakes (Figure 7-14, p. 165), (4) river systems (Figure 7-17, p. 168), and (5) inland wetlands (Figure 7-19, p. 170). The distribution of populations and species in a terrestrial or aquatic community can be vertical as well as horizontal, as shown in Figure 6-28 (p. 144) for a tropical rain forest and in Figure 7-5 (p. 156) for an ocean.

Members of the population of a particular species also may be dispersed in different patterns. For example, the population of a species may be dispersed in *clumps* (the most common pattern), *fairly uniformly*, or *randomly* (Figure 9-2, p. 199).

The physical structure within a particular type of community or ecosystem also can vary. A close look at most large terrestrial communities, ecosystems, and biomes reveals that they usually consist of a mosaic of *vegetation patches* of differing size. This leads to a combination of (1) fairly sharp edges or boundaries such as that between a forest and an open field and (2) wider and more diffuse *ecotones* or transition zones between one patch or community and another (Figure 4-10, p. 77).

Differences in the physical structure and physical properties (such as sunlight, temperature, wind, and humidity) at boundaries and in ecotones are called **edge effects**. For example, the edge area between a forest and an open field may (1) be sunnier, warmer, and drier than the forest interior and (2) have a different combination of species than the forest and field interiors.

Popular wild game animals, such as pheasants and white-tailed deer, often are more plentiful in edges and ecotones between forests and

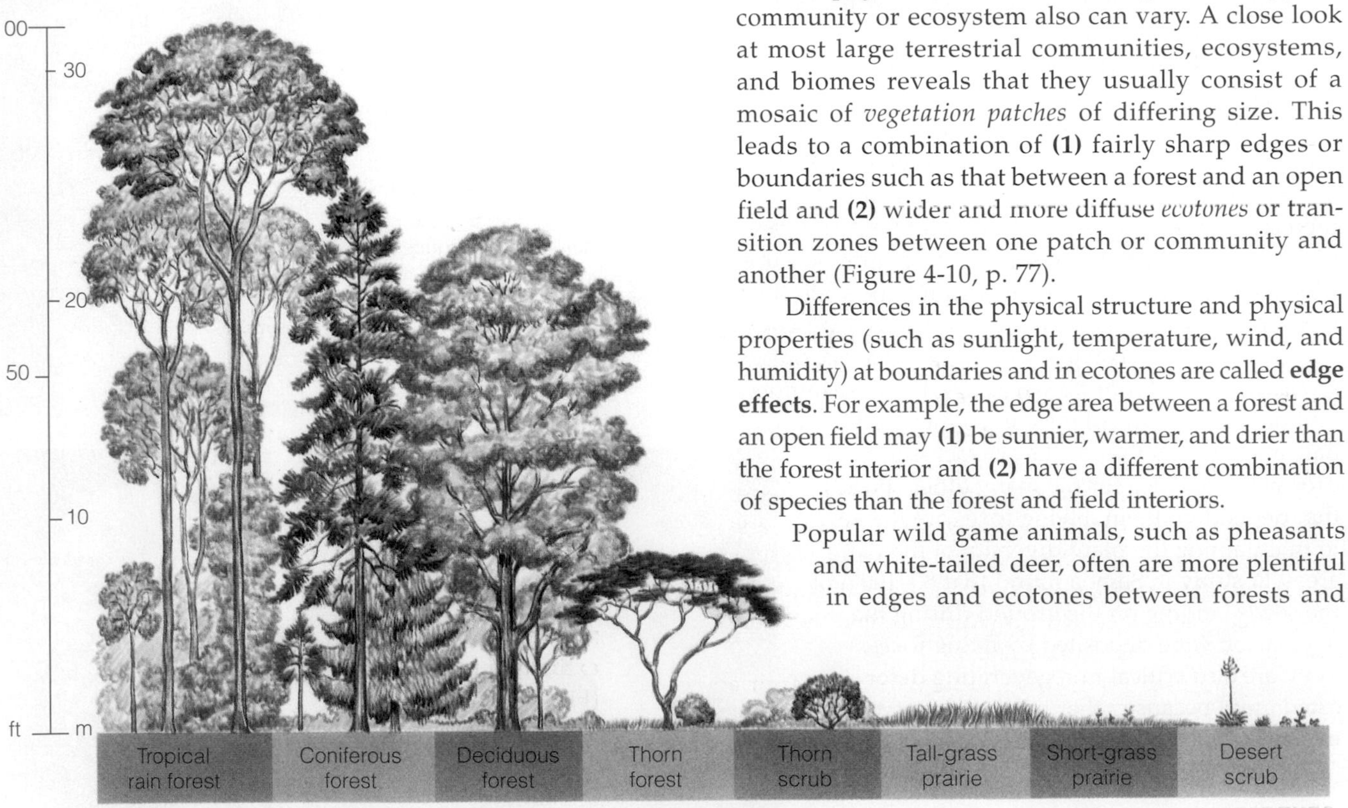

Figure 8-2 Generalized types, relative sizes, and stratification of plant species in various terrestrial communities or ecosystems.

Tropical rain forest | Coniferous forest | Deciduous forest | Thorn forest | Thorn scrub | Tall-grass prairie | Short-grass prairie | Desert scrub

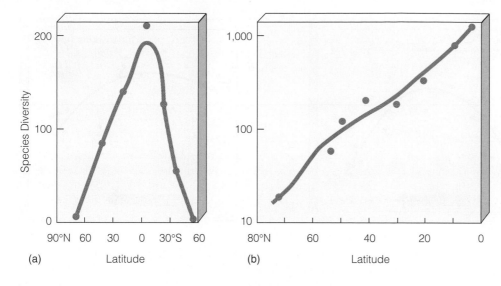

Figure 8-3 Changes in species diversity at different latitudes (distances from the equator) in terrestrial communities for **(a)** ants and **(b)** breeding birds of North and Central America. As a general rule, species diversity steadily declines as we go away from the equator toward either pole. (From *Biology: Concepts and Applications*, 4th ed. by Cecie Starr © 2000)

fields. Wild game managers sometimes create patches and edges to increase populations of such species for sport hunters.

However, the edge effects associated with habitat fragmentation can reduce the overall biodiversity of ecosystems. This occurs because the increased edge from fragmentation **(1)** makes many species more vulnerable to stresses such as predators and fire and **(2)** creates barriers that can prevent some species from colonizing new areas and finding food and mates. Conservation biologists urge that overall biodiversity be protected by **(1)** preserving large areas of habitat and **(2)** using migration corridors to link smaller habitat patches, as discussed in Section 23-6, p. 614.

Where Is Most of the World's Biodiversity Found?
Studies indicate that the most species-rich environments are **(1)** tropical forests (Figure 6-27, p. 143), **(2)** coral reefs (Figure 7-1, p. 152), **(3)** the deep sea, and **(4)** large tropical lakes.

Other places with an abundance of different species are **(1)** tropical dry habitats (deserts, shrublands, and grasslands) and **(2)** temperate shrublands with a Mediterranean climate (such as South Africa, parts of southern California, Chile, southwestern Australia, and the countries of the Mediterranean Basin).

Communities such as a tropical rain forest or a coral reef with a large number of different species (high species diversity, especially insects) generally have only a few members of each species (low species abundance).

Field investigations by ecologists have found that three major factors affect species diversity: **(1)** *latitude*

(distance from the equator) in terrestrial communities, **(2)** *depth* in aquatic systems, and **(3)** *pollution* in aquatic systems.

Such research shows that for most groups of plants and animals, species diversity on continents decreases steadily with distance from the equator toward either pole (Figure 8-3). This *latitudinal species diversity gradient* leads to the highest species diversity in tropical areas such as tropical rain forests and the lowest in polar areas such as arctic tundra. For example, the typical number of tree species per hectare (2.5 acres) is about 40–100 in a tropical forest, 10–30 in a temperate forest, and 1–5 in a northern taiga forest. Tropical Costa Rica has 205 animal species, compared to 95 in temperate France.

Major reasons for this latitudinal pattern in species diversity are as follows:

- Resource availability is higher and more reliable in the tropics than elsewhere because tropical biomes receive more sunlight and rainfall and thus have longer growing seasons and greater climate stability than biomes at higher latitudes. This leads to a variety of specialized niches (Figure 6-28, p. 144).

- It takes less time for species to evolve in equatorial communities because they have been less disturbed by advancing ice sheets and other climatic changes over the long span of geologic time.

- Higher species diversity in tropical ecosystems leads to even more species diversity as species interact and undergo coevolution (p. 109) over long periods of time.

- Tropical species have greater pressure from disease organisms and parasites because there is no winter to reduce pest populations. As a result, diseases and parasites keep any single species or group of species from dominating ecosystems. This allows a variety of species to coexist at low population densities.

- Speciation rates are higher than the background extinction rate, except during mass extinctions (Figure 5-10, p. 115).

Local variations in climate and topography also affect species richness patterns. In terrestrial communities, species diversity tends to increase with **(1)** increasing solar radiation, **(2)** increasing precipitation, **(3)** decreasing elevation, and **(4)** strong seasonal

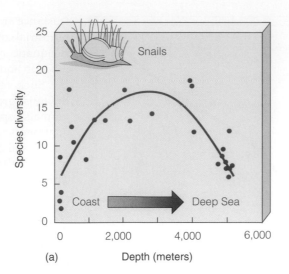

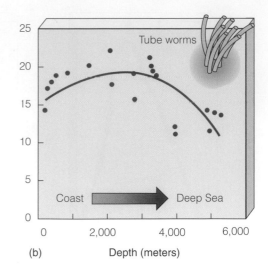

(a)   Depth (meters)

(b)   Depth (meters)

Figure 8-4 Changes in species diversity with depth in marine environments for **(a)** snails and **(b)** tube worms. As a general rule, species diversity in the ocean increases from the surface to a depth of about 2,000 meters and then decreases until the sea bottom where species diversity is usually high.

variations. In many cases, the greatest abundance of mammal species is found in areas with intermediate precipitation levels instead of in the driest or wettest habitats.

Research shows that in marine communities species diversity increases from the surface to a depth of 2,000 meters (6,600 feet) and then begins to decline with depth (Figure 8-4) until the deep sea bottom is reached, where species diversity is very high. Major reasons for this *depth-species diversity gradient* in marine communities are **(1)** greater stability in environmental conditions as one moves away from the low depth and high energy levels of the coastal zone (Figure 7-5, p. 156), **(2)** lack of nutrients in deep ocean waters below 2,000 meters, and **(3)** an abundance of nutrients and often varied habitats on the ocean floor.

A third trend is a decrease in species diversity and species abundance in aquatic systems as pollution increases and kills off or impairs the reproductivity of various aquatic species (Figure 8-5).

What Determines the Number of Species on Islands? Two factors affecting the species diversity found in an isolated ecosystem such as an island are its *size* and *degree of isolation*. In the 1960s, Robert MacArthur and Edward O. Wilson began studying communities on islands to discover why large islands tend to have more species of a certain category (such as insects, birds, or ferns) than do small islands.

To explain these differences in species diversity with island size, MacArthur and Wilson proposed what is called the **species equilibrium model** or the **theory of island biogeography**. According to this model, the number of species found on an island is determined by a balance between two factors: **(1)** the rate at which new species immigrate to the island and **(2)** the rate at which species become extinct on the island. The model predicts that at some point the rates of immigration and extinction will reach an equilibrium point (Figure 8-6a)

that determines the island's average number of different species (species diversity).

The model also predicts that immigration and extinction rates (and thus species diversity) are affected by two important features of the island: **(1)** its *size* (Figure 8-6b) and **(2)** its *distance from the nearest mainland* (Figure 8-6c). According to the model, a small island tends to have a lower species diversity than a large one for two reasons: **(1)** a small island generally has a lower immigration rate because it is a smaller target for potential colonizers, and **(2)** a small island should have a higher extinction rate because it generally has fewer resources and less diverse habitats for colonizing species.

The model also predicts that an island's distance from a mainland source of new species is important in

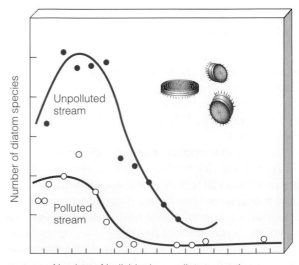

Figure 8-5 Changes in the species diversity and species abundance of diatom species in an unpolluted stream and a polluted stream. Note that both species diversity and species abundance decrease with pollution.

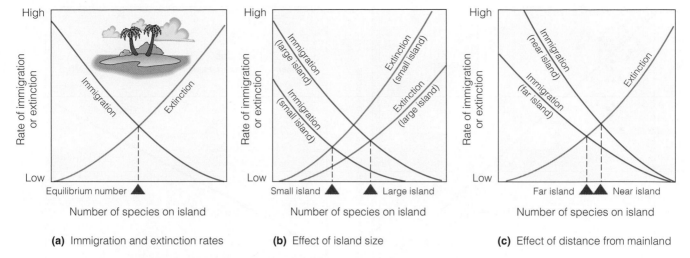

(a) Immigration and extinction rates   (b) Effect of island size   (c) Effect of distance from mainland

Figure 8-6 The species equilibrium model or theory of island biogeography, developed by Robert MacArthur and Edward O. Wilson. **(a)** The equilibrium number of species (blue triangle) on an island is determined by a balance between the immigration rate of new species and the extinction rate of species already on the island. **(b)** With time, large islands have a larger equilibrium number of species than smaller islands because of higher immigration rates and lower extinction rates on large islands. **(c)** Assuming equal extinction rates, an island near a mainland will have a larger equilibrium number of species than a more distant island because the immigration rate to a near island is higher than that to a more distant one.

determining species diversity. For two islands of about equal size and other factors, the island closest to a mainland source of immigrant species will have the higher immigration rate and thus a higher species diversity (assuming that extinction rates on both islands are about the same).

MacArthur and Wilson's original model or scientific hypothesis has been tested and supported by a series of field experiments (Figure 8-7). As a result, biologists have accepted it well enough to elevate it to the status of an important and useful scientific theory, although it may apply to a limited number of cases over short periods of time. In recent years, it has also been applied to conservation efforts to protect wildlife on land in *habitat islands* such as national parks surrounded by a sea of developed and fragmented land (Connections, p. 618).

8-2 GENERAL TYPES OF SPECIES

What Different Roles Do Various Species Play in Ecosystems? When examining ecosystems, ecologists often apply particular labels—such as *native, nonnative, indicator,* or *keystone*—to various species to clarify their ecological roles or niches (p. 110). Any given species may function as more than one of these four types in a particular ecosystem.

How Can Nonnative Species Cause Problems? Species that normally live and thrive in a particular

ecosystem are known as **native species**. Others that migrate into an ecosystem or are deliberately or accidentally introduced into an ecosystem by humans are called **nonnative species**, **exotic species**, or **alien species**. Some of these introduced species (such as crops and game species for sport hunting) are beneficial to humans, but some thrive and crowd out native species.

From a human standpoint, introduction of nonnative species can become a nightmare. In 1957, wild African bees were imported to Brazil to help increase honey production. Instead, these bees have displaced domestic honeybees and reduced the honey supply.

Since then these nonnative bee species, popularly known as "killer bees," have moved northward into Central America (killing 150 people in Mexico since 1986). They have become established in Texas (one death in 1994), Arizona (one death in 1993), New Mexico, Puerto Rico, and California and are heading north at 240 kilometers (150 miles) per year. They should be stopped eventually by cold winters in the central United States unless they can adapt genetically to cold weather.

Although they are not the killer bees portrayed in some horror movies, these bees are aggressive and unpredictable. They have killed thousands of domesticated animals and an estimated 1,000 people in the western hemisphere. Fortunately, most people not allergic to bee stings can run away. Most people killed by these honeybees have died because they fell down or became trapped and could not flee. The deliberate and accidental introduction of nonnative species is discussed in more detail on p. 565.

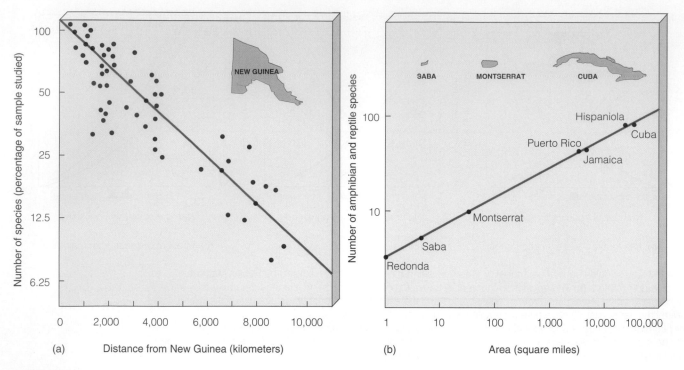

Figure 8-7 Research data supporting the theory of island biogeography with **(a)** the right graph showing that species diversity increases with island size and **(b)** the left graph showing that the species diversity of birds occupying lowland areas of South Pacific islands decreases with distance from New Guinea. (From *Biology: Concepts and Applications*, 4th ed. by Cecie Starr © 2000)

What Are Indicator Species? Species that serve as early warnings that a community or an ecosystem is being damaged are called **indicator species**. Birds are excellent biological indicators because they are found almost everywhere and respond quickly to environmental change. Research indicates that a major factor in the current decline of migratory, insect-eating songbirds in North America is habitat loss or fragmentation. The tropical forests of Latin America and the Caribbean that are winter habitats for such birds are disappearing rapidly. Their summer habitats in North America also are disappearing or are being fragmented into patches that make the birds more vulnerable to attack by predators and parasites.

The presence or absence of trout species in water at temperatures within their range of tolerance (Figure 4-14, p. 79) is an indicator of water quality because trout need clean water with high levels of dissolved oxygen. Some amphibians (frogs, toads, and salamanders), which live part of their lives in water and part on land, are also believed to be indicator species (Connections, right).

What Are Keystone Species? Not all species in a community or ecosystem contribute equally to ecosystem structure and processes. Indeed, removing some species may have little or no effect on the structure and function of an ecosystem (although identifying such species is quite difficult).

In contrast, the roles of some species in an ecosystem are much more important than their abundance or biomass suggests. Ecologists call such species **keystone species**, although this designation is controversial.* Such species play pivotal roles in the structure, function, and integrity of an ecosystem because **(1)** their strong interactions with other species affect the health and survival of these species, and **(2)** they process material out of proportion to their numbers or biomass.

Critical roles of keystone species include **(1)** pollination of flowering plant species by bees, hummingbirds, bats (Figure 8-1), and other species, **(2)** dispersion of seeds by fruit-eating animals such as bats (Figure 8-1), with the undigested seeds being scattered in their feces, **(3)** habitat modification, **(4)** predation by top carnivores to control the populations of various species, **(5)** improving the ability of plant species to obtain soil minerals and water, and **(6)** efficient recycling of animal wastes. Some species play more than one of these roles.

---

*All species play some role in their ecosystems and thus are important. Whereas some scientists consider all species equally important, others consider certain species to be more important than others in helping maintain the structure and function of ecosystems of which they are a part.

CONNECTIONS

Amphibians (frogs, toads, and salamanders) first appeared about 350 million years ago. These cold-blooded creatures range in size from a frog that can sit on your thumb to a Japanese salamander that is about 1.5 meters (5 feet) long.

Within the last two decades, populations of hundreds of the world's estimated 5,100 amphibian species (including 2,700 frog and toad species) have been vanishing or declining in almost every part of the world, even in protected wildlife reserves and parks.

In some locales, frog deformities, such as extra legs and missing legs, are occurring in unusually high numbers. This information was discovered and published by schoolchildren in Henderson, Minnesota, who were catching frogs in a farm pond.

According to the World Conservation Union, 25% of all known amphibian species are extinct, endangered, or vulnerable. The Nature Conservancy estimates that 38% of the amphibian species in the United States are endangered. In recent years, 14 amphibian species have vanished from Australia, and the golden toad is now extinct in Costa Rica.

Frogs are especially vulnerable to environmental disruption. As tadpoles they live in water and eat plants, and as adults they live mostly on land and eat insects (which can expose them to pesticides). Their eggs have no protective shells to block radiation or pollution. As adults, they take in water and air through their thin, permeable skins that can readily absorb pollutants from water, air, or soil.

Because each type of frog lives in a small, specialized habitat, they are sensitive indicators of environmental conditions in a variety of places.

Scientists have identified a variety of possible causes for amphibian declines. They include the following:

- *Loss of habitat*, especially because of the draining and filling of inland wetlands (Figure 7-19, p. 170), clear-cutting of forests, and fragmentation of habitats into pieces too small to support populations of some amphibians.

- *Prolonged drought*, which dries up breeding pools so that few tadpoles survive. Dehydration can also weaken amphibians, making them more susceptible to fatal viruses, bacteria, fungi, and parasites.

- *Pollution*. Frog eggs, tadpoles, and adults are very sensitive to many pollutants, especially **(1)** pesticides, **(2)** nitrates (mostly from runoff of commercial inorganic fertilizers and animal wastes into aquatic system), and **(3)** increased acidity of the water in lakes and ponds from acid deposition (acid rain). Exposure to such pollutants may harm amphibians' immune and endocrine systems and make them more vulnerable to bacterial infections and perhaps to a newly discovered type of skin fungus.

- *Increases in ultraviolet radiation* caused by reductions in stratospheric ozone.

- *Increased incidence of parasitism* by a flatworm (trematode), which may account for many frog deformities but not the worldwide decline of amphibians.

- *Overhunting*, especially in Asia and France, where frog legs are a delicacy.

- *Epidemic diseases* such as the chytrid fungus and iridoviruses. Pollution and other factors that weaken the immune systems of amphibians may make them more susceptible to disease organisms.

- *Immigration or introduction* of *alien predators and competitors* (such as fish) and disease organisms such as the chytrid (which may have been introduced into Australia by imported tropical fish).

In most cases, the decline or disappearance of amphibian species probably is caused by a combination of such factors. Scientists are concerned about amphibians' decline for three reasons:

- It suggests that the world's environmental health is deteriorating rapidly because amphibians generally are tough survivors and are sensitive bioindicators of changes in environmental conditions.

- Adult amphibians play important roles in the world's ecosystems. For example, amphibians eat more insects (including mosquitoes) than do birds. In some habitats, extinction of certain amphibian species could also result in extinction of other species, such as reptiles, birds, aquatic insects, fish, mammals, and other amphibians that feed on them or their larvae.

- From a human perspective amphibians represent a genetic storehouse of pharmaceutical products waiting to be discovered. Hundreds of secretions from amphibian skin have been isolated, and some of these compounds are being used as painkillers and antibiotics and in treating burns and heart attacks.

Several scientists, such as zoologist Andrew Blaustein, hypothesize that some of these causes of declines in amphibian populations may interact in a synergistic manner (p. 52) to amplify their individual effects.

As possible indicator species, amphibians may be sending us an important message. They do not need us, but we and other species need them.

Critical Thinking

On an evolutionary time scale, all species eventually become extinct. Some suggest that the widespread disappearance of amphibians is the result of natural responses to changing environmental conditions. Others contend that these losses are caused mostly by human activities and that such declines are a warning of possible danger for our own species and other species. What is your position? Why?

Beneficial *habitat modifications* by keystone species include the following:

■ Elephants push over, break, or uproot trees, creating forest openings in the savanna grasslands and woodlands of Africa. This promotes the growth of grasses and other forage plants that benefit smaller grazing species such as antelope and accelerates nutrient cycling rates.

■ Bats (p. 174) and birds regenerate deforested areas by depositing plant seeds in their droppings.

■ Beaver dams can change a fast-moving stream into a pond or lake. This attracts fish (such as bluegill), muskrats, herons, and ducks that prefer deeper, slower-moving water, as well as woodpeckers that feed on dead trees emerging from the pond. However, by felling trees to build dams beavers can also kill large expanses of terrestrial forests.

Top predator keystone species exert a stabilizing effect on their ecosystems by feeding on and regulating the populations of certain species. Examples are the wolf (p. 585), leopard, lion, alligator (Connections, right), sea otter (p. 198), and great white shark.

Have you thanked a *dung beetle* today? You should because these keystone species play a combination of vital roles in the ecosystems where they are found:

■ They rapidly remove, bury, and recycle animal wastes (dung) so that without them we would be up to our eyeballs in such waste, and many plants would be starved for nutrients.

■ They establish new plants because the dung they bury contains seeds that have passed through the digestive tracts of fruit-eating animals.

■ They churn and aerate the soil, making it more suitable for plant life.

■ They reduce populations of microorganisms that spread disease to wild and domesticated animals (including humans) because the beetle larvae feed on parasitic worms and maggots that live in the dung.

The loss of a keystone species can lead to population crashes and extinctions of other species that depend on it for certain services, a ripple or domino effect that spreads throughout an ecosystem. According to biologist Edward O. Wilson, "The loss of a keystone species is like a drill accidentally striking a power line. It causes lights to go out all over."

8-3 SPECIES INTERACTIONS: COMPETITION AND PREDATION

How Do Species Interact? An Overview When different species in an ecosystem have activities or resource requirements in common, they may interact with one another. Members of these species may be harmed by, benefit from, or be unaffected by the interaction. There are five basic types of interactions between species: **(1)** *interspecific competition*, **(2)** *predation*, **(3)** *parasitism*, **(4)** *mutualism*, and **(5)** *commensalism*.

These interactions tend to regulate the populations of species and can help them survive changes in environmental conditions, as discussed in more detail in Section 9-2, p. 202.

How Do Species Compete for Resources? Intraspecific and Interspecific Competition Competition between members of the same species for the same resources is called **intraspecific competition**. Competition between members of two or more different species for food, space, or any other limited resource is called **interspecific competition**.

Intraspecific competition can be intense because members of a particular species compete directly for the same resources. Some plants, especially in deserts, gain a competitive advantage by secreting chemicals that inhibit the growth of seedlings of their own and other species. Other plant species, such as dandelions, compete with other members of their species for living space and soil nutrients by dispersing their seeds to other sites by air (wind), water, or animals.

Another way members of the same species compete is through **territoriality**, in which organisms patrol or mark an area around their home, nesting, or major feeding site and defend it against members of their own species. Robins chase other robins away from their mating and nesting sites, and rhinos and lions use urine or scent to mark their breeding areas.

Territory size varies and is small for robins but large for species such as tigers. Two potential disadvantages of territoriality are **(1)** exclusion of many male members of a population from breeding and **(2)** large energy expenditure in defending the territory.

As long as commonly used resources are abundant, different species can share them. This allows each species to come closer to occupying the *fundamental niche* it would occupy if there were no competition from other species (Figure 5-7, p. 111).

However, most species face competition from other species for one or more limited resources (such as food, sunlight, water, soil nutrients, space, nesting sites, and good places to hide). Because of such *interspecific competition*, parts of the fundamental niches of different species overlap (Figure 5-7, p. 111). The more the niches of two species overlap, the more they compete with one another. With significant niche overlap, one of the competing species must **(1)** migrate to another area (if possible), **(2)** shift its feeding habits or behavior through natural selection and evolution (Section 5-2, p. 107), **(3)** suffer a sharp population decline, or **(4)** become extinct in that area.

Why Should We Care About Alligators?

The American alligator, North America's largest reptile, has no natural predators except humans. This species, which has been around for about 200 million years, has been able to adapt to numerous changes in the earth's environmental conditions.

This changed fairly recently when hunters began killing large numbers of these animals for **(1)** their exotic meat and **(2)** their supple belly skin, used to make shoes, belts, and pocketbooks.

Other people considered alligators to be useless and dangerous and hunted them for sport or out of hatred. Between 1950 and 1960, hunters wiped out 90% of the alligators in Louisiana, and by the 1960s the alligator population in the Florida Everglades also was near extinction.

People who say "So what?" are overlooking the alligator's important ecological role or *niche* in subtropical wetland ecosystems. Alligators dig deep depressions, or gator holes, that **(1)** collect fresh water during dry spells, **(2)** serve as refuges for aquatic life, and **(3)** supply fresh water and food for many animals.

In addition, large alligator nesting mounds provide nesting and feeding sites for herons and egrets. Alligators also eat large numbers of gar (a predatory fish) and thus help maintain populations of game fish such as bass and bream.

As alligators move from gator holes to nesting mounds, they help keep areas of open water free of invading vegetation. Without these ecosystem services, freshwater ponds and coastal wetlands in the alligator's habitat would be filled in by shrubs and trees, and dozens of species would disappear.

Some ecologists classify the North American alligator as a *keystone species* because of these important ecological roles in helping maintain the structure and function of its natural ecosystems.

In 1967, the U.S. government placed the American alligator on the endangered species list. Protected from hunters, the alligator population made a strong comeback in many areas by 1975—too strong, according to those who find alligators in their backyards and swimming pools and to duck hunters, whose retriever dogs sometimes are eaten by alligators.

In 1977, the U.S. Fish and Wildlife Service reclassified the American alligator from an *endangered* to a *threatened* species in Florida, Louisiana, and Texas, where 90% of the animals live. In 1987 this reclassification was extended to seven other states.

Alligators now number perhaps 3 million, most in Florida and Louisiana. It is generally illegal to kill members of a threatened species, but limited kills by licensed hunters are allowed in some areas of Florida, Louisiana, and South Carolina to control the population. The comeback of the American alligator from near premature extinction by overhunting is an important success story in wildlife conservation.

The increased demand for alligator meat and hides has created a booming business in alligator farms, especially in Florida. Such success reduces the need for illegal hunting of wild alligators.

Critical Thinking

Some homeowners in Florida believe that they should have the right to kill any alligator found on their property. Others argue that this should not be allowed because **(1)** alligators are a threatened species and **(2)** housing developments have invaded the habitats of alligators, not the other way around. What is your opinion on this issue? Explain.

Species compete with other species in two main ways. In **interference competition**, one species may limit another's access to some resource, regardless of its abundance, using the same types of methods found in intraspecific competition. For example, a territorial hummingbird species may defend patches of spring wildflowers from which it gets nectar by chasing away members of other hummingbird species.

In desert and grassland habitats, many plants release chemicals into the soil. These chemicals prevent the growth of competing species or reduce the rates at which their seeds germinate.

In **exploitation competition**, competing species have roughly equal access to a specific resource but differ in how fast or efficiently they exploit it. The species that can use the resource more quickly gets more of the resource and hampers the growth, reproduction, or survival of the other species.

What Is the Competitive Exclusion Principle?
Sometimes one species eliminates another species in a particular area through competition for limited resources. In 1934, Russian ecologist G. F. Gause demonstrated this effect by carrying out a laboratory experiment in which two closely related species of single-celled, bacteria-eating *Paramecium* were grown, first separately and then together in culture tubes (Figure 8-8).

The graph on the left of Figure 8-8 shows what happened when both species were grown under identical conditions in separate containers with ample supplies of food (bacteria). In this case, both species grew

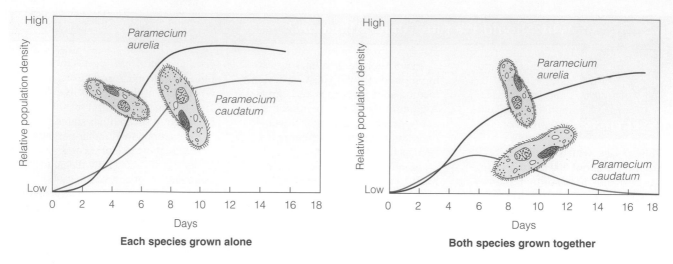

Figure 8-8 The results of G. F. Gause's classic laboratory experiment with two similar single-celled, bacteria-eating *Paramecium* species (which reproduce asexually) support the competitive exclusion principle that similar species cannot occupy the same ecological niche indefinitely.

rapidly and established stable populations. However, the smaller *Paramecium aurelia* (red curve) grew faster than the larger *Paramecium caudatum* (green curve), indicating that the former used the available food supply more efficiently than the latter. The graph on the right shows that when both species were grown together in a culture tube with a limited amount of bacteria, the smaller *Paramecium aurelia* (red curve) outmultiplied and eliminated the larger *Paramecium caudatum* (green curve).

This research, which has been supported by various laboratory and field experiments using other animal species, showed that two species that need the same resource cannot coexist indefinitely in an ecosystem in which there is not enough of that resource to meet the needs of both species. In other words, the niches of two species cannot overlap completely or significantly for very long. This finding is called the **competitive exclusion principle**.

How Have Some Species Reduced or Avoided Competition? Over a time scale long enough for evolution to occur, some species that compete for the same resources evolve adaptations that reduce or avoid competition or overlap of their fundamental niches (Figure 5-7, p. 111). One way this happens is through **resource partitioning**, the dividing up of scarce resources so that species with similar needs use them at different times, in different ways, or in different places (Figure 8-9). In effect, they evolve traits that allow them to share the wealth.

Figure 8-9 Specialized feeding niches of various bird species in a coastal wetland. Such resource partitioning reduces competition and allows sharing of limited resources.

Figure 8-10 *Resource partitioning* and *niche specialization* as a result of competition between two species. The left diagram shows the overlapping niches of two competing species. The right diagram shows that through evolution the niches of the two species become separated and more specialized (narrower) so that they avoid competing for the same resources.

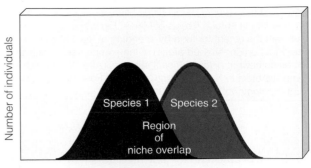

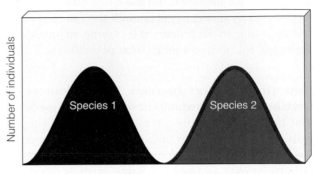

Each of the competing species occupies a *realized niche* that makes up only part of its *fundamental niche*. The result is that through evolution the fairly broad niches of two competing species (Figure 8-10, top) become more specialized (Figure 8-10, bottom). Resource partitioning through niche specialization is also found in the different layers of tropical rain forests (Figure 6-28, p. 144).

Here are some other examples of resource partitioning. When lions and leopards live in the same area, lions take mostly larger animals as prey, and leopards take smaller ones. Hawks and owls feed on similar prey, but hawks hunt during the day and owls hunt at night. Some bird species feed on the ground, whereas others seek food in trees and shrubs.

Ecologist Robert H. MacArthur studied the feeding habits of five species of warblers (small insect-eating birds) that coexist in the forests of the northeastern United States and in the adjacent area of Canada. Although they appear to be competing for the same food resources, MacArthur found that the bird species reduce competition through resource partitioning by spending at least half their time hunting for insects in different parts of trees (Figure 8-11).

On an evolutionary time scale, closely related and competing species may also partition resources and lessen competition through *character displacement*. Such species develop physical or behavioral adaptations that allow them to use different resources. In birds, for instance, bill sizes of species coexisting in the same ecosystem often differ. For example, field research has shown that one species of ground finch with a longer, thinner bill feeds on large insects, whereas another with a shorter, thicker bill feeds on seeds or small insects.

How Do Predator and Prey Species Interact? In **predation**, members of one species (the *predator*) feed directly on all or part of a living organism of another species (the *prey*). However, they do not live on or in the prey, and the prey may or may not die from the interaction. In this interaction, the predator benefits and the individual prey is clearly harmed. Together, the two kinds of organisms, such as lions (the predator or hunter) and zebras (the prey or hunted), are said to have a **predator-prey relationship**, as depicted in Figures 4-11 (p. 78), 4-12 (p. 78), 4-18 (p. 83), and 4-19 (p. 84).

At the individual level, members of the prey species are clearly harmed. However, at the population level predation can benefit the prey species because predators such as tigers and some types of sharks often kill the sick, weak, and aged members (Case Study, p. 185). Reducing the prey population gives remaining prey greater access to the available food supply. It can also improve the genetic stock of the prey population, which enhances its chances of reproductive success and long-term survival. The effects of predation on populations of predator and prey species are discussed in more detail in Section 9-2, p. 202.

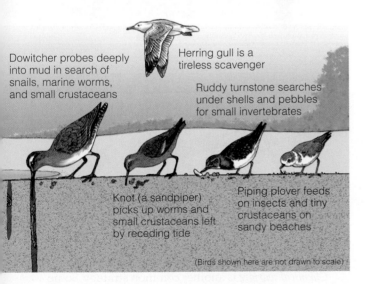

Dowitcher probes deeply into mud in search of snails, marine worms, and small crustaceans

Herring gull is a tireless scavenger

Ruddy turnstone searches under shells and pebbles for small invertebrates

Knot (a sandpiper) picks up worms and small crustaceans left by receding tide

Piping plover feeds on insects and tiny crustaceans on sandy beaches

(Birds shown here are not drawn to scale)

Figure 8-11 *Resource partitioning* of five species of common insect-eating warblers in spruce forests of Maine. Each species minimizes competition with the others for food by spending at least half its feeding time in a distinct portion (shaded areas) of the spruce trees; each also consumes somewhat different insect species. (After "Population Ecology of Some Warblers in Northeastern Coniferous Forests," by R. H. MacArthur, 1958, *Ecology*, Vol. 36, 533–36)

Some people tend to view predators with contempt. When a hawk tries to capture and feed on a rabbit, some tend to root for the rabbit. Yet the hawk (like all predators) is merely trying to get enough food to feed itself and its young; in the process, it is playing an important ecological role in controlling rabbit populations.

How Do Predators Increase Their Chances of Getting a Meal? Predators have a variety of methods that help them capture prey. *Herbivores* can simply walk, swim, or fly up to their plants they feed upon.

Carnivores feeding on mobile prey have two main options: *pursuit* and *ambush*. Some, such as the cheetah, catch prey by being able to run fast; others, such as the American bald eagle, fly and have keen eyesight; still others, such as wolves and African lions, cooperate in capturing their prey by hunting in packs.

Other predators have characteristics or strategies that enable them to hide and ambush their prey. Examples include **(1)** praying mantises (Figure 4-1, left, p. 70) sitting in flowers of a similar color and ambushing visiting insects, **(2)** white ermines (a type of weasel) and snowy owls hunting in snow-covered areas, **(3)** the alligator snapping turtle lying camouflaged on its stream-bottom habitat and dangling its worm-shaped tongue to entice fish into its powerful jaws, and **(4)** people camouflaging themselves and using traps to ambush wild game.

How Do Prey Defend Themselves Against or Avoid Predators? Species have various characteristics that enable them to avoid predators. They include **(1)** the ability to run, swim, or fly fast, **(2)** a highly developed sense of sight or smell that alerts them to the presence of predators, **(3)** protective shells (as on armadillos, which roll themselves up into an armor-plated ball, and turtles), **(4)** thick bark (giant sequoia), and **(5)** spines (porcupines) or thorns (cacti and rosebushes). Many lizards have brightly colored tails that break off when they are attacked, often giving them enough time to escape.

Other prey species use *camouflage* by having certain shapes or colors (Figure 8-12a) or the ability to change color (chameleons and cuttlefish). A frog may be almost invisible against its background (Figure 8-12b), and an arctic hare in its white winter fur blends into the snow.

Some insect species have evolved shapes that look like twigs or bird droppings on leaves.

Chemical warfare is another common strategy. Some prey species discourage predators with chemicals that are **(1)** poisonous (oleander plants), **(2)** irritating (bom-

Why Are Sharks Important Species?

CASE STUDY

The world's 350 shark species vary widely in size. The smallest is the dwarf dog shark, about the size of a large goldfish, and the largest is the whale shark, the world's largest fish, which can grow to 18 meters (60 feet) long and weigh as much as two full-grown African elephants.

Various shark species, feeding at the top of food webs, cull injured and sick animals from the ocean and thus play an important ecological role. Without such shark species the oceans would be overcrowded with dead and dying fish.

Many people, influenced by movies and popular novels, think of sharks as people-eating monsters. However, the three largest species—the whale shark, basking shark, and megamouth shark—are gentle giants that swim through the water with their mouths open, filtering out and swallowing huge quantities of *plankton* (small free-floating sea creatures).

Every year, members of a few species of shark—mostly great white, bull, tiger, gray reef, lemon, and blue—injure about 50 to 100 people worldwide and kill between 5 and 12 people. Most attacks are by great white sharks, which feed on sea lions and other marine mammals and sometimes mistake divers and surfers for their usual prey.

For every shark that injures a person, we kill at least 1 million sharks, for a total of 100 million sharks each year. Sharks are killed mostly for their fins, widely used in Asia as a soup ingredient and as a pharmaceutical cure-all and worth as much as $563 per kilogram ($256 per pound). In Hong Kong, a single bowl of shark fin soup can sell for as much as $100.

Sharks are also killed for their (1) livers, (2) meat (especially mako and thresher), (3) hides (a source of exotic, high-quality leather), and (4) jaws (especially great whites, whose jaws are worth thousands of dollars to collectors), or (5) just because we fear them. Some sharks (especially blue, mako, and oceanic whitetip) die when they are trapped as bycatch in nets or lines deployed to catch swordfish, tuna, shrimp, and other commercially important species.

Sharks also help save human lives. In addition to providing people with food, they are helping us learn how to fight cancer (which sharks almost never get), bacteria, and viruses. Their highly effective immune system is being studied because it allows wounds to heal without becoming infected. This research might help prolong your life or the life of a loved one someday.

Sharks have several natural traits that make them prone to population declines from overfishing. They (1) have only a few offspring (between 2 and 10) once every year or two, (2) take 10–24 years to reach sexual maturity and begin reproducing, and (3) have long gestation (pregnancy) periods, up to 24 months for some species.

Sharks are among the most vulnerable and least protected animals on the earth. Eight of the world's shark species, including great whites, sandtigers, and kitefins, are now considered critically endangered, endangered, or vulnerable to extinction. Of the 125 countries that commercially catch more than 100 million sharks per year, only four—Australia, Canada, New Zealand, and the United States—have implemented management plans for shark fisheries, and these plans are hard to enforce.

With more than 400 million years of evolution behind them, sharks have had a long time to get things right. Preserving their evolutionary genetic development begins with the knowledge that sharks do not need us, but we and other species need them.

Critical Thinking

After reading this information, has your attitude toward sharks changed? If so, how has it changed?

---

bardier beetles, Figure 8-12c), (3) foul smelling (skunks, skunk cabbages, and stinkbugs), or (4) bad tasting (buttercups and monarch butterflies, Figure 8-12d). Scientists have identified more than 10,000 defensive chemicals made by plants, including cocaine, caffeine, nicotine, cyanide, opium, strychnine, peyote, and rotenone (used as an insecticide).

Many bad-tasting, bad-smelling, toxic, or stinging prey species have evolved *warning coloration*, brightly colored advertising that enables experienced predators to recognize and avoid them. Examples are (1) brilliantly colored poisonous frogs (Figure 8-12e) and red-, yellow-, and black-striped coral snakes and (2) foul-tasting monarch butterflies (Figure 8-12d)

and grasshoppers. Other butterfly species, such as the nonpoisonous viceroy (Figure 8-12f), gain some protection by looking and acting like the poisonous monarch (Figure 8-12d), a protective device known as *mimicry*.

Some prey species use behavioral strategies to avoid predation. Some attempt to scare off predators by (1) puffing up (blowfish), (2) spreading their wings (peacocks), or (3) mimicking a predator (Figure 8-12h). To help fool or frighten would-be predators, some moths have wings that look likes the eyes of much larger animals (Figure 8-12g). Other prey gain some protection by living in large groups (schools of fish, herds of antelope, flocks of birds).

(a) African stoneplants

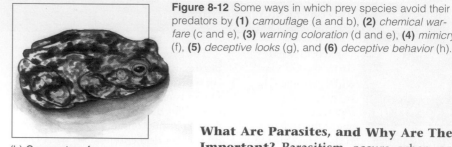

Figure 8-12 Some ways in which prey species avoid their predators by **(1)** *camouflage* (a and b), **(2)** *chemical warfare* (c and e), **(3)** *warning coloration* (d and e), **(4)** *mimicry* (f), **(5)** *deceptive looks* (g), and **(6)** *deceptive behavior* (h).

(b) Canyon tree frog

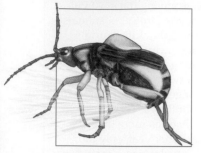

(c) Bombardier beetle

(d) Foul tasting monarch butterfly

(e) Poison dart frog

(f) Viceroy butterfly mimics monarch butterfly

(g) Hind wings of io month resemble eyes of a much larger animal.

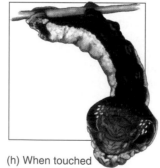

(h) When touched snake caterpillar changes shape to look like head of snake.

8-4 SYMBIOTIC SPECIES INTERACTIONS: PARASITISM, MUTUALISM, AND COMMENSALISM

What Is Symbiosis? Symbiosis is a long-lasting relationship in which species live together in an intimate association. There are three general types of symbiosis: *parasitism*, *mutualism*, and *commensalism*.

What Are Parasites, and Why Are They Important? **Parasitism** occurs when one species (the *parasite*) feeds on part of another organism (the *host*) by living on or in the host. In this symbiotic relationship, the parasite benefits and the host is harmed.

Parasitism can be viewed as a special form of predation, but unlike a conventional predator, a parasite **(1)** usually is smaller than its host (prey), **(2)** remains closely associated with, draws nourishment from, and may gradually weaken its host over time, and **(3)** rarely kills its host.

Tapeworms, disease-causing microorganisms (pathogens), and other parasites live *inside* their hosts. Other parasites, such as ticks, fleas, mosquitoes, mistletoe plants, and fungi (that cause diseases such as athlete's foot) attach themselves to the *outside* of their hosts. Some parasites move from one host to another, as fleas and ticks do; others, such as tapeworms, spend their adult lives with a single host.

From the host's point of view parasites are harmful, but parasites play important ecological roles. Collectively, the incredibly complex matrix of parasitic relationships in an ecosystem acts somewhat like glue that helps hold the species in an ecosystem together. Parasites also promote biodiversity by helping prevent some species from becoming too plentiful and eliminating other species through competition.

How Do Species Interact So That Both Species Benefit? In **mutualism** two species involved in a symbiotic relationship interact in ways that benefit both. Such benefits include **(1)** having pollen and seeds dispersed for reproduction, **(2)** being supplied with food, or **(3)** receiving protection. The *pollination* relationship between flowering plants and animals such as insects (Figure 4-1, right, p. 70), birds, and bats (p. 173) is one of the most common forms of mutualism. Examples of *nutritional mutualism* include the following:

■ *Lichens*, hardy species that can grow on trees or barren rocks, consist of colorful photosynthetic algae and chlorophyll-lacking fungi living together (Figure 17-1, p. 417). The fungi provide a home for

Figure 8-13 Two examples of mutualism. **(a)** Oxpeckers (or tickbirds) feed on the parasitic ticks that infest large, thick-skinned animals such as a black rhinoceros, and **(b)** a clownfish lives among deadly stinging sea anemones.

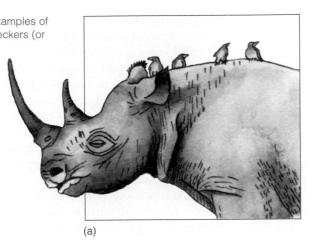

(a)

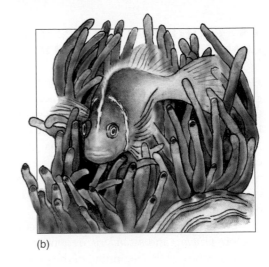

(b)

the algae, and their bodies collect and hold moisture and mineral nutrients used by both species. The algae, through photosynthesis, provide sugars as food for themselves and the fungi.

- Plants in the legume family support root nodules, where *Rhizobium* bacteria convert atmospheric nitrogen into a form usable by the plants, and the plants provide the bacteria with some simple sugars.

- Vast armies of bacteria in the digestive systems of animals break down (digest) their food. The bacteria gain a safe home with a steady food supply; the animal gains more efficient access to a large source of energy.

- Protozoans living in the guts of termites digest the wood the termites eat, and the insects use some of the resulting sugar as food.

Examples of mutualistic relationships involving *nutrition* and *protection* are as follows:

- Birds ride on the backs of large animals such as African buffalo, elephants, and rhinoceroses (Figure 8-13a). The birds remove and eat parasites from the animal's body and often make noises warning the animal when predators approach.

- Clownfish species live within sea anemones, whose tentacles sting and paralyze most fish that touch them (Figure 8-13b). The clownfish, which are not harmed by tentacles, gain protection from predators and feed on the detritus left from the meals of the anemones. The sea anemones benefit because the clownfish protect them from some of their predators.

- Minute fungi called mycorrhizae live on the roots of many plants. The fungi get nutrition from a plant's roots and in turn benefit the plant by using their myriad networks of

hairlike extensions to improve the plant's ability to extract nutrients and water from the soil.

It is tempting to think of mutualism as an example of cooperation between species, but actually it involves each species benefiting by exploiting the other.

How Do Species Interact So That One Benefits but the Other Is Not Harmed? Commensalism is a symbiotic interaction that benefits one species but neither harms nor helps the other species much, if at all. For example, redwood sorrel, a small herb, benefits from growing in the shade of tall redwood trees, with no known negative effects on the redwood trees.

Another example is the commensalistic relationship between various trees and other plants called *epiphytes* (such as some types of orchids and bromeliads) that attach themselves to the trunks or branches of large trees (Figure 8-14) in tropical and subtropical forests. These so-called air plants benefit by having a solid base on which to grow and by living in an elevated spot that gives them better access to sunlight. Their position in the tree allows them to get most of their water from the humid air and from rain collecting in their usually cupped leaves. They also absorb nutrient salts falling from the tree's upper leaves and limbs and from the dust in rainwater.

Figure 8-14 Commensalism between a white orchid (an epiphyte or air plant from the tropical forests of Latin America) that roots in the fork of a tree rather than the soil without penetrating or harming the tree. In this interaction, the epiphytes gain access to water, nutrient debris, and sunlight; the tree apparently remains unharmed unless it contains a large number of epiphytes.

8-5 ECOLOGICAL SUCCESSION: COMMUNITIES IN TRANSITION

How Do Ecosystems Respond to Change? One characteristic of all communities and ecosystems is that their structures constantly change in response to changing environmental conditions. The gradual change in species composition of a given area is called **ecological succession**. During succession some species colonize an area and their populations become more numerous, whereas populations of other species decline and even disappear.

Ecologists recognize two types of ecological succession: *primary* and *secondary*, depending on the conditions present at the beginning of the process. **Primary succession** involves the gradual establishment of biotic communities on nearly lifeless ground. In contrast, **secondary succession**, the more common type of succession, involves the reestablishment of biotic communities in an area where a biotic community is already present.

What Is Primary Succession? Establishing Life on Lifeless Ground Primary succession begins with an essentially lifeless area where there is no soil in a terrestrial ecosystem (Figure 8-15) or no bottom sediment in an aquatic ecosystem. Examples include (1) bare rock exposed by a retreating glacier or severe soil erosion, (2) newly cooled lava, (3) an abandoned highway or parking lot, or (4) a newly created shallow pond or reservoir.

Before a community of plants (producers), consumers, and decomposers can become established on land, there must be *soil*: a complex mixture of rock particles, decaying organic matter, air, water, and living

organisms. Depending mostly on the climate, it takes natural processes several hundred to several thousand years to produce fertile soil.

Soil formation begins when hardy **pioneer species** attach themselves to inhospitable patches of bare rock. Examples are wind-dispersed lichens (Figure 17-1, p. 417) and mosses, which can withstand the lack of moisture and soil nutrients and hot and cold temperature extremes found in such habitats.

These species can extract nutrients from dust in rain or snow and from bare rock. They start the soil formation process on patches of bare rock by (1) trapping wind-blown soil particles and tiny pieces of detritus, (2) producing tiny bits of organic matter, and (3) secreting mild acids that slowly fragment and break down the rock. This chemical breakdown (weathering) is hastened by physical weathering such as the fragmentation of rock when water freezes in cracks and expands.

As patches of soil build up and spread, eventually the community of lichens and mosses is replaced by a community of (1) small perennial grasses (plants that live for more than 2 years without having to reseed) and (2) herbs (ferns in tropical areas), whose seeds germinate after being blown in by the wind or carried there in the droppings of birds or on the coats of mammals.

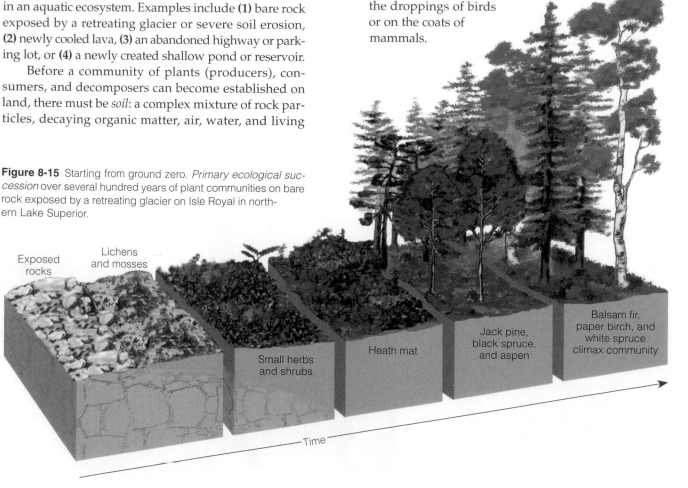

Figure 8-15 Starting from ground zero. *Primary ecological succession* over several hundred years of plant communities on bare rock exposed by a retreating glacier on Isle Royal in northern Lake Superior.

Exposed rocks

Lichens and mosses

Small herbs and shrubs

Heath mat

Jack pine, black spruce, and aspen

Balsam fir, paper birch, and white spruce climax community

Time

These **early successional plant species** (1) grow close to the ground, (2) can establish large populations quickly under harsh conditions, and (3) have short lives. Some of their roots penetrate the rock and help break it up into more soil particles, and the decay of their wastes and dead bodies adds more nutrients to the soil.

After hundreds of years the soil may be deep and fertile enough to store enough moisture and nutrients to support the growth of less hardy **midsuccessional plant species** of herbs, grasses, and low shrubs. These, in turn, are usually replaced by trees that need lots of sunlight and are adapted to the area's climate and soil.

As these tree species grow and create shade, they are replaced by **late successional plant species** (mostly trees) that can tolerate shade. Unless fire, flooding, severe erosion, tree cutting, climate change, or other natural or human processes disturb the area, what was once bare rock becomes a complex forest community (Figure 8-15).

The specific composition of pioneer, early successional, midsuccessional, and late successional communities and the rates of primary succession vary from one site to another. Generally, primary succession occurs fastest in humid tropical areas and slowest in dry polar areas.

What Is Secondary Succession? Secondary succession begins in an area where the natural community of organisms has been disturbed, removed, or destroyed but the soil or bottom sediment remains. Candidates for secondary succession include (1) abandoned farmlands, (2) burned or cut forests, (3) heavily polluted streams, and (4) land that has been dammed or flooded. Because some soil or sediment is present, new vegetation usually can begin to germinate within a few weeks. This is aided by seeds already present in soils and dispersal of seeds from nearby plants by wind and by birds and animals.

In the central (Piedmont) region of North Carolina, European settlers cleared the mature native oak and hickory forests and replanted the land with crops. Some of the land was subsequently abandoned because of erosion and loss of soil nutrients. Figure 8-16 shows how such abandoned farmland has undergone secondary succession.

Descriptions of ecological succession usually focus on changes in vegetation. However, these changes in turn affect food and shelter for various types of animals. Thus, as succession proceeds the numbers and types of animals and decomposers also change. Figure 8-17 shows some of the wildlife species likely to be found at various stages of secondary ecological succession in areas with a temperate climate.

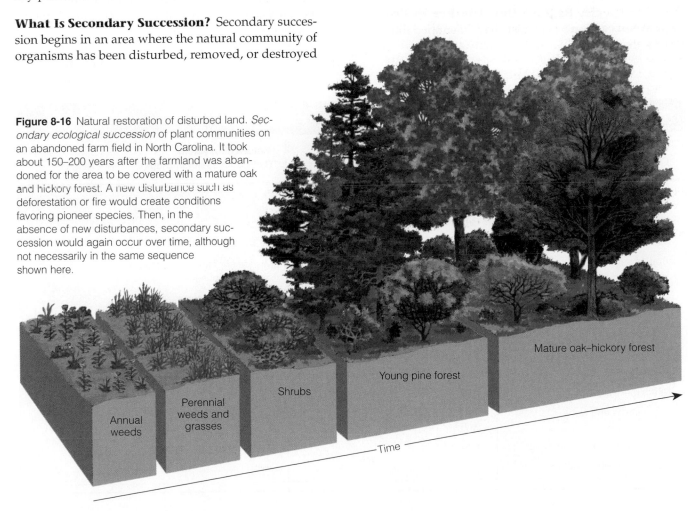

Figure 8-16 Natural restoration of disturbed land. *Secondary ecological succession* of plant communities on an abandoned farm field in North Carolina. It took about 150–200 years after the farmland was abandoned for the area to be covered with a mature oak and hickory forest. A new disturbance such as deforestation or fire would create conditions favoring pioneer species. Then, in the absence of new disturbances, secondary succession would again occur over time, although not necessarily in the same sequence shown here.

Annual weeds

Perennial weeds and grasses

Shrubs

Young pine forest

Mature oak–hickory forest

Time

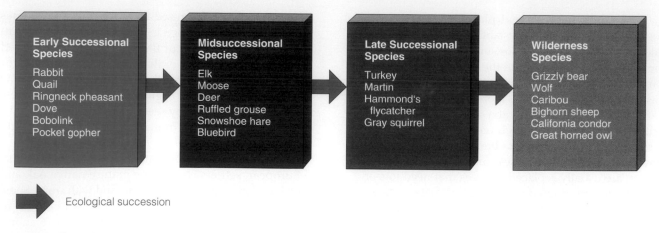

Early Successional Species
- Rabbit
- Quail
- Ringneck pheasant
- Dove
- Bobolink
- Pocket gopher

Midsuccessional Species
- Elk
- Moose
- Deer
- Ruffled grouse
- Snowshoe hare
- Bluebird

Late Successional Species
- Turkey
- Martin
- Hammond's flycatcher
- Gray squirrel

Wilderness Species
- Grizzly bear
- Wolf
- Caribou
- Bighorn sheep
- California condor
- Great horned owl

→ Ecological succession

Figure 8-17 Examples of wildlife species typically found at different stages of ecological succession in areas of the United States with a temperate climate.

Because primary and secondary succession involve changes in community structure, it is not surprising that the various stages of succession have different patterns of species diversity, trophic structure, niches, nutrient cycling, and energy flow and efficiency, as shown by the work of ecologists such as Eugene Odum (Table 8-1).

How Do Species Replace One Another in Ecological Succession? Ecologists have identified three factors that affect how and at what rate succession occurs: *facilitation*, *inhibition*, and *tolerance*.

Facilitation occurs when one set of species makes an area suitable for species with different niche requirements and is especially important in the soil-building stages of primary succession. For example, as lichens and mosses gradually build up soil on a rock in primary succession, herbs and grasses can colonize the site. Similarly, plants such as legumes add nitrogen to the soil, making it more suitable for other plants found at later stages of succession.

A common process governing secondary succession (and primary succession after soil has been built

Characteristic	Immature Ecosystem (Early Successional Stage)	Mature Ecosystem (Late Successional Stage)
Ecosystem Structure		
Plant size	Small	Large
Species diversity	Low	High
Trophic structure	Mostly producers, few decomposers	Mixture of producers, consumers, and decomposers
Ecological niches	Few, mostly generalized	Many, mostly specialized
Community organization (number of interconnecting links)	Low	High
Ecosystem Function		
Biomass	Low	High
Net primary productivity	High	Low
Food chains and webs	Simple, mostly plant → herbivore with few decomposers	Complex, dominated by decomposers
Efficiency of nutrient recycling	Low	High
Efficiency of energy use	Low	High

Table 8-1 Ecosystem Characteristics at Immature and Mature Stages of Ecological Succession

up) is *inhibition*, in which early species hinder the establishment and growth of other species. Inhibition often occurs when plants release toxic chemicals that reduce competition from other plants (interference competition). Succession then can proceed only when a fire, bulldozer, or other disturbance removes most of the inhibiting species.

In other cases, late successional plants are largely unaffected by plants at earlier stages of succession, a phenomenon known as *tolerance*. Tolerance may explain why late successional plants can thrive in mature communities without eliminating some early successional and midsuccessional plants. There is no consensus among ecologists about whether most of the stages of secondary succession occur because of inhibition, tolerance, or some combination of these two processes.

How Do Disturbances Affect Succession and Species Diversity? A **disturbance** is a change in environmental conditions that disrupts an ecosystem or community. Such disturbances can be catastrophic or gradual and caused by natural or human-caused changes (Table 8-2). At any time during primary or secondary succession, disturbances such as those listed in Table 8-2 can convert a particular stage of succession to an earlier stage.

Many people think of all environmental disturbances as harmful processes. Large catastrophic disturbances (Table 8-2) can devastate communities and ecosystems. However, many ecologists contend that in the long run some types of disturbances such as fires can be beneficial for the species diversity of communities and ecosystems. Such disturbances create new conditions that can discourage or eliminate some species but encourage others by releasing nutrients and creating unfilled niches.

For example, when a large tree falls in a tropical forest, this local disturbance increases sunlight and nutrients for growth of plants in the understory. When a log hits a rock in an intertidal zone (Figure 7-10, top, p. 160), this dislodges or kills many of the organisms that are growing on the rock and provides space for colonization of new intertidal organisms.

According to the *intermediate disturbance hypothesis*, communities that experience fairly frequent but moderate disturbances have the greatest species diversity (Figure 8-18). It is hypothesized that in such communities, moderate disturbances are large enough to create openings for colonizing species in disturbed areas but mild and infrequent enough to allow the survival of some mature species in undisturbed areas. This mixture of early and late successional plant species, along with those in between, means that the net species diversity of the whole area is greater than

Table 8-2 Changes Affecting Ecosystems	
Catastrophic*	
Natural	Drought
	Flood
	Fire
	Volcanic eruption
	Earthquake
	Hurricane or tornado
	Landslide
	Change in stream course
	Disease
Human-caused	Deforestation
	Overgrazing
	Plowing
	Erosion
	Pesticide application
	Fire
	Mining
	Toxic contamination
	Urbanization
	Water and air pollution
	Loss and degradation of wildlife habitat
Gradual*	
Natural	Climatic changes
	Immigration
	Adaption and evolution
	Ecological succession
	Disease
Human-caused	Salinization and waterlogging of soils from irrigation
	Soil compaction
	Groundwater depletion
	Water and air pollution
	Loss and degradation of wildlife habitat
	"Pests" and predator elimination
	Exotic species introduction
	Overhunting and overfishing
	Toxic contamination
	Urbanization
	Excessive tourism

*Many changes can be either catastrophic or gradual.

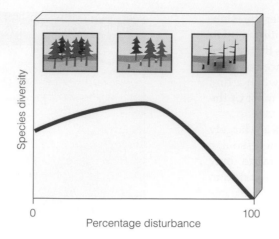

Figure 8-18 According to the *intermediate disturbance hypothesis*, moderate disturbances in communities promote greater species diversity than small or major disturbances.

Species diversity

0   Percentage disturbance   100

if the area was undisturbed or completely disrupted (Figure 8-18). Some field experiments have supported this hypothesis.

How Predictable Is Succession, and Is Nature in Balance? It is tempting to conclude that ecological succession is an orderly sequence in which each stage leads predictably to the next, more stable stage. According to this classic view, succession proceeds until an area is occupied by a generally predictable and stable type of *climax community* dominated by a few long-lived plant species and in balance with its environment. This equilibrium model of succession is what ecologists meant when they talked about the *balance of nature*.

Over the last several decades many ecologists have changed their views about balance and equilibrium in nature. When these ecologists look at a community or ecosystem, such as a young forest, they see continuous change, instability, and unpredictability instead of equilibrium, stability, and predictability.

Under the old *balance-of-nature* view, a large terrestrial community undergoing succession was viewed as eventually being covered with a predictable green blanket of climax vegetation. However, a close look at almost any ecosystem or community reveals that it consists of an ever-changing mosaic of vegetation patches at different stages of succession. These patches result from a variety of mostly unpredictable small and medium-sized disturbances (Figure 8-18).

This irregular quilt of vegetation increases the diversity of plant and animal life and provides sites where early successional species can gain a foothold. There is also a mosaic of shifting patches of plant and animal species in intertidal communities (Figure 7-10, p. 160) as a result of the random actions of waves.

Such research indicates that *we cannot predict the course of a given succession or view it as preordained progress toward an ideally adapted climax community*. Rather, succession reflects the ongoing struggle by different species **(1)** for enough light, nutrients, food, and space to survive and **(2)** to gain reproductive advantages over other species by occupying as much of their fundamental niches as possible.

This change in the way we view what is happening in nature explains why a growing number of ecologists prefer terms such as *biotic change* instead of *succession* (which implies an ordered and predictable sequence of changes). Many ecologists have also replaced the term *climax community* with terms such as *mature community* or a mosaic of *vegetation patches* at different stages of succession. Ecologists do not consider nature to be totally chaotic and unpredictable, but they have lowered their expectations in being able to make accurate predictions about the course of succession.

8-6 ECOLOGICAL STABILITY AND SUSTAINABILITY

What Is Stability? All living systems, from single-celled organisms to the biosphere, contain complex networks of negative and positive feedback loops (Section 3-2, p. 50) that interact to provide some degree of stability or sustainability over each system's expected life span.

This stability is maintained only by constant dynamic change in response to changing environmental conditions. For example, in a mature tropical rain forest, some trees die and others take their places. However, unless the forest is cut, burned, or otherwise destroyed you will still recognize it as a tropical rain forest 50 or 100 years from now.

It is useful to distinguish between three aspects of stability or sustainability in living systems:

- **Inertia**, or **persistence**: the ability of a living system to resist being disturbed or altered

- **Constancy**: the ability of a living system such as a population to keep its numbers within the limits imposed by available resources

- **Resilience**: the ability of a living system to bounce back after an external disturbance that is not too drastic

Populations, communities, and ecosystems are so complex and variable that ecologists have little understanding of how they maintain inertia, constancy, and resilience while continually responding to changes in environmental conditions.

Does Species Diversity Increase Ecosystem Stability? In the 1960s, most ecologists believed that the greater the species diversity and the accompanying web of feeding and biotic interactions in an ecosystem, the greater its stability. According to this hypothesis, an ecosystem with a diversity of species and feeding paths has more ways to respond to most environmental stresses because it does not have "all its eggs in one basket." However, most recent research indicates that there are exceptions to this intuitively appealing idea.

Of course, there is a minimum threshold of species diversity below which ecosystems cannot function; no ecosystem can function without some plants and decomposers. (Note that people and other animal consumers are not absolutely necessary for life to continue on the earth.)

Beyond this, it is difficult to know whether simple ecosystems are less stable than complex ones or to identify the threshold below which complex ecosystems fail. In part because some species play redundant roles (niches) in ecosystems, we do not know how many or which species can be eliminated before the entire ecosystem begins to lose stability or collapse.

Recent research by ecologist David Tilman and other researchers indicates that **(1)** ecosystems with more species tend to have higher net primary productivities and can be more resilient, but **(2)** the populations of individual species can fluctuate more widely in diverse ecosystems than in simpler ones. In 1999, an international team of ecologists lead by David Tilman reported that a study of eight European grasslands showed that species-rich fields had a smaller decline in net primary productivity during prolonged drought and rebounded more quickly than species-poor fields in the same area.

This supports the idea that some level of biodiversity provides insurance against catastrophe, but there is uncertainty about how much biodiversity is needed in various ecosystems. For example, some recent research suggests that average annual net primary productivity of an ecosystem reaches a peak with 10–40 producer species. Many ecosystems contain more producer species than this, but it is difficult to distinguish between those that are essential and those that are not.

Part of the problem is that ecologists disagree on how to define *stability* and *diversity*. Does an ecosystem need both high inertia and high resilience to be considered stable? Evidence suggests that some ecosystems have one of these properties but not the other. For example, tropical rain forests have high species diversity and high inertia; that is, they are resistant to significant alteration or destruction. However, once a large tract of tropical forest is severely degraded, the ecosystem's resilience sometimes is so low that the forest may not be restored. Nutrients (which are stored primarily in the vegetation, not in the soil) and other factors needed for recovery may no longer be present. Such a large-scale loss of forest cover may so change the local or regional climate that forests can no longer be supported.

Grasslands, by contrast, are **(1)** much less diverse than most forests and **(2)** have low inertia because they burn easily. However, because most of their plant matter is stored in underground roots, these ecosystems have high resilience and recover quickly. A grassland can be destroyed only if **(1)** its roots are plowed up and something else is planted in its place or **(2)** it is severely overgrazed by livestock or other herbivores.

Another difficulty is that populations, communities, and ecosystems are rarely, if ever, at equilibrium (balance). Instead, nature is in a continuing state of disturbance, fluctuation, and change.

Why Should We Bother to Protect Natural Systems? The Precautionary Principle Some developers argue that if biodiversity does not necessarily lead to increased ecological stability and if nature is mostly unpredictable, then there is no point in trying preserve and manage old-growth forests and other ecosystems. Why not **(1)** cut down diverse old-growth forests, use the timber resources, and replace the forests with tree farms, **(2)** convert the world's grasslands to cropfields, **(3)** drain and develop inland wetlands, **(4)** dump our toxic and radioactive wastes into the deep ocean, and **(5)** not worry about the premature extinction of species?

Ecologists and conservation biologists point out that there is overwhelming evidence that human disturbances (Figure 8-19) are disrupting some of the ecosystem services (Figure 4-36, p. 99) that support and sustain all life and all economies. They contend that our ignorance about the effects of our actions means that more than ever we need to make potentially damaging changes to ecosystems only with great caution.

By analogy, we know that eating too much of certain types of foods and not getting enough exercise can greatly increase the chances of heart attacks, diabetes, and other disorders. However, the exact

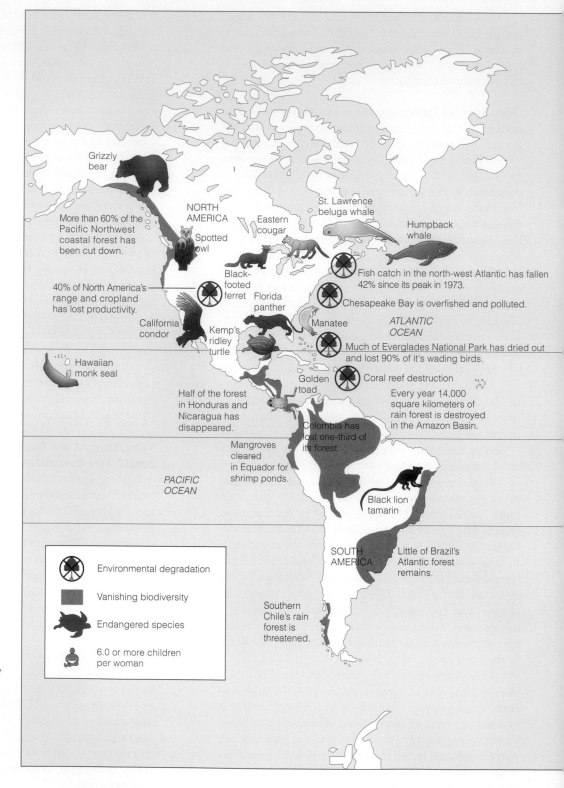

Figure 8-19 Examples of how some of the earth's natural resources are being depleted and degraded at an accelerating rate as a result of the exponential growth of the human population and resource use by humankind. (Data from The World Conservation Union, Wildlife Fund, Conservation International, United Nations, Population Reference Bureau, U.S. Fish and Wildlife Service, and Daniel Boivin)

connections between chemicals in these foods and between exercise and these health problems are largely unknown. Instead of using this uncertainty and unpredictability as an excuse to continue overeating and not exercising, the wise course is to eat better and exercise more to help *prevent* potentially serious health problems.

This approach is based on the **precautionary principle**: When there is considerable evidence that an activity raises threats of harm to human health or the environment, we should take precautionary measures to prevent harm even if some of the cause and effect relationships are not fully established scientifically. It is based on the commonsense idea behind many adages

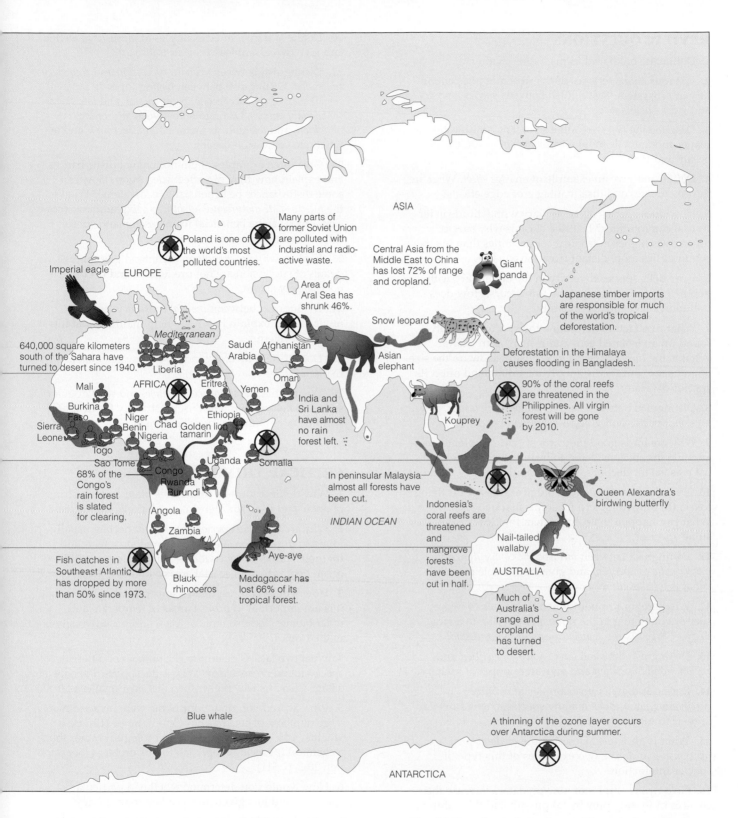

Imperial eagle

EUROPE

Poland is one of the world's most polluted countries.

Many parts of former Soviet Union are polluted with industrial and radio-active waste.

ASIA

Central Asia from the Middle East to China has lost 72% of range and cropland.

Giant panda

Area of Aral Sea has shrunk 46%.

Japanese timber imports are responsible for much of the world's tropical deforestation.

Mediterranean

Saudi Arabia

Afghanistan

Snow leopard

640,000 square kilometers south of the Sahara have turned to desert since 1940.

Liberia

Oman

Asian elephant

Deforestation in the Himalaya causes flooding in Bangladesh.

Mali

AFRICA

Eritrea

Yemen

90% of the coral reefs are threatened in the Philippines. All virgin forest will be gone by 2010.

Burkina Faso

Niger

Ethiopia

Sierra Leone

Benin

Chad

Golden lion tamarin

Kouprey

Togo

Nigeria

India and Sri Lanka have almost no rain forest left.

Sao Tome

Uganda

Somalia

68% of the Congo's rain forest is slated for clearing.

Congo Rwanda Burundi

In peninsular Malaysia almost all forests have been cut.

Queen Alexandra's birdwing butterfly

Angola

INDIAN OCEAN

Indonesia's coral reefs are threatened and mangrove forests have been cut in half.

Nail-tailed wallaby

Zambia

AUSTRALIA

Fish catches in Southeast Atlantic has dropped by more than 50% since 1973.

Black rhinoceros

Aye-aye

Madagascar has lost 66% of its tropical forest.

Much of Australia's range and cropland has turned to desert.

Blue whale

A thinning of the ozone layer occurs over Antarctica during summer.

ANTARCTICA

such as "Better safe than sorry," Look before you leap," "First, do no harm," and "Slow down for speed bumps."

In this chapter we have seen that interdependence and connectedness of species, communities, and ecosystems are essential features of life on the earth. Chapter 9 shows how populations can change when environmental conditions change.

The old idea of a static landscape, like a single musical chord sounded forever, must be abandoned, for such a landscape never existed except in our imagination. Nature undisturbed by human influence seems more like a symphony whose harmonies arise from variation and change over every interval of time.

DANIEL B. BOTKIN

REVIEW QUESTIONS

1. Define the boldfaced terms in this chapter.

2. List four characteristics of the structure of a community or ecosystem. Distinguish between *species diversity* and *species abundance*.

3. Describe the types, relative sizes, and stratification of plants and animals in **(a)** a tropical rain forest, **(b)** an ocean, and **(c)** a lake.

4. Define and give an example of an *edge effect*. What are the advantages and disadvantages of edge effects?

5. How does species diversity vary with latitude in terrestrial communities? List five reasons why species diversity is higher in tropical communities than in other communities more distant from the equator.

6. How does species diversity in the ocean vary with depth? List three reasons why this occurs. How does pollution affect aquatic diversity?

7. What two factors determine the species diversity found on an isolated ecosystem such as an island? What is the *theory of island biodiversity*? How do the size of an island and its distance from a mainland affect its species diversity?

8. Distinguish between *native, nonnative, indicator*, and *keystone* species and give an example of each.

9. Why are birds good indicator species? Explain why amphibians are considered to be indicator species and list reasons for declines in their populations.

10. Describe the keystone ecological roles of **(a)** flying foxes, **(b)** alligators, and **(c)** some shark species. What can happen in an ecosystem that loses a keystone species?

11. Distinguish between *intraspecific competition* and *interspecific competition* and give an example of each. What is *territoriality*?

12. What are four options when the niches of two species competing in the same area overlap to a large degree? What is the *competitive exclusion principle*?

13. Distinguish between *interference competition* and *exploitation competition* and give an example of each.

14. Define and give two examples of *resource partitioning*. How does it allow species to avoid overlap of their fundamental niches?

15. What is *predation*? Describe the *predator-prey relationship* and give two examples of this type of species interaction.

16. Give two examples of how predators increase their chances of finding prey by **(a)** pursuit and **(b)** ambush.

17. List six ways (adaptations) used by prey to avoid their predators and give an example of each type.

18. What is *symbiosis*? What are three types of symbiotic interactions between species?

19. Define and give two examples of *parasitism* and explain how it differs from predation. What is the ecological importance of parasitism?

20. Define and give three examples of *mutualism*. Define and give two examples of *commensalism*.

21. Distinguish between *primary succession* and *secondary succession*. Distinguish between *pioneer* (or *early successional*) *species, midsuccessional plant species*, and *late successional plant species*. Distinguish between *facilitation, inhibition*, and *tolerance* as factors that affect how and at what rate succession occurs.

22. Give three examples of environmental disturbances and explain how they can affect succession. How can some disturbances be beneficial to ecosystems? What is the *intermediate disturbance hypothesis*? Explain how occasional fires can be beneficial to succession and species in some types of ecosystems.

23. Explain why most ecologists contend that **(a)** the details of succession are not predictable and **(b)** there is no *balance of nature*.

24. Distinguish between *inertia, constancy*, and *resilience* and explain how they help maintain stability in an ecosystem.

25. Does species diversity increase ecosystem stability? Explain.

26. What is the *precautionary principle*, and why do many scientists believe that it is a useful strategy for dealing with the environmental problems we face?

CRITICAL THINKING

1. Why do deer hunters sometimes plant corn on strips of open land used for firebreaks or for telephone poles?

2. Why are **(a)** more species of trees per hectare usually found in tropical forests than in temperate forests and **(b)** more species of tube worms found at intermediate ocean depths than in coastal areas and in the deep ocean?

3. How would you respond to someone who claims that it is not important to protect areas of temperate and polar biomes because most of the world's biodiversity is in the tropics?

4. What two factors determine the number of different species found on an island? Why is the species diversity of a large island usually higher than that on a smaller island?

5. What would you do if your home were invaded by large numbers of cockroaches (Spotlight, p. 112)? See whether you can come up with an ecological rather than a chemical (pesticide) approach to this problem (See Section 20-5, p. 511).

6. How would you determine whether a particular species found in a given area is a keystone species?

7. Some butterfly species mimic other butterfly species that taste bad to predators such as birds (Figure 8-12f). Design an experiment to determine whether a particular mimic species of butterfly **(a)** tastes bad like its model (called *Müllerian mimicry* after Fritz Müller, who first recorded this phenomenon in 1878) or **(b)** does not taste bad but gains some protection by looking and acting like its bad-tasting model (called *Batesian mimicry* after

Henry W. Bates, who first described this phenomenon in the 1860s).

8. Describe how evolution can affect predator-prey relationships.

9. Science usually is defined as a search for order and predictability in nature (Section 3-1, p. 45). However, the more scientists look for order and predictability in the structure and functioning of communities and ecosystems, the more they are discovering the role of chaos and unpredictability. Does this mean that science cannot reveal very much about how nature works and that we are wasting large amounts of money, time, and talent chasing a mirage of order and predictability? Explain your answer. Closely examine the assumptions built into this view. Evaluate whether there is some useful middle ground between complete order and complete chaos.

10. How would you reply to someone who argues that **(a)** we should not worry about our effects on natural systems because succession will heal the wounds of human activities and restore the balance of nature, **(b)** if nature is unpredictable, why should we bother to preserve any natural systems, and **(c)** because there is no balance in nature and no stability in species diversity we should cut down diverse, old-growth forests and replace them with tree farms?

11. Suppose a hurricane blows down most of the trees in a forest. Timber company officials offer to salvage the fallen trees and plant a tree plantation to reduce the chances of fire and to improve the area's appearance, with an agreement that they can harvest the trees when they reach maturity and plant another tree farm. Others argue that the damaged forest should be left alone because hurricanes and other natural events are part of nature and the dead trees will serve as a source of nutrients for natural recovery through ecological succession. What do you think should be done, and why?

PROJECTS

1. Make field studies, consult research papers, and interview people to identify and evaluate **(a)** the effects of the deliberate introduction of a beneficial nonnative species into the area where you live and **(b)** the effects of the deliberate or accidental introduction of a harmful species into the area where you live.

2. Use the library or internet to find and describe two species not discussed in this textbook that are engaged in **(a)** a commensalistic interaction, **(b)** a mutualistic interaction, and **(c)** a parasite–host relationship.

3. Visit a nearby natural area and identify examples of **(a)** territoriality, **(b)** interference competition, **(c)** exploitation competition, **(d)** mutualism, and **(e)** resource partitioning.

4. Use the library or internet to list as many as possible of the parasites likely to be found in your body.

5. Visit a nearby land area such as a partially cleared or burned forest or grassland or an abandoned cropfield

and record signs of secondary ecological succession. Study the area carefully to see whether you can find patches that are at different stages of succession because of various disturbances.

6. Use the library or the internet to find bibliographic information about *Stuart L. Pimm* and *Daniel B. Botkin*, whose quotes appear at the beginning and end of this chapter.

7. Make a concept map of this chapter's major ideas, using the section heads and subheads and the key terms (in boldface). Look at the inside back cover and on the website for this book for information about concept making maps.

INTERNET STUDY RESOURCES AND RESOURCES FOR FURTHER READING AND RESEARCH

The website for this book contains helpful study aids and many ideas for further reading and research. Log on to:

http://www.brookscole.com/product/0534376975s

and click on the Chapter-by-Chapter area. Choose Chapter 8 and select a resource:

■ "Flash Cards" allows you to test your mastery of the Terms and Concepts to Remember for this chapter.

■ "Tutorial Quizzes" provides a multiple-choice practice quiz.

■ "Student Guide to InfoTrac" will lead you to Critical Thinking Projects that use InfoTrac College Edition as a research tool.

■ "References" lists the major books and articles consulted in writing this chapter.

■ "Hypercontents" takes you to an extensive list of sites with news, research, and images related to individual sections of the chapter.

INFOTRAC COLLEGE EDITION

Improve your skills with InfoTrac College Edition, a searchable online database of articles from more than 700 periodicals. Log on to:

http://www.infotrac-college.com

or access InfoTrac through the website for this book.

Try the following articles:

Ewel, J. et al. 1999. Deliberate introductions of species: research needs. *BioScience* vol. 49, no. 8, p. 619. (subject guide: native species)

Fleming, T.H. 2000. Pollination of cacti in the Sonoran Desert. *American Scientist* vol. 88, no. 5, pp. 432–437. (subject guide: pollination)

9 POPULATION DYNAMICS, CARRYING CAPACITY, AND CONSERVATION BIOLOGY*

Sea Otters: Are They Back from the Brink of Extinction?

Sea otters (*Enhydra lutris*) live in kelp forests in shallow waters along the Pacific coast (Figure 9-1a). These tool-using animals use stones to pry shellfish off rocks underwater and to break open the shells while swimming on their backs and using their bellies as a table. Each day a sea otter consumes about 25% of its weight in sea urchins (Figure 9-1b), clams, mussels, crabs, abalone, and about 40 other species of benthic organisms.

Before European settlers arrived, about 1 million sea otters lived along the Pacific coastline. By the early 1900s, the southern sea otter (*Enhydra lutris nereis*), found from Baja Mexico up the California coast, was believed to be extinct mostly because of overhunting for their warm, beautiful fur and removal by abalone fishers.

Biologists have learned that sea otters are *keystone species* (p. 178) that help keep sea urchins from depleting kelp forests in offshore waters from Alaska to southern California. Without significant predation by sea otters, sea urchin populations expand and consume much of the kelp, which in turn lowers the diversity of other plants and animals in kelp forest ecosystems.

*Paul M. Rich, associate professor of ecology and evolutionary biology and environmental studies at the University of Kansas, is coauthor of this chapter.

Kelp forests provide essential habitats for a variety of species, inhibit shore erosion, and reduce the impact of storm waves on coastlines. Thus, preventing loss of these forests has important ecological and economic benefits.

Since 1938 the population of southern sea otters has increased from about 300 to more than 2,000. Wherever sea otters have returned or have been reintroduced, formerly deforested kelp areas recover within a few years and fish populations increase. This pleases biologists but upsets commercial fishers, who argue that sea otters consume too many abalone and compete with commercial fishing.

However, during the past few years the southern sea otter population has begun to decline again for unknown reasons. One possibility is increased pollution of coastal waters. Shellfish tend to concentrate toxins that are released into coastal waters. High levels of toxins may directly kill sea otters, and exposure to lower levels over long periods may reduce their resistance to disease and parasites.

Another possibility is increased predation of the otters by killer whales. There is evidence in some areas that they may consume sea otters when there is a decline in their normal prey of Stellar sea lions and harbor seals.

The population dynamics of southern sea otter populations have helped us to better understand the ecological importance of this keystone species. *Population dynamics*, *conservation biology*, and *human impacts on natural ecosystems* are the subject of this chapter.

(a)

(b)

Figure 9-1 (a) A southern sea otter. (Jeff Foott Productions/Bruce Coleman Collection) **(b)** A sea urchin. (Jane Burton/Bruce Coleman Collection)

Through the animal and vegetable kingdoms, nature has scattered the seeds of life abroad with the most profuse and liberal hand. She has been comparatively sparing in the room and nourishment necessary to rear them.

THOMAS R. MALTHUS

This chapter addresses the following questions:

- How do populations change in size, density, and makeup in response to environmental stress?

- What is the role of predators in controlling population size?

- What different reproductive patterns do species use to enhance their survival?

- What is conservation biology?

- What impacts do human activities have on populations, communities, and ecosystems?

9-1 POPULATION DYNAMICS AND CARRYING CAPACITY

What Are the Major Characteristics of a Population? Populations are dynamic. They change in **(1)** *size* (number of individuals), **(2)** *density* (number of individuals in a certain space), **(3)** *dispersion* (spatial pattern such as clumping, uniform dispersion, or random dispersion, in which the members of a population are found in their habitat depending mostly on resource availability, Figure 9-2), and **(4)** *age distribution* (proportion of individuals of each age in a population) in response to environmental stress or changes in environmental conditions. These changes, called **population dynamics**, occur in response to environmental stress (Table 8-2, p. 191) or changes in environmental conditions.

What Limits Population Growth? Four variables— *births*, *deaths*, *immigration*, and *emigration*—govern changes in population size. A population gains individuals by birth and immigration and loses them by death and emigration:

Population change = (Births + Immigration) − (Deaths + Emigration)

These variables depend on changes in resource availability or on other environmental changes (Figure 9-3). If the number of individuals added from births and immigration equals the number lost to deaths and immigration, then there is **zero population growth**.

Populations vary in their capacity for growth, also known as the **biotic potential** of the population. The **intrinsic rate of increase (r)** is the rate at which a population would grow if it had unlimited resources. Generally, individuals in populations with a high intrinsic rate of increase **(1)** *reproduce early in life,* **(2)** *have short generation times* (the time between successive generations), **(3)** *can reproduce many times* (have a long reproductive life), and **(4)** *have many offspring each time they reproduce.*

Some species have an astounding biotic potential. For example, without any controls on its population growth, the ancestors of a single female housefly (*Musca domestica*) could total about 5.6 trillion flies within about 13 months, and within a few years there would be enough flies to cover the earth's entire surface.

However, this is not a realistic scenario because no population can grow indefinitely. In the real world, a rapidly growing population reaches some size limit imposed by a shortage of one or more limiting factors, such as light, water, space, or nutrients or too many competitors or predators. *There are always limits to population growth in nature.*

Environmental resistance consists of all the factors acting jointly to limit the growth of a population. The population size of a species in a given place and time is determined by the interplay between its biotic potential and environmental resistance (Figure 9-3).

Together biotic potential and environmental resistance determine the **carrying capacity (K)**, the number of individuals of a given species that can be sustained indefinitely in a given space (area or volume).

The intrinsic rate of increase (r) of many species depends on having a certain minimum population size, called the **minimum viable population (MVP)**. If a

Clumped
(elephants)

Uniform
(creosote bush)

Random
(dandelions)

Figure 9-2 Generalized dispersion patterns for individuals in a population throughout their habitat. The most common pattern is one in which members of a population exist in clumps throughout their habitat (left), mostly because resources usually are found in patches.

population declines below the MVP needed to support a breeding population, **(1)** certain individuals may not be able to locate mates, **(2)** genetically related individuals may interbreed and produce weak or malformed offspring, and **(3)** the genetic diversity may be too low to enable adaptation to new environmental conditions. Then the intrinsic rate of increase falls and extinction is likely.

What Is the Difference Between Exponential and Logistic Population Growth? A population that has few if any have resource limitations grows exponentially. *Exponential growth* starts out slowly and then proceeds faster and faster as the population increases. If number of individuals is plotted against time, this sequence yields a *J*-shaped exponential growth curve (Figure 9-4a).

Logistic growth involves exponential population growth when the population is small and a steady decrease in population growth with time as the population encounters environmental resistance and approaches the carrying capacity of its environment. After leveling off, a population with this type of growth typically fluctuates slightly above and below the carrying capacity. A plot of the number of individuals against time yields a sigmoid or *S*-shaped logistic growth curve (Figure 9-4b). A classic case of logistic growth involves the increase of the sheep population on the island of Tasmania, south of Australia, in the early 19th century (Figure 9-5).

What Happens if the Population Size Exceeds the Carrying Capacity? The populations of some species do not make such a smooth transition from exponential growth to logistic growth. Instead, such a population uses up its resource base and temporarily *overshoots* or exceeds the carrying capacity of its environment. This overshoot occurs because of a *reproductive time lag*, the period needed for the birth rate to fall and the death rate to rise in response to resource overconsumption.

In such cases the population suffers a *dieback* or *crash* unless the excess individuals switch to new resources or move to an area with more favorable conditions. A classic case of such a population crash

occurred when reindeer were introduced in 1910 onto a small island off the southwest coast of Alaska (Figure 9-6).

Humans are not exempt from overshoot and dieback, as shown by the tragedy on Easter Island (p. 44). Ireland also experienced a population crash after a fungus destroyed the potato crop in 1845. About 1 million people died, and 3 million people emigrated to other countries.

The earth's carrying capacity for the human species has been extended by technological, social, and other cultural changes (Figure 2-2, p. 24). We have increased food production and used large amounts of energy and matter resources to make normally uninhabitable areas of the earth habitable. However, there is growing concern about how long we will be able to keep doing this on a planet with a finite size and resources with an exponentially growing population and per capita resource use.

Carrying capacity is not a simple, fixed quantity and is affected by many factors. Examples are **(1)** com-

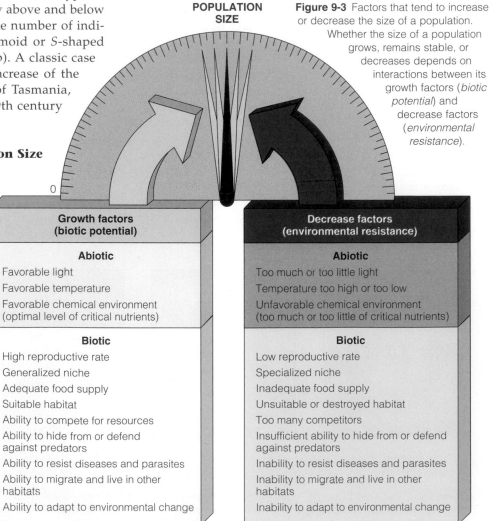

POPULATION SIZE

Figure 9-3 Factors that tend to increase or decrease the size of a population. Whether the size of a population grows, remains stable, or decreases depends on interactions between its growth factors (*biotic potential*) and decrease factors (*environmental resistance*).

Growth factors (biotic potential)	Decrease factors (environmental resistance)
Abiotic	**Abiotic**
Favorable light	Too much or too little light
Favorable temperature	Temperature too high or too low
Favorable chemical environment (optimal level of critical nutrients)	Unfavorable chemical environment (too much or too little of critical nutrients)
Biotic	**Biotic**
High reproductive rate	Low reproductive rate
Generalized niche	Specialized niche
Adequate food supply	Inadequate food supply
Suitable habitat	Unsuitable or destroyed habitat
Ability to compete for resources	Too many competitors
Ability to hide from or defend against predators	Insufficient ability to hide from or defend against predators
Ability to resist diseases and parasites	Inability to resist diseases and parasites
Ability to migrate and live in other habitats	Inability to migrate and live in other habitats
Ability to adapt to environmental change	Inability to adapt to environmental change

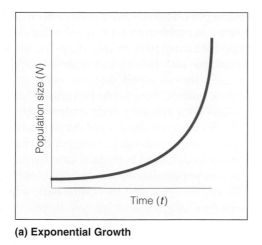

(a) Exponential Growth

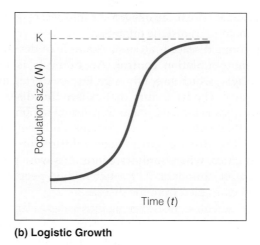

(b) Logistic Growth

Figure 9-4 Theoretical population growth curves. **(a)** *Exponential growth*, in which the population's growth rate increases with time. Exponential growth occurs when resources are not limiting and a population can grow at its *intrinsic rate of increase* (r). Exponential growth of a population cannot continue forever because eventually some factor limits population growth. **(b)** *Logistic growth*, in which the growth rate decreases as the population gets larger. With time, the population size stabilizes at or near the *carrying capacity* (K) of its environment.

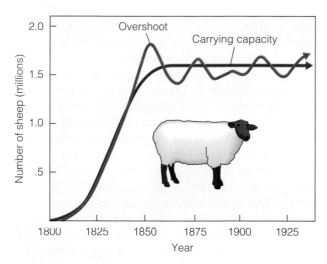
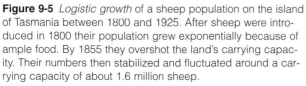

Figure 9-5 *Logistic growth* of a sheep population on the island of Tasmania between 1800 and 1925. After sheep were introduced in 1800 their population grew exponentially because of ample food. By 1855 they overshot the land's carrying capacity. Their numbers then stabilized and fluctuated around a carrying capacity of about 1.6 million sheep.

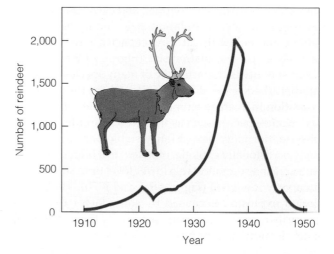

Figure 9-6 Exponential growth, overshoot, and population crash of reindeer introduced to a small island off the southwest coast of Alaska. When 26 reindeer (24 of them female) were introduced in 1910, lichens, mosses, and other food sources were plentiful. By 1935 the herd's population had soared to 2,000, overshooting the island's carrying capacity. This led to a population crash, with the herd plummeting to only 8 reindeer by 1950.

petition within and between species, **(2)** immigration and emigration, **(3)** natural and human-caused catastrophic events, and **(4)** seasonal fluctuations in the supply of food, water, hiding places, and nesting sites.

How Does Population Density Affect Population Growth? *Density-independent population controls* affect a population's size regardless of its population density. Examples include floods, hurricanes, severe drought, unseasonable weather, fire, habitat destruction

(such as clearing a forest of its trees or filling in a wetland), and pesticide spraying. For example, a severe freeze in late spring can kill many individuals in a plant population, regardless of its density.

Some limiting factors have a greater effect as a population's density increases. Examples of such *density-dependent population controls* are competition for resources, predation, parasitism, and disease. As prey populations become denser, their members compete more for limited resources. Individuals in a dense population also are more likely to be infected by parasites or contagious disease organisms.

Infectious disease is a classic example of density-dependent population control. An example is the *bubonic plague*, which swept through Europe during the 14th century. The bacterium that causes this disease normally lives in rodents. It was transferred to humans by fleas that fed on infected rodents and then bit humans. The disease spread like wildfire through crowded cities, where sanitary conditions were poor and rats were abundant. At least 25 million people in European cities died from the disease.

Health scientists are becoming increasingly alarmed about the possibility of new epidemics of common infectious diseases in crowded urban areas. The primary reason is that many common strains of disease-causing bacteria are becoming genetically resistant to most existing antibiotics (Spotlight, p. 406).

What Kinds of Population Change Curves Do We Find in Nature? In nature we find that over time species have four general types of *population fluctuations*: stable, irruptive, irregular, and cyclic (Figure 9-7). A species whose population size fluctuates slightly above and below its carrying capacity is said to have a fairly stable population size (Figures 9-5 and 9-7a). Such stability is characteristic of many species found in undisturbed tropical rain forests, where there is little variation in average temperature and rainfall.

Some species, such as the raccoon and feral house mouse, normally have a fairly stable population that may occasionally explode, or *irrupt*, to a high peak, and then crash to a more stable lower level or in some cases to a very low level (Figures 9-6 and 9-7b). The population explosion is caused by some factor that temporarily increases carrying capacity for the population, such as more food or fewer predators.

Some populations exhibit what appears to be irregular *chaotic behavior* in their changes in population size, with no recurring pattern (Figure 9-7c). Some scientists attribute such behavior to chaos in such systems. Other scientists contend that the behavior may be caused by orderly, nonchaotic behavior whose details and interactions are still poorly understood.

The fourth type consists of cyclic fluctuations of population size over a regular time period (Figure 9-7d) for poorly understood reasons. Examples are **(1)** lemmings, whose populations rise and fall every 3–4 years and **(2)** grouse, lynx, and snowshoe hare, whose populations generally rise and fall on a 10-year cycle.

9-2 THE ROLE OF PREDATION IN CONTROLLING POPULATION SIZE

Do Predators Control Population Size? The Lynx–Hare Cycle Some species that interact as predator and prey undergo cyclic changes in their numbers, with sharp increases in their numbers followed by seemingly periodic crashes (Figure 9-8). Predator–prey interactions often are blamed, but the actual causes of such *predator–prey cycles* are poorly understood.

For decades, predation has been the explanation for the correlation and time lag between the 10-year population cycles of the snowshoe hare and its predator, the Canadian lynx (Figure 9-8). According to this *top-down control* hypothesis, lynx preying on hares periodically reduce their population. The shortage of hares then reduces the lynx population, which allows the hare population to build up again. At some point the lynx population increases to take advantage of the increased supply of hares, starting the cycle again.

Recent research has cast doubt on this appealing hypothesis because snowshoe hare populations have been found to have similar 10-year boom-and-bust cycles on islands where lynx are absent. Researchers now hypothesize that the periodic crashes in the hare population may occur when large numbers of hares consume food plants (especially young shoots of willow, their prime food source) faster than they can be replenished and have a decrease in the quantity and quality of their food. Once the hare population crashes, the plants recover, and the hare population begins rising again in a hare-plant cycle. If this *bottom-up control* hypothesis is correct, instead of lynx controlling hare populations, the chang-

Figure 9-7 General types of simplified population change curves found in nature. **(a)** The population size of a species with a fairly *stable* population fluctuates slightly above and below its carrying capacity. **(b)** The populations of some species may occasionally explode, or *irrupt*, to a high peak and then crash to a more stable lower level. **(c)** The population sizes of some species change irregularly for mostly unknown reasons. **(d)** Other species undergo sharp increases in their numbers, followed by crashes over fairly regular time intervals Predators sometimes are blamed, but the actual causes of such boom-bust cycles are poorly understood.

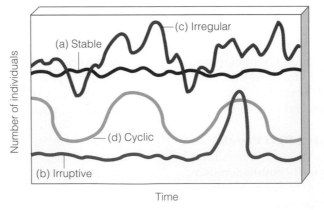

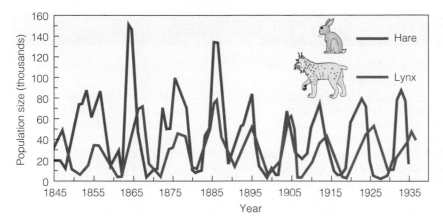

Figure 9-8 Population cycles for the snowshoe hare and Canadian lynx. At one time it was widely believed that these curves provided circumstantial evidence that these predator and prey populations regulated one another. More recent research suggests that the periodic swings in the hare population are caused by the hares themselves as they overconsume plants, die back, and then slowly recover when the plant supply is replenished. Instead of the lynx population size controlling hare population size (*top-down population control*), the rise and fall of the hare population apparently controls the size of the lynx population (*bottom-up population control*). (Data from D. A. MacLulich)

ing hare population size may be causing fluctuations in the lynx population.

These two hypotheses are not mutually exclusive. The observed cyclic population changes could be the result of a varying three-way interaction between lynx, snowshoe hares, and plants.

According to biologists, there are genuine cases of top-down control systems by predators in a number of ecosystems. Examples are **(1)** wolves controlling deer populations and moose populations (Case Study, p. 204), **(2)** large predatory fish controlling other fish populations in lakes (p. 629), **(3)** sheep and rabbits controlling plant growth in a pasture, and **(4)** sharks (Case Study, p. 185) and alligators (Connections, p. 181) controlling some fish populations.

9-3 REPRODUCTIVE PATTERNS* AND SURVIVAL

How Do Species Reproduce? Reproductive individuals in populations of all species are engaged in a struggle for genetic immortality by trying to have as many members of the next generation as possible carry their genes. Genes can be passed on to offspring by the following means:

- **Asexual reproduction**, in which all offspring are exact genetic copies (clones) of a single parent. This type of reproduction is common in single-celled species such as bacteria, in which the mother cell divides to produce two identical cells that are genetic clones or replicas of the mother cell.

---

*These are usually called *reproductive strategies*. However, the term *strategy* normally refers to a conscious plan, which biologists say wild species don't have. For this reason, it is more appropriate to use the term *reproductive patterns* or *forms of selection*. This theory of reproductive patterns was developed in 1967 by R. H. MacArthur and E. O. Wilson.

- **Sexual reproduction**, in which organisms produce offspring by combining the gametes or sex cells (such as sperm and ovum) from both parents. This produces offspring that have combinations of traits (chromosomes) from each parent. About 97% of known organisms use sexual reproduction to perpetuate their species.

Sexual reproduction has a number of ecological costs and risks, including the following:

- Females have to produce twice as many offspring to maintain the same number of young in the next generation as an asexually reproducing organism because males do not give birth.

- The chances of genetic errors and defects increase during the splitting and recombination of chromosomes.

- Mating entails costs such as time-consuming courtship and mating rituals, disease transmission, and injury inflicted by males during mating.

So if sexual reproduction has so many costs and risks, why do 97% of the earth's organisms use it, and why do we need males? One answer is that sexual reproduction provides a greater genetic diversity in the offspring (Figure 4-5, p. 73). This means that sexually reproducing organisms with "many different eggs in their genetic basket" have a greater chance of reproducing when environmental conditions change than does a brood of genetically identical clones with "only one type of egg in their genetic basket."

Another advantage of sexual reproduction is that males can gather food for the female and the young and may protect and help train the young. This is especially helpful for species such as mammals and birds that produce only a few young.

What Are Opportunist or r-Selected Species? Each species has a characteristic mode of reproduction.

CASE STUDY

Wolf and Moose Interactions on Isle Royale

For decades wildlife biologists have been studying the relationship between the moose and wolf populations on Isle Royale, an island in Lake Superior between Minnesota in the United States and Ontario in Canada (Figure 9-9).

In the early 1900s, a small herd of moose wandered across the frozen ice of Lake Superior to this island. With an abundance of food, the moose population exploded (Figure 9-9). In 1928, a wildlife biologist visiting the island successfully predicted that the large moose population would soon crash because the moose had stripped the island of most of its preferred food plants.

Sometime during the 1940s timber wolves (probably a single pair) reached Isle Royale by traveling over the ice from the Canadian mainland during winter. They reproduced and slowly grew in numbers. During winter the wolves hunt in packs and concentrate on killing the old, sick, and young moose. These individuals are the easiest to kill without undue risk because the moose is the wolf's largest and most dangerous prey. Once a target moose is selected, the wolves encircle it and try to get it to run so they can attack it from behind.

Since 1958, wildlife biologists have been tracking the populations of the two species (Figure 9-9). You might think that the wolves would have completely exterminated the moose, but instead the two species have been interacting in what appears to be an oscillating predator–prey cycle (Figure 9-9). If the wolves could drive the moose to extinction they probably would, but the moose are too formidable for this to happen.

Since 1980 the wolf population declined from a high of about 50 and has fluctuated between 12 and 25 individuals. Possible reasons for this decline are **(1)** a canine virus introduced to wolves by dogs and **(2)** a low reproduction rate because of a lack of genetic variability from inbreeding.

With the decline in wolves, the moose population rose sharply until 1995. Then it crashed from a combination of lack of food, poor reproduction, a severe winter, and a tick infestation. By 1999 the wolf population, with plenty of weakened prey, had grown to 25. If their population continues to grow, they may hold the moose numbers in check and allow damaged vegetation to recover and begin a new cycle of interactions.

Critical Thinking

What is the primary ecological lesson to be learned from the moose–wolf interaction on Isle Royale?

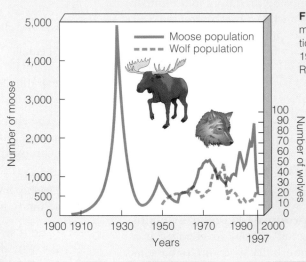

Figure 9-9 Changes in moose and wolf populations on Isle Royale from 1900 to 1999. (Data from Rolf O. Peterson, 1995)

At one extreme are species that reproduce early and put most of their energy into reproduction. They **(1)** have many (usually small) offspring each time they reproduce, **(2)** reach reproductive age rapidly, **(3)** have short generation times, **(4)** give their offspring little or no parental care or protection to help them survive, and **(5)** are short-lived (usually with a life span of less than a year). Species with this reproductive pattern overcome the massive loss of their offspring by producing so many unprotected young that a few will survive to reproduce many offspring to begin the cycle again.

Species with such a capacity for a high intrinsic rate of increase (r) are called **r-selected species** (Figure 9-10, left). Algae, bacteria, rodents, annual plants (such as dandelions), and most insects (p. 70) are examples.

Such species tend to be *opportunists*. They reproduce and disperse rapidly when conditions are favorable or when a disturbance (Table 8-2, p. 191) opens up a new habitat or niche for invasion, as in the early stages of ecological succession (Figures 8-15, p. 188, and 8-16, p. 189).

Changed environmental conditions from disturbances can allow opportunist species to gain a foothold. However, once established, their populations may crash because of changing or unfavorable environmental conditions or invasion by more competitive species. Therefore, most r-selected or opportunist species go through irregular and unstable boom–bust cycles in their population size. To survive, opportunists must continually invade new areas to compensate for being displaced by more competitive species.

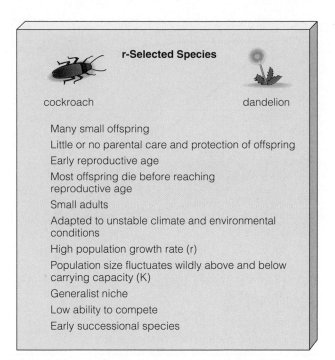

r-Selected Species

cockroach   dandelion

Many small offspring

Little or no parental care and protection of offspring

Early reproductive age

Most offspring die before reaching reproductive age

Small adults

Adapted to unstable climate and environmental conditions

High population growth rate (r)

Population size fluctuates wildly above and below carrying capacity (K)

Generalist niche

Low ability to compete

Early successional species

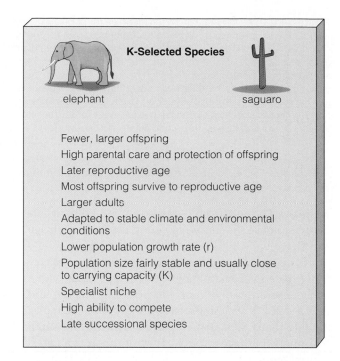

K-Selected Species

elephant   saguaro

Fewer, larger offspring

High parental care and protection of offspring

Later reproductive age

Most offspring survive to reproductive age

Larger adults

Adapted to stable climate and environmental conditions

Lower population growth rate (r)

Population size fairly stable and usually close to carrying capacity (K)

Specialist niche

High ability to compete

Late successional species

Figure 9-10 Generalized characteristics of r-selected or opportunist species and K-selected or competitor species. Many species have characteristics between these two extremes.

What Are Competitor or K-Selected Species? At the other extreme are *competitor* or **K-selected species.** These species **(1)** put fairly little energy into reproduction, **(2)** tend to reproduce late in life, **(3)** have few offspring with long generation times, and **(4)** put most of their energy into nurturing and protecting their young until they reach reproductive age.

Typically these offspring **(1)** develop inside their mothers (where they are safe), **(2)** are fairly large, **(3)** mature slowly, and **(4)** are cared for and protected by one or both parents until they reach reproductive age. This reproductive pattern results in a few big and strong individuals that can compete for resources and reproduce a few young to begin the cycle again (Figure 9-10, right).

Such species are called K-selected species because they tend to do well in competitive conditions when their population size is near the carrying capacity (K) of their environment. Their populations typically follow a logistic growth curve (Figure 9-4b, p. 210). Examples are **(1)** most large mammals (such as elephants, whales, and humans), **(2)** birds of prey, and **(3)** large and long-lived plants (such as the saguaro cactus, oak trees, redwood trees, and most tropical rain forest trees). Many K-selected species, especially those with long generation times and low reproductive rates (such as elephants, rhinoceroses, and sharks, Case Study, p. 185), are prone to extinction.

Most competitor or K-selected species thrive best in ecosystems with fairly constant environmental conditions. In contrast, opportunists thrive in habitats that have experienced disturbances (Table 8-2, p. 191) such as a tree falling, a forest fire, or the clearing of a forest or grassland for raising crops.

Many organisms have reproductive patterns between the extremes of r-selected species and K-selected species, or they change from one extreme to the other under certain environmental conditions. In agriculture we raise both r-selected species (crops) and K-selected species (livestock).

The reproductive pattern of a species may give it a temporary advantage, but *the availability of suitable habitat for individuals of a population in a particular area is what determines its ultimate population size.* Regardless of how fast a species can reproduce, there can be no more dandelions than there is dandelion habitat and no more zebras than there is zebra habitat in a particular area.

What Are Survivorship Curves? Individuals of species with different reproductive strategies tend to have different *life expectancies.* One way to represent the age structure of a population is with a **survivorship curve,** which shows the number of survivors of each age group for a particular species. There are three generalized types of survivorship curves: *late loss, early loss,* and *constant loss* (Figure 9-11).

Late loss curves are typical for K-selected species (such as elephants and humans) that produce few young and care for them until they reach reproductive age (thus reducing juvenile mortality). *Early loss* curves are typical for r-selected species (such as most annual plants and most bony fish species) with **(1)** many

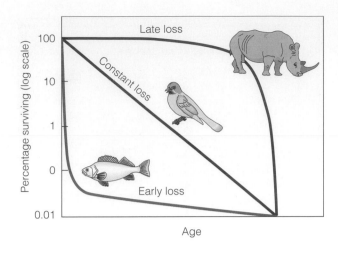

Figure 9-11 Three general survivorship curves for populations of different species, obtained by showing the percentages of the members of a population surviving at different ages. For a *late loss* population (such as elephants, rhinoceroses, and humans), there is typically high survivorship to a certain age, then high mortality. A *constant loss* population (such as many songbirds) shows a fairly constant death rate at all ages. For an *early loss* population (such as annual plants and many bony fish species), survivorship is low early in life. These generalized survivorship curves only approximate the behavior of species.

offspring, **(2)** high juvenile mortality, and **(3)** high survivorship once the surviving young reach a certain age and size.

Species with *constant loss* survivorship curves typically have intermediate reproductive patterns with a fairly constant rate of mortality in all age classes and thus a steadily declining survivorship curve. Examples include many types of songbirds, lizards, and small mammals that face a fairly constant threat from starvation, predation, and disease throughout their lives.

A table of the numbers of individuals at each age from a survivorship curve is called a *life table*. It shows the projected life expectancy and probability of death for individuals at each age. Insurance companies use life tables of human populations in various countries or regions to determine policy costs for their customers. Because life tables show that women in the United States survive an average of 7 years longer than men, a 65-year-old man normally will pay more for life insurance than a 65-year-old woman.

9-4 CONSERVATION BIOLOGY: SUSTAINING WILDLIFE POPULATIONS

What Is Conservation Biology? Conservation involves sensible and careful use of natural resources by humans. **Conservation biology** is a multidisciplinary science, originated in the 1970s, that uses the best available science to take action to preserve species and ecosystems. Its key goals are to **(1)** investigate human impacts on biodiversity and **(2)** develop practical approaches to maintaining biodiversity to ensure the continued existence of populations of wild species. This multidisciplinary science is concerned with **(1)** endangered species management (Section 22-4, p. 563), **(2)** wildlife reserves (Section 23-6, p. 614), **(3)** ecological restoration (Section 23-7, p. 623), **(4)** ecological economics (Section 26-1, p. 692), and **(5)** environmental ethics (Chapter 28 p. 740).

Conservation biology differs from *wildlife management*, which is devoted primarily to manipulating the population sizes of various animal species, especially game species prized by hunters and fishers (Section

22-6, p. 579). Conservation biology rests on three underlying principles:

- Biodiversity is necessary to all life on earth and should not be reduced by human actions.

- Humans should not cause or hasten the extinction of wildlife populations and species or disrupt vital ecological processes.

- The best way to preserve earth's biodiversity and ecological functions is to protect intact ecosystems that provide sufficient habitat for sustaining natural populations of species.

Conservation biology is based on Aldo Leopold's ethical principle that something is right when it tends to maintain the earth's life-support systems for us and other species and wrong when it does not (Section 2-5, p. 39).

Chapter 23 examines methods for sustaining the biodiversity of terrestrial ecosystems, and Chapter 24 is devoted to methods for sustaining aquatic life systems. Chapter 22 examines issues of premature extinction of terrestrial and aquatic species, with an emphasis on rare and endangered species.

How Can Conservation Biology Help Prevent Premature Extinction of Species? Conservation biology seeks answers to three questions:

- *What is the status of natural populations, and which species are in danger of extinction?*

- *What is the status of the functioning of ecosystems, and what ecosystem services (Section 4-8, p. 99, and Figure 4-36, p. 99) of value to humans and other species are we in danger of losing?*

- *What measures can we take to ensure that we maintain habitat of the quality and size needed to ensure that ecosystem functions and viable populations of wild species can be sustained?*

These are challenging questions, and the answers require extensive field research and a strong grounding in ecological theory. To understand the status of natural populations, we must **(1)** measure the current population size, **(2)** determine how population size is likely to change with time, and **(3)** determine whether existing populations are likely to be sustainable.

What Is Bioinformatics? Good conservation biology depends on good information. Increasingly, efforts of conservation biologists are focused on building computer databases that store useful information about biodiversity. Recently, a new discipline called *bioinformatics* has developed that concerns itself with providing tools for storage and access to key biological information and with building the actual databases that contain the needed biological information.

9-5 HUMAN IMPACTS ON ECOSYSTEMS: LEARNING FROM NATURE

How Have Humans Modified Natural Ecosystems? To survive and support growing numbers of people, we have greatly increased the number and area of the earth's natural systems that we have modified, cultivated, built on, or degraded (Figure 1-4, p. 8). We have used technology to alter much of the rest of nature in the following ways:

- *Fragmenting and degrading habitat.*

- *Simplifying natural ecosystems.* When we plow grasslands and clear forests, we often replace their thousands of interrelated plant and animal species with one crop or one kind of tree—called *monocultures*—or with buildings, highways, and parking lots. Then we spend a lot of time, energy, and money trying to protect such monocultures from invasion by **(1)** opportunist species of plants (weeds), **(2)** pests (mostly insects, to which a monoculture crop is like an all-you-can-eat restaurant), and **(3)** pathogens (fungi, viruses, or bacteria that harm the plants we want to grow).

- *Strengthening some populations of pest species and disease-causing bacteria by speeding up natural selection* (Figure 5-6, left, p. 110) *and causing genetic resistance through overuse of pesticides and antibiotics.*

- *Eliminating some predators.* Some ranchers want to eradicate bison or prairie dogs that compete with their sheep or cattle for grass. They also want to eliminate wolves, coyotes, eagles, and other predators that occasionally kill sheep. Big game hunters also push for elimination of predators that prey on game species.

- *Deliberately or accidentally introducing new or alien species,* some beneficial and some harmful to us and other species. In the late 1800s, several Chinese chestnut trees brought to the United States were infected with a fungus that spread to the American chestnut, once found throughout much of the eastern United States. Between 1910 and 1940, this accidentally introduced fungus almost eliminated the American chestnut.

- *Overharvesting potentially renewable resources.* Ranchers and nomadic herders sometimes allow livestock to overgraze grasslands until erosion converts these ecosystems to less productive semideserts or deserts. Farmers sometimes deplete the soil of nutrients by excessive crop growing. Fish species are overharvested. Wildlife species with economically valuable parts (such as elephant tusks, rhinoceros horns, and tiger skins) are endangered by illegal hunting (poaching).

- *Interfering with the normal chemical cycling and energy flows in ecosystems.* Soil nutrients can erode from monoculture crop fields, tree farms, construction sites, and other simplified ecosystems and can overload and disrupt other ecosystems such as lakes and coastal ecosystems. Chemicals such as chlorofluorocarbons (CFCs) released into the atmosphere can increase the amount of harmful ultraviolet energy reaching the earth by reducing ozone levels in the stratosphere. Emissions of carbon dioxide and other greenhouse gases—from burning fossil fuels and from clearing and burning forests and grasslands—can trigger global climate change by altering energy flow through the atmosphere.

To survive we must exploit and modify parts of nature. However, we are beginning to understand that any human intrusion into nature has multiple effects, most of them unpredictable (Connections, p. 208).

The challenge is to **(1)** maintain a balance between simplified, human-altered ecosystems and the neighboring, more complex natural ecosystems on which we and other life-forms depend and **(2)** slow down the rates at which we are altering nature for our purposes. If we simplify and degrade too much of the planet to meet our needs and wants, what's at risk is not the earth but our own species.

According to biodiversity expert E. O. Wilson, "If this planet were under surveillance by biologists from another world, I think they would look at us and say, 'Here is a species in the mid-stages of self-destruction.'" The evolutionary lesson to be learned from nature is that no species can get "too big for its britches," at least not for long.

Solutions: What Can We Learn from Nature About Living Sustainably? Scientific research indicates that living systems have six key features: *interdependence, diversity, resilience, adaptability, unpredictability,* and *limits.* Organisms, populations, and ecosystems are remarkably resilient when exposed to stresses (Table 8-2, p. 191) caused by natural or human-induced changes in environmental conditions. However, scientific

Malaria once infected 9 out of 10 people in North Borneo, now known as Sabah. In 1955 the World Health Organization (WHO) began spraying the island with dieldrin (a DDT relative) to kill malaria-carrying mosquitoes. The program was so successful that the dreaded disease was nearly eliminated.

However, unexpected things began to happen. The dieldrin also killed other insects, including flies and cockroaches living in houses. The islanders applauded this turn of events, but then small lizards that also lived in the houses died after gorging themselves on dieldrin-contaminated insects. Next, cats began dying after feeding on the lizards. Then, in the absence of cats, rats flourished and overran the villages. When the people became threatened by sylvatic plague carried by rat fleas, the WHO parachuted healthy cats onto the island to help control the rats.

Then the villagers' roofs began to fall in. The dieldrin had killed wasps and other insects that fed on a type of caterpillar that either avoided or was not affected by the insecticide. With most of its predators eliminated, the caterpillar population exploded, munching its way through its favorite food: the leaves used in thatched roofs.

Ultimately, this episode ended happily: Both malaria and the unexpected effects of the spraying program were brought under control. Nevertheless, the chain of unforeseen events emphasizes the unpredictability of interfering with an ecosystem. It reminds us that when we intervene in nature, we need to ask the question, "And then what?"

Critical Thinking

Do you believe that the beneficial effects of spraying pesticides on Sabah outweighed the resulting unexpected and harmful effects? Explain.

---

research indicates that environmental stresses have harmful effects on organisms, populations, and ecosystems that can affect their environmental health and long-term sustainability (Figure 9-12).

Many biologists believe that the best way for us to reduce the harmful human impacts on organisms, populations, and ecosystems and thus live more sustainably is to learn about the processes and adaptations by which nature sustains itself (Solutions, right) and then mimic these lessons from nature.

Biologists have used these lessons from nature (Solutions, right) to formulate several principles to guide us in our search for more sustainable lifestyles:

- *Our lives, lifestyles, and economies are totally dependent on the sun and the earth.*

- *Everything is connected to everything else.* The primary goal of ecology is to discover which connections in nature are the strongest, most important, and most vulnerable to disruption.

- *We can never do merely one thing.* Any human intrusion into nature has mostly unpredictable side effects. When we alter nature, we should ask the question, "And then what?"

- *We should reduce and minimize the damage we do to nature*—a prevention or precautionary strategy (p. 193)—and help heal some of the ecological wounds we have inflicted.

- *We should use care, restraint, humility, and cooperation with nature as we alter the biosphere to meet our needs and wants.*

Using such guidelines, we can create a more ecologically and economically sustainable society that lives within its ecological means by **(1)** taking no more than it needs, **(2)** using renewable resources no faster than nature replaces them, **(3)** preserving biodiversity and human cultural diversity, and **(4)** not depleting natural capital (Figure 4-36, p. 99).

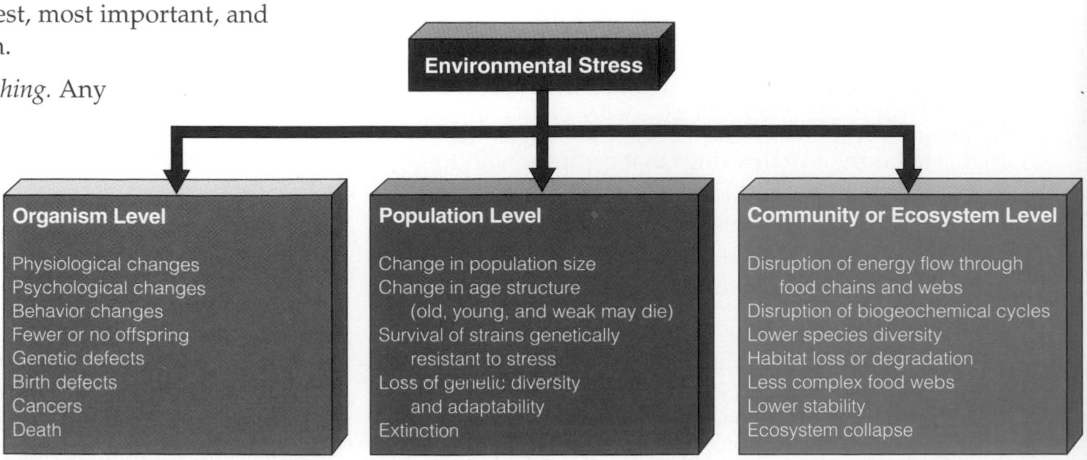

Environmental Stress

Organism Level	Population Level	Community or Ecosystem Level
Physiological changes	Change in population size	Disruption of energy flow through food chains and webs
Psychological changes	Change in age structure (old, young, and weak may die)	Disruption of biogeochemical cycles
Behavior changes	Survival of strains genetically resistant to stress	Lower species diversity
Fewer or no offspring	Loss of genetic diversity and adaptability	Habitat loss or degradation
Genetic defects	Extinction	Less complex food webs
Birth defects		Lower stability
Cancers		Ecosystem collapse
Death		

Figure 9-12 Some effects of environmental stress on organisms, populations, communities, and ecosystems.

Principles of Sustainability: Learning from Nature

SOLUTIONS

Here are four basic ecological lessons or principles of sustainability derived from observing how nature works:

- *Most ecosystems use renewable solar energy as their primary source of energy.* Thus, a sustainable society would be powered mostly by current sunlight, not ancient sunlight stored as polluting fossil fuels.

- *Ecosystems replenish nutrients and dispose of wastes by recycling chemicals.* There is almost no waste in nature, and there is no "away." The waste outputs and decomposed remains of one organism are resource inputs for other organisms.

- *Biodiversity helps maintain the sustainability and ecological functioning of ecosystems and serves as a source of adaptations to changing environmental conditions.*

- *In nature there are always limits to population growth.* The population size and growth rate of all species are controlled by their interactions with other species and with their nonliving environment.

Critical Thinking

List two ways in which human activities violate each of these four principles of sustainability. In what ways does your lifestyle violate these principles? Would you be willing to change these practices? What beneficial and harmful effects would such changes have on your lifestyle?

We cannot command nature except by obeying her.

SIR FRANCIS BACON

REVIEW QUESTIONS

1. Define the boldfaced terms in this chapter.

2. Explain how the populations of sea otters and kelp interact and why the sea otter is considered a keystone species.

3. What four factors affect population change?

4. Write an equation showing how population change is related to births, deaths, immigration, and emigration.

5. What is the *biotic potential* of a population? What are four characteristics of a population with a high *intrinsic rate of increase* (r)?

6. What are *environmental resistance* and *carrying capacity*? How do biotic potential and environmental resistance interact to determine carrying capacity? List four factors

that can alter an area's carrying capacity. What is the *minimum viable population* size of a population?

7. Distinguish between *exponential* and *logistic growth* of a population and give an example of each type.

8. How can a population overshoot its carrying capacity, and what are the consequences of doing this?

9. Distinguish between *density-dependent* and *density-independent factors* that affect a population's size and give an example of each.

10. Distinguish between stable, irruptive, irregular, and cyclic forms of population change.

11. Distinguish between *top-down control* and *bottom-up control* of a population's size. Use these concepts to describe the effects of the predator–prey interactions between the snowshoe hare and the Canadian lynx on the population of each species.

12. Describe the predator–prey interactions between wolf and moose populations on Isle Royale. Is this a *top-down* or *bottom-up* form of population control? What other factors have entered into this interaction?

13. Distinguish between *asexual reproduction* and *sexual reproduction*. What are the disadvantages and advantages of sexual reproduction?

14. List the characteristics of *r-selected* or *opportunist species* and *K-selected* or *competitor* species and give two examples of each type. Under what environmental conditions are you most likely to find **(a)** r-selected species and **(b)** K-selected species.

15. What is a *survivorship curve*, and how is it used? List three general types of survivorship curves and give an example of an organism with each type.

16. Distinguish between *conservation, conservation biology,* and *wildlife management*.

17. What are the three underlying principles of conservation biology? What three questions does conservation biology try to answer? What is *bioinformatics*, and what is its importance?

18. List seven potentially harmful ways in which humans modify natural ecosystems.

19. List four *principles of sustainability* derived from observing how natural systems are sustained.

20. List five principles we could use to help us live more sustainably.

CRITICAL THINKING

1. Why do **(a)** biotic factors that regulate population growth tend to depend on population density and **(b)** abiotic factors that regulate population tend to be independent of population density?

2. What are the advantages and disadvantages of a species undergoing **(a)** exponential growth (Figure 9-4a) and **(b)** logistic growth (Figure 9-4b)?

3. Suppose that because of disease or genetic defects from inbreeding, the wolves on Isle Royale die off.

Should we (a) intervene and import new wolves to help control the moose population or (b) let the moose population grow until it exceeds its carrying capacity and suffers another population crash? Explain.

4. Why are pest species likely to be extreme r-selected species? Why are many endangered species likely to be extreme K-selected species?

5. Why is an animal that devotes most of its energy to reproduction likely to be small and weak?

6. Given current environmental conditions, if you had a choice would you rather be an r-strategist or a K-strategist? Explain your answer. What implications does your decision have for your current lifestyle?

7. Predict the type of survivorship curve you would expect given descriptions of the following organisms:
    a. This organism is an annual plant. It lives only 1 year. During that time, it sprouts, reaches maturity, produces many wind-dispersed seeds, and dies.
    b. This organism is a mammal. It reaches maturity after 10 years. It bears one young every 2 years. The young are protected by the parents and the rest of the herd.

8. If after your death you could come back as a member of a particular type of species, what type of survivorship curve (Figure 9-11) would you like to have, and why?

9. Explain why a simplified ecosystem such as a cornfield usually is much more vulnerable to harm from insects and plant diseases than a more complex, natural ecosystem such as a grassland.

10. How has the human population generally been able to avoid environmental resistance factors that affect other populations? Is this likely to continue? Explain.

11. Explain why you agree or disagree with the five principles for living more sustainably listed on p. 208.

PROJECTS

1. Use the principles of sustainability derived from the scientific study of how nature sustains itself (Solutions, p. 209) to evaluate the sustainability of the following parts of human systems: **(a)** transportation, **(b)** cities, **(c)** agriculture, **(d)** manufacturing, **(e)** waste disposal, and **(f)** your own lifestyle. Compare your analysis with those made by other members of your class.

2. Try to trace the known and potential short- and long-term environmental impacts of the following human activities on wildlife, energy flows through ecosystems and the biosphere, and on the water, carbon, and nitrogen cycles: **(a)** driving a gasoline-burning car, **(b)** growing monoculture food crops, **(c)** cutting down diverse forests and replacing them with tree plantations, **(d)** eating meat produced by raising domesticated livestock, **(e)** increasing crop yields by applying large amounts of commercial inorganic fertilizer, **(f)** building and using a shopping center, **(g)** using air conditioning, **(h)** producing electricity by burning a fossil fuel such as coal, **(i)** producing electricity using nuclear power, **(j)** using large quantities of chemical pesticides, **(k)** using large quantities of antibiotics, **(l)** burning trash in incinerators,

(m) building sewage treatment plants and discharging the resulting effluent into nearby streams, lakes, or coastal zones, **(n)** encouraging people to live in cities, and **(o)** encouraging people to spread out and live in rural areas. Compare your analyses with those of other members of your class.

3. Use the library or the internet to find bibliographic information about *Thomas R. Malthus* and *Sir Francis Bacon*, whose quotes appear at the beginning and end of this chapter.

4. Make a concept map of this chapter's major ideas, using the section heads and subheads and the key terms (in boldface). Look at the inside back cover and on the website for this book for information about making concept maps.

INTERNET STUDY RESOURCES AND RESOURCES FOR FURTHER READING AND RESEARCH

The website for this book contains helpful study aids and many ideas for further reading and research. Log on to:

http://www.brookscole.com/product/0534376975s

and click on the Chapter-by-Chapter area. Choose Chapter 9 and select a resource:

■ "Flash Cards" allows you to test your mastery of the Terms and Concepts to Remember for this chapter.

■ "Tutorial Quizzes" provides a multiple-choice practice quiz.

■ "Student Guide to InfoTrac" will lead you to Critical Thinking Projects that use InfoTrac College Edition as a research tool.

■ "References" lists the major books and articles consulted in writing this chapter.

■ "Hypercontents" takes you to an extensive list of sites with news, research, and images related to individual sections of the chapter.

INFOTRAC COLLEGE EDITION

Improve your skills with InfoTrac College Edition, a searchable online database of articles from more than 700 periodicals. Log on to:

http://www.infotrac-college.com

or access InfoTrac through the website for this book.

Try the following articles:

Milius, S. 2000. Old lemming puzzle gets new answer. *Science News* vol. 157, no. 25, p. 395. (subject guide: lemmings)

Rapport, D.J. and W.G. Whitford. 1999. How ecosystems respond to stress. *BioScience* vol. 49, no. 3, pp. 193–203. (subject guide: ecology, environmental)

10 GEOLOGY: PROCESSES, HAZARDS, AND SOILS*

Charles Darwin Reporting from Concepción, Chile

The coastal city of Concepción, Chile, was hit by a major earthquake on February 20, 1835. Its effects were reported to the world by Charles Darwin, who sailed into Talcahuano Bay on the *HMS Beagle* soon after the earth shook.

The whole coast, said Darwin, was "strewed over with timber and furniture as if a thousand ships had

*Kenneth J. Van Dellen, professor emeritus of geology and environmental science at Macomb Community College, is co-author of this chapter.

been wrecked." Not a building was left standing, either in Concepción itself or in the port of Talcahuano; 70 other villages were destroyed, and a great wave (tsunami) nearly washed away what little was left of Talcahuano.

Quiriquina Island in the harbor was as plainly marked by the earthquake as the shore had been by the resulting tsunami. Cracks as big as a yard wide ran north and south through the ground. Enormous masses of rock had fallen from cliffs onto the beach, and the human survivors expected more to fall with the first rains.

Darwin had been miles away at Valdivia when the earthquake struck. What he felt, lying down in the woods to rest, was fairly mild: "There was no difficulty in standing upright, but the motion made me almost giddy. It lasted only two minutes, but felt much longer and affected me deeply…. Indeed, once you have felt the earth move beneath your feet, you are never quite the same again."

We live on a dynamic planet. Energy from the sun and the earth's interior and the action of water have created continents, mountains, valleys, plains, ocean basins, and soils—an ongoing process that continues to change the landscape. **Geology** is the study of the earth's dynamic history. Geologists study and analyze rocks and the features and processes of the earth's interior and surface. It is from the earth's crust that mineral resources and soil come, as do the elements that make up living organisms.

As Charles Darwin learned firsthand, the earth's dynamic processes also generate a variety of natural hazards—not just earthquakes along faults in the earth's crust (Figure 10-1) but volcanic eruptions, floods, landslides, and subsidence (sinking or collapsing) of parts of its surface.

We can best avoid or minimize risk from such geological hazards by not living in places where they pose a serious risk. However, many people live in or move to such risky areas either because they have no choice or because they believe the benefits outweigh the risks.

Figure 10-1 The San Andreas Fault as it crosses part of the Carrizo Plain between San Francisco and Los Angeles, California. This fault, which extends almost the full length of California, is responsible for earthquakes of various magnitudes. (David Parker/Photo Researchers, Inc.)

Civilization exists by geological consent, subject to change without notice.

WILL DURANT

This chapter addresses the following questions:

- What major geologic processes occur within the earth and on its surface?
- What are rocks, and how are they recycled by the rock cycle?
- What are the hazards from earthquakes and volcanic eruptions?
- What are soils, how are they eroded, and how can we reduce soil erosion?

10-1 GEOLOGIC PROCESSES

What Is the Earth's Structure? As the primitive earth cooled over eons, its interior separated into three major concentric zones, which geologists identify as the *core*, the *mantle*, and the *crust* (Figure 4-6, p. 74, and Figure 10-2). What we

know about the earth's interior comes mostly from indirect evidence: **(1)** density measurements, **(2)** seismic (earthquake) wave studies, **(3)** measurements of heat flow from the interior, **(4)** lava analyses, and **(5)** research on meteorite composition.

The earth's innermost zone, the **core** (Figure 10-2), is very hot and is made mostly of iron (with perhaps some nickel). The core has a solid inner part, surrounded by a liquid core of molten material. The outer and inner cores make up about 16% of the earth's volume and 31% of its mass.

The earth's core is surrounded by a thick, solid zone called the **mantle** (Figure 10-2). The largest zone of the earth's interior, it makes up about 82% of the earth's volume and 68% of its mass. The mantle is

rich in iron (its major constituent), silicon, oxygen, and magnesium.

Most of the mantle is solid rock, but under its rigid outermost part there is a zone of very hot, partly melted rock that flows like soft plastic. This plastic region of the mantle is called the *asthenosphere.*

The outermost and thinnest zone of the earth is called the **crust** (Figure 10-2). It consists of **(1)** the *continental crust*, which underlies the continents (including the continental shelves extending into the oceans, Figure 7-5, p. 156), and **(2)** the *oceanic crust*, which underlies the ocean basins and covers 71% of the earth's surface (Figure 10-3). Only eight elements make up 98.5% of the weight of the earth's crust (Figure 10-4).

10-2 INTERNAL AND EXTERNAL EARTH PROCESSES

What Geologic Processes Occur Within the Earth's Interior? We tend to think of the earth's crust, mantle, and core as fairly static and unchanging. However, they are constantly changed by geologic processes

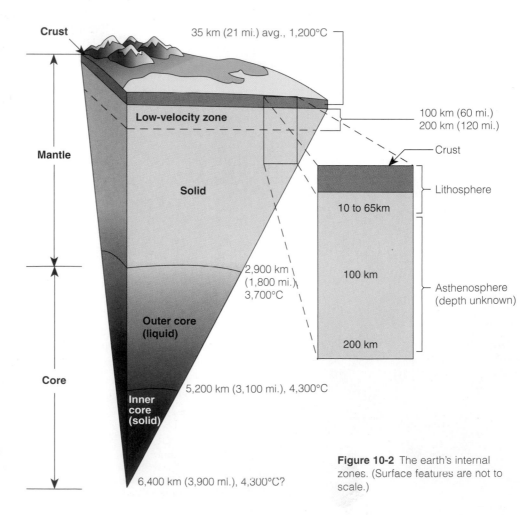

Figure 10-2 The earth's internal zones. (Surface features are not to scale.)

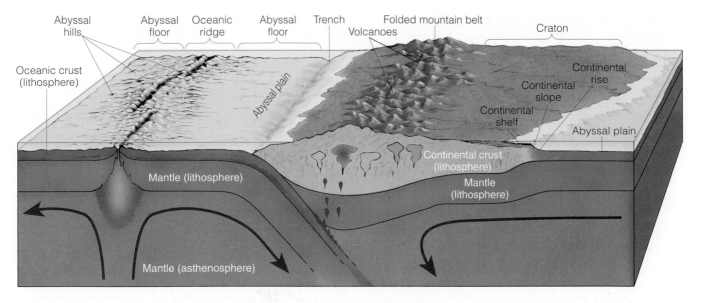

Figure 10-3 Major features of the earth's crust and upper mantle. The *lithosphere*, composed of the crust and outermost mantle, is rigid and brittle. The *asthenosphere*, a zone in the mantle, can be deformed by heat and pressure.

taking place within the earth and on the earth's surface, most over thousands to millions of years.

Geologic changes originating from the earth's interior are called *internal processes*; generally they build up the planet's surface. Heat from the earth's interior provides the energy for these processes, but gravity also plays a role.

Residual heat from the earth's formation is still being given off as the inner core cools and as the outer core cools and solidifies. Continued decay of radioactive elements in the crust, especially the continental crust, adds to the heat flow from within.

This heat from the earth's core causes much of the mantle to deform and flow slowly like heated plastic (in the same way that a red-hot iron horseshoe behaves plastically). Measurements of heat flow within the earth suggests that two kinds of movement occur in the mantle's asthenosphere:

- *Convection cells*, in which large volumes of heated rock move, following a pattern resembling convection in the atmosphere (Figure 6-7, p. 125) or a pot of boiling soup.

- *Mantle plumes*, in which mantle rock flows slowly upward in a column, like smoke from a chimney on a cold, calm morning. When the moving rock reaches the top of the plume, it moves out in a radial pattern, as if it were flowing up an umbrella through the handle and then spreading out in all directions from the tip of the umbrella to the rim.

Both convection currents and mantle plumes move upward as the heated material is displaced by denser, cooler material sinking under the influence of gravity.

What Is Plate Tectonics? A map of the earth's earthquakes and volcanoes shows that most of these phenomena occur along certain lines or belts on the earth's surface (Figure 10-5a). The areas of the earth outlined by these major belts are called **plates** (Figure 10-5b). They are about 100 kilometers (60 miles) thick and are composed of the crust and the rigid, outermost part of the mantle (above the asthenosphere)—a combination called the **lithosphere**.

These plates move constantly, supported by the slowly flowing asthenosphere like large pieces of ice floating on the surface of a lake. Some plates move faster than others, but a typical speed is about the rate at which fingernails grow.

The theory explaining the movements of the plates and the processes that occur at their boundaries is called **plate tectonics**. The

Figure 10-4 Composition by weight of the earth's crust. Various combinations of only eight elements make up the bulk of most minerals.

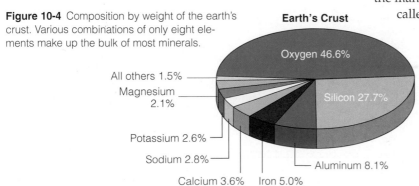

Earth's Crust

Oxygen 46.6%

Silicon 27.7%

Aluminum 8.1%

Iron 5.0%

Calcium 3.6%

Sodium 2.8%

Potassium 2.6%

Magnesium 2.1%

All others 1.5%

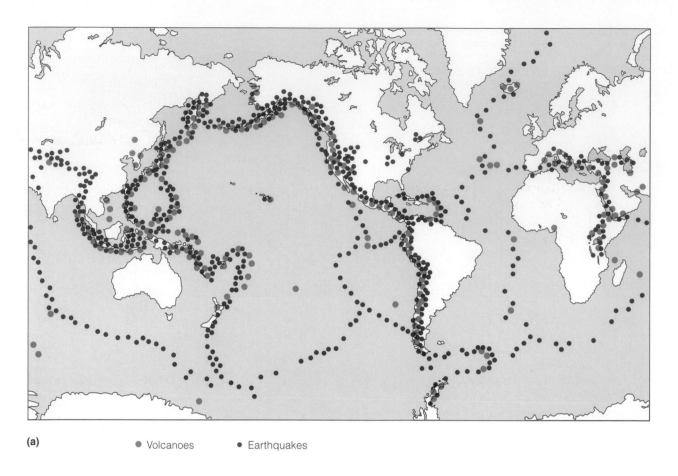

(a)   ● Volcanoes   ● Earthquakes

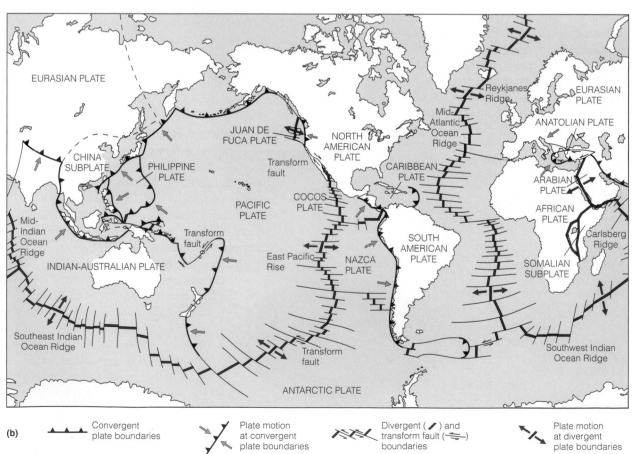

(b)   ▲▲▲ Convergent plate boundaries   Plate motion at convergent plate boundaries   Divergent (╱) and transform fault (═) boundaries   Plate motion at divergent plate boundaries

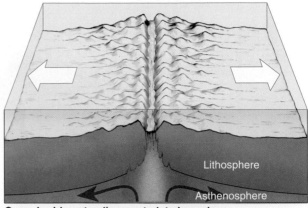
Oceanic ridge at a divergent plate boundary

Figure 10-5 (left) Earthquake and volcano sites are distributed mostly in bands along the planet's surface. **(a)** These bands correspond to the patterns for the types of lithospheric plate boundaries **(b)** shown in Figure 10-6.

concept, which became widely accepted by geologists in the 1960s, was developed from an earlier idea called *continental drift*. Throughout the earth's history, continents have split and joined as plates have drifted thousands of kilometers back and forth across the planet's surface (Figure 5-9, p. 114).

Plate motion produces mountains (including volcanoes), the oceanic ridge system, trenches, and other features of the earth's surface (Figure 10-3). Natural hazards such as volcanoes and earthquakes are likely to be found at plate boundaries (Figure 10-5a), and plate movements and interactions concentrate many of the minerals we extract and use.

The theory of plate tectonics also helps explain how certain patterns of biological evolution occurred. By reconstructing the course of continental drift over millions of years (Figure 5-9, p. 114), we can trace how life-forms migrated from one area to another when continents that are now far apart were still joined together. As the continents separated, populations became geographically and reproductively isolated and speciation occurred (Figure 5-8, p. 113).

What Types of Boundaries Occur Between the Earth's Plates? Lithospheric plates have three types of boundaries (Figure 10-6). At a **divergent plate boundary** the plates move apart in opposite directions (Figure 10-6, top), and at a **convergent plate boundary** they are pushed together by internal forces (Figures 10-5b and 10-6, middle).

At most convergent plate boundaries, oceanic lithosphere is carried downward (subducted) under the island arc or the continent at a **subduction zone**. A *trench* ordinarily forms at the boundary between the two converging plates (Figure 10-6, middle). Stresses in the plate undergoing subduction cause earthquakes at convergent plate boundaries.

The third type of plate boundary, called a **transform fault**, occurs where plates move in opposite but parallel directions along a fracture (fault) in the lithosphere (Figure 10-6, bottom). In other words, the plates slide past one another. Like the other types of plate boundaries, most transform faults are on the ocean floor. California's San Andreas Fault (Figure 10-1) is one of the exceptions.

What Geologic Processes Occur on the Earth's Surface? Erosion and Weathering Geological changes based directly or indirectly on energy from the sun and on gravity (rather than on heat in the earth's

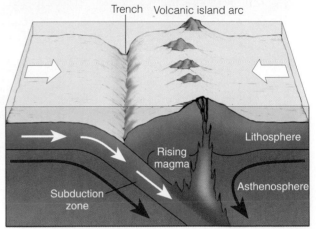
Trench and volcanic island arc at a convergent plate boundary

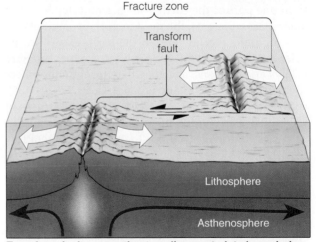
Transform fault connecting two divergent plate boundaries

Figure 10-6 Types of boundaries between the earth's lithospheric plates. All three boundary types occur both in oceans and on continents.

interior) are called *external processes*. Whereas internal processes generally build up the earth's surface, external processes tend to wear it down and produce a variety of landforms and environments formed by the buildup of eroded sediment (Figure 10-7).

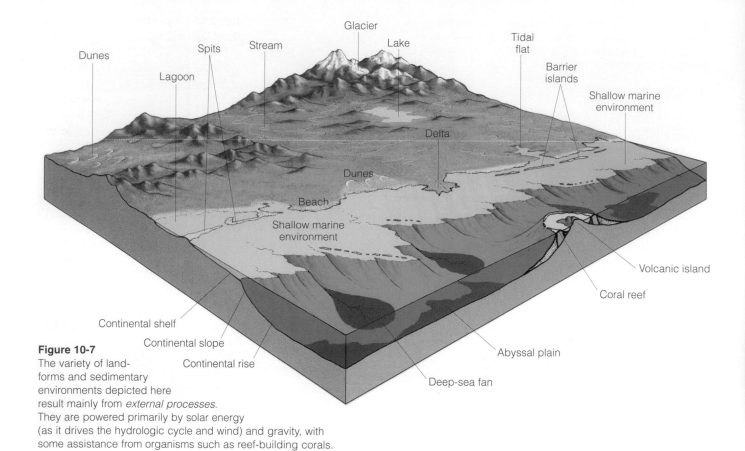

Figure 10-7
The variety of land-forms and sedimentary environments depicted here result mainly from *external processes*. They are powered primarily by solar energy (as it drives the hydrologic cycle and wind) and gravity, with some assistance from organisms such as reef-building corals.

A major external process is **erosion**. It is the process by which material is dissolved, loosened, or worn away from one part of the earth's surface and deposited in other places. Streams, the most important agent of erosion, operate everywhere on the earth except in the polar regions (Figure 7-2, p. 153). They produce ordinary valleys and canyons, and they may form deltas where streams flow into lakes and oceans (Figure 7-17, p. 168). Some erosion is caused when wind blows particles of soil from one area to another (Figure 6-1, p. 120). Human activities, particularly those that destroy vegetation, accelerate erosion, as discussed in Section 10-5.

Loosened material that can be eroded usually is produced by **weathering**, which can result from mechanical or chemical processes. In *mechanical weathering*, a large rock mass is broken into smaller fragments of the original material, similar to the results you would get by using a hammer to break a rock into small fragments. The most important agent of mechanical weathering is *frost wedging*, in which water **(1)** collects in pores and cracks of rock, **(2)** expands upon freezing, and **(3)** splits off pieces of the rock.

In *chemical weathering*, a mass of rock is decomposed by one or more chemical reactions. Most chemical weathering involves a reaction of rock material with oxygen, carbon dioxide, and moisture in the atmosphere and the ground. Weathering is responsible for soil development.

10-3 MINERALS, ROCKS, AND THE ROCK CYCLE

What Are Minerals and Rocks? The earth's crust, which is still forming in various places, is composed of minerals and rocks. It is the source of almost all the nonrenewable resources we use: fossil fuels, metallic minerals, and nonmetallic minerals (Figure 1-11, p. 11). It is also the source of soil and of the elements that make up our bodies and those of other living organisms.

A **mineral** is an element or inorganic compound that occurs naturally and is solid. It usually has a crystalline internal structure made up of an orderly, three-dimensional arrangement of atoms or ions.

Some minerals consist of a single element, such as gold, silver, diamond (carbon), and sulfur. However, most of the more than 2,000 identified minerals occur as inorganic compounds formed by various combinations of the eight elements that make up 98.5% by weight of the earth's crust (Figure 10-4). Examples are salt, mica, and quartz, all of which (along with many others) have economic importance.

Rock is any material that makes up a large, natural, continuous part of the earth's crust. Some kinds of rock, such as limestone (calcium carbonate, or $CaCO_3$) and quartzite (silicon dioxide, or SiO_2), contain only one mineral, but most rocks consist of two or more minerals.

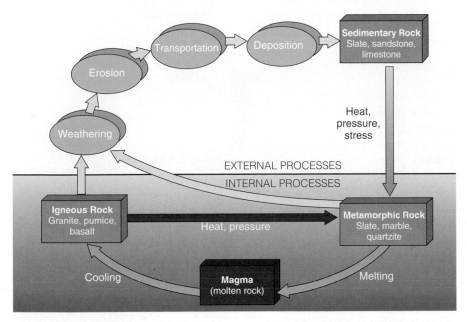

Figure 10-8 The rock cycle, the slowest of the earth's cyclic processes. The earth's materials are recycled over millions of years by three processes: melting, erosion, and metamorphism, which produce igneous, sedimentary, and metamorphic rocks. Rock of any of the three classes can be converted to rock of either of the other two classes (or can even be recycled within its own class).

What Are the Three Major Types of Rock? Based on the way it forms, rock is placed in three broad classes: igneous, sedimentary, and metamorphic. **Igneous rock** can form below or on the earth's surface when molten rock material (magma) wells up from the earth's upper mantle or deep crust, cools, and hardens into rock. Examples are granite (formed underground) and lava rock (formed above ground when molten lava cools and hardens).

Although often covered by sedimentary rocks or soil, igneous rocks form the bulk of the earth's crust. They also are the main source of many nonfuel mineral resources (Section 14-1, p. 321).

Sedimentary rock forms from sediment. Most such rocks are formed when preexisting rocks are **(1)** weathered and eroded into small pieces, **(2)** transported from their sources, and **(3)** deposited in a body of surface water. As deposited layers from weathering and erosion become buried and compacted, the resulting pressure causes their particles to bond together to form sedimentary rocks such as sandstone and shale.

Some sedimentary rocks, such as dolomite and limestone, are formed from the compacted shells, skeletons, and other remains of dead organisms. Two types of coal—lignite and bituminous coal—are sedimentary rocks derived from plant remains (Figure 14-27, p. 344).

Metamorphic rock is produced when a preexisting rock is subjected to high temperatures (which may cause it to melt partially), high pressures, chemically active fluids, or a combination of these agents. Exam-

ples are anthracite (a form of coal), slate, and marble.

What Is the Rock Cycle? Rocks are constantly exposed to various physical and chemical conditions that can change them over time. The interaction of processes that change rocks from one type to another is called the **rock cycle** (Figure 10-8).

The slowest of the earth's cyclic processes, the rock cycle recycles material over millions of years. It is responsible for concentrating the planet's nonrenewable mineral resources on which humans depend, as discussed in more detail in Section 14-1 (p. 321).

10-4 NATURAL HAZARDS: EARTHQUAKES AND VOLCANIC ERUPTIONS

What Are Earthquakes? Stress in the earth's crust can cause solid rock to deform until it suddenly fractures and shifts along the fracture, producing a *fault* (Figure 10-1). The faulting or a later abrupt movement on an existing fault causes an **earthquake**.

An earthquake has certain features and effects (Figure 10-9). When the stressed parts of the earth suddenly fracture or shift, energy is released as shock waves, which move outward from the earthquake's focus like ripples in a pool of water. The *focus* of an earthquake is the point of initial movement, and the *epicenter* is the point on the surface directly above the focus (Figure 10-9).

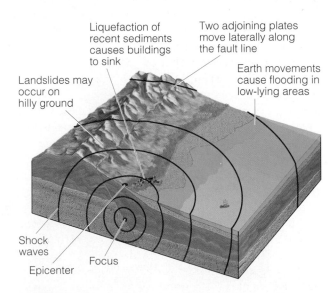

Figure 10-9 Major features and effects of an earthquake.

Figure 10-10 Expected damage from earthquakes in Canada and the contiguous United States. Except for a few regions along the Atlantic and the Gulf coasts, almost every part of the continental United States is subject to some risk from earthquakes. Several areas have a risk of moderate to major damage. This map is based on earthquake records. (Data from U.S. Geological Survey)

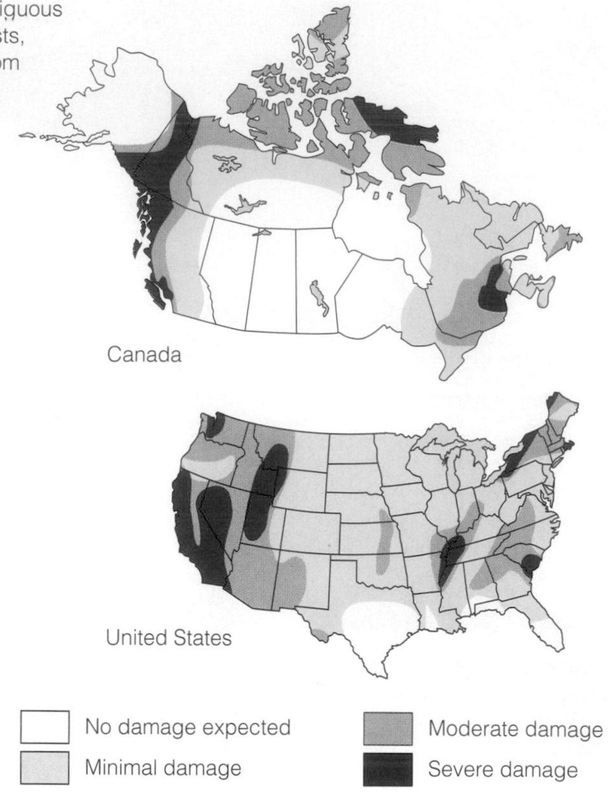

Canada

United States

| | No damage expected | | Moderate damage |
| | Minimal damage | | Severe damage |

One way to measure the severity of an earthquake is by its *magnitude* on a modified version of the Richter scale. The magnitude is a measure of the amount of energy released in the earthquake, as indicated by the amplitude (size) of the vibrations when they reach a recording instrument (seismograph). Using this approach, seismologists rate earthquakes as **(1)** *insignificant* (less than 4.0 on the Richter scale), **(2)** *minor* (4.0–4.9), **(3)** *damaging* (5.0–5.9), **(4)** *destructive* (6.0–6.9), **(5)** *major* (7.0–7.9), and **(6)** *great* (over 8.0). Each unit on the Richter scale represents an amplitude that is 10 times greater than the next smaller unit. Thus, a magnitude 5.0 earthquake is 10 times greater than a magnitude 4.0, and a magnitude 6.0 quake is 100 times greater than a magnitude 4.0 quake.

Earthquakes often have *aftershocks* that gradually decrease in frequency over a period of up to several months, and some have *foreshocks* from seconds to weeks before the main shock.

The *primary effects of earthquakes* include shaking and sometimes a permanent vertical or horizontal displacement of the ground. These effects may have serious consequences for people and for buildings, bridges, freeway overpasses, dams, and pipelines.

Secondary effects of earthquakes include rockslides, urban fires, and flooding caused by subsidence (sinking) of land. Coastal areas also can be severely damaged by large, earthquake-generated water waves, called *tsunamis* (misnamed "tidal waves," even though they have nothing to do with tides) that travel as fast as 950 kilometers (590 miles) per hour.

Solutions: How Can We Reduce Earthquake Hazards? We can reduce loss of life and property from earthquakes by **(1)** examining historical records and making geologic measurements to locate active fault zones, **(2)** making maps showing high-risk areas (Figure 10-10), **(3)** establishing building codes that regulate the placement and design of buildings in areas of high risk, and **(4)** trying to predict when and where earthquakes will occur.

Engineers know how to make homes, large buildings, bridges, and freeways more earthquake resistant. However, this can be expensive, especially if existing structures must be reinforced. And not everyone can afford to take such damage-reducing measures.

What Are Volcanoes? An active **volcano** occurs where magma (molten rock) reaches the earth's surface through a central vent or a long crack (fissure; Figure 10-11). Volcanic activity can

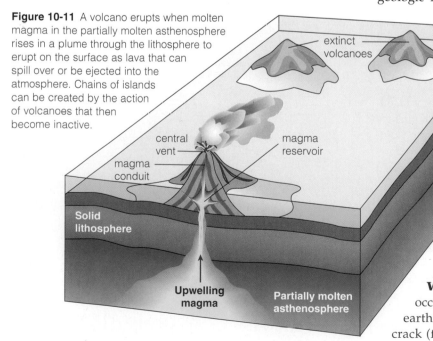

Figure 10-11 A volcano erupts when molten magma in the partially molten asthenosphere rises in a plume through the lithosphere to erupt on the surface as lava that can spill over or be ejected into the atmosphere. Chains of islands can be created by the action of volcanoes that then become inactive.

extinct volcanoes

central vent

magma reservoir

magma conduit

Solid lithosphere

Upwelling magma

Partially molten asthenosphere

The Mount St. Helens Eruption

On May 18, 1980, Mount St. Helens, in the Cascade Range near the Washington–Oregon border, erupted in what has been described as the worst volcanic disaster in U.S. history (see photos in table of contents, p. XXIV). Fifty-seven people died in the eruption, and several hundred cabins and homes were destroyed or severely damaged.

Tens of thousands of hectares of forest were obliterated, along with campgrounds and bridges. An estimated 7,000 large animals (bear, deer, elk, and mountain lions) died, as did millions of smaller animals and birds and some 11 million fish.

Salmon hatcheries were damaged and crops (including alfalfa, apples, potatoes, and wheat) were lost. Many people living in the area also lost their jobs.

On the plus side, trace elements from ash that were added to the soil may benefit agriculture in the long run. Furthermore, increased tourism to the area brought new jobs and income.

By 1990, many biologists were surprised at how fast various forms of life had begun colonizing many of the most devastated areas. This rapid recovery has taught biologists important and often surprising lessons about nature's ability to recover from what seems to be almost total devastation.

Critical Thinking

If vegetation can recover within 2 to 3 decades from the effects of a devastating volcanic eruption, why should we spend money and time on ecological restoration of highly degraded forests, grasslands, and wetlands?

---

release *ejecta* (debris ranging from large chunks of lava rock to ash that may be glowing hot), liquid lava, and gases (such as water vapor, carbon dioxide, and sulfur dioxide) into the environment (Case Study, above).

Volcanic activity is concentrated for the most part in the same areas as seismic activity (Figure 10-5a). Some volcanoes, such as those at Mount St. Helens in Washington (which erupted in 1980, Case Study, above) and Mount Pinatubo in the Philippines (which erupted in 1991), have a steep, flaring cone shape. They usually erupt explosively and eject large quantities of gases and particulate matter (soot and mineral ash) high into the troposphere.

Most of the particles of soot and ash soon fall back to the earth's surface. However, gases such as sulfur dioxide remain in the atmosphere and are converted to tiny droplets of sulfuric acid, many of which stay

above the clouds and may not be washed out by rain for up to 3 years. These tiny droplets reflect some of the sun's energy and can cool the atmosphere by as much as 0.5°C (1°F) for 1–4 years.

Other volcanic eruptions at divergent boundaries (as in Iceland) and ocean islands (such as the Hawaiian Islands) usually erupt more quietly. They involve primarily lava flows, which can cover roads and villages and ignite brush, trees, and homes.

We tend to think negatively of volcanic activity, but it also provides some benefits. One is outstanding scenery in the form of majestic mountains, some lakes (such as Crater Lake in Oregon, Figure 19-6, left p. 483), and other landforms. Perhaps the most important benefit of volcanism is the highly fertile soils produced by the weathering of lava.

Solutions: How Can We Reduce Volcano Hazards? We can reduce the loss of human life and sometimes property from volcanic eruptions by (1) land-use planning, (2) better prediction of volcanic eruptions, and (3) effective evacuation plans. The eruptive history of a volcano or volcanic center can provide some indication of where the risks are.

Scientists are also studying phenomena that precede an eruption such as (1) tilting or swelling of the cone, (2) changes in magnetic and thermal properties of the volcano, (3) changes in gas composition, and (4) increased seismic activity.

10-5 SOILS: FORMATION, EROSION, AND DEGRADATION

What Major Layers Are Found in Mature Soils? The material we call **soil** is a complex mixture of eroded rock, mineral nutrients, decaying organic matter, water, air, and billions of living organisms, most of them microscopic decomposers (Figure 10-12). Although soil is a potentially renewable resource, it is produced very slowly by the (1) weathering of rock, (2) deposit of sediments by erosion, and (3) decomposition of organic matter in dead organisms.

Mature soils are arranged in a series of zones called **soil horizons**, each with a distinct texture and composition that varies with different types of soils. A cross-sectional view of the horizons in a soil is called a **soil profile**. Most mature soils have at least three of the possible horizons (Figure 10-12).

The top layer, the *surface litter layer*, or *O horizon*, consists mostly of freshly fallen and partially decomposed leaves, twigs, animal waste, fungi, and other organic materials. Normally, it is brown or black. The *topsoil layer*, or *A horizon*, is a porous mixture of partially decomposed organic matter, called **humus**, and some inorganic mineral particles. It is usually darker and

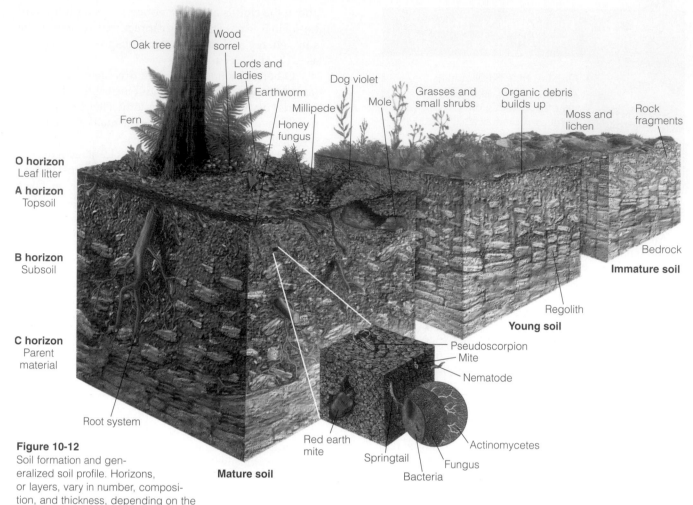

O horizon
Leaf litter

A horizon
Topsoil

B horizon
Subsoil

C horizon
Parent material

Oak tree

Wood sorrel

Lords and ladies

Dog violet

Earthworm

Mole

Grasses and small shrubs

Organic debris builds up

Rock fragments

Moss and lichen

Fern

Millipede

Honey fungus

Root system

Red earth mite

Springtail

Bacteria

Pseudoscorpion

Mite

Nematode

Actinomycetes

Fungus

Bedrock

Immature soil

Regolith

Young soil

Mature soil

Figure 10-12
Soil formation and generalized soil profile. Horizons, or layers, vary in number, composition, and thickness, depending on the type of soil. (From *Earth* by Derek Elsom, 1992. Copyright © 1992 by Marshall Editions Developments Limited. New York: Macmillan. Used by permission.)

looser than deeper layers. The roots of most plants and most of a soil's organic matter are concentrated in these two upper layers. As long as these layers are anchored by vegetation, soil stores water and releases it in a nourishing trickle instead of a devastating flood.

The two top layers of most well-developed soils teem with bacteria, fungi, earthworms, and small insects that interact in complex food webs (Figure 10-13). Bacteria and other decomposer microorganisms found by the billions in every handful of topsoil recycle the nutrients we and other land organisms need (Figure 10-14). They break down some complex organic compounds into simpler inorganic compounds soluble in water. Soil moisture carrying these dissolved nutrients is drawn up by the roots of plants and transported through stems and into leaves.

Some organic litter in the two top layers is broken down into a sticky, brown residue of partially decomposed organic material (humus). Because this humus is only slightly soluble in water, most of it stays in the

topsoil layer. A fertile soil that produces high crop yields has a thick topsoil layer with lots of humus. This helps topsoil hold water and nutrients taken up by plant roots.

The color of its topsoil tells us a lot about how useful a soil is for growing crops. For example, dark-brown or black topsoil is nitrogen-rich and high in organic matter. Gray, bright yellow, or red topsoils are low in organic matter and will need nitrogen enrichment to support most crops.

The *B horizon (subsoil)* and the *C horizon (parent material)* contain most of a soil's inorganic matter, mostly broken-down rock consisting of varying mixtures of sand, silt, clay, and gravel. The C horizon lies on a base of unweathered parent rock called *bedrock*.

The spaces, or pores, between the solid organic and inorganic particles in the upper and lower soil layers contain varying amounts of air (mostly nitrogen and oxygen gas) and water. Plant roots need oxygen for cellular respiration.

Some of the precipitation that reaches the soil percolates through the soil layers and occupies many of the soil's open spaces or pores. This downward movement of water through soil is called **infiltration**. As the water seeps down, it dissolves various soil components in

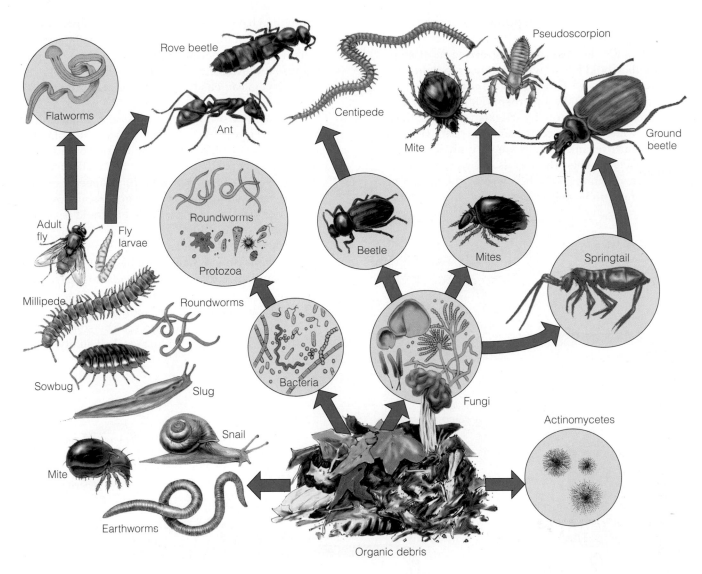

Figure 10-13 Greatly simplified food web of living organisms found in soil.

upper layers and carries them to lower layers in a process called **leaching.**

Soils develop and mature slowly. It can take 200 to 1,000 years to develop an inch (2.5 cm) of topsoil (A horizon). Five important soil types, each with a distinct profile, are shown in Figure 10-15. Most of the world's crops are grown on soils exposed when grasslands and deciduous forests are cleared.

How Do Soils Differ in Texture, Porosity, and Acidity? Soils vary in their content of **(1)** *clay* (very fine particles), **(2)** *silt* (fine particles), **(3)** *sand* (medium-size particles), and **(4)** *gravel* (coarse to very coarse particles). The relative amounts of the different sizes and types of mineral particles determine **soil texture**, as depicted in Figure 10-16, p. 224. Soils with roughly equal mixtures of clay, sand, silt, and humus are called **loams.**

To get an idea of a soil's texture, take a small amount of topsoil, moisten it, and rub it between your fingers and thumb. A gritty feel means that it contains a lot of sand. A sticky feel means a high clay content, and you should be able to roll it into a clump. Silt-laden soil feels smooth, like flour. A loam topsoil, which is best suited for plant growth, has a texture between these extremes—a crumbly, spongy feeling—with many of its particles clumped loosely together.

Soil texture helps determine **soil porosity**, a measure of the volume of pores or spaces per volume of soil and of the average distances between those spaces. Fine particles are needed for water retention and coarse ones for air spaces. A porous soil has many pores and can hold more water and air than a less porous soil. The average size of the spaces or pores in a soil determines **soil permeability**: the rate at which water and air move from upper to lower soil layers. Soil porosity is also influenced by **soil structure**: the ways in which soil particles are organized and clumped together (Figure 10-17, p. 224). Table 10-1 (p. 225) compares the main physical and chemical properties of sand, clay, silt, and loam soils.

Loams are the best soils for growing most crops because they hold lots of water, but not too tightly for plant roots to absorb. Sandy soils are easy to work, but water flows rapidly through them (Figure 10-17, left). They are useful for growing irrigated crops or those with low water needs, such as peanuts and strawberries.

The particles in clay soils are very small and easily compacted. When these soils get wet, they form large, dense clumps, which is why wet clay can be molded into bricks and pottery. Clay soils are more

porous and have a greater water-holding capacity than sandy soils, but the pore spaces are so small that these soils have a low permeability (Figure 10-17, right). Because little water can infiltrate to lower levels, the upper layers can easily become too waterlogged for most crops.

The acidity or alkalinity of a soil, as measured by its pH (Figure 3-7, p. 56), influences the uptake of soil nutrients by plants. When soils are too acidic, the acids can be partially neutralized by an alkaline substance such as lime. Because lime speeds up the decomposition of organic matter in the soil, however, manure or another organic fertilizer should be added to maintain soil fertility.

In dry regions such as much of the western and southwestern United States, rain does not leach away calcium and other alkaline compounds, so soils in such areas may be too alkaline (pH above 7.5) for some crops. Adding sulfur, which is gradually converted into sulfuric acid by soil bacteria, reduces soil alkalinity.

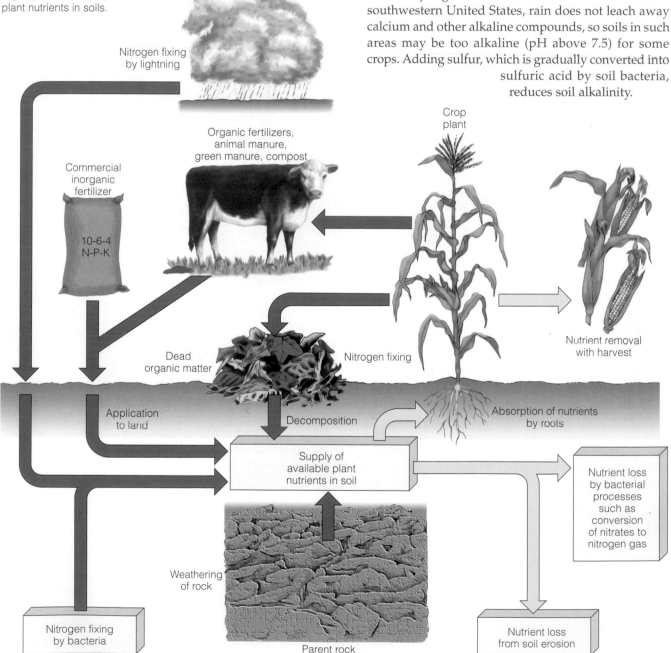

Figure 10-14 Pathways of plant nutrients in soils.

What Causes Soil Erosion? **Soil erosion** is the movement of soil components, especially surface litter and topsoil, from one place to another. It results in the buildup of sediments and sedimentary rock on land and in bodies of water (Figure 10-7). The two main agents of erosion are flowing water and wind. Some soil erosion is natural, and some is caused by human activities. In undisturbed vegetated ecosystems, the roots of plants help anchor the soil, and usually soil is not lost faster than it forms.

Farming, logging, construction, overgrazing by livestock, off-road vehicles, deliberate burning of vegetation, and other activities that destroy plant cover leave soil vulnerable to erosion. Such human activities can speed up erosion and destroy in

a few decades what nature took hundreds to thousands of years to produce.

Moving water causes most soil erosion. Soil scientists distinguish between three types of water erosion:

- *Sheet erosion* occurs when surface water moves down a slope or across a field in a wide flow and peels off fairly uniform sheets or layers of soil. Because the topsoil disappears evenly, sheet erosion may not be noticeable until much damage has been done.

- *Rill erosion* (Figure 10-18) occurs when surface water forms fast-flowing rivulets that cut small channels in the soil.

- *Gully erosion* (Figure 10-18) occurs when rivulets of fast-flowing water join together and with each succeeding rain cut the channels wider and deeper until they become ditches or gullies. Gully erosion usually happens on steep slopes where all or most vegetation has been removed.

Figure 10-15 Soil profiles of the principal soil types typically found in five different biomes.

Mosaic of closely packed pebbles, boulders

Weak humus–mineral mixture

Dry, brown to reddish-brown with variable accumulations of clay, calcium carbonate, and soluble salts

Desert Soil
(hot, dry climate)

Alkaline, dark, and rich in humus

Clay, calcium compounds

Grassland Soil
(semiarid climate)

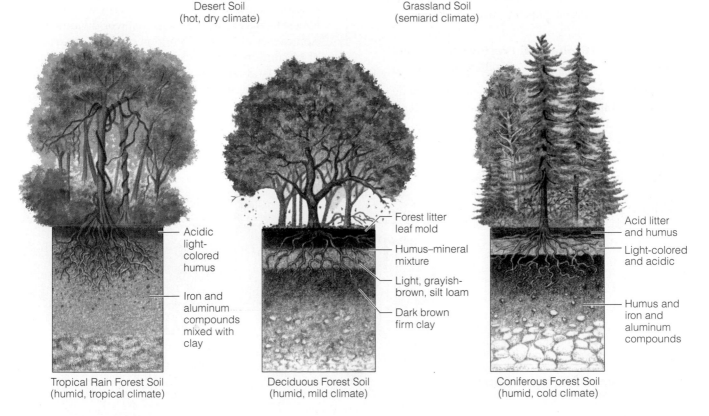

Acidic light-colored humus

Iron and aluminum compounds mixed with clay

Tropical Rain Forest Soil
(humid, tropical climate)

Forest litter leaf mold

Humus–mineral mixture

Light, grayish-brown, silt loam

Dark brown firm clay

Deciduous Forest Soil
(humid, mild climate)

Acid litter and humus

Light-colored and acidic

Humus and iron and aluminum compounds

Coniferous Forest Soil
(humid, cold climate)

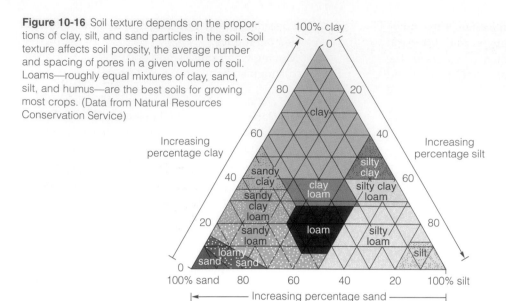

Figure 10-16 Soil texture depends on the proportions of clay, silt, and sand particles in the soil. Soil texture affects soil porosity, the average number and spacing of pores in a given volume of soil. Loams—roughly equal mixtures of clay, sand, silt, and humus—are the best soils for growing most crops. (Data from Natural Resources Conservation Service)

The two major harmful effects of soil erosion are (1) loss of soil fertility and its ability to hold water and (2) runoff of sediment that pollutes water, kills fish and shellfish, and clogs irrigation ditches, boat channels, reservoirs, and lakes.

Soil, especially topsoil, is classified as a renewable resource because it is regenerated by natural processes. However, in tropical and temperate areas it takes 200–1,000 years (depending on climate and soil type) for 2.54 centimeters (1 inch) of new topsoil to form. If topsoil erodes faster than it forms on a piece of land, the soil becomes a nonrenewable resource.

How Serious Is Global Soil Erosion? Several studies document the seriousness of soil erosion:

■ A United Nations (UN) Environment Programme survey found that topsoil is eroding faster than it forms on about one-third of the world's cropland, causing an estimated 85% of the world's land degradation from human activities (Figure 1-4, p. 8).

■ A 1992 study by the World Resources Institute and the UN Environment Programme found that soil on an area equal to the size of China and India combined had been seriously eroded since 1945

(Figure 10-19). The study also found that about 15% of land scattered across the globe was too eroded to grow crops anymore because of a combination of (1) overgrazing (35%), (2) deforestation (30%), and (3) unsustainable farming (28%). Two-thirds of these seriously degraded lands are in Asia and Africa.

■ According to a 2000 study by the Consultative Group on International Agricultural Research, (1) nearly 40% of the world's land (75% in Central America) used for agriculture is seriously degraded by erosion, salt buildup (salinization), and waterlogging, and (2) soil degradation has reduced food production on about 16% of the world's cropland.

According to Lester R. Brown (Guest Essay, p. 18), the topsoil that washes and blows into the world's streams, lakes, and oceans each year would fill a train of freight cars long enough to encircle the planet 150 times. At that rate, the world is losing about 7% of its topsoil from actual or potential cropland each decade.

The situation is worsening as many poor farmers in some developing countries plow up marginal (easily erodible) lands to survive. Soil expert David Pimentel (Guest Essay, p. 232) estimates that worldwide soil erosion causes at least $375 billion per year (an average of $42 million per hour) in (1) direct damage to agricultural lands and (2) indirect damage to waterways, infrastructure, and human health.

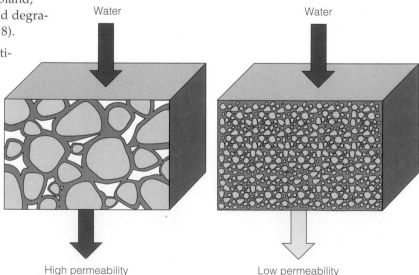

Figure 10-17 The size, shape, and degree of clumping of soil particles determine the number and volume of spaces for air and water within a soil. Soils with more pore spaces (left) contain more air and are more permeable to water flows than soils with fewer pores (right).

Table 10-1 Properties of Soils with Different Textures					
Soil texture	Nutrient-Holding Capacity	Water Infiltration Capacity	Water-Holding Capacity	Aeration	Workability
Clay	Good	Poor	Good	Poor	Poor
Silt	Medium	Medium	Medium	Medium	Medium
Sand	Poor	Good	Poor	Good	Good
Loam	Medium	Medium	Medium	Medium	Medium

How Serious Is Soil Erosion in the United States?
According to the National Resources Conservation Service, about one-third of the nation's original prime topsoil has been washed or blown into streams, lakes, and oceans, mostly as a result of overcultivation, overgrazing, and deforestation (Case Study, p. 227).

According to the U.S. Department of Agriculture (USDA), soil on cultivated land in the United States is eroding about 16 times faster than it can form. Erosion rates are even higher in heavily farmed regions, including the Great Plains, which has lost one-third or more of its topsoil in the 150 years since it was first plowed. Some of the country's most productive agricultural lands, such as those in Iowa, have lost about half their topsoil.

Because of soil conservation efforts, the USDA estimates that soil erosion in the United States decreased by about 40% between 1985 and 1997. Using these data, USDA researchers estimate that soil erosion cost the United States about $30 billion in 1997, an average loss of $3.4 million per hour.

Critics such as Pierre Crosson say that these estimates of soil erosion and damages from such erosion are exaggerated and are based on inexact models instead of field measurements of soil loss and sedimentation rates in nearby bodies of water. They point to studies by several soil scientists (1) showing that in some areas the sedimentation rate in streams was less than the amount of soil erosion predicted by models and (2) concluding that if current rates of cropland erosion in the United States continue for 100 years, crop yields will be only 3–10% less than they would be without such erosion.

However, David Pimentel (Guest Essay, p. 232) and others point out that current estimates by models and a few on-site measurements do not include all the ecological effects of soil erosion. Such effects include reductions in (1) soil depth, (2) availability of soil water for crops, and (3) soil organic matter and nutrients. When such effects are included, some soil scientists and ecologists estimate that soil erosion causes a 15–30% reduction in crop productivity.

Erosion of soil on cropland requires costly use of inorganic fertilizers to help replace lost soil nutrients. However, because fertilizers are not a substitute for fertile soil, there is a limit to the amount of fertilizer that can be applied before crop yields level off and then begin to decline.

What Is Desertification, and How Serious Is This Problem? Desertification is a process whereby the productive potential of arid or semiarid land falls by 10% or more because of human activities and climate changes. Desertification can be (1) *moderate* (with a 10–25% drop in productivity), (2) *severe* (with a 25–50%

Figure 10-18 Rill and gully erosion of vital topsoil from irrigated cropland in Arizona. (Natural Resources Conservation Service)

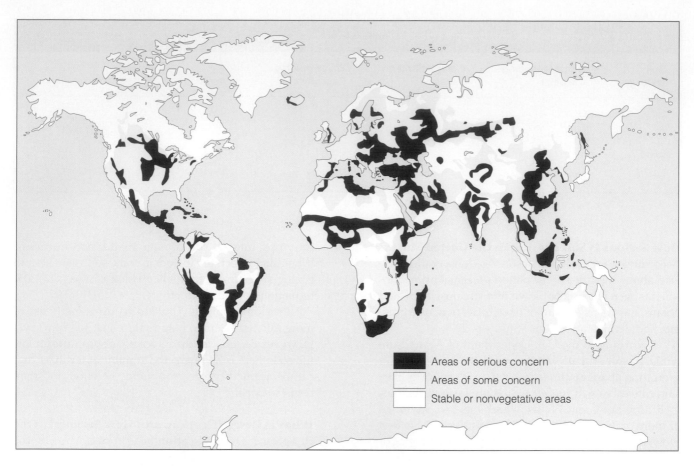

Figure 10-19 Global soil erosion. (Data from UN Environment Programme and the World Resources Institute)

drop) and **(3)** *very severe* (with a drop of 50% or more, usually creating huge gullies and sand dunes). Desertification is a serious and growing problem in many parts of the world (Figure 10-21).

Practices that leave topsoil vulnerable to desertification include **(1)** overgrazing on fragile arid and semiarid rangelands, **(2)** deforestation without reforestation, **(3)** surface mining without land reclamation, **(4)** irrigation techniques that lead to increased erosion, **(5)** salt buildup from irrigation, **(6)** farming on land with unsuitable terrain or soils, and **(7)** soil compaction by farm machinery and cattle.

Major symptoms of desertification include:

- Loss of native vegetation

- Increased erosion of the dry soil (especially by wind; Figure 6-1, p. 120)

- Salt buildup in the soil (salinization)

- Lowering of the water table as wells are dug deeper

- Reduced surface water supply as streams and ponds dry up

The consequences of desertification include **(1)** worsening drought, **(2)** famine, **(3)** declining living standards, and **(4)** swelling numbers of environmental refugees whose land is too eroded to grow crops or feed livestock.

An estimated 8.1 million square kilometers (3.1 million square miles)—an area the size of Brazil and 12 times the size of Texas—have become desertified in the past 50 years. According to a 1999 United Nations conference on desertification, **(1)** about 40% of the world's land and 70% of all drylands is suffering from the effects of desertification and **(2)** each year about 150,000 square kilometers (58,000 square miles)—an area larger than Greece—becomes desertified. This threatens the livelihoods of at least 1 billion people in 100 countries.

Solutions: How Can We Slow Desertification?
The most effective way to slow desertification is to reduce overgrazing, deforestation, and destructive forms of planting, irrigation, and mining. In addition, planting trees and grasses will anchor soil and hold water while slowing desertification and reducing the threat of global warming.

How Do Excess Salts and Water Degrade Soils?
The approximately 17% of the world's cropland that

The Dust Bowl

In the 1930s, Americans learned a harsh environmental lesson when much of the topsoil in several dry and windy midwestern states was lost through a combination of poor cultivation practices and prolonged drought.

Before settlers began grazing livestock and planting crops there in the 1870s, the deep and tangled root systems of native prairie grasses anchored the fertile topsoil firmly in place (Figure 10-15). Plowing the prairie tore up these roots, and the agricultural crops the settlers planted annually in their place had less extensive root systems.

After each harvest, the land was plowed and left bare for several months, exposing it to high winds. Overgrazing also destroyed large expanses of grass, denuding the ground. The stage was set for severe wind erosion and crop failures; all that was needed was a long drought.

Such a drought occurred between 1926 and 1934. In the 1930s, dust clouds created by hot, dry windstorms darkened the sky at midday in some areas; rabbits and birds choked to death on the dust.

During May 1934, a cloud of topsoil blown off the Great Plains traveled some 2,400 kilometers (1,500 miles) and blanketed most of the eastern United States with dust. Journalists gave the Great Plains a new name: the *Dust Bowl* (Figure 10-20).

During the "dirty thirties," large areas of cropland were stripped of topsoil and severely eroded. Thousands of displaced farm families from Oklahoma, Texas, Kansas, and Colorado migrated to California or to the industrial cities of the Midwest and East. Most found no jobs because the country was in the midst of the Great Depression.

In May 1934, Hugh Bennett of the U.S. Department of Agriculture (USDA) went before a congressional hearing in Washington to plead for new programs to protect the country's topsoil. Lawmakers took action when Great Plains dust began seeping into the hearing room.

In 1935, the United States passed the Soil Erosion Act, which established the Soil Conservation Service (SCS) as part of the USDA. With Bennett as its first head, the SCS (now called the Natural Resources Conservation Service) began promoting sound conservation practices, first in the Great Plains states and later elsewhere. Soil conservation districts were formed throughout the country, and farmers and ranchers were given technical assistance in setting up soil conservation programs.

Climate researchers see signs of a returning Dust Bowl period because of a megadrought that lasts 2 to 4 decades. By examining tree rings, archeological finds, lake sediments, and sand dunes, scientists have found that prolonged megadroughts generally hit twice a century as part of a complex drought cycle. They also found that smaller 2-year droughts strike every 20 years or so. If the earth warms as projected, the region could become even drier, and farming might have to be abandoned.

Critical Thinking

Do you think Americans learned a lesson about protecting soil as a result of the Dust Bowl in the 1930s? Explain.

Figure 10-20 The Dust Bowl of the Great Plains, where a combination of extreme drought and poor soil conservation practices led to severe wind erosion of topsoil in the 1930s. Note the connection of this area with the Ogallala Aquifer in Figure 13-17 (p. 309).

---

is irrigated produces almost 40% of the world's food. Irrigated land can produce crop yields that are two to three times greater than those from rain watering.

However, irrigation also has a downside. Most irrigation water is a dilute solution of various salts, picked up as the water flows over or through soil and rocks. Small quantities of these salts are essential nutrients for plants, but they are toxic in large amounts.

Irrigation water not absorbed into the soil evaporates, leaving behind a thin crust of dissolved salts (such as sodium chloride) in the topsoil. The accumulation of these salts, called **salinization** (Figures 10-22 and 10-23), stunts crop growth, lowers yields, and eventually kills plants and ruins the land. According to a 1995 study, severe salinization has reduced yields on 21% of the world's irrigated cropland, and another 30% has been moderately salinized. The most severe salinization occurs in Asia, especially in China, India, and Pakistan.

In the United States, salinization affects 23% of all irrigated cropland. However, the proportion is much higher in some heavily irrigated western states. This includes 66% of the irrigated land in the lower Colorado Basin and 35% of such land in California.

Precipitation can desalinate soil, but this takes thousands of years in arid and semiarid areas where irrigation is used. Salts can be flushed out of soil by

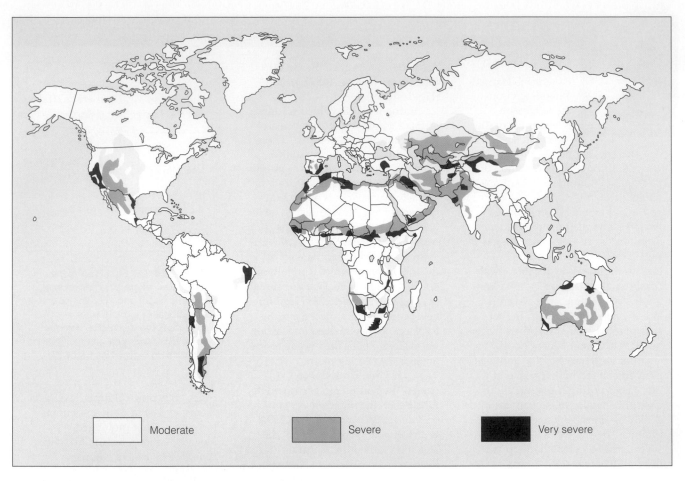

Figure 10-21 Desertification of arid and semiarid lands. (Data from UN Environmental Programme and Harold E. Drengue)

| Moderate | Severe | Very severe |

applying much more irrigation water than is needed for crop growth. However, this practice increases pumping and crop production costs and wastes enormous amounts of water.

Heavily salinized soil also can be renewed by **(1)** taking the land out of production for 2 to 5 years, **(2)** installing an underground network of perforated drainage pipes, and **(3)** flushing the soil with large quantities of low-salt water. However, this costly scheme only slows the salt buildup; it does not stop the process. Flushing salts from the soil also makes downstream irrigation water saltier unless the saline water can be drained into evaporation ponds rather than returned to the stream or canal.

Methods for reducing or bypassing the threat of salinization include **(1)** reducing the use of irrigation water (Section 13-7, p. 310), **(2)** switching to more salt-tolerant crops such as barley, cotton, sugarbeet, and semidwarf wheat, and **(3)** planting salt-loving plants (halophytes), such as saltbush, to convert heavily salinized cropland to grazing land.

There is no simple cure for salinization. As irrigation continues throughout heavily farmed parts of the world, salinization will increase and crop yields will fall.

Another problem with irrigation is **waterlogging** (Figure 10-22). Farmers often apply large amounts of irrigation water to leach salts deeper into the soil. Without adequate drainage, however, water accumulates underground and gradually raises the water table. Saline water then envelops the deep roots of plants, lowering their productivity and killing them after prolonged exposure. At least one-tenth of all irrigated land worldwide suffers from waterlogging, and the problem is getting worse.

10-6 SOLUTIONS: SOIL CONSERVATION

How Can Conservation Tillage Reduce Soil Erosion? **Soil conservation** involves reducing soil erosion and restoring soil fertility. For hundreds of years, farmers have used various methods to reduce soil erosion, most of which involve keeping the soil covered with vegetation.

In **conventional-tillage farming** the land is plowed and then the soil is broken up and smoothed to make a planting surface. In areas such as the midwestern United States, harsh winters prevent plowing just

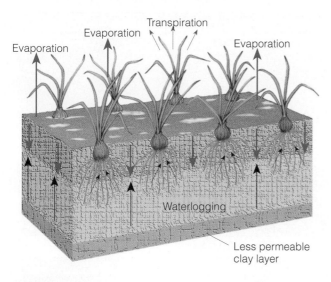

Salinization

1. Irrigation water contains small amounts of dissolved salts.

2. Evaporation and transpiration leave salts behind.

3. Salt builds up in soil.

Waterlogging

1. Precipitation and irrigation water percolate downward.

2. Water table rises.

Figure 10-22 Salinization and waterlogging of soil on irrigated land without adequate drainage lead to decreased crop yields.

before the spring growing season. Thus, cropfields often are plowed in the fall. This leaves the soil bare during the winter and early spring and makes it vulnerable to erosion.

To reduce erosion, many U.S. farmers are using **conservation-tillage farming** (either *minimum-tillage* or *no-till farming*). The idea is to disturb the soil as little as possible while planting crops. With *minimum-tillage farming*, special tillers break up and loosen the subsurface soil without turning over the topsoil, previous crop residues, and any cover vegetation. In *no-till farming*, special planting machines inject seeds, fertilizers, and weed-killers (herbicides) into slits made in the unplowed soil.

Besides reducing soil erosion, conservation tillage (1) saves fuel, (2) cuts costs, (3) holds more water in the soil, (4) keeps the soil from getting packed down, (5) allows more crops to be grown during a season (multiple cropping), (6) gives yields at least as high as those from con-

Figure 10-23 Severe salinization. Because of high evaporation, poor drainage, and severe salinization, white alkaline slats have displaced crops that once grew on this heavily irrigated land in Colorado. (Natural Resources Conservation Service)

ventional tillage, and (7) reduces the release of carbon dioxide from the soil to the air, which helps ease the threat of global warming.

At first, conservation tillage was thought to require more herbicides, but a 1990 USDA study of corn production in the United States found no real difference in levels of herbicide use between conventional and conservation tillage systems. However, no-till cultivation of corn does leave stalks. They can serve as habitats for the corn borer, which can potentially increase the use of pesticides. Crop residues also are a perfect home for a fungal disease called wheat scab, which can devastate monoculture wheat crops.

By 1998, conservation tillage was used on about 40% of U.S. cropland and is projected to be used on more than half of it by 2005. In Indiana, the Nature Conservancy is giving farmers money to buy no-till equipment in exchange for a promise to use conservation tillage for at least 3 years. The USDA estimates that using conservation tillage on 80% of U.S. cropland would reduce soil erosion by at least half. So far, the practice is not widely used in other parts of the world.

How Can Terracing, Contour Farming, Strip Cropping, and Alley Cropping Reduce Soil Erosion? Terracing can reduce soil erosion on steep slopes, each of which is converted into a series of broad, nearly level terraces that run across the land contour (Figure 10-24a). Terracing retains water for crops at each level and reduces soil erosion by controlling runoff.

In mountainous areas such as the Himalayas and the Andes, farmers traditionally built elaborate systems of terraces to grow crops. Today, however, some of these slopes are being farmed without terraces. This leaves the land too nutrient poor to grow crops or generate new forest after 10–40 years. Although most poor farmers know the risk of not terracing, many have too little time and too few workers to build terraces; they must plant crops or starve. The resultant loss of protective

(a) Terracing

(b) Contour planting and strip cropping

(c) Alley cropping

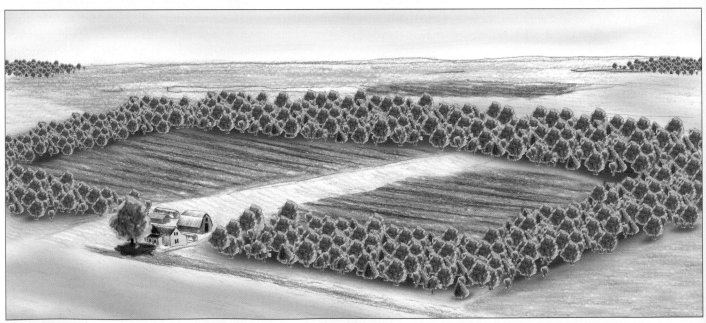

(d) Windbreaks

Figure 10-24 Soil conservation methods: **(a)** terracing, **(b)** contour planting and strip cropping, **(c)** alley cropping, and **(d)** windbreaks.

Slowing Soil Erosion in the United States

Of the world's major food-producing countries, only the United States is sharply reducing some of its soil losses through conservation tillage and government-sponsored soil conservation programs.

The 1985 Farm Act established a strategy for reducing soil erosion in the United States. In the first phase of this program, farmers are given a subsidy for taking highly erodible land out of production and replanting it with soil-saving grass or trees for 10-15 years. As of October 2000, approximately 13.6 million hectares (33.5 million acres) were in this Conservation Reserve Program (CRP).

The land in such a *conservation reserve* cannot be farmed, grazed, or cut for hay. Farmers who violate their contracts must pay back all subsidies plus interest.

According to the U.S. Department of Agriculture, since 1985 this program has cut soil losses on cropland in the United States by about 65%—a shining example of *good news*—and could eventually cut such losses as much as 80%. In 1996, Congress reauthorized the CRP until 2002.

The second phase of the program required all farmers with highly erodible land to develop government-approved 5-year soil conservation plans for their entire farms by the end of 1990. A third provision of the Farm Act authorizes the government to forgive all or part of farmers' debts to the Farmers Home Administration if they agree not to farm highly erodible cropland or wetlands for 50 years. The farmers must plant trees or grass on this land or restore it to wetland.

The 1985 Farm Act made the United States the first major food-producing country to make soil conservation a national priority. Even though these efforts to slow soil erosion are an important step, effective soil conservation is practiced on only about half of all U.S. agricultural land and on less than half of the country's most erodible cropland.

Critical Thinking

Do you believe that U.S. tax dollars should be used to pay farmers for taking highly erodible land out of production? Explain. What are the alternatives?

---

vegetation and topsoil also greatly intensifies flooding buildup of sediment in streams in valleys below.

Contour farming can reduce soil erosion by 30–50% on gently sloping land. It involves plowing and planting crops in rows across the sloped contour of the land (Figure 10-24b). Each row planted along the contour of the land acts as a small dam to help hold soil and slow water runoff.

In **strip cropping**, a row crop such as corn alternates in strips with another crop (such as a grass or a grass-legume mixture) that completely covers the soil and thus reduces erosion (Figure 10-24b). The strips of the cover crop (1) trap soil that erodes from the row crop, (2) catch and reduce water runoff, and (3) help prevent the spread of pests and plant diseases. Planting nitrogen-fixing legumes (such as soybeans or alfalfa) in some of the strips helps restore soil fertility.

Erosion also can be reduced by **alley cropping**, or **agroforestry**. It is a form of *intercropping* in which several crops are planted together in strips or alleys between trees and shrubs that can provide fruit or fuelwood (Figure 10-24c). The trees provide shade (which reduces water loss by evaporation) and help to retain and slowly release soil moisture. The tree and shrub trimmings can be used as mulch (green manure) for the crops and as fodder for livestock.

How Can Gully Reclamation, Windbreaks, and Land Classification Reduce Soil Erosion? Gully **reclamation** can restore sloping bare land on which water runoff quickly creates gullies. Ways to do this include (1) seeding small gullies with quick-growing plants such as oats, barley, and wheat for the first season, (2) building small dams at the bottoms of deep gullies to collect silt and gradually fill in the channels, (3) planting fast-growing shrubs, vines, and trees to stabilize the soil, and (4) building channels to divert water from the gully and prevent further erosion.

Long rows of trees can be planted as **windbreaks**, or **shelterbelts**, to reduce wind erosion (Figure 10-24d). Windbreaks also (1) help retain soil moisture, (2) supply some wood for fuel, and (3) provide habitats for birds, pest-eating and pollinating insects, and other animals. Many of the windbreaks planted in the upper Great Plains of the United States after the 1930s Dust Bowl disaster (Case Study, p. 227) have been cut down to make way for large irrigation systems and modern farm machinery.

Land classification can be used to identify easily erodible (marginal) land that should be neither planted in crops nor cleared of vegetation. In the United States, the National Resources Conservation Service has set up a classification system to identify types of land that are suitable or unsuitable for cultivation. Such efforts and recent farm legislation have helped reduce soil erosion in the United States (Case Study, above).

How Can We Maintain and Restore Soil Fertility? Fertilizers partially restore plant nutrients lost by erosion, crop harvesting, and leaching. Farmers can use

Land Degradation and Environmental Resources

David Pimentel

GUEST ESSAY

David Pimentel is professor of insect ecology and agricultural sciences in the College of Agriculture and Life Sciences at Cornell University. He has published more than 440 scientific papers and 20 books on environmental topics, including land degradation, agricultural pollution and energy use, biomass energy, and pesticides. He was one of the first ecologists to use an interdisciplinary, holistic approach to investigating complex environmental problems.

As the world's human population is expanding rapidly and its need for more land to produce food, fiber, and fuelwood is escalating, valuable land is being degraded through erosion and other means at an alarming rate. Soil degradation is of great concern because soil reformation is extremely slow. Globally it takes an average of 500 years (with a range of 220 to 1,000 years) to renew 2.5 centimeters (1 inch) of soil in tropical and temperate areas—a renewal rate of about 1 metric ton of topsoil per hectare of land per year. Worldwide annual erosion rates for agricultural land are about 20–100 times this average.

Erosion rates vary in different regions because of topography, rainfall, wind intensity, and the type of agricultural practices used. In China, for example, the average annual soil loss is reported to be about 30 metric tons per hectare (13 tons per acre), whereas the U.S. average is 13 metric tons per hectare (6 tons per acre). In states such as Iowa and Missouri, however, annual soil erosion averages more than 35 metric tons per hectare (16 tons per acre).

Worldwide, about 10 million hectares (25 million acres) of land—an area about the size of Virginia—are abandoned for crop production each year because of high erosion rates and waterlogging, salinization, and other forms of soil degradation. In addition, according to the UN Environment Programme, crop production becomes uneconomical on about 20 million hectares (49 million acres) each year because soil quality has been severely degraded.

Soil erosion also occurs in forestlands, but it is not as severe as that in the more exposed soil of agricultural land. Soil erosion in managed forests is a primary concern because the soil reformation rate in forests is about one-third to one-half that on agricultural land. To compound this erosion problem, at least 24 million hectares (59 million acres) of forest are being cleared each year throughout the world; most of this land is used to grow food and graze cattle.

The effects of agriculture and forestry are interrelated in other ways. Deforestation reduces fuelwood supplies and forces the poor in developing countries to substitute crop residue and manure for fuelwood. When these plant and animal wastes are burned instead of being returned to the land as ground cover and organic fertilizer, erosion is intensified and productivity of the land is decreased. These factors, in turn, increase pressure to convert more forestland into agricultural land, further intensifying soil erosion.

One reason soil erosion is not a high priority for many governments and farmers is that it usually occurs so slowly that its cumulative effects may take decades to become apparent. For example, the loss of 1 millimeter (0.04 inch) of soil is so small that it goes undetected. But over a 25-year period the loss would be 25 millimeters (1 inch), which would take about 500 years to replace by natural processes.

Besides reduced soil depth, soil erosion leads to reduced crop productivity because of losses of water, organic matter, and soil nutrients. A 50% reduction of soil organic matter on a plot of land has been found to reduce corn yields as much as 25%.

When soil erodes, vital plant nutrients such as nitrogen, phosphorus, potassium, and calcium also are lost.

---

either **organic fertilizer** from plant and animal materials or **commercial inorganic fertilizer** produced from various minerals.

Three basic types of *organic fertilizer* are animal manure, green manure, and compost. **Animal manure** includes the dung and urine of cattle, horses, poultry, and other farm animals. It improves soil structure, adds organic nitrogen, and stimulates beneficial soil bacteria and fungi.

Despite its effectiveness, the use of animal manure in the United States has decreased. Reasons for this are (1) replacement of most mixed animal-raising and crop-farming operations with separate operations for growing crops and raising animals, (2) the high costs of transporting animal manure from feedlots near urban areas to distant rural crop-growing areas, and (3) replacement of horses and other draft animals that added manure to the soil with tractors and other motorized farm machinery.

Green manure is fresh or growing green vegetation plowed into the soil to increase the organic matter and humus available to the next crop. **Compost** is a sweet-smelling, dark-brown, humuslike material that is rich in organic matter and soil nutrients. It is produced when microorganisms (mostly fungi and aerobic bacteria) in soil break down organic matter such as leaves, food wastes, paper, and wood in the presence of oxygen. Compost is a rich natural fertilizer and soil conditioner that (1) aerates soil, (2) improves its ability to retain water and nutrients, (3) helps prevent erosion, and (4) prevents nutrients from being wasted by being dumped in landfills.

Farmers, homeowners, and communities produce compost by piling up alternating layers of (1) nitrogen-rich wastes (such as grass clippings, weeds, animal

With U.S. annual cropland erosion rates of about 18 metric tons per hectare (8 tons per acre), an estimated $18 billion of plant nutrients is lost annually. Using fertilizers to replace these nutrients greatly increases the cost of crop production.

Some analysts who are unaware of the numerous and complex effects of soil erosion have falsely concluded that the damages are minor. For example, they report that soil loss causes an annual reduction in crop productivity of only 0.1–0.5% in the United States. However, we must consider all the ecological effects caused by erosion, including reductions in soil depth, availability of water for crops, and soil organic matter and nutrients. When this is done, agronomists and ecologists report a 15–30% reduction in crop productivity, leading to increased use of costly fertilizer. Because fertilizers are not a substitute for fertile soil, they can be applied only up to certain levels before crop yields begin to decline.

Reduced agricultural productivity is only one of the effects of soil erosion. In the United States, water runoff is responsible for transporting about 3 billion metric tons (3.3 billion tons) of sediment (about 60% from agricultural land) each year to waterways in the lower 48 states. Off-site damages to U.S. water storage capacity, wildlife, and navigable waterways from these sediments cost an estimated $6 billion each year. About 25% of new water storage capacity in U.S. reservoirs is built solely to compensate for sediment buildup.

When soil sediments that include pesticides and other agricultural chemicals are carried into streams, lakes, and reservoirs, fish production declines. These contaminated sediments interfere with fish spawning, increase predation on fish, and destroy fisheries in estuarine and coastal areas.

Increased erosion and water runoff on mountain slopes flood agricultural land in the valleys below, further decreasing agricultural productivity. Eroded land does not hold water very well, again decreasing crop productivity. This effect is magnified in the 80 countries (with nearly 40% of the world's population) that experience frequent droughts.

Thus, soil erosion is one of the world's critical problems, and if not slowed, it will seriously reduce agricultural and forestry production and degrade the quality of aquatic ecosystems. Solutions that are not particularly difficult often are not implemented because erosion occurs so gradually that we fail to acknowledge its cumulative effects until the damage is irreversible. Many farmers have been conditioned to believe that soil fertility losses can be remedied by applying more fertilizer or using more fossil-fuel energy.

The primary way to control soil erosion and its accompanying runoff of sediment is to maintain adequate vegetative cover on soils [by using various methods discussed in Section 10-6]. These methods also are cost-effective, especially when off-site costs of erosion are included. Scientists, policy makers, and agriculturists need to work together to implement soil and water conservation practices before much of the world's soils lose most of their productivity.

Critical Thinking

1. Some analysts contend that average soil erosion rates around the world are low and that the soil erosion problem can be solved easily with improved agricultural technology such as no-till cultivation and increased use of commercial inorganic fertilizers. Do you agree or disagree with this position? Explain.

2. What three major things do you believe elected officials should do to decrease soil erosion and the resulting water pollution by sediment in the United States or in the country where you live?

---

manure, and vegetable kitchen scraps), (2) carbon-rich plant wastes (dead leaves, hay, straw, sawdust), and (3) topsoil (Individuals Matter, p. 234). Compost (1) provides a home for microorganisms that help decompose plant and manure layers and (2) reduces the amount of plant wastes taken to landfills and incinerators.

Another form of organic fertilizer is the spores of mushrooms, puffballs, and truffles. Rapidly growing and spreading mycorrhizae fungi in the spores attach to plant roots and help them take in moisture and nutrients from the soil. Unlike typical fertilizers that must be applied every few weeks, one application of mushroom fungi lasts all year and costs just pennies per plant. The fungi also produce a bigger root system that makes plants more disease resistant.

Corn, tobacco, and cotton can deplete the topsoil of nutrients (especially nitrogen) if planted on the same land several years in a row. One way to reduce such losses is **crop rotation.** Farmers plant areas or strips with nutrient-depleting crops one year. In the next year they plant the same areas with legumes (whose root nodules add nitrogen to the soil). In addition to helping restore soil nutrients, this method (1) reduces erosion by keeping the soil covered with vegetation and (2) helps reduce crop losses to insects by presenting them with a changing target.

Can Inorganic Fertilizers Save the Soil? Today, many farmers (especially in developed countries) rely on *commercial inorganic fertilizers* containing nitrogen (as ammonium ions, nitrate ions, or urea), phosphorus (as phosphate ions), and potassium (as potassium ions). Other plant nutrients may also be present in low or trace amounts.

Fast-Track Composting

INDIVIDUALS MATTER

Compost piles must be turned over every few days for aeration to speed up decomposition. Ruth Beckner has invented a special drill bit that attaches to a cordless electric drill to inject air into a compost pile with very little physical exertion. Using her COMPOST AIR* device for about 3 minutes per day, you can create good compost in about 3 weeks instead of waiting 12–18 months for the pile to decompose naturally.

Worms, the planet's champion recyclers, can also help create compost efficiently and without any odors. Recycling coordinators in Seattle, Washington, and Toronto, Canada, have given out thousands of worm bins to homeowners.

*For information, contact Beckner & Beckner, 15 Portola Avenue, San Rafael, CA 94903 or call 1-800-58COMPOST.

Inorganic commercial fertilizers are easily transported, stored, and applied. Worldwide, their use increased about 10-fold between 1950 and 1989 but declined by 12% between 1990 and 1999. Today, the additional food they help produce feeds one of every three people in the world; without them, world food output would drop an estimated 40%.

Commercial inorganic fertilizers have some disadvantages, however. These include **(1)** not adding humus to the soil, **(2)** reducing the soil's content of organic matter and thus its ability to hold water (unless animal manure and green manure are also added to the soil), **(3)** lowering the oxygen content of soil and keeping fertilizer from being taken up as efficiently, **(4)** typically supplying only 2 or 3 of the 20 or so nutrients needed by plants, **(5)** requiring large amounts of energy for their production, transport, and application, and **(6)** releasing nitrous oxide (N_2O), a greenhouse gas that can enhance global warming, from the soil.

The widespread use of commercial inorganic fertilizers, especially on sloped land near streams and lakes, also causes water pollution as nitrate (NO_3^-) and phosphate (PO_4^{3-}) fertilizer nutrients are washed into nearby bodies of water. The resulting plant nutrient enrichment (cultural eutrophication) causes algae blooms that use up oxygen dissolved in the water, thereby killing fish, as discussed in more detail on p. 481. Rainwater seeping through the soil can also leach nitrates in commercial fertilizers into groundwater. Drinking water drawn from wells containing high levels of nitrate ions can be toxic, especially for infants.

According to soil scientists, responsibility for reducing soil erosion should not be limited to farmers. Soil erosion is also caused by timber cutting, overgrazing, mining, and urban development that is carried out without proper regard for soil conservation. Some things you can do to reduce soil erosion are listed in Appendix 6.

Below that thin layer comprising the delicate organism known as the soil is a planet as lifeless as the moon.

G. Y. JACKS AND R. O. WHYTE

REVIEW QUESTIONS

1. Define all boldfaced terms in this chapter.

2. Distinguish between the earth's *core*, *mantle*, and *crust*.

3. What are *tectonic plates*? What is the *lithosphere*? What is the *theory of plate tectonics*, and what is its importance to physical and biological processes on the earth?

4. What are the three different types of boundaries between the earth's lithospheric plates?

5. What is *erosion*, and what are its two major causes?

6. Distinguish between a *mineral* and a *rock*. Distinguish between *igneous*, *sedimentary*, and *metamorphic rock* and give two examples of each type.

7. Describe the *rock cycle* and explain its importance.

8. What is an *earthquake*, and what are its major harmful effects? List ways to reduce the hazards from earthquakes.

9. What is a *volcanic eruption*? What are some of the hazards and benefits of volcanic eruptions? Describe the effects of the Mount St. Helens volcanic eruption in the United States in 1980. List ways to reduce the hazards from volcanic eruptions.

10. What is *soil*? Distinguish between a *soil horizon* and a *soil profile*.

11. What is *humus*, and what is its importance? What does the color of topsoil tell you about how useful a soil is for growing crops?

12. Distinguish between *soil infiltration* and *leaching*. Distinguish between *soil texture*, *soil porosity*, and *soil permeability*.

13. What are *loams*, and why are they the best soils for growing most crops?

14. What are the major natural and human-related causes of soil erosion? Describe three types of soil erosion.

15. What are the major harmful effects of soil erosion?

16. How serious is soil erosion **(a)** on a global scale and **(b)** in the United States?

17. Describe the *Dust Bowl* event in the United States. Describe how the U.S. government is reducing soil erosion.

18. What is *desertification*? How serious is this problem, and what are its major causes? How can we slow desertification?

19. Distinguish between *salinization* and *waterlogging* of soils. How serious are these problems? What can be done about them?

20. What is *soil conservation*? Distinguish between *conventional-tillage farming* and *conservation-tillage farming*. What are the advantages of conservation-tillage farming?

21. Distinguish between *terracing, contour farming, strip cropping, alley cropping, gully reclamation,* and *windbreaks* as methods for reducing soil erosion.

22. Distinguish between *organic fertilizer* and *commercial inorganic fertilizer* and list the advantages of each approach for maintaining or restoring soil fertility. Distinguish between *animal manure, green manure,* and *compost* as methods for fertilizing soil. What is *crop rotation,* and why is it useful in helping maintain soil fertility?

23. List the advantages and disadvantages of using commercial inorganic fertilizers to maintain and restore soil fertility.

CRITICAL THINKING

1. List some ways, positive and negative, in which **(a)** weathering and erosion and **(b)** plate tectonics are important to you.

2. Explain what would happen if plate tectonics stopped. Explain what would happen if erosion and mass weathering stopped. If you could, would you eliminate either group of processes? Explain.

3. What might be some beneficial and harmful climatic effects of having all the current continents clustered together in one supercontinent, as was the case in the distant past (Figure 5-9, top left, p. 114)?

4. Imagine that you are an igneous rock. Act as a reporter and send in a written report on what you experience as you move through various parts of the rock cycle (Figure 10-8, p. 217). Repeat this experience, assuming in turn that you are a sedimentary rock and then a metamorphic rock.

5. In the area where you live, are you more likely to experience an earthquake or a volcanic eruption? What can you do to escape or reduce the harm if such a disaster strikes? What actions can you take when it occurs?

6. Why should everyone, not just farmers, be concerned about soil conservation?

7. What are the main advantages and disadvantages of using commercial inorganic fertilizers to restore or increase soil fertility? Why should both inorganic and organic fertilizers be used?

PROJECTS

1. Write a brief scenario describing the sequence of consequences to us and to other forms of life if the rock cycle stopped functioning.

2. Use the library or the internet to find out where earthquakes and volcanic eruptions have occurred during the past 30 years, then stick small flags on a map of the world or place dots on Figure 10-5a (p. 214). Compare their locations with the plate boundaries shown in Figure 10-5b.

3. Conduct a survey of soil erosion and soil conservation in and around your community on cropland, construction sites, mining sites, grazing land, and deforested land. Use these data to develop a plan for reducing soil erosion in your community.

4. Use the library or the internet to find bibliographic information about *Will Durant, G. Y. Jacks,* and *R. O. Whyte,* whose quotes appear at the beginning and end of this chapter.

5. Make a concept map of this chapter's major ideas, using the section heads and subheads and the key terms (in boldface). Look at the inside back cover and on the website for this book for information about making concept maps.

INTERNET STUDY RESOURCES AND RESOURCES FOR FURTHER READING AND RESEARCH

The website for this book contains helpful study aids and many ideas for further reading and research. Log on to:

http://www.brookscole.com/product/0534376975s

and click on the Chapter-by-Chapter area. Choose Chapter 10 and select a resource:

■ "Flash Cards" allows you to test your mastery of the Terms and Concepts to Remember for this chapter.

■ "Tutorial Quizzes" provides a multiple-choice practice quiz.

■ "Student Guide to InfoTrac" will lead you to Critical Thinking Projects that use InfoTrac College Edition as a research tool.

■ "References" lists the major books and articles consulted in writing this chapter.

■ "Hypercontents" takes you to an extensive list of sites with news, research, and images related to individual sections of the chapter.

INFOTRAC COLLEGE EDITION

Improve your skills with InfoTrac College Edition, a searchable online database of articles from more than 700 periodicals. Log on to:

http://www.infotrac-college.com

or access InfoTrac through the website for this book.

Try the following articles:

Dunn, S., R. Friedman, S. Baish. 2000. Coastal Erosion. *Environment* vol. 42, no. 7, p. 36–45. (subject guide: beach erosion)

Compton, J.E., and R.D. Boone. 2000. Long-term impacts of agriculture on soil carbon and nitrogen in New England forests. *Ecology* vol. 81, no. 8, pp. 2314–2325. (key words: soil carbon).

PART III

HUMAN POPULATION, RESOURCES, AND SUSTAINABILITY

I recognize the right and duty of this generation to develop and use our natural resources, but I do not recognize the right to waste them, or to rob by wasteful use, the generations that come after us.

THEODORE ROOSEVELT, 1900

11 THE HUMAN POPULATION: GROWTH, DEMOGRAPHY, AND CARRYING CAPACITY

Slowing Population Growth in Thailand

Can a country sharply reduce its population growth in only 15 years? Thailand did.

In 1971, Thailand adopted a policy to reduce its population growth (Figure 11-121, p. 249). When the program began the country's population was growing at a rate of 3.2% per year, and the average Thai family had 6.4 children.

Fifteen years later in 1986, the country's population growth rate had been cut in half to 1.6%. By 2000 the rate had fallen to 1.0%, and the average number of children per family was 1.9. Thailand's population is projected to grow from 62 million in 2000 to 72 million by 2025.

There are several reasons for this impressive feat: (1) the creativity of the government-supported family-planning program, (2) a high literacy rate among women (90%), (3) an increasing economic role for women and advances in women's rights, (4) better health care for mothers and children, (5) the openness of the Thai people to new ideas, (6) the willingness of the government to encourage and financially support family planning and to work with the private, non-profit Population and Community Development Association (PCDA), and (7) support of family planning by the country's religious leaders (95% of Thais are Buddhist).

This transition was catalyzed by the charismatic leadership of Mechai Viravidaiya, a public relations genius and former government economist who launched the PCDA in 1974 to help make family planning a national goal. PCDA workers handed out condoms at festivals, movie theaters, and even traffic jams. Between 1971 and 2000, the percentage of married women using modern birth control rose from 15% to 70%—higher than the 60% usage in developed countries and the 51% usage in developing countries.

Mechai helped establish a German-financed revolving loan plan to enable people participating in family-planning programs to install toilets and drinking water systems. Low-rate loans were offered to farmers practicing family planning. The government also offers loans to individuals from a fund that increases as their village's level of contraceptive use rises.

All is not completely rosy. Although Thailand has done well in slowing population growth and raising per capita income, it has been less successful in reducing pollution and improving public health, especially maternal health and control of AIDS and other sexually transmitted diseases.

Its capital, Bangkok, remains one of the world's most polluted and congested cities. It is plagued with notoriously high levels of traffic congestion and air pollution (Figure 11-1). The typical motorist in Bangkok spends 44 days per year sitting in traffic, costing $2.3 billion in lost work time.

Figure 11-1 This policeman in Bangkok, Thailand, is wearing a mask to reduce his intake of air polluted mainly by automobiles. Bangkok is one of the world's most car-clogged cities, with car commutes averaging 3 hours per day. Roughly one of every nine of its residents has a respiratory ailment. (Martin Harvey / Natural History Photographic Agency)

The problems to be faced are vast and complex, but come down to this: 6.1 billion people are breeding exponentially. The process of fulfilling their wants and needs is stripping earth of its biotic capacity to produce life; a climactic burst of consumption by a single species is overwhelming the skies, earth, waters, and fauna.

PAUL HAWKEN

This chapter addresses the following questions:

- How is population size affected by birth, death, fertility, and migration rates?
- How is population size affected by the percentage of males and females at each age level?
- How can population growth be slowed?
- What success have India and China had in slowing population growth?
- How can global population growth be reduced?

11-1 FACTORS AFFECTING HUMAN POPULATION SIZE

How Is Population Size Affected by Birth Rates and Death Rates? Populations grow or decline through the interplay of three factors: *births*, *deaths*, and *migration*. **Population change** is calculated by subtracting the number of people leaving a population (through death and emigration) from the number entering it (through birth and immigration) during a specified period of time (usually a year):

$$\text{Population change} = (\text{Births} + \text{Immigration}) - (\text{Deaths} + \text{Emigration})$$

When births plus immigration exceed deaths plus emigration, population increases; when the reverse is true, population declines. When these factors balance out, population size remains stable, a condition known as **zero population growth (ZPG)**.

Instead of using the total numbers of births and deaths per year, demographers use **(1)** the **birth rate**, or **crude birth rate** (the number of live births per 1,000 people in a population in a given year), and **(2)** the **death rate**, or **crude death rate** (the number of deaths per 1,000 people in a population in a given year). Figure 11-2 shows the crude birth and death rates for various groupings of countries in 2000.

Birth rates and death rates are coming down worldwide, but death rates have fallen more sharply than birth rates. As a result, there are more births than deaths; every time your heart beats 2.5 more babies are added to the world's population. At this rate, we share the earth and its resources with about 226,000 more people each day (95% of them in developing countries).

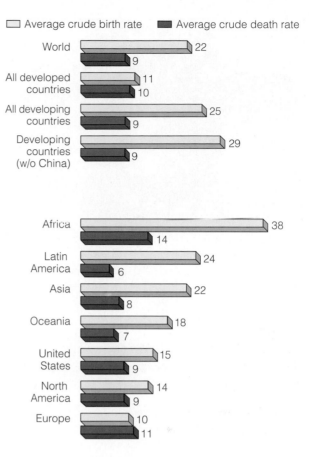

Figure 11-2 Average crude birth and death rates for various groupings of countries in 2000. (Data from Population Reference Bureau)

The rate of the world's annual population change (excluding migration) usually is expressed as a percentage:*

$$\text{Annual rate of natural population change (\%)} = \frac{\text{Birth rate} - \text{Death rate}}{1,000 \text{ people}} \times 100$$

$$= \frac{\text{Birth rate} - \text{Death rate}}{10}$$

The rate of the world's annual population growth (natural increase) dropped 39% between 1963 and 2000, from 2.2% to 1.35%. This is good news, but during the same period the population base rose by about 91%, from 3.2 billion to 6.1 billion. This drop in the rate of population increase is roughly analogous to learning that the truck heading straight at you has slowed from 100 kilometers per hour (kph) to 61 kph while its weight increased by 91%. Exponential population growth has not disappeared; it's just occurring at a slower rate.

---

*Crude birth and death rates that have not been rounded off to the nearest whole number often are used to calculate natural change; the result is then rounded off to the nearest tenth of a percent. Consequently, use of the rounded-off crude birth and death rate figures shown in Figure 11-2 will not always produce the rounded-off percentage growth figures shown in Figure 11-3.

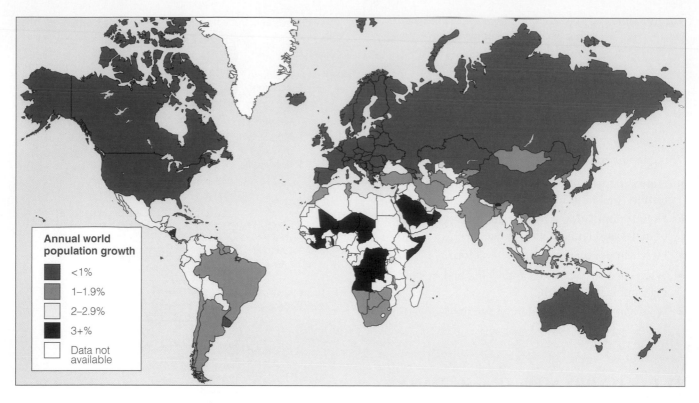

Figure 11-3 Average annual rate of population change (natural increase) in 2000. (Data from Population Reference Bureau)

However, *slower* does not mean *slow*: The world's population is still growing fast enough to double in 52 years.

Figure 11-3 presents the annual rates of population change for major parts of the world in 2000. An annual natural increase rate of 1-3% may seem small, but such exponential growth rates lead to enormous increases in population size over a 100-year period. For example, a population growing by 3% a year will increase its population 111-fold in a century.

The current annual population increase rate of 1.35% adds about 82 million people per year (6.1 billion × 1.35% = 82 million), roughly equal to adding another New York City every month, a Germany every year, and a United States every 3.4 years. Despite the drop in the rate of population growth, the larger base of population means that 82 million people were added in 2000, compared to only 69 million in 1963, when the world's population growth rate reached its peak. Figure 11-4 shows the average annual increase in the world's population from 1950 to 2000 and projected increases to 2050.

Figure 11-4 Average annual increase in the world's population, 1950–2000, and projected increase 2000–2050 (dotted line). (Data from United Nations)

In numbers of people, China (with 1.26 billion in 2000, about one of every five people in the world) and India (with 1 billion) dwarf all other countries (Figure 11-5). Together they make up 37% of the world's population. The United States, with 276 million people in 2000, has the world's third largest population but only 4.5% of the world's people. Figure 11-6 gives projected population growth in various regions between 2000 and 2025. More than 95% of this growth is projected to take place in developing countries, where acute poverty is a way of life for about 1.5 billion people.

How Have Global Fertility Rates Changed? Two types of fertility rates affect a country's population size and growth rate. The first type, **replacement-level fer-**

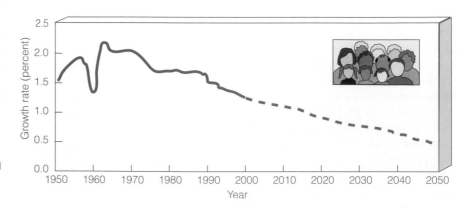

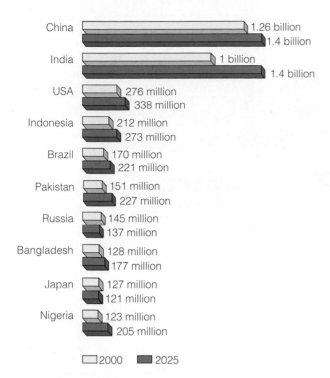

Figure 11-5 The world's 10 most populous countries in 2000, with projections of their population size in 2025. In 2000, more people lived in China than in all of Europe, Russia, North America, Japan, and Australia combined. (Data from World Bank and Population Reference Bureau)

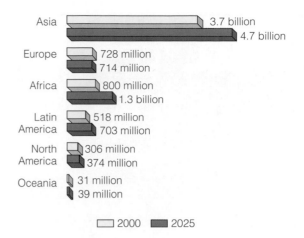

Figure 11-6 Population projections by region, 2000–2025. (Data from United Nations and Population Reference Bureau)

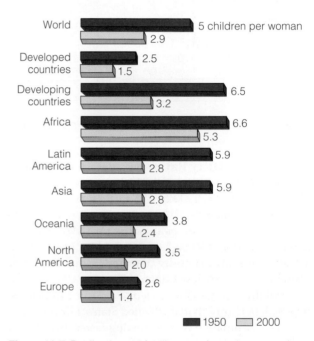

Figure 11-7 Decline in total fertility rates for various groupings of countries, 1950–2000. (Data from United Nations)

tility, is the number of children a couple must bear to replace themselves. It is slightly higher than two children per couple (2.1 in developed countries and as high as 2.5 in some developing countries in 2000), mostly because some female children die before reaching their reproductive years.

Does reaching replacement-level fertility mean an immediate halt in population growth (zero population growth)? No, because there are so many future parents already alive that if each had an average of 2.1 children and their children also had 2.1 children, the population would continue to grow for 50 years or more (assuming that death rates do not rise).

The second type of fertility rate is the **total fertility rate (TFR)**: an estimate of the average number of children a woman will have during her childbearing years if between ages 15 and 49 she bears children at the same rate as women did this year. TFRs have dropped sharply since 1950 (Figure 11-7). Although the TFR is a useful indicator of how women's actions will affect population growth this year, it is not necessarily a good indicator of their fertility decisions in future years. In 2000, 67 countries, with 44% of the world's population, had TFRs at or below the level of 2.1 children per woman. However, the average global TFR of 2.9 (3.2 in developing countries and 1.3 in developed countries) is still far above the replacement level of 2.1.

The power of exponential growth is astounding. If the world's TFR remained at its current value of 2.9 and there were no limits to human population growth, the earth's human population would reach the absurd figure of about 296 billion in only 150 years! Even if the TFR dropped to 2.5 children per woman and then remained at that level, the human population would eventually reach 28 billion people (assuming no sharp rise in death rates).

Current TFRs vary widely throughout the world (Figure 11-8), with the highest rate by far in Africa (5.3 children per woman in 2000). United Nations population projections to 2050 vary depending on the world's projected average TFR (Figure 11-9).

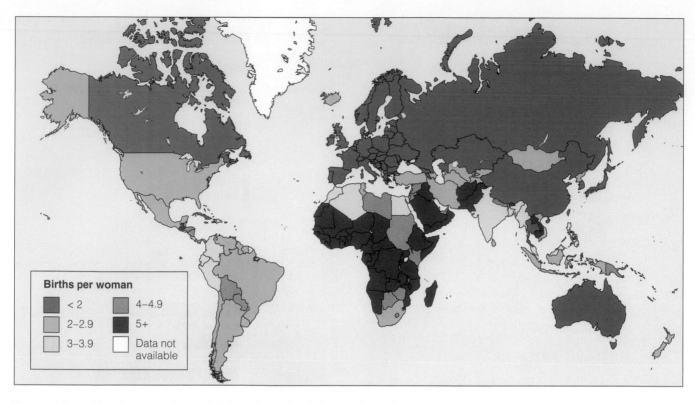

Figure 11-8 Total fertility rates in 2000. (Data from Population Reference Bureau)

Births per woman

- ■ < 2
- ■ 2–2.9
- ☐ 3–3.9
- ■ 4–4.9
- ■ 5+
- ☐ Data not available

How Have Fertility Rates Changed in the United States?

The population of the United States has grown from 76 million in 1900 to 276 million in 2000, even though the country's TFR has oscillated wildly (Figure 11-10). In 1957, the peak of the post-World War II baby boom, the TFR reached 3.7 children per woman. Since then it has generally declined, remaining at or below replacement level since 1972.

The drop in the TFR has led to a decline in the rate of population growth in the United States. However, the country's population is still growing faster than that of any other developed country and is not even close to zero population growth. In 2000, the U.S. TFR of 2.1 children per woman was made up of several different rates: **(1)** Hispanic, 2.9, **(2)** Asian and Pacific Islander, 1.9, **(3)** Native American, 2.1, **(4)** black, 2.2, and **(5)** white, 1.8.

Figure 11-9 United Nations world population projections to 2050, assuming that the world's total fertility rate is 2.5 (high), 2.0 (medium), or 1.6 (low) children per woman. (Data from United Nations)

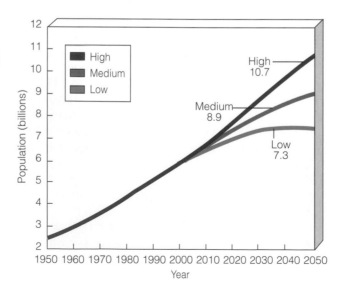

Including immigration, the U.S. population of 276 million grew by 1% in 2000—more than twice the mean rate of the world's industrialized nations. This growth added about 2.76 million people: **(1)** 1.66 million more births than deaths (accounting for about 60% of the growth), **(2)** 800,000 legal immigrants and refugees, and **(3)** an estimated 300,000 illegal immigrants.

Figure 11-11 shows U.S. birth rates between 1910 and 2000. Between 1910 and 1930, birth rates fell sharply as the country underwent industrialization and urbanization and more women got an education and began working outside the home. This shift from high birth rates to low birth rates during industrialization is called a *demographic transition.*

Birth rates remained low in the 1930s because of the Great Depression and then began rising in the 1940s during World War II. After World War II there was a sharp rise in the birth rate. This period of high birth rates between 1946 and 1964 is known as the *baby-boom period,* when 79 million people were added to the U.S. population. Between

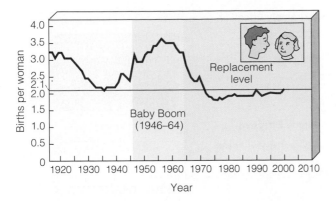

Figure 11-10 Total fertility rates for the United States between 1917 and 2000. (Data from Population Reference Bureau and U.S. Census Bureau)

1956 and 1972, birth rates began to decline as more women began working outside the home and the desired family size dropped from 4 to 2 (or no) children.

Between 1977 and 1980, a small *echo boom* in the number of births per year occurred as large number of people born during the baby boom began having children. Birth rates are projected to rise again between 2000 and 2050. According to U.S. Bureau of Census medium projections, the U.S. population will increase from 276 million to 404 million between 2000 and 2050—a 46% increase—with no stabilization on the horizon. The Census Bureau's high estimate is 507 million by 2050, an 86% increase. Between 2000 and 2100, the U.S. population is projected to more than double to 571 million people.

If immigration continues at current levels, new immigrants and their descendants would account for 80 million, or 63% of the projected 128-million increase in the U.S. medium projected population size between 2000 and 2050. Because of a high per capita rate of resource use, each addition to the U.S. population has an enormous environmental impact (Figure 1-15, p. 15).

What Factors Affect Birth Rates and Fertility Rates? Key factors affecting a country's average birth rate and TFR are the following:

- *Importance of children as a part of the labor force.* Rates tend to be higher in developing countries (especially in rural areas, where children begin working to help raise crops at an early age).

- *Urbanization.* People living in urban areas usually have better access to family-planning services and tend to have fewer children than those living in rural areas, where children are needed to perform essential tasks.

- *Cost of raising and educating children.* Rates tend to be lower in developed countries, where raising children is much more costly because children do not enter the labor force until their late teens or early 20s, or even later.

- *Educational and employment opportunities for women.* Rates tend to be low when women have access to education and paid employment outside the home. TFRs tend to decline as the female literacy rate increases. In developing countries, women with no education generally have two more children than women with a secondary school education.

- *Infant mortality rate.* In areas with low infant mortality rates, people tend to have fewer children because fewer children die at an early age.

- *Average age at marriage* (or, more precisely, the average age at which women have their first child). Women normally have fewer children when their average age at marriage is 25 or older. Birth rates tend to be much higher in countries with high birth rates among teenagers. In Africa, with the world's highest birth rates, 12% of girls ages 15 to 19 have a child, compared with 6% for the world. An estimated 70% of teenage pregnancies are unwanted.

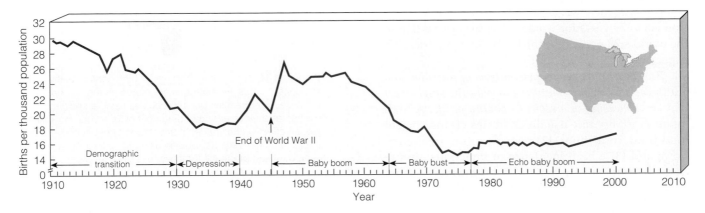

Figure 11-11 Birth rates in the United States from 1910 to 2000. (Data from U.S. Bureau of Census and U.S. Commerce Department)

- *Availability of private and public pension systems.* Pensions eliminate the need of parents to have many children to help support them in old age.

- *Availability of legal abortions.* An estimated 26 million legal abortions and 20 million illegal (and often unsafe) abortions are performed worldwide each year among the roughly 190 million pregnancies per year.

- *Availability of reliable methods of birth control* (Figure 11-12).

- *Religious beliefs, traditions, and cultural norms.* In some countries, these factors favor large families and strongly oppose abortion and some forms of birth control.

What Factors Affect Death Rates? The rapid growth of the world's population over the past 100 years is not the result of a rise in the crude birth rate. Instead, it has been caused largely by a decline in crude death rates, especially in developing countries (Figure 11-13).

More people started living longer (and fewer infants died) because of **(1)** increased food supplies and distribution, **(2)** better nutrition, **(3)** improvements in medical and public health technology (such as immunizations and antibiotics), **(4)** improvements in sanitation and personal hygiene, and **(5)** safer water supplies (which have curtailed the spread of many infectious diseases).

Two useful indicators of overall health of people in a country or region are **life expectancy** (the average number of years a newborn infant can expect to live) and the **infant mortality rate** (the number of babies out of every 1,000 born who die before their first birthday). In 2000, life expectancy averaged 75 years in developed countries and 64 years in developing countries.

The *good news* is that between 1955 and 2000, global life expectancy at birth increased from 48 years to 66 years (75 years in developed countries and 64 years in developing countries) and is projected to reach 73 by 2025. Between 1900 and 2000, life expectancy in the United States increased from 46 to 74 years for men and from 48 to 79 years for women. The *bad news* is that in the world's 38 poorest countries, mainly in Africa, life expectancy is 50 years or less.

Because it reflects the general level of nutrition and health care, infant mortality is probably the single most important measure of a society's quality of life. A high infant mortality rate usually indicates **(1)** insufficient food (undernutrition), **(2)** poor nutrition (malnutrition), and **(3)** a high incidence of infectious disease (usually from contaminated drinking water). Figure 11-14 shows the infant mortality rates in various parts of the world in 2000.

Between 1965 and 2000, the world's infant mortality rate dropped from 20 per 1,000 live births to 8 in

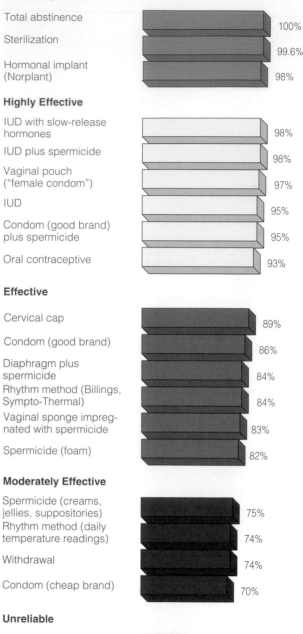

Figure 11-12 Typical effectiveness of birth control methods in the United States. Percentages are based on the number of undesired pregnancies per 100 couples using a specific method as their sole form of birth control for a year. For example, an effectiveness rating of 94% for oral contraceptives means that for every 100 women using the pill regularly for 1 year, 6 will get pregnant. Effectiveness rates tend to be lower in developing countries, primarily because of lack of education. (Data from Alan Guttmacher Institute)

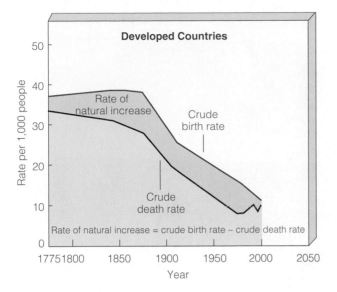

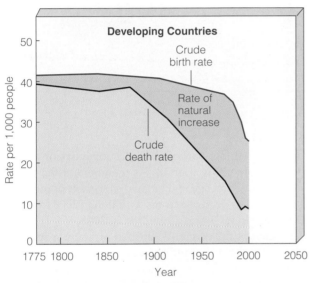

Figure 11-13 Changes in crude birth and death rates for developed and developing countries between 1775 and 2000. (Data from Population Reference Bureau and United Nations)

developed countries and from 118 to 63 in developing countries. This is an impressive achievement, but it still means that at least 8 million infants die of preventable causes during their first year of life—an average of 22,000 mostly unnecessary infant deaths per day. About 99% of these deaths occur in developing countries.

The U.S. infant mortality rate has declined steadily since 1917 (Figure 11-15). Despite this improvement, 30 countries had lower infant mortality rates than the United States in 2000. Three factors that keep the U.S. infant mortality rate higher than it could be are **(1)** inadequate health care (for poor women during pregnancy and for their babies after birth), **(2)** drug addiction among pregnant women, and **(3)** the high birth rate among teenagers.

The *good news* is that the U.S. birth rate among girls ages 15–19 in 2000 was lower than at any time since 1940. The *bad news* is that the United States has the highest teenage pregnancy rate of any industrialized country. Each year about 900,000 teenage girls become pregnant in the United States (78% of them unplanned), and about 270,000 of them have abortions. Babies born to teenagers are more likely to have low birth weights, the most important factor in infant deaths.

11-2 POPULATION AGE STRUCTURE

What Are Age Structure Diagrams? As mentioned earlier, even if the replacement-level fertility rate of 2.1 were magically achieved globally tomorrow, the world's population would keep growing for at least another 50 years, stabilizing at about 9 billion by 2054 (assuming no increase in death rates). The reason for this is a population's **age structure**: the proportion of the population (or of each sex) at each age level.

Demographers typically construct a population age structure diagram by plotting the percentages or numbers of males and females in the total population in each of three age categories: **(1)** *prereproductive* (ages 0–14), **(2)** *reproductive* (ages 15–44), and **(3)** *postreproductive* (ages 45 and up). Figure 11-16 presents generalized age structure diagrams for countries with rapid, slow, zero, and negative population growth rates.

How Does Age Structure Affect Population Growth? Any country with many people below age 15 (represented by a wide base in Figure 11-16, left) has a powerful built-in momentum to increase its population size unless death rates rise sharply. The number of births rises even if women have only one or two children because of the large number of girls who will soon be moving into their reproductive years.

In 2000, 31% of the people on the planet were under 15 years old. These 1.9 billion young people are poised to move into their prime reproductive years. In developing countries the number is even higher: 34%, compared with 19% in developed countries. This powerful force for continued population growth, mostly in developing countries, could be slowed by **(1)** an effective program to reduce birth rates or **(2)** a sharp rise in death rates.

Figure 11-17 shows the age structure in developed and developing countries in 2000. We live in a demographically divided world, as shown by demographic data in the United States, Brazil, and Nigeria (Figure 11-18, p. 248).

How Can Age Structure Diagrams Be Used to Make Population and Economic Projections? The 79-million-person increase that occurred in the U.S. population between 1946 and 1964, known as the *baby*

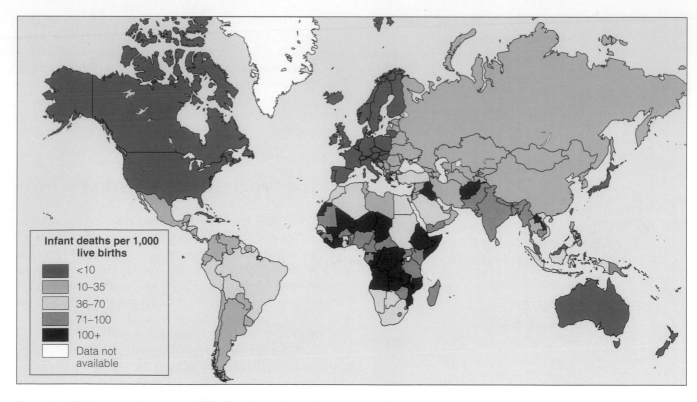

Figure 11-14 Infant mortality rates in 2000. (Data from Population Reference Bureau)

Infant deaths per 1,000 live births

- <10
- 10–35
- 36–70
- 71–100
- 100+
- Data not available

boom (Figures 11-10 and 11-11), will continue to move up through the country's age structure as the members of this group grow older (Figure 11-19).

Baby boomers now make up nearly half of all adult Americans. As a result, they dominate the population's demand for goods and services and play an increasingly important role in deciding who gets elected and what laws are passed. Baby boomers who created the youth market in their teens and 20s are now creating

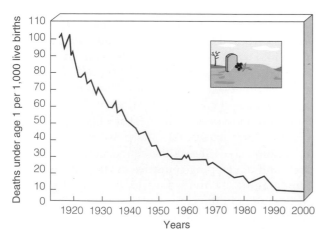

Figure 11-15 Infant mortality rates in the United States, 1917–2000. This sharp decline in infant mortality rates has led to a marked increase in average life expectancy. (Data from U.S. Census Bureau and Population Reference Bureau as presented in *Environmental Science* by Chiras, 5/E. Copyright (c) 1998 by Wadsworth.)

the 50-something market and will soon move on to create a 60-something market. As members of the baby-boom population age and begin dying, their population is projected to decrease from 79 million in 2000 to about 51 million in 2031 and 19 million in 2046.

Much of the economic burden of helping support a large number of retired baby boomers will fall on the *baby-bust generation, or generation-x*: the 44 million people born between 1965 and 1976 (when TFRs fell sharply and have remained below 2.1 since 1970; Figure 11-10). Retired baby boomers are beginning to use their political clout to force the smaller number of people in the baby-bust generation to pay higher income, health-care, and Social Security taxes. This could cause resentment and conflicts between the two generations.

The baby-bust generation is being followed by the *echo-boom generation* (Figure 11-11) consisting of about 82 million people born from 1977 to 2000. This largest generation ever should help support their baby-boom parents. This generation has grown up understanding the new digital economy and surveys show that in general they care about poverty, the environment, and global issues. Most Americans are unaware that the Social Security program has no trust fund to pay off disabled people, retirees, and survivors of retirees. Instead, Social Security taxes on workers in the current generation are used to pay off current beneficiaries. Thus, as the number of workers supporting each beneficiary declines (Figure 11-20), it will be necessary to **(1)** dramatically increase the Social Security taxes on workers, **(2)** decrease retire-

Figure 11-16 Generalized population age structure diagrams for countries with rapid (1.5-3%), slow (0.3-1.4%), zero (0-0.2%), and negative population growth rates. (Data from Population Reference Bureau)

ment benefits, or **(3)** make up the shortfall from other government funds (which could greatly increase income taxes unless government budget sources are used to fund the program).

In other respects, the baby-bust generation should have an easier time than the baby-boom generation. Fewer people will be competing for educational opportunities, jobs, and services, and labor shortages may drive up their wages, at least for jobs requiring education or technical training beyond high school. On the other hand, members of the baby-bust group may find it difficult to get job promotions as they reach middle age because members of the much larger baby boom group will occupy most upper-level positions. Many baby boomers may delay retirement because of **(1)** improved health, **(2)** the need to accumulate adequate retirement funds, or **(3)** extension of the retirement age needed to begin collecting Social Security.

From these few projections, we can see that any booms or busts in the age structure of a population create social and economic changes that ripple through a society for decades.

What Are Some Effects of Population Decline?
The populations of most of the world's countries are projected to grow throughout most of the 21st century. By 2000, however, 30 countries (most of them in Europe) with about 766 million people had roughly

Figure 11-17 Population structure by age and sex in developing countries and developed countries, 2000. In 2000, there were 1 billion young people in their prime reproductive years of 15–24 and 1.9 billion people under age 15, moving into their reproductive years. (Data from United Nations Population Division and Population Reference Bureau)

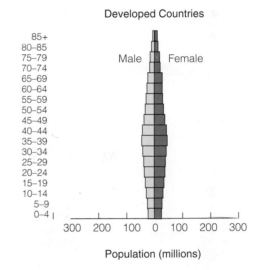

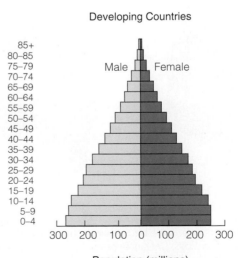

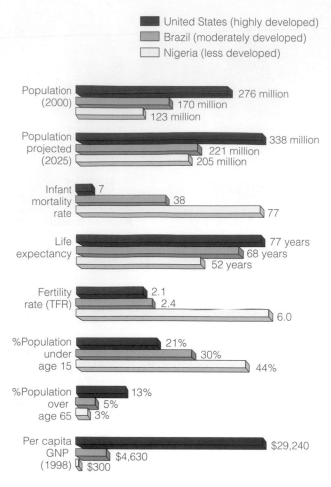

Figure 11-18 Comparison of key demographic indicators in a highly developed (United States), moderately developed (Brazil), and less developed country (Nigeria) in 2000. (Data from Population Reference Bureau)

stable populations (annual growth rates below 0.3%) or declining populations. In other words, about 13% of humanity in industrialized Europe plus Japan has achieved a stable population.

For example, in only 7 years, between 1949 and 1956, Japan (Figure 11-21) cut its birth, total fertility, and population growth rates in half. The main reason was widespread access to family planning implemented by the post-World War II U.S. occupation forces and the Japanese government.

Since 1956 these rates have declined further, mostly because of **(1)** access to family-planning services, **(2)** cramped housing, **(3)** high land prices, **(4)** late marriage ages, and **(5)** high education costs. Japan is now one of the world's least fertile and fastest-aging societies.

If this trend continues (and the country's almost negligible immigration rate does not rise), Japan's population of 127 million in 2000 is projected to fall to 100 million by 2050 and 67 million in 2100.

A much larger group of countries has reached replacement-level fertility of 2.1 children per couple, but their populations are still growing at a rate of around 1% per year because of increased numbers of young people moving into their reproductive years, increased immigration, or both. These countries, with about 40% of the world's population, include China and the United States, two of the world's most populous countries.

As the projected age structure of the world's population changes between 2000 and 2150 and the percentage of people age 60 or older increases (Figure 11-22), more and more countries will begin experiencing population declines. Because of this *global aging*,

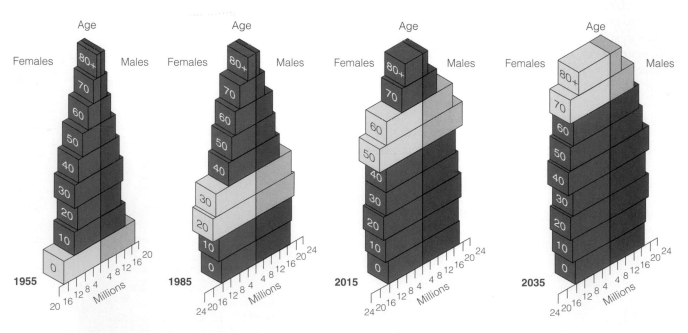

Figure 11-19 Tracking the baby-boom generation in the United States. (Data from Population Reference Bureau and U.S. Census Bureau)

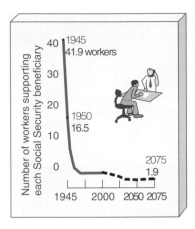

Figure 11-20 Number of workers supporting each beneficiary of the Social Security program in the United States, 1945–2075 (projected). Social Security taxes from workers in each current generation are used to pay current beneficiaries. (Data from United Nations)

in 2000 for the first time in human history people over age 60 outnumbered children under age 15 in developed countries. In 2000, working taxpayers outnumbered nonworking retired persons by 3 to 1. By 2030, this ratio will fall to 1.5 to 1 (unless official retirement ages are increased).

If population decline is gradual, its negative effects usually can be managed. However, rapid population decline, like rapid population growth, can lead to severe economic and social problems. A country undergoing rapid population decline **(1)** has a sharp rise in the proportion of older people, who consume a large share of medical care, Social Security, and other costly public services funded by working taxpayers, and **(2)** can face labor shortages unless it relies on greatly increased automation or immigration of foreign workers.

Figure 11-21 Where are Japan, Thailand, Indonesia, India, China, and Bangladesh? Some of the countries highlighted here are discussed in other chapters.

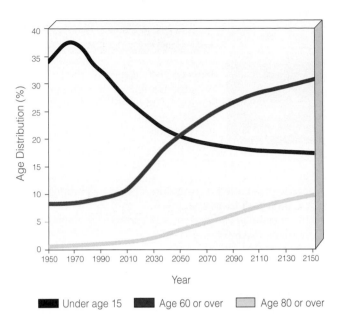

■ Under age 15   ■ Age 60 or over   □ Age 80 or over

Figure 11-22 *Global aging.* Projected percentage of world population under age 15, age 60 or over, and age 80 or over, 1950–2150, assuming the medium fertility projection shown in Figure 11-9. Between 1998 and 2050 the number of people over age 80 is projected to increase from 66 million to 370 million. The cost of supporting a much larger elderly population will place enormous strains on the world's economy. (Data from the United Nations)

11-3 SOLUTIONS: INFLUENCING POPULATION SIZE

How Is Population Size Affected by Migration? The population of an area is affected by movement of people into (immigration) and out of (emigration) that area. Most countries influence their rates of population growth to some extent by restricting immigration. Only a few countries—chiefly Canada, Australia, and the United States (Case Study, p. 250)—allow large annual increases in population from immigration.

Some migration is involuntary and involves refugees displaced by armed conflict, environmental degradation, or natural disaster. According to the International Federation of the Red Cross, in 1998 an estimated 25 million international *environmental refugees* moved from one country to another because of problems such as drought, desertification, deforestation, soil erosion, and resource shortages. The problem is worsened by the fact that emergency aid to developing countries dropped by 40% between 1993 and 1998. The number of environmental refugees could reach 50–150 million in the 21st century if global warming projections are correct.

International migration to developed countries absorbs only about 1% of the annual population growth in developing countries. Thus, population change for

Between 1820 and 1999, the United States admitted almost twice as many immigrants and refugees as all other countries combined. However, the number of legal immigrants has varied during different periods because of changes in immigration laws and rates of economic growth (Figure 11-23).

In 1999, the United States received about 800,000 legal immigrants and refugees and 300,000 illegal immigrants, together accounting for 40% of the country's population growth. The Immigration and Naturalization Service estimates that there are about 5.2 million illegal immigrants in the United States.

Currently, more than 75% of all legal immigrants live in six states: California, Florida, Illinois, New York, New Jersey, and Texas. If illegal immigrants are included, this figure rises to about 90%.

Immigrants place a tax burden on residents of such states. In California, for example, the average household pays an extra $1,200 in taxes per year because of immigrants. However, according to a 1997 study by the National Academy of Sciences, the work and taxes paid by immigrants add $1–10 billion per year to the overall U.S. economy, largely because immigration holds down wages (and thus prices) for some jobs. The study estimated that during their lifetimes immigrants pay an average of $80,000 more per person in taxes than they cost in services.

Between 1820 and 1960, most legal immigrants to

the United States came from Europe; since then, most have come from Latin America (53%) and Asia (30%). Between 2000 and 2050, the percentage of Hispanics in the U.S. population is projected to double from 12% to 24%.

In 1995, the U.S. Commission on Immigration Reform recommended reducing the number of legal immigrants and refugees to about 700,000 per year for a transition period and then to 550,000 a year. Some demographers and environmentalists go further and call for lowering the annual ceiling for legal immigrants and refugees into the United States to 300,000–450,000 or for limiting legal immigration to about 20% of annual population growth. They would accept immigrants only if they can support themselves, arguing that providing immigrants with public services turns the United States into a magnet for the world's poor.

Most of these analysts also support efforts to sharply reduce illegal immigration. However, some are concerned that a crackdown on illegal immigrants can also

lead to discrimination against legal immigrants.

Proponents argue that reducing immigration would allow the United States to stabilize its population sooner and help reduce the country's enormous environmental impact. Others oppose reducing current levels of legal immigration, arguing that **(1)** it would diminish the historical role of the United States as a place of opportunity for the world's poor and oppressed, **(2)** immigrants pay taxes and take many menial, low-paying jobs that other Americans shun, **(3)** few immigrants receive public assistance, **(4)** many immigrants open businesses and create jobs, and **(5)** according to the U.S. Census bureau, after 2020 higher immigration levels will be needed to supply enough workers as baby boomers retire.

Critical Thinking

Should the United States reduce its current level of legal immigration and tighten up on illegal immigration? Explain.

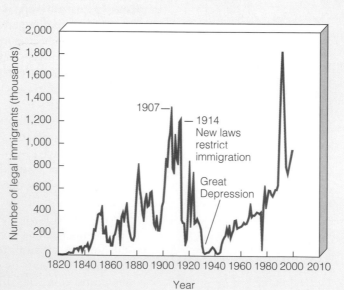

Figure 11-23 Legal immigration to the United States, 1820–1999. The large increase in immigration since 1989 resulted mostly from the Immigration Reform and Control Act of 1986, which granted legal status to illegal immigrants who could show that they had been living in the country for several years. (Data from U.S. Immigration and Naturalization Service)

most countries is determined mainly by the difference between their birth rates and death rates.

Migration within countries, especially from rural to urban areas, plays an important role in the population dynamics of cities, towns, and rural areas, as discussed on p. 659.

What Are the Pros and Cons of Reducing Births?
Currently about 93% of the world's population (and 91% of the people in developing countries) live in countries with fertility reduction programs. However, most governments spend less than 1% of their national budgets on such programs.

The projected increase of the human population from 6.1 to 9 billion or more between 2000 and 2054 raises an important question: *Can the world provide an adequate standard of living for 2.8 billion more people without causing massive environmental damage?*

There is controversy over **(1)** this question, **(2)** whether the earth is already overpopulated, and **(3)** what measures, if any, should be taken to slow population growth. To some the planet is already overpopulated (Figure 11-24), but others disagree. Some analysts, mostly economists, argue that we should encourage population growth to help stimulate economic growth.

Others believe that asking how many people the world can support is the wrong question, equivalent to asking how many cigarettes one can smoke before getting lung cancer. Instead, they say, we should be asking what the *optimum sustainable population* of the earth might be, based on the planet's *cultural carrying capacity* (Guest Essay, p. 252).

Such an optimum level would allow most people to live in reasonable comfort and freedom without impairing the ability of the planet to sustain future generations. No one knows what this optimum population might be. Some consider it a meaningless concept; some put it at 20 billion, others at 8 billion, and others as low as 2 billion.

Those who do not believe that the earth is overpopulated point out that the average life span of the world's 6.1 billion people is longer today than at any time in the past. They say that the world can support billions more people, and people are the world's most valuable resource for solving the problems we face and stimulating economic growth by becoming consumers.

Raising a family is one of the most enjoyable and rewarding things many people do. In poor countries without a social security system, raising a large family is a way for parents to have some security in their old age.

Some people view any form of population regulation as a violation of their religious beliefs, whereas others see it as an intrusion into their privacy and personal freedom. They believe that all people should be free to have as many children as they want. Some developing countries and some members of minorities in developed countries regard population control as a form of genocide to keep their numbers and power from rising.

Proponents of slowing and eventually stopping population growth point out that we fail to provide the basic necessities for one out of six people on the earth today. If we cannot (or will not) do this now, they ask, how will we be able to do this for the projected 2.8 billion more people by 2054?

Proponents of slowing population growth consider population growth in both developed and developing countries as a threat to the earth's life-support systems and the human population. They contend that if we do not sharply lower birth rates, we are deciding by default to raise death rates for humans and greatly increase environmental harm. In 1992, for example, the highly respected U.S. National Academy of Sciences and the Royal Society of London issued the following joint statement: "If current predictions of population growth and patterns of human activity on the planet remain unchanged, science and technology may not be able to prevent either irreversible degradation of the environment or continued poverty for much of the world."

Proponents of this view recognize that population growth is not the only cause of environmental and resource problems. However, they argue that adding several hundred million more people in developed countries and several billion more in developing countries can only

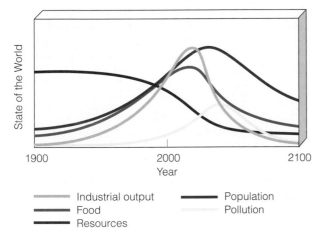

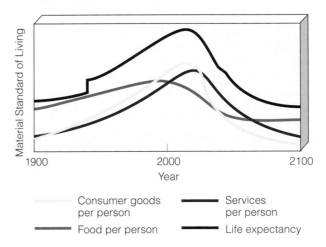

Figure 11-24 Plots of a computer model projecting what might happen if the world's population and economy continue growing exponentially at 1990 levels, assuming no major policy changes or technological innovations. This scenario projects that the world has already overshot some of its limits and that if current trends continue unchanged, we face global economic and environmental collapse sometime in the next century. (From *Beyond the Limits: Confronting Global Collapse, Envisioning a Sustainable Future*, by Donella Meadows et al., 1992. White River Junction, Vt.: Chelsea Green. Used by permission.)

Moral Implications of Cultural Carrying Capacity

Garrett Hardin

GUEST ESSAY

As longtime professor of human ecology at the University of California at Santa Barbara, Garrett Hardin made important contributions to relating ethics to biology. He has raised hard ethical questions, sometimes taken unpopular stands, and forced people to think deeply about environmental problems and their possible solutions. He is best known for his 1968 essay "The Tragedy of the Commons," which has had a significant impact on the disciplines of economics and political science and on the management of potentially renewable resources. His 17 books include Filters Against Folly: How to Survive Despite Economists, Ecologists, and the Merely Elo*quent,* Living Within Limits, *and* The Ostrich Factor: Our Population Myopia *(see Further Readings).*

For many years, Angel Island in San Francisco Bay was plagued with too many deer. A few animals transplanted there in the early 1900s lacked predators and rapidly increased to nearly 300 deer—far beyond the carrying capacity of the island. Scrawny, underfed animals tugged at the heartstrings of Californians, who carried extra food for them from the mainland to the island.

Such well-meaning charity worsened the plight of the deer. Excess animals trampled the soil, stripped the bark from small trees, and destroyed seedlings of all kinds. The net effect was to lower the island's carrying capacity, year by year, as the deer continued to multiply in a deteriorating habitat.

State game managers proposed that the excess deer be shot by skilled hunters. "How cruel!" some people protested. Then the managers proposed that coyotes be introduced onto the island. Though not big enough to kill adult deer, coyotes can kill fawns, thereby reducing the size of the herd. However, the Society for the Prevention of Cruelty to Animals was adamantly opposed to this proposal.

In the end, it was agreed that some deer would be transported to other areas suitable for deer. A total of 203 animals were caught and trucked many miles away. From the fate of a sample of animals fitted with radio collars, it was estimated that 85% of the transported deer died within a year (most of them within 2 months) from various causes: predation by coyotes, bobcats, and domestic dogs, shooting by poachers and legal hunters, and being run over by cars.

The net cost (in 1982 dollars) for relocating each animal surviving for a year was $2,876. The state refused to continue financing the program, and no volunteers stepped forward to pay future bills.

Angel Island is a microcosm of the planet as a whole. Organisms reproduce exponentially, but the environment does not increase at all. The moral is a simple ecological commandment: *Thou shalt not transgress the carrying capacity.*

Now let's examine the situation for humans. A competent physicist has placed global human carrying capacity at 50 billion, about eight times the current world population. Before you give in to the temptation to urge women to have more babies, consider what Robert Malthus said nearly 200 years ago: "There should be no more people in a country than could enjoy daily a glass of wine and piece of beef for dinner."

A diet of grain or bread and water is symbolic of minimum living standards; wine and beef are symbolic of higher living standards that make greater demands on the environment. When land that could produce plants for direct human consumption is used to grow grapes for wine or corn for cattle, more energy is expended to feed the human population. Because carrying capacity is defined as the *maximum* number of animals (humans) an area can support, using part of the area to support such cultural luxuries as wine and beef reduces the carrying capacity. This reduced capacity is called the *cultural carrying capacity*, and it is always smaller than simple carrying capacity.

Energy is the common coin of the realm for all competing demands on the environment. Energy saved by giving up a luxury can be used to produce more food staples and support more people. We could increase the simple carrying capacity of the earth by giving up any (or all) of the following "luxuries": street lighting, vacations, private cars, air conditioning, and artistic performances of all

intensify existing environmental and social problems. They call for drastic changes to prevent accelerating environmental decline and a rise in death rates (Figure 11-25).

These analysts believe that people should have the freedom to produce as many children as they want. However, such freedom would apply only if it did not reduce the quality of other people's lives now and in the future, either by impairing the earth's ability to sustain life or by causing social disruption. They point out that limiting the freedom of individuals to do anything they want to protect the freedom of other individuals is the basis of most laws in modern societies. What is your opinion on this issue?

How Can Economic Development Help Reduce Birth Rates? Demographers have examined the birth and death rates of western European countries that industrialized during the 19th century. From these data they developed a hypothesis of population change known as the **demographic transition**: As countries become industrialized, first their death rates and then their birth rates decline.

sorts. But what we consider luxuries depends on our values as individuals and societies, and values are largely matters of choice. At one extreme, we could maximize the number of human beings living at the lowest possible level of comfort. Or we could try to optimize the quality of life for a much smaller human population.

The carrying capacity of the earth is a scientific question. It may be possible to support 50 billion people at a bread-and-water level. Is that what we choose? The question, "What is the cultural carrying capacity?" requires that we debate questions of value, about which opinions differ.

An even greater difficulty must be faced. So far, we have been treating carrying capacity as a *global* issue, as if there were some global sovereignty capable of enforcing a solution on all people. But there is no global sovereignty ("one world"), nor is there any prospect of one in the foreseeable future. Thus, we must ask how some 200 nations are to coexist in a finite global environment if different sovereignties adopt different standards of living.

Consider a protected redwood forest that produces neither food for humans nor lumber for houses. Because people must travel many kilometers to visit it, the forest is a net loss in the national energy budget. However, for those fortunate enough to wander through the cathedral-like aisles beneath an evergreen vault, a redwood forest does something precious for the human spirit. But then intrudes an appeal from a distant land, where millions are starving because their population has overshot the carrying capacity; we are asked to save lives by sending food. As long as we have surpluses, we may safely indulge in the pleasures of philanthropy. But after we have run out of our surpluses, then what?

A spokesperson for the needy from that land makes a proposal: "If you would only cut down your redwood forests, you could use the lumber to build houses and then grow potatoes on the land, shipping the food to us. Since we are all passengers together on Spaceship Earth, are you not duty bound to do so? Which is more precious, trees or human beings?"

This last question may sound ethically compelling, but let's look at the consequences of assigning a preemptive and supreme value to human lives. At least 2 billion people in the world are poorer than the 34 million "legally poor" in America, and their numbers are increasing by about 1 million per year. Unless this increase is halted, sharing food and energy on the basis of need would require the sacrifice of one amenity after another in rich countries. The ultimate result of sharing would be complete poverty everywhere on the earth to maintain the earth's simple carrying capacity. Is that the best humanity can do?

To date, there has been overwhelmingly negative reaction to all proposals to make international philanthropy conditional on the cessation of population growth by overpopulated recipient nations. Foreign aid is governed by two apparently inflexible assumptions:

- The right to produce children is a universal, irrevocable right of every nation, no matter how hard it presses against the carrying capacity of its territory.

- When lives are in danger, the moral obligation of rich countries to save human lives is absolute and undeniable.

Considered separately, each of these two well-meaning doctrines might be defensible; taken together, they constitute a fatal recipe. If humanity gives maximum carrying capacity precedence over problems of cultural carrying capacity, the result will be universal poverty and environmental ruin.

Or do you see an escape from this harsh dilemma?

Critical Thinking

1. What population size would allow the world's people to have good quality of life? What do you believe is the cultural carrying capacity of the United States? Should the United States have a national policy to establish this population size as soon as possible? Explain.

2. Do you support the two principles this essay lists as the basis of foreign aid to needy countries? If not, what changes would you make in the requirements for receiving such aid?

According to this hypothesis, the transition takes place in four distinct stages (Figure 11-26). In the *preindustrial stage*, harsh living conditions lead to a high birth rate (to compensate for high infant mortality) and a high death rate. Thus, there is little population growth.

In the *transitional stage*, industrialization begins, food production rises, and health care improves. Death rates drop and birth rates remain high, so the population grows rapidly (typically 2.5–3% a year).

In the *industrial stage*, industrialization is widespread. The birth rate drops and eventually approaches the death rate. Reasons for this convergence of rates include **(1)** better access to birth control, **(2)** decline in the infant mortality rate, **(3)** increased job opportunities for women, and **(4)** the high costs of raising children who do not enter the workforce until after high school or college. Population growth continues, but at a slower and perhaps fluctuating rate, depending on economic conditions. Most developed countries are now in this third stage (Figure 11-26), and a few developing countries are entering this stage.

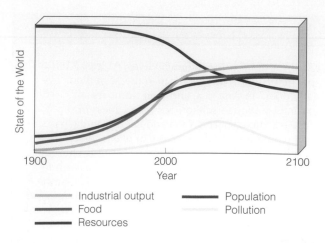
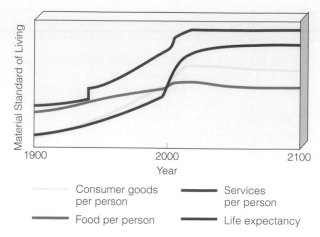

Figure 11-25 Computer-generated scenario projecting how we can avoid overshoot and collapse and make a fairly smooth transition to a sustainable future. It assumes that **(1)** technology allows us to double supplies of nonrenewable resources, double crop and timber yields, cut soil erosion in half, and double the efficiency of resource use within 20 years, **(2)** 100% effective birth control was made available to everyone by 1995, **(3)** no couple has more than two children, beginning in 1995, and **(4)** per capita industrial output is stabilized at 1990 levels. Another computer run projects that waiting until 2015 to implement these changes would lead to overshoot and collapse sometime around 2075, followed by a transition to sustainability by 2100. (From *Beyond the Limits: Confronting Global Collapse, Envisioning a Sustainable Future*, by Donella Meadows et al., 1992. White River Junction, Vt.: Chelsea Green. Used by permission.)

In the *postindustrial stage*, the birth rate declines even further, equaling the death rate and thus reaching zero population growth. Then the birth rate falls below the death rate and total population size decreases slowly. Thirty countries (most of them in Europe) containing about 13% of the world's population have entered this stage. To most population experts, the challenge is to help the remaining 85% of humanity reach this stage.

Some analysts suggest that the optimum human population (Guest Essay, p. 252) for the earth is about 2 billion people. Many observers disagree with this assessment. However, assuming that this is a desirable goal, it would take 100 years to reach that level if the world's average TFR dropped to 1.5 children per woman and remained there (assuming no sharp rise in death rates).

In most developing countries today, death rates have fallen much more than birth rates. In other words, these developing countries—mostly in Southeast Asia, Africa, and Latin America—are still in the transitional stage, halfway up the economic ladder, with high population growth rates. Some economists believe that developing countries will make the demographic transition over the next few decades without increased family-planning efforts.

However, despite encouraging declines in fertility (Figure 11-7), some population analysts fear that the still-rapid population growth in many developing countries will outstrip economic growth and overwhelm local life-support systems. This could cause many of these countries to be caught in a *demographic trap*, something that is currently happening in a number of developing countries, especially in Africa.

Analysts also point out that some of the conditions that allowed developed countries to develop are not available to many of today's developing countries. Even with large and growing populations, many developing countries **(1)** do not have enough skilled workers to produce the high-tech products needed to compete in the global economy, **(2)** lack the capital and resources needed for rapid economic development, and **(3)** since 1980 have experienced a drop in economic assistance from developed countries and a rise in their debt to such countries. Indeed, since the mid-1980s, developing countries have paid developed countries $40–50 billion a year (mostly in debt interest) more than they have received from these countries.

How Can Family Planning Help Reduce Birth and Abortion Rates and Save Lives? Family planning provides educational and clinical services that help couples choose how many children to have and when to have them. Such programs vary from culture to culture, but most provide information on birth spacing, birth control, breast-feeding, and prenatal care.

Family planning has been an important factor in increasing the proportion of married women in developing countries who use modern forms of contraception (Figure 11-27). Use of such contraceptives in developing countries has gone from 10% of married women of reproductive age in the 1960s to 52% of these

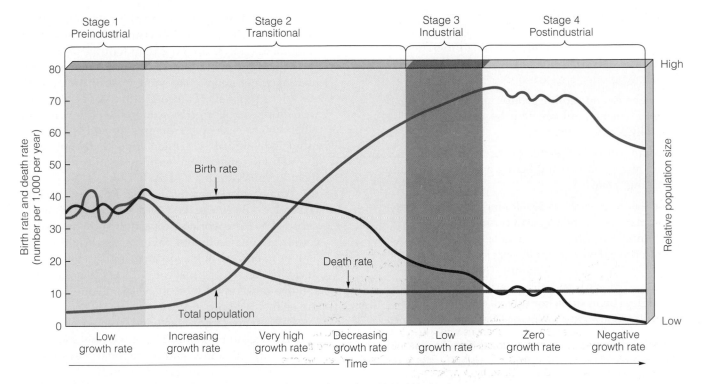

Figure 11-26 Generalized model of the demographic transition. (Data from United Nations)

women in 2000 (39% if China is excluded). Excluding China, this means that 61% of women in developing countries (90% in sub-Saharan Africa) are not using modern contraception. Research indicates that family planning is responsible for at least 55% of the drop in TFRs in developing countries, from 6 in 1960 to 3.2 in 2000 (3.7 if China is excluded).

Other advantages of family planning include the following:

- Less money is needed for children's social services.

- The number of legal and illegal abortions per year declines sharply.

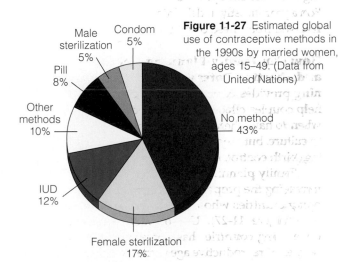

Figure 11-27 Estimated global use of contraceptive methods in the 1990s by married women, ages 15–49. (Data from United Nations)

- The risk of death from childbearing decreases. Each year at least 600,000 women die from pregnancy-related causes (an average of 1,600 deaths per day), with 99% of such deaths occurring in developing countries.

Family planning has been a significant factor in reducing birth and fertility rates in populous countries such as China (Section 11-4), Indonesia, Brazil, and Bangladesh. Family planning has also played a major role in reducing population growth in Japan, Thailand (p. 238), Mexico, South Korea, Taiwan, Iran, and several other countries with moderate to small populations. These successful programs demonstrate that population growth rates can be decreased significantly within 15-30 years.

Despite such successes, family planning has had moderate to poor results in more populous developing countries such as India, Egypt, Pakistan, and Nigeria. Results also have been poor in 79 less populous developing countries, especially in Africa and Latin America, with high or very high population growth rates.

According to United Nations studies, an estimated 300 million women in developing countries want to limit the number and determine the spacing of their children, but they lack access to services. Extending family-planning services to these women and to those who will soon be entering their reproductive years could prevent an estimated 5.8 million births a year and more than 5 million abortions a year.

Other analysts call for

- Expanding existing family planning programs to include teenagers and sexually active unmarried women, who often are excluded.

- Pro-choice and pro-life groups to join forces in greatly reducing unplanned births and abortions, especially among teenagers.

- Programs to educate men about the importance of having fewer children and taking more responsibility for raising them.

- Increased research on developing new, more effective, and more acceptable birth control methods for men. Since 1998, Chinese men have been able to put a contraceptive device about the size of a pager in their underwear for 1 hour per month. This new male contraceptive emits sperm-killing electronic pulses that can render a man sterile for up to a month. Fertility returns 2 months after the man stops using the device.

According to the United Nations, family planning could be provided in developing countries to all couples who want it for about $17 billion a year, the equivalent of less than a week's worth of worldwide military expenditures. If developed countries provided one-third of this $17 billion, each person in the developed countries would spend only $4.80 a year. This would help shrink the global TFR from 2.9 to 2.1 children and reduce the world's population by about 2.9 billion people.

How Can Empowering Women Help Reduce Birth Rates? Studies show that women tend to have fewer and healthier children and live longer when they (1) have access to education and to paying jobs outside the home and (2) live in societies in which their rights are not suppressed.

Women, roughly half of the world's population, (1) do almost all of the world's domestic work and child care, (2) provide more health care with little or no pay than all the world's organized health services combined, and (3) do more than half the work associated with growing food, gathering fuelwood, and hauling water and 80% in rural areas of Africa, Latin America, and Asia (Figure 11-28). As one Brazilian woman put it, "For poor women the only holiday is when you are asleep." Women's unpaid work, at an estimated value of $11 trillion annually (about half of the annual total global economic output), is not included in a country's GDP.

Women work two-thirds of all hours worked but (1) receive only 10% of the world's income and (2) own only 0.01% of the world's property. In most developing countries, women do not have the legal right to own land or to borrow money to increase agricultural productivity. Women also make up 70% of the world's poor and two-thirds of the more than 855 million adults who can neither read nor write.

Women are largely excluded from economic and political decision making. They hold only 14% of the world's administrative and managerial positions and occupy only 13% of parliamentary seats.

Most analysts believe that women everywhere should have full legal rights and the opportunity to become educated and earn income outside the home. This would not only slow population growth but also promote human rights and freedom. However, empowering women by seeking gender equality will require some major social changes that will be difficult to achieve in male-dominated societies.

How Can Economic Rewards and Penalties Be Used to Help Reduce Birth Rates? Some population experts argue that family planning, even coupled with economic development, cannot lower birth and fertility rates quickly enough to avoid a sharp rise in death rates in many developing countries, as is now

| 4:45 A.M. Wake, wash, and eat | 5:00 A.M.- 5:30 A.M. Walk to fields | 5:30 A.M.- 3:00 P.M. Work in fields | 3:00 P.M.- 4:00 P.M. Collect firewood | 4:00 P.M.- 5:30 P.M. Pound and grind corn | 5:30 P.M.- 6:30 P.M. Collect water | 6:30 P.M.- 8:30 P.M. Cook for family and eat | 8:30 P.M.- 9:30 P.M. Wash dishes and children | 9:30 P.M. Go to bed |

Figure 11-28 Typical workday for a woman in rural Africa. In addition to their domestic work, rural African women perform about 80% of all agricultural work. (Data from United Nations)

occurring in parts of Africa. They point to studies showing that most couples in developing countries want three or four children. This is well above the replacement-level fertility needed to bring about eventual population stabilization.

These analysts believe that we must go beyond family planning and offer economic rewards and penalties to help slow population growth. About 20 countries offer small payments to people who agree to use contraceptives or to be sterilized. However, such payments are most likely to attract people who already have all the children they want.

Some countries, including China, penalize couples who have more than one or two children by (1) raising their taxes, (2) charging other fees, or (3) eliminating income tax deductions for a couple's third child (as in Singapore, Hong Kong, and Ghana). Families who have more children than the prescribed limit may also lose health-care benefits, food allotments, and job options.

Economic rewards and penalties designed to lower birth rates work best if they (1) encourage (rather than coerce) people to have fewer children, (2) reinforce existing customs and trends toward smaller families, (3) do not penalize people who produced large families before the programs were established, and (4) increase a poor family's economic status. Once a country's population growth is out of control it may be forced to use coercive methods to prevent mass starvation and hardship, as has been the case for China (p. 258).

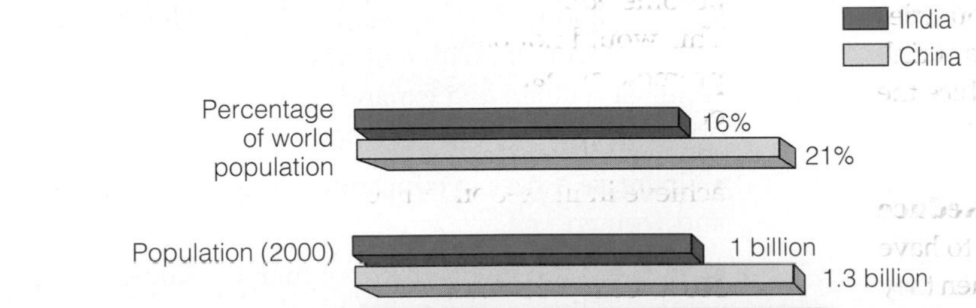

11-4 CASE STUDIES: SLOWING POPULATION GROWTH IN INDIA AND CHINA

What Success Has India Had in Controlling Its Population Growth? The world's first national family-planning program began in India (Figure 11-21) in 1952, when its population was nearly 400 million. In 2000, after 48 years of population control efforts, India was the world's second most populous country, with a population of 1 billion.

In 1952, India added 5 million people to its population; in 2000 it added 18 million—49,300 more mouths to feed each day. Figure 11-29 compares demographic data for India and China.

India's people are among the poorest in the world, with an average per capita income of about $440 a year; for 30% of the population it is less than $150 a year, or 41¢ a day. Nearly half of India's labor force is unemployed or can find only occasional work. Although India currently is self-sufficient in food grain production, about 40% of its population suffers from malnutrition, mostly because of poverty.

Some analysts fear that India's already serious malnutrition and health problems will worsen as its population continues to grow rapidly. With 16% of the world's people, India has just 2.3% of the world's land resources and 1.7% of the world's forests. About half of India's cropland is degraded as a result of soil erosion, waterlogging, salinization, overgrazing, and deforestation. About 70% of India's water is seriously polluted, and sanitation services often are inadequate.

Without its long-standing family-planning program, India's population and environmental problems would be growing even faster. Still, to its supporters the results of the program have been disappointing because of (1) poor planning, (2) bureaucratic inefficiency, (3) the low status of women (despite constitutional guarantees of equality), (4) extreme poverty, and (5) a lack of administrative and financial support.

Even though the government has provided information about the advantages of small families for years, Indian women still have an average of 3.3 children because most

- India
- China

	India	China
Percentage of world population	16%	21%
Population (2000)	1 billion	1.3 billion
Population (2025) (estimated)	1.4 billion	1.4 billion
Illiteracy (% of adults)	47%	17%
Population under age 15(%)	36%	25%
Population growth rate (%)	1.8%	0.9%
Total fertility rate	3.3 children per women (down from 5.3 in 1970)	1.8 children per women (down from 5.7 in 1972)
Infant mortality rate	72	31
Life expectancy	61 years	71 years
GNP per capita (1998)	$440	$750

Figure 11-29 Comparison of basic demographic data for India and China. (Data from United Nations and Population Reference Bureau).

couples believe they need many children to do work and care for them in old age. Because of the strong cultural preference for male children, some couples keep having children until they produce one or more boys. These factors in part explain why even though 90% of Indian couples know of at least one modern birth control method, only 43% actually use one.

What Success Has China Had in Controlling Its Population Growth? Since 1970, China (Figure 11-21) has made impressive efforts to feed its people and bring its population growth under control. Between 1972 and 2000, China cut its crude birth rate in half and cut its TFR from 5.7 to 1.8 children per woman (Figure 11-29).

In China, as in India, there is a strong preference for male children. According to some reports (denied by the government), **(1)** girls are aborted at a higher rate than boys, **(2)** some infant girls are killed, and **(3)** male children sometimes are fed better than female children.

To achieve its sharp drop in fertility, China has established the most extensive, intrusive, and strict population control program in the world. Couples are strongly urged to postpone the age at which they marry and to have no more than one child. Married couples have ready access to free sterilization, contraceptives, and abortion. Paramedics and mobile units ensure access even in rural areas.

Couples who pledge to have no more than one child receive **(1)** extra food, **(2)** larger pensions, **(3)** better housing, **(4)** free medical care, **(5)** salary bonuses, **(6)** free school tuition for their one child, and **(7)** preferential treatment in employment when their child enters the job market. Couples who break their pledge lose all the benefits. The result is that 81% of married women in China are using modern contraception. In the 1990s, China placed increased emphasis on promoting the role of women in the workforce as a way to reduce fertility.

Government officials realized in the 1960s that the only alternative to strict population control was mass starvation. China is a dictatorship, and thus, unlike India, it has been able to impose a consistent population policy throughout its society.

China's large and still growing population has an enormous environmental impact that could reduce its ability to produce enough food and threaten the health of many of its people. China has 21% of the world's population but only 7% of its fresh water and cropland, 3% of its forests, and 2% of its oil. Soil erosion in China is serious and apparently is getting worse.

Most countries prefer to avoid the coercive elements of China's program. However, other parts of this program could be used in many developing countries. Especially useful is the practice of localizing the program rather than asking people to go to distant family-planning centers. Perhaps the best lesson for other countries is to act to curb population growth before they must choose between mass starvation and coercive measures that severely restrict human freedom.

11-5 CUTTING GLOBAL POPULATION GROWTH

In 1994, the United Nations held its third once-in-a-decade Conference on Population and Development in Cairo, Egypt. One of the conference's goals was to encourage action to stabilize the world's population at 7.8 billion by 2054 instead of the projected 9 billion (Figure 11-9). The major goals of the resulting population plan, endorsed by 180 governments, are to do the following by 2015:

- Provide universal access to family-planning services and reproductive health care

- Improve the health care of infants, children, and pregnant women

- Encourage development and implementation of national population policies as part of social and economic development policies

- Bring about more equitable relationships between men and women, with emphasis on improving the status of women and expanding education and job opportunities for young women

- Increase access to education, especially for girls and women

- Increase the involvement of men in child-rearing responsibilities and family planning

- Take steps to eradicate poverty

- Reduce and eliminate unsustainable patterns of production and consumption

Most analysts also believe that government policy makers in developing and developed countries should devise policies that minimize the environmental impact of population growth. Many analysts applaud these goals, but some call them wishful thinking. Even if they wanted to, most governments could not afford to implement many of these goals.

However, the experience of Japan, Thailand (p. 238), South Korea, Taiwan, and China indicates that a country can achieve replacement-level fertility within 15–30 years. Such experience also suggests that the best way to slow population growth is a combination of investing in **(1)** family planning, **(2)** reducing poverty, and **(3)** elevating the status of women.

Our numbers expand but Earth's natural systems do not.
LESTER R. BROWN

REVIEW QUESTIONS

1. Define the boldfaced terms in this chapter.

2. How did Thailand reduce its birth rate?

3. How is population change calculated? What is *zero population growth*? What are the *crude birth rate* and the *crude death rate*? About how many people are added to the world's population each year and each day? How is the annual rate of population change calculated? What three countries have the world's largest populations?

4. Distinguish between *replacement-level fertility* and *total fertility rate*. Explain why replacement-level fertility is higher than 2. Explain why reaching replacement-level fertility does not mean an immediate halt in population growth.

5. How have fertility rates and birth rates changed in the United States since 1910? How rapidly is the U.S. population growing?

6. List five reasons why the world's death rate has declined over the past 100 years.

7. Distinguish between *life expectancy* and *infant mortality rate*. Why is the infant mortality the best measure of a society's quality of life? About how many infants die each year during their first year of life? List three factors that keep the U.S. infant mortality higher than it could be.

8. What is the *age structure* of a population? Explain why the current age structure of the world's population means that the world's population will keep growing for at least another 50 years even if replacement-level rate of 2.1 is somehow reached globally tomorrow. Draw the general shape of an age structure diagram for a country undergoing **(a)** rapid population growth, **(b)** moderate population growth, and **(c)** slow or zero population growth.

9. What percentage of population is under age 15 in **(a)** the world, **(b)** developed countries, and **(c)** developing countries? Explain how age structure diagrams can be used to make population and economic projections.

10. What are the benefits and potentially harmful effects of rapid population decline?

11. Give the major arguments for and against reducing birth rates globally.

12. What is the *demographic transition*, and what are its four phases? What factors might keep many developing countries from making the demographic transition?

13. What is *family planning*, and what are the advantages of using this approach to reduce the birth rate?

14. Explain how empowering women can help reduce birth rates. What three factors lead women to have fewer and healthier children?

15. What economic rewards and penalties have some countries used to reduce birth rates? What four conditions increase the success of using such economic rewards and penalties?

16. Briefly describe and compare the success China and India have had in reducing their birth rates. What are the major components of China's population control program?

17. List the eight goals of the current United Nations plan to stabilize the world's population at 7.8 billion by 2054, instead of the projected 9 billion.

CRITICAL THINKING

1. Why is a drop in the birth rate not necessarily a reliable indicator of future population growth?

2. Why is it rational for a poor couple in a developing country such as India to have five or six children? What changes might induce such a couple to consider their behavior irrational?

3. List what you consider to be a major local, national, and global environmental problem and describe the role of population growth in this problem.

4. Why is the replacement level higher in most developing countries than in developed countries?

5. Suppose that all women in the world began bearing children at replacement-level fertility rates of 2.1 children per woman today. Explain why this would not immediately stop global population growth. About how long would it take for population growth to stabilize?

6. What do you believe is the world's **(a)** maximum human population and **(b)** optimum human population?

7. Do you believe that the population of **(a)** your own country and **(b)** the area where you live is too high? Explain.

8. Evaluate the claims made by those opposing a reduction in births and those promoting a reduction in births, as discussed on pp. 250–252. Which position do you support, and why?

9. Explain why you agree or disagree with each of the following proposals:
   a. The number of legal immigrants and refugees allowed into the United States each year should be reduced sharply.
   b. Illegal immigration into the United States should be decreased sharply. If you agree, how would you go about achieving this?
   c. Families in the United States should be given financial incentives to have more children to prevent population decline.
   d. The United States should adopt an official policy to stabilize its population and reduce unnecessary resource waste and consumption as rapidly as possible.
   e. Everyone should have the right to have as many children as they want.

10. Some people have proposed that the earth could solve its population problem by shipping people off to space colonies, each containing about 10,000 people. Assuming that we could build such large-scale, self-sustaining space stations, how many people would have to be shipped off each day to provide living spaces for the 82 million people being added to the earth's population each year? Current space shuttles can handle about 6 to 8 passengers. If this capacity could be increased to 100 passengers per shuttle, how many shuttles would have to be launched per day to offset the 82 million people being added each year? According to your calculations, determine whether this proposal is a logical solution to the earth's population problem.

11. According to the U.S. Census Bureau and the Department of Commerce, on average, a child born in the United States in 1997 will cost $2.7 million to raise to age 18 for a high-income family, $1.5 million for a middle-income family, and $762,000 for a lower-income family. What effect do you believe this has on **(a)** the TFR in the United States and **(b)** the local, national, and global environmental impact of a typical American child based on resource use per person (Figure 1-15, p. 15)?

12. Some people believe that the most important goal is to sharply reduce the rate of population growth in developing countries, where 95% of the world's population growth is expected to take place (Figure 1-6, p. 9). Some people in developing countries agree that population growth in these countries can cause local environmental problems. However, they contend that the most serious environmental problem the world faces is disruption of the global life-support system by high levels of resource consumption per person in developed countries, which use 80% of the world's resources. What is your view on this issue? Explain.

13. Some analysts contend that we have enough food and resources for everyone but that these resources are not distributed equitably. To them, poverty, hunger, overpopulation, and environmental degradation are caused mainly by a lack of *social justice*. What is your view on this issue? Explain.

14. Why has China been more successful than India in reducing its rate of population growth? Do you agree with China's current population control policies? Explain. What alternatives, if any, would you suggest?

15. Congratulations—you have just been put in charge of the world. List the five most important features of your population policy.

PROJECTS

1. Survey members of your class to determine how many children they plan to have. Tally the results and compare them for men and women.

2. Assume that your entire class (or manageable groups of your class) is charged with coming up with a plan for halving the world's population growth rate within the next 20 years. Develop a detailed plan that would achieve this goal, including any differences between policies in developing countries and developed countries. Justify each part of your plan. Predict what problems you might face in implementing the plan, and devise strategies for dealing with these problems.

3. Prepare an age structure diagram for your community. Use the diagram to project future population growth and economic and social problems.

4. Use the library or the internet to find bibliographic information about *Theodore Roosevelt, Paul Hawken,* and *Lester R. Brown*, whose quotes appear at the beginning and end of this chapter.

5. Make a concept map of this chapter's major ideas, using the section heads and subheads and the key terms (in boldface). Look at the inside back cover and on the website for this book for information about making concept maps.

INTERNET STUDY RESOURCES AND RESOURCES FOR FURTHER READING AND RESEARCH

The website for this book contains helpful study aids and many ideas for further reading and research. Log on to:

http://www.brookscole.com/product/0534376975s

and click on the Chapter-by-Chapter area. Choose Chapter 11 and select a resource:

- "Flash Cards" allows you to test your mastery of the Terms and Concepts to Remember for this chapter.

- "Tutorial Quizzes" provides a multiple-choice practice quiz.

- "Student Guide to InfoTrac" will lead you to Critical Thinking Projects that use InfoTrac College Edition as a research tool.

- "References" lists the major books and articles consulted in writing this chapter.

- "Hypercontents" takes you to an extensive list of sites with news, research, and images related to individual sections of the chapter.

INFOTRAC COLLEGE EDITION

Improve your skills with InfoTrac College Edition, a searchable online database of articles from more than 700 periodicals. Log on to:

http://www.infotrac-college.com

or access InfoTrac through the website for this book.

Try the following articles:

Barrett, G.W. and E.P. Odum. 2000. The twenty-first century: the world at carrying capacity. *Bio-Science* vol. 50, no. 4, pp. 363–368. (subject guide: carrying capacity)

Suter, K. 1999. The Club of Rome: the global conscience. *Contemporary Review* 275(1602):1–5. (key words: Club of Rome)

12 FOOD RESOURCES

Perennial Crops on the Kansas Prairie

When you think about farms in Kansas, you probably picture seemingly endless fields of wheat or corn plowed up and planted each year. By 2040 the picture might change, thanks to pioneering research at the non-profit Land Institute near Salina, Kansas (Figure 12-1).

The institute, headed by plant geneticist Wes Jackson, is experimenting with an ecological approach to agriculture on the midwestern prairie that relies on planting a mixture of different crops planted in the same area—a growing technique called *polyculture*. The goal is to grow food crops by planting a mix of **(1)** *perennial* grasses, **(2)** legumes (a source of nitrogen fertilizer), **(3)** sunflowers, **(4)** grain crops, and **(5)** plants that provide natural insecticides in the same field.

The institute's goal is to raise food by mimicking many of the natural conditions of the prairie without losing fertile grassland soil (Figure 10-15, p. 223). Institute researchers believe that perennial polyculture can be blended with modern monoculture to reduce its harmful environmental effects.

Because these plants are perennials, the soil does not have to be plowed up and prepared each year to replant them. This takes much less labor than conventional monoculture or diversified organic farms that grow annual crops. It also reduces **(1)** soil erosion because the unplowed soil is not exposed to wind and rain, **(2)** pollution caused by chemical fertilizers and pesticides, and **(3)** the need for irrigation because the deep roots of such perennials retain more water than annuals.

Thirty years of research by the institute have shown that various mixtures of perennials

(polycultures) grown in parts of the midwestern prairie could be used as important sources of food. One such mix of perennial crops includes **(1)** *eastern gamma grass* (a warm season grass that is a relative of corn with three times as much protein as corn and twice as much as wheat), **(2)** *mammoth wildrye* (a cool season grass that is distantly related to rye, wheat, and barley), **(3)** *Illinois bundleflower* (a wild nitrogen-producing legume that can enrich the soil and whose seeds can serve as livestock feed), and **(4)** *Maximilian sunflower* (which produces seeds with as much protein as soybeans).

These discoveries may help the growing human population produce and distribute enough food to meet everyone's basic nutritional needs without degrading the soil, water, air, and biodiversity that support all food production.

Figure 12-1 The Land Institute in Salina, Kansas, is a farm, a prairie laboratory, and a school dedicated to changing the way we grow food. It advocates growing a diverse mixture of edible perennial plants to supplement traditional annual monoculture crops. (Terry Evans)

There are two spiritual dangers in not owning a farm. One is the danger of supposing that breakfast comes from the grocery, and the other that heat comes from the furnace.

ALDO LEOPOLD

This chapter addresses the following questions:

- How is the world's food produced?
- How are green-revolution and traditional methods used to raise crops?
- How much has food production increased, how serious is malnutrition, and what are the environmental effects of producing food?
- How can we increase production of (a) crops, (b) meat, and (c) fish and shellfish?
- How do government policies affect food production and food aid?
- How can we design and shift to more sustainable agricultural systems?

12-1 HOW IS FOOD PRODUCED?

What Three Systems Provide Us with Food? Some Good and Bad News Historically, humans have depended on three systems for their food supply: (1) *croplands* (mostly for producing grain), (2) *rangelands* (for producing meat from grazing livestock), and (3) *oceanic fisheries*.

Some *good news* is that since 1950 there has been a staggering increase in global food production from all three systems, mostly because of technological advances such as (1) increased use of tractors and farm machinery and high-tech fishing boats and gear, (2) inorganic chemical fertilizers, (3) irrigation, (4) pesticides, (5) high-yield varieties of wheat, rice, and corn, (6) densely populated feedlots and enclosed pens for raising cattle, pigs, and chickens, and (7) aquaculture ponds for raising some types of fish and shellfish.

To feed the 9 billion people projected by 2054, we must produce and equitably distribute more food than has been produced since agriculture began about 10,000 years ago and do this in an environmentally sustainable manner. Some analysts believe that we can continue expanding the use of industrialized agriculture to produce the necessary food.

Some *bad news* is that other analysts contend that future food production may be limited by (1) environmental degradation, (2) pollution, (3) lack of water for irrigation, (4) overgrazing by livestock, (5) overfishing, and (6) concern about loss of vital ecological services (Figure 4-36, p. 99) as human activities continue to take over or degrade more and more of the planet's *net primary productivity* (Figure 4-25, p. 88) that supports all life. The rest of this chapter is devoted to analyzing the pros and cons of the world's crop, meat, and fish production systems and how these systems can be made more sustainable.

What Plants and Animals Feed the World? Although the earth has perhaps 30,000 plant species with parts that people can eat, only 15 plant and 8 terrestrial animal species supply 90% of our food. Just three grain crops—*wheat, rice,* and *corn*—provide more than half the calories people consume. These three grains, and most other food crops, are *annuals*, whose seeds must be replanted each year.

Two-thirds of the world's people survive primarily on traditional grains (mainly rice, wheat, and corn), mostly because they cannot afford meat. As incomes rise, people consume more grain, but indirectly in the form of meat (mostly beef, pork, and chicken), eggs, milk, cheese, and other products of grain-eating domesticated livestock.

Fish and shellfish are an important source of food for about 1 billion people, mostly in Asia and in coastal areas of developing countries. However, on a global scale fish and shellfish supply less than 1% of the energy and less than 6% of the protein in the human diet.

What Are the Major Types of Food Production? All crop production involves replacing species-rich late successional communities such as mature grasslands (Figure 6-23, p. 139) and forests (Figure 6-29, p. 146) with an early successional community (Figure 8-16, p. 189) consisting of a single crop (*monoculture,* Figure 6-25, p. 141) or a mixture of crops (*polyculture*).

There are two major types of agricultural systems: industrialized and traditional. **Industrialized agriculture,** or **high-input agriculture,** uses large amounts of fossil fuel energy, water, commercial fertilizers, and pesticides to produce huge quantities of single crops (monocultures) or livestock animals for sale. Practiced on about 25% of all cropland, mostly in developed countries (Figure 12-2), high-input industrialized agriculture has spread since the mid-1960s to some developing countries.

Plantation agriculture is a form of industrialized agriculture practiced primarily in tropical developing countries. It involves growing cash crops (such as bananas, coffee, soybeans, and sugarcane, cocoa, and vegetables) on large monoculture plantations mostly for sale in developed countries.

An increasing amount of livestock production in developed countries is industrialized. Large numbers of cattle are brought to densely populated feedlots, where they are fattened up for about 4 months before slaughter. Most pigs and chickens in developed countries spend their entire lives in densely populated pens and cages and are fed mostly grain grown on cropland.

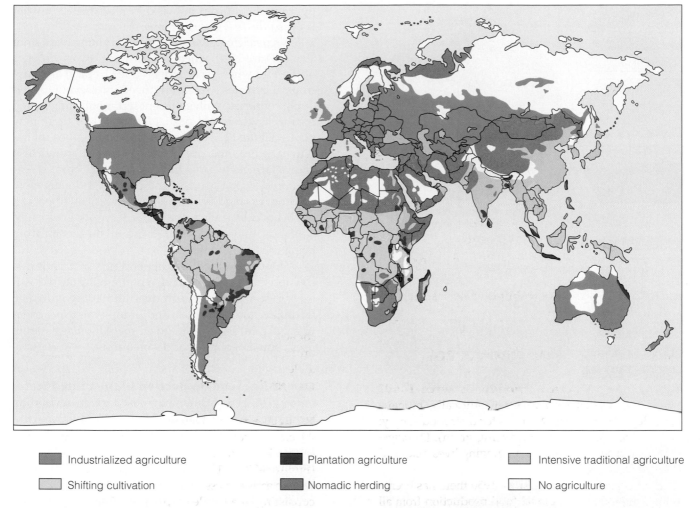

	Industrialized agriculture		Plantation agriculture		Intensive traditional agriculture
	Shifting cultivation		Nomadic herding		No agriculture

Figure 12-2 Locations of the world's principal types of food production. Excluding Antarcticta and Greenland, agricultural systems cover almost one third of the earth's land surface and account for an annual output of food worth about $1.3 trillion.

Traditional agriculture consists of two main types, which together are practiced by about 2.7 billion people (44% of the world's people) in developing countries and provide about 20% of the world's food supply. **Traditional subsistence agriculture** typically uses mostly human labor and draft animals to produce only enough crops or livestock for a farm family's survival. Examples of this very low-input type of agriculture include numerous forms of shifting cultivation in tropical forests (Figure 2-3, p. 25) and nomadic livestock herding (Figure 12-2).

In **traditional intensive agriculture**, farmers increase their inputs of human and draft labor, fertilizer, and water to get a higher yield per area of cultivated land to produce enough food to feed their families and to sell for income. Figure 12-3 shows the relative inputs of land, human and animal labor, fossil fuel energy, and financial capital needed to produce one unit of food energy in various types of food production systems.

12-2 PRODUCING FOOD BY GREEN-REVOLUTION AND TRADITIONAL TECHNIQUES

How Have Green Revolutions Increased Food Production? High-Input Monocultures in Action
Farmers can produce more food by **(1)** farming more land or **(2)** getting higher yields per unit of area from existing cropland. Since 1950 most of the increase in global food production has come from increased yields per unit of area of cropland in a process called the **green revolution**.

This process involves three steps: **(1)** developing and planting monocultures (Figure 6-25, p. 141) of selectively bred or genetically engineered high-yield varieties of key crops such as rice, wheat, and corn, **(2)** using large inputs of fertilizer, pesticides, and water on crops to produce high yields, and **(3)** increasing the intensity and frequency of cropping. This high-input approach dramatically increased crop yields in most

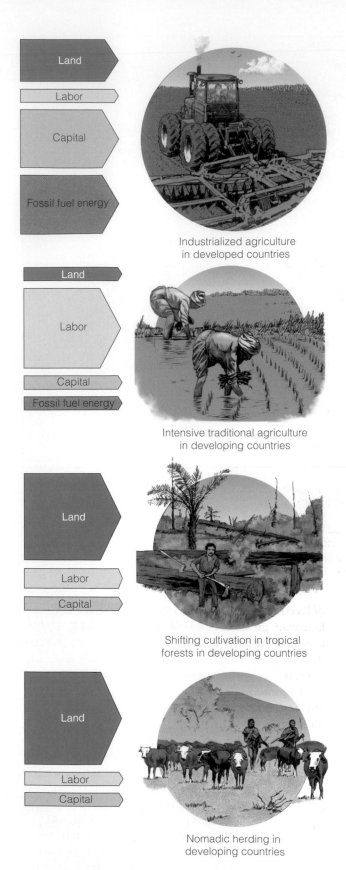

Land

Labor

Capital

Fossil fuel energy

Industrialized agriculture
in developed countries

Land

Labor

Capital

Fossil fuel energy

Intensive traditional agriculture
in developing countries

Land

Labor

Capital

Shifting cultivation in tropical
forests in developing countries

Land

Labor

Capital

Nomadic herding in
developing countries

Figure 12-3 Relative inputs of land, labor, financial capital, and fossil fuel energy in four agricultural systems. An average of 60% of the people in developing countries are involved directly in producing food, compared with only 8% in developed countries (2% in the United States).

developed countries between 1950 and 1970 in what is called the *first green revolution* (Figure 12-4).

A *second green revolution* has been taking place since 1967 (Figure 12-4), when fast-growing dwarf varieties of rice and wheat, specially bred for tropical and subtropical climates, were introduced into several developing countries. With sufficient fertile soil and enough fertilizer, water, and pesticides, yields of these new plants (Figure 12-5) can be two to five times those of traditional wheat and rice varieties. The fast growth also allows farmers to grow two or even three crops a year (multiple cropping) on the same land. Producing more food on less land is also an important way to protect biodiversity by saving large areas of forests, grasslands, wetlands, and easily eroded mountain terrain from being used to grow food.

These yield increases depend not only on fertile soil and ample water but also on high inputs of fossil fuels to run machinery, produce and apply inorganic fertilizers and pesticides, and pump water for irrigation. All told, high-input, green-revolution agriculture uses about 8% of the world's oil output.

Case Study: Food Production in the United States

Since 1940, U.S. farmers have used green-revolution techniques to more than double crop production without cultivating more land. This has kept large areas of forests, grasslands, wetlands, and easily erodible land from being converted to farmland. Figure 12-6 shows the various uses of land in the United States.

Farming has become *agribusiness* as big companies and larger family-owned farms have taken control of almost three-fourths of U.S. food production. Only about 650,000 Americans (2% of the population) are full-time farmers. However, about 9% of the population is involved in the U.S. agricultural system, from growing and processing food to distributing it and selling it at the supermarket.

In terms of total annual sales, agriculture is the biggest industry in the United States, bigger than the automotive, steel, and housing industries combined. It generates about 18% of the country's gross national product and 19% of all jobs in the private sector, employing more people than any other industry.

The U.S. agricultural system is highly productive. With only 0.3% of the world's farm labor force, U.S. farms produce about 17% of the world's grain, most of which is consumed by U.S. livestock. The United States also supplies nearly half of the world's grain exports. Most of these grain exports are used to feed livestock animals in Europe and Japan.

U.S. residents spend an average of only 10–12% of their income on food (down from 21% in 1940), compared to 18% in Japan and 40–70% in most developing countries.

This industrialization of agriculture has been made possible by the availability of cheap energy, most of it

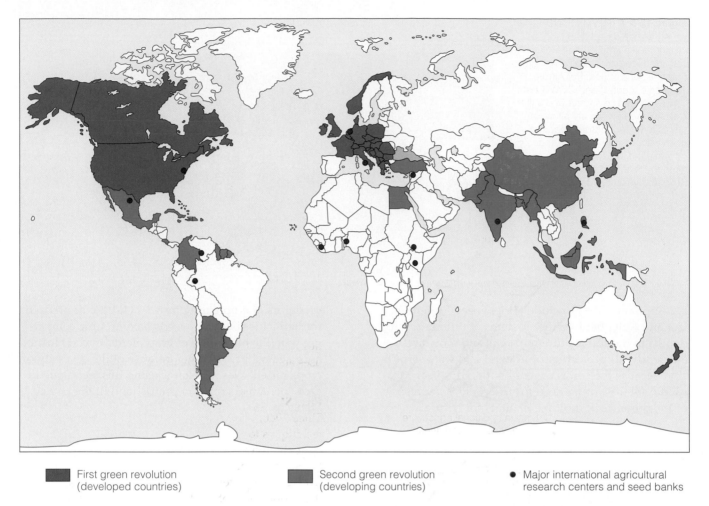

| | First green revolution (developed countries) | | Second green revolution (developing countries) | ● | Major international agricultural research centers and seed banks |

Figure 12-4 Countries whose crop yields per unit of land area increased during the two green revolutions. The first took place in developed countries between 1950 and 1970; the second has occurred since 1967 in developing countries with enough rainfall or irrigation capacity. Several agricultural research centers and gene or seed banks play a key role in developing high-yield crop varieties.

from oil. Agriculture consumes about 17% of all commercial energy in the United States each year (Figure 12-7). Most plant crops in the United States provide more food energy than the energy used to grow them. However, if we include livestock, the U.S. food production system uses about three units of fossil fuel energy to produce one unit of food energy.

Energy efficiency is much lower if we look at the whole U.S. food system. Considering the energy used to grow, store, process, package, transport, refrigerate, and cook all plant and animal food, *about 10 units of nonrenewable fossil fuel energy are needed to put 1 unit of food energy on the table.* By comparison, every unit of energy from human labor in traditional subsistence farming provides at least 1 unit of food energy and up to 10 units of food energy using traditional intensive farming.

What Growing Techniques Are Used in Traditional Agriculture? Low-Input Agrodiversity in Action Traditional farmers in developing countries today grow about 20% of the world's food on about 75%

of its cultivated land. Many traditional farmers simultaneously grow several crops on the same plot, a practice known as **interplanting**. Such crop diversity reduces the chance of losing most or all of the year's food supply to pests, bad weather, and other misfortunes.

Common interplanting strategies found throughout the world, mostly in developing countries, include the following:

- **Polyvarietal cultivation**, in which a plot is planted with several varieties of the same crop.

- **Intercropping**, in which two or more different crops are grown at the same time on a plot (for example, a carbohydrate-rich grain that uses soil nitrogen and a protein-rich legume that puts it back).

- **Agroforestry**, or **alley cropping**, in which crops and trees are planted together (Figure 10-24c, p. 230).

- **Polyculture**, a more complex form of intercropping in which many different plants maturing at various times are planted together. Advantages of this

Figure 12-5 A high-yield, semidwarf variety of rice called IR-8 (left), a part of the second green revolution, was produced by crossbreeding two parent strains of rice: PETA from Indonesia (center) and DGWG from China (right). The shorter and stiffer stalks of the new variety allow the plants to support larger heads of grain without toppling over and increase the benefit of applying more fertilizer. (International Rice Research Institute, Manila)

low-input approach include **(1)** less need for fertilizer and water because root systems at different depths in the soil capture nutrients and moisture efficiently, **(2)** protection from wind and water erosion because the soil is covered with crops year round, **(3)** little or no need for insecticides because of the presence of multiple habitats for natural predators of crop-eating insects, **(4)** little or no need for herbicides because weeds have trouble competing with the multitude of crop plants, and **(5)** insurance against bad weather because of the diversity of crops raised. Wes Jackson is using this technique to grow perennial crops on prairie land in the United States (p. 261).

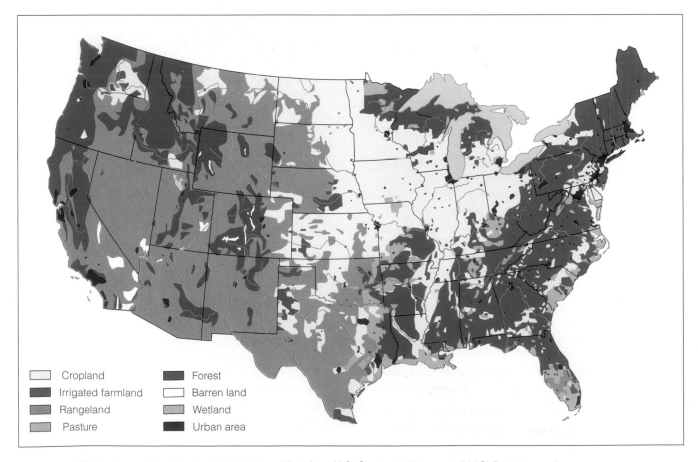

Cropland
Irrigated farmland
Rangeland
Pasture
Forest
Barren land
Wetland
Urban area

Figure 12-6 General uses of land in the United States. (Data from U.S. Geological Survey and U.S. Department of Agriculture)

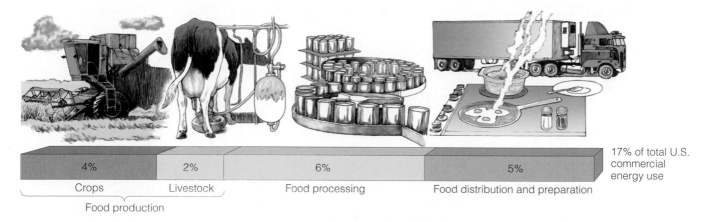

4%	2%	6%	5%	17% of total U.S. commercial energy use
Crops	Livestock	Food processing	Food distribution and preparation	

Food production

Figure 12-7 In the United States, industrialized agriculture uses about 17% of all commercial energy. On average, a piece of food eaten in the United States has traveled 2,100 kilometers (1,300 miles).

Recent ecological research on crop yields of 14 artificial ecosystems found that on average low-input polyculture (with four or five different crop species) produces higher yields per hectare of land than high-input monoculture. A 2000 field study headed by Christ Mundt at Oregon State University found that planting different varieties of rice in the same paddy greatly increased yield and sharply reduced crop losses to blast, a major fungal disease affecting rice.

12-3 FOOD PRODUCTION, NUTRITION, AND ENVIRONMENTAL EFFECTS

How Much Has Food Production Increased? Figure 12-8 illustrates the success of using high-input monoculture farming to produce food and ward off sharp rises in hunger and malnutrition. Here is some *good news*. Between 1950 and 1990, **(1)** world grain production almost tripled (Figure 12-8, left), **(2)** per capita production rose by about 36% (Figure 12-8, right), **(3)** average food prices adjusted for inflation dropped by 25%, and **(4)** the amount of food traded in the world market quadrupled.

Global meat production—mostly beef, pork, and poultry—has risen for 41 consecutive years. One indicator of the growing demand for animal protein is the ninefold increase in soybean production between 1950 and 1999. Small amounts of soybean meal are added to grain consumed by livestock and poultry to increase meat production.

Despite these impressive achievements in food production, there is some *disturbing but challenging news*:

- Population growth is outstripping food production and distribution in areas that support about 2 billion

people, especially in sub-Saharan Africa,* with extreme poverty and one of the world's highest population growth rates.

- Since 1985 global grain production has leveled off (Figure 12-8, left) and per capita grain production has declined (Figure 12-8, right). Much of the per capita decline has occurred in Africa, Russia, and the Ukraine. According to some analysts, this happened because more efficient food production has lowered the price farmers get for grain and reduced their incentive to grow more food. Other analysts attribute much of this leveling off of grain production to **(1)** limits on the amounts of water, fertilizer, and pesticides that green-revolution crops can tolerate and **(2)** loss of productivity from erosion and salinization of soil and lack of irrigation water.

- Two traditional sources of animal protein in the human diet—*rangelands* (which account for much of the world's beef and mutton production) and *fisheries*—appear to be approaching their productive limits.

- If everyone ate the diet typical of a person in a developed country, with 30–40% of the calories coming from animal products, the world's current agricultural system would support only an estimated 2.5 billion people.

How Serious Are Undernutrition and Malnutrition? According to the United Nations and the World Bank, we live in a world where one out of six people in developing countries does not get enough food and one out of seven in developed countries eats too much. People who are underfed and underweight and those who are overfed and overweight face similar health problems: **(1)** lower life expectancy, **(2)** increased susceptibility to disease and illness, and **(3)** reduced productivity and life quality.

---

*Sub-Saharan Africa includes all of Africa's countries except South Africa and the six countries north of the Sahara desert.

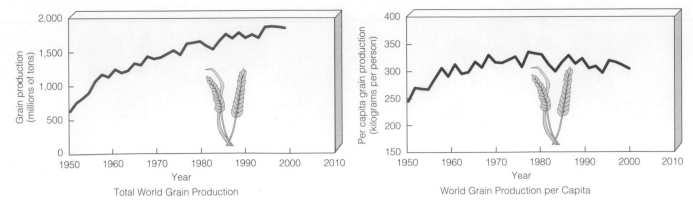

Figure 12-8 Total worldwide grain production of wheat, corn, and rice, and per capita grain production, 1950–2000. In order, the world's three largest grain-producing countries in 2000 were China, the United States, and India. (Data from U.S. Department of Agriculture, Worldwatch Institute, and UN Food and Agriculture Organization)

To maintain good health and resist disease, people need **(1)** fairly large amounts of *macronutrients* such as protein, carbohydrates, and fats, and **(2)** smaller amounts of *micronutrients* consisting of various vitamins (such as A, C, and E) and minerals (such as iron, iodine, and calcium).

People who cannot grow or buy enough food to meet their basic energy needs suffer from **undernutrition**. *Chronic undernourishment* occurs when people get less than 90% of their minimum daily calorie intake of food on a long-term basis. Such people are not starving to death, but they often do not have enough energy for an active, productive life and are more susceptible to infectious diseases.

People getting less than 80% of their minimum calorie intake are considered to be *seriously undernourished*. Children in this category are likely to **(1)** suffer from mental retardation and stunted growth and **(2)** be much more susceptible to infectious diseases (such as measles and diarrhea), which kill one child in four in developing countries.

People who are forced to live on a low-protein, high-carbohydrate diet consisting only of grains such as wheat, rice, or corn often suffer from **malnutrition**: deficiencies of protein and other key nutrients. Many of the world's desperately poor people, especially children, suffer from both undernutrition and malnutrition.

The two most common nutritional deficiency diseases are marasmus and kwashiorkor. *Marasmus* (from the Greek word *marasmos*, "to waste away") occurs when a diet is low in both calories and protein (Figure 1-9, p. 10). Most victims are either nursing infants of malnourished mothers or children who do not get enough food after being weaned from breast-feeding. If the child is treated in time with a balanced diet, most of these effects can be reversed.

Kwashiorkor (meaning "displaced child" in a West African dialect) is a severe protein deficiency occurring in infants and children ages 1–3, usually after the arrival of a new baby deprives them of breast milk. The displaced child's diet changes to grain or sweet potatoes, which provide enough calories but not enough protein. If it is caught soon enough, most of the harmful effects can be cured with a balanced diet. Otherwise, children who survive their first year or two suffer from stunted growth and mental retardation.

Here is some *good news*. Despite population growth, the estimated number of chronically malnourished people fell from 918 million in 1970 to 826 million in 1999 (96% of them in developing countries). However, according to the UN the *disturbing news* is that about one of every six people in developing countries (including one of every three children below age 5) is chronically undernourished or malnourished.

Such people are disease prone and adults are too weak to work productively or think clearly. As a result, their children also tend to be underfed, malnourished, and susceptible to disease. If these children survive to adulthood, many are locked in a malnutrition-poverty cycle (Figure 12-9) that can continue for generations.

According to the World Health Organization (WHO), each year at least 10 million people, half of them children under age 5, die prematurely from **(1)** undernutrition, **(2)** malnutrition, or **(3)** normally nonfatal diseases such as measles and diarrhea worsened by malnutrition. Such undernutrition and malnutrition have been described by the UN as a "silent and invisible global emergency with a massive impact on children" that could be prevented (Solutions, p. 270).

The most widespread micronutrient deficiencies in developing countries involve *vitamin A, iron,* and *iodine.* According to WHO, an estimated 124 million children in developing countries are deficient in vitamin A. This puts them at risk for blindness (about 500,000 cases per year) and premature death because even mild vitamin

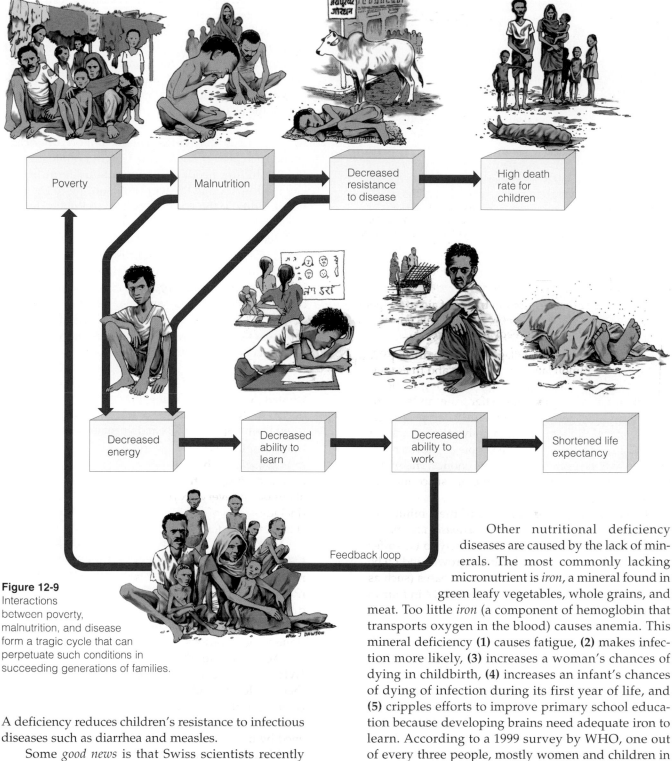

Figure 12-9
Interactions between poverty, malnutrition, and disease form a tragic cycle that can perpetuate such conditions in succeeding generations of families.

[Diagram labels:]
Poverty → Malnutrition → Decreased resistance to disease → High death rate for children

Decreased energy → Decreased ability to learn → Decreased ability to work → Shortened life expectancy

Feedback loop

A deficiency reduces children's resistance to infectious diseases such as diarrhea and measles.

Some *good news* is that Swiss scientists recently spliced genes into rice to make it rich in beta carotene, the source of vitamin A. According to the United Nations Children's Fund (UNICEF), improved vitamin A nutrition through use of this vitamin-fortified rice could prevent 1–2 million premature deaths per year among children under age 5. Scientists eventually hope to use gene splicing to fortify rice and other grains with other essential vitamins and iron.

Other nutritional deficiency diseases are caused by the lack of minerals. The most commonly lacking micronutrient is *iron*, a mineral found in green leafy vegetables, whole grains, and meat. Too little *iron* (a component of hemoglobin that transports oxygen in the blood) causes anemia. This mineral deficiency **(1)** causes fatigue, **(2)** makes infection more likely, **(3)** increases a woman's chances of dying in childbirth, **(4)** increases an infant's chances of dying of infection during its first year of life, and **(5)** cripples efforts to improve primary school education because developing brains need adequate iron to learn. According to a 1999 survey by WHO, one out of every three people, mostly women and children in tropical developing countries, suffer from iron deficiency.

Elemental *iodine*, found in seafood and crops grown in iodine-rich soils, is essential for the functioning of the thyroid gland, which produces a hormone that controls the body's rate of metabolism. Lack of iodine can cause **(1)** stunted growth, **(2)** mental retardation, and **(3)** goiter (an abnormal enlargement of the thyroid gland that can lead to deafness).

Saving Children

SOLUTIONS

According to the United Nations Children's Fund (UNICEF), "Malnourished children who live past childhood face a diminished future. They will become adults with lower physical and intellectual abilities, lower levels of productivity, and higher levels of chronic illness and disability."

UNICEF officials estimate that one-half to two-thirds of childhood deaths from nutrition-related causes could be prevented at an average annual cost of only $5-10 per child, or only 10–19¢ per week. This life-saving program would involve the following measures:

- Immunizing children against childhood diseases such as measles

- Encouraging breast-feeding

- Preventing dehydration from diarrhea by giving infants a mixture of sugar and salt in a glass of water

- Preventing blindness by giving people a vitamin A capsule twice a year at a cost of about 75¢ per person or fortifying common foods with vitamin A and other micronutrients at a cost of about 10¢ per person annually

- Providing family-planning services to help mothers space births at least 2 years apart

- Increasing education for women, with emphasis on nutrition, sterilization of drinking water, and child care

Critical Thinking

How much money (if any) would you be willing to spend each year to help implement such a program for saving children? Why has insufficient money been allocated for such a program?

Some *good news* is that programs by WHO and UNICEF to get countries to establish programs to add iodine to salt slashed the percentage of the world's people with iodine deficiency from 29% to 13% between 1994 and 1997. The *bad news* is that this still leaves at least 740 million people in developing countries with too little iodine in their diet.

How Serious Is Overnutrition? About one out of seven adults in developed countries (one out of five in the United States) suffers from **overnutrition**, a condition in which food energy intake exceeds energy use and causes excess body fat (obesity). It can be caused by too much food, too little exercise, or both.

Overnutrition is the second leading cause of preventable deaths after smoking (p. 396), mostly from coronary heart disease, cancer, stroke, and diabetes. A study of thousands of Chinese villagers indicates that the most healthful diet for humans is largely vegetarian, with only 10–15% of calories coming from fat. This is in contrast to the typical meat-based diet, in which 40% of the calories come from fat.

More than half of all adult Americans are overweight, and 20% are obese. The $36 billion Americans spend each year trying to lose weight is almost twice the $19 billion needed to eliminate undernutrition and malnutrition in the world. According to Dr. Graham Colditz at the Harvard School of Public Health, the *direct costs* (hospital stays, treatment, medicine, and doctor visits) and *indirect costs* (reduced productivity, missed workdays, and disability pensions) of obesity in the United States are about $118 billion per year, or nearly 12% of the nation's annual health-care bill.

Some nutrition analysts, including Yale University professor Kelly Brownwell, advocate discouraging consumption of unhealthful foods by:

- Regulating advertising of junk foods.

- Taxing food based on its nutrient value per calorie. Sugary and fatty foods low in nutrients would be taxed the most, and fruits and vegetables would not be taxed.

- Using the resulting tax revenues to promote healthful diets, nutrition education, and exercise programs in schools and communities.

Do We Produce Enough Food to Feed the World's People? The *good news* is that we produce more than enough food to meet the basic nutritional needs of every person on the earth today. If distributed equally, the grain currently produced worldwide is enough to give everyone a meatless subsistence diet.

The *bad news* for those not getting enough to eat is that food is not distributed equally among the world's people because of differences in **(1)** soil, **(2)** climate, **(3)** political and economic power, and **(4)** average per capita income throughout the world (Figure 1-5, p. 9).

Most agricultural experts agree that *the principal cause of hunger and malnutrition is and will continue to be poverty*, which prevents poor people from growing or buying enough food regardless of how much is available. For example, according to the UN, in the 1990s nearly 80% of all malnourished children lived in countries with food surpluses. Ways to decrease hunger and malnutrition by reducing poverty are discussed in Section 26-6 (p. 706).

Biodiversity Loss

Loss and degradation of habitat from clearing grasslands and forests and draining wetlands

Fish kills from pesticide runoff

Killing of wild predators to protect livestock

Loss of genetic diversity from replacing thousands of wild crop strains with a few monoculture strains

Soil

Erosion

Loss of fertility

Salinization

Waterlogging

Desertification

Air Pollution

Greenhouse gas emissions from fossil fuel use

Other air pollutants from fossil fuel use

Pollution from pesticide sprays

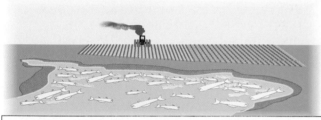

Water

Aquifer depletion

Increased runoff and flooding from land cleared to grow crops

Sediment pollution from erosion

Fish kills from pesticide runoff

Surface and groundwater pollution from pesticides and fertilizers

Overfertilization of lakes and slow-moving rivers from runoff of nitrates and phosphates from fertilizers, livestock wastes, and food processing wastes

Human Health

Nitrates in drinking water

Pesticide residues in drinking water, food, and air

Contamination of drinking and swimming water with disease organisms from livestock wastes

Bacterial contamination of meat

Figure 12-10 Major environmental effects of food production.

What Are the Environmental Effects of Producing Food? Agriculture has significant harmful impacts on air, soil, water, and biodiversity resources (Figure 12-10). David Pimentel (Guest Essay, p. 232) has estimated that the harmful environmental costs not included in the prices of food in the United States are $150–200 billion per year.

According to Norman Myers (Guest Essay, p. 117), the future ability to produce more food will be limited by a combination of (1) soil erosion (Figure 10-19, p. 226), (2) desertification (Figure 10-21, p. 228), (3) salinization and waterlogging of irrigated lands (Figure 10-22,

p. 229), (4) water deficits and droughts (Section 13-8, p. 314), (5) loss of wild species that provide the genetic resources for new foods and improved forms of existing foods, and (6) the projected effects of global warming (Section 18-4, p. 458). According to a 2000 study by the UN-affiliated International Food Policy Research Institute, nearly 40% of the world's cropland is seriously degraded (including 75% in Central America, 20% in Africa, and 11% in Asia). There is growing concern that such environmental factors may limit food production in India and China (Case Study, p. 272), the world's two most populous countries.

CASE STUDY

Since 1970 China has made significant progress in feeding its people and slowing its rate of population growth (p. 258). Despite such efforts, China is projected to add about 275 million people between 2000 and 2030—about as many people as live in the United States today.

There is growing concern that crop yields may not be able to keep up with demand. A basic problem is that with 21% of the world's people, China has only 7% of the world's cropland and fresh water, 3% of its forests, and 2% of its oil.

China depends on irrigated land to produce 70% of the grain for its population. According to World-watch Institute projections, China's grain production is likely to fall by at least 20% between 1990 and 2030, mostly because of water shortages, degraded cropland, and diversion of water from cropland to cities. According to the UN Development Program, more than 400 of China's 640 largest cities are short of water, and 100 of them are very short.

A growing amount of China's cropland is irrigated by withdrawing groundwater from aquifers faster than they are naturally recharged. As these groundwater deficits worsen, Chinese farmers will be forced to (1) take land out of irrigation, (2) switch to less thirsty crops, or (3) irrigate more efficiently.

Even if China's currently booming economy resulted in no increases in meat consumption, this 20% projected drop in grain production would mean that by 2030 China would need to import more than all of the world's current grain exports (roughly half from the United States).

However, if the increased demand for meat led to a rise in per capita grain consumption equal to the current level in Taiwan (one-half the current U.S. level), China would need to import more than the entire current grain output of the United States.

The Worldwatch Institute warns that if either of these scenarios is correct, no country or combination of countries has the potential to supply even a small fraction of China's potential food supply deficit. This is not even taking into account the huge grain deficits that are projected in other parts of the world by 2030, especially in Africa and India.

However, according to a 1997 study by the International Food Policy Institute, China should be able to feed its population and begin exporting grain again by 2020 if the government invests in (1) expanding irrigation, (2) using more water-efficient forms of irrigation (p. 311), and (3) increasing agricultural research. Recent satellite surveys also show that China has far more potential cropland than previously thought.

China's potential food problems and their need to import a large share of the world's grain exports could lead to higher grain prices. Low-income countries needing to import grain could be priced out of the market. This could lead to social unrest and possibly riots by hungry urban dwellers.

Thus, China's potential water and food shortages are linked to global military security. This explains why the U.S. intelligence community closely monitors China's water and food situation.

Critical Thinking

If the scenarios about China's future dependence on greatly increased food imports are accurate, how might this affect (a) world food prices and (b) your life. What actions, if any, do you suggest for dealing with this potential problem?

12-4 INCREASING WORLD CROP PRODUCTION

How Can Crossbreeding Be Used to Develop Genetically Improved Crop Strains? For centuries farmers and scientists have used traditional methods of *crossbreeding* through artificial selection to develop genetically improved varieties of crop strains and livestock (Spotlight, p. 108). For example, two crop plants such as a variety of a pear and of an apple can be crossbred with the goal of producing a pear with a more reddish color (Figure 12-11). This is repeated for a number of generations until the desired trait in the pear predominates.

Agricultural experts expect most future increases in food yields per hectare on existing cropland to result from improved crossbred strains of plants and from expansion of green-revolution technology to new parts of the world. For example, in 1994 crop scientists announced that they had used crossbreeding to develop new strains of corn that can increase crop yields up to 40% in regions plagued by droughts and acidic soils. In addition, a new strain of rice developed by the International Rice Research Institute could increase rice yields by as much as 20%.

However, (1) traditional crossbreeding is a slow process, typically taking 15 years or more to produce a commercially valuable new variety, (2) it can combine traits only from species that are close to one another genetically, and (3) a crossbred variety is useful for only about 5-10 years before its effectiveness is reduced by pests and diseases.

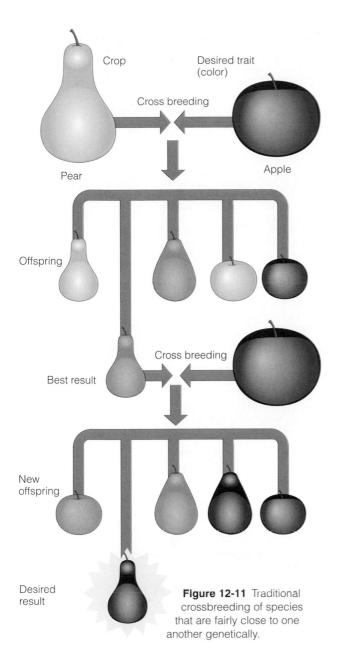

Figure 12-11 Traditional crossbreeding of species that are fairly close to one another genetically.

Labels in figure: Crop / Desired trait (color) / Cross breeding / Pear / Apple / Offspring / Best result / Cross breeding / New offspring / Desired result

How Can Genetic Engineering Be Used to Develop Genetically Improved Crop Strains? Scientists are working to create new green revolutions—actually *gene revolutions*—by using genetic engineering and other forms of biotechnology to develop new genetically improved strains of crops. **Genetic engineering** or **gene splicing** of food crops refers to the insertion of an alien gene into a commercially valuable plant (or animal) to give it new beneficial genetic traits. Such organisms are called **genetically modified organisms (GMOs)**.

Compared to traditional crossbreeding, gene splicing **(1)** takes about half as much time to develop a new crop or animal variety, **(2)** cuts costs, and **(3)** allows the insertion of genes from almost any other organism into crop or animal cells. Figure 12-12 outlines the steps involved in developing a genetically modified or transgenic plant. Scientists are also using *advanced tissue culture techniques* to produce only the desired parts of a plant such as its oils or fruits.

Ready or not, the world is entering the *age of genetic engineering*. Nearly two-thirds of the food products on U.S. supermarket shelves contain genetically engineered crops, and the proportion is increasing rapidly. None of these foods are labeled as being genetically modified.

This rapidly developing technology excites many scientists and investors. They see it as a way to produce, patent, and sell high-yield crops and livestock with more protein and key vitamins and greater resistance to diseases, pests, frost, and drought. However, there is much controversy over the rapidly growing use of genetically modified crops (Pro/Con, p. 275).

There is also controversy about the legal patenting of genetically modified life-forms. Proponents of the patent system argue that developing a new genetically modified strain can be costly and that investors deserve an equitable return on their investment capital.

However, farmers in many developing countries cannot afford to pay the patent fees and many have refused to respect patent claims on GMOs. Some developing countries oppose paying patent fees for the use of GMOs when the new organisms have been engineered from genes of natural species coming from their forests and other ecosystems. In addition to not having to paying patent fees, they feel that they should receive just compensation for use of their natural genetic resources.

Many developing countries also do not have the money to develop their own biotechnology industries. However, some nonprofit organizations are developing genetically modified crop strains and other GMOs and making them available to developing counties at no charge. Also, some international aid agencies are supporting biotechnology research designed to benefit developing countries.

Can We Continue Expanding the Green Revolution? Many analysts believe that it is possible to produce enough food to feed the 8 billion people projected by 2025 through new advances in gene-splicing technology and by spreading the use of new and existing high-yield, green-revolution techniques. Other analysts point to the following factors that have limited the success of the green and gene revolutions to date and may continue to do so:

- Without huge amounts of fertilizer and water, most green-revolution crop varieties produce yields

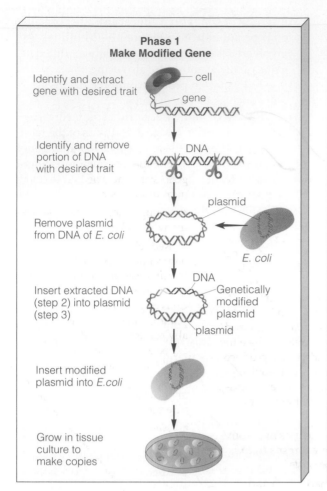

Phase 1
Make Modified Gene

Identify and extract gene with desired trait — cell / gene

Identify and remove portion of DNA with desired trait — DNA

Remove plasmid from DNA of *E. coli* — plasmid / *E. coli*

Insert extracted DNA (step 2) into plasmid (step 3) — DNA / Genetically modified plasmid / plasmid

Insert modified plasmid into *E.coli*

Grow in tissue culture to make copies

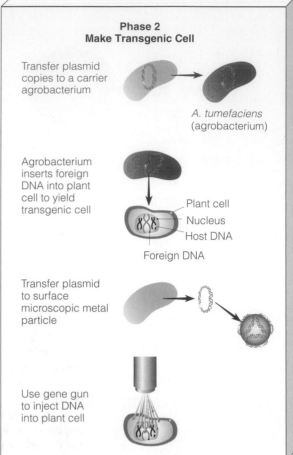

Phase 2
Make Transgenic Cell

Transfer plasmid copies to a carrier agrobacterium — *A. tumefaciens* (agrobacterium)

Agrobacterium inserts foreign DNA into plant cell to yield transgenic cell — Plant cell / Nucleus / Host DNA / Foreign DNA

Transfer plasmid to surface microscopic metal particle

Use gene gun to inject DNA into plant cell

that are no higher (and are sometimes lower) than those from traditional strains; this is why the second green revolution has not spread to many arid and semiarid areas such as much of Africa and Australia (Figure 12-4).

■ Green-revolution and genetically engineered crop strains and their needed inputs of water, fertilizer, and pesticides cost too much for most subsistence farmers in developing countries.

■ Continuing to increase inputs of fertilizer, water, and pesticides eventually produces no additional increase in crop yields.

■ Grain yields per hectare are still increasing in many parts of the world, but at a much slower rate. For example, grain yields rose about 2.1% a year between 1950 and 1990 but dropped to about 1% per year between 1990 and 2000.

■ According to Indian economist Vandana Shiva, overall gains in crop yields from new green- and gene-revolution varieties may be much lower than claimed. The reason is that the yields are based on comparisons between the output per hectare of old and new *monoculture* varieties rather than between the even higher yields per hectare for *polyculture*

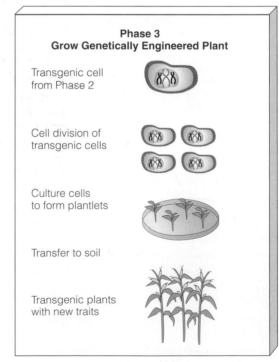

Phase 3
Grow Genetically Engineered Plant

Transgenic cell from Phase 2

Cell division of transgenic cells

Culture cells to form plantlets

Transfer to soil

Transgenic plants with new traits

Figure 12-12 Steps in genetically modifying a plant.

Genetically Modified Food

There is growing controversy over the rapidly increasing use of *genetically modified food* (GMF). Such food is seen as a potentially sustainable way to solve world food problems by its producers and called potentially dangerous "Frankenfood" by its critics in Europe.

If you live in the United States you probably had some GMF for breakfast, lunch, and dinner. Gene transfer has been used in more than 60 different crops and animals to develop **(1)** strawberries and tomatoes that resist frost because they contain "antifreeze" genes from fish, **(2)** tomatoes that stay fresh and tasty longer because of flounder genes, **(3)** potatoes that resist disease because they contain chicken genes, **(4)** smaller cows that produce more milk, and **(5)** varieties of corn, cotton, and soybeans whose cells produce a natural insecticide and can tolerate large doses of herbicides to reduce competition from weeds.

The rapidly growing agricultural biotech (ag biotech) industry, led by Monsanto, Dow, and DuPont in the United States, sees genetically modified crops as the basis of new *biotechnology revolution* that will supply more food with better nutritional values and fewer environmentally harmful effects. They envision crops genetically modified to:

- Thrive on less fertilizer and water.
- Make their own nitrogen fertilizer.
- Be more resistant to insects and plant diseases.
- Grow well in slightly salty soils.
- Withstand drought.
- Grow faster.
- Provide better nutrition.
- Reduce spoilage and improve flavor.
- Tolerate higher levels of herbicides (currently the most widespread use of transgenics). This could increase plant productivity and reduce soil erosion and water pollution by allowing increased use of conservation tillage techniques that reduce or eliminate the need to plow the soil (p. 229).

- Release pesticides (currently the second most widespread use of transgenics) and reduce the need for repeated spraying of conventional toxic pesticides. For example, a common, naturally occurring bacterium, *Bacillus thuringiensis* (*Bt*), produces a substance that is toxic to some types of moths and butterflies that in the caterpillar stage of their life cycle eat and damage the leaves of crops. *Bt* bacterial genes have been implanted into crops such as corn, potato, and cotton so that their cells produce the toxin. Because spiders also produce toxins that kill insects, researchers are planning to insert the appropriate spider genes into the cells of suitable crops.

- Reduce food allergies by removing the genes for allergens from plants such as peanuts.

- Improve human health and reduce health-care costs by transferring the genetic code of vaccines for diarrhea (recently transferred to a potato), hepatitis B (p. 408), and other widespread diseases in developing countries to crops.

However, critics and a 2000 study by the U.S. National Academy of Sciences warn of a number of potential unintended consequences from widespread use of genetically modified crops and foods, including the following:

- Causing unexpected mutations in plants, which can create new and higher levels of toxins in foods. In 1989, a genetically engineered form of the food supplement tryptophan produced toxic contaminants that killed 37 people, permanently injured 1,500 others, and caused 5,000 others to become ill.

- Introducing new allergens into foods. A few years ago researchers found that a Brazil nut gene spliced into soybean genes induced allergies in people sensitive to Brazil nuts.

- Lowering nutritional value of some foods. A recent study found that levels of compounds (phytoestrogens) thought to protect against heart disease and possibly some types of cancer were lower in genetically modified soybeans than in traditional strains.

- Doing little so far to help feed the world's hungry.

- Reducing the effectiveness of a natural pesticide (*Bacillus thuringiensis* or *Bt*) widely used in small doses by organic farmers. Cells of genetically modified corn, potato, and cotton crops continually release *Bt* into the soil. This is expected to cause rapid development of genetic resistance to *Bt* and deprive organic farmers of an important tool for making agriculture more sustainable.

- Allowing genes, such as those increasing resistance to herbicides and plant viruses, to be passed on through pollination to nearby wild relatives, especially weeds. Such hybrid superweeds, impervious to herbicides and plant viruses, could outgrow crop plants, lower crop yields, and disrupt ecosystems. Recent experiments found that in some crops foreign genes move easily and quickly from crops to wild relatives.

- Increasing herbicide use on herbicide-resistant crop varieties and raising weed-control costs for farmers.

- Increasing the costs of raising food because of the high costs of creating and patenting GMFs and complying with stricter regulation of such crops.

- Causing rapid evolution of pesticide-resistant insects, creating new plant diseases, and harming beneficial insects. For example, in 1999 researchers at Cornell University reported the results of a laboratory

(continued)

experiment in which 44% of the caterpillar (larval) forms of monarch butterflies (Figure 4-4, p. 73) died or developed abnormally after eating milkweed dusted with pollen from a widely planted bioengineered form of corn containing *Bt* genes. A 2000 field experiment by Iowa State researchers supported this laboratory study. In 1998, a Swiss study found that beneficial insects called lacewings died more quickly if they fed on corn borers reared on *Bt* corn. In Scotland, a toxicologist reported that rats eating potatoes containing insect-resistant genes and proteins suffered damaged immune systems, shrunken brains, and growth problems.

■ Introducing chemicals into genetically modified plants that might kill predators and parasites of insect pests and thus decrease nature's natural biological control of certain pests (Figure 4-36, p. 99).

■ Introducing chemicals into the soil from the decomposition of fallen leaves from genetically modified plants that may change the nutrient uptake into plants (Figure 10-14, p. 222) or kill beneficial organisms living in the soil (Figure 10-13, p. 221).

■ Causing irreversible and mostly unpredictable genetic and ecological effects by **(1)** mutating, **(2)** changing their form and behavior, **(3)** migrating or having their seeds transferred to other areas by insects, birds, and wind, and **(4)** altering the genetic traits of existing wild species. Unlike defective cars and other products, these modified organisms cannot be recalled.

■ Reducing the world's genetic diversity needed to supply the genetic raw materials for future green and gene revolutions by increasing dependence on a narrow range of bioengineered crops (Spotlight, p. 278). We have achieved unprecedented skill in moving genes around, but only nature can manufacture genes. Reducing the world's genetic biodiversity could also hinder the development of more sustainable forms of agriculture based on **(1)** planting a variety of crops (polyculture) and **(2)** using a diverse ecological approach (agroecology) for producing food instead of a narrow chemical approach.

Proponents of GMFs argue that **(1)** the potential benefits far outweigh the potential harmful effects, **(2)** environmental risks are exaggerated, and **(3)** some of the potential problems can be eliminated by using a new technique, called *chimeraplasty*, which involves inserting a chemical instruction that attaches to a gene and alters it to give desired genetic traits.

Critics recognize the potential benefits of genetically modified crops but say that we know far too little about the potential harm from widespread use of such crops to human health and ecosystems. They call for **(1)** more controlled field experiments, **(2)** more research and long-term safety testing to better understand the risks, **(3)** stricter regulation of this rapidly developing technology, and **(4)** mandatory labeling of such foods to provide consumers with a choice (as is now required in Japan, Europe, South Korea,

Canada, Australia, and New Zealand).

Makers of GMFs insist that their tests show that such foods are safe and oppose such labeling, fearing that consumers might interpret the labels as indicating that their products are unsafe. Critics contend that without labeling public health agencies will be unable to trace any health problems back to their source.

Growing numbers of food manufacturers and retailers, first in Europe and now in the United States, have stopped selling GMFs. Because of rising costs and growing public resistance, some U.S. farmers who embraced genetically modified crops are switching back to nongentically modified varieties. Insurance companies have refused to cover liability for harm caused by genetically modified organisms.

Because of the decline in consumer confidence about GMFs, financial analysts from Wall Street in the United States to Deutsche Bank in Germany have declared GMFs a risky investment. As a result, some major food biotech companies have put their biotech crop operations on a back burner or are getting out of the business.

Critical Thinking

1. Are you for or against the widespread use of transgenic crops? Explain. What are the alternatives?

2. What government controls, if any, do you believe should be applied to the development and use of genetically engineered organisms? How would you enforce any restrictions?

cropping systems and the new monoculture varieties that often replace them.

■ Yields may start dropping for a number of reasons: **(1)** the soil erodes, loses fertility, and becomes salty and waterlogged (Figure 10-22, p. 229), **(2)** underground and surface water supplies become depleted and pol-

luted with pesticides and nitrates from fertilizers, and **(3)** populations of rapidly breeding pests develop genetic immunity to widely used pesticides (p. 506).

■ Increased loss of biodiversity can limit the genetic raw material needed for future green and gene revolutions (Spotlight, p. 278).

Will People Try New Foods?
Some analysts recommend greatly increased cultivation of less widely known plants to supplement or replace such staples as wheat, rice, and corn. One of many possibilities is the *winged bean*, a protein-rich legume now common only in New Guinea and Southeast Asia. This fast-growing plant produces so many different edible parts that it has been called a supermarket on a stalk. It also needs little fertilizer because of nitrogen-fixing nodules in its roots.

Insects—called *microlivestock*—are also important potential sources of protein, vitamins, and minerals in many parts of the world (Figure 12-13). There are about 1,500 edible insect species, including black ant larvae (served in tacos in Mexico) and giant waterbugs (crushed into vegetable dip in Thailand). Two basic problems are **(1)** getting farmers to take the financial risk of cultivating new types of food crops and **(2)** convincing consumers to try new foods.

Some plant scientists believe we should rely more on polycultures of perennial crops, which are better adapted to regional soil and climate conditions than most annual crops (p. 261). Using perennials would also **(1)** eliminate the need to till soil and replant seeds each year, **(2)** greatly reduce energy use, **(3)** save water, and **(4)** reduce soil erosion and water pollution from eroded sediment.

Is Irrigating More Land the Answer?
Currently, about 40% of the world's food production comes from irrigated land. Irrigated land produces about 70% of the grain harvest in China and 50% in India (compared to 15% in the United States, Figure 12-6).

According to the UN, about 60% of the extra food needed to sustain a world population of about 8 billion by 2028 must come from irrigated agriculture. Between 1950 and 1978, the world's area of irrigated cropland increased almost threefold. However, since 1978 the amount of irrigated land per person has been falling and is projected to fall much more between 2000 and 2050 (Figure 12-14). Reasons for this downward trend include the following:

- World population has grown faster than irrigated agriculture since 1978.

- Chronic water shortages could affect 2.8 billion people by 2025 (Section 13-3, p. 299).

- Water is being pumped too rapidly from aquifers in many of the world's food-growing areas (Section 13-6, p. 304)

Figure 12-13 Insects are important food items in many parts of the world. *Mopani*—emperor moth caterpillars—are among several insects eaten in South Africa. However, this food is so popular that the caterpillars (known as mopane worms) are being overharvested. Kalahari Desert dwellers eat cockroaches, lightly toasted butterflies are a favorite food in Bali, and French-fried ants are sold on the streets of Bogota, Colombia. Most of these insects are 58–78% protein by weight—three to four times as protein-rich as beef, fish, or eggs. (Anthony Bannister/Natural History Photographic Agency)

- Irrigation water is used inefficiently (p. 311).

- Crop productivity is decreased by soil salinization (Figure 10-22, p. 229) on irrigated cropland.

- Increasing urbanization puts city dwellers and farmers in competition for limited water supplies.

- Water supplies in many food-growing areas may be disrupted by global warming (p. 458).

The majority of the world's farmers don't have enough money to irrigate their crops. Thus, rainfall is the source of water for 83% of the world's cropland.

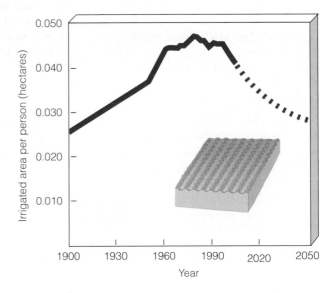

Figure 12-14 World irrigated area of cropland per person, 1900–2000, with projections to 2050. (Data from United Nations Food and Agriculture Organization, U.S. Census Bureau, and the Worldwatch Institute)

Shrinking the World's Genetic Plant Library

The UN Food and Agriculture Organization estimates that between 1900 and 2000 two-thirds of all seeds planted in developing countries were of uniform strains. Such genetic uniformity increases the vulnerability of food crops to pests, diseases, and harsh weather. Many biologists argue that this decreased variability, plus growing extinction rates of plant species, can limit the genetic raw material available to support future green and gene revolutions.

For example, in the mid-1970s, a valuable wild corn species, the only known perennial strain of corn, was barely saved from extinction. When this strain was discovered in south central Mexico, only a few thousand stalks survived in three tiny patches that were about to be cleared by squatter cultivators and commercial loggers.

Crossbreeding this perennial strain with commercial varieties could reduce the need for yearly plowing and sowing, which would reduce soil erosion and save water and energy. Even more important, this wild corn has a built-in genetic resistance to four of the eight major corn viruses, and it grows better in cooler and damper habitats than established commercial strains. Overall, the economic benefits of cultivating this barely rescued wild plant could total several billion dollars per year.

Wild varieties of the world's most important plants can be collected and stored in (1) gene or seed banks, (2) agricultural research centers, and (3) botanical gardens. However, space and money severely limit the number of species that can be preserved.

Other limitations include (1) inability to successfully store seeds of many plants (such as potatoes), (2) irreversible loss of stored seeds because of power failures, fires, or unintentional disposal, (3) death of stored seeds unless they are periodically planted (germinated) and then stored again, and (4) difficulty in reintroducing stored plants and seeds into changed habitats because they do not evolve during storage.

Critical Thinking

What are the major advantages and disadvantages of relying on a shrinking number of crop varieties? Why do seed companies favor this approach?

Key methods for using water more sustainably in crop production are (1) increasing irrigation efficiency (p. 311 and Solutions, p. 312), (2) shifting to crops that need less water (for example, depending less on water-thirsty rice and sugarcane and more on wheat and sorghum), and (3) withdrawing water from aquifers no faster than they are replenished.

Is Cultivating More Land the Answer?

Theoretically, the world's cropland could be more than doubled by clearing tropical forests and irrigating arid land (Figure 12-15). However, much of this is marginal land, where cultivation is unlikely to be sustainable.

Most of the land cleared in rain forests has nutrient-poor soils (Figure 10-15, p. 223) and cannot support crop growth for more than a couple of years. In addition, potential cropland in savanna and other semiarid land in Africa cannot be used for farming or livestock grazing because of the presence of 22 species of the tsetse fly, which transmits a protozoan parasite that causes incurable sleeping sickness in humans and a fatal disease in livestock.

Some researchers hope to develop new methods of intensive cultivation in tropical areas. However, other scientists argue that it makes more ecological and economic sense to combine ancient methods of shifting cultivation (Figure 2-3, p. 25) with various forms of polyculture.

Much of the world's potentially cultivable land lies in dry areas, especially in Australia and Africa. Large-scale irrigation in these areas would (1) require large, expensive dam projects with a mixture of beneficial and harmful impacts (Figure 13-9, p. 301), (2) use large

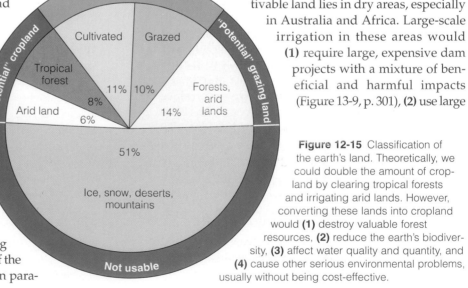

Figure 12-15 Classification of the earth's land. Theoretically, we could double the amount of cropland by clearing tropical forests and irrigating arid lands. However, converting these lands into cropland would (1) destroy valuable forest resources, (2) reduce the earth's biodiversity, (3) affect water quality and quantity, and (4) cause other serious environmental problems, usually without being cost-effective.

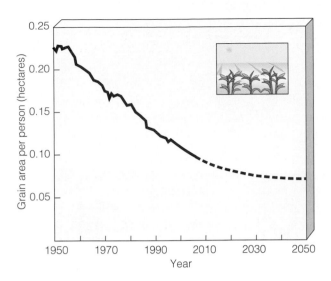

Figure 12-16 Average grain area per person worldwide, 1950–2000, with projections to 2030. (Data from U.S. Department of Agriculture and Worldwatch Institute)

inputs of fossil fuel to pump water long distances, **(3)** deplete groundwater supplies by removing water faster than it is replenished, and **(4)** require expensive efforts to prevent erosion, groundwater contamination, salinization, and waterlogging, all of which reduce crop productivity.

Thus, much of the world's new cropland that could be developed would be on land that is marginal for raising crops and requires expensive inputs of fertilizer, water, and energy. Furthermore, these potential increases in cropland would not offset the projected loss of almost one-third of today's cultivated cropland caused by erosion, overgrazing, waterlogging, salinization, and urbanization.

Such expansion in cropland would also reduce wildlife habitats and thus the world's biodiversity. According to the UN Food and Agriculture Organization (FAO), cultivating all potential cropland in developing countries would reduce forests, woodlands, and permanent pasture by 47%. Clearing these forests would also release a huge amount of carbon dioxide into the atmosphere and accelerate projected global warming, which would cause shifts in the areas where crops could be grown.

Thus, *many analysts believe that significant expansion of cropland is unlikely over the next few decades.* If this assessment is correct, the world's grainland area per person, which dropped by half between 1950 and 1998 (because population growth grew seven times faster than grainland expansion), is expected to decline further (Figure 12-16). A key question is whether crop yields per area of cropland can be increased enough to offset the shrinkage of cropland per person and keep up with projected population growth.

Can We Grow More Food in Urban Areas? Food experts believe that people in urban areas could live more sustainably and save money by growing food in empty lots, on rooftops and balconies, and in their own backyards and by raising fish in tanks and sewage lagoons. Currently, about 800 million urban gardens provide about 15% of the world's food. More food could be grown in urban areas. A study by the UN Center for Human Settlements estimated that up to 50% of the total area in many cities in developing countries is vacant public land that could be used to produce food.

12-5 PRODUCING MORE MEAT

What Are Rangeland and Pasture? About 40% of the earth's ice-free land is **rangeland** that supplies forage or vegetation for grazing (grass-eating) and browsing (shrub-eating) animals. Most rangelands are unfenced, open-range grasslands and woodlands in arid and semiarid areas that are too dry to grow crops without irrigation.

The world's rangelands make up the second land-based human food system and are also used to produce leather and wool. About 3.3 billion cattle and sheep graze on about 42% of the world's rangeland. Much of the rest is too dry, cold, or remote from population centers to support large numbers of livestock. About 29% of the total U.S. land area is rangeland, most of it short-grass prairie in the arid and semiarid western half of the country (Figure 12-6). Livestock also graze in **pastures**: managed grasslands or enclosed meadows usually planted with domesticated grasses or other forage (Figure 12-6).

Livestock can be raised **(1)** on *open ranges* where rainfall is low but fairly regular and **(2)** by *nomadic herding* where rainfall is sparse and irregular and herders must move livestock to find ample grass and prevent overgrazing (Figure 12-17).

What Is the Ecology of Rangeland Plants? Rangeland grasses are hearty species that are adapted to withstand drought, occasional fires, and moderate grazing by livestock and other herbivores. Most rangeland grasses have deep and complex root systems (Figure 10-15, p. 223) that **(1)** help anchor the plants, **(2)** extract underground water so plants can withstand drought, and **(3)** store nutrients so plants can grow again after a drought or fire.

Blades of rangeland grass grow from the base, not the tip. Thus, as long as only its upper half—called its *metabolic reserve*—is eaten and its lower half remains, rangeland grass is a renewable resource that can be grazed again and again (Figure 12-18, top). The exposed metabolic reserve of a grass plant is where

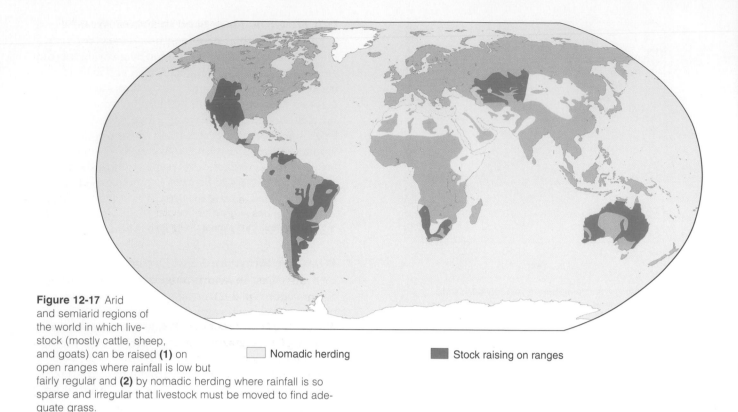

Figure 12-17 Arid and semiarid regions of the world in which livestock (mostly cattle, sheep, and goats) can be raised **(1)** on open ranges where rainfall is low but fairly regular and **(2)** by nomadic herding where rainfall is so sparse and irregular that livestock must be moved to find adequate grass.

☐ Nomadic herding   ■ Stock raising on ranges

photosynthesis takes place to provide food for the deep roots of rangeland grasses. If all or most of the lower half (metabolic reserve) of the plant is eaten, the plant is weakened and can die (Figure 12-18, bottom).

Rangeland plants vary in their ability to recover from stresses such as fire, drought, and excessive grazing (Figure 12-19). When native plant species disappear from a rangeland, weedy invaders move in and gradually decrease the nutritional value of the available forage (Figure 12-19, bottom).

How Is Meat Produced, and What Are Its Environmental Consequences? Meat and meat products are good sources of high-quality protein. Between 1950 and 1998, world meat production increased fourfold and per capita meat production rose by 29%. According to Worldwatch Institute estimates, the world's meat production is likely to more than double between 2000 and 2050 as affluence rises in middle-income developing countries and people begin consuming more meat.

Most of the world's cattle and sheep and other grazing livestock are raised on rangeland by open grazing or nomadic herding (Figure 12-17). However, an increasing amount of livestock production in developed countries is industrialized, with animals living in densely populated factorylike farms and being fattened for slaughter by feeding on grain grown on cropland or meal produced from fish.

In the United States most production of cattle, pigs, and poultry is concentrated in increasingly large, factorylike production facilities in only a few states (Figure 12-20). This arrangement concentrates pollution problems such as **(1)** foul odors, **(2)** water pollution

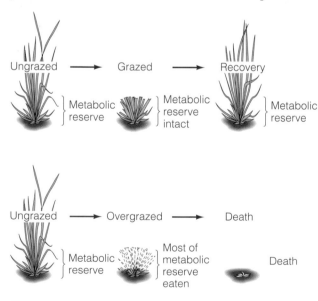

Figure 12-18 Rangeland grasses grow from the bottom up and are renewable as long as the bottom half of the plant (metabolic reserve), where photosynthesis takes place, is not eaten (top). If the metabolic reserve is eaten, the plant is weakened and can die (bottom) (From *Environmental Science*, 5/E by Chiras, p.208. Copyright (c) 1998 by Wadsworth.)

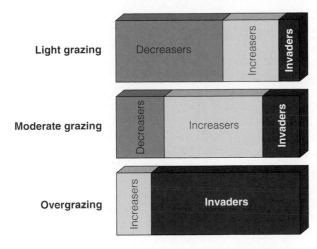

Light grazing — Decreasers / Increasers / Invaders

Moderate grazing — Decreasers / Increasers / Invaders

Overgrazing — Increasers / Invaders

Figure 12-19 Effects of various degrees of grazing on the relative amounts of three major types of grassland plants. *Decreasers* are grass species that decline in abundance with moderate grazing; *increasers* are those that increase with moderate to heavy grazing pressure. *Invaders* are plants that colonize an area because of overgrazing or other changes in rangeland conditions.

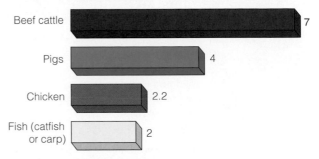

Kilograms of grain needed per kilogram of body weight

Beef cattle — 7
Pigs — 4
Chicken — 2.2
Fish (catfish or carp) — 2

Figure 12-21 Efficiency of converting grain into animal protein. Data in kilograms of grain per kilogram of body weight added. (Data from U.S. Department of Agriculture)

when lagoons storing animal wastes collapse or are flooded (as occurred in many pig farm operations in North Carolina in 1999 after Hurricane Floyd p.486), **(3)** and contamination of drinking water wells by nitrates from animal wastes.

Livestock and fish vary widely in the efficiency with which they convert grain into animal protein (Figure 12-21). A more sustainable form of agriculture would involve shifting to more grain-efficient sources of animal protein, such as poultry. Livestock produc-

tion also has an enormous environmental impact (Connections, p. 282).

What Are the Effects of Overgrazing and Undergrazing? Overgrazing occurs when too many animals graze for too long and exceed the carrying capacity of a grassland area. Most overgrazing is caused by excessive numbers of domestic livestock feeding for too long in a particular area, especially rangelands with a fairly stable climate (Figure 12-17, green areas).

Such overgrazing **(1)** lowers the net primary productivity of grassland vegetation (Figure 4-25, p. 88), **(2)** reduces grass cover and exposes the soil to erosion by water and wind (Figure 12-22, left), **(3)** compacts the soil (which diminishes its capacity to hold water), **(4)** enhances invasion of exposed land by woody shrubs such as mesquite and prickly pear cactus (Figure 12-19, bottom), and **(5)** is a major cause of desertification (Figure 10-21, p. 228).

Some grassland can suffer from **undergrazing**, where absence of grazing for long periods (at least 5 years) can reduce the net primary productivity of grassland vegetation and grass cover. Moderate grazing of such areas removes accumulation of standing dead material and stimulates new biomass production.

Undergrazing is more likely in *arid* areas (such as sub-Saharan Africa and parts of the Mediterranean area) with erratic rainfall and wide, largely unpredictable swings in plant productivity. These are mostly areas where livestock are raised by nomadic herding (Figure 12-17, yellow areas).

In recent years, nomadic herding has been reduced by **(1)** wars, **(2)** travel restrictions, **(3)** population growth, **(4)** increased crop growing and urbanization, and **(5)** the drilling of fairly inexpensive tubewells (narrow wells drilled into an aquifer to provide

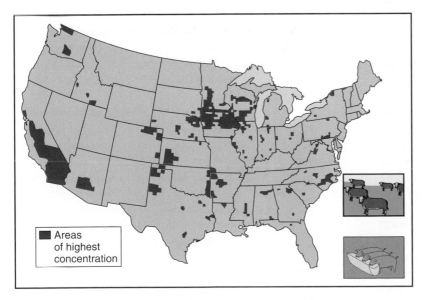

Areas of highest concentration

Figure 12-20 Large-scale, factorylike production of livestock—mostly cattle, pigs, and poultry—in the United States and the resulting pollution problems are concentrated mostly in a few states. (Data from the U.S. Department of Agriculture)

Some Environmental Consequences of Meat Production

The meat-based diet of affluent people in developed and developing countries has the following environmental effects:

- More than half of the world's cropland (19% in the United States) is used to produce livestock feed grain (mostly field corn, sorghum, and soybeans).

- Livestock and fish raised for food consume about 34% of the world's grain production (70% of U.S. production).

- Livestock use more than half the water withdrawn from rivers and aquifers each year, mostly to irrigate crops fed to livestock and to wash away manure from crowded livestock pens and feedlots.

- Manure washing off the land or leaking from lagoons used to store animal wastes is a significant source of water pollution that kills fish by depleting dissolved oxygen (pp. 478-9).

- About 14% of U.S. topsoil loss is directly associated with livestock grazing.

- Overgrazing of sparse vegetation and trampling of the soil by too many livestock is the major cause of desertification in arid and semiarid areas (Figure 10-21, p. 228).

- Cattle belch out 12–15% of all the methane (a greenhouse gas) released into the atmosphere (Figure 18-4, p. 450).

- Some of the nitrogen in commercial inorganic fertilizer used to grow livestock feed is converted to nitrous oxide, another greenhouse gas (Figure 18-4, p. 450).

- Livestock in the United States produce 20 times more waste (manure) than is produced by the country's human population. Each year this manure could fill an imaginary train of 6.7 million boxcars circling the earth almost 20 times. Only about half of this nutrient-rich livestock waste is recycled to the soil.

Some environmentalists have called for reducing livestock production (especially cattle) to reduce its environmental effects and to feed more people. This would decrease the environmental impact of livestock production, but it would not free up much land or grain to feed more of the world's hungry people.

Cattle and sheep that graze on rangeland use a resource (grass) that humans cannot eat, and most of this land is not suitable for growing crops. Moreover, because of poverty, insufficient economic aid, and the nature of global economic and food distribution systems, very little if any additional grain grown on land used to raise livestock or livestock feed would reach the world's hungry people.

Critical Thinking

Would you be willing to eat less meat or not eat any meat? Explain.

---

water for livestock). However, the resulting reduced movement of livestock can lead to loss of grass cover through overgrazing in areas around the wells and through undergrazing in other areas.

What Is the Condition of the World's Rangelands? *Range condition* usually is classified as **(1)** *excellent* (containing more than 75% of its potential forage production), **(2)** *good* (51–75%), **(3)** *fair* (26–50%), and **(4)** *poor* (0–25%). Limited data from surveys in various countries indicate that overgrazing by livestock has caused as much as 20% of the world's rangeland to lose productivity, mostly by desertification (Figure 10-21, p. 228).

Most of the rangeland in the United States is in the West (Figure 12-17). About 60% is privately owned and the rest is public land managed by the Bureau of Land Management (BLM) and the U.S. Forest Service. Only about 2% of the 120 million cattle and 10% of the 20 million sheep raised in the United States graze on public rangelands.

In 1990 (the latest data available), the BLM and the General Accounting Office rated 68% of nonarctic U.S.

public rangeland as being in unsatisfactory (fair or poor) condition, compared with 84% in 1936. This is a great improvement, but there is still a long way to go. Conservation biologists and some range experts also point out that surveys of U.S. rangeland condition neglect the damage livestock inflict on vital riparian zones (Spotlight, right).

How Can Rangelands Be Managed Sustainably to Produce More Meat? The primary goal of sustainable rangeland management is to maximize livestock productivity without overgrazing or undergrazing rangeland vegetation. *Rangeland management* methods include **(1)** controlling the number, types, and distribution of livestock grazing on land, **(2)** deferred grazing, and **(3)** rangeland restoration and improvement.

The most widely used method for sustainable rangeland management is controlling the number of grazing animals and the duration of their grazing in a given area so that the carrying capacity of the area is not exceeded. However, determining the carrying capacity of a range site is difficult and costly, and carrying

According to some wildlife and rangeland experts, estimates of rangeland condition do not take into account severe damage to heavily grazed thin strips of lush vegetation along streams called **riparian zones**. Because cattle need lots of water, they congregate near riparian zones and feed there until the riparian vegetation is destroyed by trampling and overgrazing (Figure 12-23, left).

These ecologically important zones **(1)** help prevent floods, **(2)** help keep streams from drying out during droughts by storing and releasing water slowly from spring runoff and summer storms, and **(3)** provide habitats, food, water, and shade for wildlife in the arid and semiarid Western lands. Studies indicate that 65–75% of the wildlife in the Western United States is totally dependent on riparian habitats.

Riparian areas can be restored by using fencing to restrict access to degraded areas and by developing off-stream watering sites for livestock (Figure 12-23, right). Sometimes protected areas can recover in a few years.

Critical Thinking

Do you believe that riparian zones on public rangelands in the United States should receive stronger protection? Explain. If so, how would you see that such protection is provided?

---

capacity varies with factors such as **(1)** climatic conditions (especially drought), **(2)** past grazing use, **(3)** soil type, **(4)** invasions by new species, **(5)** kinds of grazing animals, and **(6)** intensity of grazing.

Another way to make maximum use of rangeland vegetation is to graze small herds of animals of different species so that all vegetation is used and none is overgrazed. For example, cattle and sheep graze on grasses and herbaceous plants, goats browse on low woody shrubs, and camels feed on large woody plants and tree leaves.

Livestock tend to aggregate around natural water sources and stock ponds. As a result, areas around water sources tend to be overgrazed and other areas can be undergrazed. Managers can prevent this and help promote more uniform use of rangeland by **(1)** fencing off damaged rangeland and riparian zones (Figure 12-23), **(2)** moving livestock from one grazing area to another, **(3)** providing supplemental feed at selected sites, and **(4)** situating water holes and tanks and salt blocks in strategic places.

One method used to help sustain rangeland grasses is *deferred grazing* (Figure 12-24). With this approach, livestock are rotated on a regular schedule from one fenced area to another so that over a 6-year period each area is protected from grazing for 2 years to allow recovery. However, protecting areas too long (more than 5 years) can lead to undergrazing.

A more expensive and less widely used method of rangeland management involves suppressing the growth of unwanted plants (mostly invaders, Figure 12-19, bottom) by herbicide spraying, mechanical removal, or controlled burning. A cheaper way to discourage unwanted vegetation is controlled, short-term trampling by large numbers of livestock.

Replanting barren areas with native grass seeds and applying fertilizer can increase growth of desirable vegetation and reduce soil erosion. Reseeding is an excellent way to restore severely degraded rangeland, but it is expensive.

Figure 12-22 Rangeland: overgrazed (left) and lightly grazed (right). (USDA, Natural Resources Conservation Service)

Figure 12-23 **Figure 12-23** Cattle on a riparian zone of a public rangeland along Arizona's San Pedro River (left) in the mid-1980s just before this section of waterway was protected by banning domestic livestock grazing for 15 years, eliminating sand and gravel operations and water pumping rights in nearby areas, and limiting access by off-highway vehicles. The photo on the right shows the recovery of this riparian area at the same time of year after 10 years of protection. (Bureau of Land Management)

Analysts expect most future increases in meat production to come from feedlots rather than rangelands. This will increase pressure on the world's grain supply because feedlot livestock consume grain produced on cropland instead of feeding on natural grasses (Figure 12-21).

Pasture A					
First year	Second year	Third year	Fourth year	Fifth year	Sixth year
Deferred	Grazed last	Grazed second	Grazed first	Grazed first	Grazed second

Pasture B					
First year	Second year	Third year	Fourth year	Fifth year	Sixth year
Grazed first	Grazed second	Deferred	Grazed last	Grazed second	Grazed first

Pasture C					
First year	Second year	Third year	Fourth year	Fifth year	Sixth year
Grazed second	Grazed first	Grazed first	Grazed second	Deferred	Grazed last

Figure 12-24 Deferred grazing plan in which during a 6-year rotation cycle each fenced in field gets nearly a 2-year rest from grazing. (From *Environmental Science*, 5/E by Chiras, 208. Copyright (c) 1998 by Wadsworth.)

12-6 CATCHING AND RAISING MORE FISH

How Are Fish and Shellfish Harvested? The world's third major food-producing system consists of **fisheries**: concentrations of particular aquatic species suitable for commercial harvesting in a given ocean area or inland body of water. Some commercially important marine species of fish and shellfish are shown in Figure 12-25. Most commercial fishing involves catching six major groups of fish and invertebrates:

- *Herring*, *sardines*, and *anchovies* (Figure 12-25). These small, pelagic fish feed on plankton near the surface in coastal waters and form large schools containing up to several billion individuals.

- *Salmon* species that live in northern temperate waters of the Atlantic and Pacific Oceans and are valued for their size, taste, and texture. Although these pelagic species do not form schools, they aggregate in large numbers and are easily captured when they migrate back to their home freshwater streams to breed.

- The *cod family*, consisting of about 55 species including pollack, haddock, and hake (Figure 12-25). These demersal fish form large schools and feed on the bottom, cruising mostly for crabs, mollusks, and small fish.

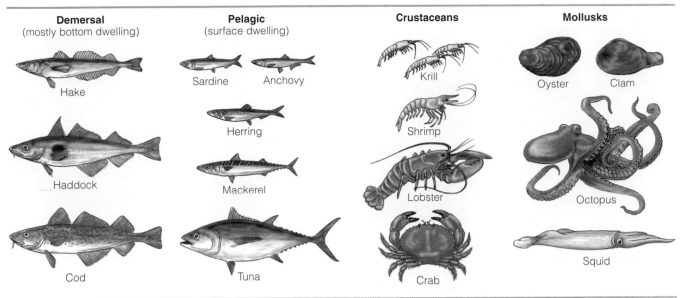

Fish		Shellfish	
Demersal (mostly bottom dwelling)	**Pelagic** (surface dwelling)	**Crustaceans**	**Mollusks**

Hake

Sardine   Anchovy

Krill

Oyster   Clam

Haddock

Herring

Shrimp

Mackerel

Lobster

Octopus

Cod

Tuna

Crab

Squid

Figure 12-25 Some major types of commercially harvested marine fish and shellfish.

- *Mackerels* and *tunas* (Figure 12-25), consisting of about 50 species that bring high prices. These species are top carnivores that hunt in high-speed schools for smaller fish and sometimes squid in the surface waters of the open sea.

- *Flatfish* species such as plaice, flounder, sole, and halibut. These demersal species spend most of their life on the ocean bottom in fairly shallow water on the continental shelf.

- *Invertebrate mollusks* and *crustaceans* such as shrimp, craps, and lobsters (Figure 12-25) that command the highest market prices.

Six countries—Japan, the former Soviet Union, Peru, Chile, China, and the United States—account for more than 50% of the annual global marine catch. Today the commercial fishing industry is dominated by industrial fishing fleets using satellite positioning equipment, sonar, massive nets, spotter planes, and factory ships that can process and freeze their catches. Various methods for harvesting fish are shown in Figure 12-26. Some fishing boats, called *trawlers*, catch demersal fish and shellfish (especially shrimp) by dragging a funnel-shaped net held open at the neck along the ocean bottom, scraping up everything that lies on it and often destroying bottom habitats. Newer trawling nets are large enough to swallow 12 jumbo jets in a single gulp, and even larger ones are on the way. The large mesh of the net allows most small fish to escape but can capture and kill other species such as seals and endangered and threatened sea turtles. Only the large fish are

kept, with most of the fish and other aquatic species (called *bycatch*) thrown back into the ocean either dead or dying. The bycatch from trawling can be enormous. For commercially valuable shrimp, the bycatch can be up to eight times the weight of the shrimp that are kept.

Pelagic species such as tuna, which feed in schools near the surface or in shallow areas, often are caught by *purse-seine fishing* (Figure 12-26). Once located, a tuna school is surrounded by a large purse-seine net, which is closed like a drawstring purse to trap the fish. Nets used to capture yellowfin tuna in the eastern tropical Pacific Ocean have killed large numbers of dolphins, which swim on the surface above schools of the tuna.

Another widely used method for catching open-ocean fish species such as swordfish, tuna, and sharks (Case Study, p. 185) is *longlining*, in which fishing vessels put out lines up to 130 kilometers (80 miles) long, hung with thousands of baited hooks (Figure 12-26). Longlines also hook pilot whales, dolphins, endangered sea turtles, and sea-feeding albatross.

A serious threat to marine biodiversity (Section 24-2, p. 632) is *drift-net fishing* (Figure 12-26). Each net hangs as much as 15 meters (50 feet) below the surface and is up to 55 kilometers (34 miles) long. Almost anything that comes in contact with these nearly invisible "curtains of death" becomes entangled. They can lead to overfishing of the desired species and trap and kill large quantities of unwanted fish and marine mammals (such as dolphins, porpoises, and seals), marine turtles, and seabirds.

In 1990, the UN General Assembly declared a moratorium on the use of drift nets longer than 2.5 kilometers (1.6 miles) in international waters after December 31,

Figure 12-26 Major commercial fishing methods used to harvest various marine species.

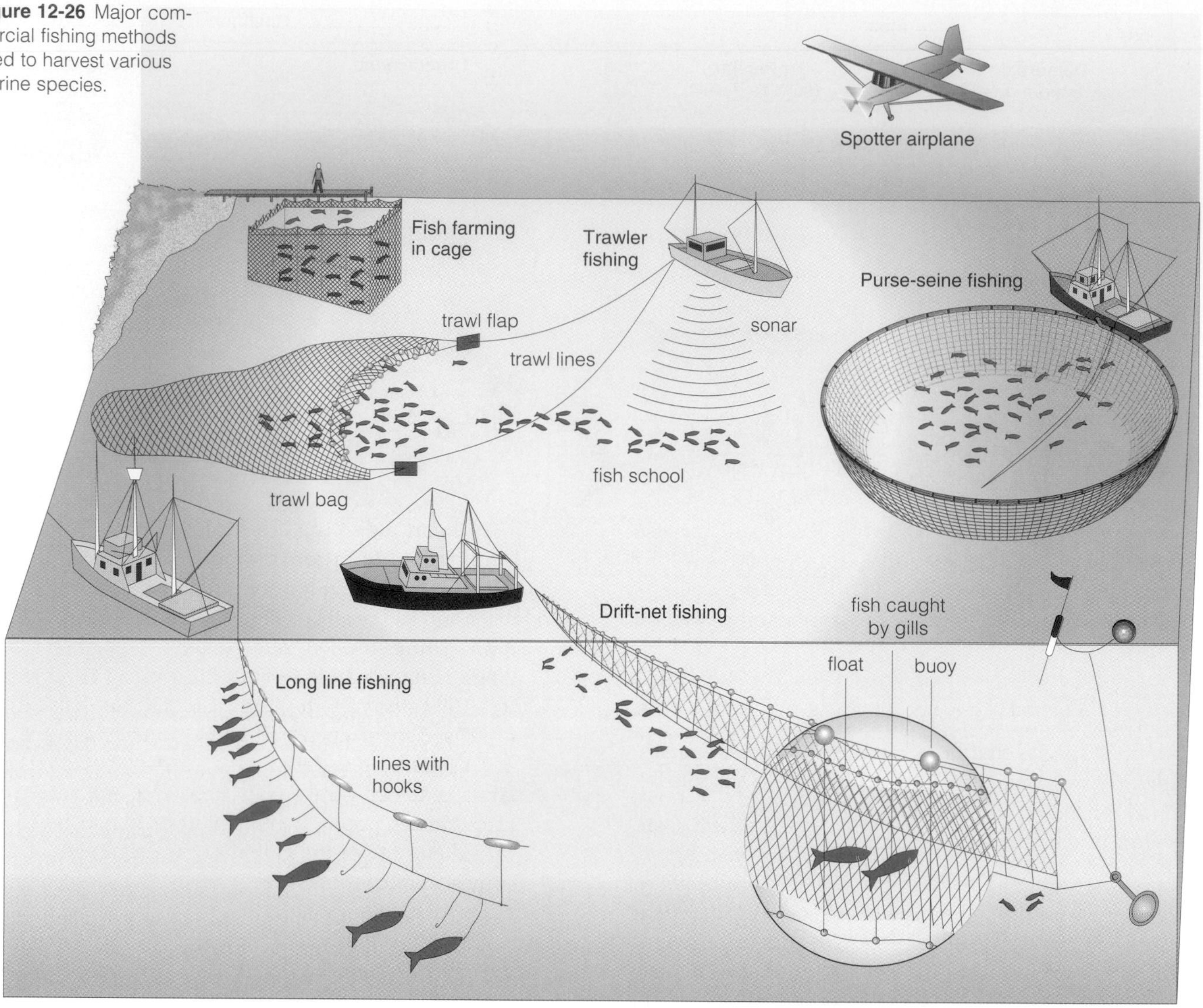

Spotter airplane

Fish farming in cage

Trawler fishing

sonar

Purse-seine fishing

trawl flap

trawl lines

trawl bag

fish school

Drift-net fishing

fish caught by gills

float

buoy

Long line fishing

lines with hooks

1992, to help reduce overfishing and reduction of marine biodiversity. This has sharply reduced this harvesting technique, but **(1)** compliance is voluntary, **(2)** it is difficult to monitor fishing fleets over vast ocean areas, and **(3)** the decrease has led to increased use of longlines, which often have similar effects on marine wildlife.

About 65% of the annual commercial catch of fish and shellfish comes from the ocean. About 99% of this catch is taken from plankton-rich coastal waters, but this vital coastal zone is being disrupted and polluted (p. 158, and Section 19-4, p. 488).

The remainder of the annual catch comes from using **(1)** aquaculture to raise fish in ponds and underwater cages (25%) and **(2)** from inland freshwater fishing from lakes, rivers, reservoirs, and ponds (10%). About one-third of the world fish harvest is used as animal feed, fish meal, and oils.

Can We Harvest More Fish and Shellfish?
Between 1950 and 1998, **(1)** the annual commercial fish

catch (marine plus freshwater harvest) increased 4.8-fold (Figure 12-27, left) and **(2)** the per capita seafood catch more than doubled (Figure 12-27, right).

However, because the human population has grown faster than the annual catch, the per capita commercial fish catch has been falling since 1989 and may continue to decline because of overfishing, pollution, habitat loss, and population growth.

Connections: How Are Overfishing and Habitat Degradation Affecting Fish Harvests? Fish are potentially renewable resources as long as the annual harvest leaves enough breeding stock to renew the species for the next year. Ideally, an annual **sustainable yield**—the size of the annual catch that could be harvested indefinitely without a decline in the population of a species—should be established for each species to avoid depleting the stock.

However, determining sustainable yields is difficult. Estimating mobile aquatic populations is not easy,

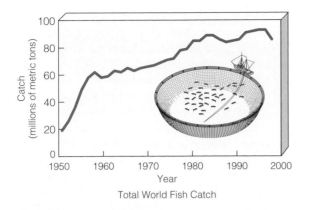

Total World Fish Catch

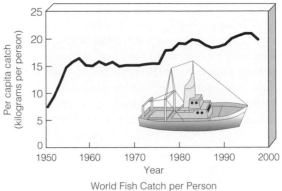

World Fish Catch per Person

Figure 12-27 World fish catch and world fish catch per person, 1950-98. Worldwide per capita fish catch did not rise much between 1968 and 1989 and has dropped since then. (Data from UN Food and Agriculture Organization and Worldwatch Institute)

and sustainable yields shift from year to year because of changes in climate, pollution, and other factors. Furthermore, sustainably harvesting the entire annual surplus of one species may severely reduce the population of other species that rely on it for food.

Overfishing is the taking of so many fish that too little breeding stock is left to maintain numbers; that is, overfishing is a harvest that exceeds a species' sustainable yield. Prolonged overfishing leads to **commercial extinction** when the population of a species declines to the point at which it is no longer profitable to hunt for them. Fishing fleets then move to a new species or a new region, hoping that the overfished species will recover eventually. However, some species may not recover from overfishing. For example, catches of orange roughy in South Pacific waters plummeted by 70% in only 6 years. Because members of this long-lived species need at least 25 years to mature to their breeding age, they have been overfished to the point at which recovery may not be possible.

Fisheries are also depleted by high levels of *bycatch*, the nontarget fish that are caught and then thrown back into the sea, usually dead or dying. Nearly one-fourth of the annual global fish catch is bycatch that depletes marine biodiversity and does not provide food.

According to the UN Food and Agriculture Organization, *11 of the world's 15 most important oceanic fishing areas have either reached or exceeded their sustainable productivity limits.* As a result, about 75% of the world's commercial oceanic fish stocks are in decline (28%) are being fished at their biological limit (47%). Canada has closed down its cod fishery on the Grand Banks and much of salmon fishery in southern British Columbia, putting more than 50,000 people out of work. The United States has closed most of its bottom fishing on the Georges Bank between Cape Cod and Newfoundland and has depleted its salmon fishery region in Washington and Oregon. Because of overfishing, there are more than 100 disputes over rights to marine fisheries between countries.

According to the U.S. National Fish and Wildlife Foundation, 14 major commercial fish species in U.S. waters (accounting for one-fifth of the world's annual catch and half of all U.S. stocks) are so depleted that even if all fishing stopped immediately it would take up to 20 years for stocks to recover (Figure 12-28). According to a 1999 study by the Commerce Department, 98 out of 127 species taken in U.S. marine waters are overfished.

Degradation, destruction, and pollution of wetlands, estuaries, coral reefs, salt marshes, and mangroves also threaten populations of fish and shellfish. An estimated 80–90% of the global commercial marine catch comes from coastal waters within 320 kilometers (200 miles) of the shoreline.

The possibility of global climate change over the next 50–100 years (Section 18-2, p. 450) is also a serious

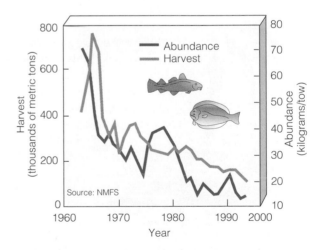

Figure 12-28 The harvest of groundfishes (yellowtail flounder, haddock, and cod) in the Georges Bank off the coast of New England in the North Atlantic, once one of the world's most productive fishing grounds, has declined sharply since 1965. Stocks dropped to such low levels that since December 1994 the National Marine Fisheries Services has banned fishing of these species in the Georges Bank. (Data from U.S. National Marine Fisheries Service)

threat to the global fish catch. Warmer ocean water can degrade or destroy highly productive coral reefs (p. 152) and enhance the harmful effects of habitat degradation and pollution on fish populations. The thinning of the ozone layer (Section 18-6, p. 465) can also damage surface-dwelling marine species by leading to increased penetration of harmful UV radiation into ocean waters. Fish populations also are affected by shorter-term climatic changes such as El Niño-Southern Oscillations (ENSOs, Figure 6-10, p. 127).

Is Aquaculture the Answer? **Aquaculture**, in which fish and shellfish are raised for food, supplies about 25% of the world's commercial fish harvest. Aquacul-ture production increased 3.3-fold between 1984 and 1998 and by 2005 may account for one-third of the world's fish harvest. China is the world leader in aquaculture (producing about 62% of the world's output), followed by India and Japan.

There are two basic types of aquaculture. **Fish farming** involves cultivating fish in a controlled environment, often a pond or tank, and harvesting them when they reach the desired size. **Fish ranching** involves **(1)** holding anadromous species such as salmon (that live part of their lives in fresh water and part in salt water) in captivity for the first few years of their lives (usually in fenced-in areas or floating cages in coastal lagoons and estuaries), **(2)** releasing them,

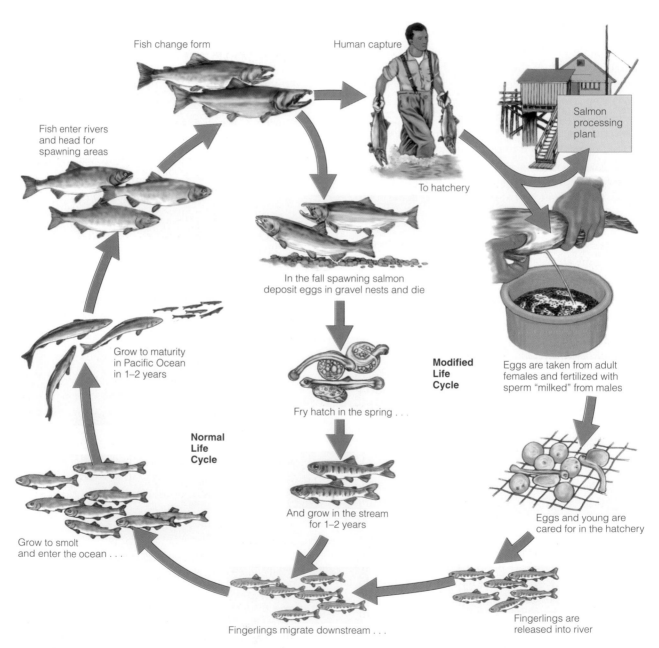

Figure 12-29 Normal life cycle of wild salmon (left) and human-modified life cycle of hatchery-raised salmon (right). Salmon spend part of their lives in fresh water and part in salt water.

and then **(3)** harvesting the adults when they return to spawn (Figure 12-29).

Species cultivated in developing countries (mostly by inland aquaculture) include carp, tilapia, milkfish, clams, and oysters, all of which feed on phytoplankton and other aquatic plants and thus eat low on the food chain. In developed countries and some rapidly developing countries in Asia, aquaculture is used mostly to **(1)** stock lakes and streams with game fish or **(2)** to raise expensive fish and shellfish such as oysters, catfish, crayfish, rainbow trout, shrimp, and salmon. Aquaculture now produces **(1)** 90% of all oysters, **(2)** 40% of all salmon (75% in the United States), **(3)** 30% of the shrimp and prawns (50% in the United States), and **(4)** 65% of freshwater fish sold in the global marketplace. Many of these species are fed grain or fish meal.

Figure 12-30 lists the major advantages and disadvantages of aquaculture. Some analysts project that freshwater and saltwater aquaculture production could double during the next 10 years. Other analysts warn that the harmful environmental effects of aquaculture (Figure 12-30) could limit future production.

Aquaculture has been promoted as a way to boost the global seafood harvest while taking the pressure off of the world's overharvested marine fisheries. However, a 2000 study by Stanford University researcher Rosamond Taylor found that increased use of aquaculture has stressed marine fisheries by **(1)** raising the demand for some ocean fish such as anchovies that are ground into fish meal that is fed to some aquaculture species and **(2)** creating vast amounts of animal waste that have fouled coastal areas, which are important sources of fish and shellfish.

Ways to use fisheries and aquaculture more sustainably and to sustain marine biodiversity are discussed in Section 24-3 (p. 636) and Section 24-4 (p. 647). Nearly 1 billion people worldwide, most of them in Asia, rely on fish as their primary source of protein. However, even under the most optimistic projections, harvesting and raising more fish and shellfish will not increase world food supplies significantly because fish and shellfish supply only about 1% of the energy and 6% of the protein in the human diet.

12-7 AGRICULTURAL POLICY AND FOOD AID

How Do Government Agricultural Policies Affect Food Production? Agriculture is a financially risky business. Whether farmers have a good year or a bad year depends on factors over which they have little control: weather, crop prices, crop pests and diseases, interest rates, and the global market. Because of the need for reliable food supplies despite fluctuations in these factors, most governments provide various forms of assistance to farmers and consumers.

One approach is to *keep food prices artificially low.* This makes consumers happy but means that farmers may not be able to make a living. Many governments in developing countries keep food prices in cities lower than in the countryside to prevent political unrest. However, lower food prices in cities encourages rural people migrate to urban areas. This can aggravate urban problems and unemployment and increase the chances of political unrest.

A second approach is to *give farmers subsidies to keep them in business and encourage them to increase food production.* In developed countries, government price supports and other subsidies for agriculture total more than $300 billion per year (including $100 billion per year in the United States). If government subsidies are too generous and the weather is good, farmers may produce more food than can be sold. The resulting surplus depresses food prices, which reduces the financial incentive for farmers in developing countries to increase domestic food production. Moreover, the taxes paid by citizens in developed countries to provide agricultural subsidies can offset the lower food prices they enjoy.

A third policy is to *eliminate most or all price controls and subsidies.* This allows free market competition to be the primary factor determining food prices and thus the amount of food produced. Some analysts call for gradually phasing out all government price controls and subsidies and

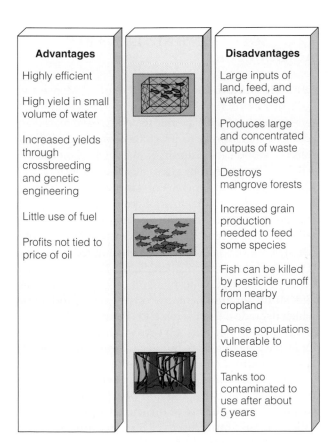

Advantages		Disadvantages
Highly efficient		Large inputs of land, feed, and water needed
High yield in small volume of water		Produces large and concentrated outputs of waste
Increased yields through crossbreeding and genetic engineering		Destroys mangrove forests
Little use of fuel		Increased grain production needed to feed some species
Profits not tied to price of oil		Fish can be killed by pesticide runoff from nearby cropland
		Dense populations vulnerable to disease
		Tanks too contaminated to use after about 5 years

Figure 12-30 Advantages and disadvantages of aquaculture.

Most people view international food aid as a humanitarian effort to prevent people from dying prematurely. However, some analysts contend that giving food to starving people in countries with high population growth rates does more harm than good in the long run. By not helping people grow their own food, they argue, food relief can condemn even greater numbers to premature death from starvation and disease in the future.

Biologist Garrett Hardin (Guest Essay, p. 252) suggests that we use the concept of *lifeboat ethics* to decide which countries get food relief. His basic premise is that there are already too many people in the lifeboat we call earth.

Thus, if food relief is given to countries that are not reducing their populations, the effect is to add more people to an already overcrowded lifeboat. According to Hardin, sooner or later the overloaded boat will sink and most of the passengers will drown.

Large amounts of food relief can also **(1)** depress local food prices and decrease food production, **(2)** stimulate mass migration from farms to already overburdened cities, and **(3)** discourage governments from investing in the rural agricultural development needed to enable farmers to grow enough food for the population.

Another problem is that much food relief does not reach hunger victims because some of the food **(1)** rots or is devoured by pests before it can be distributed because of inadequate storage systems, **(2)** is stolen and sold by officials, and **(3)** is sometimes given to officials as bribes.

Providing food and other forms of aid during famines can also be risky and costly. For example, supplying starving people in Somalia in 1992 required a UN peacekeeping force that probably cost at least 10 times more than the food that was distributed.

Most critics are not against providing aid. However, they believe that such aid should help countries control population growth and grow enough food to feed their populations by using more sustainable agricultural methods (Figure 12-31). Temporary food relief, they believe, should be given only when there is a complete breakdown of an area's food supply because of natural disaster.

Critical Thinking

Is sending food to famine victims helpful or harmful in the **(a)** short run and **(b)** long run? Explain. Would you attach any conditions to providing such aid? Explain.

---

letting farmers and fishers respond to market demand. However, these analysts urge that any phaseout of farm and fishery subsidies should be coupled with increased aid for the poor and the lower middle class, who would suffer the most from any increase in food prices.

Many environmentalists believe that instead of eliminating all subsidies, we should use them to reward farmers and ranchers who protect the soil, conserve water, reforest degraded land, protect and restore wetlands, conserve wildlife, and practice more sustainable agriculture and fishing. What do you think?

The growing emphasis on global free trade in agricultural products can also cause a loss of land and increased malnutrition for the rural poor in developing countries. Current trade agreements, such as the Agreement on Agriculture under the World Trade Organization (WTO, p. 734), allow industrial farmers in Europe and North America to sell surpluses of government-subsidized grain and other crops cheaply in developing countries. This undercuts local farmers, forces many of them off the land, and can increase poverty and malnutrition.

Another way in which governments and private organizations deal with a lack of food production and hunger is through food aid (Spotlight, above).

12-8 SOLUTIONS: SUSTAINABLE AGRICULTURE

What Is Sustainable Agriculture? Evidence suggests that two of the world's major food production systems, rangelands and oceanic fisheries, have reached or are near their limits to produce more food. If these trends continue, most future growth in food production will have to come from increasing the amount of land used to grow crops, increasing crop yields, or both.

However, the total area of cropland is unlikely to expand because of a lack of affordable and environmentally sustainable land. In addition, increasing the yields per area of existing cropland may be limited because of **(1)** a lack of water for irrigation (Figure 12-14), **(2)** reduced genetic diversity, **(3)** a leveling off of yields per hectare, and **(4)** the environmental effects of food production, which degrade existing cropland (Figure 12-10).

If these projections are correct, then the two main tools in trying to reduce hunger, malnutrition, poverty, and the harmful environmental effects of agriculture will be **(1)** slowing population growth (Section 11-3, p. 249) and **(2)** developing and phasing in systems of **sustainable agriculture** or **low-input agriculture** (also called **organic farming**) over the next three decades,

Gordon Conway, president of the Rockefeller Foundation, calls this a new *doubly green revolution* that will (1) increase crop yields in an environmentally sustainable manner and (2) benefit the poor more directly than the first two green revolutions.

Instead of using synthetic pesticides and artificial fertilizers, low-input or organic farming relies on ecological interactions to increase yields, reduce pest losses, and build soil fertility. Under U.S. Department of Agriculture regulations, foods labeled *organic* (1) must consist of at least 95% organically produced ingredients, (2) cannot include genetically engineered foods, (3) cannot be irradiated (p. 514), and (4) cannot be grown on soil fertilized by sewage sludge (p. 495). Organic meat and poultry must be fed 100% organically grown feed and be given access to outdoor range or pasture. Use of antibiotics is prohibited.

Figure 12-31 lists the major components of more sustainable, low-input agriculture. Studies have shown that low-input organic farming (1) produces equivalent yields with lower carbon dioxide emissions, (2) uses about 50% less energy than conventional farming, (3) improves soil fertility, and (4) generally is more profitable for the farmer than high-input farming.

Most proponents of more sustainable agriculture are not opposed to high-yield agriculture. Instead, they see it as vital for protecting the earth's biodiversity by reducing the need to cultivate new and often marginal land. They call for using environmentally sustainable forms of both high-yield polyculture and high-yield monoculture for growing crops. They also believe that current agricultural research and economic incentives should be redirected to encourage increases in yield per hectare without depleting or degrading soil, water, and biodiversity.

Can We Make the Transition to More Sustainable Agriculture? A growing number agricultural analysts believe that over the next 30 years we must make a transition from unsustainable and environmentally harmful agriculture (Figure 12-10) to more sustainable forms of agriculture (Figure 12-31).

In developed countries, even a partial shift to more environmentally sustainable food production will not be easy. It will be opposed by (1) agribusiness, (2) successful farmers with large investments in unsustainable forms of industrialized agriculture, (3) specialized farmers unwilling to learn the more demanding art of farming sustainably, and (4) many consumers unwilling or unable to pay higher prices for food when more of agriculture's harmful environmental and health costs are included in the market prices of food.

To help farmers make the transition to more sustainable agriculture, analysts call for:

- Greatly increasing research on sustainable agriculture and improving human nutrition. Despite grow-

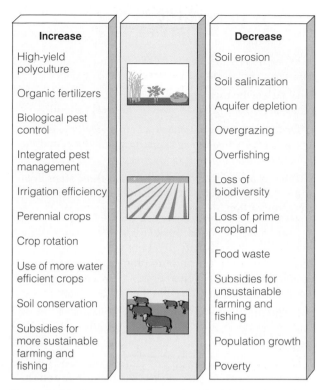

Figure 12-31 Components of more sustainable, low-throughput agriculture.

ing demand for organic foods, less than 0.2% of the U.S. Department of Agriculture's annual research budget is devoted to organic cultivation methods.

- Setting up demonstration projects throughout each country so that farmers can see how sustainable agricultural systems work

- Increasing agricultural aid to developing countries (which, when adjusted for inflation, declined by about 50% between 1982 and 1999), with emphasis on developing more sustainable, low-input agriculture

- Establishing training programs in sustainable agriculture for farmers and government agricultural officials and encouraging the creation of college curricula in sustainable agriculture and human nutrition

Sustainable agriculture involves applying the four principles of sustainability (Solutions, p. 209) to producing food. The goal is to feed the world's people while sustaining and restoring the earth's natural capital. Some actions you can take to help promote more sustainable agriculture are listed in Appendix 6.

The need to bring birth rates well below death rates, increase food production while protecting the environment, and distribute food to all who need it is the greatest challenge our species has ever faced.

PAUL AND ANNE EHRLICH

REVIEW QUESTIONS

1. Define the boldfaced terms in this chapter.

2. What are *perennial crops*? What advantages do they have over conventional annual crops?

3. What are the three main systems that supply our food? What three crops provide most of the world's food?

4. Distinguish between *industrialized agriculture, plantation agriculture, traditional subsistence agriculture,* and *traditional intensive agriculture.*

5. What is a *green revolution*, and what three steps does it involve? Distinguish between the first and second green revolutions.

6. Explain how producing more food on less land can help protect biodiversity.

7. Describe the nature and importance of the agricultural industry in the United States. How energy-efficient is industrialized agriculture in the United States?

8. Distinguish between *interplanting, polyvarietal cultivation, intercropping, agroforestry,* and *polyculture.* What are the major advantages and disadvantages of polyculture?

9. List four pieces of good news about world food production since 1950? List four pieces of disturbing and challenging news about present and future food production.

10. Distinguish between *undernutrition, malnutrition,* and *overnutrition.* What are the two most common nutritional deficiency diseases, and how do they differ? What are the effects of deficiencies of (a) vitamin A, (b) iron, and (c) iodine?

11. About how many chronically malnourished people are there in the world? About how many people die each year from undernutrition, malnutrition, or diseases worsened by malnutrition? List six major ways to reduce sickness and premature death of children from malnutrition. What is the primary cause of hunger in the world?

12. List the major harmful environmental effects of producing food.

13. What are China's major food problems? How could lower food production in China affect the rest of the world?

14. Describe how genetically improved crop strains are developed by (a) *crossbreeding* and (b) *genetic engineering.*

15. List the pros and cons of growing more food by (a) increasing crop yields through green revolutions, (b) genetically modifying crops and foods, (c) getting people to switch to new types of foods, (d) irrigating more cropland, (e) cultivating more land, and (f) growing more food in urban areas. List seven factors that could limit greatly increased food production through green-revolution and genetic engineering techniques. Why should we be concerned about loss of genetic diversity in agricultural crop strains? List three ways to use water more sustainably in crop production.

16. What is *rangeland*? Explain how rangeland grass can be a renewable resource for livestock and wild herbivores. How is meat produced in developed and developing countries? What are the advantages and

disadvantages of raising livestock in factorylike production facilities?

17. Distinguish between *overgrazing* and *undergrazing.* What are three major effects of overgrazing? Describe the general condition of rangelands throughout the world and in the United States. Describe three ways to manage rangelands more sustainably for meat production. What are *riparian areas*, and why are they important?

18. What are some of the harmful environmental effects of meat production? What are some of the advantages of a meat-based diet?

19. What are *fisheries*? Distinguish between *trawling, purse-seine, longlining,* and *drift-net* methods for harvesting fish. Describe trends in the world's total fish catch and per capita fish catch since 1950 and explain why the per capita catch is expected to decline.

20. What is the *sustainable yield* of a fishery? Distinguish between *overfishing* and *commercial extinction* of a fish species. About what percentage of the world's fish stocks are in decline because of overfishing and pollution?

21. Distinguish between *fish farming* and *fish ranching.* What are the pros and cons of aquaculture?

22. List three types of policy governments use to affect food production. List the pros and cons of each approach.

23. List the pros and cons of international food aid.

24. What is *sustainable agriculture*, why is it important, and what are its major components?

CRITICAL THINKING

1. Summarize the major economic and ecological advantages and limitations of each of the following proposals for increasing world food supplies and reducing hunger over the next 30 years: (a) cultivating more land by clearing tropical forests and irrigating arid lands, (b) catching more fish in the open sea, (c) producing more fish and shellfish with aquaculture, and (d) increasing the yield per area of cropland.

2. Why should it matter to people in developed countries that many people in developing countries are malnourished and hungry? What are the three most important things that you believe should be done to reduce hunger (a) in the country where you live and (b) in the world?

3. Some people argue that starving people could get enough food by eating nonconventional plants and insects; others point out that most starving people do not know what plants and insects are safe to eat and cannot take a chance on experimenting when even the slightest illness could kill them. If you had no money to grow or buy food, would you collect and eat protein-rich grasshoppers, moths, beetles, or other insects?

4. What could happen to energy-intensive agriculture in the United States and other industrialized countries if world oil prices rose sharply?

5. Some have suggested that some rangelands could be used to raise wild grazing animals for meat instead of

conventional livestock. Others consider it unethical to raise and kill wild herbivores for food. What do you think? Explain.

6. Should all price supports and other government subsidies paid to farmers be eliminated? Explain. Try to consult one or more farmers in answering this question.

7. What are the economic and ecological advantages and disadvantages of relying on a small number of genetic varieties of major crops and livestock?

8. What are the major economic and ecological advantages and disadvantages of genetically modified food? Are you for or against widespread use of such food? Explain.

9. What are the major economic and ecological advantages and disadvantages of relying more on perennial food crops (p. 261)? Why will conventional seed companies be opposed to relying more on perennial crop varieties?

10. Suppose you live near a coastal area and a company wants to use a fairly large area of coastal marshland for an aquaculture operation. If you were an elected local official, would you support or oppose such a project? Explain. What safeguards or regulations would you impose on the operation?

11. Should governments phase in agricultural tax breaks and subsidies to encourage farmers to switch to more sustainable farming? Explain.

12. How could knowledge about secondary ecological succession (Figure 8-16, p. 189) be used to make agriculture more sustainable?

PROJECTS

1. If possible, visit both a conventional industrialized farm and an organic or low-input farm. Compare **(a)** soil erosion and other forms of land degradation, **(b)** use and costs of energy, **(c)** use and costs of pesticides and inorganic fertilizer, **(d)** use and costs of natural pest control and organic fertilizer, **(e)** yields per hectare for the same crops, and **(f)** overall profit per hectare for the same crops.

2. Evaluate cattle grazing on private and public rangeland and pastures in your local area. Try to document any harmful environmental impacts. Have the economic benefits to the community outweighed any harmful environmental effects?

3. Gather information from your local planning office to determine how much cropland in your area has been lost to urbanization since 1980. What policies, if any, do your state and local community have for promoting cropland preservation?

4. Try to gather data evaluating the harmful environmental effects of nearby agriculture on your local community. What things are being done to reduce these effects?

5. Use the library or the internet to find out what controls now exist on genetically engineered organisms in the United States (or the country where you live) and how well such controls are enforced.

6. Make a survey in the nearest urban area to estimate what percentage of the food is grown by urban dwellers.

Survey unused land and use it to estimate how much it could contribute to urban food production. Use these data to draw up a plan for increasing urban food production and present it to city officials.

7. Use the library or the internet to find bibliographic information about *Aldo Leopold* and *Paul* and *Anne Ehrlich*, whose quotes appear at the beginning and end of this chapter.

8. Make a concept map of this chapter's major ideas, using the section heads and subheads and the key terms (in boldface). Look at the inside back cover and on the website for this book for information about making concept maps.

INTERNET STUDY RESOURCES AND RESOURCES FOR FURTHER READING AND RESEARCH

The website for this book contains helpful study aids and many ideas for further reading and research. Log on to:

http://www.brookscole.com/product/0534376975s

and click on the Chapter-by-Chapter area. Choose Chapter 12 and select a resource:

■ "Flash Cards" allows you to test your mastery of the Terms and Concepts to Remember for this chapter.

■ "Tutorial Quizzes" provides a multiple-choice practice quiz.

■ "Student Guide to InfoTrac" will lead you to Critical Thinking Projects that use InfoTrac College Edition as a research tool.

■ "References" lists the major books and articles consulted in writing this chapter.

■ "Hypercontents" takes you to an extensive list of sites with news, research, and images related to individual sections of the chapter.

INFOTRAC COLLEGE EDITION

Improve your skills with InfoTrac College Edition, a searchable online database of articles from more than 700 periodicals. Log on to:

http://www.infotrac-college.com

or access InfoTrac through the website for this book.

Try the following articles:

Shepherd, V. 2000. Down on this farm the times they are a-changin' (organic farming). *Smithsonian*, July 2000, p. 64+. (subject guide: organic farming)

Pauly, D., V. Christensen, R. Froese, and M.L. Palomares. 2000. Fishing down aquatic food webs. *American Scientist* vol. 88, no. 1, pp. 46–50. (subject guide: aquatic food webs)

13 WATER RESOURCES

Water Wars in the Middle East

If there is another war between countries in the Middle East, it could be fought over water, not oil. Most water in this dry region comes from three shared river basins: the Nile, Jordan, and Tigris-Euphrates (Figure 13-1). Water in much of this arid region is already in short supply.

Ethiopia, which controls the headwaters that feed 86% of the Nile's flow, plans to divert more of this water; so does Sudan. This could reduce the amount of water available to water-short Egypt, whose terrain is desert except for a green area of irrigated cropland running down its middle along the Nile and its delta (see part opening photo on p. 237). Between 2000 and 2025, Egypt's population is expected to increase from 67 million to 97 million, greatly increasing the demand for already scarce water.

Egypt's options are to (1) go to war with Sudan and Ethiopia to obtain more water, (2) cut population growth, (3) improve irrigation efficiency, (4) spend $2 billion to build the world's longest concrete canal and pump water out of Lake Nasser (the reservoir created from the Nile by the Aswan High Dam) and create more irrigated farmland in the middle of the desert, (5) import more grain to reduce the need for irrigation water, (6) work out water-sharing agreements with other countries, or (7) suffer the harsh human and economic consequences.

The Jordan Basin is by far the most water-short region, with fierce competition for its water between Jordan, Syria, Palestine (Gaza and the West Bank), and Israel (Figure 13-1). The com-

bined populations of these already water-short countries are projected to increase from 31 million to 51 million between 2000 and 2025.

Syria plans to build dams and withdraw more water from the Jordan River, decreasing the downstream water supply for Jordan and Israel. Israel warns that it will consider destroying the largest dam that Syria plans to build.

Turkey, located at the headwaters of the Tigris and Euphrates rivers, controls how much water flows downstream to Syria and Iraq before emptying into the Persian Gulf (Figure 13-1). Turkey is building 24 dams along the upper Tigris and Euphrates rivers to (1) generate huge quantities of electricity, (2) irrigate a large area of land, (3) boost the region's income fivefold, and (4) generate about 3.5 million jobs for its 65 million people.

If completed, these dams will reduce the flow of water downstream to Syria and Iraq by up to 35% in normal years and much more in dry years. Syria also plans to build a large dam along the Euphrates River to divert water arriving from Turkey, which will leave little water for Iraq and possibly lead to a war between Syria and Iraq.

Clearly, water distribution will be a key issue in any peace talks in this region. Resolving these problems will require a combination of (1) regional cooperation in allocating water supplies, (2) slowed population growth, (3) improved efficiency in water use, and (4) increased water prices to encourage water conservation and improve irrigation efficiency.

As fresh water becomes scarcer, access to water resources will be a major factor in determining the economic, environmental, and military security of a growing number of countries. According to Lester Brown (Guest Essay, p. 18) and Christopher Flavin of the Worldwatch Institute, "The spreading scarcity of fresh water may be the most underestimated resource issue facing the world as it enters the new millennium."

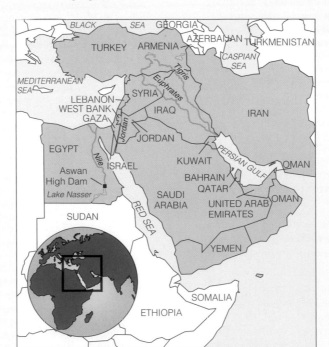

Figure 13-1 The Middle East, whose countries have some of the highest population growth rates in the world. Because of the dry climate, food production depends heavily on irrigation. Existing conflicts between countries in this region over access to water may soon overshadow both long-standing religious and ethnic clashes and attempts to take over valuable oil supplies.

Our liquid planet glows like a soft blue sapphire in the hard-edged darkness of space. There is nothing else like it in the solar system. It is because of water.

JOHN TODD

This chapter addresses the following questions:

- What are water's unique physical properties?

- How much fresh water is available to us, and how much of it are we using?

- What causes freshwater shortages, and what can be done about this problem?

- What are the pros and cons of using dams and reservoirs to supply more water?

- What are the pros and cons of transferring large amounts of water from one place to another?

- What are the pros and cons of withdrawing groundwater and converting salt water to fresh water?

- How can we waste less water?

- What are the causes of flooding, and what can be done to reduce the risk of flooding and flood damage?

- How can we use the earth's water more sustainably?

13-1 WATER'S IMPORTANCE AND UNIQUE PROPERTIES

Why Is Water So Important? We live on the water planet, with a precious film of water—most of it salt water—covering about 71% of the earth's surface (Figure 7-4, p. 156). Water is the basis of life and has two characteristics that distinguish it from other resources:

- *No plant or animal can survive without it.* All organisms are made up mostly of water; a tree is about 60% water by weight, and most animals are about 50-65% water.

- *There are no substitutes for most of its uses.*

Each of us needs only about a dozen cupfuls of water per day to survive, but huge amounts of water are needed to supply us with food, shelter, and our other needs and wants. Water also plays a key role in (1) sculpting the earth's surface, (2) moderating climate, and (3) diluting pollutants.

What Are Some Important Properties of Water? Water is a remarkable substance with a unique combination of properties:

- *There are strong forces of attraction (called hydrogen bonds, Appendix 3, Figure 4) between molecules of water.* These attractive forces are the major factor determining water's unique properties.

- *Water exists as a liquid over a wide temperature range because of the strong forces of attraction between water molecules.* Its high boiling point of 100°C (212°F) and low freezing point of 0°C (32°F) mean that water remains a liquid in most climates on the earth.

- *Liquid water changes temperature very slowly because it can store a large amount of heat without a large change in temperature.* This high heat capacity (1) helps protect living organisms from temperature fluctuations, (2) moderates the earth's climate, and (3) makes water an excellent coolant for car engines, power plants, and heat-producing industrial processes.

- *It takes a lot of heat to evaporate liquid water because of the strong forces of attraction between its molecules.* Water absorbs large amounts of heat as it changes into water vapor and releases this heat as the vapor condenses back to liquid water. This is a primary factor in distributing heat throughout the world (Figure 6-7, p. 125) and thus plays an important role in determining the climates of various areas. This property also makes water evaporation an effective cooling process, which is why you feel cooler when perspiration or bathwater evaporates from your skin.

- *Liquid water can dissolve a variety of compounds.* This enables it to (1) carry dissolved nutrients into the tissues of living organisms, (2) flush waste products out of those tissues, (3) serve as an all-purpose cleanser, and (4) help remove and dilute the water-soluble wastes of civilization. Water's superiority as a solvent also means that water-soluble wastes pollute it easily.

- *Water molecules can break down (ionize) into hydrogen ions (H^+) and hydroxide ions (OH^-), which help maintain a balance between acids and bases in cells, as measured by the pH of water solutions* (Figure 3-7, p. 56).

- *Water filters out wavelengths of ultraviolet radiation* (Figure 3-10, p. 58) *that would harm some aquatic organisms.*

- *The strong attractive forces between the molecules of liquid water cause its surface to contract (high surface tension) and to adhere to and coat a solid (high wetting ability).* These cohesive forces pull water molecules at the surface layer together so strongly that it can support small insects. The combination of high surface and wetting ability allow water to rise through a plant from the roots to the leaves (capillary action).

- *Unlike most liquids, water expands when it freezes.* This means that ice has a lower density (mass per unit of volume) than liquid water. Thus ice floats on water. Without this property, lakes and streams in cold climates would freeze solid and eliminate most of their current forms of aquatic life. Because water expands upon freezing, it can also break pipes, crack engine blocks (which is why we use antifreeze), and break up streets and fracture rocks (thus forming soil).

Water is the lifeblood of the biosphere. It connects us to one another, to other forms of life, and to the entire planet. Despite its importance, water is one of our most poorly managed resources. We waste it and pollute it. We also charge too little for making it available, thus encouraging still greater waste and pollution of this renewable resource, for which there is no substitute.

13-2 SUPPLY, RENEWAL, AND USE OF WATER RESOURCES

How Much Fresh Water Is Available? Only a tiny fraction of the planet's abundant water is available to us as fresh water (Figure 13-2). About 97.4 % by volume is found in the oceans and is too salty for drinking, irrigation, or industry (except as a coolant).

Most of the remaining 2.6% that is fresh water is locked up in ice caps or glaciers or is in groundwater too deep or salty to be used.

Thus, only about 0.014% of the earth's total volume of water is easily available to us as soil moisture, usable groundwater, water vapor, and lakes and streams (Figure 13-2). If the world's water supply were only 100 liters (26 gallons), our usable supply of fresh water would be only about 0.003 liter (one-half teaspoon).

Fortunately, the available fresh water amounts to a generous supply. Moreover, this water is continuously collected, purified, recycled, and distributed in the solar-powered *hydrologic cycle* (Figure 4-28, p. 90) as long as we do not **(1)** overload it with slowly degradable and nondegradable wastes or **(2)** withdraw it from underground supplies faster than it is replenished. Unfortunately, we are doing both.

Differences in average annual precipitation divide the world's countries and people into water haves and have-nots. For example, Canada, with only 0.5% of the world's population, has 20% of the world's fresh water, whereas China, with 21% of the world's people, has only 7% of the supply.

As population, irrigation, and industrialization increase, water shortages in already water-short regions will intensify and heighten tensions between and within countries (p. 294).

Global warming is projected to alter the global hydrological cycle (Section 18-4, p. 458). A warmer atmosphere will increase global rates of evaporation, shift precipitation patterns, and disrupt water supplies in unpredictable ways. Some areas will get more precipitation and some less. River flows will change. Monsoons and hurricanes are likely to intensify, and the average sea level will rise from thermal expansion of the oceans and partial melting of ice caps and mountain glaciers. The basic problem is that scientists cannot predict precisely when, where, and to what degree such changes will occur.

What Is Surface Water? The fresh water we use first arrives as the result of precipitation. Precipitation that does not infiltrate the ground or return to the atmosphere by evaporation (including transpiration) is called **surface runoff** that flows into streams, lakes, wetlands, and reservoirs.

About two-thirds of the world's annual runoff is lost by seasonal floods and is not available for human use. The remaining one-third is **reliable runoff** that generally can be counted on as a stable source of water from year to year.

A **watershed**, also called a **drainage basin**, is a region from which water drains into a stream, lake, reservoir, wetland, or other body of water.

What Is Groundwater? Some precipitation infiltrates the ground and percolates downward through voids (pores, fractures, crevices, and other spaces) in soil and rock (Figure 13-3). The water in these voids is called **groundwater**.

The bedrock that supports the earth's soils and rocks cannot be penetrated by water, except where it has been fractured. Groundwater accumulates in the pores of rock and soil above this impermeable barrier. The voids in the layer of soil and rock above this barrier are completely filled with water, making up what is called the **zone of saturation.**

The **water table** is located at the top of the zone of saturation. It falls in dry weather and rises in wet weather. An unsaturated zone, or **zone of aeration**, lies above the water table. In this zone, pores of rock and soil contain air and may be moist but not saturated with water.

Porous, water-saturated layers of sand, gravel, or bedrock through which groundwater flows are called **aquifers** (Figure 13-3). Any area of land through which water passes downward or laterally into an aquifer is called a **recharge area**. Aquifers are replenished naturally by precipitation that percolates downward through soil and rock in what is called **natural recharge**, but some are recharged from the side by *lateral recharge*.

Aquifers are not underground pools or streams of flowing water, except where water flows through

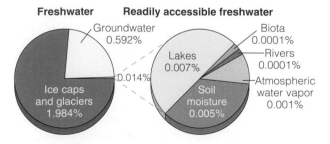

Figure 13-2 The planet's water budget. Only a tiny fraction by volume of the world's water supply is fresh water available for human use.

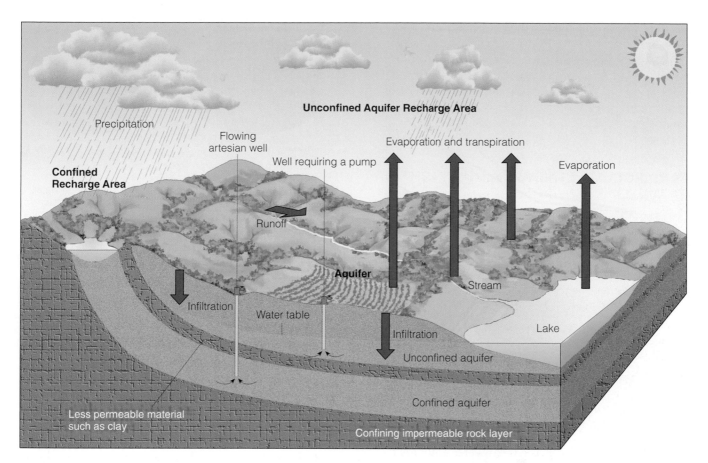

Figure 13-3 The groundwater system. An *unconfined aquifer* is an aquifer with a water table. A *confined aquifer* is bounded above and below by less permeable beds of rock. Groundwater in this type of aquifer is confined under pressure.

large fractures in bedrock and through some caverns. Instead, aquifers are like large, elongated sponges through which groundwater seeps.

Groundwater moves from the *recharge area* through an aquifer and out to a *discharge area* (well, spring, lake, geyser, stream, or ocean) as part of the hydrologic cycle (Figure 4-28, p. 90). Groundwater normally moves from points of high elevation and pressure to points of lower elevation and pressure. This movement is quite slow, typically only a meter or so (about 3 feet) per year and rarely more than 0.3 meter (1 foot) per day.

Some aquifers get very little (if any) recharge and on a human time scale are nonrenewable resources. They are often found fairly deep underground and were formed tens of thousands of years ago. Withdrawals from such aquifers amount to *water mining* that, if kept up, will deplete these ancient deposits of water.

How Much of the World's Reliable Supply of Water Are We Using? About 20% of all the water running to the sea each year is in rivers too remote to supply cities and farming regions. Downpours of rain that cannot be collected make up another 50% of the total global runoff. This leaves about 30% of the total global runoff for human use.

Since 1900, global water use has increased about ninefold and per capita use has quadrupled, with irrigation accounting for the largest increase in water use (Figure 13-4). As a result, humans now withdraw about 35% of the world's reliable runoff. At least another 20% of this runoff is left in streams to transport goods by boats, dilute pollution, and sustain fisheries.

Thus, we are directly or indirectly using more than half of the world's reliable runoff. Because of increased population growth and economic development, global withdrawal rates of surface water are projected to at least double in the next two decades and exceed the reliable surface runoff in a growing number of areas.

Nature's water delivery does not match up with the distribution of much of the world's population. For example, Asia, with 61% of the world's people, has only 36% of the earth's reliable annual runoff. On the other hand, South America, with 26% of the earth's reliable runoff, has only 8% of the world's people.

However, in some areas the largest rivers (which carry most of the runoff) are far from agricultural and population centers where the water is needed. For example, about 60% of South America's runoff flows through the Amazon River in remote areas where few people live because of infertile soil and inhospitable conditions.

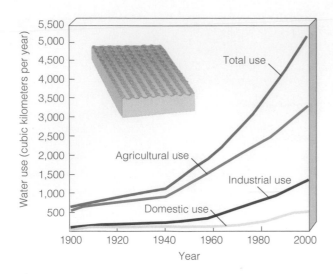

Figure 13-4 Global water use, 1900–2000. Between 2000 and 2054, the world's population is expected to increase by about 3 billion people and greatly increase the demand for water. (Data from World Commission on Water Use in the 21st Century)

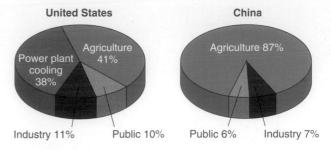

Figure 13-5 Water use in the United States and China. The United States has the world's highest per capita use of water, amounting to an average of 4,800 liters (1,280 gallons) per person per day. Between 1980 and 1999, total water use in the United States decreased by 10% despite a 17% increase in population, mostly because of more efficient irrigation. (Data from Worldwatch Institute and World Resources Institute)

How Do We Use the World's Fresh Water? Uses of withdrawn water vary from one region to another and from one country to another (Figure 13-5). Worldwide, about 70% of all water withdrawn each year from rivers, lakes, and aquifers is used to irrigate 17% of the world's cropland.

Industry uses about 20% of the water withdrawn each year, and cities and residences use the remaining 10%. Industrial water use ranges from about 70% in Germany and several other industrialized European countries to about 5–7% in less industrialized countries such as India, Egypt, and China (Figure 13-5). According to the Worldwatch Institute, on average, a ton of water used in industry generates goods or services worth about $14,000—about 70 times more than the economic values from using the same amount of water to grow grain.

Agriculture and manufacturing use large amounts of water (Figure 13-6). Much of this water could be used more efficiently and some could be reused (Section 13-7, p. 310).

Case Study: Freshwater Resources in the United States Although the United States has plenty of fresh water, much of it is in the wrong place at the wrong time or is contaminated by agricultural and industrial practices. The eastern states usually have ample precipitation, whereas many western states have too little (Figure 13-7, top).

In the East, the largest uses for water are for energy production, cooling, and manufacturing. The largest use by far in the West is for irrigation (which accounts for about 85% of all water use).

In many parts of the eastern United States the most serious water problems are **(1)** flooding, **(2)** occasional urban shortages, and **(3)** pollution. For example, the 3 million residents of Long Island, New York, get most of their water from an aquifer that is becoming severely contaminated.

The major water problem in the arid and semiarid areas of the western half of the country is a shortage of runoff, caused by **(1)** low precipitation (Figure 13-7, top), **(2)** high evaporation, and **(3)** recurring prolonged drought. Water tables in many areas are dropping rapidly as farmers and cities deplete aquifers faster than they are recharged.

Records indicate that the United States experiences a prolonged drought cycle about every 30 years, with severe dry periods in the 1870s, 1900s, 1930s, and 1970s. If this pattern continues, the Unites States could be entering another drought cycle.

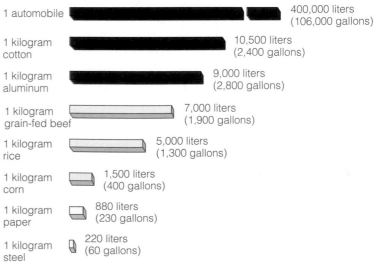

Figure 13-6 Amount of water needed to produce some common agricultural (green) and manufactured (red) products. (Data from U.S. Geological Survey)

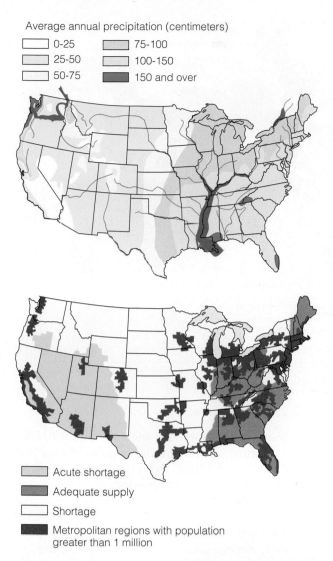

Average annual precipitation (centimeters)

☐ 0-25	▨ 75-100		
▨ 25-50	▨ 100-150		
☐ 50-75	■ 150 and over		

Acute shortage

Adequate supply

Shortage

Metropolitan regions with population greater than 1 million

Figure 13-7 Average annual precipitation and major rivers (top) and water deficit regions in the continental United States and their proximity to metropolitan areas having populations greater than 1 million (bottom). (Data from U.S. Water Resources Council and U.S. Geological Survey)

In the United States, many major urban centers (especially those in the West and Midwest) are located in areas that do not have enough water (Figure 13-7, bottom). Water experts project that conflicts over water supplies within and between states will intensify as more industries and people migrate west and compete with farmers for scarce water. These shortages could worsen in some areas if climate warms as a result of an enhanced greenhouse effect.

13-3   TOO LITTLE WATER

What Causes Freshwater Shortages? According to water expert Malin Falkenmark, there are four causes of water scarcity: **(1)** a *dry climate* (Figure 6-4, p. 124), **(2)** *drought* (a period of 21 days or longer in which precipitation is at least 70% lower and evaporation is higher than normal), **(3)** *desiccation* (drying of the soil because of such activities as deforestation and overgrazing by livestock), and **(4)** *water stress* (low per capita availability of water caused by increasing numbers of people relying on limited levels of runoff).

Figure 13-8 shows the degree of stress on the world's major river systems, based on a comparison of the water available with the amount of water used by humans. A country is said to be *water stressed* when the volume of its reliable runoff per person drops to below about 1,700 cubic meters (60,000 cubic feet).

According to the United Nations and the World Bank, about 500 million people live in 34 Asian, African, and Middle Eastern countries suffering from water stress, and this could increase to almost 3 billion people in 50 countries by 2025. Water shortages in Africa are expected to be especially acute because **(1)** 67% of its land is dry (40%) or is desert (27%) and **(2)** its population is expected to increase from 800 million to 1.3 billion between 2000 and 2025.

All but 2 of the world's 34 water-stressed countries must import about one-fourth of the world's total grain exports to supplement their food production. As population grows and water scarcity increases, such grain exports are expected to increase. This could **(1)** heighten competition for the world's grain exports between countries such as China (Case Study, p. 272) and other developing countries in Asia, Africa, and the Middle East, **(2)** raise grain prices, and **(3)** increase hunger and malnutrition in developing countries that cannot afford to pay higher prices for grain.

Since the 1970s, water scarcity intensified by prolonged drought has killed more than 24,000 people per year and created millions of environmental refugees. According to the United Nations, in 1998 about 25 million people had to flee their homes because of water shortages, pollution, and flooding in their river basins. By 2025, the number of such environmental refugees could quadruple. In water-short rural areas in developing countries, many women and children must walk long distances each day, carrying heavy jars or cans, to get a meager and sometimes contaminated supply of water.

Evidence of water stress is seen in the damming and draining of rivers to supply water for irrigation and cities and dropping water tables in some of the world's major food-producing areas. According to a 1999 report by the World Commission on Water in the 21st Century (funded by the United Nations and World Bank), half of the world's major rivers (Figure 7-2, p. 153) are going dry part of the year or are seriously polluted.

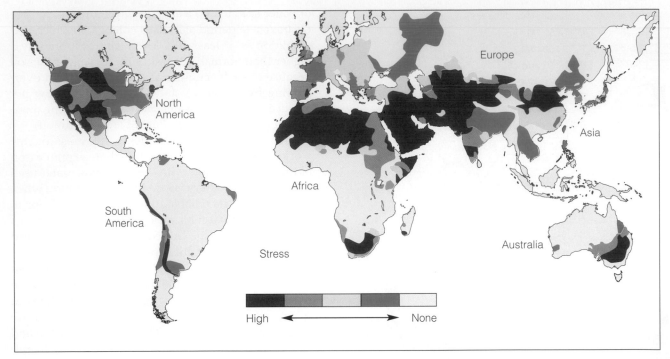

Figure 13-8 Stress on the world's major river basins, based on a comparison of the amount of water available with the amount used by humans. (Data from World Commission on Water Use in the 21st Century)

A number of environmental, political, and economic analysts believe that *access to water resources, already a key foreign policy and environmental security issue for water-short countries, will become even more important over the next 10-20 years.* Two countries share almost 150 of the world's 214 major river systems (57 of them in Africa), and another 50 are shared by 3 to 10 countries. Some 40% of the world's population already clashes over water, especially in the Middle East (p. 294).

How Can We Increase Freshwater Supplies? There are five ways to increase the supply of fresh water in a particular area: **(1)** Build dams and reservoirs to store runoff, **(2)** bring in surface water from another area, **(3)** withdraw groundwater, **(4)** convert salt water to fresh water (desalination), and **(5)** improve the efficiency of water use.

In developed countries, people tend to live where the climate is favorable and then bring in water from another watershed. In developing countries, most people (especially the rural poor) must settle where the water is and try to capture and use as much precipitation as they can.

13-4 USING DAMS AND RESERVOIRS TO SUPPLY MORE WATER

What Are the Pros and Cons of Large Dams and Reservoirs? Large dams and reservoirs have benefits

and drawbacks (Figure 13-9 and Case Study, p. 302). The main purpose of dams and large reservoirs is to capture and store runoff and release it as needed for **(1)** controlling floods, **(2)** producing hydroelectric power, and **(3)** supplying water for irrigation and for towns and cities. Reservoirs also provide recreational activities such as swimming, fishing, and boating.

Many rivers resemble an elaborate plumbing system, with multiple dams used to control the timing and flow of water, like water from a faucet. The goal of this engineering approach is to capture and use as much of a river's flow as possible. This has worked, with the world's dams increasing the annual runoff available for human use by nearly one-third.

However, a series of dams on a river, especially in arid areas, can reduce downstream flow to a trickle and prevent it from reaching the sea as a part of the hydrologic cycle. Major rivers that run dry and do not reach the sea anymore during the dry season include the **(1)** Colorado in the southwestern United States (Case Study, p. 303), **(2)** Yellow in northern China, **(3)** Nile in the Middle East (Figure 13-1), **(4)** Ganges and Indus in South Asia, and **(5)** Amu Darya and Syr Darya in five counties that once were part of the Soviet Union (Case Study, p. 305).

This engineering approach to river management often impairs the important ecological services rivers provide (Figure 13-11, p. 304). Ideally, in developing a dam and reservoir the human needs for water should be balanced with preserving a river's ecological services. However, achieving this goal is difficult because

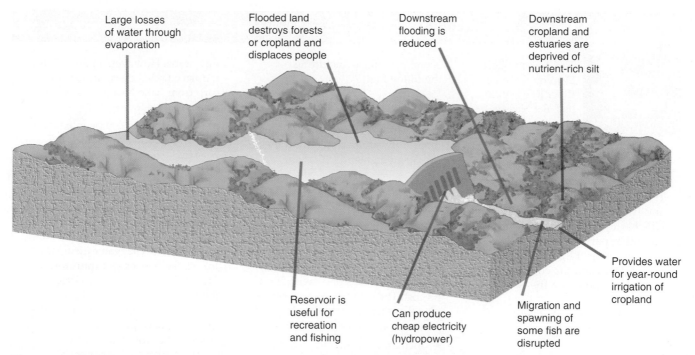

Large losses of water through evaporation

Flooded land destroys forests or cropland and displaces people

Downstream flooding is reduced

Downstream cropland and estuaries are deprived of nutrient-rich silt

Provides water for year-round irrigation of cropland

Reservoir is useful for recreation and fishing

Can produce cheap electricity (hydropower)

Migration and spawning of some fish are disrupted

Figure 13-9 Main advantages (green) and disadvantages (orange) of large dams and reservoirs. The world's 45,000 large dams now impound about 14% of the world's runoff.

the amount of available and usable water from a river varies with **(1)** the time of year, **(2)** conditions such as drought and higher-than-normal precipitation, **(3)** pollution loads, **(4)** habitat needs of aquatic life, and **(5)** the values people place on wildlife, fisheries, and recreational use of river basins. One approach is to develop computer models to **(1)** take such factors into account and **(2)** estimate the minimum amount of water needed to satisfy basic human and ecological needs.

Case Study: China's Three Gorges Dam When completed, China's Three Gorges project on the mountainous upper reaches of the Yangtze River will be the world's largest hydroelectric dam and reservoir. This superdam, with the electric output of 20 large coal-burning or nuclear power plants, will **(1)** supply power to industries and to 150 million Chinese, **(2)** help China reduce its dependence on coal, which causes severe air pollution and releases enormous amounts of the greenhouse gas carbon dioxide into the atmosphere, and **(3)** hold back the Yangtze River's floodwaters, which have killed more than 500,000 people during the past 100 years.

According to Chinese officials, the 400 million people living in the Yangtze River Valley who will benefit from the dam and the hundreds of thousands of lives that will be saved from reduced flooding far exceed the 1.9 million people who will be relocated. In 1998 alone, flooding of the Yangtze killed 4,000 people, dislocated 223 million people, flooded a large area of cropland, and cost $36 billion.

This is more than the projected $17–30 billion cost of the 20-year Yangtze project scheduled for completion by 2009. However, estimates by independent experts put the completion date at around 2019 and the cost at $70 billion or more.

Critics point out that the severity of flooding has greatly increased during the past few decades because 85% of the forest cover in the Yangtze basin has been cleared for timber and agriculture. This loss of water-absorbing cover allows more rapid flow of water across the land's surface, which increases flooding and silting of the river with eroded soil. In 1998, Chinese officials **(1)** banned logging in the upper Yangtze watershed, **(2)** started additional land reclamation projects in the river's floodplain, and **(3)** earmarked $2 billion to reforest the watershed.

Critics point to a number of drawbacks for the Yangtze dam project. Forming the gigantic 596-kilometer-long (370-mile-long) reservoir behind the dam will **(1)** flood large areas of productive farmland and scenic land, 100 towns, and 2 cities (each with about 100,000 people) and **(2)** displace about 1.9 million people from their homes.

Critics also charge that the dam and reservoir will **(1)** radically change the region's entire ecosystem, **(2)** increase water pollution because of the reduced river flow, **(3)** expose half a million people to severe flooding if the reservoir fills up with sediment and overflows (especially if the reservoir is kept filled at a high level, as planned, to provide maximum hydroelectric power), **(4)** reduce annual deposits of nutrient-rich sediments

Egypt's Aswan High Dam: Blessing or Disaster?

The Aswan High Dam on the Nile River in Egypt and its Lake Nasser reservoir (Figure 13-1) demonstrate the mix of advantages and disadvantages of such projects.

The dam was built in the 1960s to provide flood control and irrigation water for the lower Nile basin and electricity for Cairo and other parts of Egypt. All cropland in arid Egypt must be irrigated, and the country is totally dependent on the Nile for this water.

The dam's major benefits include the following:

- Supplying about one-third of Egypt's electrical power.

- Storing and releasing water for irrigation. This saved Egypt's rice and cotton crops during severe droughts in the 1970s and 1980s and helped avert massive famines. To many Egyptians, this more than paid for the cost of the dam.

- Increasing food production by allowing year-round irrigation of nearly 3.3 million hectares (8.2 million acres) of land in the lower Nile basin.

- Providing flood control for the lower Nile basin.

- Providing such benefits for tens of millions of Egyptians, which Egyptian officials say greatly exceeds the harm of uprooting 125,000 people when Lake Nasser was flooded.

However, the dam has also produced harmful ecological and economic effects, including the following:

- Ending the yearly flooding that for thousands of years had fertilized the Nile's floodplain with silt, most of it washed down from the Ethiopian highlands. Now the river's silt accumulates behind the dam, filling Lake Nasser and eventually making the dam useless.

- Necessitating the use of commercial fertilizer on cropland in the Nile Delta basin at an annual cost of more than $100 million to make up for plant nutrients once available at no cost. The country's new fertilizer plants use up much of the electrical power produced by the dam.

- Increasing salinization because there is no natural annual flooding to flush salts from the irrigated soil. This has offset about three-fourths of the gain in food production from new land irrigated by water from the reservoir.

- Eliminating 94% of Nile water that once reached the Mediterranean Sea each year and upsetting the ecology of waters near the mouth of the Nile.

- Eliminating the annual sediment discharge where the Nile reaches the sea. This has caused the coastal delta to erode and advance inland and has reduced productivity on large areas of agricultural land.

- Eradicating most of Egypt's sardine, mackerel, shrimp, and lobster fishing industries because nutrient-rich silt no longer reaches the river's mouth. This has led to losses of approximately 30,000 jobs, millions of dollars annually, and an important source of protein for Egyptians. However, a new fishing industry taking bass, catfish, and carp from Lake Nasser has offset some of these losses.

- Uprooting 125,000 people when land was flooded to create Lake Nasser.

In 1997, the Egyptian government (1) opened a new canal to transfer Nile water under the Suez Canal to irrigate land in the Sinai desert and (2) began a 20-year project to divert Nile water upstream from Lake Nasser and transport it hundreds of kilometers to irrigate land in Egypt's southwestern desert and open up land for human settlement to relieve population pressures in the crowded Nile valley.

These plans are threatened because about 80% of the water flowing into Lake Nasser comes from Ethiopia (Figure 13-1), which plans to build dams for irrigation and hydropower projects. This may not leave enough water in the Nile to support Egypt's plans for increased irrigation and could increase tension between the two countries.

Some analysts believe that the dam's benefits outweigh its economic and ecological costs. Other analysts consider it an economic and ecological disaster. Time will tell who is right.

Critical Thinking

1. Do you believe that the benefits of the Aswan High Dam outweigh its drawbacks? Explain.

2. List two principles for designing and building large dams based on lessons from the Aswan High Dam.

---

below the dam, and (5) promote saltwater intrusion into drinking water supplies near the mouth of the river.

13-5 TRANSFERRING WATER FROM ONE PLACE TO ANOTHER

What Are the Pros and Cons of Large-Scale Water Transfers? Tunnels, aqueducts, and underground pipes can transfer stream runoff collected by dams and reservoirs from water-rich areas to water-poor areas. Although such transfers have benefits, they also create environmental problems (Case Study, p. 305). Indeed, most of the world's dam projects and large-scale water transfers illustrate the important ecological principle that *you cannot do just one thing*.

One of the world's largest watershed transfer projects is the *California Water Project*. In California, the

The Colorado River flows 2,300 kilometers (1,400 miles) from the mountains of central Colorado to the Mexican border and eventually to the Gulf of California (Figure 13-10). During the past 50 years, this once free-flowing river has been tamed by a gigantic plumbing system consisting of **(1)** 14 major dams and reservoirs (Figure 13-10), **(2)** hundreds of smaller dams, and **(3)** a network of aqueducts and canals that supply water to farmers, ranchers, and cities.

This has economically transformed the southwestern United States. Today, this domesticated river provides **(1)** electricity (from hydroelectric plants at major dams), **(2)** water for more than 25 million people in seven states, **(3)** water used to produce about 15% of the nation's produce and livestock, and **(4)** a multi-billion-dollar recreation industry of whitewater rafting, boating, fishing, camping, and hiking enjoyed by more than 15 million people a year.

Take away this tamed river and **(1)** Las Vegas, Nevada, would be a mostly uninhabited desert area, **(2)** San Diego, California (which gets 70% of its water from the Colorado), could not support its present population, and **(3)** California's Imperial Valley (which grows a major portion of the nation's vegetables) would consist mostly of cactus and mesquite plants.

However, four major problems are associated with use of this river's water:

- The Colorado River basin includes some of the driest lands in the United States and Mexico (Figure 13-7, bottom).

- Legal pacts in 1922 and 1944 allocated more water to the states in the river's *upper basin* (Wyoming, Utah, Colorado, and New Mexico) and *lower basin* (Arizona, Nevada, and California; Figure 13-10) and to Mexico than now flows through the river, even in years without a drought.

- Because of so many withdrawals, the river rarely makes it to the Gulf of California. Instead, it fizzles into a trickle that disappears into the Mexican desert or (in drought years) the Arizona desert. This **(1)** threatens the survival of species that spawn in such rivers, **(2)** destroys estuaries that serve as breeding grounds for numerous aquatic species, and **(3)** increases saltwater contamination of aquifers near coasts.

There are increasing legal battles over how much of the river's limited water can be withdrawn and used by cities, farmers, ranchers, and Native Americans (who as senior owners of water rights dating back to the mid-1880s have the law on their side and have been winning legal battles to withdraw more water). Environmentalists have also initiated mostly unsuccessful attempts to keep more of the river wild by not building so many large dams and removing some of the existing dams to help protect the river's ecological services (Figure 13-11).

Traditionally, about 80% of the water withdrawn from the Colorado has been used to irrigate crops and raise cattle because ranchers and farmers (after Native Americans) got there first and established legal rights to use a certain amount of water each year. This large-scale use of water for agriculture was made possible because the government paid for the dams and reservoirs and under long-term contracts has supplied many of the farmers and ranchers with water at a very low price. This has led to **(1)** inefficient use of irrigation water and **(2)** growing crops such as rice, cotton, and alfalfa (for cattle feed) that need a lot of water.

Some cities (such as Tucson, Arizona, and Colorado Springs, Colorado) have been buying up the legal water rights of nearby farmers and ranchers. Others are paying farmers to install less wasteful irrigation systems (Figure 13-19, p. 311) so that more water will be available to support urban areas. It is estimated that improving overall irrigation efficiency by about 10–15% would provide enough water to support projected urban growth in the areas served by the river to 2020.

There is also controversy over water quality, especially in the lower basin. As more water is withdrawn, the remaining flow gets saltier, largely because of evaporation and because water withdrawn to irrigate fields trickles back to the river laden with salts from the soil.

These controversies illustrate the problems that governments and people in semiarid regions with shared river systems face as population and economic growth place increasing demands on limited supplies of surface water.

Critical Thinking

1. What are the pros and cons of reducing or eliminating government subsidies that provide farmers, ranchers, and cities with cheap water from the Colorado River?

2. If the legal system allowed it, put the following users in order of how much water you would allocate to them from the Colorado River: farmers, ranchers, cities, Native Americans, and Mexico. Explain your choices.

---

basic water problem is that 75% of the population lives south of Sacramento but 75% of the state's rain occurs north of Sacramento.

The California Water Project uses a maze of giant dams, pumps, and aqueducts to transport water from water-rich northern California to heavily populated areas and to arid and semiarid agricultural regions, mostly in southern California (Figure 13-13, p. 306).

For decades, northern and southern Californians have been feuding over how the state's water should

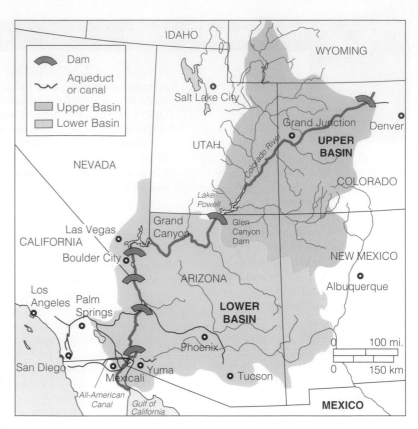

Figure 13-10 The Colorado River basin. The area drained by this basin is equal to more than one-twelfth of the land area of the lower 48 states.

be allocated under this project. Southern Californians say they need more water from the north to support Los Angeles, San Diego, and other growing urban areas and to grow more crops. Agriculture uses 74% of the water withdrawn in California, much of it for water-thirsty crops.

Opponents in the north say that sending more water south would **(1)** degrade the Sacramento River, **(2)** threaten fisheries, and **(3)** reduce the flushing action that helps clean San Francisco Bay of pollutants. They also argue that much of the water sent south is wasted unnecessarily and that making irrigation just 10% more efficient would provide enough water for domestic and industrial uses in southern California. However, if water supplies in northern California and in the Colorado River basin (Figure 13-10) were to drop sharply because of global warming, the amount of water delivered by the huge distribution system would plummet.

Pumping out more groundwater is not the answer because groundwater is already being withdrawn faster than it is replenished throughout much of California. To most analysts, quicker and cheaper solutions are **(1)** improving irrigation efficiency (Figure 13-20, p. 313) and **(2)** allowing farmers to sell their legal rights to withdraw certain amounts of water from rivers.

Case Study: The James Bay Watershed Transfer Project Another major watershed transfer project is Canada's James Bay project. It is a $60-billion, 50-year scheme to harness the wild rivers that flow into Que-

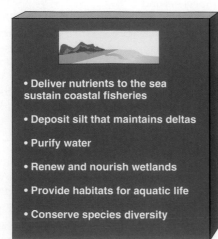

Figure 13-11 Some ecological services provided by rivers. Currently, the services are given little or no monetary value when the costs and benefits of dam and reservoir projects are assessed. According to environmental economists, attaching even crudely estimated monetary values to these ecosystem services would help sustain them.

- Deliver nutrients to the sea sustain coastal fisheries
- Deposit silt that maintains deltas
- Purify water
- Renew and nourish wetlands
- Provide habitats for aquatic life
- Conserve species diversity

bec's James and Hudson Bays to produce electric power for Canadian and U.S. consumers (Figure 13-14, p. 307). If completed, this megaproject would **(1)** construct 600 dams and dikes that will reverse or alter the flow of 19 giant rivers covering a watershed three times the size of New York State, **(2)** flood an area of boreal forest and tundra equal in area to Washington State or Germany, and **(3)** displace thousands of indigenous Cree and Inuit, who for 5,000 years have lived off James Bay by subsistence hunting, fishing, and trapping.

After 20 years, the $16-billion phase I has been completed. The second and much larger phase was postponed indefinitely in 1994 because of **(1)** an excess of power generated, **(2)** opposition by the Cree (whose ancestral hunting grounds would have been flooded) and Canadian and U.S. environmentalists, and **(3)** New York State's cancellation of two contacts to buy electricity produced by phase II.

13-6 TAPPING GROUNDWATER AND CONVERTING SALT WATER TO FRESH WATER

What Are the Pros and Cons of Withdrawing Groundwater? Pumping groundwater from aquifers has several advantages over tapping more erratic flows from streams. Groundwater **(1)** can be removed as needed year-round, **(2)** is not lost by evaporation, and

The Aral Sea Water Transfer Disaster

The shrinking of the Aral Sea (Figure 13-12) is a result of a large-scale water transfer project in an area of the former Soviet Union with the driest climate in central Asia. Since 1960, enormous amounts of irrigation water have been diverted from the inland Aral Sea and its two feeder rivers to irrigate cotton, vegetable, fruit, and rice crops and create area one of the world's largest irrigated areas. The irrigation canal, the world's longest, stretches over 1,300 kilometers (800 miles).

This water diversion project (coupled with droughts) has caused a regional ecological, economic, and health disaster, described by one former Soviet official as "ten times worse than the 1986 Chernobyl nuclear power-plant accident" (p. 350).

Since 1960, as more and more water from the feeder rivers has been diverted to irrigate crops, **(1)** the sea's salinity has tripled, **(2)** its surface area has shrunk by 54% (Figure 13-12), **(3)** its volume has decreased by 75%, **(4)** its two supply rivers have become mere trickles, and **(5)** about 36,000 square kilometers (14,000 square miles) of former lake bottom has become a human-made desert covered with glistening white salt. The process continues, and within 10–20 years the once enormous Aral Sea may break up into three small brine lakes.

Twenty of the area's 24 native fish species have become extinct. This has devastated the area's fishing industry, which once provided work for more than 60,000 people. Fishing villages and boats once on the sea's coastline now are in the middle of a salt desert and have been abandoned.

Wetlands have shrunk by 85% and, combined with high levels of pollution from agricultural chemicals, have greatly reduced waterfowl populations. Roughly half the area's bird and mammal species have disappeared.

Winds pick up the salty dust that encrusts the lake's now-exposed bed and blow it onto fields as far as 300 kilometers (190 miles) away. As the salt spreads, it kills wildlife, crops, and other vegetation and pollutes water.

To raise yields, farmers have increased inputs of herbicides, insecticides, fertilizers, and irrigation water on some crops. Many of these chemicals have percolated

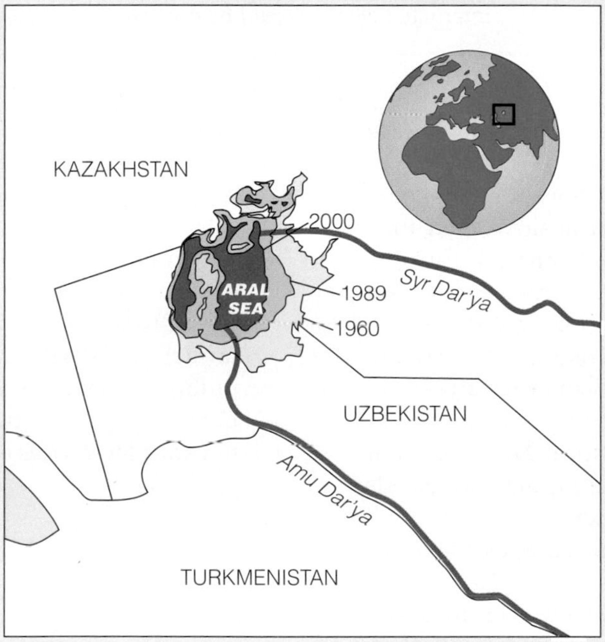

Figure 13-12 Once the world's fourth-largest freshwater lake, the Aral Sea has been shrinking and getting saltier since 1960 because most of the water from the rivers that replenish it has been diverted to grow cotton and food crops. As the lake shrinks, it leaves behind a salty desert, economic ruin, increasing health problems, and severe ecological disruption.

downward and accumulated to dangerous levels in the groundwater, from which most of the region's drinking water comes. Rising water tables from excessive irrigation have brought these chemicals near the surface and contaminated drinking water supplies. The lower

river flows have also concentrated salts, pesticides, and other toxic chemicals, making surface water supplies hazardous to drink.

The Aral Sea basin may have one of the world's worst salinization problems. This situation is getting worse because of positive feedback loops (p. 51). Irrigators apply more water to their fields during the season when crops are not grown to flush the accumulated salts out of the root zone before planting the next crop. This extra input of irrigation water **(1)** increases water use, **(2)** further shrinks the Aral Sea and increases the salt blowing onto cropland, and **(3)** adds more salt from irrigation water to the soil. Eventually this runaway feedback system could cause the collapse of most agriculture in this area.

Conversion of much of the Aral Sea to a salt desert has also affected the area's climate. The once-huge sea acted as a thermal buffer that moderated the heat of summer and the extreme cold of winter. Now there is less rain, summers are hotter and drier, winters are colder, and the growing season is shorter. This, coupled with severe salinization of almost a third of the area's cropland, has caused crop yields to drop 20–50%.

More water could be withdrawn to flush out and lessen the area's acute salt problem. However, Russian scientists estimate that freeing up this much water would mean retiring about half of the area's irrigated cropland, an unthinkable solution considering the region's already dire economic conditions.

Winds whip up fertilizer and pesticide residues from the poisoned agricultural land and salt from the bare floor of the shriveled Aral Sea. The combination of toxic

(continued)

dust, salt, and contaminated water has caused serious health problems for a growing number of the 58 million people living in the Aral Sea's watershed, and the area's population is expected to increase to 83 million by 2025. Such problems include abnormally high rates of **(1)** infant mortality, **(2)** tuberculosis, **(3)** anemia, **(4)** respiratory illness (one of the world's highest), **(5)** eye diseases (from salt dust), **(6)** throat cancer, **(7)** kidney and liver diseases (especially cancers), **(8)** arthritic diseases, **(9)** typhoid fever, and **(10)** hepatitis.

Can the Aral Sea be saved and can the area's serious ecological and human health problems be reduced? Between 1999 and 2002, the United Nations and the World Bank plan to spend $600 million to **(1)** purify

drinking water, **(2)** upgrade irrigation and drainage systems to improve irrigation efficiency, flush salts from croplands, and boost crop productivity, and **(3)** construct wetlands and artificial lakes to help restore aquatic vegetation, wildlife, and fisheries. However, this process will take decades and will not prevent the shrinkage of the Aral Sea into a few brine lakes.

In 1994, the presidents of the five countries in the Aral Sea basin developed a regional water management plan to address the area's dire water, ecological, and health problems. This plan **(1)** acknowledges that the region's current agricultural practices are unsustainable and **(2)** confirms that principles of international law should be used to

allocate water between the five countries. However, it does not discourage most of these countries from expanding irrigated land to help support their declining economies.

This agreement is a milestone in cooperative management of a large shared water basin. However, the water use, agricultural, population, economic, and social reforms needed to achieve more sustainable use of the Aral Sea basin will be very difficult to implement economically and politically.

Critical Thinking

What ecological and economic lessons can we learn from the Aral Sea tragedy?

---

(3) usually is less expensive to develop than surface water systems.

Aquifers provide drinking water for at least one-fourth of the planet's people. In Asia alone, more than 1 billion people depend on groundwater for drinking. Aquifers supply more than half of India's irrigation water and 90% of Bangladesh's drinking water.

In the United States, about half of the drinking water (96% in rural areas and 20% in urban areas) and 43% of irrigation water is pumped from aquifers. In Florida, Hawaii, Idaho, Mississippi, Nebraska, and New Mexico, more than 90% of the population depends on groundwater for drinking water.

However, overuse of groundwater can cause or intensify several problems: **(1)** *water table lowering* (Figure 13-15), **(2)** *aquifer depletion* (Figure 13-16, top), **(3)** *aquifer subsidence* (sinking of land when groundwater is withdrawn, Figure 13-16, bottom), **(4)** *intrusion of salt water into aquifers*, **(5)** *drawing of chemical contamination in groundwater toward wells*, and **(6)** *reduced stream flow* because of diminished flows of groundwater into streams. Groundwater can also become contaminated by industrial and agricultural activities, septic tanks, and other sources, as discussed in Section 19-3, p. 485.

Excessive water removal from rivers and its consequences are fairly easy to see, but aquifer depletion is hidden from view. According to Worldwatch Institute estimates, unsustainable depletion of groundwater is being used to produce about 10% of the world's annual grain harvest. This example of the tragedy of the commons (Connections, p. 12) is expected to increase as irrigated areas are expanded to help feed the 2 billion more people projected to join the ranks of humanity between 2000 and 2028.

In addition to limiting future food production, overpumping aquifers is increasing the gap between the rich and poor in some areas. As water tables drop, farmers must **(1)** drill deeper wells, **(2)** buy larger pumps to bring the water to the surface, and **(3)** use more electricity to run the pumps. Poor farmers cannot afford to do this and often end up losing their land

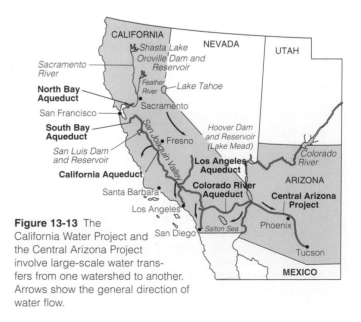

Figure 13-13 The California Water Project and the Central Arizona Project involve large-scale water transfers from one watershed to another. Arrows show the general direction of water flow.

and either working for richer farmers or hoping to survive by migrating to cities.

Currently, *groundwater in the United States is being withdrawn at four times its replacement rate.* The most serious overdrafts are occurring **(1)** in parts of the huge Ogallala Aquifer, extending from southern South Dakota to central Texas (Case Study, p. 309) and **(2)** in parts of the arid southwestern United States (Figure 13-16, top), especially California's Central Valley, which supplies about half the country's vegetables and fruits.

Aquifer depletion is also a problem in **(1)** Saudi Arabia, **(2)** central and northern China, **(3)** northwest and southern India (where one-fourth of the country's grain is being produced by unsustainable groundwater withdrawal), **(4)** northern Africa (especially Libya and Tunisia), **(5)** southern Europe, **(6)** the Middle East, and **(7)** parts of Mexico, Thailand, and Pakistan. Many cities are also overexploiting groundwater. Portions of Mexico City, Mexico, and Bangkok, Thailand, are sinking as geologic formations compact and subside after groundwater is removed.

When fresh water from an aquifer near a coast is withdrawn faster than it is recharged, salt water intrudes into the aquifer (Figure 13-17). Such intrusion can contaminate the drinking water of many towns and cities along coastal areas.

Ideally, *water should not be withdrawn from an aquifer faster than it is replaced.* However, this is costly to monitor and difficult to enforce because of **(1)** laws and customs that allow users unlimited water extraction from aquifers underlying their land and **(2)** the tragedy of the commons (Connections, p. 12).

Figure 13-14 If completed, the James Bay project in northern Quebec will alter or reverse the flow of 19 major rivers and flood an area the size of Washington State to produce hydropower for consumers in Quebec and the United States, especially in New York State. Phase I of this 50-year project is completed. In 1994, phase II was postponed indefinitely because of **(1)** a surplus of electricity and **(2)** opposition by environmentalists and the indigenous Cree, whose ancestral hunting grounds would have been flooded.

Ways to prevent or slow groundwater depletion include **(1)** controlling population growth, **(2)** not planting water-intensive crops such as cotton and sugarcane in dry areas, **(3)** shifting to crops that need less water in dry areas, **(4)** developing crop strains that need less water and are more resistant to heat stress, and **(5)** wasting less irrigation water (p. 310).

How Useful Is Desalination? Removing dissolved salts from ocean water or from brackish (slightly salty) groundwater, called **desalination**, is another way to increase supplies of fresh water. The two most widely used methods are **(1)** *distillation*, which involves heating salt water until it evaporates (and leaves behind salts in solid form) and condenses as fresh water, and **(2)** *reverse osmosis* in which salt water is pumped at high pressure through a thin membrane whose pores allow water molecules, but not dissolved salts, to pass through.

About 11,100 desalination plants in 120 countries (especially in the arid Middle East and parts of North

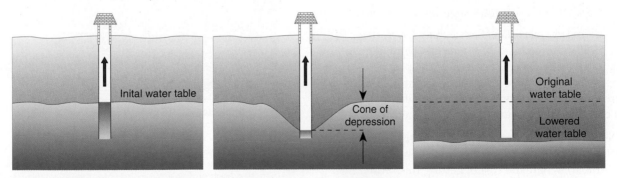

Figure 13-15 Lowering of the water table when a well is drilled into an aquifer (left). A cone of depression (middle) in the water table forms if groundwater is pumped to the surface faster than it can flow through the aquifer to the well. If this excessive water removal continues, the water table falls (right).

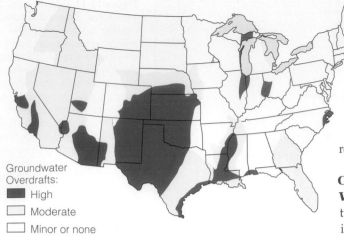

Groundwater
Overdrafts:
■ High
▨ Moderate
□ Minor or none

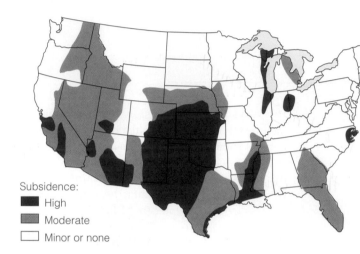

Subsidence:
■ High
▨ Moderate
□ Minor or none

Figure 13-16 Areas of greatest aquifer depletion and groundwater contamination (top) and ground subsidence (bottom) in the continental United States. Aquifer depletion is also high in Hawaii and Puerto Rico (not shown on map). (Data from U.S. Water Resources Council and U.S. Geological Survey)

Africa) meet less than 0.15% of the world's water needs. Desalination would have to increase 33-fold just to supply 5% of current world water use.

This is unlikely because desalination has two major disadvantages:

- *It is expensive because it takes large amounts of energy.*

- *It produces large quantities of waste water (brine) containing high levels of salt and other minerals.* Dumping the concentrated brine into the ocean near the plants increases the local salt concentration and threatens food resources in estuary waters, and dumping it on land could contaminate groundwater and surface water.

Desalination can provide fresh water for coastal cities in arid countries (such as sparsely populated Saudi Arabia), where the cost of getting fresh water by any method is high. Scientists are working to develop new membranes for reverse osmosis that can separate water from salt more efficiently and under less pressure. If successful, this could help bring down the cost of using desalinization to produce drinking water. However, desalinated water probably will never be cheap enough to irrigate conventional crops or meet much of the world's demand for fresh water unless **(1)** affordable solar-powered distillation plants can be developed and **(2)** someone can figure out what to do with the resulting mountains of salt.

Can Cloud Seeding and Towing Icebergs Improve Water Supplies? For decades several countries, particularly the United States, have been experimenting with seeding clouds with tiny particles of chemicals (such as silver iodide). The particles form water condensation nuclei and thus produce more rain over dry regions and more snow over mountains.

However, cloud seeding **(1)** is not useful in very dry areas, where it is most needed, because rain clouds rarely are available there and **(2)** would introduce large amounts of the cloud-seeding chemicals into soil and water systems, possibly harming people, wildlife, and agricultural productivity.

Another obstacle to cloud seeding is legal disputes over the ownership of water in clouds. During the 1977 drought in the western United States, the attorney general of Idaho accused officials in neighboring Washington of "cloud rustling" and threatened to file suit in federal court.

There also have been proposals to tow massive icebergs to arid coastal areas (such as Saudi Arabia and southern California) and then to pump the fresh water from the melting bergs ashore. However, the technology for doing this is not available and the costs may be too high, especially for water-short developing countries.

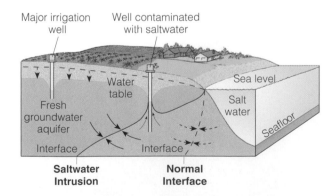

Figure 13-17 Saltwater intrusion along a coastal region. When the water table is lowered, the normal interface (dotted line) between fresh and saline groundwater moves inland (solid line), making coastal drinking water supplies unusable. (Data from U.S. Geological Survey)

Mining Groundwater: The Shrinking Ogallala Aquifer

Large amounts of water have been pumped from the Ogallala, the world's largest known aquifer (Figure 13-18). This has helped transform vast areas of arid high plains prairie land into one of the largest and most productive agricultural regions in the United States.

Mostly because of irrigated farming, this region produces 20% of U.S. agricultural output (including 40% of its feedlot beef), valued at $32 billion per year. This has brought prosperity to many farmers and merchants in this region, but the hidden environmental and economic cost has been increasing aquifer depletion in some areas.

Although this aquifer is gigantic, it is essentially a nonrenewable aquifer (stored during the retreat of the last ice age about 15,000–30,000 years ago) with an extremely slow recharge rate. In some areas, water is being pumped out of the aquifer 8–10 times faster than the aquifer's natural recharge rate.

The northernmost states (Wyoming, North Dakota, South Dakota, and parts of Colorado) still have ample supplies. However, supplies in parts of the southern states, where the aquifer is thinner (Figure 13-18) are being depleted rapidly, with about two-thirds of the aquifer's depletion taking place in the Texas High Plains.

Water experts project that at the current rate of withdrawal, one-fourth of the aquifer's original supply will be depleted by 2020 and much sooner in areas where it is shallow. It will take thousands of years to replenish the aquifer.

Government subsidies designed to increase crop production also increase depletion of the Ogallala by (1) encouraging the growth of water-thirsty cotton in the lower basin, (2) providing crop-disaster payments, and (3) providing tax breaks in the form of groundwater depletion allowances, with larger breaks for heavier groundwater use.

Depletion of this essentially non-renewable water resource can be delayed if farmers (1) use more efficient forms of irrigation (Figure 13-19), (2) switch to crops that need less water, or (3) irrigate less land. Many farmers in northwest Texas have reduced their water use by 20-25% by switching to more efficient irrigation technologies. Total irrigated area has been declining in five of the seven states using this aquifer because of rising drilling and pumping costs as the water table drops.

Cities using this groundwater can also implement policies and technologies to reduce their water use and waste. People enjoying the benefits of this aquifer can pitch in by installing water-saving toilets and showerheads and converting their lawns to plants that can survive in an arid climate with little watering (Figure 13-21, p. 313).

Critical Thinking

1. What are the pros and cons of giving government subsidies to farmers and ranchers using water withdrawn from the Ogallala to grow crops and raise livestock that need large amounts of irrigation water? How do you benefit from such subsidies?

2. Should these subsidies be reduced or eliminated and replaced with subsidies that encourage farmers to use more efficient forms of irrigation and switch to crops that need less water? Explain.

WYOMING
SOUTH DAKOTA
NEBRASKA
KANSAS
COLORADO
NEW MEXICO
OKLAHOMA
TEXAS

Miles
0   100

0   160
Kilomiters

Saturated thickness of Ogallala Aquifer

Less than 61 meters (200 ft.)

61–183 meters (200–600 ft.)

More than 183 meters (600 ft.)
(as much as 370 meters or 1,200 ft. in places)

Figure 13-18 The Ogallala is the world's largest known aquifer. If the water in this aquifer were above ground, it could cover all 50 states with 0.5 meter (1.5 feet) of water. Water withdrawn from this aquifer is used to grow crops, raise cattle, and provide cities and industries with water. As a result, this aquifer, which is renewed very slowly, is being depleted (especially at its thin southern end in parts of Texas, New Mexico, Oklahoma, and Kansas). (Data from U.S. Geological Survey)

13-7 USING WATER MORE EFFICIENTLY

Why Do We Waste So Much Water? Mohamed El-Ashry of the World Resources Institute estimates that *65–70% of the water people use throughout the world is lost through evaporation, leaks, and other losses.* The United States, the world's largest user of water, does slightly better but still loses about 50% of the water it withdraws. El-Ashry believes that it is economically and technically feasible to reduce such water losses to 15%, thereby meeting most of the world's water needs for the foreseeable future.

This will involve greatly increased use of water-saving technologies and practices that do more with less water. This will also **(1)** decrease the burden on wastewater plants, **(2)** reduce the need for expensive dams and water transfer projects that destroy wildlife habitats and displace people, **(4)** slow depletion of groundwater aquifers, and **(5)** save energy and money.

According to water resource experts, three major causes of water waste are:

■ *Government subsidies of water supply projects that create artificially low water prices.* According to Sandra

Postel, "By heavily subsidizing water, governments give out the false message that it is abundant and can afford to be wasted—even as rivers are drying up, aquifers are being depleted, fisheries are collapsing, and species are going extinct." However, farmers, industries, and others benefiting from water subsidies argue that such subsidies **(1)** promote settlement and agricultural production in arid and semiarid areas, **(2)** stimulate local economies, and **(3)** help lower prices of food and manufactured goods for consumers.

■ *Water laws* (Spotlight, below).

■ *Fragmented watershed management.* The Chicago, Illinois, metropolitan area, for example, has 349 water supply systems divided among some 2,000 local units of government over a six-county area. By contrast, in England and Wales the British Water Act of 1973 replaced more than 1,600 agencies with 10 regional water authorities based on natural watershed boundaries. Each water authority owns, finances, and manages all water supply, water pollution control, and waste treatment facilities in its region. Each authority is managed by elected local officials and a smaller number of officials appointed by the national government.

SPOTLIGHT | Water Rights in the United States

Laws regulating access to and use of surface water differ in the eastern and western parts of the United States. In most of the East, water use is based on the *doctrine of riparian rights.*

Basically, this system of water law gives anyone whose land adjoins a flowing stream the right to use water from the stream as long as some is left for downstream landowners. However, as population and water-intensive land uses grow, there is often too little water to meet the needs of all the people along a stream.

In the arid and semiarid West, the riparian system does not work because large amounts of water are needed in areas far from major surface water sources. In most of this region, water use is regulated by the *principle of prior appropriation.*

In this first-come, first-served approach, the first user of water

from a stream establishes a legal right for continued use of the amount originally withdrawn. If there is a shortage, later users are cut off in order until there is enough water to satisfy the demands of the earlier users. Some states have a combination of riparian and prior appropriation water rights.

Most groundwater use is based on *common law*, which holds that subsurface water belongs to whoever owns the land above such water. This allows landowners to withdraw as much groundwater as they want.

When many users tap the same aquifer, that aquifer becomes a common-property resource. Because the largest users have little incentive to conserve, this can deplete the aquifer for everyone and create a tragedy of the commons (Connections, p. 12).

A system of legally protected water rights allows individuals owning rights to sell, trade, or lease them to make money, ease water short-

ages, or protect the ecosystem services of rivers (Figure 13-11). For example, some water-short cities in the western United States are paying nearby farmers to install more efficient irrigation methods in exchange for the water the farmers save.

Private organizations and government agencies are also buying up water rights and using them to help restore aquatic environments by returning the water to rivers and wetlands. However, water markets must be regulated to avoid excessive water prices (especially for the poor) and inequalities in water distribution.

Critical Thinking

What are the advantages and disadvantages of **(a)** the principles of riparian rights and prior appropriation for access to surface water and **(b)** the common law approach to groundwater use in the United States? If you disagree with these approaches, how would divide up water rights?

Solutions: How Can We Waste Less Water in Irrigation? Globally, only about 40% of the water used reaches crops. Most irrigation systems distribute water from a groundwater well or a surface water source and allow it to flow by gravity through unlined ditches in cropfields so that the water can be absorbed by crops (Figure 13-19, left). This flood irrigation method **(1)** delivers far more water than needed for crop growth and **(2)** typically allows only 60% of the water to reach crops because of evaporation, seepage, and runoff.

Water waste can be reduced by using more efficient irrigation systems such as:

- *Center-pivot low-pressure sprinklers* (Figure 13-19, right), which typically allow 80% of the water input to reach crops and reduce water use over conventional gravity flow systems by 25%.

- *Low-energy precision application (LEPA) sprinklers.* This form of center-pivot irrigation allows 90-95% of the water input to reach crops by spraying it closer to the ground and in larger droplets than the center-pivot, low-pressure system. LEPA sprinklers use 20–30% less energy than low-pressure sprinklers and typically use 37% less water than conventional gravity flow systems.

- *Using surge or time-controlled valves on conventional gravity flow irrigation systems* (Figure 13-19, left). These valves send water down irrigation ditches in pulses instead of in a continuous stream. This can raise irrigation efficiency to 80% and reduce water use by 25%.

- *Using soil moisture detectors to water crops only when they need it.* For example, some farmers in Texas bury a $1 cube of gypsum, the size of a lump of sugar, at the root zone of crops. Wires embedded in the gypsum are run back to a small, portable meter that indicates soil moisture. Farmers using this technique can use 33–66% less irrigation water with no change in crop yields.

- *Drip irrigation systems* (Figure 13-19, center, and Solutions, p. 312), which can raise water efficiency to 90-95% and reduce water use by 37–70%.

Other ways to reduce water waste in irrigating crops are listed in Figure 13-20. Since 1950, water-short Israel has used many of these techniques to slash irrigation water waste by about 84% while irrigating 44% more land. Israel now treats and reuses 65% of its municipal sewage water for crop production and plans to increase this to 80% by 2025. The government also has **(1)** gradually removed most government water subsidies to raise the price of irrigation water to one of the highest in the world, **(2)** imports most of its water-intensive wheat and meat, and **(3)** concentrates on growing fruits, vegetables, and flowers that need less water.

Many of the world's poor farmers cannot afford to use most of the modern technological methods for increasing irrigation and irrigation efficiency. Such farmers increase irrigation by using small-scale and low-cost traditional technologies such as **(1)** pedal-powered treadle pumps to move water through irrigation ditches (widely used in Bangladesh), **(2)** animal-powered irrigation pumps, **(3)** buckets with holes for drip

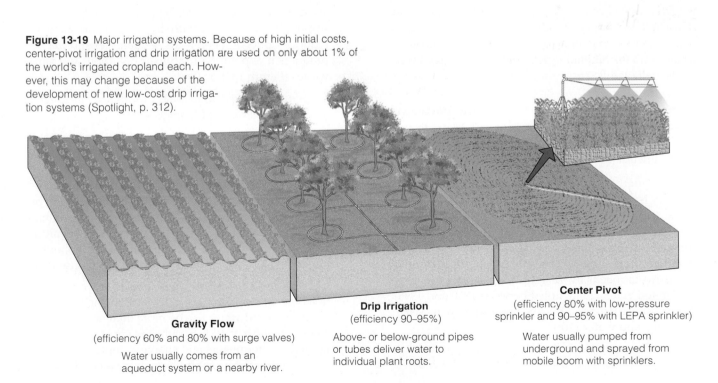

Figure 13-19 Major irrigation systems. Because of high initial costs, center-pivot irrigation and drip irrigation are used on only about 1% of the world's irrigated cropland each. However, this may change because of the development of new low-cost drip irrigation systems (Spotlight, p. 312).

Gravity Flow
(efficiency 60% and 80% with surge valves)

Water usually comes from an aqueduct system or a nearby river.

Drip Irrigation
(efficiency 90–95%)

Above- or below-ground pipes or tubes deliver water to individual plant roots.

Center Pivot
(efficiency 80% with low-pressure sprinkler and 90–95% with LEPA sprinkler)

Water usually pumped from underground and sprayed from mobile boom with sprinklers.

SOLUTIONS

The Promise of Drip Irrigation

The development of inexpensive, weather-resistant, and flexible plastic tubing after World War II paved the way for use of a new form of micro-irrigation called *drip irrigation* (Figure 13-19, middle). It consists of a network of perforated plastic tubing, installed at or below the ground surface. The small holes or emitters in the tubing deliver drops of water at a slow and steady rate close to plant roots.

This technique, developed in Israel in the 1960s and now used by half the country's farmers, has a number of advantages, including the following:

- *Adaptability*. The tubing system can easily be fitted to match the patterns of crops in a field and left in place or moved to different locations.

- *Efficiency*, with 90–95% of the water input reaching crops.

- *Lower operating costs* because 37-70% less energy is needed to

pump this water at low pressure and less labor is needed to move sprinkler systems.

- *Ability to apply fertilizer solutions in precise amounts*, which reduces fertilizer use and waste, salinization, and water pollution from fertilizer runoff.

- *A increase in crop yields of 20–90%* by getting more crop growth per drop.

- *Healthier plants and higher yields* because plants are neither under-watered nor overwatered.

Despite these advantages, drip irrigation is used on less than 1% of the world's irrigated area. The main reason is that the capital cost of conventional drip irrigation systems is too high for most poor farmers and for use on low-value row crops. However, drip irrigation is economically feasible for high-profit fruit, vegetable, and orchard crops and for home gardens.

Some *good news* is that the capital cost of a new drip irrigation system is one-tenth as much per

hectare as conventional drip systems. This system has **(1)** simple holes instead of emitters, **(2)** cloth filters instead of costly filtration equipment, and **(3)** better portability so that each drip line can water 10 rows of crops instead of 1.

After promising field trials, the company developing this system and the World Bank have been working together to market the system in India and other dry and water-scarce areas.

These low-cost drip irrigation systems could bring about a revolution in more sustainable irrigated agriculture that would **(1)** increase food yields, **(2)** reduce water use and waste, and **(3)** lessen some of the environmental problems associated with agriculture (Figure 12-10, p. 271).

Critical Thinking

Should governments provide subsidies to farmers who use drip irrigation based on how much water they save? Explain.

---

irrigation, **(4)** check dams, ponds, and tanks to collect rainwater for irrigation, **(5)** terracing (Figure 10-24a, p. 230) to reduce water loss on crops grown on steep terrain, and **(6)** cultivating seasonally waterlogged wetlands, delta lands, and valley bottoms.

Solutions: How Can We Waste Less Water in Industry, Homes, and Businesses? Ways to use water more efficiently in industries, homes, and businesses include the following:

- *Redesigning manufacturing processes*. A paper mill in Hadera, Israel, uses one-tenth as much water as most of the world's other paper mills, and a German paper plant nearly eliminated water use by completely recycling and purifying its water. Manufacturing aluminum from recycled scrap rather than virgin ore can reduce water needs by 97%. Although water use in the United States quadrupled between 1950 and 1998, industrial use dropped by 20% because of increased efficiency and water reuse.

- *Replacing green lawns in arid and semiarid regions with vegetation adapted to a dry climate*. This form of

landscaping, called *xeriscaping* (pronounced "ZER-i-scaping"), reduces water use by 30–85% and sharply reduces inputs of labor, fertilizer, and fuel and the production of polluted runoff, air pollution, and yard wastes (Figure 13-21).

- *Using drip irrigation to water gardens and other vegetation around homes and businesses*.

- *Fixing leaks in water mains, pipes, toilets, and faucets*. Leaks waste about half of the water supply in many cities in developing countries and 20–35% of water withdrawn from public supplies in the United States and the United Kingdom. In water-short Cairo, Egypt, people often wade across streets ankle deep in water because of leaky water pipes. Leaks from toilet valves, dripping faucets, and aging pipes account for about one-tenth of the water used in typical U.S. household.

- *Using water meters to monitor and charge for municipal water use*. In Boulder, Colorado, introducing water meters reduced water use by more than one-third. About one-fifth of all U.S. public water systems do not have water meters and charge a single low rate for almost unlimited use of high-quality water. Many

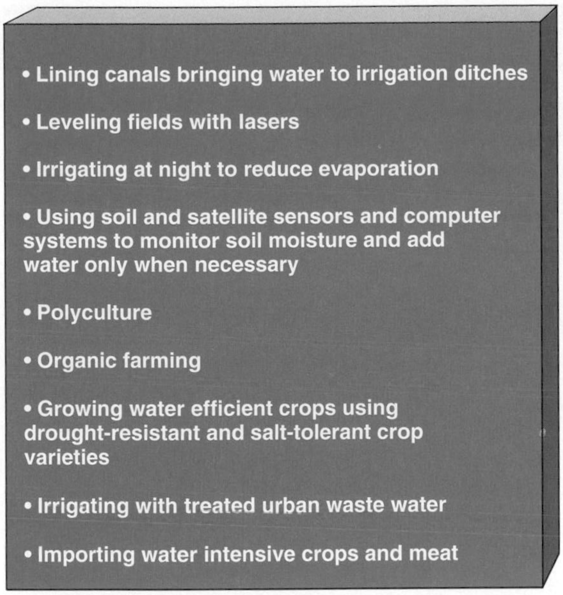

- Lining canals bringing water to irrigation ditches

- Leveling fields with lasers

- Irrigating at night to reduce evaporation

- Using soil and satellite sensors and computer systems to monitor soil moisture and add water only when necessary

- Polyculture

- Organic farming

- Growing water efficient crops using drought-resistant and salt-tolerant crop varieties

- Irrigating with treated urban waste water

- Importing water intensive crops and meat

Figure 13-20 Methods for reducing water waste in irrigation.

apartment dwellers have little incentive to conserve water because their water use is included in their rent.

■ *Having ordinances requiring water conservation in water-short cities.* Because of such ordinances, the desert city of Tucson, Arizona, consumes half as much water per person as Las Vegas, a desert city with even less rainfall and less emphasis on water conservation (Spotlight, p. 314).

■ *Requiring or encouraging use of water-saving toilets and showerheads.* Since 1994 all new toilets sold in the United States must use no more than 6 liters (1.6 gallons) per flush, and similar laws have been passed in Mexico and in Ontario, Canada. A low-flow showerhead costing about $20 saves about $34-56 per year in water heating costs. Audits conducted by students in Brown University's environmental studies program showed that the school could save $44,000 a year by using low-flow showerheads in dormitories. A California water utility cut per capita water use 40% in 1 year by giving rebates for customers switching to water-saving toilets and showerheads.

■ *Using washing machines that load from the front instead of the top.* Such machines **(1)** use 40–75% less water, **(2)** make clothes last longer because they are not agitated, and **(3)** save money.

■ *Reusing gray water from bathtubs, showers, bathroom sinks, and clothes washers for irrigating lawns and nonedible plants and raising fish.* About 50-75% of the water used by a typical house could be reused as gray water. In the United States, California has become the first state to legalize reuse of gray water to irrigate landscapes. About 65% of the wastewater in Israel is reused.

■ *Installing or leasing systems that purify and completely recycle wastewater from houses, apartments, or office buildings.* In Tokyo, Japan, all the water used in Mitsubishi's 60-story office building is purified for reuse by an automated recycling system.

■ *Collecting and using rainwater for flushing toilets, irrigating gardens, watering lawns, and putting out fires.* In Tokyo, Japan, large tanks on top of 579 city buildings capture and use rainwater.

■ *Reducing personal water use and waste* by actions such as those listed in Appendix 6.

Two decades of experience with droughts in northern California has shown that water demand can be cut by more than 50% for homes, 60% for parks, and 20% for businesses without economic hardships. Reducing water use and waste can also save money. Between 1987 and 1998, Boston, Massachusetts, reduced total water demand 24% by **(1)** fixing leaky pipes, **(2)** installing water-saving fixtures, and **(3)** educating the public about how to save water. This allowed the city to avoid diverting two large rivers to supply more water, which would have cost two to three times more than its water conservation program.

Raising the price of water for domestic and industrial consumers (as Israel has done) is one way to reduce wasteful water use. You might think that charging more for water supplied by public water systems would hurt the poor. Instead, this usually lowers the

Figure 13-21 Xeriscaping in Berkeley, California. This technique can reduce water use by as much as 80% by landscaping with rocks and plants that need little water and are adapted to the growing conditions in arid and semiarid areas. (©Mark E. Gibson/Visuals Unlimited)

SPOTLIGHT

Running Out of Water in Las Vegas, Nevada

Las Vegas, Nevada, located in the Mojave Desert, is an artificial aquatic wonderland of large trees, green lawns and golf courses, waterfalls, and swimming pools. Residents in this arid area have to apply 3 meters (10 feet) of water on a lawn each year to keep it green.

Las Vegas is one of the fastest-growing cities in the United States, with its population doubling from 550,000 in 1985 to 1.2 million in 1998. It is estimated that the city uses more water per person than any city in the world.

Tucson Arizona, in the Sonora Desert, gets only 30 centimeters (12 inches) of rainfall a year, but Las Vegas averages only 10 centimeters (4 inches). Tucson, now a model of water conservation, began a strict water conservation program in 1976, including raising water rates 500% for some residents.

In contrast, Las Vegas only recently started to encourage water conservation by **(1)** raising water rates sharply (but they are still less than half those in Tucson) and **(2)** encouraging replacement of lawns with rocks and native plants that survive on little water (Figure 13-21).

Water experts project that even if these recent water conservation efforts are successful, the city will run short of water by 2007.

Critical Thinking

1. If you were an elected official in charge of Las Vegas, what three things would you do to improve water conservation? What might be the political implications of instituting such a program?

2. If water shortages by 2007 limit the growth of the area's population, would you consider this good or bad? Explain.

cost of water for the poor because most are paying 10 to 12 times more per liter of water to buy it from private water vendors than citizens receiving often purer water from public systems.

13-8 TOO MUCH WATER

What Are the Causes and Effects of Flooding? Natural flooding by streams is caused primarily by heavy rain or rapid melting of snow. This causes water in a stream to overflow its normal channel and flood the adjacent area, called a **floodplain** (Figure 13-22). Floodplains, which include highly productive wetlands, help **(1)** provide natural flood and erosion control, **(2)** maintain high water quality, and **(3)** recharge groundwater.

People have settled on floodplains since the beginnings of agriculture. They have many advantages, including **(1)** fertile soil, **(2)** ample water for irrigation, **(3)** flat land suitable for crops, buildings, highways, and railroads, and **(4)** use of nearby rivers for transportation and recreation. In the United States, 10 million households and businesses with property valued at $1 trillion exist in flood-prone areas.

Floods are a natural phenomenon and have several

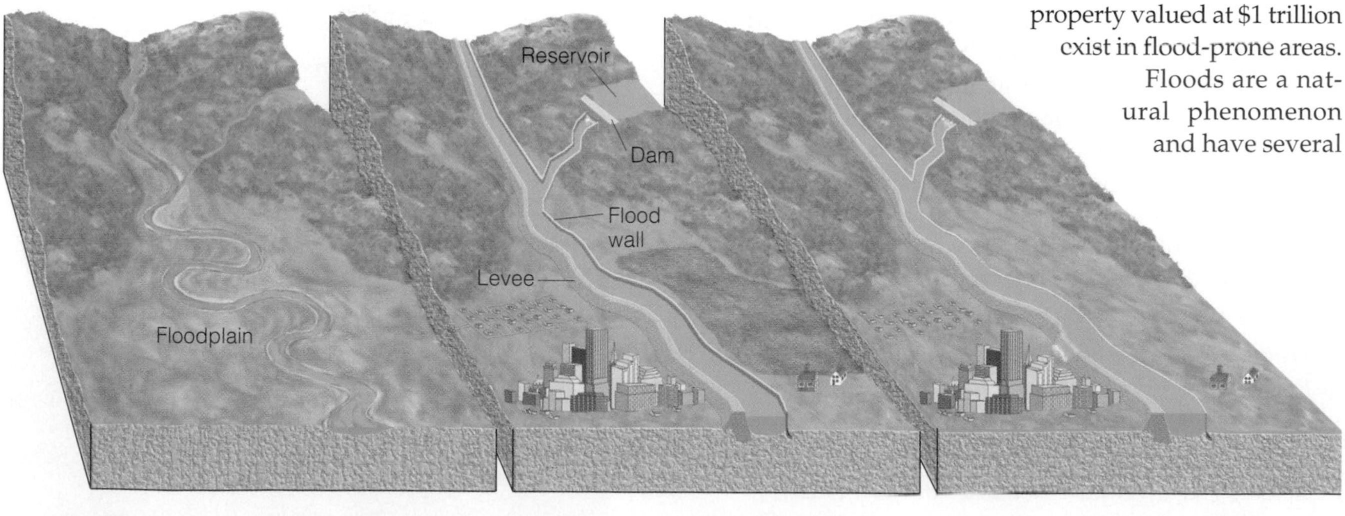

Figure 13-22 Land in a natural floodplain (left) often is flooded after prolonged rains. When the floodwaters recede, deposits of silt are left behind, creating a nutrient-rich soil. To reduce the threat of flooding (and thus allow people to live in floodplains), rivers have been **(1)** dammed to create reservoirs that store and release water as needed, **(2)** narrowed and straightened (channelization, middle), and **(3)** equipped with protective levees and walls (middle). These alterations can give a false sense of security to floodplain dwellers living in high-risk areas. In the long run, such measures can greatly increase flood damage because they can be overwhelmed by prolonged rains (right), as happened in the midwestern United States during the summer of 1993 (Figure 13-23, right).

Living Dangerously on Floodplains in Bangladesh

CONNECTIONS

Bangladesh (Figure 11-21, p. 249) is one of the world's **(1)** most densely populated countries, with 128 million people packed into an area roughly the size of Wisconsin, and **(2)** poorest countries, with an average per capita GNP of about $350, or 96¢ per day.

The people of Bangladesh benefit from moderate annual flooding during the summer monsoon season. They depend on these seasonal floodwaters to **(1)** grow rice and **(2)** help maintain soil fertility in the delta basin by receiving an annual deposit of eroded Himalayan soil.

However, excessive flooding can be disastrous. In the past, great floods occurred every 50 years or so, but since the 1970s they have come about every 4 years.

Bangladesh's increased flood problems begin in the Himalayan watershed. A combination of rapid population growth, deforestation, overgrazing, and unsustainable farming on steep, easily erodible mountain slopes has greatly dimin-

ished the soil's ability to absorb water. Instead of being absorbed and released slowly, water from the monsoon rains runs off the denuded Himalayan foothills, carrying vital topsoil with it (Figure 13-24).

This runoff, combined with heavier-than-normal monsoon rains, has increased the severity of flooding along Himalayan rivers and downstream in Bangladesh. For example, a disastrous flood in 1998 **(1)** covered two-thirds of Bangladesh's land area for 9 months, **(2)** leveled 2 million homes, **(3)** drowned at least 2,000 people, **(4)** left 30 million people homeless, **(5)** destroyed more than one-fourth of the country's crops, which caused thousands of people to die of starvation, and **(6)** caused at least $3.4 billion in damages.

Living on a coastal floodplain also carries dangers from storm surges and cyclones. Since 1961, 17 devastating cyclones have slammed into Bangladesh. In 1970, as many as 1 million people drowned in one storm and another surge killed an estimated 140,000 people in 1991.

In their struggle to survive, the poor in Bangladesh have cleared many of the country's coastal mangrove forests (Figure 7-6, p. 157) for fuelwood, farming, and aquaculture ponds for raising shrimp. This has led to more severe flooding because these coastal wetlands shelter Bangladesh's low-lying coastal areas from storm surges and cyclones. Damages and deaths from cyclones in areas of Bangladesh still protected by mangrove forests have been much lower than in areas where the forests have been cleared.

Critical Thinking

1. Bangladesh's population is growing rapidly and is expected to increase from 128 million to 177 million between 2000 and 2025. How could slowing its rate of population growth help reduce poverty and the harmful impacts of excessive flooding?

2. How could reforestation in the upstream countries of Bhutan, China, India, and Nepal reduce flooding in those countries and in Bangladesh?

---

benefits. They **(1)** provide the world's most productive farmland because they are regularly covered with nutrient-rich silt left after floodwaters recede, **(2)** recharge groundwater, and **(3)** refill wetlands.

However, each year floods kills thousands of people and causes tens of billions of dollars in property damage (Connections, above). During the spring and summer of 1993, heavy rain soaked the upper Midwest of the United States, causing the Mississippi River and its tributaries to flood nine states (Figure 13-23). This massive flood, considered by many experts to be the worst in U.S. history, **(1)** killed 50 people, **(2)** destroyed 70,000 homes, **(3)** damaged large areas of cropland, and **(4)** caused more than $12 billion in damages to homes, business, and crops. Billions more are being spent to repair the hundreds of levees and embankments that were damaged or destroyed.

Floods, like droughts, usually are considered natural disasters, but since the 1960s human activities have contributed to the sharp rise in flood deaths and damages. Three ways humans increase the severity of flood damage are by **(1)** removing water-absorbing vegetation, especially on hillsides (Figure 13-24), **(2)** draining

wetlands that absorb floodwaters and reduce the severity of flooding, and **(3)** living on floodplains. In developed countries, people deliberately settle on floodplains and then expect dams, levees, and other devices to protect them from floodwaters. However, when heavier-than-normal rains occur these devices do not work (Figure 13-22, right). In many developing countries, the poor have little choice but to try to survive in flood-prone areas (Connections, above).

Urbanization also increases flooding by replacing water-absorbing vegetation, soil, and wetlands with highways, parking lots, and buildings that cause rapid runoff of rainwater. If sea levels rise during the next century, as projected, many low-lying coastal cities, wetlands, and croplands will be under water.

Solutions: How Can We Reduce Flood Risks? One controversial way of reducing flooding is *channelization*, in which a section of a stream is deepened, widened, or straightened to allow more rapid runoff (Figure 13-22, middle). Channelization can reduce upstream flooding, but the increased flow of water can also increase upstream bank erosion and downstream flooding and

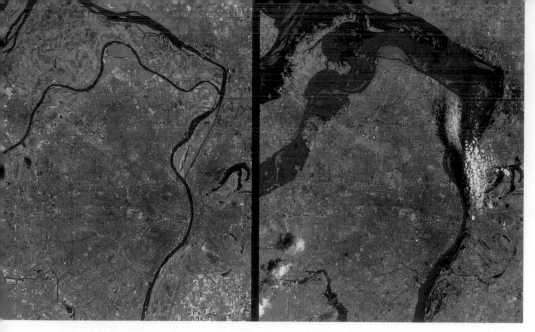

Figure 13-23 The satellite image on the left shows the area around St. Louis, Missouri, before the massive flood on July 4, 1988 (left), and the image on the right shows the same area on July 18, 1993, after the flood. Note the large increase in the flooded blue area (right). (Earth Satellite Corporation)

sediment deposition. Channelization also reduces habitats for aquatic wildlife by **(1)** removing bank vegetation, **(2)** increasing stream velocity, and **(3)** eliminating resting and hiding places for fish and other aquatic life.

Another approach is to build *artificial levees* along stream banks to reduce the chances of water overflow-ing into nearby floodplains. These walls may be permanent or temporary (such as sandbags placed when a flood is imminent).

Levees, like channelization, contain and speed up stream flow but increase the water's capacity for doing damage downstream. They also do not protect against unusually high and powerful floodwaters, as occurred in 1993 when two-thirds of the levees built along the Mississippi River were damaged or destroyed (Figure 13-23, right).

A *flood control dam* built across a stream can reduce flooding by storing water in a reservoir and releasing it gradually (Figure 13-9). However, this can reduce floods from prolonged rains only if the reservoir water level is kept low. Dam operators find it more profitable to keep water levels high for producing electricity and supplying irrigation water. As a result, after prolonged rains the reservoir can overflow, or operators may release large volumes of water to prevent overflow, thereby worsening the severity of flooding downstream.

Some flood control dams have failed, causing sudden, catastrophic flooding and threatening lives, property, and wildlife. According to the Federal Emergency Management Agency, the United States has about 1,300

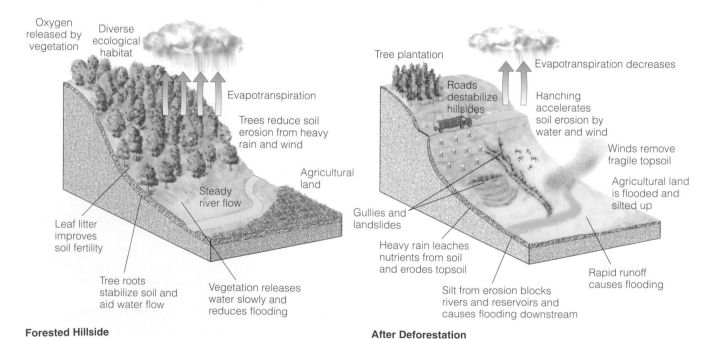

Forested Hillside

After Deforestation

Figure 13-24 A hillside before and after deforestation. Once a hillside has been deforested for timber and fuelwood, livestock grazing, or unsustainable farming, water from precipitation **(1)** rushes down the denuded slopes, **(2)** erodes precious topsoil, and **(3)** floods downstream areas. A 3,000-year-old Chinese proverb says, "To protect your rivers, protect your mountains."

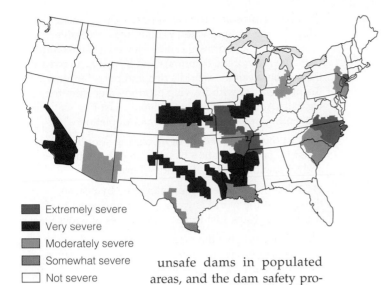

- Extremely severe
- Very severe
- Moderately severe
- Somewhat severe
- Not severe

unsafe dams in populated areas, and the dam safety programs of most states are inadequate.

Since 1934 the U.S. Army Corps of Engineers has spent almost $250 billion on channelization, levees, and dams, which the corps estimates has saved at least $380 billion in damages. Even so, the average financial losses from flood damage (adjusted for inflation) in the United States more than doubled between 1934 and 1999, partly because government-subsidized insurance programs have encouraged people to live in flood-prone areas.

After the devastating 1993 flood, the president of the United States commissioned a Flood Plain Management Task Force to reassess national flood policy. In its 1994 report, the task force recommended that **(1)** some towns built along rivers should be moved back so the floodplains can be restored to their natural condition and usefulness and **(2)** there should be less reliance on levees. Germany is bulldozing some of its levees to allow parts of floodplains to flood regularly.

From a human safety and damage viewpoint, *floodplain management* is considered the best approach. The first step is to construct a *flood frequency curve* (based on historical records and an examination of vegetation) to determine how often on average a flood of a certain size occurs in a particular area. This can be used to develop a generalized map (Figure 13-25) and more localized maps of flood-prone areas. Flood frequency data and maps of flood-prone areas do not tell us when floods will occur, but they give a general idea of how often and where floods might occur, based on an area's history.

Using these data, a plan is developed to **(1)** prohibit certain types of buildings or activities in high-risk flood zones, **(2)** elevate or otherwise floodproof buildings that are allowed on legally defined floodplains, and **(3)** construct a floodway that allows floodwater to flow through the community with minimal damage. Floodplain management based on thousands of years of experience can be summed up in one idea: *Sooner or later the river (or the ocean) always wins.*

13-9 SOLUTIONS: ACHIEVING A MORE SUSTAINABLE WATER FUTURE

Sustainable water use is based on the commonsense principle stated in an old Inca proverb: "The frog does not drink up the pond in which it lives." Figure 13-26 lists ways to implement this principle.

The challenge in developing such a *blue revolution* is to implement a mix of strategies built around **(1)** irrigating crops more efficiently, **(2)** using water-saving technologies in industries and homes, and **(3)** improving and integrating management of water basins and groundwater supplies.

Accomplishing such a revolution in water use and management will be difficult and controversial. However, water experts contend that not developing such strategies will eventually lead to **(1)** economic and health problems, **(2)** increased environmental degradation and loss of biodiversity, **(3)** heightened tensions and perhaps armed conflicts over water supplies, **(4)** larger

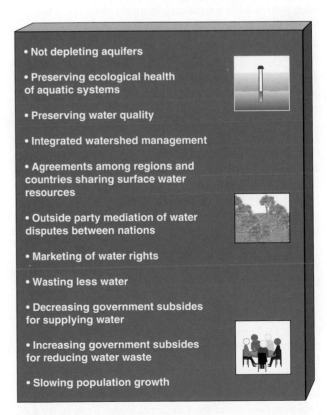

- Not depleting aquifers
- Preserving ecological health of aquatic systems
- Preserving water quality
- Integrated watershed management
- Agreements among regions and countries sharing surface water resources
- Outside party mediation of water disputes between nations
- Marketing of water rights
- Wasting less water
- Decreasing government subsides for supplying water
- Increasing government subsides for reducing water waste
- Slowing population growth

Figure 13-26 Methods for achieving more sustainable use of the earth's water resources.

numbers of environmental refugees, and **(5)** threats to national and global military, economic, and environmental security.

It is not until the well runs dry that we know the worth of water.

BENJAMIN FRANKLIN

REVIEW QUESTIONS

1. Define the boldfaced terms in this chapter.

2. Explain why there is a danger of water wars in the Middle East.

3. What two characteristics distinguish water (and air) from other resources?

4. List nine unique properties of water and explain the importance of each property.

5. What percentage of the earth's total volume of water is available for use by us? How might global warming alter the hydrologic cycle?

6. Distinguish between *surface runoff, reliable runoff, watershed, groundwater, zone of saturation, water table, aquifer, recharge area,* and *natural recharge.* Explain how the water in some aquifers can be depleted.

7. Since 1900 how much has the total use and per capita use of water by humans increased? About what percentage of the world's reliable surface runoff is used by humanity?

8. About what percentage of the water we withdraw each year is used for **(a)** irrigation, **(b)** industry, and **(c)** residences and cities?

9. What are the major water uses and problems of **(a)** the eastern United States and **(b)** the western United States?

10. List four causes of water scarcity. What is *water stress*? About how many people in the world live in countries suffering from water stress, and how many people are expected to face such shortages by 2025? How can growth in the number of water-stressed countries affect **(a)** global grain exports and prices, **(b)** hunger and malnutrition in developing countries, **(c)** the number of environmental refugees, and **(d)** the daily workload of many poor women and children?

11. List five ways to increase the supply of fresh water in a particular area.

12. List six ecological services provided by rivers. What physical and economic factors make it difficult to protect such services?

13. List the major pros and cons of building large dams and reservoirs to supply fresh water. List the pros and cons of **(a)** Egypt's Aswan dam project, **(b)** building numerous dams along the Colorado River basin, and **(c)** China's Three Gorges dam project.

14. List the major pros and cons of supplying water by transferring it from one watershed to another. List the pros and cons of **(a)** the California Water Project, **(b)** the James Bay project in Canada, and **(c)** the Aral Sea water transfer project in Central Asia.

15. List the pros and cons of supplying more water by withdrawing groundwater. Explain why excessive groundwater withdrawal can be viewed as an example of the tragedy of the commons and how it can increase the gap between the world's rich and poor. Summarize the problems of withdrawing groundwater from the Ogallala Aquifer in the United States. List five ways for preventing or slowing groundwater depletion.

16. List the pros and cons of increasing supplies of fresh water by **(a)** desalination of salt water, **(b)** cloud seeding, and **(c)** towing icebergs to water-short areas.

17. What percentage of the water used by people throughout the world is wasted? List five benefits of conserving water. List three major causes of water waste.

18. Define and give the pros and cons of **(a)** the *doctrine of riparian rights,* **(b)** the *principle of prior appropriation* used to govern legal rights to surface water, and **(c)** the *common law* approach used to govern legal rights to groundwater in the United States. How can such rights promote **(a)** water waste and **(b)** water conservation?

19. List ways to reduce water waste in **(a)** irrigation and **(b)** industry, homes, and businesses. List six advantages of *drip irrigation* and explain how it could help bring about a revolution in improving water efficiency. Explain why raising water prices can economically benefit the urban poor in many areas.

20. What is a *floodplain*? What three major services are provided by a floodplain? List four reasons why so many people live on floodplains. List the major benefits and disadvantages of floods. List four ways that humans increase the damages from floods. Describe the nature and causes of the flooding problems in Bangladesh.

21. List the pros and cons of trying to reduce flood risks by **(a)** stream channelization, **(b)** artificial levees and embankments, **(c)** dams, and **(d)** floodplain management.

22. List eleven ways to use the world's water more sustainably and five disadvantages of not implementing such strategies.

CRITICAL THINKING

1. What would happen to your body if suddenly your water molecules no longer formed hydrogen bonds with one another (Appendix 3, Figure 4)?

2. What would happen to plant life if the planet's water molecules did not form hydrogen bonds with one another?

3. How do human activities increase the harmful effects of prolonged drought? How can we reduce these effects?

4. Explain how dams and reservoirs can cause more flood damage than they prevent. Should all proposed large dam and reservoir projects be scrapped? Explain.

5. Do you believe that the projected benefits of China's massive Three Gorges dam and reservoir project on the Yangtze River will outweigh its potential drawbacks? Explain. What are the alternatives?

6. What role does population growth play in water supply problems?

7. What are the pros and cons of **(a)** gradually phasing out government subsidies of irrigation projects in the western United States (or in the country where you live) to increase water conservation and **(b)** providing government subsidies to farmers for improving irrigation efficiency?

8. Should the prices of water for all uses be raised sharply to include more of its environmental costs and to encourage water conservation? Explain. What harmful and beneficial effects might this have on **(a)** business and jobs, **(b)** your lifestyle and the lifestyles of any children or grandchildren you might have, **(c)** the poor, and **(d)** the environment?

9. Should we use up slowly renewable underground water supplies such as the Ogallala aquifer (Figure 13-18) or save them for future generations? Explain.

10. Critics contend that the current water system for supplying water for use in homes and business in developed countries should be changed to a system *that matches the quality of the water to its use.* They argue that it is too costly and wasteful to have all water supplied for domestic use pure enough to drink when most of this water is used for bathing, washing clothes, and flushing toilets that do not need such high-grade water. They recommend that two grades of water be supplied for such purposes. Others recommend encouraging homes and businesses to install or lease systems for recycling all domestic wastewater to solve this problem. What are the pros and cons of each of these approaches? Which one, if either, do you favor? Explain.

11. Calculate how many liters and gallons of water are wasted in 1 month by a toilet that leaks two drops of water per second (1 liter of water equals about 3,500 drops and 1 liter equals 0.265 gallon).

12. List five major ways to conserve water for personal use (see Appendix 6). Which, if any, of these practices do you now use or intend to use?

13. How do human activities contribute to flooding and flood damage? How can these effects be reduced?

PROJECTS

1. In your community,
    a. What are the major sources of the water supply?
    b. How is water use divided between agricultural, industrial, power plant cooling, and public uses? Who are the biggest consumers of water?
    c. What has happened to water prices during the past 20 years? Are they too low to encourage water conservation and reuse?
    d. What water supply problems are projected?
    e. How is water being wasted?

2. Use the library or the internet to discover **(a)** which industries in the country where you live use the most water and **(b)** which industries have done the most to improve water use efficiency in the last 20 years.

3. Develop a water conservation plan for your school and submit it to school officials.

4. Consult with local officials to identify any floodplain areas in your community. Develop a map showing these areas and the types of activities (such as housing, manufacturing, roads, and recreational use) found on these lands.

5. Use the library or the internet to find bibliographic information about *John Todd* and *Benjamin Franklin*, whose quotes appear at the beginning and end of this chapter.

6. Make a concept map of this chapter's major ideas, using the section heads and subheads and the key terms (in boldface). Look at the inside back cover and on the website of this book for information about making concept maps.

INTERNET STUDY RESOURCES AND RESOURCES FOR FURTHER READING AND RESEARCH

The website for this book contains helpful study aids and many ideas for further reading and research. Log on to:

http://www.brookscole.com/product/0534376975s

and click on the Chapter-by-Chapter area. Choose Chapter 13 and select a resource:

- "Flash Cards" allows you to test your mastery of the Terms and Concepts to Remember for this chapter.

- "Tutorial Quizzes" provides a multiple-choice practice quiz.

- "Student Guide to InfoTrac" will lead you to Critical Thinking Projects that use InfoTrac College Edition as a research tool.

- "References" lists the major books and articles consulted in writing this chapter.

- "Hypercontents" takes you to an extensive list of sites with news, research, and images related to individual sections of the chapter.

INFOTRAC COLLEGE EDITION

Improve your skills with InfoTrac College Edition, a searchable online database of articles from more than 700 periodicals. Log on to:

http://www.infotrac-college.com

or access InfoTrac through the website for this book.

Try the following articles:

Barlow, M. 1999. We are running out of water. *The Ecologist* vol. 29, no. 3, pp. 182–183. (subject guide: water, shortage)

D.M. Rosenberg; P. McCully; C.M. Pringle. 2000. Global-scale environmental effects of hydrological alterations: introduction. *BioScience* vol. 50, no. 9, pp. 746–751. (subject guide: dams)

Controversy over the General Mining Law of 1872

Some people have gotten rich by using the little-known General Mining Law of 1872. This law was designed to (1) encourage mineral exploration and mining of gold, silver, copper, zinc, nickel, uranium, and other *hardrock minerals* on U.S. public lands and (2) help develop the then largely unpopulated West.

Under this 1872 law, a person or corporation has been able to assume legal ownership of parcels of land on essentially all U.S. public land except parks and wilderness areas by *patenting* it. This involves (1) declaring their belief that the land contains valuable hardrock minerals, (2) spending $500 to improve the land for mineral development, (3) filing a claim, (4) paying an annual fee of $100 to maintain the claim, and (5) if desired, paying the federal government $6–12 per hectare ($2.50–5.00 an acre) for the land. Once purchased, the land can be used, leased, or sold for essentially any purpose.

So far, public lands containing an estimated $240–385 billion (adjusted for inflation) of publicly owned mineral resources have been transferred to private interests at 1872 prices. Domestic and foreign mining companies operating under this law remove mineral resources worth about $2–3 billion per year on once-public land they bought at very low prices. Coal, oil, and natural gas companies pay an 8–16% royalty on the net value of fossil fuels they remove from public lands, but hardrock mining companies pay no royalties on the minerals they extract from such lands.

There is also no provision in the 1872 law requiring mining companies to pay for environmental cleanup of any damage they cause to these lands and nearby water resources. It is estimated that cleanup costs for land and streams (Figure 14-1) damaged by 557,000 abandoned hardrock mines and open pits (mostly in the West) will cost U.S. taxpayers $33–72 billion.

Mining companies defend the 1872 law. They argue that they must invest large sums of money (often $100 million or more) to locate and develop an ore site before any profits are made from mining hardrock minerals. They also point out that their mining operations (1) provide high paying jobs to miners, (2) supply vital resources for industry, (3) stimulate the national and local economies, (4) reduce trade deficits, and (5) save American consumers money on products produced from such minerals.

Mining companies also contend that (1) less than 0.25% of U.S. public lands have been have been transferred to private ownership, and (2) paying royalties on their profits and requiring them to pay cleanup costs would force them to move their mining operations to other countries.

For decades, environmentalists have been trying, without success, to have this law revised to protect taxpayers and the environment by (1) permanently banning the patenting (sale) of public lands but allowing 20-year leases of designated public land for hardrock mining, (2) requiring mining companies to pay an 8–12% royalty on the *net* value of all minerals removed from public land, and (3) making mining companies legally and financially responsible for environmental cleanup and restoration of each site or charging them an additional mining fee to help pay for cleanup costs. Canada, Australia, South Africa, and other countries that are major extractors of hardrock minerals have laws with such requirements.

Figure 14-1 Bear Trap Creek in Montana is one example of how gold mining can contaminate water with highly toxic cyanide or mercury used to extract gold from its ore. Air and water also convert the sulfur in gold ore to sulfuric acid, which releases toxic metals such as cadmium and copper into streams and groundwater. (Bryan Peterson)

Mineral resources are the building blocks on which modern society depends. Knowledge of their physical nature and origins, the web they weave between all aspects of human society and the physical earth can lay the foundations for a sustainable society.

ANN DORR

This chapter addresses the following questions:

- What are nonrenewable mineral resources, and how are they formed?

- How do we find and extract nonrenewable mineral and energy resources from the earth's crust?

- What are the environmental effects of extracting and using mineral resources?

- How fast are nonfuel mineral supplies being used up?

- How can we increase supplies of key nonfuel minerals?

- How should we evaluate energy alternatives?

- What are the advantages and disadvantages of oil?

- What are the advantages and disadvantages of natural gas?

- What are the advantages and disadvantages of coal?

- What are the advantages and disadvantages of conventional nuclear fission, breeder nuclear fission, and nuclear fusion?

14-1 NATURE AND FORMATION OF MINERAL RESOURCES

What Are Mineral Resources? A mineral resource is a concentration of naturally occurring material in or on the earth's crust that can be extracted and processed into useful materials at an affordable cost. Over millions to billions of years the earth's internal and external geologic processes (Section 10-2, p. 212, and Figure 10-8, p. 217) have produced numerous nonfuel mineral resources and energy resources. Because they take so long to produce, they are classified as *nonrenewable resources.*

We know how to find and extract more than 100 nonrenewable minerals from the earth's crust. They include **(1)** *metallic mineral resources* (iron, copper, aluminum), **(2)** *nonmetallic mineral resources* (salt, gypsum, clay, sand, phosphates, water, and soil), and **(3)** *energy resources* (coal, oil, natural gas, and uranium).

Ore is rock containing enough of one or more metallic minerals to be mined profitably. We convert about 40 metals extracted from ores into many everyday items that we either **(1)** use and discard (Figure 3-19, p. 66) or **(2)** learn to reuse, recycle, or use less wastefully (Figure 3-20, p. 67).

The U.S. Geological Survey (USGS) divides nonrenewable mineral resources into two broad categories: *identified* and *undiscovered* (Figure 14-2). **Identified resources** are deposits of a nonrenewable mineral resource that have a *known* location, quantity, and quality, or deposits based on direct geological evidence and measurements. **Undiscovered resources** are potential supplies of a nonrenewable mineral resource that are assumed to exist on the basis of geologic knowledge and theory (although specific locations, quality, and amounts are unknown).

Reserves are identified resources from which a usable nonrenewable mineral can be extracted profitably at current prices. **Other resources** are identified and undiscovered resources not classified as reserves.

Most published estimates of the supply of a given nonrenewable resource refer to *reserves.* Reserves can increase **(1)** when new deposits are found or **(2)** when price increases or improved mining technology make it profitable to extract deposits that previously were too expensive to extract. Theoretically, all of the *other resources* could eventually be converted to reserves, but this is highly unlikely.

How Do Ores Form from Magma? Ores form as a result of several internal and external geologic processes.

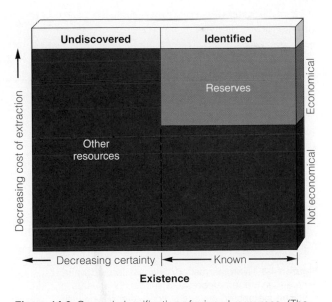

Figure 14-2 General classification of mineral resources. (The area shown for each class does not represent its relative abundance.) In theory, all mineral resources classified as *other resources* could become reserves because of rising mineral prices or improved mineral location and extraction technology. In practice, geologists expect only a fraction of other resources to become reserves.

Plate tectonics (Figure 10-5b, p. 214) **(1)** shapes the earth's crust as the earth's plates collide, retreat, and slide across one another at the boundaries between them (Figure 10-6, p. 215) and **(2)** determines where the earth's richest mineral deposits form.

One way this happens is when movement of the earth's plates allows *magma* (molten rock) to flow up into the earth's crust at divergent and convergent plate boundaries (Figure 10-6, p. 215). As this magma cools, it crystallizes into various layers of mineral-containing igneous rocks.

The most common way that ore deposits are formed is through *hydrothermal processes*. When two tectonic plates retreat from one another, gaps created in the earth's crust fill with upwelling magma and seawater. The seawater seeping into these cracks becomes super-heated and dissolves metals from rock or magma. As these metal-bearing solutions cool, their dissolved minerals cool and form *hydrothermal ore deposits*.

Sometimes these hydrothermal solutions form rich veins of ore in fractured rock on land or soak into rock pores near veins to form less concentrated ore deposits. Other hydrothermal ore deposits form on the seafloor as dissolved minerals precipitate out and sink to the seabed, where they are often buried under sediment.

Hydrothermal ore deposits also occur when upwelling magma solidifies into chimney-shaped *black smokers* (Figure 14-3) in volcanically active regions of the ocean floor near spreading oceanic ridges (Figure 10-3, p. 213). These smokers are miniature volcanoes that shoot jets of hot, black, mineral-rich water through vents in solidified magma on the seafloor. As the hot water comes into contact with cold seawater, black particles of metal sulfides precipitate out and accumulate as chimneylike structures near the hot water vents (Figure 14-3).

Because they are so rich in nutrients (especially sulfur), these hydrothermal deposits on the dark ocean floor support colonies of bacteria that produce food through *chemosynthesis* and support a variety of animals such as red tube worms, clams, crustaceans, and other forms of marine life (Figure 14-3). These ore deposits formed in underwater volcanic environments are especially rich in copper, lead, zinc, silver, gold, and other minerals.

Another potential source of metals from the ocean floor is *manganese nodules* that cover about

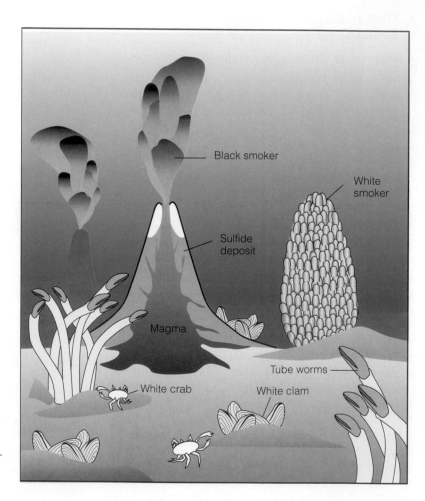

Figure 14-3 Hydrothermal ore deposits formed when mineral-rich, superheated water shoots out of vents in solidified magma on the ocean floor. After mixing with cold seawater, black particles of metal ore precipitate out and build up as chimneylike ore deposits around the vents. A variety of organisms, supported by bacteria producing food by chemosynthesis, exist in the dark ocean around the black smokers.

25–50% of the Pacific Ocean floor. These cherry- to potato-sized rocks contain 30–40% manganese by weight and small amounts of other important metals such as iron, copper, and nickel. Geologists believe that these ore modules crystallized from hot solutions arising from volcanic activity at midoceanic ridges, perhaps by black smokers.

How Do Ores and Other Minerals Form from Sedimentary and Weathering Processes? As sediments settle they can form ore deposits by *sedimentary sorting* and *precipitation*. For example, many streams carry a mixture of silt, sand, gravel, and occasional small grains of gold. When the stream current slows down these minerals settle out on the basis of their density. Because gold is denser than any other mineral, it falls to the bottom of a stream first in this *sedimentary sorting* process. Over time fairly rich deposits of settled gold particles, called *placer deposits*, concentrate near bedrock or coarse gravel in streams. Miners use pans or sieves to scoop up such streambed material and sort out the gold particles.

In deserts, groundwater flows can form lakes with no outlets to the sea. Some of the mineral-containing groundwater flowing into these lakes (or into enclosed seas) evaporates. This causes the concentrations of the dissolved salts to increase to the point where they precipitate to form *evaporite mineral deposits*. Important minerals formed this way include table salt, borax, and sodium carbonate.

Ore deposits can also form because of the *weathering* of rock by water in areas with high rainfall. Torrents of water dissolve and remove most soluble metal ions from rock and soil near the earth's surface. This weathering process leaves ions of insoluble compounds in the soil to form *residual deposits* of metal ores such as iron and aluminum (bauxite ore).

14-2 FINDING AND REMOVING NONRENEWABLE MINERAL RESOURCES

How Are Buried Mineral Deposits Found? Mining companies use several methods to find promising mineral deposits. They include:

- Using aerial photos and satellite images to reveal protruding rock formations (outcrops) associated with certain minerals.

- Using planes equipped with *radiation measuring equipment* to detect deposits of radioactive metals such as uranium and with a *magnetometer* to measure changes in the earth's magnetic field caused by magnetic minerals such as iron ore.

- Using a *gravimeter* to measure differences in gravity because the density of an ore deposit usually differs from that of the surrounding rock.

- Drilling a deep well and extracting core samples.

- Putting sensors in existing wells to detect electrical resistance or radioactivity to pinpoint the location of oil and natural gas.

- Making *seismic surveys* on land and at sea by detonating explosive charges that send shock waves to get information about the makeup of buried rock layers.

- Performing *chemical analysis* of water and plants to detect deposits of underground minerals that have leached into nearby bodies of water or have been absorbed by plant tissues.

After suitable mineral deposits are located, several different types of mining techniques are used to remove the deposits, depending on their location and type. Shallow deposits are removed by **surface mining** (Figure 14-4), and deep deposits are removed by **subsurface mining** (Figure 14-5).

In surface mining, mechanized equipment strips away the **overburden** of soil and rock and usually discards it as waste material called **spoil**. In the United States, surface mining extracts about 90% of the nonfuel mineral and rock resources and about 60% of the coal by weight.

The type of surface mining used depends on the resource being sought and on local topography. Methods include the following:

- **Open-pit mining** (Figure 14-4a), in which machines dig holes and remove ores such as iron and copper and sand, gravel, and stone (such as limestone and marble).

- **Dredging** (Figure 14-4b), in which chain buckets and draglines scrape up underwater mineral deposits.

- **Area strip mining** (Figure 14-4c), used where the terrain is fairly flat. An earthmover strips away the overburden, and a power shovel digs a cut to remove the mineral deposit. After the mineral is removed, the trench is filled with overburden and a new cut is made parallel to the previous one. The process is repeated over the entire site. If the land is not restored, area strip mining leaves a wavy series of highly erodible hills of rubble called *spoil banks*.

- **Contour strip mining** (Figure 14-4d), used on hilly or mountainous terrain. A power shovel cuts a series

(a) Open Pit Mine

(b) Dredging

(c) Area Strip Mining

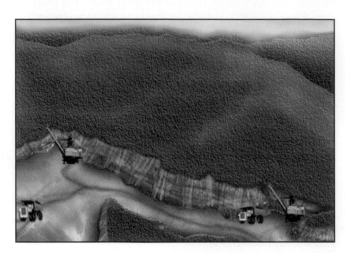

(d) Contour Strip Mining

Figure 14-4 Major mining methods used to extract surface deposits of solid mineral and energy resources. (National Archives/EPA Documerica)

of terraces into the side of a hill. An earthmover removes the overburden, and a power shovel extracts the coal, with the overburden from each new terrace dumped onto the one below. Unless the land is restored, a wall of dirt is left in front of a highly erodible bank of soil and rock called a *highwall*.

Surface-mined land can be restored (except in arid and semiarid areas), but this is expensive and is not done in many countries. In the United States, the Surface Mining Control and Reclamation Act of 1977 requires mining companies to restore most surface-mined land so that it can be used for the same purpose as it was before it was mined.

The law also levied a tax on mining companies to restore land that was disturbed by surface mining before the law was passed. However, more than 6,000 abandoned coal and metal mines, covering an area

of about the size of the state of Virginia, have not been restored. An even larger area of abandoned rock quarries and gravel and sand mines has not been reclaimed.

Subsurface mining (Figure 14-5) is used to remove coal and various metal ores that are too deep to be extracted by surface mining. Miners **(1)** dig a deep vertical shaft, **(2)** blast subsurface tunnels and chambers to get to the deposit, and **(3)** use machinery to remove the ore or coal and transport it to the surface.

Subsurface mining disturbs less than one-tenth as much land as surface mining and usually produces less waste material. However, it leaves much of the resource in the ground and is more dangerous and expensive than surface mining. Hazards include **(1)** collapse of roofs and walls of underground mines, **(2)** explosions of dust and natural gas, and **(3)** lung diseases caused by prolonged inhalation of mining dust.

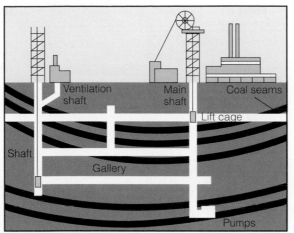

Figure 14-5 Major mining methods used to extract underground deposits of solid mineral and energy resources. **(a)** Mine shafts and tunnels are dug and blasted out. **(b)** In *room-and-pillar* mining, machinery is used to gouge out coal and load it onto a shuttle car in one operation, and pillars of coal are left to support the mine roof. **(c)** In *longwall coal mining*, movable steel props support the roof and cutting machines shear off the coal onto a conveyor belt. As the mining proceeds, roof supports are moved forward and the roof behind is allowed to fall.

(a) Underground Coal Mine

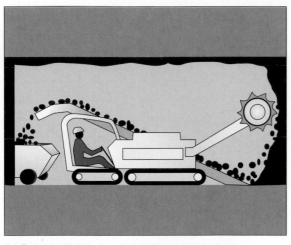

(b) Room-and-pillar

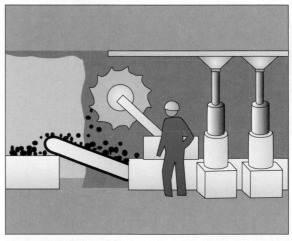

(c) Longwall Mining of Coal

14-3 ENVIRONMENTAL EFFECTS OF EXTRACTING, PROCESSING, AND USING MINERAL RESOURCES

What Are the Environmental Impacts of Using Mineral Resources? The mining, processing, and use of mineral resources takes enormous amounts of energy and often causes land disturbance, soil erosion, and air and water pollution (Figure 14-6).

Mining can affect the environment in several ways, including:

- Scarring and disruption of the land surface (Figure 14-4).

- Collapse or subsidence of land above underground mines, which can cause houses to tilt, sewer lines to crack, gas mains to break, and groundwater systems to be disrupted.

- Wind- or water-caused erosion of toxin-laced mining wastes.

- Acid mine drainage, when rainwater seeping through a mine or mine wastes **(1)** carries sulfuric acid (H_2SO_4, produced when aerobic bacteria act on iron sulfide minerals in spoil) to nearby streams and groundwater (Figure 14-7), **(2)** contaminates water supplies, and **(3)** destroys aquatic life.

- Emissions of toxic chemicals into the atmosphere. In the United States, the mining industry produces more toxic emissions than any other industry (accounting for 48% of such emissions in 1998).

- Exposure of wildlife to toxic mining wastes stored in holding ponds and leakage of toxic wastes from such ponds (Figure 14-1).

Figure 14-8 shows the typical life cycle of metal resource. Ore extracted from the earth's crust typically

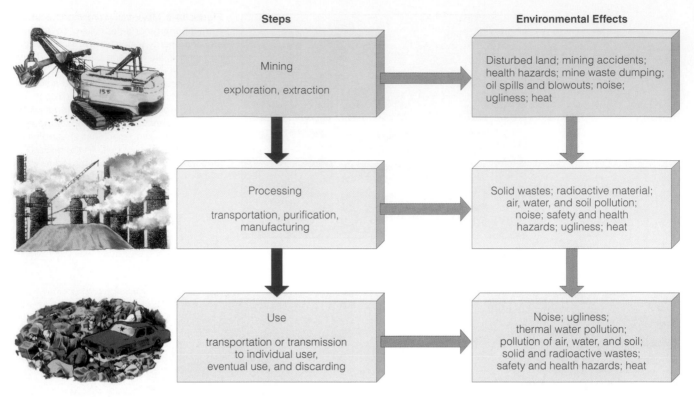

Steps
Environmental Effects
Mining
exploration, extraction
Disturbed land; mining accidents; health hazards; mine waste dumping; oil spills and blowouts; noise; ugliness; heat
Processing
transportation, purification, manufacturing
Solid wastes; radioactive material; air, water, and soil pollution; noise; safety and health hazards; ugliness; heat
Use
transportation or transmission to individual user, eventual use, and discarding
Noise; ugliness; thermal water pollution; pollution of air, water, and soil; solid and radioactive wastes; safety and health hazards; heat

Figure 14-6 Some harmful environmental effects of extracting, processing, and using nonrenewable mineral and energy resources. The energy used to carry out each step causes additional pollution and environmental degradation.

has two components: **(1)** the *ore mineral* containing the desired metal and **(2)** waste material called *gangue*. Removing the gangue from ores produces piles of waste called *tailings***.** Particles of toxic metals blown or leached from tailings by rainfall can contaminate surface water and groundwater.

Most ores consist of one or more compounds of the desired metal. After gangue has been removed, **smelting** is used to separate the metal from the other elements in the ore mineral. Without effective pollution control equipment, smelters emit enormous quantities of air pollutants, which damage vegetation and soils in the surrounding area. It can take decades for vegetation in such areas to be restored by a combination of secondary ecological succession (Figure 8-16, p. 189) and expensive restoration efforts.

Subsurface Mine Opening
Surface Mine
Runoff of sediment
Acid drainage from reaction of mineral or ore with water
Spoil banks
Leaching may carry acids into soil and groundwater supplies
Percolation to groundwater
Leaching of toxic metals and other compounds from mine spoil

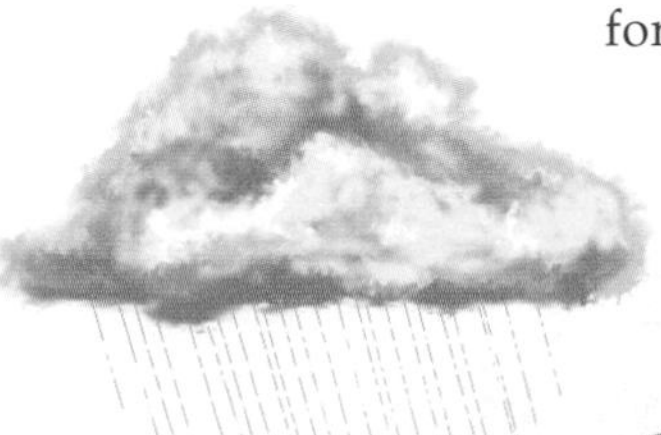

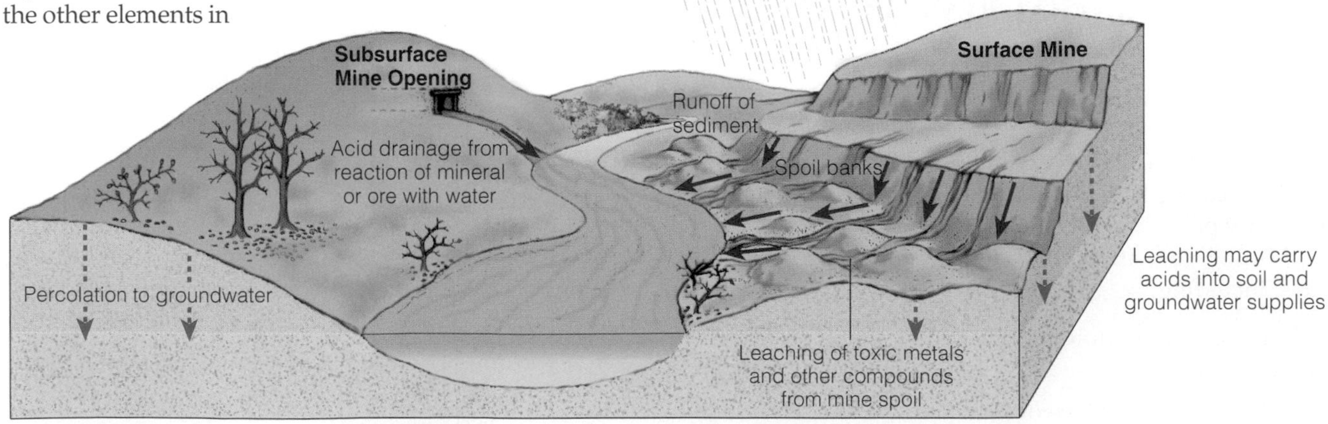

Figure 14-7 Pollution and degradation of a stream and groundwater by runoff of acids—called *acid mine drainage*—and by toxic chemicals from surface and subsurface mining. These substances can kill fish and other aquatic life. Acid mine drainage has damaged more than 19,000 kilometers (12,000 miles) of streams in the United States, mostly in Appalachia and the West.

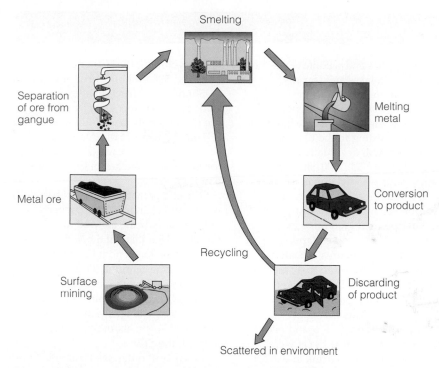

Smelting

Separation
of ore from
gangue

Melting
metal

Metal ore

Conversion
to product

Recycling

Surface
mining

Discarding
of product

Scattered in environment

Figure 14-8 Typical life cycle of a metal resource. Each step in this process uses energy and produces some pollution and waste heat.

Smelters also cause water pollution and produce liquid and solid hazardous wastes that must be disposed of safely. Some companies are using improved technology to (1) reduce pollution from smelting, (2) lower production costs, (3) save costly cleanup bills, and (4) decrease liability for damages.

Once the pure metal has been produced by smelting, it is usually melted and converted to desired products, which are then used and discarded or recycled (Figure 14-8).

Are There Environmental Limits to Resource Extraction and Use?
Some environmentalists and resource experts believe that the greatest danger from the world's continually increasing levels of resource consumption is not the exhaustion of nonrenewable resources but the environmental damage caused by their extraction, processing, and conversion to products (Figure 14-6).

The mineral industry accounts for 5–10% of world energy use, making it a major contributor to air and water pollution and to emissions of greenhouse gases such as carbon dioxide. The environmental impacts from mining an ore are affected by its percentage of metal content, or *grade* (Case Study, p. 328). Usually more accessible and higher-grade ores are exploited first. As they are depleted, it takes more money, energy, water, and other materials to exploit lower-grade ores, and land disruption, mining waste, and pollution increase accordingly.

Currently, most of the harmful environmental costs of mining and processing minerals are not included in the prices for processed metals and consumer products produced from such metals. This gives mining companies and manufacturers little incentive to reduce resource waste and pollution because they can pass many of the harmful environmental costs of their production on to society and future generations.

14-4 SUPPLIES OF MINERAL RESOURCES

Will There Be Enough Mineral Resources?
The future supply of nonrenewable minerals depends on two factors: (1) the actual or potential supply and (2) the rate at which that supply is used. We never completely run out of any mineral. However, a mineral becomes *economically depleted* when it costs more to find, extract, transport, and process the remaining deposit than it is worth. At that point, there are five choices: (1) Recycle or reuse existing supplies, (2) waste less, (3) use less, (4) find a substitute, or (5) do without.

Depletion time is the time it takes to use up a certain proportion (usually 80%) of the reserves of a mineral at a given rate of use (Figure 1-12, p. 13). When experts disagree about depletion times, they are often using different assumptions about supply and rate of use (Figure 14-9).

A traditional measure of the projected availability of nonrenewable resources is the **reserve-to-production ratio**: the number of years that proven reserves of a particular nonrenewable mineral will last at current annual production rates. Reserve estimates are continually changing because new deposits often are discovered, and new mining and processing can allow some of the minerals classified as other resources (Figure 14-2) to be converted to reserves. Under these circumstances, the reserve-to-production ratio is the best available projection of the current estimated supply and its estimated depletion time.

The shortest depletion time assumes no recycling or reuse and no increase in reserves (curve A, Figure 14-9). A longer depletion time assumes that recycling will stretch existing reserves and that better mining technology, higher prices, and new discoveries will increase reserves (curve B, Figure 14-9). An even longer depletion time assumes that new discoveries will further expand reserves and that recycling, reuse, and

CASE STUDY

Gold miners typically remove ore equal to the weight of 50 automobiles to extract an amount of gold that would fit inside your clenched fist. Most newlyweds would be surprised to know that about 5.5 metric tons (6 tons) of mining waste was created to make their two gold wedding rings.

The solid waste remaining after gold is extracted from its ore is left piled near the mine sites and can pollute the air, surface water (Figures 14-1 and 14-7), and groundwater. In South Africa, each metric ton of gold that is mined on average kills 1 worker and seriously injures 11 others.

In Australia and North America, a mining technology called *cyanide heap leaching* is cheap enough to allow mining companies to level entire mountains containing very low-grade gold ore. To extract the gold, miners spray a cyanide solution (which reacts with gold) onto huge open-air piles of crushed ore. They then **(1)** collect the solution in leach beds and overflow ponds, **(2)** recirculate it a number of times, and **(3)** extract gold from it.

A problem is that cyanide is extremely toxic to birds and mammals drawn to cyanide solution collection ponds as a source of water. Cyanide leach pads and collection ponds can also leak or overflow, posing threats to underground drinking water supplies and wildlife (especially fish) in lakes and streams.

Special liners beneath the ore heaps and in the collection ponds can prevent leaks, but some have failed. According to the EPA, all such liners will eventually leak.

On January 30, 2000, snow and heavy rains washed out an earthen dam containing an aboveground cyanide leach pond at an Australian-operated gold mine in Romania. The dam's collapse released large amounts of water laced with cyanide and toxic metals into the Tisza and Danube rivers flowing through parts of Romania, Hungary, and Yugoslavia.

Several hundred thousand people living along these rivers were told not to fish or to drink or extract water from affected rivers or from wells along the rivers. Food industries and paper mills were shut down. Thousands of fish and other forms of aquatic life were killed. The accident could have been prevented if the mine had installed a stronger containment dam and a backup collection pond to prevent leakage into nearby surface water.

Some gold mining companies take care to avoid environmental damage, but other companies do not. A glaring example is the Summitville gold mine site in the San Juan Mountains of southern Colorado. A Canadian company used the 1872 mining law (p. 320) to buy the land from the federal government for a pittance, spent $1 million developing the site, and then removed $98 million worth of gold.

Shoddy construction allowed acids and toxic metals to leak from the site and poison a 27-kilometer (17-mile) stretch of the Alamosa River, the source of irrigation water for farms and ranches in Colorado's San Luis Valley.

The company then declared bankruptcy and abandoned the property, but only after being allowed to retrieve $2.5 million of the $7.5 million reclamation bond it had posted with the state. Summitville is now a Superfund site, with the EPA spending $40,000 a day to contain the site's toxic wastes. Ultimately, the EPA expects to spend about $120 million to finish the cleanup.

Since 1980 millions of miners have streamed into tropical forests and other areas in search of gold. These small-scale miners use destructive mining techniques such as **(1)** digging large pits by hand, **(2)** river dredging, and **(3)** hydraulic mining (a technique, outlawed in the United States, in which water jets wash entire hillsides into sluice boxes).

Highly toxic mercury usually is used to extract the gold from the other materials. In the process, much of the mercury ends up contaminating water supplies and fish consumed by people.

Critical Thinking

What regulations, if any, would you impose on gold mining operations?

---

reduced consumption will extend supplies (curve C, Figure 14-9). Finding a substitute for a resource leads to a new set of depletion curves for the new resource.

Case Study: Mineral Resources in the United States Three countries—the United States, Germany, and Russia—with only 8% of the world's population consume about 75% of the world's most widely used metals. The United States, with 4.5% of the world's population, uses about 20% of the world's metal production and 25% of the fossil fuels produced each year.

Each year Americans directly or indirectly consume about 11 metric tons (12 tons) of nonfuel miner-als and another 26 metric tons (24 tons) of wood, paper, chemicals, plastics, fossil fuels, and other resources each year. Researchers at the University of British Columbia estimate that trying to sustain the entire world at an American (or Canadian) level of resource use would take the land area of three planets the size of the earth (Figure 1-10, p. 11).

No single nation is self-sufficient in all key minerals because the earth's geological processes have not distributed the world's mineral resources evenly. Five nations—the United States, Canada, Russia, South Africa, and Australia—supply most of the nonrenewable mineral resources used by modern societies.

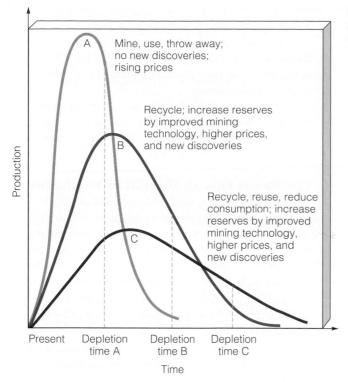

A — Mine, use, throw away; no new discoveries; rising prices

B — Recycle; increase reserves by improved mining technology, higher prices, and new discoveries

C — Recycle, reuse, reduce consumption; increase reserves by improved mining technology, higher prices, and new discoveries

Production

Present | Depletion time A | Depletion time B | Depletion time C

Time

Figure 14-9 Depletion curves for a nonrenewable resource (such as aluminum or copper) using three sets of assumptions. Dashed vertical lines represent times when 80% depletion occurs.

Currently, the United States depends on 25 other countries for more than half of its mineral resources (Figure 14-10). Some of these minerals are imported because they are used faster than they can be produced from domestic supplies. Others are imported because foreign ore deposits are of a higher grade and cheaper to extract than remaining U.S. reserves.

How Does Economics Affect Mineral Resource Supplies? Geologic processes determine the quantity and location of a mineral resource in the earth's crust, but economics determines what part of the known supply is used.

According to standard economic theory, in a competitive free market a plentiful mineral resource is cheap when its supply exceeds demand. However, when a resource becomes scarce its price rises. This can (1) encourage exploration for new deposits, (2) stimulate development of better mining technology, (3) make it profitable to mine lower-grade ores, and (4) encourage resource conservation.

However, according to some economists this theory may no longer apply to most developed countries. One reason is that industry and government in such countries control the supply, demand, and prices of minerals to such a large extent that a truly competitive free market does not exist (p. 691).

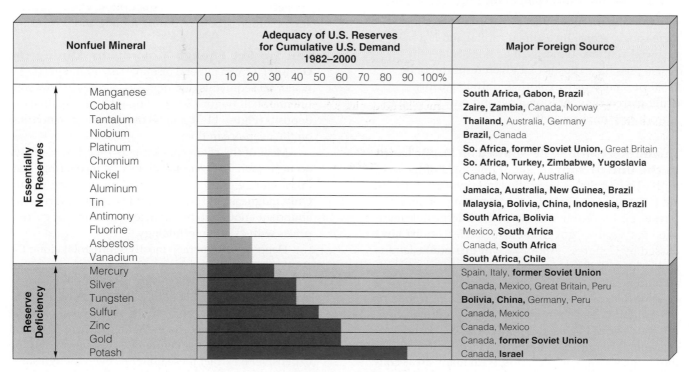

Nonfuel Mineral	Adequacy of U.S. Reserves for Cumulative U.S. Demand 1982–2000	Major Foreign Source
	0 10 20 30 40 50 60 70 80 90 100%	
Essentially No Reserves — Manganese		**South Africa, Gabon, Brazil**
Cobalt		**Zaire, Zambia,** Canada, Norway
Tantalum		**Thailand,** Australia, Germany
Niobium		**Brazil,** Canada
Platinum		**So. Africa, former Soviet Union,** Great Britain
Chromium		**So. Africa, Turkey, Zimbabwe, Yugoslavia**
Nickel		Canada, Norway, Australia
Aluminum		**Jamaica, Australia, New Guinea, Brazil**
Tin		**Malaysia, Bolivia, China, Indonesia, Brazil**
Antimony		**South Africa, Bolivia**
Fluorine		Mexico, **South Africa**
Asbestos		Canada, **South Africa**
Vanadium		**South Africa, Chile**
Reserve Deficiency — Mercury		Spain, Italy, **former Soviet Union**
Silver		Canada, Mexico, Great Britain, Peru
Tungsten		**Bolivia, China,** Germany, Peru
Sulfur		Canada, Mexico
Zinc		Canada, Mexico
Gold		Canada, **former Soviet Union**
Potash		Canada, **Israel**

Figure 14-10 Evaluation of the supply of selected nonfuel minerals in the United States and major foreign sources of these minerals in 2000. Foreign sources subject to potential interruption of supply by political, economic, or military disruption are shown in boldface. Almost all U.S. supplies of four important minerals—manganese, cobalt, platinum, and chromium—are imported from potentially unstable countries. (Data from U.S. Geological Survey and U.S. Bureau of Mines)

Most mineral prices are low because countries subsidize development of their domestic mineral resources to help promote economic growth and national security. In the United States, for instance, mining companies (1) get depletion allowances amounting to 5–22% of their gross income (depending on the mineral) and (2) can deduct much of their costs for finding and developing mineral deposits. In addition, hardrock mining companies operating in the United States can buy public land at 1872 prices and pay no royalties to the government on the minerals they extract (p. 320).

Between 1982 and 2000, U.S. mining companies received more than $6 billion in government subsidies. Critics argue that taxing rather than subsidizing the extraction of nonfuel mineral resources would (1) provide governments with revenue, (2) create incentives for more efficient resource use, (3) promote waste reduction and pollution prevention, and (4) encourage recycling and reuse.

Mining company representatives say they need subsidies and low taxes to (1) keep the prices of minerals low for consumers and (2) encourage them not to move their mining operations to other countries without such taxes and less stringent mining regulations.

Two other economic problems also hinder the development of new supplies of mineral resources:

- Mineral scarcity does not raise the market price of products very much because the cost of mineral resources is only a small part of the final cost of goods.

- Exploring for new mineral resources takes lots of increasingly scarce investment capital and is a risky financial venture. Typically, if geologists identify 10,000 possible deposits of a given resource, only 1,000 sites are worth exploring; only 100 justify drilling, trenching, or tunneling; and only 1 becomes a producing mine or well.

Should More Mining Be Allowed on Public Lands in the United States? About one-third of the land in the United States is public land owned jointly by all U.S. citizens. This land, consisting of national forests, parks, resource lands, and wilderness, is managed by various government agencies under laws passed by Congress. About 72% of this public land is in Alaska and 22% is in western states (where 60% of all land is public land).

For decades, resource developers, environmentalists, and conservationists have argued over how this land should be used (Spotlight, p. 32). Extractors of mineral resources complain that three-fourths of the country's vast public lands, with many areas containing rich deposits of mineral resources, are off limits to mining.

In recent decades, they have stepped up efforts, in what is known as the Sagebrush Rebellion (p. 35) and

the anti-environmental Wise-Use movement (p. 589), to have Congress (1) open up most of these lands to mineral development or (2) return ownership of many of these lands to states and allow state legislatures to determine how they should be used or sell these lands to private companies.

Conservation biologists and environmentalists strongly oppose such efforts. They argue that this would increase environmental degradation and decrease biodiversity (Section 23-1, p. 586).

Can We Get Enough Minerals by Mining Lower-Grade Ores? Some analysts contend that all we need to do to increase supplies of a mineral is to extract lower grades of ore. They point to new earth-moving equipment, improved techniques for removing impurities, and other technological advances in mineral extraction and processing during the past few decades.

In 1900, for instance, the average copper ore mined in the United States was about 5% copper by weight. Today it is 0.5%, and copper costs less (adjusted for inflation). New methods of mineral extraction may allow even lower-grade ores of some metals to be used (Solutions, right).

However, the mining of lower-grade ores can be limited by (1) increased cost of mining and processing larger volumes of ore, (2) availability of fresh water needed to mine and process some minerals (especially in arid and semiarid areas), and (3) the environmental impact of the increased land disruption, waste material, and pollution produced during mining and processing.

Can We Get Enough Minerals by Mining the Oceans? Ocean mineral resources are found in (1) seawater, (2) sediments and deposits on the shallow continental shelf (Figure 7-5, p. 156), (3) hydrothermal ore deposits (Figure 14-3), and (4) manganese-rich nodules on the deep-ocean floor.

Most of the chemical elements found in seawater occur in such low concentrations that recovering them takes more energy and money than they are worth. Only magnesium, bromine, and sodium chloride are abundant enough to be extracted profitably at current prices with existing technology.

Deposits of minerals (mostly sediments) along the continental shelf and near shorelines are significant sources of sand, gravel, phosphates, sulfur, tin, copper, iron, tungsten, silver, titanium, platinum, and diamonds.

Rich deposits of gold, silver, zinc, and copper are found as sulfide deposits in the deep ocean floor and around black smokers (Figure 14-3). Currently, it is too expensive to extract these minerals even though some of these deposits contain large concentrations of important metals.

Manganese-rich modules that are found on the deep ocean floor at various sites may be a future source

Mining with Microbes

SOLUTIONS

One way to improve mining technology is the use of microorganisms for in-place (in situ, pronounced "in-SEE-two") mining. This biological approach to mining would (1) remove desired metals from ores while leaving the surrounding environment undisturbed, (2) reduce air pollution associated with the smelting of metal ores, and (3) reduce water pollution associated with using hazardous chemicals such as cyanides and mercury to extract gold (Case Study, p. 328).

Once a commercially viable ore deposit has been identified, wells are drilled into it and the ore is fractured. Then the ore is inoculated with natural or genetically engineered bacteria to extract the desired metal. Next the well is flooded with water, which is pumped to the surface, where the desired metal is removed. Then the water is recycled.

This technique permits economical extraction from low-grade ores, which are increasingly being used as high-grade ores are depleted. Since 1958, the copper industry has been using natural strains of the bacterium *Thiobacillus ferroxidans* to remove copper from low-grade copper ore. Currently, more than 30% of all copper produced worldwide, worth more than $1 billion a year, comes from such *biomining*. If naturally occurring bacteria cannot be found to extract a particular metal, genetic engineering techniques could be used to produce such bacteria.

However, microbiological ore processing is slow. It can take decades to remove the same amount of material that conventional methods can remove within months or years. So far, biological mining methods are economically feasible only with low-grade ore (such as gold and copper) for which conventional techniques are too expensive.

Critical Thinking

1. If you had a large sum of money to invest, would you invest it in the microbiological mining of aluminum ore? Explain.

2. Are the ecological and human health risks from using genetically engineered organisms to mine metals likely to be higher or lower than those of genetically engineered organisms in food (Pro/Con, p. 275)? Explain.

of manganese and other key metals. They might be sucked up from the ocean floor by giant vacuum pipes or scooped up by buckets on a continuous cable operated by a mining ship.

Currently, these nodules and resource-rich mineral beds in international waters are not being developed

because of (1) high costs and (2) squabbles over who owns these common-property resources and how any profits from extracting them should be distributed among the world's nations. Some of these issues may be resolved by the international Law of the Sea Treaty, which recently went into effect.

Some environmentalists believe that seabed mining probably would cause less environmental harm than mining on land. However, they are concerned that removing seabed mineral deposits and dumping back unwanted material will (1) stir up ocean sediments, (2) destroy seafloor organisms, and (3) have potentially harmful effects on poorly understood ocean food webs and marine biodiversity (Chapter 24).

Can We Find Substitutes for Scarce Nonrenewable Mineral Resources? The Materials Revolution Some analysts believe that even if supplies of key minerals become very expensive or scarce, human ingenuity will find substitutes. They point to the current *materials revolution* in which silicon and new materials, particularly ceramics and plastics, are being developed and used as replacements for metals.

Ceramics have many advantages over conventional metals. They are harder, stronger, lighter, and longer lasting than many metals and they withstand intense heat and do not corrode. Within a few decades we may have high-temperature ceramic superconductors in which electricity flows without resistance. Such a development may lead to faster computers, more efficient power transmission, and affordable electromagnets for propelling magnetic levitation trains.

Plastics also have advantages over many metals. High-strength plastics and composite materials strengthened by lightweight carbon and glass fibers are likely to transform the automobile and aerospace industries. They cost less to produce than metals because they take less energy, do not need painting, and can be molded into any shape. New plastics and gels also are being developed to provide superinsulation without taking up much space. One new plastic can withstand extremely high temperatures and is not even affected by exposure to the most intense laser beams.

A few decades ago iron oxide (ordinary rust) was used only as a paint pigment. Now it is applied to computer diskettes to preserve digital information and to plastic tapes to store audio and video information.

Substitutes undoubtedly can be found for many scarce mineral resources. However, finding substitutes for some key materials may be difficult or impossible. Examples are helium, phosphorus for phosphate fertilizers, manganese for making steel, and copper for wiring motors and generators.

In addition, some substitutes are inferior to the minerals they replace. For example, aluminum could replace copper in electrical wiring. However, producing aluminum takes much more energy than producing copper, and aluminum wiring is a greater fire hazard than copper wiring.

14-5 EVALUATING ENERGY RESOURCES

What Types of Energy Do We Use? *Some 99% of the energy used to heat the earth and all of our buildings comes directly from the sun.* Without this direct input of essentially inexhaustible solar energy, the earth's average temperature would be −240°C (−400°F), and life as we know it would not exist.

This incoming solar energy is based on the nuclear fusion of hydrogen atoms that make up the sun's mass (Figure 3-17, p. 64). Thus, life on earth is made possible by a gigantic nuclear fusion reactor safely located about 150 million kilometers (93 million miles) away.

This direct input of solar energy also produces several other *indirect forms of renewable solar energy*: **(1)** wind, **(2)** falling and flowing water (hydropower), and **(3)** biomass (solar energy converted to chemical energy stored in chemical bonds of organic compounds in trees and other plants).

Commercial energy sold in the marketplace makes up the remaining 1% of the energy we use to supplement the earth's direct input of solar energy. Most commercial energy comes from extracting and burning nonrenewable mineral resources obtained from the earth's crust, primarily carbon-containing fossil fuels (petroleum, natural gas, and coal; Figure 14-11).

Over the past 60,000 years, major cultural changes (Section 2-1, p. 24) and technological advances (Figure 2-2, p. 24) have greatly increased energy use per person (Figure 14-12). As a result of such advances, about 81% of the commercial energy consumed in the world comes from *nonrenewable* energy resources (75% from fossil fuels and 6% from nuclear power; Figure 14-13, left).

In developing countries, the most important supplemental source of renewable energy is from *biomass* (mostly fuelwood and charcoal made from fuelwood). It is the main source of energy for heating and cooking for roughly half the world's population. Within a few decades, one-fourth of the world's population in developed countries may face an oil shortage. However, half the world's people in developing countries already face a fuelwood shortage and cannot afford to use fossil fuels.

Energy resources come and go as a result of shortages, cost, and new technologies. For example, during the 20th century there were major shifts in the world's major sources of energy. Specifically, the use of **(1)** *coal* dropped from 55% to 22%, **(2)** *oil* increased from 2% to 30%, **(3)** *natural gas* rose from 1% to 23%, **(4)** *nuclear energy* increased from 0% to 6%, and **(5)** *renewable energy* (mostly wood and flowing water) decreased from 42% to 19%.

The United States is the world's largest energy user. With only 4.5% of the population, it uses 24% of the world's commercial energy. In contrast, India,

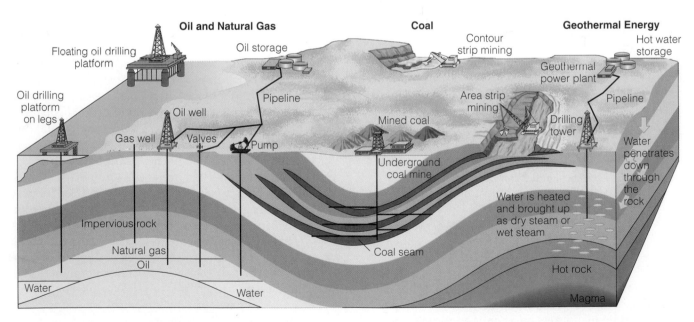

Figure 14-11 Important nonrenewable energy resources that can be removed from the earth's crust are coal, oil, natural gas, and some forms of geothermal energy. Nonrenewable uranium ore is also extracted from the earth's crust and then processed to increase its concentration of uranium-235, which can be used as a fuel in nuclear reactors to produce electricity.

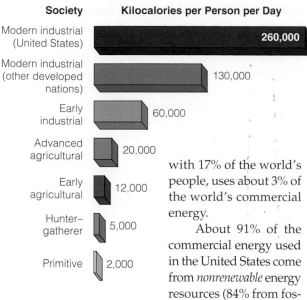

Society	Kilocalories per Person per Day
Modern industrial (United States)	260,000
Modern industrial (other developed nations)	130,000
Early industrial	60,000
Advanced agricultural	20,000
Early agricultural	12,000
Hunter–gatherer	5,000
Primitive	2,000

Figure 14-12 Average direct and indirect daily energy use per person at various stages of human cultural development. A typical citizen in a modern industrial society uses 65 to 130 times as much energy per day as a hunter-gatherer.

with 17% of the world's people, uses about 3% of the world's commercial energy.

About 91% of the commercial energy used in the United States come from *nonrenewable* energy resources (84% from fossil fuels and 7% from nuclear power, Figure 14-13, right). Energy use per person in the United States and Canada is about twice as high as in Japan, Germany, France, and the United Kingdom and at least 100 times as high as that in China and India. According to Maurice Strong, secretary general of the 1992 United Nations (UN) Earth Summit, "Typical citizens of advanced industrialized nations each consume as much energy in six months as typical citizens in developing countries consume during their entire life."

How Should We Evaluate Energy Resources? The world's current dependence on nonrenewable fossil fuels (Figure 14-13, left) is the primary cause of air and water pollution, land disruption, and greenhouse gas emissions. Moreover, affordable oil, the most widely used energy resource in developed countries, may be *economically depleted* within 42–93 years and gradually replaced by other energy resources.

For many analysts, the key problem is not running out of oil or other fossil fuels. Instead, it is reducing the disruptive economic and ecological effects of projected global warming (Section 18-4, p. 458) by **(1)** reducing energy waste and **(2)** shifting to a mix of new energy resources that produce little or no carbon dioxide and other greenhouse gases.

What is our best immediate energy option? The general consensus is to cut out unnecessary energy waste by improving energy efficiency. This means using energy technologies such as insulation, high-mileage cars, and efficient light bulbs and appliances that give us heat, cooling, light, and motion using less energy, as discussed in Section 15-1, p. 359 and 15-2, p. 362.

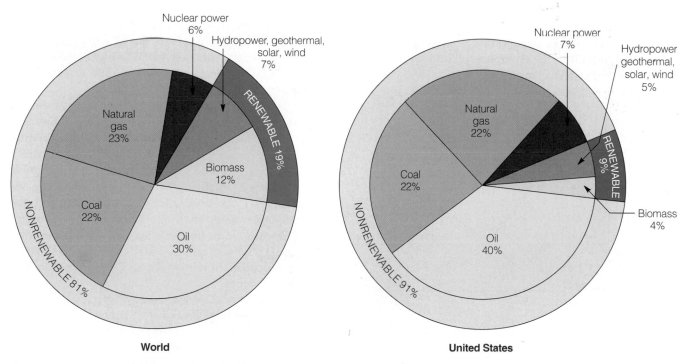

Figure 14-13 Commercial energy use by source for the world and the United States in 1999. Commercial energy amounts to only 1% of the energy used in the world; the other 99% is direct solar energy received from the sun and is not sold in the marketplace. (Data from U.S. Department of Energy, British Petroleum, and Worldwatch Institute)

There is disagreement about our next best energy option. Some say we should get much more of the energy we need from a mix renewable energy sources such as the (1) sun, (2) wind, (3) flowing water (hydropower), (4) biomass, (5) hydrogen gas, and (6) heat stored in the earth's interior (geothermal energy), as discussed in Chapter 15.

Others say we should burn more coal and synthetic liquid and gaseous fuels made from coal because it is our most plentiful energy resource. Some think we can find more oil. Many believe that fairly abundant and clean-burning natural gas is the answer, especially as a transition fuel to a new solar-hydrogen age built around improved energy efficiency and increased use of renewable energy. Some think nuclear power is the answer. These nonrenewable energy options are evaluated in this chapter.

Experience shows that it usually takes at least 50 years and huge investments to phase in new energy alternatives (Figure 14-14) to the point where they provide 10-20% of total energy use. What energy resources will we be using over the next 20 to 50 years? Making such projections involves answering the following questions for *each* energy alternative:

- How much of the energy source will be available in the near future (the next 15–25 years) and in the long term (the next 25–50 years)?

- What is this source's net energy yield?

- How much will it cost to develop, phase in, and use this energy resource?

- How will extracting, transporting, and using the energy resource affect the environment, human health, and the earth's climate?

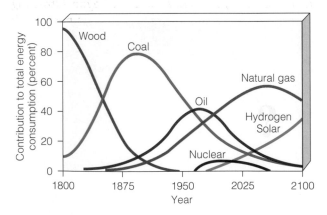

Figure 14-14 Shifts in the use of commercial energy resources in the United States since 1800, with projected changes to 2100. Shifts from wood to coal and then from coal to oil and natural gas have each taken about 50 years. (Data from U.S. Department of Energy)

What Is Net Energy? The Only Energy That Really Counts It takes energy to get energy. For example, oil must be (1) found, (2) pumped up from beneath the ground or ocean floor, (3) transferred to a refinery and converted to useful fuels (such as gasoline, diesel fuel, and heating oil), (4) transported to users, and (5) burned in furnaces and cars before it is useful to us. Each of these steps uses energy, and the second law of energy (p. 65) tells us that each time we use energy to perform a task, some of it is always wasted and is degraded to low-quality energy.

The usable amount of *high-quality energy* (Figure 3-11, p. 59) available from a given quantity of an energy resource is its **net energy**: the total amount of energy available from an energy resource minus the energy needed to find, extract, process, and get that energy to consumers. It is calculated by estimating the total energy available from the resource over its lifetime minus the amount of energy (1) used (the first law of energy), (2) automatically wasted (the second law of energy), and (3) unnecessarily wasted in finding, processing, concentrating, and transporting the useful energy to users.

Net energy is like your net spendable income (your wages minus taxes and other deductions). For example, suppose that for every 10 units of energy in oil in the ground we have to use and waste 8 units of energy to find, extract, process, and transport the oil to users. Then we have only 2 units of *useful energy* available from every 10 units of energy in the oil.

We can express net energy as the ratio of useful energy produced to the useful energy used to produce it. In the example just given, the *net energy ratio* would be 10/8, or approximately 1.25. The higher the ratio, the greater the net energy. When the ratio is less than 1, there is a net energy loss.

Figure 14-15 shows estimated net energy ratios for various types of space heating, high-temperature heat for industrial processes, and transportation. Currently, oil has a high net energy ratio because much of it comes from large, accessible deposits such as those in the Middle East. When those sources are depleted, the net energy ratio of oil will decline and prices will rise. Then more money and high-quality fossil-fuel energy will be needed to find, process, and deliver new oil from (1) widely dispersed small deposits, (2) deposits buried deep in the earth's crust, or (3) deposits located in remote areas.

Conventional nuclear energy has a low net energy ratio because large amounts of energy are needed to (1) extract and process uranium ore, (2) convert it into a usable nuclear fuel, (3) build and operate nuclear power plants, (4) dismantle the highly radioactive plants after their 15–40 years of useful life, and (5) store the resulting highly radioactive wastes for thousands of years.

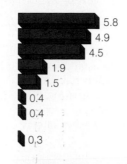

Space Heating

Passive solar	5.8
Natural gas	4.9
Oil	4.5
Active solar	1.9
Coal gasification	1.5
Electric resistance heating (coal-fired plant)	0.4
Electric resistance heating (natural-gas-fired plant)	0.4
Electric resistance heating (nuclear plant)	0.3

High-Temperature Industrial Heat

Surface-mined coal	28.2
Underground-mined coal	25.8
Natural gas	4.9
Oil	4.7
Coal gasification	1.5
Direct solar (highly concentrated by mirrors, heliostats, or other devices)	0.9

Transportation

Natural gas	4.9
Gasoline (refined crude oil)	4.1
Biofuel (ethyl alcohol)	1.9
Coal liquefaction	1.4
Oil shale	1.2

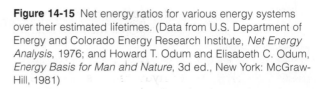

Figure 14-15 Net energy ratios for various energy systems over their estimated lifetimes. (Data from U.S. Department of Energy and Colorado Energy Research Institute, *Net Energy Analysis*, 1976; and Howard T. Odum and Elisabeth C. Odum, *Energy Basis for Man and Nature*, 3d ed., New York: McGraw-Hill, 1981)

14-6 OIL

What Is Crude Oil, and How Is It Extracted and Processed? **Petroleum**, or **crude oil** (oil as it comes out of the ground), is a thick liquid consisting of hundreds of combustible hydrocarbons along with small amounts of sulfur, oxygen, and nitrogen impurities. This fossil fuel was produced by the decomposition of dead organic matter from plants and animals that were buried under lake and ocean sediments hundreds of millions of years ago and subjected to high temperatures and pressures over millions of years as part of the carbon cycle (Figure 4-29, p. 92).

Deposits of crude oil and natural gas often are trapped together under a dome deep within the earth's crust on land or under the seafloor (Figure 14-11). Crude oil is rarely found in a large, hollow reservoir. Instead, it is dispersed in pores and cracks in underground rock formations, somewhat like water saturating a sponge.

Primary oil recovery involves drilling a well and pumping out the lighter crude oil that flows out of the rock pores by gravity into the bottom of the well. More of the heavier crude oil left in the well can be removed by **secondary oil recovery**. This involves (1) pumping water under high pressure into a nearby well to force some of the remaining heavy crude oil through rock pores to the original well, (2) pumping the oil and water mixture to the surface, (3) separating out the

heavy oil, and (4) reusing the water to recover more heavy oil.

On average, producers get only about 35% of the oil out of an oil deposit by primary and secondary recovery. The well is then abandoned because the *heavy crude oil* that remains is too difficult or expensive to recover.

As oil prices rise, it can become economical to remove about 10–25% of this remaining heavy oil by **tertiary oil recovery**. In one method, superheated steam or carbon dioxide is injected into a well to force out additional heavy crude. Another approach is to (1) inject detergent to dissolve some of the remaining heavy oil, (2) pump the mixture to the surface, (3) separate out the oil, and (4) reuse the detergent to recover more oil. However, the net energy yield for such recovered oil is lower because it takes the energy in one-third of a barrel of refined oil to retrieve each barrel of heavy crude oil.

Drilling for oil causes only moderate damage to the earth's land because the wells occupy fairly little land area. However, oil companies have begun extracting petroleum from ecologically fragile environments such as the Arctic tundra in Alaska and the ocean floor (Figure 14-11). Despite great care, accidents occurring during drilling and extraction of the crude oil from the ocean floor have released large amounts of the oil into ocean waters, harming or killing marine life and disrupting marine ecosystems. Ocean waters and coastlines have also been polluted by accidents involving oil tankers (Case Study, p. 336).

Once crude oil has been extracted, it is transported to a *refinery* by pipeline, truck, or ship (oil tanker). There it is heated and distilled in gigantic columns to separate it into components with different boiling points (Figure 14-16). Some of the products of oil distillation, called **petrochemicals**, are used as raw materials in industrial organic chemicals, pesticides, plastics, synthetic fibers, paints, medicines, and many other products.

The Exxon Valdez Oil Spill

Crude oil from Alaska's North Slope fields near Prudhoe Bay (Figure 14-17, p. 338) is carried by pipeline to the port of Valdez and then shipped by tanker to the West Coast. On March 24, 1989, the *Exxon Valdez*, a tanker more than three football fields long, went off course in a 16-kilometer-wide (12-mile-wide) channel in Prince William Sound near Valdez, Alaska. It hit submerged rocks, creating the worst oil spill ever in U.S. waters.

The rapidly spreading oil slick coated more than 1,600 kilometers (1,000 miles) of shoreline, almost the length of the U.S. shoreline between New Jersey and South Carolina. The full loss of wildlife will never be known because most of the dead animals sank and decomposed without being counted. By 1999, 10 years after the accident, only 2 of 22 damaged populations (the bald eagle and river otters) had recovered, 12 were gradually recovering, and the other 8 (including orcas and harbor seals) had not recovered.

Exxon spent $2.2 billion directly on the cleanup, but some of the cleanup methods did more harm than good. For example, the use of high-pressure jets of hot water to clean beaches killed coastal plants and animals that had survived the spill. As a result, a year after the spill the oiled sites had recovered more rapidly than the washed sites.

In 1991, Exxon pleaded guilty to federal felony and misdemeanor charges and agreed to pay the federal government and the state of Alaska $1 billion in fines and civil damages. In 1994, a jury awarded members of the fishing industry, landowners, and other Alaska residents $5 billion in punitive damages and penalties, a decision the courts upheld in 2000.

Studies show that others must share the blame for this tragedy. State officials had been lax in monitoring the preparedness of oil companies for dealing with oil spill, and the U.S. Coast Guard did not effectively monitor tanker traffic because of inadequate radar equipment and personnel. Americans who drive fuel-inefficient cars and do not insist that we use and waste less oil also share part of the blame.

This roughly $8.5-billion accident (one-fourth of it paid for by U.S. taxpayers because of tax write-offs by Exxon) might have been prevented if Exxon had spent only $22.5 million to fit the tanker with a double hull (which it still has not done). In the early 1970s, interior secretary Rogers Morton told Congress that all oil tankers using Alaskan waters would have double hulls, but under pressure from oil companies the requirement was dropped.

Today, most cargo-carrying ships have double hulls, but only 15% of oil tankers have such hulls. Legislation passed since the spill requires all new tankers to have double hulls and all existing large single-hulled oil tankers to be phased out by 2015. However, the oil industry is working to weaken these and other requirements.

This spill highlighted the importance of pollution prevention and the advantages of shifting to improving energy efficiency and renewable energy (Chapter 15) to reduce dependence on oil. Even with the best technology and a fast response by well-trained people, scientists estimate that no more than 12–15% of the oil from a major spill can be recovered.

Critical Thinking

Congratulations: You have just been put in charge of preventing oil spills from tankers throughout the world. What are the three most important things you would do to accomplish your mission?

Who Has the World's Oil Supplies? The 11 countries that make up the Organization of Petroleum Exporting Countries (OPEC)* have 67% of the world's crude oil reserves, which explains why OPEC is expected to have long-term control over world oil supplies and prices.

About 64% of the world's crude oil reserves are found in the Middle East. Saudi Arabia, with 26%, has by far the largest proportion of the world's crude oil reserves, followed by Iraq, Kuwait, Iran, and United Arab Emirates, each with 9–10%.

The remaining 36% of global crude oil reserves are found in (1) Latin America (14%, with 7% in Venezuela and 5% in Mexico), (2) Africa (7%), (3) the former Soviet Union (6%), (4) Asia (4%, with 3% in China), (5) the United States (2.3%), and (6) Europe (2%).

Geologists believe that the Middle East also contains the majority of the world's undiscovered oil. However, geologists expect to find some additional oil in China, Mexico, and several other developing countries.

Figure 14-17 shows the locations of the major known deposits of fossil fuels (oil, natural gas, and coal) in the United States and Canada and ocean areas where more crude oil and natural gas might be found.

The United States uses nearly 30% of the crude oil extracted worldwide each year (68% of it for trans-

---

*OPEC was formed in 1960 so that developing countries with much of the world's known and projected oil supplies could get a higher price for this resource. Today its members are Algeria, Indonesia, Iran, Iraq, Kuwait, Libya, Nigeria, Qatar, Saudi Arabia, United Arab Emirates, and Venezuela.

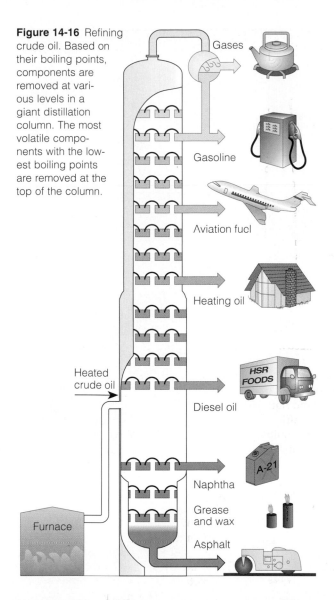

Figure 14-16 Refining crude oil. Based on their boiling points, components are removed at various levels in a giant distillation column. The most volatile components with the lowest boiling points are removed at the top of the column.

Gases

Gasoline

Aviation fuel

Heating oil

Heated crude oil

Diesel oil

Naphtha

Grease and wax

Asphalt

Furnace

portation), mostly because oil is an abundant and convenient fuel to use and is cheap (Figure 14-18). In 1999, the United States imported about 52% of the oil it used (up from 36% in 1973 during the OPEC oil embargo), mostly because of declining domestic oil reserves and increased oil use. According to the Department of Energy, the United States could be importing 64% or more of the oil it uses by 2020.

This dependence on imported oil (about half of it from OPEC countries) and the possibility of much higher oil prices within 10–20 years could drain the United States, Japan, and other major oil-importing nations of vast amounts of money. Economists warn that this could lead to severe inflation and widespread economic recession, perhaps even a major depression.

How Long Will Oil Supplies Last, and What Are the Pros and Cons of Oil? Since the first oil well was drilled in Pennsylvania in 1859, oil use has been growing exponentially at about 4–5% per year, except during the Great Depression of the 1930s and the global economic recession of the 1980s. As a result, **(1)** production of the world's estimated oil reserves is expected to peak between 2010 and 2030 (Figure 14-19, top), and **(2)** production of estimated U.S. reserves peaked in 1975 (Figure 14-19, bottom).

Identified global reserves of oil should last about 53 years at the current usage rate, and 42 years if usage increases as projected by about 2% per year. Undiscovered oil that is thought to exist might add another 20–40 years to global oil supplies, probably at higher prices. Thus, *known and projected supplies of oil are projected to be 80% depleted within 42–93 years depending on the annual rate of use.*

At today's consumption rate, U.S. oil reserves should last about 15–24 years (Figure 14-19, bottom) and 10–15 years if consumption increases as projected. However, potential reserves might yield an additional 24 years of production. Thus, *U.S. oil supplies are projected to be 80% depleted within 15–48 years, depending on the annual rate of use.*

Some analysts contend that rising oil prices (when oil consumption exceeds oil production) will stimulate exploration and lead to enough new reserves to meet future demand through the next century or longer. Other analysts argue that such optimistic projections about future oil supplies ignore the consequences of the high (2–5% per year) exponential growth in oil consumption.

Even assuming that we continue to use crude oil at the current rate, **(1)** Saudi Arabia, with the largest known crude oil reserves, could supply all the world's oil needs for only about 10 years, and **(2)** the estimated reserves under Alaska's North Slope (the largest ever found in North America) would meet current world demand for only 6 months or U.S. demand for 3 years. In short, just to keep on using oil at the *current rate*, we must discover and add to global oil reserves the equivalent of a new Saudi Arabian supply *every 10 years.*

According to the U.S. Geological Survey, global discovery of large oil fields peaked in 1962 and has been declining since then. Oil production in Alaska peaked in 1988 and has been declining since then. There is controversy over whether to allow new oil and gas exploration and development in Alaska's Arctic National Wildlife Refuge (Pro/Con, p. 343).

Burning any fossil fuel releases carbon dioxide into the atmosphere and thus can promote global warming. Figure 14-20 compares the relative amounts of CO_2 emitted per unit of energy by the major fossil fuels and nuclear power. Figure 14-21 lists the pros and cons of using conventional crude oil as an energy resource.

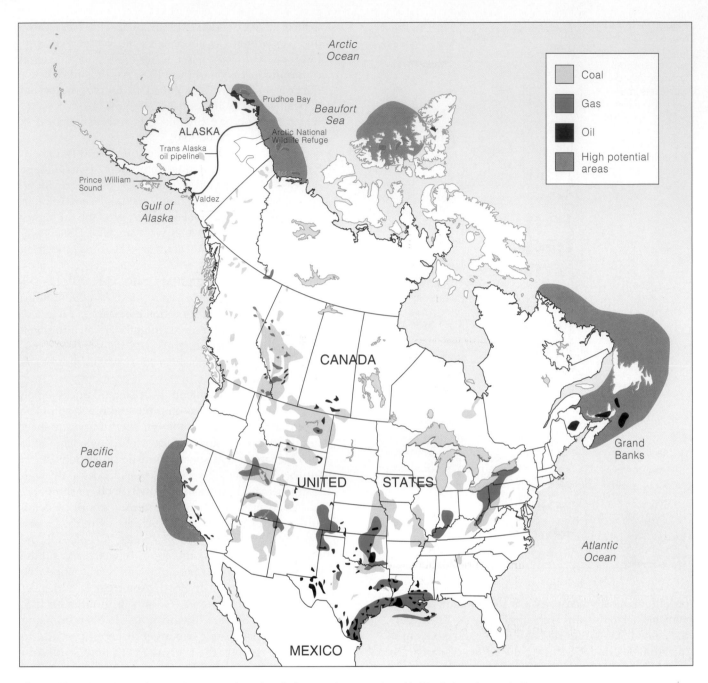

Figure 14-17 Locations of the major known deposits of oil, natural gas, and coal in North America and off-shore areas where more crude oil and natural gas might be found. Geologists do not expect to find very much new oil and natural gas in North America. (Data from Council on Environmental Quality and U.S. Geological Survey)

How Useful Are Heavy Oils from Oil Shale and Tar Sands?
Oil shale is a fine-grained sedimentary rock (Figure 14-22, left) containing solid combustible organic material called **kerogen.** This material can be distilled from oil shale by heating it in a large container (Figure 14-23) to yield **shale oil** (Figure 14-22, right, p. 340). Before the thick shale oil can be sent by pipeline to a refinery, it must be heated to increase its flow rate and processed to remove sulfur, nitrogen, and other impurities.

The shale oil potentially recoverable from U.S. deposits, mostly on federal lands in Colorado, Utah, and Wyoming, could probably meet the country's crude oil demand for about 40 years at current use levels. Canada, China, and Russia also have large oil shale

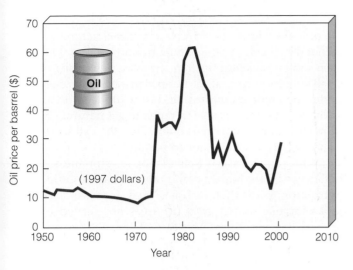

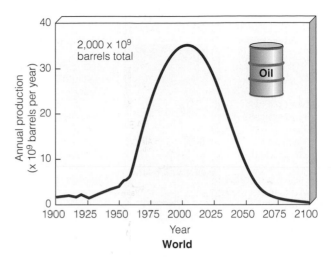

World

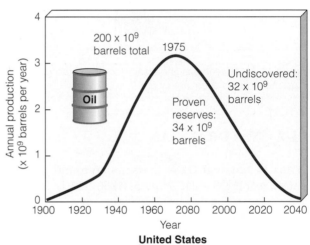

United States

Figure 14-18 Inflation-adjusted price of oil, 1950–2000. When adjusted for inflation, oil costs about the same as it did in 1975. Although low oil prices have stimulated economic growth, they have discouraged improvements in energy efficiency and increased use of renewable energy resources. (Data from U.S. Department of Energy and Department of Commerce)

deposits. Estimated potential global supplies of shale oil are 200 times larger than estimated global supplies of conventional oil. However, most deposits of oil shale are of such low grade that it takes more energy and money to mine and convert the kerogen to crude oil than the resulting fuel is worth.

Tar sand (or oil sand) is a mixture of clay, sand, water, and a combustible organic material called **bitumen** (a thick, high-sulfur heavy oil). Most deposits are too deep underground to be mined at a profit, but some deposits are close enough to the earth's surface to be removed by surface mining. The bitumen is removed, purified, and chemically upgraded into a synthetic crude oil suitable for refining (Figure 14-24).

The world's largest known deposits of tar sands, the Athabasca Tar Sands, lie in northern Alberta, Canada. About 10% of these deposits lie close enough to the surface to be surface mined. Currently, these deposits supply about 21% of Canada's oil needs. These deposits could supply all of Canada's projected oil needs for about 33 years at its current consumption rate, but they would last the world only about 2 years. Other large deposits of tar sands are in Utah, Venezuela, Colombia, and Russia.

Figure 14-25 (p. 342) lists the pros and cons of using heavy oil from oil shale and tar sand as energy resources. Because of low net energy yields and the high costs needed to develop and process them, neither of these resources is expected to provide much of the world's energy in the foreseeable future.

Figure 14-19 Petroleum production curves for the world (top) and the United States (bottom). (Data from U.S. Geological Survey)

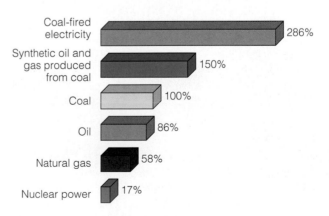

Figure 14-20 Carbon dioxide emissions per unit of energy produced by various fuels, expressed as percentages of emissions produced by coal.

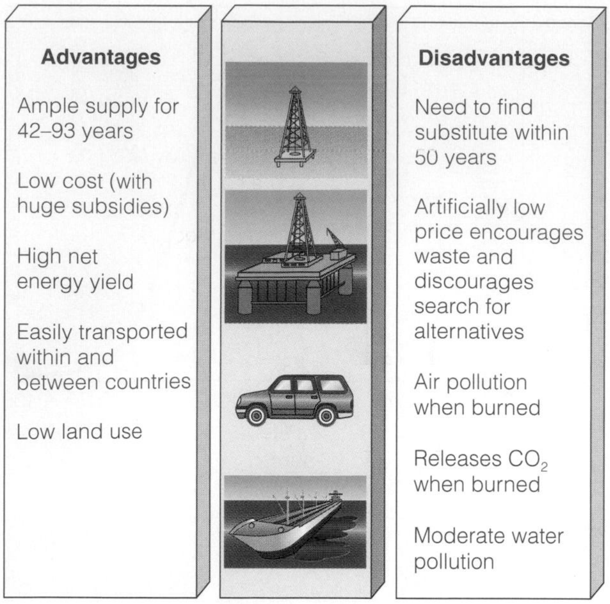

Advantages		Disadvantages
Ample supply for 42–93 years		Need to find substitute within 50 years
Low cost (with huge subsidies)		Artificially low price encourages waste and discourages search for alternatives
High net energy yield		
Easily transported within and between countries		Air pollution when burned
Low land use		Releases CO$_2$ when burned
		Moderate water pollution

Figure 14-21 Advantages and disadvantages of using conventional oil as an energy resource.

14-7 NATURAL GAS

What Is Natural Gas? In its underground gaseous state, **natural gas** is a mixture of **(1)** 50–90% by volume of methane (CH$_4$), the simplest hydrocarbon, **(2)** smaller amounts of heavier gaseous hydrocarbons such as ethane (C$_2$H$_6$), propane (C$_3$H$_8$), and butane (C$_4$H$_{10}$), and **(3)** small amounts of highly toxic hydrogen sulfide (H$_2$S), a by-product of naturally occurring sulfur in the earth.

Figure 14-22 Oil shale and the shale oil extracted from it. Big U.S. oil shale projects have been canceled because of excessive cost. (U.S. Department of Energy)

Conventional natural gas lies above most reservoirs of crude oil (Figure 14-11). Unconventional natural gas is found by itself in other underground sources. One such source is methane hydrate, which is composed of small bubbles of natural gas trapped in ice crystals deep under the arctic permafrost and beneath deep ocean sediments. So far it costs too much to get natural gas from such unconventional sources, but the extraction technology is being developed rapidly.

When a natural gas field is tapped, propane and butane gases are liquefied and removed as **liquefied petroleum gas (LPG)**. LPG is stored in pressurized tanks for use mostly in rural areas not served by natural gas pipelines. The rest of the gas (mostly methane) is **(1)** dried to remove water vapor, **(2)** cleansed of poisonous hydrogen sulfide and other impurities, and **(3)** pumped into pressurized pipelines for distribution. At a very low temperature of −184°C (−300°F), natural gas can be converted to **liquefied natural gas (LNG)**. This highly flammable liquid can then be shipped to other countries in refrigerated tanker ships.

Who Has the World's Natural Gas Supplies? Russia and Kazakhstan have about 42% of the world's natural gas reserves. Other countries with large known natural gas reserves are Iran (15%), Qatar (5%), Saudi Arabia (4%), Algeria (4%), the United States (3%), Nigeria (3%), and Venezuela (3%).

Geologists expect to find more natural gas, especially in unexplored developing countries. Most U.S. natural gas reserves are located in the same places as crude oil (Figures 14-11 and 14-17).

How Long Will Natural Gas Supplies Last? The outlook for natural gas supplies is much better than for oil. At the current consumption rate, known reserves and undiscovered, potential reserves of conventional natural gas are expected to last the world for 125 years and the United States for 65–80 years.

It is estimated that conventional supplies of natural gas, plus unconventional supplies available at higher prices, will last at least 200 years at the current consumption rate and 80 years if usage rates rise 2% per year. Thus, global supplies of conventional and unconventional natural gas should last 205–325 years, depending on how rapidly natural gas is used.

What Is the Future of Natural Gas? Figure 14-26 lists the pros and cons of using natural gas as an energy resource. In fairly new combined-cycle natural gas systems, natural gas is burned in combustion turbines, which are essentially giant jet engines bolted to the ground. This system can produce electricity more efficiently and cheaply than burning coal or oil or using nuclear power.

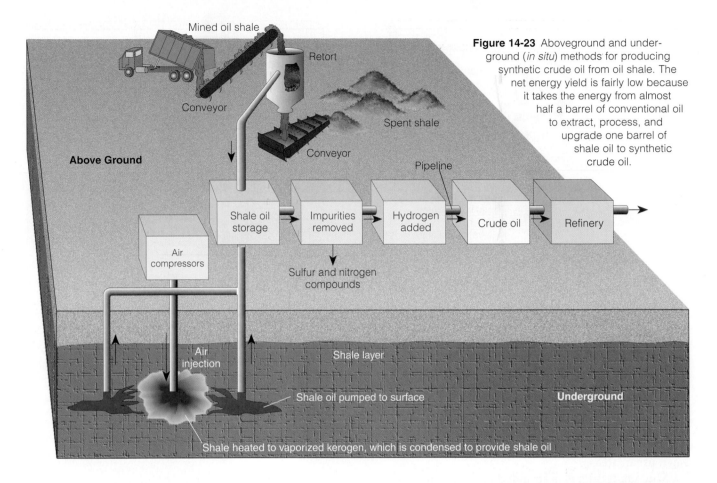

Figure 14-23 Aboveground and underground (*in situ*) methods for producing synthetic crude oil from oil shale. The net energy yield is fairly low because it takes the energy from almost half a barrel of conventional oil to extract, process, and upgrade one barrel of shale oil to synthetic crude oil.

In 2000, General Electric Co. and the U.S. Department of Energy announced development of a combined-cycle natural gas electric power plant that is 60% efficient, compared with the 32–40% efficiency of today's coal, gas, oil, and nuclear power plants. In addition to using less fuel, this new system produces much less carbon dioxide and smog-causing nitrogen oxides than coal-burning power plants.

According to the U.S. Department of Energy, most of the more than 200 new power plants to be built in the United States between 2000 and 2015 will use natural gas turbine systems. These systems can also provide backup power for solar energy and wind power systems.

Smaller combined-cycle natural gas units being developed could supply all of the heat and electricity needs of an apartment building or office building. This will be aided by a catalytic combustion device made by Catalytica (a small company in Mountain View, California) that eliminates nearly all air pollution from

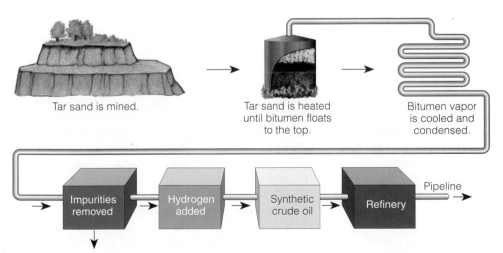

Tar sand is mined.

Tar sand is heated until bitumen floats to the top.

Bitumen vapor is cooled and condensed.

Figure 14-24 Generalized summary of how synthetic crude oil is produced from tar sand. The net energy yield is low because it takes the energy in almost one-half a barrel of conventional oil to extract and process one barrel of bitumen and upgrade it to synthetic crude oil.

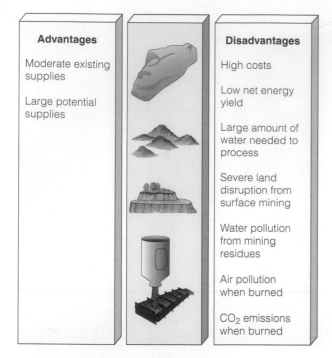

Advantages		Disadvantages
Moderate existing supplies		High costs
Large potential supplies		Low net energy yield
		Large amount of water needed to process
		Severe land disruption from surface mining
		Water pollution from mining residues
		Air pollution when burned
		CO_2 emissions when burned

Figure 14-25 Advantages and disadvantages of using heavy oils from oil shale and tar sand as energy resources.

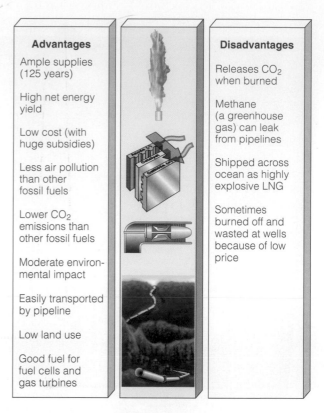

Advantages		Disadvantages
Ample supplies (125 years)		Releases CO_2 when burned
High net energy yield		Methane (a greenhouse gas) can leak from pipelines
Low cost (with huge subsidies)		Shipped across ocean as highly explosive LNG
Less air pollution than other fossil fuels		Sometimes burned off and wasted at wells because of low price
Lower CO_2 emissions than other fossil fuels		
Moderate environmental impact		
Easily transported by pipeline		
Low land use		
Good fuel for fuel cells and gas turbines		

Figure 14-26 Advantages and disadvantages of using conventional natural gas as an energy resource.

gas-fired electric generators, making them suitable for use in urban areas.

Much natural gas is stranded at wells and costs too much to transport to users. Currently this stranded gas is burned off (flared) at natural gas wells. In addition to wasting natural gas, this adds carbon dioxide to the atmosphere.

This may change. In 1999, Daimler-Chrysler and Syntropetroleum of Oklahoma announced development of a chemical process to convert stranded natural gas at well sites to a clean-burning fuel for use in diesel, fuel cell, and electric hybrid vehicles.

Because of its advantages over oil, coal, and nuclear energy, some analysts see natural gas as the best fuel to help make the transition to improved energy efficiency and greater use of renewable energy over the next 50 years.

14-8   COAL

What Is Coal, and How Is It Extracted and Processed? Coal is a solid fossil fuel that is mostly combustible carbon (20–98% depending on the type), with much smaller amounts of water and sulfur and mercury and trace amounts of radioactive materials found in the earth. Most coal deposits were formed during the Carboniferous period (360 and 285 million years ago), when swamps covered large areas of low-

lying land. Coal formed in several stages (Figure 14-27) as buried plant remains in these swamps were subjected to intense heat and pressure over many millions of years as part of the carbon cycle (Figure 4-29, p. 92).

As coal ages, its carbon content increases and its water content decreases (Figure 14-27). Anthracite, which is about 98% carbon, is the most desirable type of coal because of its high heat content and low sulfur content. However, because it takes much longer to form, it is less common and therefore more expensive.

Some coal is extracted underground (Figure 14-5) by miners working in tunnels and shafts. Such mining is one of the world's most dangerous occupations because of accidents and black lung disease (caused by prolonged inhalation of coal dust particles). When coal lies close to the earth's surface it is extracted by *area strip mining* (Figure 14-4c) on flat terrain and *contour strip mining* (Figure 14-4d) on hilly or mountainous terrain.

After coal is removed it is transported (usually by train) to a processing plant, where it is broken up, crushed, and then washed to remove impurities. The coal is then dried and shipped (again usually by train) to users, mostly power plants and industrial plants.

How Is Coal Used, and Where Are the Largest Supplies? Coal provides about 22% of the commercial energy used in the world and the United States. It is

Should Oil and Gas Development Be Allowed in the Arctic National Wildlife Refuge?

The Arctic National Wildlife Refuge on Alaska's North Slope (Figure 14-17, p. 338) contains more than one-fifth of all the land in the U.S. National Wildlife Refuge System and has been called the crown jewel of the system. The refuge's coastal plain, its most biologically productive part, is the only stretch of Alaska's arctic coastline not open to oil and gas development.

The Alaskan National Interest Lands Conservation Act of 1989 requires specific authorization from Congress before drilling or other development can take place on this coastal plain. For years, U.S. oil companies have been lobbying Congress to grant them permission to carry out exploratory drilling in the coastal plain because they believe that this area might contain oil and natural gas deposits.

Oil company officials argue that such exploration and development of any oil and natural gas they find is needed to **(1)** increase U.S. oil and natural gas supplies and **(2)** reduce U.S. dependence on oil imports. They also point out that they **(1)** already have a large industrial complex of roads, pipelines, storage tanks and other facilities in nearby Prudhoe Bay to support oil production from the refuge, **(2)** have developed Alaska's Prudhoe Bay oil fields without significant harm to wildlife, **(3)** seek to open to oil and gas development only 1.5% of the total area of coastal plain region, and **(4)** can develop this area in an environmentally responsible manner with little lasting environmental impact.

Environmentalists and many biologists oppose this proposal and urge Congress to designate the entire coastal plain as wilderness. They contend that **(1)** the Department of Interior estimates that there is only a 19% chance of finding as much oil there as the United States consumes every 6–30 months, **(2)** improvements in energy efficiency would save far more oil at a much lower cost (Sections 15-1 and 15-2),

(3) potential degradation of any portion of this irreplaceable wildlife area is not worth the risk, **(4)** the huge oil spill from the tanker *Exxon Valdez* in Alaska's Prince William Sound (Case Study, p. 336) casts doubt on the claim of environmentally responsible development, **(5)** a 1988 study by U.S. Fish and Wildlife Service revealed that oil drilling at Prudhoe Bay has caused much more air and water pollution than was anticipated before drilling began in 1972, **(6)** a 1995 study by the Department of the Interior concluded that long-lasting ecological harm would be caused by oil drilling in the refuge's fragile tundra ecosystem, and **(7)** it is not financially feasible to restore damaged areas in the Arctic to their natural states (about the only point that both environmentalists and oil company officials agree on).

Critical Thinking

Do you believe that oil companies should be allowed to explore and remove oil from this wildlife refuge? Why or why not?

---

used to generate 62% of the world's electricity and make 75% of its steel. The remainder of the world's electricity is supplied by renewable hydroelectric dams (20%), nonrenewable nuclear energy (17%), and other renewable alternatives such as wind, solar, and geo-thermal energy (1%).

Countries getting more than half of the energy they use from coal are South Africa (78%), China (73%), Poland (68%), India (57%), Kazakhstan (54%), and the Czech Republic (51%). About 66% of the world's proven coal reserves and 85% of the estimated undiscovered coal deposits are located in the United States (with 24% of global reserves, Figure 14-17), Russia, China, and India. China is the world's largest user of coal, followed closely by the United States.

In the United States, coal is burned to generate about 52% of the country's electricity. The rest of the electricity produced in the United States comes from **(1)** nuclear power (20%), **(2)** natural gas (14%), **(3)** renewable energy (12%, with 10% of this from hydropower), and **(4)** oil (2%).

How Long Will Coal Supplies Last? Coal is the world's most abundant fossil fuel. *Identified* world reserves of coal should last at least 225 years at the current usage rate, and 65 years if usage rises 2% per year. The world's *unidentified* coal reserves are projected to last about 900 years at the current consumption rate and 149 years if the usage rate increases 2% per year. Thus, *identified and unidentified supplies of coal could last the world for 965–1,125 years, depending on the rate of usage.*

China, with 11% of the world's reserves, has enough coal to last 300 years at its current rate of consumption. Identified U.S. coal reserves should last about 300 years at the current consumption rate, and unidentified U.S. coal resources could extend those supplies for perhaps 100 years, at a higher cost.

What Is the Future of Coal? Figure 14-28 lists the pros and cons of using coal as an energy resource. Coal is very abundant, but it has the highest environmental impact of any fossil fuel from **(1)** land disturbance (Figures 14-4 and 14-6), **(2)** air pollution, **(3)** CO_2

Figure 14-27 Stages in coal formation over millions of years. Peat is a soil material made of moist, partially decomposed organic matter. Lignite and bituminous coal are sedimentary rocks, whereas anthracite is a metamorphic rock.

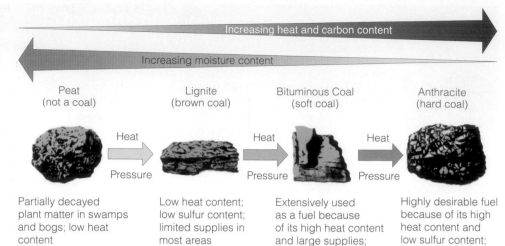

Increasing heat and carbon content

Increasing moisture content

| Peat (not a coal) | Lignite (brown coal) | Bituminous Coal (soft coal) | Anthracite (hard coal) |

Heat / Pressure → Heat / Pressure → Heat / Pressure →

Partially decayed plant matter in swamps and bogs; low heat content

Low heat content; low sulfur content; limited supplies in most areas

Extensively used as a fuel because of its high heat content and large supplies; normally has a high sulfur content

Highly desirable fuel because of its high heat content and low sulfur content; supplies are limited in most areas

emissions (Figure 14-20, accounting for about 43% of the world's annual emissions), and **(4)** water pollution (Figure 14-7).

In the United States, electric power generation (mostly from coal) is the second largest producer of toxic emissions. According to a 1998 Environmental Protection Agency study, the most threatening material produced by coal-burning power plants consists of particles of toxic mercury.

According to a 2000 report by the National Academy of Science, an estimated 60,000 babies a year may be born with neurological damage because of exposure to mercury in pregnant women who have consumed mercury-laden fish. Burning coal also releases thousands of times more radioactive particles into the

atmosphere per unit of energy produced than does a normally operating nuclear power plant. In addition, mining coal releases large amounts of methane gas, another potent greenhouse gas.

Each year in the United States alone, air pollutants from coal burning **(1)** kill thousands of people (with estimates ranging from 65,000 to 200,000), **(2)** cause at least 50,000 cases of respiratory disease, and **(3)** result in several billion dollars of property damage. However, new ways, such as *fluidized-bed combustion* (Figure 14-29), have been developed to burn coal more cleanly and efficiently and may be phased in over the next several decades.

Many analysts project a decline in coal use over the next 40–50 years because of **(1)** its high CO_2 emissions (Figure 14-20) and harmful health effects and **(2)** the availability of safer and cheaper ways to produce electricity such as burning natural gas in combined-cycle gas turbines (now the fuel of choice for power generation in industrial nations) and wind energy (p. 381).

What Are the Pros and Cons of Converting Solid Coal into Gaseous and Liquid Fuels? Solid coal can be converted into **synthetic natural gas (SNG)** by **coal gasification** (Figure 14-30) or into a liquid fuel such as methanol or synthetic gasoline by **coal liquefaction**. Figure 14-31 lists the pros and cons of using these *synfuels* produced from coal. Most analysts expect synfuels to play only a minor role as an energy resource in the next 30–50 years.

14-9 NUCLEAR ENERGY

How Does a Nuclear Fission Reactor Work? To evaluate the pros and cons of nuclear power, we must know first how a conventional nuclear power plant and its accompanying nuclear fuel cycle work. In a nuclear

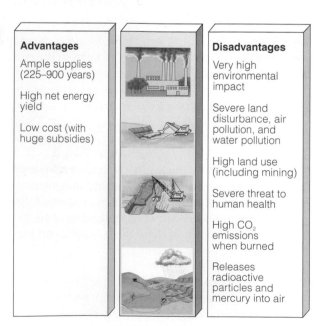

Advantages	Disadvantages
Ample supplies (225–900 years)	Very high environmental impact
High net energy yield	Severe land disturbance, air pollution, and water pollution
Low cost (with huge subsidies)	High land use (including mining)
	Severe threat to human health
	High CO_2 emissions when burned
	Releases radioactive particles and mercury into air

Figure 14-28 Advantages and disadvantages of using coal as an energy resource.

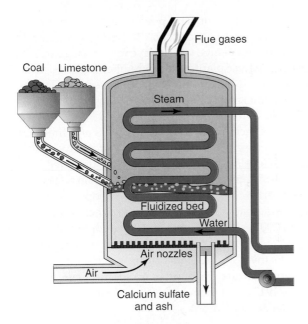

Figure 14-29 Fluidized-bed coal combustion. A stream of hot air is blown into a boiler to suspend a mixture of powdered coal and crushed limestone. This method **(1)** removes most of the sulfur dioxide, **(2)** sharply reduces emissions of nitrogen oxides, and **(3)** burns the coal more efficiently and cheaply than conventional combustion methods.

fission chain reaction, neutrons split the nuclei of atoms such as uranium-235 and plutonium-239 and release energy mostly as high-temperature heat (Figure 3-16, p. 64). In the reactor of a nuclear power plant, the rate of fission is controlled and the heat generated is used to produce high-pressure steam, which spins turbines that generate electricity.

Light-water reactors (LWRs) like the one diagrammed in Figure 14-32 produce about 85% of the world's

nuclear-generated electricity (100% in the United States). An LWR has the following key parts:

- *Core* containing 35,000-50,000 long, thin fuel rods, each packed with fuel pellets (each about one-third the size of a cigarette). Each pellet contains the energy equivalent of 0.9 metric tons (1 ton) of coal.

- *Uranium oxide fuel* consisting of about 97% nonfissionable uranium-238 and 3% fissionable uranium-235. To create a suitable fuel, the concentration of uranium-235 in the ore is increased (enriched) from 0.7% (its natural concentration in uranium ore) to 3% by removing some of the uranium-238.

- *Control rods*, which are moved in and out of the reactor core to absorb neutrons and thus regulate the rate of fission and amount of power the reactor produces.

- *Moderator*, which slows down the neutrons emitted by the fission process so that the chain reaction can be kept going. This is a material such as **(1)** liquid water (75% of the world's reactors, called *pressurized water reactors*, Figure 14-32), **(2)** solid graphite (20% of reactors), or **(3)** heavy water (D_2O, 5% of reactors). Graphite-moderated reactors can also produce fissionable plutonium-239 for nuclear weapons.

- *Coolant*, usually water, which circulates through the reactor's core to remove heat (to keep fuel rods and other materials from melting) and produce steam for generating electricity.

Nuclear power plants, each with one or more reactors, are only one part of the nuclear fuel cycle (Figure 14-33). In evaluating the safety and economic feasibility of nuclear power, we need to look at this entire cycle, not just the nuclear plant itself.

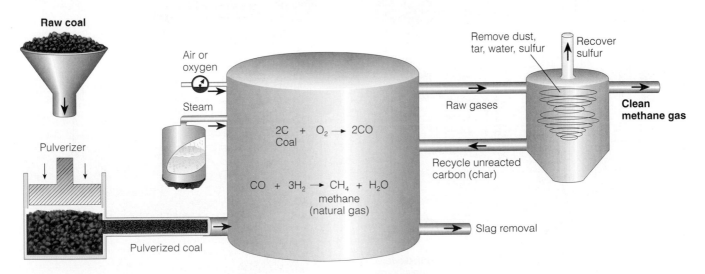

Figure 14-30 Coal gasification. Generalized view of one method for converting solid coal into synthetic natural gas (methane).

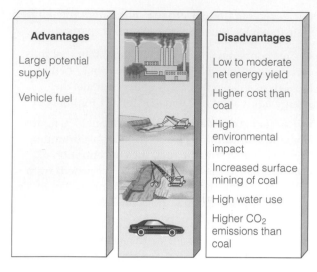

Advantages		Disadvantages
Large potential supply		Low to moderate net energy yield
Vehicle fuel		Higher cost than coal
		High environmental impact
		Increased surface mining of coal
		High water use
		Higher CO_2 emissions than coal

Figure 14-31 Advantages and disadvantages of using synthetic natural gas (SNG) and liquid synfuels produced from coal.

Accidents at nuclear power plants get most of the publicity, but they can also happen at nuclear fuel reprocessing plants, which have fewer safeguards than nuclear power plants. On September 30, 1999, an accident caused by human error occurred in a Japanese nuclear fuel reprocessing plant in village 110 kilometers (70 miles) from Tokyo.

This accident exposed 49 workers and firefighters (who were sent to the plant without protective clothing) to high levels of radioactivity as they attempted to stop an out-of-control chain reaction. Radiation released into the nearby community exposed 7 people to high levels of radioactivity and forced the evacuation of more than 200 residents. Some 300,000 people within a 10-kilometer (6-mile) radius of the plant were told to stay inside their homes. According to France's Nuclear Safety Institute, 59 similar accidents have occurred since 1945 (33 of them in the United States and 19 in the former Soviet Union).

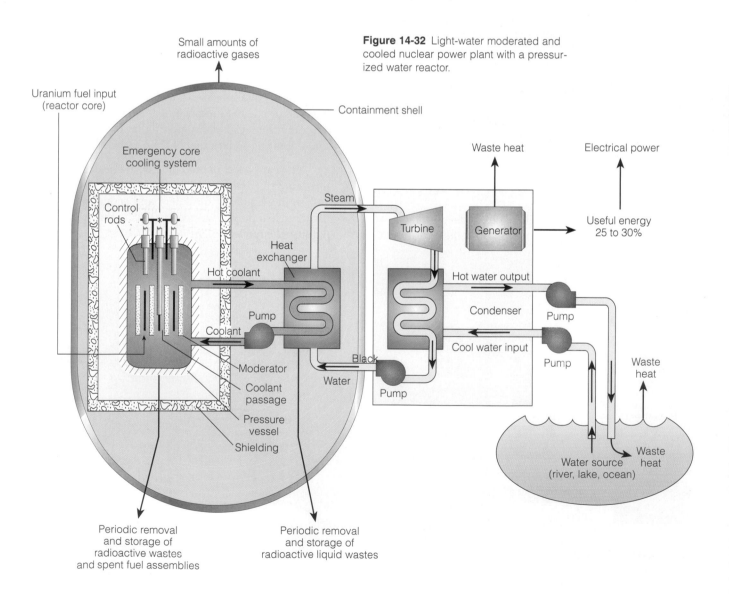

Figure 14-32 Light-water moderated and cooled nuclear power plant with a pressurized water reactor.

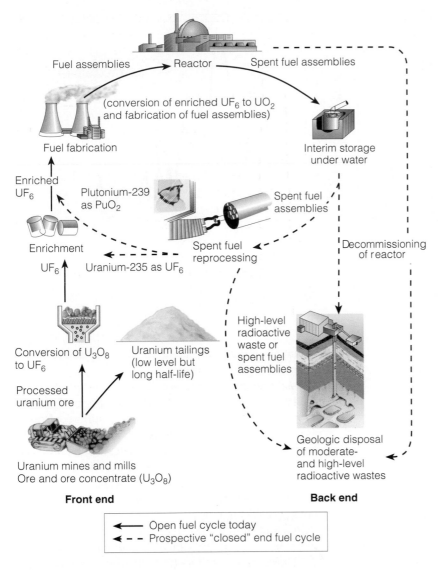

Fuel assemblies → Reactor → Spent fuel assemblies

(conversion of enriched UF_6 to UO_2 and fabrication of fuel assemblies)

Fuel fabrication

Interim storage under water

Enriched UF_6

Plutonium-239 as PuO_2

Spent fuel assemblies

Enrichment

UF_6

Uranium-235 as UF_6

Spent fuel reprocessing

Decommissioning of reactor

Conversion of U_3O_8 to UF_6

Uranium tailings (low level but long half-life)

Processed uranium ore

High-level radioactive waste or spent fuel assemblies

Uranium mines and mills
Ore and ore concentrate (U_3O_8)

Geologic disposal of moderate- and high-level radioactive wastes

Front end

Back end

← Open fuel cycle today
◄- - Prospective "closed" end fuel cycle

Figure 14-33 The nuclear fuel cycle.

What Happened to Nuclear Power? Studies indicate that U.S. utility companies began developing nuclear power plants in the late 1950s for three reasons:

- The Atomic Energy Commission (which had the conflicting roles of promoting and regulating nuclear power) promised utility executives that nuclear power would produce electricity at a much lower cost than coal and other alternatives. Indeed, president Dwight D. Eisenhower declared in a 1953 speech that nuclear power would be "too cheap to meter."

- The government paid about one-fourth of the cost of building the first group of commercial reactors and guaranteed that there would be no cost overruns.

- After insurance companies refused to insure nuclear power, Congress passed the Price-Anderson Act to protect the U.S. nuclear industry and utilities from significant liability to the general public in case of accidents.

In the 1950s, researchers predicted that by the year 2000, at least 1,800 nuclear power plants would supply 21% of the world's commercial energy (25% in the United States), and most of the world's electricity.

However, after almost 50 years of development, enormous government subsidies, and an investment of $2 trillion, these goals have not been met. Instead,

- By 1999, 435 commercial nuclear reactors in 32 countries were producing 6% of the world's commercial energy and 17% of its electricity.

- Since 1989 the growth in electricity production from nuclear power has essentially leveled off and its capacity is expected to decline between 2003 and 2020 as existing plants wear out and are retired or become too costly to run (Figure 14-34, left).

- Since 1975 the number and capacity of new nuclear reactors has declined sharply (Figure 14-34, right), with only 36 reactors currently under construction (26 of them in Asia).

- Germany (with 31% of its electricity from nuclear power) and Sweden (with 47%) plan to phase out nuclear power over the next 20 to 30 years.

- Public opposition to nuclear power has been so intense in Japan (with 36% of its electricity from nuclear power) that only two sites for new nuclear plants have been approved since 1979.

- No new nuclear power plants have been ordered in the United States since 1978, and all 120 plants ordered since 1973 have been canceled. In 1999, the 103 licensed commercial nuclear power plants in the United States generated about 20% of the country's electricity. This percentage is expected to decline over the next two decades as existing plants wear out and are retired (decommissioned) or become too costly to run.

China is not following this global trend in the declining use of nuclear power. It plans to build more than 50 new plants by 2020 as a way to reduce its dependence on highly polluting coal. France, which gets 75% of its electricity from nuclear power, is also a strong advocate of nuclear power. However, in 1999 it placed a moratorium on further construction of nuclear power plants.

The major reasons for the failure of nuclear power to grow as projected are **(1)** multi-billion-dollar construction cost overruns, **(2)** stricter government

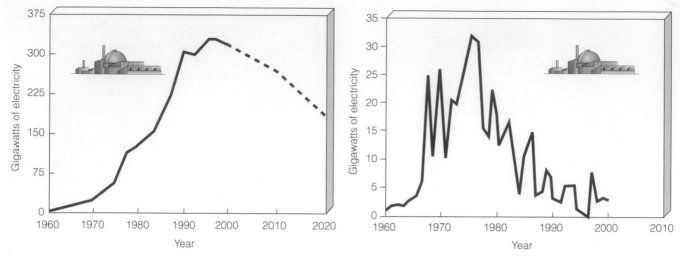

Figure 14-34 Global nuclear power generating capacity peaked in the 1990s and is projected to decline (left) because of the **(1)** sharp decline in construction of new nuclear reactors since 1975 (right) and **(2)** retirement of many of the world's existing reactors as they wear out or cost too much to run. (Data from U.S. Department of Energy and the Worldwatch Institute).

safety regulations, **(3)** high operating costs, **(4)** more malfunctions than expected, **(5)** poor management, **(6)** public concerns about safety after the Chernobyl (p. 350) and Three Mile Island (p. 351) accidents, and **(7)** investor concerns about the economic feasibility of nuclear power.

What Are the Pros and Cons of Nuclear Power?
Figure 14-35 lists the major advantages and disadvantages of nuclear power. Using nuclear power to produce electricity has some important advantages over coal-burning power plants (Figure 14-36).

How Safe Are Nuclear Power Plants and Other Nuclear Facilities?
Because of the built-in safety features, the risk of exposure to radioactivity from nuclear power plants in the United States and most other developed countries is *extremely low*. However, a partial or complete meltdown or explosion is possible, as accidents at the Chernobyl (Case Study, p. 350) nuclear power plant in Ukraine and the Three Mile Island plant in Pennsylvania (Case Study, p. 351) have taught us.

The U.S. Nuclear Regulatory Commission (NRC) estimates that there is a 15–45% chance of a complete core meltdown at a U.S. reactor during the next 20 years. The NRC also found that 39 U.S. reactors have an 80% chance of containment shell failure from a meltdown or an explosion of gases inside the containment structures.

Nuclear scientists and government officials throughout the world urge the shutdown of 35 poorly designed and poorly operated nuclear reactors in some republics of the former Soviet Union and in eastern Europe. However, without considerable economic aid from the world's developed countries, it is unlikely that these

potentially dangerous plants will be closed and replaced with safer nuclear or nonnuclear alternatives.

All French pressurized water nuclear reactors are built using a cost-saving standardized design. However, plants with this generic design have suffered from a growing number of faulty steam generator problems and cracked reactor vessel heads, which must be replaced at great cost to minimize the chances of serious accidents.

In the United States, there is widespread public distrust in the ability of the NRC and the Department of Energy (DOE) to enforce nuclear safety in commercial (NRC) and military (DOE) nuclear facilities. Congressional hearings in 1987 uncovered evidence that high-level NRC staff members **(1)** destroyed documents and obstructed investigations of criminal wrongdoing by utilities, **(2)** suggested ways utilities could evade commission regulations, and **(3)** provided utilities and their contractors with advance notice of "surprise" inspections. In 1996, George Galatis, a respected senior nuclear engineer, said, "I believe in nuclear power but after seeing the NRC in action I'm convinced a serious accident is not just likely, but inevitable….They're asleep at the wheel."

The nuclear power industry claims that nuclear power plants in the United States have not killed anyone and cause far less environmental harm than coal-burning plants (Figure 14-36). However, according to U.S. National Academy of Sciences estimates, U.S. nuclear plants cause 6,000 premature deaths and 3,700 serious genetic defects each year. If correct, this annual death toll is much smaller than the 65,000 to 200,000 deaths per year caused by coal-burning plants in the United States. However, critics point out that the estimated annual deaths from both of these types of plants

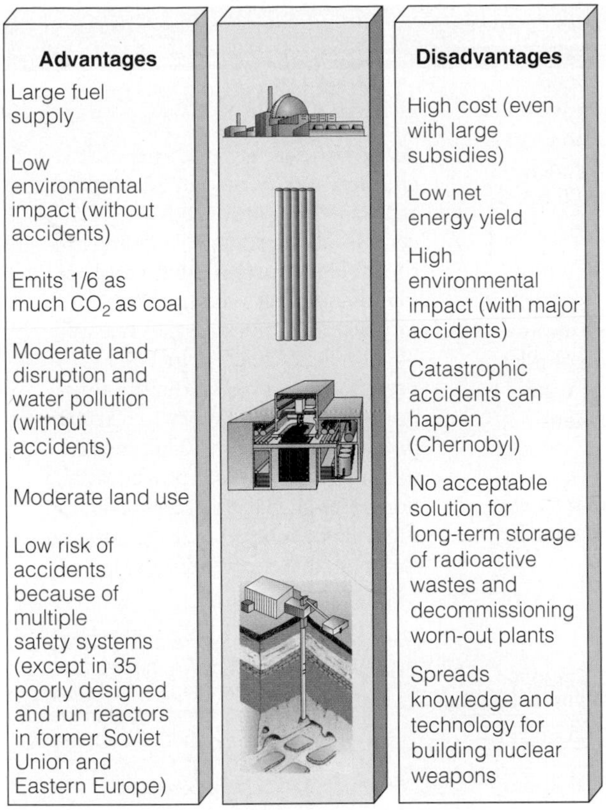

Advantages			Disadvantages
Large fuel supply	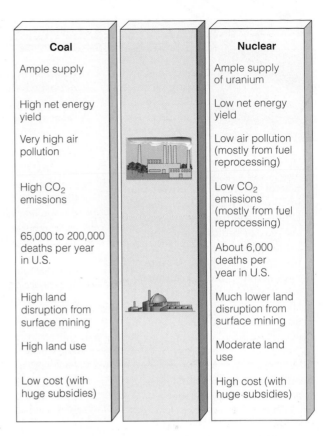		High cost (even with large subsidies)
Low environmental impact (without accidents)			Low net energy yield
Emits 1/6 as much CO_2 as coal			High environmental impact (with major accidents)
Moderate land disruption and water pollution (without accidents)			Catastrophic accidents can happen (Chernobyl)
Moderate land use			No acceptable solution for long-term storage of radioactive wastes and decommissioning worn-out plants
Low risk of accidents because of multiple safety systems (except in 35 poorly designed and run reactors in former Soviet Union and Eastern Europe)			Spreads knowledge and technology for building nuclear weapons

Figure 14-35 Advantages and disadvantages of using nuclear power to produce electricity. This evaluation includes the entire nuclear fuel cycle (Figure 14-33).

Coal		Nuclear
Ample supply		Ample supply of uranium
High net energy yield		Low net energy yield
Very high air pollution		Low air pollution (mostly from fuel reprocessing)
High CO_2 emissions		Low CO_2 emissions (mostly from fuel reprocessing)
65,000 to 200,000 deaths per year in U.S.		About 6,000 deaths per year in U.S.
High land disruption from surface mining		Much lower land disruption from surface mining
High land use		Moderate land use
Low cost (with huge subsidies)		High cost (with huge subsidies)

Figure 14-36 Comparison of the risks of using nuclear power and coal-burning plants to produce electricity.

are unacceptable, given the availability of much less harmful alternatives.

What Do We Do with Low-Level Radioactive Waste? Each part of the nuclear fuel cycle (Figure 14-33) produces low-level and high-level solid, liquid, and gaseous radioactive wastes with various half-lives (Table 3-2, p. 62). Wastes classified as *low-level radioactive wastes* give off small amounts of ionizing radiation and must be stored safely for 100–500 years before decaying to safe levels.

From the 1940s to 1970, most low-level radioactive waste produced in the United States (and most other countries) was put into steel drums and dumped into the ocean; the United Kingdom and Pakistan still dispose of their low-level radioactive wastes in this way.

Today, low-level waste materials from commercial nuclear power plants, hospitals, universities, industries, and other producers in the United States are put in steel drums and shipped to the two remaining regional landfills run by federal and state governments. Attempts to build new regional dumps for low-level radioactive waste using improved technology (Figure 14-38) have met with fierce public opposition.

In 2000 the NRC was evaluating a controversial plan that would allow nuclear power plant operators to sell their soils contaminated with low-level radioactive waste to construction companies, farmers, golf courses, and for use in home landscaping projects, athletic fields, and playgrounds.

What Should We Do with High-Level Radioactive Waste? *High-level radioactive wastes* give off large amounts of ionizing radiation (Figure 3-12, p. 62) for a short time and small amounts for a long time. Such wastes must be stored safely for at least 10,000 years and about 240,000 years if plutonium-239 is not removed by reprocessing (Figure 3-13, p. 62). Most high-level radioactive wastes are spent fuel rods from commercial nuclear power plants (now being stored in pools of water at plant sites) and an assortment of wastes from plants that produce plutonium and tritium for nuclear weapons.

After 50 years of research, scientists still do not agree on whether there is any safe method of storing these wastes. Some scientists believe that the long-term safe storage or disposal of high-level radioactive wastes is technically possible. Others disagree, pointing out

The Chernobyl Nuclear Power-Plant Accident

Chernobyl is a chilling word recognized around the globe as the site of a major nuclear disaster (Figure 14-37). On April 26, 1986, a series of explosions in one of the reactors in a nuclear power plant in Ukraine (then part of the Soviet Union) blew the massive roof off the reactor building and flung radioactive debris and dust high into the atmosphere. A huge radioactive cloud spread over much of Belarus, Russia, the Ukraine, and other parts of Europe and eventually encircled the planet.

Here are some consequences of this disaster, caused by poor reactor design and human error:

■ In 1998, the Ukrainian Health Ministry put the official death toll from the accident at 3,576. However, Greenpeace Ukraine estimates that by 1995 the total death toll from the accident was about 32,000.

■ According to the United Nations, almost 400,000 people were forced to leave their homes, probably never to return. Most were not evacuated until at least 10 days after the accident.

■ According to a United Nations study, some 160,000 square kilometers (62,000 square miles)—about the size of the state of Florida—of the former Soviet Union remain contaminated with radioactivity.

■ More than half a million people were exposed to dangerous levels of radioactivity. In Belarus,

where 70% of the radiation was deposited, the World Health Organization says that thyroid cancer rates among children are 100 times preaccident levels.

■ Government officials say that the total cost of the accident will reach at least $358 billion, many times more than the value of all the nuclear electricity that was ever generated in the former Soviet Union.

Chernobyl taught us that *a major nuclear accident anywhere is a nuclear accident everywhere.*

Critical Thinking

After a nuclear accident, exposed children and adults can take iodine tablets to help prevent thyroid cancer. The stable isotope of iodine in the tablets saturates the thyroid gland and blocks the uptake of radioactive iodine isotopes released by a nuclear accident. In 1997, French officials began distributing potassium iodide tablets to 600,000 people living within 10 kilometers (6 miles) of its nuclear power installations. Has such action been taken in the country where you live? Are you for or against doing this? Explain.

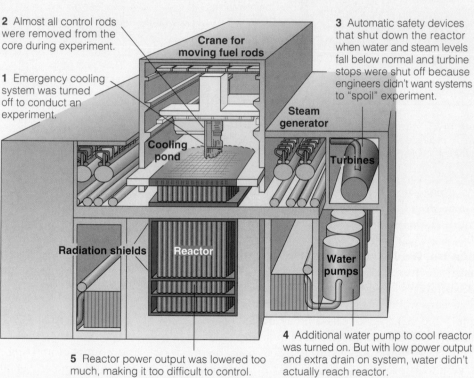

2 Almost all control rods were removed from the core during experiment.

1 Emergency cooling system was turned off to conduct an experiment.

Crane for moving fuel rods

3 Automatic safety devices that shut down the reactor when water and steam levels fall below normal and turbine stops were shut off because engineers didn't want systems to "spoil" experiment.

Steam generator

Cooling pond

Turbines

Radiation shields

Reactor

Water pumps

4 Additional water pump to cool reactor was turned on. But with low power output and extra drain on system, water didn't actually reach reactor.

5 Reactor power output was lowered too much, making it too difficult to control.

Figure 14-37 Major events leading to the Chernobyl nuclear power-plant accident on April 26, 1986, in the former Soviet Union. The accident happened because **(1)** engineers turned off most of the reactor's automatic safety and warning systems (to keep them from interfering with an unauthorized safety experiment), **(2)** the safety design of the reactor was inadequate (there was no secondary containment shell, as in Western-style reactors, and no emergency core cooling system to prevent reactor core meltdown), and **(3)** a design flaw led to unstable operation at low power. After the reactor exploded, crews exposed themselves to lethal levels of radiation to put out fires and encase the shattered reactor in a hastily constructed concrete tomb. This 20-story concrete tomb is sagging and is full of holes that allow water to seep in and radioactive dust to drift out. Building a new tomb for the reactor will cost at least $1.5 billion—money the Ukrainian government does not have.

The Three Mile Island Nuclear Power-Plant Accident

On March 29, 1979, the number 2 reactor at the Three Mile Island (TMI) nuclear plant near Harrisburg, Pennsylvania, lost its coolant water because of a series of mechanical failures and human operator errors not anticipated in safety studies.

The reactor's core became partially uncovered and about 50% of it melted and fell to the bottom of the reactor. Unknown amounts of radioactive materials escaped into the atmosphere, 50,000 people were evacuated, and another 50,000 fled the area on their own.

Most analysts say that the estimated radiation released was not enough to cause deaths or cancers. However, there is controversy over how much radiation was released and there have been higher than normal increases in certain cancers among people who lived near the plant.

Partial cleanup of the damaged TMI reactor, lawsuits, and payment of damage claims have cost $1.2 billion so far, almost twice the reactor's $700-million construction cost. Banks and other lending institutions have little interest in financing new U.S. power plants because the TMI accident showed that utility companies could lose more than $1 billion in equipment and cleanup costs, even without any established harmful effects on public health.

Critical Thinking

According to the nuclear power industry, the TMI accident showed that its safety systems work because no deaths were known to have been caused by the accident. Other analysts dispute this claim and argue that the accident was a wake-up call about the potential dangers of nuclear power plants that led to tighter and better safety regulations. What is your position? Explain.

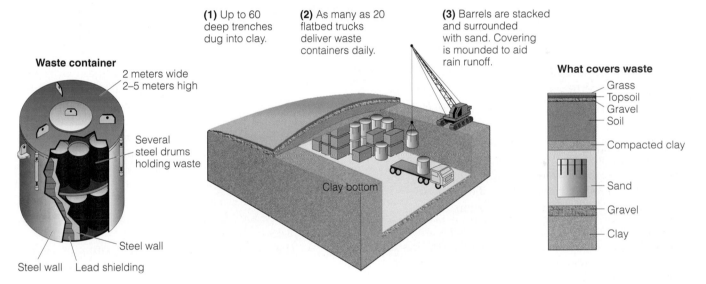

(1) Up to 60 deep trenches dug into clay.

(2) As many as 20 flatbed trucks deliver waste containers daily.

(3) Barrels are stacked and surrounded with sand. Covering is mounded to aid rain runoff.

Waste container

2 meters wide
2–5 meters high

Several steel drums holding waste

Steel wall

Steel wall   Lead shielding

Clay bottom

What covers waste

Grass
Topsoil
Gravel
Soil

Compacted clay

Sand

Gravel

Clay

Figure 14-38 Proposed design of a state-of-the-art low-level radioactive waste landfill. According to the EPA, all landfills eventually leak. The best-designed landfills may not leak for several decades, after which the companies running them would no longer be liable for leaks and problems. Then any problems would be passed on to taxpayers and future generations. (Data from U.S. Atomic Industrial Forum)

that it is impossible to demonstrate that any method will work for the 10,000–240,000 years of fail-safe storage needed for such wastes.

Here are some of the proposed methods and their possible drawbacks:

▪ *Bury it deep underground* (Figure 14-39, p. 352 and Case Study, p. 353). This favored strategy is under study by all countries producing nuclear waste.

However, according to a 1990 report by the U.S. National Academy of Sciences, "Use of geological information to pretend to be able to make very accurate predictions of long-term site behavior is scientifically unsound."

▪ *Shoot it into space or into the sun.* Costs would be very high, and a launch accident, such as the explosion of the space shuttle *Challenger*, could disperse high-level radioactive wastes over large areas of the earth's surface. This strategy has been abandoned for now.

▪ *Bury it under the antarctic ice sheet or the Greenland ice cap.* The long-term stability of the ice sheets is not known. They could be destabilized by heat from the wastes, and retrieving the wastes would be difficult

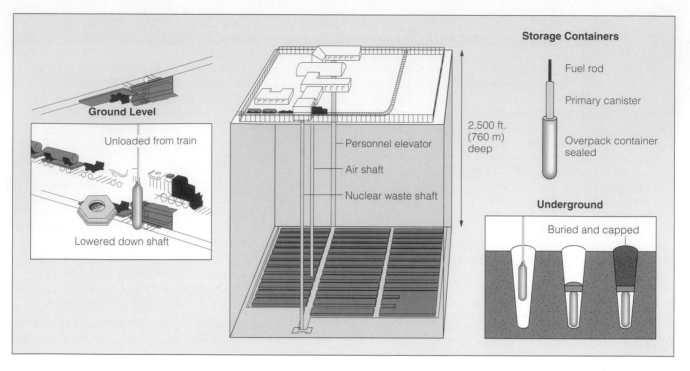

Figure 14-39 Proposed general design for deep-underground permanent storage of high-level radioactive wastes from commercial nuclear power plants in the United States. (U.S. Department of Energy)

or impossible if the method failed. This strategy is prohibited by international law.

- *Dump it into descending subduction zones in the deep ocean* (Figure 10-6, middle, p. 215). Wastes eventually might be spewed out somewhere else by volcanic activity, containers might leak and contaminate the ocean before being carried downward, and retrieval would be impossible if the method did not work. This strategy is prohibited by international law.

- *Bury it in thick deposits of mud on the deep ocean floor in areas that tests show have been geologically stable for 65 million years.* The waste containers would eventually corrode and release their radioactive contents. This approach is prohibited by international law.

- *Change it into harmless, or less harmful, isotopes.* Currently there is no way to do this. Even if a method were developed, **(1)** costs probably would be very high and **(2)** the resulting toxic materials and low-level (but very long-lived) radioactive wastes would still need to be disposed of safely.

How Widespread Are Contaminated Radioactive Sites? In 1992, the EPA estimated that as many as 45,000 sites in the United States may be contaminated with radioactive materials, 20,000 of them belonging to the DOE and the Department of Defense. According to the DOE, it will cost taxpayers at least $230 billion over the next 75 years to clean up these facilities (with some analysts estimating that the price will range from $400 to $900 billion). According to a 2000 DOE report, more than two-thirds of 144 highly contaminated sites used to produce nuclear weapons will never be completely cleaned up and will have to be protected and monitored for centuries. Some critics question spending so much money on a problem that is ranked by scientific advisers to the EPA as a low-risk ecological problem and not among the top high-risk human health problems (Figure 16-13, p. 411).

The radioactive contamination situation in the United States pales in comparison to the post-Cold War legacy of nuclear waste and contamination in the republics of the former Soviet Union. Land in various parts of the former Soviet Union is dotted with **(1)** areas severely contaminated by nuclear accidents, **(2)** 25 operating nuclear power plants with flawed and unsafe designs, **(3)** nuclear-waste dump sites, **(4)** radioactive waste-processing plants, **(5)** contaminated nuclear test sites, and **(6)** coastal waters where nuclear wastes and retired nuclear-powered submarines were dumped.

A 1957 explosion of a nuclear waste storage tank at Mayak, a plutonium production facility in southern Russia, spewed 2.5 times as much radiation into the atmosphere as the Chernobyl accident. Because of radioactive contamination, **(1)** no one can live within about 2,600 square kilometers (1,000 square miles) of the surrounding area, and **(2)** nearby Lake Karachay

Underground Storage of High-Level Radioactive Wastes in the United States

In 1985, the U.S. Department of Energy (DOE) announced plans to build a repository for underground storage of high-level radioactive wastes from commercial nuclear reactors on federal land in the Yucca Mountain desert region, 160 kilometers (100 miles) northwest of Las Vegas, Nevada. The facility (Figure 14-39), which is expected to cost about $43 billion to build (financed by a tax on nuclear power), is scheduled to open by 2010.

It may never open, mostly because (1) rock fractures may allow water to leak into the site and corrode radioactive waste storage casks, (2) there is an active volcano nearby, and (3) there are 36 active earthquake faults on the site. In 1998, Jerry Szymanski, formerly the DOE's top geologist at Yucca Mountain and now an outspoken opponent of the site, said that if water flooded the site it could cause an explosion so large that "Chernobyl would be small potatoes."

However, in 1999 scientists working on the site released a preliminary report (1) finding nothing so far to disqualify the site and (2) contending that it could safely store high-level radioactive wastes for several hundred thousand years.

Regardless of the storage method, most citizens strongly oppose the location of low-level or high-level nuclear waste disposal facilities anywhere near them. Even if the problem is technically solvable, it may be politically unacceptable.

Critical Thinking

1. What do you believe should be done with high-level radioactive wastes? Explain.

2. Would you favor having high-level nuclear waste transported by truck or train through the area where you live to the Yucca Mountain site? Explain.

radioactive materials in high-level nuclear waste storage facilities (Figure 14-39), (2) putting up a physical barrier and setting up full-time security for 30–100 years before the plant is dismantled, or (3) enclosing the entire plant in a tomb that must last for several thousand years.

At least 228 large commercial reactors worldwide (20 in the United States) are scheduled for retirement between 2000 and 2012. By 2030 all U.S. reactors will have to be retired, based on the life of their current operating licenses, and many may be retired early for safety or financial reasons.

U.S. utilities have been setting aside funds for decommissioning more than 100 reactors, but to date these funds fall far short of the estimated costs. If sufficient funds are not available for decommissioning, the balance of the costs could be passed along to ratepayers and taxpayers.

Some utility companies have proposed extending the life of current reactors to 60 years. By 2000 five U.S. plants had obtained 20-year licence extensions and 30 other plants may apply for such extensions by 2005. However, opponents contend that this could increase the risk of nuclear accidents in aging reactors.

What Is the Connection Between Nuclear Reactors and the Spread of Nuclear Weapons? Currently, 60 countries—1 of every 3 in the world—have nuclear weapons or the knowledge and ability to build them. Information and fuel needed to build these nuclear weapons has come mostly from the research and commercial nuclear reactors that the United States and 14 other countries have been giving away and selling in the international marketplace for decades.

Chances of nuclear war have been decreased by the recent dismantling of thousands of Russian and American nuclear warheads. However, this greatly increases the amount of bomb-grade plutonium-239 removed from these weapons that must be kept from use in nuclear weapons.

Can We Afford Nuclear Power? Experience has shown that nuclear power is an extremely expensive way to boil water to produce electricity, even when huge government subsidies partially shield it from free-market competition with other energy sources.

Costs rose dramatically in the 1970s and 1980s because of unanticipated safety problems and stricter regulations after the Three Mile Island and Chernobyl accidents. In 1995, the World Bank said that nuclear power is too costly and risky and *The Economist* says of nuclear power plants that "not one, anywhere in the world, makes commercial sense."

Forbes business magazine has called the failure of the U.S. nuclear power program "the largest managerial disaster in U.S. business history, involving $1 trillion in

(which was a dump site for the facility's radioactive wastes between 1949 and 1967) is so radioactive that standing on its shores for about an hour would be fatal.

What Can We Do with Worn-Out Nuclear Plants?
After approximately 15–40 years of operation, a nuclear reactor becomes dangerously contaminated with radioactive materials, and many of its parts are worn out. Then the plant must be *decommissioned* or retired (the last step in the nuclear fuel cycle, Figure 14-33) by (1) dismantling it and storing its large volume of highly

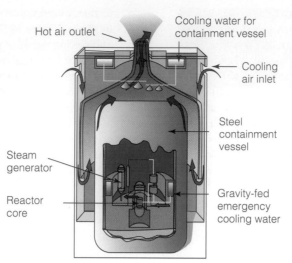

Figure 14-40 One of several new and smaller reactor designs that nuclear engineers say should improve the safety of nuclear power and reduce its costs. This reactor has a *passive safety design* that **(1)** makes it almost impossible for the reactor to go beyond acceptable levels of power and **(2)** in the case of an accident allows cooling water to flow into the reactor by gravity instead of using a complex system of pumps to bring it in. Other designs reduce chances of a runaway chain reaction by encapsulating uranium fuels in tiny, heat-resistant ceramic spheres instead of packing large numbers of fuel pellets into long metal rods. (Data from Westinghouse)

Labels on figure: Hot air outlet · Cooling water for containment vessel · Cooling air inlet · Steel containment vessel · Steam generator · Reactor core · Gravity-fed emergency cooling water

wasted investment and $10 billion in direct losses to stockholders." To financial analysts and utility company executives in most countries, the issue is no longer whether it is economical to build and operate nuclear power plants but whether it makes economic sense to continue operating many of the plants already built.

The U.S. nuclear industry hopes to persuade the federal government and utility companies to build hundreds of smaller second-generation plants using standardized designs, which they claim are safer and can be built more quickly (in 3–6 years) (Figure 14-40). These *advanced light-water reactors (ALWRs)* have built-in *passive safety features* designed to make explosions or the release of radioactive emissions almost impossible. However, according to *Nucleonics Week*, an important nuclear industry publication, "Experts are flatly unconvinced that safety has been achieved—or even substantially increased—by the new designs."

Is Breeder Nuclear Fission a Feasible Alternative? Some nuclear power proponents urge the development and widespread use of **breeder nuclear fission reactors**, which generate more nuclear fuel than they consume by converting nonfissionable uranium-238 into fissionable plutonium-239. Because breeders would use more than 99% of the uranium in ore deposits, the world's known uranium reserves would last at least 1,000 years, and perhaps several thousand years.

However, if the safety system of a breeder reactor fails, the reactor could lose some of its liquid sodium coolant, which ignites when exposed to air and reacts explosively if it comes into contact with water. This could cause a runaway fission chain reaction and perhaps a nuclear explosion powerful enough to blast open the containment building and release a cloud of highly radioactive gases and particulate matter. Leaks of flammable liquid sodium can also cause fires, as has

happened with all experimental breeder reactors built so far.

In addition, existing experimental breeder reactors produce plutonium so slowly that it would take 100–200 years for them to produce enough plutonium to fuel a significant number of other breeder reactors. In 1994, the United States ended government-supported research for breeder technology after providing about $9 billion in research and development funding.

In December 1986, France opened a commercial-size breeder reactor. It was so expensive to build and operate that after spending $13 billion the government closed it permanently in 1998. Because of this experience, other countries have abandoned their plans to build full-size commercial breeder reactors.

Is Nuclear Fusion a Feasible Alternative? Scientists hope that controlled nuclear fusion will provide an almost limitless source of high-temperature heat and electricity. Research has focused on the D-T nuclear fusion reaction, in which two isotopes of hydrogen—deuterium (D) and tritium (T)—fuse at about 100 million degrees (Figure 3-17, p. 64).

After 50 years of research and huge expenditures of mostly government funds, controlled nuclear fusion is still in the laboratory stage. None of the approaches tested so far have produced more energy than they use. In 1989, two chemists claimed to have achieved deuterium-deuterium (D-D) nuclear fusion at room temperature using a simple apparatus, but subsequent experiments have not substantiated their claim.

If researchers can eventually get more energy out of nuclear fusion than they put in, the next step would be to build a small fusion reactor and then scale it up to commercial size, an extremely difficult engineering problem. The estimated cost of a commercial fusion reactor is several times that of a comparable conventional fission reactor.

Proponents contend that with greatly increased federal funding, a commercial nuclear fusion power plant might be built by 2030. However, many energy experts do not expect nuclear fusion to be a significant energy source until 2100, if then.

What Should Be the Future of Nuclear Power in the United States? Since 1948 nuclear energy has received about 65% of all federal energy research and development funds in the United States. Some analysts call for phasing out all or most government subsidies and tax breaks for nuclear power and using such funds to subsidize and accelerate the development of promising newer energy technologies such as (1) natural gas turbines, (2) improving energy efficiency, (3) forms of renewable energy such as wind, solar cells, and hydrogen (Chapter 15), and (4) smaller and more flexible *micropower* turbines and fuel cells (Section 15-9, p. 389).

To these analysts, nuclear power is a complex, expensive, inflexible, and centralized way to produce electricity. They believe that it is a technology whose time has passed in a world where electricity will increasingly be provided by small, decentralized, easily expandable power plants such as natural gas turbines, wind turbines, arrays of solar cells, and fuel cells (Section 15-9, p. 389). According to investors and World Bank economic analysts, nuclear power simply cannot compete in today's increasingly open and unregulated energy market.

Proponents of nuclear power argue that governments should continue funding research and development and pilot plant testing of potentially safer and cheaper reactor designs (Figure 14-40) along with breeder fission and nuclear fusion. They argue that we need to keep these nuclear options available for use in the future if natural gas turbines, improved energy efficiency, hydrogen-powered fuel cells, wind turbines, and other renewable energy options fail to (1) keep up with electricity demands and (2) reduce CO_2 emissions to acceptable levels.

We are embarked on the beginning of the last days of the Age of Oil.

MIKE BOWLIN (CEO ARCO OIL)

REVIEW QUESTIONS

1. Define the boldfaced terms in this chapter.

2. Describe the basic features of the 1872 Mining Law in the United States and list its pros and cons.

3. Distinguish between a *mineral resource* and an *ore*. Distinguish between *identified resources*, *undiscovered resources*, *reserves*, and *other resources*. Give two factors that can increase the reserves of a mineral resource.

4. Describe how mineral deposits are formed by *magma flows*, *hydrothermal processes*, *sedimentary sorting*, *precipitation*, and *rock weathering*.

5. List seven ways mining companies use to find mineral deposits. Distinguish between **(a)** *overburden* and *spoil*, **(b)** *surface mining* and *subsurface mining*, and **(c)** *open-pit mining*, *dredging*, *area strip mining*, and *contour strip mining*.

6. List five major environmental impacts of mining, processing, and using mineral resources. Describe the life cycle of a metal resource. Distinguish between *ore mineral*, *gangue*, and *tailings*. What is *smelting*, and what are

its major environmental impacts? What are three environmental limits associated with extracting and using mineral resources?

7. What is *economic depletion* of a mineral resource? When such depletion occurs, what five choices are available? Distinguish between *depletion time* and the *reserve-to-production ratio* for a mineral resource. How can the estimated depletion time for a resource be extended?

8. What five countries supply most of the nonrenewable mineral resources used by modern societies?

9. Explain how a competitive free market should increase supplies of a scarce mineral resource and list three reasons why this idea may not be effective under today's economic conditions.

10. List the pros and cons of allowing more mineral exploration and mining on public lands in the United States.

11. List the pros and cons of increasing supplies of mineral resources by **(a)** mining lower-grade ores, **(b)** mining the oceans, and **(c)** finding substitutes for scarce nonrenewable mineral resources.

12. List the pros and cons of **(a)** current methods used to mine gold in developed and developing countries and **(b)** using bacteria to extract metals from ores.

13. Where does most of the energy we use come from? What percentage of the energy we use comes from nonrenewable and from renewable energy in **(a)** the world and **(b)** the United States?

14. How long does it usually take to phase in a new energy alternative to the point where it accounts for 10–20% of total energy use? What four questions should we try to answer about each energy resource?

15. What is *net energy*, and why is it important in evaluating an energy resource?

16. What is *petroleum* or *crude oil*? Distinguish between *primary*, *secondary*, and *tertiary recovery* of crude oil. How is oil extracted from the earth's crust? Describe the nature of and the effects of the oil spill by the tanker *Exxon Valdez*.

17. What happens to crude oil at a refinery? What are *petrochemicals*?

18. Who has most of the world's oil reserves? What percentage of the world's oil reserves is found in the United States? What percentage of the world's oil does the United States use? What percentage of the oil used in the United States is imported? How long are known and projected supplies of conventional oil expected to last in **(a)** the world and **(b)** the United States?

19. What are the pros and cons of using oil as an energy resource?

20. What are *oil shale* and *kerogen*, and how can these substances be converted to shale oil? What is *tar sand*, and how can it be converted to synthetic crude oil? What are the pros and cons of using heavy oil from shale oil and tar sand as energy resources?

21. What is *natural gas*? Who has most of the world's reserves of natural gas? What is *methane hydrate*? Distinguish between *liquefied petroleum gas (LPG)* and *liquefied natural gas (LNG)*.

22. How long are known and projected supplies of natural gas expected to last in **(a)** the world and **(b)** the United States? What is a *combined-cycle natural gas turbine system*, and why is its use rising rapidly? What are the pros and cons of using natural gas as an energy resource?

23. What is *coal*, and how is it formed? Distinguish between *lignite, bituminous*, and *anthracite* coal. How is coal extracted from the earth's crust? How is coal used? What four countries have the largest reserves of coal?

24. How long are known and projected supplies of coal expected to last **(a)** in the world and **(b)** in the United States?

25. What are the pros and cons of using coal as an energy resource? What are the pros and cons of converting solid coal into gaseous and liquid fuels?

26. Describe how a *nuclear fission reactor* works. What are the five major components of a *light-water nuclear reactor*, and what role does each play? What is the *nuclear fuel cycle*?

27. List three reasons why commercial nuclear power plants were developed in the United States after World War II. What five factors have contributed to the leveling off of the use of nuclear power plants to produce electricity?

28. List the major advantages and disadvantages of using conventional nuclear fission to produce electricity. Compare the advantages and disadvantages of using nuclear power and burning coal to produce electricity.

29. How safe are nuclear power plants? Describe **(a)** the Chernobyl nuclear power plant in the former Soviet Union and **(b)** the Three Mile Island nuclear power-plant accident in Pennsylvania.

30. What is being done with *low-level radioactive waste* produced by the nuclear fuel cycle? What are the options for dealing with *high-level radioactive waste*? List the pros and cons of the proposed site for storing high-level nuclear wastes at Yucca Mountain in Nevada. How widespread are contaminated radioactive waste sites in **(a)** the United States and **(b)** the former Soviet Union?

31. What are the three options for retiring (decommissioning) nuclear power plants?

32. What is the relationship between the development of commercial nuclear power and the spread of nuclear weapons throughout much of the world?

33. Why are the costs of nuclear power so high? What is the likely financial future of nuclear power?

34. What are the pros and cons of using **(a)** breeder nuclear fission and **(b)** nuclear fusion as an energy resource? What are the pros and cons of continuing large-scale government subsidies for research and development of conventional nuclear power, breeder fission, and nuclear fusion?

CRITICAL THINKING

1. Use the second law of energy (p. 65) to analyze the scientific and economic feasibility of each of the following processes: **(a)** extracting most minerals dissolved in seawater, **(b)** recycling mineral products such as aluminum drink cans that have been discarded and widely dispersed, **(c)** mining increasingly lower-grade deposits of minerals, **(d)** using inexhaustible solar energy to mine minerals, and **(e)** continuing to mine, use, and recycle minerals at increasing rates.

2. Explain why you support or oppose each of the following proposals concerning extraction of hardrock minerals on public land in the United States: **(a)** not granting title to public lands in the United States for actual or claimed hardrock mining, **(b)** requiring mining companies to pay a royalty of 8–12% on the net revenues they earn from hardrock minerals they extract from public lands, **(c)** making hardrock mining companies responsible for restoring the land and cleaning up environmental damage caused by their activities.

3. Why does extracting and minerals from low-grade ores cause more environmental damage than extracting them from high-grade ores?

4. List the pros and cons of including the environmental and health costs caused by mining, processing, and producing mineral resources (Figure 14-6) in the prices of metals to manufacturers and in the prices of consumer products. Do you favor including environmental costs in the prices of products? Explain. How would you institute such a policy?

5. Just to continue using oil at the current rate (not the projected higher exponential increase in its annual use) we must discover and add to global oil reserves the equivalent of a new Saudi Arabia supply (the world's largest) *every 10 years*. Do you believe this is possible? If not, what effects might this have on your life and on the life of a child or grandchild you might have?

6. If you were trying to find new deposits of oil, would you search primarily in formations of igneous rock, sedimentary rock, or metamorphic rock? Explain.

7. List five things you can do to reduce your dependence on oil and resources such as gasoline and most plastics derived from oil (see Appendix 6). Which of these things do you actually plan to do?

8. The United States now imports more than half the oil it uses and could be importing 64% of its oil by 2020. Explain why you are for or against continuing to increase oil imports. What do you believe are the five best ways to reduce dependence on oil imports?

9. Explain why you agree or disagree with the following proposals by various energy analysts: **(a)** To solve present and future U.S. energy problems, all we need to do

is find and develop more domestic supplies of oil, natural gas, and coal and increase dependence on nuclear power, **(b)** a heavy federal tax should be placed on gasoline and imported oil used in the United States (or the country where you live), and **(c)** by 2020, the United States should phase out all nuclear power plants.

10. Explain why you agree or disagree with each of the following proposals made by the U.S. nuclear power industry: **(a)** using government subsidies to build a large number of new, better-designed nuclear fission power plants (Figure 14-40) to reduce dependence on imported oil and slow down projected global warming, **(b)** restoring government subsidies to develop a breeder nuclear fission reactor program, and **(c)** greatly increasing current federal subsidies for developing nuclear fusion.

11. If you had to choose, would you rather live next door to a coal-fired power plant or a nuclear plant? Explain.

12. Should the United States and other developed countries provide economic and technical aid for closing 35 poorly designed and poorly operated nuclear reactors in some republics of the former Soviet Union and in eastern Europe? Explain.

PROJECTS

1. What mineral resources are extracted in your local area? What mining methods are used, and what have been their environmental impacts? How has mining these resources benefited the local economy?

2. How is the electricity in your community produced? How has the cost of that electricity changed since 1970 compared to general inflation?

3. Write a two-page scenario of what your life might be like without oil. Compare and discuss the scenarios developed by members of your class.

4. Use the library or the internet to find bibliographic information about *Ann Dorr* and *Mike Bowlin*, whose quotes appear at the beginning and end of this chapter.

5. Make a concept map of this chapter's major ideas, using the section heads and subheads and the key terms (in boldface). Look at the inside back cover and on the website for this book for information about making concept maps.

INTERNET STUDY RESOURCES AND RESOURCES FOR FURTHER READING AND RESEARCH

The website for this book contains helpful study aids and many ideas for further reading and research. Log on to:

http://www.brookscole.com/product/0534376975s

and click on the Chapter-by-Chapter area. Choose Chapter 14 and select a resource:

- "Flash Cards" allows you to test your mastery of the Terms and Concepts to Remember for this chapter.

- "Tutorial Quizzes" provides a multiple-choice practice quiz.

- "Student Guide to InfoTrac" will lead you to Critical Thinking Projects that use InfoTrac College Edition as a research tool.

- "References" lists the major books and articles consulted in writing this chapter.

- "Hypercontents" takes you to an extensive list of sites with news, research, and images related to individual sections of the chapter.

INFOTRAC COLLEGE EDITION

Improve your skills with InfoTrac College Edition, a searchable online database of articles from more than 700 periodicals. Log on to:

http://www.infotrac-college.com

or access InfoTrac through the website for this book.

Try the following articles:

Campbell, C.J. 2000. Depletion and denial: the final years of oil supplies. *USA Today (Magazine)* vol. 129, no. 2666, pp. 18–20. (subject guide: petroleum conservation)

Radetzki, M. 2000. Coal or nuclear in new power stations: the political economy of an undesirable but necessary choice. *The Energy Journal* vol. 21, no. 1, pp. 135–147. (subject guide: coal, nuclear)

15 ENERGY EFFICIENCY AND RENEWABLE ENERGY

The Coming Energy-Efficiency and Renewable-Energy Revolution

Energy experts Hunter and Amory Lovins (Guest Essay, p. 361) have built a large, passively heated, superinsulated, partially earth-sheltered home and office in Snowmass, Colorado (Figure 15-1), where winter temperatures can drop to −40°C (−40°F).

This structure, which also houses the research center for the Rocky Mountain Institute, an office used by 40 people, gets 99% of its space and water heating and 95% of its daytime lighting from the sun. It uses one-tenth the usual amount of electricity for a structure of its size.

Today's superinsulating windows mean that a house can have large numbers of windows without much heat loss in cold weather or heat gain in hot weather. Thinner insulation material now being developed will allow roofs and walls to be insulated far better than in today's best superinsulated houses.

A small but growing number of people in developed and developing countries are getting their electricity from *solar cells* that convert sunlight directly into electricity. They can be attached like shingles to a roof or applied to window glass as a coating. Solar-cell prices are high but are falling rapidly.

Many scientists and executives of oil and automobile companies believe we are in the beginning stages of a *solar-hydrogen revolution*. Electricity produced by large banks of solar cells or farms of wind turbines could be passed through water to make hydrogen gas (H_2), which could be used to fuel vehicles, industries, and buildings. Another solution is to burn hydrogen in energy-efficient *fuel cells* that produce electricity to run cars and heat buildings and water.

Burning hydrogen produces water vapor, small amounts of controllable nitrogen oxides, and no carbon dioxide. Thus, shifting to hydrogen as our primary energy resource during the 21st century would eliminate most of the world's air pollution and greatly slow projected global warming.

These are only a few of the components of the *energy-efficiency and renewable-energy revolution* that many analysts believe will help us make the transition to more sustainable societies over the next 40–50 years.

Figure 15-1 The Rocky Mountain Institute in Colorado. This facility is a home and a center for the study of energy efficiency and sustainable use of energy and other resources. It is also an example of energy-efficient passive solar design. (Robert Millman/Rocky Mountain Institute)

If the United States wants to save a lot of oil and money and increase national security, there are two simple ways to do it: Stop driving Petropigs and stop living in energy sieves.
AMORY B. LOVINS

This chapter addresses the following questions:

- What are the advantages and disadvantages of improving energy efficiency?
- What are the advantages and disadvantages of using solar energy to heat buildings and water and to produce electricity?
- What are the advantages and disadvantages of using flowing water and solar energy stored as heat in water to produce electricity?
- What are the advantages and disadvantages of using wind to produce electricity?
- What are the advantages and disadvantages of burning plant material (biomass) to heat buildings and water, produce electricity, and propel vehicles (biofuels)?
- What are the advantages and disadvantages of producing hydrogen gas and using it to produce electricity, heat buildings and water, and propel vehicles?
- What are the advantages and disadvantages of extracting heat from the earth's interior (geothermal energy)?
- What are the advantages and disadvantages of using smaller, decentralized micropower sources to heat buildings and water, produce electricity, and propel vehicles?
- How can we make a transition to a more sustainable energy future?

15-1 THE IMPORTANCE OF IMPROVING ENERGY EFFICIENCY

What Is Energy Efficiency? Doing More with Less **Energy efficiency** is the percentage of total energy input into an energy conversion device or system that does useful work and is not converted to low-quality, essentially useless heat. Improving the energy efficiency of a car motor, home heating system, or other energy conversion device involves using less energy to do more useful work.

You may be surprised to learn that *84% of all commercial energy used in the United States is wasted* (Figure 15-2). About 41% of this energy is wasted automatically because of the degradation of energy quality imposed by the second law of energy (Section 3-7, p. 65). However, about 43% is wasted unnecessarily, mostly by **(1)** using fuel-wasting motor vehicles, furnaces, and other devices

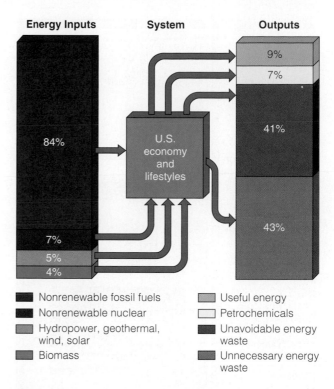

Figure 15-2 Flow of commercial energy through the U.S. economy. Note that only 16% of all commercial energy used in the United States ends up performing useful tasks or being converted to petrochemicals; the rest is either automatically and unavoidably wasted because of the second law of energy (41%) or wasted unnecessarily (43%).

and **(2)** by living and working in leaky, poorly insulated, poorly designed buildings (Guest Essay, p. 361).

According to the Department of Energy, the United States unnecessarily wastes as much energy as two-thirds of the world's population consumes. Improvements in energy efficiency since the OPEC oil embargo in 1973 have cut U.S. energy bills by $275 billion a year. However, unnecessary energy waste still costs the United States about $300 billion per year (an average of $570,000 per minute)—more than the $271 billion military budget in 2000.

Other developed countries also waste large amounts of energy, but many such as Japan, Germany, Sweden, and Denmark are about twice as energy-efficient as the United States. However, the world's most energy-inefficient countries are in the developing world. Reducing energy waste has a number of economic and environmental advantages (Figure 15-3).

The energy conversion devices we use vary in their energy efficiencies (Figure 15-4). We can save energy and money by buying more energy-efficient cars, lighting, heating systems, water heaters, air conditioners, and appliances. Some energy-efficient models may cost more initially, but in the long run they usually save money by having a lower **life cycle cost**: initial cost plus lifetime operating costs.

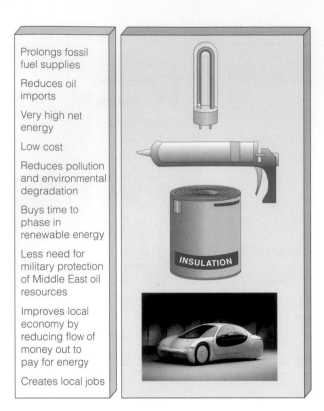

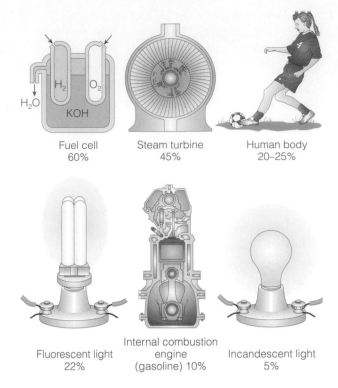

Fuel cell
60%

Steam turbine
45%

Human body
20–25%

Fluorescent light
22%

Internal combustion
engine
(gasoline) 10%

Incandescent light
5%

Figure 15-4 Energy efficiency of some common energy conversion devices.

Figure 15-3 Advantages of reducing energy waste. Global improvements in energy efficiency could save the world about $1 trillion per year

The net energy efficiency of the entire energy delivery process for a space heater, water heater, or car is determined by the efficiency of each step in the energy conversion process. For example, the sequence of energy-using (and energy-wasting) steps involved in using electricity produced from fossil or nuclear fuels is

Extraction → Transportation → Processing →
Transportation to power plant → Electric generation →
Transmission → End use

Figure 15-5 shows the net energy efficiency for heating two well-insulated homes: **(1)** one with electricity produced at a nuclear power plant, transported by wire to the home, and converted to heat (electric resistance heating), and **(2)** the other heated passively, with an input of direct solar energy through high-efficiency windows facing the sun, with heat stored in heat-absorbing materials for slow release.

This analysis shows that the process of **(1)** converting the high-quality energy in nuclear fuel to high-quality heat at several thousand degrees in the power plant, **(2)** converting this heat to high-quality electricity, **(3)** transmitting the electricity to users, and **(4)** using the electricity to provide low-quality heat for warming a house to only about 20°C (68°F) is very wasteful

of high-quality energy (Figure 3-11, p. 59). Burning coal or any fossil fuel at a power plant to supply electricity for heating water or space is also inefficient. It is much less wasteful to **(1)** collect solar energy from the environment, **(2)** store the resulting heat in heat-absorbing materials, and, **(3)** if necessary, use a small backup heating system to raise its temperature slightly to provide space heating or household hot water.

Figure 15-5 shows that one way to waste less energy (and money) is not using high-quality energy to do a job that can be done with lower-quality energy (Figure 3-11, p. 59). This helps explain why using electricity to heat a house (resistance heating) costs about three times more per unit of energy than using a heat pump (useful in warm to moderate climates only) and about twice as much as heating with oil or with an efficient natural gas furnace.

Perhaps the three least efficient energy-using devices in widespread use today are **(1)** incandescent light bulbs (which waste 95% of the energy input), **(2)** vehicles with internal combustion engines (which waste 86–90% of the energy in their fuel), and **(3)** nuclear power plants producing electricity for space heating or water heating (which waste 86% of the energy in their nuclear fuel, and probably 92% when the energy needed to deal with radioactive wastes and

Technology Is the Answer (But What Was the Question?)

Amory B. Lovins

GUEST ESSAY

Physicist and energy consultant Amory B. Lovins is one of the world's most respected experts on energy strategy. In 1989, he received the Delphi Prize for environmental work; in 1990 the Wall Street Journal *named him one of the 39 people most likely to change the course of business in the 1990s. He is research director at Rocky Mountain Institute, a nonprofit resource policy center that he and his wife, Hunter, founded in Snowmass, Colorado, in 1982. He has served as a consultant to more than 200 utilities, private industries, and international organizations, and to many national, state, and local governments. He is active in energy affairs in more than 35 countries and has published several hundred papers and a dozen books on energy strategies and policies.*

It is fashionable to suppose that we're running out of energy and ask how can we get more of it. However, the more important questions are **(1)** How much energy do we need and **(2)** What are the cheapest and least environmentally harmful ways to meet these needs?

How much energy it takes to make steel, run a car, or keep ourselves comfortable in our houses depends on how cleverly we use energy. It is now cheaper, for example, to double the efficiency of most industrial electric motor drive systems than to fuel existing power plants to make electricity. (Just this one saving can more than replace the entire U.S. nuclear power program.) We know how to make lights five times as efficient as those currently in use and how to make household appliances that give us the same work as now but use one-fifth as much energy (saving money in the process).

Ten automakers have made good-sized, peppy, safe prototype cars averaging 29–59 kilometers per liter (67–138 miles per gallon), and within a decade automakers could have cars getting 64–128 kpl (150–300 mpg) on the road if consumers demanded such cars. We know today how to make new buildings (and many old ones) so heat-tight (but still well ventilated) that they need essentially no outside energy to maintain comfort year-round, even in severe climates. In fact, I live and work in one [Figure 15-1].

These energy-saving measures are all cheaper than going out and getting more energy. However, the old view of the energy problem included a worse mistake than forgetting to ask how much energy we needed: It sought more energy, in any form, from any source, at any price, as if all kinds of energy were alike.

Just as there are different kinds of food, so there are many different forms of energy whose different prices and qualities suit them to different uses [Figure 3-11, p. 59]. After all, there is no demand for energy as such; nobody wants raw kilowatt-hours or barrels of sticky black goo. People instead want energy services: comfort, light, mobility, hot showers, cold beverages, and the ability to cook food and make cement. In developing energy resources we should start by asking, "What tasks do we want energy for, and what amount, type, and source of energy will do each task most cheaply?"

Electricity is a particularly high-quality, expensive form of energy. An average kilowatt-hour delivered in the United States in 1999 was priced at about 6.7¢, equivalent to buying the heat content of oil costing $111 per barrel—over three times the average world price of crude oil in 2000. No new nuclear plants are being built in the United States (and only a few are being built in other parts of the world). One reason is that the average cost of electricity from the last nuclear power plant built in the United States is about 13.5¢ per kilowatt-hour, equivalent on a heat basis to buying oil at about $216 per barrel.

Such costly energy might be worthwhile if it were used only for the premium tasks that need it, such as lights, motors, electronics, and smelters. However, those special uses—only 8% of all delivered U.S. energy needs—are already met twice over by today's power stations. Two-fifths of the electricity used in the United States is for uneconomic, low-grade uses such as water heating, space heating, and air conditioning. Yet no matter how efficiently we use electricity (even with heat pumps), we can never get our money's worth on these applications.

Thus, *supplying more electricity is irrelevant to the energy problem we have.* Even though electricity accounts for almost all the federal energy research-and-development budget and at least half the national energy investment, it is the wrong kind of energy to meet the nation's needs economically. Arguing about what kind of new power station to build—coal, nuclear, or solar—is like shopping for the best buy in antique Chippendale chairs to burn in your stove, or for expensive brandy to put in your car's gas tank.

The real question is, "What is the cheapest way to do low-temperature heating and cooling?" The answer is weather-stripping, insulation, heat exchangers, greenhouses, superwindows (which have as much insulating value as the outside wall of a typical house), roof overhangs, trees, and so on. These measures generally cost about 0.5¢–2¢ per kilowatt-hour, the lowest-cost way by far to supply energy.

If we need more electricity, we should get it from the cheapest sources first. In approximate order of increasing price, these include:

- Converting to efficient lighting equipment. This would save the United States electricity equal to the output of 120 large power plants, plus $30 billion a year in fuel and maintenance costs.

- Using more efficient electric motors to save up to half the energy used by motor systems. This would save

(continued)

electricity equal to the output of another 150 large power plants and repay the cost in about a year.

- Displacing the electricity now used for water heating and for space heating and cooling with good architecture, weatherization, insulation, and mostly passive solar techniques.

- Improving the energy efficiency of appliances, smelters, and the like.

Just these four measures can quadruple U.S. electrical efficiency, making it possible to run today's economy with no changes in lifestyles and using no power plants, whether old or new or fueled with oil, gas, coal, uranium, or solar energy. We would need only the present hydroelectric capacity, readily available small-scale hydroelectric projects, and a modest amount of wind power.

If we still wanted more electricity, the next cheapest sources would include (1) cogenerating electricity and heat in industrial plants and power plants, (2) using low-temperature heat engines run by industrial waste heat or by solar ponds, (3) filling empty turbine bays and upgrading equipment in existing big dams, (4) using modern wind machines or small-scale hydroelectric turbines in good sites, (5) using combined-cycle natural-gas turbines, and perhaps (6) using recently developed more efficient solar cells when their price is reduced by mass production.

It is only after we have exhausted all these cheaper opportunities that we would even consider building a new central power station of any kind—the slowest and costliest known way to get more electricity (or to save oil).

To emphasize the importance of starting with energy end uses rather than energy sources, consider a story from France. In the mid-1970s, energy conservation planners in the French government found that their biggest need for energy was to heat buildings and that even with good heat pumps, electricity would be the costliest way to do this. So they had a fight with their government-owned and -run utility company; they won, and electric heating was supposed to be discouraged or even phased out because it was so wasteful of money and fuel.

Meanwhile, down the street, the energy supply planners (who were far more numerous and influential in the French government) said, "Look at all that nasty imported oil coming into our country. We must replace that oil with some other source of energy. Voilà! Nuclear reactors can give us energy, so we'll build them all over the country." However, they paid little attention to who would use that extra energy and no attention to relative prices.

Thus, these two groups of the French energy establishment went on with their respective solutions to two different, indeed contradictory, French energy problems: *more energy of any kind* versus *the right kind to do each task in the most inexpensive way*. It was only in 1979 that these conflicting perceptions collided. The supply-side planners suddenly realized that the only thing they would be able to sell all that nuclear electricity for would be electric heating, which they had just agreed not to do.

Every industrial country is in this embarrassing position. Supply-oriented planners think the problem boils down to whether to build coal or nuclear power stations (or both). Energy-use planners realize that *no* kind of new power station can be an economic way to meet the needs for using electricity to provide low- and high-temperature heat and for the vehicular liquid fuels that are 92% of our energy problem.

So if we want to provide energy services at the lowest cost, we need to begin by determining what we need the energy for!

Critical Thinking

1. The author argues that building more nuclear, coal, or other electrical power plants to supply electricity for the United States is unnecessary and wasteful of energy and money. List your reasons for agreeing or disagreeing with this viewpoint.

2. Explain why you agree or disagree that increasing the supply of energy, instead of improving energy efficiency, is the wrong answer to our energy problems.

---

retired nuclear plants is included). Energy experts call for us to replace these devices or greatly improve their energy efficiency over the next few decades.

Coal-burning plants are also big energy wasters. About 34% of the energy in coal burned in a typical electric power plant is used to produced electricity, and the remaining 66% ends up as waste heat that flows into the environment. As a result, U.S. coal-burning power plants throw away as much heat as all the energy used by Japan, the world's second largest economy.

15-2 WAYS TO IMPROVE ENERGY EFFICIENCY

How Can We Use Waste Heat? Could we save energy by recycling energy? No. The second law of energy tells us that we cannot recycle energy. However, we can slow the rate at which waste heat flows into the environment when high-quality energy is degraded.

For a house, the best way to do this is to (1) insulate it thoroughly, (2) eliminate air leaks (Figure 15-6),

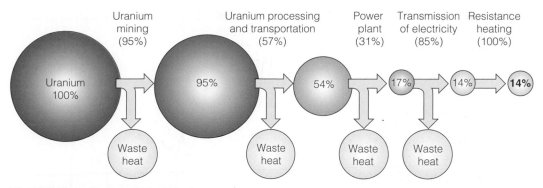

Electricity from Nuclear Power Plant

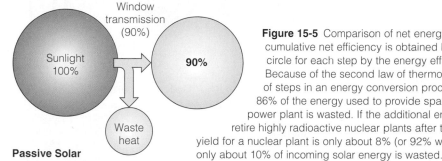

Passive Solar

Figure 15-5 Comparison of net energy efficiency for two types of space heating. The cumulative net efficiency is obtained by multiplying the percentage shown inside the circle for each step by the energy efficiency for that step (shown in parentheses). Because of the second law of thermodynamics, in most cases the greater the number of steps in an energy conversion process, the lower its net energy efficiency. About 86% of the energy used to provide space heating by electricity produced at a nuclear power plant is wasted. If the additional energy needed to deal with nuclear wastes and to retire highly radioactive nuclear plants after their useful life is included, then the net energy yield for a nuclear plant is only about 8% (or 92% waste). By contrast, with passive solar heating, only about 10% of incoming solar energy is wasted.

and **(3)** equip it with an air-to-air heat exchanger to prevent buildup of indoor air pollutants.

In office buildings and stores, waste heat from lights, computers, and other machines can be collected and distributed to reduce heating bills during cold weather. During hot weather, the collected heat can be vented outdoors to reduce cooling bills.

How Can We Save Energy in Industry? Three important ways to save energy and money in industry are:

- **Cogeneration**, or *combined heat and power (CHP)* systems, in which two useful forms of energy (such as steam and electricity) are produced from the same fuel source. These systems have an efficiency of up to 80%

Figure 15-6 An infrared photo (thermogram) showing heat loss (red, white, and orange) around the windows, doors, roofs, and foundations of houses and stores in Plymouth, Michigan. Many homes and buildings in the United States (and in most other countries) are so full of leaks that their heat loss in cold weather and heat gain in hot weather are equivalent to having a large window-size hole in the wall of the house. (VANSCAN® Continuous Mobile Thermogram by Daedalus Enterprises, Inc.)

(compared to about 30–40% for coal-fired boilers and nuclear power plants) and emit two-thirds less carbon dioxide per unit of energy produced than conventional coal-fired boilers. Cogeneration has been widely used in western Europe for years, and its use in the United States and China is growing. In Germany, small cogeneration units that run on natural gas or liquefied petroleum gas (LPG) supply restaurants, apartment buildings, and houses with all their energy. In 6–8 years, they pay for themselves in saved fuel and electricity.

- *Replacing energy-wasting electric motors.* Running electric motors (mostly in industry) consumes about half of all electricity produced in the United States. Most of these motors are inefficient because they run only at full speed with their output throttled to match the task. Each year a heavily used electric motor consumes 10 times its purchase cost in electricity—equivalent to using $200,000 worth of gasoline each year to fuel a $20,000 car. The costs of replacing such motors with new adjustable-speed drive motors would be paid back in about 1 year and save an amount of energy equal to that generated by 150 large (1,000-megawatt) power plants.

- *Switching to high-efficiency lighting* (Guest Essay, p. 361).

How Can We Save Energy in Transportation?

According to most energy analysts, the best way to save energy (especially oil) and money in transportation is to *increase the fuel efficiency of motor vehicles*.

Between 1973 and 1985, the average fuel efficiency doubled for new American cars and rose 37% for all passenger cars on the road because of government-mandated standards, called the Corporate Average Fuel Economy (CAFE) standards. CAFE regulations require new cars to meet certain average fuel efficiency standards, averaged over all cars produced.

Actually, between 1985 and 1999 the average fuel efficiency of new vehicles in the United States fell from 11 kilometers per liter (25.9 miles per gallon) to 10 kilometers per liter (23.8 miles per gallon) because of the popularity of (1) sport utility vehicles (SUVs), minivans, and light trucks (subject to much lower mileage standards than cars) and (2) larger, less efficient autos. According to the EPA, increasing average fuel economy by 1.3 hpl (3 mpg) would (1) save $25 billion a year in fuel costs, (2) reduce CO_2 emissions, and (3) save 1 million barrels of oil per day. However, automakers have successfully opposed any increase in the CAFE standards and are lobbying Congress to eliminate them.

Since 1985, at least 10 automobile companies have developed prototype cars with fuel efficiencies of 30–60 kilometers per liter (70–140 miles per gallon). These cars are (1) manufactured with light and strong materials, (2) meet current safety and air pollution standards,

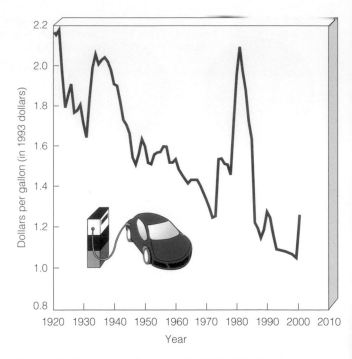

Figure 15-7 Real price of gasoline (in 1993 dollars) in the United States, 1920-2000. The 212 million motor vehicles in the United States use about 40% of the world's gasoline. Gasoline is one of the cheapest items American consumers buy. (U.S. Department of Energy)

carry four or five passengers, and (3) accelerate as rapidly as most current models.

If such cars were mass-produced, their slightly higher costs would be more than offset by their fuel savings. The problem is that there is little consumer interest in fuel-efficient cars mostly because (1) the inflation-adjusted price of gasoline today in the United States is low despite increases in gasoline prices in 2000 (Figure 15-7), and (2) two-thirds of consumers prefer SUVs and other large, inefficient vehicles.

Another way to save energy is to *shift to more energy-efficient ways to move people* (Figure 15-8) *and*

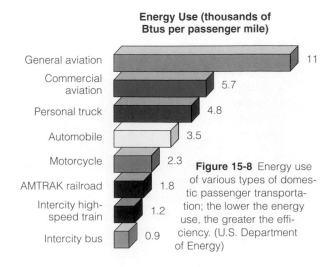

Figure 15-8 Energy use of various types of domestic passenger transportation; the lower the energy use, the greater the efficiency. (U.S. Department of Energy)

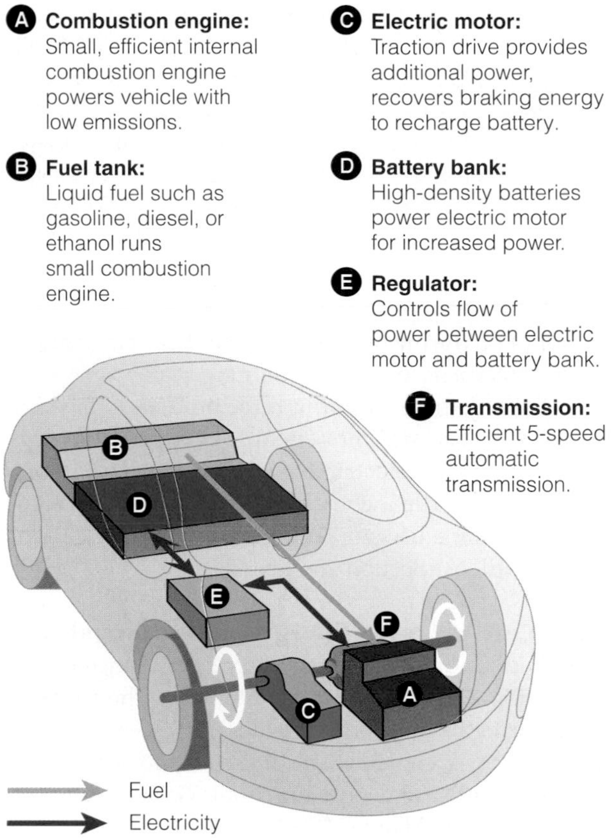

Figure 15-9 General features of a car powered by a *hybrid gas-electric engine*. A small internal combustion engine recharges the batteries, thus reducing the need for heavy banks of batteries and solving the problem of the limited range of conventional electric cars. The bodies of future models of such cars probably will be made of lightweight composite plastics that **(1)** offer more protection in crashes, **(2)** do not need to be painted, **(3)** do not rust, **(4)** can be recycled, and **(5)** have fewer parts than conventional cars. (Concept information from DaimlerChrysler, Ford, Honda, and Toyota)

A **Combustion engine:** Small, efficient internal combustion engine powers vehicle with low emissions.

B **Fuel tank:** Liquid fuel such as gasoline, diesel, or ethanol runs small combustion engine.

C **Electric motor:** Traction drive provides additional power, recovers braking energy to recharge battery.

D **Battery bank:** High-density batteries power electric motor for increased power.

E **Regulator:** Controls flow of power between electric motor and battery bank.

F **Transmission:** Efficient 5-speed automatic transmission.

→ Fuel

→ Electricity

freight. More freight could be shifted from trucks and planes to more energy-efficient trains and ships. New freight transport trucks could be made 50% more fuel-efficient than today's conventional trucks through use of improved aerodynamic design, turbocharged diesel engines, and radial tires.

Are Electric Cars the Answer? Conventional battery-powered *electric cars* are extremely quiet, need little maintenance, and can accelerate rapidly. The cars themselves produce no air pollution, but using coal and nuclear power plants to produce the electricity needed to recharge their batteries produces air pollution and nuclear wastes—something called *elsewhere pollution*. If solar cells or wind turbines could be used to recharge the car batteries, CO_2 and other air pollution emissions would be almost eliminated.

On the negative side, today's electric cars **(1)** can travel only 81–161 kilometers (50–100 miles) before needing a 3- to 8-hour recharge (although a new device may reduce recharge time to 10–20 minutes), and **(2)** batteries must be replaced about every 48,000 kilometers (30,000 miles) at a cost of at least $2,000. This plus the costs for daily recharging and a buying a recharger ($700–3,500) means that today's electric cars have twice the operating cost of gasoline-powered cars. Because of high costs and a lack of consumer interest, in 1999 major car companies abandoned their production of electric cars.

Are Hybrid and Fuel Cell Cars the Answer? There is growing interest in developing *superefficient cars* that could eventually get 34–128 kilometers per liter (80–300 miles per gallon). One type of highly efficient car uses **(1)** a small *hybrid electric-internal combustion engine* that runs on gasoline or some other liquid fuel and **(2)** a small battery (recharged by the internal combustion engine) to provide the energy needed for acceleration and hill climbing (Figure 15-9). In 1999, Toyota and Honda began selling the first generation of fuel-efficient hybrid engine cars.

Another type of superefficient car is an electric vehicle that uses *fuel cells* (Figure 15-10). Fuel cells consist of two electrodes immersed in a solution (electrolyte) that conducts electricity. They produce electricity by combining hydrogen and oxygen ions, typically from hydrogen and oxygen gas supplied as fuel for the cell (Figure 15-10). Such cars are 50–60% efficient (compared to 10–14% efficiency for gasoline-powered vehicles). As a result, fuel cell cars running on hydrogen should get 37–47 kilometers per liter (87–108 miles per gallon).

Most major automobile companies have developed prototype fuel-cell cars and hope to begin marketing them by 2004 using two different approaches. Some will be fueled by methanol (CH_3OH), a liquid usually produced from natural gas. A *liquid reformer system* in the engine extracts the hydrogen from methanol for use in the fuel cell. Drivers would fill up with methanol at pumps in conventional filing stations. A new process developed in 2000 allows fuel cells to produce hydrogen directly from hydrocarbons such as natural gas, gasoline, or diesel fuel. Instead of using complicated liquid fuel reformers, General Motors and Honda use a chemical process to store hydrogen fuel as a *solid metal hydride* compound that can be heated as needed to provide H_2.

How Can Electric Bicycles Reduce Energy Use and Waste? For urban trips, some people may begin using *electric bicycles*, now being sold by several companies such as EV Global Motors (founded by former Chrysler Corporation head Lee Iacocca). These bicycles are powered by a small electric motor and cost about $1,100. They **(1)** travel at up to 30 kilometers per hour (18 miles per hour), **(2)** go about 48 kilometers (30 miles) without pedaling on a full electric charge, and **(3)** produce

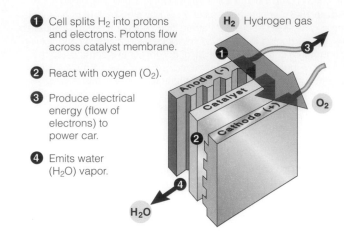

Figure 15-10 General features of an electric car powered by a *fuel cell* running on hydrogen gas. Such cars **(1)** will be almost pollution-free, emitting only water vapor and small amounts of nitrogen oxides and no CO_2, and **(2)** should get at least twice the mileage of comparable gasoline-powered cars. Several automobile companies have developed prototypes and are working to get costs down and improve hydrogen fuel storage systems. Early models could be on the road by 2004. (Concept information from DaimlerChrysler, Ford, Ballard, Toyota, and Honda)

❶ Cell splits H_2 into protons and electrons. Protons flow across catalyst membrane.

❷ React with oxygen (O_2).

❸ Produce electrical energy (flow of electrons) to power car.

❹ Emits water (H_2O) vapor.

Ⓐ **Fuel cell stack:** Hydrogen and oxygen combine chemically to produce electricity.

Ⓑ **Fuel tank:** Hydrogen gas or liquid or solid metal hydride stored on board or made from gasoline or methanol.

Ⓒ **Turbo compressor:** Sends pressurized air to fuel cell.

Ⓓ **Traction inverter:** Module converts DC electricity from fuel cell to AC for use in electric motor.

Ⓔ **Electric motor / transaxle:** Converts electrical energy to mechanical energy to turn wheels.

→ Fuel
→ Electricity

no pollution during operation (and only a small amount for the electricity used in recharging them).

Globally there are about three times as many bicycles as cars because most people cannot afford cars. Increased use of electric bicycles and electric motor scooters in middle-income developing countries could reduce gasoline use, air pollution, CO_2 emissions, and urban gridlock (Section 25-3, p. 670).

How Can We Save Energy in Buildings? Most energy in residential and commercial buildings is used for heating, air conditioning, and lighting. The 110-story, twin-towered World Trade Center in New York City is a monument to energy waste: It uses as much electricity as a city of 100,000 people for about 53,000 employees.

In contrast, Atlanta's 13-story Georgia Power Company building uses 60% less energy than conventional office buildings of the same size. The largest surface of the building faces south to capture solar energy. Each floor extends out over the one below it, blocking out the higher summer sun to reduce air conditioning costs but allowing warming by the lower winter sun. Energy-efficient lights focus on desks rather than illuminating entire rooms. The Georgia Power model and other existing cost-effective commercial building technologies could **(1)** reduce energy use by 75% in buildings, **(2)** cut carbon dioxide emissions in half, and **(3)** in the United States save more than $130 billion per year in energy bills.

There are a number of ways to improve the energy efficiency of buildings, some of them discussed in the opening of this chapter (p. 358). One is to build more *superinsulated houses* (Figure 15-11). Such houses typically cost 5% more to build than conventional houses of the same size. However, this extra cost is paid back by energy savings within about 5 years and can save a homeowner $50,000–100,000 over a 40-year period.

Since the mid-1980s there has been growing interest in building superinsulated houses called *strawbale houses** with walls consisting of compacted bales of cer-

---

*Strawbale houses and barns were first built in Nebraska the early 1900s because no trees were available. Some of these durable structures are still in use today. For information on strawbale houses, see Steen et al., *The Straw Bale House* (White River Junction, Vt.: Chelsea Green, 1994); and GreenFire Institute, 1509 Queen Anne Ave. North, #606, Seattle, WA 98109, 206-284-7470.

tain types of straw (available at a low cost almost everywhere) covered with plaster or adobe (Figure 15-12). By 2000, there were more than 1,200 such homes built or under construction in the United States (Guest Essay, p. 377). Using straw, an *annually* renewable agricultural residue often burned as a waste product, for the walls reduces the need for wood and thus slows deforestation. The main problem is getting banks and other moneylenders to recognize the potential of this and other unconventional types of housing and provide homeowners with construction loans.

Another way to save energy is to *use the most energy-efficient ways to heat houses* (Figure 15-13). The

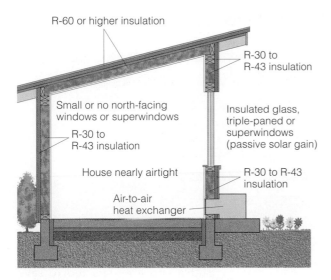

R-60 or higher insulation

R-30 to R-43 insulation

Small or no north-facing windows or superwindows

Insulated glass, triple-paned or superwindows (passive solar gain)

R-30 to R-43 insulation

House nearly airtight

R-30 to R-43 insulation

Air-to-air heat exchanger

Figure 15-11 Major features of a superinsulated house. Such a house is so heavily insulated and so airtight that it can be warmed by heat from direct sunlight, appliances, and human bodies, with little or no need for a backup heating system. An air-to-air heat exchanger prevents buildup of indoor air pollution.

most energy-efficient ways to heat space are **(1)** a superinsulated house, **(2)** passive solar heating, **(3)** heat pumps in warm climates (but not in cold climates because at low temperatures they automatically switch to costly electric resistance heating), and **(4)** a high-efficiency (85–98%) natural gas furnace. The most wasteful and expensive way is to use electric resistance heating with the electricity produced by a coal-fired or nuclear power plant**.**

The energy efficiency of existing houses and buildings can be improved significantly by adding insulation, plugging leaks, and installing energy-saving windows and lighting. About one-third of heated air in U.S. homes and buildings escapes through closed windows and holes and cracks (Figure 15-6)—equal to the energy in all the oil flowing through the Alaska pipeline every year. During hot weather these windows and cracks also let heat in, increasing the use of air conditioning.

Replacing all windows in the United States with low-E (low-emissivity) windows would cut these expensive losses by two-thirds and reduce CO_2 emissions. Widely available superinsulating windows insulate as well as 8–12 sheets of glass. Although they cost 10–15% more than double-glazed windows. this cost is paid back rapidly by the energy they save. Even better windows will reach the market soon.

Simply wrapping a water heater in a $20 insulating jacket can save a homeowner $45 a year and reduce CO_2 emissions. Leaky ducts allow 20–30% of a home's heating and cooling energy to escape and draw unwanted moisture and heat into the home. Careful sealing can reduce this loss. Some designs for new homes keep the ducts inside the home's thermal envelope so that escaping hot or cool air leaks into the living space.

An energy-efficient way to heat hot water for washing and bathing is to use tankless instant water heaters (about the size of bookcase loudspeakers) fired by natural gas or LPG. These devices, widely used in many parts of Europe, heat the water instantly as it flows through a small burner chamber and provide hot water only when it is needed. A well-insulated, conventional natural gas or LPG water heater is fairly efficient. However, all conventional natural gas and electric resistance heaters waste energy by keeping a large tank of water hot

Figure 15-12 An energy-efficient, environmentally healthy, and affordable Victorian-style strawbale house designed and built by Alison Gannett in Crested Butte, Colorado. The left photo is during construction, and the right photo is the completed house. Depending on the thickness of the bales, plastered strawbale walls have an insulating value of R-35 to R-60, compared to R-12 to R-19 in a conventional house. (The R-value is a measure of resistance to heat flow.) (Alison Gannett)

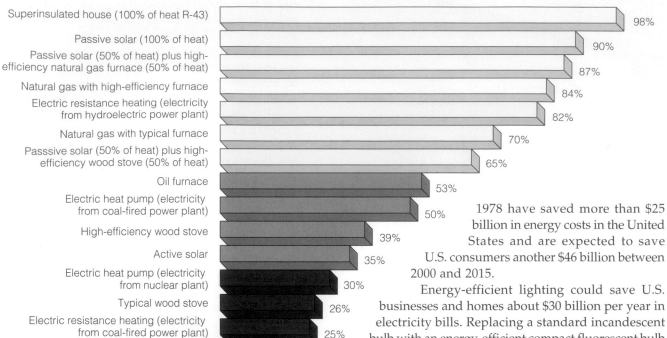

Superinsulated house (100% of heat R-43) — 98%

Passive solar (100% of heat) — 90%

Passive solar (50% of heat) plus high-efficiency natural gas furnace (50% of heat) — 87%

Natural gas with high-efficiency furnace — 84%

Electric resistance heating (electricity from hydroelectric power plant) — 82%

Natural gas with typical furnace — 70%

Passsive solar (50% of heat) plus high-efficiency wood stove (50% of heat) — 65%

Oil furnace — 53%

Electric heat pump (electricity from coal-fired power plant) — 50%

High-efficiency wood stove — 39%

Active solar — 35%

Electric heat pump (electricity from nuclear plant) — 30%

Typical wood stove — 26%

Electric resistance heating (electricity from coal-fired power plant) — 25%

Electric resistance heating (electricity from nuclear plant) — 14%

Figure 15-13 Net energy efficiencies for various ways to heat an enclosed space such as a house. (Data from Howard T. Odum)

all day and night and can run out after a long shower or two.

Using electricity produced by any type of power plant is the most inefficient and expensive way to heat water for washing and bathing. A $425 electric water heater can cost $5,900 in energy over its 15-year life, compared to about $4,000 for a comparable natural gas water heater over the same period.

Setting higher energy-efficiency standards for new buildings would also save energy. Building codes could require that all new houses use 60–80% less energy than conventional houses of the same size, as has been done in Davis, California (p. 658). Because of tough energy-efficiency standards, the average Swedish home consumes about one-third as much energy as the average American home of the same size.

Another way to save energy is to *buy the most energy-efficient appliances and lights.** Federal energy efficiency standards set for more than 20 appliances—including a tripling of refrigerator efficiency—since

*Each year the American Council for an Energy-Efficient Economy (ACEEE) publishes a list of the most energy-efficient major appliances mass-produced for the U.S. market. A copy can be obtained from the council at 1001 Connecticut Ave. NW, Suite 801, Washington, DC 20036. Each year they also publish *A Consumer Guide to Home Energy Savings*, available in bookstores or from the ACEEE.

1978 have saved more than $25 billion in energy costs in the United States and are expected to save U.S. consumers another $46 billion between 2000 and 2015.

Energy-efficient lighting could save U.S. businesses and homes about $30 billion per year in electricity bills. Replacing a standard incandescent bulb with an energy-efficient compact fluorescent bulb (Figure 15-14) saves about $48–70 per bulb over its 10-year life. Thus, replacing 25 incandescent bulbs in a house or building with energy-efficient fluorescent bulbs saves $1,250–1,750. Students in Brown University's environmental studies program showed that the school could save more than $40,000 per year just by replacing the incandescent light bulbs in exit signs with compact fluorescent bulbs.

According to the Alliance to Save Energy, if every U.S. household replaced four 100-watt incandescent light bulbs that burn 4 hours or more per day with 23-watt fluorescent bulbs, the reduced air pollution would equal that from 7 million cars. Despite their advantages, less than 10% of U.S. homes use these energy- and money-saving bulbs, which pay for themselves in 2–4 years. Studies show that such low use has resulted from **(1)** the high initial cost of the bulbs, **(2)** lack of information and education about life-cycle costing, **(3)** initial lack of suitable light fixtures for these larger bulbs (light fixtures and smaller bulbs are now available), and **(4)** dissatisfaction with the color and intensity of the light produced compared to incandescent bulbs (corrected with newer bulbs).

The energy saved from this fairly small use of fluorescent bulbs has largely been wiped out by sales of cheap halogen torchière lamps. Halogen bulbs are **(1)** very inefficient, **(2)** emit large amounts of heat that increase air-conditioning bills, and **(3)** have caused fires. More efficient and safer incandescent versions of the lamps are now available.

If all U.S. households used the most efficient frost-free refrigerator now available, 18 large (1,000-megawatt) power plants could close. Microwave ovens can cut electricity use for cooking by 25–50% (but not if used

CONNECTIONS

According to a study by Forrester Research, between 1998 and 2003, **(1)** consumer e-commerce is expected to grow from $7.8 billion to $108 billion, and **(2)** business e-commerce is expected to rise from $43 billion to $1 trillion.

In addition to revolutionizing business practices around the world, the internet is having a positive environmental impact. According to a 2000 report by researchers at the Center for Energy and Climate Solutions, increasing use of the internet is saving energy by:

- Allowing more employees to work at home and allowing highly mobile workers to be assigned flexible office space only when they are in the office.

- Saving energy and material resources by allowing a reduction in retail, manufacturing, warehouse, and commercial office space. Some online companies keep no merchandise in warehouses and have it shipped to customers directly from manufacturers. Many manufactur-

ers are using the internet to sell their goods directly to consumers and other businesses. For example, Home Depot uses information technology and the internet to move 85% of its merchandise directly from manufacturers to their stores.

- Using less energy to get products from sellers to consumers. Sending a package by overnight air uses about 40% less fuel than driving round-trip to the mall to get the product. Shipping by rail saves even more energy.

- Not being a major energy user because the internet draws heavily on the existing communication infrastructure. The average personal computer and monitor use only 150 watts of power, and this is dropping because of increased use of more energy-efficient models and laptops.

- Reducing the energy, paper, and materials used to produce, package, and market consumer items such as computer software and music CDs by allowing them to be downloaded.

- Reducing energy used in transporting goods by using internet-based systems to auction off empty

shipping space on trucks, aircraft, and trains.

The resulting savings in energy consumption also reduce air pollution and carbon dioxide emissions by reducing fossil fuel use. By 2010, the projected energy savings from increased business on the internet should reduce carbon dioxide emissions by an amount equivalent to that from 170 large (1,000,000-kilowatt) coal-burning power plants.

Paper production (a highly energy- and resource-intensive industry) is expected to decrease as **(1)** consumers use the internet to download software and view magazines, newspapers, research articles, telephone directories, encyclopedias, and books, **(2)** consumers and businesses send more e-mail and less mail and reports by envelope and packages, and **(3)** easily updated online catalogs replace paper catalogs.

Critical Thinking

What types of environmental harm might be increased by greatly expanded use of the internet to sell more and more goods in the global marketplace?

for defrosting food). Clothes dryers with moisture sensors cut energy use by 15%, and front-loading washers use 50% less energy than top-loading models but cost about the same. New microwave clothes dryers soon to be available will use 15% less energy than con-

ventional electric dryers and 28% less energy than gas units. Increased use of the internet for business and shopping transactions reduces energy use and decreases emissions of carbon dioxide and other air pollutants (Connections, above).

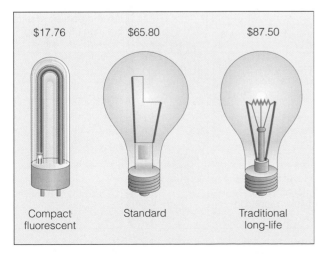

$17.76 — Compact fluorescent

$65.80 — Standard

$87.50 — Traditional long-life

Figure 15-14 Cost of electricity for comparable light bulbs used for 10,000 hours. Because conventional incandescent bulbs are only 5% efficient and last only 1,500 hours, they waste enormous amounts of energy and money and add to the heat load of houses during hot weather. Socket-type fluorescent lights use one-fourth as much electricity as conventional bulbs. Although these bulbs cost $6–15 per bulb, they last up to 100,000 hours (60–70 times longer than conventional incandescent bulbs and 25 times longer than halogen bulbs), saving a lot of money (compared with less efficient incandescent and halogen bulbs) over their long life. Recently, smaller compact fluorescent bulbs (costing about $6) that fit into ordinary light fixtures and ones that can be dimmed have been developed. (Data from Electric Power Research Institute)

Why Aren't We Doing More to Reduce Energy Waste? With such an impressive array of benefits (Figure 15-3), why isn't there more emphasis on improving energy efficiency? The major reasons are as follows:

- A glut of low-cost fossil fuels (Figure 14-18, p. 339, and Figure 15-7). As long as energy is cheap, people are more likely to waste it and not make investments in improving energy efficiency.

- Lack of sufficient tax breaks and other economic incentives for consumers and businesses to invest in improving energy efficiency.

- Lack of information about the availability of energy-saving devices and the amount of money such items can save consumers by using *life cycle cost* analysis.

15-3 USING SOLAR ENERGY TO PROVIDE HEAT AND ELECTRICITY

What Are the Major Advantages and Disadvantages of Solar Energy? In 1999, renewable energy provided about 9% of the commercial energy used in the United States (Figure 14-13, p. 333). About 4% of this energy came from hydropower, 3% from biomass and geothermal, and 2% from wind and direct solar energy. California—with the world's sixth largest economy—gets 12% of its electricity from renewable wind, geothermal, solar, and biomass sources.

This could change over the next 20–50 years. In 1994, Shell International Petroleum in London projected that renewable energy (especially using wind and solar cells to produce electricity and to produce hydrogen for fuel cells) could account for 50% of world energy production by 2050.

Figure 15-15 lists some of the advantages and disadvantages of making a shift to greatly increased use of direct solar energy and indirect forms of solar energy such as wind. Like fossil fuels and nuclear power (Chapter 14), each renewable energy alternative has a mix of advantages and disadvantages, as discussed in the remainder of this chapter.

How Can We Use Solar Energy to Heat Houses and Water? Buildings and water can be heated by solar energy using two methods: passive and active (Figure 15-16). A **passive solar heating system** absorbs and stores heat from the sun directly within a structure (Figures 15-1, 15-16 (left), and 15-17 and Guest Essay, p. 377).

Figure 15-15 Major advantages and disadvantages of using direct and indirect solar energy systems to produce heat and electricity. Specific advantages and disadvantages of different direct and indirect solar and other renewable energy systems are discussed in this chapter.

Energy-efficient windows, greenhouses, and sunspaces face the sun to collect solar energy by direct gain. Walls and floors of concrete, adobe, brick, stone, salt-treated timber, and water in 55-gallon drums store much of the collected solar energy as heat and release it slowly throughout the day and night. A small backup heating system such as a vented natural gas or propane heater may be used but is not necessary in many climates.

Engineer and builder Michael Sykes has designed and built several versions of a solar envelope house that is heated and cooled passively by solar energy, with the energy stored and slowly released by massive timbers and the earth beneath the house (Figure 15-18). The front and back of this house are double walls of heavy timber impregnated with salt to increase the wood's ability to store heat. The space between the two walls and the basement forms a convection loop or envelope around the inner shell of the house. In summer, roof vents release heated air from the convection loop throughout the day; at night, these roof vents, with the aid of a fan, draw air into the loop, passively cooling the house. The interior temperature of the house typically stays within 2° of 21°C (70°F) year-round without any conventional cooling or heating system. In cold or cloudy climates, a small wood stove or vented natural gas heater in the basement can be used as a backup to heat the air in the convection loop.

Advantages

Save money (wind)

Reduce air pollution (99% less than coal)

Greatly reduce CO_2 emissions

Reduce dependence on imported oil

Last as long as coal and nuclear plants (30–40 years)

Land use less than for coal

Low land use with new solar cell and window glass system

Backup and storage devices available (such as gas turbines, batteries, and flywheels)

Backup need reduced by distributing and storing solar-produced hydrogen gas

Disadvantages

Making solar cells produces toxic chemicals

Solar systems last only 30–40 years

Take large amounts of land because of diffuse nature of sunlight

Can damage fragile desert ecosystems used to collect solar energy

Need backup systems at night and during cloudy and rainy weather

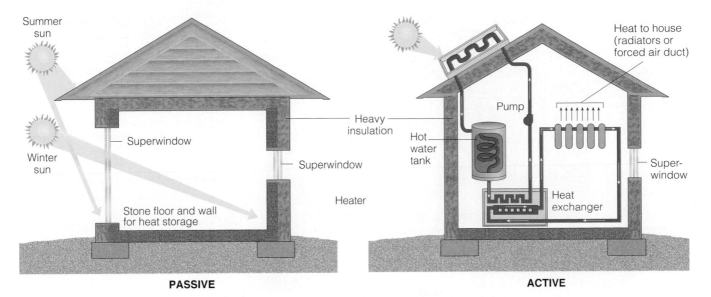

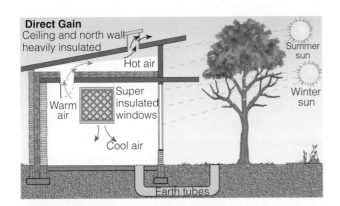

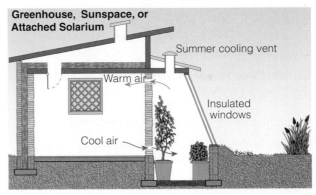

Figure 15-16 Passive and active solar heating for a home.

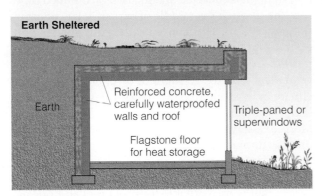

Figure 15-17 Three examples of passive solar design for houses.

On a life cycle cost basis, good passive solar and superinsulated design is the cheapest way to heat a home or small building in regions where ample sunlight is available during daylight hours (Figure 15-19). Such a system usually adds 5–10% to the construction cost, but the life cycle cost of operating such a house is 30–40% lower. The typical payback time for passive solar features is 3–7 years.

In an **active solar heating system**, collectors absorb solar energy, and a fan or a pump supplies part of a building's space-heating or water-heating needs (Figure 15-16, right). Several connected collectors usually are mounted on the roof with an unobstructed exposure to the sun. Some of the heat can be used directly, and the rest can be stored in insulated tanks containing rocks, water, or a heat-absorbing chemical for release as needed. Active solar collectors can also supply hot water.

Figure 15-20 lists the major advantages and disadvantages of using passive or active solar energy for heating buildings. As architects, developers, and home buyers become more aware of the monetary and aesthetic values of good passive solar design, its use in new homes should increase in areas with ample sunlight. However, passive solar cannot be used to heat existing homes and buildings **(1)** not oriented to receive sunlight and **(2)** whose access to sunlight is blocked by other buildings and structures. Most analysts do not

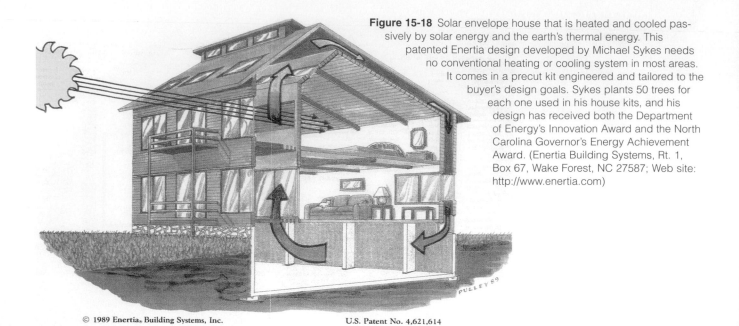

Figure 15-18 Solar envelope house that is heated and cooled passively by solar energy and the earth's thermal energy. This patented Enertia design developed by Michael Sykes needs no conventional heating or cooling system in most areas. It comes in a precut kit engineered and tailored to the buyer's design goals. Sykes plants 50 trees for each one used in his house kits, and his design has received both the Department of Energy's Innovation Award and the North Carolina Governor's Energy Achievement Award. (Enertia Building Systems, Rt. 1, Box 67, Wake Forest, NC 27587; Web site: http://www.enertia.com)

© 1989 Enertia® Building Systems, Inc.   U.S. Patent No. 4,621,614

expect widespread use of active solar collectors for heating houses because of high costs, maintenance, and unappealing appearance.

How Can We Cool Houses Naturally? Ways to make a building cooler include:

- Using superinsulation and superinsulating windows.

- Blocking the high summer sun with deciduous trees, window overhangs, or awnings (Figure 15-17, top left).

- Using windows and fans to take advantage of breezes and keep air moving.

- Suspending reflective insulating foil in an attic to block heat from radiating down into the house.

- Placing plastic (PVC) *earth tubes* 3–6 meters (10–20 feet) underground where the earth is cool year-round and using a tiny fan to pipe cool and partially dehumidified air into an energy-efficient house (Figure 15-17 top left).*

- Using solar-powered evaporative air conditioners (which work well only in dry climates and cost too much for residential use).

How Can We Use Solar Energy to Generate High-Temperature Heat and Electricity? Several so-called *solar thermal systems* collect and transform radiant energy from the sun into high-temperature thermal energy (heat), which can be used directly or converted to electricity (Figure 15-21). In one such *cen-*

*They work. I used them in a passively heated and cooled office and home for 15 years. People allergic to pollen and molds should add an air purification system, but this is also necessary with a conventional cooling system.

tral receiver system, called a *power tower*, huge arrays of computer-controlled mirrors called *heliostats* track the sun and focus sunlight on a central heat collection tower (Figure 15-21a).

A government-subsidized power tower system, called Solar Two, began operating in the California desert in 1996. However, this experimental plant **(1)** cost about eight times more to build than a coal-fired plant, **(2)** produced electricity at about twice the cost of a coal-fired plant, and **(3)** was shut down in 1999.

In a *solar thermal plant* or *distributed receiver system*, sunlight is collected and focused on oil-filled pipes running through the middle of curved solar collectors (Figure 15-21b). This concentrated sunlight can generate temperatures high enough for industrial processes or for producing steam to run turbines and generate electricity. At night or on cloudy days, high-efficiency combined-cycle natural gas turbines can supply backup electricity as needed. In California's Mojave Desert, such a solar thermal system with a natural gas turbine backup system produced power much more cheaply than nuclear power plants. However, the company went bankrupt, partly because of a lack of tax breaks that were available for fossil fuel and nuclear power plants.

Another type of distributed receiver system uses *parabolic dish collectors* (that look somewhat like TV satellite dishes) instead of parabolic troughs. These collectors can track the sun along two axes and generally are more efficient than troughs. A pilot plant is being built in northern Australia. The U.S. Department of Energy projects that early in this century parabolic dishes with a natural gas turbine backup should be able to produce electrical power costing about the same as that from coal-burning plants.

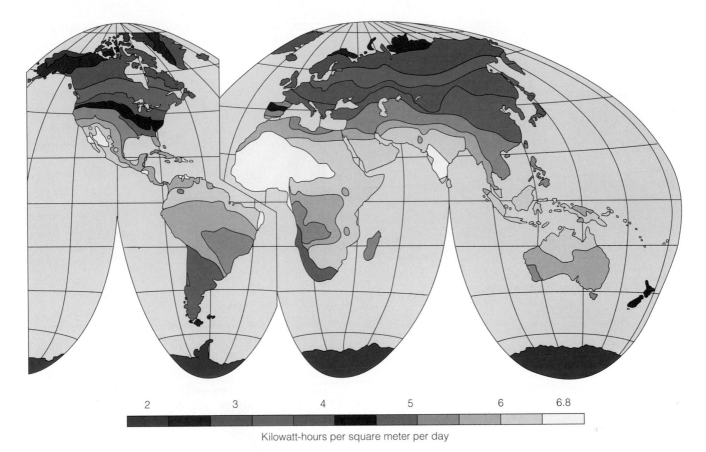

Figure 15-19 Map of global solar energy availability. Areas with more than 3.5 kilowatt-hours per square meter per day (see scale) are good candidates for passive and active solar heating systems and use of solar cells to produce electricity. (Data from U.S. Department of Energy)

Another approach for intensifying incoming solar energy about 80,000 times is a *nonimaging optical solar concentrator*. With this technology, the sun's rays are allowed to scramble instead of being focused on a par-

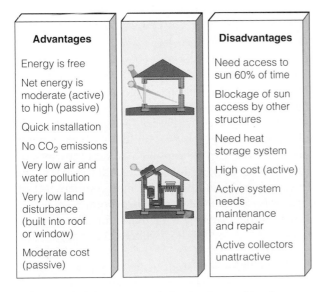

Figure 15-20 Advantages and disadvantages of heating a house with passive or active solar energy.

ticular point (Figure 15-21c). Because of its high efficiency and ability to generate extremely high temperatures, nonimaging concentrators may make solar energy practical for widespread industrial and commercial use within 10-20 years.

Inexpensive solar cookers can focus and concentrate sunlight and cook food, especially in rural villages in sunny developing countries. They can be made by fitting an insulated box big enough to hold three or four pots with a transparent, removable top (Figure 15-21d). Solar cookers reduce deforestation for fuelwood, the time and labor needed to collect firewood, and indoor air pollution from smoky fires.

Figure 15-22 lists the advantages and disadvantages of concentrating solar energy to produce high-temperature heat or electricity. Most analysts do not expect widespread use of such technologies over the next few decades because of their high costs and the availability of much cheaper ways to produce electricity such as combined-cycle natural gas turbines and wind turbines.

How Can We Produce Electricity with Solar Cells? Solar energy can be converted directly into electrical energy by **photovoltaic (PV) cells,** commonly

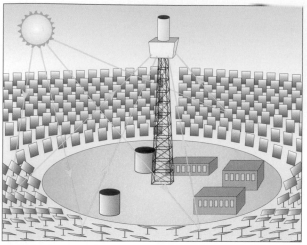

(a) Solar Power Tower

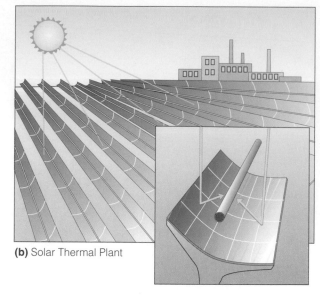

(b) Solar Thermal Plant

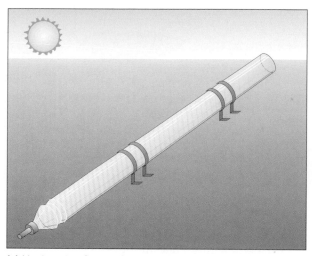

(c) Nonimaging Optical Solar Concentrator

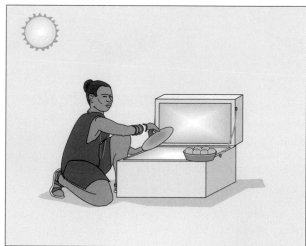

(d) Solar Cooker

Figure 15-21 Several ways to collect and concentrate solar energy to produce high-temperature heat and electricity. Because of their high costs (except for solar ovens), such systems are not expected to provide much of the world's energy.

called **solar cells** (Figure 15-23). A solar cell is a transparent wafer that contains a *semiconductor* material with a thickness ranging from less than that of a human hair to a sheet of paper. Sunlight energizes and causes electrons in the semiconductor to flow, creating an electrical current.

Because a single solar cell produces only a tiny amount of electricity, many cells are wired together in modular panels to produce the amount of electricity needed. The direct current (DC) electricity produced can be **(1)** stored in batteries and used directly or **(2)** converted to conventional alternating-current (AC) electricity by a separate inverter or an inverter built into the cells (Guest Essay, p. 377).

Traditional-looking solar-cell roof shingles and photovoltaic panels that resemble metal roofs (developed in Japan) reduce the cost of solar-cell installations by saving on roof costs (Figure 15-23). With this technology, the roof becomes a building's power plant. Solar cells also can be deployed along highways, on bridges, over parking lots, on or under bridges, and atop municipal buildings. A German company is testing a *solar-electric window* that incorporates solar cells into a semitransparent glazing that simultaneously generates electricity and provides filtered light during daylight hours.

Researchers envision using easily expandable banks of solar cells to **(1)** provide electricity for rural vil-

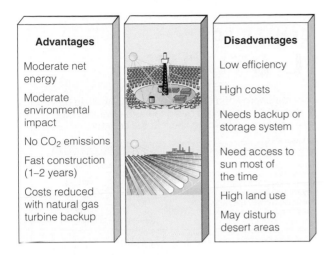

Advantages	Disadvantages
Moderate net energy	Low efficiency
Moderate environmental impact	High costs
No CO_2 emissions	Needs backup or storage system
Fast construction (1–2 years)	Need access to sun most of the time
Costs reduced with natural gas turbine backup	High land use
	May disturb desert areas

Figure 15-22 Advantages and disadvantages of using solar energy to generate high-temperature heat and electricity.

lages in developing countries, **(2)** produce electricity at a small power plant, using combined-cycle natural gas turbines to provide backup power when the sun is not shining, and **(3)** convert water to hydrogen gas that can be distributed to energy users by pipeline, as natural gas is. Researchers are developing flywheels, improved deep-cycle batteries, and supercapacitors to store solar (or wind) power for later use as needed. One promising system is to **(1)** use rooftop solar cells to produce hydrogen when the sun is shining, **(2)** store the hydrogen, and **(3)** use it in a fuel cell to provide electricity and heat as needed.

In New York City, the Durst Building at 4 Times Square incorporates green design on a grand scale. This 48-story skyscraper uses **(1)** low-E insulating windows, **(2)** energy-efficient lighting, **(3)** PV panels built into the walls along its south and east sides, **(4)** two large fuel cells in the basement that provide hot water, supplement daytime electricity, and provide all the building's electricity at night, **(5)** recycled building materials, and **(6)** separate waste chutes to facilitate recycling by tenants. This design is projected to produce 40% fewer greenhouse gas emissions than a traditionally powered building of the same volume.

In 1998, the 350-room Mauna Lani Bay Hotel, a luxury resort on the Kona-Kohala coast of the island of Hawaii, covered its roof with solar cells, which act as a 100-kilowatt power station.

Oberlin College in Ohio recently built a new environmental studies center incorporating major elements of *green design*, including use of passive solar energy and solar cells (Figure 15-24). David Orr (head of Oberlin's environmental studies program, Guest Essay, p. 683) designed the building with the help of more than 250 students, faculty, and townspeople.

In 1999, researchers announced the development of a new breed of solar cells that are thinner than a human hair and can be created with materials costing only a few pennies. Instead of silicon, they consist of an ultrathin layer of a semiconductor compound (copper indium diselenide) deposited on a material such as glass. Mass production and technological advances could bring costs down as low as 4¢ per kilowatt-hour by 2015—below the cost of all other ways to produce electricity. (A kilowatt-hour of electricity can light a 100-watt light bulb all night or run a typical hairdryer for about 1 hour.)

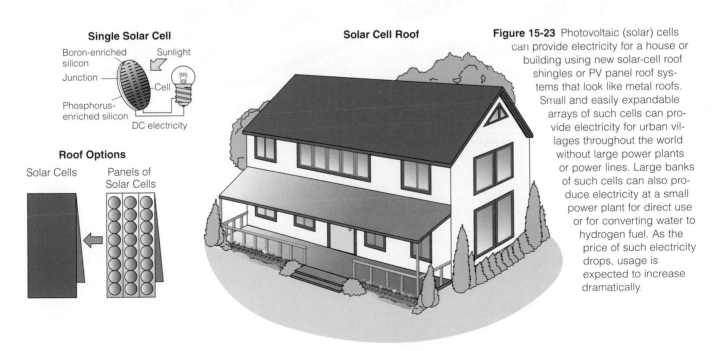

Figure 15-23 Photovoltaic (solar) cells can provide electricity for a house or building using new solar-cell roof shingles or PV panel roof systems that look like metal roofs. Small and easily expandable arrays of such cells can provide electricity for urban villages throughout the world without large power plants or power lines. Large banks of such cells can also produce electricity at a small power plant for direct use or for converting water to hydrogen fuel. As the price of such electricity drops, usage is expected to increase dramatically.

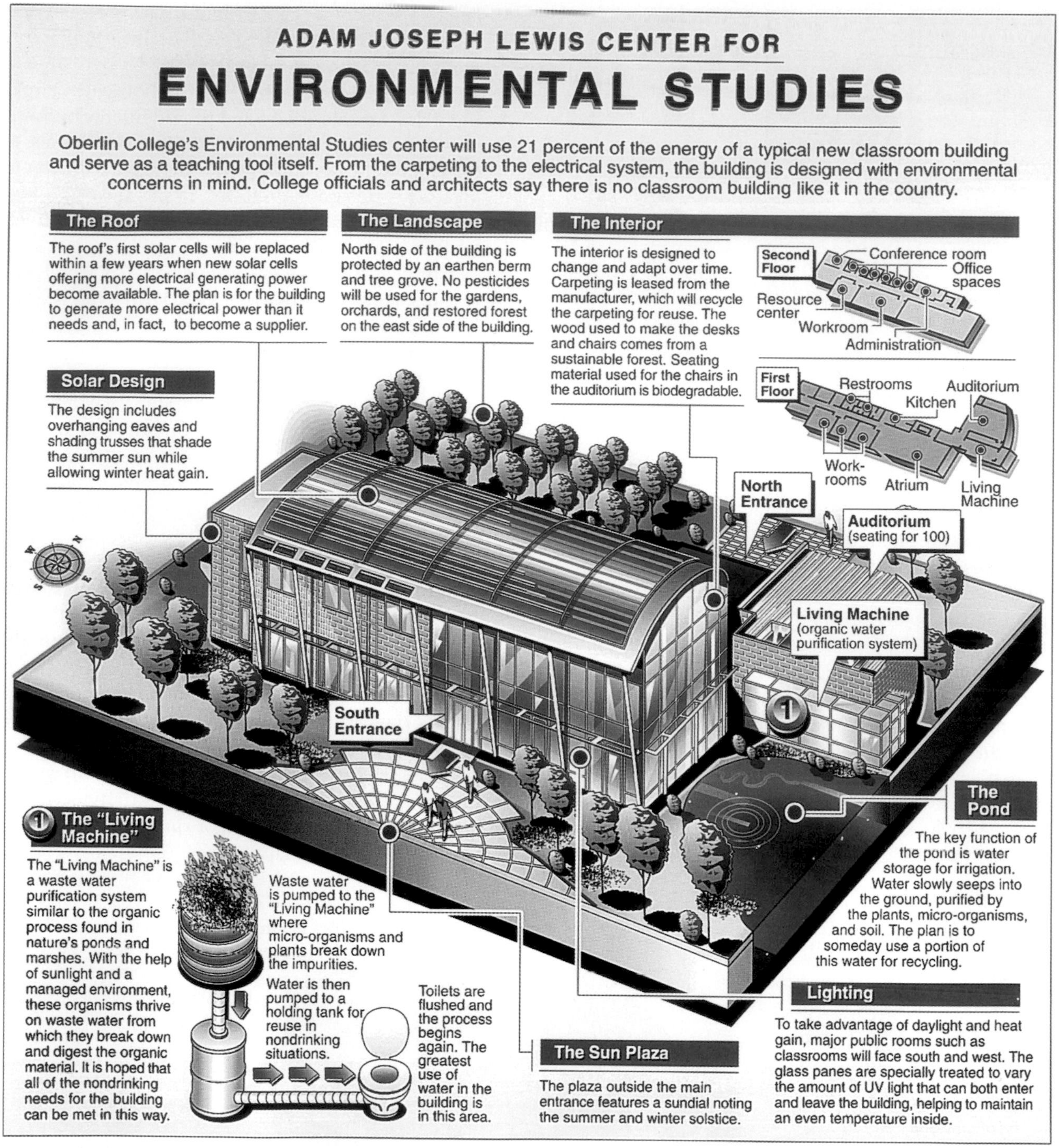

ADAM JOSEPH LEWIS CENTER FOR

ENVIRONMENTAL STUDIES

Oberlin College's Environmental Studies center will use 21 percent of the energy of a typical new classroom building and serve as a teaching tool itself. From the carpeting to the electrical system, the building is designed with environmental concerns in mind. College officials and architects say there is no classroom building like it in the country.

The Roof

The roof's first solar cells will be replaced within a few years when new solar cells offering more electrical generating power become available. The plan is for the building to generate more electrical power than it needs and, in fact, to become a supplier.

Solar Design

The design includes overhanging eaves and shading trusses that shade the summer sun while allowing winter heat gain.

The Landscape

North side of the building is protected by an earthen berm and tree grove. No pesticides will be used for the gardens, orchards, and restored forest on the east side of the building.

The Interior

The interior is designed to change and adapt over time. Carpeting is leased from the manufacturer, which will recycle the carpeting for reuse. The wood used to make the desks and chairs comes from a sustainable forest. Seating material used for the chairs in the auditorium is biodegradable.

Second Floor — Conference room — Office spaces
Resource center — Workroom — Administration

First Floor — Restrooms — Auditorium — Kitchen — Workrooms — Atrium — Living Machine

North Entrance

Auditorium (seating for 100)

Living Machine (organic water purification system)

South Entrance

① The "Living Machine"

The "Living Machine" is a waste water purification system similar to the organic process found in nature's ponds and marshes. With the help of sunlight and a managed environment, these organisms thrive on waste water from which they break down and digest the organic material. It is hoped that all of the nondrinking needs for the building can be met in this way.

Waste water is pumped to the "Living Machine" where micro-organisms and plants break down the impurities.

Water is then pumped to a holding tank for reuse in nondrinking situations.

Toilets are flushed and the process begins again. The greatest use of water in the building is in this area.

The Sun Plaza

The plaza outside the main entrance features a sundial noting the summer and winter solstice.

The Pond

The key function of the pond is water storage for irrigation. Water slowly seeps into the ground, purified by the plants, micro-organisms, and soil. The plan is to someday use a portion of this water for recycling.

Lighting

To take advantage of daylight and heat gain, major public rooms such as classrooms will face south and west. The glass panes are specially treated to vary the amount of UV light that can both enter and leave the building, helping to maintain an even temperature inside.

Figure 15-24 Major elements of green design in the Adam Joseph Lewis Center for Environmental Studies at Oberlin College in Ohio. There were no waste products as a result of the construction and no use of toxic materials. The building is the brainchild of David Orr (head of Oberlin's Environmental Studies program) and a number of students, faculty, and townspeople. In North Carolina, Catawba College's Center for the Environment is also a model of green design. (Illustration by James Owen. Courtesy of Oberlin College)

In 2000, researchers were working to produce electricity from solar cells made of microscopic quantum rods 1 nanometer (1 billionth of a meter) thick and 1–10 nanometers long. They contain tiny microscopic crystals, each consisting of clusters of 100–100,000 molecules of cadmium selenide, a semiconducting material.

Researchers hope to produce these large quantities of quantum rods to make these solar cells.

Solar cells are an ideal technology for providing electricity to 2 billion people in rural areas in most developing countries who have no electricity. With financing from the World Bank, India (the world's number-one market

Living Lightly on the Earth at Round Mountain Organics

Nancy Wicks

Nancy Wicks is an ecopioneer trying to live her ideals. She grew up in a small town in Iowa and did undergraduate studies in a village in Nepal. Both of these life experiences inspired her to live more sustainably by creating Round Mountain Organics, an organic garden at an altitude of 1.6 kilometers (8,500 feet) in the Rocky Mountains near Crested Butte, Colorado. Nancy lives in a passive solar strawbale house, which is powered by the wind and sun. She's a fanatic reuser, recycler, and composter. She received the "Sustainable Business of the Year, 2000" award from the High Country Citizen's Alliance.

After studying in Nepal, where sustainability is a do or die situation, I have tried incorporate as many sustainable practices into my life as possible at my house and organic garden business called Round Mountain Organics. This includes **(1)** being a member of a buying co-op (where buying in bulk not only saves re-sources but also saves money), **(2)** reusing everything from plastic and paper bags to trays and pots for garden plants, and **(3)** composting food waste (which saves money on trash bills and fertilizes the soil).

After moving onto the land that is now home to Round Mountain Organics, I spent 4 years planning and building an octagonal strawbale house with a stucco exterior—the first such house to be built in the county. I chose to build with straw because **(1)** straw is a natural building material and renewable resource, **(2)** there is a surplus of straw after harvesting grains such as wheat (which I used), oats, barley, and rice, **(3)** its insulation value of R-54 comes in handy when you live in an area where winter temperatures can dip to 40°F below zero, and **(4)** they are easy to build (with only 1 week needed to put up the strawbale walls). Since the 1980s strawbale houses have also been built in Arizona and New Mexico to beat the heat.

I used a passive solar design by orienting the house to the south to take advantage of Colorado's abundant sunshine. During the day the insulated window covers are drawn up and the sun shines onto flagstone tiles covering a cement slab that stores and releases heat slowly to keep the house comfortably warm or cool regardless of outside conditions. At night the window covers are let down to hold the heat in.

Because I live in one of the world's sunniest places, I decided not to get hooked up to the electrical grid and instead get my electricity from a small wind turbine and panels of solar cells. The electricity is stored in a bank of 12 batteries and an inverter converts the stored direct current (DC) electricity to ordinary 120-volt alternating current (AC).

If it's cloudy and not windy for a couple of days (which is rare), I fire up a small gas generator to charge the batteries. I also use some propane to provide hot water with a small on-demand water heater. Only a small pilot light stays lit until the hot water faucet is turned on. Then a large flame is ignited that the water is piped through. This way, I do not use energy to keep a tank of water hot around the clock.

I use many energy-saving devices. They include compact florescent light bulbs, an oversized pressure tank so the well pump does not have to kick on every time the faucet turns on, and a superinsulated energy-efficient DC refrigerator.

I use organic gardening to grow flowers, herbs, and vegetables for my own use and for sale to local residents and restaurants. I incorporate some pioneering organic gardening techniques such as Rudolph Steiner's biodynamics (developed in 1924) and Bill Mollison's permaculture (developed in 1978).

Insect pests are picked off by hand and beneficial insects such as ladybugs are used to eat harmful insects such as aphids. Compost, aged animal manure, and cover crops that are plowed in as green manure are used to add nutrients to the soil. Crop rotation is used so as not to deplete the soil of nutrients.

Cold frames (a type of mini-greenhouse) are used to extend the growing season to 150 days in the cold climate where there are only about 90 days without a killing frost. My latest endeavor is building a passively heated strawbale greenhouse to provide the community with fresh salad greens, herbs, and flowers all winter long. The chicken coop is in the northeast corner of the greenhouse, with the heat given off by the chickens helping to warm the greenhouse.

My next venture is to start a nonprofit, Round Mountain Sustainable Living Institute, to educate people on how they can live in harmony with the earth.

Critical Thinking

Would you like to live a lifestyle similar to that of Nancy Wicks? Explain. Why do you think more people do not try to live more sustainably, as she does? List three ways to help encourage people to adopt such a lifestyle.

---

for solar cells) is installing solar-cell systems in 38,000 villages, and Zimbabwe is bringing solar electricity to 2,500 villages. Houston-based Enron Corporation announced plans to build large, grid-connected solar photovoltaic power plants in the desert regions of China, India, and the United States. In 1999, Shell Oil (which owns two PV companies) launched a solar electrification project in South Africa that will provide 50,000 homes in impoverished areas with solar panels and batteries.

In 1998, the U.S. Department of Energy launched a program to encourage utilities, the solar industry, and governments to work together to install solar-cell

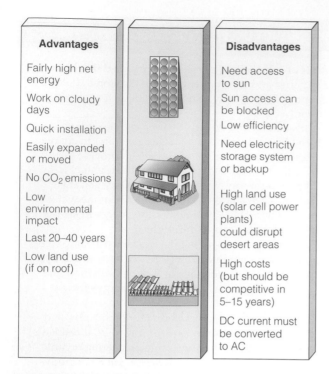

Figure 15-25 Advantages and disadvantages of using solar cells to produce electricity.

energy systems on 1 million roofs in the United States by 2010. Similar government-supported programs are expected to install 100,000 solar-cell roofs in Japan and Germany and 10,000 in Italy. Such programs can stimulate the mass production of solar cells that could reduce their costs by 75% or more.

Figure 15-25 lists the advantages and disadvantages of solar cells. With an aggressive program, analysts project that solar cells could supply 17% of the world's electricity by 2020—as much as nuclear power does today—at a lower cost and much lower risk. With a strong push from governments and private investors, by 2050 solar cells could provide as much as 25% of the world's electricity and at least 35% of the electricity in

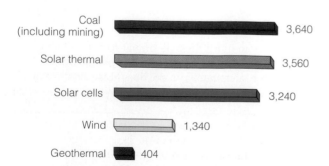

Figure 15-26 Approximate land use of various systems for producing electricity in the United States. Numbers give the land occupied in square meters per gigawatt-hour of electricity produced for 30 years. (Data from Worldwatch Institute)

the United States. If such projections are correct, the production, sale, and installation of solar cells (already a $2-billion-per-year business) could become one of the world's largest and fastest-growing businesses.

Critics of solar energy contend that producing electricity using large banks of solar cells, solar thermal plants (Figure 15-21b), and wind farms (photo, table of contents, p. xxiii, and Figure 15-29, p. 381) uses too much land. However, these three ways of producing electricity use less land per unit of electricity produced than coal (including the land disrupted from coal mining), the most widely used method for producing electricity (Figure 15-26).

15-4 PRODUCING ELECTRICITY FROM MOVING WATER AND FROM HEAT STORED IN WATER

How Can We Produce Electricity Using Hydropower Plants? Electricity can be produced from flowing water by:

- *Large-scale hydropower*, in which a high dam is built across a large river to create a reservoir (Figure 13-9, p. 301). Some of the water stored in the reservoir is allowed to flow through huge pipes at controlled rates, spinning turbines and producing electricity.

- *Small-scale hydropower*, in which a low dam with no reservoir (or only a small one) is built across a small stream and the stream's flow of water is used to spin turbines and produce electricity.

- *Pumped-storage hydropower*, in which pumps using surplus electricity from a conventional power plant pump water from a lake or a reservoir to another reservoir at a higher elevation. When more electricity is needed, water in the upper reservoir is released, flows through turbines, and generates electricity on its return to the lower reservoir.

Hydropower supplies about **(1)** 6% of the world's total commercial energy (4% in the United States) and **(2)** 20% of the world's electricity (10% in the United States but about 63% of the power used along the West Coast). Hydropower supplies about 99% of the electricity in Norway, 75% in New Zealand, 50% in developing countries, and 25% in China.

According to a 1999 study by Resources for the Future, the average cost for hydropower in the United is 4¢ per kilowatt-hour, compared to **(1)** 3.5–6¢ for natural gas, **(2)** 3.5–5¢ for wind (down from 40¢ in 1980), **(3)** 4–7¢ for geothermal, **(4)** 5–6¢ for coal, **(5)** 10–21¢ for nuclear power, and **(6)** 20¢ using solar cells (although these costs are expected to fall sharply over the next 10–20 years).

Figure 15-27 lists the advantages and disadvantages of using large-scale hydropower plants to produce electricity. According to the United Nations, only about 13% of the world's technically exploitable potential for hydropower has been developed, with much of this untapped potential in South Asia (especially China, p. 301), South America, and parts of the former Soviet Union. However, because of increasing concern about the environmental and social consequences of large dams (Figure 13-9, p. 301), there has been growing pressure on the World Bank and other development agencies to stop funding new large-scale hydropower projects. In 2000, the World Commission on Dams published a study indicating that hydropower is a major emitter of greenhouse gases. This is because reservoirs that power the dams can trap rotting vegetation, which can emit greenhouse gases such as CO_2 and CH_4.

Small-scale hydropower projects eliminate most of the harmful environmental effects of large-scale projects, but they can (1) threaten recreational activities and aquatic life, (2) disrupt the flow of wild and scenic rivers, and (3) destroy wetlands. In addition, their electrical output can vary with seasonal changes in stream flow.

Is Producing Electricity from Tides and Waves a Useful Option? Twice a day in high and low tides,

Advantages	Disadvantages
Moderate to high net energy	High construction costs
High efficiency (80%)	High environmental impact
Low-cost electricity	High CO_2 emissions from biomass decay in shallow tropical reservoirs
Long life span	
No CO_2 emissions during operation	Floods natural areas
May provide flood control below dam	Converts land habitat to lake habitat
Provides water for year-round irrigation	Danger of collapse
	Uproots people
	Decreases fish harvest below dam
	Decreases flow of natural fertilizer (silt) to land below dam

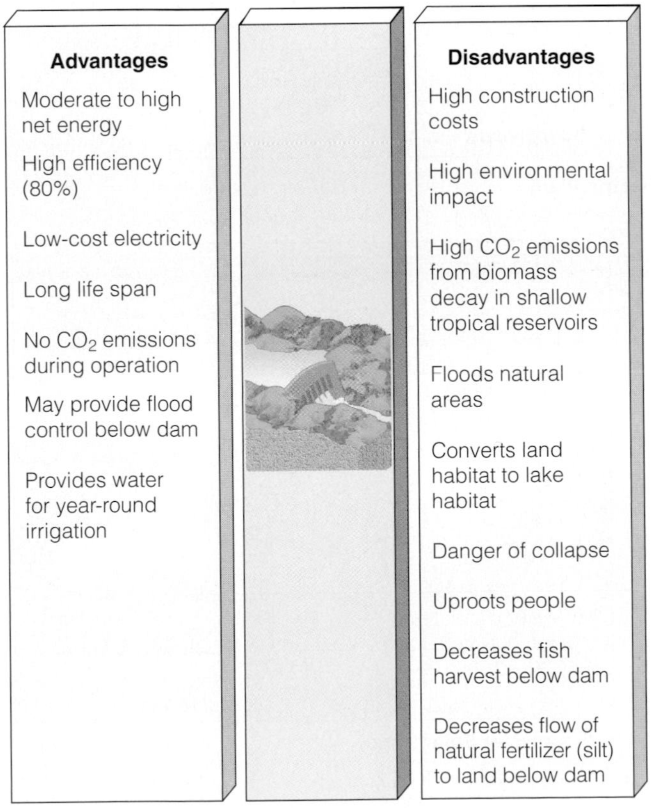

Figure 15-27 Advantages and disadvantages of using large dams and reservoirs to produce electricity.

water that flows into and out of coastal bays and estuaries can spin turbines to produce electricity (Figure 15-28a). Two large tidal energy facilities are currently operating, one at La Rance in France and the other in Canada's Bay of Fundy. However, most analysts expect tidal power to make only a tiny contribution to world electricity supplies. There are few suitable sites, and construction costs are high.

The kinetic energy in ocean waves, created primarily by wind, is another potential source of electricity (Figure 15-28b). Most analysts expect wave power to make little contribution to world electricity production, except in a few coastal areas with the right conditions (such as western England). Construction costs are moderate to high and the net energy yield is moderate, but equipment can be damaged or destroyed by saltwater corrosion and severe storms.

How Can We Produce Electricity from Heat Stored in Water? Japan and the United States have been evaluating the use of the large temperature differences (between the cold, deep waters and the sun-warmed surface waters) of tropical oceans for producing electricity. If economically feasible, this would be done in *ocean thermal energy conversion* (OTEC) plants anchored to the bottom of tropical oceans in suitable sites (Figure 15-28c). However, most energy analysts believe that the large-scale extraction of energy from ocean thermal gradients may never compete economically with other energy alternatives.

Saline solar ponds, usually located near inland saline seas or lakes in areas with ample sunlight, can be used to produce electricity (Figure 15-28d). Heat accumulated during the day in the denser bottom layer can be used to produce steam that spins turbines, generating electricity. A small experimental saline solar pond power plant on the shore of the Israeli side of the Dead Sea operated for several years but was closed in 1989 because of high operating costs.

Freshwater solar ponds can be used to heat water and space (Figure 15-28e). A shallow hole is dug and lined with concrete. A number of large, black plastic bags, each filled with several centimeters of water, are placed in the hole and then covered with fiberglass insulation panels. The panels let sunlight in but keep most of the heat stored in the water during the daytime from being lost to the atmosphere. When the water in the bags has reached its peak temperature in the afternoon, a computer turns on pumps to transfer hot water from the bags to large, insulated tanks for distribution.

Saline and freshwater solar ponds use no energy storage and backup systems, emit no air pollution, and have a moderate net energy yield. Freshwater solar ponds can be built in almost any sunny area and have moderate construction and operating costs.

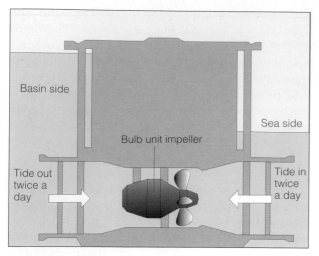

a. Tidal Power Plant

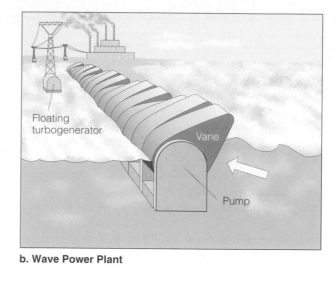

b. Wave Power Plant

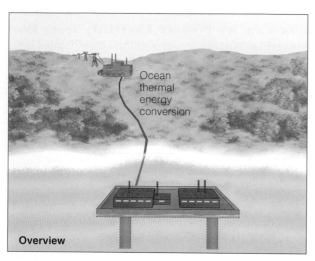

c. Ocean Thermal Electric Plant

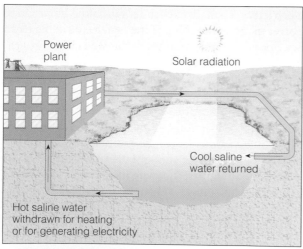

d. Saline Water Solar Pond

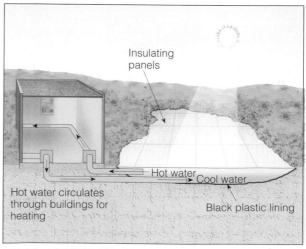

e. Freshwater Solar Pond

Figure 15-28 Ways to produce electricity from moving water and to tap into solar energy stored in water as heat. None of these systems are expected to be significant new sources of energy in the near future.

15-5 PRODUCING ELECTRICITY FROM WIND

How Rapidly Has the Use of Wind Power Grown?

Wind power is the world's fastest-growing energy resource (with an average growth of 22% per year during the 1990s and more than $2 billion in sales of wind turbines in 2000). In 2000, wind turbines (Figure 15-29) worldwide produced almost 15,000 megawatts of electricity, enough to meet the needs of 5.2 million homes. This was more than three times the capacity in 1995 and 1,300 times the capacity in 1980.

Despite its rapid growth, wind power produced only about 1% of the energy used in the United States in 2000 because it is still in its infancy. However, the U.S. Department of Energy has launched a program designed to have 5% of the country's energy produced by the wind by 2020.

How Much Does It Cost to Produce Electricity from Wind?

Between 1980 and 2000, the price of electricity produced by wind in the United States fell from 40¢ to 3.5–5¢ per kilowatt-hour, about the same as for new gas- and coal-fired power plants. According to the Department of Energy, with increased government funding for research and development and tax credits, the cost could fall to 2.5¢ within 3–5 years, making it the country's cheapest way to produce electricity.

What Areas Have the Greatest Potential for Wind Power?

Figure 15-30 shows the potential areas for use of wind power in the United States. The U.S. Department of Energy calls the Midwestern United States the "Saudi Arabia of wind." The Dakotas and Texas alone have enough wind resources to meet all the nation's electricity needs. Sizable wind farm projects are being developed in 12 states, with the world's largest single wind project now being developed in Iowa. Individuals can also use small wind turbines to supply some or all of their electricity (Guest Essay, p. 377).

Currently, California has the largest number of wind turbines, which produce about 1.5% of the state's electricity—enough to power about 300,000 homes. However, California ranks 17th among states with the best future potential for wind power because most of its best sites have already been developed.

The global potential for wind power has barely been tapped. Inland China's Inner Mongolia has enough wind resources to provide all the country's electricity, and England also has an enormous potential supply of wind resources. Denmark (with wind generating more than 8% of its electricity) is the world's largest user of wind and producer of wind turbines. Wind power also is being developed rapidly in Germany (the world's third largest user of wind power), Spain, and India (the world's number-two market for wind energy).

In the long run, electricity from large wind farms in remote areas might be used to make hydrogen gas from water during off-peak periods. The hydrogen could then be fed into a pipeline and storage system.

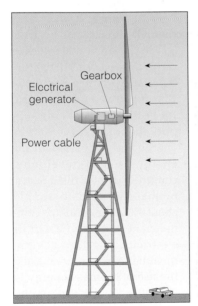

Wind Turbine

Wind Farm

Figure 15-29 Wind turbines can be used to produce electricity individually or in clusters called wind farms.

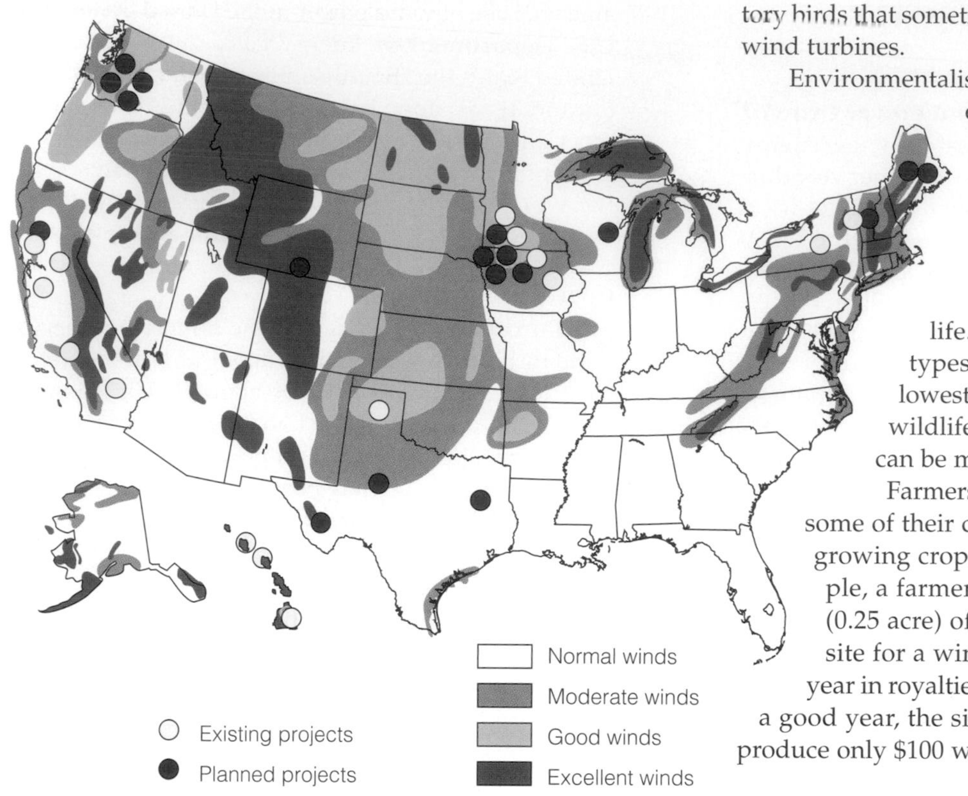

tory birds that sometimes get caught in the blades of the wind turbines.

Environmentalists and wind turbine manufacturers are working on this problem.

Oil spills, air pollution, water pollution, and release of toxic wastes from use of fossil fuels such as coal and oil have also killed enormous numbers of birds, fish, and other forms of wildlife. The key questions are (1) which types of energy resources lead to the lowest loss of wildlife and (2) how loss of wildlife from use of any energy resource can be minimized.

Farmers can boost their income by leasing some of their cropland for wind turbines, while growing crops around the turbines. For example, a farmer in Iowa who leases 0.10 hectare (0.25 acre) of cropland to the local utility as a site for a wind turbine typically gets $2,000 a year in royalties from the electricity produced. In a good year, the site occupied by the turbine could produce only $100 worth of corn.

Figure 15-30 Potential for use of wind power in the United States. In principle, all the power needs of the United States could be provided by exploiting the wind potential of just three states: North Dakota, South Dakota, and Texas. (Data from U.S. Department of Energy)

○	Existing projects
●	Planned projects

□	Normal winds
▨	Moderate winds
▨	Good winds
■	Excellent winds

What Are the Major Advantages and Disadvantages of Wind Power?

Figure 15-31 lists the advantages and disadvantages of using wind to produce electricity. Some environmentalists and other critics have pointed out that wind turbines have been responsible for the death of approximately 10,000 predatory birds (such as some types of hawks, kestrels, vultures, and eagles) in the United States over the past 20 years. The problem is that wind turbine towers attract bird prey, which attract preda-

Figure 15-31 Advantages and disadvantages of using wind to produce electricity. Wind power experts project that by 2050 wind power could supply more than 10% of the world's electricity and 10–25% of the electricity used in the United States.

Advantages	Disadvantages
Moderate to high net energy	Steady winds needed
High efficiency	Backup systems when needed winds are low
Moderate capital cost	High land use for wind farm
Low electricity cost (and falling)	Visual pollution
Very low environmental impact	Noise when located near populated areas
No CO$_2$ emissions	May interfere in flights of migratory birds and kill birds of prey
Quick construction	
Easily expanded	

15-6 PRODUCING ENERGY FROM BIOMASS

How Useful Is Burning Solid Biomass? Biomass is plant materials and animal wastes used as sources of energy. Biomass comes in many forms and can be burned directly as a solid fuel or converted into gaseous or liquid **biofuels** (Figure 15-32).

Most biomass is burned (1) directly for heating, cooking and industrial processes or (2) indirectly to drive turbines and produce electricity. Burning wood and manure for heating and cooking supplies about 12% of the world's energy and about 30% of the energy used in developing countries. In the United States, biomass is used to supply about 4% of the country's commercial energy and 2% of its electricity. The U.S. government has a goal of increasing the use of biomass energy to 9% of the country's total commercial energy by 2010.

Almost 70% of the people living in developing countries heat their homes and cook their food by burning

Figure 15-32 Principal types of biomass fuel.

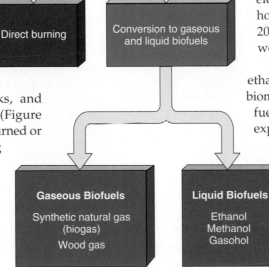

Solid Biomass Fuels
Wood logs and pellets
Charcoal
Agricultural waste
(stalks and other plant debris)
Timbering wastes
(branches, treetops, and wood chips)
Animal wastes (dung)
Aquatic plants (kelp and water hyacinths)
Urban wastes (paper, cardboard, and other combustible materials)

Direct burning

Conversion to gaseous and liquid biofuels

Gaseous Biofuels
Synthetic natural gas (biogas)
Wood gas

Liquid Biofuels
Ethanol
Methanol
Gasohol

wood or charcoal. However, about 2.7 billion people in these countries cannot find (or are too poor to buy) enough fuelwood to meet their needs.

One way to produce biomass fuel is to plant, harvest, and burn large numbers of fast-growing **(1)** trees (especially cottonwoods, poplars, sycamores, willows, and leucaenas), **(2)** shrubs, **(3)** perennial grasses (such as switchgrass), and **(4)** water hyacinths in *biomass plantations*.

In agricultural areas, *crop residues* (such as sugarcane residues, rice husks, cotton stalks, and coconut shells) and *animal manure* (Figure 23-22, p. 612) can be collected and burned or converted into biofuels. Burning *bagasse*—the residue left after sugarcane harvesting and processing—supplies about 10% of the electricity in Hawaii and in Brazil. According to a 1999 study by the Union of Concerned Scientists, energy crops and crop wastes from the Midwest alone could theoretically provide about 16% of the electricity used in the United States, without irrigation and without competing with food crops for land.

Some ecologists argue that it makes more sense to use animal manure as a fertilizer and crop residues to feed livestock, retard soil erosion, and fertilize the soil. Figure 15-33 lists the general advantages and disadvantages of burning solid biomass as a fuel. One problem is that burning biomass produces carbon dioxide. However, if the rate of use of biomass does not exceed the rate at which it is replenished by new plant growth (which takes up CO_2), there is no net increase in CO_2 emissions.

Is Producing Gaseous and Liquid Fuels from Solid Biomass a Useful Option? Bacteria and various chemical processes can convert some forms of biomass into gaseous and liquid biofuels (Figure 15-32). Examples include **(1)** *biogas*, a mixture of 60% methane and 40% carbon dioxide, **(2)** *liquid ethanol* (ethyl, or

grain, alcohol), and **(3)** and *liquid methanol* (methyl, or wood alcohol).

In China, anaerobic bacteria in more than 6 million *biogas digesters* convert plant and animal wastes into methane fuel for heating and cooking. These simple devices can be built for about $50 including labor. After the biogas has been separated, the solid residue is used as fertilizer on food crops or, if contaminated, on trees. When they work, biogas digesters are very efficient. However, they are also slow and unpredictable, a problem that could be corrected with development of more reliable models.

According to the U.S. Department of Energy, gasifying biomass and burning it in advanced combustion turbines could produce electricity costing 4.5¢ per kilowatt-hour by 2010. Shell Oil projects that by 2010 biomass could provide 5% of the world's electricity (worth $20 billion).

Some analysts believe that liquid ethanol and methanol produced from biomass could replace gasoline and diesel fuel when oil becomes too scarce and expensive. *Ethanol* can be made from sugar and grain crops (sugarcane, sugar beets, sorghum, sunflowers, and corn) by fermentation and distillation. Gasoline mixed with 10-23% pure ethanol makes *gasohol*, which can be burned in conventional gasoline engines and is sold as super unleaded or ethanol-enriched gasoline.

Another alcohol, *methanol*, is made mostly from natural gas but also can be produced at a higher cost from wood, wood wastes, agricultural wastes (such as corncobs), sewage sludge, garbage, and coal. Some of the first generation of cars using hydrogen-powered fuel cells will use reformers to convert methanol to hydrogen. The advantages and disadvantages of using ethanol, methanol, and several other fuels as alternatives to gasoline are summarized in Table 15-1, p. 385. According to a 1997 analysis by David Pimentel and two other researchers, "Large-scale biofuel production is not an alternative to the current use of oil and is not even an advisable option to cover a significant fraction of it."

Projections about the future role of biomass energy vary widely, with this resource providing 14–40% of the world's energy by 2050 depending on assumptions about **(1)** biomass gasification, **(2)** use in energy-efficient fuel cells, and **(3)** establishment of biomass plantations on degraded forestland and idle cropland.

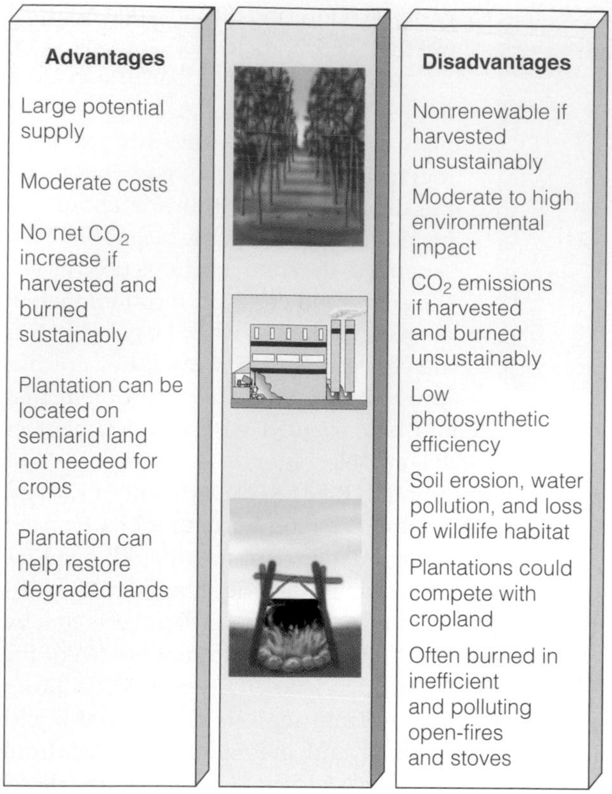

Advantages		Disadvantages
Large potential supply		Nonrenewable if harvested unsustainably
Moderate costs		Moderate to high environmental impact
No net CO_2 increase if harvested and burned sustainably		CO_2 emissions if harvested and burned unsustainably
Plantation can be located on semiarid land not needed for crops		Low photosynthetic efficiency
		Soil erosion, water pollution, and loss of wildlife habitat
Plantation can help restore degraded lands		Plantations could compete with cropland
		Often burned in inefficient and polluting open-fires and stoves

Figure 15-33 General advantages and disadvantages of burning solid biomass as a fuel.

Whether such a projection becomes reality depends on **(1)** the availability of large areas of productive land and adequate water (resources that may be in short supply in coming decades), **(2)** the ability to minimize the harmful environmental effects of large-scale biomass production, and **(3)** whether biomass is used sustainably so that there is no net increase in CO_2 emissions.

15-7 THE SOLAR-HYDROGEN REVOLUTION

What Can We Use to Replace Oil? Good-Bye Oil and Smog, Hello Hydrogen When oil is gone (or when what's left costs too much to use), what will we use to fuel vehicles, industry, and buildings? Many scientists and executives of major oil companies and automobile companies say the fuel of the future is hydrogen gas (H_2) (Table 15-1, right).

When hydrogen gas burns in air, it combines with oxygen gas in the air and produces nonpolluting water vapor* and some nitrogen oxides (produced because

---

*Water vapor is a potent greenhouse gas. However, because there is already so much of it in the atmosphere, human additions of this gas are insignificant.

air, which is 78% nitrogen, is used to burn hydrogen). This eliminates most of the air pollution problems we face today and greatly reduces the threats from global warming by emitting no carbon dioxide.

There is very little hydrogen gas (H_2) around. Instead, it is combined with other elements in compounds such as water, most organic compounds such as methane (CH_4) and methanol (CH_3OH), and those found in oil and gasoline. Methods for producing hydrogen include the following:

- *Reforming*, in which chemical processes are used to separate hydrogen from carbon atoms in organic compounds such as methane (CH_4) or methanol (CH_3OH).

- *Electrolysis of water*, in which direct electrical current is passed through water to convert its molecules into gaseous hydrogen and oxygen (Figure 15-34).

- *Photoelectrolysis*, in which electricity produced by solar cells (Figure 15-23) is used to split water molecules into gaseous hydrogen and oxygen. In 2000, researchers at the Isreal Institute of Technology created a combined *photovoltaic–photoelectrochemical cell* that uses sunlight to split water into hydrogen and oxygen at an efficiency of 18.3%. The researchers believe that such systems are capable of reaching a conversion efficiency of over 30%.

- *Coal gasification* (Figure 14-30, p. 345).

- *Biomass gasification*, in which wood chips and agricultural wastes are superheated to turn them into hydrogen and other gases.

- *Thermolysis*, in which high temperatures (up to 3000°C) are used to split water molecules into H_2 and O_2.

- *Biological production*, in which some types of algae and bacteria use sunlight to produce hydrogen under certain conditions (Spotlight, p. 387).

What Is the Catch? If you think using hydrogen as an energy source sounds too good to be true, you're right. Several problems must be solved to make hydrogen one of our primary energy resources, but scientists are making rapid progress in finding solutions to these problems.

One problem is that it takes energy (and thus money) to produce this fuel. We could burn coal to produce high-temperature heat or use electricity from coal-burning and nuclear power plants to split water and produce hydrogen. However, this subjects us to the harmful environmental effects associated with using these fuels (Chapter 14), and it costs more than the hydrogen fuel is worth.

Most proponents of using hydrogen gas believe that if we are to get its very low pollution benefits, the

Table 15-1 Evaluation of Alternatives to Gasoline

Advantages	Disadvantages
Compressed Natural Gas	
Fairly abundant, inexpensive domestic and global supplies	Large fuel tank needed; one-fourth the range
Low hydrocarbon, CO, and CO_2 emissions	Expensive engine modification needed ($2,000)
Vehicle development advanced; well suited for fleet vehicles	New filling stations needed
Reduced engine maintenance	Nonrenewable resource
Electricity	
Renewable if not generated from fossil fuels or nuclear power	Limited range and power
Zero vehicle emissions	Batteries expensive
Electric grid in place	Slow refueling (6–8 hours)
Efficient and quiet	Power plant emissions if generated from coal or oil
Reformulated Gasoline (Oxygenated Fuel)	
No new filling stations needed	Nonrenewable resource
Low to moderate CO emissions reduction	Dependence on imported oil perpetuated
No engine modification needed	No CO_2 emission reduction
	Higher cost
	Groundwater contaminated by leakage and spills (especially by MTBE, a possible human carcinogen)
	No longer needed because of improved emission control system
A-55 (55% water, 45% naphtha)	
Can be sold in conventional filling station	Not yet widely available
Much lower emissions of nitrogen oxide and particulates than diesel fuel	Independent tests needed to verify pollution reduction claims
Cannot explode or catch fire	Refineries may limit supply or drive up price of less-profitable naphtha
Lower cost (25–50%)	Large amounts of water needed to produce
Naphtha produces 90% less pollution at refineries than gasoline or diesel fuel	
Low-cost engine modification ($300 for cars, $1000 for trucks and buses)	
Modified engine can run A-55, gasoline of diesel	
Methanol	
High octane	Large fuel tank needed; one-half the range
Reduction of CO_2 emissions (total amount depends on method of production)	Corrosive to metal, rubber, plastic
Reduced total air pollution (30–40%)	Increased emissions of potentially carcinogenic formaldehyde
	High CO_2 emissions if generated by coal
	High capital cost to produce
	Hard to start in cold weather
Ethanol	
High octane	Large fuel tank needed; lower range
Reduction of CO_2 emissions (total amount depends on distillation process and efficiency of crop growing)	Much higher cost
Reduction of CO emissions	Corn supply limited
Potentially renewable	Competition with food growing for cropland
	Smog formation possible
	Corrosive
	Hard to start in cold weather
Solar–Hydrogen	
Renewable if produced using solar energy	Nonrenewable if generated by fossil fuels or nuclear power
Lower flammability than gasoline	Large fuel tank needed
Virtually emission-free	No distribution system in place
No emissions of CO_2	Engine redesign needed
Nontoxic	Currently expensive

Figure 15-34 Hydrogen gas as an energy source. Producing hydrogen gas takes electricity, heat, or solar energy to decompose water, thus leading to a negative net energy yield. However, hydrogen is a clean-burning fuel that can replace oil, other fossil fuels, and nuclear energy. Using solar energy (probably solar cells and wind turbines) to produce hydrogen from water also could eliminate most air pollution and greatly reduce the threat of global warming.

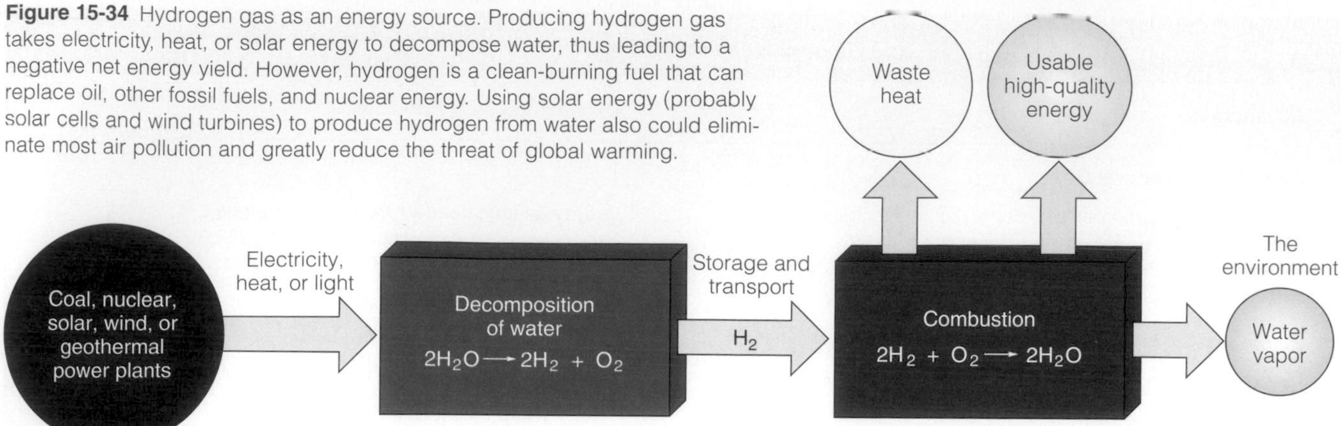

energy to produce the gas from water must come from the sun, probably **(1)** in the form of electricity generated by sources such as hydropower, solar cells, solar thermal power plants (Figure 15-21b, p. 374), and wind farms and **(2)** perhaps eventually from bacteria and algae (Spotlight, right).

If scientists and engineers can learn how to use sunlight to decompose water cheaply enough, they will set in motion a *solar-hydrogen revolution* over the next 50 years and change the world as much as the agricultural and industrial revolutions did. Currently, using solar energy to produce hydrogen gas is too costly, but the costs of using solar energy to produce electricity are coming down. The goal of the Department of Energy is to have the cost of hydrogen equal to that of natural gas by 2030.

The first widespread use of hydrogen probably will be to combine it with gasoline, ethanol, methanol, and natural gas to increase fuel performance and reduce air pollution. Adding just 5% hydrogen to gasoline can reduce emissions of nitrogen oxides (NO and NO_2) by 30–40%.

Next, blends of natural gas and hydrogen produced by solar-cell or thermal solar power plants (in sunny, mostly desert areas) or wind turbines could be mixed with natural gas or carried alone in modified natural gas pipelines to users. Researchers estimate that using pipelines to transport hydrogen long distances should cost only about one-fourth as much as transmitting electricity the same distance.

Once produced, hydrogen must be stored for use in cars, furnaces, air conditioners, or fuel cells. Hydrogen can be stored:

- In *compressed gas storage tanks*. The technology is available, but the costs of tanks and compression are high, and tanks are too heavy for use in motor vehicles.

- As *liquid hydrogen*. Condensing hydrogen gas into more dense liquid form allows a larger quantity of

hydrogen to be stored and transported. However, this conversion takes a large input of energy and is costly.

- As *solid metal hydride compounds*, which when heated decompose and release hydrogen gas. This is a safe and efficient way to store hydrogen, but an input of energy is needed to release the hydrogen.

- By absorption on *activated charcoal*, which when heated releases hydrogen gas. Like hydrides, this is a safe and efficient way to store hydrogen, but an input of energy is needed to release the hydrogen.

- Inside *glass microspheres*. Currently, tiny glass spheres are being developed for this purpose.

Unlike gasoline, metal hydrides, charcoal powders, and glass microspheres containing hydrogen will not explode or burn if a vehicle's tank is ruptured in an accident. However, it's difficult to store enough hydrogen gas in a car as a compressed gas, liquid, or a solid for it to run very far, a problem similar to the one the electric car faces. Scientists and engineers are seeking solutions to this problem.

Another possibility is to power a car with a *fuel cell* (Figures 15-4 and 15-10) in which hydrogen and oxygen gas combine to produce electrical current. Fuel cells produce no air pollution and have energy efficiencies of 65–95%, several times the efficiency of conventional gasoline-powered engines and electric cars.

A number of prototype fuel-cell systems for cars, buses, homes, and buildings are being tested and evaluated. All major automobile companies have developed fuel-cell cars (Figure 15-10) and hope to begin marketing them by 2004. In 1999, DaimlerChrysler and Shell announced plans to turn the tiny country of Iceland into the world's first "hydrogen economy," eventually replacing the gasoline and diesel engines on all its cars, buses, and fishing vessels with hydrogen.

Two U.S. companies are developing residential fuel cells that use methane to produce hydrogen fuel for the cell. These units are about the size of a dishwasher,

Producing Hydrogen from Green Algae Found in Pond Scum

SPOTLIGHT

In a few decades we may be able to use large-scale cultures of green algae to produce hydrogen gas. This simple plant grows all over the world and is commonly found in pond scum.

When living in ordinary air and sunlight, green algae carry out photosynthesis like other plants and produce carbohydrates and oxygen gas. However, in 2000, Tasios Melis, a researcher at the University of California, Berkeley, found a way to make these algae produce bubbles of hydrogen rather than oxygen.

First, he grew cultures of hundreds of billions of the algae in the normal way with plenty of sunlight, nutrients, and water. Then he cut off their supply of a two key nutrients: sulfur and oxygen. Within 20 hours, the plant cells underwent a metabolic change and switched from an oxygen-producing to a hydrogen-producing metabolism, allowing the researcher to collect hydrogen gas bubbling from the culture.

Melis believes that he can increase the efficiency of this hydrogen-producing process tenfold. If so, sometime in the future a biological hydrogen plant might cycle an algae-water mixture through a system of clear tubes exposed to sunlight to produce hydrogen. The gene responsible for producing the hydrogen might even be transferred to other plants to produce hydrogen.

Critical Thinking

What might be some ecological problems related to the widespread use of this method for producing hydrogen?

should cost less than $4,000, and could be on the market by 2002. A single unit could supply all the heating, cooling, cooking, refrigeration, and electrical needs of a home and provide hydrogen fuel for one or more cars at an affordable price.

Figure 15-35 lists the pros and cons of using hydrogen as an energy resource. The Department of Energy has a goal of hydrogen energy providing 10% of all U.S. energy consumption by 2025. Even if this is only partially accomplished, it could greatly reduce emissions of CO_2 and other air pollutants and decrease U.S. dependence on oil imports.

Advantages
Can be produced from water
Low environmental impact
No CO_2 emissions
Good substitute for oil
Competitive price if environmental and social costs are included in cost comparisons
Easier to store than electricity
Safer than gasoline and natural gas
High efficiency (65–95%) in fuel cells
Disadvantages
Not found in nature
Energy is needed to produce fuel
Negative net energy
High costs (but expected to come down)
Short driving range for current fuel cell cars

Figure 15-35 Advantages and disadvantages of using hydrogen as a fuel for vehicles and for providing heat and electricity.

15-8 GEOTHERMAL ENERGY

How Can We Tap the Earth's Internal Heat? Going Underground **Geothermal energy** is energy extracted from the earth's internal heat. Under the earth's crust, there is a layer of hot and molten rock called magma in the earth's mantle (Figure 10-3, p. 213). Because it is less dense than the surrounding rock, magma rises slowly toward the earth's crust, carrying heat from below.

Sometimes the hot magma reaches the earth's surface as lava. However, most magma remains below earth's crust, heating nearby rock and groundwater. Some of this hot geothermal water travels up through *geysers*. However, most of it remains deep underground, trapped in cracks and porous rock. This natural collection of hot water is called a *geothermal reservoir*.

Heat, mostly from the radioactive decay of naturally radioactive elements, is continually transferred to underground reservoirs of **(1)** *dry steam* (steam with no water droplets), **(2)** *wet steam* (a mixture of steam and water droplets), and **(3)** *hot water* trapped in fractured or porous rock at various places in earth's crust.

If such geothermal reservoirs are close to the surface, wells can be drilled to extract the dry steam, wet steam (Figure 15-36), or hot water. This thermal energy can be used to heat homes and buildings and to produce electricity.

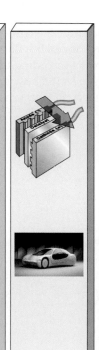

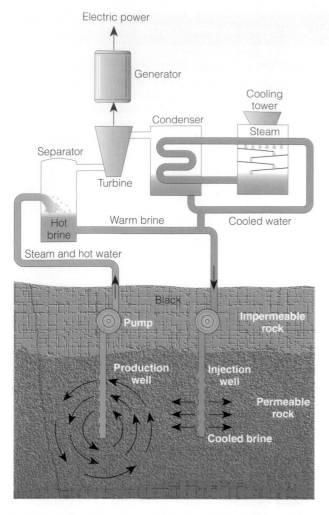

Figure 15-36 Tapping the earth's heat or geothermal energy in the form of wet steam to produce electricity.

Utah. The Philippines is the world's second largest user, followed by Mexico. In 1999, Santa Monica, California, became the first city in the world to get all its electricity from geothermal energy.

The world's largest operating geothermal system, called *The Geysers*, extracts energy from a dry steam reservoir north of San Francisco, California. Electricity production at this site began in 1960. Now the area contains 26 power plants producing enough power for a city of 1.3 million people. However, heat is being withdrawn from this geothermal site about 80 times faster than it is being replenished, converting this potentially renewable resource to a nonrenewable source of energy.

Geothermal water has been used to heat homes and buildings in Klamath Falls, Oregon, and Boise, Idaho, for more than a century. In Iceland, every building is heated by hot spring water.

Three other virtually nondepletable sources of geothermal energy are **(1)** *molten rock* (magma), **(2)** *hot dry-rock zones*, where molten rock that has penetrated the earth's crust heats subsurface rock to high temperatures, and **(3)** low- to moderate-temperature *warm-rock reservoir deposits*, which could be used to preheat water and run heat pumps for space heating and air conditioning. Research is being carried out in several countries to see whether hot dry-rock zones, which can be found almost anywhere about 8–10 kilometers (5–6 miles) below the earth's surface, can provide affordable geothermal energy.

Figure 15-38 lists the pros and cons of using geothermal energy. Currently, the cost of tapping geothermal energy is too high for all but the most concentrated and accessible sources. According to the U.S. Geothermal Energy Association, at best geothermal energy could meet 5% of all U.S. energy needs over the next several decades.

However, these geothermal reservoirs can be depleted if heat is removed faster than natural processes renew it. Thus, geothermal resources can be nonrenewable on a human time scale, but the potential supply is so vast that it is usually classified as a renewable energy resource.

Figure 15-37 shows the locations of the world's known global reservoirs of moderate- to high-temperature geothermal energy. Currently, about 22 countries (most of them in the developing world) are extracting energy from geothermal sites to produce about 1% of the world's electricity.

The United States accounts for 38% of the 7,000 megawatts of geothermal electricity generated worldwide, with most of the favorable sites in California, Hawaii, Nevada, and

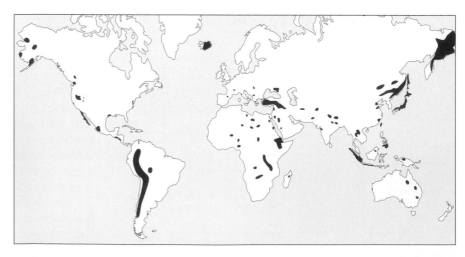

Figure 15-37 Known global reservoirs of moderate- to high-temperature geothermal energy. (Data from Canadian Geothermal Resources Council)

Advantages		Disadvantages
Very high efficiency		Scarcity of suitable sites
Moderate net energy at accessible sites		Depleted if used too rapidly
Lower CO_2 emissions than fossil fuels		CO_2 emissions
Low cost at favorable sites		Moderate to high local air pollution
Low land use		Noise and odor (H_2S)
Low land disturbance		
Moderate environmental impact		

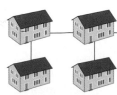

Figure 15-38 Advantages and disadvantages of using geothermal energy for space heating and to produce electricity or high-temperature heat for industrial processes.

15-9 ENTERING THE AGE OF DECENTRALIZED MICROPOWER

What Is Micropower? According to Chuck Linderman, director of energy supply policy for the Edison Electric Institute, the era of big central power-plant systems (Figure 15-39) is over. According to a growing number of energy and financial analysts, countries, companies, and investors trying to preserve or build large, centralized coal-burning and nuclear power plants may find themselves saddled with expensive technological dinosaurs.

Most energy analysts believe that the chief feature of electricity production over the next few decades is significant *decentralization* to dispersed, small-scale power systems (Figure 15-40). These **micropower systems** generate 1–10,000 kilowatts of power. This shift from centralized *macropower* to dispersed *micropower* is analogous to the computer industry's shift from large, cen-

tralized mainframes to increasingly smaller, widely dispersed PCs, laptops, and handheld computers.

In a special August 1999 issue of *Business Week* titled "21 Ideas for the 21st Century," the drastic downsizing of power-producing systems to micropower plants headed the list. Increasing amounts of venture capital and entrepreneurial talent in small startup energy companies and in large companies (such as General Electric, BP, and Shell Oil) are moving into this rapidly growing investment opportunity.

Between the mid-1960s and the mid-1980s, the average size of a new utility power station in the United States fell from 1,000,000 kilowatts (1 megawatt) to 600,000 kilowatts. Then between the mid-1980s and 1998 it fell again to an average of 210,000 kilowatts, the size of a typical natural gas combined-cycle power station.

This trend of power-plant downsizing is continuing with increased use of **(1)** moderate-size industrial cogeneration plants (50,000 kilowatts) and **(2)** energy-efficient, natural gas-burning gas generators (microturbines) for commercial buildings and residences (5–10,000 kilowatts). These microturbines are tiny jet engines that use heat released by combustion to spin a shaft that spins a high-speed generator. They can be fitted with lean burn engines and catalytic converters to reduce air pollution (mostly nitrogen oxides) and mufflers and soundproofing to reduce noise. Like central air conditioning units, they are serviced regularly by professionals.

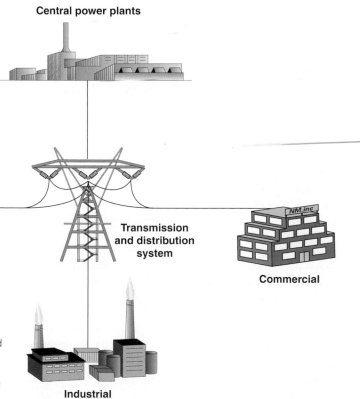

Central power plants

Transmission and distribution system

Residential

Commercial

Industrial

Figure 15-39 Centralized power system in which electricity produced mainly by a fairly small number of **(1)** large coal-burning and nuclear power plants (producing 600,000 to 1 million kilowatts of power) and **(2)** natural-gas turbines (producing about 200,000 kilowatts of power) is distributed by a system of high-voltage wires to users.

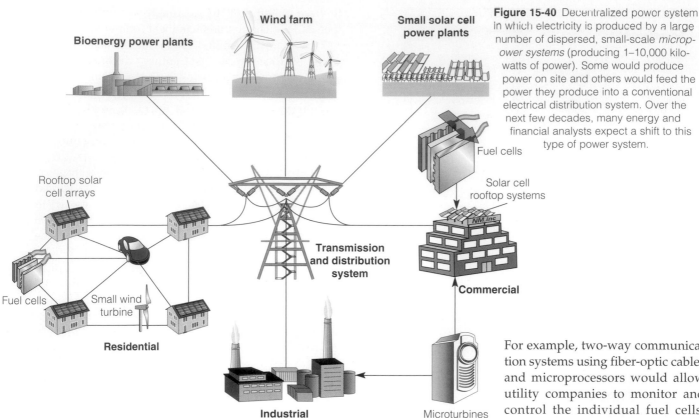

Figure 15-40 Decentralized power system in which electricity is produced by a large number of dispersed, small-scale *micropower systems* (producing 1–10,000 kilowatts of power). Some would produce power on site and others would feed the power they produce into a conventional electrical distribution system. Over the next few decades, many energy and financial analysts expect a shift to this type of power system.

This downsizing will be accelerated by the switch to increasingly smaller and more dispersed micropower systems such as **(1)** *wind turbines* (1–3,000 kilowatts), **(2)** low-cost *microturbines* for businesses (25–300 kilowatts), **(3)** energy-efficient *Stirling engines* (0.1–100 kilowatts), **(4)** efficient, quiet, reliable, low-maintenance *fuel cells* (1–10,000 kilowatts), and **(5)** quiet, reliable, low-maintenance household *solar panels and solar roofs* (1–1,000 kilowatts, Figure 15-23). Figure 15-41 lists some of the advantages of decentralized micropower systems (Figure 15-40) over traditional macropower systems (Figure 15-39).

How Can Decentralized Micropower Systems Be Managed? Some electric utility officials believe that a decentralized micropower system controlled mostly by customers instead of a centralized control system can lead to chaos. However, a growing number of energy analysts argue that by using multiple feedback loops integrated through a communication network, a decentralized, dispersed micropower system can be more resilient than a centralized system.

They envision using modern communication technologies to integrate centralized and decentralized power systems. As with the internet, individuals and businesses can use such a system to make their own power choices using a common set of connection rules or protocols.

For example, two-way communication systems using fiber-optic cables and microprocessors would allow utility companies to monitor and control the individual fuel cells, solar-cell rooftop systems, wind turbines, cogenerators, air conditioners, and water heaters of their customers. These communication systems could **(1)** turn such devices on or off (to save energy), **(2)** withdraw and buy excess power from customers, and **(3)** sell the power as needed. Controlled by computers, such an integrated system could maximize its overall energy efficiency and respond immediately to any problems.

How Rapidly Can the Transition to Micropower Be Made? No one knows how fast the transition to micropower systems can be made. However, much of the technology has already been developed, and investment capital is flowing rapidly into development and use of such systems. The potential for financial gain by companies and investors is huge, with a $10-trillion market projected for the global energy supply between 2000 and 2020. With the right economic stimulus and government policies, some analysts believe that micropower systems could dominate global markets for new power within 5–10 years.

Decentralized micropower systems could allow 2 billion people in isolated villages in developing countries to leapfrog over more expensive centralized power systems. Such villages could be powered by micropower systems such as **(1)** wind turbines and solar cells backed by fuel cells running on hydrogen produced from renewable sources and **(2)** Stirling engines and microturbines burning locally available biomass resources.

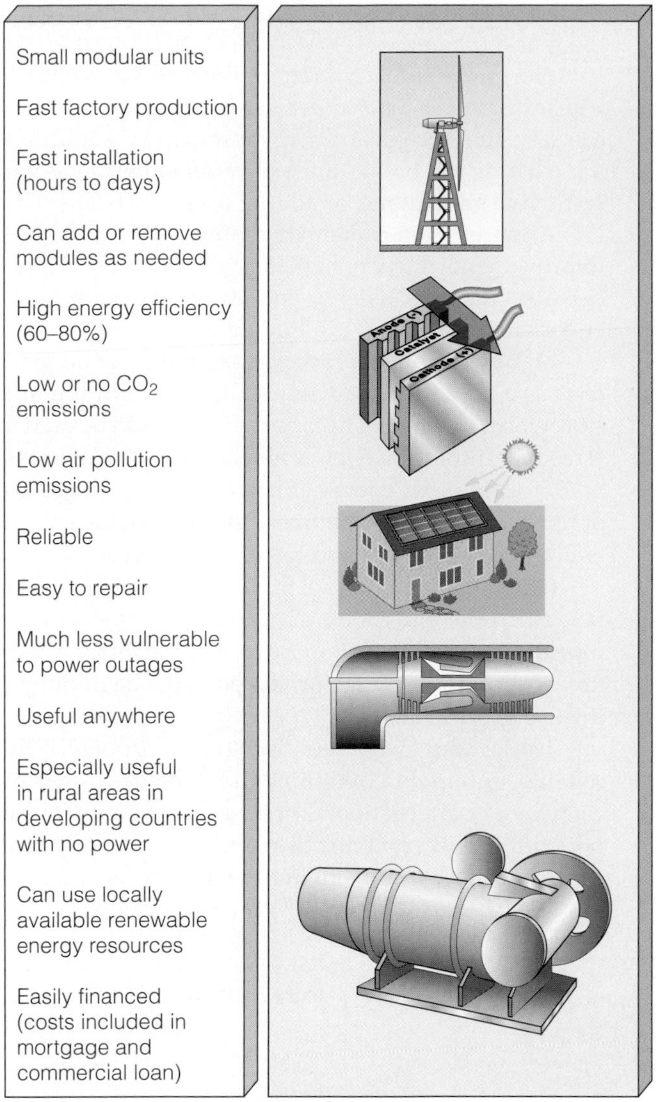

- Small modular units

- Fast factory production

- Fast installation (hours to days)

- Can add or remove modules as needed

- High energy efficiency (60–80%)

- Low or no CO_2 emissions

- Low air pollution emissions

- Reliable

- Easy to repair

- Much less vulnerable to power outages

- Useful anywhere

- Especially useful in rural areas in developing countries with no power

- Can use locally available renewable energy resources

- Easily financed (costs included in mortgage and commercial loan)

Figure 15-41 Some advantages of micropower systems.

15-10 SOLUTIONS: A SUSTAINABLE ENERGY STRATEGY

What Are the Best Energy Alternatives? We have a variety of nonrenewable and renewable energy resources, each with certain advantages and disadvantages. Many scientists and energy experts who have evaluated these energy alternatives have come to the following general conclusions:

- *There will be a shift from centralized macropower systems* (Figure 15-39) *to smaller, decentralized micropower systems* (Figures 15-40 and 15-41).

- *The best alternatives are a combination of improved energy efficiency and using natural gas as a fuel to make the transition to increased use of a variety of small-scale, decentralized, locally available renewable energy resources.*

- *Because there is not enough financial capital to develop all energy alternatives, governments and private companies must carefully choose which alternatives to support.*

- *Over the next 50 years the choice is not between using nonrenewable fossil fuels and various types of renewable energy.* Because of their supplies and low prices, fossil fuels will continue to be used in large quantities. The key questions are **(1)** how can we reduce the harmful environmental impacts of widespread fossil fuel use (especially to reduce air pollution and slow projected global warming) and **(2)** what roles can improving energy efficiency and depending more on some forms of renewable energy play in achieving these goals?

What Role Does Economics Play in Energy Resource Use? To most analysts the key to making a shift to more sustainable energy resources and societies is not technology but economics and politics. Governments use three basic economic and political strategies to help stimulate or dampen the short-term and long-term use of a particular energy resource.

The first approach is *allowing all energy resources to compete in a free market without government interference.* Free-market economists believe that letting the marketplace decide is the most effective way to develop future energy alternatives.

However, this is difficult to accomplish because of well-entrenched government intervention into the marketplace in the form of subsidies, taxes, and regulations. Another potential problem is that the emphasis on short-term profits for investors can inhibit development of new energy resources. During their 20- to 40-year development and phase-in period, new energy alternatives can rarely compete economically with established energy resources and face a *chicken-and-egg dilemma.* There must be enough orders for companies to invest in mass production facilities that will bring the price down, but most users will not place orders until the price comes down. Unless private companies invest large amounts of capital in long-term research and development, government support is needed during this period.

However, unless it is hindered by government regulations, private enterprise may accelerate the transition from *centralized macropower systems* to *decentralized micropower systems* because of the potential for huge profits.

The second approach is *trying to keep energy prices artificially low to encourage use of selected energy resources.* This is done mostly by **(1)** providing research and development subsidies and tax breaks and **(2)** enacting regulations that help stimulate the development and use of energy resources receiving such support.

For example, most governments have helped stimulate the development of fossil fuels and nuclear power by supporting research and development, providing subsides, and enacting favorable regulations for more than 50 years. Since the mid-1970s some governments have also been subsidizing energy efficiency and renewable energy technologies, but usually at much lower rates than fossil fuels and nuclear power.

Critics argue that subsidies and tax breaks do not work because **(1)** they amount to only about 1–2% of the total energy economy and **(2)** after 20 years of government subsidies and tax breaks, solar and wind power provide only about 1% of the energy used in the United States and the world and are too costly.

Supporters of such subsidies, tax breaks, and favorable regulations for energy efficiency and renewable energy resources argue that **(1)** they have not been in place very long compared to those supporting fossil fuels and nuclear energy, and **(2)** they have been too low and too erratic to help these new industries reach a take-off point and achieve lower prices through mass production. For example, subsidies and tax breaks put into place in the mid-1970s for energy conservation and renewable energy in the United States were sharply reduced in the 1980s and early 1990s before being increased somewhat since 1995.

Even with favorable treatment, some energy alternatives (such as nuclear power) may not be able to compete economically (p. 353). Proponents of increased support of renewable energy resources argue that they should be helped to reach the take-off point to find out how well they can compete on their own in the marketplace.

The third approach is *keeping energy prices artificially high to discourage use of an energy resource.* Governments can raise the price of an energy resource by withdrawing existing tax breaks and other subsidies, enacting restrictive regulations, or adding taxes on its use. This **(1)** increases government revenues, **(2)** encourages improvements in energy efficiency, **(3)** reduces dependence on imported energy, and **(4)** decreases use of an energy resource that has a limited future supply.

Many economists favor *increasing taxes on fossil fuels* as a way to reduce air and water pollution and slow global warming. The tax revenues would be used to **(1)** reduce income taxes on wages and profits (Solutions, p. 707), **(2)** improve energy efficiency, **(3)** encourage use of renewable energy resources, and **(4)** provide energy assistance to the poor and lower middle class.

Other analysts criticize this *output approach,* which taxes carbon emissions at the end of the energy pipeline. Instead, they propose an *input approach,* which taxes each unit of energy produced at the beginning of the energy pipeline.

These input taxes would be **(1)** proportional to the relative amount of environmental damage caused by each energy alternative, **(2)** phased in over a 10-year period to avoid economic disruption, and **(3)** adjusted downward as harmful environmental effects decrease.

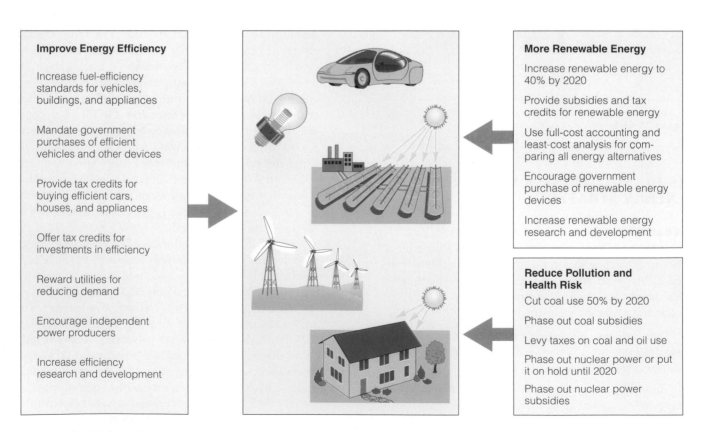

Improve Energy Efficiency

Increase fuel-efficiency standards for vehicles, buildings, and appliances

Mandate government purchases of efficient vehicles and other devices

Provide tax credits for buying efficient cars, houses, and appliances

Offer tax credits for investments in efficiency

Reward utilities for reducing demand

Encourage independent power producers

Increase efficiency research and development

More Renewable Energy

Increase renewable energy to 40% by 2020

Provide subsidies and tax credits for renewable energy

Use full-cost accounting and least-cost analysis for comparing all energy alternatives

Encourage government purchase of renewable energy devices

Increase renewable energy research and development

Reduce Pollution and Health Risk

Cut coal use 50% by 2020

Phase out coal subsidies

Levy taxes on coal and oil use

Phase out nuclear power or put it on hold until 2020

Phase out nuclear power subsidies

Figure 15-42 Solutions. Suggestions various analysts have made to help make the transition to a more sustainable energy future.

Based on current environmental data, such taxes would be **(1)** zero for energy efficiency, **(2)** fairly low for most forms of renewable energy, **(3)** slightly more for natural gas and more environmentally harmful forms of renewable energy, and **(4)** highest for nonrenewable oil, coal, and nuclear power.

How Can We Develop a More Sustainable Energy Future? Figure 15-42 lists a variety of strategies analysts have suggested for making the transition to a more sustainable energy future over the next few decades.

Energy experts estimate that implementing policies such as those shown in Figure 15-42 over the next 20–30 years could **(1)** save money, **(2)** create a net gain in jobs, **(3)** reduce greenhouse gas emissions, and **(4)** sharply reduce air and water pollution. Some actions you can take to promote a more sustainable energy future are listed in Appendix 6.

The move to micropower may accelerate the evolution to a carbon-free hydrogen economy, as proliferating fuel cells create a growing demand for hydrogen as an energy carrier.

SETH DUNN AND CHRISTOPHER FLAVIN

REVIEW QUESTIONS

1. Define the boldfaced terms in this chapter.

2. What is *energy efficiency*? How much of the energy used in the United States is wasted? What percentage of this is wasted because of the second law of energy, and what percentage is wasted unnecessarily? What is *life cycle cost*? What are three of the least efficient energy-using devices?

3. Explain why we cannot recycle energy. List three ways to slow down the flow of heat from **(a)** a house and **(b)** an office building.

4. What are the advantages of saving energy?

5. What is *cogeneration*, and how efficient is it compared with producing electricity by a conventional coal-burning or nuclear power plant? List two other ways to save energy in industry.

6. What do most experts believe is the best way to save energy in transportation?

7. List the pros and cons of using **(a)** battery-powered electric cars, **(b)** hybrid cars, **(c)** fuel-cell cars, and **(d)** electric bicycles.

8. Describe how we can save energy in homes by using **(a)** superinsulated houses and **(b)** strawbale houses. What are the four most efficient ways to heat a house? Describe ways to make an existing house more energy efficient. What are the most efficient and least efficient ways to heat water for washing and bathing? List the pros and cons of switching from inefficient incandescent and halogen light bulbs to efficient compact fluorescent light bulbs.

9. Describe how the internet can save energy and help reduce carbon dioxide emissions.

10. List three reasons why there is little emphasis on saving energy in the United States, despite its important benefits.

11. What are the major advantages and disadvantages of relying more on direct and indirect renewable energy from the sun?

12. Distinguish between a *passive solar heating system* and an *active solar heating system* and list the pros and cons of each system.

13. Describe three ways to cool houses naturally.

14. Distinguish between the following solar systems used to generate high-temperature heat and electricity: **(a)** power tower, **(b)** solar thermal plant, **(c)** parabolic dish collection system, **(d)** nonimaging optical solar concentrator, and **(e)** solar stoves. List the advantages and disadvantages of concentrating solar energy to produce high-temperature heat or electricity.

15. What is a *solar cell*? List the advantages and disadvantages of using solar cells to produce electricity.

16. Distinguish between *large-scale hydropower*, *small-scale-hydropower*, and *pumped-storage hydropower* systems. List the advantages and disadvantages of using hydropower to produce electricity.

17. List the advantages and disadvantages of using the following systems for storing heat in water to produce electricity: **(a)** ocean thermal energy conversion (OTEC), **(b)** saline solar ponds, and **(c)** freshwater solar ponds.

18. List the advantages and disadvantages of using wind to produce electricity. What parts of the world have the greatest potential for using wind resources?

19. List the advantages and disadvantages of **(a)** burning solid biomass as a source of energy and **(b)** producing gaseous and liquid fuels from solid biomass.

20. What is the *solar-hydrogen revolution*? Describe seven ways to produce hydrogen and five ways to store hydrogen. What is a *fuel cell*, and what are the advantages and disadvantages of using this technology? List the advantages and disadvantages of using hydrogen as a source of energy.

21. What is *geothermal energy*? Describe three types of geothermal reservoirs. List the advantages and disadvantages of using geothermal energy to produce heat and electricity.

22. What is *micropower*, and what are its advantages over macropower electricity systems? Describe five types of micropower systems.

23. What four conclusions have energy experts reached about possible future energy alternatives?

24. Summarize the three different economic approaches that can be used to stimulate or dampen the use of a particular energy resource. List the pros and cons of each approach.

25. What are major ways to help make the transition to a more sustainable energy future?

CRITICAL THINKING

1. A homebuilder installs electric baseboard heat and claims that "it's the cheapest and cleanest way to go." Apply your understanding of the second law of energy (thermodynamics) to evaluate his claim.

2. Someone tells you that we can save energy by recycling it. How would you respond?

3. Should the Corporate Average Fuel Economy (CAFE) standards for motor vehicles used in the United States be increased, left at 1985 levels (the current situation), or eliminated? Explain. Should the CAFE standards for light trucks, vans, and sport utility vehicles be increased to the same level as for cars? Explain. List the positive and negative effects on your health and lifestyle if CAFE standards are **(a)** increased or **(b)** eliminated.

4. What are the five most important things an individual can do to save energy at home and in transportation (see Appendix 6)? Which, if any, of these do you currently do? Which, if any, do you plan to do?

5. Congratulations. You have just won $150,000 to build a house of your choice anywhere in the country. What type of house would you build? Where would you locate it? What types of materials would you use? What types of materials would you *not* use? How would you heat and cool your house? How would you heat your water? Considering fuel and energy efficiency, what sort of lighting, stove, refrigerator, washer, and dryer would you use? Which of these appliances could you do without?

6. Explain why you agree or disagree with the following proposals by various energy analysts: **(a)** Federal subsidies for all energy alternatives should be eliminated so that all energy choices can compete in a true free-market system, **(b)** all government tax breaks and other subsidies for conventional fuels (oil, natural gas, coal), synthetic natural gas and oil, and nuclear power (fission and fusion) should be removed and replaced with subsidies and tax breaks for improving energy efficiency and developing solar, wind, geothermal, and biomass energy alternatives, and **(c)** development of solar and wind energy should be left to private enterprise and receive little or no help from the federal government, but nuclear energy and fossil fuels should continue to receive large federal subsidies.

7. Explain why you agree or disagree with the proposals suggested in Figure 15-42 (p. 392) as ways to promote a more sustainable energy future.

8. Congratulations. You have just been put in charge of the world. List the five most important features of your energy policy.

PROJECTS

1. Make a study of energy use in your school and use the findings to develop an energy-efficiency improvement program. Present your plan to school officials.

2. Learn how easy it is to produce hydrogen gas from water using a battery, some wire for two electrodes, and a dish of water. Hook a wire to each of the poles of the battery, immerse the electrodes in the water, and observe bubbles of hydrogen gas being produced at the negative electrode and bubbles of oxygen at the positive electrode. Carefully add a small amount of battery acid to the water and notice that this increases the rate of hydrogen production.

3. Use the library or the internet to find bibliographic information about *Amory Lovins, Seth Dunn,* and *Christopher Flavin,* whose quotes appear at the beginning and end of this chapter.

4. Make a concept map of this chapter's major ideas, using the section heads and subheads and the key terms (in boldface). Look at the inside back cover and on the website for this book for information about making concept maps.

INTERNET STUDY RESOURCES AND RESOURCES FOR FURTHER READING AND RESEARCH

The website for this book contains helpful study aids and many ideas for further reading and research. Log on to:

http://www.brookscole.com/product/0534376975s

and click on the Chapter-by-Chapter area. Choose Chapter 15 and select a resource:

- "Flash Cards" allows you to test your mastery of the Terms and Concepts to Remember for this chapter.

- "Tutorial Quizzes" provides a multiple-choice practice quiz.

- "Student Guide to InfoTrac" will lead you to Critical Thinking Projects that use InfoTrac College Edition as a research tool.

- "References" lists the major books and articles consulted in writing this chapter.

- "Hypercontents" takes you to an extensive list of sites with news, research, and images related to individual sections of the chapter.

INFOTRAC COLLEGE EDITION

Improve your skills with InfoTrac College Edition, a searchable online database of articles from more than 700 periodicals. Log on to:

http://www.infotrac-college.com

or access InfoTrac through the website for this book.

Try the following articles:

Rosentreter, R. 2000. Oil, profits, and the question of alternative energy. *The Humanist* vol. 60, no. 5, pp. 8–13. (subject guide: alternative energy)

Bond, M. 2000. Solar energy: seeing the light (industry overview). *Geographical* vol. 72, no. 11, pp. 28–31. (subject guide: solar energy)

PART IV

ENVIRONMENTAL QUALITY AND POLLUTION

In our every deliberation, we must consider the impact of our decisions on the next seven generations.

IROQUOIS CONFEDERATION, 18TH CENTURY

16 RISK, TOXICOLOGY, AND HUMAN HEALTH

The Big Killer

What is roughly the diameter of a 30-caliber bullet, can be bought almost anywhere, is highly addictive, and kills about 11,000 people every day, or 460 per hour? It's a cigarette. *Cigarette smoking is the single most preventable major cause of death and suffering among adults.*

The World Health Organization (WHO) estimates that each year tobacco contributes to the premature deaths of at least 4 million people from heart disease, lung cancer, other cancers, bronchitis, emphysema, and stroke. The annual death toll from smoking-related diseases is projected to reach 10 million by 2030 (70% of them in developing countries)—an average of about 27,400 preventable deaths per day. In China alone, health experts project that 2 billion people will die prematurely each year from tobacco-related causes by 2020.

Smoking kills about 431,000 Americans per year, an average of 1,180 deaths per day (Figure 16-1). This death toll is roughly equivalent to three fully loaded jumbo (400-passenger) jets crashing every day with no survivors. Smoking causes more deaths each year in the United States than do all illegal drugs, alcohol (the second most harmful legal drug after nicotine),

automobile accidents, suicide, and homicide combined (Figure 16-1).

According to a 1998 study, secondhand smoke (inhaled by nonsmokers) causes 30,000–60,000 premature deaths per year in the United States. Each year, parental smoking prematurely kills an estimated 6,000 children and causes 5.4 million serious child ailments in the United States.

The overwhelming consensus in the scientific community is that the nicotine (and probably the acetaldehyde) inhaled in tobacco smoke is highly addictive. Only 1 in 10 people who try to quit smoking succeed—about the same relapse rate as for recovering alcoholics and those addicted to heroin or crack cocaine. A British government study showed that adolescents who smoke more than one cigarette have an 85% chance of becoming smokers.

Worldwide, the cost of treating smoking-related illnesses is estimated at $200 billion a year, and in the United States $70–100 billion a year is spent on **(1)** medical bills, **(2)** increased insurance costs, **(3)** disability, **(4)** lost earnings and productivity because of illness, and **(5)** property damage from smoking-caused fires. This is an average of $3–4 per pack of cigarettes sold in the United States.

Many health experts urge that a $2–4 federal tax be added to the price of a pack of cigarettes in the United States. In England a pack of cigarettes costs about $5, versus $2 in the United States. Such a tax would mean that the users of cigarettes (and other tobacco products), not the rest of society, would pay a much greater share of the health, economic, and social costs associated with their smoking: a *user-pays* approach.

Other suggestions for reducing the death toll and health effects of smoking in the United States include **(1)** banning all cigarette advertising, **(2)** forbidding the sale of cigarettes and other tobacco products to anyone under 21 (with strict penalties for violators), **(3)** banning all cigarette vending machines, **(4)** classifying nicotine as an addictive and dangerous drug (and placing its use in tobacco or other products under the jurisdiction of the Food and Drug Administration), **(5)** eliminating all federal subsidies and tax breaks to U.S. tobacco farmers and tobacco companies, and **(6)** using cigarette tax income to finance a massive antitobacco advertising and education program.

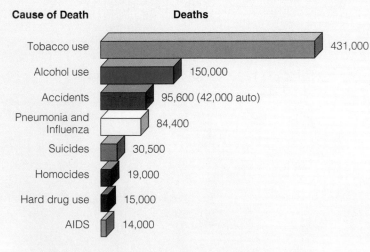

Cause of Death | **Deaths**

- Tobacco use — 431,000
- Alcohol use — 150,000
- Accidents — 95,600 (42,000 auto)
- Pneumonia and Influenza — 84,400
- Suicides — 30,500
- Homocides — 19,000
- Hard drug use — 15,000
- AIDS — 14,000

Figure 16-1 Annual deaths in the United States from tobacco use and other causes. Smoking is by far the nation's leading cause of preventable death, causing more premature deaths each year than all the other categories in this figure combined. (Data from National Center for Health Statistics)

The dose makes the poison.

PARACELSUS, 1540

This chapter addresses the following questions:

- What types of hazards do people face?

- What is toxicology, and how do scientists measure toxicity?

- What chemical hazards do people face, and how can they be measured?

- What types of disease (biological hazards) threaten people in developing countries and developed countries?

- How can risks be estimated, managed, and reduced?

16-1 RISK, PROBABILITY, AND HAZARDS

What Is Risk? **Risk** is the possibility of suffering harm from a hazard that can cause injury, disease, economic loss, or environmental damage. Risk is expressed in terms of **probability**: a mathematical statement about how likely it is that some event or effect will occur. In these terms, **risk** is defined as the probability of exposure times the probability of harm (Risk = Exposure × Harm).

Probability often is stated in terms such as "The lifetime probability of developing cancer from exposure to a certain chemical is 1 in 1 million." This means that one of every 1 million people exposed to the chemical at a specified average daily dosage will develop cancer over a typical lifetime (usually considered to be 70 years).

It is important to distinguish between *possibility* and *probability*. Saying that an event or effect is possible means that it could occur—a very inexact statement. Probability describes in mathematical terms how likely it is that the possible event or effect will occur and how likely it is to cause harm.

How Are Risks Assessed and Managed? Nothing we do is completely safe or entirely free from potential harm. Thus, individuals and government regulators have to assess the risk from a particular hazard to determine whether there is a *low risk* or a *high risk* of harm.

Risk assessment involves **(1)** identifying a real or potential hazard ("What is the hazard?"), **(2)** determining the probability of its occurrence ("How likely is the event?"), and **(3)** assessing the severity of its health, environmental, economic, and social impact ("How much damage is it likely to cause?"; Figure 16-2).

Figure 16-2 Risk assessment and risk management. These are important, difficult, and controversial processes.

This is a complex, difficult, and controversial process. For example, assessing the risk of exposure to a toxic chemical involves estimating **(1)** the number of people or other organisms exposed, **(2)** the level and duration of exposure, and **(3)** other possible contributing factors such as age, health, sex, personal habits, and interactions with other chemicals.

After a risk has been assessed, the next step is **risk management**, in which people make decisions about **(1)** how serious it is compared to other risks (*comparative risk analysis*), **(2)** how much (if at all) the risk should be reduced, **(3)** how such risk reduction can be accomplished, and **(4)** how much money should be devoted to reducing the risk to an acceptable level (Figure 16-2). This is even more difficult and controversial than risk assessment because of a lack of information and the economic, health, and political implications of such decisions.

What Are the Major Types of Hazards? The various kinds of hazards we face can be categorized as follows:

- *Cultural hazards* such as unsafe working conditions, smoking (left), poor diet, drugs, drinking, driving, criminal assault, unsafe sex, and poverty. Some *good news* is that between 1900 and 2000, deaths from industrial accidents in the United States decreased from about 35,000 per year to 6,100 per year despite an almost fourfold increase in population.

- *Chemical hazards* from harmful chemicals in the air (Chapter 17), water (Chapter 19), soil (p. 227), and food (p. 271). The bodies of most human beings contain small amounts of about 500 synthetic organic chemicals—whose health effects are mostly unknown—that did not exist in 1920.

- *Physical hazards* such as ionizing radiation (p. 62), fire, earthquake (p. 217), volcanic eruption (p. 218), flood (p. 314), tornadoes, and hurricanes.

- *Biological hazards* from pathogens (bacteria, viruses, and parasites), pollen and other allergens, and animals such as bees and poisonous snakes.

According to a 1998 study by Cornell University scientist David Pimentel (Guest Essay, p. 232), environmental factors such as malnutrition, smoking, cooking fires, skin cancer, exposures to pesticides and other hazardous chemicals, and air and water pollution contribute to about 40% of the world's annual deaths.

16-2 TOXICOLOGY

What Determines Whether a Chemical Is Harmful? Dose and Response Toxicity is a measure of how harmful a substance is. Whether a chemical (or other agent such as ionizing radiation) is harmful depends on several factors. One is the **dosage**, the amount of a potentially harmful substance that a person has ingested, inhaled, or absorbed through the skin. Whether a chemical is harmful depends on **(1)** the size of the dosage over a certain period of time, **(2)** how often an exposure occurs, **(3)** who is exposed (adult or child, for example), **(4)** how well the body's detoxification systems (liver, lungs, and kidneys) work, and **(5)** genetic makeup that determines an individual's sensitivity to a particular toxin (Figure 16-3).

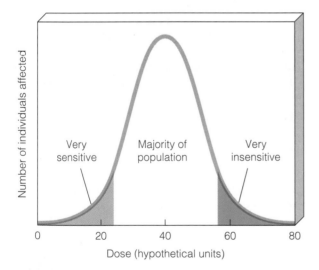

Figure 16-3 Typical variations in sensitivity to a toxic chemical within a population, mostly because of differences in genetic makeup. Some individuals in a population are very sensitive to small does of a toxin (left), and others are very insensitive (right). Most people fall between these two extremes (middle).

This genetic variation in individual responses to exposure to various toxins raises a difficult ethical, political, and economic question. When regulating levels of a toxin in the environment, should the allowed level be set to protect **(1)** the most sensitive individuals (at great cost) or **(2)** the average person?

The harm caused by a substance can also be affected by:

- *Solubility. Water-soluble toxins* (which are often inorganic compounds) can move throughout the environment and get into water supplies. *Oil- or fat-soluble toxins* (which are usually organic compounds) can accumulate in body tissues and cells.

- *Persistence.* Many chemicals, such as plastics, chlorofluorocarbons (CFCs), chlorinated hydrocarbons, and plastics, are widely used because of their persistence or resistance to breakdown. However, this persistence also means that they can have long-lasting effects on the health of wildlife and people.

- **Bioaccumulation**, in which some molecules are absorbed and stored in specific organs or tissues at levels higher than normally would be expected.

- **Biomagnification**, in which the levels of some toxins in the environment are magnified as they pass through food chains and webs (Figure 16-4). Examples of chemicals that can be biomagnified include long-lived, fat-soluble organic compounds such as **(1)** the pesticide DDT, **(2)** PCBs (oily chemicals used in electrical transformers), and **(3)** some radioactive isotopes (such as strontium-90, Table 3-2, p. 62). Stored in body fat, such chemicals can be passed along to offspring during gestation or egg laying and as mothers nurse their young.

- *Chemical interactions* that can decrease or multiply the harmful effects of a toxin. An *antagonistic interaction* can reduce the harmful response. For example, vitamins E and A apparently interact to reduce the body's response to some carcinogens. A *synergistic interaction* (p. 52) multiplies harmful effects. For example, workers exposed to asbestos increase their chances of getting lung cancer 20-fold. However, asbestos workers who also smoke have a 400-fold increase in lung-cancer rates.

The type and amount of health damage that result from exposure to a chemical or other agent are called the **response**. An *acute effect* is an immediate or rapid harmful reaction to an exposure; it can range from dizziness or a rash to death. A *chronic effect* is a permanent or long-lasting consequence (kidney or liver damage, for example) of exposure to a harmful substance.

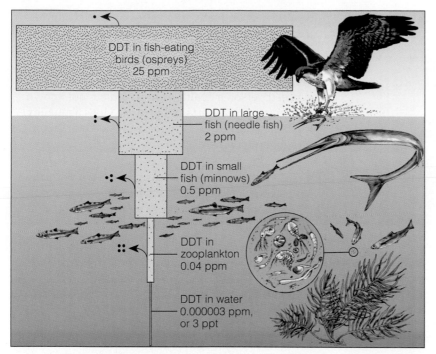

Figure 16-4 *Bioaccumulation* and *biomagnification*. DDT is a fat-soluble chemical that can bioaccumulate in the fatty tissues of animals. In a food chain or food web, the accumulated concentrations of DDT can be biologically magnified in the bodies of animals at each higher trophic level. This diagram shows that the concentration of DDT in the fatty tissues of organisms was biomagnified about 10 million times in this food chain in an estuary near Long Island Sound. If each phytoplankton organism in such a food chain takes up from the water and retains one unit of DDT, a small fish eating thousands of zooplankton (which feed on the phytoplankton) will store thousands of units of DDT in its fatty tissue. Then each large fish that eats 10 of the smaller fish will ingest and store tens of thousands of units, and each bird (or human) that eats several large fish will ingest hundreds of thousands of units. Dots represent DDT, and arrows show small losses of DDT through respiration and excretion.

Should We Be Concerned About Trace Levels of Toxic Chemicals in the Environment and in Our Bodies? The answer is that it depends on the chemical and its concentration. The detection of trace amounts of a chemical in air, water, or food does not necessarily mean that it is there at a level harmful to most people or to wildlife.

A basic concept of toxicology is that any synthetic or natural chemical (even water) can be harmful if ingested in a large enough quantity. Drinking 100 cups of strong coffee one after another would expose most people to a lethal dosage of caffeine. Similarly, downing 100 tablets of aspirin or 1 liter (1.1 quarts) of pure alcohol (ethanol) would kill most people.

The critical question is how much exposure to a particular toxic chemical causes a harmful response. This is the meaning of the quote by German scientist Paracelsus about the dose making the poison (found at the beginning of this chapter).

Most chemicals have some safe or threshold level of exposure below which their harmful effects are insignificant because:

- The human body has mechanisms for breaking down (usually by enzymes found in the liver), diluting, or excreting small amounts of most toxins to keep them from reaching harmful levels.

- Individual cells have enzymes that can repair damage to DNA and protein molecules.

- Cells in some parts of the body (such as the skin and linings of the gastrointestinal tract, lungs, and blood vessels) reproduce fast enough to replace damaged cells. However, such high rates of cell reproduction can sometimes be altered by exposure to ionizing radiation and certain chemicals so that cell growth accelerates and creates a nonmalignant or malignant (cancerous) tumor.

Some people have the mistaken idea that all natural chemicals are safe and all synthetic chemicals are harmful. In fact, many synthetic chemicals are quite safe if used as intended, and many natural chemicals are deadly. For example, the average person is far more likely to be killed by aflatoxin in peanut butter than by lightning or a shark. However, the chance of dying from eating several spoonfuls of peanut butter a day is quite small.

In addition, the ability of chemists to detect increasingly small amounts of potentially toxic chemicals in air, water, and food can give the false impression that dangers from toxic chemicals are increasing. In 1980, chemists could routinely detect concentrations of substance in parts per million (ppm) Table 3-1, p. 61. By 1990, chemists could detect parts per billion (ppb) and today they can detect concentrations of parts per trillion (ppt) and in some cases parts per quadrillion (ppq).

What Is a Poison? Legally, a **poison** is a chemical that has an LD_{50} of 50 milligrams or less per kilogram of body weight. The LD_{50} is the **median lethal dose**: the amount of a chemical received in one dose that kills exactly 50% of the animals (usually rats and mice) in a test population (usually 60–200 laboratory animals) within a 14-day period (Figure 16-5).

Chemicals vary widely in their toxicity (Table 16-1). Some poisons can cause serious harm or death after a single acute exposure at extremely low dosages.

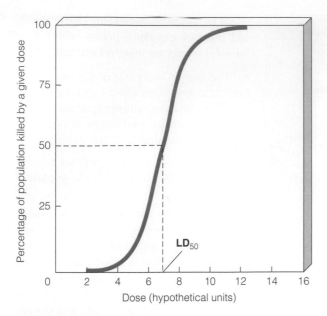

Figure 16-5 Hypothetical dose–response curve showing determination of the LD$_{50}$, the dosage of a specific chemical that kills 50% of the animals in a test group.

Others cause such harm only at such huge dosages that it is nearly impossible to get enough into the body. Most chemicals fall between these two extremes.

What Methods Do Scientists Use to Determine Toxicity? Scientists use three methods to determine the level at which a substance poses a health threat:

- *Case reports* (usually made by physicians) provide information about people suffering some adverse

health effect or death after exposure to a chemical. Such information often involves accidental poisonings, drug overdoses, homicides, or suicide attempts. Most case reports are not a reliable source for determining toxicity because the actual dosage and the exposed person's health status often are not known. However, such reports can provide clues about environmental hazards and suggest the need for laboratory investigations.

- *Laboratory investigations* (usually on test animals) are used to determine **(1)** toxicity, **(2)** residence time, **(3)** what parts of the body are affected, and **(4)** sometimes how the harm takes place.

- **Epidemiology** ("ep-i-deem-ee-OL-oh-gee") in populations of humans exposed to certain chemicals or diseases is used to find out why some people get sick and others do not.

How Are Laboratory Experiments Used to Determine Toxicity? Acute toxicity and chronic toxicity usually are determined by exposing a population of live laboratory animals (especially mice and rats, which are small and prolific and can be housed inexpensively in large numbers) to measured doses of a specific substance under controlled conditions. Animal tests take 2–5 years and cost $200,000 to $2 million per substance tested.

Animal welfare groups want to limit or ban all use of test animals (or to ensure that experimental animals are treated in the most humane manner possible). More humane methods for carrying out toxicity tests include

Table 16-1 Toxicity Ratings and Average Lethal Doses for Humans

Toxicity Rating	LD$_{50}$ (milligrams per kilogram of body weight)*	Average Lethal Dose†	Examples
Supertoxic	Less than 0.01	Less than 1 drop	Nerve gases, botulism toxin, mushroom toxins, dioxin (TCDD)
Extremely Toxic	Less than 5	Less than 7 drops	Potassium cyanide, heroin, atropine, parathion, nicotine
Very Toxic	5–50	7 drops to 1 teaspoon	Mercury salts, morphine, codeine
Toxic	50–500	1 teaspoon to 1 ounce	Lead salts, DDT, sodium hydroxide, sodium fluoride, sulfuric acid, caffeine, carbon tetrachloride
Moderately Toxic	500–5,000	1 ounce to 1 pint	Methyl (wood) alcohol, ether, phenobarbital, amphetamines (speed), kerosene, aspirin
Slightly toxic	5,000–15,000	1 pint to 1 quart	Ethyl alcohol, Lysol, soaps
Essentially nontoxic	15,000 or greater	More than 1 quart	Water, glycerin, table sugar

*Dosage that kills 50% of individuals exposed
†Amounts of substances that are liquids at room temperature when given to a 70.4-kilogram (155-Pound) human

using **(1)** bacteria, **(2)** cell and tissue cultures, and **(3)** chicken egg membranes. In 1999, scientists developed a cheaper and much more sensitive way to determine toxicity by measuring changes in the electrical properties of individual animal cells every quarter-second 24 hours a day.

These alternatives can greatly decrease the use of animals for testing toxicity. However, scientists point out that some animal testing is needed because the alternative methods cannot adequately mimic the complex biochemical interactions of a live animal.

Acute toxicity tests are run to develop a **dose–response curve**, which shows the effects of various dosages of a toxic agent on a group of test organisms (Figure 16-6). Such tests are *controlled experiments* in which the effects of the chemical on a *test group* are compared with the responses of a *control group* of organisms not exposed to the chemical. Care is taken to ensure that organisms in each group are as identical as possible in age, health status, and genetic makeup and that they are exposed to the same environmental conditions.

Fairly high dosages are used to reduce the number of test animals needed, obtain results quickly, and lower costs. Otherwise, tests would have to be run on millions of laboratory animals for many years, and manufacturers could not afford to test most chemicals. For the same reasons, the results of high-dose exposures usually are extrapolated to low-dose levels using mathematical models. Then the extrapolated low-dose results on the test organisms are extrapolated to humans to estimate LD_{50} values for acute toxicity (Table 16-1).

According to the *nonthreshold dose–response model* (Figure 16-6, left), any dosage of a toxic chemical or ionizing radiation causes harm that increases with the

dosage. Many chemicals that cause birth defects or cancers show this kind of response.

With the *threshold dose–response model* (Figure 16-6, right) there is a threshold dosage before any detectable harmful effects occur, presumably because the body can repair the damage caused by low dosages of some substances. It is extremely difficult to establish which of these models applies at low dosages. To be on the safe side, the nonthreshold dose–response model often is assumed.

Some scientists challenge the validity of extrapolating data from test animals to humans because human physiology and metabolism often are different from those of the test animals. Also, different species of test animals can react differently to the same toxin because of differences in body size, physiology, metabolism, and toxin sensitivity (Figure 16-3). Other scientists counter that such tests and models work fairly well (especially for revealing cancer risks) when the correct experimental animal is chosen or when a chemical is toxic to several different test animal species.

How Is Epidemiology Used to Determine Toxicity? In an *epidemiological study*, the health of people exposed to a particular toxic agent or disease organism (the experimental group) is compared with the health of another group of statistically similar people not exposed to these conditions (the control group). The goal of such studies to establish a strong, moderate, weak, or no statistical association between a hazard and a health problem.

Three major limitations of epidemiology are that **(1)** too few people have been exposed to sufficiently high levels of many toxic agents to allow detection of statistically significant differences, **(2)** conclusively linking an observed effect with exposure to a particular hazard is very difficult because people are exposed to many different toxic agents and disease-causing factors throughout their lives, and **(3)** it cannot be used to evaluate hazards from new technologies, substances, or diseases to which people have not been exposed.

How Valid Are Estimates of Toxicity? As we have seen, all methods for estimating toxicity levels and risks have serious limitations. However, they are all we have. To take this uncertainty into account and minimize harm, standards for allowed exposure to toxic substances and ionizing radiation typically are set at levels 1/100 or even 1/1,000 of the estimated harmful levels.

Despite their many limitations, carefully conducted and evaluated toxicity studies are important sources of information we can use to understand dose–response effects and to estimate and set exposure standards. However, citizens, lawmakers, and regulatory officials must recognize the huge uncertainties and guesswork involved in all such studies.

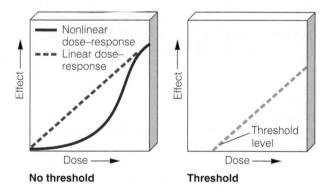

Figure 16-6 Hypothetical dose–response curves. The linear and nonlinear curves in the left graph show that exposure to any dosage of a chemical or ionizing radiation has a harmful effect that increases with the dosage. The curve on the right shows that a harmful effect occurs only when the dosage exceeds a certain *threshold level*. There is much uncertainty about which of these models applies to various harmful agents because of the difficulty in estimating the response to very low dosages. (Adapted from *Environmental Science*, 5/E by Chiras, p. 352, fig. 19.12. Copyright ©1998 by Wadsworth)

16-3 CHEMICAL HAZARDS

What Are Toxic and Hazardous Chemicals? **Toxic chemicals** generally are defined as substances that are fatal to more than 50% of test animals (LD_{50}) at given concentrations. **Hazardous chemicals** cause harm by (1) being flammable or explosive, (2) irritating or damaging the skin or lungs (strong acidic or alkaline substances such as oven cleaners), (3) interfering with or preventing oxygen uptake and distribution (asphyxiants such as carbon monoxide and hydrogen sulfide), or (4) inducing allergic reactions of the immune system (allergens).

What Are Mutagens? **Mutagens** are agents, such as chemicals and ionizing radiation, that cause random *mutations*, or changes in the DNA molecules found in cells. Mutations in a sperm or egg cell can be passed on to future generations and cause diseases such as manic depression, cystic fibrosis, hemophilia, sickle-cell anemia, Down's syndrome, and some types of cancer. Mutations in other cells are not inherited but may cause harmful effects.

Most mutations are harmless, probably because all organisms have biochemical repair mechanisms that can correct mistakes or changes in the DNA code. In addition, some mutations play a vital role in evolution (p. 107). There is no agreement on the best ways to test substances for genetic damage in humans.

What Are Teratogens? **Teratogens** are chemicals, radiation, or viruses that cause birth defects while the human embryo is growing and developing during pregnancy, especially during the first 3 months. Chemicals known to cause birth defects in laboratory animals include PCBs, thalidomide, steroid hormones, and heavy metals such as arsenic, cadmium, lead (p. 540), and mercury.

What Are Carcinogens? According to the WHO, environmental and lifestyle factors play a key role in causing or promoting up to 80% of all cancers. Major sources of carcinogens are cigarette smoke (30–40% of cancers), diet (20–40%), occupational exposure (5–15%), and environmental pollutants (1–10%). Inherited genetic factors and certain viruses cause about 10–20% of all cancers.

Carcinogens are chemicals, radiation, or viruses that cause or promote the growth of a malignant (cancerous) tumor, in which certain cells multiply uncontrollably. Many cancerous tumors spread by **metastasis** when malignant cells break off from tumors and travel in body fluids to other parts of the body. There, they start new tumors, making treatment much more difficult.

Because there are more than 100 types of cancer (depending on the types of cells involved), there are many different causes. These include genetic predisposition, viral infections, and exposure to various mutagens and carcinogens.

Typically, 10–40 years may elapse between the initial exposure to a carcinogen and the appearance of detectable symptoms. Partly because of this time lag, many healthy teenagers and young adults have trouble believing that their smoking (p. 396), drinking, eating, and other lifestyle habits today could lead to some form of cancer before they reach age 50.

How Can Chemicals Harm the Immune, Nervous, and Endocrine Systems? Since the 1970s there has been a growing body of research on wildlife and laboratory animals and epidemiological studies of humans indicating that long-term (often low-level) exposure to various toxic chemicals in the environment can disrupt the body's immune, nervous, and endocrine systems.

The *immune system* consists of specialized cells and tissues that protect the body against disease and harmful substances by forming antibodies to invading agents and rendering them harmless. Viruses such as the human immunodeficiency virus (HIV), ionizing radiation, malnutrition, and some synthetic chemicals can weaken the human immune system. This can leave the body wide open to attacks by allergens, infectious bacteria, viruses, and protozoans. Recent studies of laboratory animals and wildlife as well as epidemiological studies of humans (especially in developing countries) have linked immune system suppression to several widely used pesticides.

Synthetic chemicals in the environment threaten the human *nervous system* (brain, spinal cord, and peripheral nerves). Many poisons are *neurotoxins*, which attack nerve cells (neurons). Examples are (1) chlorinated hydrocarbons (DDT, PCBs, dioxins), (2) organophosphate pesticides (Table 20-1, p. 504), (3) formaldehyde, (4) various compounds of arsenic, mercury, lead, and cadmium, and (5) widely used industrial solvents such as trichloroethylene (TCE), toluene, and xylene.

The *endocrine system* is a complex network of glands and hormones that regulates many of the body's functions. Each type of hormone has a specific molecular shape that allows it to attach only to certain cell receptors (Figure 16-7, left). Once bonded together, the hormone and its receptor molecule move to the cell's nucleus to execute the chemical message carried by the hormone.

The endocrine glands release extremely small amounts of *hormones* into the bloodstream that act as natural chemical messengers to control body functions such as sexual reproduction, growth, development, and behavior in humans and other animals. These naturally occurring hormones have profound effects on the human nervous, reproductive, and immune systems. There is concern that human exposure to low levels of synthetic chemicals, known as *hormonally active agents* (HAAs), can mimic and disrupt the effects of natural hormones (Connections, p. 404).

Why Do We Know So Little About the Harmful Effects of Chemicals? According to risk assessment expert Joseph V. Rodricks, "Toxicologists know a great deal about a few chemicals, a little about many, and next to nothing about most." The U.S. National Academy of Sciences estimates that only about 10% of at least 75,000 chemicals in commercial use have been thoroughly screened for toxicity, and only 2% have been adequately tested to determine whether they are carcinogens, teratogens, or mutagens. Hardly any of the chemicals in commercial use have been screened for damage to the nervous, endocrine, and immune systems.

Each year about 1,000 new synthetic chemicals are introduced into the marketplace, with little knowledge about their potentially harmful effects. Currently, federal and state governments do not regulate about 99.5% of the commercially used chemicals in the United States. There are three major reasons for this lack of information and regulation.

- Under existing laws most chemicals are considered innocent until proven guilty. No one is required to investigate whether they are harmful.

- There are not enough funds, personnel, facilities, and test animals to provide such information for more than a small fraction of the many chemicals we encounter in our daily lives.

- It is too difficult and expensive to analyze the combined effects of multiple exposures to various chemicals and the possible interactions of such chemicals. For example, just studying the possible different three-chemical interactions of the 500 most widely used industrial chemicals would take 20.7 million experiments—a physical and financial impossibility.

What Is the Precautionary Approach? The difficulty and expense of getting information about the harmful effects of chemicals are one reason an increasing number of scientists and health officials are pushing for much greater emphasis on *pollution prevention*. This strategy greatly reduces the need for statistically uncertain and controversial toxicity studies and exposure standards. It also reduces the risk posed by potentially hazardous chemicals and products and their possible but poorly understood multiple interactions.

This approach is based on the **precautionary principle**. According to this concept, when there is considerable scientific uncertainty about potentially serious harm from chemicals or technologies, decision makers should act to prevent harm to humans and the environment. It is based on familiar axioms: Look before you leap, better safe than sorry, and an ounce of prevention is worth a pound of cure.

Under this approach, those proposing to introduce a new chemical or technology would bear the burden

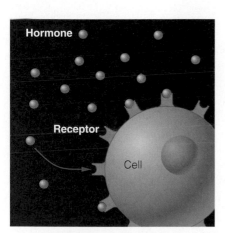

Normal Hormone Process

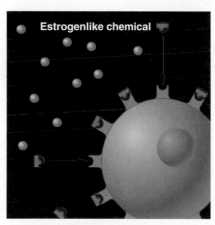

Hormone Mimic

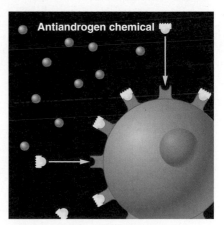

Hormone Blocker

Figure 16-7 Hormones are molecules that act as messengers in the endocrine system to regulate various bodily processes, including reproduction, growth, and development. Each type of hormone has a unique molecular shape that allows it to attach to specially shaped receptors on the surface of or inside cells and transmit its chemical message (left). Molecules of certain pesticides and other molecules have shapes similar to those of natural hormones. Some of these hormone impostors, called *hormone mimics*, disrupt the endocrine system by attaching to estrogen receptor molecules (center). Others, called *hormone blockers*, prevent natural hormones such as androgens (male sex hormones) from attaching to their receptors (right). Some pollutants called *thyroid disrupters* may disrupt hormones released by thyroid glands and cause growth and weight disorders and brain and behavioral disorders.

Are Hormone Disrupters and Mimics a Health Threat?

CONNECTIONS

Over the last 15 years experts from a number of disciplines have been piecing together field studies on wildlife, studies on laboratory animals, and epidemiological studies of human populations. This analysis suggests that a variety of human-made chemicals can act as *hormone* or *endocrine disrupters*, known as *hormonally active agents (HAAs)*. By 1998, 60 endocrine disrupters had been identified, and with further testing and screening the list of HAAs could reach several hundred.

Some, called *hormone mimics*, are estrogenlike chemicals that disrupt the endocrine system by attaching to estrogen receptor molecules (Figure 16-7, center).

Others, called *hormone blockers*, disrupt the endocrine system by preventing natural hormones such as androgens (male sex hormones) from attaching to their receptors (Figure 16-7, right). There is also growing concern about pollutants that can act as *thyroid disrupters* and cause growth, weight, brain, and behavioral disorders.

Most natural hormones are broken down or excreted. However, many synthetic hormone impostors are stable, fat-soluble compounds whose concentrations can be biomagnified as they move through food chains and webs (Figure 16-4). Thus, they can pose a special threat to humans and other carnivores dining at the top of food webs.

Numerous wildlife and laboratory studies reveal various possible effects of estrogen mimics and hormone blockers (HAAs). Here are a few of many examples:

- Ranch minks fed Lake Michigan fish contaminated with endocrine disrupters such as DDT and PCBs failed to reproduce.
- Exposure to PCBs has reduced penis size in some test animals and in 118 boys born to women who were exposed to a PCB spill in Taiwan in 1979.
- A 1999 study by Michigan State University zoologists found that female rats exposed to PCBs were reluctant to mate, raising the possibility that such contaminants could cause low sex drives in women.
- In 1973, estrogen mimics called PBBs accidentally got into cattle feed in Michigan, and from there into beef. Pregnant women who ate the beef (and whose breast milk had high levels of PBBs) had sons with undersized penises and malformed testicles.
- During the past 50 years there have been dramatic increases in testicular and prostate cancer in humans almost everywhere.
- Average sperm counts among men in the United States and Europe have declined by 50% during the past 60 years.

In 1999, the U.S. National Academy of Sciences released a report based on a 4-year review of the scientific literature on hormone disrupters (HAAs). The panel concluded that far too little is known about the effects of such chemicals to come to a definitive conclusion about their effects on humans.

Scientists on this panel called for greatly increased research to (1) verify current frontier science findings and (2) determine whether low levels of most hormone-disrupting chemicals in the environment pose a threat to the human population. However, the report also concluded that at present the at least 75,000 industrial chemicals in commercial use cannot be tested to determine whether they are hormone disrupters because the necessary tests do not exist.

If such research (which will take decades) shows that exposure to small amounts of hormone disrupters is harmful to humans and some forms of wildlife, the only reasonable choice may be to prevent such chemicals from reaching the environment. This will be a difficult and controversial economic and political decision.

Some health scientists believe that we should begin sharply reducing the use of potential hormone disrupters now because they meet the two requirements of the *precautionary principle*: great scientific uncertainty and a reasonable suspicion of harm.

Critical Thinking

1. Do you consider the possible threat from hormone disrupters a problem that could affect you or any child you might have? Explain.

2. Do you believe that the precautionary approach should be used to deal with this problem while more definite research is carried out over the next two decades? Explain. What harmful effects could using this approach have on the economy and on your lifestyle? Do such effects outweigh the risks? Explain.

---

of establishing its safety. In other words, new chemicals and technologies would be assumed to be guilty until proven innocent. Manufacturers and businesses contend that doing this would make it too expensive and almost impossible to introduce any new chemical or technology.

16-4 BIOLOGICAL HAZARDS: DISEASE IN DEVELOPED AND DEVELOPING COUNTRIES

What Are Nontransmissible Diseases? A **nontransmissible disease** is not caused by living organisms and does not spread from one person to another. Examples are cardiovascular (heart and blood vessel) disorders, most cancers, diabetes, bronchitis, emphysema, and malnutrition. Such diseases typically have multiple (and often unknown) causes and tend to develop slowly and progressively over time.

What Are Transmissible Diseases? A *transmissible disease* is caused by a living organism (such as a bacterium, virus, protozoa, or parasite) and can be spread from one person to another. These infectious agents are called *pathogens* and are spread by air, water, food, body fluids, some insects, and other nonhuman carriers called *vectors*.

Typically, a *bacterium* is a one-celled microorganism capable of replicating itself by simple cell division. A *virus* is a microscopic, noncellular infectious agent. Its DNA contains instructions for making more viruses, but it has no apparatus to do so. To replicate, a virus must invade a host cell and take over the cell's DNA to create a factory for producing more viruses.

According to the WHO and UNICEF, every year in developing countries at least 2 million children under age 5 die of mostly preventable infectious diseases—an average of at least 5,500 premature deaths per day. About 80% of all illnesses in developing countries are caused by waterborne infectious diseases (such as diarrhea, hepatitis, typhoid fever, and cholera), mainly from unsafe drinking water and inadequate sanitation systems.

Antibiotics have greatly reduced the incidence of infectious disease caused by bacteria. However, their widespread use and misuse have increased the genetic resistance of many disease-causing bacteria, which can reproduce rapidly (Spotlight, p. 406).

Worldwide, infectious diseases cause about one of every four deaths each year. According to the WHO, the world's seven deadliest infectious diseases are **(1)** *acute respiratory infections*, mostly pneumonia and flu (caused by bacteria and viruses and killing about 3.7 million people per year), **(2)** *HIV/AIDS* (a viral disease, 2.6 million), **(3)** *diarrheal diseases* (caused by bacteria and viruses, 2.5 million), **(4)** *tuberculosis* (TB, a bacterial disease, 2 million; Case Study, p. 407), **(5)** *malaria* (caused by parasitic protozoa, 1.5 million), **(6)** *measles* (a viral disease, 1 million), and **(7)** *hepatitis B* (a viral disease, 1 million).

How Rapidly Are Viral Diseases Spreading? Viral diseases include **(1)** *influenza* or *flu* (transmitted by the bodily fluids or airborne emissions of an infected person), **(2)** *Ebola* (transmitted by the blood or other body fluids of an infected person), **(3)** *rabies* (transmitted by dogs, coyotes, raccoons, skunks, and bats), and **(4)** *HIV/AIDS*. Viruses, like bacteria, can genetically adapt rapidly to different conditions.

Although health officials worry about the emergence of new viral diseases (such those caused by Ebola viruses), they recognize that the greatest virus health threat to humans is the emergence of new, very virulent stains of influenza. Flu viruses move through the air and are highly contagious. During 1918 and 1919, a flu epidemic infected more than half the world's population and killed 20–30 million people (including about 500,000 in the United States). Today, flu kills about 1 million people per year (20,000 of them in the United States).

Sex can be dangerous to one's health. Every day an estimated 114 million acts of sexual intercourse take place in the world. Some 919,000 of these acts lead to conception and about 360,000 pass on a bacterium or virus that causes a sexually transmitted disease (STD). In the United States, STDs infect about 15.3 million people each year.

According to a 1998 report by the American Social Health Association, at least one in every three sexually active people in the United States will contract an STD by the age of 24. Some of these diseases can cause infertility in men and women. Others can cause warts and genital cancers or, in the case of HIV, death.

There is a growing threat from the spread of *acquired immune deficiency syndrome* or *AIDS*, which is caused by HIV. The virus itself is not deadly, but it kills immune cells and leaves the body defenseless against infectious bacteria and other viruses. HIV can be transmitted **(1)** during unprotected sexual activity, **(2)** from one intravenous drug user to another through shared needles, **(3)** from an infected mother to an infant during birth, and **(4)** by exposure to infected blood.

According to the WHO, in December 2000 some 36.1 million people (22 million of them in sub-Saharan Africa and 1.2 million of them children under age 15) were infected with HIV. During 2000, 5.4 million people (80% of them in Africa and Asia) were newly infected with HIV—an average of 15,300 new infections per day.

Within about 7–10 years, 95% of those with HIV develop AIDS. This long incubation period means that infected people often spread the virus for several years without knowing that they are infected. So far, there is no cure for AIDS, although drugs may help some infected people live longer (if they can afford the treatment).

By January 2000, about 19 million people (3.6 million of them children under age 15 and 420,000 people in the United States) had died of AIDS-related diseases. About 2.6 million of these deaths occurred in 1999 (1.8 million of them in Africa). In the 29 African countries hit hardest by AIDS, the average life expectancy at birth is 7 years less than it would have been without the presence of AIDS. According to the WHO, between 2000 and 2010 AIDS probably will kill as many people as all the wars in the 20th century combined.

Are We Losing the War Against Infectious Bacteria?

There is growing and alarming evidence that we may be losing our war against infectious bacterial diseases because bacteria are among the earth's ultimate survivors. When a colony of bacteria is dosed with an antibiotic such as penicillin, most of the bacteria are killed.

However, a few have mutant genes that make them immune to the drug. Through natural selection (Figure 5-6, left, p. 110), a single mutant can pass such traits on to most of its offspring, which can amount to 16,777,216 in only 24 hours.

Each time this strain of bacterium is exposed to penicillin or some other antibiotic, a larger proportion of its offspring are genetically resistant to the drug. The rapid multiplication of resistant bacteria in a victim is made easier because the antibiotics also wipe out their bacterial competitors.

Even worse, bacteria can become genetically resistant to antibiotics they have never been exposed to. When a resistant and a nonresistant bacterium touch one another (say, on a hospital bedsheet or in a human stomach), they can exchange a small loop of DNA called a plasmid, thereby transferring genetic resistance from one organism to another.

The incredible genetic adaptability of bacteria is one reason the world faces a potentially serious rise in the incidence of some infectious bacterial diseases once controlled by antibiotics. Other factors also play a key role, including (1) spread of bacteria (some beneficial and some harmful) around the globe by human travel and the trade of goods, (2) overuse of antibiotics by doctors, often at the insistence of their patients, (3) failure of many

patients to take all of their prescribed antibiotics, which promotes bacterial resistance, (4) availability of antibiotics in many countries without prescriptions, (5) overuse of pesticides (p. 508), which increases populations of pesticide-resistant insects and other carriers of bacterial diseases, and (6) widespread use of antibiotics in the livestock and dairy industries to control disease in livestock animals (about 20% of their use) and to promote animal growth (about 80% of their use).

The result of these factors acting together is that every major disease-causing bacterium now has strains that resist at least one of the roughly 160 antibiotics we use to treat bacterial infections. In 1998, health officials were alarmed to learn of the existence of a strain of bubonic plague in Madagascar that is resistant to multiple antibiotics.

In 2000, officials at the U.S. Centers for Disease Control and Prevention estimated that about 2.2 million people (most with a weakened immunity system) a year get sick and at least 88,000 die from infectious diseases they pick up in U.S. hospitals, nursing homes, or home health-care settings. Most of these infections are caused by (1) contaminated catheters, intravenous lines, and breathing tubes and (2) failure of doctors and other health-care personnel to carefully wash their hands with water or with waterless alcohol-based antimicrobial hand rubs and frequently change their latex gloves. Patients can reduce such infections by asking any doctor or health-care worker coming into their room, "Did you wash your hands?" or "Did you change your gloves?"

There is growing controversy over the widespread use of antibiotics to increase the growth rate of

about 80% of the livestock animals raised each year in the United States, mainly cattle, pigs, and poultry. Each year this use accounts for about 40% of all antibiotics used in the United States and for more than half the global production of antibiotics.

The European Union, the World Health Organization, the American Public Health Association, and the U.S. Centers for Disease Control and Prevention all favor the immediate phaseout of all antibiotics used to promote growth in livestock animals that are the same as or closely related to antibiotics used in humans. Several European countries have imposed such bans, and since 1986 Sweden has banned all use of antibiotics for growth promotion in livestock.

A 1999 study by the U.S. National Academy of Sciences concluded that (1) bacteria that resist antibiotics can be passed from food animals to people, (2) not enough is known to determine the risks this poses to human health, and (3) phasing out antibiotics used as growth promoters would cost about $5–10 per person annually in higher meat and fish prices.

Critical Thinking

1. What role, if any, have you played in the increase in genetic resistance of bacteria to widely used antibiotics? List three ways to reduce this threat.

2. Do you believe that the use of the same antibiotics to treat human illness and to fatten livestock should be banned in the United States (or the country where you live)? Explain. Would you favor using small amounts of such antibiotics to treat disease in livestock?

Once a viral infection starts, it is much harder to fight than infections by bacteria and protozoans. Only a few antiviral drugs exist because most drugs that will kill a virus also harm the cells of its host. Treating viral infections (such as colds, flu, and most

mild coughs and sore throats) with antibiotics is useless and increases genetic resistance in disease-causing bacteria (Spotlight, above).

Medicine's only effective weapons against viruses are vaccines that stimulate the body's immune system

The Global Tuberculosis Epidemic

Since 1990 one of the world's most underreported stories has been the rapid spread of tuberculosis (TB). According to the World Health Organization, this highly infectious bacterial disease kills about 2 million people and infects about 8 million people per year (Figure 16-8). It is the leading cause of death among women of reproductive age. In India, where nearly half the population is infected with TB, the disease kills a half a million people each year.

The bacterium causing TB infection moves from person to person, mainly in airborne droplets produced by coughing, sneezing, singing, or even talking. Whereas TB kills about 2 million people per year, highly publicized Ebola viruses, which are hard to transmit, have killed about 1,000 people over the past 20 years—an average of 50 people per year.

About one of every three people in the world is infected with the TB bacillus. During their lifetime about 5–10% of these people will become sick or infectious with active TB, especially when their immune system is weakened. Left untreated, each person with active TB will infect 10–15 other people. Because many of the infected people do not appear to be sick and about half of them do not even know they are infected, this serious health prob-

lem has been called a *silent global epidemic*.

Until this resurgence, the incidence of TB had fallen sharply (except among the poor). This occurred mostly because of **(1)** prevention programs (such as X-ray screening and detection of people with active TB) and **(2)** treatment with antibiotics, which began in the 1940s. Before

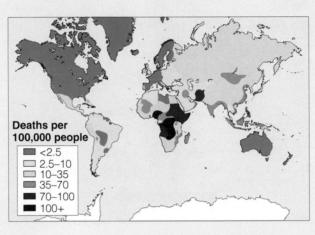

Deaths per 100,000 people
- <2.5
- 2.5–10
- 10–35
- 35–70
- 70–100
- 100+

Figure 16-8 The current global tuberculosis epidemic. This easily transmitted disease is spreading rapidly and now kills about 2 million people a year. (Data from World Health Organization)

antibiotics, the spread of the disease was controlled by isolating patients with active TB in sanitariums until they died or recovered.

Major reasons for the recent increase in TB are **(1)** poor TB screening and control programs (especially in developing countries, where about 95% of the new cases occur), **(2)** development of strains of the tuberculosis bacterium that are genetically resistant to almost all effective

antibiotics (typically leading to mortality rates of more than 50%), **(3)** population growth and increased urbanization (which increase contacts among people), **(4)** poverty, and **(5)** the spread of AIDS, which greatly weakens the immune system and allows TB bacteria to multiply.

Slowing the spread of the disease involves early identification and treatment of people with active TB, usually those with a chronic cough. Treatment with a combination of four inexpensive drugs can cure 90% of those with active TB. However, to be successful the drugs must be taken every day for 6 to 8 months. Because the symptoms disappear after a few weeks, many patients think they are cured and stop taking the drugs. This allows the disease to recur in a hard-to-treat form. It then spreads to other people, and drug-resistant strains of TB bacteria develop.

According to the World Health Organization, a worldwide campaign to help control TB would cost about $360 million to help save at least 20 million lives during the next decade.

Critical Thinking

Before you read this report, were you aware of the serious global TB epidemic? Why do you think that this important story has gotten so little media attention compared to other diseases that cause many fewer deaths per year?

---

to produce antibodies to ward off viral infections. Immunization with vaccines has helped tame viral diseases such as smallpox, polio, rabies, influenza, measles, and hepatitis B (Solutions, p. 408).

Connections: What Factors Can Affect the Spread of Transmissible Diseases? Outbreaks of infectious diseases often occur because of a change in

the physical, social, or biological environment of disease reservoirs, carrier vectors, or exposure to new host populations. Important factors include the following:

- *Increased international air travel*, which can rapidly spread diseases such as flu, measles, cholera, yellow fever, and TB. Between 1960 and 2000, international air travel increased from about 50 million to 500 million people per year.

Producing an Edible Hepatitis B Vaccine

SOLUTIONS

Worldwide, about 300 million people (most in developing countries) are infected with liver-damaging hepatitis B. Each year about 1 million people die from this disease and there are about 1 million new infections.

A vaccine for hepatitis B has been developed from bioengineered yeast. However, it **(1)** costs about $100 per person, **(2)** requires three shots over a 6-month period, and **(3)** must be refrigerated continuously before use. This makes it impractical to carry out mass immunizations for hepatitis B in developing countries.

Researchers are developing a much cheaper and easily administered vaccine by inserting DNA extracted from the hepatitis B virus into the cells of leaves of plants such as potatoes (Figure 16-9). Mice eating raw potatoes from the plant developed immunity to hepatitis B. Researchers hope to introduce the vaccine into bananas, which might cost as little as 5¢ per dose to produce.

Critical Thinking

What might be some disadvantages of introducing hepatitis B vaccine into the genes of food plants? Do you believe that the advantages of using this vaccine in food outweigh its disadvantages? Explain.

(1) DNA of hepatitis B virus put into potato leaf.

Hepatitis B virus

(2) Potato plant grown from leaf has the hepatitis B virus in it's cells.

Potato leaf

Potato plant

Potato

Virus protein

(3) Potatoes produce protein containing hepatitis B virus.

(4) Mouse eats potato and develops immunity to hepatitis B virus.

Figure 16-9 Using genetic engineering to produce food crops that contain a vaccine for hepatitis B.

- *Migration to urban areas*, which increases the probability of infection from diseases such as TB (Case Study, p. 407), cholera, and sexually transmitted diseases.

- *Migration to uninhabited rural areas and deforestation in tropical developing countries*, which can expose people to new diseases and disease vectors such as malaria, sleeping sickness, and yellow fever.

- *Migration to suburbs in developed countries*. For example, as more people have moved to wooded suburbs in the eastern United States, they have come into greater contact with ticks infested with bacteria that cause Lyme disease, which causes fever, lethargy, and (sometimes) long-lasting arthritis.

- *Hunger and malnutrition* (p. 267), which increase the number of children killed by infectious diseases such as measles and diarrhea.

- *Increased rice cultivation* in flooded fields and paddies, which creates ideal breeding grounds for mosquitoes and other insects that transmit diseases to humans.

- *Global warming*, which is leading to the spread of tropical infectious diseases such as malaria (Figure 16-10), yellow fever, and dengue fever (called "breakbone fever" by those whose experience the excruciat-

ing pain it causes in joints) to temperate areas. A 2000 study by researchers at the Johns Hopkins School of Public Health found that each 1°C (1.8°F) rise in temperature causes an 8% increase in diarrhea in children under age 5 in developing countries.

- *High winds or hurricanes*, which can transfer infectious organisms and carriers of disease (such as insects) from tropical to temperate areas.

- *Accidental introduction of insect vectors*. The Asian tiger mosquito is a vector for dengue fever, yellow fever, and other viruses. In 1985, it was brought accidentally to the United States inside used tires shipped from Asia. Today, this mosquito species has become established in Hawaii and the southeastern United States and has begun extending its range north toward Chicago and Washington, D.C.

- *Flooding*, which **(1)** often contaminates water supplies with raw sewage and **(2)** creates areas of standing water and moist soil that are ideal breeding grounds for mosquitoes and other insects that spread infectious diseases.

Case Study: Malaria, a Protozoal Disease About 45% of the world's population live in tropical and subtropical regions in which malaria is present (Figure

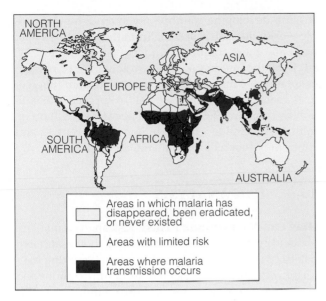

Figure 16-10 Worldwide distribution of malaria. About 45% of the world's current population live in areas in which malaria is present, with the disease killing at least 1.5 million people a year. If the world becomes warmer, as projected by current climate models, by 2046 malaria could affect 60% of the world's population. (Data from the World Health Organization)

Legend:
- Areas in which malaria has disappeared, been eradicated, or never existed
- Areas with limited risk
- Areas where malaria transmission occurs

16-10). Currently, an estimated 300–500 million people are infected with malaria parasites worldwide, and there are 270–480 million new cases each year.

Malaria's symptoms come and go and include fever and chills, anemia, an enlarged spleen, severe abdominal pain and headaches, extreme weakness, and greater susceptibility to other diseases. The disease kills about 1.5 million people each year, more than half of them children under age 5.

Malaria is caused by four species of protozoa of the genus *Plasmodium*. Most cases of the disease are transmitted when an uninfected female of any one of 60 species of *Anopheles* mosquito **(1)** bites an infected person, **(2)** ingests blood that contains the parasite, and **(3)** later bites an uninfected person (Figure 16-11). When this happens, *Plasmodium* parasites move out of the mosquito and into the human's bloodstream, multiply in the liver, and then enter blood cells to continue multiplying. Malaria also can be transmitted by blood transfusions or by sharing needles.

The malaria cycle repeats itself until immunity develops, treatment is given, or the victim dies. Over the course of human history, malarial protozoa probably have killed more people than all the wars ever fought.

During the 1950s and 1960s, the spread of malaria was sharply curtailed by **(1)** draining swamplands and marshes, **(2)** spraying breeding areas with insecticides, and **(3)** using drugs to kill the parasites in the bloodstream. Since 1970, however, malaria has come roaring back. Most species of the malaria-carrying *Anopheles* mosquito have become genetically resistant to most insecticides. Worse, the *Plasmodium* parasites have become genetically resistant to common anti-malarial drugs.

Researchers are working to develop new anti-malarial drugs, vaccines, and biological controls for *Anopheles* mosquitoes. However, such approaches are underfunded and have proved more difficult than originally thought. Researchers also are studying the feasibility of altering the genetic makeup of mosquitoes so that they cannot carry and transmit the parasite to humans.

According to health experts, prevention is the best approach to slowing the spread of malaria. Methods include **(1)** increasing water flow in irrigation systems to prevent mosquito larvae from developing (an expensive and wasteful use of water), **(2)** using mosquito nets dipped in a nontoxic insecticide (permethrin) in windows and doors of homes, **(3)** cultivating fish that feed on mosquito larvae (biological control), **(4)** clearing vegetation around houses, **(5)** planting trees that soak up water in low-lying marsh areas where mosquitoes thrive (a method that can degrade or destroy ecologically important wetlands), and **(6)** using zinc and vitamin A supplements to boost children's resistance to malaria.

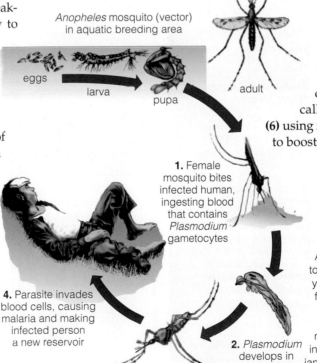

Anopheles mosquito (vector) in aquatic breeding area

eggs
larva
pupa
adult

1. Female mosquito bites infected human, ingesting blood that contains *Plasmodium* gametocytes

2. *Plasmodium* develops in mosquito

3. Mosquito injects *Plasmodium* sporozoites into human host

4. Parasite invades blood cells, causing malaria and making infected person a new reservoir

Figure 16-11 The life cycle of malaria. This life cycle of *Plasmodium* circulates from mosquito to human and back to mosquito. Although various species of mosquitoes carry diseases (such as malaria, yellow fever, encephalitis, and dengue fever to humans and heartworm to dogs), mosquitoes play important ecological roles. Their eggs are a major food source for fish, various insects, and frogs and other amphibians, and adult mosquitoes are an important source of food for bats, spiders, and many species of insects and birds.

What Are the Major Diseases in Developed Countries? As a country industrializes, it usually makes an *epidemiological transition*. The infectious diseases of childhood become less important, and the chronic diseases of adulthood (heart disease and stroke, cancer, and respiratory conditions) become more important in causing mortality. In 1999, for example, infectious and parasitic diseases were responsible for 43% of all deaths in developing countries but only 1% in developed countries.

In 1999, about 31% of the deaths in the United States were from heart attacks, 23% from cancer, 7% from strokes, 7% from infectious diseases (mostly pneumonia, influenza, and AIDS), 4% from accidents (half from automobile accidents), 3% from diabetes, and 1% each from suicides, kidney diseases, and liver diseases. Almost two-thirds of these deaths resulted from chronic diseases (such as heart disease, strokes, and cancer) that take a long time to develop and have multiple causes. In 1980, infectious diseases ranked as America's fifth leading killer. By 1999, they were tied with stroke as the third leading cause of death, representing a partial reversal of the epidemiologic transition in the United States.

According to the U.S. Department of Health and Human Services, about 95% of the money spent on health care in the United States each year is used to *treat* rather than to *prevent* disease—one reason why health-care costs are so high and continue to climb. Health experts estimate that changing harmful lifestyle factors could prevent **(1)** 40–70% of all premature deaths, **(2)** one-third of all cases of acute disability, and **(3)** two-thirds of all cases of chronic disability in the United States.

How Can We Reduce Infectious Diseases in Developing Countries? Figure 16-12 lists ways that health scientists and public health officials suggest for preventing or reducing the incidence of infectious diseases that affect humanity. They also call for increased emphasis on preventive health care in developing countries (Solutions, p. 413).

The WHO estimates that only 2% of the world's global research and development funds are devoted to infectious dis-

eases in developing countries, even though more people worldwide suffer and die from these diseases than from all others combined. Indeed, according to 2000 study by the International Federation of Red Cross, an estimated 150 million people have died from tuberculosis, malaria, and AIDS since 1945, compared to 23 million in wars. In 1995, the world spent $864 billion on military protection and $15 billion on preventing and controlling tuberculosis, malaria, and AIDS.

16-5   RISK ANALYSIS

How Can We Estimate Risks? Risk analysis involves **(1)** identifying hazards and evaluating their associated risks (*risk assessment*, Figure 16-2, left), **(2)** ranking risks (*comparative risk analysis*), **(3)** determining options and making decisions about reducing or eliminating risks (*risk management*, Figure 16-2, right), and **(4)** informing decision makers and the public about risks (*risk communication*).

Statistical probabilities based on past experience, animal testing and other tests, and epidemiological studies are used to estimate risks from older technologies and chemicals (Section 16-1). To evaluate new technologies and products, risk evaluators use more uncertain statistical probabilities, based on models rather than actual experience.

The left side of Figure 16-13 is an example of *comparative risk analysis*, summarizing the greatest ecological and health risks identified by a panel of scientists acting as advisers to the U.S. Environmental Protection Agency. Note the difference between the comparison of relative risk by scientists (Figure 16-13, left) and the general public (Figure 16-13, right). These differences result largely from failure of professional risk evaluators to communicate the nature of risks and their relative importance to the public, teachers, and members of the media. Some risk experts contend that much of our risk education is based on often misleading media reports on the latest risk scare of the week (based mainly on frontier science) that do not put such risks in perspective.

Increase research on tropical diseases and vaccines

Reduce poverty

Decrease malnutrition

Improve drinking water quality

Reduce unnecessary use of antibiotics

Educate people to take all of an antibiotic prescription

Reduce antibiotic use to promote livestock growth

Careful hand washing by all medical personnel

Slow global warming to reduce spread of tropical diseases to temperate areas

Increase preventative health care

Figure 16-12 Solutions. Ways to prevent or reduce the incidence of infectious diseases.

Scientists (Not in rank order in each category)	Citizens (In rank order)
High-Risk Health Problems • Indoor air pollution • Outdoor air pollution • Worker exposure to industrial or farm chemicals • Pollutants in drinking water • Pesticide residues on food • Toxic chemicals in consumer products	**High-Risk Problems** • Hazardous waste sites • Industrial water pollution • Occupational exposure to chemicals • Oil spills • Stratospheric ozone depletion • Nuclear power-plant accidents • Industrial accidents releasing pollutants • Radioactive wastes • Air pollution from factories • Leaking underground tanks
High-Risk Ecological Problems • Global climate change • Stratospheric ozone depletion • Wildlife habitat alteration and destruction • Species extinction and loss of biodiversity	
Medium-Risk Ecological Problems • Acid deposition • Pesticides • Airborne toxic chemicals • Toxic chemicals, nutrients, and sediment in surface waters	**Medium-Risk Problems** • Coastal water contamination • Solid waste and litter • Pesticide risks to farm workers • Water pollution from sewage plants
Low-Risk Ecological Problems • Oil spills • Groundwater pollution • Radioactive isotopes • Acid runoff to surface waters • Thermal pollution	**Low-Risk Problems** • Air pollution from vehicles • Pesticide residues in foods • Global climate change • Drinking water contamination

Figure 16-13 Comparative risk analysis of the most serious ecological and health problems according to scientists acting as advisers to the U.S. Environmental Protection Agency (left column). Risks in each of these categories are not listed in rank order. The right side of this figure represents polls showing how U.S. citizens rank the ecological and health risks they perceive as being the most serious. Why do you think there is such a great difference between the ranking by risk experts and by the general public? (Data from Science Advisory Board, *Reducing Risks*, Washington, D.C.: Environmental Protection Agency, 1990)

Once a risk assessment has been completed, decision makers much decide what level of risk is acceptable. Figure 16-14 shows four methods used to determine the acceptability of a risk. The most widely used method is *cost-benefit analysis*, which attempts to determine whether the estimated short- and long-term risks or costs of using a particular technology or chemical outweigh its the estimated short- and long-term benefits.

What Are the Greatest Risks People Face? The greatest risks many people face today are rarely dramatic enough to make the daily news. In terms of reduced life span from malnutrition, exposure to disease-causing organisms and dangerous chemicals, and lack of basic health care, *the greatest risk by far is poverty* (Figure 16-15, p. 414).

After the health risks associated with poverty, the greatest risks of premature death are mostly the result of voluntary choices people make about their lifestyles (Figures 16-1 and 16-15).

By far the best ways to reduce one's risk of premature death and serious health risks are to (1) not smoke, (2) avoid excess sunlight (which ages skin and causes skin cancer), (3) not drink alcohol or drink only in moderation (no more than two drinks in a single day), (4) reduce consumption of foods containing cholesterol and saturated fats, (5) eat a variety of fruits and vegetables, (6) exercise regularly, (7) lose excess weight, and (8) for those who can afford a car, drive as safely as possible in a vehicle with the best available safety equipment.

However, we have little or no control over some factors that can influence our vulnerability to some risks. For example, we cannot control (1) our gender, (2) the genes we inherited from our parents, and (3) our social and psychological environment during early childhood.

How Can We Estimate Risks for Technological Systems? The more complex a technological system and the more people needed to design and run it, the more difficult it is to estimate the risks. The overall reliability of any technological system (expressed as a percentage) is the product of two factors:

System reliability (%) = Technology reliability × Human reliability

With careful design, quality control, maintenance, and monitoring, a highly complex system such as a nuclear power plant or space shuttle can achieve a high degree of technology reliability. However, human reliability usually is much lower than technology reliability and is almost impossible to predict: To err is human.

Suppose that the technology reliability of a nuclear power plant is 95% (0.95) and that human reliability is 75% (0.75). Then the overall system reliability is 71% (0.95 × 0.75 = 0.71 = 71%). Even if we could make the technology 100% reliable (1.0), the overall system reliability would still be only 75% (1.0 × 0.75 = 0.75 =75%). The crucial dependence of even the most carefully designed systems on unpredictable human reliability helps explain essentially "impossible" tragedies such as the Chernobyl (p. 350) nuclear power-plant accident and the explosion of the space shuttle *Challenger*.

One way to make a system more foolproof or fail-safe is to move more of the potentially fallible elements from the human side to the technical side. However, **(1)** chance events such as a lightning bolt can knock out automatic control system, **(2)** no machine or computer program can completely replace human judgment, **(3)** the parts in any automated control system are manufactured, assembled, tested, certified, and maintained by fallible human beings, and **(4)** computer software programs used to monitor and control complex systems can also contain human error or can be deliberately modified by computer viruses to malfunction.

What Are the Limitations of Risk Analysis? Here are some of the key questions involved in evaluating risk analysis:

- How reliable are risk assessment data and models?

- Who profits from allowing certain levels of harmful chemicals into the environment, and who suffers? Who decides this?

- Should estimates emphasize short-term risks, or should more weight be put on long-term risks? Who should make this decision?

- Should the primary goal of risk analysis be to determine how much risk is acceptable (the current approach) or to figure out how to do the least damage (a prevention approach)?

- Who should do a particular risk analysis, and who should review the results? A government agency? Independent scientists? The public?

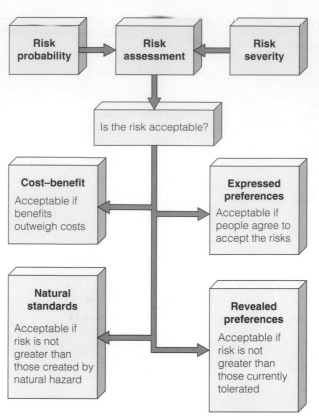

Figure 16-14 Methods for determining the acceptability of a risk after a risk assessment has been made. Cost–benefit analysis is the most widely used method. (Adapted from *Environmental Science*, 5/E by Chiras, p. 351. Copyright ©1998 by Wadsworth)

- Should cumulative effects of various risks be considered, or should risks be considered separately, as is usually done? Suppose a pesticide is found to have an annual risk of killing 1 person in 1 million from cancer, the current EPA limit. Cumulatively, however, effects from 40 such pesticides might kill 40, or 400, of every 1 million people. Is this acceptable and to whom?

- How widespread is each risk?

- Should risk levels be higher for workers (as is almost always the case) than for the general public? What say should workers and their families have in this decision? According to government estimates, the exposure of workers to toxic chemicals in the United States causes 50,000–70,000 deaths (at least half from cancer) and 350,000 new cases of illness per year. The situation is much worse in developing countries, with more than 1 million work-related deaths occurring worldwide each year. Is this a necessary cost of doing business?

- How much risk is acceptable and to whom? According to the National Academy of Sciences, exposure to toxic chemicals is responsible for 2–4% of the 521,000 cancer deaths in the United States; this amounts to 10,400–20,800 premature cancer deaths per year. Is this acceptable or unacceptable and to whom?

Proponents contend that risk analysis is a useful way to **(1)** organize and analyze available scientific information, **(2)** identify significant hazards, **(3)** focus on areas that warrant more research, **(4)** help regulators decide how money for reducing risks should be allocated, and **(5)** stimulate people to make more informed decisions about health and environmental goals and priorities.

However, critics point out that results of risk analysis are very uncertain. For example, a recent study documented the significant uncertainties involved in even simple risk analysis. Eleven European governments established 11 different teams of their best scientists and engineers (including those from private companies) to assess the hazards and risks from a small plant storing only one hazardous chemical (ammonia). The 11 teams, consisting of world-class experts analyzing this

Improving Health Care in Developing Countries

With adequate funding, the health of people in developing countries and the poor in developed countries can be improved dramatically, quickly, and cheaply by providing the following forms of mostly preventive health care:

- Better nutrition, prenatal care, and birth assistance for pregnant women. At least 585,000 women in developing countries die each year of mostly preventable causes related to pregnancy and childbirth, compared with about 6,000 in developed countries. According to the World Health Organization, the majority of these deaths in low-income countries could be prevented for as little as $2 per woman per year.

- Better nutrition for children.

- Greatly improved postnatal care (including promotion of breast-feeding) to reduce infant mortality.

Breast-fed babies get natural immunity to many diseases from antibodies in their mothers' milk.

- Immunization against the world's five largest preventable infectious diseases: tetanus, measles, diphtheria, typhoid fever, and polio. Since 1971 the percentage of children in developing countries immunized against these diseases has increased from 10% to 84%, saving about 10 million lives a year.

- Oral rehydration therapy for victims of diarrheal diseases, which cause about one-fourth of all deaths of children under age 5. A simple solution of boiled water, salt, and sugar or rice, at a cost of only a few cents per person, can prevent death from dehydration. According to the British medical journal *Lancet*, this simple treatment is "the most important medical advance of the 20th century."

- Careful and selective use of antibiotics for infections (Spotlight, p. 406).

- Clean drinking water and sanitation facilities for the one-third of the world's population that lacks them.

According to the World Health Organization, extending such primary health care to all the world's people would cost an additional $10 billion per year, about 4% of what the world spends every year on cigarettes or devotes every 4 days to military spending. The cost of this program is about $1 per child.

Critical Thinking

1. Do you believe that developed countries should foot at least half the bill for implementing such proposals? What economic and environmental benefits would this provide for developed countries?

2. How many dollars per year of your taxes would you be willing to spend for such a preventive health program in developing countries?

---

very simple system, disagreed with one another on fundamental points and varied in their assessments of the hazards by a factor of 25,000.

Such built-in uncertainty in risk analysis is analogous to a radar device that can detect a car speeding at 160 kilometers (100 miles) per hour but can tell us only that the car is traveling somewhere between 0.16 kilometer (0.1 mile) per hour and 160,000 kilometers (100,000 miles) per hour. Such inherent uncertainty explains why regulators setting human exposure levels for toxic substances usually divide the best results by 100 to 1,000 to provide the public with a margin of safety.

According to critics, the main decision-making tool we should rely on is not to find out how much risk is acceptable—which is mostly a political question. Instead, it should be to find out the least damaging reasonable alternatives by asking, "Which alternative will bring sufficient benefits and minimize damage to humans and to the earth?" In other words, the emphasis should be on *alternative assessment* not *risk assessment*.

How Should Risks Be Managed? Risk management includes the administrative, political, and economic actions taken to decide whether and how to reduce a particular societal risk to a certain level and at what cost.

Risk management involves deciding:

- The reliability of the risk analysis for each risk.

- Which risks should be given the highest priority.

- How much risk is acceptable (Figure 16-14).

- How much it will cost to reduce each risk to an acceptable level.

- How limited funds should be spent to provide the greatest benefit.

- How the risk management plan will be monitored, enforced, and communicated to the public.

Each step in this process involves making value judgments and weighing trade-offs to find some reasonable compromise among conflicting political, economic, health, and environmental interests.

How Well Do We Perceive Risks? How much risk is acceptable? Studies indicate that if the chance of death from a chemical or activity is less than 1 in 100,000, most people are not likely to be worried enough to change their ways.

However, most of us do poorly in assessing the relative risks from the hazards that surround us (Figure 16-13 and 16-15). Also, many people deny or shrug off

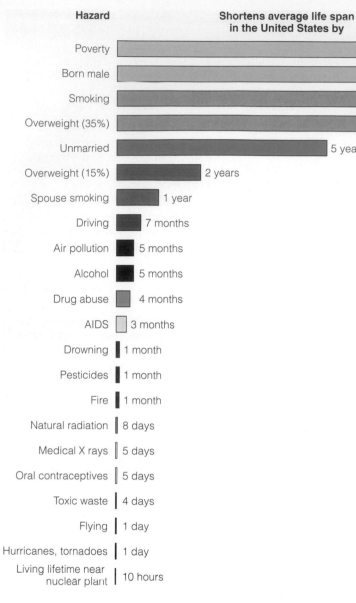

Hazard	Shortens average life span in the United States by
Poverty	7–10 years
Born male	7.5 years
Smoking	6 years
Overweight (35%)	6 years
Unmarried	5 years
Overweight (15%)	2 years
Spouse smoking	1 year
Driving	7 months
Air pollution	5 months
Alcohol	5 months
Drug abuse	4 months
AIDS	3 months
Drowning	1 month
Pesticides	1 month
Fire	1 month
Natural radiation	8 days
Medical X rays	5 days
Oral contraceptives	5 days
Toxic waste	4 days
Flying	1 day
Hurricanes, tornadoes	1 day
Living lifetime near nuclear plant	10 hours

Figure 16-15 Comparison of risks people face, expressed in terms of shorter average life span. After poverty, the greatest risks people face result mostly from voluntary choices they make about their lifestyles. These are only generalized relative estimates. Individual response to some of these risks can vary with factors such as **(1)** genetic variation (Figure 16-3), **(2)** family medical history, **(3)** emotional makeup, **(4)** stress, and **(5)** social ties and support. (Data from Bernard L. Cohen)

the high-risk chances of dying (or injury) from voluntary activities they enjoy, such as **(1)** motorcycling (1 death in 50 participants), **(2)** smoking (1 in 300 participants by age 65 for a pack-a-day smoker), **(3)** hanggliding (1 in 1,250), and **(4)** driving (1 in 2,500 without a seatbelt and 1 in 5,000 with a seatbelt).

Yet some of these same people may be terrified about the possibility of dying from a **(1)** commercial airplane crash (1 in 4.6 million), **(2)** train crash (1 in 20 million), **(3)** snakebite (1 in 36 million), **(4)** shark attack (1 in 300

million), or **(5)** exposure to trichloroethylene (TCE) in drinking water at the trace levels allowed by the EPA (1 in 2 billion).

Being bombarded with news about people killed or harmed by various hazards distorts our sense of risk. However, *the most important good news each year is that about 99.1% of the people on the earth did not die.* Despite the greatly increased use of synthetic chemicals in food production and processing, the general health and average life expectancy of people in the United States (and most developed countries) have increased during the past 50 years.

Our perceptions of risk and our responses to perceived risks often have little to do with how risky the experts say something is (Figures 16-13 and 16-15). The public generally sees a technology or a product as being riskier than experts do when:

- *It is new or complex rather than familiar.* Examples include genetic engineering (Pro/Con, p. 275) or nuclear power, as opposed to large dams or coal-fired power plants (Figure 14-36, p. 349).

- *It is perceived as being mostly involuntary.* Examples include nuclear power plants or food additives, as opposed to driving or smoking.

- *It is viewed as unnecessary rather than as beneficial or necessary.* Examples might include using chlorofluorocarbon (CFC) propellants in aerosol spray cans or using food additives that increase sales appeal, as opposed to cars or aspirin.

- *Its use involves a large, well-publicized death toll from a single catastrophic accident rather than the same or an even larger death toll spread out over a longer time.* Examples might include a severe nuclear power plant accident (p. 350), an industrial explosion, or a plane crash, as opposed to coal-burning power plants, automobiles, or smoking.

- *Its use involves unfair distribution of the risks.* Citizens are outraged when government officials decide to put a hazardous-waste landfill or incinerator in or near their neighborhood, even when the decision is based on risk analysis. This is usually seen as politics, not science. Residents will not be satisfied by estimates that the lifetime risks of cancer death from the facility are not greater than, say, 1 in 100,000. Living near the facility means that they, not the 99,999 people living farther away, have a much higher risk of dying from cancer by having this risk involuntarily imposed on them.

- *The people affected are not involved in the decision-making process from start to finish.*

- *Its use does not involve a sincere search for and evaluation of alternatives.* People who believe that their lives and the lives of their families are being threatened want to know what the alternatives are and which alternative causes the least harm to them and the earth.

Better education and communication about the nature of risks will help bring the public's perceptions of various risks closer to those of professional risk evaluators. However, such education will not eliminate the emotional, cultural, and ethical factors that decision makers must take into account in determining the acceptability of a particular risk and evaluating the possible alternatives.

The burden of proof imposed on individuals, companies, and institutions should be to show that pollution prevention options have been thoroughly examined, evaluated, and used before lesser options are chosen.

JOEL HIRSCHORN

REVIEW QUESTIONS

1. Define the boldfaced terms in this chapter.

2. What human activity kills the largest number of people each year? List six ways to help reduce the harmful effects of smoking.

3. What are *risk* and *probability*? Distinguish between *risk assessment* and *risk management*.

4. List two (a) cultural hazards, (b) chemical hazards, (c) physical hazards, and (d) biological hazards.

5. What is *toxicity*? Distinguish between *dosage* and *response* for a potentially harmful substance. List five factors that determine whether a chemical is harmful. Distinguish between *bioaccumulation* and *biomagnification*. List three mechanisms by which the human body can reduce the harmful effects of most harmful chemicals.

6. What is a *poison*? What is an LD_{50}? List three methods used to determine toxicity and list the limitations of each method. Describe how laboratory tests are used to determine toxicity. What is a *dose–response curve*? Distinguish between a *linear dose–response curve* and a *threshold dose–response curve*.

7. Distinguish between *toxic chemicals* and *hazardous chemicals*. Distinguish between *mutagens*, *teratogens*, and *carcinogens* and give one example of each.

8. Distinguish between the *immune system*, *nervous system*, and *endocrine system* and give an example of something that causes harm to each system. What are *hormone disrupters* and *hormone mimics*? List two examples of such chemicals.

9. About what percentage of the 75,000 chemicals in commercial use in the United States have been screened (a) to assess toxicity, (b) to determine whether they are carcinogens, teratogens, or mutagens, and (c) to determine whether they damage the nervous, endocrine, or immune systems? What percentage of the commercially used chemicals in the United States do federal and state governments regulate?

10. List three reasons for the lack of information about the potentially harmful effects of most chemicals in commercial use. Distinguish between the regulation strategy and the pollution prevention strategy for protecting the public from potentially harmful chemicals. What is the *precautionary principle*? Why is it rarely used?

11. Distinguish between *nontransmissible* and *transmissible diseases* and give two examples of each type. About how many children die each year in developing countries from mostly preventable infectious diseases? What are the seven deadliest infectious diseases in order of the number of deaths they cause each year?

12. What are the major types of diseases in (a) developing countries, (b) developed countries, and (c) the United States? What is an *epidemiologic transition*?

13. How do infectious bacteria become resistant to antibiotics? List seven factors that have led to an increase in infectious diseases that cannot be controlled by most antibiotics.

14. What is the best way to treat (a) a bacterial disease and (b) a viral disease? List two examples of each type of disease.

15. What causes tuberculosis, and how is it transmitted? About how many people died from TB during the past year? List five reasons for the increase in TB infections in recent years. How can the spread of this bacterial infectious disease be slowed?

16. Distinguish between *HIV* and *AIDS*. List four ways in which HIV can be transmitted. About how many people in the world are (a) infected with HIV and (b) have died of AIDS? During the past year, about how many people were infected with HIV and how many people died of AIDS? List ways to prevent the spread of this viral infectious disease.

17. List 10 factors that can affect the spread of infectious diseases.

18. What causes malaria? About how many people die from malaria each year? List six ways to help prevent this protozoal infectious disease.

19. List 10 ways to prevent or reduce the incidence of infectious diseases throughout the world. List seven major ways to improve health care in developing countries. About how much would implementing these measures cost per year?

20. What is *risk analysis*? What are the major limitations of risk analysis?

21. List five of the greatest risks people face in terms of reduced life span. List eight ways to reduce your risk of premature death and serious health problems. How can we estimate the risks from technological systems?

22. What is *risk management*? What six questions do risk managers try to answer? About what percentage of the people on the earth die each year? List seven reasons that lead people to perceive that certain risks are greater than experts say they are.

CRITICAL THINKING

1. Explain why you agree or disagree with the proposals made by health officials for reducing the death toll and other harmful effects of smoking listed on p. 396.

2. Do you think chemicals should be regulated based on their effects on the nervous, immune, and endocrine systems? Explain.

3. Should we have zero pollution levels for all hazardous chemicals? Explain.

4. Do you believe that health and safety standards in the workplace should be strengthened and enforced more vigorously, even if this causes a loss of jobs when companies transfer operations to countries with weaker standards? Explain.

5. Evaluate the following statements:
   a. We should not get so worked up about exposure to toxic chemicals because almost any chemical can cause some harm at a large enough dosage.
   b. We should not worry so much about exposure to toxic chemicals because through genetic adaptation we can develop immunity to such chemicals.
   c. We should not worry so much about exposure to toxic chemicals because we can use genetic engineering to reduce or eliminate such problems.

6. How can changes in the age structure of a human population increase the spread of infectious diseases? How can the spread of infectious diseases affect the age structure of human populations?

7. Should pollution levels be set to protect the most sensitive people in a population (Figure 16-3, left) or the average person (Figure 16-3, middle)? Explain.

8. What are the five major risks you face from your lifestyle, where you live, and what you do for a living? Which of these risks are voluntary and which are involuntary? List the five most important things you can do to reduce these risks. Which of these things do you actually plan to do?

9. How would you answer each of the questions raised about (a) risk analysis on p. 412, and (b) risk assessment and risk management on p. 413? Explain each of your answers.

PROJECTS

1. Assume that members of your class (or small manageable groups in your class) have been appointed to a technology risk-benefit assessment board. As a group, decide why you would approve or disapprove of widespread use of each of the following: (a) drugs to retard aging, (b) electrical or chemical devices that would stimulate the brain to eliminate anxiety, fear, unhappiness, and aggression, and (c) genetic engineering to produce people with superior intelligence and strength.

2. Use the library or the internet to find recent articles describing the rise of genetic resistance of disease-causing bacteria to commonly used antibiotics. Evaluate the evidence and claims in these articles.

3. Pick a specific viral disease and use the library or internet to find out about (a) how it spreads, (b) its effects, (c) strategies for controlling its spread, and (d) possible treatments.

4. Use the library or the internet to find bibliographic information about *Paracelsus* and *Joel Hirschorn*, whose quotes appear at the beginning and end of this chapter.

5. Make a concept map of this chapter's major ideas, using the section heads and subheads and the key terms (in boldface). Look at the inside back cover and on the website for this book for information about making concept maps.

INTERNET STUDY RESOURCES AND RESOURCES FOR FURTHER READING AND RESEARCH

The website for this book contains helpful study aids and many ideas for further reading and research. Log on to:

http://www.brookscole.com/product/0534376975s

and click on the Chapter-by-Chapter area. Choose Chapter 16 and select a resource:

- "Flash Cards" allows you to test your mastery of the Terms and Concepts to Remember for this chapter.

- "Tutorial Quizzes" provides a multiple-choice practice quiz.

- "Student Guide to InfoTrac" will lead you to Critical Thinking Projects that use InfoTrac College Edition as a research tool.

- "References" lists the major books and articles consulted in writing this chapter.

- "Hypercontents" takes you to an extensive list of sites with news, research, and images related to individual sections of the chapter.

INFOTRAC COLLEGE EDITION

Improve your skills with InfoTrac College Edition, a searchable online database of articles from more than 700 periodicals. Log on to:

http://www.infotrac-college.com

or access InfoTrac through the website for this book.

Try the following articles:

Gregory, R. 2000. Using stakeholder values to make smarter environmental decisions. *Environment* vol. 42, no. 5, pp. 34–41. (subject guide: risk assessment)

Hunter, B.T. 2000. New alternatives in safety testing: testing product safety without using animals. *Consumers' Research Magazine* vol. 83, no. 5, pp. 26–30. (toxicology, technique)

17 AIR AND AIR POLLUTION

When Is a Lichen Like a Canary?

Nineteenth-century coal miners took canaries with them into the mines—not for their songs, but for the moment when they stopped singing. Then the miners knew it was time to get out of the mine because the air contained methane, which could ignite and explode.

Today we use sophisticated equipment to monitor air quality, but living things such as lichens (Figure 17-1) still can warn us of bad air. A lichen consists of a fungus and an alga living together, usually in a mutually beneficial (mutualistic) partnership.

These hearty pioneer species are good air pollution detectors because they are always absorbing air as a source of nourishment. Certain lichen species are sensitive to specific air-polluting chemicals. Old man's beard (*Usnea trichodea*) (Figure 17-1, right) and yellow *Evernia* lichens, for example, sicken or die in the presence of too much sulfur dioxide.

Because lichens are widespread, long-lived, and anchored in place, they can also help track pollution to its source. The scientist who discovered sulfur dioxide pollution on Isle Royale in Lake Superior (Case Study, p. 204), where no car or smokestack had ever intruded, used *Evernia* lichens to point the finger northward to coal-burning facilities at Thunder Bay, Canada.

Radioactive particles spewed into the atmosphere by the Chernobyl nuclear power-plant disaster (p. 350) fell to the ground over much of northern Scandinavia and were absorbed by lichens that carpet much of Lapland. The area's Saami people depend on reindeer meat for food, and the reindeer feed on lichens. After Chernobyl more than 70,000 reindeer had to be killed and the meat discarded because it was too radioactive to eat. Scientists helped the Saami identify which of the remaining reindeer to move by analyzing lichens (which absorbed some of the radioactive fallout) to pinpoint the most contaminated areas.

Last but not least, lichens can replace electronic monitoring stations that cost more than $100,000 each. This is not so much a triumph of nature over technology as a partnership between the two, for technicians use highly sophisticated methods to analyze lichens for pollution and measure their rates of photosynthesis.

We all must breathe air from a global atmospheric commons in which air currents and winds can transport some pollutants long distances. Thus, air pollution anywhere is a potential threat elsewhere. Lichens can alert us to the danger, but as with all forms of pollution, the best solution is prevention.

Figure 17-1 Red and yellow crustose lichens growing on slate rock in the foothills of the Sierra Nevada near Merced, California (left), and *Usnea trichodea* lichen growing on a branch of a larch tree in Gifford Pinchot National Park, Washington (right). The vulnerability of various lichen species to specific air pollutants can help researchers detect levels of these pollutants and track down their sources. (Left, Kenneth W. Fink/Ardea, London; right, Milton Rand/Tom Stack & Associates)

I thought I saw a blue jay this morning. But the smog was so bad that it turned out to be a cardinal holding its breath.

MICHAEL J. COHEN

This chapter addresses the following questions:

- What layers are found in the atmosphere?

- What are the major outdoor air pollutants, and where do they come from?

- What are two types of smog?

- What is acid deposition, and how can it be reduced?

- What are the harmful effects of air pollutants?

- How can we prevent and control air pollution?

17-1 THE ATMOSPHERE

What Is the Troposphere? Weather Breeder We live at the bottom of a sea of air called the **atmosphere**. This sea of life-sustaining gases surrounding the earth is divided into several spherical layers (Figure 17-2). Each layer is characterized by abrupt changes in temperature, the result of differences in the absorption of incoming solar energy.

About 75–80% of the mass of the earth's air is found in the atmosphere's innermost layer, the **troposphere**, which extends only about 17 kilometers (11 miles) above sea level at the equator and about 8 kilometers (5 miles) over the poles. If the earth were the size of an apple, this lower layer containing the air we breathe would be no thicker than the apple's skin. This thin and turbulent layer of rising and falling air currents and winds is the planet's weather breeder.

Take a deep breath. About 99% of the volume of the air you inhaled consisted of two gases: nitrogen (78%) and oxygen (21%). The remainder consisted of **(1)** water vapor (varying from 0.01% at the frigid poles to 4% in the humid tropics), **(2)** slightly less than 1% argon (Ar), **(3)** 0.037% carbon dioxide (CO_2), and **(4)** trace amounts of several other gases.

During several billion years of chemical and biological evolution, the composition of the earth's atmosphere has varied (Spotlight, p. 106). A turning point occurred about 2.3–2.7 billion years ago when photosynthesizing cyanobacteria began adding oxygen to the troposphere.

The fact that oxygen now makes up about 21% of the volume of the earth's atmosphere is fortunate for us and other species that survive by aerobic respiration (p. 80). Flammability calculations indicate that if oxygen levels were to rise above 25%, the earth would be an inferno. Essentially all vegetation and other organic material would burn. If levels were to fall below 15%, even the driest material would not burn.

Both the average pressure exerted by the gases in the atmosphere and their average density (mass of gases per unit volume) decrease with altitude. Temperature also declines with altitude in the troposphere but abruptly begins to rise at the top of this zone, called the *tropopause* (Figure 17-2). This temperature change limits mixing between the troposphere and upper layers of the atmosphere.

What Is the Stratosphere? Earth's Global Sunscreen The atmosphere's second layer is the **stratosphere**, which extends from about 17 to 48 kilometers (11 to 30 miles) above the earth's surface (Figure 17-2). Although the stratosphere contains less matter than the troposphere, its composition is similar, with two notable exceptions: **(1)** Its volume of water vapor is about 1/1,000 as much, and **(2)** its concentration of ozone is much higher (Figure 17-3).

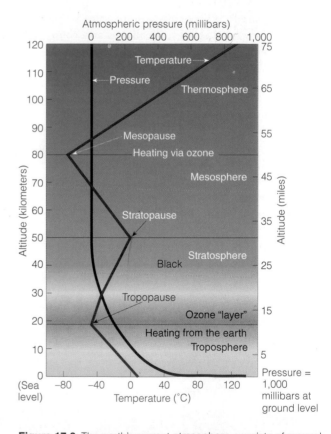

Figure 17-2 The earth's current atmosphere consists of several layers. The average temperature of atmosphere varies with altitude (red line). The average temperature of the atmosphere at the earth's surface is determined by a combination of **(1)** *natural heating* by incoming sunlight and certain greenhouse gases that release absorbed energy as heat into the lower troposphere (the natural *greenhouse effect*, Figure 6-13, p. 128) and **(2)** *natural* cooling by surface evaporation of water and convection processes that transfer heat to higher altitudes and latitudes (Figure 6-7, p. 125, and Figure 6-8, p. 126). Most UV radiation from the sun is absorbed by ozone (O_3), which is found primarily in the stratosphere in the *ozone layer* between 17 and 26 kilometers (10 to 16 miles) above sea level.

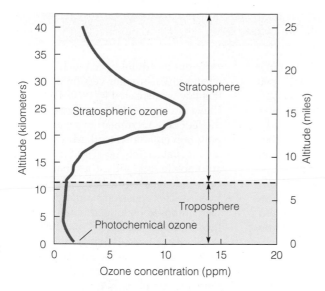

Figure 17-3 Average distribution and concentrations of ozone in the troposphere and stratosphere. *Beneficial ozone* that forms in the stratosphere protects life on earth by filtering out of the incoming harmful ultraviolet radiation emitted by the sun. *Harmful* or *photochemical* ozone forms in the troposphere when various air pollutants undergo chemical reactions under the influence of sunlight. Ozone in this portion of the atmosphere near the earth's surface damages plants, humans, and some materials such as rubber. (Data from NOAA)

Stratospheric ozone is produced when some of the oxygen molecules there interact with ultraviolet (UV) radiation emitted by the sun. This "global sunscreen" of ozone in the stratosphere keeps about 95% of the sun's harmful UV radiation from reaching the earth's surface. This UV filter (1) allows humans and other forms of life to exist on land, (2) helps protect humans from sunburn, skin and eye cancer, cataracts, and damage to the immune system, and (3) prevents much of the oxygen in the troposphere from being converted to photochemical ozone, a harmful air pollutant (Figure 17-3).

Thus, certain levels of *stratospheric ozone* are essential to the continued existence of most life on the earth. On the other hand, very low levels of *tropospheric* or *photochemical ozone* damage plants and materials such as rubber and have harmful effects on our respiratory systems (Figure 17-3). There is much evidence that some human activities are (1) *decreasing* the amount of ozone in the stratosphere and (2) *increasing* the amount of ozone in the troposphere.

What Are Two Important Global Processes Taking Place in the Atmosphere? Two *natural processes* that play crucial roles in the earth's climate and biodiversity are the (1) *greenhouse effect* (Figure 6-13, p. 128), which helps heat the troposphere and the earth's surface, and (2) *ozone shield* in the stratosphere (Figure 17-2), which filters out most of the sun's UV radiation. Both of these natural processes are necessary for life as we

know it. However, there is much evidence that chemicals we add to the atmosphere can (1) enhance the natural greenhouse effect and lead to *global warming* Section 18-2, p. 450 and (2) reduce the concentration of ozone in the stratosphere, an effect called *ozone depletion*, as discussed in Section 18-6 (p. 465).

How Are We Disrupting the Earth's Gaseous Biogeochemical Cycles? The earth's biogeochemical cycles (Section 4-6, p. 87) work well as long as we do not disrupt them by (1) overloading them with essential chemicals at certain points or (2) removing too many vital chemicals at other points. There is ample evidence that our activities are disrupting the

- *Carbon cycle* (Figure 4-29, p. 92) by adding about one-fourth as much CO_2 to the troposphere (mostly from burning fossil fuels and clearing forests faster than they grow back) as the rest of nature. These human inputs of CO_2 have the potential to warm the earth's atmosphere and alter global climate and food-producing regions (Sections 18-3 and 18-4).

- *Nitrogen cycle* (Figure 4-30, p. 94) by releasing three times more nitrogen oxides (NO, NO_2, and N_2O) and gaseous ammonia (NH_3) into the troposphere (mostly from burning fossil fuels and using nitrogen fertilizers) than do natural processes. Most of the nitrogen oxides emitted into the troposphere are converted to nitric acid vapor (HNO_3) and acid-forming nitrate salts. When these compounds dissolve in water and return to the earth's surface, they increase the acidity (Figure 3-7, p. 56) of soils, streams, and lakes and can harm plant and animal life.

- *Sulfur cycle* (Figure 4-33, p. 97) by releasing about twice as much sulfur dioxide (SO_2) into the troposphere (mostly from petroleum refining and burning of coal and oil) as do natural sources. Much of this sulfur dioxide is converted to sulfuric acid (H_2SO_4) and sulfate salts that return to the earth's surface and increase the acidity of soils, streams, and lakes and harm plant and animal life.

- Biogeochemical cycles of toxic metals such as *arsenic, cadmium,* and *lead* by injecting into the troposphere about twice as much arsenic as the rest of nature, 7 times as much cadmium, and 27 times as much lead.

17-2 OUTDOOR AIR POLLUTION

What Are the Major Types and Sources of Air Pollution? **Air pollution** is the presence of one or more chemicals in the atmosphere in sufficient quantities and duration to cause harm to humans, other forms of life, and materials. The effects of airborne pollutants range from annoying to lethal. Air pollution is not new (Spotlight, p. 420).

SPOTLIGHT

Air Pollution in the Past: The Bad Old Days

Modern civilization did not invent air pollution. It probably began when humans discovered fire and used it to burn wood in poorly ventilated caves.

In A.D. 61, the Roman author Seneca wrote of the "stink, soot, and heavy air" from the burning of wood. During the Middle Ages a haze of wood smoke hung over densely packed urban areas.

The industrial revolution brought even worse air pollution as coal was burned to power factories and heat homes. In 1273, King Edward I of England banned the burning of coal in London (except by blacksmiths) to reduce air pollution. However, the ban was ignored because of a scarcity of fuelwood. More than 500 years later, the English poet Shelley observed, "Hell must be much like London, a smoky and populous city."

By the 1850s, London had become well known for its "pea-soup" fog, consisting of a mixture of coal smoke and fog that blanketed the city. In 1880, a prolonged coal fog killed an estimated 2,200 people. Another in 1911 killed more than 1,100 Londoners. The authors of a report on this disaster coined the word *smog* for the deadly mixture of smoke and fog that enveloped the city.

In 1952, an even worse yellow fog lasted for 5 days and killed 4,000 Londoners, prompting Parliament to pass the Clean Air Act of 1956. Additional air pollution disasters in 1956, 1957, and 1962 killed 2,500 more people. Because of

strong air pollution laws, London's air today is much cleaner, and "pea soup" fogs are a thing of the past.

The industrial revolution, powered by coal-burning factories and homes, brought air pollution to the United States. Large industrial cities such as Pittsburgh, Pennsylvania, and St. Louis, Missouri, were known for their smoky air. By the 1940s, the air over some cities was so polluted that people had to use their automobile headlights during the day.

The first documented air pollution disaster in the United States occurred during October 1948, at the town of Donora in Pennsylvania's Monongahela River Valley south of Pittsburgh. Pollutants from the area's industries became trapped in a fog that stagnated over the valley for 5 days. After several days the fog was so dense that people could not see well enough to drive, even at noon with their headlights on. About 7,000 of the town's 14,000 inhabitants became sick, and 22 of them died. This killer fog resulted from a combination of mountainous terrain surrounding the valley and weather conditions that trapped and concentrated deadly pollutants emitted by the community's steel mill, zinc smelter, and sulfuric acid plant.

In 1963, high concentrations of air pollutants accumulated in the air over New York City, killing about 300 people and injuring thousands. Other episodes in New York, Los Angeles, and other large cities in the 1960s led to much stronger air-pollution control programs in the 1970s.

Congress passed the original version of the Clean Air Act in

1963, but it did not have much effect until a stronger version of this law was enacted in 1970. The Clean Air Act of 1970 empowered the federal government to set air pollution emission standards (with an adequate safety margin) for automobiles and industries that each state was required to enforce. Even stricter emission standards were imposed by amendments to the Clean Air Act in 1977 and 1990.

Mostly as a result of these laws and actions by states and local areas, the United States has not had any more Donora or New York City incidents. The *good news* is that in most places smog has decreased and rain has become less acidic. Despite significant population growth and economic growth, emissions of the six most common (criteria) outdoor air pollutants decreased by 31% between 1970 and 1998.

However, the *bad news* is that much more can be done to continue improving air quality in the United States. According to the EPA, **(1)** about 62 million Americans are exposed to levels of one or more of the six most common outdoor air pollutants that exceed federal emission standards, and **(2)** air pollution in the United States causes the premature deaths of 150,000–350,000 people per year (most from exposure to indoor air pollution).

Critical Thinking

Explain why you agree or disagree with the statement that air pollution in the United States should not be a major concern because of the significant progress in reducing outdoor air pollution since 1970.

Table 17-1 lists the major classes of pollutants commonly found in outdoor (ambient) air. Such air pollutants come from both natural sources and human (anthropogenic) activities. Examples of natural sources include **(1)** dust and other forms of suspended particulate matter from wind storms and soil (Figure 6-1, p. 120), **(2)** sulfur oxides and particulate matter from volcanoes, **(3)** carbon oxides, nitrogen oxides, and particulates from forest fires, **(4)** hydrocarbons and pollen from live plants, **(5)** methane and hydrogen sulfide from decaying plants, and **(6)** salt particulates from the sea. Most natural sources of air pollution are spread out and, except for those from volcanic eruptions and some forest fires, rarely reach harmful levels.

Table 17-1 Major Classes of Air Pollutants	
Class	**Examples**
Carbon oxides	Carbon monoxide (CO) and carbon dioxide (CO_2)
Sulfur oxides	Sulfur dioxide (SO_2) and sulfur trioxide (SO_3)
Nitrogen oxides	Nitric oxide (NO), nitrogen dioxide (NO_2), nitrous oxide (N_2O) (NO and NO_2 often are lumped together and labeled NO_x)
Volatile organic compounds (VOCs)	Methane (CH_4), propane (C_3H_8), chlorofluorocarbons (CFCs)
Suspended particulate matter (SPM)	Solid particles (dust, soot, asbestos, lead, nitrate, and sulfate salts), liquid droplets (sulfuric acid, PCBs, dioxins, and pesticides)
Photochemical oxidants	Ozone (O_3), peroxyacyl nitrates (PANs), hydrogen peroxide (H_2O_2), aldehydes
Radioactive substances	Radon-222, iodine-131, strontium-90, plutonium-239 (Table 3-2, p. 62)
Hazardous air pollutants (HAPs), which cause health effects such as cancer, birth defects, and nervous system problems	Carbon tetrachloride (CCl_4), methyl chloride (CH_3Cl), chloroform ($CHCl_3$), benzene (C_6H_6), ethylene dibromide ($C_2H_2Br_2$), formaldehyde (CH_2O_2)

Most outdoor pollutants in urban areas enter the atmosphere from the burning of fossil fuels in power plants and factories (*stationary sources*) and motor vehicles (*mobile sources*). According to the World Bank and the United Nations, **(1)** burning coal to produce electricity is responsible for 67% of SO_2, 36% of CO_2, 33% of mercury, and 28% of NO_x emissions each year in the United States and **(2)** vehicle emissions account for 75% of CO, 33% of CO_2, and 44% of NO_x in U.S. urban air. In car-clogged cities such as Los Angeles, California; São Paulo, Brazil; Bangkok, Thailand (Figure 11-1, p. 238); Rome, Italy; and Mexico City, Mexico (Case Study, p. 660), motor vehicles are responsible for 80–88% of the air pollution.

Primary pollutants are those emitted directly into the troposphere in a potentially harmful form. While in the troposphere, some of these primary pollutants may react with one another or with the basic components of air to form new pollutants, called **secondary pollutants** (Figure 17-4).

With their large concentrations of cars and factories, cities normally have higher air pollution levels than rural areas. However, prevailing winds can spread long-lived primary and secondary air pollutants emitted in urban and industrial areas to the countryside and to other downwind urban areas.

Indoor pollutants come from **(1)** infiltration of polluted outside air and **(2)** various chemicals used or produced inside buildings, as discussed in more detail in Section 17-5. Risk analysis experts rate indoor and outdoor air pollution as high-risk human health problems (Figure 16-13, left, p. 411).

According to the World Health Organization (WHO), more than 1.1 billion people—one of every five—live in urban areas where the air is unhealthy to breathe. Most live in densely populated cities in developing countries where air pollution control laws do not exist or are poorly enforced.

In the United States (and in most other countries), government-mandated standards set maximum allowable atmospheric concentrations for six *criteria air pollutants* commonly found in outdoor air (Table 17-2).

17-3 PHOTOCHEMICAL AND INDUSTRIAL SMOG

What Is Photochemical Smog? Brown-Air Smog? Any chemical reaction activated by light is called a *photochemical reaction*. Air pollution known as **photochemical smog** is a mixture of primary and secondary pollutants formed under the influence of sunlight (Figure 17-5, p. 424). The resulting mixture of more than 100 chemicals is dominated by *ozone*, a highly reactive gas that harms most living organisms (Figure 17-3).

Here is a simplified version of the complex chemistry of photochemical smog formation. It begins when nitrogen and oxygen in air react at the high temperatures found inside automobile engines and the boilers in coal-burning power and industrial plants to produce colorless nitric oxide ($N_2 + O_2 \longrightarrow$ 2NO). Once in the troposphere, the nitric oxide slowly reacts with oxygen to form nitrogen dioxide, a yellowish-brown gas with a choking odor ($2NO + O_2 \longrightarrow 2NO_2$). The NO_2

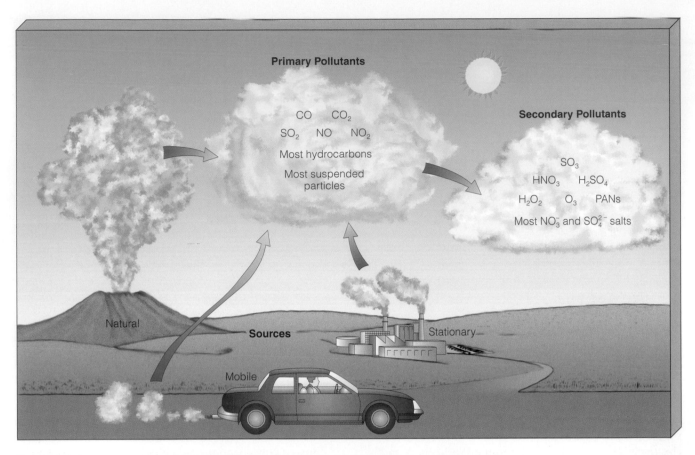

Figure 17-4 Sources and types of air pollutants. Human inputs of air pollutants may come from *mobile sources* (such as cars) and *stationary sources* (such as industrial and power plants). Some *primary air pollutants* may react with one another or with other chemicals in the air to form *secondary air pollutants*.

Table 17-2 Common Criteria Air Pollutants in the United States

CARBON MONOXIDE (CO)

Description: Colorless, odorless gas that is poisonous to air-breathing animals; forms during the incomplete combustion of carbon-containing fuels ($C + O_2 \longrightarrow 2CO$).

Major human sources: Cigarette smoking (p. 396), incomplete burning of fossil fuels. About 77% (95% in cities) comes from motor vehicle exhaust.

Health effects: Reacts with hemoglobin in red blood cells and reduces the ability of blood to bring oxygen to body cells and tissues. This impairs perception and thinking; slows reflexes; causes headaches, drowsiness, dizziness, and nausea; can trigger heart attacks and angina; damages the development of fetuses and young children; and aggravates chronic bronchitis, emphysema, and anemia. At high levels it causes collapse, coma, irreversible brain cell damage, and death.

NITROGEN DIOXIDE (NO₂)

Description: Reddish-brown, irritating gas that gives photochemical smog its brownish color; in the atmosphere can be converted to nitric acid (HNO_3), a major component of acid deposition.

Major human sources: Fossil fuel burning in motor vehicles (49%) and power and industrial plants (49%).

Health effects: Lung irritation and damage; aggravates asthma and chronic bronchitis; increases susceptibility to respiratory infections such as the flu and common colds (especially in young children and older adults).

Environmental effects: Reduces visibility; acid deposition of HNO_3 can damage trees, soils, and aquatic life in lakes.

Property damage: HNO_3 can corrode metals and eat away stone on buildings, statues, and monuments; NO_2 can damage fabrics.

SULFUR DIOXIDE (SO₂)

Description: Colorless, irritating; forms mostly from the combustion of sulfur-containing fossil fuels such as coal and oil ($S + O_2 \longrightarrow SO_2$); in the atmosphere can be converted to sulfuric acid (H_2SO_4), a major component of acid deposition.

Major human sources: Coal burning in power plants (88%) and industrial processes (10%).

Health effects: Breathing problems for healthy people; severe restriction of airways in people with asthma; chronic exposure can cause a permanent condition similar to bronchitis. According to the World Health Organization, at least 625 million people are exposed to unsafe levels of sulfur dioxide from fossil fuel burning.

Environmental effects: Reduces visibility; acid deposition of H_2SO_4 can damage trees, soils, and aquatic life in lakes.

Property damage: SO_2 and H_2SO can corrode metals

is responsible for the brownish haze that hangs over many cities during the afternoons of sunny days, explaining why photochemical smog sometimes is called *brown-air smog*.

Some of the NO_2 reacts with water vapor in the atmosphere to form nitric acid vapor and nitric oxide ($3NO_2 + H_2O \longrightarrow 2HNO_3 + NO$). When the remaining NO_2 is exposed to ultraviolet radiation from the sun, some of it is converted to nitric oxide and oxygen atoms ($NO_2 + UV\ radiation \longrightarrow NO + O$). The highly reactive oxygen atoms then react with O_2 to produce ozone ($O_2 + O \longrightarrow O_3$). Both the oxygen atoms and ozone then react with volatile organic compounds (mostly hydrocarbons released by vegetation, vehicles, gas stations, oil refineries, and dry cleaners) to produce aldehydes. In addition, hydrocarbons, oxygen, and nitrogen dioxide react to produce peroxyacyl nitrates, or PANs (hydrocarbons + O_2 + $NO_2 \longrightarrow$ PANs).

Collectively, NO_2, O_3, and PANs are called *photochemical oxidants* because they can react with and oxidize certain compounds in the atmosphere (or inside your lungs) that normally are not oxidized. Mere traces of these photochemical oxidants (especially ozone) and aldehydes in photochemical smog can irritate the respiratory tract and damage crops and trees.

The hotter the day, the higher the levels of ozone and other components of photochemical smog. As traffic increases in the morning, levels of NO_x and unburned hydrocarbons rise and begin reacting in the presence of sunlight to produce photochemical smog. On a sunny day the photochemical smog (dominated by O_3) builds up to peak levels by early afternoon, irritating people's eyes and respiratory tracts (Figure 17-6).

All modern cities have photochemical smog, but it is much more common in cities with sunny, warm, dry climates and lots of motor vehicles. Examples of such cities are Los Angeles, California; Denver, Colorado; and Salt Lake City, Utah in the United States, as well as Sydney, Australia; Mexico City, Mexico; and São Paulo and Buenos Aires in Brazil. According to a 1999 article in *Geophysical Research Letters*, if 400 million Chinese drivers drive cars by 2050 as projected, the resulting photochemical smog could cover the entire western Pacific in ozone, extending to the United States.

High levels of ozone have also been measured above the tropical forests of the Amazon basin, where cars are not the culprit. In this case, hydrocarbons naturally released by the trees and other vegetation can interact with natural levels of NO_2 and sunlight to produce ozone. In 2000, researchers at the University of California at Irvine found that atmospheric particles of sea salt (NaCl) can be converted by sunlight into chlorine atoms. When these highly reactive Cl atoms come in contact with pollutants emitted from fossil fuel combustion they can form ground-level ozone.

What Is Industrial Smog? Gray-Air Smog Fifty years ago cities such as London, England, and Chicago and Pittsburgh in the United States burned large amounts of coal and heavy oil (which contain sulfur impurities; Figure 14-27, p. 344) in power plants and factories and for space heating. During winter, people

---

and eat away stone on buildings, statues, and monuments; SO_2 can damage paint, paper, and leather.

SUSPENDED PARTICULATE MATTER (SPM)

Description: Variety of particles and droplets (aerosols) small and light enough to remain suspended in atmosphere for short periods (large particles) to long periods (small particles; Figure 17-7); we see these particles as smoke, dust, and haze.

Major human sources: Burning coal in power and industrial plants (40%), burning diesel and other fuels in vehicles (17%), agriculture (plowing, burning off fields), unpaved roads, construction.

Health effects: Nose and throat irritation, lung damage, and bronchitis; aggravates bronchitis and asthma; causes early death; toxic particulates (such as lead, cadmium, PCBs, and dioxins) can cause mutations, reproductive problems, or cancer.

Environmental effects: Reduces visibility; acid deposition of H_2SO_4 droplets can damage trees, soils, and aquatic life in lakes.

Property damage: Corrodes metal; soils and discolors buildings, clothes, fabrics, and paints.

OZONE (O_3)

Description: Highly reactive, irritating gas with an unpleasant odor that forms in the troposphere as a major component of photochemical smog (Figures 17-5 and 17-6).

Major human sources: Chemical reaction with volatile organic compounds (VOCs, emitted mostly by cars and industries) and nitrogen oxides to form photochemical smog (Figure 17-5).

Health effects: Breathing problems; coughing; eye, nose, and throat irritation; aggravates chronic diseases such as asthma, bronchitis, emphysema, and heart disease; reduces resistance to colds and pneumonia; may speed up lung tissue aging.

Environmental effects: Ozone can damage plants and trees; smog can reduce visibility.

Property damage: Damages rubber, fabrics, and paints.

LEAD

Description: Solid toxic metal and its compounds, emitted into the atmosphere as particulate matter.

Major human sources: Paint (old houses), smelters (metal refineries), lead manufacture, storage batteries, leaded gasoline (being phased out in developed countries).

Health effects: Accumulates in the body; brain and other nervous system damage and mental retardation (especially in children); digestive and other health problems; some lead-containing chemicals cause cancer in test animals.

Environmental effects: Can harm wildlife. Data from U.S. Environmental Protection Agency.

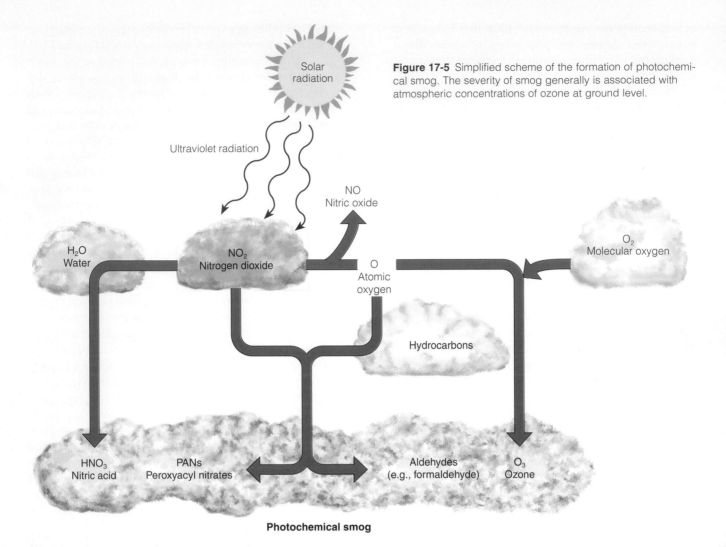

Figure 17-5 Simplified scheme of the formation of photochemical smog. The severity of smog generally is associated with atmospheric concentrations of ozone at ground level.

Photochemical smog

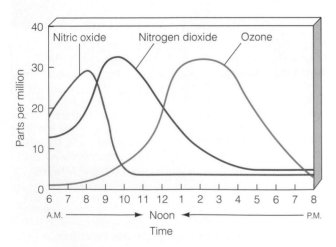

Figure 17-6 Typical daily pattern of changes in concentrations of air pollutants that lead to development of photochemical smog in a city such as Los Angeles, California.(From *Environmental Science*, 5/E by Chiras, p. 363. Copyright ©2000 by Wadsworth)

carbon dioxide (C + O_2 ⟶ CO_2) and carbon monoxide (2C + O_2 ⟶ 2CO). Some of the unburned carbon also ends up in the atmosphere as suspended particulate matter (soot).

The sulfur compounds in coal and oil also react with oxygen to produce sulfur dioxide, a colorless, suffocating gas (S + O_2 ⟶ SO_2). Sulfur dioxide also is emitted into the troposphere when metal sulfide ores (such as lead sulfide, PbS) are roasted or smelted to convert the metal ore to the free metal (Figure 14-8, p. 327).

In the troposphere, some of the sulfur dioxide reacts with oxygen to form sulfur trioxide ($2SO_2$ + O_2 ⟶ $2SO_3$), which then reacts with water vapor in the air to produce tiny suspended droplets of sulfuric acid (SO_3 + H_2O ⟶ H_2SO_4). Some of these

in such cities were exposed to **industrial smog** consisting mostly of **(1)** sulfur dioxide, **(2)** suspended droplets of sulfuric acid (formed from some of the sulfur dioxide, Figure 17-4), and **(3)** a variety of suspended solid particles and droplets (called aerosols; Figure 17-7).

The chemistry of industrial smog is fairly simple. When burned, the carbon in coal and oil is converted to

droplets react with ammonia in the atmosphere to form solid particles of ammonium sulfate ($2NH_3 + H_2SO_4 \longrightarrow (NH_4)_2SO_4$). The tiny suspended particles of such salts and carbon (soot) give the resulting industrial smog a gray color, explaining why it is sometimes called *gray-air smog.*

Urban industrial smog is rarely a problem today in most developed countries because coal and heavy oil are burned only in large boilers with reasonably good pollution control or with tall smokestacks (which transfer the pollutants to downwind areas). However, industrial smog is a problem in industrialized urban areas of China, India, Ukraine, and some eastern European countries, where large quantities of coal are burned with inadequate pollution controls.

For example, the world's 10 most air-polluted cities are found in China (9 cities) and India (1 city), mostly because of heavy use of coal. An estimated 800 million of China's 1.2 billion people burn coal in their homes for cooking and heating, often in unvented stoves that fill their homes with toxic fumes. Globally, indoor air pollution from such coal burning accounts for about 2.8 million of the estimated 3 million people who die prematurely each year from air pollution.

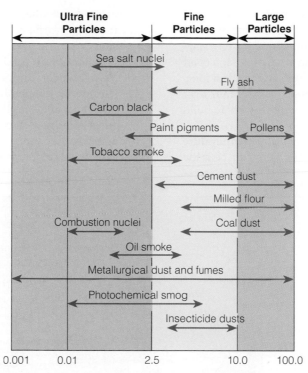

Figure 17-7 Suspended particulate matter consists of particles of solid matter and droplets of liquid that are small and light enough to remain suspended in the atmosphere for short periods (large particles) to long periods (small particles). Suspended particles are found in a wide variety of types and sizes, ranging from 0.001 micrometer to 100 micrometers (a micrometer, or micron, is one millionth of a meter, or about 0.00004 inches). Since 1987, the EPA has focused on *fine particles* smaller than 10 microns (known as *PM-10*). In 1997, the agency began focusing on reducing emissions of *ultrafine particles* with diameters less than 2.5 microns (known as *PM-2.5*) because these particles are small enough to reach the lower part of human lungs and contribute to respiratory diseases.

Wind and rain can help cleanse urban air, but because of the law of conservation of matter (p. 60), the pollutants do not disappear. They are blown somewhere else or are deposited from the sky onto surface waters, soil, and buildings.

In addition, a 2000 study of clean and dirty clouds over three countries by atmospheric scientist Daniel Rosenfield indicated that air pollution can reduce the rain and snowfall that help cleanse the air of pollutants. He found that clean clouds hold large droplets of water that can easily be converted into rain or snow. However, dirty clouds contain small water droplets that are less likely to bump into each other and stick together in large enough clumps to form air-cleansing raindrops and snowflakes.

Three factors that can increase air pollution are:

- *Urban buildings* that can slow wind speed and reduce dilution and removal of pollutants.

- *Hills and mountains* that tend to reduce the flow of air in valleys below them and allow pollutant levels to build up at ground level.

- *High temperatures* that promote the chemical reactions that lead to photochemical smog formation.

What Are Temperature Inversions? During daylight the sun warms the air near the earth's surface. Normally, this warm air and most of the pollutants it contains rises to mix with the cooler air above it. This mixing of warm and cold air creates turbulence, which disperses the pollutants.

Under certain atmospheric conditions, however, a layer of warm air can lie atop a layer of cooler air nearer the ground, a situation known as a **temperature inversion**. Because the cooler air is denser than the warmer air above it, the air near the surface does not rise and mix with the air above it. Pollutants can concentrate in this stagnant layer of cool air near the ground.

There are two types of temperature inversions. One, called a **subsidence temperature inversion**,

What Factors Influence the Formation of Photochemical and Industrial Smog? The frequency and severity of smog in an area depend on **(1)** local climate and topography, **(2)** population density, **(3)** the amount of industry, and **(4)** the fuels used in industry, heating, and transportation.

Air pollution can be reduced by:

- *Rain and snow*, which help cleanse the air of pollutants. This explains why cities with dry climates are more prone to photochemical smog than ones with wet climates.

- *Winds*, which help sweep pollutants away, dilute pollutants by mixing them with cleaner air, and bring in fresh air.

occurs when a large mass of warm air moves into a region at a high altitude and floats over a mass of colder air near the ground. This keeps the air over a city stagnant and prevents vertical mixing and dispersion of air pollutants. Normally such conditions do not last long, but sometimes warmer air masses can remain over cooler air below for days and allow pollutants to build up to harmful levels.

The second type, called a **radiation temperature inversion**, typically occurs at night as the air near the ground cools faster than the air above it. As the sun rises and warms the earth's surface, a radiation inversion normally disappears by noon and disperses the pollutants built up during the night.

However, under certain conditions, radiation or subsidence temperature inversions can last for several days and allow pollutants to build up to dangerous concentrations. Areas with two types of topography and weather conditions are especially susceptible to prolonged temperature inversions (Figure 17-8).

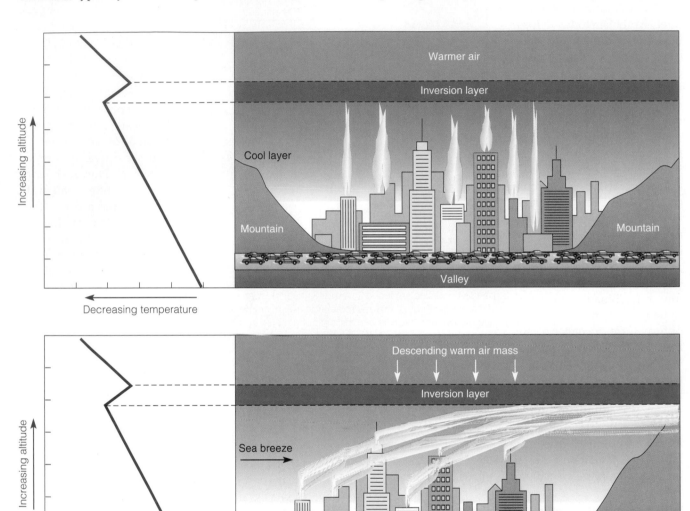

Figure 17-8 Topography and weather conditions that can create more frequent and prolonged *radiation temperature inversions*, in which a layer of warm air sits atop a cooler layer of air near the ground. In such cases, pollutant concentrations in the air near the earth's surface can build up to harmful levels. The left figure shows how air pollutants can build up in the air near the ground in a valley surrounded by mountains. The right figure shows how frequent and prolonged radiation temperature inversions can occur in an area (such as Los Angeles, California) with a sunny climate, light winds, mountains on three sides, and the ocean on the other. The layer of descending warm air (right) prevents ascending air currents from dispersing and diluting pollutants from the cooler air near the ground. Because of their topography, Los Angeles in the United States and Mexico City in Mexico have frequent thermal inversions, many of them prolonged during the summer.

One such area is a town or city located in a valley surrounded by mountains that experiences cloudy and cold weather during part of the year (Figure 17-8, top). In such cases, the surrounding mountains block out the winter sun needed to reverse the nightly radiation temperature inversion. As long as these stagnant conditions persist, concentrations of pollutants in the valley below will build up to harmful and even lethal concentrations. This is what happened during the 1948 air pollution disaster in the valley town of Donora, Pennsylvania (Spotlight, p. 420).

A city with several million people and motor vehicles in an area with a (1) sunny climate, (2) light winds, (3) mountains on three sides, and (4) the ocean on the other has ideal conditions for photochemical smog worsened by frequent subsidence thermal inversions (Figure 17-8, bottom). This describes California's heavily populated Los Angeles basin, which has prolonged subsidence temperature inversions at least half of the year, mostly during the warm summer and fall months. High-pressure air off the coast of California much of the year creates a descending warm air mass that sits atop an air mass below that is cooled by the nearby ocean. When this subsidence thermal inversion persists throughout the day, the surrounding mountains prevent the polluted surface air from being blown away by sea breezes (Figure 17-8, bottom).

17-4 REGIONAL OUTDOOR AIR POLLUTION FROM ACID DEPOSITION

What Is an Acid, and How Is Acidity Measured? An **acid** is any chemical that releases hydrogen ions (H^+) when dissolved in water. Examples of acids include hydrochloric acid (HCl), nitric acid (HNO_3), sulfuric acid (H_2SO_4), and carbonic acid (H_2CO_3). The higher the concentration of hydrogen ions in a solution, the more acidic the solution.

A numerical scale of pH values is commonly used to express hydrogen ion concentrations. The lower the pH of a water solution, the greater its acidity. Solutions with pH values less than 7 are *acidic*, and those with pH values greater than 7 are *alkaline* or *basic* (Figure 3-7, p. 56).

Each whole number change on the pH scale represents a tenfold change in the concentration of hydrogen ions in a water solution. For example, a solution with a pH of 4 is 10 times as acidic (has 10 times as many H^+ ions per unit of volume) as one with a pH of 5 and 100 times as acidic as a solution with a pH of 6. How many times as acidic is a solution with a pH of 2 as a solution with a pH of 6?

Water and carbon dioxide in the atmosphere react to produce a weakly acidic solution of carbonic acid ($H_2O + CO_2 \longrightarrow H_2CO_3$). As a result, natural rainfall has a pH of about 5.6 (Figure 3-7, p. 56).

What Is Acid Deposition? Most coal-burning power plants, ore smelters, and other industrial plants in developed countries use tall smokestacks to emit sulfur dioxide, suspended particles, and nitrogen oxides above the inversion layer (Figure 17-8), where mixing, dilution, and removal by wind are more effective. Thus, tall smokestacks reduce *local* air pollution but increase *regional* air pollution downwind.

The primary pollutants, sulfur dioxide and nitrogen oxides, emitted into the atmosphere above the inversion layer are transported as much as 1,000 kilometers (600 miles) by prevailing winds. During their trip, they form secondary pollutants such as nitric acid vapor, droplets of sulfuric acid, and particles of acid-forming sulfate and nitrate salts (Figure 17-4).

These acidic substances remain in the atmosphere for 2-14 days, depending mostly on prevailing winds, precipitation, and other weather patterns. During this period they descend to the earth's surface in two forms: (1) *wet deposition* (as acidic rain, snow, fog, and cloud vapor with a pH less than 5.6) and (2) *dry deposition* (as acidic particles). The resulting mixture is called **acid deposition** (Figure 17-9), sometimes called *acid rain*. Most dry deposition occurs within about 2–3 days fairly near the emission sources, whereas most wet deposition occurs takes place in 4-14 days in more distant downwind areas (Figure 17-9).

What Areas Are Most Affected by Acid Deposition? Acid deposition is a regional problem in the eastern United States (Figure 17-10) and in other parts of the world (Figure 17-11). Most of these regions are downwind from coal-burning power plants, smelters, or factories or are major urban areas with large numbers of motor vehicles.

In the United States, coal-burning power and industrial plants in the Ohio Valley (Figure 17-10) emit the largest quantities of sulfur dioxide and other acidic pollutants. Mostly as a result of these emissions, typical precipitation in the eastern United States has a pH of 4.2 to 4.7 (Figure 17-10), 10 or more times the acidity of natural precipitation with a pH of 5.6. Some mountaintop forests in the eastern United States and east of Los Angeles, California, are bathed in fog and dews as acidic as lemon juice, with a pH of 2.3—about 1,000 times the acidity of normal precipitation.

A **buffer** is a substance that can react with hydrogen ions in a solution and thus hold the pH fairly constant. Calcium carbonate ($CaCO_3$), or limestone, is a natural buffer that protects many soils and lakes from acid precipitation. Hydrogen ions in acids percolating through soils derived from limestone rock are neutralized (buffered) when they react with $CaCO_3$ ($CaCO_3 + 2H^+ \longrightarrow Ca^{2+} + CO_2 + H_2O$).

The areas most sensitive to acid deposition are (1) those containing thin, acidic soils derived mostly from granitic rock without such natural buffering

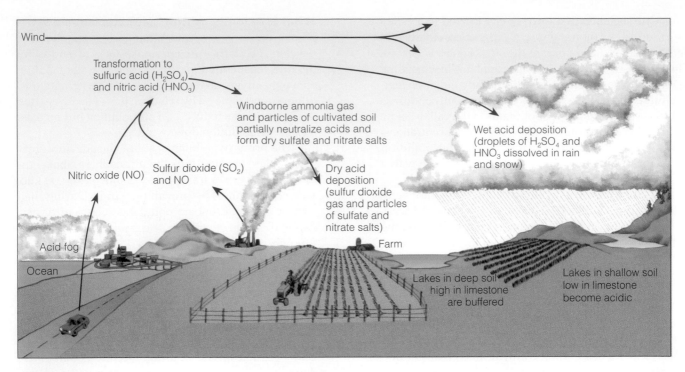

Figure 17-9 Acid deposition, which consists of rain, snow, dust, or gas with a pH lower than 5.6, is commonly called acid rain. Soils and lakes vary in their ability to buffer or remove excess acidity.

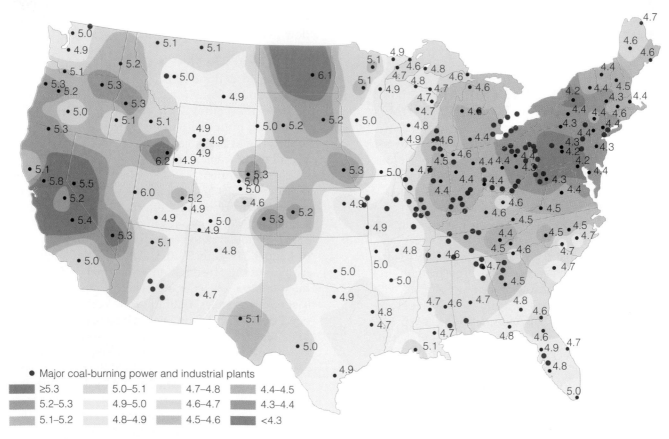

- Major coal-burning power and industrial plants

≥5.3	5.0–5.1	4.7–4.8	4.4–4.5
5.2–5.3	4.9–5.0	4.6–4.7	4.3–4.4
5.1–5.2	4.8–4.9	4.5–4.6	<4.3

Figure 17-10 pH values from field measurements in 48 states in 1998. Red dots show major sources of sulfur dioxide (SO_2) emissions, mostly large coal-burning power plants. In the eastern United States, the primary component of acid deposition is H_2SO_4 (formed from SO_2 emitted by coal-burning plants). In the western United States, HNO_3 predominates (formed mostly from NO_x emissions from motor vehicles). (National Atmospheric Deposition Program [NRSP-3]/National Trends Network, 1998. NADP Program Office, Illinois State Water Survey, 2204 Griffith Dr., Champaign, IL 61820)

(Figure 17-11, green and most red areas) and **(2)** those in which the buffering capacity of soils has been depleted by decades of acid deposition (some red areas in Figure 17-11).

Many acid-producing chemicals generated by power plants, factories, smelters, and cars in one country are exported to other countries by prevailing winds. For example, studies show that some acid deposition in:

- Norway, Switzerland, Austria, Sweden, the Netherlands, and Finland is blown to those countries from industrialized areas of western Europe (especially the United Kingdom and Germany) and eastern Europe.

- Southeastern Canada (Figure 17-11) can be traced to SO_2 and other emissions from the Ohio Valley of the United States (Figure 17-10).

- The eastern United States has been traced to emissions from two large metal smelters in southeastern Canada.

- Japan and North and South Korea comes from China.

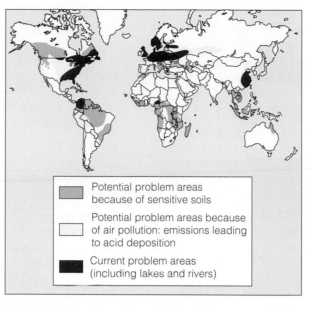

Figure 17-11 Regions where acid deposition is now a problem (red) and regions with the potential to develop this problem (yellow). Such regions have **(1)** large inputs of air pollution (mostly from power plants, industrial plants, and ore smelters, red dots in Figure 17-10) or **(2)** sensitive areas with soils and bedrock that cannot neutralize (buffer) inputs of acidic compounds (green areas and most red areas). (Data from World Resources Institute and U.S. Environmental Agency)

The worst acid deposition is in Asia, especially in China, which gets 73% of its energy from burning coal. Acid deposition already affects more than one-fourth of China's land area and could increase over the next two decades because of projected increases in SO_2 and NO_x emissions. This could **(1)** overwhelm many of the country's soils, **(2)** reduce crop productivity (Case Study, p. 272), and **(3)** hinder efforts to restore deforested areas. Acid deposition is also a growing problem in eastern Europe, Russia, Nigeria, Venezuela, and Colombia (Figure 17-11).

In 1999, researchers at California's Scripps Institution of Oceanography found a thick, brown haze of air pollution covering much of the Indian Ocean during winter (Figure 17-12). The haze-covered area is about the size of the United States and rises as high as 3,000 meters (10,000 feet). During the late spring and summer, when prevailing winds reverse, some of the haze

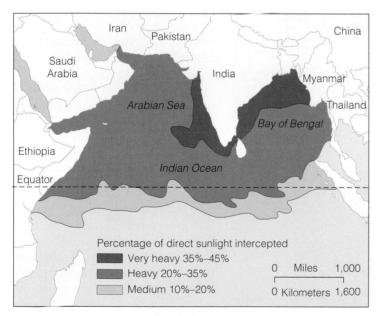

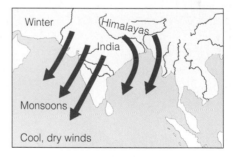

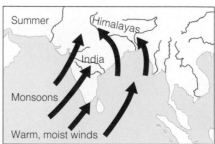

Figure 17-12 A thick, brown haze of air pollution occurs over much of the Indian Ocean during winter. This air pollution comes from burning fossil fuels for transportation and industry in China, Southeast Asia, and India. These pollutants are blown out over the ocean during the winter monsoon season when prevailing winds flow down from the Himalayas and out to sea. In the late spring and summer, the winds reverse. This can blow some of the haze back across the land, where it can combine with monsoon rains and fall as acid deposition. (Data from Veerabhadran Ramanathan, Scripps Institution of Oceanography)

can be blown back onto the land, where it can combine with monsoon rains and fall as acid deposition.

What Are the Effects of Acid Deposition on Human Health, Materials, and the Economy?
Acid deposition:

- Contributes to human respiratory diseases such as bronchitis and asthma, as discussed in Section 17-6.

- Can leach toxic metals such as lead and copper from water pipes into drinking water.

- Damages statues, buildings, metals, and car finishes. Limestone and marble (which is a form of limestone) are especially susceptible because they dissolve even in weak acid solutions ($CaCO_3 + 2H^+ \rightarrow Ca^{2+} + CO_2 + H_2O$).

- Decreases atmospheric visibility (mostly because of sulfate particles).

- Lowers profits and causes job losses because of lower productivity in fisheries, forests, and farms.

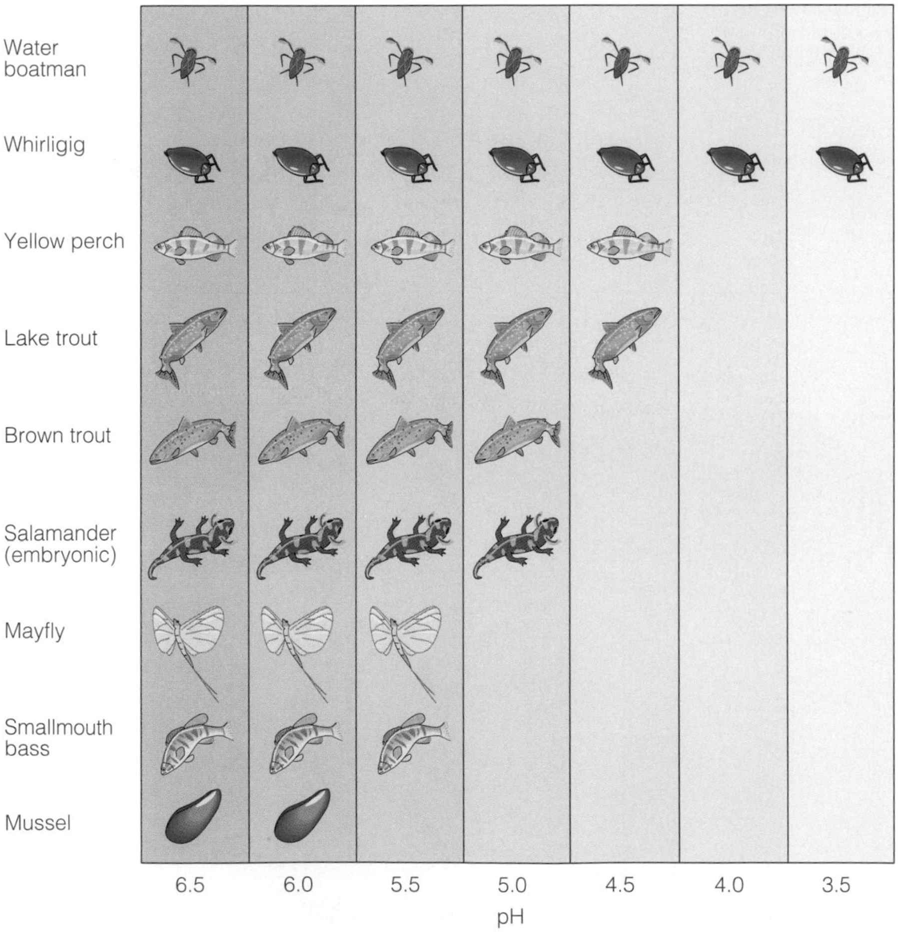

What Are the Effects of Acid Deposition on Aquatic Ecosystems?
Most freshwater lakes, streams, and ponds have a natural pH in the range of 6 to 8. Acid deposition has many harmful ecological effects when the pH of most aquatic systems falls below 6 and especially below 5 (Figure 17-13).

Here are some effects of increased acidity on aquatic systems:

- As the pH approaches 5, less desirable species of plankton and mosses may begin to invade, and populations of fish such as smallmouth bass disappear (Figure 17-13).

- Below a pH of 5, (1) fish populations begin to disappear (Figure 17-13), (2) the bottom is covered with undecayed material, and (3) mosses may dominate nearshore areas.

- Below a pH of 4.5, the water is essentially devoid of fish (Figure 17-13).

- Aluminum ions (Al^{3+}) attached to minerals in nearby soil can be released into lakes, where they can kill many kinds of fish by stimulating excessive mucus formation. This asphyxiates the fish by clogging their gills.

- Toxic mercury emitted by coal-burning plants can contaminate lakes, streams, and fish in downwind areas. Small aquatic organisms can transform inorganic mercury compounds deposited in lakes by air deposition or leached from nearby soils into highly toxic *methylmercury*. Methylmercury is soluble in the fatty tissue of fish. Thus, it can contaminate fish and be biomagnified to higher concentrations in aquatic food chains and webs (Figure 16-4, p. 399) and poison humans, birds, marine mammals, and other wildlife. Humans who eat the contaminated fish can suffer permanent kidney failure, tremors, severe brain damage, and death.

Much of the damage to aquatic life in sensitive areas with little buffering capacity (Figure 17-11) is a result of *acid shock*. This is caused by the sudden runoff of large amounts of highly acidic water and aluminum ions into lakes and streams, when snow melts in the spring or after unusually heavy rains.

Because of excess acidity,

- In Norway and Sweden, at least 16,000 lakes contain no fish, and

Figure 17-13 Fish and other aquatic organisms vary in their sensitivity to acidity. The figure shows the lowest pH (highest acidity) at which the various species can survive. Note that the greatest effects occur when the pH drops below 5.5. (From *Environmental Science*, 5/E by Chiras, p. 393. Copyright ©2000 by Wadsworth)

52,000 more lakes have lost most of their acid-neutralizing capacity.

- In Canada, some 14,000 acidified lakes contain few if any fish, and some fish populations in 150,000 more lakes are declining because of increased acidity.

- In the United States, about 9,000 lakes (most in the northeast and upper Midwest, Figure 17-10) are threatened with excess acidity, one-third of them seriously. According to the Environmental Protection Agency's National Surface Water Survey (NSWS), acid deposition causes about 75% of the acidic lakes and about 50% of the acidic streams in the United States.

What Are the Effects of Acid Deposition on Plants and Soil Chemistry? Acid deposition (often along with other air pollutants such as ozone) can harm forests and crops, especially when the soil pH falls below 5.1. Some of this injury is direct. For example, acidic substances can damage the protective waxy surface of leaves and needles. This lowers the resistance of plants to disease and allows nutrients to be leached from leaves and needles. Acid deposition can also impair germination of the seeds of trees such as spruce.

Most of the damage to forests and crops is indirect and is caused by chemical interactions in forest and cropland soils. At first, acid deposition can stimulate tree and plant growth by adding nitrogen and sulfur nutrients to the soil. In time, however, continued inputs of these acidic chemicals can

- Leach essential plant nutrients such as calcium and magnesium salts from soils. This reduces plant productivity and the ability of the soils to buffer or neutralize acids.

- Release aluminum ions (Al^{3+}) attached to insoluble soil compounds, which can hinder uptake and use of soil nutrients and water by plants.

- Dissolve insoluble soil compounds and release ions of metals such as lead, cadmium, and mercury that can be absorbed by plants and are highly toxic to plants and animals.

- Promote the growth of acid-loving mosses that kill trees by (1) holding so much water that feeder roots are drowned, (2) eliminating air from the soil, and (3) killing mycorrhizal fungi that help tree roots absorb nutrients.

- Weaken trees and other plants so they become more susceptible to other types of damage such as (1) severe cold, (2) diseases, (3) insect attacks, (4) drought, and (5) harmful mosses.

Many of these factors can interact synergistically to make matters worse (Connections, below).

Hidden Synergistic Interactions and Forest Decline

CONNECTIONS

Loss of calcium (and magnesium) from soils can lead to several interacting, cascading, and synergistic effects. Acting over several decades, these mostly hidden interactions can escalate forest damage and decline through a positive feedback mechanism (p. 51).

Reduced calcium decreases plant productivity and the ability of the soil to neutralize (buffer) incoming acids. This in turn causes a sharp rise in soil acidification, which further depletes soil nutrients.

The increase in acidity also releases aluminum ions from the soil. This reduces the ability of tree roots to absorb remaining calcium and other soil nutrients and further decreases tree productivity.

Increased acidity can also cause a decline in earthworms, beetles, and other soil organisms that decompose leaves and woody debris and recycle calcium and other soil nutrients (Figure 10-13, p. 221). This slows down replenishment of calcium and other plant nutrients, further aggravating forest damage and decline.

It can take decades to hundreds or thousands of years for soil development to replenish nutrients leached out by acid deposition. Thus, losses in plant productivity could continue for many decades even when air-pollution control programs reduce emissions of sulfur dioxide and nitrogen oxides.

There are other interactions. Nitrogen is often a limiting nutrient that determines the amount of plant growth for many species in terrestrial ecosystems. Until recently scientists expected some of the acid-caused loss in forest productivity to be offset by increased productivity from the larger input of nitric acid and nitrate salts from acid deposition.

However, things are not that simple. A higher nitrogen input increases the growth of certain trees and plants. However, too much nitrogen can (1) inhibit the ability of plant roots to absorb soil nutrients and (2) reduce a tree's ability to withstand cold weather. The harmful effects from excess nitrogen and increased acidity can interact synergistically to amplify plant productivity losses and tree death.

These interacting changes in soil chemistry can take place underground and remain hidden for years until at some threshold point most of a forest's dominant tree species begin dying.

Critical Thinking

Why should we be concerned about forest damage and decline resulting from prolonged exposure to acid deposition and other air pollutants?

Mountaintop forests are the terrestrial areas hardest hit by acid deposition because **(1)** they tend to have thin soils without much buffering capacity and **(2)** trees on mountaintops (especially conifers such as red spruce and balsam fir that keep their leaves year-round) are bathed almost continuously in very acidic fog and clouds.

Many interacting factors can decrease the health of trees (Figure 17-14 and Connections, p. 431). Any one of these stresses may be the final cause of severe tree damage or death. However, the underlying cause often is years of exposure to an atmospheric cocktail of air pollutants and soil overloaded with acids, which weakens trees and makes them more vulnerable to other stresses.

For example, scientists have analyzed decades of data on nutrient losses from soils at Yale University's Hubbard Brook Experimental Forest in New Hampshire. They found that since 1960 acid deposition has leached out more than 50% of the available calcium in the generally alkaline (and thus buffering) soils found in this research forest. As a result, there has been no net tree growth since the 1980s.

How Serious Is Acid Deposition in the United States? Since 1980 the federal government has sponsored the National Acid Precipitation Assessment Program (NAPAP) to **(1)** coordinate government acid deposition research, **(2)** assess the costs, benefits, and effectiveness of the country's acid deposition legislation and control program, and **(3)** report its findings to Congress. In its 1998 report to Congress, the NAPAP provided both good and bad news.

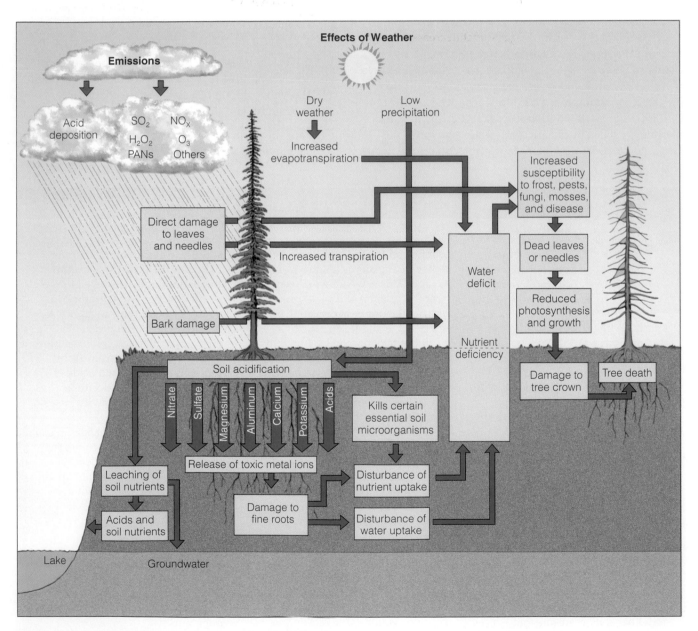

Figure 17-14 Air pollutants are one of several interacting stresses that can damage, weaken, or kill trees.

The *good news* is that:

- Reductions in SO_2 and NO_x emissions from previously unregulated coal-fired power plants required by Phase I of the 1990 amendments to the Clean Air Act have been met. Although SO_2 emissions dropped between 1990 and 1999, NO_x emissions have remained roughly the same (mostly because of emissions from motor vehicles). Phase II (which began in 2000) will require further reduction in both pollutants from a larger number of sources and will be evaluated in a 2002 report.

- There have been declines in the acidity and sulfate concentrations in precipitation in the Midwest and Northeast.

- Dissolved concentrations of sulfate (a major cause of acidification) have dropped in monitored lakes and streams.

- Epidemiological studies suggest that decreased emissions of SO_2, NO_x, and particulate matter (especially fine particles) could reduce premature deaths from cardiovascular and respiratory causes.

- The magnitude of human health and visibility benefits from SO_2 reductions alone could exceed the costs of complying with the 1990 amendments to the Clean Air Act (but the report provides no cost-benefit calculations).

Some *bad news* is that:

- Deposition of sulfur and nitrogen compounds has caused adverse impacts on highly sensitive forest ecosystems in the United States.

- A scientific study of the carefully monitored Calhoun Experimental Forest in South Carolina found a 10-fold increase in acidity (a one pH unit drop) in the first 36 centimeters (14 inches) of soil and half that amount in the next 36 centimeters (14 inches) between 1962 and 1990.

- Adverse impacts from acidic deposition could develop in other forest ecosystems through chronic, long-term exposure.

- Atmospheric nitrate concentrations have not decreased.

- Although sulfate levels in many lakes have declined, there has been no decline in the acidity or acid-neutralizing capacity of sensitive aquatic systems, the key to a recovery of aquatic life.

In 1999, a team of 23 scientists (led by aquatic biologist John Stoddard) evaluated data on changes in acidity conditions in 205 lakes and streams across North America and Europe between 1980 and 1995. The *good news* is that **(1)** in seven of the eight regions studied, the concentration of dissolved sulfate declined dramatically, and **(2)** in Europe the drop in sulfate levels led to a decrease in acidity.

The *bad news* is that in four of the five regions studied in North America, there has been no drop in acidity despite dramatic drops in dissolved sulfate concentrations. According to the researchers, the soils around these lakes and deposits on their bottoms have lost much of their natural buffering capacity because acid deposition has leached out calcium and magnesium compounds. In such cases, it can take decades or centuries for this buffering capacity to be restored by weathering of underlying bedrock.

These findings were confirmed by a 2000 study by the U.S. General Accounting Office, which reported a 26% drop in sulfur deposition and a 2% increase in nitrogen deposition in the eastern states between 1989 and 1998. According to this report, "Acid rain is having a tremendously damaging effect on the Northeast."

In other words, the 1990 amendments to Clean Air Act have helped reduce some of the impacts of acid deposition in the United States. However, there is still a long way to go. According to the General Accounting Office and the EPA, the additional cuts in SO_2 and NO_x emissions required by Phase II of the 1990 U.S. Clean Air Act will help, but sensitive lakes and soils probably will not recover without further cuts in the emissions of these two gases.

Solutions: What Can Be Done to Reduce Acid Deposition? Figure 17-15 summarizes ways to reduce acid deposition. According to most scientists studying acid deposition, the best solutions are *prevention approaches* that reduce or eliminate emissions of SO_2, NO_x, and particulates, as discussed in Section 17-7.

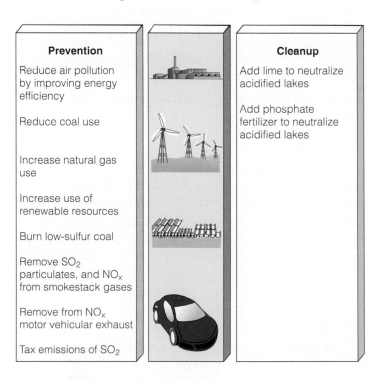

Prevention

Reduce air pollution by improving energy efficiency

Reduce coal use

Increase natural gas use

Increase use of renewable resources

Burn low-sulfur coal

Remove SO_2 particulates, and NO_x from smokestack gases

Remove NO_x from motor vehicular exhaust

Tax emissions of SO_2

Cleanup

Add lime to neutralize acidified lakes

Add phosphate fertilizer to neutralize acidified lakes

Figure 17-15 Solutions: methods for reducing acid deposition.

Controlling acid deposition is a difficult political problem because (1) the people and ecosystems it affects often are quite distant from those who cause the problem, (2) countries with large supplies of coal (such as China, India, Russia, and the United States) have a strong incentive to use it as a major energy resource, and (3) owners of coal-burning power plants say that the costs of adding air-pollution reducing equipment, using low-sulfur coal, or removing sulfur from coal are too high.

Large amounts of limestone or lime can be used to neutralize acidified lakes or surrounding soil—the only cleanup approach now being used. However, there are several problems with liming.

- It is an expensive and temporary remedy that usually must be repeated annually. Using lime to reduce excess acidity in U.S. lakes would cost at least $8 billion per year.

- It can kill some types of plankton and aquatic plants and can harm wetland plants that need acidic water.

- It is difficult to know how much of the lime to put where (in the water or at selected places on the ground).

Recently researchers in England found that adding a small amount of phosphate fertilizer can neutralize excess acidity in a lake. However, the effectiveness of this approach is still being evaluated.

17-5   INDOOR AIR POLLUTION

What Are the Types and Sources of Indoor Air Pollution? If you are reading this book indoors, you may be inhaling more air pollutants with each breath than if you were outside (Figure 17-16). According to EPA studies, in the United States

- Levels of 11 common pollutants are generally two to five times higher inside homes and commercial buildings than outdoors and as much as 100 times higher in some cases.

- Levels of fine particles (Figure 17-7), which can contain toxins and metals such as lead and cadmium, can be as much as 60% higher indoors than outdoors.

- Concentrations of several pesticides (such as chlordane), approved for outdoor use only, were 10 times greater inside than outside monitored homes (some coming from pesticide dust tracked in on shoes).

- Pollution levels inside cars in traffic-clogged U.S. urban areas can be up to 18 times higher than those outside the vehicles.

The health risks from exposure to such chemicals are magnified because people typically spend 70–98% of their time indoors or inside vehicles. In 1990, the EPA placed indoor air pollution at the top of the list of 18 sources of cancer risk, and it is rated by risk analysis scientists as a high-risk health problem for humans (Figure 16-13, left, p. 411). According to the EPA, more than 3,000 cases of cancer per year in the United States may be caused by exposure to indoor air pollutants. At greatest risk are (1) smokers, (2) infants and children under age 5, (3) the old, (4) the sick, (5) pregnant women, (6) people with respiratory or heart problems, and (7) factory workers.

Danish and U.S. EPA studies have linked pollutants found in buildings to dizziness, headaches, coughing, sneezing, nausea, burning eyes, chronic fatigue, and flu-like symptoms, known as the *sick building syndrome*. New buildings are more commonly "sick" than old ones because of reduced air exchange (to save energy) and chemicals released from new carpeting and furniture. According to the EPA, at least 17% of the 4 million commercial buildings in the United States are considered "sick" (including EPA headquarters). Indoor air pollution in the United States costs an estimated $100 billion per year in absenteeism, reduced productivity, and health-care costs. Mostly because of differences in genetic makeup, some individuals can be acutely sensitive to one or a number of indoor air pollutants (Figure 16-3, p. 398).

According to the EPA and public health officials, the three most dangerous indoor air pollutants are (1) cigarette smoke (p. 396), (2) formaldehyde, and (3) radioactive radon-222 gas. Worker exposure to asbestos fibers in mines and in factories making asbestos material is also a serious indoor air pollution problem, especially in developing countries. A number of research studies on laboratory animals have also identified tiny fibers of *fiberglass* as a widespread and potentially potent carcinogen in indoor air.

The chemical that causes most people difficulty is *formaldehyde*, a colorless, extremely irritating gas widely used to manufacture common household materials. As many as 20 million Americans suffer from chronic breathing problems, dizziness, rash, headaches, sore throat, sinus and eye irritation, wheezing, and nausea caused by daily exposure to low levels of formaldehyde emitted (outgassed) from common household materials. These include (1) building materials (such as plywood, particleboard, paneling, and high-gloss wood used in floors and cabinets), (2) furniture, (3) drapes, (4) upholstery, (5) adhesives in carpeting and wallpaper, (6) urethane-formaldehyde insulation, (7) fingernail hardener, and (8) wrinkle-free coating on permanent-press clothing (Figure 17-16). The EPA estimates that as many as 1 of every 5,000 people who live in manufactured homes for more than 10 years will develop cancer from formaldehyde exposure.

In developing countries, the burning of wood, dung, crop residues, and coal in open fires or in unvented or poorly vented stoves for cooking and heat-

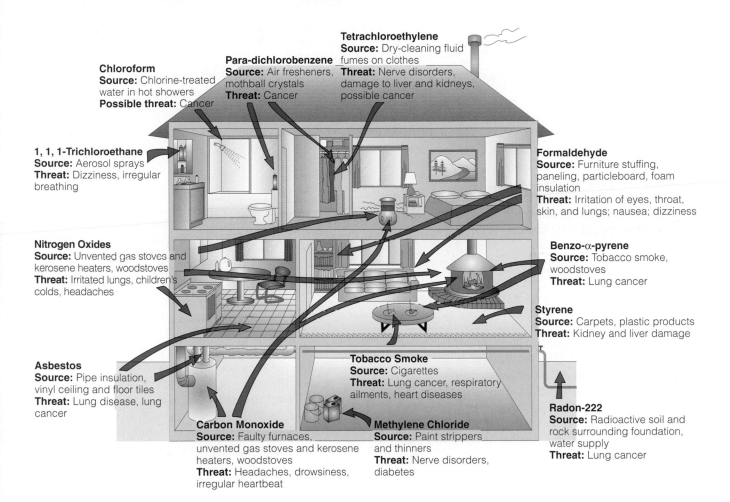

Chloroform
Source: Chlorine-treated water in hot showers
Possible threat: Cancer

Para-dichlorobenzene
Source: Air fresheners, mothball crystals
Threat: Cancer

Tetrachloroethylene
Source: Dry-cleaning fluid fumes on clothes
Threat: Nerve disorders, damage to liver and kidneys, possible cancer

1, 1, 1-Trichloroethane
Source: Aerosol sprays
Threat: Dizziness, irregular breathing

Formaldehyde
Source: Furniture stuffing, paneling, particleboard, foam insulation
Threat: Irritation of eyes, throat, skin, and lungs; nausea; dizziness

Nitrogen Oxides
Source: Unvented gas stoves and kerosene heaters, woodstoves
Threat: Irritated lungs, children's colds, headaches

Benzo-α-pyrene
Source: Tobacco smoke, woodstoves
Threat: Lung cancer

Styrene
Source: Carpets, plastic products
Threat: Kidney and liver damage

Asbestos
Source: Pipe insulation, vinyl ceiling and floor tiles
Threat: Lung disease, lung cancer

Tobacco Smoke
Source: Cigarettes
Threat: Lung cancer, respiratory ailments, heart diseases

Carbon Monoxide
Source: Faulty furnaces, unvented gas stoves and kerosene heaters, woodstoves
Threat: Headaches, drowsiness, irregular heartbeat

Methylene Chloride
Source: Paint strippers and thinners
Threat: Nerve disorders, diabetes

Radon-222
Source: Radioactive soil and rock surrounding foundation, water supply
Threat: Lung cancer

Figure 17-16 Some important indoor air pollutants. (Data from U.S. Environmental Protection Agency)

ing exposes inhabitants (especially women and young children) to very high levels of particulate air pollution.

Case Study: Is Your Home Contaminated with Radon Gas? Radon-222 is naturally occurring radioactive gas that you cannot see, taste, or smell. It is produced by the radioactive decay of uranium-238. Small amounts of uranium-238 are found in most soil and rock, but this isotope is much more concentrated in underground deposits of minerals such as uranium, phosphate, granite, and shale.

When radon gas from such deposits seeps upward through the soil and is released outdoors, it disperses quickly in the atmosphere and decays to harmless levels. However, radon gas can enter buildings above such deposits through **(1)** cracks in foundations and walls, **(2)** openings around sump pumps and drains, and **(3)** hollow concrete blocks (Figure 17-17) and build up to high levels, especially in unventilated lower levels of homes and buildings.

Radon-222 gas quickly decays into solid particles of other radioactive elements that, if inhaled, expose lung tissue to a large amount of ionizing radiation from alpha particles (Figure 3-12, p. 62). This can damage lung tissue and lead to lung cancer over the course of a 70-year lifetime. Your chances of getting lung cancer from radon depend mostly on **(1)** how much radon is in your home, **(2)** the amount of time you spend in your home, and **(3)** whether you are a smoker or have ever smoked.

In 1998, the National Academy of Science estimated that prolonged exposure for a lifetime of 70 years to low levels of radon or radon acting together with smoking is responsible for 15,000 to 22,000 (or 12%) of the lung cancer deaths each year in the United States. This makes radon the second leading cause of lung cancer after smoking (p. 396). Most of the deaths are among smokers or former smokers, with about 2,100 to 2,900 among nonsmokers.

These estimates are based on assuming that **(1)** there is no safe threshold dose for radon exposure (Figure 16-6, left, p. 401) and that **(2)** the incidence of the lung cancer in uranium miners exposed to high levels of radon in mines can be extrapolated to estimate lung cancer deaths for people in homes exposed to much lower levels of radon. Some scientists question these

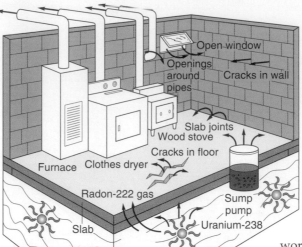

Outlet vents for furnaces and dryers

Open window

Openings around pipes

Cracks in wall

Slab joints

Wood stove

Cracks in floor

Furnace

Clothes dryer

Radon-222 gas

Sump pump

Slab

Uranium-238

Radium-222

Soil

Figure 17-17 Sources and paths of entry for indoor radon-222 gas. (Data from U.S. Environmental Protection Agency)

assumptions and say that these estimates are too high. They also point to the contradictory results of several epidemiological studies on the risks of lung cancer from radon exposure.

EPA indoor radon surveys suggest that 4–5 million U.S. homes may have annual radon levels above 4 picocuries per liter of air* and that 50,000–100,000 homes may have levels above 20 picocuries per liter. If the 4 picocuries per liter standard is adopted (as proposed by the EPA), the cost of testing and correcting the problem could run about $50 billion, with a 15–20% reduction in radon-related deaths. Some researchers argue that it makes more sense to spend perhaps only $500 million to find and fix homes and buildings with radon levels above 20 picocuries per liter until more reliable data are available on the threat from exposure to lower levels of radon.

Because radon hot spots can occur almost anywhere, it is impossible to know which buildings have unsafe levels of radon without conducting tests. In 1988, the EPA and the U.S. Surgeon General's Office recommended that everyone living in a detached house, a town house, a mobile home, or on the first three floors of an apartment building test for radon. Ideally, radon levels should be monitored continuously in the main living areas (not basements or crawl spaces) for 2 months to a year. By 2000, only about 6% of U.S. households had conducted radon tests (most lasting only 2 to 7 days and costing $20–100 per home).

If testing reveals an unacceptable level, homeowners can consult the free EPA publication *Radon Reduction Methods* for ways to reduce radon levels and health risks. According to the EPA, radon control could add $350–500 to the cost of a new home, and correcting a radon problem in an existing house could run $800–2,500.

Case Study: What Should Be Done About Asbestos?

Asbestos is a name given to several different fibrous forms of silicate minerals. For decades it has been widely used as a building material and for large water pipelines because of its strength, flexibility, and

*A *picocurie* is a trillionth of a curie, which is the amount of radioactivity emitted by a gram of radium.

low cost compared to competing materials. Unless completely sealed within a product, asbestos crumbles easily into a dust of fibers tiny enough to become suspended in the air and inhaled deep into the lungs, where they remain for many years.

Prolonged exposure to asbestos fibers can cause (1) *asbestosis* (a chronic, sometimes fatal disease that makes breathing very difficult and was recognized as a hazard among asbestos workers as early as 1924), (2) *lung cancer*, and (3) *mesothelioma* (a fatal cancer of the lung's lining). Epidemiological studies have shown that smokers exposed to asbestos fibers have a much greater chance of dying from lung cancer than do nonsmokers exposed to such fibers.

Most of these diseases occur in workers exposed for years to high levels of asbestos fibers. This group includes asbestos miners, insulators, pipefitters, shipyard employees, and workers in asbestos-producing factories. Family members who breathe asbestos dust brought home in the clothes and hair of asbestos workers also have higher than expected cancer rates, as do people who live near asbestos manufacturing plants with inadequate control of asbestos emissions.

According to health officials, asbestos fibers caused the premature cancer deaths of almost 172,000 asbestos workers in the United States between 1967 and 2000, the worst occupational health disaster of the 20th century. An additional 119,000 premature cancer deaths among U.S. asbestos workers are predicted between 2000 and 2025, mostly from workers exposed to unsafe conditions before working conditions were improved.

After being swamped with health claims from workers, most U.S. asbestos manufacturing companies either declared bankruptcy or moved their operations to other countries (such as Mexico and Brazil) with weaker environmental laws and lax enforcement.

Since 1980 the focus in the United States has shifted to the possible health effects of low levels of asbestos fibers in buildings on the general public. Between 1900 and 1984, asbestos was sprayed on ceilings and walls of schools and other public and private buildings in the United States for fireproofing, soundproofing, insulation of heaters and pipes, and wall and ceiling decoration. The EPA banned those uses in 1984.

In 1989, the EPA ordered a ban on almost all remaining uses of asbestos (such as brake linings, roofing shingles, and water pipes) in the United States by

1997. Representatives of the asbestos industry in the United States and Canada (which now produces most of the asbestos used in the United States) challenged the ban in court. They contended that with proper precautions these asbestos products can be used safely and that the costs of the ban outweigh the benefits. In 1991, a federal appeals court overturned the 1989 EPA ban.

In 1979, the EPA recommended removing existing asbestos in one of every seven commercial and public buildings in the United States (including 30,000 schools), at a cost of at least $100 billion. However, risk analysis and scientific studies indicate that the risk from indoor exposure to asbestos fibers (even in buildings rich in asbestos materials) is extremely low and is about one-tenth the risk from breathing asbestos fibers found in outdoor air from natural sources (wind and erosion).

Critics argued that removing asbestos from buildings is a waste of money, with each life saved costing $100–500 million. They called for sealing, wrapping, and other forms of containment instead of removal, except where asbestos has been damaged or disturbed. Indeed, improper or unnecessary removal can release more asbestos fibers than does sealing off asbestos that is not crumbling. In 1990, the EPA agreed with this and reversed its earlier policy. In 1998, chemists developed a foam that lets building owners treat asbestos-containing fireproofing material without removing it.

According to health experts, the major health risk from asbestos today is occurring among asbestos miners and workers in developing countries (especially Russia, China, Brazil, India, and Thailand), where the use of asbestos as an inexpensive building material is growing rapidly. Experts say that 90% of the worker deaths from asbestos exposure can be prevented by (1) using a good-fitting facemask, (2) wetting asbestos to control dust, and (3) changing clothes before and after handling asbestos.

17-6 EFFECTS OF AIR POLLUTION ON LIVING ORGANISMS AND MATERIALS

How Does the Human Respiratory System Help Protect Us from Air Pollution? Table 17-2 (p. 422) listed the major health effects from the six most common (criteria) outdoor air pollutants. Your respiratory system (Figure 17-18) has a number of mechanisms that help protect you from such air pollution. They include:

- Hairs in your nose that filter out large particles.

- Sticky mucus in the lining of your upper respiratory tract that captures smaller (but not the smallest) particles and dissolves some gaseous pollutants.

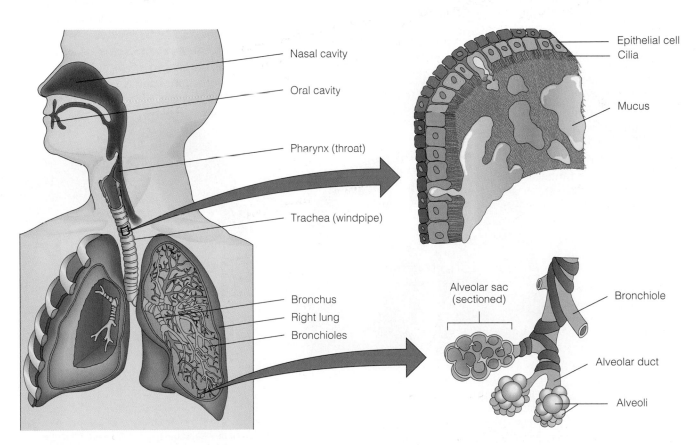

Figure 17-18 Major components of the human respiratory system.

- Sneezing and coughing that expel contaminated air and mucus when pollutants irritate your respiratory system.

- Hundreds of thousands of tiny, mucus-coated hair-like structures called *cilia* that line your upper respiratory tract. They continually wave back and forth and transport mucus and the pollutants they trap to your throat (where they are swallowed or expelled).

Years of smoking and exposure to air pollutants can overload or break down these natural defenses. This can cause or contribute to respiratory diseases such as **(1)** *lung cancer*, **(2)** *asthma* (typically an allergic reaction causing sudden episodes of muscle spasms in the bronchial walls, resulting in acute shortness of breath), **(3)** *chronic bronchitis* (persistent inflammation and damage to the cells lining the bronchi and bronchioles, causing mucus buildup, painful coughing, and shortness of breath), and **(4)** *emphysema* (irreversible damage to air sacs or alveoli leading to abnormal dilation of air spaces, loss of lung elasticity, and acute shortness of breath (Figure 17-19). Older adults, infants, pregnant women, and people with heart disease, asthma, or other respiratory diseases are especially vulnerable to air pollution.

How Many People Die Prematurely from Air Pollution? It is difficult to estimate by risk analysis how many people die prematurely from respiratory or cardiac problems caused or aggravated by air pollution because people are exposed to so many different pollutants over their lifetimes.

In the United States, estimates of annual deaths related to outdoor air pollution range from 65,000 to 200,000 (most from exposure to fine particles, Spotlight, right). If indoor air pollution is included, estimated annual deaths from air pollution in the United States range from 150,000 to 350,000 people—equivalent to 1–2 fully loaded 400-passenger jumbo jets crashing *each day* with no survivors.

Millions more become ill and lose work time. A 2000 study by state and local air pollution officials estimated that each year more than 125,000 Americans (120,000 of them in urban areas) get cancer from breathing

diesel fumes from buses, trucks, and other diesel engines. According to the EPA and the American Lung Association, air pollution in the United States costs at least $150 billion annually in health care and lost work productivity, with $100 billion of that caused by indoor air pollution.

According to a 1999 study by Australia's Commonwealth Science Council, worldwide at least 3 million people (most of them in Asia) die prematurely each year from the effects of air pollution—an average of 8,200 deaths per day. About 2.8 million of these deaths are from *indoor* air pollution, and 200,000 are from *outdoor* pollution (Figure 17-20).

Most people who live in large cities in developing countries breathe air that is the equivalent of smoking 2–3 packs of cigarettes a day. According to WHO estimates, up to 700,000 premature deaths per year worldwide could be prevented in developing countries if three pollutants—suspended particulate matter, carbon monoxide, and lead—were brought down to safer levels.

How Are Plants and Aquatic Systems Damaged by Air Pollutants? Figure 17-14 (p. 432) summarizes some of the direct and indirect effects of acidic deposition, ozone, and other air pollutants on trees, and the effects of acid deposition on trees and plants were discussed on p. 431.

The effects of exposure to a mix of air pollutants may not become visible for several decades, when large numbers of trees suddenly begin dying because of depletion of soil nutrients and increased susceptibility to pests, diseases, fungi, and drought (Connections, p. 431). This phenomenon, known as *Waldsterben* (for-

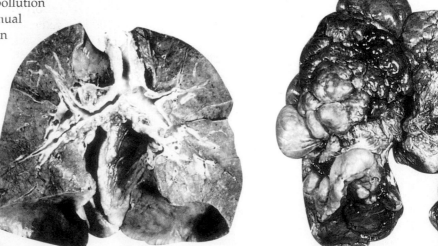

Figure 17-19 Normal human lungs (left) and the lungs of a person who died of emphysema (right). Prolonged smoking and exposure to air pollutants can cause emphysema in anyone, but about 2% of emphysema cases result from a defective gene that reduces the elasticity of the air sacs in the lungs. Anyone with this hereditary condition, for which testing is available, should not smoke and should not live or work in a highly polluted area. (O. Auerbach/Visuals Unlimited)

Health Dangers from Fine Particles

SPOTLIGHT

Research indicates that invisible particles—especially *fine particles* with diameters less than 10 microns (PM-10) and *ultrafine particles* with diameters less than 2.5 microns (PM-2.5)—pose a significant health hazard. Such particles are emitted by incinerators, motor vehicles, radial tires, wind erosion, wood-burning fireplaces, and power and industrial plants (Figure 17-7).

Such tiny particles **(1)** are not effectively captured by modern air-pollution control equipment, **(2)** are small enough to penetrate the respiratory system's natural defenses against air pollution (Figure 17-18), and **(3)** can bring with them droplets or other particles of toxic or cancer-causing pollutants that become attached to their surfaces.

Once they are lodged deep within the lungs, these fine particles can cause chronic irritation that

can **(1)** trigger asthma attacks, **(2)** aggravate other lung diseases, **(3)** cause lung cancer, and **(4)** interfere with the blood's ability to take in oxygen and release CO_2. This strains the heart, increasing the risk of death from heart disease.

Several recent studies of air pollution in U.S. cities have indicated that fine and ultrafine particles prematurely kill 65,000–200,000 Americans each year. There is no known threshold level below which the harmful effects of fine particles disappear.

Exposure to particulate air pollution is much worse in most developing countries, where urban air quality has generally deteriorated. The World Bank estimates that if particulate levels were reduced globally to WHO guidelines, 300,000–700,000 premature deaths per year could be prevented.

In 1997, the EPA announced stricter emission standards for ultrafine particles with diameters less than 2.5 microns (PM-2.5). The

EPA estimates the cost of implementing the standards at $7 billion per year, with the resulting health and other benefits estimated at $120 billion per year.

According to industry officials, the new standard is based on flimsy scientific evidence, and its implementation will cost $200 billion per year. EPA officials say that their review of the scientific evidence—one of the most exhaustive ones ever undertaken by the agency—supports the need for the new standard for ultrafine particles. Furthermore, a 2000 study by the Health Effects Institute of 90 large American cities confirmed the link between fine and ultrafine particles and higher rates of death and disease.

Critical Thinking

Are you for or against the stricter standard for emissions of ultrafine particles? Explain.

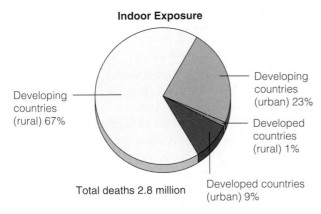

Indoor Exposure

Developing countries (rural) 67%

Developing countries (urban) 23%

Developed countries (rural) 1%

Developed countries (urban) 9%

Total deaths 2.8 million

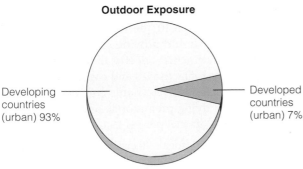

Outdoor Exposure

Developing countries (urban) 93%

Developed countries (urban) 7%

Total deaths 0.2 million

est death), has turned whole forests of spruce, fir, and beech into stump-studded meadows and mountainsides (see photo on p. 395). It is estimated that air pollution has been a key factor in reducing the overall productivity of European forests by about 16% and causing damage valued at roughly $30 billion per year.

Forest diebacks have also occurred in the United States. The most seriously affected areas are high-elevation spruce trees that populate the ridges of the Appalachian Mountains from Maine to Georgia, including the Shenandoah and Great Smoky Mountain national parks.

Air pollution, mostly by ozone, also threatens some crops—especially corn, wheat, and soybeans, the three most important U.S. crops—and is reducing U.S. food production by 5–10%. In the United States, estimates of agricultural losses as a result of air pollution (mostly by ozone) range from $2 to $6 billion per year, with an estimated $1 billion of damages in California alone. The effects of high acidity (low pH) on aquatic life were discussed on p. 430.

Figure 17-20 Estimated premature deaths per year caused by indoor and outdoor air pollution in developing and developed countries. (Data from Australia's Commonwealth Science Council, 2000)

Table 17-3 Harmful Effects of Air Pollution on Materials

Material	Effects	Principal Air Pollutants
Stone and concrete	Surface erosion, discoloration, soiling	Sulfur dioxide, sulfuric acid, nitric acid, particulate matter
Metals	Corrosion, tarnishing, loss of strength	Sulfur dioxide, sulfuric acid, nitric acid, particulate matter, hydrogen sulfide
Ceramics and glass	Surface erosion	Hydrogen fluoride, particulate matter
Paints	Surface erosion, discoloration, soiling	Sulfur dioxide, hydrogen sulfide, ozone, particulate matter
Paper	Embrittlement, discoloration	Sulfur dioxide
Rubber	Cracking, loss of strength	Ozone
Leather	Surface deterioration, loss of strength	Sulfur dioxide
Textiles	Deterioration, fading, soiling	Sulfur dioxide, nitrogen dioxide, ozone, particulate matter

What Are the Harmful Effects of Air Pollutants on Materials? Each year, air pollutants cause billions of dollars in damage to various materials we use (Table 17-3). The fallout of soot and grit on buildings, cars, and clothing requires costly cleaning. Air pollutants break down exterior paint on cars and houses, and they deteriorate roofing materials. Irreplaceable marble statues, historic buildings, and stained glass windows around the world have been pitted, gouged, and discolored by air pollutants. The EPA estimates damage to buildings in the United States from acid deposition alone at $5 billion per year.

17-7 SOLUTIONS: PREVENTING AND REDUCING AIR POLLUTION

How Have Laws Been Used to Reduce Air Pollution in the United States? The U.S. Congress passed Clean Air Acts in 1970, 1977, and 1990. These laws use a *command-and-control* approach in which the federal government establishes air pollution regulations that are enforced by each state and by major cities.

Congress directed the EPA to establish *national ambient air quality standards (NAAQS)* for six outdoor criteria pollutants (Table 17-2). The EPA regulates these chemicals by using *criteria* developed from risk assessment methods (Section 16-2, p. 398) to set maximum permissible levels in outdoor air.

One limit, called a *primary standard*, is set to protect human health, and another called a *secondary standard*, is intended to prevent environmental and property damage. Each standard specifies the maximum allowable level, averaged over a specific period, for a certain pollutant in outdoor (ambient) air. A geographic area that meets or does better than the primary standard for a particular pollutant is called an *attainment area*, and one that does not meet the primary standard is called a *nonattainment area*.

The EPA has also established national emission standards for more than 100 different toxic air pollutants that are known to cause or suspected of causing cancer or other adverse health effects.

Here is some *good news*. According to the EPA,

- Between 1970 and 1998, national total emissions of the six criteria pollutants declined 31%, while U.S. population increased 31%, gross domestic product increased 114%, and vehicle miles traveled rose 127%.

- Between 1978 and 1998, mean concentrations of the six criteria air pollutants in the troposphere decreased by **(1)** 97% for lead, **(2)** 60% for carbon monoxide, **(3)** 58% for sulfur dioxide, **(4)** 30% for ground-level ozone, **(5)** 25% for suspended particulate matter (10 micrometers or less in diameter), and **(6)** 2% for nitrogen dioxide.

- The mean estimated human health and environmental benefits from air pollution regulations between 1970 and 1990 amounted to $6.8 trillion, compared to $436 million (in 1990 dollars) spent to implement all federal, state, and local air pollution regulations. Thus, the net economic benefit of the Clean Air Act between 1970 and 1990 was $6.4 trillion. During this 20-year period, the act prevented an estimated 1.6 million premature deaths and 300 million cases of respiratory disease.

- Between 1990 and 2010, the 1990 amendments to the Clean Air Act should provide four times more health and environmental benefits than the estimated compliance costs to industries and consumers. By 2010, the 1990 amendments should prevent 23,000 Americans from dying prematurely and avert 1,700,000 asthma attacks per year.

- A U.S. Court of Appeals decision in 2000 will allow implementation of a 1998 regulation requiring 32 older coal-fired plants in 10 states (mostly in the Ohio Valley and Midwest, Figure 17-10) to meet the same air pollution emission standards as new coal-burning plants. According to the EPA, this will have the same effect as taking 26 million cars off the road and will bring cleaner, safer air to more than 138 million people living in the eastern half of the United States.

Here is some *bad news*.

- Between 1970 and 1998, emissions of nitrogen oxides (NO_x) increased 11%.

- Despite continued improvements in air quality, in 1999 approximately 62 million people lived in 130 nonattainment areas with air that did not meet the primary standards for one or more of the six criteria pollutants.

How Could U.S. Air Pollution Laws Be Improved?
The Clean Air Act of 1990 was an important step in the right direction, but many environmentalists point to the following deficiencies in this law:

- *Continuing to rely mostly on pollution cleanup rather than prevention.* In the United States, the air pollutant with the largest drop (97% between 1970 and 1998) in its atmospheric level was lead, which was virtually banned in gasoline.

- *Failing to increase fuel efficiency standards for cars and light trucks.* According to environmental scientists, this would reduce air pollution more quickly and effectively than any other method and would save consumers enormous amounts of money (p. 364).

- *Not adequately regulating emissions from inefficient, two-cycle gasoline engines used in devices such as lawnmowers, leaf blowers, chain saws, and personal marine engines used to power jet skis, outboard motors, and personal watercraft.* According to recent studies by the California Air Resources Board, **(1)** a 1-hour ride on a typical jet ski creates more air pollution than the average U.S. car does in a year, **(2)** operating a 100-horsepower marine engine for 7 hours emits more air pollutants than a new car driven 160,000 kilometers (100,000 miles), and **(3)** each year the fuel and oil spilled by the 14 million small marine engines in the United States is 15 times the amount spilled by the *Exxon Valdez* oil tanker (Case Study, p. 336).

- *Doing too little to reduce emissions of carbon dioxide and other greenhouse gases* (Section 18-5, p. 460).

Executives of companies affected by implementing such policies strongly oppose such changes in air pollution laws. They claim that implementing such changes would cost too much, harm economic growth, and cost jobs.

Proponents contend that history has shown that almost all industry estimates of implementing various air pollution control standards in the United States were many times the actual cost of implementation. In addition, implementing such standards has helped increase economic growth and create jobs by stimulating companies to develop new technologies for reducing air pollution emissions. Many of these technologies are sold in the international marketplace.

Should We Use the Marketplace to Reduce Pollution? To help reduce SO_2 emissions, the Clean Air Act of 1990 allows an *emissions trading policy*, which enables the 110 most polluting power plants in 21 states (primarily in the Midwest and East, Figure 17-10) to buy and sell SO_2 pollution rights.

Each year a power plant is given a certain number of pollution credits or rights that allow it to emit a certain amount of SO_2. A utility that emits less SO_2 than its limit receives more pollution credits. It can use these credits **(1)** to avoid reductions in SO_2 emissions from some of its other facilities, **(2)** bank them for future plant expansions, or **(3)** sell them to other utilities, private citizens, or environmental groups. Proponents of this system argue that it allows the marketplace to determine the cheapest, most efficient way to get the job done instead of having the government dictate how to control pollution.

Some environmentalists see this market approach as an improvement over the current regulatory approach, as long as it achieves net reduction in SO_2 pollution. This would be done by limiting the total number of credits and gradually lowering the annual number of credits, something that is not required by the 1990 amendments to the Clean Air Act.

Some environmentalists contend that marketing pollution rights allows utilities with older, dirtier power plants to buy their way out and keep on emitting unacceptable levels of SO_2. They also warn that this approach creates incentives to cheat. Air quality regulation is based largely on self-reporting of emissions, and pollution monitoring is incomplete and imprecise. Thus, sellers of permits will benefit by understating their reductions (to get more permits), and permit buyers will benefit by underreporting emissions (to reduce their permit purchases).

Here is some *good news*. Between 1994 and 1997, the emission trading system helped reduce SO_2 emissions in the United States by 30%. The cost of doing this was less than one-tenth the cost projected by industry because this market-based system motivated companies to reduce emissions in more efficient ways.

In 1997, the EPA proposed a voluntary emissions trading program involving smog-forming nitrogen oxides (NO_x) for 22 eastern states and the District of Columbia. Emissions trading may also be implemented for particulate emissions and volatile organic compounds.

How Can We Reduce Outdoor Air Pollution? Figure 17-21 summarizes ways to reduce emissions of sulfur oxides, nitrogen oxides, and particulate matter from stationary sources (such as electric power plants and industrial plants that burn coal). Until recently, emphasis has been on dispersing and diluting the pollutants

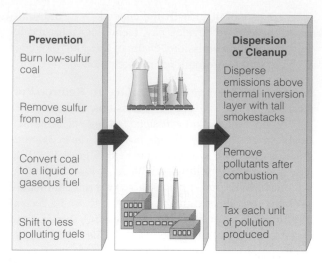

Figure 17-21 Solutions: methods for reducing emissions of sulfur oxides, nitrogen oxides, and particulate matter from stationary sources such as coal-burning electric power plants and industrial plants.

lution is to get older, high-polluting vehicles off the road. According to EPA estimates, 10% of the vehicles on the road in the United States emit 50–70% of the pollutants. A problem is that many old cars are owned by people who cannot afford to buy a newer car. One suggestion would be to pay people to take their old cars off the road, which would result in huge savings in health and air-pollution control costs.

California's South Coast Air Quality Management District Council developed a drastic and controversial program to produce an 80% reduction in ozone, photochemical smog, and other major air pollutants in the Los Angeles area by 2009. This plan would **(1)** sharply reduce use of gasoline-burning engines over two decades by converting cars, trucks, buses, chain saws, outboard motors, and lawnmowers to run on electricity or alternative fuels, **(2)** substantially raise parking fees and assess high fees for families owning more than one car, **(3)** require gas stations to use a hydrocarbon vapor recovery system on gas pumps and to sell alternative

by using tall smokestacks or adding equipment that removes some of the particulate pollutants after they are produced (Figure 17-22). However, under the sulfur reduction requirements of the 1990 amendments to the Clean Air Act, more utilities are switching to low-sulfur coal to reduce SO_2 emissions. Environmentalists call for taxes on air pollutant emissions and greater emphasis on prevention.

Figure 17-23 lists ways to reduce emissions from motor vehicles, the primary culprits in producing photochemical smog. Use of alternative vehicle fuels to reduce air pollution is evaluated in Table 15-1, p. 385.

An important way to make significant reductions in air pol-

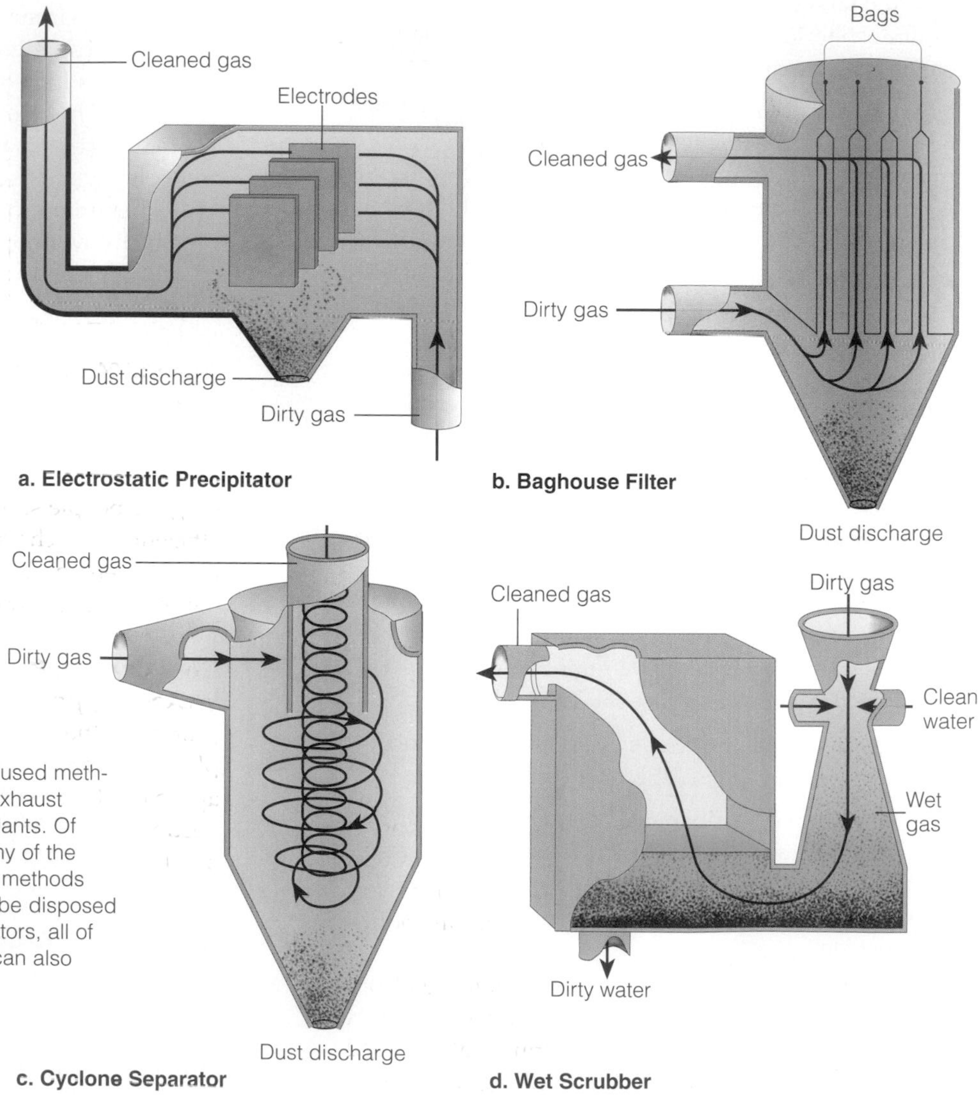

a. Electrostatic Precipitator

Cleaned gas
Electrodes
Dust discharge
Dirty gas

b. Baghouse Filter

Bags
Cleaned gas
Dirty gas
Dust discharge

Figure 17-22 Solutions: four commonly used methods for removing particulates from the exhaust gases of electric power and industrial plants. Of these, only baghouse filters remove many of the more hazardous fine particles. All these methods produce hazardous materials that must be disposed of safely, and except for cyclone separators, all of them are expensive. The wet scrubber can also reduce sulfur dioxide emissions.

c. Cyclone Separator

Cleaned gas
Dirty gas
Dust discharge

d. Wet Scrubber

Cleaned gas
Dirty gas
Clean water
Wet gas
Dirty water

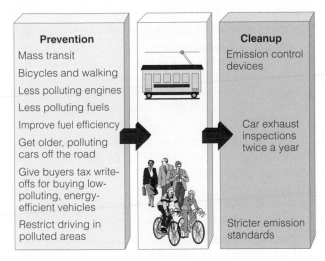

Prevention		Cleanup
Mass transit		Emission control devices
Bicycles and walking		
Less polluting engines		
Less polluting fuels		
Improve fuel efficiency		Car exhaust inspections twice a year
Get older, polluting cars off the road		
Give buyers tax write-offs for buying low-polluting, energy-efficient vehicles		
Restrict driving in polluted areas		Stricter emission standards

Figure 17-23 Solutions: methods for reducing emissions from motor vehicles.

fuels (Table 15-1, p. 385), **(4)** strictly control or relocate industrial plants and businesses that release large quantities of hydrocarbons and other pollutants, and **(5)** find substitutes for or ban consumer products that release hydrocarbons, including aerosol propellants, paints, household cleaners, and barbecue starter fluids.

Here is some *good news*. Since the 1960s, Tokyo, Japan (with a current population of about 28 million), has implemented a strict air-pollution control program that has sharply reduced levels of sulfur dioxide, carbon monoxide, and ozone. During the past 30 years outdoor air quality in most western European cities has also improved. The *bad news* is that outdoor air quality has remained about the same or has gotten worse in most rapidly growing urban areas in developing countries.

How Can We Reduce Indoor Air Pollution? In the United States indoor air pollution poses a much greater health risk for many people than outdoor air pollution. Yet the EPA spends about $500 million per year fighting outdoor air pollution and only about $13 million a year on indoor air pollution.

To reduce indoor air pollution, it is not necessary to impose indoor air quality standards and monitor the more than 100 million homes and buildings in the United States. Instead, air pollution experts suggest that indoor air pollution can be reduced by several means (Figure 17-24). Another possibility for cleaner indoor air in high-rise buildings is rooftop greenhouses through which building air can be circulated. Some actions you can take to reduce your exposure to indoor air pollutants are listed in Appendix 6.

In developing countries, indoor air pollution from open fires and leaky and inefficient stoves that burn wood, charcoal, or coal (and the resulting high levels of respiratory illnesses) could be reduced if governments

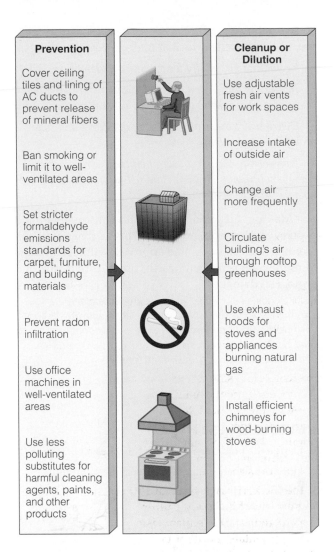

Prevention		Cleanup or Dilution
Cover ceiling tiles and lining of AC ducts to prevent release of mineral fibers		Use adjustable fresh air vents for work spaces
Ban smoking or limit it to well-ventilated areas		Increase intake of outside air
Set stricter formaldehyde emissions standards for carpet, furniture, and building materials		Change air more frequently
		Circulate building's air through rooftop greenhouses
Prevent radon infiltration		Use exhaust hoods for stoves and appliances burning natural gas
Use office machines in well-ventilated areas		
Use less polluting substitutes for harmful cleaning agents, paints, and other products		Install efficient chimneys for wood-burning stoves

Figure 17-24 Solutions: ways to prevent and reduce indoor air pollution.

(1) gave people simple stoves that burn biofuels more efficiently (which would also reduce deforestation) and that are vented outside or **(2)** provided them with simple solar cookers (Figure 15-21d, p. 374).

How Can We Protect the Atmosphere? An Integrated Approach Environmentalists believe that protecting the atmosphere, and thus the health of people and many other organisms, will take a global approach that integrates many different strategies. Suggestions for doing this over the next 40–50 years include the following:

- *Putting more emphasis on pollution prevention*
- *Improving energy efficiency*
- *Reducing use of fossil fuels (especially coal and oil)*
- *Increasing use of renewable energy*
- *Slowing population growth*

- *Integrating air pollution, water pollution, energy, land-use, population, economic, and trade policies*
- *Regulating air quality for an entire region or airshed*
- *Taxing the production of air pollutants and using the revenue to reduce taxes on income and wealth*
- *Distributing cheap and efficient cookstoves and solar cookstoves in developing countries*
- *Transferring the latest energy-efficiency, renewable-energy, pollution prevention, and pollution control technologies to developing countries*

Making such changes will be controversial and expensive. However, proponents argue that not implementing such an integrated approach will cost far more in money, poor human health, premature death, and ecological damage.

Turning the corner on air pollution requires moving beyond patchwork, end-of-pipe approaches to confront pollution at its sources. This will mean reorienting energy, transportation, and industrial structures toward prevention.

HILARY F. FRENCH

REVIEW QUESTIONS

1. Define the boldfaced terms in this chapter.

2. How can lichens be used to detect air pollutants?

3. Briefly describe the history of air pollution in Europe and the United States.

4. Distinguish between *atmosphere*, *troposphere*, and *stratosphere*. What key role does the stratosphere play in maintaining life on the earth?

5. Distinguish between the *greenhouse effect* and *the ozone shield* and explain the importance of these natural processes in sustaining life on the earth.

6. Explain how human activities are disrupting the carbon, nitrogen, and sulfur biogeochemical cycles.

7. Distinguish between *air pollution*, *primary air pollutants*, and *secondary air pollutants*. List the major classes of pollutants found in outdoor air. Distinguish between *stationary* and *mobile* sources of pollution for outdoor air. What are the two major sources of indoor air pollution?

8. List the six *criteria* air pollutants regulated in the United States (and in most developed countries). For each of these pollutants, summarize its major human sources and health effects.

9. What is *photochemical smog*, and how does it form? What is *industrial smog*, and how does it form?

10. List major factors that can (a) reduce air pollution and (b) increase air pollution. What is a *temperature inversion*, and what are its harmful effects? Distinguish between a *subsidence thermal inversion* and a *radiation tem-perature inversion*. What types of places are most likely to suffer from prolonged inversions of each type?

11. Distinguish between *acid deposition*, *wet deposition*, and *dry deposition*. What areas tend to be affected by acid deposition? What is a *buffer*, and what types of geologic areas can neutralize or buffer some inputs of acidic chemicals? What two types of areas are most sensitive to acid deposition?

12. What are the major harmful effects of acid deposition on (a) human health, (b) materials, (c) soils, (d) aquatic life, (e) trees and other plants, and (f) some forms of wildlife?

13. Describe hidden synergistic interactions that can accelerate forest damage and decline by air pollution.

14. Summarize the major *good news* and *bad news* about acid deposition in the United States.

15. List seven ways to prevent acid deposition. What are the advantages and disadvantages of liming acidified lakes to reduce the effects of acid deposition?

16. How serious is indoor air pollution, and what are some of its sources? What is the *sick-building syndrome*? According to the EPA, what are the three most dangerous indoor air pollutants in the United States? What is the most dangerous indoor air pollutant in most developing countries?

17. Summarize the problem of indoor pollution from (a) formaldehyde, (b) radioactive radon gas, and (c) asbestos fibers.

18. List four defenses your body has against air pollution. What are the major harmful health effects of (a) carbon monoxide, (b) suspended particulate matter, (c) sulfur dioxide, (d) nitrogen oxides, and (e) ozone (See Table 17-2, p. 422)? Describe the health dangers from inhaling fine particles.

19. About how many people die prematurely each year from exposure to air pollutants in (a) the United States and (b) the world? What percentage of these deaths occurs in developing countries? What percentage of these deaths is the result of indoor air pollution?

20. What is the Clean Air Act, and how has it helped reduce outdoor air pollution in the United States? Distinguish between *national ambient air quality standards*, *primary standards*, and *secondary standards*. Distinguish between *attainment* and *nonattainment areas*.

21. Summarize the major *good news* and *bad news* about the effectiveness of the Clean Air Act in reducing outdoor air pollution in the United States. According to environmentalists, what are four weaknesses of the current Clean Air Act in the United States?

22. What is an *emission trading policy*, and what are the pros and cons of using this approach to help reduce air pollution?

23. List the major prevention and cleanup methods for dealing with air pollution from (a) emissions of sulfur oxides, nitrogen oxides, and particulate matter from sta-

tionary sources, **(b)** automobile emissions, **(c)** indoor air pollution in developed countries, and **(d)** indoor air pollution in developing countries.

24. What are ten components of an integrated approach for dealing with air pollution?

CRITICAL THINKING

1. Evaluate the pros and cons of the following statement: "Because we have not proven absolutely that anyone has died or suffered serious disease from nitrogen oxides, current federal emission standards for this pollutant should be relaxed."

2. Identify climate and topographic factors in your local community that **(a)** intensify air pollution and **(b)** help reduce air pollution.

3. Should all tall smokestacks be banned? Explain.

4. Explain how sulfur in coal can contribute to the acidity of rainwater.

5. Suppose you become trapped in your car during a snowstorm and you have your engine and heater running to keep warm. Explain why you should roll down one of your windows just a little.

6. Why are most severe air pollution episodes associated with subsidence temperature inversions rather than radiation temperature inversions?

7. Evaluate your exposure to some or all of the indoor air pollutants in Figure 17-16 in your school, workplace, and home. Come up with a plan for reducing your exposure to these pollutants.

8. Should annual government-held auctions of marketable trading permits be used as a primary way of controlling and reducing air pollution? Explain. What conditions, if any, would you put on this approach?

9. Do you agree or disagree with the possible weaknesses of the U.S. Clean Air Act listed on p. 441? Defend each of your choices. Can you identify other weaknesses?

PROJECTS

1. Have buildings at your school been tested for radon? If so, what were the results? What has been done about areas with unacceptable levels? If this testing has not been done, talk with school officials about having it done.

2. Write 1- to 2-page scenarios speculating about what types of species would survive and evolve if the concentration of oxygen in the earth's atmosphere **(a)** increased to 30% and **(b)** decreased to 15%.

3. Use the library or the internet to find bibliographic information about *Michael J. Cohen* and *Hilary French*,

whose quotes appear at the beginning and end of this chapter.

4. Make a concept map of this chapter's major ideas, using the section heads and subheads and the key terms (in boldface). Look at the inside back cover and on the website for this book for information about making concept maps.

INTERNET STUDY RESOURCES AND RESOURCES FOR FURTHER READING AND RESEARCH

The website for this book contains helpful study aids and many ideas for further reading and research. Log on to:

http://www.brookscole.com/product/0534376975s

and click on the Chapter-by-Chapter area. Choose Chapter 17 and select a resource:

- "Flash Cards" allows you to test your mastery of the Terms and Concepts to Remember for this chapter.

- "Tutorial Quizzes" provides a multiple-choice practice quiz.

- "Student Guide to InfoTrac" will lead you to Critical Thinking Projects that use InfoTrac College Edition as a research tool.

- "References" lists the major books and articles consulted in writing this chapter.

- "Hypercontents" takes you to an extensive list of sites with news, research, and images related to individual sections of the chapter.

INFOTRAC COLLEGE EDITION

Improve your skills with InfoTrac College Edition, a searchable online database of articles from more than 700 periodicals. Log on to:

http://www.infotrac-college.com

or access InfoTrac through the website for this book.

Try the following articles:

Dale, J. 2000. On the road to cleaner air. *State Legislatures* vol. 26, no. 3, pp. 12–17. (subject guide: air pollution, management)

Jaret, P. 2000. Experts are turning up some surprising sources of indoor air pollution. *National Wildlife* Feb-March 2000. (subject guide: indoor air quality)

18 CLIMATE CHANGE AND OZONE LOSS

A.D. 2060: Green Times on Planet Earth

Mary Wilkins sat in the living room of the solar-powered earth-sheltered house (Figure 18-1) she shared with her daughter Jane and her family. It was July 4, 2060: Independence Day.

She began putting the finishing touches on her grandchildren's costumes for this afternoon's pageant in Rachel Carson Park. It would honor earth heroes who began the Age of Ecology in the 20th century, as well as those who continued this tradition in the 21st century.

She was delighted that her 12-year-old grandchild Jeffrey had been chosen to play Aldo Leopold, who in the late 1940s began urging people to work with the earth (p. 39). Her pride swelled when her 10-year-old grandchild Lynn was chosen to play Rachel Carson, who in the 1960s alerted us to threats from increasing exposure to pesticides and other harmful chemicals (Individuals Matter, p. 36). Her neighbor's son Manuel had been chosen to play biologist Edward O. Wilson, who in the last third of the 20th century alerted us to the need to preserve the earth's biodiversity.

Even in her most idealistic dreams, she had never guessed that she would see the loss of global biodiversity slowed to a trickle. Most air pollution gradually disappeared when energy from the sun (p. 370), wind (p. 381), and hydrogen (p. 384) replaced most use of fossil fuels. Most food was grown by sustainable agriculture (Figure 12-31, p. 291).

Preventing pollution and reducing waste had become important, money-saving priorities for businesses and households based on the four Rs of resource consumption: *reduce, reuse, recycle* and *refuse*. Walking and bicycling had increased in cities and towns designed as vibrant communities for people instead of cars (p.681). Low-polluting and safe ecocars got 128 kilometers per liter (300 miles per gallon), and most places had efficient mass transportation.

World population stabilized at 8 billion in 2028 and then begun a slow decline. The threat of climate change from atmospheric warming enhanced by human activities lessened as the use of fossil fuels declined. International treaties enacted in the 1990s effectively banned the chemicals that had begun depleting ozone in the stratosphere. By 2050, ozone levels in the stratosphere had returned to 1980 levels.

Two hours later, she, her daughter Jane, and her son-in-law Gene watched with pride as 40 beautiful children honored the leaders of the Age of Ecology. At the end, Lynn stepped forward and said, "Today we have honored many earth heroes, but the real heroes are the ordinary people in this audience and around the world who worked to help sustain the earth's life-support systems for us and other species. Thank you, Grandma, Mom, Dad, and everyone here for giving us such a wonderful gift. We promise to leave the earth even better for our children and grandchildren and all living creatures."

Figure 18-1 An earth-sheltered house in Will County, Illinois, in the United States. About 13,000 families across the United States have built such houses. Mary Wilkins's fictional house in 2060 could be similar to this one. (Pat Armstrong/Visuals Unlimited)

We are embarked on the most colossal ecological experiments of all time—doubling the concentration in the atmosphere of an entire planet of one of its most important gases—and we really have little idea of what might happen.

PAUL A. COLINVAUX

This chapter addresses the following questions:

- How has the earth's climate changed in the past?
- How might the earth's climate change in the future?
- What factors can affect changes in the earth's average temperature?
- What are some possible effects of climate change from a warmer earth?
- What can we do to slow or adapt to projected climate change caused by natural processes, human activities, or both?
- Are human activities depleting ozone in the stratosphere, and why should we care?
- What can we do to slow and eventually reverse ozone depletion in the stratosphere caused by human activities?

18-1 PAST CLIMATE CHANGE AND THE NATURAL GREENHOUSE EFFECT

How Have the Earth's Temperature and Climate Changed in the Past? Climate change is neither new nor unusual. The earth's average surface temperature and climate have been changing throughout the world's 4.7-billion-year history, sometimes gradually (over hundreds to millions of years) and at other times fairly quickly (over a few decades).

Figure 18-2 shows how the *estimated* average global temperature of the atmosphere near the earth's surface has changed during four time scales in the past: **(1)** 900,000 years, **(2)** 22,000 years, **(3)** 1,000 years, and **(4)** 130 years.

Over the past 900,000 years the average temperature of the atmosphere near the earth's surface has undergone prolonged periods of *global cooling* and *global warming* (Figure 18-2, top). During each cold period, thick glacial ice covered much of the earth's surface for about 100,000 years. Each of these periods was followed by a warmer interglacial period lasting 10,000–12,500 years, during which most of the ice melted. During the past 10,000 years we have had the good fortune to live in an interglacial period with a fairly stable climate and average global surface temperature, compared to many of the more dramatic changes in the past.

What Is the Greenhouse Effect, and Why Is It Important to Us? For the earth and its entire

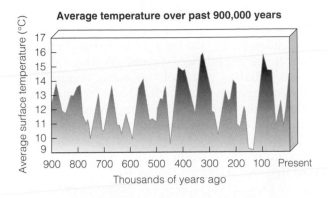

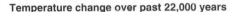

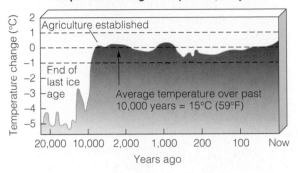

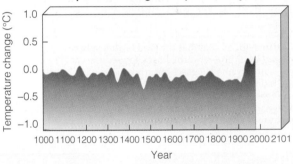

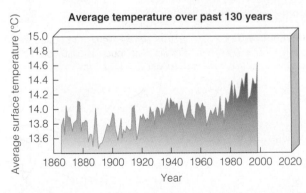

Figure 18-2 Estimated changes in the average global temperature of the atmosphere near the earth's surface over different periods of time. Past temperature changes are estimated by analysis of **(1)** plankton and isotopes in ocean sediments, **(2)** ice cores from ancient glaciers, **(3)** temperature measurements at different depths in boreholes drilled deep into the earth's surface, **(4)** pollen from lake bottoms and bogs, **(5)** tree rings, **(6)** historical records, and **(7)** temperature measurements (since 1860). (Data from Goddard Institute for Space Studies, Intergovernmental Panel on Climate Change, National Academy of Sciences, National Aeronautics and Space Agency, National Center for Atmospheric Research, and National Oceanic and Atmospheric Administration)

atmosphere (or any system) to remain at a constant temperature, incoming solar energy (visible light that warms the earth) must be balanced by an equal amount of outgoing energy (longer-wavelength, lower-energy, infrared radiation that cools the earth; Figure 3-10, p. 58, and Figure 6-13, p. 128). Although the overall average temperature of the atmosphere is constant, the average temperature at various altitudes varies (Figure 17-2, p. 418, red line).

In addition to incoming sunlight, a natural process called the *greenhouse effect* (Figure 6-13, p. 128) warms the earth's lower troposphere and surface. Recall that it occurs because molecules of certain atmospheric gases, called *greenhouse gases* (primarily water vapor and carbon dioxide), warm the earth's surface. They do this by absorbing some of the infrared radiation (heat) radiated by the earth's surface. This causes their molecules to vibrate and transform the absorbed energy into longer-wavelength infrared radiation (heat) in the troposphere (Figure 6-13, p. 128).*

Actually, the atmosphere does not behave like a real greenhouse (or a car with its windows closed). The primary reason air heats up in a greenhouse (or car interior) is that the closed windows keep the air from being carried away by convection to the outside air. By contrast, heat released by molecules of greenhouse

---

*It is incorrect to say that that the earth's atmosphere *traps* heat or *reradiates* heat it has absorbed back toward the earth's surface. Molecules of greenhouse gases absorb various wavelengths of infrared radiation and transform them into infrared radiation with different (longer) wavelengths. Because the originally absorbed wavelengths of infrared radiation no longer exist, it is incorrect to say that they have been trapped or reradiated.

gases in the atmosphere is spread through the atmosphere by convection. A more scientifically accurate term for this phenomenon would be *tropospheric heating effect*. However, the term *greenhouse effect* is widely used and accepted.

Swedish chemist Svante Arrhenius first recognized this natural tropospheric heating effect in 1896. Since then it has been confirmed by numerous laboratory experiments and measurements of atmospheric temperatures at different altitudes and is one of the most widely accepted theories in the atmospheric sciences.

If the natural greenhouse effect acted by itself, the average surface temperature of the earth would be about 54°C (130°F). However, a *natural cooling process* also takes place at the earth's surface. This occurs when **(1)** large quantities of heat are absorbed by the evaporation of liquid surface water and **(2)** the water vapor molecules rise, condense to form droplets in clouds, and release their stored heat higher in the atmosphere (Figure 6-7, p. 125). The combined effects of these natural heating and cooling effects means that the earth's average surface temperature is about 15°C (59°F) instead of a frigid −18°C (0°F).

The two greenhouse gases with the largest concentrations in the atmosphere are **(1)** water vapor controlled by the hydrologic cycle (Figure 4-28, p. 90) and **(2)** carbon dioxide controlled by the carbon cycle (Figure 4-29, p. 92). Other greenhouse gases present in lower concentrations include methane (CH_4), nitrous oxide (N_2O), CFCs, SF_6, and SF_5CF_3. Inputs of these gases into the atmosphere can come from **(1)** natural sources and **(2)** human activities (Table 18-1), except for CFCs, SF_6, and SF_5CF_3, which come only from human sources.

Table 18-1 Greenhouse Gases from Human Activities

Greenhouse Gas	Human Sources	Average Time in the Troposphere	Relative Warming Potential (compared to CO_2)**
Carbon dioxide (CO_2)	Fossil fuel burning, especially coal (70–75%), deforestation, and plant burning (20–25%)	50-500 years	1
Methane (CH_4)	Rice paddies, guts of cattle and termites, landfills, coal production, coal seams, and natural gas leaks from oil and gas production and pipelines	9–15 years	24
Nitrous oxide (N_2O)	Fossil fuel burning, fertilizers, livestock wastes, and nylon production	120 years	360
Chlorofluorocarbons (CFCs)*	Air conditioners, refrigerators, plastic foams	11–20 years (65–110 years in the stratosphere)	1,500–7,000

*CFC use is being phased out, but it will take 50–100 years for the ozone layer to recover.
**The relative warming potential for SF_5CF_3 is 18,000 times that for CO_2 and its molecules last for about 1,000 years in the troposphere.

Figure 18-3 Estimated long-term variations in average global temperature of the atmosphere near the earth's surface and average tropospheric carbon dioxide levels over the past 160,000 years. These CO_2 levels were obtained by inserting metal tubes deep into antarctic glaciers, removing the ice, and analyzing bubbles of ancient air trapped in ice at various depths throughout the past. The rough correlation between tropospheric CO_2 levels and temperature shown in these estimates based on ice core data suggests a connection between these two variables, although no definitive causal link has been established. In 1999, the world's deepest ice core sample revealed a similar correlation between air temperatures and the greenhouse gases CO_2 and CH_4 going back for 460,000 years. (Data from Intergovernmental Panel on Climate Change and National Center for Atmospheric Research)

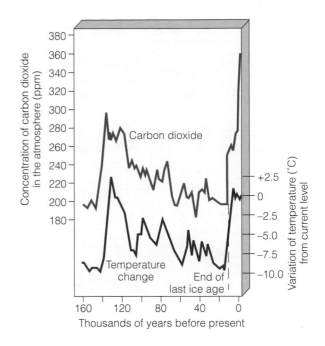

Analysis of gases in bubbles trapped at various depths in ancient glacial ice shows that over the past 160,000 years levels of tropospheric water vapor (the dominant greenhouse gas) have remained fairly constant. However, during most of this period, CO_2 levels have fluctuated between 190 and 290 parts per million. These estimated changes in tropospheric CO_2 levels correlate fairly closely with estimated variations in the atmosphere's average global temperature near the earth's surface during the past 160,000 years (Figure 18-3).

Have Human Activities Influenced the Earth's Climate? Since the beginning of the industrial revolution around 1750 (and especially since 1950) there has been a sharp rise in **(1)** the use of fossil fuels, which release large amounts of the greenhouse gases CO_2 and CH_4 into the troposphere, **(2)** deforestation and clearing and burning of grasslands to raise crops, which release CO_2 and N_2O into the atmosphere, and **(3)** cultivation of rice in paddies and use of inorganic fertilizers, which release N_2O into the troposphere.

Figure 18-4 shows that since 1860 there has been a sharp rise in the concentrations of the greenhouse gases CO_2, CH_4, and N_2O, with especially sharp increases since 1950. Based on evidence about past changes in atmospheric CO_2 concentrations and atmospheric temperatures (Figure 18-3), it is hypothesized that increased inputs of CO_2 and other greenhouse gases from human activities (Figure 18-4) could enhance the earth's natural greenhouse effect and raise the average global temperature of the atmosphere near the earth's surface. This enhanced greenhouse effect usually is called **global warming** and should not be confused with the different problem of ozone depletion (Table 18-2 and Section 18-6, p. 465).

Analysis of ice core samples, temperature measurements at different levels in several hundred boreholes in the earth's surface, and atmospheric temperature measurements show that

- The concentration of CO_2 in the troposphere is higher than it has been in the past 420,000 years and is rising by about 0.5% a year.

- The 20th century was the hottest century in the past 1,000 years (Figure 18-2).

- Since 1860, the average global temperature of the troposphere near the earth's surface has risen 0.6–0.7°C (1.1–1.3°F), with most of this increase taking place since 1946 (Figure 18-2).

- The 15 warmest years on record since 1860 have occurred in the past two decades (Figure 18-2), with the 5 hottest years occurring in the 1990s (in order 1998, 1997, 1995, 1990, and 1999). A significant portion of the sharp temperature rise in 1997 and 1998 (most of it over the tropical oceans) could have been the result of the strong 1997–98 El Niño-Southern Oscillation (ENSO; Figure 6-11, p. 127).

Other observed signs of a warmer troposphere during recent decades include **(1)** increased temperatures and melting of ice caps and floating ice at the earth's poles (Connections, p. 454), **(2)** retreat of some glaciers on the tops of mountains in the Alps, Andes, Himalayas, and northern Cascades of Washington, **(3)** northward migration of some warm-climate fish and trees, and **(4)** bleaching of coral reefs (Figure 7-1, top, p. 152) in tropical areas with warmer water.

Do these data indicate that our dramatic increase in fossil fuel use, agriculture, and deforestation since 1850 and especially since 1950 have had an influence on the earth's ever-changing temperature and climate? The simple answer ranges from *maybe* to *probably*, but we cannot be sure because of our limited knowledge about how earth's complex climate system works.

Here are several possibilities:

- Some or most of the almost 0.7°C rise since 1860 could result from normal but still poorly understood fluctuations in the average global temperature built into the earth's climate system.

■ The observed global warming since 1860 could be a **(1)** natural climate change trend or **(2)** a natural climate change trend enhanced by human activities.

■ Such warming of the earth's surface could accelerate and last for decades to hundreds of years, or it

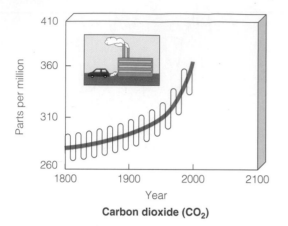

Carbon dioxide (CO₂)

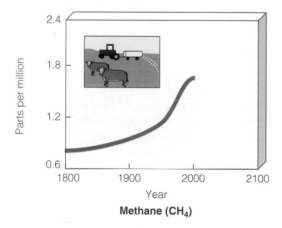

Methane (CH₄)

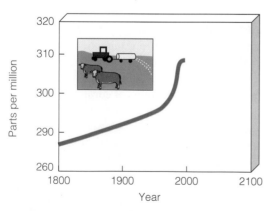

Figure 18-4 Increases in average concentrations of the greenhouse gases carbon dioxide, methane, and nitrous oxide in the troposphere between 1860 and 1999. (Data from Intergovernmental Panel on Climate Change, National Center for Atmospheric Research, and World Resources Institute)

could be temporary, with temperatures leveling off or even dropping sometime in the future.

It is clear that during the past 200 years human activities have been changing the chemical composition of the atmosphere more rapidly than it has changed at any time during the last 10,000 years. What is not clear is how this global experiment might affect the earth's climate and life-support systems. Because there is only one earth, scientists warn that this is a potentially dangerous experiment.

Regardless of the cause, significant climate change caused by atmospheric warming or cooling over several decades to a hundred years has important implications for human life, wildlife, and the world's economies. Such rapid climate change can **(1)** affect the availability of water resources by altering rates of evaporation and precipitation, **(2)** shift areas where crops can be grown, **(3)** change average sea levels, and **(4)** alter the structure and location of the world's biomes (Figure 6-16, p. 131), as discussed in more detail in Section 18-4.

18-2 PROJECTING FUTURE CHANGES IN THE EARTH'S CLIMATE

What Might Happen if Greenhouse Gas Levels Keep Rising? It is projected that there will be significant increases in the emissions of CO_2, CH_4, and N_2O during the 21st century (Figure 18-5) and that such increases could enhance the earth's natural greenhouse effect. These projected increases in greenhouse gases raise three important questions:

■ When, where, and how much might the global average temperature and climate change in the future?

■ What beneficial and harmful effects might projected natural climate change or human-influenced climate change have on humans, other species, and economies?

■ What can we do to minimize or adapt to the harmful effects of fairly rapid climate change, whether caused by natural factors, human factors, or both?

How Do Scientists Model Climate Changes? Computer Models as Crystal Balls To project the effects of increases in greenhouse gases (Figure 18-5) on average global temperature and the earth's climate, scientists develop *mathematical models* of such systems and run them on supercomputers (Spotlight, p. 452). Most climate models are derived from models developed for weather forecasting, called *general circulation models* (GCMs). These models attempt to provide a three-dimensional representation of how energy, air masses, and moisture flow through the atmosphere, based on

Table 18-2 Major Characteristics of Global Warming and Ozone Depletion		
Characteristic	**Global Warming**	**Ozone Depletion**
Region of atmosphere involved	Troposphere.	Stratosphere.
Major substances involved	CO_2, CH_4, N_2O (greenhouse gases).	O_3, O_2, chlorofluorocarbons (CFCs).
Interaction with radiation	Molecules of greenhouse gases absorb infared (IR) radiation from the earth's surface, vibrate, and release longer-wavelength IR radiation (heat) into the lower troposphere. This natural greenhouse effect helps warm the lower troposphere (Figure 6-13, p. 128).	About 95% of incoming ultraviolet (UV) radiation from the sun is absorbed by O_3 molecules in the stratosphere and does not reach the earth's surface (Figure 4-18, p. 75).
Nature of problem	Increasing concentrations of greenhouse gases in the troposphere from burning fossil fuels, deforestation, and agriculture could be enhancing the natural greenhouse effect and raising the earth's average surface temperature.	CFCs and other ozone-depleting chemicals released into the troposphere by human activities have made their way to the stratosphere, where they decrease O_3 concentration. This can allow more harmful UV radiation to reach the earth's surface.
Possible consequences	Changes in climate, agricultural productivity, water supplies, and sea level.	Increased incidence of skin cancer, eye cataracts, and immune system suppression and damage to crops and phytoplankton.
Possible responses	Decrease fossil fuel use and deforestation.	Eliminate CFCs and other ozone-depleting chemicals and find acceptable substitutes.

the laws of physics and general air circulation patterns between the earth's warm equator and cold poles (Figure 6-6, p. 125). The latest atmospheric climate models also include **(1)** ocean circulation models (Figure 6-4, p. 124), **(2)** interactions between the atmosphere and the ocean, **(3)** changes in solar output, **(4)** inputs of aerosols (tiny particles and droplets) into the atmosphere by volcanoes and pollution from human activities, and **(5)** a more realistic but still limited estimate of the effect of cloudiness. When this is done, the projected values for the atmosphere's average global surface temperature closely follow the measured values between 1860 and 1999 (Figure 18-7).

Greatly improved climate models and measurements have increased our ability to predict future global changes in the earth's climate and average surface temperature. However, there are still significant scientific uncertainties, as discussed in Section 18-3.

What Is the Scientific Consensus About Future Climate Change and Its Effects? The Intergovernmental Panel on Climate Change (IPCC) is a network of about 2,500 of the world's leading climate experts from 70 nations established by the United Nations and the World Meteorological Organization to study climate change. In 1990, 1995, and 2000, the IPCC published major reports evaluating the available evidence

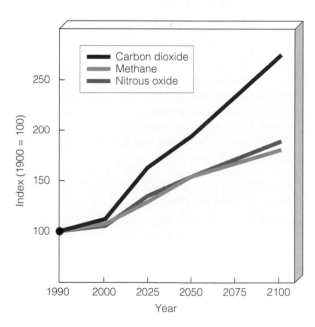

Figure 18-5 Projected emissions of three important greenhouse gases as a result of human activities, 1990–2100. (Data from United Nations Food and Agriculture Organization, 1997)

about past changes in global temperatures (Figure 18-2) and climate models projecting future changes in average global temperatures and climate. The U.S. National Academy of Sciences and the American

SPOTLIGHT

How Do Climate Models Work?

Here is a crude representation of how current three-dimensional general circulation models (GCM) of global climate are built.

- Simulate the earth's atmosphere mathematically by covering the earth's surface with 9–64 stacked layers of gigantic cells or boxes, each several hundred kilometers on a side and about 3 kilometers (2 miles) high (Figure 18-6).

- Assign an *initial condition* (starting value) for each variable to each box in the stacked layers. These variables include (1) temperature, (2) air pressure, (3) wind speed and direction, and (4) concentrations of key chemicals such as H_2O and CO_2.

- Develop a set of equations describing the expected flow of various types of energy and matter in and out of the box-filled atmosphere. These inputs, called *boundary conditions*, include variables such as (1) solar radiation, (2) precipitation, (3) heat radiated by the earth, (4) cloudiness, (5) interactions between the atmosphere and oceans, (6) greenhouse gases, and (7) air pollutants (especially aerosols).

- Develop an elaborate set of mathematical equations that connect the cells so that when the value of any of these input variables changes in one cell, the values of variables in surrounding cells change in a realistic way.

- Run the model on a supercomputer to (1) simulate changes in average global temperature and climate in the past (to verify the model) and (2) project future changes.

Such models can provide us with scenarios of what *could* happen based on various assumptions and data fed into each model. How well the results correspond to the real world depends on (1) the assumptions of the model based on current knowledge about the systems making up the earth, oceans, and atmosphere, (2) the accuracy of the data used, (3) magnification of tiny errors over time, (4) factors in the earth's climate system that amplify (positive feedback) or dampen (negative feedback) changes in average global temperatures (Section 18-3), and (5) the effects of totally unexpected or unpredictable events (chaos).

Current models reproduce fairly accurately the changes in the earth's average global temperature that have taken since 1865 (Figure 18-7). The models can also help us evaluate the possible effects of various ways of slowing or even halting any global warming projected by the models.

Although mathematical climate models are improving rapidly, they still have many limitations because of a lack of understanding of the complexity of the earth's climate system. The models also cannot make reliable projections of the nature and rate of climate change in different regions of the world. However, mathematical models are the most useful tools we have for understanding and projecting the behavior of complex systems such as global climate (p. 50).

Critical Thinking

How much should we rely on mathematical climate models to help us evaluate the possible effects of human activities on the earth's climate and make decisions about how to prevent the global warming the models project? What are the alternatives?

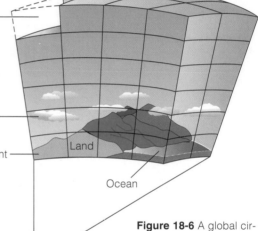

Figure 18-6 A global circulation model (GCM) of climate (and weather) divides the earth's atmosphere into large numbers of gigantic boxes or cells stacked many layers high. The laws of physics and our understanding of global air circulation patterns (Figure 6-8, p. 126) are used to describe numerically what happens to major variables affecting climate in each cell and from one cell to another.

Geophysical Union have also evaluated possible future climate changes.

According to the IPCC's 1995 report,

- "The balance of evidence suggests a discernable human influence on climate over the last 50 years."

- The earth's mean surface temperature is likely to warm by 1–3.5°C (1.8–6.3°F) between 2000 and 2100.

According to the IPCC's 2000 report,

- "There is stronger evidence" on the human influence on climate and "it is likely that (greenhouse gases from human acivities) have contributed substantially to the observed warning over the last 50 years."

- The earth's mean surface temperature is likely to increase 1.4–5.8°C (2.5–10.4°F) between 2000 and

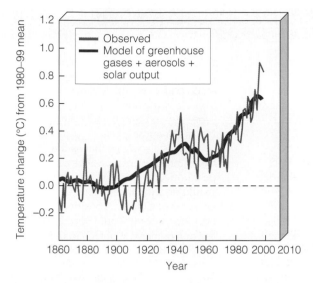

Figure 18-7 Comparison of measured changes in the average atmospheric temperature near the earth's surface (blue line) with those projected by the latest climate models between 1860 and 1999 (red line). (Data from Tom L. Wigley, Pew Center on Global Climate Change)

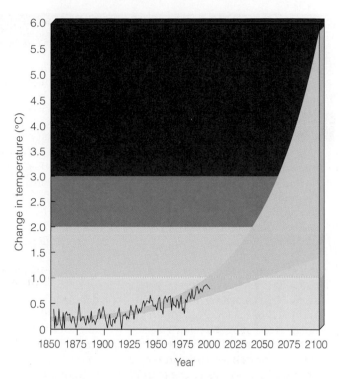

Figure 18-8 Comparison of measured changes in the average temperature of the atmosphere at the earth's surface between 1860 and 1999 and the projected range of temperature increase during the 21st century. Current models indicate that the average global temperature at the earth's surface will rise by 1.4–5.8°C (2.5–10.4°F) sometime during the 21st century. Because of the uncertainty involved in current climate models, the projected warming shown here could be *overestimated* or *underestimated* by a factor of two. (Data from U.S. National Academy of Sciences, National Center for Atmospheric Research, and Intergovernmental Panel on Climate Change)

2100 (Figure 18-8). The report defines "likely" as a chance of 66–99%.

- This new projection of temperature increases is higher than in 1995 mostly because the latest climate models assume that developed and developing nations (especially China and India) will cut emissions of sulfur dioxide produced by burning coal in electric power and industrial plants during the 21st century. SO_2 and aerosols (tiny liquid droplets and solid particles) formed from it in the atmosphere produce a haze that has a cooling influence on the lower atmosphere. Thus, lowering atmospheric levels of these chemicals to reduce their harmful impacts on human health, crops, and forests (Section 17-6, p. 437) can enhance the natural greenhouse effect.

However, a small minority of atmospheric scientists inside and outside the IPCC contends that

- There is still insufficient knowledge about natural climate variables that could change the assessment (up or down), as discussed in Section 18-3.

- Computer models used to predict climate change are improving but are still not very reliable.

IPCC and other climate scientists holding the consensus view

- Agree that we need to support greatly increased research on **(1)** how the climate system works and **(2)** improving climate models to help narrow the range of uncertainty in projected temperature (Figure 18-8).

- Point out that the biggest concern is not just a temperature increase, but how rapidly it rises. They warn that warming or cooling of the earth's surface by even 1°C (1.8°F) per century **(1)** would be faster than any temperature change occuring in the last 10,000 years, since the beginning of agriculture (Figure 18-2), and **(2)** could cause serious disruptions of the earth's ecosystems and human societies (Section 18-4).

18-3 FACTORS AFFECTING CHANGES IN THE EARTH'S AVERAGE TEMPERATURE

Will the Earth Continue to Get Warmer? The average temperature at the earth's surface has increased significantly during the past 20 years (Figure 18-2). Will this trend continue?

The answer is that *we do not know.* Scientists have identified a number of natural and human-influenced factors that might *amplify* (positive feedback) or *dampen*

Early Warnings from the Earth's Ice

CONNECTIONS

As the atmosphere warms, it causes more convection that transfers surplus heat from equatorial to polar areas (Figure 6-6, p. 125). Thus, temperature increases tend to be greater at the earth's poles than at the midlatitudes. This explains why the earth's frigid poles are regarded as early warning sentinels of global warming.

The sea ice floating at the Arctic is large enough to cover the continental United States and is highly sensitive to changes in the air above and the ocean below. News from the Arctic is alarming. Between 1968 and 1999, surface temperatures at nine stations north of the arctic circle rose by about 6.1°C (11°F)—more than 10 times the average global temperature increase during this period.

The bright color of the floating sea ice in the Arctic Ocean helps cool the earth by reflecting 80% of the sunlight it receives back into space. If this ice melted, the darker Arctic Ocean would become a heat collector, absorbing 80% of its input of

sunlight, and drastically affect global climate, especially in the northern hemisphere.

Satellite data and measurements by surface boats and submarines indicate that the summer Arctic Ocean ice **(1)** shrunk in area by about 3% between 1978 and 1998 (a loss in area equivalent to the size of Texas) and **(2)** thinned by 42% between 1958 and 1998. If the ice continues to disappear, the Arctic Ocean will warm rapidly and melt more ice.

It is not known whether this shrinkage and thinning of arctic sea ice is the result of natural polar climate fluctuations or global warming caused by increases in greenhouse gases. According to climate models, if global warming is the culprit the entire ice pack eventually will disappear.

Because it is floating, large-scale melting of Arctic Ocean ice will not raise global sea levels (just as an ice cube in a glass of water does not raise the water level when it melts). However, widespread melting would greatly amplify warming of the Arctic region. This could reroute

warm ocean currents (Figure 6-4, p. 124) and weather patterns further south and cause significant cooling in parts of the northern hemisphere, especially in Europe and eastern North America.

Studies also show that as the frozen north warms and thaws, peat buried in the arctic tundra soil would decay and release large amounts of CO_2. This positive feedback mechanism would speed up the rate of increase of CO_2 in the atmosphere and amplify global warming. According to climate researchers at San Diego State University (California), since 1982 the arctic tundra has warmed so much that it has been giving off more carbon dioxide than it absorbs—a positive feedback loop that could accelerate tropospheric warming.

The news from the Antarctic is also disturbing. The huge antarctic ice cap, which is almost twice the area of Australia, contains 70% of the world's fresh water and 90% of the world's highly reflective ice, which helps cool the earth.

The near-freezing meltwater that runs off the cap, along with water

(negative feedback) changes in the earth's average surface temperature. These factors could influence how much and how fast temperatures might climb or drop and what the effects might be in various areas.

How Might Changes in Solar Output Affect the Earth's Temperatures? Solar output varies by about 0.1% over the 18-year and 22-year sunspot cycles and over 80-year and other much longer cycles. These up-and-down changes in solar output can temporarily warm or cool the earth and thus affect the projections of climate models.

Currently, atmospheric scientists have not been able to identify a mechanism related to changes in solar output that could account for more than 25–50% of the atmospheric warming between 1900 and 1999. However, according to a hypothesis by climate scientists Paal Beckke and Henrik Svensmark, changes in the sun's exterior magnetic field could affect the atmospheric temperature by altering the amount of cosmic rays striking the earth. These rays can convert neutral molecules of gases in the atmosphere into charged ions that can affect cloud formation.

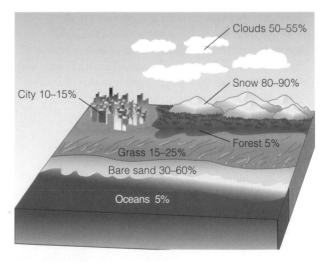

Figure 18-9 The albedo, or reflectivity of incoming solar energy, of different parts of the earth's surface varies greatly. (Data from NOAA)

It is hypothesized that

- An increase in cosmic rays because of a decrease in the sun's exterior magnetic field may help cool the

from melting floating icebergs, falls to the ocean floor and surges northward. This surge affects deep-sea circulation (Figure 18-10), which in turn affects the climate in various parts of the world. A major meltdown of the Antarctic ice cap would not only affect these ocean currents but also raise the average global sea level.

Since 1947, the average temperature of the Antarctic Peninsula has risen by about 2.8°C (5°F) in summer and 5.6°C (10°F) in winter. As a result, huge pieces of ice shelf—some as large as the state of Delaware—have begun breaking off (calving) the peninsula's eastern shore.

The adjacent western antarctic ice sheet—the size of Texas and Colorado combined—has been melting slowly for decades. The breakup and eventual collapse of this ice sheet appears to be part of an ongoing natural cycle initiated by the melting of northern hemispheric ice sheets at the end of the last glacial period, when average sea levels began their huge rise (Figure 18-11).

However, this breakup could speed up if other natural processes or human activities continue to warm the atmosphere and oceans. A partial melting or breakdown of the western antarctic ice sheet would raise global sea levels by as much as 0.9 meters (3 feet) by 2100. Some scientists are considering the possibility that a much larger area of this sheet could be gone by 2100. If that occurred, the average global sea level would rise by at least 5–6 meters (15–20 feet) and severely flood many of the world's low-lying coastal cities, wetlands, and islands.

For some scientists, the most disturbing news comes from Greenland, a hugh island three times the size of Texas found more than 14,500 kilometers (9,000 miles) north of Antarctica. A 2000 study of ice cores found that the glaciers of Greenland are more likely to melt because they are closer to the equator than the west antarctic ice sheet. If this occurred, as it did in a previous interglacial warm period 110,000–130,000 years ago (Figure 18-11), average sea levels would rise by 4-meters (13–20 feet). This influx of fresh water would cause flooding in the world's low-lying coastal areas and cities and could shut off currents such as the Gulf Stream and North Atlantic, which keep Europe warmer than it would otherwise be.

These changes in ice cover at the earth's poles may result from natural climate cycles, human activities, or a combination of both. In any case, they can have serious long-term implications for wildlife and human economies.

One comedian jokes that he plans to buy land in Kansas because it will probably become valuable beachfront property. Another boasts that she is not worried because she lives in a houseboat—the "Noah strategy."

Critical Thinking

What difference might it make in your life and in that of any child you might have if human activities play an important role in continued warming of the earth's polar regions and Greenland? What three important things could you do to help slow down such possible warming?

---

earth's surface by forming more thick, umbrella-like low clouds.

- A decrease in cosmic rays because of an increase in the sun's exterior magnetic field (as has been observed since 1901) may help warm the earth's surface by forming more high, thin clouds.

However, according to Kevin Treberth of the U.S. National Center for Atmospheric Research,

- There is very good evidence that the amount of cooling, low-cloud cover has increased throughout much of the world in recent decades–the opposite of what would be expected according to the cosmic ray hypothesis with an increase in the sun's exterior magnetic field.

How Might Changes in the Earth's Reflectivity Affect Atmospheric Temperatures? Different parts of the earth's surface vary in their **albedo**, or ability to reflect light (Figure 18-9). White or shiny surfaces such as snow, ice, and sand reflect most of the sunlight that hits them. This reflected energy does not heat the earth (Figure 4-8, p. 75).

The albedo of the planet is not constant and changes as a result of natural and human-induced changes in the earth's surface. For example, albedo increases when polar ice caps expand during glacial periods and decreases when they melt and expose less reflective land and ocean surfaces (Connections, left). Satellite measurements of how albedo varies throughout the world and projections of how this might change have been incorporated into current climate models.

How Might the Oceans Affect Climate? Currently, the oceans help moderate the earth's average surface temperature by naturally removing about 29% of the excess CO_2 we pump into the atmosphere as part of the global carbon cycle (Figure 4-29, p. 92). We do not know whether the oceans can absorb more CO_2.

The oceans also affect global climate by absorbing heat from the atmosphere and transferring some of it to the deep ocean, where it is stored temporarily. It is hypothesized that this absorption of heat by the oceans apparently has delayed part of the warming of the

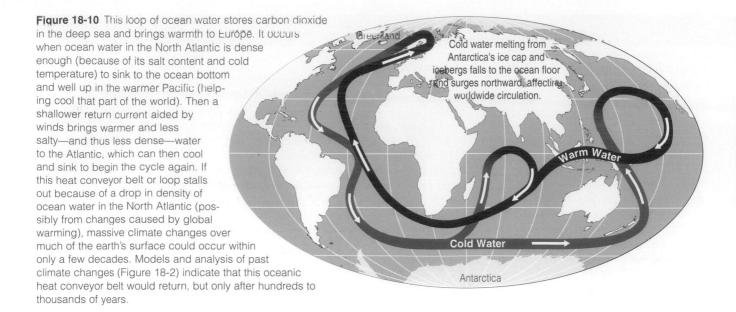

Figure 18-10 This loop of ocean water stores carbon dioxide in the deep sea and brings warmth to Europe. It occurs when ocean water in the North Atlantic is dense enough (because of its salt content and cold temperature) to sink to the ocean bottom and well up in the warmer Pacific (helping cool that part of the world). Then a shallower return current aided by winds brings warmer and less salty—and thus less dense—water to the Atlantic, which can then cool and sink to begin the cycle again. If this heat conveyor belt or loop stalls out because of a drop in density of ocean water in the North Atlantic (possibly from changes caused by global warming), massive climate changes over much of the earth's surface could occur within only a few decades. Models and analysis of past climate changes (Figure 18-2) indicate that this oceanic heat conveyor belt would return, but only after hundreds to thousands of years.

earth's atmosphere, as climate modelers had already projected and taken into account. However, over the next few decades it is projected that some of this stored heat probably will be released back to the atmosphere and could amplify global warming.

There is also concern that global warming could disrupt ocean currents (Figure 6-4, p. 124), which are driven largely by differences in water density and winds. Connected deep and surface currents act like a gigantic conveyor belt to transfer heat from one place to another and store carbon dioxide and heat in the deep sea (Figure 18-10). There is concern that an influx of fresh water from thawing ice in the Arctic and the Antarctic (Connections, p. 454) might slow or disrupt this conveyer belt and slow or halt the amount of heat it brings to the North Atlantic region. If this loop stalls out, evidence from past clim-ate changes indicates that this could trigger atmosphere temperature changes of more than 5°C (9°F) over periods as short as 40 years.

Changes in average sea level affect (1) the amount of heat and CO_2 that can be stored in the ocean and (2) changes in the earth's biomes (Figure 6-16, p. 131). Figure 18-11 shows estimated changes in the earth's average sea level over the last 250,000 years based on data obtained from cores drilled in the ocean floor. The climax of the earth's last glacial period was about 18,000 years ago. Since that time the earth's average sea level has risen about 125 meters (410 feet) as most glaciers covering the earth's land surface melted during the current interglacial period. According to the 2000 IPCC report, the rate of sea level rise is now faster than at any other time during the past 1,000 years.

There is concern that accelerated global warming from natural or human-related factors will enhance

glacial melting and raise sea levels slightly during the 21st century and possibly much more if there is a significant melting of ice sheets in Greenland (Connections, p. 455). Even a slight increase in sea level would flood many coastal wetlands, low-lying islands, and cities.

How Do Water Vapor Content and Clouds Affect Climate? Warmer temperatures increase evaporation of surface water and create more clouds. These additional clouds could have (1) a warming effect by absorbing and releasing heat into the troposphere or (2) a cooling effect by reflecting sunlight back into space.

The net result of these two opposing effects depends on (1) whether it is day or night and (2) the type (thin

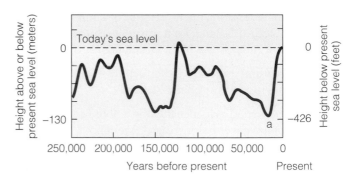

Figure 18-11 Changes in average sea level over the past 250,000 years based on data from cores removed from the ocean floor. The rise and fall of sea level are caused largely by the coming and going of glacial periods (ice ages). As glaciers melted and retreated since the peak of the last glacial period about 18,000 years ago, the earth's average sea level has risen about 125 meters (410 feet) and changed the earth's coastal zones (Figure 7-5, p. 156) significantly. (Adapted from *Oceanography: An Invitation to Marine Science*, 3d ed., by Tom Garrison ©1998. Reprinted by permission of Brooks/Cole, a division of Thomson Leaning. Fax 800-730-2215)

or thick) and altitude of clouds. Scientists do not know which of these factors might predominate or how cloud types and heights might vary in different parts of the world as a result of global warming. The effects of clouds have been included in recent climate models, but there is still much uncertainty about these effects on climate.

In 1999, researchers at the University of Colorado reported that the wispy condensation trails left behind by jet planes may have a greater impact on the earth's climate than scientists had thought. Using infrared satellite images, they found that jet contrails expand and turn into large cirrus clouds that tend to release heat into the upper atmosphere. If these preliminary results are confirmed, jet planes emissions could be responsible for as much as half of the atmospheric warming in the northern hemisphere.

How Might Air Pollution Affect Climate? Projected global warming might be partially offset by *aerosols* (tiny droplets and solid particles, Figure 17-7, p. 425) of various air pollutants released or formed in the atmosphere by volcanic eruptions and human activities. It is hypothesized that higher levels of aerosols attract enough water molecules to form condensation nuclei, leading to increased cloud formation.

Some of the resulting clouds have a high albedo and reflect more incoming sunlight back into space during daytime. This could help counteract the heating effects of increased greenhouse gases.

Nights would be warmer because the clouds would still be there and prevent some of the heat stored in the earth's surface (land and water) during the day from being radiated into space. These pollutants may explain why most of the recent warming in the northern hemisphere occurs at night.

However, these interactions are complex. Pollutants in the lower troposphere can either warm or cool the air, depending on the reflectivity of the underlying surface (Figure 18-9).

According to scientists, it is unlikely that air pollutants will counteract projected global warming very much in the next 50 years because:

■ Aerosols fall back to the earth or are washed out of the atmosphere within weeks or months, whereas CO_2 and other greenhouse gases remain in the atmosphere for decades to several hundred years.

■ Aerosols are major components of acid deposition (Figure 17-9, p. 428), which can slow forest growth and weaken or kill trees (Figure 17-14, p. 432). This reduces the ability of growing trees to absorb some of the CO_2 we are putting into the atmosphere and could accelerate atmospheric warming.

■ Aerosol inputs into the atmosphere are being reduced (Section 17-7, p. 440) because they kill or sicken large numbers of people each year (p. 438) and damage food crops, trees, and other forms of vegetation (Section 17-4, p. 427).

How Might Increased CO_2 Levels Affect Photosynthesis and Methane Emissions? Some studies suggest that more CO_2 in the atmosphere could increase the rate of photosynthesis in areas with adequate amounts of water and other soil nutrients. This would remove more CO_2 from the atmosphere and help slow atmospheric warming.

However, this effect would slow as the plants reach maturity and take up less CO_2. In addition, when the plants die and are decomposed or burned, the carbon they stored is returned to the atmosphere as carbon dioxide.

Other studies suggest that **(1)** this effect varies with different types of plants and in different climate zones and **(2)** much of any increased plant growth could be offset by plant-eating insects that breed more rapidly and weeds that grow more rapidly and year-round in warmer temperatures.

Atmospheric warming could be accelerated by increased release of methane (a potent greenhouse gas, Table 18-1) from:

■ Bogs and other wetlands.

■ Icelike compounds called methane hydrates trapped beneath the Artic permafrost and sediments on the floor of the Artic Ocean. Large amounts of methane would be released into the trosphere if **(1)** the blanket of permafrost in tundra soils melts and the Arctic Ocean warms considerably or **(2)** methane hydrates on the ocean floor are extracted to increase natural gas supplies (a possibility now under study in the United States).

How Rapidly Could Climate Shift? If moderate change takes place gradually over several hundred years, people in areas with unfavorable climate changes may be able to adapt to the new conditions. However, if the projected global temperature change takes place over several decades, we may not have enough time (and money) to **(1)** switch food-growing regions, **(2)** relocate people from low-lying coastal areas, and **(3)** build elaborate systems of dikes and levees to help protect the large portion of the world's population living near coastal areas. This could lead to large numbers of premature deaths from lack of food and social and economic chaos, especially in developing countries, which would not have the money needed to make such investments. Such rapid changes would also reduce the earth's biodiversity because many species could not move or adapt.

Recent data from analyses of ice cores and deep-sea sediments suggest that average temperatures during

Agriculture
- Shifts in food-growing areas
- Changes in crop yields
- Increased irrigation demands
- Increased pests, crop diseases, and weeds in warmer areas

Water Resources
- Changes in water supply
- Decreased water quality
- Increased drought
- Increased flooding

Forests
- Changes in forest composition and locations
- Disappearance of some forests
- Increased fires from drying
- Loss of wildlife habitat and species

Biodiversity
- Extinction of some plant and animal species
- Loss of habitats
- Disruption of aquatic life

Sea Level and Coastal Areas
- Rising sea levels
- Flooding of low-lying islands and coastal cities
- Flooding of coastal estuaries, wetlands, and coral reefs
- Beach erosion
- Disruption of coastal fisheries
- Contamination of coastal aquifers with salt water

Weather Extremes
- Prolonged heat waves and droughts
- Increased flooding
- More intense hurricanes, typhoons, tornadoes, and violent storms

Human Population
- Increased deaths
- More environmental refugees
- Increased migration

Human Health
- Increased deaths from heat and disease
- Disruption of food and water supplies
- Spread of tropical diseases to temperate areas
- Increased respiratory disease
- Increased water pollution from coastal flooding

the warm interglacial period that began about 125,000 years ago (Figure 18-2) varied as much as 10°C (18°F) in only a decade or two and that such warming and cooling periods each lasted 1,000 years or more. If these findings are correct and also apply to the current interglacial period, fairly small rises in greenhouse gas concentrations could trigger rapid up-and-down shifts in the earth's average surface temperatures.

As a result of uncertainties in climate models and the factors discussed in this section, climate scientists estimate that *projections from current climate models about atmospheric warming and rises in average sea levels during the next 50-100 years could be half (the best-case scenario) or twice (the worst-case scenario) the current projections (Figure 18-8).* In any event, possible climate change and its effects are likely to be erratic and mostly unpredictable.

18-4 SOME POSSIBLE EFFECTS OF A WARMER WORLD

Why Should We Worry if the Earth's Temperature Rises a Few Degrees? So what's the big deal? Why should we worry about a possible rise of only a few degrees in the earth's average surface temperature? We often have that much change between May and July, or even between yesterday and today.

The key point is that we are not talking about normal swings in *local weather* but a projected *global* change in *climate*—weather averaged over decades, centuries, and millennia.

What Are Some Possible Effects of Atmospheric Warming? A warmer global climate could have a number of harmful and beneficial effects (Figure 18-12)

depending on where one lives. If the earth's surface warms, climate models project that:

- Some places will get hotter and others colder.

- Some areas will be drier and others wetter. For example, climate change is projected to increase water scarcity in the Middle East (p. 294).

- Warmer soil, especially at high latitudes, will speed up plant decomposition and release more CO_2.

- Wetter areas will experience more intense rainfall, which will increase soil erosion and flooding.

- Weather extremes such as (1) heat waves, (2) prolonged droughts (from increased evaporation and drier soils), (3) flooding (from more intense rainfall), and (4) violent storms are projected to increase in frequency and severity. In 1998 alone, weather-related disasters (1) killed an estimated 32,000 people, (2) displaced 300 million people (more than the population of the United States), and (3) caused record-high damages of $92 billion worldwide—

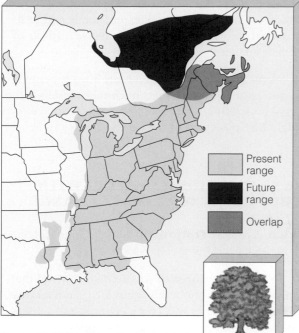

Figure 18-13 Possible effects of global warming on the geographic range of beech trees based on archeological evidence and computer models. According to one projection, if CO_2 emissions doubled between 1990 and 2050, beech trees (now common throughout the eastern United States) would survive only in a greatly reduced range in northern Maine and southeastern Canada. This is only one of a number of tree species whose geographic ranges could be changed drastically by increased atmospheric warming. (Data from Margaret B. Davis and Catherine Zabinski, University of Minnesota)

Present range

Future range

Overlap

much greater than the losses for the entire decade of the 1980s.

- The largest temperature increases will take place at the earth's poles and probably cause more melting of floating ice and glaciers (Connections, p. 454). This could **(1)** decrease the earth's ability to reflect incoming sunlight (albedo, Figure 18-9), **(2)** amplify global warming, **(3)** reduce the amount of tundra available to Arctic species, **(4)** threaten many of the world's distinctive mammals (such as arctic fox, polar bears, and bowhead whales) found only in the Arctic and migratory birds that spend part of each year in the Arctic, and **(5)** threaten some species unique to the Antarctic (Figure 4-19 and Connections, right).

- According to a 2000 study by the World Wildlife Fund, projected global warming could fundamentally alter one-third of the world's wildlife habitats and 70% of those in high northern altitudes by 2100.

- For each 1°C (1.8°F) rise in the earth's average temperature, climate belts in midlatitude regions would shift toward the earth's poles by 100–150 kilometers (60–90 miles) or upward 150 meters (500 feet) in altitude. Such shifts could change areas where crops could be grown and affect the makeup and

Global Warming, Kirtland's Warblers, and Adélie Penguins

Global warming may sharply reduce populations of some species, especially those with specialized niches. For example, Kirtland's warbler, an endangered bird that nests exclusively under young jack pines in northern Michigan, probably would become extinct. The reason is that if northern Michigan ends up with less rain and more heat, jack pines there probably will die off in the next 30–90 years.

Another example involves Adélie penguins (Figure 4-19, p. 84) that live in the western Antarctic, which has warmed up by several degrees since 1947 (Connections, p. 454). During the past 22 years, University of Montana ecologist William Fraserhas has observed a 40% drop in the population of this penguin species.

He suggests that sea ice melted by warmer temperatures has reduced populations of shrimplike krill (Figure 4-19, p. 84) that are the Adélie's favorite food. Apparently populations of marine plankton that the krill eat have been dropping, perhaps because of **(1)** melting of the sea ice and **(2)** increased ultraviolet radiation because of drastic drops in concentrations of protective ozone in the stratosphere above Antarctica for several months each year (Figure 18-17, p. 467). In addition, the penguins' ability to produce young may be decreased by increased spring snowfall (caused by warmer, moisture-laden air) that buries the Adélie's eggs under snowbanks.

Fraser and other scientists believe that the plight of Adélie penguins is an early warning of changes other parts of the earth could experience on a much larger scale if projected global warming takes place.

Critical Thinking

Identify another species with a specialized niche whose population size could be sharply reduced because of projected global warming.

location of at least one-third of today's forests. Northward expansion of crop-growing regions from the midwestern United States into Canada would be limited by the thinner and less fertile soils there. However, food production could be increased by northward expansion of crop-growing areas in parts of northern Asia with fertile soils.

- Current models cannot predict whether higher crop yields in some areas would compensate for lower yields in others (such as a 20–40% drop in rice yields). However, there is general agreement that food production would diminish in some of the world's poorest countries, especially in Africa.

- Tree species whose seeds are spread by wind may not be able to migrate fast enough to keep up with climate shifts and would die out (Figure 18-13). Poor soils may also limit the rate at which tree species can spread northward.

- Forest diebacks from climate change and from greatly increased wildfires in areas where the climate became drier would release large amounts of CO_2 into the environment. This could accelerate atmospheric warming and affect timber production, wildlife habitat, and recreational activities.

- Changes in the structure and location of wildlife habitats could cause extinction of plant and animal species that could not migrate to new areas and threaten species with specialized niches (Connections, p. 459).

- Shifts in regional climate would threaten many parks, wildlife reserves, wilderness areas, wetlands, and coral reefs, wiping out many current efforts to stem the loss of biodiversity.

- Fish would die as (1) temperatures soared in streams and lakes, (2) dissolved oxygen levels dropped, and (3) lower water levels concentrated pesticides and other pollutants.

- Ocean currents may shift and change (Figure 18-10). This could cause sharp temperature drops in areas such as northern Europe and Japan where climates are more temperate because of warm-water currents (such as the Gulf Stream and Kuroshio, Figure 6-4, p. 124).

- Global sea levels will rise, mainly because water expands slightly when heated. During the 20th century, the average global sea level rose 10–25 centimeters (4–10 inches) and is projected to rise by 15–95 centimeters (6–36 inches) during the 21st century, with a best estimate of 49 centimeters (19 inches) by 2100. Even a modest rise in sea level of about 95 centimeters (36 inches) would (1) threaten half of the world's coastal estuaries, wetlands, and coral reefs and disrupt marine fisheries, (2) cause severe beach erosion (especially along the U.S. East Coast), (3) flood coastal regions and put an estimated 300 million people living in 30 of the world's largest cities directly at risk, (4) flood agricultural lowlands and deltas in parts of Bangladesh, India, and China, where much of the world's rice is grown, and (5) submerge some low-lying islands in the Pacific and Caribbean. If warming at the poles caused increased melting of land-based ice sheets, sea levels would rise much more (Connections, p. 454).

- The largest burden will fall on developing nations, which do not have the economic and technological ability to adapt to the adverse impacts of climate change.

Some of these general trends are already beginning to take place. However, current climate models do not agree on specific regions or areas where such effects might take place and how long the effects might last.

18-5 SOLUTIONS: DEALING WITH THE THREAT OF CLIMATE CHANGE

What Are Our Options? There are four schools of thought concerning global warming.

- *Do nothing.* A dozen or so scientists contend that climate change from human activities is not a threat, and a few popular press commentators and writers even claim that global warming is a hoax.

- *Do more research before acting.* A second group of scientists and economists point to the considerable uncertainty about climate change and its effects. They call for more research before making such far-reaching economic and political decisions as phasing out fossil fuels and sharply reducing deforestation.

- *Act now to reduce the risks from climate change.* A third group of scientists and economists urge us to adopt a *precautionary strategy.* They believe that when dealing with risky and far-reaching environmental problems such as climate change, the safest course is to take informed preventive action *before* there is overwhelming scientific knowledge to justify acting. In 1997, 2,700 economists led by 8 Nobel laureates declared, "As economists, we believe that global climate change carries with it significant environmental, economic, social, and geopolitical risks and that preventive steps are justified."

- *Act now as part of a no-regrets strategy.* Scientists and economists supporting this approach say that we should take the key actions needed to slow projected atmospheric warming even if it is not a serious threat because such actions lead to other important environmental, health, and economic benefits (Solutions, right). For example, a reduction in the combustion of fossil fuels, especially coal, will lead to sharp reductions in air pollution that harms and prematurely kills large numbers of people, lowers food and timber productivity, and decreases biodiversity.

Those who favor doing nothing or waiting before acting point out that there is a 50% chance that we are *overestimating* the impact of rising greenhouse gases.

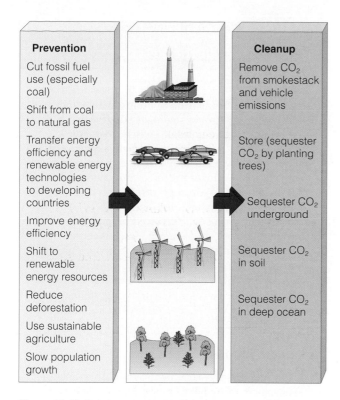

Prevention		Cleanup
Cut fossil fuel use (especially coal)		Remove CO_2 from smokestack and vehicle emissions
Shift from coal to natural gas		
Transfer energy efficiency and renewable energy technologies to developing countries		Store (sequester CO_2 by planting trees)
Improve energy efficiency		Sequester CO_2 underground
Shift to renewable energy resources		Sequester CO_2 in soil
Reduce deforestation		Sequester CO_2 in deep ocean
Use sustainable agriculture		
Slow population growth		

Figure 18-14 Solutions: methods for slowing projected atmospheric warming during the 21st century.

However, those urging action now point out that there is also a 50% chance that we are *underestimating* such effects.

How Can We Reduce the Threat of Climate Change from Human Activities?

Figure 18-14 presents a variety of prevention and cleanup solutions analysts have suggested for slowing climate change from increased greenhouse gas emissions. Gradually implementing such solutions over the next 20–30 years could simultaneously reduce the threats from global warming, air pollution, deforestation, and biodiversity loss.

According to some widely publicized economic models, reducing carbon dioxide emissions will be too costly. Other economists criticize these models as being unrealistic and too gloomy because they (1) do not include the huge cost savings from implementing many of the strategies listed in Figure 18-14 (Solutions, right) and (2) underestimate the ability of the marketplace to act rapidly when there is money to be made from reducing greenhouse gas emissions.

According to a number of economic studies, implementing the strategies listed in Figure 18-14 would (1) boost the global economy, (2) provide much-needed jobs (especially in developing countries with large numbers of unemployed and underemployed people), and (3) cost much less than trying to deal with the harmful effects of these problems. In other words, the

SOLUTIONS

Energy Efficiency to the Rescue

According to energy expert Amory Lovins (Guest Essay, p. 361), *the major remedies for slowing possible global warming are things we should be doing already even if there were no threat of global warming.* Lovins argues that we should (1) waste less energy, (2) reduce air pollution by cutting down on our use of fossil fuels and switching to renewable forms of energy, and (3) harvest trees more sustainably.

According to Lovins, improving energy efficiency (Section 15-2, p. 362) would be the fastest, cheapest, and surest way to slash emissions of CO_2 and most other air pollutants within two decades, using existing technology. Lovins estimates that increased energy efficiency would also save the world up to $1 trillion per year in reduced energy costs—as much as the annual global military budget—and $300 billion per year in the United States.

Lovins also warns that using fossil fuels such as natural gas (CH_4) as a source of hydrogen to power fuel cells (Figure 15-10, p. 366) instead of hydrogen derived from H_2O (Figure 15-34 p. 386) could increase the threat of global warming. The reason is that stripping hydrogen from carbon-based fossil fuels leaves behind carbon dioxide, which would probably be vented into the atmosphere.

Using energy more efficiently would also (1) reduce pollution, (2) help protect biodiversity, (3) deter arguments between governments about how CO_2 reductions should be divided up and enforced, (4) make the world's supplies of fossil fuel last longer, (5) reduce international tensions over who gets the dwindling oil supplies, and (6) allow more time to phase in renewable energy sources. A growing number of companies have gotten the message and are saving money and energy while reducing their carbon dioxide emissions and selling their emission reductions in the global marketplace.

Critical Thinking

1. Do you agree that improving energy efficiency should be done regardless of its impact on the threat of global warming as part of a no-regrets strategy? Explain.

2. Why do you think there has been little emphasis on improving energy efficiency?

earth's climate, air, and biodiversity could be protected *at a profit* for most of the world's major businesses.

According to a 1999 study by the Tellus Institute in Cambridge, Massachusetts, if the United States adopted aggressive policies to boost energy efficiency, increase use of renewable energy, and reduce CO_2 emissions, by 2010 this would:

- Save $43 billion per year in energy costs.

- Decrease jobs in the fossil fuel industry but result in a net increase of 870,000 new jobs.

- Reduce U.S. energy use by 18% and electricity use by 30%.

- Cut SO_2 emissions by 50% and emissions of nitrogen oxides by 25%.

- Reduce U.S. CO_2 emissions by 14% below 1990 levels, twice the reduction specified in the 1997 Kyoto climate treaty.

How Can Government Regulation be Used to Reduce or Prevent Greenhouse Gas Emissions?

Governments could significantly reduce CO_2 emissions over a decade by:

- Phasing out government subsidies for coal and oil (about $300 billion per year globally and $21 billion annually in the United States).

- Retaining government subsidies for natural gas, which could help us make the 40- to 50-year transition to an age of energy efficiency and renewable energy. When burned, natural gas emits only half as much CO_2 per unit of energy as coal (Figure 14-20, p. 339), and it emits far smaller amounts of most other air pollutants.

- Increasing emphasis on reducing emissions of methane, which absorbs and releases about 20 times as much heat per molecule as CO_2 (Table 18-1). Ways to reduce methane imputs include capturing and burning methane gas released by landfills (Figure 21-12, p. 537) and reducing leaks from tanks, pipelines, and other natural gas handling facilities.

- Agreeing to global and national limits on greenhouse gas emissions and encouraging industries and countries to meet these limits by selling and trading greenhouse gas emission permits in the marketplace. This approach also stimulates companies to develop new technologies that reduce emissions of greenhouse gases and increase their profits.

- Phasing in (1) output-based *carbon taxes* on each unit of CO_2 emitted by fossil fuels (especially coal and gasoline) or (2) input-based *energy taxes* on each unit of fossil fuel (especially coal and gasoline) that is burned. Costa Rica has enacted a 15% carbon tax and uses a third of the revenues to finance tree-planting projects by farmers.

- Decreasing taxes on income, labor, and profits to match any increases in consumption taxes on carbon emissions or fossil fuel use (Connections, p. 53, and Solutions, p. 707).

Organic Farming to the Rescue

SOLUTIONS

Recent studies by the Rodale Institute in Pennsylvania and other scientists suggest that a wholesale switch to organic farming could help slow global warming. They have shown that crops grown with organic fertilizer produce equivalent yields but much less CO_2 through respiration than crops grown with commercial fertilizer.

CO_2 outputs also are reduced because organic farming uses 50% less energy than conventional farming methods. Farmers can also reduce greenhouse emissions by adopting well-known soil conservation methods such as (1) using cultivation tillage to reduce or eliminate plowing, (2) using cover crops in winter, and (3) preserving buffer strips of trees along riverbanks.

Critical Thinking

List three things governments could do to encourage farmers to switch to more sustainable organic farming and soil conservation methods.

- Increasing government subsidies for energy-efficiency and renewable-energy technologies to help speed up the switch to these alternatives.

- Funding the transfer of energy efficiency and renewable energy technologies from developed countries to developing countries. Increasing the current tax on each international currency transaction by a quarter of a penny could finance this technology transfer. This would generate revenues of $200–300 billion a year.

- Increasing use of nuclear power (Section 14-9, p. 344) because it produces only about one-sixth as much CO_2 per unit of electricity as coal (Figure 14-20, p. 339). Other analysts oppose this because of (1) the danger of large-scale releases of highly radioactive materials from nuclear power-plant accidents (p. 350), (2) its very high cost (p. 353), and (3) the problem of how to store radioactive wastes safely for thousands of years. (p. 351).

- Using government regulations and subsidies to reduce deforestation (p. 609) and encourage increased use of sustainable agriculture (Figure 12-31, p. 291, and Solutions, above).

- Establishing national policies and funding international efforts to slow population growth (p. 258). If we cut per capita greenhouse gas emissions in half but world population doubles, we're back where we started.

How Can the Marketplace be Used to Reduce or Prevent Greenhouse Gas Emissions? Another approach is to agree to global and national limits on greenhouse gas emissions and encourage industries and countries to meet these limits by selling and trading greenhouse gas emission permits in the marketplace. This approach also stimulates companies to develop new technologies that reduce emissions of greenhouse gases and increase their profits. In the United States, this market approach has been used to reduce SO_2 emissions ahead of target goals at a small fraction of the projected cost (p. 441).

This market-based approach is included in the ongoing negotiations over implementation of the climate change treaty developed in 1997 in Kyoto, Japan. Countries or businesses that reduce their emissions of greenhouse gases could make money by selling their reductions to other countries and business. Ways to earn such emission credits include **(1)** improving energy efficiency (Section 15-2, p. 362), **(2)** adopting certain farming, ranching, and soil-building and conservation practices that emit less CO_2, N_2O, and CH_4 (Solutions, left), **(3)** switching from coal to natural gas, **(4)** switching from coal and other fossil fuels to forms of renewable energy such as solar, wind, hydrogen, and geothermal (Chapter 15), and **(5)** sequestering (removing) CO_2 from the atmosphere by reforestation or injecting it into the deep ocean or secure underground reservoirs. For example, a coal-burning power plant in Ohio could earn credits to allow some CO_2 emissions by financing a CO_2-removing tree planting project in Oregon or Costa Rica.

Some analysts believe that this market-based approach is more politically and economically feasible than relying primarily on government regulation to impose taxes on carbon emissions or each unit of fossil fuel used. According to these analysts, government regulation is needed to establish clear guidelines, goals, and quotas, but companies should be free to meet these goals in any way that works and makes money. They agree with the advice given 2000 years ago by Lao-tzu: "Govern a country as you would fry a small fish: Don't poke at it too much."

Since 1992, the United States and other countries have been arguing over how to **(1)** reduce CO_2 emissions by about 5.2% (7% in the United States) below 1990 level by 2010 and **(2)** implement carbon trading under the 1997 Kyoto climate change treaty. Meanwhile, an increasing number of private companies have leapfrogged over such negotiations. They are making money by trading in **(1)** carbon dioxide emission reductions and **(2)** removal (sequestering) of CO_2 from the atmosphere by trees or storage in deep underground reservoirs or the deep ocean.

Whereas the United States has reluctantly agreed to cut its greenhouse emissions to 7% below their 1990 level by 2010, several major companies have made money by sharply cutting their greenhouse gas emissions. For example, by 2000 DuPont had cut its emissions of CO_2 by about 50% from their 1991 level and saved more than $6 in net costs for each ton of avoided carbon emissions. In addition, under the 1997 Kyoto treaty, these cuts will allow DuPont to earn marketable emission credits that could someday contribute billions to the its net earnings.

However, there is controversy over the emission trading scheme in the Kyoto treaty because it allowed credits for marketable emission rights based on reductions that had already been achieved, which critics called "hot air" emission rights. The trading of emission rights is based on sound economic theory and some success in reducing air pollution. However, there is concern that trading "hot air" rights could undermine the legitimacy of the system.

How Can We Remove CO_2 from the Atmosphere? Scientists are evaluating several ways to remove CO_2 from the atmosphere or from smokestacks and store (sequester) it in **(1)** immature trees, **(2)** plants that store it in the soil, **(3)** deep underground reservoirs, and **(4)** the deep ocean.

One way to remove CO_2 from the atmosphere temporarily would be to *plant trees* over an area equivalent to the size of Australia in a massive global reforestation program. Such a program would also help restore degraded lands. However, the rate of removal of CO_2 from the atmosphere by photosynthesis decreases as trees mature and grow at a slower pace. In addition, trees release their stored CO_2 back into the atmosphere when they die and decompose or if they catch fire. Studies suggest that a global reforestation program (requiring each person in the world to plant and tend to an average of 1,000 trees every year) would offset only about 3 years of our current CO_2 emissions from burning fossil fuels.

Agricultural scientists are investigating the use of plants such as switchgrass that could remove CO_2 from the air and deposit it in the soil. Farmers could make money by receiving sequestering payments from power companies to grow such plants on land that is not suitable for ordinary crops. This would also reduce soil erosion and water pollution. The mature crops could be harvested and sold to power companies as a biomass fuel, and the field could be replanted.

Other sequestering approaches include collecting CO_2 from smokestacks (and natural gas wells) and **(1)** *pumping it deep underground* into unminable coal seams and abandoned oil fields (as is currently done to push up more oil, p. 335), or **(2)** *injecting it into the deep ocean*. In one type of ocean sequestering suggested by oceanographer Peter Brewer, the collected CO_2 would be injected deep enough into the sea that it would form heavy CO_2 icebergs that could sit on the ocean floor undisturbed for centuries.

In another approach, CO_2 emitted by power plants would be dissolved in seawater to form a solution of carbonic acid ($CO_2 + H_2O \longrightarrow H_2CO_3$). This solution would then be reacted with a solution of pulverized carbonate minerals such as limestone ($CaCO_3$) to produce a solution containing bicarbonate ions (HCO_3^-) that could be released into the ocean. This would mimic the removal of carbon dioxide as bicarbonates in ocean sediments by the natural carbon cycle (Figure 4-29, p. 92). It is estimated that this approach would cost only about one-fourth as much direct injection of CO_2 into the deep sea.

However, any method of underground or deep sea sequestration would take a costly investment in materials, transportation of the CO_2 (presumably by pipeline) to storage sites, and infrastructure. In addition, injecting large quantities of CO_2 into the ocean could upset the global carbon cycle and some forms of deep-sea life in unpredictable ways. Another problem is that current methods can remove only about 30% of the CO_2 from smokestack emissions, and using them would double the cost of electricity.

Can Technofixes Save Us? Some scientists have suggested various technofixes for reducing the threat of possible global warming, including **(1)** adding iron to the oceans to stimulate the growth of marine algae (which could remove more CO_2 through photosynthesis), **(2)** unfurling gigantic foil-surfaced sun mirrors in space or placing such mirrors on about 50,000 orbiting satellites to reduce solar input, **(3)** releasing trillions of reflective balloons filled with cheap helium into the atmosphere, and **(4)** injecting sunlight-reflecting sulfate particulates or firing sulfur dioxide cannonballs into the stratosphere to cool the earth's surface (this would turn the sky white and increase depletion of stratospheric ozone).

Many of these costly schemes might not work, and most probably would produce unpredictable short- and long-term harmful environmental effects. Moreover, once started, those that work could never be stopped without a renewed rise in CO_2 levels. Instead of spending huge sums of money on such schemes, many scientists believe it would be more effective and less expensive to improve energy efficiency (Solutions, p. 461) and shift to renewable forms of energy that do not produce carbon dioxide (Chapter 15).

What Has Been Done to Reduce Greenhouse Gas Emissions? At the 1992 Earth Summit in Rio de Janeiro, Brazil, 106 nations approved a Convention on Climate Change in which developed countries committed themselves to reducing their emissions of CO_2 and other greenhouse gases to 1990 levels by the year 2000. However, the convention did not *require* countries to reach this goal, and most countries did not achieve this goal.

In December 1997, more than 2,200 delegates from 161 nations met in Kyoto, Japan, to negotiate a new treaty to help slow global warming. The resulting treaty would **(1)** require 38 developed countries to cut greenhouse emissions to an average of about 5.2% below 1990 levels between 2008 to 2012, **(2)** not require developing countries to make any cuts in their greenhouse gas emissions, and **(3)** allow emissions trading.

Some analysts praise the Kyoto agreement as a small but important step in dealing with the problem of global warming and hope that the conditions of the treaty will be strengthened in future negotiating sessions. According to computer models, the 5.1% reduction goal of the Kyoto Protocol would shave only about $0.06°C$ ($0.1°F$) off the $0.7–1.7°C$ ($1–3°F$) temperature rise projected by 2060.

There is also controversy over what role developing countries should take in reducing their CO_2 emissions. Developing countries were not required to reduce their emissions in the first phase of the treaty because these countries argued that

- Developed countries should be the first to reduce their CO_2 emissions because of their higher total and per capita emissions. For example, **(1)** average per capita CO_2 emissions in developed countries are about six times higher than in developing countries, **(2)** the average American is responsible for nearly eight times as much CO_2 emissions per person as the average Chinese, and **(3)** each year the average American is responsible for the emission of about 18 metric tons (20 tons) of CO_2 into the atmosphere—roughly equal to the mass of four elephants.

- They are just beginning to expand some of their economies and should be entitled to some increases in CO_2 emissions, as the developed countries had during the early stages of their economic growth and development.

By 2001, the U.S. Congress had not ratified the treaty, mostly because of **(1)** its failure to require emission reductions from developing countries and **(2)** intensive lobbying by a coalition of coal, oil, steel, chemical, and automobile companies opposed to the treaty who argued that it would have a devastating impact on the U.S. economy and workers. By 2001, only 18 countries had agreed to be bound by the treaty, far fewer than the 55 countries needed to give it legal force.

During 2000, some of the major oil and automobile companies dropped out of the Global Climate Coalition that opposed the treaty and most other action on reducing possible climate change from global warming. Their CEOs **(1)** indicated that global warming was a potential risk that should be addressed and **(2)** agreed with most economists that dealing with this problem should stimulate the U.S. economy and create many new jobs.

Some analysts suggest that the stalemate between developed and developing counties over reducing greenhouse gas emissions might be eased by (1) giving developing countries a 10-year grace period before they are required to meet specified reductions and (2) setting up an international fund financed by developed countries to transfer energy-efficiency and renewable-energy technologies to developing countries. These two strategies were used to develop an international treaty that is gradually reducing inputs of ozone-depleting chemicals into the stratosphere (p. 472 and Figure 18-22, p. 473).

How Can We Prepare for Possible Global Warming? According to latest global climate models, an immediate 60% reduction in current global CO_2 emissions would be needed to stabilize concentrations of CO_2 in the air at their present levels. Such a change is extremely unlikely on political and economic grounds because it would take rapid, massive changes in industrial processes, energy sources, transportation options, and individual lifestyles. This strategy would not even be considered unless there were dramatic and incontestable evidence that human activities were changing climate in ways that would devastate economies, human health, and biodiversity.

Without a sense of urgency, many (perhaps most) of the actions climate experts have recommended for slowing atmospheric warming (Figure 18-14) either will not be done or will be done too slowly. As a result, a growing number of analysts suggest that we should also begin to prepare for the possible harmful effects of long-term atmospheric warming and climate change (Figure 18-12). Figure 18-15 shows some ways to do this.

Implementing the key measures for slowing or adapting to climate change listed in Figures 18-14 and 18-15 will be very costly. However, a number of studies indicate that in the long run the savings would greatly exceed the costs. Some actions you can take to reduce the threat of global warming are listed in Appendix 6.

18-6 OZONE DEPLETION IN THE STRATOSPHERE

What Is the Threat from Ozone Depletion? A layer of ozone in the lower stratosphere (Figure 17-2, p. 418, and Figure 17-3, p. 419) keeps about 95% of the sun's harmful ultraviolet (UV) radiation from reaching the earth's surface. This UV filter (1) allows humans and other forms of life to exist on land, (2) helps protect humans from sunburn, skin cancer, eye cataracts, and damage to the immune system, and (3) prevents much of the oxygen in the troposphere from being converted to photochemical ozone, a harmful air pollutant (Figure 17-5, p. 424, and Table 18-2).

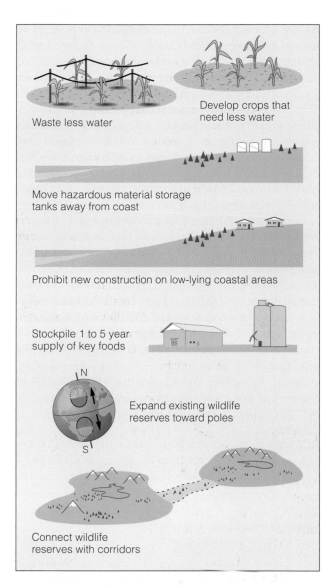

Figure 18-15 Solutions: ways to prepare for the possible long-term effects of projected climate change from increased atmospheric temperatures.

Measuring instruments on balloons, aircraft, and satellites clearly show seasonal depletion (thinning) of ozone concentrations in the stratosphere above Antarctica and the Arctic. Similar measurements reveal a lower overall thinning everywhere except over the tropics. Based on these measurements and chemical models, the overwhelming consensus of researchers in this field is that ozone depletion (thinning) in the stratosphere is a serious long-term threat to (1) humans, (2) many other animals, and (3) the sunlight-driven primary producers (mostly plants) that support the earth's food chains and webs.

What Causes Ozone Depletion? From Dream Chemicals to Nightmare Chemicals This situation started when Thomas Midgley, Jr., a General Motors chemist, discovered the first **chlorofluorocarbon (CFC)**

in 1930, and chemists made similar compounds to create a family of highly useful CFCs. The two most widely used are CFC-11 (trichlorofluoromethane, CCl_3F) and CFC-12 (dichlorodifluoromethane, CCl_2F_2), known by their trade name as Freons.

These chemically stable (nonreactive), odorless, nonflammable, nontoxic, and noncorrosive compounds seemed to be dream chemicals. Cheap to make, they became popular as (1) coolants in air conditioners and refrigerators (replacing toxic sulfur dioxide and ammonia), (2) propellants in aerosol spray cans, (3) cleaners for electronic parts such as computer chips, (4) sterilants for hospital instruments, (5) fumigants for granaries and ship cargo holds, and (6) bubbles in plastic foam used for insulation and packaging. Between 1960 and the early 1990s, CFC production rose sharply.

However, CFCs were too good to be true. In 1974, calculations by two University of California-Irvine chemists, Sherwood Rowland and Mario Molina, indicated that CFCs were lowering the average concentration of ozone in the stratosphere. They shocked both the scientific community and the $28-billion-per-year CFC industry by calling for an immediate ban of CFCs in spray cans (for which substitutes were available).

According to Rowland and Molina,

- Large quantities of CFCs were being released into the troposphere mostly from (1) the use of CFCs as propellants in spray cans, (2) leaks from refrigeration and air conditioning equipment, and (3) the production and burning of plastic foam products.

- CFCs remain in the troposphere because they are insoluble in water and are chemically unreactive.

- Over 11–20 years they rise into the stratosphere mostly through convection, random drift, and the turbulent mixing of air in the troposphere.

- In the stratosphere, the CFC molecules break down under the influence of high-energy UV radiation. This releases highly reactive chlorine atoms, which speed up the breakdown of very reactive ozone (O_3) into O_2 and O in a cyclic chain of chemical reactions (Figure 18-16). This causes ozone in various parts of the stratosphere to be destroyed faster than it is formed.

- Each CFC molecule can last in the stratosphere for 65-385 years (depending on its type), with the most widely used CFCs lasting 75–111 years. During that time, each chlorine atom released from these molecules can convert up to 100,000 molecules of O_3 to O_2.

According to Rowland and Molina's calculations and later models and atmospheric measurements of CFCs in the stratosphere, these dream molecules have turned into a nightmare of global ozone destroyers.

The CFC industry (led by the DuPont Company), a powerful, well-funded adversary with a lot of profits and jobs at stake, attacked Rowland and Molina's calculations and conclusions. However, they held their

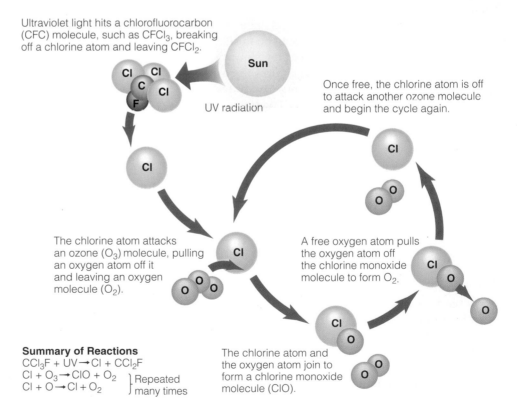

Figure 18-16 A simplified summary of how chlorofluorocarbons (CFCs) and other chlorine-containing compounds destroy ozone in the stratosphere. Note that chlorine atoms are continuously regenerated as they react with ozone. Thus, they act as *catalysts*, chemicals that speed up chemical reactions without being used up by the reaction. Bromine atoms released from bromine-containing compounds that reach the stratosphere also destroy ozone by a similar mechanism.

Ultraviolet light hits a chlorofluorocarbon (CFC) molecule, such as $CFCl_3$, breaking off a chlorine atom and leaving $CFCl_2$.

Sun

UV radiation

Once free, the chlorine atom is off to attack another ozone molecule and begin the cycle again.

The chlorine atom attacks an ozone (O_3) molecule, pulling an oxygen atom off it and leaving an oxygen molecule (O_2).

A free oxygen atom pulls the oxygen atom off the chlorine monoxide molecule to form O_2.

The chlorine atom and the oxygen atom join to form a chlorine monoxide molecule (ClO).

Summary of Reactions
$CCl_3F + UV \rightarrow Cl + CCl_2F$
$Cl + O_3 \rightarrow ClO + O_2$ } Repeated
$Cl + O \rightarrow Cl + O_2$ } many times

ground, expanded their research, and explained the meaning of their calculations to other scientists, elected officials, and the media. It was not until 1988—14 years after Rowland and Molina's study—that DuPont officials acknowledged that CFCs were depleting the ozone layer and agreed to stop producing them once they found substitutes. In 1995, Rowland and Molina received the Nobel prize in chemistry for their work.

What Other Chemicals Deplete Stratospheric Ozone? Other ozone-depleting compounds (ODCs) include:

- *Halons* and *HBFCs*, used in fire extinguishers.

- *Methyl bromide* (CH_3Br), a widely used fumigant.

- *Carbon tetrachloride* (CCl_4), a cheap, highly toxic solvent.

- *Methyl chloroform*, or 1,1,1-trichloroethane ($C_2H_3Cl_3$), used as a cleaning solvent for clothes and metals and as a propellant in more than 160 consumer products such as correction fluid, dry-cleaning sprays, spray adhesives, and other aerosols.

- *Hydrogen chloride* (HCl), emitted into the stratosphere by U.S. space shuttles.

The oceans and occasional volcanic eruptions also release chlorine compounds into the troposphere. However, most of these chlorine compounds do not make it to the stratosphere because they easily dissolve in water and are washed out of the troposphere in rain. Measurements and models indicate that 75–85% of the observed ozone losses in the stratosphere since 1976 are the result of ODCs released into the atmosphere by human activities beginning in the 1950s.

Why Is There Seasonal Thinning of Ozone over the Poles? In 1984, researchers analyzing satellite data discovered that 40-50% of the ozone in the upper stratosphere over Antarctica was being destroyed dur-

ing the antarctic spring and early summer (September–December), when sunlight returned after the dark antarctic winter. When they reexamined earlier data they found that the seasonal loss of ozone above Antarctica did not change much between 1956 and 1976, but that such losses had increased significantly since 1976 (Figure 18-17). The data showing these large losses had not been noticed because a computer had been programmed to ignore large drops in ozone concentration on the assumption that they were errors.

In retrospect, the drops since 1976 were understandable because it takes 11–20 years for CFCs to reach the stratosphere. Thus, the large amounts produced in the 1960s would be having an impact on ozone in the stratosphere around the mid-1970s, and the even larger amounts produced in the 1970s reached the stratosphere in the 1990s. When would you expect CFCs produced in the 1980s (with 1989 being the peak production year) and in the early 1990s to begin reaching the stratosphere?

Figure 18-18 shows the seasonal variation of ozone over Antarctica during 1997. The observed seasonal loss during the Antarctic summer has been incorrectly called an *ozone hole*. A more accurate term is *ozone thinning* because the ozone depletion varies with altitude (Figure 18-18) and location.

The total area of the atmosphere above Antarctica that suffers from ozone thinning during the peak season varies from year to year (Figure 18-19). In 2000, seasonal ozone thinning above Antarctica was the largest ever and covered an area three times the size of the continental United States.

Measurements indicate that CFCs are the primary culprits. Each sunless winter, steady winds blow in a

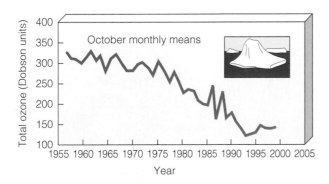

Figure 18-17 Mean total level of ozone for October over the Halley Bay measuring station in Antarctica, 1956–1999. (Data from British Antarctic Survey and World Meteorological Organization)

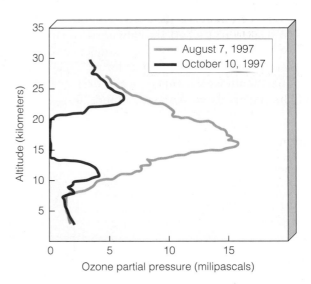

Figure 18-18 Variation of ozone level with altitude over Antarctica during 1997. Note the severe depletion of ozone during October (during the Antarctic summer) and its return to more normal levels in August (during the Antarctic winter). (Data from National Oceanic and Atmospheric Administration)

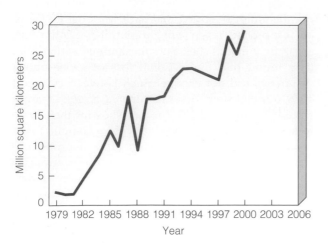

Figure 18-19 Variation in the average area of ozone thinning above Antarctica between September 7 and October 10 from 1979 to 1999. (Data from National Oceanic and Atmospheric Administration)

circular pattern over the earth's poles. This creates a *polar vortex*: a huge swirling mass of very cold air that is isolated from the rest of the atmosphere until the sun returns a few months later.

When water droplets in clouds enter this circling stream of extremely frigid air, they form tiny ice crystals. The surfaces of these ice crystals collect CFCs and other ODCs in the stratosphere and speed up (catalyze) the chemical reactions that release Cl atoms and ClO. Instead of entering a chain reaction of ozone destruction (Figure 18-16), the ClO atoms combine with one another to form Cl_2O_2 molecules. In the dark of winter, the Cl_2O_2 molecules cannot react with ozone, so they accumulate in the polar vortex.

When sunlight and the antarctic spring return 2–3 months later (in October), the light breaks up the stored Cl_2O_2 molecules, releasing large numbers of Cl atoms and initiating the catalyzed chlorine cycle (Figure 18-16). Within weeks, this typically destroys 40–50% of the ozone above Antarctica (and 100% in some

places). The returning sunlight **(1)** gradually melts the ice crystals, **(2)** breaks up the vortex of trapped polar air, and **(3)** allows it to begin mixing again with the rest of the atmosphere. Then new ozone forms over Antarctica until the next dark winter.

When the vortex breaks up, huge masses of ozone-depleted air above Antarctica flow northward and linger for a few weeks over parts of Australia, New Zealand, South America, and South Africa. This raises biologically damaging UV-B levels in these areas by 3–10%, and in some years by as much as 20%.

In 1988, scientists discovered that similar but less severe ozone thinning occurs over the Arctic during the arctic spring and early summer (February–June), with a seasonal ozone loss of 11–38% (compared to a typical 50% loss above Antarctica). However, in 1997 and 1999, the average seasonal loss over the Arctic was about 60%. When this mass of air above the Arctic breaks up each spring, large masses of ozone-depleted air flow south to linger over parts of Europe, North America, and Asia. According to a 1998 model developed by National Aeronautics and Space Agency (NASA) scientists at the Goddard Institute for Space Studies, ozone depletion over the Antarctic and Arctic will be at its worst between 2010 and 2019 (Figure 18-20).

Why Should We Be Worried About Ozone Depletion? Life in the Ultraviolet Zone Why should we care about ozone loss? From a human standpoint the answer is that with less ozone in the stratosphere, more biologically damaging UV-A and UV-B radiation will reach the earth's surface. This will give humans **(1)** worse sunburns, **(2)** more eye cataracts (a clouding of the eye's lens that reduces vision and can cause blindness if not corrected), and **(3)** more skin cancers (Connections, right). The number of new cases of skin cancer in the United States has quadrupled over the past 20 years, resulting in nearly 10,000 deaths per year.

According to UN Environment Programme estimates, the additional UV-B radiation reaching the

Figure 18-20 Projected total ozone loss, averaged over 2010–2019, during September for the Antarctic (left) and during March for the Arctic. According to the model used to make these projections, during this period the severity of ozone loss over the Arctic may approach that over the Antarctic. Dark red represents ozone depletion of 54% or more; light blue, 18–30%; and dark blue, 6-12%. (Data from NASA Goddard Institute for Space Studies)

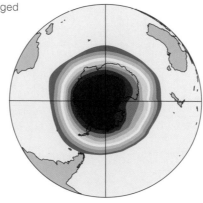

Antarctic

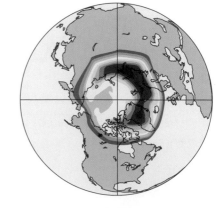

Arctic

The Cancer You Are Most Likely to Get

CONNECTIONS

Research indicates that years of exposure to UV-B ionizing radiation in sunlight is the primary cause of *squamous cell* (Figure 18-21, left) and *basal cell* (Figure 18-21, center) *skin cancers*. Together these two types make up 95% of all skin cancers. Typically there is a 15- to 40-year lag between excessive exposure to UV-B and development of these cancers.

Caucasian children and adolescents who get only a single severe sunburn double their chances of getting these two types of cancers. Some 90–95% of these types of skin cancer can be cured if detected early enough, although their removal may leave disfiguring scars. These cancers kill only 1–2% of their victims, but this still amounts to about 2,300 deaths in the United States each year.

A third type of skin cancer, *malignant melanoma* (Figure 18-21, right), occurs in pigmented areas such as moles anywhere on the body's surface. Within a few months, this type of cancer can spread to other organs. It kills about one-fourth of its victims (most under age 40) within 5 years, despite surgery, chemotherapy, and radiation treatments. Each year it kills about 100,000 people (including 7,300 Americans in 1998), mostly Caucasians. It can be cured if detected early enough, but recent studies show that some melanoma survivors have a recurrence more than 15 years later.

Recent evidence suggests that about 90% of sunlight's melanoma-causing effect may come from exposure to UV-A and 10% from UV-B. Some sunscreens do not protect from UV-A, and tanning booth lights emit mostly UV-A.

Evidence indicates that people (especially Caucasians) who get three or more blistering sunburns before age 20 are five times more likely to develop malignant melanoma than those who have never had severe sunburns. About 10% of those who get malignant melanoma have an inherited gene that makes them especially susceptible to the disease.

To protect yourself, the safest course is to stay out of the sun (especially between 10 A.M. and 3 P.M., when UV levels are highest) and avoid tanning parlors. When you are in the sun, **(1)** wear tightly woven protective clothing, **(2)** a wide-brimmed hat, and **(3)** sunglasses that protect against UV-A and UV-B radiation (ordinary sunglasses may actually harm your eyes by dilating your pupils so that more UV radiation strikes the retina).

Because UV rays can penetrate clouds, overcast skies do not protect you; neither does shade, because UV rays can reflect off sand, snow, water, or patio floors. People who take antibiotics and women who take birth control pills are more susceptible to UV damage.

Use a sunscreen that offers protection against both UV-A and UV-B and has a protection factor of 15 or more (25 if you have light skin). Apply to all exposed skin

and reapply it after swimming or excessive perspiration. Most people do not realize that the protection factors for sunscreens are based on using one full ounce of the product—far more than most people apply. Some people are also increasing their risk of skin cancer by falsely assuming that sunscreens allow them to spend more time in the sun.

Children who use a sunscreen with a protection factor of 15 every time they are in the sun from age 1 to age 18 decrease their chance of getting skin cancer by 80%. Babies under a year old should not be exposed to the sun at all and should not have sunscreens applied until they are at least 6 months old.

Become familiar with your moles and examine your skin at least once a month. The warning signs of skin cancer are **(1)** a change in the size, shape, or color of a mole or wart (the major sign of malignant melanoma, which must be treated quickly), **(2)** sudden appearance of dark spots on the skin, or **(3)** a sore that keeps oozing, bleeding, and crusting over but does not heal. Be alert for precancerous growths (reddish-brown spots with a scaly crust). If you observe any of these signs, consult a doctor immediately.

Critical Thinking

What precautions, if any, do you take to reduce your chances of getting skin cancer from exposure to sunlight? Explain why you do or do not take such precautions.

earth's surface resulting from an annual 10% loss of global ozone (already a strong possibility within a few years) could lead to **(1)** 300,000 additional cases of squamous cell cancer (Figure 18-21, left) and basal cell cancer (Figure 18-21, center) worldwide each year, **(2)** 4,500–9,000 additional cases of potentially fatal malignant melanoma (Figure 18-21, right) each year, and **(3)** 1.5 million new cases of cataracts (which account for more than half of the world's 25-35 million cases of blindness) each year.

Other effects of increased UV exposure are:

- Immune system suppression, which makes the body more susceptible to infectious diseases and some forms of cancer.

- An increase in damaging acid deposition (Figure 17-9, p. 428) and eye-burning photochemical smog in the troposphere (Figure 17-5, p. 424).

- Lower yields of key crops such as corn, rice, cotton, soybeans, beans, peas, sorghum, and wheat, with

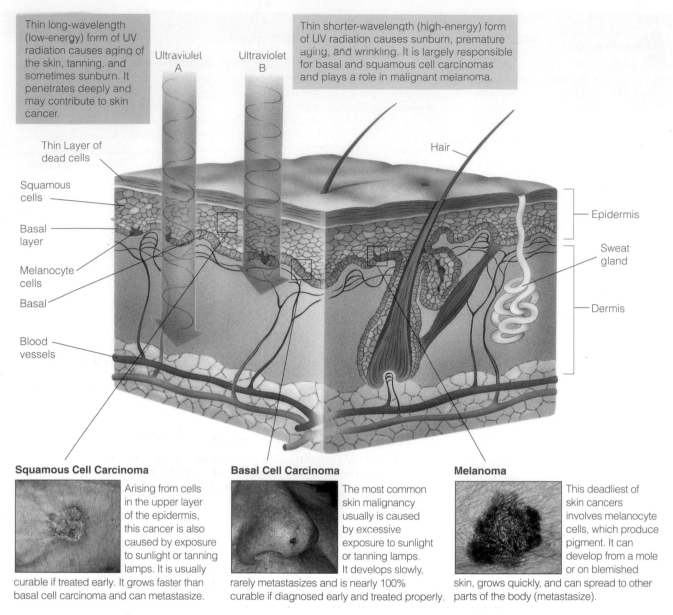

Thin long-wavelength (low-energy) form of UV radiation causes aging of the skin, tanning, and sometimes sunburn. It penetrates deeply and may contribute to skin cancer.

Thin shorter-wavelength (high-energy) form of UV radiation causes sunburn, premature aging, and wrinkling. It is largely responsible for basal and squamous cell carcinomas and plays a role in malignant melanoma.

Ultraviolet A

Ultraviolet B

Hair

Thin Layer of dead cells

Squamous cells

Basal layer

Melanocyte cells

Basal

Blood vessels

Epidermis

Sweat gland

Dermis

Squamous Cell Carcinoma

Arising from cells in the upper layer of the epidermis, this cancer is also caused by exposure to sunlight or tanning lamps. It is usually curable if treated early. It grows faster than basal cell carcinoma and can metastasize.

Basal Cell Carcinoma

The most common skin malignancy usually is caused by excessive exposure to sunlight or tanning lamps. It develops slowly, rarely metastasizes and is nearly 100% curable if diagnosed early and treated properly.

Melanoma

This deadliest of skin cancers involves melanocyte cells, which produce pigment. It can develop from a mole or on blemished skin, grows quickly, and can spread to other parts of the body (metastasize).

Figure 18-21 Structure of the human skin and the relationships between ultraviolet (UV-A and UV-B) radiation and the three types of skin cancer. The incidence of these types of cancer is rising, mostly because more fair-skinned people have increased their exposure to sunlight by moving to areas with sunnier climates and spending more of their leisure time exposed to sunlight. If ozone-destroying chemicals continue to reduce stratospheric ozone levels, incidence of these types of cancers is expected to rise. (The Skin Cancer Foundation)

estimated losses totaling $2.5 billion per year in the United States before the mid-21st century.

- Decline in forest productivity of the many tree species sensitive to UV-B radiation. This could reduce CO_2 uptake and enhance global warming.

- Increased breakdown and degradation of materials such as various types of paints, plastics, and outdoor materials. Such damage could cost billions of dollars per year.

- Reduction in the productivity of surface-dwelling phytoplankton, which could **(1)** upset aquatic food webs, **(2)** decrease yields of seafood eaten by humans, and **(3)** possibly accelerate global warming by decreasing the oceanic uptake of CO_2.

Humans can make cultural adaptations to increased UV-B radiation by staying out of the sun, protecting their skin with clothing, and applying sunscreens (Connections, p. 469). However, UV-sensitive plants and other animals that help support us and other forms of

Table 18-3 CFC Substitutes

Types	Pros	Cons
Hydrochlorofluorocarbons (HCFCs)	Break down faster (2–20 years). Pose about 90% less danger to ozone layer. Can be used in aerosol sprays, refrigeration, air conditioning, foam, and cleaning agents.	Are greenhouse gases. Will still deplete ozone, especially if used in large quantities. Health effects largely unknown. Expensive. May lower energy efficiency of appliances. Can be degraded to trifluoroacetate (TFA), which can inhibit plant growth in wetlands.
Hydrofluorocarbons (HFCs)	Break down faster (2–20 years). Do not contain ozone-destroying chlorine. Can be used in aerosol sprays, refrigeration, air conditioning, and insulating foam.	Are potent greenhouse gases. Expensive. Safety questions about flammability and toxicity still unresolved. Can be degraded to TFA, which can inhibit plant growth in wetlands. May lower energy efficiency of appliances. Production of HFC-134a, a refrigerant substitute, yields an equal amount of methyl chloroform, a serious ozone depleter.
Hydrocarbons (HCs) such as propane and butane	Cheap and readily available. Can be used in aerosol sprays, refrigeration, foam, and cleaning agents. Not patentable. Can be made locally in developing countries.	Can be flammable and poisonous if released. Some increase in ground-level air pollution.
Ammonia	Simple alternative for refrigerators; widely used before CFCs.	Toxic if inhaled. Must be handled carefully.
Water and steam	Effective for some cleaning operations and for sterilizing medical instruments.	Creates polluted water that must be treated. Wastes water unless the used water is cleaned up and reused.
Terpenes (from the rinds of lemons and other citrus fruits)	Effective for cleaning electronic parts.	None.
Helium	Effective coolant for refrigerators, freezers, and air conditioners.	This rare gas may become scarce if use is widespread, but very little coolant is needed per appliance.

life cannot make such changes except through the long process of biological evolution.

18-7 SOLUTIONS: PROTECTING THE OZONE LAYER

How Can We Protect the Ozone Layer? The scientific consensus of researchers in this field is that we should immediately stop producing all ozone-depleting chemicals. Even with immediate action, the models indicate that it will take about 50 years for the ozone layer to return to 1980 levels and about 100 years for recovery to pre-1950 levels.

Substitutes are already available for most uses of CFCs (Table 18-3), and others are being developed (Individuals Matter, p. 472).

To a growing number of scientists, hydrocarbons (HCs) such as propane and butane are a better way to reduce ozone depletion while doing little to increase global warming. Developing countries could use HC refrigerator technology to leapfrog ahead of industrialized countries without having to invest in costly HFC and HCFC technologies that will have to be phased out within a few decades because they are potent greenhouse gases (Table 18-3). This approach is also less costly because HCs cannot be patented and can be manufactured locally, reducing the need to import expensive HFCs and HCFCs.

Can Technofixes Save Us? What about a quick fix from technology, so that we can keep on using CFCs? One proposed scheme for removing ozone from the

Ray Turner and His Refrigerator

INDIVIDUALS MATTER

Ray Turner, an aerospace manager at Hughes Aircraft in California, made an important low-tech, ozone-saving discovery by using his head—and his refrigerator. His concern for the environment led him to look for a cheap and simple substitute for the CFCs used as cleaning agents for removing films of oxidation from the electronic circuit boards manufactured at his plant.

He started by looking in his refrigerator. He decided to put drops of various substances on a corroded penny to see whether any of them removed the film of oxidation. Then he used his soldering gun to see whether solder would stick to the surface of the penny, indicating that the film had been cleaned off.

First, he tried vinegar. No luck. Then he tried some ground-up lemon peel, also a failure. Next he tried a drop of lemon juice and watched as the solder took hold. The rest, as they say, is history.

Today, Hughes Aircraft uses inexpensive citrus-based solvents that are CFC-free to clean circuit boards. This new cleaning technique has reduced circuit board defects by about 75% at Hughes. And Turner got a hefty bonus. Now other companies, such as AT&T, clean computer boards and chips using acidic chemicals extracted from cantaloupes, peaches, and plums. Maybe you can find a solution to an environmental problem in your refrigerator, grocery, drugstore, or backyard.

atmosphere is to launch a fleet of 20–30 football-field-long, radio-controlled blimps into the stratosphere above Antarctica. Hanging from each blimp would be a huge curtain of electrical wires that would inject negatively charged electrons into the stratosphere. The electrons would convert ozone-destroying chlorine atoms (Cl) in the stratosphere to chloride ions ($Cl + e \longrightarrow Cl^-$) that would not react with ozone. A second suspended sheet of positively charged wires could be used to attract the negatively charged ions. However, atmospheric chemist Ralph Ciecerone believes that this plan will not work because other chemical species in the stratosphere snatch electrons more readily than does chlorine. This scheme could also have unpredictable side effects on atmospheric chemistry.

Others have suggested using tens of thousands of lasers to blast CFCs out of the atmosphere before they can reach the stratosphere. However, enormous amounts of energy would be needed to do this, and decades of research would be needed to perfect the types of lasers needed. Moreover, we cannot predict the possible effects of such powerful laser blasts on climate, birds, or planes.

What Is Being Done to Reduce Ozone Depletion?
Some Hopeful Progress In 1987, 36 nations meeting in Montreal developed a treaty, commonly known as the *Montreal Protocol*, to cut emissions of CFCs (but not other ozone depleters) into the atmosphere by about 35% between 1989 and 2000. This agreement required developing countries to freeze their CFC consumption in mid-1999 and phase it out altogether by 2010. By 2000, developing countries as a group were ahead of this schedule.

However, measurements and dramatic computer-generated color images made after 1989 showed an increasing seasonal loss of ozone over the Antarctic (Figure 18-17) that covered larger areas (Figure 18-19). Spurred on by this bad news, representatives of 93 countries met in London in 1990 and in Copenhagen in 1992 and adopted a new protocol accelerating the phaseout of key ozone-depleting chemicals.

At the 1990 and 1992 meetings, the developed countries agreed to establish an international fund to reimburse developing countries for various costs of complying with the international protocol. An EPA study indicated that the costs of doing this were very low compared to the costs of a more severely damaged ozone layer. By 2000, industrial countries had contributed more than $1 billion to the fund to finance some 3,000 projects in 116 developing countries.

The landmark international agreements reached so far and signed by more than 170 countries are important examples of global cooperation in response to a serious global environmental problem. Because of these agreements, CFC production fell by 93% between 1989 (its peak production year) and 1999. Global production of carbon tetrachloride and methyl chloroform has also dropped sharply, but production of methyl bromide, halons (70% produced by China), and HCFCs continues to rise.

Methyl bromide is widely used **(1)** as a fumigant to control certain soil-dwelling pathogens for more than 100 crops worldwide and **(2)** to control insects in homes and food-processing facilities. Under the 1992 Copenhagen Protocol, its use is to be phased out in the United States by 2005. U.S. Department of Agriculture scientists have found two alternatives for methyl bromide: **(1)** a naturally occurring, nontoxic, non-ozone-depleting chemical (benzaldehyde) in peaches and several other fruits that can be used as a soil fumigant and **(2)** diatomaceous earth (the skeletal remains of free-floating diatoms found in the ocean) for controlling insects in homes and food-processing facilities.

There is also the problem of how to destroy CFCs removed from old refrigerators, air conditioners, and other devices. In 1998, researchers at Yale University extracted a chemical from rhubarb leaves and heated it to a high temperature in a tube to transform CFCs into table salt, carbon, and sodium fluoride, which is used in some toothpastes.

According to a 1998 World Meteorological Organization (WMO) study by 350 scientists, the ozone layer will

- Continue to be depleted for several decades because of the 11- to 20-year time lag between when ODCs are released into the stratosphere and when they reach the stratosphere and their persistence for decades in the stratosphere. For example, according to WMO study, long-lasting ODCs already found in the stratosphere will cause ozone losses of 10–30% over the heavily populated northern hemisphere in the next few decades.

- Return to 1980 levels by about 2050 and to 1950 levels by about 2100 (Figure 18-22), assuming the international agreements are followed and there are no major volcanic eruptions (which can temporarily deplete stratospheric ozone). Without the 1992 international agreement, ozone depletion would be a much more serious threat (Figure 18-22).

One piece of *disturbing news* in the 1998 WMO study is that ozone depletion in the stratosphere has been cooling the troposphere. This has helped offset or disguise as much as 30% of the projected global warming from our emissions of greenhouse gases. Thus, restoring the ozone layer could lead to an increase in global warming.

The *good news* is that the ozone treaty has set an important precedent for global cooperation and action to avert potential global disaster. Nations and companies agreed to work together to solve this problem because:

- There was convincing and dramatic scientific evidence of a serious problem.

- CFCs were produced by a small number of international companies.

- The certainty that CFC sales would decline unleashed the economic and creative resources of the private sector to find even more profitable substitute chemicals.

International cooperation in dealing with projected atmospheric warming is much more difficult because:

- We lack clear-cut and dramatic evidence that there is a serious problem.

- Greenhouse gas emissions result from the actions of hundreds of different large and politically powerful industries (such as coal, oil, chemicals, automobiles, and steel) and billions of consumers.

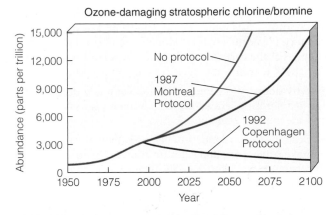

Figure 18-22 Projected concentrations of ozone-depleting chemicals (ODCs) in the stratosphere under three scenarios: **(1)** no action, **(2)** the 1987 Montreal Protocol, and **(3)** the 1992 Copenhagen Protocol. (Adapted from *Environmental Science*, 5/E by Chiras, fig 21.5. Copyright ©1998 by Wadsworth)

- Reducing greenhouse gas emissions will take far-reaching changes within most industries and lifestyle changes for billions of consumers.

Thus, reducing and adapting to the threat of global warming is a difficult political and scientific challenge. However, it can be done over the next few decades (Figures 18-14 and 18-15). Moreover, numerous economic studies show that meeting this challenge will **(1)** save huge amounts of money, **(2)** create many jobs, and **(3)** be highly profitable for many of the world's major businesses compared to doing nothing or waiting to act. As this major industrial transition takes place, analysts point out that the economic rewards will go to the companies and countries that lead the way.

The atmosphere is the key symbol of global interdependence. If we can't solve some of our problems in the face of threats to this global commons, then I can't be very optimistic about the future of the world.

MARGARET MEAD

REVIEW QUESTIONS

1. Define the boldfaced terms in this chapter.

2. Summarize briefly how the earth's climate has changed over the past 900,000 years and over the past 130 years. Distinguish between *glacial* and *interglacial* periods.

3. How do scientists get information about past changes in the earth's climate?

4. What three factors determine the average temperature of the atmosphere near the earth's surface?

5. What is the earth's *natural greenhouse effect*? How widely is this theory accepted? What are the two major greenhouse gases? Describe the *natural cooling process* that takes place near the earth's surface.

6. List three human activities that increase the input of greenhouse gases into the troposphere and could enhance the earth's natural greenhouse effect. Describe changes at the earth's poles and in Greenland that indicate that the troposphere has warmed in recent decades.

7. Why does rapid climate change over a few decades to 100 years pose a serious threat to human life, wildlife, and the world's economies?

8. Describe how scientists develop mathematical models to make projections about future climate change. According to the latest models, between 2000 and 2100 about how much increase is projected for **(a)** the global average temperature and **(b)** global sea levels?

9. Explain how each of the following factors might enhance or dampen projected global warming: **(a)** changes in solar output, **(b)** changes in the earth's reflectivity (albedo), **(c)** the oceans, **(d)** water vapor content and clouds, **(e)** air pollution, and **(f)** effects of increased carbon dioxide levels on photosynthesis and methane emissions. How rapidly might climates shift? What is the range of error in current climate change projections?

10. Why should we worry about a possible rise of only one to a few degrees in the average temperature at the earth's surface?

11. Explain how atmospheric warming might affect each of the following: **(a)** food production, **(b)** water supplies, **(c)** forests, **(d)** biodiversity, **(e)** sea levels, **(f)** weather extremes, **(g)** human health, **(h)** developing countries, and **(i)** species such as Kirtland's warblers and Adélie penguins.

12. What are the four schools of thought about what should be done about global warming?

13. According to climate scientists, by how much will we need to reduce the current level of fossil fuel use to stabilize carbon dioxide emissions at their current level?

14. List six prevention methods and four cleanup methods for slowing climate change from increased greenhouse gas emissions.

15. Explain how improving energy efficiency and relying more on organic farming could help reduce greenhouse gas emissions.

16. Summarize progress made in developing an international treaty to help reduce greenhouse gas emissions. Describe the controversy over the role developing nations should play in an international treaty to reduce greenhouse gas emissions.

17. List seven ways that we might prepare for and adjust to the harmful effects of global warming.

18. What is stratospheric *ozone depletion* and how serious is this problem? What types of chemicals cause ozone depletion? How do these chemicals cause such depletion?

19. Explain how seasonal ozone thinning occurs each year over the earth's poles.

20. What are the major harmful effects of ozone depletion on **(a)** human health, **(b)** crop yields, **(c)** forest productivity, **(d)** materials such as plastics and paints, and **(e)** productivity of plankton?

21. Distinguish between *squamous cell skin cancer*, *basal cell skin cancer*, and *malignant melanoma*. List ways that you can reduce your chances of getting skin cancer.

22. If all ozone-depleting chemicals were banned now, about how long would it take for average concentrations of ozone in the stratosphere to return to levels in **(a)** 1980 and **(b)** 1950?

23. Summarize the progress that has been made in reducing the threat of ozone depletion and explain the importance of such efforts.

24. List three factors that helped countries agree to an international treaty to phase out ozone-depleting chemicals. List three reasons why getting countries to develop an international treaty to reduce greenhouse gas emissions is a much more difficult problem.

CRITICAL THINKING

1. In preparation for the 1992 UN Conference on the Human Environment in Rio de Janeiro, president George Bush's top economic adviser gave an address in Williamsburg, Virginia, to representatives of governments from a number of countries. He told his audience not to worry about global warming because the average temperature increases scientists are predicting were much less than the temperature increase he experienced in coming from Washington, D.C., to Williamsburg. What is the fundamental flaw in this reasoning?

2. What changes might occur in the global hydrologic cycle (Figure 4-28, p. 90) if the atmosphere were to experience significant **(a)** warming or **(b)** cooling? Explain.

3. What effect should clearing forests and converting them to grasslands and crops have on the earth's **(a)** reflectivity (albedo) and **(b)** average surface temperature? Explain.

4. Which of the four schools of thought about what should be done about possible global warming (p. 460) do you favor? Explain.

5. Explain why you agree or disagree with each of the proposals listed in **(a)** Figure 18-14 (p. 461) for slowing down emissions of greenhouse gases into the atmosphere and **(b)** Figure 18-15 (p. 465) for preparing for the effects of global warming. What might be the harmful effects on your life of *not* carrying out these actions?

6. Why are most developing countries more reluctant to reduce their use of fossil fuels than many developed countries? Do you agree with the developing countries' argument that developed countries should bear the brunt of reducing CO_2 emissions because they produce much more of these emissions than developing countries?

7. Explain why you agree or disagree with **(a)** giving developing countries a 10-year grace period before they are required to meet specified reductions of greenhouse gases under an international climate change treaty and **(b)** setting up an international fund financed by developed countries to transfer energy-efficiency and renewable-energy technologies to developing countries (with the fund financed by increasing the current tax on each international currency transaction by a quarter of a penny).

8. Explain why developed nations are able to adapt to problems caused by increased global warming more easily than developing countries. Should developed countries provide significant aid to help developing countries adapt to the harmful effects of global warming? Explain.

9. What consumption patterns and other features of your lifestyle directly add greenhouse gases to the atmosphere? Which, if any, of these things would you be willing to give up to slow projected global warming and reduce other forms of air pollution?

10. Explain how ozone can be beneficial or harmful to people depending on where it is located in the atmosphere.

11. You have been diagnosed with a treatable non-melanoma form of skin cancer. Explain why you could increase your chances of getting more of such cancers within 15–40 years by moving to Australia or Florida.

PROJECTS

1. As a class, conduct a poll of students at your school to determine **(a)** whether they understand the difference between global warming of the troposphere and ozone depletion in the stratosphere and **(b)** whether they believe that global warming from an enhanced greenhouse effect is a very serious problem, a moderately serious problem, or of little concern. Tally the results to see whether there are differences related to year in school, political leaning (liberal, conservative, independent), or sex of poll participants.

2. As a class, conduct a poll of students at your school to determine whether they believe that stratospheric ozone depletion is a very serious problem, a moderately serious problem, or of little concern. Tally the results to see whether there are differences related to year in school, political leaning (liberal, conservative, independent), or sex of poll participants.

3. Write a 1- to 2-page scenario of what your life could be like 60 years from now if nations, companies, and individuals do not take steps to reduce projected global warming caused at least partly by human activities. Contrast your scenario with the positive scenario at the opening of this chapter. Compare and critique scenarios written by different members of your class.

4. Use the library or the internet to find bibliographic information about *Paul A. Colinvaux* and *Margaret Mead*, whose quotes appear at the beginning and end of this chapter.

5. Make a concept map of this chapter's major ideas, using the section heads and subheads and the key terms (in boldface). Look at the inside back cover and on the website for this book for information about making concept maps.

INTERNET STUDY RESOURCES AND RESOURCES FOR FURTHER READING AND RESEARCH

The website for this book contains helpful study aids and many ideas for further reading and research. Log on to:

http://www.brookscole.com/product/0534376975s

and click on the Chapter-by-Chapter area. Choose Chapter 18 and select a resource:

- "Flash Cards" allows you to test your mastery of the Terms and Concepts to Remember for this chapter.

- "Tutorial Quizzes" provides a multiple-choice practice quiz.

- "Student Guide to InfoTrac" will lead you to Critical Thinking Projects that use InfoTrac College Edition as a research tool.

- "References" lists the major books and articles consulted in writing this chapter.

- "Hypercontents" takes you to an extensive list of sites with news, research, and images related to individual sections of the chapter.

INFOTRAC COLLEGE EDITION

Improve your skills with InfoTrac College Edition, a searchable online database of articles from more than 700 periodicals. Log on to:

http://www.infotrac-college.com

or access InfoTrac through the website for this book.

Try the following articles:

Shaver, G.R. et al. 2000. Global warming and terrestrial ecosystems: a conceptual framework for analysis. *BioScience* vol. 50, no. 10, p. 871–882. (subject guide: global warming, research)

Danis, F. 2000. Article explains ozone loss over south pole. *Ozone Depletion Network Online Today* Nov 16, 2000. (subject guide: ozone layer depletion)

19 WATER POLLUTION

Learning Nature's Ways to Purify Sewage

Some communities and individuals are seeking better ways to purify sewage by working with nature. Ecologist John Todd designs, builds, and operates innovative ecological wastewater treatment systems called *living machines* (Figure 19-1). They look like aquatic botanical gardens and are powered by the sun in greenhouses or outdoors, depending on the climate.

This ecological purification process begins when sewage flows into a passive solar greenhouse or outdoor site containing rows of large open tanks populated by an increasingly complex series of organisms. In the first set of tanks, algae and microorganisms decompose organic wastes into nutrients that are taken up by aquatic plants such as water hyacinths, cattails, and bulrushes.

In these tanks, algae and microorganisms decompose wastes into nutrients absorbed by the plants. The decomposition is speeded up by sunlight.

After flowing though several of these natural purification tanks, algae and organic waste are filtered out as the water passes through an artificial marsh of sand, gravel, and bulrush plants. Some of the plants also absorb (sequester) toxic metals such as lead and mercury and secrete natural antibiotic compounds that kill pathogens.

Next the water flows into engineered ecosystems in aquarium tanks, where snails and zooplankton consume microorganisms and are in turn consumed by crayfish, tilapia, and other fish that can be eaten or sold as bait. After 10 days, the clear water flows into a second artificial marsh for final filtering and cleansing.

The water can be made pure enough to drink by using ultraviolet light or passing the water through an ozone generator, usually immersed out of sight in an attractive pond or wetland habitat. The chief by-products of such living machines are ornamental plants, trees, and baitfish that can be sold. Operating costs are about the same as for a conventional sewage treatment plant.

This technology is sold by Living Technologies, which has installed 30 living machine purification systems in the United States and six other countries, including the system for the innovative environmental science building at Oberlin College in Ohio (Figure 15-24, p. 376). Some communities and industries are working with nature by using natural and artificial wetlands to purify waste water, as discussed on p. 496.

Water pollution is related to air pollution, land-use practices, climate change, and the number of people, farms, and industries producing sewage. These connections explain why solving water pollution problems should be integrated with air pollution, energy, land-use, climate, and population policies that emphasize pollution prevention. Otherwise, we will continue shifting potential pollutants from one part of the biosphere to another until threshold levels of damage are exceeded as more people, farms, and industries produce more wastes.

Figure 19-1 Ecological wastewater purification by a *living machine*. At the Providence, Rhode Island, Solar Sewage Treatment Plant, biologist John Todd demonstrates how ecological waste engineering in a greenhouse can be used to purify wastewater in an ecological process he invented. Todd and others are conducting research to perfect solar-aquatic sewage treatment systems based on working with nature. (Ocean Arks International)

> *Today everybody is downwind or downstream from somebody else.*
>
> WILLIAM RUCKELSHAUS

This chapter addresses the following questions:

- What pollutes water, where do the pollutants come from, and what effects do they have?

- What are the major water pollution problems of streams and lakes?

- How is groundwater polluted, and what can be done to prevent such pollution?

- What are the major water pollution problems of oceans?

- How can we prevent and reduce surface-water pollution?

- How safe is drinking water, and how can it be made safer?

19-1 TYPES AND SOURCES OF WATER POLLUTION

What Are the Major Water Pollutants, and How Do We Detect Them? Water pollution is any chemical, biological, or physical change in water quality that has a harmful effect on living organisms or makes water unsuitable for desired uses. Table 19-1 lists the major classes of water pollutants along with their major human sources and harmful effects. Table 19-2 lists some common diseases that can be transmitted to humans through drinking water contaminated with infectious agents.

Various methods are used to determine water quality. A good indicator of water quality in terms of infectious agents is the number of colonies of *coliform bacteria* present in a 100-milliliter (0.1-quart) sample of water. The World Health Organization recommends a coliform bacteria count of 0 colonies per 100 milliliters for drinking water, and the U.S. Environmental Protection

---

Table 19-1 Major Categories of Water Pollutants States

INFECTIOUS AGENTS

Examples: Bacteria, viruses, protozoa, and parasitic worms
Major Human Sources: Human and animal wastes
Harmful Effects: Disease (Table 19-2)

OXYGEN-DEMANDING WASTES

Examples: Organic waste such as animal manure and plant debris that can be decomposed by aerobic (oxygen-requiring) bacteria
Major Human Sources: Sewage, animal feedlots, paper mills, and food processing facilities
Harmful Effects: Large populations of bacteria decomposing these wastes can degrade water quality by depleting water of dissolved oxygen. This causes fish and other forms of oxygen-consuming aquatic life to die.

INORGANIC CHEMICALS

Examples: Water-soluble (1) acids, (2) compounds of toxic metals such as lead (Pb), arsenic (As), and selenium (Se), and (3) salts such as NaCl in ocean water and fluorides (F^-) found in some soils

Major Human Sources: Surface runoff, industrial effluents, and household cleansers
Harmful Effects: Can (1) make freshwater unusable for drinking or irrigation, (2) cause skin cancers and crippling spinal and neck damage (F^-), (3) damage the nervous system, liver, and kidneys (Pb and As), (4) harm fish and other aquatic life, (5) lower crop yields, and (6) accelerate corrosion of metals exposed to such water.

ORGANIC CHEMICALS

Examples: Oil, gasoline, plastics, pesticides, cleaning solvents, detergents
Major Human Sources: Industrial effluents, household cleansers, surface runoff from farms and yards
Harmful Effects: can (1) threaten human health by causing nervous system damage (some pesticides), reproductive disorders (some solvents) and some cancers (gasoline, oil, and some solvents) and (2) harm fish and wildlife

PLANT NUTRIENTS

Examples: Water-soluble compounds containing nitrate (NO_3^-), phosphate (PO_4^{3-}), and ammonium (NH_4^+) ions
Major Human Sources: Sewage, manure, and runoff of agricultural and urban fertilizers
Harmful Effects: Can cause excessive growth of algae and other aquatic plants which die, decay, deplete water of dissolved oxygen, and kill fish. Drinking water with excessive levels of nitrates lowers the oxygen-carrying capacity of the blood and can kill unborn children and infants ("blue-baby syndrome").

SEDIMENT

Examples: Soil, silt
Major Human Sources: Land erosion
Harmful Effects: Can (1) cloud water and reduce photosynthesis, (2) disrupt aquatic food webs, (3) carry pesticides, bacteria, and other harmful substances, (4) settle out and destroy feeding and spawning grounds of fish, and (5) clog and fill lakes, artificial reservoirs, stream channels, and harbors.

RADIOACTIVE MATERIALS

Examples: Radioactive isotopes of iodine, radon, uranium, cesium, and thorium
Major Human Sources: Nuclear power plants, mining and processing of uranium and other ores, nuclear weapons production, natural sources
Harmful Effects: Genetic mutations, miscarriages, birth defects, and certain cancers

HEAT (THERMAL POLLUTION)

Examples: Excessive heat
Major Human Sources: Water cooling of electric power plants (Figure 14-32, p. 346) and some types of industrial plants. Almost half of all water withdrawn in the United States each year is for cooling electric power plants.
Harmful Effects: Lowers dissolved oxygen levels and makes aquatic organisms more vulnerable to disease, parasites, and toxic chemicals. When a power plant first opens or shuts down for repair, fish and other organisms adapted to a particular temperature range (Figure 4-14, p. 79) can be killed by the abrupt change in water temperature—known as *thermal shock*.

Table 19-2 Common Diseases Transmitted to Humans Through Contaminated Drinking Water

Type of Organism	Disease	Effects
Bacteria	Typhoid fever	Diarrhea, severe vomiting, enlarged spleen, inflamed intestine; often fatal if untreated
	Cholera	Diarrhea, severe vomiting, dehydration; often fatal if untreated
	Bacterial dysentery	Diarrhea; rarely fatal except in infants without proper treatment
	Enteritis	Severe stomach pain, nausea, vomiting; rarely fatal
Viruses	Infectious hepatitis	Fever, severe headache, loss of appetite, abdominal pain, jaundice, enlarged liver; rarely fatal but may cause permanent liver damage
Parasitic protozoa	Amoebic dysentery	Severe diarrhea, headache, abdominal pain, chills, fever; if not treated can cause liver abscess, bowel perforation, and death
	Giardiasis	Diarrhea, abdominal cramps, flatulence, belching, fatigue
Parasitic worms	Schistosomiasis	Abdominal pain, skin rash, anemia, chronic fatigue, and chronic general ill health

Agency (EPA) recommends a maximum level for swimming water of 200 colonies per 100 milliliters.

Water pollution from oxygen-demanding wastes and plant nutrients can be determined by measuring the level of dissolved oxygen (Figure 19-2). The quantity of oxygen-demanding wastes in water can be determined by measuring the **biological oxygen demand (BOD)**: the amount of dissolved oxygen needed by aerobic decomposers to break down the organic materials in a certain volume of water over a 5-day incubation period at 20°C (68°F).

Chemical analysis is used to determine the presence and concentrations of most inorganic and organic chemicals that pollute water. Living organisms can also be used as *indicator species* to monitor water pollution. For example, the tissues of *filter-feeding mussels*, harvested from the sediments of coastal waters, can be analyzed for the presence of various industrial chemicals, toxic metals (such as mercury and lead), and pesticides. Aquatic plants such as cattails can be removed and analyzed to determine pollution in areas contaminated with fuels, solvents, and other organic chemicals.

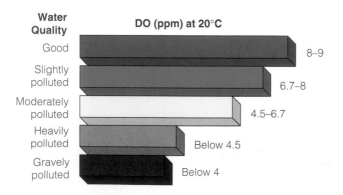

Figure 19-2 Water quality and dissolved oxygen (DO) content in parts per million (ppm) at 20°C (68°F). Only a few fish species can survive in water with less than 4 ppm of dissolved oxygen at this temperature.

What Are Point and Nonpoint Sources of Water Pollution? **Point sources** discharge pollutants at specific locations through pipes, ditches, or sewers into bodies of surface water. Examples include (1) factories, (2) sewage treatment plants (which remove some but not all pollutants), (3) active and abandoned underground mines, and (4) oil tankers. Because point sources are at specific places, they are fairly easy to identify, monitor, and regulate. In developed countries many industrial discharges are strictly controlled, whereas in most developing countries such discharges are largely uncontrolled.

Nonpoint sources are sources that cannot be traced to any single site of discharge. They are usually large land areas or airsheds that pollute water by runoff, subsurface flow, or deposition from the atmosphere. Examples include runoff of chemicals into surface water from croplands, livestock feedlots, logged forests, urban streets, lawns, golf courses, and parking lots and acid deposition (Figure 17-9, p. 428).

Nonpoint pollution from agriculture includes sediments, inorganic fertilizers, manure, salts dissolved in irrigation water, and pesticides. In the United States, such pollution is responsible for an estimated 64% of the total mass of pollutants entering streams and 57% of those entering lakes. Little progress has been made in controlling nonpoint water pollution because of the difficulty and expense of identifying and controlling discharges from so many diffuse sources.

19-2 POLLUTION OF STREAMS AND LAKES

What Are the Pollution Problems of Streams? Flowing streams, including large ones called *rivers*, can recover rapidly from degradable, oxygen-demanding wastes and excess heat by a combination of dilution and bacterial decay. This natural recovery process works as long as (1) streams are not overloaded with these pollutants and (2) their flow is not reduced by drought, damming, or diversion for agriculture and industry. However, these natural dilution and biodegradation processes do not eliminate slowly degradable and nondegradable pollutants.

In a flowing stream, the breakdown of degradable wastes by bacteria depletes dissolved oxygen, which reduces or eliminates populations of organisms with high oxygen requirements until the stream is cleansed of wastes. The depth and width of the resulting *oxygen sag curve* (Figure 19-3), and thus the time and distance needed for a stream to recover, depend on the volume of incoming degradable wastes and the stream's (1) volume, (2) flow rate, (3) temperature, and (4) pH level (Figure 3-7, p. 56). Similar oxygen sag curves can be plotted when heated water from industrial and power plants is discharged into streams.

What Progress Have We Made in Reducing Stream Pollution? Requiring cities to withdraw their drinking water downstream rather than upstream (as is done now) would improve water quality dramatically. Then each city would be forced to clean up its own waste outputs rather than passing them downstream. However, upstream users, who already have the use of fairly clean water without high cleanup costs, fight this pollution prevention approach.

Here is some *good news*. Water pollution control laws enacted in the 1970s have greatly increased the number and quality of wastewater treatment plants in the United States and many other developed countries. Laws have also required industries to reduce or eliminate point-source discharges into surface waters. These efforts have enabled the United States to hold the line against increased pollution of most of its streams by disease-causing agents and oxygen-demanding wastes. This is an impressive accomplishment given the rise in economic activity and population since the laws were passed.

One success story is the cleanup of Ohio's Cuyahoga River, which was so polluted that in 1959 and again in 1969 it caught fire and burned for several days as it flowed through Cleveland, Ohio. The highly publicized image of this burning river prompted city, state, and federal officials to (1) enact laws limiting the discharge of industrial wastes into the river and sewage systems and (2) appropriate funds to upgrade sewage treatment facilities. Today the river has made a comeback and is widely used by boaters and anglers.

Pollution control laws passed since 1970 have also led to improvements in dissolved oxygen content in

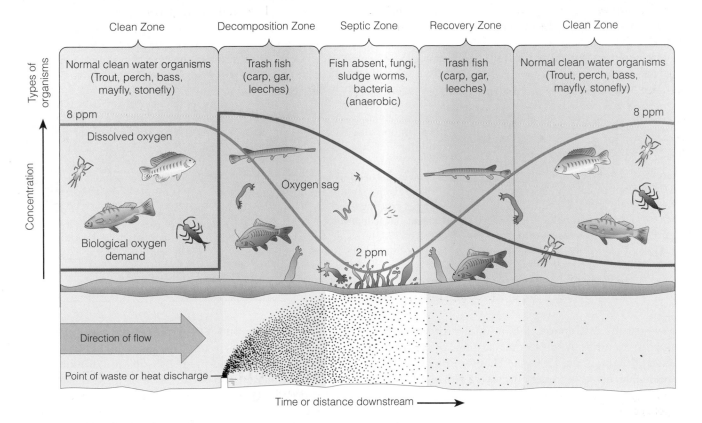

Figure 19-3 Dilution and decay of degradable, oxygen-demanding wastes and heat, showing the oxygen sag curve (orange) and the curve of oxygen demand (blue). Depending on flow rates and the amount of pollutants, streams recover from oxygen-demanding wastes and heat if they are given enough time and are not overloaded.

Tracking Down the *Pfiesteria* Cell from Hell

JoAnn M. Burkholder is professor of aquatic biology and marine science at North Carolina State University. She knows what it is to be sickened by a newly identified fish-killing microbe and to experience the political heat when you go public with your research to alert people about a potentially serious health threat.

In 1986, she investigated why a colleague's laboratory research fish were dying mysteriously and discovered that the culprit was a new microbe so tiny that dozens could fit on the head of a pin. She and her codiscoverer named it *Pfiesteria piscida* (pronounced "fee-STEER-e-ah pis-kuh-SEED-uh"), but some biologists call it the cell from hell.

She and her colleagues discovered that this complex microscopic organism could behave as both a plant and an animal and assume at least 24 guises in its lifetime. Without suitable prey, the microbe can masquerade as a plant or lie dormant for years. Then under certain conditions these microbes can change from alga eaters into fish-killing dinoflagellates that **(1)** release neurotoxins that stun fish in rivers and coastal estuaries and **(2)** usually kills them within 10 minutes to several hours.

The neurotoxin can also form an aerosol above the water. In 1993, Burkholder and her chief research aid experienced nausea, burning eyes and cramps, weakness, slow-healing sores, difficulty breathing, and severe loss of memory and mental powers from breathing toxic fumes released in tanks of fish dying from *Pfiesteria* attacks. They eventually recovered but still cannot exercise strenuously without severe shortness of breath and the onset of respiratory illness.

Since then more than 100 researchers, fishermen, and water-skiers in North Carolina, Virginia, and Maryland have experienced one or more of these symptoms when exposed to water or air contaminated by *Pfiesteria* toxins. In 1995, Burkholder and her colleagues detected a second species of fish-eating *Pfiesteria* in North Carolina's New River after a major spill from a hog-waste lagoon. These single-cell organisms live in waters from the Chesapeake Bay to the Gulf Coast of Florida and Alabama and each year cause more than $60 million in losses to U.S. fisheries and tourism.

Through lab and field research, Burkholder developed evidence that connected outbreaks or blooms of *Pfiesteria* with excessive levels of nitrogen (as nitrates) and phosphorus (as phosphates) in rivers and estuaries. High levels of such nutrients are found in runoff from fertilized croplands, industrial development, and feedlots (especially those used to raise hogs and chickens) into rivers flowing into coastal estuaries.

In 1991, Burkholder went public with her findings and urged North Carolina state legislators to put curbs on hog farming and enact much tougher laws to reduce the flow of nutrients and other pollutants into the state's rivers. Hog farmers, developers, farming interests, fishing industry officials (worried about whether it is safe to eat fish and shellfish from affected rivers and estuaries), tourist industry officials (alarmed about a negative image of the state's huge coastal recreational industry), and some state officials reacted negatively to her political activism. Some challenged her character and competence and accused her of using the results of preliminary research to push for questionable policies. She also received some anonymous death threats.

Burkholder has not backed down and continues to criticize state health officials and legislators for not taking her concerns about public health seriously enough. Under the glare of state and national publicity,* the state now supports research on the problem and since 1997 has had a moratorium on construction new hog farms (due to expire in July 2001).

Research by other scientists has confirmed the link between *Pfiesteria* outbreaks and nutrient overloading of rivers. Since 1997, a number of federal and state environmental, health, and agricultural agencies have set up a coordinated research effort to learn more about what triggers outbreaks of the organism and how they affect humans and other organisms.

*For a popularized description of her research and political battle to alert the public and elected officials to the dangers posed by this microbe, see Rodney Barker's *And the Waters Turned to Blood: The Ultimate Biological Threat* (New York: Simon & Schuster, 1997).

---

many streams in Canada, Japan, and most western European countries. A spectacular cleanup has occurred in Great Britain. In the 1950s the Thames River was little more than a flowing anaerobic sewer. However, after more than 40 years of effort and hundreds of millions of dollars spent by British taxpayers and private industry, the Thames has made a remarkable recovery. Commercial fishing is thriving, and many species of waterfowl and wading birds have returned to their former feeding grounds.

What Is the Bad News About Stream Pollution?
Despite progress in improving stream quality in most developed countries, large fish kills and drinking water contamination still occur. Most of these disasters are caused by **(1)** accidental or deliberate releases of toxic inorganic and organic chemicals by industries or mines (Case Study, p. 328), **(2)** malfunctioning sewage treatment plants, and **(3)** nonpoint runoff of pesticides and nutrients (eroded soil, fertilizer, and animal waste) from cropland or animal feedlots (Individuals Matter, above).

Available data indicate that stream pollution from discharges of sewage and industrial wastes is a serious and growing problem in most developing countries, where waste treatment is practically nonexistent. Numerous streams in the former Soviet Union and in eastern European countries are severely polluted. Currently, more than two-thirds of India's water resources are polluted with industrial wastes and sewage. Of the 78 streams monitored in China, 54 are seriously polluted with untreated sewage and industrial wastes. About 20% of China's rivers are too polluted to use for irrigation, and 80% of them are so degraded they no longer support fish. In Latin America and Africa, most streams passing through urban or industrial areas are severely polluted.

What Are the Pollution Problems of Lakes? In lakes, reservoirs, and ponds, dilution often is less effective than in streams because:

- Lakes and reservoirs often contain stratified layers (Figure 7-14, p. 165) that undergo little vertical mixing.

- They have little flow. For example, the flushing and changing of water in lakes and large artificial

reservoirs can take from 1 to 100 years, compared with several days to several weeks for streams.

- Ponds contain small volumes of water.

As a result, lakes, reservoirs, and ponds are more vulnerable than streams to contamination by plant nutrients, oil, pesticides, and toxic substances such as lead, mercury, and selenium. These contaminants can destroy both bottom life and fish and birds that feed on contaminated aquatic organisms.

Many toxic chemicals and acids (Figure 17-9, p. 428) also enter lakes and reservoirs from the atmosphere. Concentrations of some chemicals, such as DDT (Figure 16-4, p. 399), PCBs (Figure 19-4), some radioactive isotopes, and some mercury compounds can be biologically magnified as they pass through food webs in lakes.

Lakes receive inputs of nutrients and silt from the surrounding land basin as a result of natural erosion and runoff. This natural nutrient enrichment of lakes is called **eutrophication**. Over time, some lakes become more eutrophic (Figure 7-15, bottom, p. 166), but others do not because of differences in the surrounding drainage basin.

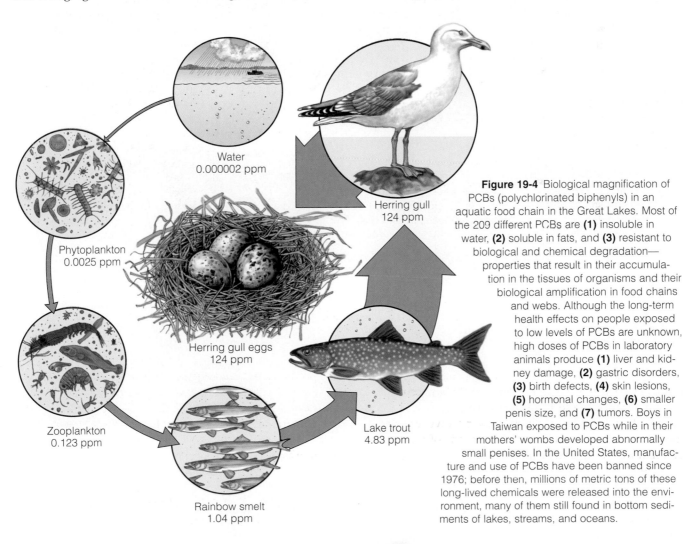

Water
0.000002 ppm

Phytoplankton
0.0025 ppm

Zooplankton
0.123 ppm

Herring gull eggs
124 ppm

Herring gull
124 ppm

Lake trout
4.83 ppm

Rainbow smelt
1.04 ppm

Figure 19-4 Biological magnification of PCBs (polychlorinated biphenyls) in an aquatic food chain in the Great Lakes. Most of the 209 different PCBs are **(1)** insoluble in water, **(2)** soluble in fats, and **(3)** resistant to biological and chemical degradation—properties that result in their accumulation in the tissues of organisms and their biological amplification in food chains and webs. Although the long-term health effects on people exposed to low levels of PCBs are unknown, high doses of PCBs in laboratory animals produce **(1)** liver and kidney damage, **(2)** gastric disorders, **(3)** birth defects, **(4)** skin lesions, **(5)** hormonal changes, **(6)** smaller penis size, and **(7)** tumors. Boys in Taiwan exposed to PCBs while in their mothers' wombs developed abnormally small penises. In the United States, manufacture and use of PCBs have been banned since 1976; before then, millions of metric tons of these long-lived chemicals were released into the environment, many of them still found in bottom sediments of lakes, streams, and oceans.

Near urban or agricultural areas, human activities can greatly accelerate the input of plant nutrients to a lake, which results in a process known as **cultural eutrophication**. Such a change is caused mostly by nitrate- and phosphate-containing effluents from **(1)** sewage treatment plants, **(2)** runoff of fertilizers and animal wastes, and **(3)** accelerated erosion of nutrient-rich topsoil (Figure 19-5).

This excessive input of plant nutrients does not poison lakes and ponds, and such lakes are not by any means "dead," as is sometimes reported. Instead, the high nutrient input overnourishes the plant life in such lakes and causes populations of algae and cyanobacteria to explode (Figure 19-6, right). This disrupts **(1)** their nitrogen (Figure 4-30, p. 94) and phosphorus (Figure 4-32, p. 96) cycles and **(2)** the distribution and types of organisms found in such bodies of water.

During hot weather or drought, this nutrient overload produces dense growths of organisms such as algae, cyanobacteria (Figure 19-6, right), water hyacinths, and duckweed. Dissolved oxygen (in both the surface layer of water near the shore and in the bottom layer) is depleted when large masses of algae die, fall to the bottom, and are decomposed by aerobic bacteria. This oxygen depletion can kill fish and other aerobic aquatic animals. If excess nutrients continue to flow into a lake, anaerobic bacteria take over and produce gaseous decomposition products such as smelly, highly toxic hydrogen sulfide and flammable methane.

About one-third of the 100,000 medium to large lakes and about 85% of the large lakes near major population centers in the United States suffer from some degree of cultural eutrophication. One-fourth of China's lakes are classified as eutrophic.

Ways to *prevent* or reduce cultural eutrophication include **(1)** advanced waste treatment (Section 19-5), **(2)** bans or limits on phosphates in household detergents and other cleaning agents, and **(3)** soil conservation and land-use control to reduce nutrient runoff.

Major *cleanup methods* are **(1)** removing excess weeds, **(2)** controlling undesirable plant growth with herbicides and algicides, and **(3)** pumping air through

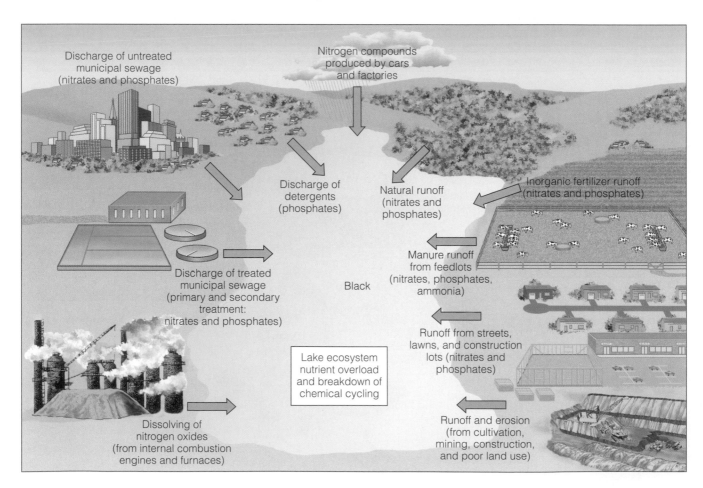

Figure 19-5 Principal sources of nutrient overload causing *cultural eutrophication* in lakes. The amount of nutrients from each source varies according to the types and amounts of human activities occurring in each airshed and watershed. Levels of dissolved oxygen (Figure 19-2) drop when enlarged populations of algae and plants (stimulated by increased nutrient input) die and are decomposed by aerobic bacteria. Lowered oxygen levels can kill fish and other aquatic life and reduce the aesthetic and recreational value of the lake.

lakes and reservoirs to avoid oxygen depletion (an expensive and energy-intensive method). As usual, pollution prevention is more effective and usually cheaper in the long run than cleanup.

Case Study: Chemical Pollution in the Great Lakes The five interconnected Great Lakes (Figure 19-7) contain at least 95% of the fresh surface water in the United States and 20% of the world's fresh surface water. The Great Lakes basin is home for about 38 million people—about 30% of the Canadian population and 14% of the U.S. population. Its huge drainage basin receives inputs of water and pollutants from seven states and much of Ontario in Canada (Figure 19-7).

Despite their enormous size, these lakes are vulnerable to pollution from point and nonpoint sources because less than 1% of the water entering the Great Lakes flows out to the St. Lawrence River each year. In addition to land runoff, these lakes receive large quantities of acids, pesticides, and other toxic chemicals by deposition from the atmosphere (often blown in from hundreds or thousands of kilometers away).

By the 1960s, many areas of the Great Lakes were suffering from severe cultural eutrophication, huge fish kills, and contamination from bacteria and a variety of toxic industrial wastes. The impact on Lake Erie was particularly intense because it is the shallowest of the Great Lakes and it has the highest concentrations of people and industrial activity along its shores. Many bathing beaches had to be closed, and by 1970 the lake had lost nearly all its native fish (Figure 19-8, left).

Here is some *good news*. Since 1972, a $20-billion Great Lakes pollution control program has been carried out jointly by Canada and the United States. This joint program has **(1)** significantly decreased levels of phosphates, coliform bacteria, and many toxic industrial chemicals, **(2)** decreased algae blooms, **(3)** increased dissolved oxygen levels and sport and commercial fishing catches (Figure 19-8, right), and **(4)** allowed most swimming beaches to reopen.

These improvements occurred mainly because of **(1)** new or upgraded sewage treatment plants, **(2)** better treatment of industrial wastes, and **(3)** banning of phosphate detergents, household cleaners, and water conditioners.

Here is some *bad news*:

- Less than 3% of the lakes' shoreline is clean enough for swimming or for supplying drinking water.

- Nonpoint land runoff of pesticides and fertilizers from urban sprawl has surpassed industrial pollution as the greatest threat to the lakes.

- Forty-three toxic hot spots (Figure 19-7) are still heavily polluted.

- Atmospheric deposition of pesticides, mercury from coal-burning plants, and other toxic chemicals from as far as Mexico and Russia account for an estimated 50% of the input of toxic compounds.

Figure 19-6 The effect of nutrient enrichment on a lake. Crater Lake in Oregon (left) is an example an oligotrophic lake (Figure 7-15, top, p. 166) that is low in nutrients. Because of a low density of plankton, its water is quite clear. The lake on the right, found in western New York, is a eutrophic lake (Figure 7-15, bottom, p. 166). Because of an excess of plant nutrients, its surface is covered with mats of algae and cyanobacteria. (Left, Jack Carey; right, W. A. Bannazewski, Visuals Unlimited)

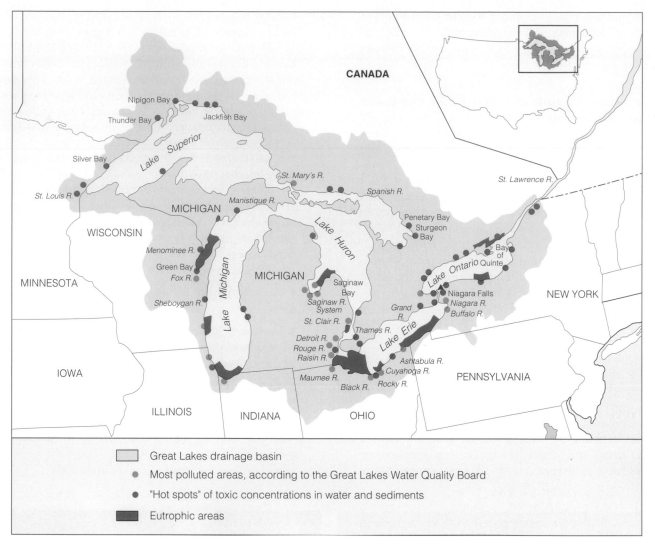

Great Lakes drainage basin

● Most polluted areas, according to the Great Lakes Water Quality Board

● "Hot spots" of toxic concentrations in water and sediments

Eutrophic areas

Figure 19-7 The Great Lakes basin and the locations of some of its water quality problems. The Great Lakes region is dotted with several hundred abandoned toxic waste sites that are listed by the EPA as Superfund sites to receive cleanup priority (p. 543). (Data from Environmental Protection Agency)

- Toxic chemicals such as PCBs have **(1)** built up in food chains and webs (Figure 19-4), **(2)** contaminated many types of sport fish, and **(3)** depleted populations of birds, river otters, and other animals feeding on contaminated fish.

- A survey by Wisconsin biologists revealed that one fish in four taken from the Great Lakes is unsafe for human consumption.

In 1991, the U.S. government passed a law requiring accelerated cleanup of the lakes, especially 43 toxic hot spots, and an immediate reduction in air pollutant emissions in the region. However, a lack of federal and state funds may delay meeting these goals.

Some environmentalists call for a ban on **(1)** the use of chlorine as a bleach in the pulp and paper industry around the Great Lakes, **(2)** all new incinerators in the

area, and **(3)** discharge into the lakes of 70 toxic chemicals that threaten human health and wildlife. Officials of these industries strongly oppose such bans.

Connections: How Might Projected Climate Change Affect Water Pollution? Global warming (Section 18-4, p. 458) can affect water quality by putting more water vapor into the atmosphere and altering existing patterns of precipitation. As a result, some areas will get much more precipitation and others will experience more prolonged drought.

A moisture-laden atmosphere generates more intense downpours, which can flush more harmful chemicals, plant nutrients, and microorganisms into waterways. Massive flooding, whether caused by normal weather patterns or human impacts on climate, often spreads disease-carrying pathogens (Table 19-2)

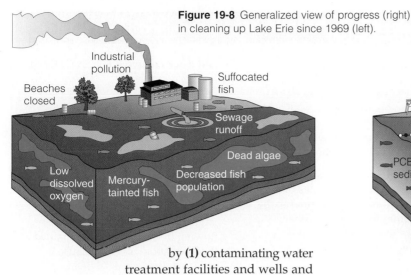

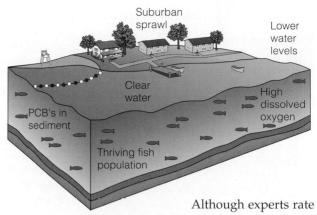

Figure 19-8 Generalized view of progress (right) in cleaning up Lake Erie since 1969 (left).

by **(1)** contaminating water treatment facilities and wells and **(2)** causing the overflow of animal waste lagoons and combined sewer lines (Case Study, p. 486). Warmer water temperatures can also increase the rate at which harmful microorganisms such as *Salmonella* grow.

Waterborne infectious diseases can also increase in countries experiencing prolonged drought as a result of climate change. In drought conditions, disease can spread more rapidly because people lack enough water to stay clean.

19-3 GROUNDWATER POLLUTION AND ITS PREVENTION

Why Is Groundwater Pollution Such a Serious Problem? According to many scientists, a serious threat to human health is the out-of-sight pollution of groundwater, a prime source of water for drinking and irrigation. Groundwater supplies about 75% of the drinking water in Europe, 51% in the United States, 32% in Asia, and 29% in Latin America. In selected parts of the world, essentially all drinking and irrigation water is pumped up from aquifers.

Studies indicate that groundwater pollution comes from numerous sources as we dump more of our wastes into **(1)** storage lagoons, **(2)** septic tanks, **(3)** landfills, **(4)** hazardous waste dumps, and **(5)** deep injection wells and store gasoline, oil, solvents, and hazardous wastes in metal underground tanks that after 20–40 years can corrode and leak (Figure 19-9). Groundwater is also contaminated by people who dump or spill gasoline, oil, and paint thinners and other organic solvents onto the ground.

Groundwater in aquifers is easy to deplete (Case Study, p. 309) and pollute because much of it is renewed very slowly. For example, the average recycling time for groundwater is about 1,400 years, compared to only 20 days for river water.

Although experts rate groundwater pollution as a low-risk ecological problem, they consider pollutants in drinking water (much of it from groundwater) a high-risk health problem (Figure 16-13, left, p. 411). Human health risks come mostly from groundwater contaminated with **(1)** petrochemicals (such as gasoline and oil), **(2)** organic solvents (such as trichloroethylene, or TCE), **(3)** pesticides, **(4)** arsenic (As), **(5)** lead (Pb), and **(6)** fluoride (F⁻; Table 19-1).

When groundwater becomes contaminated, it cannot cleanse itself of *degradable wastes* as flowing surface water does (Figure 19-3) because **(1)** it flows so slowly (usually less than 0.3 meters or 1 foot per day) that contaminants are not diluted and dispersed effectively, **(2)** it has much smaller populations of decomposing bacteria, and **(3)** its cold temperature slows down the chemical reactions that decompose wastes. This means that it can take hundreds to thousands of years for contaminated groundwater to cleanse itself of *degradable* wastes, and on a human time scale *nondegradable wastes* (such as toxic lead, arsenic, and fluoride) are there permanently.

What Is the Extent of Groundwater Pollution? The answer is that *we do not know* because few countries go to the great expense of locating, tracking, and testing aquifers. However, studies that have been made by scientists in scattered parts of the world are alarming. Here are a few of the findings:

- Up to 25% of usable groundwater in the United States is contaminated (and in some areas as much as 75%). In New Jersey, for example, every major aquifer is contaminated.

- The EPA has documented groundwater contamination by 74 pesticides in 38 states, mostly in farming regions. In California, pesticides contaminate the drinking water of more than 1 million people. Contamination of groundwater by pesticides has also been found in farming regions of western Europe, south Asia, and Latin America.

Hurricanes, Hog Farms, and Water Pollution in Eastern North Carolina

With 10 million animals statewide, North Carolina is the nation's second largest hog producer, after Iowa. Ninety-percent of the state's 2,400 hog farms in this $1.9-billion-per-year business are located in a portion of eastern North Carolina. The area is also a large producer of poultry.

In 1999, three hurricanes dropped more than 0.9 meter (3 feet) of rain in less than 2 months on eastern North Carolina, with the largest coming from Hurricane Floyd in September. The resulting massive flooding:

- Knocked out electricity in at least 24 municipal sewage treatment plants, causing large spills of raw sewage into coastal rivers.

- Covered more than 50 hog and poultry waste lagoons, collapsed five lagoons, and released large amounts of animal waste into nearby streams, rivers, residential areas, and water supplies—a potential danger that biologist JoAnn M. Burkholder (Individuals Matter, p. 480) had warned about in 1991.

- Required hog and poultry farmers to pump large volumes of diluted animal wastes onto nearby fields from hundreds of other lagoons to keep them from overflowing or collapsing.

- Caused structural damage to a number of the area's 2,000 other animal waste lagoons that could make them vulnerable to future flooding.

- Polluted houses and wells, broke water mains, and threatened water supplies with bacteria, viruses, and parasites. After Floyd, a health crisis was largely averted because National Guard helicopters and trucks delivered clean water to several counties.

The stormwater (1) became mixed with raw human and animal sewage, junkyard waste, leaking underground gasoline tanks and oil drums, and pesticide-laced sediments from farm fields and (2) poured down rivers into Pamlico Sound, the nation's second largest estuary. Water and wastes flowing into this estuary remained there for about a year because it only has three small outlets to the Atlantic Ocean.

Critical Thinking

What three things could the government of North Carolina do to help reduce water pollution and other harmful effects from flooding caused by future hurricanes and large storms?

- According to the EPA and the U.S. Geological Survey, about 45% of *municipal* groundwater supplies in the United States are contaminated with one or more organic chemicals.

- A 2000 U.S. Geological Survey study found that about 42 million Americans use groundwater that is subject to contamination by volatile organic compounds (VOCs; Figure 19-10)

- An EPA survey found that (1) one-third of 26,000 industrial waste ponds and lagoons in the United States have no liners to prevent toxic liquid wastes from seeping into aquifers, and (2) one-third of these sites are within 1.6 kilometers (1 mile) of a drinking water well.

- The EPA estimates that at least 1 million underground tanks storing gasoline, diesel fuel, home heating oil, and toxic solvents are leaking their contents into groundwater in the United States. A slow leak of just 4 liters (1 gallon) of gasoline per day can contaminate the water supply for 50,000 people. During this century, scientists expect many of the millions of such tanks installed around the world in recent decades to corrode, leak, contaminate groundwater, and become a major global health problem.

- Determining the extent of a leak from an underground tank can cost $25,000–250,000, and cleanup costs range from $10,000 to more than $250,000. If the chemical reaches an aquifer, effective cleanup is rarely possible. Replacing a leaking tank adds an additional $10,000–60,000. In California's Silicon Valley, where electronics industries use underground tanks to store a variety of waste solvents, local authorities found leaks from 85% of the tanks they inspected.

- About 60% of the liquid hazardous waste in the United States is disposed of by injections into deep underground wells (Figure 19-9). Although these wastes are injected below the deepest sources of drinking water, some of the injection pipes can leak, and some of the wastes have entered aquifers used for drinking water in parts of Texas, Florida, Oklahoma, and Ohio.

- Groundwater contamination by toxic *arsenic (As)* can occur when tubewells are drilled in areas where the soils are naturally rich in arsenic (such as India's state of West Bengal and parts of Bangladesh, where 35–77 million people are drinking water with arsenic levels 5–100 times the World Health Organization limit).

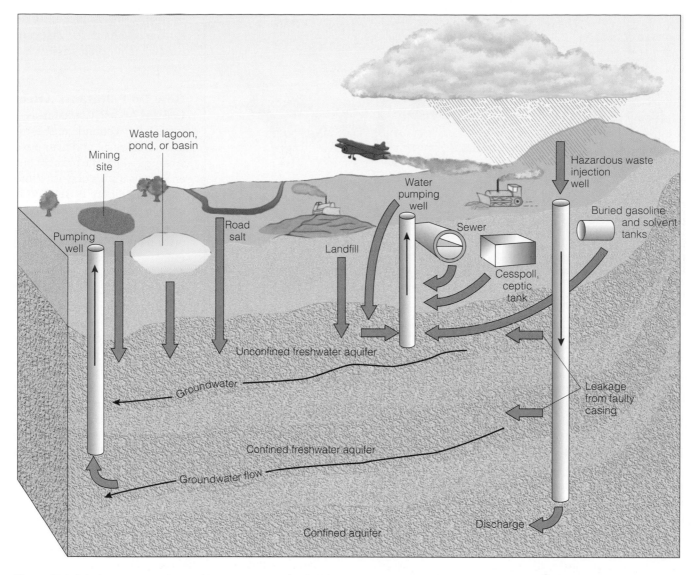

Figure 19-9 Principal sources of groundwater contamination in the United States.

- According to the World Health Organization, an estimated 70 million people in northern China and 30 million in northwestern India drink groundwater contaminated with high levels of naturally occurring *fluoride (F^-)*. This can cause crippling backbone and neck damage and a variety of dental problems.

- In coastal areas, excessive pumping of water from aquifers can lead to contamination of drinking water by saltwater intrusion (Figure 13-17, p. 308).

Solutions: How Can We Protect Groundwater?
Contaminated aquifers are almost impossible to clean because of their **(1)** enormous volume, **(2)** inaccessibility, and **(3)** slow movement. Pumping polluted groundwater to the surface, cleaning it up, and returning it to the aquifer is extremely expensive. Moreover,

recent attempts to pump and treat contaminated aquifers indicate that it may take 50–1,000 years of continuous pumping before all the contamination is forced to the surface and drinking water quality is achieved. Thus, *preventing contamination is considered the only effective way to protect groundwater resources.* Ways to do this include:

- *Monitoring aquifers near landfills and underground tanks.*

- *Requiring leak detection systems for underground tanks used to store hazardous liquids.*

- *Banning or more strictly regulating disposal of hazardous wastes in deep injection wells and landfills.*

- *Storing hazardous liquids above ground in tanks with systems that detect and collect leaking liquids.*

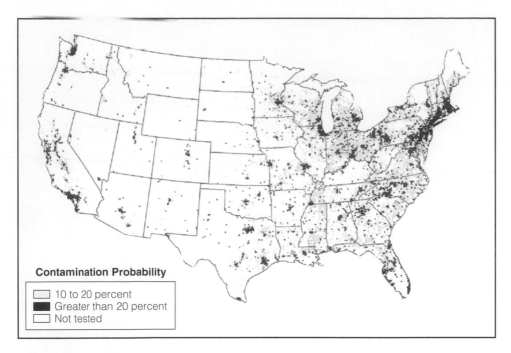

Figure 19-10 Probability of drinking water contamination by volatile organic compounds (VOCs) in the 48 contiguous states. John Zorgoski, who headed the study, urges people who live in areas with red or yellow dots to have their well and municipal water tested for VOCs. (Data from the U.S. Geological Survey)

19-4 OCEAN POLLUTION

How Much Pollution Can the Oceans Tolerate?
The oceans are the ultimate sink for much of the waste matter we produce, as summarized in the African proverb, "Water may flow in a thousand channels, but it all returns to the sea."

Oceans can dilute, disperse, and degrade large amounts of raw sewage, sewage sludge, oil, and some types of degradable industrial waste, especially in deep-water areas. Some forms of marine life have also proved to be more resilient than originally expected. This has led some analysts to suggest that it is safer to dump sewage sludge and most other hazardous wastes into the deep ocean than to bury them on land or burn them in incinerators.

Other scientists disagree, pointing out that we know less about the deep ocean than we do about outer space. They add that dumping waste in the ocean would delay urgently needed pollution prevention and promote further degradation of this vital part of the earth's life-support system.

There is increasing agreement that we must greatly increase funding to help us understand more about how the oceans work and how to help sustain their natural processes and biodiversity (Chapter 24). According to oceanographer Sylvia Earle, "We've made the investment to venture into the skies, and it has paid off mightily. We've neglected the oceans, and it has cost

us dearly. This is the time to do for the oceans in the 21st century what our predecessors did for space."

How Do Pollutants Affect Coastal Areas? Coastal areas—especially wetlands and estuaries, coral reefs, and mangrove swamps—bear the brunt of our enormous inputs of wastes into the ocean (Figure 19-11). This is not surprising because **(1)** about 40% the world's population lives on or within 100 kilometers (160 miles) of the coast, **(2)** 14 of the world's 15 largest metropolitan areas, each with 10 million people or more, are near coastal waters, and **(3)** coastal populations are growing more rapidly than the global population.

In most coastal developing countries (and in some coastal developed countries), municipal sewage and industrial wastes are dumped into the sea without treatment. The most polluted seas lie off the densely populated coasts of Bangladesh, India, Pakistan, Indonesia, Malaysia, Thailand, and the Philippines. About 85% of the sewage from large cities along the Mediterranean Sea, which has a coastal population of 200 million people during tourist season, is discharged into the sea untreated. This causes widespread beach pollution and shellfish contamination.

Most U.S. harbors and bays are badly polluted by municipal sewage, industrial wastes, and oil. Scuba divers talk of swimming through clouds of half-dissolved feces and of bay and harbor bottoms covered with toxic sediment known as black mayonnaise. They see lobsters and crabs covered with mysterious burn holes and fish with cancerous sores and rotting fins. Each year at least one-third of the area of U.S. coastal waters around the lower 48 states is closed to shellfish harvesting because of pollution and habitat disruption.

New studies show that vast colonies of human viruses from raw sewage, effluents from sewage treatment plants (which do not remove viruses), and leaking septic tanks migrate into coastal waters. One study found that about one-fourth of the people using coastal beaches develop ear infections, sore throats and eyes, respiratory disease, or gastrointestinal disease.

Runoff of sewage and agricultural wastes into coastal waters and acid deposition from the atmosphere (Figure 17-9, p. 428) introduce large quantities of nitrate

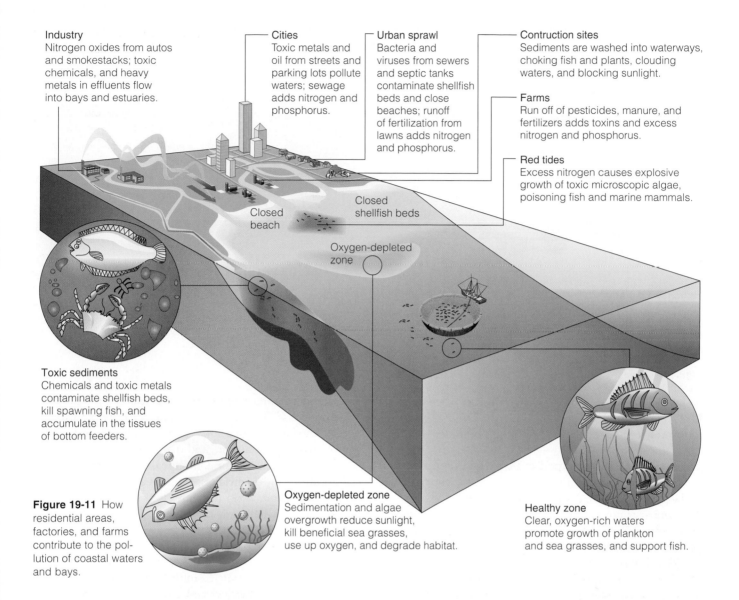

Industry
Nitrogen oxides from autos and smokestacks; toxic chemicals, and heavy metals in effluents flow into bays and estuaries.

Cities
Toxic metals and oil from streets and parking lots pollute waters; sewage adds nitrogen and phosphorus.

Urban sprawl
Bacteria and viruses from sewers and septic tanks contaminate shellfish beds and close beaches; runoff of fertilization from lawns adds nitrogen and phosphorus.

Contruction sites
Sediments are washed into waterways, choking fish and plants, clouding waters, and blocking sunlight.

Farms
Run off of pesticides, manure, and fertilizers adds toxins and excess nitrogen and phosphorus.

Red tides
Excess nitrogen causes explosive growth of toxic microscopic algae, poisoning fish and marine mammals.

Closed beach

Closed shellfish beds

Oxygen-depleted zone

Toxic sediments
Chemicals and toxic metals contaminate shellfish beds, kill spawning fish, and accumulate in the tissues of bottom feeders.

Figure 19-11 How residential areas, factories, and farms contribute to the pollution of coastal waters and bays.

Oxygen-depleted zone
Sedimentation and algae overgrowth reduce sunlight, kill beneficial sea grasses, use up oxygen, and degrade habitat.

Healthy zone
Clear, oxygen-rich waters promote growth of plankton and sea grasses, and support fish.

(NO_3^-) and phosphate (PO^{3-}) plant nutrients, which can cause explosive growth of harmful algae.

These algal blooms—called red, brown, or green tides, depending on their color—can **(1)** release waterborne and airborne toxins that damage fisheries, **(2)** kill some fish-eating birds, **(3)** reduce tourism, and **(4)** poison seafood. When the algae die and decompose, coastal waters are depleted of oxygen, and a variety of marine species die.

According to a 2000 report by the U.S. National Academy of Sciences, fish and other marine life are being killed in more than a third of the nation's coastal areas from algae blooms caused by runoff of excess plant nutrients. Each year some 61 large *oxygen-depleted zones* (sometimes inaccurately called a dead zones) form in the world's coastal waters and in landlocked seas such as the Baltic and Black because of excessive fertilizer inputs. In these zones, much of the aquatic life dies or moves elsewhere.

The biggest such zone in U.S. waters and the third largest in the world forms every summer in a narrow stretch of the Gulf of Mexico (Figure 19-12). For half the year, this zone in the Gulf of Mexico is rich with shrimp, fish, and many other forms of aquatic life. However, the spring snowmelt and rains flush nitrogen and phosphorus fertilizers from farms and cities throughout the Mississippi River's giant watershed, which covers more than 40% of the continental United States from Montana to western New York (Figure 19-12). As a result, this zone becomes overfertilized, depleted of oxygen, and devoid of fish and many other creatures for about 6 months a year. By summer, the oxygen-depleted zone covers about 20,700 square kilometers (8,000 square miles).

In the Baltic Sea, excessive cultural eutrophication has killed almost all bottom-dwelling animal life over an area of about 109,000 square kilometers (42,000 square miles)—about the size of Guatemala or Tennessee. In 1998, a deadly red tide appeared on the

Figure 19-12 A large zone of oxygen-depleted water forms for half of the year in the Gulf of Mexico as a result of oxygen-depleting algal blooms. It is created by huge inputs of nitrate (NO_3^-) and phosphate (PO_4^{3-}) plant nutrients from the massive Mississippi River Basin.

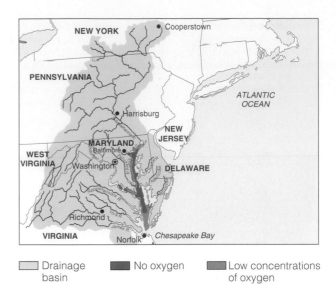

☐ Drainage basin   ■ No oxygen   ■ Low concentrations of oxygen

Figure 19-13 Chesapeake Bay, the largest estuary in the United States, is severely degraded as a result of water pollution from point and nonpoint sources in six states and from deposition of air pollutants.

China coast, where none has been recorded before. Within a few hours, it wiped out fish farms, leaving thousands of tons of rotting fish.

Case Study: The Chesapeake Bay The Chesapeake Bay, the largest estuary in the United States, is in trouble because of human activities. Between 1940 and 2000, the number of people living in the Chesapeake Bay area grew from 3.7 million to 17 million, and within a few years its population may reach 18 million.

The estuary receives wastes from point and nonpoint sources scattered throughout a huge drainage basin that includes 9 large rivers and 141 smaller streams and creeks in parts of six states (Figure 19-13). The bay has become a huge pollution sink because **(1)** it is quite shallow and **(2)** only 1% of the waste entering it is flushed into the Atlantic Ocean.

Phosphate and nitrate levels have risen sharply in many parts of the bay, causing algae blooms and oxygen depletion (Figure 19-13). Studies have shown that point sources, primarily sewage treatment plants, contribute about 60% by weight of the phosphates. Nonpoint sources—mostly runoff from urban, suburban, and agricultural land and deposition from the atmosphere—are the origins of about 60% by weight of the nitrates.

Air pollutants account for nearly 40% of the nitrogen, 90% of the mercury (mostly from older coal-burning power plants), and 95% of the lead entering the estuary. Large quantities of pesticides run off cropland and urban lawns, and industries discharge large amounts of toxic wastes, often in violation of their discharge permits. Commercial harvests of oysters (Solutions, p. 492), crabs, and several important fish have fallen sharply since 1960 because of a combination of overfishing, pollution, and disease.

In the 1980s, the Chesapeake Bay Program, the country's most ambitious attempt at integrated coastal management, was implemented. Results have been impressive. Between 1985 and 1993, phosphorus levels declined 16% and nitrogen levels dropped 7%—a significant achievement given the increasing population in the watershed and the fact that nearly 40% of the nitrogen inputs come from the atmosphere.

Reaching the declared goal of a 40% reduction in nutrient levels and a significant improvement in habitat water quality throughout the bay will be especially difficult because the area's population is expected to grow by 25% between 1995 and 2020. So far, however, the Chesapeake Bay Program shows what can be done when diverse interested parties work together to achieve goals that benefit both wildlife and people.

What Pollutants Do We Dump into the Ocean? Industrial waste dumping off U.S. coasts has stopped, although it still occurs in a number of other developed countries and some developing countries. However, barges and ships still legally dump large quantities of **dredge spoils** (materials, often laden with toxic metals, scraped from the bottoms of harbors and rivers to maintain shipping channels) at 110 sites off the Atlantic, Pacific, and Gulf coasts.

In addition, many countries dump into the ocean large quantities of sewage **sludge**: a gooey mixture of toxic chemicals, infectious agents, and settled solids removed from wastewater at sewage treatment plants. Since 1992, this practice has been banned in the United States.

Fifty countries with at least 80% of the world's merchant fleet have agreed not to dump sewage and garbage at sea, but this agreement is difficult to enforce and often is violated. Most ship owners save money by dumping wastes at sea and risk only small fines if they are caught. Each year, as many as 2 million seabirds and more than 100,000 marine mammals (including whales, seals, dolphins, and sea lions) die when they ingest or become entangled in fishing nets, ropes, and other debris dumped into the sea and discarded on beaches.

Under the London Dumping Convention of 1972, 100 countries agreed not to dump highly toxic pollutants and high-level radioactive wastes in the open sea beyond the boundaries of their national jurisdictions. Since 1983, these same nations have observed a moratorium on the dumping of low-level radioactive wastes at sea, which in 1994 became a permanent ban. However, France, Great Britain, Russia, China, and Belgium may legally exempt themselves from this ban. In 1992, it was learned that for decades the former Soviet Union had been dumping large quantities of high- and low-level radioactive wastes into the Arctic Ocean and its tributaries.

What Are the Effects of Oil on Ocean Ecosystems? *Crude petroleum* (oil as it comes out of the ground) and *refined petroleum* (fuel oil, gasoline, and other processed petroleum products, Figure 14-16, p. 337) are accidentally or deliberately released into the environment from a number of sources.

Tanker accidents (Case Study, p. 336) and blowouts at offshore drilling rigs (when oil escapes under high pressure from a borehole in the ocean floor) get most of the publicity. However, more oil is released **(1)** during normal operation of offshore wells, **(2)** from washing tankers and releasing the oily water, and **(3)** from pipeline and storage tank leaks.

Natural oil seeps also release large amounts of oil into the ocean at some sites, but most ocean oil pollution comes from activities on land. Almost half (some experts estimate 90%) of the oil reaching the oceans is waste oil dumped, spilled, or leaked onto the land or into sewers by cities, industries, and individuals changing their own motor oil. Worldwide, about 10% of the oil that reaches the ocean comes from the atmosphere, mostly from smoke emitted by oil fires.

The effects of oil on ocean ecosystems depend on a number of factors: **(1)** type of oil (crude or refined), **(2)** amount released, **(3)** distance of release from shore, **(4)** time of year, **(5)** weather conditions, **(6)** average water temperature, and **(7)** ocean currents.

Volatile organic hydrocarbons in oil immediately kill a number of aquatic organisms, especially in their vulnerable larval forms. Some other chemicals form tar-like globs that float on the surface and coats the feathers of birds (especially diving birds) and the fur of marine mammals. This destroys their natural insulation and buoyancy, causing many of them to drown or die of exposure from loss of body heat. Heavy oil components that sink to the ocean floor or wash into estuaries can smother bottom-dwelling organisms such as crabs, oysters, mussels, and clams or make them unfit for human consumption. Some oil spills have killed reef corals.

Research shows that most (but not all) forms of marine life recover from exposure to large amounts of *crude oil* within 3 years. However, recovery from exposure to *refined oil*, especially in estuaries, can take 10 years or longer. The effects of spills in cold waters and in shallow enclosed gulfs and bays generally last longer (Case Study, p. 336).

Oil slicks that wash onto beaches can have a serious economic impact on coastal residents, who lose income from fishing and tourist activities. Oil-polluted beaches washed by strong waves or currents become clean after about a year, but beaches in sheltered areas remain contaminated for several years. Estuaries and salt marshes suffer the most and longest-lasting damage. Despite their localized harmful effects, experts rate oil spills as a low-risk ecological problem (Figure 16-13, left, p. 411).

How Can Oil Spills Be Cleaned Up? If they are not too large, oils spills can be partially cleaned up by mechanical, chemical, fire, and natural methods. *Mechanical methods* include using **(1)** floating booms to contain the oil spill or keep it from reaching sensitive areas, **(2)** skimmer boats to vacuum up some of the oil into collection barges, and **(3)** absorbent pads or large mesh pillows filled with feathers or hair to soak up oil on beaches or in waters too shallow for skimmer boats.

Chemical methods include the use of **(1)** coagulating agents to cause floating oil to clump together for easier pickup or to sink to the bottom, where it usually does less harm, and **(2)** dispersing agents to break up oil slicks. However, these agents can also damage some types of organisms. Fire can burn off floating oil, but crude oil is hard to ignite, and this approach produces air pollution. In time, the natural action of wind and waves mixes or emulsifies oil with water (like emulsified salad dressing), and bacteria biodegrade some of the oil.

These methods remove only part of the oil, and none work well on a large spill. Scientists estimate that no more than 12–15% of the oil from a major spill can be recovered. This explains why preventing oil pollution is the most effective and in the long run the least

Bring Back the Oysters

SOLUTIONS

A number of scientists and environmentalists are looking for ways to rebuild the Chesapeake Bay's once huge population of the eastern oyster as a way to help clean up the water. Oysters are filter feeders that vacuum up the algae and nutrient-laden suspended silt that cause many of the Chesapeake's problems.

According to aquatic biologist Roger Newell, the bay's once prodigious oyster population served as a natural water purifier by filtering the bay's entire volume of water every 3 or 4 days. However, overharvesting and two parasitic oyster diseases have reduced the oyster population to about 1% of its historic high. As a result, today's oyster population needs about a year to filter the bay's water.

Computer models project that increasing the oyster population to only 10% of its historic high would improve water quality and spur the growth of underwater sea grass. Methods for restoring the bay's oyster population include **(1)** developing disease-resistant oyster stocks, **(2)** seeding protected oyster beds with large, older oysters presumed to have some disease resistance, **(3)** dumping hundreds of millions of oyster shells on dozens of historic reef areas with the goal of resurrecting some of the old oyster breeding reefs, **(4)** trying to find a reef-building substitute to hasten reef reconstruction, **(5)** setting aside 20–25% of the bay's oyster beds as sanctuaries to protect stocks from overfishing, and **(6)** greatly increasing funds for research and implementation of such a program.

Critical Thinking

Should more of the scarce funds for reducing pollution and degradation of the Chesapeake Bay be used to increase the oyster population? Explain.

---

prevent and control air pollution because an estimated 33% of all pollutants entering the ocean worldwide comes from air emissions from land-based sources.

Analysts have suggested the following measures to prevent and reduce excessive pollution of coastal waters:

Prevention

- *Use separate sewage and storm runoff lines in coastal urban areas.* Otherwise, excessive rainfall causes lines carrying sewage and storm runoff to overflow and release raw sewage into coastal waters.

- *Discourage or ban ocean dumping of sludge and hazardous dredged materials.*

- *Protect sensitive and ecologically valuable coastal areas from development, oil drilling, and oil shipping.*

- *Regulate coastal development.*

- *Require double hulls for all oil tankers.*

- *Recycle used oil.*

Cleanup

- *Improve oil spill cleanup capabilities.*

- *Require at least secondary treatment of coastal sewage, or use wetlands, solar aquatic (p. 476), or other treatment methods.*

19-5   SOLUTIONS: PREVENTING AND REDUCING SURFACE WATER POLLUTION

What Can We Do About Water Pollution from Nonpoint Sources? Ways to help control nonpoint water pollution, most of it from agriculture, include the following:

- Reduce fertilizer runoff into surface waters and leaching into aquifers by using slow-release fertilizer and using none on steeply sloped land.

- Reduce the need for fertilizer by alternating plantings between row crops and soybeans or other nitrogen-fixing plants.

- Plant buffer zones of vegetation between cultivated fields and nearby surface water.

- Reduce pesticide runoff by applying pesticides only when needed and using biological control or integrated pest management (p. 515).

- Control runoff and infiltration of manure from animal feedlots by **(1)** improving manure control, **(2)** planting buffers, and **(3)** not locating feedlots and animal waste on steeply sloped land near surface water and in flood zones.

---

costly approach, as revealed by the large spill from *Exxon Valdez* oil tanker in 1989 (Case Study, p. 336). However, a 1998 study by a National Aeronautics and Space Agency (NASA) scientist concluded that mesh pillows filled with 640 metric tons (700 tons) of hair could have soaked up the entire *Exxon Valdez* oil spill in a weeks, saving much of the $2 billion Exxon spent to capture only about 12% of the oil spilled.

Solutions: How Can We Protect Coastal Waters? The key to protecting oceans is to reduce the flow of pollution from the land and from streams emptying into the ocean. Such efforts must be integrated with efforts to

- Reduce soil erosion and flooding by reforesting critical watersheds.

Under the U.S. Clean Water Act, states are required to protect watersheds from nonpoint water pollution from farms and forests. States are supposed to designate waterways impaired by such pollution and develop a plan to curtail the pollution. However, a 2000 study by the National Wildlife Federation shows that 38 of the 50 states have done little to address nonpoint water pollution under this federal law.

What Can We Do About Water Pollution from Point Sources? The Legal Approach According to Sandra Postel, director of the Global Water Policy Project, most cities in developing countries discharge 80–90% of their untreated sewage directly into rivers, streams, and lakes, which are used for drinking water, bathing, and washing clothes.

In Latin America, less than 2% of urban sewage is treated. In China, only 4.5% of municipal wastewater and 17% of industrial discharge is treated. In India, only about 209 of its 3,100 largest cities have partial waste treatment plants, and only 8 have modern waste treatment plants.

In developed countries, most wastes from point sources are purified to varying degrees. The Federal Water Pollution Control Act of 1972 (renamed the Clean Water Act when it was amended in 1977) and the 1987 Water Quality Act form the basis of U.S. efforts to control pollution of the country's surface waters.

In 1995, the EPA developed a *discharge trading policy* designed to use market forces to reduce water pollution (as has been done with sulfur dioxide for air pollution control, p. 441). The policy would allow a water pollution source, such as an industrial plant or a sewage treatment plant, to sell credits for its excess reductions to another facility that cannot reduce its discharges as cheaply.

Here is some *good news*. The Clean Water Act of 1972 led to the following improvements in U.S. water quality between 1972 and 1998: **(1)** The percentage of U.S. rivers and lakes tested that are fishable and swimmable increased from 36% to 62%, **(2)** the amount of topsoil lost through agricultural runoff was cut by about 1.1 billion metric tons (1 billion tons) annually, **(3)** the proportion of the U.S. population served by sewage treatment plants increased from 32% to 74%, and **(4)** annual wetland losses decreased by 83%.

Here is some *bad news*: **(1)** About 44% of lakes, 38% of rivers (up from 26% in 1984), and 32% of estuaries in the United States are still unsafe for fishing, swimming, and other recreational uses, **(2)** of 27,300 surface waters analyzed in 1998, 12–17% were impaired by sediments, excessive nutrients, pathogens, and low oxygen

levels, **(3)** hog, poultry, and cattle farm runoff pollutes 70% of U.S. rivers, **(4)** despite aggressive monitoring nearly 110,000 metric tons (100,000 tons) of toxic industrial wastes are illegally dumped into U.S. rivers each year, **(5)** fish caught in more than 1,400 different waterways are unsafe to eat because of high levels of pesticides and other toxic substances, **(6)** antiquated sewage systems in 1,100 cities still dump poorly treated sewage into streams, lakes, and coastal waters, **(7)** less than 2% of the country's 5.8 million kilometers (3.6 million miles) of streams are healthy enough to be considered high quality, **(8)** 40% of the country's surface and groundwater is unsafe for human use, and **(9)** a 2000 study by Friends of the Earth and the Environmental Working Group found that officials in all but seven states have allowed industrial water pollution permits to expire, effectively giving discharging industries an unlimited license to pollute.

Some environmentalists call for the Clean Water Act to be strengthened by **(1)** increasing funding and authority to control nonpoint sources of pollution, **(2)** increasing monitoring of state programs to see that pollution permits are not allowed to expire, **(3)** strengthening programs to prevent and control toxic water pollution, **(4)** providing more funding and authority for integrated watershed and airshed planning to protect groundwater and surface water from contamination, **(5)** permitting states with good environmental records to take over parts of the clean water program, under looser federal control, and **(6)** expanding the rights of citizens to bring lawsuits to ensure that water pollution laws are enforced.

Many people oppose these proposals, contending that the Clean Water Act's regulations are already too restrictive and costly. Farmers and developers **(1)** see the law as a curb on their rights as property owners to fill in wetlands and **(2)** believe that they should be compensated for property value losses because of federal wetland protection regulations (p. 589). State and local officials want more discretion in testing for and meeting water quality standards. They argue that in many communities it is unnecessary and too expensive to test for all the water pollutants required by federal law.

What Can We Do About Water Pollution from Point Sources? The Technological Approach As population, urbanization, and industrialization grow, the volume of wastewater needing treatment will increase enormously. In rural and suburban areas with suitable soils, sewage from each house usually is discharged into a **septic tank** (Figure 19-14). About 25% of all homes in the United States are served by septic tanks, which should be cleaned out every 3–5 years by a reputable contractor so that they will not contribute to groundwater pollution.

Figure 19-14 Septic tank system used for disposal of domestic sewage and wastewater in rural and suburban areas. This system traps greases and large solids and discharges the remaining wastes over a large drainage field. As these wastes percolate downward, the soil filters out some potential pollutants, and soil bacteria decompose biodegradable materials. To be effective, septic tank systems must be **(1)** properly installed in soils with adequate drainage, **(2)** not placed too close together or too near well sites, and **(3)** pumped out when the settling tank becomes full.

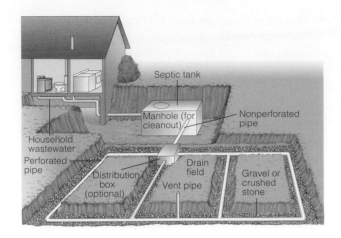

In urban areas, most waterborne wastes from homes, businesses, factories, and storm runoff flow through a network of sewer pipes to wastewater treatment plants. Some cities have separate lines for stormwater runoff, but in 1,200 U.S. cities the lines for these two systems are combined because it is cheaper. When rains cause combined sewer systems to overflow, they discharge untreated sewage directly into surface waters.

When sewage reaches a treatment plant, it can undergo up to three levels of purification. **Primary sewage treatment** is a *mechanical* process that uses screens to filter out debris such as sticks, stones, and rags and allows suspended solids to settle out as sludge in a settling tank (Figure 19-15). Improved primary treatment uses chemically treated polymers to remove suspended solids more thoroughly. By itself, primary treatment removes about 60% of the suspended solids and 30% of the oxygen-demanding organic wastes from sewage but removes no phosphates, nitrates, salts, radioisotopes, and pesticides.

Secondary sewage treatment is a *biological* process in which aerobic bacteria are used to remove up to 90% of biodegradable, oxygen-demanding organic wastes (Figure 19-15). Some treatment plants use *trickling filters*, in which aerobic bacteria degrade sewage as it seeps through a bed of crushed stones covered with bacteria and protozoa. Others use an *activated sludge process*, in which the sewage is pumped into a large tank and mixed for several hours with bacteria-rich sludge and air bubbles to facilitate degradation by microorganisms. The water then goes to a sedimentation tank, where most of the suspended solids and microorganisms settle out as sludge. The sludge produced by primary or secondary treatment is broken down in an anaerobic

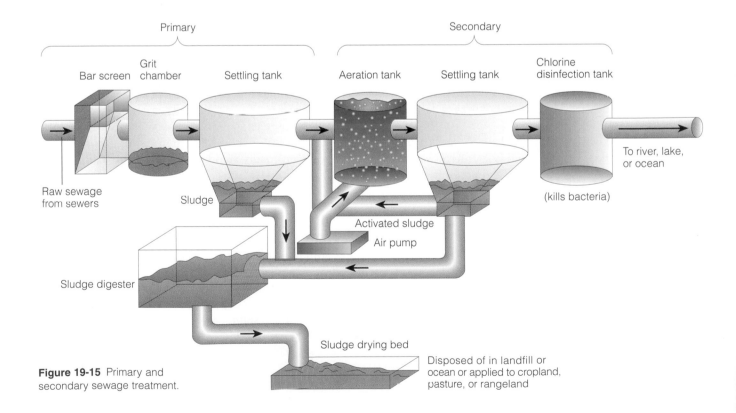

Figure 19-15 Primary and secondary sewage treatment.

digester and **(1)** incinerated, **(2)** dumped into the ocean or a landfill, or **(3)** applied to land as fertilizer.

A combination of primary and secondary treatment (Figure 19-15) removes about **(1)** 97% by weight of the suspended solids, **(2)** 95–97% of the oxygen-demanding organic wastes, **(3)** 70% of most toxic metal compounds and nonpersistent synthetic organic chemicals, **(4)** 70% of the phosphorus (mostly as phosphates), **(5)** 50% of the nitrogen (mostly as nitrates), and **(6)** 5% of dissolved salts. Almost no long-lived radioactive isotopes or persistent organic substances such as pesticides are removed.

As a result of the Clean Water Act, most U.S. cities have combined primary and secondary sewage treatment plants (Figure 19-15). However, government studies have found that **(1)** at least two-thirds of these plants have violated water pollution regulations, **(2)** 500 cities have failed to meet federal standards for sewage treatment plants, and **(3)** 34 East Coast cities simply screen out large floating objects from their sewage before discharging it into coastal waters.

Advanced sewage treatment is a series of specialized chemical and physical processes that remove specific pollutants left in the water after primary and

secondary treatment (Figure 19-16). Advanced treatment is rarely used because such plants typically cost twice as much to build and four times as much to operate as secondary plants.

Before water is discharged after primary, secondary, or advanced treatment, it is bleached (to remove water coloration) and disinfected (to kill disease-carrying bacteria and some but not all viruses). The usual method for doing this is *chlorination*. However, chlorine can react with organic materials in water to form small amounts of chlorinated hydrocarbons, some of which cause cancers in test animals and may damage the human nervous, immune, and endocrine systems (Connections, p. 404). Use of other disinfectants, such as ozone and ultraviolet light, is increasing, but they cost more than chlorination and are not as long lasting.

What Should We Do with Sewage Sludge? Sewage treatment produces a toxic, gooey *sludge*. In the United States, about **(1)** 9% by weight is converted to compost for use as a soil conditioner and **(2)** 36% is applied to farmland, forests, golf courses, cemeteries, parkland, highway medians, and degraded land as fertilizer. The remaining 55% is **(1)** dumped in conventional landfills

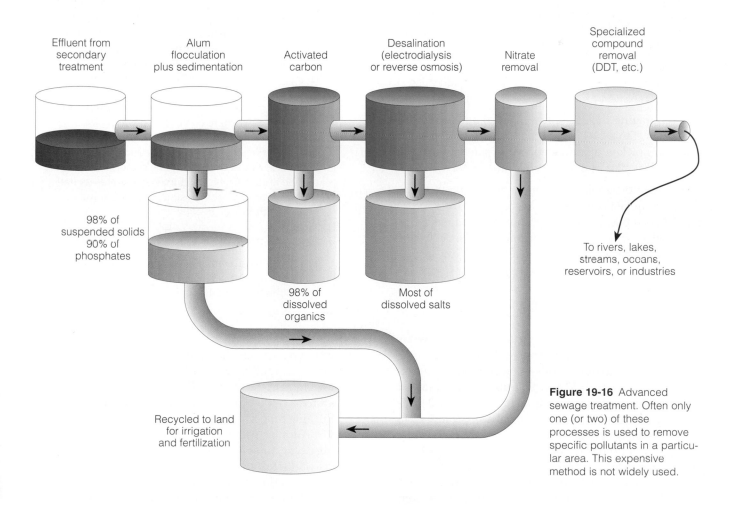

Figure 19-16 Advanced sewage treatment. Often only one (or two) of these processes is used to remove specific pollutants in a particular area. This expensive method is not widely used.

(where it can contaminate groundwater) or **(2)** incinerated (which can pollute the air with traces of toxic chemicals and produces a toxic ash, which usually is buried in landfills that EPA experts say will leak eventually).

From an environmental standpoint, it is desirable to recycle the plant nutrients in sewage sludge to the soil on land not used to grow food crops. As long as harmful bacteria and toxic chemicals are not present or are removed, sludge can also be used to fertilize land used for food crops or livestock. However, removing bacteria (usually by heating), toxic metals, and organic chemicals is expensive and is rarely done in the United States.

A growing number of health problems and lawsuits have resulted from use of sludge to fertilize crops in the United States. To protect consumers and avoid lawsuits, some food packers such as DelMonte and Heinz have banned produce grown on farms using sludge as a fertilizer. Also, farm credit bureaus are refusing to finance farms that use sludge because of the financial risks from contaminated soils.

Environmental scientist Peter Montague (Guest Essay, p. 526) believes that the entire sewage treatment approach should be redesigned with the goal of preventing toxic and hazardous waste from reaching sewage treatment plants. This would be accomplished by:

- Requiring industries and businesses to remove all toxic and hazardous wastes from water sent to municipal sewage treatment plants.

- Encouraging or requiring industries to reduce or eliminate toxic chemical use and waste through clean production (Guest Essay, p. 526, and Solutions, p. 525).

- Encouraging the use of less harmful household chemicals (Table 21-1, p. 521) and having harmful chemicals picked up and disposed of safely on a regular basis as part of garbage collection services.

- Having households, apartment buildings, and offices eliminate sewage outputs by switching to modern composting toilet systems that are installed, maintained, and managed by professionals. Such systems would be cheaper than current sewage systems because they do not require vast systems of underground pipes connected to centralized sewage treatment plants and conserve large amounts of water.

How Can We Treat Sewage by Working with Nature? Some communities and individuals are seeking better ways to purify contaminated water by working with nature (p. 476 and Solutions, above). Mark Nelson, who spent 2 years inside the Biosphere 2 facility in Arizona (p. 740), has developed a small, low-tech, and inexpensive artificial wetland system to treat raw sewage from hotels, restaurants, and homes in developing countries (Figure 19-17). This *wastewater garden* system removes 99.9% of fecal coliform bacteria and more than 80% of the nitrates and phosphates from

Using Wetlands to Treat Sewage

SOLUTIONS

Waste treatment is one of the important ecological services provided by wetlands. More than 600 cities, towns, and industries in the United States now use natural and artificial wetlands to treat sewage as a low-tech, low-cost alternative to expensive waste treatment plants.

Some communities have created artificial wetlands to treat their water, as the residents of Arcata, California, did, led by Humboldt State University professors Robert Gearheart and George Allen.

In this coastal town of 17,000, some 63 hectares (155 acres) of wetlands has been created between the town and the adjacent Humboldt Bay. The marshes, developed on land that was once a dump, act as an inexpensive, natural waste treatment plant. The project cost less than half the estimated cost of a conventional treatment plant.

Here's how it works: First, sewage is held in sedimentation tanks, where the solids settle out as sludge that is removed and processed for use as fertilizer. The liquid is pumped into oxidation ponds, where remaining wastes are broken down by bacteria. After a month or so, the water is released into the artificial marshes, where it is further filtered and cleansed by plants and bacteria.

Although the water is clean enough to be discharged directly into the bay, state law requires that it first be chlorinated. The town chlorinates the water and then dechlorinates it before sending it into the bay, where oyster beds thrive.

The marshes and lagoons also serve as an Audubon Society bird sanctuary and provide habitats for thousands of otters, seabirds, and marine animals. The town even celebrates its natural sewage treatment system with an annual "Flush with Pride" festival.

Critical Thinking

List some possible drawbacks to creating artificial wetlands to treat sewage. Do these drawbacks outweigh the benefits? Explain.

incoming sewage that in most developing countries is often dumped untreated into the ocean or into shallow holes in the ground. The water flowing out of such systems can be used to irrigate gardens or fields or for flushing toilets and thus helps save water.

Another promising new technology for processing domestic sewage in developing countries is a *double-vault treatment system*. In this approach, human wastes and organic kitchen wastes are deposited in one chamber and later moved to a second chamber,

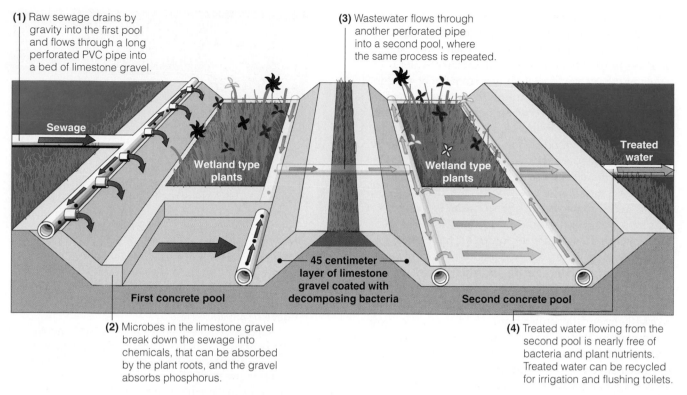

(1) Raw sewage drains by gravity into the first pool and flows through a long perforated PVC pipe into a bed of limestone gravel.

(3) Wastewater flows through another perforated pipe into a second pool, where the same process is repeated.

Sewage

Wetland type plants

Wetland type plants

Treated water

45 centimeter layer of limestone gravel coated with decomposing bacteria

First concrete pool

Second concrete pool

(2) Microbes in the limestone gravel break down the sewage into chemicals, that can be absorbed by the plant roots, and the gravel absorbs phosphorus.

(4) Treated water flowing from the second pool is nearly free of bacteria and plant nutrients. Treated water can be recycled for irrigation and flushing toilets.

Figure 19-17 *Wastewater garden.* This small, artificial, gravity-fed wetland system for treating sewage from hotels, restaurants, and homes in developing countries was developed by the creators of the Biosphere 2 facility in Arizona (p. 740). This wastewater treatment system uses only 1.9–3.8 square meters (20–30 square feet) of space per person.

where they are allowed to compost for several months. Solar heating and bacteria convert the waste in the second chamber into a safe and odorless soil conditioner that can be used on household gardens or sold to nearby farms. When the soil conditioner is removed, the wastes stored into the first holding chamber are transferred to the composting chamber. Such dry composting units use no water and are small enough to serve one or two families.

Other double-vault units are neighborhood mini-plants that biologically process the wet or water-flushed wastes of up to 1,000 people. These systems separate gray water from sewage and percolate the water through a bed of sand and gravel until it is pure enough to be used on gardens or to irrigate flowers, grass, or trees. Such systems typically cost one-seventh as much as a conventional sewer and waste treatment plant system and can produce fertilizer that can be sold to recoup the startup costs.

In addition to *living machine* sewage treatment systems (p. 476), scientists at Living Technologies have developed neighborhood-level *sewage walls* that would run along the length of a residential block. Sewage would be channeled through a series of four terraced planters that progressively filter and purify the waste. Each planter would be capped with glass to allow use of sunlight and contain the bacteria and plants best suited for the various stages of treatment. The resulting

effluent could be used on local gardens, and the plants could be harvested periodically and converted to compost for use on neighborhood gardens.

Another approach is to use wastewater to grow forests. While the trees are growing, this approach can also remove carbon dioxide from the atmosphere and help slow global warming.

19-6 DRINKING WATER QUALITY

Is the Water Safe to Drink? About one-fourth of people in developing countries do not have access to clean drinking water. In China, an estimated 700 million people drink contaminated water, and only 6 of China's 27 largest cities provide drinking water that meets government standards. Contaminated drinking water is considered a key factor in the doubling of liver disease and cancer deaths in China since 1970. In Russia, half of all tap water is unfit to drink, and a third of the aquifers are too contaminated for drinking purposes. About 290 million Africans—about equal to the entire U.S. population—do not have access to safe drinking water.

In many poor villages in developing countries, people get their water from **(1)** shallow groundwater wells that are easily contaminated, **(2)** nearby polluted river water, or **(3)** mudholes used by both animals and humans.

In most urban slums in developing countries, drinking water is not pumped in or the poor there cannot afford a house connection. Such poor urban dwellers must either **(1)** drink contaminated water from rivers or other sources or **(2)** buy it from street vendors at an average cost of 12 times more per liter than middle-class families pay for water piped to their houses. A 1999 study by the World Commission on Water for the 21st Century found that much of the water sold by urban street vendors is drawn from polluted rivers or other contaminated sources.

The United Nations estimates that it would cost about $25 billion a year over 8–10 years to bring low-cost safe water and sanitation to the 1.4 billion people—one of every four—who do not have access to clean drinking water. These expenditures could prevent many of the 5 million deaths (including 2 million children under age 5) and 3.4 billion cases of illness caused each year by unsafe water. Currently, the world is spending only about $8 billion a year on clean water efforts. The $17-billion shortfall is about equal to what people in Europe and the United States spend each year on pet food or about what the world spends every 8 days for military purposes. Researchers are trying to find cheap and simple ways to purify drinking water in developing nations (Individuals Matter, right).

How Is Drinking Water Purified? Treatment of water for drinking by city dwellers is much like wastewater treatment. Areas that depend on surface water usually store it in a reservoir for several days to improve clarity and taste by allowing the dissolved oxygen content to increase and suspended matter to settle out. The water is then pumped to a purification plant, where it is treated to meet government drinking water standards. Usually the water is run through sand filters and activated charcoal before it is disinfected. In areas with very pure groundwater sources, little treatment is necessary.

How Is the Quality of Drinking Water Protected? About 54 countries, most of them in North America and Europe, have safe drinking water standards. The U.S. Safe Drinking Water Act of 1974 requires the EPA to establish national drinking water standards, called *maximum contaminant levels*, for any pollutants that may have adverse effects on human health.

Privately owned wells are not required to meet federal drinking water standards, primarily because of the costs of testing each well regularly (at least $1,000) and opposition to mandatory testing and compliance by some homeowners.

It is difficult to estimate how many people in the United States get sick or die each year from drinking

INDIVIDUALS MATTER

Using UV Light, Horseradish, and Slimes to Purify Water

Ashok J. Gadgil, a physicist at California's Lawrence Berkeley National Laboratory, and his colleagues have recently developed a simple device that uses ultraviolet light to kill disease-causing organisms in drinking water. When water from a well or hand pump (in a village or household in a developing country) is passed through this tabletop system, UV radiation from a mercury vapor lamp zaps germs in the water.

This $300 device weighs only 7 kilograms (15 pounds) and can disinfect 57 liters (15 gallons) of water per minute at a very low cost. It draws only 40 watts of power, supplied by solar cells, and can run unsupervised in remote areas of developing countries.

Recently, Pennsylvania State soil biochemists discovered that chopped horseradish mixed with hydrogen peroxide (H_2O_2) helps rid contaminated water of organic pollutants called *phenols*. The horseradish contains an enzyme that speeds up the breakdown of the phenols.

Judith Bender and Peter Phillips at Clark Atlanta University in Georgia have found a way to use slime produced by cyanobacteria to decompose chlorinated hydrocarbons that contaminate drinking water. Within 3 weeks, a slimy, floating bacterial mat can surround and decompose a glob of toxic chlordane (a pesticide banned in the United States as a suspected carcinogen). Such slime mats can also remove lead, copper, chromium, cadmium, selenium, and other toxic metals from water.

contaminated water. Estimates include **(1)** 7 million illnesses and 1,200 deaths per year according to a 1994 study by the Natural Resources Defense Council, **(2)** 1 million illnesses and about 900 deaths according to a 1997 report by the Centers for Disease Control and Prevention, and **(3)** 239,000 illnesses and 50 deaths according to a 1999 EPA study.

In April 1993, a dramatic incident occurred when residents of Milwaukee, Wisconsin, were told that water from their taps was unsafe to drink. Before this health crisis was over, 111 people had died and 403,000 people had become sick as a result of exposure to *Cryptosporidium*, a parasitic organism. Milwaukee spent an estimated $54 million to deal with this outbreak. In May 1994, another outbreak of this parasite killed 19 and sickened more than 100 people in Las Vegas, Nevada.

After these incidents, the EPA stepped up efforts to require public water systems to detect and prevent contamination by disease-causing pathogens such as *E. coli* and *Cryptosporidium* in drinking water obtained from surface sources such as lakes and rivers. In 2000, the EPA proposed new rules that would require public water systems getting all or part of their water from underground aquifers to monitor more closely for such organisms and use disinfectants if such threats are found. According to the EPA, these new regulations are expected to prevent more than 115,000 illnesses a year.

In 1999, an EPA audit found that communities and states were not reporting about 88% of all violations of the Safe Drinking Water Act to the government database used to alert consumers and trigger legal actions when water systems do not meet health standards. These violations range from missed water quality tests to contamination problems.

According to the Natural Resources Defense Council (NRDC), U.S. drinking water supplies could be made safer at a cost of only about $30 a year per household. However, Congress is being pressured by water-polluting industries to weaken the Safe Drinking Water Act by (1) eliminating national tests of drinking water, (2) eliminating the requirement that the media be advised of emergency water health violations and that water system officials notify their customers of such violations, (3) allowing states to give drinking water systems a permanent right to violate the standard for a given contaminant if the provider claims it cannot afford to comply, and (4) eliminating the requirement that water systems use affordable, feasible technology to remove cancer-causing contaminants.

Environmentalists call for the U.S. Safe Drinking Water Act to be strengthened by (1) improving treatment by combining at least half of the 50,000 water systems that serve fewer than 3,300 people each with larger ones nearby, (2) strengthening and enforcing public notification requirements about violations of drinking water standards, and (3) banning all lead in new plumbing pipes, faucets, and fixtures (current law allows fixtures with up to 10% lead to be sold as lead-free).

Is Bottled Water the Answer? Despite some problems, experts say that the United States has some of the world's cleanest drinking water. Yet about half of all Americans worry about getting sick from tap water contaminants, and many drink bottled water or install expensive water purification systems. Studies indicate that many of these consumers are being ripped off and in some cases may end up drinking water that is dirtier than water they can get from their taps.

An estimated one-third of the bottled water purchased in the United States is contaminated with bacteria. To be safe, consumers purchasing bottled water should determine whether the bottler belongs to the International Bottled Water Association (IBWA) and adheres to its testing requirements.* Some companies pay $2,500 annually to obtain more stringent certification by the National Sanitation Foundation, an independent agency that tests for 200 chemical and biological contaminants.

Before drinking expensive bottled water and buying costly home water purifiers, health officials suggest that consumers have their water tested by local health authorities or private labs (not companies trying to sell water purification equipment) to (1) identify what contaminants, if any, must be removed and (2) recommend the type of purification needed to remove such contaminants. Independent experts contend that unless tests show otherwise, for most urban and suburban Americans served by large municipal drinking water systems, home water treatment systems are not worth the expense and maintenance hassles.

Buyers should carefully check out companies selling water purification equipment and be wary of claims that the EPA has approved a treatment device. Although the EPA does *register* such devices, it neither tests nor approves them.

How Can We Reduce Water Pollution? An Integrated Approach According to environmentalists, a sustainable approach to dealing with water pollution requires that we shift our emphasis from pollution cleanup to pollution prevention by (1) *reducing* the toxicity or volume of pollutants (for example, replacing organic solvent-based inks and paints with water-based materials), (2) *reusing* wastewater instead of discharging it (for example, reusing treated wastewater for irrigation), and (3) *recycling* pollutants (for example, cleaning up and recycling contaminated solvents for reuse) instead of discharging them.

To make such a shift, we need to accept that the environment—air, water, soil, and life—is an interconnected whole. Without an integrated approach to all forms of pollution, environmentalists argue that we will continue to shift environmental problems from one part of the environment to another. Some actions you can take to help reduce water pollution are listed in Appendix 6.

It is a hard truth to swallow, but nature does not care if we live or die. We cannot survive without the oceans, for example, but they can do just fine without us.

ROGER ROSENBLATT

---

*Check for the IBWA seal of approval on the bottle or contact the International Bottled Water Association (113 North Henry Street, Alexandria, VA 22314; phone: 703-683-5213) for a member list.

REVIEW QUESTIONS

1. Define the boldfaced terms in this chapter.

2. Describe John Todd's *living machines* used to purify sewage.

3. What is *water pollution*? Describe how a *coliform bacteria count*, measurement of *biological oxygen demand*, and *biological indicators* can be used to determine water quality. What are eight types of water pollutants, and what are the major sources and effects of each type?

4. Distinguish between *point* and *nonpoint sources of water pollution* and give two examples of each type. Which type is easier to control? Why?

5. What are the major water pollution problems of streams? Explain how streams can handle some loads of biodegradable wastes, and explain the limitations of this approach.

6. Summarize the good and bad news about attempts to prevent or control stream pollution.

7. What are the major water pollution problems of lakes? Distinguish between *eutrophication* and *cultural eutrophication*. What are the major causes of cultural eutrophication? List three methods for **(a)** preventing cultural eutrophication and **(b)** cleaning up cultural eutrophication.

8. Summarize the good and bad news about attempts to reduce water pollution in the Great Lakes.

9. Explain how climate change from projected global warming can decrease the quality of surface water.

10. List five major sources of groundwater contamination. List three reasons why groundwater pollution is such a serious problem.

11. List five examples indicating the seriousness of groundwater pollution. List four ways to prevent groundwater contamination.

12. List the major pollution problems of the oceans. Why are most of these problems found in coastal areas?

13. Describe the ocean pollution problems caused by large inputs of plant nutrients from river systems.

14. Summarize the major pollution problems of the Chesapeake Bay in the United States and the progress made in dealing with these problems.

15. Distinguish between ocean pollution from *dredge spoils* and from *sewage sludge*. Distinguish between *crude petroleum* and *refined petroleum* and summarize the major water pollution problems caused by oil. What are the three major sources of oil pollution in the world's oceans? List six ways to help prevent oil pollution of the oceans and two ways to help clean up such pollution.

16. List six ways to help prevent water pollution from nonpoint sources. Why has there been so little emphasis on dealing with this problem?

17. Explain how the United States and most developed countries have reduced water pollution from point sources by enacting laws, and summarize the good and bad news about such efforts. List six ways in which environmentalists believe water pollution control laws in the United States should be strengthened, and list two reasons why there is opposition to such changes.

18. Distinguish between *septic tanks*, *primary sewage treatment*, *secondary sewage treatment*, and *advanced sewage treatment* as ways to reduce water pollution. List ways to deal with the sludge produced by waste treatment methods. What are the pros and cons of each approach?

19. Describe three ways to treat sewage based on working with nature.

20. What percentage of the world's people does not have access to clean drinking water? About how many children under age 5 die each year from infectious diseases caused by drinking contaminated water?

21. How is drinking water purified? How is the quality of drinking water protected in the United States? How successful have these efforts been? List three ways in which environmentalists believe the U.S. Safe Drinking Water Act should be strengthened and four ways in which opponents believe it should be weakened. List the pros and cons of drinking bottled water.

22. List three ways to shift the emphasis from cleanup to prevention of water pollution.

CRITICAL THINKING

1. Why is dilution not always the solution to water pollution? Give examples and conditions for which this solution is or is not applicable.

2. How can a stream cleanse itself of oxygen-demanding wastes? Under what conditions will this natural cleansing system fail?

3. Which of the eight categories of pollutants listed in Table 19-1 are most likely to originate from **(a)** point sources and **(b)** nonpoint sources?

4. A large number of fish are found floating dead on a lake during the summer. You are called to determine the cause of the fish kill. What reason would you suggest for the kill? What measurements would you make to verify your hypothesis?

5. Explain why a number of communities around the Great Lakes have banned the use of phosphate-containing detergents in recent years.

6. Should injection of liquid hazardous wastes into deep wells below drinking water aquifers (Figure 19-9) be banned? Explain. What are the alternatives?

7. Explain why some health officials project that groundwater pollution could be one of the world's most serious health problems between 2050 and 2100.

8. Should all dumping of wastes and untreated sewage in the ocean be banned? Explain. If so, where would you put the wastes instead? What exceptions would you permit, and why? How would you enforce such regulations?

9. Your town (Town B) is located on a river between towns A and C. What are the rights and responsibilities

of upstream communities to downstream communities? Should sewage and industrial wastes be dumped at the upstream end of a community that generates them?

10. Congratulations. You have been placed in charge of sharply reducing nonpoint water pollution throughout the world. What are the three most important things that you would do?

11. Congratulations. You have been placed in charge of sharply reducing groundwater pollution throughout the world. What are the three most important things you would do?

PROJECTS

1. In your community,
  a. What are the principal nonpoint sources of contamination of surface water and groundwater?
  b. What is the source of drinking water?
  c. How is drinking water treated?
  d. How many times during each of the past 5 years have levels of tested contaminants violated federal standards? Was the public notified about the violations?
  e. Is fishing prohibited in any lakes or rivers in your region because of pollution? Are people warned about this?
  f. Is groundwater contamination a problem? If so, where, and what has been done about the problem?
  g. Is there a vulnerable aquifer or critical recharge zone that should be protected to ensure the quality of groundwater? Is your local government aware of this? What action (if any) has it taken?

2. Are storm drains and sanitary sewers combined or separate in your area? Are there plans to reduce pollution from stormwater runoff? If not, make an economic evaluation of the costs and benefits of developing separate storm drains and sanitary sewers, and then present your findings to local officials.

3. Arrange a class or individual tour of a sewage treatment plant in your community. Compare the processes it uses with those shown in Figure 19-15. What happens to the sludge produced by this plant? What improvements, if any, would you suggest for this plant?

4. Use library research, the internet, and user interviews to evaluate the relative effectiveness and costs of home water purification devices. Determine the type or types of water pollutants each device removes and the effectiveness of this process.

5. Use the library or the internet to find bibliographic information about *William Ruckelshaus* and *Roger Rosenblatt*, whose quotes appear at the beginning and end of this chapter.

6. Make a concept map of this chapter's major ideas, using the section heads and subheads and the key terms (in boldface). Look at the inside back cover and on the website for this book for information about making concept maps.

INTERNET STUDY RESOURCES AND RESOURCES FOR FURTHER READING AND RESEARCH

The website for this book contains helpful study aids and many ideas for further reading and research. Log on to:

  http://www.brookscole.com/product/0534376975s

and click on the Chapter-by-Chapter area. Choose Chapter 19 and select a resource:

■ "Flash Cards" allows you to test your mastery of the Terms and Concepts to Remember for this chapter.

■ "Tutorial Quizzes" provides a multiple-choice practice quiz.

■ "Student Guide to InfoTrac" will lead you to Critical Thinking Projects that use InfoTrac College Edition as a research tool.

■ "References" lists the major books and articles consulted in writing this chapter.

■ "Hypercontents" takes you to an extensive list of sites with news, research, and images related to individual sections of the chapter.

INFOTRAC COLLEGE EDITION

Improve your skills with InfoTrac College Edition, a searchable online database of articles from more than 700 periodicals. Log on to:

  http://www.infotrac-college.com

or access InfoTrac through the website for this book.

Try the following articles:

1999. Watershed management is key to improving America's water resources. *Journal of Environmental Health* vol. 61, no. 9, pp. 43–44. (subject guide: water quality, standards)

Kreeger, K. 2000. Down on the fish farm: developing effluent standards for aquaculture. *BioScience* vol. 50, no. 11, pp. 949–953. (keywords: aquaculture, effluent)

20 PESTICIDES AND PEST CONTROL

Along Came a Spider

Since about 10,000 years ago, when agriculture began, we have been competing with insect pests for the food we grow. Today we are not much closer to winning this competition than we were then. The reasons are the astounding ability of insect pests to multiply (p. 70) and to rapidly develop genetic resistance to poisons we throw at them through directional natural selection (Figure 5-6, left, p. 110).

Some Chinese farmers recently decided to switch from a chemical to a biological strategy to help control insect pests. Instead of spraying their rice and cotton fields with poisons, they build little straw huts around the fields in the fall.

These farmers are encouraging insects' worst enemy, one that has hunted them for millions of years: *spiders* (Figure 20-1). The little huts are for hibernating spiders. Protected from the worst of the cold by the huts, far more of the hibernating spiders become active in the spring. Ravenous after their winter fast, they scuttle off into the fields to stalk their insect prey.

Even without human help, the world's 30,000 known species of spiders kill far more insects every year than insecticides do. A typical acre of meadow or woods contains an estimated 50,000 to 2 million spiders, each devouring hundreds of insects per year.

Entomologist Willard H. Whitcomb found that leaving strips of weeds around cotton and soybean fields provides the kind of undergrowth favored by insect-eating wolf spiders (Figure 20-1, left). He also sings the praises of a type of banana spider, which lives in warm climates and can keep a house clear of cockroaches (Spotlight, p. 112).

In Maine, Daniel Jennings of the U.S. Forest Service uses spiders to help control the spruce budworm, which devastates the Northeast's spruce and fir forests. Spiders also attack the much-feared gypsy moth, which destroys tree foliage.

The idea of encouraging populations of spiders in fields, forests, and even houses scares some people because spiders have bad reputations. From a human standpoint, however, spiders are helpful and mostly harmless creatures.

A few spider species, such as the black widow, the brown recluse, and eastern Australia's Sydney funnel web, are dangerous to people. However, most spider species, including the ferocious-looking wolf spider (Figure 20-1, left), do not harm humans. Even the giant tarantula rarely bites people, and its venom is too weak to harm us or other large mammals. As we seek new ways to coexist with the insect rulers of the planet (p. 70), we should be sure that spiders are in our corner.

This chapter looks first at the pros and cons of the conventional chemical approach to pest control based on using synthetic chemical pesticides. Then it discusses the pros and conse of a variety of biological and ecological alternatives for controlling pest populations.

Figure 20-1 Spiders are insects' worst enemies. Most spiders, such as the crab spider (below) and the wolf spider (left), found in many parts of the world, are harmless to humans. (Left, James C. Cokendolpher; right, Dan Kline/Visuals Unlimited)

A weed is a plant whose virtues have not yet been discovered.
RALPH WALDO EMERSON

This chapter addresses the following questions:

- What are pesticides, and what types are used?
- What are the pros and cons of using chemicals to kill insects and weeds?
- How well is pesticide use regulated in the United States?
- What alternatives are there to using conventional pesticides, and what are the advantages and disadvantages of each alternative?

20-1 PESTICIDES: TYPES AND USES

How Does Nature Keep Pest Populations Under Control? A **pest** is any species that (1) competes with us for food, (2) invades lawns and gardens, (3) destroys wood in houses, (4) spreads disease, or (5) is simply a nuisance. In natural ecosystems and many polyculture agroecosystems (p. 265), *natural enemies* (predators, parasites, and disease organisms) control the populations of 50–90% of pest species as part of the earth's natural capital and help keep any one species from taking over for very long.

When we replace polyculture agriculture with monoculture agriculture and spray fields with massive doses of pesticides, we upset many of these natural population checks and balances. Then we must devise ways to protect our monoculture crops, tree farms, and lawns from insects and other pests that nature once controlled at no charge.

To help control pest organisms, we have developed a variety of **pesticides** (or *biocides*): chemicals to kill organisms we consider undesirable. Common types of pesticides include (1) *insecticides* (insect-killers), (2) *herbicides* (weed-killers), (3) *fungicides* (fungus-killers), (4) *nematocides* (roundworm-killers), and (5) *rodenticides* (rat- and mouse-killers).

We did not invent the use of chemicals to repel or kill other species; plants have been producing chemicals to ward off or poison herbivores that feed on them for about 225 million years. This is a never-ending, ever-changing process: Herbivores overcome various plant defenses through natural selection; then the plants use natural selection to develop new defenses. The result of these dynamic interactions between predator and prey species is what biologists call *coevolution* (p. 109).

What Was the First Generation of Pesticides and Repellents? As the human population grew and agriculture spread, people began looking for ways to protect their crops, mostly by using chemicals to kill or repel insect pests. Sulfur was used as an insecticide well before 500 B.C.; by the 1400s, toxic compounds of arsenic, lead, and mercury were being applied to crops as insecticides. Farmers abandoned this approach in the late 1920s when the increasing number of human poisonings and fatalities prompted a search for less toxic substitutes. However, traces of these nondegradable toxic metal compounds are still being taken up by tobacco, vegetables, and other crops grown on soil dosed with them long ago.

In the 1600s, nicotine sulfate, extracted from tobacco leaves, came into use as an insecticide. In the mid-1800s two more natural pesticides were introduced: (1) *pyrethrum*, obtained from the heads of chrysanthemum flowers, and (2) *rotenone*, from the roots of various tropical forest legumes. These *first-generation pesticides* were mainly natural substances, chemicals borrowed from plants that had been defending themselves from insects for eons.

In addition to protecting crops, people have used chemicals (produced by plants) to (1) repel or kill insects in their households, yards, and gardens, (2) save money, and (3) reduce potential health hazards associated with using some commercial insecticides. Natural chemicals or methods to repel or kill common pests such as ants, mosquitoes, cockroaches, flies, and fleas and control weeds are listed in Appendix 6.

What Is the Second Generation of Pesticides? A major pest control revolution began in 1939, when entomologist Paul Müller discovered that DDT (dichlorodiphenyltrichloroethane), a chemical known since 1874, was a potent insecticide. DDT, the first of the so-called *second-generation pesticides*, soon became the world's most-used pesticide, and Mueller received the Nobel Prize in 1948 for his discovery. Since 1945 chemists have developed hundreds of synthetic organic chemicals for use as pesticides.

Since 1950 pesticide use has increased more than 50-fold, and most of today's pesticides are more than 10 times as toxic as those used in the 1950s. Worldwide, about 2.3 million metric tons (2.5 million tons) of second-generation pesticides are used yearly. About 75% of these chemicals are used in developed countries, but use in developing countries is soaring.

In the United States, about 630 different biologically active (pest-killing) ingredients and about 1,820 inert (inactive) ingredients are mixed to make some 25,000 different pesticide products. About 25% of pesticide use in the United States is for ridding houses, gardens, lawns, parks, playing fields, swimming pools, and golf courses of pests. According to the U.S. Environmental Protection Agency (EPA), the average lawn in the United States is doused with 10 times more synthetic pesticides per hectare than U.S. cropland.

The EPA estimates that 84% of U.S. homes use pesticide products such as bait boxes, pest strips, bug bombs, flea collars, and pesticide pet shampoos and weed killers for lawns and gardens. Each year, more than 250,000 people in the United States become ill because of household use of pesticides, and such pesticides are a major source of accidental poisonings and deaths for children under age 5.

Some pesticides, called *broad-spectrum agents*, are toxic to many species; others, called *selective* or *narrow-spectrum agents*, are effective against a narrowly defined group of organisms. Pesticides vary in their *persistence*, the length of time they remain deadly in the environment (Table 20-1). In 1962, biologist Rachel Carson warned against relying on synthetic organic chemicals to kill insects and other species we deem pests (Individuals Matter, p. 36).

20-2 THE CASE FOR PESTICIDES

Proponents of conventional chemical pesticides contend that their benefits outweigh their harmful effects. Here are some of the major benefits of conventional pesticides:

- *They save human lives.* Since 1945, DDT and other chlorinated hydrocarbon and organophosphate insecticides have probably prevented the premature deaths of at least 7 million people from insect-transmitted diseases such as (1) malaria (carried by the *Anopheles* mosquito; Figure 16-11, p. 409), (2) bubonic plague (rat fleas), (3) typhus (body lice and fleas), and (4) sleeping sickness (tsetse fly).

- *They increase food supplies and lower food costs.* About 55% of the world's potential human food supply is lost to pests before (35%) or after (20%) harvest. Pests before and after harvest destroy an estimated 37% of the potential U.S. food supply; insects cause 13% of these losses, plant pathogens 12%, and weeds 12%. As a result, food production for humans and livestock worth at least $65 million a year is lost to pests. Without pesticides, these losses would be worse, and food prices would rise. Figure 20-2 shows five of the most common insect pests in the United States and their ranges.

Table 20-1 Major Types of Pesticides

Type	Examples	Persistence	Biologically Magnified?
Insecticides			
Chlorinated hydrocarbons	DDT, aldrin, dieldrin, toxaphene, lindane, chlordane, methoxychlor, mirex	High (2–15 years)	Yes (Figure 16-4, p. 399)
Organophosphates	Malathion, parathion, diazinon, TEPP, DDVP, mevingphos	Low to moderate (1–2 weeks), but some can last several years	No
Carbamates	Aldicarb, carbaryl (Sevin), propoxur, maneb, zineb	Low (days to weeks)	No
Botanicals	Rotenone, pyrethrum, and camphor extracted from plants, synthetic pyrethroids (variations of pyrethrum) and rotenoids (variations of rotenone)	Low (days to weeks)	No
Microbotanicals	Various bacteria, fungi, protozoa	Low (days to weeks)	No
Herbicides			
Contact chemicals	Atrazine, simazine, paraquat	Low (days to weeks)	No
Systemic chemicals	2,4-D, 2,4,5-T, Silvex, diruon, daminozide (Alar), alachlor (Lasso), glyphosate (Roundup)	Mostly low (days to weeks)	No
Soil sterilants	Tribualin, diphenamid, dalapon, butylate	Low (days)	No
Fungicides			
Various chemicals	Captan, pentachlorphenol, zeneb, methyl bromide, carbon bisulfide	Most low (days)	No
Fumigants			
Various chemicals	Carbon tetrachloride, ethylene dibromide, methyl bromide	Mostly high	Yes (for most)

- *They work faster and better than alternatives.* Pesticides can **(1)** control most pests quickly and at a reasonable cost, **(2)** have a long shelf life, **(3)** are easily shipped and applied, and **(4)** are safe when handled properly. When genetic resistance occurs, farmers can use stronger doses or switch to other pesticides.

- *When used properly, their health risks are insignificant compared with their benefits.* According to Elizabeth Whelan, director of the American Council on Science and Health (ACSH), which presents the position of the pesticide industry, "The reality is that pesticides, when used in the approved regulatory manner, pose no risk to either farm workers or consumers." Pesticide company officials and scientists consider media reports describing pesticide health scares to be distorted science and irresponsible reporting.

- *Newer pesticides are safer and more effective than many older pesticides.* Greater use is being made of botanicals and microbotanicals (Table 20-1), derived originally from plants, that are safer to users and less damaging to the environment. Genetic engineering is also being used to develop pest-resistant crop strains and genetically altered crops that produce pesticides (Pro/Con, p. 275).

- *Many new pesticides are used at very low rates per unit area compared to older products.* For example, application amounts per hectare for many new herbicides are $\frac{1}{100}$ the rates for older ones.

Scientists continue to search for the ideal pest-killing chemical, which would:

- *Kill only the target pest*
- *Harm no other species*
- *Disappear or break down into something harmless after doing its job*
- *Not cause genetic resistance in target organisms*
- *Be more cost-effective than doing nothing*

The search continues, but so far no known natural or synthetic pesticide chemical meets all or even most of these criteria. Conventional pesticides have disadvantages (Section 20-3, right), but so do the alternatives (as discussed in Section 20-5, p. 511). In each case, the question is whether the advantages outweigh the disadvantages.

20-3 THE CASE AGAINST PESTICIDES

How Do Pesticides Cause Genetic Resistance?
Opponents of widespread pesticide use believe that their harmful effects outweigh their benefits. The biggest problem is the development of *genetic resistance* to pesticides by pest organisms. Insects breed rapidly

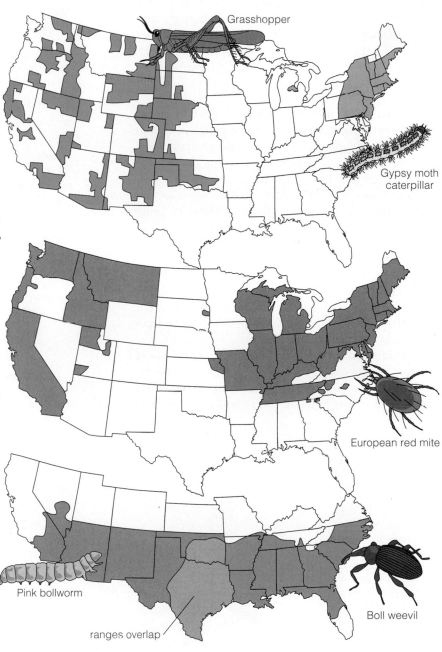

Figure 20-2 Geographic range of five major pests in the lower 48 states of the United States. (Data from U.S. Department of Agriculture)

(Figure 20-3), and within 5–10 years (much sooner in tropical areas) they can develop immunity to pesticides through directional natural selection (Figure 5-6, left, p. 110) and come back stronger than before (see cartoon). Weeds and plant disease organisms also develop genetic resistance, but more slowly.

Since 1945, about 1000 major pest species of insects and mites, weeds, plant diseases, and rodents (mostly rats) have developed genetic resistance to one or more pesticides (Figure 20-4). At least 17 insect pest species are resistant to all major classes of insecticides, and several fungal plant diseases are immune to most widely used fungicides. DDT is now banned for agricultural use in at least 86 countries, but is still widely used for mosquito control in many developing countries. Because of genetic resistance, many insecticides (such as DDT) no longer protect people from insect-transmitted diseases, such as malaria (Figure 16-11, p. 409), in some parts of the world.

How Can Pesticides Kill Natural Pest Enemies and Create New Pests? Another problem is that broad-spectrum insecticides kill natural predators and parasites that may have been maintaining the population of a pest species at a reasonable level. With wolf spiders (Figure 20-1, left), wasps, predatory beetles, and other natural enemies out of the way, the population of a rapidly reproducing insect pest species can rebound and even get larger within days or weeks after initially being controlled.

Wiping out natural predators can also unleash new pests whose populations its predators had previously held in check, causing other unexpected effects (Connections, p. 208). Currently 100 of the 300 most destructive insect pests in the United States were secondary pests that became major pests after widespread use of insecticides. For example, after the use of pesticides the European red mite (Figure 20-2) became an important pest on apple trees in the northeastern United States.

What Is the Pesticide Treadmill? When genetic resistance develops, pesticide sales representatives usually recommend **(1)** more frequent applications, **(2)** larger doses, or **(3)** a switch to new chemicals to keep the resistant species under control. This can put farmers on a *pesticide treadmill*, whereby they may pay more and more for a pest control program that often becomes less and less effective as genetic resistance increases. Some biologists also warn that increasing dependence on genetically altered crops (transgenics) can put farmers on a similar *genetic treadmill* (Pro/Con, p. 275). Pesticide companies say that *pesticide treadmill* and *genetic treadmill* are highly emotional terms that ignore the important benefits of pesticides (p. 504).

David Pimentel (Guest Essay, p. 232), an expert in insect ecology, has evaluated data from more than 300 agricultural scientists and economists and come to the following conclusions:

- Although the use of synthetic pesticides has increased 33-fold since 1942, it is estimated that more of the U.S. food supply is lost to pests today (37%) than in the 1940s (31%). Losses attributed to insects almost doubled (from 7% to 13%) despite a 10-fold increase in the use of synthetic insecticides.

- The estimated environmental, health, and social costs of pesticide use in the United States range from $4 to $10 billion per year. The International Food Policy Research Institute puts the estimate much higher, at $100–200 billion per year, or $5–10 in damages for every dollar spent on pesticides.

- Alternative pest control practices (Section 20-5) could halve the use of chemical pesticides on 40 major U.S. crops without reducing crop yields.

Figure 20-3 A boll weevil, just one example of an insect capable of rapid breeding. In the cotton fields of the southern United States, these insects lay thousands of eggs, producing a new generation every 21 days and as many as six generations in a single growing season. Attempts to control the cotton boll weevil account for at least 25% of insecticide use in the United States. For example, it typically takes about 114 grams (one-quarter pound) of pesticides to make one cotton T-shirt. Some farmers are increasing their use of natural predators and other biological methods to control this major pest. (U.S. Department of Agriculture)

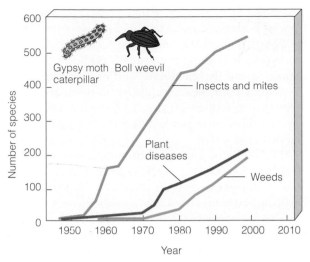

Figure 20-4 Rise of genetic resistance to pesticides, 1945–98. (Data from U.S. Department of Agriculture and the Worldwide Institute)

Numerous studies and experience show that pesticide use can be reduced sharply without reducing yields, and in some cases yields even increase. Over the past few years, Sweden has cut pesticide use in half with almost no decrease in the harvest. Campbell Soup uses no pesticides on tomatoes it grows in Mexico, and yields have not dropped. After a 65% cut in pesticide use on rice in Indonesia, yields increased by 15%.

Where Do Pesticides Go? *Pesticides do not stay put.* According to the U.S. Department of Agriculture (USDA), no more than 2% (and often less than 0.1%) of the insecticides applied to crops by aerial spraying (Figure 20-5) or ground spraying reaches the target pests; less than 5% of herbicides applied to crops reaches the target weeds.

Pesticides that miss their target pests end up in the air, surface water, groundwater, bottom sediments, food, and nontarget organisms, including humans and wildlife. Pesticide waste and mobility from ground spraying can be reduced by **(1)** using ground sprayers that suck in and recover spray in the atmosphere, **(2)** covering booms that spray pesticides to reduce drift, and **(3)** laying ropelike wicks along the ground to deliver herbicides directly to weeds and reduce herbicide use by 90%. Crops that have been genetically altered to release small amounts of pesticides directly to pests can help overcome this problem but can also increase the rate of genetic resistance to such pesticides (Pro/Con, p. 275).

How Can Pesticides Harm Wildlife? During the 1950s and 1960s, populations of fish-eating birds such as the osprey, cormorant, brown pelican, and bald eagle plummeted. Research indicated that a chemical derived from DDT, when biologically magnified in food webs (Figure 16-4, p. 399), made the birds' eggshells so fragile that they could not reproduce successfully. Also hard-hit were such predatory birds as the prairie falcon, sparrow hawk, and peregrine falcon (Figure 20-6), which help control rabbits, ground squirrels, and other crop-eaters. Since the U.S. ban on DDT in 1972, most of these species have made a comeback.

According to the USDA and the U.S. Fish and Wildlife Service, each year in the United States pesticides applied to cropland:

- Wipe about 20% of U.S. honeybee colonies and damage another 15%, costing farmers at least $200 million per year from reduced pollination of vital crops.

- Kill more than 67 million birds.

- Kill 6–14 million fish when they wash off cropland into surface waters.

Figure 20-5 A crop duster spraying an insecticide on grapevines south of Fresno, California. Aircraft apply about 25% of the pesticides used on U.S. cropland, but only 0.1-2% of these insecticides actually reach the target pests. To compensate for the drift of pesticides from target to nontarget areas, aircraft apply up to 30% more pesticide than ground-based application does. (National Archives/EPA Documerica)

- Menace about 20% of the endangered and threatened species in the United States

How Can Pesticides Threaten Human Health?

DDT and related persistent chlorinated hydrocarbon pesticides have been banned in the United States and most developed countries and have been largely replaced by nonpersistent organophosphate pesticides (Table 20-1). Although the organophosphates are much less persistent, most are highly toxic to humans and other mammals.

Because of their high toxicity and the more frequent applications needed to control pests, organophosphates are a significant hazard for agricultural workers and others exposed to these pesticides. It is estimated that organophosphates are responsible for about 70% of human pesticide poisonings and deaths.

According to the World Health Organization (WHO) and the UN Environment Programme (UNEP), an estimated 3 million agricultural workers in developing countries are seriously poisoned by pesticides each year, resulting in an estimated 180,000 deaths—an average of 490 premature deaths per day. Health officials believe that the actual number of pesticide-related illnesses and deaths among the world's farm workers probably is greatly underestimated because of (1) poor records, (2) lack of doctors and disease reporting in rural areas, and (3) faulty diagnoses.

Farm workers in developing countries are especially vulnerable to pesticide poisoning because three-quarters of all pesticides are applied by hand. In addition, (1) educational levels are low, (2) warning labels often are vague, nonexistent, or written in languages that farm workers cannot read, (3) pesticide regulations are lax or nonexistent, and (4) use of protective equipment is rare (especially in the hot and humid tropics). Workers' clothing also spreads pesticide contamination to family members, and it is common in developing countries for children to work or play in fields treated with pesticides. To make matters worse, some families reuse pesticide containers to store food and drinking water.

In the United States, health officials estimate that (1) at least 300,000 farm workers suffer from pesticide-related illnesses each year, (2) at least 30,000 of these pesticide poisonings are acute, and (3) typically 25 farm workers die every year from such poisonings. Each year another 110,000 Americans, mostly children, get sick from misuse or unsafe storage of pesticides in the home, and about 20 die.

According to scientific literature reviewed by the EPA,

- Approximately 165 of the active ingredients approved for use in U.S. pesticide products are known or suspected human carcinogens. By June 2000, use of only 41 of these chemicals had been banned by the EPA or discontinued voluntarily by manufacturers.

- A study of Missouri children revealed a statistically significant correlation between childhood brain cancer and use of various pesticides in the home, including (1) flea and tick collars, (2) no-pest strips, and (3) chemicals used to control pests such as roaches, ants, spiders, mosquitoes, and termites.

- Children whose homes contained pest strips (with dichlorovos) faced 2.5–3 times the risk of leukemia as children whose homes did not contain such strips.

- In 2000, EPA scientists published a report indicating that atrazine (widely used as a weed killer by farmers growing corn, sorghum, citrus fruits, and other crops) could cause uterine, prostate, and breast cancer in humans and disrupt reproductive development.

- Recently, Swedish medical researchers found that exposure to glyphosate (the active ingredient in the widely used weed killer Roundup) nearly tripled a person's chances of developing a type of cancer known as non-Hodgkin's lymphoma.

- Children whose yards were treated with pesticides (mostly herbicides) were four times more likely to suffer from cancers of muscle and connective tissues than children whose yards were not treated.

Pesticide company officials and scientists say that these studies are only preliminary and have not been verified.

Some scientists are becoming increasingly concerned about possible (1) genetic mutations, (2) birth defects, (3) nervous system disorders (especially behav-

Figure 20-6 Until 1994, the peregrine falcon was listed as an endangered species in the United States, mostly because DDT caused their young to die before hatching because the eggshells were too thin to protect them. Only about 120 peregrine falcons were left in the lower 48 states in 1975; today there are about 3,400, most of them bred in captivity and then released into the wild in a $2.7-million-per-year recovery program. Because of the success of this program, the peregrine falcon in the United States has been removed from the list of endangered and threatened species. (Hans Reinhard/Bruce Coleman Collection)

A Black Day in Bhopal

December 2, 1984, will long be a black day for India because on that date the world's worst industrial accident occurred at a Union Carbide pesticide plant in Bhopal, India.

Some 36 metric tons (40 tons) of highly toxic methyl isocyanate (MIC) gas, used to produce carbamate pesticides, erupted from an underground storage tank after water leaked in through faulty valves and corroded pipes, causing an explosive chemical reaction. Once in the atmosphere, some of the toxic MIC was converted to even more deadly hydrogen cyanide gas.

The toxic cloud of gas settled over about 78 square kilometers (30 square miles), exposing up to 600,000 people. Many of these people were illegal squatters living near the plant because they had no other place to go.

According to Indian officials, at least 6,000 people (some say 7,000–16,000, based on the sales of shrouds and cremation wood) were killed. An international team of medical specialists (the International Medical Commission on Bhopal) estimated in 1996 that 50,000–60,000 people sustained permanent injuries such as blindness, lung damage, and neurological problems.

The economic damage from the accident was estimated at $4.1 billion. Indian officials claim that Union Carbide probably could have prevented the tragedy by spending no more than $1 million to upgrade plant equipment and improve safety. However, Union Carbide officials contend that the accident was the result of sabotage by a disgruntled Indian employee.

After the accident, Union Carbide reduced the corporation's liability risks for compensating victims by selling off a portion of its assets and giving much of the profits to its shareholders in the form of special dividends. In 1994, Union Carbide sold its holdings in India.

In 1989, Union Carbide agreed to pay an out-of-court settlement of $470 million to compensate the victims without admitting any guilt or negligence concerning the accident. The company also spent $100 million to build a hospital for the victims. By 2000, most victims with injuries had received $600 in compensation, and families of victims who died had received less than $3,000.

On December 2, 1999—the 15th anniversary of the disaster—survivors and families of people killed by the accident filed a lawsuit in a New York U.S. district court charging the Union Carbide Company and its former chief executive officer with violating international law and the fundamental human rights of the victims and survivors of the Bhopal plant accident.

The lawsuit alleges (1) that Union Carbide "demonstrated a reckless and depraved indifference to human life in the design, operation, and maintenance" of the Union Carbide facility in Bhopal and (2) "that the defendants are liable for fraud and civil contempt for their total failure to comply with the lawful orders of the courts of both the United States and India."

Critical Thinking

Why do you think Union Carbide's stock price rose after the company agreed to pay a $470 million out-of-court settlement to compensate the victims of the Bhopal accident?

---

ioral disorders), and (4) effects on the immune and endocrine systems from long-term exposure to low levels of various pesticides (Connections, p. 404). Very little research has been conducted on these potentially serious threats to human health.

Accidents and unsafe practices in pesticide manufacturing plants can expose workers, their families, and sometimes the general public to harmful levels of pesticides or toxic chemicals used in their manufacture (Case Study, above).

20-4 PESTICIDE REGULATION IN THE UNITED STATES

Is the Public Adequately Protected? The Federal Insecticide, Fungicide, and Rodenticide Act (FIFRA), established by Congress in 1947 and amended in 1972, requires that all commercial pesticides be approved by the EPA for general or restricted use. Pesticide companies must evaluate the biologically active ingredients in their products for toxicity to animals (and by extrapolation, to humans, p. 400). Then, EPA officials review these data before a pesticide can be registered for use. When a pesticide is legally approved for use on fruits or vegetables, the EPA sets a *tolerance level* specifying the amount of toxic pesticide residue that can legally remain on the crop when the consumer eats it.

Between 1972 and 2000, the EPA banned or severely restricted the use of 56 active pesticide ingredients. The banned chemicals include (1) most chlorinated hydrocarbon insecticides, (2) several carbamates and organophosphates, and (3) the systemic herbicides 2,4,5-T and Silvex (Table 20-1). However, banned or unregistered pesticides may be manufactured in the United States and exported to other countries (Connections, p. 510).

What Goes Around Can Come Around

U.S. pesticide companies can make and export to other countries pesticides that have been banned or severely restricted—or never even approved—in the United States. Between 1992 and 1999, U.S. exports of such pesticides (most to developing countries) averaged more than 27 metric tons (30 tons) per day. During this period U.S. chemical companies increased their exports of domestically banned pesticides by more than 18% and their exports of unregistered pesticides by more than 40%. Other industrial countries also export banned and unapproved pesticides.

However, in what environmentalists call a *circle of poison*, residues of some of these banned or unapproved chemicals exported to other countries can return to the exporting countries on imported food. Persistent pesticides such as DDT can also be carried by winds from other countries to the United States. A number of documented cases have linked pesticides exported from industrialized countries to health disasters in developing countries.

Many developing countries now have large stocks of unused and obsolete pesticides, many of them bought from manufacturers and donated under foreign aid programs. The UN Food and Agriculture Organization (FAO) estimates that developing countries now have more than 100,000 metric tons (110,000) tons of such chemicals.

In many cases, these chemicals are stored out in the open in drums that are leaking and corroding and contaminating soils, groundwater, and drinking and irrigation water. In Africa alone, the FAO estimated that it would cost $80–100 million to dispose of stocks of these chemicals.

Environmentalists have urged Congress—without success—to ban such exports. Supporters of pesticide exports argue that **(1)** such sales increase economic growth and provide jobs, **(2)** if the United States did not export pesticides, other countries would, and **(3)** banned pesticides are exported only with the consent of the importing countries.

The FAO publishes a "red alert" list of more than 50 pesticides banned in five or more countries. This list is designed to help developing countries become more aware of dangerous pesticides before agreeing to import such chemicals.

In 1998, more than 50 countries met to finalize an international treaty that requires exporting countries to have informed consent from importing counties for exports of 22 pesticides and 5 industrial chemicals. More than 100 countries also began negotiations in 1998 to develop an international agreement to ban or phase out the use of 12 especially hazardous persistent organic pollutants (9 of them persistent chlorinated hydrocarbon pesticides).

Critical Thinking

Should U.S. companies be allowed to export pesticides that have been banned, severely restricted, or not approved for use in the United States? Explain.

FIFRA required the EPA to reevaluate the more than 600 active ingredients approved for use in pre-1972 pesticide products to determine whether any of them caused cancer, birth defects, or other health risks. In the late 1970s, the EPA discovered that a now-defunct laboratory falsified data used to support registrations for more than 200 pesticide active ingredients, which still have not been fully reevaluated. By 2000, 28 years after Congress ordered EPA to evaluate these chemicals, less than 10% of these 600 active ingredients had been evaluated fully. The EPA contends that it has not been able to complete this evaluation because of the difficulty and expense of determining the health effects of chemicals (Section 16-2, p. 398).

According to studies by National Academy of Sciences,

- Federal laws regulating pesticide use in the United States are inadequate and poorly enforced by the EPA, Food and Drug Administration (FDA), and USDA.

- Exposure to pesticide residues in food causes 4,000–20,000 cases of cancer per year in the United States. Because roughly 50% of people with cancer die prematurely, this amounts to about 2,000–10,000 premature deaths per year in the United States from exposure to legally allowed pesticide residues in foods.

- Up to 98% of the potential risk of developing cancer from pesticide residues on food grown in the United States would be eliminated if EPA standards were as strict for pre-1972 pesticides as they are for later ones.

Representatives from pesticide companies dispute these findings. Indeed, the food industry denies that anyone in the United States has ever been harmed by eating food that has been grown using pesticides for the past 50 years.

Since 1987, the EPA has been evaluating the potential health effects of the 1,820 so-called *inert* ingredients (such as chlorinated hydrocarbon solvents) that make up 80–99% by weight of pesticide products. It has labeled 100 of them "of known or potential toxicological concern" but has not yet banned their use.

The National Academy of Sciences and environmentalists point out that FIFRA:

- Allows the EPA to leave inadequately tested pesticides on the market.

- Allows the EPA to license new chemicals without full health and safety data.

- Gives the EPA unlimited time to remove a chemical, even when its health and environmental risks are shown to outweigh its economic benefits.

- Has built-in appeals and other procedures that often keep a dangerous chemical on the market for up to 10 years.

- Is the only major environmental statute that does not allow citizens to sue the EPA for not enforcing the law.

A 1993 study of pesticide safety by the U.S. National Academy of Sciences urged the government to do the following things:

- Make human health the primary consideration for setting limits on pesticide levels allowed in food.

- Collect more and better data on pesticide exposure for different groups, including farm workers, adults, and children.

- Develop new and better test procedures for evaluating the toxicity of pesticides, especially for children.

- Consider cumulative exposures of all pesticides in food and water, especially for children, instead of basing regulations on exposure to a single pesticide.

Some *good news* is that progress has been made with the passage of the 1996 Food Quality Protection Act (FQPA), which:

- Requires new standards for pesticide tolerance levels in foods, based on a reasonable certainty of no harm to human health (defined for cancer as producing no more than one additional cancer per million people exposed to a certain pesticide over a lifetime).

- Requires manufacturers to demonstrate that the active ingredients in new pesticide products are safe for infants and children.

- Allows the EPA to apply an additional 10-fold safety factor to pesticide tolerance levels to protect infants and children.

- Requires the EPA to consider exposure to more than one pesticide when setting pesticide tolerance levels.

- Requires the EPA to develop rules for a program to screen all active and inactive ingredients for their estrogenic and endocrine effects by 1999 (which had not been achieved by the end of 2000).

Environmentalists believe that this law should be strengthened to **(1)** help prevent contamination of groundwater by pesticides (Section 19-3, p. 485), **(2)** improve the safety of farm workers who are exposed to high levels of pesticides, and **(3)** allow citizens to sue the EPA for not enforcing the law.

Pesticide control laws in the United States may have some weaknesses, but most other countries (especially developing countries) have little protection against the potentially harmful effects of pesticides.

20-5   OTHER WAYS TO CONTROL PESTS

What Should Be the Primary Goal of Pest Control? In most cases, the primary goal of spraying with conventional pesticides is to eradicate pests in the area affected. However, critics say that the primary goal of any pest control strategy should be to reduce crop damage to an economically tolerable level (Figure 20-7). The point at which the economic losses caused by pest

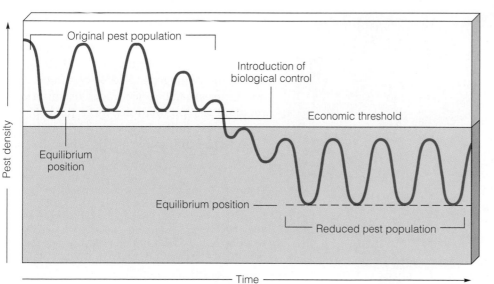

Figure 20-7 According to environmentalists, the primary goal of integrated pest management should be to keep each pest population just below the size at which it causes economic loss.

damage outweigh the cost of applying a pesticide is called the **economic threshold** (Figure 20-7). Because of the risk of increased genetic resistance and other problems (Section 20-3), continuing to spray beyond the economic threshold can make matters worse and cost more than it is worth.

The problem is determining when the economic threshold has been reached. This involves careful monitoring of cropfields to assess crop damage and determine pest populations (often by using traps baited with chemicals that attract the key pests).

Many farmers do not want to bother doing this and instead are likely to use additional *insurance spraying* to be on the safe side. One method used to reduce unnecessary insurance spraying is the purchase of *pest-loss insurance*. It pays farmers for losses caused by pests and is usually cheaper than using excess pesticides.

Another source of increased pesticide use is *cosmetic spaying*. Extra pesticides are used because consumers often buy only the best-looking fruits and vegetables even though there is nothing wrong with blemished ones. The only solution to this problem is consumer education.

How Can Cultivation Practices Help Control Pests? The following cultivation practices can be used to help reduce pest damage:

- Changing the type of crop planted in a field each year (crop rotation).

- Planting rows of hedges or trees around fields to hinder insect invasions and provide habitats for their natural enemies (with the added benefit of reduced soil erosion, Figure 10-24d, p. 230).

- Adjusting planting times so that major insect pests either starve or get eaten by their natural predators.

- Growing crops in areas where their major pests do not exist (Figure 20-2).

- Planting trap crops to lure pests away from the main crop.

- Switching from vulnerable monocultures to intercropping, agroforestry, and polyculture, which use plant diversity to reduce losses to pests (p. 265).

- Plowing under or burning diseased or infected plants and stalks and other crop residues that harbor pests that remain in cropfields after harvesting.

- Using plastic that degrades slowly in sunlight to keep weeds from sprouting between crop rows.

- Using vacuum machines to gently remove harmful bugs from plants.

With the rise of industrial agriculture and the greatly increased use of synthetic pesticides over the last 45 years, many farmers in developed countries no longer use many of these traditional cultivation methods for controlling pest populations. However, as the problems and expense of using pesticides have risen since 1980, more farmers are returning to these methods.

How Can Genetically Resistant Plants and Crops Help Lower Pest Losses? Plants and animals that are genetically resistant to certain pest insects, fungi, and diseases can be developed. However, resistant varieties usually take a long time (10–20 years) and lots of money to develop by conventional crossbreeding methods (Figure 12-11, p. 273). Moreover, insects and plant diseases can develop new strains that attack the once-resistant varieties (Figure 20-4), forcing scientists to continually develop new resistant strains.

Genetic engineering is now helping to speed up this process through the development of transgenic crops (Figure 20-8). However, there is controversy over the increasing used of genetically modified plants and foods (Pro/Con, p. 275).

How Can Natural Enemies Help Control Pests? Predators (Figure 20-1 and 20-9), parasites, and pathogens (disease-causing bacteria and viruses) can be encouraged or imported to regulate pest populations. More than 300 **biological pest control** projects worldwide have been successful, especially in China (p. 502).

Figure 20-8 The results of one example of using genetic engineering to reduce pest damage. Both tomato plants were exposed to destructive caterpillars. The normal plant's leaves are almost gone (left), whereas the genetically altered plant (right) shows little damage. (Monsanto)

In the United States, natural enemies have been used to control more than 70 insect pests and have saved farmers an average of $25 for every $1 invested.

Here are some examples of biological control:

- In Nigeria, crop-duster planes release parasitic wasps instead of pesticides to fight the cassava mealybug; farmers get a $178 return for every $1 they spend on the wasps.

- Rabbits in Australia have been controlled by an infectious virus.

- Within 2 years after its introduction, the vedalia beetle had eliminated the cottony-cushion scale from citrus orchards in the United States.

The *good news* is that biological pest control

- Focuses on selected target species

- Is nontoxic to other species

- Can be self-perpetuating and save large amounts of money once a population of effective natural predators or parasites is established

- Minimizes genetic resistance because pest and predator species interact and change together (coevolution)

The *bad news* is that biological control agents

- Can take years of research to understand how a particular pest interacts with its various enemies and to choose the best agent

- Cannot always be mass-produced

- Often are slower acting and more difficult to apply than conventional pesticides

- Must be protected from pesticides sprayed in nearby fields

- Can multiply and in some cases attack desirable insect species or other unintended hosts and become pests themselves

How Can Biopesticides Help Control Pests? Botanicals such as synthetic pyrethroids (Table 20-1) are an increasingly popular pest control method. Microbes also are being drafted for insect wars, especially by organic farmers. For example, *Bacillus thuringensis (Bt)* toxin is a registered pesticide sold commercially as a dry powder. Each of the thousands of strains of this common soil bacterium kills a specific pest. Various strains of *Bt* are used by almost all organic farmers as a nonchemical pesticide. The *bad news* for organic farmers is that genetic resistance is developing to some *Bt* toxins, especially because *Bt* genes have been transferred to widely planted genetically al-tered crops such as corn, potatoes, and soybeans (Pro/Con, p. 275).

Figure 20-9 Biological pest control: An adult convergent ladybug (right) is consuming an aphid (left). (Peter J. Bryant/Biological Photo Service)

How Can Insect Birth Control and Genetic Engineering Help Control Pests? Males of some insect pest species can be **(1)** raised in the laboratory, **(2)** sterilized by radiation or chemicals, and **(3)** released into an infested area to mate unsuccessfully with fertile wild females. Males are sterilized rather than females because the male insects mate several times, whereas the females mate only once.

The USDA used the sterile male approach to essentially eliminate the screwworm fly, a major livestock pest (Figure 20-10), from the southeastern states between 1962 and 1971. This has saved the cattle industry an estimated $300 million a year. The sterile male technique also was used successfully to **(1)** control the Mediterranean fruit fly (medfly) during a 1990 outbreak

Figure 20-10 Infestation of a steer by screwworm fly larvae in Texas. An adult steer can be killed in 10 days by thousands of maggots feeding on a single wound. (U.S. Department of Agriculture)

in California and (2) to eradicate the tsetse fly (which transmits sleeping sickness) from the island of Zanzibar.

Problems with the sterile male approach include (1) high costs, (2) difficulties in knowing the mating times and behaviors of each target insect, (3) the large number of sterile males needed, (4) the few species for which this strategy works, and (5) the need to release sterile males continually to prevent resurgence.

In 2000, researchers at Oxford University in London reported laboratory findings that releasing insects genetically engineered to carry a lethal gene could control the population of a target insect. Further tests are needed to evaluate the effectiveness and costs of using this approach in the field.

How Can Sex Attractants and Hormones Help Control Pests? In many insect species, a female that is ready to mate releases a minute amount (typically about one-millionth of a gram) of a chemical sex attractant called a *pheromone*. Whether extracted from insects or synthesized in the laboratory, pheromones can lure pests into traps or attract their natural predators into cropfields (usually the more effective approach). More than 50 companies worldwide sell about 250 pheromones to control pests (Figure 20-11).

These chemicals (1) attract only one species, (2) work in trace amounts, (3) have little chance of causing genetic resistance, and (4) are not harmful to nontarget species. However, it is costly and time-consuming to identify, isolate, and produce the specific sex attractant for each pest or predator.

Each step in the insect life cycle is regulated by the timely natural release of *juvenile hormones* (JH) and *molting hormones* (MH; Figure 20-12). These chemicals, which can be extracted from insects or synthesized in the laboratory, can disrupt an insect's normal life cycle, causing the insect to fail to reach maturity and reproduce (Figure 20-13).

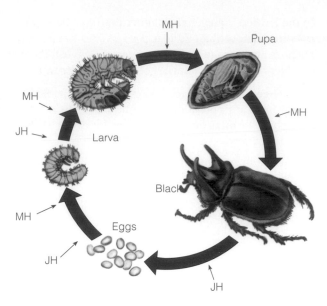

Figure 20-12 For normal insect growth, development, and reproduction to occur, certain juvenile hormones (JH) and molting hormones (MH) must be present at genetically determined stages in the insect's life cycle. If applied at the proper time, synthetic hormones disrupt the life cycles of insect pests and help control their populations.

Insect hormones have the same advantages as sex attractants. However, they (1) take weeks to kill an insect, (2) often are ineffective with large infestations of insects, (3) sometimes break down before they can act, (4) must be applied at exactly the right time in the target insect's life cycle, (5) can sometimes affect the target's predators and other nonpest species, (6) can kill crustaceans if they get into aquatic ecosystems, and (7) are difficult and costly to produce.

How Can Hot Water Be Used to Kill Pests? Recently some farmers have begun using a machine that sprays boiling water on crops to kill weeds and insects. So far, the system has worked well on cotton, alfalfa, and potato fields and in citrus groves in Florida, where the machine was invented. The cost is roughly equal to that of using chemical pesticides.

How Can Exposing Foods to Ionizing Radiation Help Control Pests? Exposing foods after harvest to high-energy gamma radiation (Figure 3-12, p. 62) can kill some pests. Such *food irradiation* extends food shelf life and kills (1) insects, (2) parasitic worms (such as trichinae in pork), and (3) bacteria (such as salmonella, which infects at least 51,000 Americans and kills 2,000 each year, and *E. coli*, which infects more than 20,000 Americans and kills about 250 each year). A food does not become radioactive when it is irradiated, just as exposure to X-rays does not make the body radioactive.

According to the U.S. FDA and the WHO, more than 2,000 studies over three decades show that foods

Figure 20-11 Pheromones can help control populations of pests, such as the red scale mites that have infested this lemon grown in Florida. (Agricultural Research Service/USDA)

exposed to low doses of ionizing radiation are safe for human consumption. Currently, 37 countries (8 of them in western Europe) allow irradiation of one or more food items. Proponents of this technology contend that the potential benefits of food irradiation greatly exceed the risks because it is likely to lower health hazards to people and reduce the need for pesticides and some food additives.

Critics of food irradiation:

- Are concerned that irradiating food forms trace amounts of certain chemicals called free radicals, some of which have caused cancer in laboratory animals. Proponents reply that free radicals are normal components of food and are also produced when foods are fried or broiled.

- Point out that we do not know the long-term health effects of eating irradiated food.

- Warn that current levels of irradiation do not destroy botulinum spores, but they do destroy the bacteria that give off the rotten odor warning us of their presence.

- Contend that consumers want fresh, wholesome food, not old, possibly less nutritious food made to appear fresh and healthy by irradiation:

In the United States, the FDA has approved the use of irradiation on poultry, pork, beef, spices, and all domestic fruits and vegetables to delay sprouting, kill insect pests, and slow ripening. However, New York, New Jersey, and Maine have prohibited the sale and distribution of irradiated food, as have Germany, Austria, Denmark, Sweden, Switzerland, Sudan, Singapore, Australia, and New Zealand.

How Can Integrated Pest Management Help Control Pests? An increasing number of pest control experts and farmers believe that the best way to control crop pests is a carefully designed **integrated pest management (IPM)** program. In this approach, each crop and its pests are evaluated as parts of an ecological system. Then a control program is developed that includes a mix of cultivation (p. 512), biological (p. 513), and chemical methods applied in proper sequence and with the proper timing.

The overall aim of IPM is not to eradicate pest populations but to reduce crop damage to an economically tolerable level (Figure 20-7). Fields are carefully monitored, often by *field scouts*. Such scouts are trained in monitoring pest populations (usually by sampling numbers with traps baited with sex attractants) to determine whether pest populations exceed the economic threshold.

When a damaging level of pests is reached, farmers first use biological methods (natural predators, parasites, and disease organisms) and cultivation controls, including vacuuming up harmful bugs. Small amounts

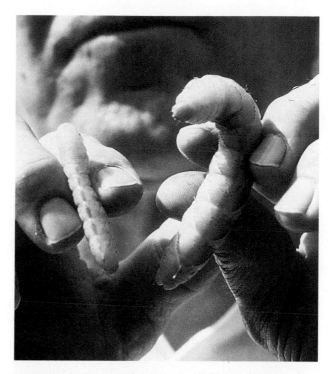

Figure 20-13 A use of hormones to prevent insects from maturing completely, making it impossible for them to reproduce. The stunted tobacco hornworm (left) was fed a compound that prevents production of molting hormones; a normal hornworm is shown on the right. (Agricultural Research Service/USDA)

of insecticides (mostly botanicals or microbotanicals) are applied only as a last resort, and different chemicals are used to slow development of genetic resistance and to avoid killing predators of pest species.

In 1986, the Indonesian government **(1)** banned the use of 57 of the 66 pesticides used on rice, **(2)** phased out pesticide subsidies over a 2-year period, and **(3)** used some of the money to help launch a nationwide program to switch to IPM, including a major farmer education program. The results were dramatic: Between 1987 and 1992, **(1)** pesticide use dropped by 65%, **(2)** rice production rose by 15%, and **(3)** more than 250,000 farmers were trained in IPM techniques. By 1993, the program had saved the Indonesian government more than $1.2 billion (most from elimination of $120-million-per-year pesticide subsidies)—more than enough to fund its IPM program. According to an FAO study of seven Asian countries, rice farmers using IPM were able to cut pesticide use by an average of 46%, while increasing rice yields by an average of 10%.

The experiences of various countries show that a well-designed IPM program can **(1)** reduce pesticide use and pest control costs by 50–90%, **(2)** reduce preharvest pest-induced crop losses by 50%, **(3)** improve crop yields, **(4)** reduce inputs of fertilizer and irrigation water, and **(5)** slow the development of genetic resistance because pests are assaulted less often and with lower doses of pesticides. Thus, IPM is an important form of pollution prevention that reduces risks to wildlife and human health. Consumers Union estimates that if all U.S. farmers practiced IPM by 2020, public health risks from pesticides would drop by 75%.

In Massachusetts, apple growers using IPM cut pesticide use by 43% in 10 years. In one region of Costa Rica, use of IPM on banana plantations has eliminated all pesticide use. In Brazil, IPM has reduced pesticide use on soybeans by 90%.

Recently, the United Nations Food and Agriculture Organization (FAO), UNEP, World Bank, and United Nations Development Program (UNDP) joined forces to cosponsor an international IPM facility. Its mission is to establish networks of researchers, agricultural extension services, and farmers to provide information about how IPM works and its economic and ecological benefits.

Despite its promise, IPM, like any other form of pest control, has some disadvantages. It requires expert knowledge about each pest situation, it is slower acting than conventional pesticides, and methods developed for a crop in one area might not apply to areas with even slightly different growing conditions. Although long-term costs typically are lower than those of using conventional pesticides, initial costs may be higher.

Another problem is that some IPM programs are not integrated. Instead, they use pesticide consultants to monitor crops only to determine when pesticides should be used, without integrating the use of cultural and biological control programs. Critics say that these monitored pesticide management programs should not be called IPM programs.

Widespread use of IPM is hindered by government subsidies of conventional chemical pesticides and by opposition from agricultural chemical companies, whose pesticide sales would drop sharply. In addition, farmers get most of their information about pest control from pesticide salespeople (and in the United States from USDA county farm agents). Most of these advisers do not have adequate training in IPM.

A 1996 study by the National Academy of Sciences recommended that the United States shift from chemically based approaches to ecologically based pest management approaches. A growing number of scientists urge the USDA to promote IPM in the United States by (1) adding a 2% sales tax on pesticides and using the revenue to fund IPM research and education, (2) setting up a federally supported IPM demonstration project on at least one farm in every county, (3) training USDA field personnel and county farm agents in IPM so that they can help farmers use this alternative, (4) providing federal and state subsidies, and perhaps government-backed crop insurance, to farmers who use IPM or other approved alternatives to pesticides, and (5) gradually phasing out subsidies to farmers who depend almost entirely on pesticides as effective IPM methods

are developed for major pest species. Some actions you can take to reduce your use of and exposure to pesticides are listed in Appendix 6.

We need to recognize that pest control is basically an ecological, not a chemical problem.

ROBERT L. RUDD

REVIEW QUESTIONS

1. Define the boldfaced terms in this chapter.

2. How can spiders be useful in controlling insect pests?

3. What is a *pest*? How are species we consider pests controlled in nature? What happens to this natural pest control service when we simplify ecosystems? How have plants evolved to protect themselves from herbivores?

4. Distinguish between *insecticides, herbicides, fungicides, nematocides,* and *rodenticides*. Distinguish between *first-generation pesticides* and *second-generation pesticides* used by people to combat pests, and give two examples of each type. Distinguish between *broad-spectrum* and *narrow-spectrum* pesticides.

5. List six benefits of pesticide use. List five traits of the ideal pest-killing chemical.

6. Summarize the harmful effects of pesticides in terms of (a) development of genetic resistance among pests, (b) killing of natural predators and parasites that help control pest species, (c) creation of new pest species, (d) migration of pesticides into the environment, (e) effects on wildlife, and (f) effects on the health of pesticide and farm workers and the general public.

7. What are the *pesticide treadmill* and the *genetic treadmill,* and how are they perpetuated? What are the alternatives?

8. How are pesticides regulated in the United States? List major weaknesses in these laws. According to the U.S. National Academy of Sciences, what three things should be done to minimize the harmful effects of pesticides?

9. Describe the events that took place at the Dow pesticide plant in Bhopal, India.

10. With respect to pesticides, what is the circle of poison? How might it affect you?

11. List nine cultivation practices that can be used to help control pests. Why are these methods not widely used today?

12. List the pros and cons of the following pest control methods: (a) development of plants and crops that are genetically resistant to pests, (b) use of crops genetically modified to tolerate higher doses of herbicides or to release pesticides (transgenics), (c) biological control, (d) biopesticides, (e) insect birth control through sterilization, (f) insect sex attractants, (g) insect hormones (pheromones), (h) spraying with hot water, (i) food irradiation, and (j) integrated pest management (IPM).

CRITICAL THINKING

1. Overall, do you think that the benefits of pesticides outweigh their disadvantages? Explain your position.

2. Should DDT and other pesticides be banned from use in malaria control efforts throughout the world? Explain. What are the alternatives?

3. Do you agree or disagree that because DDT and the other banned chlorinated hydrocarbon pesticides pose no demonstrable threat to human health and have saved millions of lives, they should again be approved for use in the United States? Explain.

4. If increased mosquito populations threatened you with malaria, would you spray DDT in your yard and inside your home to reduce the risk? Explain. What are the alternatives?

5. List the advantages and disadvantages of the switch from persistent chlorinated hydrocarbon pesticides (such as DDT) to nonpersistent organophosphate pesticides. On balance, have the benefits of this switch outweighed the disadvantages? Explain.

6. Explain how widespread use of a pesticide can **(a)** increase the damage done by a particular pest and **(b)** create new pest organisms.

7. Explain why biological pest control often is more successful on a small island than on a continent.

8. Do you believe that farmers should be given economic incentives for switching to IPM? Explain your position.

9. If IPM and other alternatives to conventional pesticides are considered to be safer and more effective in the long run than conventional pesticides, why are these methods not being more widely used by farmers?

10. Should certain types of foods be irradiated to help control disease organisms and increase shelf life? Explain. If so, should such foods be required to carry a clear label stating that they have been irradiated? Explain.

11. What changes, if any, do you believe should be made in the Federal Insecticide, Fungicide, and Rodenticide Act and the Food Quality Protection Act that regulate pesticide use in the United States?

12. Congratulations. You have just been placed in charge of pest control for the entire world. What are the three most important components of your global pest control strategy?

PROJECTS

1. How are bugs and weeds controlled in **(a)** your yard and garden, **(b)** the grounds of your school, and **(c)** public school grounds, parks, and playgrounds in your community?

2. Make a survey of all pesticides used in or around your home. Compare the results for your entire class.

3. Some research shows that although many people agree that we need to make greater use of alternatives to conventional pesticides for controlling pests, when they are faced with an actual infestation from insects or

rodents the first thing they do is spray with pesticides. Survey members of your class and other groups to help determine the validity of these research findings.

4. Use the library or the internet to find bibliographic information about *Ralph Waldo Emerson* and *Robert L. Rudd*, whose quotes appear at the beginning and end of this chapter.

5. Make a concept map of this chapter's major ideas, using the section heads and subheads and the key terms (in boldface). Look at the inside back cover and on the website for this book for information about making concept maps.

INTERNET STUDY RESOURCES AND RESOURCES FOR FURTHER READING AND RESEARCH

The website for this book contains helpful study aids and many ideas for further reading and research. Log on to:

> http://www.brookscole.com/product/0534376975s

and click on the Chapter-by-Chapter area. Choose Chapter 20 and select a resource:

- "Flash Cards" allows you to test your mastery of the Terms and Concepts to Remember for this chapter.

- "Tutorial Quizzes" provides a multiple-choice practice quiz.

- "Student Guide to InfoTrac" will lead you to Critical Thinking Projects that use InfoTrac College Edition as a research tool.

- "References" lists the major books and articles consulted in writing this chapter.

- "Hypercontents" takes you to an extensive list of sites with news, research, and images related to individual sections of the chapter.

INFOTRAC COLLEGE EDITION

Improve your skills with InfoTrac College Edition, a searchable online database of articles from more than 700 periodicals. Log on to:

> http://www.infotrac-college.com

or access InfoTrac through the website for this book.

Try the following articles:

Blindauer, K.M., R.J. Jackson, M. McGeehin, C. Pertowski, C. Rubin. 1999. Environmental pesticide illness and injury: the need for a national surveillance system. *Journal of Environmental Health* vol. 61, no. 10, pp. 9–14. (subject guide: pesticide, surveillance)

Hargrove, T.R. 1999. Wrangling over refuge. *American Scientist* vol. 86, no. 1, pp. 24–25. (subject guide: biopesticides)

21 SOLID AND HAZARDOUS WASTE

There Is No "Away": Love Canal

Between 1942 and 1953, Hooker Chemicals and Plastics (owned by OxyChem since 1968) sealed chemical wastes containing at least 200 different chemicals into steel drums and dumped them into an old canal excavation (called Love Canal after its builder, William Love) near Niagara Falls, New York.

In 1953, Hooker Chemicals filled the canal, covered it with clay and topsoil, and sold it to the Niagara Falls school board for $1. The company inserted in the deed a disclaimer denying legal liability for any injury caused by the wastes. In 1957, Hooker warned the school board not to disturb the clay cap because of the possible danger from toxic wastes.

By 1959 an elementary school, playing fields, and 949 homes had been built in the 10-square-block Love Canal area (Figure 21-1). Roads and sewer lines crisscrossed the dump site, some of them disrupting the clay cap covering the wastes. An expressway built at one end of the dump in the 1960s blocked groundwater from migrating to the Niagara River. This created a "bathtub effect" that allowed contaminated groundwater and rainwater to build up and overflow the disrupted cap.

Residents began complaining to city officials in 1976 about chemical smells and chemical burns their children received playing in the canal area, but these complaints were ignored. In 1977, chemicals began leaking from the badly corroded steel drums into storm sewers, gardens, basements of homes next to the canal, and the school playground.

In 1978, after much media publicity and pressure from residents led by Lois Gibbs (a mother galvanized into action as she watched her children come down with one illness after another; Guest Essay, p. 523), the state acted. It closed the school and arranged for the 239 homes closest to the dump to be evacuated, purchased, and destroyed.

Two years later, after protests from families still living fairly close to the landfill, president Jimmy Carter (1) declared Love Canal a federal disaster area, (2) had the remaining families relocated, and (3) offered federal funds to buy 564 more homes. Residents of all but 72 of the homes moved out. Some residents who remained claim that the entire problem was exaggerated by other residents, environmentalists, and the media.

After more than 15 years of court cases, OxyChem agreed in 1994 to (1) pay a $98-million settlement to New York State and (2) be responsible for all future treatment of wastes and wastewater at the Love Canal site. In 1999, the company also agreed to reimburse the federal government and New York State $7.1 million for the Love Canal cleanup.

Because of the difficulty in linking exposure to a variety of chemicals to specific health effects (Section 16-2, p. 398), the long-term health effects of exposure to hazardous chemicals on Love Canal residents remain unknown and controversial. However, for the rest of their lives the evacuated families will worry about the possible effects of the chemicals on themselves and their children and grandchildren.

The dump site has been covered with a new clay cap and surrounded by a drainage system that pumps leaking wastes to a new treatment plant. In June 1990, the U.S. Environmental Protection Agency (EPA) declared the area (renamed Black Creek Village) safe and allowed state officials to begin selling 234 of the remaining houses at 10–20% below market value. Most of the houses have now been sold. Buyers must sign an agreement stating that New York State and the federal government make no guarantees or representations about the safety of living in these homes.

The Love Canal incident is a vivid reminder that (1) we can never really throw anything away, (2) wastes do not stay put, and (3) preventing pollution is much safer and cheaper than trying to clean it up.

Figure 21-1 The Love Canal housing development near Niagara Falls, New York, was built near a hazardous-waste dump site. The photo shows the area when it was abandoned in 1980. In 1990, the EPA allowed people to buy some of the remaining houses and move back into the area. (New York State Department of Environmental Conservation)

Solid wastes are only raw materials we're too stupid to use.
Arthur C. Clarke

This chapter addresses the following questions:

- What are solid waste and hazardous waste, and how much of each type do we produce?

- What can we do to reduce, reuse, and recycle solid waste and hazardous waste?

- What are we doing to recycle aluminum, paper, and plastics?

- What are the advantages and disadvantages of burning or burying wastes?

- What can we do to reduce exposure to lead, hazardous chlorine compounds, and dioxins?

- How is hazardous waste regulated in the United States?

- How can we make the transition to a more sustainable low-waste society?

21-1 WASTING RESOURCES

What Is Solid Waste, and How Much Is Produced? The United States, with only 4.5% of the world's population, produces about 33% of the world's **solid waste**: any unwanted or discarded material that is not a liquid or a gas. Each year, the United States generates about 11 billion metric tons (12 billion tons) of solid waste—an average of 40 metric tons (97,000 pounds) per person. About 98.5% of this solid waste comes from **(1)** mining (Section 14-3, p. 325, and Case Study, p. 328), **(2)** oil and natural gas production (Section 14-6, p. 335), **(3)** agriculture (Figure 12-10, p. 271), and **(4)** industrial activities used to produce goods and services for consumers (Figure 21-2).

The remaining 1.5% of solid waste produced in the United States is **municipal solid waste** (MSW) from homes and businesses in or near urban areas. The amount of MSW, often called *garbage*, currently produced in the United States each year amounts to about 200 million metric tons (440 billion pounds)—almost twice as much as in 1970. This is enough waste to fill a bumper-to-bumper convoy of garbage trucks encircling the globe almost eight times.

This amounts to an average of 730 kilograms (1,600 pounds) per person in the United States—the world's highest per capita solid waste production—and many times the rate in developing countries. After the United States, the five other countries with the highest amounts of per capita solid waste per year, in order, are Australia, Canada, Switzerland, France, and Norway. Figure 21-3 shows what happened to the MSW produced in the United States during 1999.

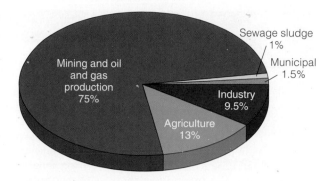

Figure 21-2 Sources of the estimated 11 billion metric tons (12 billion tons) of solid waste produced each year in the United States. Mining, agricultural, and industrial activities produce 55 times as much solid waste as household activities. (Data from U.S. Environmental Protection Agency and U.S. Bureau of Mines)

What Does It Mean to Live in a High-Waste Society? Here are a few of the solid wastes U.S. consumers throw away:

- Enough aluminum to rebuild the country's entire commercial airline fleet every 3 months

- Enough tires each year to encircle the planet almost three times

- About 18 billion disposable diapers per year, which if linked end to end would reach to the moon and back seven times

- About 2 billion disposable razors, 19 million computers, and 8 million television sets each year

- Some 8.6 million metric tons (17 billion pounds) of polystyrene peanuts used to protect items during shipping

- Used carpet that would cover 7,800 square kilometers (3,000 square miles)

- About 1.7 billion metric tons (3.7 trillion pounds) of construction waste per year—an average of 6 metric tons (13,200 pounds) per person

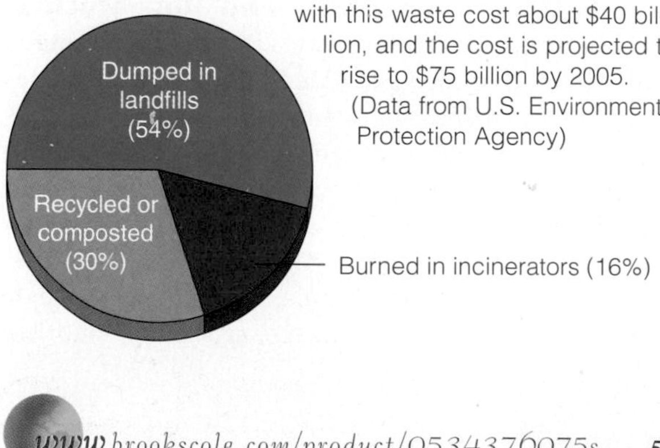

Figure 21-3 Fate of municipal solid waste produced in the United States in 1999. The recycling rate in 1999 was more than six times the rate in 1970. Dealing with this waste cost about $40 billion, and the cost is projected to rise to $75 billion by 2005. (Data from U.S. Environmental Protection Agency)

Dumped in landfills (54%)

Recycled or composted (30%)

Burned in incinerators (16%)

Mining and Smelting Waste in Montana

Environmental laws passed since the 1970s (Appendix 5) require mining companies and metal smelters operating in the United States to reduce their inputs of air and water into the environment and to stabilize the mining wastes they produce. This has helped, but these operations still produce large quantities of solid waste and pollutants (p. 325 and Case Study, p. 328).

Mining and smelting operations before these laws were enacted have produced a legacy of toxic waste sites. The largest complex of such sites consists of a huge copper mine near Butte, Montana, and a large copper smelter in Anaconda, Montana. Mining began in Butte the 1800s and continued at a high rate until the early 1900s.

Typically, the ore extracted contained as little as 0.3% by weight

copper. The remaining 99.7% of the rock that was waste was left in piles near the mine mouths and smelters. Today, the Butte-Anaconda area has about 25 square kilometers (10 square miles) of mine spoils and smelter wastes in piles averaging about 15 meters (49 feet) high—about as tall as a five-story building.

Acid wastes (Figure 14-7, p. 326) and toxic metals from these piles have polluted the area's soils, groundwater, and streams and killed aquatic life and stunted plant growth. Windblown dust from these spoils and smelter wastes has contaminated entire counties with toxic metal dust.

In the 1980s, arsenic levels in wells in Miltown, a small town on the Clark Fork River about 200 kilometers (120 miles) downstream from Butte, were so high that town officials closed all wells and had to build an expensive alternative drinking water system.

A partial cleanup costing hundreds of millions of dollars is under way. According to the Clark Ford Foundation, a local environmental organization, it will take about $1 billion to fully clean up the Butte mine site, the Anaconda smelter site, and the Clark Ford drainage basin.

Similar toxic legacies from mining are being created in many of the world's developing countries where there are few or poorly enforced environmental regulations.

Critical Thinking

Should mining and smelting companies or U.S. taxpayers be responsible for cleaning up wastes from mines and smelters that were operating before stricter environmental laws were enacted in the 1970s? Explain.

- About 2.5 million nonreturnable plastic bottles every hour

- About 670,000 metric tons (1.5 billion pounds) of edible food per year

- Enough office paper to build a 3.5-meter (11-foot) high wall from New York City to San Francisco, California

- About 19 billion catalogs (an average of 69 per American) per year

- Some 186 billion pieces of junk mail (an average of 675 per American) each year, about 45% of which is thrown in the trash unopened

This is only part of the 1.5% of all solid waste labeled "municipal" in Figure 21-2.

What Is Hazardous Waste, and How Much Is Produced? In the United States, **hazardous waste** is legally defined as any discarded solid or liquid material that (1) contains one or more of 39 toxic, carcinogenic, mutagenic, or teratogenic compounds (Section 16-3, p. 402) at levels that exceed established limits (including many solvents, pesticides, and paint strippers), (2) catches fire easily (gasoline, paints, and solvents), (3) is reactive or unstable enough to explode or release toxic fumes (acids, bases, ammonia, chlorine

bleach), or (4) is capable of corroding metal containers such as tanks, drums, and barrels (industrial cleaning agents and oven and drain cleaners).

This narrow official definition of hazardous wastes (mandated by Congress) does *not* include the following materials: (1) radioactive wastes (p. 349), (2) hazardous and toxic materials discarded by households (Table 21-1), (3) mining wastes (Case Study, above, and Case Study, p. 328), (4) oil- and gas-drilling wastes (routinely discharged into surface waters or dumped into unlined pits and landfills), (5) liquid waste containing organic hydrocarbon compounds (80% of all liquid hazardous waste), (6) cement kiln dust, produced when liquid hazardous wastes are burned in a cement kiln, and (7) wastes from the thousands of small businesses and factories that generate less than 100 kilograms (220 pounds) of hazardous waste per month.

As a result, *hazardous-waste laws do not regulate 95% of the country's hazardous waste.* In most other countries, especially developing countries, little, if any, of the hazardous waste is regulated.

Including all categories, the EPA estimates that at least 5.5 billion metric tons (12 trillion pounds) of hazardous waste are produced each year in the United States—an average of 20 metric tons (44,000 pounds) per person. This amounts to about 75% of the world's hazardous waste.

Table 21-1 Common Household Toxic and Hazardous Materials

Cleaning Products

Disinfectants

Drain, toilet, and window cleaners

Oven cleaners

Bleach and ammonia

Cleaning solvents and spot removers

Septic tank cleaners

Paint and Building Products

Latex and oil-based paints

Paint thinners, solvents, and strippers

Stains, varnishes, and lacquers

Wood preservatives

Acids for etching and rust removal

Asphalt and roof tar

Gardening and Pest Control Products

Pesticide sprays and dusts

Weed killers

Ant and rodent killers

Flea powder

Automotive Products

Gasoline

Used motor oil

Antifreeze

Battery acid

Solvents

Brake and transmission fluid

Rust inhibitor and rust remover

General Products

Dry cell batteries (mercury and cadmium)

Artists' paints and inks

Glues and cements

21-2 PRODUCING LESS WASTE AND POLLUTION

What Are Our Options? There are two ways to deal with the solid and hazardous waste we create: **(1)** *waste management* and **(2)** *pollution (waste) prevention.* Waste management is a *high-waste approach* (Figure 3-19, p. 66) that views waste production as an unavoidable product of economic growth. It attempts to manage the

resulting wastes in ways that reduce environmental harm, mostly by **(1)** burying them, **(2)** burning them, or **(3)** shipping them off to another state or country. In effect, it transfers solid and hazardous waste from one part of the environment to another.

Preventing pollution and waste is a *low-waste approach* that recognizes that there is no "away" and views most solid and hazardous waste either **(1)** as potential resources (that we should be recycling, composting, or reusing) or **(2)** as harmful substances that we should not be using in the first place (Figure 3-20, p. 67, and Figures 21-4 and 21-5). This approach focuses on discouraging waste production and encouraging waste prevention (Guest Essay, p. 523).

According to the U.S. National Academy of Sciences (Figures 21-4 and 21-5), the low-waste approach should have the following hierarchy of goals: **(1)** *reduce* waste and pollution, **(2)** *reuse* as many things as possible, **(3)** *recycle and compost* as much waste as possible, **(4)** *chemically or biologically treat or incinerate* waste that cannot be reduced, reused, recycled, or composted, and **(5)** *bury* what is left in state-of-the-art landfills or aboveground vaults after the first four goals have been met.

Scientists estimate that in a low-waste society 60–80% of the solid and hazardous waste produced could be eliminated through *reduction, reuse, recycling* (including composting), and *redesign* of manufacturing processes and

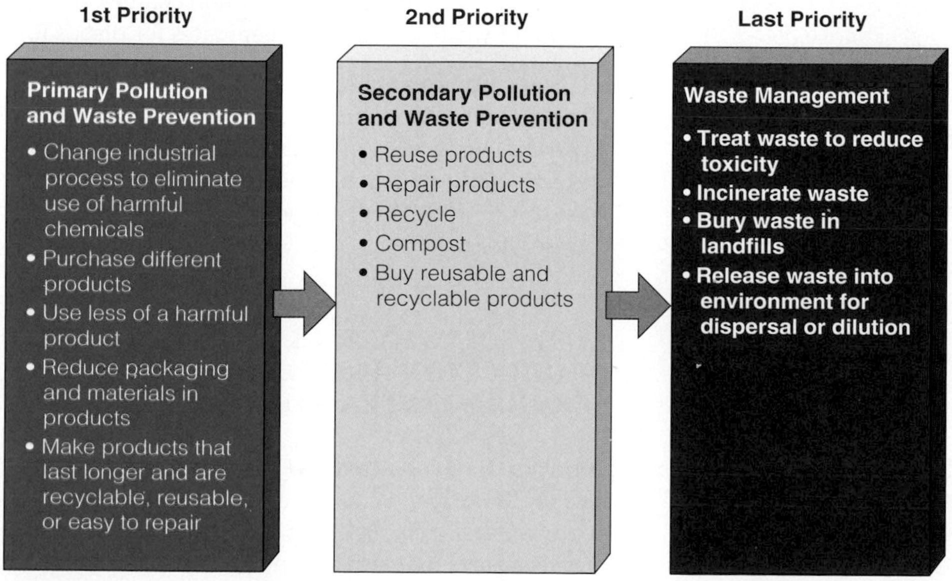

1st Priority

Primary Pollution and Waste Prevention

- Change industrial process to eliminate use of harmful chemicals
- Purchase different products
- Use less of a harmful product
- Reduce packaging and materials in products
- Make products that last longer and are recyclable, reusable, or easy to repair

2nd Priority

Secondary Pollution and Waste Prevention

- Reuse products
- Repair products
- Recycle
- Compost
- Buy reusable and recyclable products

Last Priority

Waste Management

- Treat waste to reduce toxicity
- Incinerate waste
- Bury waste in landfills
- Release waste into environment for dispersal or dilution

Figure 21-4 Solutions: priorities suggested by prominent scientists for dealing with material use and solid waste. To date, these priorities have not been followed in the United States (and in most other countries). Instead, most efforts are devoted to waste management (bury it or burn it). (U.S. Environmental Protection Agency and U.S. National Academy of Scientists)

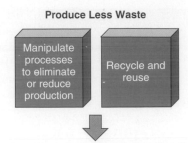

Figure 21-5 Solutions: priorities suggested by prominent scientists for dealing with hazardous waste. To date, these priorities have not been followed in the United States (and in most other countries). (U.S. National Academy of Scientists)

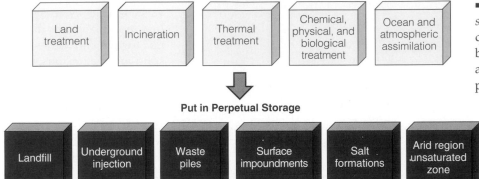

Produce Less Waste

Manipulate processes to eliminate or reduce production

Recycle and reuse

Convert to Less Hazardous or Nonhazardous Substances

Land treatment | Incineration | Thermal treatment | Chemical, physical, and biological treatment | Ocean and atmospheric assimilation

Put in Perpetual Storage

Landfill | Underground injection | Waste piles | Surface impoundments | Salt formations | Arid region unsaturated zone

buildings. Currently, the order of priorities shown in Figures 21-4 and 21-5 for dealing with solid and hazardous wastes is reversed in the United States (and in most other countries).

Solutions: How Can We Reduce Waste and Pollution? Ways to reduce resource use, waste, and pollution (Figure 3-20, p. 67) include:

■ *Decreasing consumption.* Before any purchase this involves asking questions such as Do I really need this? Can I buy it secondhand? Can I borrow or lease it?

■ *Doing more with less* by *redesigning manufacturing processes and products to use less material and energy per unit of output of goods and services* (Solutions, p. 525, and Guest Essay, p. 526).

■ *Redesigning manufacturing processes to produce less waste and pollution.* Most toxic organic solvents can be recycled within plants or replaced with water-based or citrus-based solvents (Individuals Matter, p. 472). Hydrogen peroxide can be used instead of toxic chlorine to bleach paper and other materials. Clothes can be washed without water by using detergents dissolved in liquid CO_2. Microwave drying or a CO_2-based process can replace dry cleaning with toxic organic solvents such as perchloroethylene (PERC).

■ *Developing products that are easy to repair, reuse, remanufacture, compost, or recycle.* Several European auto manufacturers design their cars for easy disassembly and for reuse and recycling of up to 80% of their parts (75% in the United States). In the United States, more than 73,000 remanufacturing firms (including IBM and Xerox), employing more than 480,000 people, generate

revenues of about $60 billion per year. Xerox's latest photocopier, with every part reusable or recyclable for easy remanufacturing, should eventually save the company $1 billion in manufacturing costs.

■ *Designing products to last longer.* Tires are now being produced have an average life of 97,000 kilometers (60,000 miles), but researchers believe this could be extended to at least 160,000 kilometers (100,000 miles).

■ *Eliminating or reducing unnecessary packaging.* Here are some key questions that environmentalists believe designers, manufacturers, and consumers should ask about packaging: Is it necessary? Can it use fewer materials? Can it be reused? Are the resources that went into it nonrenewable or renewable? Does it contain the highest feasible amount of consumer-discarded (postconsumer) recycled material? Is it designed to be recycled easily? Can it be incinerated without producing harmful air pollutants or a toxic ash? Can it be buried and decomposed in a landfill without producing chemicals that can contaminate groundwater? Objects can now be protected during shipping by replacing plastic peanuts with thin plastic bags filled with compressed air and made of 100% recyclable polyethylene. After use, the bags can be deflated and recycled or, with the proper equipment, reinflated and reused.

■ *Using trash taxes* to reduce waste and using the revenues to reduce taxes on income and wealth (Connections, p. 53, and Solutions, p. 707), as a number of European countries have done. A related *pay-as-you-throw* system reduces solid waste and encourages recycling by basing garbage collection charges on the amount of waste a household generates for disposal.

Some firms and cities are leading the way by eliminating almost all their waste. The city of Canberra, Australia, has a goal of achieving no waste by 2010, and the Netherlands has set a goal of reducing its waste by 70–90%. What are your community and your country doing?

21-3 SOLUTIONS: CLEANER PRODUCTION AND SELLING SERVICES INSTEAD OF THINGS

What Is the Ecoindustrial Revolution, and What Are Its Benefits? Some analysts urge us to bring about an *ecoindustrial revolution* over the next 50 years as a way to help achieve industrial, economic, and environmental

We Have Been Asking the Wrong Questions About Wastes

Lois Marie Gibbs

GUEST ESSAY

In 1977, Lois Marie Gibbs was a young housewife with two children living near the Love Canal toxic dump site. She had never engaged in any sort of political action until her children began experiencing unexplained illnesses and she learned that toxic chemicals were oozing from the dump site into many of the area's yards and basements. Then she organized her neighborhood and became the president and major strategist for the Love Canal Homeowners Association. This grassroots political action brought hazardous-waste issues to national prominence and spurred passage of the federal Superfund legislation to help clean up abandoned hazardous-waste sites. Lois Gibbs then moved to Washington, D.C., and formed Citizens' Clearinghouse for Hazardous Wastes (renamed the Center for Health, Environment, and Justice), an organization that has helped more than 7,000 community organizations protect themselves from hazardous wastes. Her story is told in her autobiography, Love Canal: My Story *(State University of New York Press, 1982). She was also the subject of a CBS movie,* Lois Gibbs: The Love Canal, *which aired in 1982. Her latest book is* Dying from Dioxin *(Boston: South End Press, 1995).*

Just about everyone knows our environment is in danger. One of the most serious threats is the massive amount of waste we put into the air, water, and ground every year. All across the United States and around the world are thousands of places that have been, and continue to be, polluted by toxic chemicals, radioactive waste, and just plain garbage.

For generations, the main question people have asked is "Where do we put all this waste? It's got to go somewhere." That is the wrong question, as has been shown by a series of experiments in waste disposal and by the simple fact that there is no "away" in "throwaway."

We tried dumping our waste in the oceans. That does not work. We tried injecting it into deep, underground wells. That does not work. We've been trying to build landfills that do not leak, but according to the EPA all landfills eventually leak. We've been trying to get rid of waste by burning it in high-tech incinerators; that only produces different types of pollution, such as air pollution and toxic ash. Even recycling, which is a very good thing to do, suffers from the same problem as all the other methods: It addresses waste *after* it has been produced.

For many years, people have been assuming, "It's got to go somewhere," but now many people, especially young people, are starting to ask why. Why do we produce so much waste? Why do we need products and services that have so many toxic by-products? Why can't industry change the way it makes things so that it stops producing so much waste?

When you start asking these questions, you start getting answers that lead to *pollution prevention* and *waste reduction* instead of *pollution control* and *waste management*. People, young and old, who care about pollution prevention are challenging companies to stop making products with gases that reduce ozone in the ozone layer [Section 18-6, p. 465] and contribute to the threatening possibility of global warming [Section 18-4, p. 458]. They are asking why so many goods are wrapped in excessive, throwaway packaging. They are challenging companies that sell pesticides, cleaning fluids, batteries, and other hazardous products to either remove the toxins from those products or take them back for recovery or recycling rather than disposing of them in the environment. They are demanding alternatives to throwaway materials in general.

Since 1988, hundreds of student groups have contacted my organization to get help and advice in taking these effective types of actions. Oregon students took legal action to get rid of cups and plates made from bleached paper because the paper contains the deadly poison dioxin. As a result the school systems switched to reusable cups, plates, and utensils.

Dozens of student groups have joined with local grassroots organizations to get toxic-waste sites cleaned up or to stop construction of new toxic-waste sites, radioactive-waste sites, or waste incinerators.

Waste issues are not simply environmental issues; they are also tied up with our economy, which is geared to producing and then disposing of waste. *Somebody* is making money from every scrap of waste and has a vested interest in keeping things the way they are.

Waste issues are also issues of *justice* and *fairness*. There is a lot of debate between industry officials and environmentalists, especially those in federal and state environmental agencies, about so-called acceptable risk [Section 16-5, p. 410]. These industry officials and environmentalists decide the degree of people's exposure to toxic chemicals but do not ask the people who will actually be exposed how they feel about it.

Risk analysts often say, "There's only a one in a million chance of increased death from this toxic chemical." That may be true, but suppose I took a pistol and went to the

(continued)

sustainability. The goals of this emerging concept of *cleaner production* (Guest Essay, p. 526), or *industrial ecology*, are to redesign all industrial products and processes integrate them into **(1)** an essentially closed system of cyclical material flows (Figure 3-20, p. 67) or **(2)** a network in which the wastes of one manufacturer become raw materials for another, and companies take back packaging and used products from consumers for reuse, recycling, repair, or remanufacturing.

In effect, companies would mimic natural chemical cycles (Section 4-6, p. 87) and interact in complex *resource exchange webs* similar to food webs in natural

edge of your neighborhood and began shooting into it. There's probably only a one in a million chance that I'd hit somebody, but would you issue me a license to do that?

As long as we do not stand up for our rights and demand that "bullets" in the form of hazardous chemicals not be "fired" in our neighborhoods, we are giving environmental regulators and waste producers a license to kill a certain number of us without our even being consulted.

From my personal experience, I know that decisions made to dump wastes at Love Canal [p. 518] and in thousands of other places were not made purely on the basis of the best available scientific knowledge. The same holds true for decisions about how to manage the wastes we produce today and how to produce less waste.

We live in a world that is shaped by decisions based on money and power. If you really want to understand what's behind any given environmental issue, the first question you should ask is, "Who stands to profit from this?" Then ask, "Who is going to pay the price?" You can then identify both sides of the issue and decide whether you want to be part of the problem or part of the solution.

Critical Thinking

1. Do you believe that we should put primary emphasis on pollution prevention and waste reduction? Explain.

2. What changes would you be willing to make in your own lifestyle to prevent pollution and reduce waste?

ecosystems. A prototype of this industrial ecosystem concept exists in Kalundborg, Denmark, where **(1)** a coal-fired power plant, **(2)** an oil refinery, **(3)** a sulfuric acid producer, **(4)** a sheetrock plant, **(5)** a pharmaceutical plant, **(6)** a cement manufacturer, **(7)** local farms, **(8)** horticulture greenhouses, **(9)** a fish farm, and **(10)** nearby homes are working together to save money by exchanging and converting their wastes into resources for one another.

For example, instead of cooling waste steam and releasing it as hot water into the nearby sea, the coal-fired plant supplies **(1)** steam to the oil refinery and pharmaceutical plant and **(2)** heat to warm greenhouses, area homes, and water in the fish farm. Surplus natural gas from the oil refinery is sold to the power plant and the sheetrock plant, and sulfur removed from oil by the refinery is sold to the sulfuric acid manufacturer.

Sulfur dioxide removed from power plant emissions in the form of calcium sulfate is sold to the sheetrock plant, and fly ash removed from the emissions is sold to the cement manufacturer, which uses it to build roads. Farmers fertilize their fields with waste from the fish farms and high-nutrient sludge produced by the fermentation plant.

These resource exchanges, which took about a decade to negotiate and develop, were done primarily to save money. However, these examples of industrial cooperation also benefit the environment by reducing resource use, saving energy, and reducing solid waste and pollution.

A similar arrangement in Fiji links together a chicken-raising operation, a mushroom farm, a brewery, fish farms, hydroponic gardens, and a methane gas production unit. Since 1993, Halifax (Nova Scotia) and more than 20 U.S. cities—including Baltimore (Maryland), Rochester (New York), Chattanooga (Tennessee, p. 684),

and the Brownsville/Matamoros region along the Texas-Mexico border—have announced plans to build ecoindustrial parks similar to the one in Kalundborg.

These important forms of *biomimicry*, which eliminate most waste, also provide economic benefits to businesses by:

- Reducing the costs of controlling pollution and complying with pollution regulations.

- Improving the health and safety of workers by reducing exposure to toxic and hazardous material (and thus reducing company health-care insurance expenses).

- Reducing future legal liability for toxic and hazardous wastes.

- Stimulating companies to come up with new, environmentally beneficial chemicals, processes, and products that can be sold worldwide.

- Giving companies a better image among consumers based on results rather than on public relations campaigns.

In 1975, the Minnesota Mining and Manufacturing Company (3M), which makes 60,000 different products in 100 manufacturing plants, began a Pollution Prevention Pays (3P) program. Mostly through employee-run projects it **(1)** redesigned equipment and processes, **(2)** used fewer hazardous raw materials, **(3)** identified hazardous chemical outputs (and recycled or sold them as raw materials to other companies), and **(4)** began making more nonpolluting products.

Forced to comply with new regulations to reduce solvent emissions by 90%, 3M scientists found a way to avoid the use of solvents altogether by coating products with water-based solutions. This innovation gave the company a temporary advantage over its

Doing More with Less: Increasing Resource Productivity

During the last few decades there has been a *design revolution* that has allowed businesses to use less material and energy per unit of goods and services, mostly by **(1)** finding substitutes for products that use less material and **(2)** redesigning or improving products so that they take less material and energy to produce.

Today, a single fiber-optic cable containing about 65 grams (3 ounces) of silica (SiO_2) can carry many times more electronic messages than a comparable length of cable containing 0.9 metric tons (1 ton) of copper. Increased use of wireless communication will greatly reduce the use of wires of any type.

Paper documents such as product catalogs, phone directories, technical reference manuals, and parts directories can be accessed on CD-ROMs or at various internet sites, saving millions of dollars and tons of paper. All the phone books in the United States can be put on about three CD-ROMs, and a single DVD-ROM could hold all the world's phone numbers.

A skyscraper built today includes about 35% less steel than the same building built in the 1960s because of the use of lighter-weight but higher-strength steel. Use of such steel and replacement of many steel parts with light-weight plastics and composite materials has **(1)** reduced the weight of cars by about 25% without compromising performance and safety, **(2)** increased fuel efficiency, and **(3)** reduced the average weight per unit of appliances such as stoves, washers, dryers, air conditioners, TV sets, and computers.

Conventional lumber is being replaced by engineered structural beams and joists and wall framing (studs) made by compressing and gluing wood wastes or by gluing many layers of wood together. Such products make roof and floor supports in houses so rigid that no internal load-bearing walls are needed. This **(1)** allows more flexible and useful living space, **(2)** reduces the wood needed for internal walls by at least 70%, and **(3)** doubles the amount of space for insulation (paid for by saved wood and lumber), which saves energy and allows smaller and less costly heating and cooling systems.

Since the mid-1970s **(1)** the thickness of plastic grocery bags has been reduced by 70% without sacrificing strength, **(2)** plastic milk jugs weigh 40% less, **(3)** aluminum drink cans contain one-third less aluminum, **(4)** steel cans are 60% lighter, **(5)** disposable diapers contain 50% less paper pulp, and **(6)** plastic frozen food bags weigh 89% less.

Because of increased population and per capita consumption, the total amount of municipal solid waste (p. 519) in the United States continues to grow. However, improvements in material efficiency (dematerialization) and increased recycling and composting have helped cut the rate of growth of such waste in half since 1990.

These improvements in resource productivity are important, but according to some analysts they can be greatly increased through a new *resource productivity revolution*. In their 1999 book *Natural Capitalism*, Paul Hawken (Guest Essay, p. 6), Amory Lovins (Guest Essay, p. 361), and Hunter Lovins contend that we already have the knowledge and technology to greatly increase *resource productivity* by getting 75–90% more work or service from each unit of material resources we use. They provide numerous examples of how this can be done and is being done in developed nations without diminishing the quantity or quality of the services people want. According to these analysts, such a revolution in *resource productivity* by doing more with less would

- Sharply decrease the depletion and degradation of the earth's natural capital that supports all economies (Guest Essay, p. 6).

- Give companies and countries making such improvements a huge competitive advantage in the global marketplace.

- Help reduce unemployment and poverty by making it more profitable to employ people. This is especially important in developing countries, where human labor is an abundant and underused economic resource.

- Not be as difficult as it might seem because the current use of matter and energy resources to fuel economic growth is extremely inefficient and amounts to throwing money away. According to a 1989 report by the National Academy of Sciences, only about 6% of the vast flows of materials through the U.S. economy actually end up in products. In addition, about 43% of the energy flowing through the U.S. economy is unnecessarily wasted (Figure 15-2, p. 359).

To these analysts, the only major impediments to such an economic and ecological revolution are laws, policies, taxes, and subsidies that continue to reward inefficient resource use.

Critical Thinking

1. Identify three items not listed above in which less material is used per item.

2. Do you believe that it is possible for resource waste to be decreased by 75–90% within the next 20 years? Explain. What might be some disadvantages of making such a shift? Do you believe that such disadvantages outweigh the advantages? Explain.

Cleaner Production: A New Environmentalism for the 21st Century

Peter Montague

Peter Montague is director of the Environmental Research Foundation in Washington, D.C., which studies and informs the public about environmental problems and the technologies and policies that might help solve them. He has served as project administrator of a hazardous waste research program at Princeton University and has taught courses in environmental impact analysis at the University of New Mexico. He is the coauthor of two books on toxic heavy metals in the natural environment and is editor of Rachel's Environment and Health Biweekly, *an informative newsletter on environmental problems and solutions.*

GUEST ESSAY

Environmentalism as we have known it for the last 30 years is dead. The environmentalism of the 1970s advocated strict numerical controls on releases into the environment of *dangerous wastes* (any unwanted or uncontrolled materials that can harm living things or disrupt ecosystems).

However, after several decades of effort by government regulatory agencies and concerned citizens (the environmental movement), most dangerous chemicals are not regulated in any way. Even the few that are covered by regulations have not been adequately controlled.

In short, the *pollution management* approach to environmental protection has failed; *pollution prevention* is our only hope. An ounce of prevention really is worth a pound of cure.

Here, in list form, is the situation facing environmentalists today:

■ *All waste disposal—landfilling, incineration, deep-well injection—is polluting because "disposal" means dispersal into the environment.* Once wastes are created, they cannot be contained or controlled because of the scientific laws of matter conservation [p. 60] and energy conservation [p. 65]. The old environmentalism failed to recognize this important truth and thus squandered enormous resources trying to achieve the impossible.

■ *The inevitable result of our reliance on waste treatment and disposal systems has been an unrelenting buildup of toxic synthetic materials in humans and other forms of life worldwide.* For example, breast milk of women in industrialized countries is so contaminated with pesticides and industrial hydrocarbons that if human milk were bottled and sold commercially, it could be banned by the U.S. Food and Drug Administration as unsafe for human consumption. If a whale beaches itself on U.S. shores and dies, its body must be treated as "hazardous waste" because whales contain concentrations of PCBs [polychlorinated biphenyls; Figure 19-4, p. 481], legally defined as hazardous.

■ *The ability of humans and other life forms to adapt to changes in their environment is strictly limited by the genetic code each form of life inherits.* Continued contamination at a rate hundreds of times faster than we can adapt will subject humans to increasingly widespread sickness and degradation of the species and could ultimately lead to extinction.

■ *Damage to humans (and other life forms) is abundantly documented.* Birds, fish, and humans in industrialized countries are enduring steadily rising levels of cancer, genetic mutations, and damage to their nervous, immune, and hormonal systems [Connections, p. 404] as a result of pollution.

If we will but look, the handwriting is on the wall everywhere.

To deal with these problems, industrial societies must abandon their reliance on waste treatment and disposal and on the regulatory system of numerical standards created to manage the damage that results from relying on waste disposal instead of waste prevention. We must quickly move the industrialized and industrializing countries to new technical approaches accompanied by new industrial goals: *clean production* or zero-discharge systems.

The concept of clean production involves industrial systems that avoid or eliminate dangerous wastes and dangerous products and minimize the use and waste of raw materials, water, and energy. Goods manufactured in a clean production process must not damage natural ecosystems throughout their entire life cycle, including (1) raw material selection, extraction, and processing, (2) product conceptualization, design, manufacture, and assembly, (3) material transport during all phases, (4) industrial and household usage, and (5) reintroduction of the product into industrial systems or into the environment when it no longer serves a useful function.

Clean production does not rely on *end-of-pipe* pollution controls such as filters or scrubbers [Figure 17-22, p. 442] or chemical, physical, or biological treatment [Figure 19-15, p. 494]. Measures that (1) pretend to reduce the volume of waste by incineration or concentration, (2) mask the hazard by dilution, or (3) transfer pollutants from one environmental medium to another are also excluded from the concept of clean production.

A new industrial pattern, and thus a new environmentalism, is emerging. Human survival and life quality depend on our willingness to make and pay for the changes needed to shift to this cleaner form of industrial production.

Critical Thinking

1. Some environmentalists point to the successes of the *pollution management* approach to environmental protection practiced during the past 30 years and do not agree that it has failed. What is your position? Explain.

2. List three undesirable economic, health, consumption, and lifestyle changes you might experience as a consequence of putting much greater emphasis on pollution prevention.

competitors and shortened the time of such products to market because the water-based product did not have to go through the EPA approval process for solvent-based coatings.

By 1998, 3M's overall waste production was down by one-third, emissions of air pollutants per unit of production were reduced by 70%, and the company had saved over $750 million in waste disposal and materials costs. During the 1990s a growing number of companies adopted similar pollution prevention programs that convert their manufacturing processes into essentially closed-loop systems.

Cleaner production systems are based on the idea that a *key principle of industrial design should be to avoid a problem in the first place*. Such systems:

- Do not introduce toxic chemicals into the environment.

- Are energy efficient (Section 15-2, p. 362).

- Produce products that **(a)** are durable and reusable, **(b)** are easy to dismantle, repair, and rebuild, and **(c)** use minimal packaging that can easily be reused, recycled, or composted.

- Are nonpolluting throughout their entire life cycle.

- Help preserve biodiversity and cultural diversity.

What Is a Service Flow Economy, and What Are Its Advantages? In the mid-1980s, German chemist Michael Braungart and Swiss industry analyst Walter Stahel independently proposed a new economic model that would provide profits while greatly reducing resource use and waste. Their idea involves shifting from our current *material flow economy* (Figure 3-19, p. 66) to a *service flow economy* over the next few decades. Instead of buying most goods outright, customers would lease or rent the *services* such goods provide.

For a fee (often on a monthly basis), companies would provide all the materials and products needed to supply and maintain their services and would take back all materials at the end of their useful lives in what Stahel called a "cradle-to-cradle" process. With such a service flow or product stewardship economy, a product produced by a manufacturer remains as an asset that yields more profit if it **(1)** uses the minimum amount of materials, **(2)** lasts as long as possible, **(3)** is easy to maintain, repair, remanufacture, reuse, or recycle, and **(4)** provides customers with the services they want instead of trying to keep selling them newer models of outmoded products.

A service flow economy would also increase employment because it takes more people to service and remanufacture equipment than it does to extract virgin materials and use them to fabricate products that are often thrown away and replaced with newer models.

This shift is already under way:

- Since 1992 the Xerox Corporation has been leasing most of its copy machines as part of its mission to provide *document services* instead of selling photocopiers. The company replaces or upgrades copier cartridges and other parts in its modularly designed copiers at no extra cost to the customer. When the service contract expires, Xerox takes the machine back for reuse or remanufacture and has a goal of sending no material to landfills or incinerators. To save money, machines are built to order and are designed to use recycled paper, have few parts, be energy efficient, and emit as little noise, heat, ozone, and copier chemicals as possible.

- Ray Anderson, CEO of a large carpet tile company, plans to lease rather than sell carpet (Individuals Matter, p. 528).

- For years, 160 firms, called *chauffagistes*, have been providing 10 million buildings in metropolitan France with heat. These firms provide *warmth services* by contracting to keep a client's space within a specified temperature during certain hours at a designated cost.

- Carrier, the world's leading maker of air-conditioning equipment, now sells leases to provide its customers with *cooling services*. Carrier also teams up with other service providers to install superwindows and more efficient lighting and make other energy-efficiency upgrades that reduce the cooling needs of its customers. Carrier makes money doing this by having to install less or even no air conditioning equipment.

- Dow and several other chemical companies are doing a booming business in leasing organic solvents (mostly used to remove grease from surfaces), photographic developing chemicals, and dyes and pigments. In this *chemical service* business, the company **(1)** delivers the chemicals, **(2)** helps the client set up a recovery system, **(3)** takes away the recovered chemicals, and **(4)** delivers new chemicals as needed.

As the service flow economy matures, companies may interact in a global network of service interactions resembling a food web. For example, many service companies may **(1)** lease all their office furniture and equipment, manufacturing equipment, and buildings as needed anywhere in the world and **(2)** hire other service firms on a contract basis to provide all their research and development, manufacturing, marketing, order-taking, delivery, accounting, and product maintenance needs. By not having large amounts of capital tied up in fixed equipment and personnel, such service companies could adapt more rapidly, flexibly and effectively to changes in technology and market forces.

Ray Anderson

Ray Anderson is CEO of Interface, an Atlanta-based company that makes carpet tiles. The company has 26 factories in 6 countries, customers in 110 countries, and more than $1 billion in annual sales.

Ray changed the way he viewed the world and his business after reading Paul Hawken's book *The Ecology of Commerce* (Guest Essay, p. 6). In 1994, he announced plans to develop the nation's first totally sustainable green corporation.

He has implemented hundreds of projects with the goals of (1) zero waste, (2) greatly reduced energy use, and (3) eventually zero use of fossil fuels by relying on renewable solar energy. By 1999, the company had reduced resource waste by almost 30% and reduced energy waste enough to save $100 million. One of Interface's factories in California is being run on solar cells to produce the world's first solar-made carpet.

To achieve the goal of zero waste, Anderson plans to stop selling carpet and lease it as a way to control recycling. For a monthly fee, the company will install, clean, and inspect the carpet on a monthly basis, repair worn carpet tiles overnight, and recycle worn-out tiles into new carpeting. As Anderson puts it, "We want to harvest yesterday's carpets and recycle them with zero scrap going to the landfill and zero emissions into the ecosystem—and run the whole thing on sunlight."

DuPont and several other chemical companies have developed processes to remove the nylon and plastic PVC fibers in carpet and recycle it into other lower-quality use products (open-loop recycling).

Interface has gone further and developed a new polymer material, called Solenium, that (1) when worn out can be completely recycled back into new carpet tiles (more desirable closed-loop recycling), (2) does not mildew, (3) is highly stain resistant, and (4) is easily cleaned with water. Making this material takes fewer steps and produces 99.7% less waste than making normal carpet, and the material lasts about four times longer than conventional carpet.

The company also has plans to install and lease a raised-floor system that goes beneath its carpet tiles and integrate this with cooling and heating services provided by other service companies.

Anderson is one of a growing number of business leaders committed to a finding a more economically and ecologically sustainable way of doing business while still making a profit for stockholders. Since he instituted his earth-friendly policies, the company's share price and earnings have increased, and he says he's having a blast.

21-4 REUSE

What Are the Advantages of Refillable Containers? *Reuse* is a form of waste reduction that (1) extends resource supplies, (2) keeps high-quality matter resources from being reduced to low-matter-quality waste (Figure 3-9, p. 57), and (3) reduces energy use (Figure 21-6) and pollution even more than recycling.

Two examples of reuse are refillable glass beverage bottles and refillable soft drink bottles made of polyethylene terephthalate (PET) plastic. Unlike throwaway and recyclable cans and bottles, refillable beverage bottles create local jobs related to their collection and refilling. Moreover, studies by Coca-Cola and PepsiCo of Canada show that their soft drinks in 0.5-liter (16-ounce) throwaway bottles cost one-third less in refillable bottles.

In 1964, 89% of all soft drinks and 50% of all beer in the United States were sold in refillable glass bottles. Today such bottles make up only about 7% of the beer and soft drink market, and only 10 states have refillable glass bottles. The disappearance of most local bottling companies has led to a loss of local jobs, income, and tax revenues. Some call for reinstatement of this bottling reuse system in the United States; others say it is not practical because the system of collections and returns has been dismantled. What do you think?

Denmark has led the way by banning all beverage containers that cannot be reused. To encourage use of refillable glass bottles, Ecuador has a refundable beverage container deposit fee that is 50% of the cost of the drink. In Finland, 95% of the soft drink, beer, wine, and spirits containers are refillable, and in Germany, 73% are refillable.

Other examples of reusable items are:

- Metal or plastic lunchboxes.

- Plastic containers for storing lunchbox items and refrigerator leftovers instead of using throwaway plastic wrap and aluminum foil.

- Cloth shopping bags (Solutions, right).

- Shipping pallets made of recycled plastic waste instead of throwaway wood pallets (400 million per year in the United States). In 1991, Toyota shifted entirely to reusable shipping containers. A similar move by the Xerox Corporation saves the company $2–5 million per year.

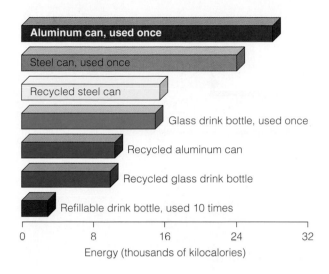

Figure 21-6 Energy consumption for different types of 35-milli-liter (12-fluid-ounce) beverage containers. (Data from Argonne National Laboratory)

- *Tool libraries* (such as those in Berkeley, California, and Takoma Park, Maryland) where people can check out a variety of power and hand tools.

- Photocopiers that strip off old toner so that a sheet of paper can be reused up to 10 times.

- E-paper, a flexible and cordless computer screen being developed by Xerox, that **(1)** looks like a sheet of paper, **(2)** uses no energy for storing or viewing writing or images, and **(3)** can be electronically written and rewritten at least a million times, making it equivalent to more than a million sheets of paper.

21-5 RECYCLING

What Are the Two Types of Recycling? Recycling has a number of benefits to people and the environment (Figure 21-7). There are two types of recycling for materials such as glass, metals, paper, and plastics:

- *Primary*, or *closed-loop*, *recycling*, in which wastes discarded by consumers (*postconsumer wastes*) are recycled to produce new products of the same type (such as newspaper into newspaper and aluminum cans into aluminum cans). This reduces use of virgin resources and pollution and saves energy (Figure 21-6).

- *Secondary*, or *open-loop*, *recycling*, in which waste materials are converted into different and usually lower-quality products.

Primary recycling reduces the amount of virgin materials in a product by 20–90%, whereas secondary recycling reduces virgin material by 25% at most.

Environmentalists urge us not to be misled by labels claiming that paper and plastic bags or other items are recyclable. Just about anything is in theory recyclable. What counts is **(1)** whether an item is actually recycled and **(2)** whether we complete the recycling loop by buying products using the maximum feasible content of postconsumer recycled materials.

Case Study: Recycling Municipal Solid Waste in the United States A material that is a good candidate for recycling is **(1)** easily isolated from other wastes, **(2)** available in large quantities in a fairly uniform form, and **(3)** valuable. In 1999, about 30% of U.S. MSW was recycled or composted (Solutions, p. 531)—the highest

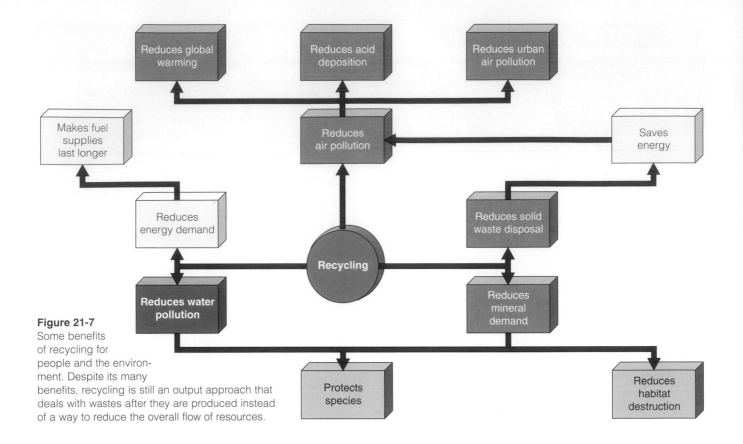

Figure 21-7
Some benefits of recycling for people and the environment. Despite its many benefits, recycling is still an output approach that deals with wastes after they are produced instead of a way to reduce the overall flow of resources.

rate of any industrialized country. The United States has more than 8,800 municipal curbside recycling programs serving 51% of the population.

These programs recycle **(1)** 98% of the steel used in cars, **(2)** 96% of car batteries, **(3)** 74% of aluminum cans, **(4)** 70% of lead, **(5)** 43% of wastepaper and paperboard, **(6)** 40% of yard waste, and **(7)** 27% of glass containers. Recent pilot studies in several U.S. communities show that 60–80% recycling and composting rates are possible.

Studies show that one of the best ways to encourage recycling is a *pay-as-you-throw* program that bases garbage collection charges on the amount of waste a household generates for disposal; materials sorted out for recycling are hauled away free. Currently, more than 2,800 communities in North America have such systems. At least 11 U.S. cities have 45–60% recycling rates.

Is Centralized Recycling of Mixed Solid Waste the Answer? Large-scale recycling can be accomplished by collecting mixed urban waste and transporting it to centralized *materials-recovery facilities (MRFs)*. There, machines shred and automatically separate the mixed waste to recover valuable materials for sale to manufacturers as raw materials (Figure 21-8). The remaining paper, plastics, and other combustible wastes are recycled or burned to produce steam or electricity to run the recovery plant or to sell to nearby

industries or homes. Ash from the incinerator is buried in a landfill.

Currently, there are more than 225 MRFs in the United States. However, such plants **(1)** are expensive to build, operate, and maintain (which is why some have been shut down), **(2)** can emit toxic air pollutants if not operated properly, and **(3)** produce a toxic ash that must be disposed of safely. MRFs also must have a large input of garbage to make them financially successful. Thus, their owners have a vested interest in increased *throughput* of matter and energy resources to produce more trash—the reverse of what prominent scientists believe we should be doing (Figure 21-4).

Is Separating Solid Wastes for Recycling the Answer? Many solid-waste experts argue that it makes more sense economically and environmentally for households and businesses to separate trash into recyclable and reusable categories (such as glass, paper, metals, certain types of plastics, and compostable materials). Then compartmentalized city collection trucks, private haulers, or volunteer recycling organizations pick up the segregated wastes and sell them to scrap dealers, compost plants, and manufacturers. Another alternative (especially in less populated areas) is to establish a network of drop-off centers, buyback centers, and deposit-refund programs in which people

Using Composting to Recycle Biodegradable Wastes

Biodegradable organic waste such as paper, food scraps, and lawn waste is a valuable resource that can composted to produce plant nutrients that can be recycled to the soil. Biodegradable wastes make up about 35% by weight of the MSW output in the United States and could be converted to compost (p. 232).

However, only about 5% of the MSW in the United States is composted (compared to 17% in France and 10% in Switzerland). Some cities in Austria, Belgium, Denmark, Germany, Luxembourg, and Switzerland recover and compost more than 85% of their biodegradable wastes.

Individuals can compost such wastes in backyard bins (Individu-als Matter, p. 234) or indoor containers. They can also be collected and composted in centralized community facilities, as is done in many western European countries.

The resulting compost can be used as an organic soil fertilizer or conditioner, as topsoil, or as a landfill cover. Compost also can be used to help restore eroded soil on hillsides and along highways, strip-mined land, overgrazed areas, and eroded cropland.

To be successful, a large-scale composting program must (1) overcome siting problems (few people want to live near a giant compost pile or plant), (2) control odors, and (3) exclude toxic materials that can contaminate the compost and make it unsuitable for use as a fertilizer on crops and lawns.

Three ways to control or reduce odors for large-scale composting operations are (1) enclosing the facilities and filtering the air inside (but residents near large composting plants still complain of unacceptable odors), (2) creating municipal compost operations near existing landfills or at other isolated sites, and (3) decomposing biodegradable wastes in a closed metal container in which air is recirculated to give precise control of available oxygen and temperature (a technique that has been used successfully in the Netherlands for 20 years).

Critical Thinking

How would you go about increasing the rate of composting of biodegradable wastes in your community?

---

deliver and either sell or donate their separated recyclable materials.

This source separation approach (1) produces little air and water pollution, (2) has low startup costs and moderate operating costs, (3) saves more energy and provides more jobs per unit of material than MRFs, landfills, and incinerators, (4) yields cleaner and usually more valuable recyclables, and (5) educates people about the need for waste reduction, reuse, and recycling.

Aluminum and paper separated out for recycling are worth a lot of money. As a result, in a growing number of cities people steal these materials—from curbside containers set out by residents and from unprotected recycling drop-off centers—and sell them. This undermines municipal recycling programs by lowering the income available from selling these high-value materials.

Does Recycling Make Economic Sense? The answer is yes and no, depending on different ways of looking at the economic and environmental costs and benefits of recycling. Critics contend that recycling (1) has become almost a religion that is above criticism regardless of how much it costs communities, (2) does not make sense if it costs more to recycle materials than to send them to a landfill or incinerator (as is the case in some areas), (3) often is not needed to save landfill space because many areas in the United States are not run-ning out of landfill space, and (4) may make economic sense for valuable and easy-to-recycle materials (such as aluminum, paper, and steel) but not for cheap or plentiful resources (such as glass from silica) and most plastics (which are expensive to recycle).

However, recycling proponents argue that:

- Recycling programs should not be judged on whether they pay for themselves any more than are conventional garbage disposal systems based on land burial or incineration.

- The primary benefit of recycling is not a reduction in the use of landfills and incinerators but the other important benefits it provides for people and the environment (Figure 21-7).

- Studies show that the net economic, health, and environmental benefits of recycling (Figure 21-7) far outweigh the costs.

- Programs with (1) high recycling rates, (2) a single pickup system (for both materials to be recycled and garbage that cannot be recycled) instead of a more expensive dual collection system, and (3) a pay-as-you-throw system tend to make money and have higher recycling rates.

Why Don't We Have More Reuse and Recycling? Three factors that hinder recycling (and reuse) are

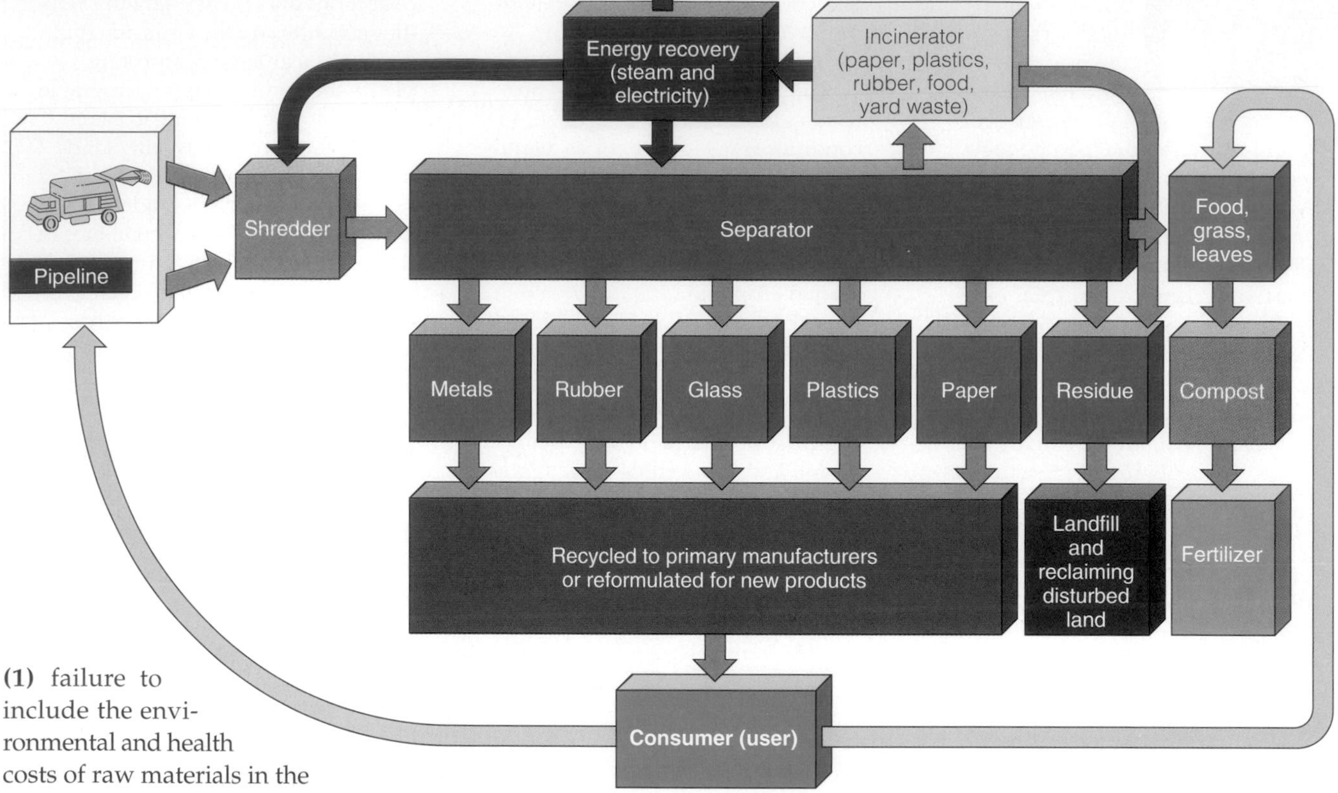

Figure 21-8 Schematic of a generalized materials-recovery facility used to sort mixed wastes for recycling and burning to produce energy. Because such plants need high volumes of trash to be economical, they discourage reuse and waste reduction.

(1) failure to include the environmental and health costs of raw materials in the market prices of consumer items, (2) more tax breaks and subsidies for resource-extracting industries ($2.6 billion per year in the United States) than for recycling and reuse industries, and (3) lack of large, steady markets for recycled materials.

Analysts have suggested that these obstacles can be overcome by:

- Taxing virgin resources and phasing out subsidies for extraction of virgin resources.

- Lowering or eliminating taxes on recycled materials based on postconsumer waste content.

- Providing subsidies for reuse and postconsumer waste recycling.

- Greatly increasing use of the *pay-as-you-throw* system.

- Encouraging or requiring government purchases of recycled products to help increase demand and lower prices. The U.S. government, which accounts for 2% of all paper bought in the country, has mandated that its purchases have a minimum of 30% postconsumer recycled paper content.

- Viewing landfilling and incineration of solid wastes as last resorts to be used only for wastes that cannot be reused, composted, or recycled (Figure 21-4).

- Requiring labels on all products listing preconsumer and postconsumer recycled content.

Countries such as Germany are leading the way in recycling, reuse, and waste reduction (Case Study, right).

21-6 CASE STUDIES: RECYCLING ALUMINUM, WASTEPAPER, AND PLASTICS

How Much Aluminum Is Being Recycled? The United States gets 60% of its aluminum from virgin ore. Recycling the remaining 40% produces 95% less air pollution and 97% less water pollution and uses 95% less energy than mining and processing aluminum ore. In 1999, 74% of aluminum beverage cans produced in the United States were recycled. Despite this progress, about 26% of the 105 billion aluminum cans produced in 1999 in the United States were thrown away. Laid end to end, these cans would wrap around the planet more than 120 times.

Recycling aluminum cans is great, but many environmentalists believe that these cans are an example of an unnecessary item that could be replaced by more

Recycling, Reuse, and Waste Reduction in Germany

In 1991, Germany enacted the world's toughest packaging law, which was designed to **(1)** reduce the amount of waste being landfilled or incinerated, **(2)** reduce waste production, and **(3)** recycle or reuse 65% of the nation's packaging by 1995. Product distributors must take back their boxes and other containers for reuse or recycling, and incineration of packaging is not allowed.

To implement the system, more than 600 German manufacturers and distributors pay fees to a nonprofit service company they formed to collect, sort, and reprocess the packaging discards (coded with a green dot) of member firms.

The system has been so successful that in 1997, 86% of all packaging in Germany was recycled (up from 12% in 1992). Plastic packaging has lost one-third of its market share to glass and cardboard, and four out of five German manufacturers have reduced their use of packaging by 17%. About 75% of all produce shipped through Germany is boxed in leased, reusable shipping crates.

Inspired by the German system, more than 28 countries have enacted *take-back* laws for packaging materials, 16 have done so for batteries, and 12 are considering such legislation for electronic items. In 1998, Taiwan started a take-back program for computers, televisions, and large home appliances. In the Netherlands, all packaging waste is banned from landfills.

Critical Thinking

What might be some long-term disadvantages of Germany's program? Do you believe that a similar program should be put into place in the country where you live? Explain.

---

energy-efficient (Figure 21-6) and less polluting refillable glass or PET plastic bottles—a switch from recycling to reuse that also creates local jobs and stimulates local economies. One way to encourage this change would be to place a heavy tax on nonrefillable containers and no tax on reusable beverage containers, as is done in at least nine countries.

How Is Paper Made, and How Much Is Being Recycled? With about 27% of the world's population, the United States, Japan, and China use about 71% of the world' paper. The amount of paper produced in the United States in 1999 would make a pile reaching to the moon and back more than eight times. Per capita use of paper and paperboard in the United States is more than two times that of other industrial countries as a whole and 19 times that of developing countries (56 times that of Africa).

It takes about 3 metric tons (6,600 pounds) of trees and 89 metric tons (196,000 pounds) of other resources such as water and chemicals to make 1 metric ton (2,200 pounds) of paper. The pulp and paper industry **(1)** is the world's fifth largest industrial user of energy and **(2)** uses more water to produce a metric ton of product than any other industry.

To make paper, trees are stripped of their bark and chipped into small pieces. A chemical such as caustic soda (sodium hydroxide) or a mechanical grinder is then used to separate or pulp the wood in the chips to a soft mush or pulp of cellulose fibers. The brownish pulp is **(1)** rinsed and washed several times (which yields large volumes of pulping liquor that are either treated or in many developing countries dumped directly into waterways), **(2)** often bleached to make it white (typically using toxic chlorine or chlorine compounds), and **(3)** pressed into a thin sheet and dried.

Recycling paper involves removing its ink, glue, and coating and reconverting it to pulp that is pressed again into new paper. The perception that recycled paper is weak, coarse, flecked and too expensive is no longer true.

Globally, about **(1)** 50% of the fiber used to make paper comes from virgin wood, **(2)** 43% comes from recycled fibers, and **(3)** 7% comes from nonwood fibers such as wheat straw and hemp. In 1999, the United States recycled about 43% of its wastepaper (up from 25% in 1989) and 70% of its corrugated cardboard containers. At least 10 other countries recycle 50-96% of their wastepaper and paperboard, with a global recycling rate of 43%. Despite a 43% recycling rate, the amount of paper thrown away each year in the United States is more than all of the paper consumed in China (where the recycling rate is only 27%).

Recycling paper **(1)** does not involve cutting new trees, **(2)** saves energy because it takes 30–64% less energy to produce the same weight of recycled paper as to make the paper from trees, **(3)** reduces air pollution from pulp mills by 74–95%, **(4)** lowers water pollution by 35%, **(5)** helps prevent groundwater contamination by toxic ink left after paper rots in landfills over a 30- to 60-year period, **(6)** conserves large quantities of water, **(7)** takes little or no bleaching because the fibers recycled from white paper have already been bleached, **(8)** can save landfill space, **(9)** creates five times more jobs than harvesting trees for pulp, and **(10)** can save money.

Currently, there are several zero-effluent paper mills in the world that use only recycled paper and reuse water, chemicals, and other materials in a closed-loop system that eliminates any releases of pollutants to the air, water, or land. Some mills are experimenting with *biopulping*, in which local fungi are used to break down the wood chips into cellulose fiber. This reduces energy use, improves the water quality of effluents, and increases paper strength.

Chlorine (Cl_2) and chlorine compounds (such as chlorine dioxide, ClO_2) used to bleach at least 40% of the world's paper (1) corrode processing equipment, (2) are hazardous for workers, (3) are hard to recover and reuse, and (4) are extremely harmful when released to the environment. A growing number of paper mills (mostly in Europe) have replaced chlorine-based bleaching chemicals with oxygen-based chemicals such as hydrogen peroxide (H_2O_2) or ozone (O_3). Such pollution prevention processes (1) nearly eliminate the release of air pollutants (including highly toxic chlorine-containing dioxin, p. 542), (2) use less water and energy, (3) allow reuse of the water many times, (4) reduce the treatment needed for water that is discharged, and (5) save money. Worldwide, about 6% of the world's pulp (27% in Scandinavia but only 1% in the United States and Canada) is bleached without using chlorine.

Buying recycled paper products can save trees and energy and reduce pollution, but it does not necessarily reduce solid waste. Only products made from *postconsumer waste*—waste intercepted on its way from consumer to the landfill or incinerator—does that. The Bank of America recycles about 61% of its paper, saving about $500,000 a year, and 75% of the paper it purchases has some postconsumer recycled content. The U.S. Postal Service, United Parcel Service, and Airborne Express use more than 80% postconsumer wastepaper for boxes and paperboard.

Most recycled paper is made from *preconsumer waste*: scraps and cuttings recovered from paper and printing plants. Because paper manufacturers have always recycled this waste, it has never contributed to landfill problems. Now this paper is labeled "recycled" as a marketing ploy, giving the false

impression that people who buy such products (often at higher prices) are helping to reduce solid waste. Most "recycled" paper has no more than 50% recycled fibers, with only 10% from postconsumer waste. Environmentalists propose that governments require companies to use labels giving postconsumer recycled content and indicating whether the paper was bleached with chlorine or by a chlorine-free process.

Is It Feasible to Recycle Plastics? Plastics are various types of polymer molecules made by chemically linking monomer molecules (petrochemicals) produced mostly from oil and natural gas (Figure 21-9). Plastics now account for about 9% by weight and 22% by volume of MSW in the United States and about 60% of the debris found on U.S. beaches.

Currently, only about 7% by weight of all plastic wastes and 10% of plastic containers in the United States are recycled, mostly because

- Plastics are very difficult to isolate from other wastes because they (1) occur in so many different and often difficult to identify forms of resins, (2) sometimes consist of composites or laminated layers of different plastics, and (3) contain stabilizers and other chemicals that must be removed before recycling.

- Recovering individual plastic resins does not yield much material because only small amounts of any given resin are used per product.

- The price of oil is so low (Figure 14-18, p. 339) that the cost of virgin plastic resins (except for PET, used mostly in plastic drink bottles) is about 40% lower than that of recycled resins.

Thus, mandating that plastic products contain a certain amount of recycled plastic resins is unlikely to work and could hinder the use of plastics in reducing the resource content and weight (Solutions, p. 525) of many widely used items such as plastic bags, bottles, and other containers.

In 1998, the Daimler Chrysler Corporation built an experimental automobile with a body made from plastic derived entirely from recycled PET bottles and expects to sell such vehicles within a

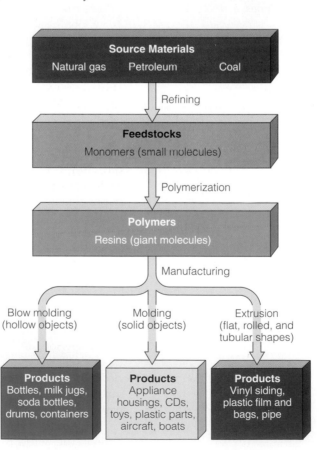

Figure 21-9 How plastics are made. (Adapted from the Society of the Plastics Industry)

decade. Chrysler says the plastic cars will be cheaper and lighter and get better gas mileage. In 2000, Cargill and Dow Chemical announced a joint venture to develop a facility in Nebraska to make a polymer resin from renewable resources, such as corn, that can be used in items such as clothing, food containers, and home and office furnishings and eventually in bottles and foams.

Environmentalists recognize the beneficial qualities of plastics: **(1)** durability (in products such as car and machine parts, carpeting, toys, furniture, reusable tubs and containers, and refillable bottles), **(2)** light weight, **(3)** unbreakability (compared to glass), and **(4)** in some cases reusability as containers.

On the other hand,

- The plastics industry is among the leading producers of hazardous waste.

- In landfills, toxic cadmium and lead compounds used as binders, colorants, and heat stabilizers can leach out of plastics into groundwater and surface water.

- Most plastics are nondegradable or take 200-400 years to degrade.

- Some widespread uses of plastics—especially excessive and often unnecessary single-use packaging and throwaway beverage and food containers—could be sharply reduced and replaced with less harmful and less wasteful alternatives.

21-7 DETOXIFYING, BURNING, BURYING, AND EXPORTING WASTES

How Can Hazardous Waste Be Detoxified? Denmark has the world's most comprehensive and effective hazardous-waste detoxification program. Each Danish municipality has at least one facility that accepts paints, solvents, and other hazardous wastes from households. Hazardous and toxic waste from industries is delivered to 21 transfer stations throughout the country. All waste is then transferred to a large treatment facility, where about 75% of it is detoxified and what's left is buried in a carefully designed and monitored landfill.

Some consider biological treatment of hazardous waste, or *bioremediation*, to be the wave of the future for cleaning up some types of toxic and hazardous waste. In this process, microorganisms (usually natural or genetically engineered bacteria) and enzymes are used to destroy toxic or hazardous substances or convert them to harmless compounds. This approach mimics nature by using decomposers to recycle matter but speeds up the process by using catalysts and altering the temperature, pH, moisture, oxygen, and nutrients.

Contaminated soil and water are treated on site (*in situ*), eliminating the need to remove and transport them to a cleanup facility.

Bioremediation is effective for a number of organic wastes, including pesticides, gasoline, diesel fuel, polychlorinated biphenyls (PCBs), and organic solvents. However, it does not appear to work very well for **(1)** toxic metals, **(2)** highly concentrated chemical wastes, or **(3)** complete digestion of some complex mixtures of toxic chemicals.

Another biological way to treat hazardous wastes is *phytoremediation*, which involves using natural or genetically engineered plants to filter and remove contaminants, again mimicking nature. Selected plants can be used to clean up soil and water contaminated with chemicals such as pesticides, organic solvents, radioactive metals, and toxic metals such as lead and mercury.

For example, fast-growing poplar trees can be planted at contaminated industrial or other sites. The trees act like straws to suck up contamination from soil and groundwater and either safely store the chemicals in their tissues or metabolize them into harmless compounds released into the atmosphere. Field tests show that special strains of sunflower plants can remove 95% of lead and other toxic contaminants from soil and water within 24 hours. The mulberry bush is effective on industrial sludge, mustard plants absorb lead, and canola plants soak up excess selenium in the soil. Planting vegetation on a site also reduces soil erosion by water and wind. Plants are also used to remove contaminants in sewage treatment systems that work with nature (Figure 19-1, p. 476, and Figure 19-17, p. 497).

Phytoremediation **(1)** is inexpensive, **(2)** does not involve heavy machinery that produces air pollution, and **(3)** can reduce the amount of material dumped in landfills. On the other hand, **(1)** it is often slow (it can take several growing seasons to clean a site), **(2)** it is effective only at depths that plant roots can reach, and **(3)** in some cases animals may feed on pollutant-containing leaves.

Is Burning Solid and Hazardous Waste the Answer? In the United States, about 20% of the mixed trash in municipal solid waste in the United States is combusted in about 170 *mass-burn incinerators* (Figure 21-10). About 80% of the hazardous waste is burned in 172 commercial incinerators, cement kilns, and lightweight aggregate kilns, and the other 20% is combusted in industrial boilers and other types of industrial furnaces. In Japan—which has little space for landfills—most hazardous waste and MSW is burned in more than 3,800 highly polluting incinerators, which emit large amounts of toxic dioxins. By 2002, U.S. facilities burning hazardous waste must meet more stringent air pollution standards to reduce emissions of hazardous air pollutants such as dioxin (p. 542), lead (p. 540), and

Figure 21-10 Schematic of a waste-to-energy incinerator with pollution controls that burns mixed solid waste and recovers some of the energy to produce steam used for heating or producing electricity. (Adapted from EPA, *Let's Reduce and Recycle*)

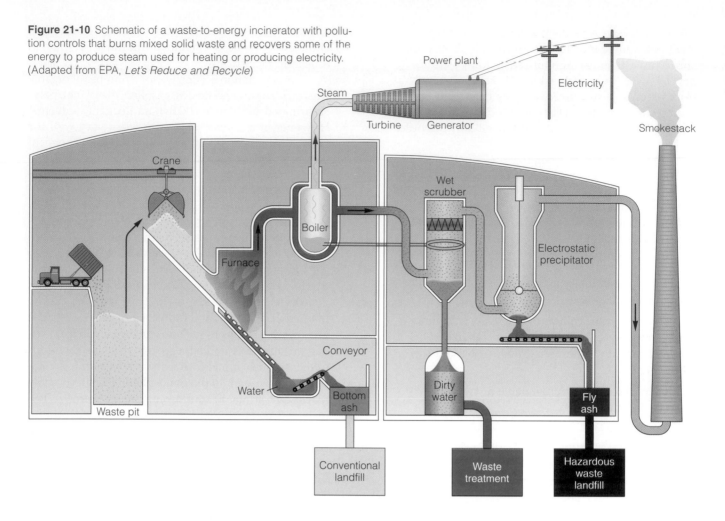

mercury along with other criteria air pollutants (Table 17-2, p. 422). Figure 21-11 lists the pros and cons of using incinerators to burn solid and hazardous waste.

Since 1985, there has been a decrease in the use of incineration for treating wastes in some parts of the world because of high costs, health threats from air pollution, and intense citizen opposition. For example, **(1)** Sweden banned the construction of new incinerators in 1985, **(2)** Rhode Island and West Virginia banned solid-waste incineration in 1992, **(3)** several solid-waste incinerators in the United States have been shut down because of excessive costs and pollution, **(4)** more than 280 new incinerator projects have been blocked, delayed, or canceled in the United States since 1985, **(5)** in 1999, the Philippines became the first country to ban all waste incineration, followed by Costa Rica, and **(6)** a growing number of hospitals are destroying infectious material by using steam heat and pressure in autoclaves (which are cheaper to run than incinerators) and other non-incinerator methods for treating medical wastes.

Is Land Disposal of Solid Waste the Answer? Currently, about 54% by weight of the MSW in the United States is buried in sanitary landfills (compared to 90% in the United Kingdom, 80% in Canada, 15% in Japan, and 12% in Switzerland). In a **sanitary landfill**, solid wastes are spread out in thin layers, compacted, and covered daily with a fresh layer of clay or plastic foam.

Modern state-of-the-art landfills on geologically suitable sites are lined with clay and plastic before

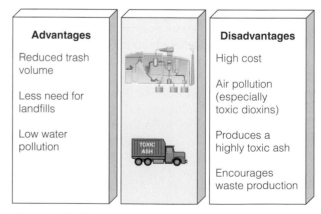

Advantages		Disadvantages
Reduced trash volume		High cost
Less need for landfills		Air pollution (especially toxic dioxins)
Low water pollution		Produces a highly toxic ash
		Encourages waste production

Figure 21-11 Advantages and disadvantages of incinerating solid and hazardous waste.

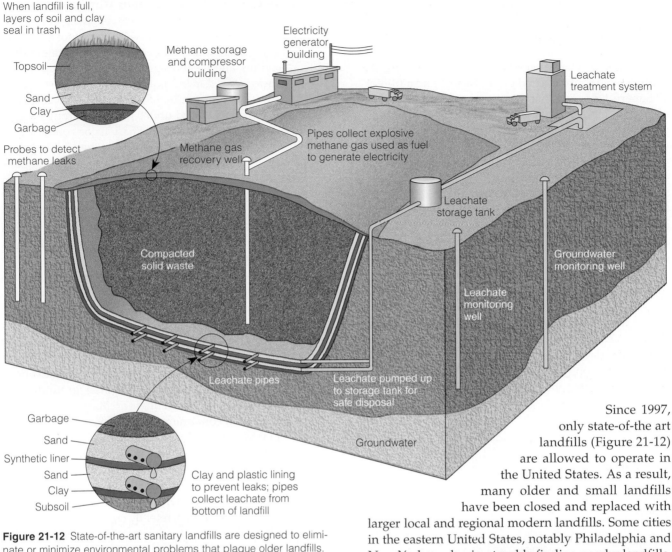

When landfill is full, layers of soil and clay seal in trash

Topsoil

Sand

Clay

Garbage

Probes to detect methane leaks

Methane storage and compressor building

Electricity generator building

Methane gas recovery well

Pipes collect explosive methane gas used as fuel to generate electricity

Leachate treatment system

Leachate storage tank

Leachate storage tank

Compacted solid waste

Groundwater monitoring well

Leachate monitoring well

Leachate pipes

Leachate pumped up to storage tank for safe disposal

Groundwater

Garbage

Sand

Synthetic liner

Sand

Clay

Subsoil

Clay and plastic lining to prevent leaks; pipes collect leachate from bottom of landfill

Figure 21-12 State-of-the-art sanitary landfills are designed to eliminate or minimize environmental problems that plague older landfills. Even such state-of-the-art landfills are expected to leak eventually, passing both the effects of contamination and cleanup costs on to future generations.

being filled with garbage (Figure 21-12). The bottom is covered with a second impermeable liner, usually made of several layers of clay, thick plastic, and sand. This liner collects *leachate* (rainwater contaminated as it percolates through the solid waste) and is intended to prevent its leakage into groundwater. Collected leachate is pumped from the bottom of the landfill, stored in tanks, and sent to a regular sewage treatment plant or an on-site treatment plant. When full, the landfill is covered with clay, sand, gravel, and topsoil to prevent water from seeping in. Several wells are drilled around the landfill to monitor any leakage of leachate into nearby groundwater. Figure 21-13 lists the pros and cons of using sanitary landfills for disposal of solid wastes.

Since 1997, only state-of-the art landfills (Figure 21-12) are allowed to operate in the United States. As a result, many older and small landfills have been closed and replaced with larger local and regional modern landfills. Some cities in the eastern United States, notably Philadelphia and New York, are having trouble finding nearby landfills. However, there is no shortage of landfill space in most parts of the United States.

These new landfills are equipped with a connected network of vent pipes to collect landfill gas (consisting mostly of two greenhouse gases, methane and carbon dioxide) released by the underground (anaerobic) decomposition of wastes. The methane is filtered out and burned in small gas turbines to produce steam or electricity (Figure 21-12) for nearby facilities (Individuals Matter, p. 541) or sold to utilities.

However, thousands of older and abandoned landfills do not have such systems and will emit methane and carbon dioxide, both potent greenhouse gases, for decades.

A 1998 study by the New York State Department of Health found that women living near solid-waste landfills where toxic gases are escaping have a 400% greater risk of bladder cancer or leukemia (cancer of blood-forming cells). Several other studies of landfills in 339 U.S. counties, Canada, and Germany found increased incidence of leukemias and cancers of the bladder, liver, prostate, lung, and cervix.

Contamination of groundwater and nearby surface water by leachate from unlined and lined older landfills is a serious problem. Some 86% of older U.S. landfills studied have contaminated groundwater, and a fifth of all Superfund hazardous-waste sites (p. 543) are former municipal landfills that will cost billions of dollars to clean up.

Modern double-lined landfills (Figure 21-12) delay the release of toxic leachate into groundwater below landfills but do not prevent it. These landfills are designed to accept waste for 10–40 years, and current EPA regulations require owners to maintain and monitor landfills for at least 30 years after they are closed. However, they could begin to leak after this period, passing the health risks and costs of contamination to future generations.

According to G. Fred Lee, an experienced landfill consultant, the best solution to the leachate problem is to apply clean water to landfills continuously and then collect and treat the resulting leachate in carefully designed and monitored facilities. He contends that after 10–20 years of such washing, little potential for groundwater pollution should remain. This wetting would also hasten the breakdown of wastes and thus allow old landfills to be dug out and used again.

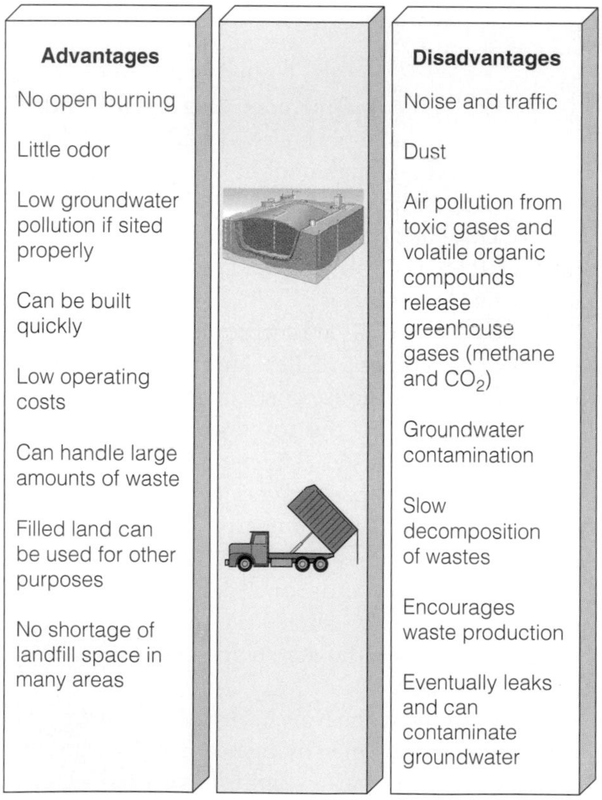

Is Land Disposal of Hazardous Waste the Answer?

Hazardous waste in the United States is disposed of on land in (1) deep underground wells, (2) surface impoundments such as ponds, pits, or lagoons, (3) state-of-the-art landfills, and (4) aboveground storage facilities.

In *deep-well disposal*, liquid hazardous wastes (such as cleaning solutions, metals, cyanides, and corrosive solutions) are pumped under pressure through a pipe into dry, porous geologic formations or zones of rock far beneath aquifers tapped for drinking and irrigation water (Figure 19-9, p. 487). In theory, these liquids soak into the porous rock material and are isolated from overlying groundwater by essentially impermeable layers of rock. Figure 21-14 lists the pros and cons of deep-well disposal of liquid hazardous wastes. Many scientists believe that current regulations for deep-well disposal are inadequate and need to be improved.

Surface impoundments are excavated depressions such as ponds, pits, or lagoons into which liquid hazardous wastes are drained and stored (Figure 19-9, p. 487). As water evaporates, the waste settles and becomes more concentrated. Figure 21-15 lists the pros and cons of this method. EPA studies found that 70% of these storage basins in the United States have no liners, and as many as 90% may threaten groundwater. According to the EPA, all liners are likely to leak eventually and can contaminate groundwater.

Liquid and solid hazardous waste also can be put into drums or other containers and buried in carefully designed and monitored *secure hazardous-waste landfills* (Figure 21-16). Sweden goes further and buries its concentrated hazardous wastes in underground vaults made of reinforced concrete. By contrast, in the United Kingdom, most hazardous wastes are mixed with

Advantages		Disadvantages
No open burning		Noise and traffic
Little odor		Dust
Low groundwater pollution if sited properly		Air pollution from toxic gases and volatile organic compounds release greenhouse gases (methane and CO_2)
Can be built quickly		
Low operating costs		
Can handle large amounts of waste		Groundwater contamination
Filled land can be used for other purposes		Slow decomposition of wastes
No shortage of landfill space in many areas		Encourages waste production
		Eventually leaks and can contaminate groundwater

Figure 21-13 Advantages and disadvantages of using sanitary landfills to dispose of solid waste.

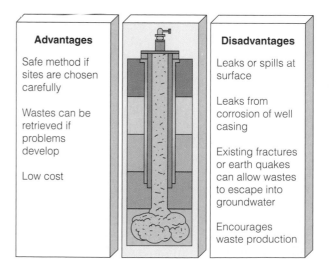

Advantages		Disadvantages
Safe method if sites are chosen carefully		Leaks or spills at surface
Wastes can be retrieved if problems develop		Leaks from corrosion of well casing
Low cost		Existing fractures or earth quakes can allow wastes to escape into groundwater
		Encourages waste production

Figure 21-14 Advantages and disadvantages of injecting liquid hazardous wastes into deep underground wells.

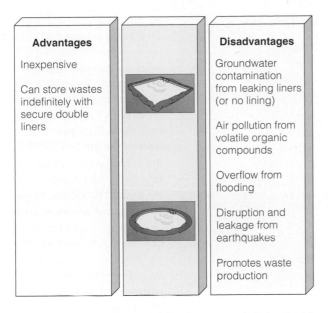

Advantages		Disadvantages
Inexpensive		Groundwater contamination from leaking liners (or no lining)
Can store wastes indefinitely with secure double liners		Air pollution from volatile organic compounds
		Overflow from flooding
		Disruption and leakage from earthquakes
		Promotes waste production

Figure 21-15 Advantages and disadvantages of storing liquid hazardous wastes in surface impoundments.

household garbage and stored in hundreds of conventional landfills throughout the country.

With proper design and maintenance, secure landfills can effectively store hazardous wastes. However, **(1)** this method is expensive, **(2)** burrowing animals may make holes in the clay cap, **(3)** the liners are likely to leak eventually, and **(4)** it discourages hazardous waste reduction.

Hazardous wastes also can be stored in carefully designed *aboveground buildings* (Figure 21-17). These two-story buildings **(1)** are built of reinforced concrete

Figure 21-16 Secure hazardous-waste landfill.

Gas vent · Topsoil · Earth · Sand · Plastic cover · Impervious clay cap

Clay cap
Bulk waste
Impervious clay
Earth
Water table
Groundwater

Leak detection system
Reactive wastes in drums
Groundwater monitering well

Double leachate collection system · Plastic double liner

to prevent damage by storms and hurricanes and to help contain any leakage and **(2)** use fans and filters to create a negative air pressure to prevent the release of toxic gases.

The first floor contains no wastes but has inspection walkways so people can easily check for leaks from the upper story. Any leakage is collected, treated, solidified, and returned to the storage building. Earthquakes could damage such structures, but this is also a potential problem for all other methods for storing hazardous wastes.

There is also growing concern about accidents during some of the more than 500,000 shipments of hazardous wastes (mostly to landfills and incinerators in trucks or by train) in the United States each year. On average, trucks and trains carrying hazardous materials are involved in about 13,000 accidents per year in the United States. In a typical year, these accidents kill about 100 people, cause more than 10,000 injuries, and require evacuation of more than 500,000 people. Most communities do not have the equipment and trained personnel to deal with hazardous-waste spills.

Is Exporting Hazardous Waste the Answer?
Some hazardous-waste producers in the United States and several other industrialized countries have been getting rid of some of these wastes by legally (or illegally) shipping it to other countries, especially developing countries.

Waste disposal firms can charge high prices for picking up hazardous wastes. If they can then dispose of them (legally or illegally) at low costs, they pocket huge profits.

In 1989, countries met and drew up the Basel Convention on Hazardous Waste. It requires exporters to get approval from the recipient nation before a shipment of hazardous wastes can be sent but did not ban trade in such wastes. By 1994, more than 100 developing countries had unilaterally decided to ban hazardous waste imports.

In 1995, the Basel convention was strengthened to ban all hazardous waste exports from developed countries to developing countries. If ratified by enough countries and enforced, this ban on hazardous-waste exports will help. However, this would not end illegal trade in these wastes because the potential profits are much too great. To most environmental scientists, the only real solution to the hazardous-waste problem is to produce as little as possible in the first place (Figure 21-5, p. 522).

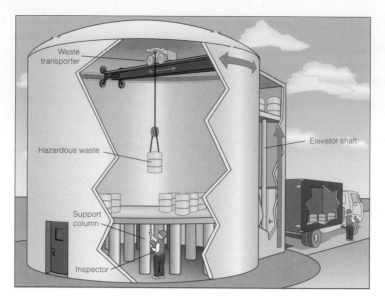

Figure 21-17 Above-ground hazardous-waste storage facilities can help keep wastes from contaminating groundwater. So far, there has been little use and evaluation of this method. Some scientists and engineers believe that with careful design this would be a safer way to deal with hazardous wastes than incinerators, deep wells, surface impoundments, and secure landfills.

Labels in figure: Waste transporter · Hazardous waste · Support column · Inspector · Elevator shaft

21-8 CASE STUDIES: LEAD, DIOXINS, AND CHLORINE

How Can We Reduce Exposure to Lead? Each year, 12,000-16,000 American children under age 9 are treated for acute lead poisoning, and about 200 die. About 30% of the survivors suffer from palsy, partial paralysis, blindness, and mental retardation.

Lead can also cause damage at levels far below those that cause acute lead poisoning, especially in children and unborn fetuses. Research indicates that children under age 6 and unborn fetuses with fairly low blood levels of lead are especially vulnerable to (1) nervous system impairment, (2) a lowered IQ (by 4-7 points), (3) a shortened attention span, (4) hyperactivity, (5) hearing damage, and (6) various behavior disorders. A 5-point drop in IQ—the main effect of lead poisoning—puts 50% more children in the low IQ (80 or less) category and reduces the number of children with high IQs.

It is *great news* that between 1976 and 1995 the percentage of U.S. children ages 1–5 with lead levels above the current blood level standard dropped from 85% to 4.4%, preventing at least 9 million childhood lead poisonings. The primary reason is that government regulations banned leaded gasoline in 1976 (with a gradual phaseout by 1986) and lead-based paints in 1970 (but illegal use continued until about 1978).

In 1923, researchers at General Motors concluded that ethyl alcohol (ethanol) was a better and safer chemical to add to gasoline to reduce engine knock than tetraethyl lead (TEL). However, because TEL could be patented and ethanol could not, General Motors executives decided it would be more profitable to use TEL—a decision opposed by Alice Hamilton (p. 36). As a result, 6.4 million metric tons (7 million tons) of TEL were added to gasoline between 1923 and 1986, and

68 million American children were exposed to lead. Because lead is nondegradable, this 63-year legacy of leaded gasoline use in the United States left a residue of about 5.9 million metric tons (13 billion pounds) of lead in the environment in the form of a fine, toxic dust, much of which is still moving around in soil and house dust. Leaded gasoline is still being used in many other countries, with winds depositing some particles of this lead onto the surface waters and land of countries throughout the world.

In addition, most of the estimated 4.5 million metric tons (10 billion pounds) of lead used in paint between 1889 and 1979 in the United States remains in homes and buildings, slowly deteriorating into toxic dust and chips that toddlers can breathe or pick up with their hands and transfer to their mouths.

Some *bad news* is that even with the encouraging drop in average blood levels of lead, the U.S. Centers for Disease Control and Prevention estimate that at least 890,000 U.S. children still have unsafe levels of lead from exposure to a number of sources (Figure 21-18). In addition, a 1993 study by the National Academy of Sciences and numerous other studies indicate that there is no safe level of lead in children's blood. This means that several million children in the United States (and an estimated 130-200 million children throughout the world) may be suffering from reduced mental capacity and other harmful effects of lead poisoning.

Health scientists have proposed a number of ways to help protect children from lead poisoning:

- *Testing all children for blood levels of lead by age 1.* A 1999 study by the U.S. General Accounting Office found that most states were not testing young children on Medicaid for lead poisoning, as required by a 1989 law. Each test costs $10 or less and is paid for by the federal Medicaid program.

- *Banning incineration of solid and hazardous waste or greatly increasing current pollution control standards for old and new incinerators.*

- *Phasing out leaded gasoline worldwide over the next decade.*

- *Testing older housing and buildings for leaded paint and lead dust and removing this hazard.* According to government estimates, 83% of U.S. homes (and 86% of public housing units) built between 1950 and 1978

Using Landfill Methane to Heat a School

In January 1997, Pattonville High School in Maryland Heights, Missouri (near St. Louis), became the first public school to use methane gas produced by a nearby landfill as a source of heat for the school's 117 classrooms and 2 gymnasiums.

Most students at the high school looked at the nearby landfill and saw garbage. In 1993, however, members of the school's Ecology Club saw it as an environmental opportunity.

After doing their homework, they convinced school officials to build a pipeline to transfer waste methane gas from the landfill to the school and to rebuild two basement heating system boilers to burn the methane to heat the school. The school is expected to recoup its investment by 2003 and save money thereafter. In addition, students at the school are getting a lesson in applied environmental science by studying the operation and design of the system.

Critical Thinking

This project may not have wide application because most schools are not near landfills. Are there any renewable energy resources (Chapter 15) available at your school site that could be used to provide all or part of the school's heat or electricity? If so, consider evaluating such resources and presenting a plan to school officials.

contain lead-based paint. A 1996 study showed that reducing lead hazards in public housing units would cost $458 million but yield $1.5 billion in health and other benefits for about 1.7 million children and save taxpayers about $1 billion. So far, Congress has not allocated funds to carry out this prevention strategy.

- *Banning all lead solder in plumbing pipes and fixtures and in food cans.* The EPA estimates that 15% of U.S. households (most built before 1988) have lead in pipes, faucets, and well pumps. Homeowners can reduce this threat by running the tap for 1–2 minutes each morning to flush out the pipes.

- *Removing lead from piping and other parts of municipal drinking water systems within 10 years.*

- *Washing fresh fruits and vegetables and hands thoroughly to remove particles of lead dust.*

- *Testing ceramicware used to serve food for lead glazing.*

- *Banning imports of lead-laden polyvinyl chloride (PVC) miniblinds.*

Doing most of these things will cost an estimated $50 billion in the United States. However, health officials say the alternative is to keep poisoning and mentally handicapping millions of children. In 2000, President Clinton launched a program with the goal of eliminating childhood lead poisoning by 2010.

Although the threat from lead has been reduced in the United States, this is not the case in developing countries. The World Health Organization (WHO) estimates that (1) 130-200 million children around the world are at risk from lead, and (2) 15-18 million children in developing countries suffer permanent brain damage because of lead poisoning, with 90% of the lead coming from leaded gasoline.

Blood levels of lead are especially high in India, Bangladesh, China, Egypt, Russia, and some African countries. For example, recent studies indicate that 64% of the children in New Delhi, India, and 65–100% of those in Shanghai, China, have unsafe blood levels of lead.

What Should We Do About Chlorine? Modern society depends heavily on chlorine (Cl_2) and chlorine-containing compounds. However, some of the around 11,000 chlorine-containing organic compounds we produce (1) are persistent, (2) accumulate in body fat, and (3) are harmful to human health (according to animal and other toxicity studies; Section 16-2, p. 398).

In 1993, the Governing Council of the American Public Health Association (APHA), a major U.S. scientific and medical association, unanimously approved a statement urging the American chemical industry to phase out use of chlorine except for (1) producing some pharmaceuticals and (2) disinfecting public water supplies within 10–20 years.

The three largest uses of chlorine are (1) plastics (mostly PVC), (2) solvents, and (3) paper and pulp bleaching. Together, they account for about 80% of all chlorine use. Most of these uses have substitutes and could be phased out, as a small but growing number of companies are doing. Nonchlorine plastics and other materials could replace most uses of PVCs, and PVC incineration could be prohibited.

Studies estimate that 60% of the chlorinated organic solvents currently used could be phased out over a decade and replaced with less harmful, affordable substitutes such as (1) soap and water, (2) steam cleaning, (3) citrus-based solvents (Individuals Matter, p. 472), and (4) physical cleaning (including elbow grease and blasting with plastic beads and a pressurized solution of water and baking soda). Such a phaseout has already begun, not just for environmental reasons but also to reduce legal liability for possible effects of such chemicals on workers and consumers and to save money.

Using chlorine to bleach wood pulp and paper could be replaced with less harmful processes that rely on oxygen, ozone, and hydrogen peroxide, as several

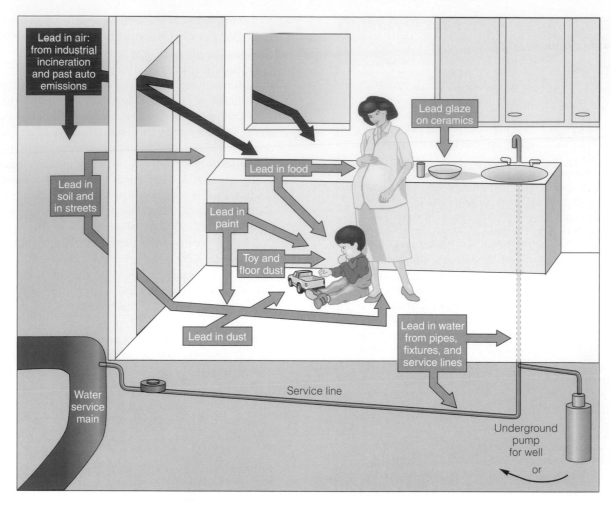

Figure 21-18 Sources of lead exposure for children and fetuses.

paper companies are doing. By the early 1990s, the German paper industry had achieved totally chlorine-free paper production (which is 30% cheaper than processes using chlorine); the rest of Europe is following Germany's lead.

About 5% of the chlorine produced in the United States is used to purify drinking water (1%) and wastewater from sewage treatment plants (4%). Much of this could be replaced gradually with ozone and other nonchlorine purification processes. However, drinking water may still need to be purified by chlorination because it kills harmful microorganisms in water lines.

The remaining 35% of chlorine is used to manufacture a variety of organic and inorganic chemicals, many of them for small and specialized uses. Each of these chemicals could be evaluated to determine which are essential (many pharmaceuticals) and which might be replaced by affordable and environmentally acceptable substitutes.

If given proper financial incentives and 10–20 years to develop and phase in substitutes, chemical producers could profit from sales of chlorine substitutes and

protect themselves from adverse publicity, lawsuits, and cleanup costs.

How Dangerous Are Dioxins? *Dioxins* are a family of more than 75 different chlorinated hydrocarbon compounds formed (along with toxic furans) as unwanted by-products in high-temperature chemical reactions involving chlorine and hydrocarbons. Exposing chlorine-containing compounds to high temperature creates conditions that can produce dioxins and related compounds called furans. Worldwide, incineration of municipal and medical wastes accounts for about 70% of dioxin and furan releases to the atmosphere. Other sources include (1) residential wood-burning fireplaces, (2) coal-fired power plants and metal smelting and refining facilities, (3) wood pulp and paper mills, and (4) sludge from municipal wastewater treatment plants.

One dioxin compound, TCDD, is the most harmful and the most widely studied. Dioxins such as TCDD are persistent chemicals that linger in the environment for decades, especially in soil and human fat tissue.

More than 90% of human exposure to dioxin comes from eating fatty foods and animal products such as meat, fish, milk, cheese, and ice cream. In the United States, 46 states have warned consumers not to eat local fish because of dioxin contamination.

There is growing concern about the harmful health effects of dioxins on humans. For example,

- A 2000 draft report of a comprehensive, EPA-sponsored, 6-year, $4-million review by more than 100 scientists around the world concluded that (1) TCDD should be classified as a definite human carcinogen, and other dioxin compounds should be classified as likely human carcinogens, especially for people who eat large amounts of fatty meats and dairy products, (2) the most powerful effect of exposure to low levels of dioxin by the general public is disruption of the reproductive, endocrine, and immune systems (Connections, p. 404), and (3) very small levels of dioxin released into the environment can cause serious damage to certain wildlife species. An independent panel of scientists reviewing the EPA draft report agreed with most of the findings but concluded that the agency had overestimated the cancer risk from dioxin exposure.

- A 1988 study by a group of German scientists concluded that dioxins may be responsible for 12% of human cancers in industrialized countries. If this estimate is correct, dioxins are responsible for about 120,000 cancers each year in the United States. Health scientists extrapolating the results of the 2000 EPA draft study estimate that an average 365 cancer deaths per year (1 per day) in the United States are caused by dioxin.

- A 2000 study by Kyle Steenland, a researcher with the National Institute for Occupational Safety and Health, found that (1) chemical workers exposed to high dioxin levels over a long time increase their chances of dying from all types of cancer by 60%, and (2) there is no significant risk of cancer for the general public exposed to low levels of dioxin.

Industries producing the chemical say that the dangers of long-term exposure to low levels are overestimated. Some environmental scientists argue that it will take decades to resolve these issues and that meanwhile we should adopt stricter regulations to reduce dioxin emissions as a *precautionary* strategy.

According to these scientists, the best way to sharply reduce dioxin emissions in the United States (and elsewhere) would be to (1) eliminate the chlorinated hydrocarbon compounds that produce dioxins from hazardous wastes burned in incinerators, iron ore sintering plants, and cement kilns and (2) use nonchlorine methods to bleach paper. In 1999, India banned the burning of PVC in medical waste generators.

21-9 HAZARDOUS-WASTE REGULATION IN THE UNITED STATES

What Is the Resource Conservation and Recovery Act? In 1976, the U.S. Congress passed the Resource Conservation and Recovery Act (RCRA, pronounced "RICK-ra") and amended it in 1984. This law requires (1) the EPA to identify hazardous wastes and set standards for their management by states, (2) firms that store, treat, or dispose of more than 100 kilograms (220 pounds) of hazardous wastes per month to have a permit stating how such wastes are to be managed, and (3) permit holders to use a cradle-to-grave system to keep track of waste transferred from point of origin to approved off-site disposal facilities.

What Is the Superfund Act? In 1980, the U.S. Congress passed the Comprehensive Environmental Response, Compensation, and Liability Act, commonly known as the *Superfund* program. Through taxes on chemical raw materials, this law (plus later amendments) provides a fund for (1) identifying abandoned hazardous-waste dump sites (p. 518 and Spotlight, p. 544) and underground tanks leaking toxic chemicals, (2) protecting and if necessary cleaning up groundwater near such sites, (3) cleaning up the sites, and, (4) when they can be found, requiring the responsible parties to pay for the cleanup.

Once the EPA identifies a site, the following procedure is used:

- The groundwater around the site is tested for contamination. If these tests reveal no immediate threat to human health, nothing more may be done.

- If there is a threat to human health, immediate steps are taken to isolate and stabilize the wastes and protect the public. This may include (1) digging a deep trench and installing a concrete containment dike around the site and (2) covering the site with impervious layers of plastic and clay to prevent further infiltration and leaching of wastes into nearby groundwater.

- The worst sites that represent an immediate and severe threat to human health are put on a *National Priorities List (NPL)* and scheduled for total cleanup using the most cost-effective cleanup method. Cleanup methods include (1) removing and treating any chemical wastes stored in drums or other containers, (2) excavating contaminated soil and burning it in an on-site incinerator or kiln, and (3) cleaning up contaminated soil by bioremediation or phytoremediation (p. 535). So far, site cleanup site takes an average of 12 years at a mean cost of $30 million.

To keep taxpayers from footing most of the bill, cleanups are based on the *polluter-pays principle*. The

Using Honeybees to Detect Toxic Pollutants

Honeybees are being used to detect the presence of toxic and radioactive chemicals in concentrations as low as parts per trillion. On their forays from a hive, bees pick up water, nectar, pollen, suspended particulate matter, volatile organic compounds, and radioactive material that contaminate the air near the sites they visit.

The bees then bring these materials back to the hive. There they fan the air vigorously with their wings to regulate the hive's temperature. This releases and circulates pollutants they picked up into the air inside the hive.

Scientists have put portable hives, each containing 7,000 to 15,000 bees, near known or suspected hazardous waste sites. A small copper tube attached to the side of each hive pumps air out. Portable equipment is used to analyze the air for toxic and radioactive materials.

To find out where the bees have gone to forage for food, a botanist uses a microscope to examine pollen grains to determine what kinds of plants they come from. These data can be correlated with the plants found in an area up to 1.6 kilometers (1 mile) from each hive. This allows scientists to develop maps of toxicity levels and hot spots on large tracts of land.

This approach is much cheaper than setting up a number of air pollution monitors around mines, hazardous waste dumps, and other sources of toxic and radioactive pollutants. By 2000, these *biological indicator* species had helped locate and track toxic pollutants and radioactive material at about 30 sites across the United States.

Critical Thinking

Because honeybees can pick up toxic pollutants anywhere they go, should honey from all beehives be tested for such pollutants before it is placed on the market? Explain.

EPA is charged with **(1)** finding the parties responsible for each site, **(2)** ordering them to pay for the entire cleanup, and **(3)** suing them if they do not. When the EPA can find no responsible party, it draws money out of the Superfund for cleanup. The fund has grown from about $300 million per year in the 1980s to about $2 billion per year in the 1990s.

To implement the polluter-pays principle, the Superfund legislation considers all polluters of a site to be subject to strict, joint, and several liability. This controversial strategy means that each individual polluter (regardless of its contribution) can be held liable for the entire cost of cleaning up a site if the other parties cannot be found or have gone bankrupt. So far only about 25% of the cleanup costs have been paid from the Superfund.

Since 1981 about 1,300 sites have been placed on a National Priority List for cleanup because they pose a real or potential threat to nearby populations. Emergency cleanup had been carried out at almost all sites, and wastes on more than half the sites have been contained and stabilized to prevent leakage. Between 1981 and 2000, 750 sites were cleaned up and removed from the list at a cost of more than $300 billion. Remaining sites are in various stages of cleanup. Attempts are being made to find ways to improve the Superfund Act without seriously weakening it (Solutions, right).

The former U.S. Office of Technology Assessment and the Waste Management Research Institute estimate that the Superfund list could eventually include at least 10,000 priority sites, with cleanup costs of up to $1 trillion, not counting legal fees. Other studies project only 2,000 sites with estimated cleanup costs of $200–300 billion over the next 30 years.

Numerous other sites containing hazardous and toxic wastes are not covered by the Superfund law. According to estimates by the responsible government agencies,

- Cleaning up toxic military dumps will cost $100–200 billion and take at least 30 years.

- Cleaning up 113 Department of Energy sites used to make nuclear weapons will cost $168–212 billion and take 70 years.

- Cleaning up 300,000 active and abandoned mine sites, 200,000 sites leased for oil and gas drilling, 3,000 landfills, and 4,200 leaking underground storage tanks on public lands under the Department of Interior will cost at least $100 billion.

These estimated costs of $568 billion to $1.5 trillion are only for cleanup to prevent future damage and do not include the health and ecological costs associated with such wastes. Some environmentalists and economists cite this as compelling evidence that preventing pollution costs much less than cleaning it up.

What Are Brownfields? *Brownfields* are industrial and commercial sites that have been abandoned and in most cases contaminated. Examples include empty factories, abandoned junkyards, older landfills, and boarded-up gas stations. There are at least 450,000 such sites in the United States, many of them in economically distressed inner city areas.

Many of these sites could be cleaned up and developed. However, these efforts have been hampered by concerns of urban planners, developers, and banks that lend money about legal liability and lawsuits from real

Since the Superfund program began, polluters and their insurance companies have been working hard to do away with the polluter-pays principle at the heart of the program and make it mostly a *public-pays* approach.

This strategy, which is working, has three components: **(1)** Deny responsibility (stonewall) to tie up the EPA in expensive legal suits for years, **(2)** sue local governments and small businesses to make them responsible for cleanup, both as a delay tactic and to turn local governments and small businesses into opponents of Superfund's strict liability requirements, and **(3)** mount a public relations campaign declaring that toxic dumps pose little threat, that cleanup is too expensive compared to the risks involved, and that Superfund is unfair and is wasteful and ineffective.

The strict joint and several liability of the Superfund law may seem unfair to some, but the authors of the legislation argued that any other liability scheme would not work. If the EPA had to identify and bring an enforcement action against every party liable for a dump site, the agency would be overwhelmed with lawsuits, and the pace of cleanup would be still slower. Administrative costs would also rise sharply, causing the program to become more of a

public-pays program favored by the polluters.

The EPA also points out that the strict polluter-pays principle in the Superfund Act has been effective in making illegal dump sites virtually a thing of the past—an important form of pollution prevention for the future. It has also forced waste producers who are fearful of future liability claims to reduce their production of such waste and to recycle or reuse much more of what they generate.

The program has also been criticized because of the **(1)** slow progress and high costs caused by overly stringent EPA cleanup standards and **(2)** lack of involvement in cleanup decisions by local government and people living near Superfund sites. Critics say that it has taken 30 years and more than $300 billion to stabilize the roughly 1,300 sites on the list but clean up only about half of them. According to the EPA, the delaying tactics just described have caused much of the delay and high costs.

Critics and defenders have been looking for ways to improve the program. Proposals include:

- Reducing lawsuits and speeding cleanup by setting up an $8-billion Environmental Insurance Resolution Fund funded by insurers for a 10-year period. Companies found liable for cleanup of wastes they disposed of before 1986 would be able to use money from this fund

rather than going to their individual insurance companies.

- Exempting innocent individuals, small businesses, and municipalities from Superfund liability.

- Involving local governments and people living near Superfund sites in cleanup decisions from start to finish.

Most experts agree that the Superfund's ultimate goal of permanent cleanup of the most dangerous sites should be maintained. However, many argue that sites on the National Priorities List should be ranked in three general categories: **(1)** sites needing immediate full cleanup, **(2)** sites considered to pose a serious hazard but located far from concentrations of people or endangered ecosystems (these sites would receive emergency cleanup and then be isolated by barriers and signs, with more complete cleanup to come later), and **(3)** lower-risk sites needing only stabilization (capping and containment) and monitoring.

Critical Thinking

Do you support the *polluter-pays principle* used in the original Superfund legislation? Defend your position. Use the library or internet to find out whether there have been any recent changes in the Superfund law. If so, evaluate these changes and decide which of them you support. Defend your position.

---

or perceived past contamination of such sites. To remove some of these obstacles,

- In 1996, Congress changed the law to limit the liability of banks that funded brownfield evaluation and development.

- Insurance companies have developed new policies that protect developers from any lawsuits they incur while cleaning up brownfields.

- Almost half of the states have passed brownfield laws that relax cleanup regulations for brownfield sites destined for commercial or industrial use but not for sites to be used for housing.

- Since 1995 the EPA has removed 30,000 brownfield sites that pose no hazard to human health from the Superfund database, relieving developers of the costs and red tape of getting the EPA to approve such sites for development.

- Based on three decades of assessing and cleaning up Superfund sites, new technology is available to make assessment and cleanup of brownfield sites cheaper, quicker, and more reliable.

As a result, a number of these sites are being redeveloped as part of urban revitalization. For example, a former industrial site in downtown Bridgeport,

Connecticut, has been converted into a new 5,500-seat baseball stadium that will eventually house a new museum and an indoor ice-skating rink. The city used a seed grant from the federal government's brownfield program to help attract a developer. The developer then invested $11 million in cleanup and construction, and the city and state added another $3 million. In Portland, Oregon, an industrial site has been recycled into a bustling and financially successful mixed-use development called RiverPlace.

21-10 SOLUTIONS: ACHIEVING A LOW-WASTE SOCIETY

What Is the Role of Grassroots Action? Bottom-Up Change In the United States, local citizens have worked together to prevent hundreds of incinerators, landfills, or treatment plants for hazardous and radioactive wastes from being built in or near their communities.

Most members of such groups recognize that health risks from incinerators and landfills, when averaged over the entire country, are quite low. However, they also know that the risks for the people near these facilities are much higher. These people, not the rest of the population, are the ones whose health, lives, and property values are being threatened.

Manufacturers and waste industry officials point out that something must be done with the toxic and hazardous wastes produced to provide people with certain goods and services. They contend that if local citizens adopt a "not in my backyard" (NIMBY) approach, the waste still ends up in someone's backyard.

Many citizens do not accept this argument. To them, the best way to deal with most toxic or hazardous wastes is to produce much less of them, as suggested by the National Academy of Sciences (Figure 21-5, p. 522, and Guest Essay, p. 523). For such materials, their goal is "not in anyone's backyard" (NIABY) or "not on planet Earth" (NOPE). What do you think?

What Can Be Done at the International Level? Between 1989 and 1994, an international treaty to limit transfer of hazardous waste from one country to another was developed (p. 539). In 2000, delegates from 122 countries completed a global treaty to control 12 *persistent organic pollutants* (POPs), which will go into effect when ratified by 50 countries.

These widely used toxic chemicals are insoluble in water and soluble in fatty tissues. This allows them to be concentrated in the fatty tissues of humans and other organisms feeding at high trophic levels in food webs to levels hundreds of thousand times higher than in the general environment (Figure 16-4, p. 399, and Figure 19-4, p. 481). They can also be transported long distances by wind and water.

The list of 12 chemicals, called the *dirty dozen*, includes DDT and 8 other chlorine-containing persistent pesticides, PCBs, dioxins, and furans. The goals of the treaty are to (1) ban or phase out use of these chemicals and (2) detoxify or isolate stockpiles of such chemicals in warehouses and dumps. About 25 countries will still be allowed to use DDT to combat malaria until safer alternatives are available. Developed nations will provide developing nations with $150 million per year to help them switch to safer alternatives for the 12 POPs.

How Can We Make the Transition to a Low-Waste Society? According to physicist Albert Einstein, "A clever person solves a problem, a wise person avoids it." To prevent pollution and reduce waste, many environmental scientists urge us to understand and live by four key principles: (1) Everything is connected, (2) there is no "away" for the wastes we produce, (3) dilution is not always the solution to pollution, and (4) the best and cheapest way to deal with waste and pollution is to produce less of them and then reuse and recycle most of the materials we use (Figures 21-4 and 21-5).

Visible signs of the *cleaner production* (p. 522 and Guest Essay, p. 526), increased *resource productivity* (Solutions, p. 525), and *service flow* businesses (p. 527) are emerging. Ecoindustrial parks are being built and planned. Architects and builders are creating structures that capture sunlight to provide heat and electricity, purify their own wastewater, and provide conditions that improve the productivity, health, and morale of people living or working in them (Figure 15-24, p. 376). Energy-wasting centralized power plants are beginning to be replaced by smaller-scale renewable micropower plants (p. 389).

Such revolutions start off slowly but can increase rapidly as their economic, ecological, and health advantages become more apparent to investors, business leaders, elected officials, and citizens. Some actions you can take to reduce your production of solid waste and hazardous waste and help bring about such revolutions are listed in Appendix 6.

Nearly all environmental and social harm is caused by the uneconomically wasteful use of human and natural resources.
PAUL HAWKEN

REVIEW QUESTIONS

1. Define the boldfaced terms in this chapter.

2. Summarize what happened at Love Canal and the lessons learned from this environmental problem.

3. Distinguish between *solid waste* and *municipal solid waste*. What are the major sources of solid waste in the United States? What happens to municipal solid waste in the United States? Describe the results of mining and smelting waste in Montana. Give five examples of solid waste thrown away in the United States.

4. What is *hazardous waste*? Hazardous-waste laws regulate what percentage of the overall hazardous waste produced in the United States?

5. Distinguish between the *high-waste* and *low-waste* approaches to solid and hazardous waste management. According to the U.S. National Academy of Sciences, what should be the five goals of solid and hazardous waste management in order of their importance? According to scientists, what percentage of solid and hazardous waste produced could be eliminated through a combination of waste reduction, reuse, and recycling (including composting)?

6. What does Lois Gibbs think we do with solid and hazardous waste, and what major questions does she think we should ask about such wastes?

7. List seven ways to reduce waste and pollution. What is *resource productivity*? List five ways in which resource productivity has been improved. By how much do some experts think we can improve resource productivity? List four advantages of greatly increasing resource productivity.

8. What is *cleaner production*? Describe the resource exchange system used in Denmark and the pollution prevention program implemented by the Minnesota Mining and Manufacturing (3M) company in the United States. List five economic benefits of cleaner production.

9. Explain why Peter Montague believes that the regulation and standards approach used to control pollution during the past 30 years has failed and what he thinks should replace this approach.

10. What is a *service flow economy*? What are four economic advantages of such an economy for businesses and consumers? List four examples of how a service economy is being implemented. Describe how Ray Anderson is developing a carpet tile service economy business.

11. What is *reuse*, and what are three advantages of using this approach for waste reduction? List five examples of reuse.

12. Distinguish between *primary (closed-loop)* and *secondary (open-loop) recycling*. What is *compost*, and how is it used as a way to deal with solid waste? About what percentage of the municipal solid waste produced in the United States is recycled and composted, and what percentage do experts believe could be recycled and composted? What is a *pay-as-you-throw* garbage collection program, and how does it encourage recycling?

13. Distinguish between the *centralized recycling* of mixed solid waste and *consumer separation* of solid waste, and list the pros and cons of each approach to recycling.

14. List the pros and cons of recycling. What three factors hinder recycling and reuse? List seven ways to encourage more recycling and reuse. Describe Germany's approach to recycling.

15. Summarize the recycling of **(a)** aluminum, **(b)** wastepaper, and **(c)** plastics.

16. Describe how paper is made and how paper is recycled. List nine benefits from recycling paper. Distinguish between *preconsumer* and *postconsumer paper waste*.

17. List three reasons why so few plastics are recycled. List four advantages and four disadvantages of using plastics.

18. Describe Denmark's hazardous-waste detoxification program. What are *bioremediation* and *phytoremediation*?

19. Describe the major components of a *mass burn incinerator* and a *sanitary landfill*. List the pros and cons of dealing with solid and hazardous waste by **(a)** burning it in incinerators and **(b)** burying it in sanitary landfills.

20. List the pros and cons of storing hazardous wastes in **(a)** deep underground wells, **(b)** surface impoundments, **(c)** secure landfills, and **(d)** aboveground buildings, and **(e)** exporting solid and hazardous waste to other areas or countries. Describe what is being done at the international level about the exporting of hazardous wastes from one country to another.

21. Describe the hazards of exposure to lead and list eight ways to reduce such exposure.

22. What are three major problems with many chlorine-containing compounds? What are the three major uses of chlorine? How can we reduce exposure to harmful chlorine-containing compounds?

23. What are *dioxins*, how are they produced, what harm can they cause, and how can we reduce exposure to these hazardous chemicals?

24. How is the Resource Conservation and Recovery Act used to deal with the problem of hazardous wastes in the United States?

25. What is the Superfund Act, and what are its strengths and weaknesses? List three ways to improve this act.

26. What are *brownfields*, and what has been done to help redevelop such sites in the United States?

27. Describe international efforts to control use of 12 persistent organic pollutants.

28. List four principles that can be used as guidelines for making the transition to a low-waste society.

CRITICAL THINKING

1. Explain why you support or oppose requiring that **(a)** all beverage containers be reusable, **(b)** all households and businesses sort recyclable materials into separate containers for curbside pickup, **(c)** garbage-collecting systems implement the pay-as-you-throw approach, and **(d)** consumers pay for plastic or paper bags at grocery and other stores to encourage the use of reusable shopping bags.

2. Use the second law of energy (thermodynamics) to explain why a *properly designed* source-separation recycling program takes less energy and produces less pollution than a centralized program that collects mixed waste over a large area and hauls it to a centralized facility where workers or machinery separate the wastes for recycling.

3. What short- and long-term disadvantages (if any) might an ecoindustrial revolution based on cleaner production (p. 526) bring? Do you believe that it will be possible to phase in such a revolution in the country where you live over the next two to three decades? Explain. What are the three most important strategies for doing this?

4. Explain why some businesses participating in an exchange and chemical-cycling network (p. 524) might produce large amounts of waste for use as resources elsewhere rather than redesigning their manufacturing processes to reduce waste production.

5. What short- and long-term disadvantages (if any) might there be in shifting to a service flow economy (p. 527)? Do you believe that it will be possible to phase in such a shift in the country where you live over the next two to three decades? Explain. What are the three most important strategies for doing this?

6. Would you oppose having a hazardous-waste landfill, waste treatment plant, deep-injection well, or incinerator in your community? Explain. If you oppose these disposal facilities, how do you believe the hazardous waste generated in your community and your state should be managed?

7. Give your reasons for agreeing or disagreeing with each of the following proposals for dealing with hazardous waste:

    a. Reducing the production of hazardous waste and encouraging recycling and reuse of hazardous materials by charging producers a tax or fee for each unit of waste generated

    b. Banning all land disposal and incineration of hazardous waste to encourage recycling, reuse, and treatment and to protect air, water, and soil from contamination

    c. Providing low-interest loans, tax breaks, and other financial incentives to encourage industries producing hazardous waste to reduce, recycle, reuse, treat, and destroy such waste

    d. Banning the shipment of hazardous waste from one country to another

8. Congratulations! You have just been put in charge of bringing about a cleaner production, resource productivity, and service flow economic revolution throughout the world over the next 20 years. List the five most important components of your strategy.

PROJECTS

1. For 1 week, keep a list of the solid waste you throw away. What percentage of this waste consists of materials that could be recycled, reused, or burned for energy? What percentage of the items could you have done without in the first place? Tally and compare the results for your entire class.

2. What percentage of the municipal solid waste in your community is **(a)** landfilled, **(b)** incinerated, **(c)** composted, and **(d)** recycled? What technology is used in local landfills and incinerators? What leakage and pollution problems have local landfills or incinerators had? Does your community have a recycling program? Is it voluntary or mandatory? Does it have curbside collection? Drop-off centers? Buyback centers?

3. What hazardous wastes are produced **(a)** at your school and **(b)** in your community? What happens to these wastes?

4. Use the library or the internet to find bibliographic information about *Arthur C. Clarke* and *Paul Hawken*, whose quotes appear at the beginning and end of this chapter.

5. Make a concept map of this chapter's major ideas, using the section heads and subheads and the key terms (in boldface). Look at the inside back cover and on the website for this book for information about making concept maps.

INTERNET STUDY RESOURCES AND RESOURCES FOR FURTHER READING AND RESEARCH

The website for this book contains helpful study aids and many ideas for further reading and research. Log on to:

http://www.brookscole.com/product/0534376975s

and click on the Chapter-by-Chapter area. Choose Chapter 21 and select a resource:

- "Flash Cards" allows you to test your mastery of the Terms and Concepts to Remember for this chapter.

- "Tutorial Quizzes" provides a multiple-choice practice quiz.

- "Student Guide to InfoTrac" will lead you to Critical Thinking Projects that use InfoTrac College Edition as a research tool.

- "References" lists the major books and articles consulted in writing this chapter.

- "Hypercontents" takes you to an extensive list of sites with news, research, and images related to individual sections of the chapter.

INFOTRAC COLLEGE EDITION

Improve your skills with InfoTrac College Edition, a searchable online database of articles from more than 700 periodicals. Log on to:

http://www.infotrac-college.com

or access InfoTrac through the website for this book.

Try the following articles:

    Gardner, G., and P. Sampat. 1999. Making things last: reinventing our material culture. *The Futurist* vol. 33, no. 5. pp. 24–28. (keywords: recycling, economics)

    O'Neill, K. 2000. Managing hazardous waste: the global challenge. *Environment* vol. 42, no. 3, pp. 43–44. (keywords: Basel Convention)

PART V

BIODIVERSITY, LAND USE, AND CONSERVATION

Only when the last tree has died and the last river poisoned and the last fish has been caught will we realize that we cannot eat money.
19TH-CENTURY CREE INDIAN SAYING

The Passenger Pigeon: Gone Forever

In the early 1800s, bird expert Alexander Wilson watched a single migrating flock of an estimated 2 million passenger pigeons darken the sky for more than 4 hours. By 1914 the passenger pigeon (Figure 22-1) had disappeared forever.

How could a species that was once the most common bird in North America become extinct in only a few decades? The answer is humans. The main reasons for the extinction of this species were (1) uncontrolled commercial hunting and (2) loss of the bird's habitat and food supply as forests were cleared to make room for farms and cities.

Passenger pigeons were good to eat, their feathers made good pillows, and their bones were widely used for fertilizer. They were easy to kill because they flew in gigantic flocks and nested in long, narrow colonies.

Commercial hunters would capture one pigeon alive, sew its eyes shut, and tie it to a perch called a stool. Soon a curious flock would land beside this "stool pigeon," and the birds would then be shot or ensnared by nets that might trap more than 1,000 birds at once.

Beginning in 1858, passenger pigeon hunting became a big business. Shotguns, traps, artillery, and even dynamite were used. Burning grass or sulfur below their roosts sometimes suffocated birds. Live birds were used as targets in shooting galleries. In 1878, one professional pigeon trapper made $60,000 by killing 3 million birds at their nesting grounds near Petoskey, Michigan.

By the early 1880s, only a few thousand birds were left. At that point, recovery of the species was doomed because the females laid only one egg per nest. On March 24, 1900, a young boy in Ohio shot the last known wild passenger pigeon. The last passenger pigeon on earth, a hen named Martha after Martha Washington, died in the Cincinnati Zoo in 1914. Her stuffed body is now on view at the National Museum of Natural History in Washington, D.C.

Eventually, all species become extinct or evolve into new species. However, biologists estimate that every day 3–200 species become *prematurely extinct* primarily because of human activities. Studies indicate that this rate of loss of biodiversity is increasing as the human population (1) grows, (2) consumes more resources, (3) disturbs more of the earth's land, and (4) uses more of the earth's net plant productivity that supports all species.

Ways to deal with this loss of biodiversity include (1) protecting terrestrial species from premature extinction, discussed in this chapter, and (2) sustaining the terrestrial ecosystems where the world's species live, discussed in Chapter 23. Chapter 24 focuses on sustaining aquatic biodiversity.

Figure 22-1 Passenger pigeons, extinct in the wild since 1900. The last known passenger pigeon died in the Cincinnati Zoo in 1914. (John James Audubon/The New York Historical Society)

The last word in ignorance is the person who says of an animal or plant: "What good is it?"...If the land mechanism as a whole is good, then every part of it is good, whether we understand it or not....Harmony with land is like harmony with a friend; you cannot cherish his right hand and chop off his left.

ALDO LEOPOLD

This chapter addresses the following questions:

- How have human activities affected the earth's biodiversity?
- Are human activities causing a new mass extinction?
- Why should we care about biodiversity?
- What human activities endanger wildlife?
- How can we prevent premature extinction of species?
- How can we manage game animals more sustainably?

22-1 HUMAN IMPACTS ON BIODIVERSITY

How Have Human Activities Affected Global Biodiversity? Factors that tend to *increase* biodiversity are **(1)** a physically diverse habitat, **(2)** moderate environmental disturbance (Figure 8-18, p. 192), **(3)** small variations in environmental conditions such as nutrient supply, precipitation, and temperature, **(4)** the middle stages of succession (Figure 8-15, p. 188, and Figure 8-16, p. 189), and **(5)** evolution (Section 5-2, p. 107).

Factors that tend to *decrease* biodiversity are **(1)** environmental stress, **(2)** large environmental disturbance, **(3)** extreme environmental conditions, **(4)** severe limitation of an essential nutrient, habitat, or other resource, **(5)** introduction of an alien species, and **(6)** geographic isolation. Many of these factors occur because of human activities.

Figure 22-2 summarizes the major connections between human activities and the earth's biodiversity. Here are some examples of how human activities have decreased and degraded the earth's biodiversity.

- Humans have taken over, disturbed, or degraded 40–50% of the earth's land surface (Figure 1-4, p. 8), especially by filing in wetlands and converting grasslands and forests to cropfields and urban areas. Such activities are projected to disturb or degrade another third of the planet's land surface over the next 100 years.

- Humans use, waste, or destroy about **(1)** 27% of the earth's total potential net primary productivity and **(2)** 40% of the net primary productivity of the planet's terrestrial ecosystems (Figure 4-26, p. 88), and this resource use is expected to increase.

- According to a 1999 report by the World Wide Fund for Nature and several other organizations,

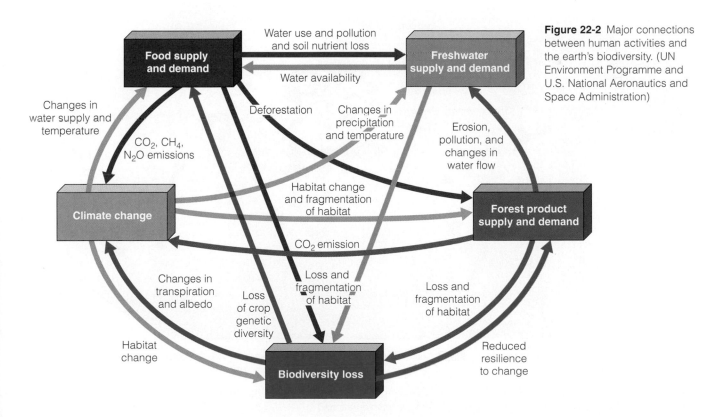

Figure 22-2 Major connections between human activities and the earth's biodiversity. (UN Environment Programme and U.S. National Aeronautics and Space Administration)

*www.*brookscole.com/product/0534376975s   **551**

between 1970 and 1995, the earth lost one-third of its natural capital.

■ Biologists estimate that the current global extinction rate of species is 100 to 1,000 times what it would be without human-induced changes. According to biologist Peter Raven, the human-caused extinction rate may climb to 10,000 times the natural rate of extinction during the 21st century.

■ A 2000 joint study by the World Conservation Union and Conservation International and a 1999 study by the World Wildlife Fund found that **(1)** 34% of the world's fish species (51% of freshwater species), **(2)** 25% of amphibians (Connections, p. 179), **(3)** 24% of mammals, **(4)** 20% of reptiles, **(5)** 14% of plants, and **(6)** 12% of bird species are under threat of extinction.

■ A comprehensive 2000 study by the World Conservation Union identified 11,046 plants and animals at risk of becoming extinct.

These threats to the world's biodiversity are projected to increase sharply by 2018 (Figure 22-3). In 2000, a team of researchers led by Osvaldo Sala developed a computer model to project global biodiversity loss scenarios for the world by 2100 (Connections, right).

How Have Human Activities Affected Biodiversity in the United States? Numerous studies have shown sharp declines in the biodiversity of the United States over the past 200 years.

■ According to a 2000 survey by the Nature Conservancy and the Association for Biodiversity Information, **(1)** at least 539 species have become extinct since 1600 (mostly because of loss of habitat), and **(2)** about 33% of the 21,000 animal and plant species in the United States are vulnerable to premature extinction, mostly because of human activities (Figure 22-4).

■ The International Union for Conservation (IUCN) estimates that 29% of the known plant species in the United States are threatened with extinction.

■ A 1999 study by the U.S. Geological Survey found that **(1)** 95–98% of the virgin and 99% of the virgin eastern deciduous forests in the lower 48 states have been destroyed since 1620, **(2)** 96% of the virgin southeastern coastal plain longleaf pine forests are gone, **(3)** 98% of tallgrass prairie in the Midwest and Great Plains has disappeared, **(4)** 99% of California's native grassland, 91% of its wetlands, and 85% of its redwood forests are gone, **(5)** 90% of Hawaii's dry forests and grasslands has disappeared, **(6)** 81% of the nation's fish communities have been disturbed by human activities, and **(7)** more than half of the country's original wetlands have been destroyed.

How Serious Is Biodiversity Loss? Some analysts say that global and national estimates of biodiversity

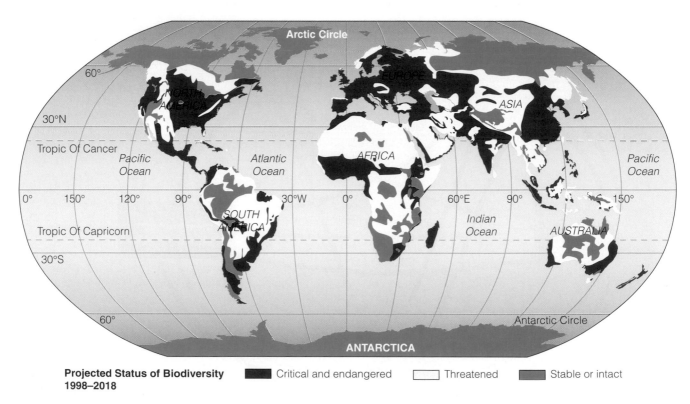

Projected Status of Biodiversity 1998–2018 ■ Critical and endangered ☐ Threatened ■ Stable or intact

Figure 22-3 Projected status of the earth's biodiversity between 1998 and 2018. (Data from World Resources Institute, World Conservation Monitoring Center, and Conservation International)

Global Biodiversity in 2100

In 2000, an international team of 19 researchers identified five factors that have the greatest effect on biodiversity loss and degradation. They used computer modeling to project scenarios for changes in biodiversity for 10 different terrestrial biomes and freshwater aquatic systems in the year 2100.

According to this study, biodiversity is affected strongly by changes in **(1)** land use, **(2)** climate (Figure 18-8, p. 453), **(3)** nitrogen deposition and acid deposition (Figure 17-11, p. 429), **(4)** biotic exchanges involving the deliberate or accidental movement of nonnative species from one ecosystem to another, and **(5)** levels of atmospheric carbon dioxide (Figure 18-5, p. 451).

According to their model, during the 21st century:

■ Mediterranean and grassland ecosystems are likely to experience the greatest proportional change in biodiversity because they are affected by all five factors, and arctic, alpine, and desert ecosystems are expected to have only moderate change in biodiversity.

■ Northern temperate ecosystems are likely to experience the least additional change in biodiversity because humans have already extensively altered them.

■ Land-use changes such as conversion of temperate grasslands and tropical forests into cropland are expected to have the greatest effect on the biodiversity of terrestrial ecosystems and freshwater ecosystems.

■ Invasive species are expected to pose the greatest threat to freshwater ecosystems (especially lakes) and Mediterranean and southern temperate forests.

■ Climate change is expected to have the greatest effect on biodiversity at high latitudes (arctic and boreal biomes) and the least effect in tropical areas.

■ Nitrogen deposition poses the greatest threat to the biodiversity of forests near northern cities in temperate boreal zones and the smallest threat in deserts and tropical forests (Figure 17-11, p. 429).

■ Global freshwater biodiversity (Section 24-6, p. 652) is expected to decline at far greater rates than terrestrial biodiversity.

Critical Thinking

1. Based on the results of this study, during the 21st century **(a)** will changes in biodiversity where you live be low, moderate, or high, and **(b)** which of the five major factors are likely to have the greatest effect on biodiversity where you live?

2. How do your answers to the preceding question correlate with the results of the study projecting changes in global biodiversity by 2018 (shown in Figure 22-3)?

---

loss and degradation are exaggerated. They point out that:

■ We do not know how many species there are, with estimates ranging from 5–100 million and a best guess of 12–14 million.

■ We have identified and named only about 1.4–1.8 million species and know a fair amount about only one-third of these species.

■ We do not know how important various species are to the functioning and sustainability of their ecosystems and to humans because we understand the detailed roles and interactions of only a small number of the world's known species.

■ It is very difficult to document that even a widely studied species has become extinct.

■ Estimates of species extinction rates are based on **(1)** models such as the theory of island biogeography (Figure 8-6, p. 177, and Figure 8-7, p. 178), **(2)** changes in species diversity at different latitudes (Figure 8-3, p. 175), **(3)** statistical sampling mostly in tropical forests, and **(4)** differing assumptions about the earth's total number of species (5–14 million), the proportion of these species that are found in tropical

forests, and the rate at which tropical forests are being cleared (0.8–2% per year). Because of these variables, biodiversity researchers estimate that extinction rates range from 1,000 to 75,000 species per year (a loss of about 3–200 species per day).

Scientists agree that their estimates of extinction rates are based on inadequate data and sampling, and they are continuously striving to get better data. However, they point out that:

■ This is all the information we have, and such estimates can give us an idea of general trends in biodiversity loss.

■ Even if the best available estimates of extinction rates are 100–1,000 times higher than the actual rates, there has been a serious loss and degradation of global biodiversity.

■ Because of projected increases in population and resource consumption per person, the rate and extent of biodiversity loss and degradation are likely to increase.

■ Arguing over the numbers and waiting to get better data should not be used as excuses for not implementing a *precautionary strategy* to reduce the threats

Figure 22-4 The state of U.S. species. Recently scientists surveyed the status of about 21,000 of the estimated 205,000 plant and animal species in the United States. The *good news* is that about two-thirds of the species surveyed apparently are not in danger of extinction. The *bad news* is that about one-third of these species are in trouble. Life-forms with the highest levels of risk are flowering plants and freshwater aquatic species such as mussels (70% of species in danger), crayfishes (more than half of species in danger), amphibians (Connections, p. 179), and fishes. (Data from the Nature Conservancy)

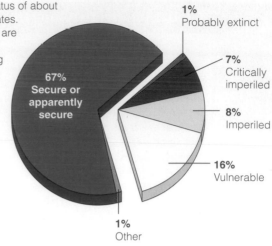

to human health and economies and to the world's ecosystems and wildlife from this potentially serious global ecological problem.

Should Conservation Efforts Focus on Sustaining Species or Ecosystems? Figure 22-5 outlines the goals, strategies, and tactics for **(1)** preventing the premature extinction of species, as discussed in this chapter, and **(2)** preserving and restoring the ecosystems and aquatic systems that provide habitats and resources for the world's species, as discussed in Chapter 23 for terrestrial species and Chapter 24 for aquatic species.

The consensus among most conservation biologists is that we need to protect both species and ecosystems. The ecosystem protection approach focuses on ensuring that enough land and aquatic areas are protected to provide habitat for the majority of terrestrial and aquatic wild species.

By contrast, a species-by-species approach involves **(1)** identifying which species are at greatest risk of becoming extinct from human activities, **(2)** gaining a detailed understanding of such species, and **(3)** focusing efforts on protecting them. However, much of the world's biodiversity is not known, especially in the species-rich tropics. Even in the temperate zone our knowledge of populations of wild organisms often is limited. Thus, even though a species-by-species approach may be desirable, many biologists argue that an ecosystem approach often is more practical and more likely to succeed.

22-2 SPECIES EXTINCTION

What Are Three Types of Species Extinction? Biologists distinguish between three levels of species extinction:

- *Local extinction*: A species is no longer found in an area it once inhabited but is still found elsewhere in the world. Most local extinctions involve losses of one or more populations of species.

- *Ecological extinction*: There are so few members of a species left that it can no longer play its ecological roles in the biological communities where it is found.

- *Biological extinction*: A species is no longer found anywhere on the earth. Biological extinction is forever

and is an irreversible loss of a unique pool of genes and individuals that evolution took millions of years to produce (Figure 22-6).

Some biologists believe that loss of *local populations* of key species may be a better indicator of biodiversity loss than species extinction because these local populations provide most of the ecological services of an area. According to a 1997 study by biologist Jennifer Hughes and her colleagues at Stanford University,

- There are 3.1 billion populations of the world's estimated 14 million species.

- If species diversity in tropical forests is declining by roughly 9,000–26,000 species per year (25–71 species per day), then an estimated 16 million populations per year (44,000 per day) are being lost in tropical forests alone.

- This loss of local populations is tens of thousands of times greater than the estimated rate of species loss of 3 to 200 per day.

What Are Endangered and Threatened Species? Species heading toward biological extinction are classified as either *endangered* or *threatened* (Figure 22-7). An **endangered species** has so few individual survivors that the species could soon become extinct over all or most of its natural range. A **threatened** or **vulnerable species** is still abundant in its natural range but is declining in numbers and is likely to become endangered in the near future. Endangered and threatened species are ecological smoke alarms.

Biodiversity researchers are also concerned about **rare species** that **(1)** have naturally small numbers of individuals often because of limited geographic ranges or low population densities or **(2)** have been locally depleted by human activities. Although these species may not be in immediate danger of extinction, their small numbers make them vulnerable to extinction.

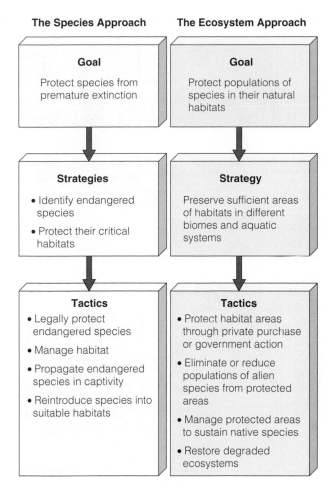

The Species Approach

Goal
Protect species from premature extinction

↓

Strategies
- Identify endangered species
- Protect their critical habitats

↓

Tactics
- Legally protect endangered species
- Manage habitat
- Propagate endangered species in captivity
- Reintroduce species into suitable habitats

The Ecosystem Approach

Goal
Protect populations of species in their natural habitats

↓

Strategy
Preserve sufficient areas of habitats in different biomes and aquatic systems

↓

Tactics
- Protect habitat areas through private purchase or government action
- Eliminate or reduce populations of alien species from protected areas
- Manage protected areas to sustain native species
- Restore degraded ecosystems

Figure 22-5 Goal, strategies, and tactics for protecting biodiversity.

Some species have characteristics that make them more vulnerable than others to biological extinction (Figure 22-8, p. 558, Connections, p. 181, and Case Study, p. 559).

How Do Biologists Estimate Extinction Rates?

Biodiversity researchers often use the well-established fact that the number of species present increases with the size of an area (Figure 22-9, p. 558) to estimate the number of extinctions resulting from habitat destruction. This *species-area relationship* suggests that, on average, a 90% loss of habitat causes the extinction of about 50% of the species living in that habitat.

For example, suppose we assume that (1) about 50% of the world's existing terrestrial species live in tropical forests and (2) about one-third of the remaining tropical forests will be cut or burned during the next few decades. The species-area relationship suggests that about 1 million species in these forests will become extinct during this period.

How Do Biologists Estimate Extinction Risks?

Biodiversity researchers use models to estimate the *risk* that a particular species will become endangered or extinct within a certain time. These models take into account factors such as (1) trends in population size, (2) changes in habitat availability, and (3) changes in the genetic variability within populations of a species.

One model used to estimate population's risk of extinction is a **population viability analysis (PVA)**. It is a *risk assessment* using mathematical and statistical methods to predict the probability that a population will persist for a certain number of generations based on current population size and habitat conditions. A PVA may take into account factors such as a population's (1) resource needs and availability, (2) vulnerable stages in its life cycle, (3) genetic variability, (4) interactions with other species (Section 8-3, p. 180, and Section 8-4, p. 186), (5) reproductive rates and population dynamics (Section 9-3, p. 203), (6) effects of habitat loss, fragmentation, and degradation, and (7) responses to changing environmental conditions.

Population variability analysis often includes the concept of **minimum viable population (MVP)**: an estimate of the smallest number of individuals necessary to ensure the survival of a population in a region for a specified time period, typically ranging from decades to 100 years. Most population viability analyses

Passenger pigeon   Great Auk   Dodo   Dusky seaside sparrow   Aepyornis (Madagascar)

Figure 22-6 Some animal species that have become prematurely extinct largely because of human activities, mostly habitat destruction and overhunting.

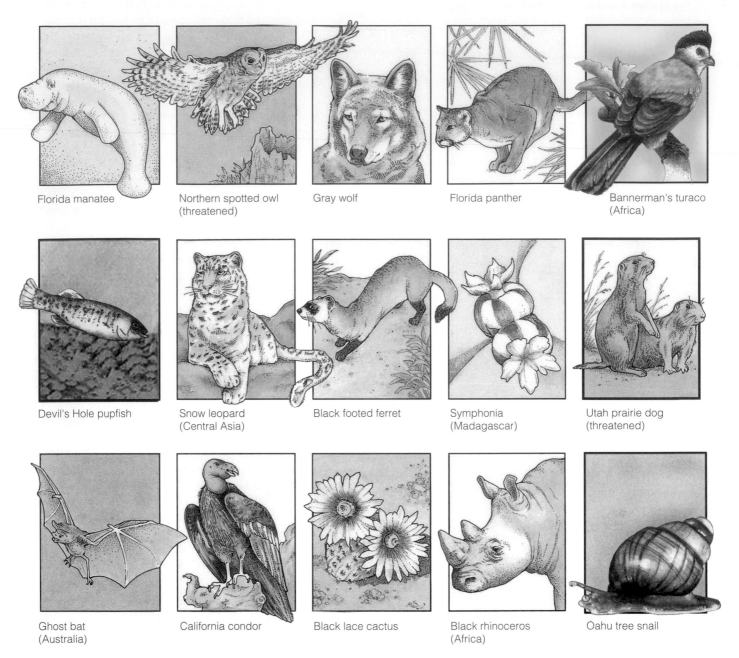

Figure 22-7 Species that are endangered or threatened with premature extinction largely because of human activities. Almost 30,000 of the world's species and 1,200 of those in the United States are officially listed as being in danger of becoming extinct. Most biologists believe that the actual number of species at risk is much larger.

Labels within figure:

Florida manatee
Northern spotted owl (threatened)
Gray wolf
Florida panther
Bannerman's turaco (Africa)

Devil's Hole pupfish
Snow leopard (Central Asia)
Black footed ferret
Symphonia (Madagascar)
Utah prairie dog (threatened)

Ghost bat (Australia)
California condor
Black lace cactus
Black rhinoceros (Africa)
Oahu tree snail

indicate that at least a few thousand individuals are needed to ensure the survival of most species for more than a few decades.

Once a minimum viable population size has been estimated for a species, an estimate is made of its **minimum dynamic area (MDA)**: the minimum area of suitable habitat needed to maintain the minimum viable population. It is estimated by studying the **(1)** home range size for individuals of a species, **(2)** colonies of endangered species, and **(3)** availability of nearby individuals and populations for rebuilding depleted populations.

Studies suggest that MDAs of 100-1,000 square kilometers (39–390 square miles) are needed to maintain

Grizzly bear
(threatened)

Arabian oryx
(Middle East)

White top pitcher plant

Kirtland's warblers

African elephant
(Africa)

Mojave desert tortoise
(threatened)

Swallowtail butterfly

Humpback chub

Golden lion tamarin
(Brazil)

Siberian tiger
(Siberia)

West Virginia spring
salamander

Giant panda
(China)

Knowlton cactus

Whooping crane

Blue whale

Mountain gorilla
(Africa)

Swamp pink

Pine barrens
tree-frog (male)

Hawksbill sea turtle

El Segundo blue butterfly

Characteristic	Examples
Low reproductive rate (K-strategist)	Blue whale, giant panda, rhinoceros
Specialized niche	Blue whale, giant panda, Everglades kite
Narrow distribution	Many island species, elephant seal, desert pupfish
Feeds at high trophic level	Bengal tiger, bald eagle, grizzly bear
Fixed migratory patterns	Blue whale, whooping crane, sea turtles
Rare	Many island species, African violet, some orchids
Commercially valuable	Snow leopard, tiger, elephant, rhinoceros, rare plants and birds
Large territories	California condor, grizzly bear, Florida panther

Figure 22-8 Characteristics of species that are prone to extinction.

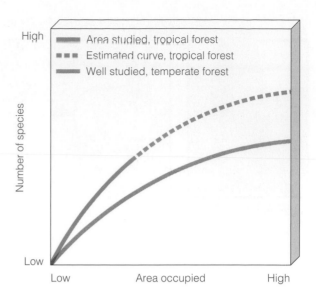

Figure 22-9 A species-area curve showing how the number of species increases with habitat size. The slope of such curves varies from one habitat to another. However, the basic pattern of an initial rapid rise in the number of species followed by a leveling off in the number of species has been found to occur regardless of habitat.

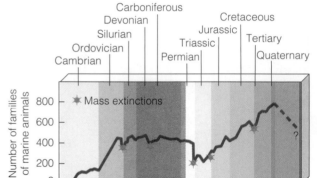

Figure 22-10 Fossils and radioactive dating indicate that five major mass extinctions have taken place in the past 500 million years. Many scientists believe that we are now in the midst of a sixth mass extinction, caused primarily by human activities.

populations of many small mammals. However, an estimated MDA of 49,000 square kilometers (18,900 square miles) is needed to sustain a population of 50 grizzly bears (which have a large home range), and a population of 100 bears would need an area of 98,000 square kilometers (37,800 square miles).

How Does Background Extinction Differ from Mass Extinction? Evolutionary biologists estimate that more than 99.9% of all the species that have ever existed are now extinct because of a combination of background and mass extinctions.

Each year, a small number of species become extinct naturally at a low rate, a phenomenon called the **background** or **natural rate of extinction**. According to the fossil record, the earth's background extinction rate is about 3–14 species per year if the earth has 14 million species and about 1–5 species per year if there are 5 million species.

In contrast, **mass extinction** is a rise in extinction rates above the background level. It is a catastrophic, widespread (often global) event in which large groups of existing species (perhaps 25–70%) are wiped out. Most mass extinctions are believed to result from one or a combination of global climate changes that kill many species and leave behind those able to adapt to the new conditions.

Fossils and geological evidence indicate that the earth's species have experienced five great mass extinc-

tions (20–60 million years apart) during the past 500 million years in which large numbers of species became extinct each year for tens of thousands to millions of years (Figure 5-10, p. 115, and Figure 22-10). Evidence also shows that these mass extinctions were followed by other periods, called *adaptive radiations*, when the diversity of life increased and spread for 10 million years or more (Figure 5-10, p. 115, and Figure 5-11, p. 116). The last mass extinction took place about 65 million

Bats Are Getting a Bad Rap

CASE STUDY

Despite their variety (950 known species) and worldwide distribution, bats—the only mammals that can fly—have certain traits that make them vulnerable to extinction. They reproduce slowly, and many bat species that live in huge colonies in caves and abandoned mines become vulnerable to destruction when people block the passageways or disturb their hibernation.

Bats play important ecological roles. About 70% of all bat species feed on crop-damaging nocturnal insects and other pest species such as mosquitoes, making them the primary control agents for such insects.

In some tropical forests and on many tropical islands, pollen-eating bats pollinate flowers, and fruit-eating bats distribute plants throughout tropical forests by excreting undigested seeds (p. 173). As *keystone species*, they are vital for maintaining plant biodiversity and for regenerating large areas of tropical forest cleared by human activities. If you enjoy bananas, cashews, dates, figs, avocados, or mangos, you can thank bats.

Many people mistakenly view bats as fearsome, filthy, aggressive, rabies-carrying bloodsuckers. However, most bat species are harmless to people, livestock, and crops. In the United States, only 10 people have died of bat-transmitted disease in four decades of recordkeeping; more Americans die each year from falling coconuts.

Because of unwarranted fears of bats and lack of knowledge about their vital ecological roles, several bat species have been driven to extinction. Currently, 26% of the world's bat species, including the ghost bat (Figure 22-7), are listed as endangered or threatened. Conservation biologists urge us to view bats as valuable allies, not as enemies.

Critical Thinking

How do you feel about bats? What would you do if you found bats flying around your yard at night?

---

years ago, when the dinosaurs became extinct after thriving for 140 million years.

Why Do Most Biologists Believe There Is a New Mass Extinction Crisis? According to a 1998 survey, 70% of the biologists polled believe that we are in the midst of a mass extinction and that this loss of species will pose a major threat to human health and economies in the 21st century. In 1999, Peter Raven, president of the International Botanical Congress, reported, "We are predicting the extinction of about two-thirds of all mammal, butterfly, and plant species by the end of the 21st century."

To these biologists, we are not heeding Aldo Leopold's (Section 2-5, p. 39) warning about preserving species as we tinker with the earth: "To keep every cog and wheel is the first precaution of intelligent tinkering."

As mentioned earlier, biologists estimate that currently 1,000–73,000 species become extinct each year (on average, 3–200 species per day) depending on the models, assumptions, and data used. Assuming that there are 5–14 million species on the earth, this is thousands of times the estimated natural background extinction rate of 1–14 species per year and about 1 million times the earth's natural speciation rate.

Even if this estimate of the erosion of biodiversity is as much as 1,000 times too high, the current extinction rate is many times the estimated background extinction rate. Scientists expect this extinction rate to accelerate as the human population grows and takes over more of the planet's wildlife habitats and net primary productivity (Figure 22-3).

Mass extinctions occurred long before humans evolved. However, biologists point out two important differences between the current mass extinction most biologists believe we are bringing about and those of the past:

- It is taking place in only a few decades, rather than over thousands to millions of years (Figure 5-10, p. 115).

- We are eliminating or degrading many biologically diverse environments (such as tropical forests, tropical coral reefs, wetlands, and estuaries) that in the past have served as evolutionary centers for the 5- to 10-million-year recovery of biodiversity after a mass extinction (Figure 5-11, p. 116).

In the words of Norman Myers (Guest Essay, p. 117), "Within just a few human generations, we shall—in the absence of greatly expanded conservation efforts—impoverish the biosphere to an extent that will persist for at least 200,000 human generations or twenty times longer than the period since humans emerged as a species." However, a few analysts question whether there is an extinction crisis (Spotlight, p. 560).

22-3 WHY SHOULD WE CARE ABOUT BIODIVERSITY?

Why Preserve Wild Species and Ecosystems? What is biodiversity good for? If all species eventually become extinct, why should we worry about losing a few more because of our activities? Does it matter that **(1)** the passenger pigeon (Figure 22-1), **(2)** the great auk (Figure 22-6), **(3)** the green sea turtle (photo on p. 43), **(4)** the 30–50

Is There Really an Extinction Crisis?

SPOTLIGHT

Some social scientists and a few biologists question the existence of a human-caused extinction crisis. Some of these critics also point out that biologists contending that we are in the midst of a human-caused extinction crisis are making the questionable assumption that any loss of habitat also means a net loss in species, usually in some proportion to the amount of habitat lost (Figure 22-9).

For example, when an old-growth forest is cleared, a number of species (many of them with specialized niches) are lost. However, a cleared area that returns by secondary ecological succession to a second-growth forest can still support some of the original species. There may be a net loss of species, but not as high as projected by some estimates.

Biologists use area-size models and field data to estimate current and future extinction rates. Using such approaches, they estimate the annual loss of tropical forest habitat at about 1.8% per year. Edward O. Wilson and several tropical biologists who have counted species in patches of tropical forest before and after destruction or degradation estimate

that this 1.8% loss in habitat results in roughly a 0.5% loss of species.

Do such estimates add up to an extinction crisis? Let us assume, as Wilson and many other biologists do, that a loss of 1 million species over 50–150 years (a very short time in evolutionary history) represents an extinction crisis, with an extinction rate comparable to that during the last mass extinction 65 million years ago (Figure 22-10).

If we assume the global decline in species to be 0.5% per year, then we will lose 25,000 species per year if there are 5 million species and 70,000 per year if there are 14 million species. If these assumptions are correct, we will lose 1 million species in 40 years if there are 5 million species and in 14 years with 100 million species.

Let's assume, however, that the estimate of 0.5% species loss per year is too high for the earth as a whole because of replacement of some species by ecological succession or other factors. If it is 0.25% per year, then we will lose 1 million species in 80 years with 5 million species and in 29 years with 14 million species. Even if we halve the estimated species loss again, to 0.125% per year, we can still lose 1 million species within 57 to 160 years, enough to qualify the situa-

tion as an extinction crisis. According to biodiversity expert Edward O. Wilson, "Clearly, we are in the midst of one of the great extinction spasms of geological history."

Mathematical models recently developed by ecologists indicate a time lag of several generations between habitat loss and extinction, primarily because habitat loss also removes potential colonization sites for the emergence of new species. If these models are correct, biologists may be greatly underestimating the magnitude of current and projected biotic impoverishment.

Biologists do not contend that their estimates of extinction rates are precise enough to make firm predictions. Instead, they argue that there is ample evidence that we are destroying and degrading wildlife habitats at an increasing rate and that our actions certainly lead to a significant loss of species, even though the number and rate vary in different parts of the world.

Critical Thinking

Do you believe that we are in the midst of a sixth mass extinction caused mostly by human activities? Explain. If so, list five ways in which you contribute to this loss of biodiversity.

---

remaining Florida panthers (photo on p. 549), or **(5)** some unknown plant or insect in a tropical forest becomes prematurely extinct because of human activities?

Because ecosystems are constantly changing in response to changing environmental conditions (Section 8-6, p. 192), why should we try to preserve ecosystems? Does it matter that tropical forests, grasslands, wetlands, coral reefs, and others systems making up the earth's ecological diversity are being destroyed or degraded by human activities?

Conservation biologists and ecologists contend that the answer to these questions is *yes* because wild species, natural ecosystems, and the earth's overall biodiversity have two types of value (Figure 22-11):

- **Instrumental value** because of their usefulness to us.

- **Intrinsic value** because they exist, regardless of whether they have any usefulness to us.

What Are the Instrumental Values of Biodiversity? The genes, species, and ecosystems that make up the earth's biodiversity have several types of instrumental or human-centered values that can be classified as **(1)** *utilitarian* or *use* and **(2)** *nonutilitarian* or *nonuse* (Figure 22-11).

Examples of utilitarian values involve the following uses of nature:

Economic Goods

- Species provide food, fuel, fiber, lumber, paper, medicine, and many other useful products.

- About 90% of today's food crops were domesticated from wild tropical plants (Figure 12-5, p. 266), and the genetic diversity of existing wild plant species is needed to develop future crop strains.

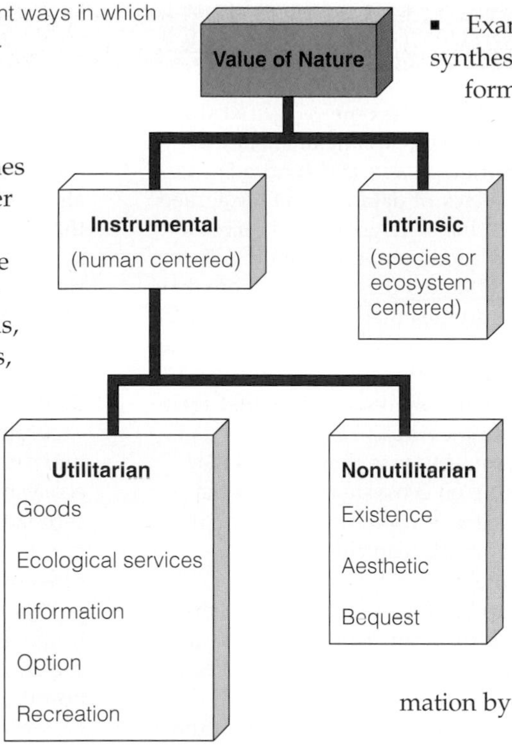

Figure 22-11 Different ways in which humans value nature.

Value of Nature

Instrumental (human centered)

Intrinsic (species or ecosystem centered)

Utilitarian
Goods
Ecological services
Information
Option
Recreation

Nonutilitarian
Existence
Aesthetic
Bequest

- At least 40% of all medicines (worth at least $200 billion per year) and 80% of the top 150 prescription drugs used in the United States were originally derived from living organisms, with 74% derived from plants, mostly from tropical developing countries (Figure 22-12). Yet less than 0.1% of the world's known plant species have been screened by chemists for the presence of compounds that might help fight disease.

Ecological Services

- The ecological services provided by wild species and ecosystems are key factors in sustaining the earth's biodiversity and ecological functions that support human economies and human health.

- Examples of ecological services include photosynthesis, pollination of crops and other plants, soil formation and maintenance, nutrient recycling, pest control, climate regulation, moderation of weather extremes, flood control, drinking and irrigation water, waste decomposition, absorption and detoxification of pollutants, and clean air and water (Figure 4-36, p. 99).

Information

- The genetic information in species allows them to adapt to changing environmental conditions and to form new species that provide us with ecological services and goods.

- The genetic information in the genes of species used in genetic engineering (Pro/Con, p. 275).

- We obtain educational or scientific information by studying genes, species, and ecosystems.

Options

- People would be willing to pay in advance to preserve the option of directly using a resource such as a tree, an elephant, a forest, or clean air in the future.

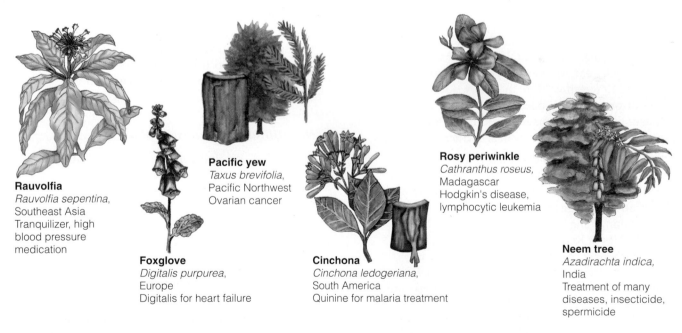

Rauvolfia
Rauvolfia sepentina, Southeast Asia
Tranquilizer, high blood pressure medication

Foxglove
Digitalis purpurea, Europe
Digitalis for heart failure

Pacific yew
Taxus brevifolia, Pacific Northwest
Ovarian cancer

Cinchona
Cinchona ledogeriana, South America
Quinine for malaria treatment

Rosy periwinkle
Cathranthus roseus, Madagascar
Hodgkin's disease, lymphocytic leukemia

Neem tree
Azadirachta indica, India
Treatment of many diseases, insecticide, spermicide

Figure 22-12 Nature's pharmacy. Parts of these and a number of other plants and animals (many of them found in tropical forests) are used to treat a variety of human ailments and diseases. About 70% of the 3,000 plants identified by the National Cancer Institute as sources of cancer-fighting chemicals come from tropical forests. Despite their economic and health potential, fewer than 1% of the estimated 125,000 flowering plant species in tropical forests (and a mere 1,100 of the world's 266,000 known plant species) have been examined for their medicinal properties. Many of these tropical plant species are likely to become extinct before we can study them.

What Is the Economic Worth of the Earth's Ecological Services?

In 1997, a team of 13 ecologists, economists, and geographers attempted to estimate how much the earth's natural ecological services (Figure 4-36, p. 99) are worth in monetary terms. According to this crude appraisal led by ecological economist Robert Costanza of the University of Maryland, the economic value of income from the earth's natural capital is at least $36 trillion per year—close to the annual gross world product of $39 trillion. Thus, each year nature's natural income provides us with free goods and services worth somewhere between $2,600 and $8,900 per person. To provide an annual natural income of $36 trillion per year, the world's natural capital would have a value of at least $500 trillion—an average of about $82,000 for each person on earth.

To make these estimates, the researchers divided the earth's surface into 16 biomes (Figure 6-16, p. 131) and aquatic life zones (they omitted deserts and tundra because of a lack of data). Then they agreed on a list of 17 goods and services provided by nature (Figure 4-36, p. 99) and sifted through more than 100 studies that attempted to put a dollar value on such services in the 16 different types of ecosystems.

Some analysts believe that such estimates are misleading and dangerous because they put a dollar value on ecosystem services that have an infinite value because they are irreplaceable. In the 1970s, economist E. F. Schumacher warned that "to undertake to measure the immeasurable is absurd" and is a "pretense that everything has a price."

The researchers admit that their estimates rely on many assumptions and omissions and could easily be too low by a factor of 10 to 1 million or more. For example, their calculations included only estimates of the ecosystem services themselves, not the natural capital that generates them, and omitted the value of nonrenewable minerals and fuels. They also recognize that as the supply of ecosystem services declines their value will rise sharply and that such services can be viewed as having an infinite value.

However, they contend that their estimates are much more accurate than the *zero* value the market usually assigns to these ecosystem services. They hope such estimates will call people's attention to the fact that the earth's ecosystem services (Figure 4-36, p. 99) are absolutely essential for all humans and their economies (Figure 26-7, p. 694) and that their economic value is huge.

Critical Thinking

Should we put a price on nature's services? Explain. What are the alternatives?

Recreation

- We value recreational pleasure provided by wild plants and animals and natural ecosystems. For example, each year Americans spend more than three times more to watch wildlife than they do to watch movies or professional sporting events. Examples of such uses that do not consume wildlife include nature photography, nature walks, and birdwatching.

- Another normally nonconsumptive use of nature is wildlife tourism, or *ecotourism*. It generates at least $500 billion worldwide, and perhaps twice as much. Conservation biologist Michael Soulé estimates that one male lion living to age 7 generates $515,000 in tourist dollars in Kenya but only $1,000 if killed for its skin. Similarly, over a lifetime of 60 years, a Kenyan elephant is worth about $1 million in ecotourist revenue—many times more than its tusks are worth when sold illegally for their ivory.

Ideally, ecotourism (1) should not cause ecological damage, (2) should provide income for local people to motivate them to preserve wildlife, and (3) should provide funds for the purchase and maintenance of wildlife preserves and conservation programs. However, most nature tourism does not meet these goals and excessive and unregulated ecotourism can destroy or degrade fragile areas and promote premature species extinction. Also, typically less than 1% of the money earned from ecotours ends up as income for local people or for wildlife conservation. Appendix 6 lists some guidelines for evaluating ecotours.

Nonutilitarian (Nonuse) Values

- *Existence:* We find value in knowing that something such as a redwood forest, wilderness, or an endangered species exists, even if we will never see it or get direct use from it.

- *Aesthetic*: We find value in a resource such as a tree, forest, wild species, or a vista because of its beauty.

- *Bequest*: People are willing to pay to protect some forms of natural capital for use by future generations.

What Is the Total Economic Value of the Earth's Biodiversity? Recently, ecologists and economists have attempted to place a monetary value on the instrumental ecosystem services provided by earth's natural resources (Spotlight, above). They hope that such estimates of the total instrumental value or wealth of nature will **(1)** alert people to the value of these free services to our lives and economies and **(2)** help slow down the unsustainable use and degradation of many of the world's ecosystems and species.

What Is the Intrinsic Value of Biodiversity? Some people believe that all wild species and ecosystems and the world's overall biodiversity have *intrinsic value*. They are said to have an inherent right to exist that is unrelated to their usefulness to humans.

According to this view, we have an ethical responsibility to **(1)** protect species from becoming prematurely extinct as a result of human activities and **(2)** prevent the degradation of the world's ecosystems and its overall biodiversity, as discussed in more detail in Section 28-2, p. 742. Biologist Edward O. Wilson believes that most people feel obligated to protect other species and the earth's biodiversity because most humans seem to have a natural affinity for nature (Connections, right).

Some people distinguish between the survival rights of plants and those of animals, mostly for practical reasons. Poet Alan Watts once said that he was a vegetarian "because cows scream louder than carrots." Other people distinguish between various types of species. For example, they might think little about killing a mosquito, cockroach (Spotlight, p. 112), or rat or ridding the world of disease-causing bacteria.

Some proponents go further and assert that each individual organism, not just each species, has a right to survive without human interference. Others apply this to individuals of some species but not to those of other species. Unless they are strict vegetarians, for example, some people might see no harm in having others kill domesticated animals in slaughterhouses to provide them with meat, leather, and other products. However, these same people might deplore the killing of wild animals such as deer, squirrels, or rabbits.

Others emphasize the importance of preserving the whole spectrum of biodiversity by protecting entire ecosystems rather than individual species or organisms, as discussed in Chapter 23.

22-4 CAUSES OF DEPLETION AND PREMATURE EXTINCTION OF WILD SPECIES

What Are the Causes of Wildlife Depletion and Premature Extinction? The basic underlying causes of population reduction and premature extinction of wildlife are:

- Human population growth (Figure 1-1, p. 2, and Figure 1-6, p. 9).

- Economic systems and policies that fail to value the environment and its ecological services (Figure

Biophilia

CONNECTIONS

Biologist Edward O. Wilson contends that because of the billions of years of biological connections leading to the evolution of the human species (Figure 5-4, p. 105), we have an inherent affinity for the natural world, a phenomenon he calls *biophilia* (love of life).

Evidence of this natural affinity for life is seen in the preference most people have for almost any natural scene over one from an urban environment. Given a choice, most people prefer to live in an area where they can see water, grassland, or a forest. More people visit zoos and aquariums than attend all professional sporting events combined.

In the 1970s, I was touring the space center at Cape Canaveral in Florida. During our bus ride the tour guide pointed out each of the abandoned multimillion-dollar launch sites and gave a brief history of each launch. Most of us were utterly bored. Suddenly people started rushing to the front of the bus and staring out the window with great excitement. What they were looking at was a baby alligator—a dramatic example of how *biophilia* can triumph over *technophilia*.

Critical Thinking

1. Do you have an affinity for wildlife and wild ecosystems (biophilia)? If so, how do you display this love of wildlife in your daily actions? What patterns of your consumption help destroy and degrade wildlife?

2. Some critics contend that there is no verifiable genetic evidence of Wilson's biophilia concept and view this idea as opinion, not science. If you have an affinity for wildlife, do you believe that it is inherent in your genes or that it is culturally conditioned? Does it matter? Explain.

4-36, p. 99), thereby promoting unsustainable exploitation.

- Greater per capita resource use as a result of increasing affluence and economic growth (Figure 1-7, p. 9, and Figure 1-15, p. 15).

- Increasing appropriation of the earth's net primary productivity (Figure 4-25, p. 88), which supports all life for human use.

- Poverty, which pushes many of the poor in developing countries to cut forests, grow crops on marginal land, overgraze grasslands, deplete fish species, and kill endangered animals for their valuable furs, tusks, or other parts.

These underlying causes lead to other more direct causes of the endangerment and premature extinction of wild species (Figure 22-13).

What Is the Role of Habitat Loss, Degradation, and Fragmentation? Biologists agree that the greatest threat to wild species is habitat loss (Figure 22-14), degradation, and fragmentation. According to biodiversity researchers, tropical deforestation (Section 23-4, p. 604) is the greatest eliminator of species, followed by **(1)** destruction of coral reefs and wetlands (p. 152 and p. 161), **(2)** plowing of grasslands (Figure 6-25, p. 141), and **(3)** pollution of freshwater (Section 19-2, p. 478) and marine habitats (Section 19-4, p. 488). Globally, temperate biomes have been affected more by habitat disturbance, degradation, and fragmentation than have tropical biomes because of widespread development in temperate developed countries over the past 200 years (Figure 22-15).

According to the Nature Conservancy, the major types of habitat disturbance threatening endangered species in the United States are in order of importance **(1)** agriculture, **(2)** commercial development, **(3)** water development, **(4)** outdoor recreation including off-road vehicles, **(5)** livestock grazing, and **(6)** pollution.

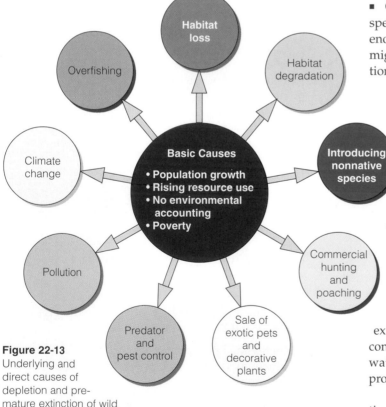

Figure 22-13
Underlying and direct causes of depletion and premature extinction of wild species. The two biggest direct causes of wildlife depletion and premature extinction are **(1)** habitat loss, fragmentation, and degradation and **(2)** deliberate or accidental introduction of nonnative species into ecosystems.

Island species, many of them *endemic species* found nowhere else on earth, are especially vulnerable to extinction. The theory of island biogeography (Figure 8-6, p. 177) has been used to predict the number and percentage of species that would become extinct when habitats on islands are destroyed, degraded, or fragmented. For example, many continuously forested landscapes have been converted into a quilt of small patches of trees criss-crossed with roads, subdivisions, cropfields, and other forms of human land use. This model has been extended from islands to national parks, tropical rain forests, lakes, and nature reserves, which can be viewed as *habitat islands* in an inhospitable sea of human-altered habitat.

Habitat fragmentation occurs when a large, continuous area of habitat is reduced in area and divided into a patchwork of isolated areas or fragments. The three main problems caused by habitat fragmentation are:

- A decrease in the sustainable population size for many species when an existing population is divided into two or more isolated subpopulations.

- Increased surface area or edge, which makes some species more vulnerable to **(1)** predators, **(2)** competition from nonnative and pest species, **(3)** wind, and **(4)** fire.

- Creation of barriers that limit the ability of some species to disperse and colonize new areas, find enough to eat, and find mates. Some species of migrating birds face loss, degradation, or fragmentation of their seasonal habitats (Case Study, p. 567).

Types of species that are vulnerable to local and regional extinction because of habitat fragmentation are those that **(1)** are rare, **(2)** need to roam unhindered over a large area, **(3)** cannot rebuild their population because of a low reproductive capacity, **(4)** have specialized niches (habitat or resource requirements), and **(5)** are sought by people for furs, food, medicines, or other uses.

What Harm Is Caused by Nonnative Species?
After habitat loss and degradation, deliberate or accidental introduction of nonnative species into ecosystems is the biggest cause of animal and plant extinctions. The nonnative invaders arrive from other continents as **(1)** stowaways on aircraft, **(2)** in the bilge water of tankers, and **(3)** as hitchhikers on imported products such as wooden packing crates.

The United States is home for about 50,000 nonnative species. According to a 2000 study by David Pimentel, damages and pest control costs for unwanted species amount to an estimated $137 billion per year—an average loss of $16 million per hour (Figure 22-16).

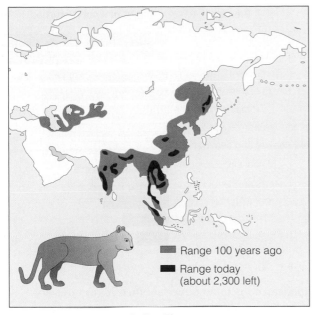

Indian Tiger

Range 100 years ago
Range today
(about 2,300 left)

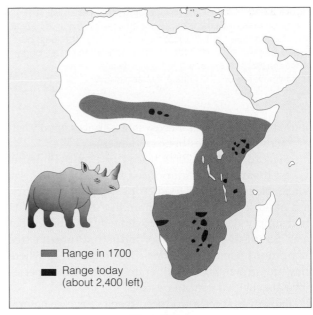

Black Rhino

Range in 1700
Range today
(about 2,400 left)

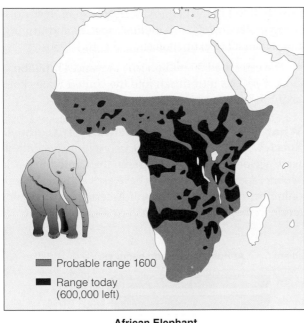

African Elephant

Probable range 1600
Range today
(600,000 left)

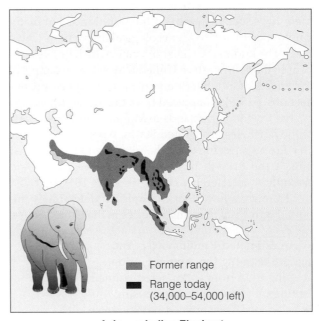

Asian or Indian Elephant

Former range
Range today
(34,000–54,000 left)

Figure 22-14 Reductions in the ranges of four wildlife species, mostly the result of habitat loss and hunting. What will happen to these and millions of other species when the world's human population doubles and per capita resource consumption rises sharply in the next few decades? (Data from International Union for the Conservation of Nature and World Wildlife Fund)

These are conservative estimates because of **(1)** inadequate economic data on many nonnative species and **(2)** inability to assign accurate monetary values to losses in biodiversity, ecosystem services, and aesthetics. If such costs were included, the overall annual damage from nonnative species would be much higher.

According to the U.S. Wildlife Service, 49% of the more than 1,200 endangered and threatened species in the United States (and 95% of those in Hawaii) are threatened by nonnative species.

What Is the Role of Deliberately Introduced Species? Deliberate introduction of nonnative species can be beneficial or harmful depending on the species and where they are introduced. We depend heavily on

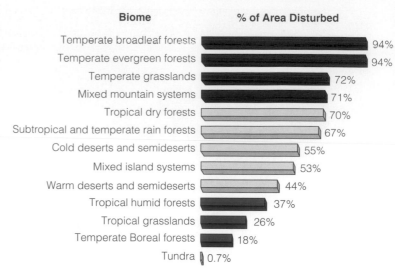

Figure 22-15 Habitat disturbance by biome. Generally, temperate biomes (red) have experienced greater levels of habitat disturbance than have tropical biomes (yellow). The least disturbed biomes are boreal forests and arctic biomes, although boreal forests are expected to be increasingly disturbed by logging. (Data from the Nature Conservancy)

Biome	% of Area Disturbed
Temperate broadleaf forests	94%
Temperate evergreen forests	94%
Temperate grasslands	72%
Mixed mountain systems	71%
Tropical dry forests	70%
Subtropical and temperate rain forests	67%
Cold deserts and semideserts	55%
Mixed island systems	53%
Warm deserts and semideserts	44%
Tropical humid forests	37%
Tropical grasslands	26%
Temperate Boreal forests	18%
Tundra	0.7%

nonnative organisms for ecosystem services, food, shelter, medicine, and aesthetic enjoyment.

According to a 2000 study by David Pimentel (Guest Essay, p. 232), introduced species such as corn, wheat, rice, and other food crops, and cattle, poultry, and other livestock supply more than 98% of the U.S. food supply at a value of approximately $800 billion per year. Similarly, about 85% of industrial forestry tree plantations use species nonnative to the areas where they are grown. Some deliberately introduced species have also helped control pests.

However, some introduced species have no natural predators, competitors, parasites, or pathogens to control their numbers in their new habitats. This can allow them to reduce or wipe out the populations of many native species and trigger ecological disruptions (Table 22-1). The kudzu ("CUD-zoo") vine, which grows rampant in the southeastern United States, is an example of the effects of a deliberately introduced species (Connections, p. 569). Knapweed from the Republic of Georgia has invaded grasslands in Montana and other parts of the northwestern United States. Apparently it gains an edge by releasing toxic chemicals from its roots that inhibit the growth of competing native plants—an example of interference competition (p. 181).

Here are a few other examples of deliberately imported terrestrial nonnative species in the United States:

- An estimated 1 million *wild (feral) pigs* are roaming the state of Florida, hogging food from endangered animals, rooting up farm fields, and causing traffic accidents. They arrived with Spanish conquistadors more than 450 years ago. Game and wildlife officials have had little success in controlling their numbers with hunting and trapping and say there is no way to stop them.

- *Brazilian pepper*, once sold as a landscape ornamental plant, now infests more than 40,000 hectares (100,000 acres) of Everglades National Park.

- The Australian *melaleuca tree*, introduced in south Florida in 1906, was planted as windbreaks and fence rows. Now it has invaded 217,000 hectares (536,000 acres) of the forest and grassland ecosystems of the Florida Ever-

glades and is taking over another 30 hectares (74 acres) every day.

- In Hawaii, the carnivorous *rosy tree snail* was imported to kill the giant African tree snail. Instead, it began devouring local mollusk species and driving more than 50 to extinction since the mid-1950s.

- The estimated 30 million *feral cats* and 41 million *outdoor pet cats* introduced into the United States kill about 568 million birds per year.

What Is the Role of Accidentally Introduced Species? In the late 1930s, extremely aggressive fire ants were accidentally introduced into the United States in Mobile, Alabama. They may have arrived on shiploads of lumber imported from South America or by hitching a ride in the soil-containing ballast of cargo ships.

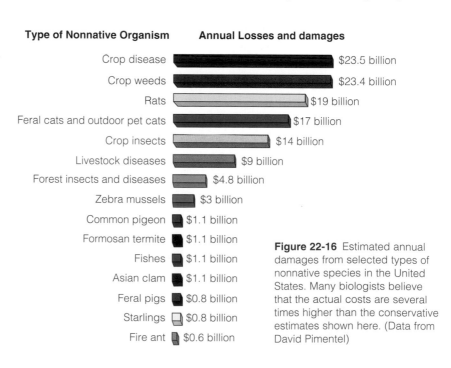

Type of Nonnative Organism	Annual Losses and damages
Crop disease	$23.5 billion
Crop weeds	$23.4 billion
Rats	$19 billion
Feral cats and outdoor pet cats	$17 billion
Crop insects	$14 billion
Livestock diseases	$9 billion
Forest insects and diseases	$4.8 billion
Zebra mussels	$3 billion
Common pigeon	$1.1 billion
Formosan termite	$1.1 billion
Fishes	$1.1 billion
Asian clam	$1.1 billion
Feral pigs	$0.8 billion
Starlings	$0.8 billion
Fire ant	$0.6 billion

Figure 22-16 Estimated annual damages from selected types of nonnative species in the United States. Many biologists believe that the actual costs are several times higher than the conservative estimates shown here. (Data from David Pimentel)

The Plight of Migrating Birds

Migrating bird species face a double habitat problem. Nearly half of the 700 U.S. bird species spend two-thirds of the year in the tropical forests of Central or South America or on Caribbean islands. During the summer they return to North America to breed.

A U.S. Fish and Wildlife study showed that between 1978 and 1987, populations of 44 of the 62 surveyed species of insect-eating, migratory songbirds in North America declined; 20 species experienced drops of 25–45%.

Researchers have identified several possible culprits:

- Logging of tropical forests in the birds' winter habitats.

- Fragmentation of their summer forest and grassland habitats in North America. The intrusion of farms, freeways, and suburbs breaks forests into patches. This makes it easier for **(1)** predators to feast on the eggs and the young of migrant songbirds, **(2)** parasitic cowbirds to lay their eggs in the nests of various songbird species and have those birds raise their young for them, and **(3)** invasions of nonnative shrubs to give nesting birds less protection from predators. In Texas, an environmental group is paying ranchers to set traps for parasitic cowbirds to decrease their harmful effects on songbirds.

- Deaths of at least 4 million migrating songbirds each year when they fly into TV, radio, and phone towers.

However, there is no universal trend in North American songbird populations. Evidence indicates that some species are generally declining, whereas populations of other species are declining in some areas and increasing in other areas.

Approximately 68% of the world's 9,600 known bird species are declining in numbers (58%) or are threatened with extinction (12%), mostly because of habitat loss and fragmentation. Conservation biologists view this decline of bird species as an early warning of the greater loss of biodiversity to come. Birds are excellent environmental indicators because they **(1)** live in every climate and biome, **(2)** respond quickly to environmental changes in their habitats, and **(3)** are easy to track and count.

In addition to serving as indicator species, birds play important ecological roles such as **(1)** helping control populations of insects (including the spruce budworm, gypsy moth, and tent caterpillar, which decimate many tree species) and rodents, **(2)** pollinating a wide variety of flowering plants, and **(3)** spreading plants throughout their habitats by consuming plant seeds and excreting them in their droppings.

Critical Thinking

What three things would you do to help prevent the decline in populations of migrating birds?

Without natural predators, these ants have spread rapidly by land and water (they can float) throughout the South, from Texas to Florida and as far north as Tennessee and Virginia (Figure 22-18, p. 570) and are also found in Puerto Rico. Recently they have also invaded California and New Mexico.

Wherever the fire ant has gone, up to 90% of native ant populations have been sharply reduced or wiped out. Their extremely painful stings have also killed deer fawn, birds, livestock, pets, and at least 80 people allergic to their venom. They have also **(1)** invaded cars and caused accidents by attacking drivers, **(2)** made crop-fields unplowable, **(3)** disrupted phone service and electrical power, **(4)** caused some fires by chewing through underground cables, and **(5)** cost the United States an estimated $600 million per year.

Widespread pesticide spraying in the 1950s and 1960s temporarily reduced populations of fire ants. In the end, however, this chemical warfare hastened the advance of the rapidly multiplying fire ant by **(1)** reducing populations of many native ant species and **(2)** promoting development of genetic resistance to heavily used pesticides in the rapidly multiplying fire ants.

Researchers at the U.S. Department of Agriculture are experimenting with use of biological control to reduce fire ant plantations. These include use of:

- A tiny parasitic Brazilian fly that lays its eggs on the fire ant's body. After the eggs hatch, the larvae eat their way through the ant's head.

- A pathogen (*Thelohania solenopsae*) imported from South America that infects fire ant colonies and generally can wipe out a colony within 9–18 months.

Before widespread use of these biological control agents, however, researchers must be sure they will not cause problems for native ant species or become pests themselves.

About 50 years ago the *brown tree snake* arrived accidentally on the island of Guam in the cargo of a military plane. Since then this highly aggressive, nocturnal, climbing snake with no natural enemies has eliminated or decimated bird and lizard species on Guam. It has also snatched chickens and pets from yards, attacked babies asleep in cribs, and shorted out power lines. In some areas of Guam, more than 1,900 of

Table 22-1 Damage Caused by Species Imported into the United States

Name	Origin	Mode of Transport	Type of Damage
Mammals			
European wild boar	Russia	Intentionally imported (1912), escaped captivity	Destroys habitat by rooting; damages crops
Nutria (cat-sized rodent)	Argentina	Intentionally imported (1940), escaped captivity	Alters marsh ecology; damages levees and earth dams; destroys crops
Birds			
European starling	Europe	Intentionally released (1890)	Competes with native songbirds; damages crops; transmits swine diseases; causes airport nuisance
House sparrow	England	Intentionally released by Brooklyn Institute (1853)	Damages crops; displaces native songbirds
Fish			
Carp	Germany	Intentionally released (1877)	Displaces native fish; uproots water plants; lowers waterfowl populations
Sea lamprey	North Atlantic Ocean	Entered Great Lakes via Welland Canal (1829)	Wiped out lake trout, lake whitefish, and sturgeon in Great Lakes
Walking catfish	Thailand	Imported into Florida	Destroys bass, bluegill, and other fish
Insects			
Argentine fire ant	Argentina	Probably entered via coffee shipments from Brazil (1918)	Damages crops; destroys native ant species
Camphor scale insect	Japan	Accidentally imported on nursery stock (1920s)	Damaged nearly 200 plant species in Louisiana, Texas, and Alabama
Japanese beetle	Japan	Accidentally imported on irises or azaleas (1911)	Defoliates more than 250 species of trees and other plants, including many of commercial importance
Plants			
Water hyacinth	Central America	Accidentally introduced (1882)	Clogs waterways; shades out other aquatic vegetation
Chestnut blight (fungus)	Asia	Accidentally imported on nursery plants (1900)	Killed nearly all eastern U.S. chestnut trees; disturbed forest ecology
Dutch elm disease (fungus)	Europe	Accidentally imported on infected elm timber used for veneers (1930)	Killed millions of elms; disturbed forest ecology

*Source: Modified from Biological Conservation by David W. Ehrenfeld, 1970. Holt, Rinehart & Winston, Inc.

these snakes per square kilometer (5,000 per square mile) have been counted.

This light-sensitive snake can select an airplane wheel or outgoing cargo as a hiding place during daylight hours and thus spread to other areas such as Hawaii, which is home for 41% of all endangered bird species in the United States. Hawaii has a lot to lose from an invasion by this snake. Another harmful invader is the Formosan termite (Case Study, p. 571).

In 1985, *Tiger mosquitoes*, which breed in scrap tires, arrived in the United States in a Japanese ship carrying tires to a Houston, Texas, recapping plant. Since

then they have spread to 25 states and to parts of Africa and South America. Aggressive biters, these mosquitoes can transmit 18 potentially fatal tropical viruses, including dengue fever, yellow fever, and forms of encephalitis. University of Kentucky scientists have flown a radar-equipped plane over the Ohio River Valley to find the Asian tiger mosquito. Radar cannot detect the mosquitoes, but it can spot hidden mounds of scrap tires, where the insects breed.

Accidentally introduced species can also create trade controversies. In 1998, the United States banned the import of goods from China in untreated wooden

Deliberate Introduction of the Kudzu Vine

In the 1930s, the *kudzu vine* was imported from Japan and planted in the southeastern United States to help control soil erosion. It does control erosion, but it is so prolific and difficult to kill that it engulfs hillsides, trees, abandoned houses and cars, stream banks, patches of forest, and anything else in its path (Figure 22-17).

This vine, sometimes called "the vine that ate the South," has spread throughout much of the southern United States and could spread as far north as the Great Lakes by 2040 if projected global warming occurs.

Although kudzu is considered a menace in the United States, Asians use a powdered kudzu starch in beverages, gourmet confections, and herbal remedies for a range of diseases. A Japanese firm has built a large kudzu farm and processing plant in Alabama and ships the extracted starch to Japan.

In an ironic twist, kudzu—which can engulf and kill trees—could eventually help save trees from loggers. Research at Georgia Institute of Technology indicates that kudzu may be used as a source of tree-free paper.

Critical Thinking

On balance, do you think the potential beneficial uses of the kudzu vine outweigh its harmful ecological effects? Explain.

Figure 22-17 Kudzu taking over a field, trees, and a sign in South Carolina. This vine can grow 0.3 meter (1 foot) per day and is now found from East Texas to Florida and as far north as southeastern Pennsylvania and Illinois. Kudzu was deliberately introduced into the United States for erosion control, but it can't be stopped by being dug up or burned. Grazing by goats and repeated doses of herbicides can destroy it, but goats and herbicides also destroy other plants, and herbicides can contaminate water supplies. Recently, scientists have found a common fungus (*Myrithecium verrucaria*) that can kill kudzu within a few hours, apparently without harming other plants. (Angela Lax/Photo Researchers, Inc.)

packing crates. Such crates were the primary culprits in the recent invasion of the voracious *Asian long-horned beetle*, which poses a major threat to U.S. hardwood trees. China has complained that the ban is an unfair trade barrier and hopes to get the World Trade Organization (p. 733) to overturn the ban.

What Can Be Done to Reduce the Threat from Nonnative Species? Once a nonnative species has become established in an ecosystem, its wholesale removal is almost impossible—somewhat like trying to get smoke back into a chimney or trying to unscramble an egg. Thus, the best way to limit the harmful impacts of nonnative species is to prevent them from being introduced and becoming established. This can be done by:

- Identifying major characteristics that allow species to become successful invaders and the types of communities that are vulnerable to invaders and using this information to screen out potentially harmful invaders (Figure 22-19).

- Stepping up inspections of goods coming into a country.

- Identifying major harmful invader species and passing international laws banning their transfer from one country to another (as is now done for endangered species).

- Requiring ships to discharge their ballast (bilge) water and replace it with saltwater at sea before entering ports and to sterilize such water.

For invaders that get across borders, the best strategies are to **(1)** set up a monitoring network to spot invaders early and **(2)** once an invader is detected, mount a rapid, coordinated response to eliminate it or keep it

1918

2000

from spreading. Methods for controlling the spread of new invading species and trying to reduce the spread of entrenched species include the following:

- *Mechanical control* involving physically removing the species. This can be expensive and labor intensive.

- *Chemical control* by using pesticides. This can sometimes affect nontarget organisms (including humans) and cause genetic resistance in the invader species (p. 505).

- *Biological control* by introducing a natural predator, parasite, or disease organism from the invading species' natural habitat. This can lead to environmentally sound control with minimal expense, but some control agents do not last, and some can become invasive species themselves.

- *Ecological control*, in which environmental factors such as fire and water flow can be manipulated to give native species a competitive advantage over invader species.

In 1999, President Bill Clinton issued an executive order allocating $28 million to combat species invasions into the United States and creating an interagency coun-cil to produce a plan to mobilize the federal government to defend against invasions by nonnative species.

What Is the Role of Commercial Hunting and Poaching? The international trade in wild plants and animals brings in $10–20 billion a year. It is estimated that at least one-fourth of this trade involves the illegal sale of endangered and threatened species or their parts. Organized crime has moved into illegal wildlife smuggling because of the huge profits involved (second only to drug smuggling). At least two-thirds of all live animals smuggled around the world die in transit.

Worldwide, some 622 species of animals and plants face extinction, mostly because of illegal trade. To poachers, a live mountain gorilla is worth $150,000, a gyrfalcon $120,000, a chimpanzee $50,000, an Imperial Amazon macaw $30,000, and rhinoceros horn (Figure 22-20) as much as $28,600 per kilogram ($13,000 per pound). Excessive hunting brought the American bison to near extinction (p. 23).

In much of West Africa, wildlife in the form of bushmeat is an important source of protein for many local people (Figure 22-21). The growing bushmeat trade also generates revenues of more than $150 million per year and can endanger species such as the gorilla.

In 1950, there were an estimated 100,000 tigers in the world. Despite international protection, today only about 6,000 tigers are left (about 4,000 of them in India), mostly because of habitat loss and poaching for their furs and bones. Bengal tigers are at risk because a tiger fur sells for $100,000 in Tokyo. With the body parts of a single tiger worth more than $10,000, it is not surprising that illegal hunting has skyrocketed, especially in India. Without emergency action, there may be few or no tigers left in the wild within 20 years.

As more species become endangered, the demand for them on the black market soars, hastening their chances of premature extinction from poaching. Most poachers are not caught, and the money to be made far outweighs the risk of fines and the much smaller risk of imprisonment.

Case Study: How Should We Protect Elephants from Extinction? Habitat loss (Figure 22-14) and legal and illegal trade in elephant ivory (Figure 22-22) have reduced wild African elephant numbers from 2.5 million in 1970 to about 500,000 today (plus about 300 in zoos).

Since 1989, this decline has been slowed by an international ban on the sale of ivory from African elephants, but things are not quite that simple. First, the

Characteristics of Successful Invader Species	Characteristics of Ecosystems Vulnerable to Invader Species
• High reproductive rate, short generation time (r-selected species)	• Similar climate to habitat of invader
• Pioneer species	• Absence of predators on invading species
• Long lived	• Early successional species
• High dispersal rate	• Low diversity of native species
• Release growth-inhibiting chemicals into soil	• Absence of fire
• Generalists	• Disturbed by human activities
• High genetic variability	

Figure 22-19 Some general characteristics of successful invader species and communities vulnerable to invading species.

The Termite from Hell

Forget killer bees and fire ants. The homeowner's nightmare is the Formosan termite. It is the most voracious, aggressive, and prolific of more than 2,000 known termite species.

These termites invaded the Hawaiian Islands by 1900 and probably arrived on the mainland of United States during or soon after World War II in wooden packing materials on military cargo ships that docked in southern ports such as New Orleans, Louisiana, and Houston, Texas.

They can quickly munch through wooden beams and plywood and can even chew through brick. Their huge colonies, which can contain up to 10 million insects (10 times as many as a nest of native termites), can be found under houses or in their attics.

Over the past decade, the Formosan termite has caused more damage in New Orleans, Louisiana, than hurricanes, floods, and tornadoes combined. In New Orleans almost every building in every neighborhood has been struck, and the famous French Quarter has one of the most concentrated infestations in the world.

Once confined to Louisiana, they have invaded at least a dozen other states, including Alabama, Florida, Mississippi, North and South Carolina, Texas, and California. They cause at least $1 billion in damage each year.

Most currently used pesticides do not work on these termites. In New Orleans, the U.S. Department of Agriculture is using a variety of techniques all at once in an attempt to control the species in a heavily infested 15-block area of the French Quarter. They hope to develop techniques for dealing with these invaders elsewhere.

Critical Thinking

What important ecological roles do termites play in nature? If the Formosan termite and other termite species could be eradicated (a highly unlikely possibility), would you favor doing this? Explain.

---

ban on elephant ivory sales has increased the killing of other species with ivory parts such as bull walruses (for their ivory tusks, worth $500–1,500 a pair) and hippos (for their ivory teeth, worth $70 per kilogram). Second, an increase in elephant populations in areas where their habitat has shrunk has resulted in widespread destruction of vegetation by these animals. This in turn reduces the niches available for other wild species.

Some wildlife conservationists and leaders of several southern African nations have called for a partial lifting of the elephant ivory ban in areas where elephant populations are not endangered. Ivory from the sustainable culling of elephants in these areas would be marketed in a way that certifies that it was obtained legally. Most of the profits would go to local people and for wildlife conservation.

In 1997, the Convention on International Trade in Endangered Species (CITES) voted to give Zimbabwe, Botswana, and Namibia one-time limited permits to sell ivory obtained from culling elephant herds and from natural deaths to Japan under strict monitoring. The $5 million raised from the sales is being used for wildlife protection and rural development programs. These three countries and South Africa have proposed that they be allowed to cull and sell ivory from their elephant herds annually.

Some wildlife conservationists oppose such sales. They believe that the best way to protect African elephants from extinction is to continue the ivory ban, which discourages poaching by keeping the price of ivory down. They point out that no international monitoring program to detect increased poaching was set up and that there has been an increase in elephant poaching since 1999, when the three African countries

Figure 22-20 Rhinoceros horns are carved into ornate dagger handles that sell for $500–12,000 in Yemen and other parts of the Middle East. In China and other parts of Asia, powdered rhino horn is used for medicinal purposes, particularly as a proven fever reducer and occasionally as an alleged aphrodisiac. All five rhinoceros species are threatened with extinction because of poachers (who kill them for their horns) and habitat loss. Between 1973 and 1998 the population of African black rhinos (Figure 22-14) dropped from approximately 63,000 to about 2,400. (R. F. Porter/Ardea London)

Figure 22-21 Bushmeat, such as this gorilla head, is consumed as a source of protein by local people in parts of West Africa and sold in the national and international marketplace. You can find bushmeat on the menu in Cameroon and the Congo in West Africa as well as in Paris, France, and Brussels, Belgium. (©Karl Ammann, Biosynergy Institute)

were allowed to sell culled ivory. What do you think should be done?

What Is the Role of Predators and Pest Control?

People try to exterminate species that compete with them for food and game animals. For example, U.S. fruit farmers exterminated the Carolina parakeet around 1914 because it fed on fruit crops. The species was easy prey because when one member of a flock was shot, the rest of the birds hovered over its body, making themselves easy targets.

African farmers kill large numbers of elephants to keep them from trampling and eating food crops. Many ranchers, farmers, and hunters in the United States support the killing of coyotes, wolves, and other species that can prey on livestock and on species prized by game hunters.

Since 1929, U.S. ranchers and government agencies have poisoned 99% of North America's prairie dogs because horses and cattle sometimes step into the burrows and break

their legs. This has also nearly wiped out the endangered black-footed ferret (Figure 22-7, about 600 left in the wild), which preyed on the prairie dog.

What Is the Role of the Market for Exotic Pets and Decorative Plants? The global legal and illegal trade in wild species for use as pets is a huge and very profitable business. However, for every live animal captured and sold in the pet market, an estimated 50 other animals are killed.

About 25 million U.S. households have exotic birds as pets, 85% of them imported. More than 60 bird species, mostly parrots, are endangered or threatened because of this wild bird trade. According to the U.S. Fish and Wildlife Service, collectors of exotic birds may pay $10,000 for a threatened hyacinth macaw smuggled out of Brazil; however, during its lifetime, a single macaw left in the wild might yield as much as $165,000 in tourist income. A 1992 study suggested that keeping a pet bird indoors for more than 10 years doubles a person's chances of getting lung cancer from inhaling tiny particles of bird dander.

Other wild species whose populations are depleted because of the pet trade include amphibians, reptiles, mammals, and tropical fish (taken mostly from the coral reefs of Indonesia and the Philippines). Divers catch tropical fish by using plastic squeeze bottles of cyanide to stun them. For each fish caught alive, many more die. In addition, the cyanide solution kills the coral animals that create the reef, which is a center for marine biodiversity (Figure 7-1, p. 152, and Figure 7-13, p. 163).

Some exotic plants, especially orchids and cacti, are endangered because they are gathered (often illegally) and sold to collectors to decorate houses, offices, and landscapes. A collector may pay $5,000 for a single rare orchid, and a single rare mature crested saguaro cactus can earn cactus rustlers as much as $15,000.

What Are the Roles of Climate Change and Pollution? Most natural climate changes in the past have taken place over long periods of time (Figure 18-2, p. 447), which gave species more time to adapt and evolve to these changing environmental conditions. However, there is concern that human activities such as emissions of greenhouse gases and deforestation can bring about rapid climate change over several decades (Figure 18-8, p. 453). If these projected climate changes take place, **(1)** many wild species may not have enough time to adapt or migrate (Figure 18-13, p. 459), and

Figure 22-22 Vultures are feeding on this elephant carcass, poached in Tanzania for its ivory tusks, which are used to make jewelry, piano keys, ornamental carvings, and art objects. (E. R. Degginger)

(2) some wildlife in well-protected and well-managed terrestrial reserves and ocean sanctuaries could be depleted within a few decades.

According to a 2000 study by the World Wildlife Fund, global warming could increase extinction by altering one-third of the world's wildlife habitats by 2100. This includes 70% of the habitat in high latitude arctic and boreal biomes.

Toxic chemicals, such as pesticides (Figure 16-4, p. 399), degrade wildlife habitats and kill some terrestrial plants and animals and aquatic species (p. 480, p. 484, and p. 507). According to the U.S. Fish and Wildlife Service, pesticides menace about 20% of the endangered and threatened species in the United States.

What Role Does Loss of Genetic Diversity Play?
Loss of genetic diversity limits the ability of populations and species to survive by reducing their ability to reproduce and adapt to changing conditions. This is an especially serious problem for small populations, which can lose genetic diversity by:

- The *founder effect*, in which the limited genetic diversity of a few individuals founding a population may not be enough to sustain the population.

- *Inbreeding* among closely related individuals of a population.

- A *demographic bottleneck*, which occurs when only a few individuals survive to perpetuate a population after a catastrophe.

- *Genetic drift*, caused by unequal reproductive success, in which some individuals breed more than others and their genes eventually dominate the population.

22-5 SOLUTIONS: PROTECTING WILD SPECIES FROM DEPLETION AND EXTINCTION

How Can Bioinformatics Help Protect Biodiversity? To protect biodiversity, we need fundamental information about the biology and ecology of wild species. In particular, information is needed about **(1)** species names, **(2)** descriptions, **(3)** distributions, **(4)** status of populations, **(5)** habitat requirements, and **(6)** interactions with other species (Section 8-3, p. 180, and Section 8-4, p. 186).

Bioinformatics is the applied science of managing, analyzing, and communicating biological information. It involves **(1)** *building computer databases* to organize and store useful biological information, **(2)** *developing computer tools to find, visualize, and analyze* the information, and **(3)** *communicating* the information, especially using the internet. Bioinformatics is being applied to

many aspects of biology, ranging from storing DNA sequences to storing names, descriptions, and locations of collections of biological organisms in museums (Spotlight, p. 574).

How Can International Treaties Help Protect Endangered Species? Several international treaties and conventions help protect endangered or threatened wild species. One of the most far-reaching is the 1975 Convention on International Trade in Endangered Species (CITES). This treaty, now signed by 152 countries, lists **(1)** more than 800 species that cannot be commercially traded as live specimens or wildlife products because they are in danger of extinction and **(2)** 29,000 other species whose international trade is monitored and regulated because they are at risk of becoming threatened.

However, the effects of this treaty are limited because **(1)** enforcement is difficult and spotty, **(2)** convicted violators often pay only small fines, **(3)** member countries can exempt themselves from protecting any listed species, and **(4)** much of highly profitable illegal trade in wildlife and wildlife products goes on in countries that have not signed the treaty.

The *Convention on Biological Diversity* (CBD) came out of the 1992 Rio Earth Summit. This treaty, which has been ratified by 172 countries (but not the United States), legally binds signatory governments to reversing the global decline of biological diversity.

Under the convention, countries are supposed to accomplish this goal by **(1)** adopting national biodiversity protection strategies and action, **(2)** establishing nationwide systems of protected areas, **(3)** restoring degraded habitats, **(4)** conserving endangered and threatened species and ecosystems, **(5)** making sure that biological resources are used in ecologically sustainable ways, **(6)** ensuring the safe use and application of biotechnology products, and **(7)** seeing that the benefits from biological diversity (such as new medicines) are shared equitably.

The CBD is by far the world's most comprehensive treaty. However, its implementation has proceeded slowly because of failure of some key countries, such as the United States, to ratify the treaty. Also, converting its worthy goals into tangible progress is difficult. In addition, the treaty contains no severe penalties or other enforcement mechanisms.

How Can National Laws Help Protect Endangered Species? The United States controls imports and exports of endangered wildlife and wildlife products through two important laws. The *Lacey Act of 1900* prohibits transporting live or dead wild animals or their parts across state borders without a federal permit. The *Endangered Species Act of 1973* (amended in 1982 and 1988) makes it illegal for Americans to import

SPOTLIGHT

A new environmental buzzword is *bioinformatics*. By providing new computer technologies for organizing, managing, accessing, and analyzing information about biodiversity, bioinformatics is transforming the way biologists and ecologists work.

Traditionally, natural history museums, arboreta, botanical gardens, herbaria, and zoos have been much more than places where the public could view exhibits and learn about nature. They have also **(1)** housed research institutions dedicated to exploring the natural world, **(2)** served as repositories of *biological collections*, and **(3)** been centers for the study of *systematics*, or taxonomy, the branch of biology that deals with descriptions, names, classifications, and evolutionary relationships of the earth's organisms.

In recent years, the role of these institutions and the importance of systematics have been underappreciated. Some have even viewed systematics and biological collecting as old-fashioned, and a shortage of qualified systematists has resulted. However, this is changing because of a growing appreciation for the need for fundamental biodiversity information.

As a result, **(1)** most institutions that house biological collections are busily entering information about their collections into computerized databases, and **(2)** most systematists are building databases about the characteristics of organisms they study.

One example is Species 2000, an internet-based global research project that has the goal of providing information about all known species of plants, animals, fungi, and microbes on the earth as the baseline data set for studies of global biodiversity. Users worldwide will be able to verify the scientific name, status, and classification of any known species via the Species Locator, which will provide online access to authoritative species data drawn from an array of participating databases.

Another example is the proposed Global Biological Information Facility, which would put more than 350 years of information on biological resources currently housed in the world's museums on the internet. A Web site would list each of the world's named species, with links to information about each species, literature references, and which museums house specimens.

Critical Thinking

Explain why some developers and extractors of resources on public land oppose the development of databases such as Species 2000 and national databases of the species found in a particular country.

---

or trade in any product made from an endangered or threatened species unless it is used for an approved scientific purpose or to enhance the survival of the species.

The Endangered Species Act (ESA) authorizes the National Marine Fisheries Service (NMFS) to identify and list endangered and threatened ocean species; the U.S. Fish and Wildlife Service (USFWS) identifies and lists all other endangered and threatened species. These species cannot be hunted, killed, collected, or injured in the United States.

Any decision by either agency to add or remove a species from the list must be based on biology only, not on economic or political considerations. However, economic factors can be used in deciding whether and how to protect endangered habitat and in developing recovery plans for listed species. The act also forbids federal agencies to carry out, fund, or authorize projects that would jeopardize an endangered or threatened species or destroy or modify the critical habitat it needs to survive. On private lands, fines and even jail sentences can be imposed to ensure protection of the habitats of endangered species.

Between 1973 and 2000, the number of U.S. species on the official endangered and threatened list increased from 92 to more than 1,200 species (about 60% of them plants and 40% animals). About 78% of these species are listed as endangered and 22% as threatened. Each year about 50-85 species are added to the list.

According to a 2000 study by the Nature Conservancy, about 33% of the country's species are at risk of extinction, and 15% of them are at high risk. This amounts to about 30,000 species, compared to the only 1,200 species currently protected under the ESA. The study found that many of the country's rarest and most imperiled species are concentrated in a few hot spots (Figure 22-23).

The ESA generally requires the secretary of the interior to designate and protect the *critical habitat* needed for the survival and recovery of each listed species. By June 2000, however, only 124 designated critical habitats had been established.

Getting listed is only half the battle. Next, the USFWS or the NMFS is supposed to prepare a plan to help the species recover. By 2000, final recovery plans had been developed and approved for about 75% of the endangered or threatened U.S. species, but about half of those plans exist only on paper.

The ESA requires that all commercial shipments of wildlife and wildlife products enter or leave the country through one of nine designated ports. Few ille-

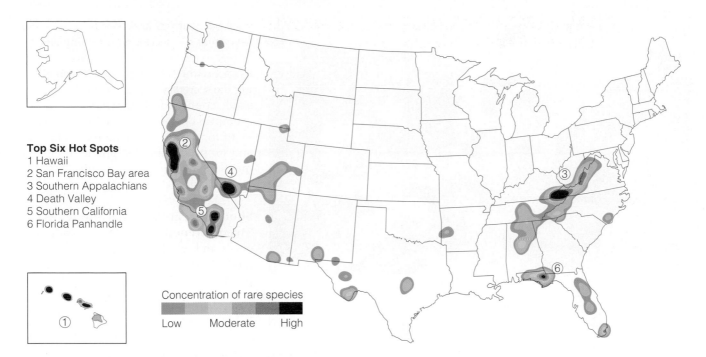

Top Six Hot Spots
1 Hawaii
2 San Francisco Bay area
3 Southern Appalachians
4 Death Valley
5 Southern California
6 Florida Panhandle

Concentration of rare species

Low   Moderate   High

Figure 22-23 Biodiversity hot spots in the United States. This map shows areas that contain the largest concentrations of rare and potentially endangered species. (Data from State Natural Heritage Programs, the Nature Conservancy, and Association for Biodiversity Information)

gal shipments are confiscated (Figure 22-24) because the 60 USFWS inspectors can examine only about one-fourth of at least 90,000 shipments that enter and leave the United States each year. Even if caught, many violators are not prosecuted, and convicted violators often pay only a small fine.

How Can Private Landowners Be Encouraged to Protect Endangered Species? One problem with the ESA is that it has encouraged some developers, timber companies, and other private landowners to avoid government regulation under the ESA by managing their land to reduce its use by endangered species. The National Association for Homebuilders has published practical tips for developers and other landowners to avoid ESA issues. Suggestions include **(1)** planting crops, **(2)** plowing fields between crops to prevent native vegetation and endangered species from occupying the fields, **(3)** clearing forests, and **(4)** burning or managing vegetation to make it unsuitable for local endangered species.

In 1982, Congress amended the ESA to allow the secretary of the interior to use *habitat conservation plans* (HCPs) to strike a compromise between the interests of private landowners and the interests of endangered and threatened wildlife with the goal of not reducing the recovery chances of a protected species. With an HCP, landowners, developers, or loggers are allowed to destroy some critical habitat or kill all or a certain

number of endangered or threatened species on private land in exchange for taking steps to protect that species. Such protective measures might include **(1)** setting aside a part of the species' habitat as a preserve, **(2)** paying to relocate the species to another suitable habitat, or **(3)** paying money to have the government buy suitable habitat elsewhere. Once the plan is approved it cannot be changed, even if new data show that the plan is inadequate to protect a species and help it recover. By June 2000, almost 300 HCPs had been approved, covering more than 8 million hectares (20 million acres).

This approach is supported by some wildlife conservationists because it can head off use of evasive techniques and reduce political pressure to seriously weaken or eliminate the ESA. However, there is growing concern that such plans are being developed too quickly, without enough scientific evaluation of the effects of such plans on a species' recovery. Suggestions for improving HCPs include **(1)** developing scientific standards for such plans, **(2)** having plans reviewed by a scientific advisory committee, and **(3)** requiring more compensating efforts by private landowners as a buffer when important data such as population trends or assessment of the impacts on the affected species are not known.

In 1999, the USFWS approved two new approaches that remove barriers in the ESA that have kept private landowners from protecting threatened or endangered species on their property:

Figure 22-24 Confiscated products made from endangered species. Because of a scarcity of funds and inspectors, probably no more than one-tenth of the illegal wildlife trade in the United States is discovered. The situation is even worse in most other countries. (Steve Hillebrand/U.S. Fish and Wildlife Service)

- *Safe harbor agreements* in which landowners voluntarily agree to take specified steps to restore, improve, or maintain habitat for threatened or endangered species located on their land. In return, landowners get **(1)** technical help from local conservation agencies, **(2)** government assurances that the land, water, or other natural resources involved will not face future restrictions once the agreement is over, and **(3)** assurances that after the agreement has expired landowners can return the property to its original condition without penalty.

- Voluntary *candidate conservation agreements* in which landowners agree to take specific steps to help conserve a species whose population is declining but is not yet listed as endangered or threatened. Participating landowners receive technical help and assurances that no additional resource use restrictions will be imposed on the land covered by the agreement if the species is listed as endangered or threatened in the future.

Should the Endangered Species Act Be Weakened? Opponents of the ESA contend that it has not worked and has caused severe economic losses by hindering development on private land. Since 1995 there have been efforts to weaken the ESA by **(1)** making protection of endangered species on private land voluntary, **(2)** having the government pay landowners if it forces them to stop using part of their land to protect endangered species (p. 589), **(3)** making it harder and more expensive to list newly endangered species by requiring government wildlife officials to navigate through a

series of hearings and peer-review panels, **(4)** giving the secretary of the interior the power to permit a listed species to become extinct without trying to save it, **(5)** allowing the secretary of the interior to give any state, county, or landowner permanent exemption from the law, with no requirement for public notification or comment, and **(6)** prohibiting the public from commenting on or bringing lawsuits to change poorly designed HCPs.

Should the Endangered Species Act Be Strengthened? Most conservation biologists and wildlife scientists contend that the ESA has not been a failure (Spotlight, p. 578). They also refute the charge that the ESA has caused severe economic losses.

- Between 1979 and 1992, only 69 (0.05%) of more than 145,000 projects evaluated by the USFWS were blocked or canceled as a result of the ESA.

- The act does allow for economic concerns. By law, a decision to list a species must be based solely on science. However, once a species is listed, economic considerations can be weighed against species protection in protecting critical habitat and designing and implementing recovery plans.

- The act also allows a special Cabinet-level panel, called the "God Squad," to exempt any federal project from having to comply with the act if the economic costs are too high.

- The act allows the government to issue permits and exemptions to landowners with listed species living on their property and to use HCPs to bargain with private landowners.

A study by the National Academy of Sciences recommended changes to make the ESA more scientifically sound and effective by **(1)** greatly increasing the meager funding for implementing the act, **(2)** developing recovery plans more quickly, **(3)** developing guidelines to avoid provisions that are scientifically or economically unsound and that spell out which actions are likely to harm recovery, and **(4)** establishing a core of "survival habitat" as a temporary emergency measure when a species is first listed that could support the species for 25–50 years.

Most biologists and wildlife conservationists believe that we should develop a new system to protect and sustain biological diversity and ecosystem function based on three principles: **(1)** Find out what species and ecosystems we have, **(2)** locate and protect the most endangered ecosystems and species (Connections, p. 618), and **(3)** give private landowners financial carrots (tax breaks and write-offs), technical help, and assurances of no future requirements (safe harbor and candidate conservation agreements) for helping protect endangered species and ecosystems.

Should We Try to Protect All Endangered and Threatened Species? Because of limited funds, information, and trained personnel, only a few endangered and threatened species can be saved. Some analysts suggest that the limited funds available for preserving threatened and endangered wildlife be concentrated on species that **(1)** have the best chance for survival, **(2)** have the most ecological value to an ecosystem, and **(3)** are potentially useful for agriculture, medicine, or industry.

New research suggests that the health of an ecosystem should be judged not on the basis of sheer numbers of species but on which species **(1)** play keystone roles and **(2)** are tolerant to environmental change such as acid deposition, climate change, and toxins. Researchers say preserving 10 key and adaptable species in an ecosystem is likely to be better than preserving 10,000 weak ones. In addition to protecting *keystone species*, biologists call for protection of keystone resources such as **(1)** salt licks and mineral pools that provide essential minerals for wildlife, **(2)** deep pools in streams and springs that serve as refuges for fish and other aquatic species during dry periods, and **(3)** hollow tree trunks used as breeding sites and homes for many bird and mammal species.

Others oppose selective protection of species on ethical grounds or contend that we don't have enough biological information to make such evaluations. Proponents argue that **(1)** in effect we are already deciding by default which species to save and that **(2)** despite limited knowledge, the selective approach is more effective and a better use of limited funds than the current one.

How Can Wildlife Refuges and Other Protected Areas Help Protect Endangered Species? Since 1903, when President Theodore Roosevelt established the first U.S. federal wildlife refuge at Pelican Island, Florida, the National Wildlife Refuge System has grown to 522 refuges. Some 34 million Americans visit these refuges each year to hunt, fish, hike, or watch birds and other wildlife.

More than three-fourths of the refuges are wetlands for protecting migratory waterfowl. About 20% of the U.S. endangered and threatened species have habitats in the refuge system, and some refuges have been set aside for specific endangered species. These have helped Florida's key deer, the brown pelican, and the trumpeter swan to recover.

Conservation biologists urge the establishment of more refuges for endangered plants. They also urge Congress and state legislatures to allow abandoned military lands that contain significant wildlife habitat to become national or state wildlife refuges.

According to a General Accounting Office study, activities considered harmful to wildlife occur in nearly 60% of the nation's wildlife refuges. There is much controversy over whether to allow oil and gas development in Alaska's Arctic National Wildlife Refuge (Pro/Con, p. 343).

Can Gene Banks, Botanical Gardens, and Farms Help Save Most Endangered Species? Gene or seed banks are used preserve genetic information and endangered plant species by storing their seeds in refrigerated, low-humidity environments (Spotlight, p. 278). Most of the world's 50 seed banks have focused on storing seeds of the approximately 100 plant species that provide about 90% of the food consumed by humans. However, some banks are devoting more attention to storing seeds for a wider range of species that may be threatened with extinction or a loss of genetic diversity. The U.S. National Genetic Resource Program consists of a network of seed banks that store 450,000 seed samples from 8,000 species.

Scientists urge that many more such banks be established, especially in developing countries. However, some species cannot be preserved in gene banks, and maintaining the banks is very expensive.

The world's 1,600 botanical gardens and arboreta contain about 4 million living plants, representing about 80,000 species or approximately 30% of the world's known plant species. The world's largest botanical garden is the Royal Botanical Gardens of England at Kew. It contains an estimated 25,000 species of living plants, or about 10% of the world's total. About 2,700 of these species are listed as threatened.

Botanical gardens are increasingly focusing on cultivation of rare and endangered plant species. In the United States, the Center for Plant Conservation has coordinated efforts among 28 botanical gardens to store nearly 600 endangered U.S. plant species and to propagate and reintroduce some of them into the wild. In addition, botanical gardens help educate an estimated 150 million visitors a year about the need for plant conservation. However, these sanctuaries have too little storage capacity and too little funding to preserve most of the world's rare and threatened plants.

Pressure can be taken off some endangered species by raising individuals on farms for commercial sale. One example is the use of farms in Florida to raise alligators for their meat and hides (p. 181).

Another example is *butterfly farming* in Papua New Guinea, where many butterfly species are threatened by habitation destruction and fragmentation, commercial overexploitation, and environmental degradation. In 1966, the government banned the collection and trading of seven species of threatened birdwing butterflies and established several butterfly reserves.

In 1974, the government established a butterfly farm program, with all the proceeds going to the people who tend the farms. Today the country has more than 500 butterfly farms.

A typical butterfly farm is surrounded by a thick hedge of *Poinsettia* and *Hibiscus vines* that keep pigs out, attract adult butterflies, and provide them with nectar. *Aristolochia* vines are grown on the branches of trees inside the farm to serve as food for the caterpillar larvae. Local people tend the caterpillars and collect the butterflies for sale when they emerge from the pupae. Some wild butterflies are still collected in Papua New Guinea, but the farm-raised specimens reduce the pressure on the wild species.

Can Zoos Help Protect Most Endangered Species?
Zoos, game parks, and animal research centers are increasingly being used to preserve some individuals of critically endangered animal species, with the long-term goal of reintroducing the species into protected wild habitats.

Two techniques for preserving endangered species are egg pulling and captive breeding. *Egg pulling* involves collecting wild eggs laid by critically endangered bird species and then hatching them in zoos or research centers. In *captive breeding*, some or all of the wild individuals of a critically endangered species are captured for breeding in captivity, with the aim of reintroducing the offspring into the wild.

Other techniques for increasing the populations of captive species include **(1)** artificial insemination, **(2)** surgical implantation of eggs of one species into a surrogate mother of another species (embryo transfer), **(3)** use of incubators, **(4)** cross-fostering, in which the young of a rare species are raised by parents of a similar species, **(5)** using computer databases of the family lineages of species in zoos and DNA analysis to match individuals for mating and prevent genetic erosion through inbreeding and **(6)** genetic cloning.

Captive breeding programs at zoos in Phoenix, Arizona, and San Diego and Los Angeles, California, temporarily saved the nearly extinct Arabian oryx (Figure 22-7), a large antelope that once lived throughout the Middle East. By 1960, it had been hunted nearly to extinction in the wild (with only about 48 left) by people riding in jeeps and helicopters and wielding rifles and machine guns. Between 1980 and 1996, the return of oryx bred in captivity to protected habitats in the Middle East allowed the population to reach about 400. However, since 1996 poachers have been capturing these animals for sale to private zoos. As a result, since 1996 the wild population has been reduced to about 100, and the species is again facing extinction.

Accomplishments of the Endangered Species Act

SPOTLIGHT

Critics of the ESA call it an expensive failure because only a few species have been removed from the endangered list. Most biologists strongly disagree that the act has been a failure, for several reasons. *First*, species are listed only when they are already in serious danger of extinction. This is like setting up a poorly funded hospital emergency room that takes only the most desperate cases, often with little hope for recovery, and then saying it should be shut down because it has not saved enough patients.

Second, it takes decades for most species to become endangered or threatened. Thus, it should not be surprising that it usually takes decades to bring a species in critical condition back to the point where it can be removed from the list. Expecting the ESA (which has been in existence only since 1973) to quickly repair the biological depletion of many decades is unrealistic.

The most important measure of the law's success is that the conditions of almost 40% of the listed species are stable or improving. A hospital emergency room taking only the most desperate cases yet stabilizing or improving the condition of 40% of its patients would be considered an astounding success.

Third, the federal endangered species budget was only $93 million in 1999 (up from $23 million in 1993)—about one-third the cost of one C-17 transport plane. The amount spent by the federal government to help protect endangered species amounts to only about 34¢ a year per U.S. citizen, and that represents a threefold increase since 1993.

To most biologists, it's amazing that so much has been accomplished in stabilizing or improving the condition of almost 40% of the listed species on a shoestring budget.

Critical Thinking

Explain why you agree or disagree each of the proposals made on p. 576 to **(a)** weaken the ESA and **(b)** strengthen the act.

In 2000, five endangered golden lion tamarin monkeys (Figure 22-7) raised in captivity at the Brookfield Zoo near Chicago were introduced into the wild in a forest reserve in Rio de Janeiro, Brazil. The project aims to have 2,000 golden lion tamarins living in the wild by 2025. Population studies suggest that this is the minimum number needed to ensure sufficient genetic diversity for the future survival of the species.

Before release, many captive-raised animals and birds need extensive training to help them learn how to find food and shelter, avoid predators, and interact in social groups. Because released populations often are small and thus vulnerable, scientists often try to estimate the *minimum viable population* needed for an introduced species to survive.

Endangered species now being bred in captivity in the United States and returned to the wild include the peregrine falcon (Figure 20-6, p. 508, removed from the endangered species list in 1999) and the black-footed ferret (Figure 22-7). However, most reintroductions fail because of (1) lack of suitable habitat, (2) inability of individuals bred in captivity to survive in the wild, or (3) renewed overhunting or capture of some returned species (such as the Arabian oryx).

When the money is available, scientists can gain information about the numbers, movement, and behavior of natural or released populations of animals by using radio transmitters and satellite tracking or setting up video cameras throughout the habitat of a species.

Ways to establish new populations of rare and endangered plants at suitable sites include (1) promoting seed dispersal by wind, animals, water, and manual seeding, (2) planting wild-collected or greenhouse-grown seedlings or adult plants, and (3) controlling competing plant species and herbivore populations until the new populations become established.

Efforts to maintain populations of endangered species in zoos and research centers are limited by lack of space and money. The captive population of each species must number 100 to 500 individuals to avoid extinction through accident, disease, or loss of genetic diversity through inbreeding. Recent genetic research indicates that 10,000 or more individuals are needed for an endangered species to maintain its capacity for biological evolution.

It is estimated that today's zoos and research centers have space to preserve healthy and sustainable populations of only 925 of the 2,000 large vertebrate species that could vanish from the planet. According to one estimate, if all of the space in U.S. zoos were used for captive breeding, only about 100 species of large animals could be sustained on a long-term basis.

In many cases, the economic costs of captive breeding are much greater than protecting species in the wild. For example, conservationist Nigel Leader-Williams estimates that the annual cost of protecting endangered black rhinos (Figure 22-14) in the wild is about one-third the cost of maintaining them in a zoo.

Instead of seeing zoos as Noah's Arks, some critics see them as prisons for once-wild animals. They also contend that zoos foster the notion that we do not need to preserve large numbers of wild species in their natural habitats.

Whether one agrees or disagrees with this position, conservation biologists point out that zoos and botanical gardens are not a biologically or economically feasible solution for most of the world's current endangered species and the much larger number expected to become endangered over the next few decades.

22-6 WILDLIFE MANAGEMENT

How Can Wildlife Populations Be Managed? **Wildlife management** involves manipulating wildlife populations (especially game species) and their habitats for their welfare and for human benefit. The *wildlife management approach* manages game species for sustained yield by (1) using laws to regulate hunting and fishing, (2) establishing harvest quotas, (3) developing population management plans, (4) improving wildlife habitat, and (5) using international treaties to protect migrating game species such as waterfowl.

In the United States, funds for state game management programs come from the sale of hunting and fishing licenses and from federal taxes on hunting and fishing equipment. Two-thirds of the states also have provisions on state income tax returns that allow individuals to contribute money to state wildlife programs.

Only 10% of all government wildlife dollars are spent to study or benefit nongame species, which make up nearly 90% of the country's wildlife species. Since the passage of the Wildlife Restoration Act in 1937 there has been a large increase in the number of game animals (such as white-tailed deer, wild turkeys, Rocky Mountain elk, and pronghorn antelope) sought by many sport hunters.

The first step in wildlife management is to decide which species are to be managed in a particular area. Ecologists and conservation biologists emphasize preserving biodiversity, wildlife conservationists are concerned about endangered species, birdwatchers want the greatest diversity of bird species, and hunters want large populations of game species. In the United States and other developed countries, most wildlife management is devoted to producing surpluses of game animals and birds for hunters.

After goals have been set, the wildlife manager must develop a management plan. Ideally, this is based on the principles of (1) ecological succession (Figure 8-15, p. 188, and Figure 8-16, p. 189), (2) wildlife population dynamics (Figure 9-3, p. 200, and Figure 9-7, p. 202), (3) an understanding of the cover, food, water, space, and other habitat needs of each species to be managed, and (4) the *maximum sustained yield (MSY)* of a population, in which harvested individuals are not removed faster than they can be replaced through reproduction. The manager must also consider the number of potential hunters, their likely success rates, and the regulations for preventing excessive hunting.

This information is difficult, expensive, and time-consuming to obtain. It involves much educated guesswork and trial and error, which is why wildlife management is as much an art as a science. Management plans must also be sensitive to political pressures from conflicting groups and to budget constraints.

Estimating maximum sustained yields is difficult because (1) the necessary data on population size and breeding potential often are not available, (2) the estimated yields vary with environmental factors such as climate change, drought, and availability of sufficient habitat and other resources, and (3) estimates must take into account how particular harvesting levels affect other species or ecosystem properties.

How Can Vegetation and Water Supplies Be Manipulated to Manage Wildlife?
Wildlife managers can encourage the growth of plant species that are the preferred food and cover for a particular animal species by controlling the ecological succession of the vegetation in an area.

Wildlife species can be classified into four types, according to the stage of ecological succession at which they are most likely to be found: early successional, midsuccessional, late successional, and wilderness (Figure 8-17, p. 190):

- *Early successional species* find food and cover in weedy pioneer plants that invade an area that has been disturbed, whether by human activities or natural phenomena (fires, volcanic activity, or glaciation).

- *Midsuccessional species* are found around abandoned croplands and partially open areas created by (1) logging of small timber stands, (2) controlled burning, and (3) clearing of vegetation for roads, houses, firebreaks, oil and gas pipelines, and electrical transmission lines. Such openings of the forest canopy (1) promote the growth of vegetation favored by midsuccessional mammal and bird species and (2) increase the amount of edge habitat, where two communities such as a forest and a field come together. This transition zone allows animals such as deer to feed on vegetation in clearings and then quickly escape to cover in the nearby forest.

- *Late successional species* rely on old-growth and mature forests to produce the food and cover on which they depend. These animals need the protection of moderate-sized, old-growth forest refuges.

- *Wilderness species* flourish only in undisturbed, mature vegetation communities such as large areas of old-growth forests, tundra, grasslands, and deserts. They can survive only in large wilderness areas and wildlife refuges.

Various types of habitat management can be used to attract a desired species and encourage its population growth. Examples are (1) planting seeds, (2) transplanting certain types of vegetation, (3) building artificial nests, and (4) deliberately setting controlled, low-level ground fires to help control vegetation. Wildlife managers often create or improve ponds and lakes in wildlife refuges to provide water, food, and habitat for waterfowl and other wild animals.

How Useful Is Sport Hunting in Managing Wildlife Populations?
Most developed countries use sport hunting laws to manage populations of game animals. Licensed hunters are allowed to hunt only during certain portions of the year to protect animals during their mating season. Limits are set on the size, number, and sex of animals that can be killed, as well as on the number of hunters allowed in a given area.

Close control of sport hunting is difficult. Accurate data on game populations may not exist and may cost too much to get. People in communities near hunting areas, who benefit from money spent by hunters, may seek to have hunting quotas raised.

There is controversy over sport hunting. Proponents argue that without carefully regulated sport hunting, deer and other large game animals exceed the carrying capacity of their habitats and destroy vegetation they and other species need. For example, between 1900 and 2000, the estimated population of white-tailed deer in the United States increased from 500,000 to 25–30 million. Many of these deer (1) invade subdivisions and eat shrubs and home gardens, (2) raid farmers' fields and orchards, (3) threaten rare plants and animals in some areas, (4) help spread Lyme disease, carried by deer ticks to humans, and (5) are involved in more than 500,000 deer-vehicle collisions each year that kill more Americans than any other wild animal, injure thousands, and cost at least $1.1 billion. Wildlife biologist William Porter estimates that without hunting, the U.S. deer population would be five times its current level and cause extensive ecological damage.

According to its proponents, sport hunting also (1) provides recreational pleasure for millions of people (15 million in the United States), (2) stimulates local economies, and (3) provides money through sales of hunting licenses and taxes on firearms and ammunition (more than $1.7 billion since 1937) that is used to buy, restore, and maintain wildlife habitats and to support wildlife research in the United States. Environmental groups such as the Sierra Club and Defenders of Wildlife support carefully controlled sport hunting as a way to preserve biological diversity by helping to prevent depletion of other native species of plants and animals.

However some individuals and groups, including the Humane Society, oppose sport hunting. They argue that **(1)** it inflicts unnecessary pain and suffering on wild animals (most of which are not killed to supply food humans need for survival), and **(2)** game managers create a surplus of game animals by deliberately eliminating their natural predators (such as wolves) and then claim that the surplus must be harvested by hunters to prevent habitat degradation or starvation of the game. Instead of eliminating natural predators, say opponents, wildlife managers should reintroduce them to reduce the need for sport hunting.

Supporters of hunting point out that **(1)** populations of many game species (such as deer) are so large that predators such as wolves cannot control them, and **(2)** because most wildlife habitats are fragmented, introducing predators can lead to the loss of nearby livestock. However, critics of hunting contend that deer (which make up only 2% of the 200 million animals hunters kill in the United States each year) are being used as a smokescreen argument to allow killing of many other game species that do not threaten vegetation. What do you think?

How Can Populations of Migratory Waterfowl Be Managed? Migratory birds—including ducks, geese, swans, and many songbirds (Case Study, p. 567)—make north–south journeys from one habitat to another each year, usually to find food, suitable cli-

mate, and other conditions necessary for reproduction. Such bird species use many different north-south routes called **flyways**, but only about 15 are considered major routes (Figure 22-25).

Some countries along such flyways have entered into agreements and treaties to protect crucial habitats needed by such species, both along their migration routes and at each end of their journeys. However, the populations of migrating snow geese have risen to the point where they are causing serious ecological damage (Connections, p. 582).

Wildlife officials manage waterfowl by regulating hunting, protecting existing habitats, and developing new habitats, including artificial nesting sites, ponds, and nesting islands. More than 75% of the federal wildlife refuges in the United States are wetlands used by migratory birds. Local and state agencies and private conservation groups such as Ducks Unlimited, the Audubon Society, and the Nature Conservancy (Solutions, p. 616) have also established waterfowl refuges.

Since 1934 the Migratory Bird Hunting and Conservation Stamp Act has required waterfowl hunters to buy a duck stamp each season they hunt. Revenue from these sales goes into a fund to buy land and easements for the benefit of waterfowl.

In this chapter, we have seen that protecting the terrestrial species that make up part of the earth's

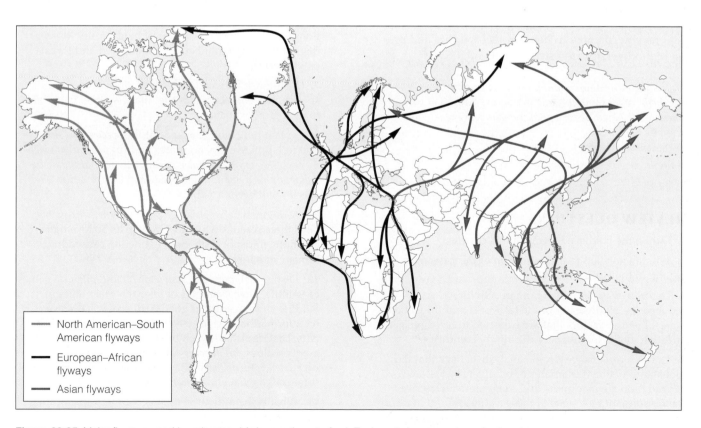

Figure 22-25 Major flyways used by migratory birds, mostly waterfowl. Each route has a number of subroutes.

What Should Be Done About Snow Geese?

CONNECTIONS

The populations of the lesser snow goose and Ross's geese have grown so large that they threaten their arctic tundra and wetlands habitats. According to the U.S. Fish and Wildlife Service, the populations of these species rose from about 800,000 in the 1960s to about 5 billion by 1999. Hunters kill about 500,000 of these birds annually, but their population is still growing at a rate of about 250,000 birds per year.

Their massive numbers have converted vast tracts of arctic tundra in Canada and Alaska to highly saline, bare soil where few plants can grow, creating what has been called a botanical desert. These geese have also destroyed about one-third of the salt marshes along the Hudson and

James Bays and have heavily damaged another third.

These geese have thrived because each year they can gorge themselves on the abundant grains in the U.S. agricultural heartland from the Great Plains to the Gulf Coast. As a result, they arrive at their winter breeding grounds in better condition, leading to higher rates of reproduction.

Two options have been suggested for dealing with this problem. One is to let nature takes its course. Eventually the geese will destroy their winter breeding ground and their population will crash. However, this may take decades and lead to widespread ecosystem degradation.

Another strategy is to allow and encourage hunters to kill more of these birds each year. In 1999, the

U.S. Fish and Wildlife Service announced plans to allow 24 states to ease hunting restrictions on the species, hoping to at least double the number killed. Though endorsed by many conservation and hunting groups, this plan was denounced by the Humane Society of the United States, which is bringing a lawsuit saying that the expanded hunt violates laws protecting migratory birds.

This problem, which connects high agricultural productivity with overpopulation of a species and ecosystem degradation, is another reminder that we can never do one thing in nature.

Critical Thinking

What do you think should be done about snow goose overpopulation? Explain.

---

biodiversity from premature extinction is a difficult, controversial, and challenging responsibility. Some actions you can take to help protect wildlife and preserve biodiversity are listed in Appendix 6.

A greening of the human mind must precede the greening of the earth. A green mind is one that cares, saves, and shares. These are the qualities essential for conserving biological diversity now and forever.
M. S. SWAMINATHAN

REVIEW QUESTIONS

1. Define the boldfaced terms in this chapter.

2. What factors led to the extinction of the passenger pigeon in the United States?

3. Describe ways in which human activities have reduced the biodiversity of the **(a)** world and **(b)** United States. List five factors that are expected to have a major effect on global biodiversity during this century.

4. List five reasons why some analysts believe that the global and national estimates of biodiversity loss and degradation are inaccurate and exaggerated. Describe the response of biologists to these claims.

5. List the major goals, strategies, and tactics for **(a)** preventing the premature extinction of species and **(b)** preserving and restoring ecosystems and aquatic systems.

6. Distinguish between *local*, *ecological*, and *biological* extinction of a species. Explain why loss of local populations may be a better indicator of biodiversity loss and degradation than numbers of endangered and threatened species.

7. Distinguish between *endangered*, *threatened*, and *rare* species. List characteristics that make species vulnerable to biological extinction.

8. Explain how biologists use **(a)** the *species-area relationship* to estimate extinction rates and **(b)** *population viability analysis*, *minimum viable population*, and *minimum dynamic area* to estimate the risk that a population of a species will become extinct.

9. Distinguish between *background* and *mass extinction*. Give three reasons why most biologists believe that we are in the midst of sixth mass extinction caused by human activities.

10. Distinguish between *instrumental* and *intrinsic* values of wildlife. Give examples of instrumental values of wildlife as sources of **(a)** goods, **(b)** ecological services, **(c)** information, and **(d)** psychological and aesthetic benefits. Distinguish between the following types of instrumental values: **(a)** *direct use*, **(b)** *option*, **(c)** *indirect use*, **(d)** *existence*, **(e)** *aesthetic*, **(f)** *educational*, and **(g)** *bequest*. List some *consumptive* and *nonconsumptive* values of wildlife.

11. What is the estimated total economic value of the earth's biodiversity?

12. What is the *intrinsic value* of biodiversity? What is *biophilia*?

13. What are five underlying causes of the population reduction and extinction of wild species? Describe how each of the following factors contributes to the premature biological extinction of species and give an example of a species affected by each factor: **(a)** habitat loss and degradation, **(b)** habitat fragmentation, **(c)** deliberately introduced nonnative species, **(d)** accidentally introduced nonnative species, **(e)** commercial hunting and illegal hunting (poaching), **(f)** predator and pest control, **(g)** the legal and illegal market for exotic pets and decorative plants, **(h)** climate change and pollution, and **(i)** loss of genetic diversity.

14. Describe ways to reduce the threat from nonnative species.

15. Why are many island species especially vulnerable to extinction? Why are some migrating birds vulnerable to premature extinction? What are some important ecological roles of bats, and why are they vulnerable to extinction?

16. What is *bioinformatics*, and how can it be used to help protect biodiversity?

17. List the benefits and limitations of protecting species using **(a)** the Convention on International Trade in Endangered Species (CITES) and **(b)** the Endangered Species Act (ESA) in the United States.

18. Give reasons why you believe that the Endangered Species Act has been a failure or a success. List measures that would strengthen and weaken the ESA in the United States.

19. Distinguish between *habitat conservation plans*, *safe harbor agreements*, and *candidate conservation plans* used as ways to help implement the Endangered Species Act.

20. What are the pros and cons of trying to protect all endangered and threatened species? List three guidelines that could be used to decide which species to protect.

21. Summarize the pros and cons of using the following to help protect endangered species: **(a)** wildlife refuges, **(b)** gene banks and botanical gardens, and **(c)** zoos and animal research centers. Distinguish between *egg pulling* and *captive breeding*.

22. What is *wildlife management*, and what are the major steps in managing wildlife? What are the major limitations of wildlife management?

23. Describe how each of the following methods can be used to help manage wildlife, and list the limitations of each approach: **(a)** manipulating vegetation and water supplies, **(b)** sport hunting, and **(c)** controlling populations of migratory waterfowl. List the pros and cons of sport hunting.

CRITICAL THINKING

1. How is biodiversity affected by **(a)** population growth, **(b)** poverty, and **(c)** climate change?

2. Discuss your gut-level reaction to the following statement: "Eventually all species become extinct. Thus, it does not really matter that the passenger pigeon is extinct and that the blue whale, the whooping crane, the California condor, and the world's remaining species of rhinoceros and tigers are endangered mostly because of human activities." Be honest about your reaction, and give arguments for your position.

3. (a) Do you accept the ethical position that each *species* has the inherent right to survive without human interference, regardless of whether it serves any useful purpose for humans? Explain. Would you extend this right to the *Anopheles* mosquito, which transmits malaria? What about infectious bacteria? **(b)** Do you believe that each *individual* of an animal species has an inherent right to survive? Explain. Would you extend such rights to individual plants and microorganisms? What about tigers that kill people? Explain.

4. Explain why you agree or disagree with **(a)** using animals for research, **(b)** keeping animals captive in a zoo, and **(c)** killing surplus animals produced by a captive breeding program in a zoo when no suitable habitat is available for their release.

5. Is the current mass extinction that scientists believe is taking place necessarily bad? Explain.

6. Your lawn and house are invaded by fire ants, which can cause painful bites. What would you do?

7. Which of the following statements best describes your feelings toward wildlife: **(a)** As long as it stays in its space, wildlife is OK, **(b)** as long as I do not need its space, wildlife is OK, **(c)** I have the right to use wildlife habitat to meet my own needs, **(d)** when you've seen one redwood tree, fox, elephant, or some other form of wildlife you've seen them all, so lock up a few of each species in a zoo or wildlife park and do not worry about protecting the rest, and **(e)** wildlife should be protected.

8. List your three favorite species. Examine why they are your favorites. Are they cute and cuddly-looking, like the giant panda and the koala? Do they have humanlike qualities, like apes or penguins that walk upright? Are they large, like elephants or blue whales? Are they beautiful, like tigers and monarch butterflies? Are any of them plants? Are any of them species such as bats, sharks, snakes, or spiders that most people are afraid of? Are any of them microorganisms that help keep you alive? Reflect on what your choice of favorite species tells you about your attitudes toward most wildlife.

9. Environmental groups in a heavily forested state want to restrict logging in some areas to save the habitat of an endangered squirrel. Timber company officials argue that the well-being of one type of squirrel is not as important as the well-being of the many families affected if the restriction causes them to lay off thousands of workers. If you had the power to decide this issue, what would you do and why? Can you come up with a compromise?

10. If you lived in the suburbs and deer invaded your yard and ate your shrubs and vegetables, what would you do?

11. We named ourselves the wise (*sapiens*) species. Do you believe that we are a wise species? Explain. If not, what are the three most important things you believe we should do to live up to the name we have given ourselves?

12. Recently scientists have begun using gene-transfer techniques to clone some endangered species as a way to help prevent their extinction. However, some conservationists are concerned that this could divert limited money and attention from what endangered animals need most: protected habitats and protection from poaching. What is your opinion on this issue? Explain.

13. Congratulations. You have been put in charge of preventing the premature extinction of the world's existing species from human activities. What would be the three major components of your program to accomplish this goal?

PROJECTS

1. Make a log of your own consumption of all products for a single day. Relate your level and types of consumption to the decline of wildlife species and the increased destruction, degradation, and fragmentation of wildlife habitats in the United States and in tropical forests.

2. Identify examples of habitat destruction or degradation in your community that have had harmful effects on the populations of various wild plant and animal species. Develop a management plan for rehabilitating these habitats and species.

3. Choose a particular endangered animal or plant species and use the library or the internet to find out what is being done to protect it from extinction. Develop a protection plan for this species.

4. Use the library or the internet to find bibliographic information about *Aldo Leopold* and *M. S. Swaminathan*, whose quotes appear at the beginning and end of this chapter.

5. Make a concept map of this chapter's major ideas, using the section heads and subheads and the key terms (in boldface). Look at the inside back cover and on the website for this book for information about making concept maps.

INTERNET STUDY RESOURCES AND RESOURCES FOR FURTHER READING AND RESEARCH

The website for this book contains helpful study aids and many ideas for further reading and research. Log on to:

http://www.brookscole.com/product/0534376975s

and click on the Chapter-by-Chapter area. Choose Chapter 22 and select a resource:

- "Flash Cards" allows you to test your mastery of the Terms and Concepts to Remember for this chapter.

- "Tutorial Quizzes" provides a multiple-choice practice quiz.

- "Student Guide to InfoTrac" will lead you to Critical Thinking Projects that use InfoTrac College Edition as a research tool.

- "References" lists the major books and articles consulted in writing this chapter.

- "Hypercontents" takes you to an extensive list of sites with news, research, and images related to individual sections of the chapter.

INFOTRAC COLLEGE EDITION

Improve your skills with InfoTrac College Edition, a searchable online database of articles from more than 700 periodicals. Log on to:

http://www.infotrac-college.com

or access InfoTrac through the website for this book.

Try the following articles:

Mattoon, A. 2000. Amphibia Fading. *World Watch* vol. 13, no. 3, p. 12–23. (keywords: endandered species, amphibia)

Wilson, E.O. 2000. Vanishing before our eyes. *Time* vol. 155, no. 17, p28+. (subject guide: extinction)

23 SUSTAINING TERRESTRIAL BIODIVERSITY: THE ECOSYSTEM APPROACH

Who's Afraid of the Big Gray Wolf?

At one time, the gray wolf (Figure 23-1) ranged over most of North America. Between 1850 and 1900, however, an estimated 2 million wolves were shot, trapped, and poisoned by ranchers, hunters, and government employees. The idea was to make the West and the Great Plains safe for livestock and for big game animals prized by hunters.

This strategy worked. When the Endangered Species Act (p. 573) was passed in 1973, there were only about 400–500 gray wolves in the lower 48 states, primarily in Minnesota and Michigan. In 1974, the gray wolf was listed as endangered in all 48 lower states except Minnesota. Alaska was not included because it had 6,000–8,000 gray wolves.

With such protection, the gray wolf population has increased to about 3,000 in seven states (Arizona, Idaho, Michigan, Minnesota, New Mexico, Wisconsin, and Wyoming). As a result, the U.S. Fish and Wildlife Service (USFWS) proposed in 2000 that the gray wolf be moved from the endangered to the threatened category in these states.

Ecologists now recognize the important role this keystone predator species once played in parts of the West and the Great Plains. These wolves (1) culled herds of bison, elk, caribou, and mule deer, (2) kept down coyote populations, and (3) provided uneaten meat for scavengers such as ravens, bald eagles, ermines, and foxes.

In recent years, herds of elk, moose, and antelope have proliferated, devastating some of the area's vegetation, increasing erosion, and threatening the niches of other wildlife species. Reintroducing a keystone species such as the gray wolf into a terrestrial ecosystem is one way to help sustain its biodiversity and prevent environmental degradation.

In 1987, the USFWS proposed reintroducing gray wolves into the Yellowstone ecosystem, an idea that brought outraged protests. Some objections came from ranchers who feared the wolves would attack their cattle and sheep; one enraged rancher said that the idea was "like reintroducing smallpox." Other protests came from (1) hunters who feared that the wolves would kill too many big game animals and (2) miners and loggers who worried that the government would force them to cease operations on wolf-populated federal lands.

Since 1995, federal wildlife officials have caught gray wolves in Canada and relocated them in Yellowstone National Park and northern Idaho. By 2000, the population of these wolves had grown by natural reproduction to around 300.

National Park Service officials trap or shoot wolves that kill livestock outside park areas. In addition, a private fund established by Defenders of Wildlife pays ranchers for sheep or cattle verified to have been killed by gray wolves that have wandered out of the Yellowstone ecosystem.

Some ranchers and hunters say that they'll take care of the wolves quietly—what they call the "shoot, shovel, and shut up" solution. Meanwhile, there are continuing efforts in Congress to eliminate the program and its funding. Ranchers and hunters are also trying to block efforts by the USFWS to reintroduce an experimental population of grizzly bears—North America's largest carnivore—into two large protected wilderness areas in the Bitterroot Mountains of Montana and Idaho.

Forests, grasslands, parks, wilderness, and other storehouses of terrestrial biodiversity are coming under increasing pressure from population growth and economic development. Biodiversity researchers and protectors urge us to use these renewable natural resources more sustainably, as discussed in this chapter.

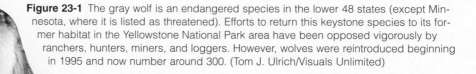

Figure 23-1 The gray wolf is an endangered species in the lower 48 states (except Minnesota, where it is listed as threatened). Efforts to return this keystone species to its former habitat in the Yellowstone National Park area have been opposed vigorously by ranchers, hunters, miners, and loggers. However, wolves were reintroduced beginning in 1995 and now number around 300. (Tom J. Ulrich/Visuals Unlimited)

Forests precede civilizations, deserts follow them.

FRANCOIS-AUGUSTE-RENÉ DE CHATEAUBRIAND

This chapter addresses the following questions:

- How is land used in the world and in the United States?

- What are the major types of public lands in the United States, and how are they used?

- Why are forest resources important, and how are they used and managed in the world and in the United States?

- Are tropical deforestation and fuelwood shortages serious problems? If so, what can we do about them?

- What problems do parks face, and how should we manage them?

- How should we establish, design, protect, and manage nature reserves?

- What is ecological restoration, and why is it important?

23-1 LAND USE IN THE WORLD AND THE UNITED STATES

How Is the World's Land Used? The land that covers about 29% of the earth's surface (Figure 7-4, p. 156) is used for several major purposes (Figure 23-2). About 39% of the world's land is used by humans for livestock, crops, and urban areas. The remaining 61% of the world's land consists of (1) forests and wetlands that are used to varying degrees by humans and (2) desert, tundra, rock, ice-covered land, and steep mountain terrain that is not widely used because it is unsuitable for

development. Human activities have disturbed about two-thirds of the earth's inhabitable land area (Figure 1-4, p. 8).

How Is Land Used in the United States and Canada? Figure 23-3 shows how land is used (top) and owned (bottom) in the United States. About 48% of U.S. land is used for livestock, crops, and urban areas. The remaining 52% is used to varying degrees by humans.

No nation has set aside as much of its land—about 42%—for public use, resource extraction, enjoyment, and wildlife as has the United States (Figure 23-3, top). Roughly 35% of the country's land belongs to every American and is managed for them by the federal government (Figure 23-3, bottom). About 73% of this federal public land is in Alaska, and another 22% is in the western states (where 60% of all land is public land). Each year there are about 2 billion visits to these national parks and forests, wildlife refuges, and other public lands.

In Canada, about 46% of the land is forest (94% of it owned and managed by provincial governments), 6% is cropland, 2% is rangeland, and 1% is urban. Most of the rest of the land (30%) is tundra and wetland.

What Are the Major Types of U.S. Public Lands? Federal public lands are classified as multiple-use lands, moderately restricted-use lands, and restricted-use lands.

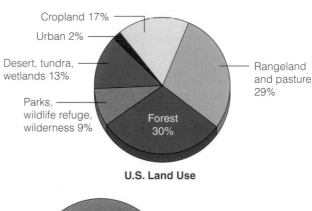

U.S. Land Use

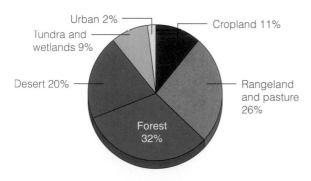

Figure 23-2 How the world's land is used. Excluding uninhabitable areas of rock, ice, desert, and steep mountain terrain, only about 27% of the planet's land area remains undisturbed by human activities (Figure 1-4, p. 8). (Data from UN Food and Agriculture Organization)

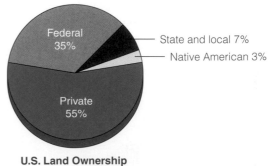

U.S. Land Ownership

Figure 23-3 How land is used (top) and owned (bottom) in the United States. (Data from U.S. Department of Agriculture)

Multiple-Use Lands

- The 155 forests (Figure 23-4) and 20 grasslands of the *National Forest System* are managed by the U.S. Forest Service. These forests are used for logging, mining, livestock grazing, farming, oil and gas extraction, recreation, sport hunting, sport and commercial fishing, and conservation of watershed, soil, and wildlife resources.

Off-road vehicles usually are restricted to designated routes.

- *National Resource Lands* in the western states and Alaska are managed by the Bureau of Land Management (BLM). The emphasis is on providing a secure domestic supply of energy and strategic minerals and preserving rangelands for livestock grazing under a permit system.

Figure 23-4 National forests, national parks, and wildlife refuges managed by the U.S. federal government. These and other public lands are owned jointly by all U.S. citizens. (Data from U.S. Geological Survey)

■ National parks and preserves   ■ National forests   ■ National wildlife refuges

Moderately Restricted-Use Lands

- The 522 *National Wildlife Refuges* (Figure 23-4) are managed by the U.S. Fish and Wildlife Service (USFWS). Most refuges protect habitats and breeding areas for waterfowl and big game to provide a harvestable supply for hunters; a few protect endangered species from extinction. Sport hunting, trapping, sport and commercial fishing, oil and gas development, mining, logging, grazing, some military activities, and farming are permitted as long as the Department of the Interior finds such uses compatible with the purposes of each unit.

Restricted-Use Lands

- The 379 units of the *National Park System* include 55 major parks (mostly in the West; Figure 23-4) and 324 national recreation areas, monuments, memorials, battlefields, historic sites, parkways, trails, rivers, seashores, and lakeshores managed by the National Park Service. National parks may be used only for camping, hiking, sport fishing, and boating. Motor vehicles are permitted only on roads, although off-road vehicles are permitted in some parks. In national recreation areas, these same activities, plus sport hunting, mining, and oil and gas drilling, are allowed.

- The 630 roadless areas of the *National Wilderness Preservation System*, which lie within the national parks, national wildlife refuges, and national forests, are managed by the National Park Service (42%), Forest Service (33%), USFWS (20%), and BLM (5%). These areas are open only for recreational activities such as hiking, sport fishing, camping, nonmotorized boating, and, in some areas, sport hunting and horseback riding. Roads, motorized vehicles, logging, livestock grazing, mining, commercial activities, and buildings are banned, except when they predate the wilderness designation.

How Should U.S. Public Lands Be Managed? Federal public lands contain at least 20% of the country's oil reserves, 30% of its natural gas reserves, 40% of its coal reserves, and 40% of its commercial forests, as well as large amounts of hard-rock minerals (p. 320 and Figure 23-5). Because of the resources they contain, there has been intense controversy over how public lands should be used and managed (Spotlight, p. 32). Most conservation biologists and ecological economists and many free-market economists believe that the following four principles should govern use of public land:

- Protecting biodiversity, wildlife habitats, and the ecological functioning of public land ecosystems should be the primary goal.

- No one should be given subsidies or tax breaks for using or extracting resources on public lands. In

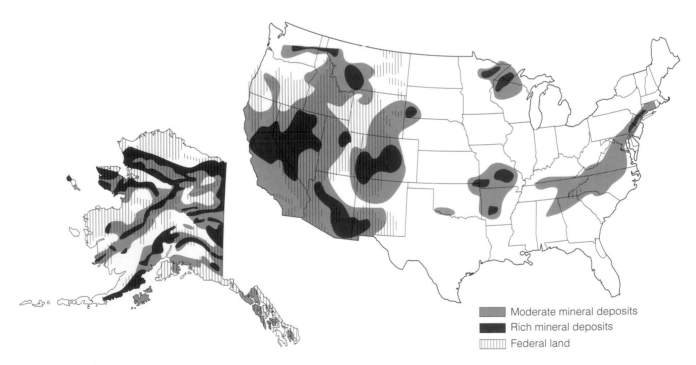

Figure 23-5 Mineral deposits on federal public lands in the United States. Minerals found on public lands include **(1)** energy resources such as oil, oil shale, natural gas, coal, and uranium and **(2)** metallic minerals such as copper, cobalt, nickel, titanium, and platinum. Public lands also contain valuable forest and grassland resources. There is much conflict over how these publicly owned resources should be used. (Data from U.S. Geological Survey)

recent years, the government has given more than $1 billion a year in subsidies to mining (p. 320), logging, and grazing interests using U.S. public lands.

- The American people deserve fair compensation for the use of their property.

- All users or extractors of resources on public lands should be fully responsible for any environmental damage they cause.

Most of these proposals are based on Aldo Leopold's land use ethic (Section 2-5, p. 39).

There is strong opposition to these ideas. In the late 1970s, a coalition of ranchers, miners, loggers, developers, farmers, and some elected officials (including president Ronald Reagan, p. 35) launched a political campaign known as the *sagebrush rebellion* to convert public lands and their resources to private ownership. Until this can be done, their goal is to severely weaken the power of the federal government to regulate how public lands are used by **(1)** slashing federal funding for public land administration and **(2)** weakening or eliminating federal laws that regulate use of public lands.

Members of this group view public lands primarily in terms of their **(1)** usefulness in providing mineral, forest, grazing, and other resources (Figure 23-5) and **(2)** ability to increase short-term economic growth.

Since 1988, several hundred local and regional grassroots groups in the United States have formed a national anti-environmental coalition called the *Wise-Use movement* to change the way in which public lands in the United States are used. Much of their money comes from real estate developers and from timber, mining, oil, coal, ranching, and off-road vehicle interests.

According to Ron Arnold, who played a key role in setting up this movement, the specific goals of the Wise-Use movement are to:

- *Have the government sell resource-rich public lands to private enterprise.*

- *Cut all old-growth forests in the national forests and replace them with tree plantations.*

- *Open all national parks, national wildlife refuges, and wilderness areas to oil drilling, mining, off-road vehicles, and commercial development.*

- *Do away with the National Park Service and launch a 20-year construction program of new concessions and theme parks run by private firms in the national parks.*

- *Continue mining on public lands under the provisions of the 1872 Mining Law (p. 320), which allows mining interests to pay no royalties to taxpayers for hard-rock minerals they remove and to buy public lands for a pittance.*

- *Modify the Endangered Species Act so that economic factors override preservation of endangered and threatened species.*

- *Eliminate government restrictions on development of wetlands found on public lands.*

- *Recognize private property rights to mining claims, water, grazing permits, and timber contracts on public lands and prevent fees for these activities from being increased.*

- *Provide civil penalties against anyone who legally challenges economic action or development on federal lands.*

- *Allow pro-industry (Wise-Use) groups or individuals to sue as "harmed parties" on behalf of corporations threatened by environmentalists.*

Most anti-environmental groups either affiliated with or generally supportive of the Wise-Use movement and its offshoot, the *Alliance for America* (formed in 1991), use environmentally friendly-sounding names.*

Case Study: The Takings and Property Rights Controversy Since 1991, the Wise-Use movement has helped spearhead a legal tactic called the *takings and property rights movement* to paralyze government regulation of public lands. One approach, called the *county movement*, has involved lobbying state legislatures to pass laws that allow county zoning, land-use plans, and environmental ordinances to take precedence over any federal laws that affect private property rights on public land within county borders.

If successful, such legislation at the state level would severely limit the ability of the federal government to enforce grazing, logging, mining, wildlife protection, zoning, environmental, and health laws on public lands. Government regulatory agencies would be **(1)** tied up in lengthy court cases, **(2)** required to make complex evaluations of possible financial losses from restrictions on the use of public land, and **(3)** forced to compensate parties for any losses incurred.

A different controversy in the takings and property rights issue focuses on government action that affects the use of *private* land. It involves the controversial issue of whether federal and state governments must compensate property owners when federal laws or regulations **(1)** limit how the owners can use their property or **(2)** decrease its financial value—something called *regulatory taking*.

The Fifth Amendment of the U.S. Constitution gives the government the power, known as *eminent*

---

*For lists and information about these oranizations, see *The Greenpeace Guide to Anti-Environmental* organizations (1993, Odonian Press), *Masks of Deception: Corporate Front Groups in America* by Mark Megalli and Andy Friedman (1993, *Essential information*, P.O. Box 19367, Washington, D.C. 20036), *Let the People Judge: A Reader on the Wise-Use Movement* (Washington, D.C.: Island Press, 1993), *The War Against the Greens* (San Francisco: Sierra Club Books, 1994) and *A Pocket Guide to Environmental Bad Guys* by James Ridgeway et al (1999, Thunder's Mouth Press). Also see the Anti-Environmental Myth Web site at http://members.aol.com/jimn469897/myths.htm.

domain, to force a citizen to sell property needed for a public good. If your land is needed for a road, for example, the government can take your land, but must pay its fair market value in exchange for your loss of its use.

The current controversy is over whether the Constitution requires the government to compensate you if instead of taking your property (a *physical taking*) it reduces its value by not allowing you to do certain things with it (a *regulatory taking*). For example, you might not be allowed to (1) build on some or all of your property because it is a wetland protected by law or (2) harvest trees on all or part of your land because it is a habitat for an endangered species.

The problem of regulatory takings is a complicated and highly controversial private property issue. Most people feel that they should be compensated for losses in property values as a result of government regulation. The problem is that requiring compensation for regulatory takings could cripple the financial ability of state and federal government to protect the public good in enforcing environmental, land use, health, and safety laws by imposing massive costs.

This would undermine the long-standing rule of law in the United States that landowners must not use their land in any way that creates a public or private nuisance (in other words, harms neighbors or the public). Indeed, most government land-use, environmental, health, zoning, and safety regulations are designed to protect the community from harmful actions by individual property owners.

Environmentalists warn that requiring compensation for regulatory takings (1) will do little to protect the private property rights and values of ordinary landowners and citizens and (2) could lower their property values because of increased pollution and environmental degradation. Instead, they contend that government compensation for regulatory takings will mostly benefit the 5% of the nation's largest private landowners (mostly timber and mining companies, agribusinesses, energy companies, and developers) who own almost three-quarters of all privately owned land in the United States.

The controversy over regulatory takings is a continuation of the classic conflict over two types of freedoms: (1) the right to be protected by law from the damaging actions of others and (2) the right to do as one pleases without undue government interference. Achieving a balance between these often conflicting types of individual rights is a difficult problem that governments have been wrestling with for centuries.

Since 1991, members of Congress have attached takings amendments to a wide variety of major pieces of legislation. By January 2001, these amendments had been dropped in committee negotiations or vetoed by president Bill Clinton.

Case Study: Livestock and U.S. Public Rangeland

About 23,000 U.S. ranchers hold permits to graze about 2% of the country's cattle and 10% of the sheep on BLM and Forest Service rangelands in 16 western states. Permit holders pay the federal government a grazing fee for this privilege.

Since 1981, grazing fees on public rangeland have been set by Congress at one-fourth to one-tenth the going rate for comparable private land. This means that taxpayers give the roughly 1% of U.S. ranchers with federal grazing permits subsidies amounting to $20–150 million a year, depending on how the costs of federal rangeland management are calculated. In 1998, for example, the BLM and U.S. Forest Service spent at least $75 million on administering and maintaining the federal grazing program but took in only $20 million in grazing fees.

The public subsidy does not end with low grazing fees. Other government costs include water pipelines, stock ponds, weed control, livestock predator control, clearing of undesirable vegetation, grass planting, erosion, and biodiversity loss. When these costs are added, it is estimated that taxpayers are giving 23,000 ranchers an annual subsidy of about $2 billion (an average of $69,000 per rancher) to produce 2% of the country's beef and 10% of its mutton. It is not surprising that politically influential permit holders have fought hard to block any change in this system.

Three common positions on the use of public rangelands are as follows:

- Phase out commercial grazing of livestock on western public lands over the next 10–15 years because the water-poor western rangeland (Figure 13-7, p. 299) is not a good place to raise cattle and sheep, which need a lot of water and can overgraze rangeland (Figure 12-22, left, p. 283) and degrade riparian areas (Figure 12-23, left, p. 284).

- Get government out of the management of public rangeland and allow it to be (1) managed by grazing boards made up of ranchers exempt from most environmental laws or (2) sold to private ranching or development interests.

- Get progressive ranchers (Individuals Matter, right), federal land managers, and environmentalists to work together to (1) develop sustainable ways to manage public rangelands and (2) help keep ranchers using public and private lands in business so that rangeland is not destroyed by being converted to houses, condos, small "ranchettes," and tourist attractions.

Although their numbers are still small, some coalitions of environmentalists, federal range managers, and ranchers are working together to develop more sustainable ways of using public rangeland such as:

- *Banning or strictly limiting grazing on riparian areas* (Figure 12-23, p. 284, and Individuals Matter, right).

The Eco-Rancher

Wyoming rancher Jack Turnell is one of a new breed of cow-puncher who gets along with environmentalists. He talks about riparian ecology and biodiversity as fluently as he talks about cattle: "I guess I have learned how to bridge the gap between the environmentalists, the bureaucracies, and the ranching industry."

Turnell grazes cattle on his 32,000-hectare (80,000-acre) ranch south of Cody, Wyoming, and on 16,000 hectares (40,000 acres) of Forest Service land on which he has grazing rights. For the first decade after he took over the ranch, he raised cows the conventional way. Since then, he's made some changes.

Turnell disagrees with the proposals by some environmentalists to raise grazing fees and remove sheep and cattle from public rangeland. He believes that if ranchers are kicked off the public range, ranches like his will be sold to developers and chopped up into vacation sites and homes ("ranchettes"), destroying the range for wildlife and livestock alike.

At the same time, he believes that ranches can be operated in more ecologically sustainable ways. To demonstrate this, Turnell (1) began rotating his cows away from the riparian areas (Figure 12-23, p. 284), (2) gave up most uses of fertilizers and pesticides, and (3) crossed his Hereford and Angus cows with a French breed that tends to congregate less around water. Most of his ranching decisions are made in consultation with range and wildlife scientists, and changes in range condition are carefully monitored with photographs.

The results have been impressive. Riparian areas on the ranch and Forest Service land are lined with willows and other plant life. This has provided lush habitat for an expanding population of wildlife, including pronghorn antelope, deer, moose, elk, bear, and mountain lions. In addition, this eco-rancher makes more money because the higher-quality grass puts more meat on his cattle.

■ *Banning or strictly limiting grazing on rangeland where it endangers habitats of threatened and endangered species, causes desertification (Figure 10-21, p. 228), or is otherwise ecologically unsustainable.*

■ *Supporting endangered ranchers and rangeland by setting up Grassbanks. Ranchers needing to rest areas of public or private rangeland to allow recovery can graze their cattle on nearby private or public ranch land in a Grassbank without paying leasing fees. In return, ranchers using such land must place a legally binding conservation easement on equivalent areas of their private ranch land*

that bars the land from being subdivided or otherwise used for development.

■ *Allowing individuals and environmental groups to purchase grazing permits and not use the land for grazing.* This practice was upheld by a 2000 ruling by the U.S. Supreme Court.

So far, sagebrush rebel and Wise-use western ranchers with grazing permits have wielded enough political power to see that most such cooperative measures are not implemented by elected officials and government land management agencies.

23-2 MANAGING AND SUSTAINING FORESTS

What Is the Ecological Importance of Forests?
Forests provide numerous ecological services by:

■ Supporting food webs and energy flow (Section 4-4, p. 82) and cycling nutrients (Section 4-6, p. 87) in forest ecosystems.

■ Acting as giant sponges that (1) slow runoff, (2) hold water that recharges springs, streams, and groundwater, (3) regulate the flow of water from mountain highlands to croplands and urban areas, and (4) reduce the amount of sediment washing into streams, lakes, and reservoirs by reducing soil erosion (Figure 13-24, p. 316).

■ Influencing local and regional climate. For example, 50–80% of the moisture in the air above tropical forests comes from trees via transpiration and evaporation. If large areas of these forests are cleared, average annual precipitation drops and the region's climate gets hotter and drier. This process eventually can convert a diverse tropical forest into a sparse grassland or even a desert.

■ Providing numerous habitats and niches for wildlife species (Figure 6-28, p. 144, and Figure 8-11, p. 184). The world's forests provide habitats for 50–90% of all terrestrial species such as mammals, birds, amphibians, and reptiles. Many of these forest-dwelling species are threatened by extinction, primarily because of habitat loss and fragmentation, and this biodiversity loss is projected to increase sharply over the next 50 years.

■ Purifying the air.

What Is the Economic Importance of Forests?
People use trees harvested from forests for a number of purposes that are worth more than $400 billion per year.

Economic uses of forests include:

■ Fuelwood, accounting for about 50% of the wood removed from forests. Limbs, cut-up logs, bark, wood pellets, and wood converted to charcoal are burned as a fuel (1) in small stoves or open fires to provide heat and cook food for about 2 billion people, mostly in developing countries, and (2) to fire boilers, which create steam or generate electricity, mostly in developed countries.

■ Industrial timber and roundwood (unprocessed logs) used to make lumber, plywood, particleboard, chipboard, and veneer. This accounts for about 30% of the wood removed from forests.

■ Pulp to make paper (p. 533), accounting for about 20% of the annual wood harvest.

■ Medicines extracted from plants, bark, and some animals (Figure 22-12, p. 561).

■ Mining, livestock grazing, and recreation.

What Are the Major Types of Forests? There are three general types of forests, based primarily on climate: *tropical, temperate,* and *polar (boreal)* (Figure 6-17, p. 132), described in detail in Section 6-5 (p. 142). Currently, forests with 50% or more tree cover occupy about 32% of the earth's land surface (Figure 6-16, p. 131, and photo on p. 1).

Old-growth forests (sometimes called *frontier forests*) are uncut forests or regenerated forests that have not been seriously disturbed by human activities or natural disasters for at least several hundred years. Old-growth forests provide ecological niches for a multitude of wildlife species (Figure 6-28, p. 144, and Figure 23-6). These forests also have large numbers of standing dead trees (snags) and fallen logs, which provide habitats for a variety of species. Decay of this dead vegetation returns plant nutrients to the soil (Figure 4-15, p. 81).

Second-growth forests are stands of trees resulting from secondary ecological succession (Figure 8-16, p. 189) after the trees in an area have been removed by (1) human activities such as clear-cutting for timber or conversion to cropland or (2) natural forces such as fire, hurricanes, or volcanic eruption.

Some old-growth and second-growth forests have been cleared and converted to **tree farms** or **plantations** (Figure 6-31, p. 148). Such farms are managed tracts with uniformly aged trees of one species that are harvested by clear-cutting as soon as they become commercially valuable. Then they are replanted and clear-cut again on regular cycles (Figure 23-7).

Tree farms or plantations are attempts to produce timber, pulpwood, and plantation crops (such as bananas, rubber, cocoa, and coffee) more efficiently by setting back secondary ecological succession (Figure 8-16, p. 189) to create greatly simplified ecosystems. This is the same approach used to produce more food per hectare by planting monoculture crops. Although tree farms reduce biodiversity where they are planted, they can help preserve biodiversity elsewhere by

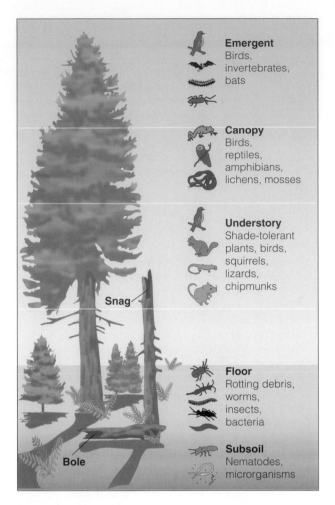

Figure 23-6 Layers of biodiversity in an old-growth Douglas fir forest in the Pacific Northwest. There is much controversy over whether to clear-cut such forests on public lands in the United States and Canada for their valuable timber or to preserve them for their biodiversity and important ecosystem services.

reducing pressures to cut down old-growth and second-growth forests.

Currently, industrial tree plantations occupy about 0.5% of the world's forested land and supply 22% of of the world's timber and pulp. Pulpwood plantations, 80% of them in South America and the Asia-Pacific region, supply about 16% of the world's fiber used to make paper.

What Is Happening to the World's Forests? Forests are renewable resources as long as the rate of cutting and degradation does not exceed the rate of regrowth. However, according to the World Resources Institute, during the past 8,000 years human activities have reduced the earth's original forest cover by about 46%, much of it removed in the last three decades.

The area of many temperate forests in North America and Europe is increasing slightly because of reforestation from secondary ecological succession (Figure 8-16, p. 189) on cleared forest areas and abandonment of marginal croplands (Figure 23-8, green). However,

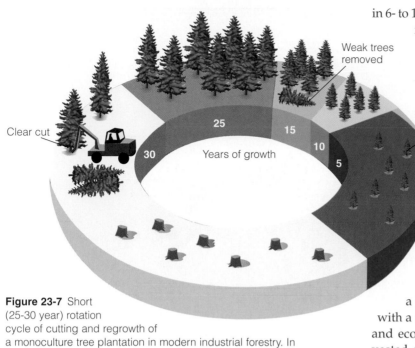

Figure 23-7 Short (25-30 year) rotation cycle of cutting and regrowth of a monoculture tree plantation in modern industrial forestry. In tropical countries, where trees can go more rapidly year-round, the rotation cycle can be 6–10 years.

Labels on figure: Clear cut · 30 · 25 · Years of growth · 15 · Weak trees removed · 10 · 5 · Seedlings planted

forests are being cut or fragmented and degraded faster than they can regenerate in many parts the world, especially in tropical areas of Latin America, Africa, and Asia (Figure 23-8, red, orange, and yellow). A 2000 report by the World Resources Institute estimated that the world's forests are shrinking by at least 140,000 square kilometers (54,000 square miles) per year, with 90% of this loss occurring in tropical countries (Figure 23-8).

Deforestation can provide land for crops, livestock, or urban growth. However, it also **(1)** decreases the overall net primary productivity (Figure 4-25, p. 88) of the cleared area, **(2)** reduces the stock of nutrients once stored in the trees and leaf litter, **(3)** diminishes biodiversity, **(4)** makes the soil more prone to erosion and drying, **(5)** increases the rate of runoff of water and soil nutrients from the land (Case Study, p. 95, and Figure 4-31, p. 95), and **(6)** reduces the uptake of CO_2, which is the major contributor to projected global warming (Figure 18-8, p. 453), and adds CO_2 to the atmosphere if the cleared trees are burned or allowed to decay.

What Are the Major Types of Forest Management? The total volume of wood produced by a particular stand of forest varies as it goes through different stages of growth and ecological succession (Figure 23-9). If the goal is to produce fuelwood or fiber for paper production in the shortest time, the forest usually is harvested on a short rotation cycle (Figure 23-7), well before the volume of wood produced peaks (point A in Figure 23-9). Typically, pulpwood plantations are harvested

in 6- to 10-year rotations in the tropics and 20- to 30-year rotations in temperate regions (Figure 23-7).

Harvesting at point B in Figure 23-9 gives the maximum yield of wood per unit of time. If the goal is high-quality wood for fine furniture or veneer, managers use longer rotations to develop larger, older-growth trees (point C in Figure 23-9), whose rate of growth has leveled off and is much lower than that of young trees.

There are two basic forest management systems: even-aged and uneven-aged. With **even-aged management**, trees in a given stand are maintained at about the same age and size. In this approach, sometimes called *industrial forestry*, a biologically diverse natural forest is replaced with a simplified *tree farm* of one or two fast-growing and economically desirable species that can be harvested every 10–100 years, depending on the species (Figure 23-7). Crossbreeding and genetic engineering can improve the quality and quantity of tree farm wood. Currently, about 54% of the pulpwood used to make paper (p. 533) comes from second-growth forests, 29% from pulpwood plantations (although this share is growing rapidly), and 17% from old-growth forests (mostly boreal forests in Canada and Russia).

However, decades of experience with intensive even-aged management of German forests has shown that in a forest in which almost all the trees are repeatedly cut and removed, the soil is depleted of nutrients. Trees then have little defense against stresses such as drought, diseases, pests, and air pollutants (Figure 17-14, p. 432).

In **uneven-aged management**, a variety of tree species in a given stand are maintained at many ages and sizes to foster natural regeneration. Here the goals are **(1)** biological diversity, **(2)** long-term sustainable production of high-quality timber (Figure 23-9, point C), **(3)** a moderate economic return, and **(4)** multiple use of the forest for timber, wildlife, watershed protection, and recreation. Mature trees are cut selectively, and the removal of all trees is used only on small patches of species that benefit from such a practice.

The rate of economic return owners expect on a forest asset is the key factor determining whether most will use short-term, even-aged management or longer-term, uneven-aged management. Before the 1960s, forest owners figured a rate of return on their investment of 2–3%. Because the trees grew about as fast as the money invested in them, owners could afford to wait fairly long times before they harvested them, normally using uneven-aged management.

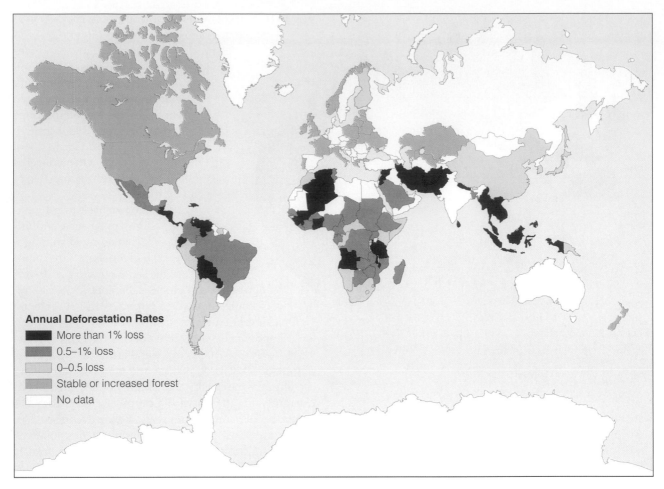

Annual Deforestation Rates

- More than 1% loss
- 0.5–1% loss
- 0–0.5 loss
- Stable or increased forest
- No data

Figure 23-8 Estimated annual changes in forest area, 1990–1995 (latest available data). Deforestation is occurring in most of the world except North America and parts of Europe. (United Nations Food and Agriculture Organization)

Since the 1960s many forest owners have been basing their management decisions on returns of almost 10%, reflecting what they could earn by putting their money into other investments. This means that they can make more money **(1)** clear-cutting diverse, uneven-aged forests, **(2)** investing the profits in something else, **(3)** growing new even-aged stands of trees as quickly as possible, **(4)** cutting them down, **(5)** reinvesting the money, and **(6)** repeating this process until the soil is exhausted.

Extensive clear-cutting can greatly increase short-term profits for forest owners but can lead to a boom-and-bust cycle for the economies of local communities. For example, in the 1980s, Champion International leveled more than 2,600 square kilometers (1,000 square miles) of forests in Montana. After liquidating much of its valuable timber in the Big Sky Country, Champion moved on, leaving behind hundreds of unemployed loggers and mill workers, economically depressed towns, and large areas of heavily logged land.

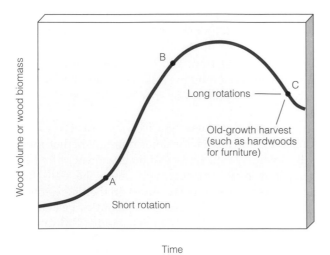

Figure 23-9 Changes in wood volume over various growth and harvest cycles in forest management. Forest management occurs over a cycle of decisions and events called a *rotation* (Figure 23-7). The most important steps in a rotation include **(1)** taking an inventory of the site, **(2)** developing a forest management plan, **(3)** building roads into the site, **(4)** preparing the site for harvest, **(5)** harvesting timber, and **(6)** regenerating and managing the site until the next harvest.

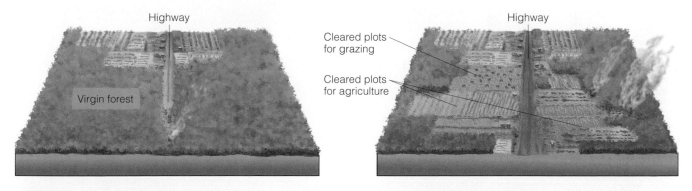

Figure 23-10 Building roads into previously inaccessible forests paves the way to fragmentation, destruction, and degradation.

How Are Trees Harvested? Before large amounts of timber can be harvested, roads must be built for access and timber removal. Even with careful design, logging roads have a number of harmful effects (Figure 23-10) such as **(1)** increased erosion and sediment runoff into waterways, **(2)** habitat fragmentation and biodiversity loss, **(3)** exposure of forests to invasion by nonnative pests, diseases, and wildlife species, and **(4)** opening of once-inaccessible forests to farmers, miners, ranchers, hunters, and off-road vehicle users. In addition, logging roads on public lands in the United States disqualify the land for protection as wilderness.

Once loggers can reach a forest, they use various methods to harvest the trees (Figure 23-11). With **selective cutting**, intermediate-aged or mature trees in an uneven-aged forest are cut singly or in small groups (group selection), creating gaps no larger than the height of the standing trees (Figure 23-11a). Selective cutting **(1)** reduces crowding, **(2)** encourages growth of younger trees, **(3)** maintains an uneven-aged stand of trees of different species, **(4)** allows natural regeneration from the surrounding trees, **(5)** can help protect the site from soil erosion and wind damage, **(6)** can be used to remove diseased trees, and **(7)** allows a forest to be used for multiple purposes.

However, much of the logging in tropical forests involves a form of selective cutting called *high grading* in which only the largest and best specimens of the most desirable species are cut and hauled out. Studies have shown that for every large tree that is felled, 16 or 17 other trees are damaged or pulled down because the canopies of trees in tropical forests usually are connected by a network of vines. Studies in Malaysia and the Amazon Basin found that selective logging of 8% of the mature trees in a tropical rain forest damaged nearly half the trees and reduced canopy coverage by 50%. This causes the forest floor to become warmer, drier, and more flammable and increases erosion of the forest's thin and usually nutrient-poor soil (Figure 10-15, p. 223).

Some tree species grow best in full or moderate sunlight in moderate to large clearings. Such sun-loving species usually are harvested by shelterwood cutting, seed-tree cutting, or clear-cutting (in which all trees on a site are removed in a single cut). **Shelterwood cutting** removes all mature trees in two or typically three cuttings over a period of about 10 years (Figure 23-11b). This method **(1)** allows natural seeding from the best seed trees, **(2)** keeps seedlings from being crowded out, **(3)** leaves a fairly natural-looking forest that can serve a variety of purposes until the final cut is made, **(4)** helps reduce soil erosion, and **(5)** provides a good habitat for some types of wildlife. Two problems are that **(1)** loggers sometimes take too many trees in the first cut, and **(2)** eventually a diverse forest is converted to a mostly monoculture tree plantation that is clear-cut.

Seed-tree cutting harvests nearly all a stand's trees in one cutting, leaving a few uniformly distributed seed-producing trees to regenerate the stand (Figure 23-11c). By allowing several species to grow at once, seed-tree cutting leaves an aesthetically pleasing forest that is useful for recreation, deer hunting, erosion control, and wildlife conservation. Leaving the best trees for seed can also lead to genetic improvement in the new stand. However, eventually the seed trees are cut and a diverse forest is converted to a monoculture tree plantation that is clear-cut.

Clear-cutting is the removal of all trees from an area in a single cutting. The clear-cut area may be a whole stand (Figure 23-11d), a strip, or a series of patches. After all trees are cut, the site usually is reforested **(1)** naturally by seed released by the harvest (or from nearby uncut trees if the clear-cut area is small) or **(2)** artificially as foresters broadcast seed over the site or plant seedlings raised in a nursery (Figure 23-7).

According to timber companies, clear-cutting **(1)** increases timber yield per hectare, **(2)** permits reforesting with genetically improved stocks of fast-growing trees, **(3)** shortens the time needed to establish a new stand of trees (Figure 23-7), **(4)** takes less skill and

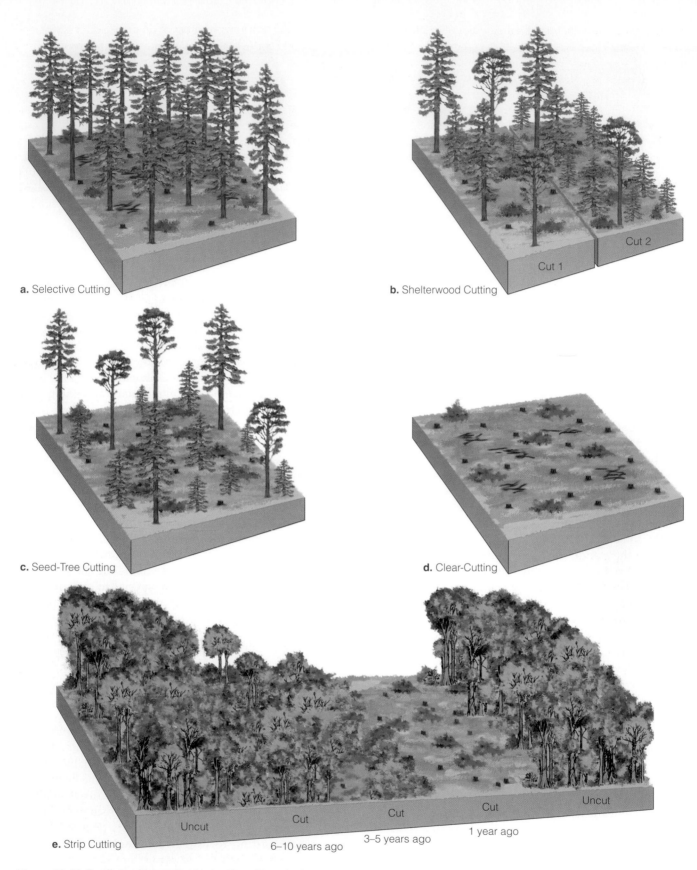

a. Selective Cutting

b. Shelterwood Cutting

Cut 1

Cut 2

c. Seed-Tree Cutting

d. Clear-Cutting

e. Strip Cutting

Uncut

Cut

6–10 years ago

Cut

3–5 years ago

Cut

1 year ago

Uncut

Figure 23-11 Tree harvesting methods. Another alternative is to preserve biodiversity and enhance the sustainability of diverse old-growth forests by not cutting them.

planning than other harvesting methods, **(5)** usually provides the maximum economic return in the shortest time, and, **(6)** if done carefully and responsibly, often is the best way to harvest tree farms and stands of some tree species that need full or moderate sunlight for growth.

On the negative side, clear-cutting **(1)** leaves moderate to large forest openings, **(2)** eliminates most recreational value for several decades, **(3)** reduces biodiversity, disrupts ecosystem processes, and destroys and fragments some wildlife habitats, **(4)** makes nearby trees more vulnerable to being blown down by windstorms, and **(5)** leads to severe soil erosion, sediment water pollution, and flooding when done on steep slopes (Figure 13-24, p. 316).

A variation of clear-cutting that can allow a sustainable timber yield without widespread destruction is **strip cutting**, in which a strip of trees is clear-cut along the contour of the land, with the corridor narrow enough to allow natural regeneration within a few years (Figure 23-11e). After regeneration, another strip is cut above the first, and so on. This allows a forest to be clear-cut in narrow strips over several decades with minimal damage. Typically each strip is not harvested again for 30–40 years.

How Can Forests Be Managed More Sustainably?
Instead of increased use of conventional short-rotation industrial forestry (Figure 23-7), biodiversity researchers and a growing number of foresters call for more sustainable forest management that:

- *Grows more timber on long rotations,* generally about 100–200 years (point C, Figure 23-9), depending on the species and soil quality.

- *Emphasizes* **(a)** *selective cutting of individual trees or small groups of most tree species* (Figure 23-11a), **(b)** *strip cutting* (Figure 23-11e) *instead of conventional clear-cutting, and* **(c)** *not clear-cutting (including seed-tree and shelterwood cutting) on land that slopes more than 15°.*

- *Minimizes fragmentation of remaining larger blocks of forest.*

- *Uses road-building and logging methods that minimize soil erosion and compaction.*

- *Leaves most standing dead trees (snags) and fallen timber to maintain diverse wildlife habitats and to be recycled as nutrients* (Figure 4-15, p. 81).

- *Has timber grown by sustainable methods certified and labeled by outside certifying groups* (Solutions, p. 598).

- *Includes the estimated ecological services provided by trees and a forest* (Spotlight, p. 562) *in estimates of their economic value.*

Another approach is to greatly increase the use of more environmentally sustainable industrial forestry. Some foresters have suggested that the United States and the world meet most of its demand for wood and wood products by growing genetically improved trees on carefully managed tree plantations (Figure 23-7).

According to one estimate, the entire world demand for forest products (except fuelwood) could be supplied by *tree plantations* occupying only about 5% of the world's forested land (compared to about 0.5% currently occupied by such plantations). If the efficiency of wood harvesting from such plantations were increased by 60–80%, all the world's wood needs (except fuelwood) could be supplied on land equal to only 0.1–0.3% of the world's current forested area. This is a land area roughly equal to that of Louisiana or Iowa.

Proponents of this approach contend that this practice would be more sustainable if it were **(1)** done with minimum use of pesticides and commercial fertilizer and **(2)** carried out mostly on already cleared and degraded marginal farmland and pastureland that has existing access roads. This would reduce pressures to cut timber in old-growth forests and help protect wildlife habitats. Currently, New Zealand meets all its domestic wood needs from plantations and has protected about 30% of its original native forest.

Conservation biologists and environmentalists agree that farming trees on plantations is preferable to harvesting the world's remaining old-growth forests and could be used to help restore degraded land and increase carbon dioxide uptake. However, they point out that this approach should be used only if:

- Harvesting of remaining old-growth forests is banned.

- Harvesting of second-growth forests is phased out over a 10- to 20-year period as the plantations are phased in.

- Plantations are established only on lands that are already truly degraded, not on any newly cleared land or existing cropland.

- All government subsidies and tax breaks for harvesting timber from old-growth and second-growth forests are phased out and replaced with subsidies for establishing tree plantations in truly degraded areas.

How Can Pathogens and Insects Affect Forests?
Insects and tree diseases can damage and kill trees in natural forests and tree plantations. Three deadly tree diseases caused by parasitic fungi that were accidentally introduced into the United States from other countries are **(1)** chestnut blight (from China), **(2)** Dutch elm disease (from Asia via Europe), and **(3)** white pine blister rust (from Europe).

Examples of insects that can have caused serious damage to certain tree species in the United States include:

- *Bark beetles* (such as the bronze birch borer and Asian longhorned beetle), which bore channels

Certifying Sustainably Grown Timber

SOLUTIONS

Collins Pine owns and manages a large area of productive timberland in northeastern California. The company's goals are to maintain ecological, economic, and social sustainability by using selective cutting (since the 1940s) to **(1)** provide a profitable harvest of logs from several tree species, **(2)** maintain jobs as the largest employer in Chester, California, and **(3)** take care of the land and its ecosystems.

All the timber the company harvests each year are certified as being sustainably produced by Scientific Certification Systems (SCS). It is part of the nonprofit Forest Stewardship Council (FSC), formed in 1993 to develop a list of environmentally sound practices that could be used to certify timber and products made from such timber.

Each year, SCS evaluates all of Collins's landholdings to ensure that **(1)** cutting has not exceeded long-term forest regeneration, **(2)** roads and harvesting systems have not caused unreasonable ecological damage, **(3)** management of soils, downed wood, and standing snags provides adequate nutrient cycling and wildlife habitat, and **(4)** the company is a good employer and a good steward of its land and water resources.

Collins does not have gates or "Keep Out" signs on its land. People in surrounding communities can use company land for fishing, hunting, and camping. The company's owners see certification as **(1)** recognition of their dedication to land stewardship and **(2)** a way to place a market value on their ecological goals.

Another forest certification success story involves the Menominee nation. Since 1890, they have been selectively harvesting mixed species and ages from their tribal reservation land near Green Bay, Wisconsin, which is the state's single largest tract of virgin forest. Lumber from the tribal forest has been certified by the Rainforest Alliance's Smart Wood Program as being harvested in an environmentally and socially responsible manner. Each year thousands of people flock to the reservation to look at a sustainable forest operation that honors wildlife and biodiversity while providing jobs and income for the local economy.

Although certified timber makes up only a small share of all timber production, demand for the product is growing rapidly and exceeds the available supply. This increased demand is being driven not only by consumer demand but also by business self-interest. Sellers of wood products know that almost half of the world's original forests are gone and that remaining forests will not last long if they are not harvested sustainably. For example,

- In 2000, Home Depot and Lowe's, two of the largest buyers and retailers of lumber products in the United States, announced that by the end of 2002 they will **(1)** no longer buy wood products that come from endangered ecosystems, **(2)** give preference to buying wood from certified forests, and **(3)** promote efficient wood use and environmentally beneficial alternatives to wood products.

- By 2000, there were more than 16 million hectares (40 million acres) of certified forests in 31 countries, and the area is doubling each year.

- In Great Britain, wood certified by the FSC accounts for one-fourth of all wood sold nationally and 75% of the forests in Sweden grow FSC-certified sustainable timber.

- In 1998, the World Bank and the World Wildlife Fund entered into a partnership to bring about a 10-fold increase in the area of certified forests by 2005.

Critical Thinking

Suppose you inherited a sustainable timber company operation similar to the one described in this box. If you switched to clear-cutting most of the land, you could **(1)** make millions in a short time, **(2)** sell the clear-cut land, and **(3)** invest the money elsewhere at a much higher rate of return than long-term sustainable timber harvesting. What would you do with your inheritance? Explain.

---

through the layer beneath the bark of spruce, fir, birch, and pine trees.

- *Spruce budworm* and *gypsy moth larvae* (accidentally introduced from Europe on 1869 and now established in 16 states), which can kill trees by eliminating the foliage they need to carry out photosynthesis and produce food.

- The *hemlock woolly adelgid* (accidentally introduced into the United States from Asia in 1924), which during the 1990s destroyed large areas of hemlock forest throughout the eastern United States. In 1996, scientists began trying to control this invader by introducing a tiny beetle from Japan that is a natural enemy of the woolly adelgid.

Ways to reduce the impact of tree diseases and insects on forests include **(1)** preserving biodiversity, **(2)** banning imported timber that might introduce harmful new pathogens or insect pests, **(3)** removing infected and infested trees or clear-cutting infected and infested areas and burning all debris, **(4)** treating diseased trees with antibiotics, **(5)** developing disease-resistant tree species, **(6)** applying pesticides (Section 20-2, p. 504), and **(7)** using integrated pest management (p. 515).

How Do Fires Affect Forest Ecosystems? Intermittent natural fires set by lightning are an important part of the ecological cycle of some types of forests. Such fires maintain the vegetation of many ecosystems at a certain stage of ecological succession (Figure 8-16, p. 189). Examples of fire-maintained communities include (1) savanna, (2) temperate grasslands (Figure 6-23, p. 139), (3) chaparral, (4) southern pine forests, (5) western forests containing giant sequoia trees, and (6) northern coniferous forest (Figure 6-30, p. 147). In such ecosystems, occasional fires burn away much of the low-lying vegetation and small trees. A burst of new vegetation follows.

Some tree species need occasional fires. The seeds of the giant sequoia, the lodgepole pine, and the jack pine, for instance, are released from cones or germinate only after being exposed to intense heat from a fire. Mature trees of such species are not harmed by low-level ground fires because of their thick, fire-resistant bark and lack of branches near the ground.

Forest ecosystems can be affected by different types of fires. Some, called *surface fires* (Figure 23-12, left), usually burn only undergrowth and leaf litter on the forest floor. These fires can kill seedlings and small trees but spare most mature trees and allow most wild animals to escape.

Occasional surface fires (1) burn away flammable ground material and help prevent more destructive fires, (2) release valuable mineral nutrients tied up in slowly decomposing litter and undergrowth, (3) increase the activity of underground nitrogen-fixing bacteria, (4) stimulate the germination of certain tree seeds, and (5) help control pathogens and insects. In addition, some wildlife species such as deer, moose, elk, muskrat, woodcock, and quail depend on occasional surface fires to maintain their habitats and to provide food in the form of vegetation that sprouts after fires.

Some extremely hot fires, called *crown fires* (Figure 23-12, right), may start on the ground but eventually burn whole trees and leap from treetop to treetop. They usually occur in forests in which no surface fires have occurred for several decades, allowing dead wood, leaves, and other flammable ground litter to build up. These rapidly burning fires can destroy most vegetation, kill wildlife, and increase soil erosion.

Sometimes surface fires go underground and burn partially decayed leaves or peat. Such *ground fires* are most common in northern peat bogs. They may smolder for days or weeks before being detected and are difficult to extinguish.

Four approaches used to protect forest resources from fire are (1) *prevention*, (2) *prescribed burning* (setting controlled ground fires to prevent buildup of flammable material), (3) *presuppression* (early detection and control of fires), and (4) *suppression* (fighting fires once they have started). Ways to prevent forest fires include (1) requiring burning permits, (2) closing all or parts of a forest to travel and camping during periods of drought and high fire danger, and (3) educating the public. The Smokey Bear educational campaign of the Forest Service and the National Advertising Council, for example, has prevented countless forest fires in the United States, saved many lives, and prevented billions of dollars in losses.

This educational program has convinced most members of the general public that all forest fires are bad and should be put out. According to ecologists, however, preventing all fires can increase the likelihood of highly destructive crown fires (Figure 23-12, right) because it allows large quantities of highly flammable underbrush and undergrowth and smaller trees to accumulate in forests that need occasional fires for regeneration. This can convert what would have been fairly harmless surface fires into fires intense enough to destroy the larger fire-resistant species needed for forest regeneration. An entirely different ecosystem may then develop on such sites.

Since 1972, U.S. Park Service policy has been to allow most lightning-caused fires to burn themselves out as long as they do not threaten human lives, park facilities, private property, or endangered wildlife. The Park Service and the Forest Service have also set controlled prescribed fires in ecosystems in which ecologists consider occasional fires to be beneficial to trees and various forms of forest life. However, great care must be taken in setting controlled fires to ensure that they do not get out of control, as occurred during the spring of 2000 in an area managed by the Park Service near Los Alamos, New Mexico.

After the fires in parts of Yellowstone National Park and many other fires that broke out in a number of national forests in the West during the summers of 1988,

Surface fire   **Crown fire**

Figure 23-12 *Surface fires* (left) usually burn undergrowth and leaf litter on a forest floor and can help prevent more destructive *crown fires* (right) by removing flammable ground material. Sometimes carefully controlled surface fires are deliberately set to prevent buildup of flammable ground material in forests.

1994, and 2000, some people called for a reversal of the federal let-it-burn policy. However, biologists contend that these fires caused more damage than they should have because the previous policy of fighting all fires in the park had allowed the buildup of flammable ground litter and small plants. Moreover, the Yellowstone fire was ecologically beneficial. Today, Yellowstone's meadows and forests have lush growths of regenerating trees and wildflowers and more diverse habitats and better food supplies for many of its natural species.

Ecologists believe that **(1)** many fires in national parks, national forests, and wilderness areas should be allowed to burn, and **(2)** carefully controlled prescribed ground fires should be set to keep down flammable ground litter in fire-adapted forests as part of the natural ecological cycle of succession and regeneration (Figure 8-16, p. 189).

Another approach is to allow commercial logging companies to thin out trees in national forests, parks, and wilderness areas to reduce the fire hazard. If done under close government supervision and strict guidelines, this could help in some areas. However, environmentalists warn that **(1)** allowing salvage logging in protected public forests and wilderness areas is an excuse to advance the agenda of logging companies to greatly increase commercial logging and road building in national forests and in roadless areas eligible for inclusion in the national wilderness systems and **(2)** without close supervision, logging companies can go beyond thinning out younger trees and remove larger trees.

How Do Air Pollution and Climate Change Threaten Forests? Forests at high elevations and those downwind from urban and industrial centers are exposed to a variety of air pollutants that can harm trees, especially conifers. Besides doing direct harm, prolonged exposure to multiple air pollutants makes trees much more vulnerable to drought, diseases, and

insects (Figure 17-14, p. 432). The solution is to reduce emissions of the offending pollutants from coal-burning power plants, industrial plants, and motor vehicles (Section 17-7, p. 440).

In coming decades, an even greater threat to forests (especially temperate and boreal forests) may come from regional climate changes brought about by global warming that can increase the threat of forest fires in some areas and cause some forest types to die out in some areas (Figure 18-13, p. 459). Ways to deal with projected global warming are discussed in Section 18-5, p. 460.

23-3 FOREST RESOURCES AND MANAGEMENT IN THE UNITED STATES

What Is the Status of Forests in the United States? Forests cover about one-third of the lower 48 United States, provide habitats for more than 80% of the country's wildlife species, and supply about two-thirds of the nation's total water runoff. Some *good news* is that today forests in the United States generally are bigger (Figure 23-8) and often healthier than they were in 1920, when the country's population was around 100 million.

However, some *discouraging news* is that:

■ Between 1620 and 1998, most of the existing old-growth forests in the lower 48 states were cut (Figure 23-13). However, many of the old-growth forests that were cleared or partially cleared between 1620 (Figure 23-13, left) and 1960 have grown back naturally through secondary ecological succession as fairly diverse second-growth (and in some cases third-growth) forests (Figure 8-16, p. 189).

■ Since the mid-1960s, an increasing area of the nation's remaining old-growth and fairly diverse second-growth forests has been clear-cut and replaced with tree plantations. Although this does not reduce

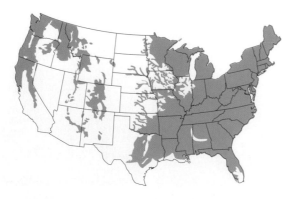

Virgin forests, 1620

Virgin forests, 1998

Figure 23-13 Vanishing old-growth forests in the United States, 1620 and 1998. Since 1960, some of these forests have been replaced with second-growth forests and tree plantations. (Data from the Wilderness Society and the U.S. Forest Service)

the nation's overall tree cover, biodiversity researchers argue that it does reduce overall forest biodiversity and the efficiency of ecosystem processes such as energy flow and chemical cycling. Others argue that tree farms help preserve overall forest biodiversity by reducing the pressure to clear-cut more diverse old-growth and second-growth forests. The basic question is, "What percentage of the nation's forests should be converted to tree farms to meet the wood and pulp needs of consumers and still preserve the overall biodiversity of the nation's forests?"

■ Since 1980 the rate at which forests and farmland are being lost to urban and suburban development has increased. According to the U.S. Department of Agriculture, between 1992 and 1997 more than 24,000 square kilometers (9,300 square miles) of forest were cleared for urban and suburban development, and the rate is increasing (Figure 25-8, p. 664).

Case Study: How Should U.S. National Forests Be Managed? The 156 national forests (Figure 23-4) managed by the U.S. Forest Service:

■ Contain about 19% of the country's forest area and supply about 3% of the nation's softwood timber (down from 15% in the 1980s).

■ Serve as grazing lands for more than 3 million cattle and sheep each year.

■ Provide about $4 billion worth of minerals, oil, and natural gas per year.

■ Contain a network of more than 612,000 kilometers (380,000 miles) of roads—equal in area to the entire U.S. interstate highway system. Most are narrow dirt roads built at taxpayers' expense for logging by private companies.

■ Provide habitats for almost 200 threatened and endangered species and hundreds of other wild species and fish.

■ Are the principal habitat for thousands of pollinator species that contribute $4–7 billion per year to U.S. agriculture.

■ Provide some of the country's cleanest drinking water for more than 60 million Americans in more than 3,400 communities. The estimated value of this water is $3.7 billion per year.

■ Contain about one-third of the country's protected wilderness area.

■ Receive more visits for recreation, hunting, and fishing than any other federal public lands, with recreational use rising sharply since 1930 (Figure 23-14, right).

The Forest Service is required by law to manage national forests according to the principles of (1) *sustained yield* (which states that potentially renewable tree resources should not be harvested or used faster than they are replenished) and (2) *multiple use* (which says that each of these forests should be managed for a variety of simultaneous uses such as sustainable timber harvesting, recreation, livestock grazing, watershed protection, and wildlife).

There is much controversy over how forest resources in the national forests should be used. Timber companies push to cut as much of this timber as possible at low prices. Biodiversity experts and environmentalists believe that (1) tree harvesting in public forests should be reduced or eliminated, (2) timber companies should

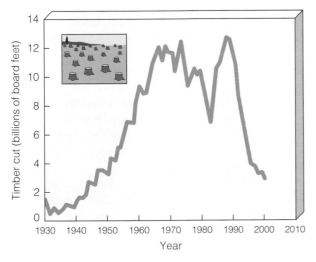
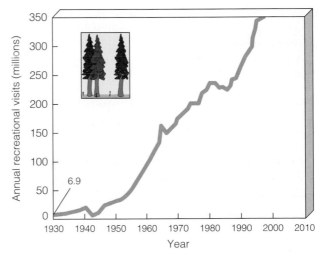

Figure 23-14 Timber harvest levels (left) and recreational visits in visitor days (right) in U.S. national forests, 1930–1999. Beginning in 1965, the Forest Service started counting visits to national forest in visitor days. One *visitor day* equals one person spending a 12-hour day in a national forest. (Data from U.S. Forest Service)

pay more for trees they harvest from national forests, and (3) national forests should be managed primarily to provide recreation and to sustain their biodiversity, water resources, and other ecological services.

Between 1930 and 1988, overall timber harvesting from national forests increased sharply (Figure 23-14, left) mostly because of (1) intense lobbying of Congress by timber company interests, (2) a law that allows the Forest Service to keep most of the money it makes on timber sales (and thus increase its budget), and (3) a 1908 law that gives counties within the boundaries of national forests 25% of the gross receipts from timber sales.

By law, the Forest Service must sell timber for no less than the cost of reforesting the land from which it was harvested. However, the cost of access roads is not included in this price and is provided as a subsidy to logging companies. Usually, the companies also get the timber itself for less than they would pay a private landowner for an equivalent amount of timber. According to the Wilderness Society, a lumber company harvesting trees in a national forest pays as little as $2 for 20-meter (65-foot), 100-year-old pine tree that will produce timber products worth thousands of dollars.

Because of such subsidies, timber sales from U.S. federal lands have turned a profit for taxpayers in only 3 of the last 100 years. Studies by the U.S. General Accounting Office show that between 1978 and 1998, national forests lost at least $7 billion from below-cost timber sales. According to a 2000 study by the National Forest Protection Alliance, logging in National Forests (1) causes more economic harm than good to local communities near such forests, (2) costs the nation's taxpayers about $1.2 billion per year in logging subsidies (an estimate confirmed by the U.S. Congressional Research Service), and (3) costs taxpayers and nearby communities many billions of dollars more each year from rivers polluted and fisheries damaged by sediments from logged lands, increased flooding, lost recreational opportunities, and degraded scenery.

Timber company officials argue that logging in national forests:

■ Is needed to satisfy the country's demand for wood.

■ Provides cheap timber that benefits consumers by keeping lumber and paper prices down.

■ Improves forest health by removing diseased trees and helping prevent forest fires.

■ Provides jobs and stimulates economic growth in nearby communities.

Environmentalists respond that:

■ Timber cutting in national forests provides only about 3% of the country's wood, and there is ample private forestland to meet the country's demand for wood (Figure 23-8).

■ Below-cost timber prices in national forests increase pressure for timber cutting in national and state forests and have little effect on the consumer costs of lumber and paper.

■ Communities relying heavily on proceeds from national forest timber sales experience severe economic slumps as the timber is depleted and timber companies move to other areas (p. 594).

■ A 1999 study by the Earth Island Institute found that if the government stopped cutting timber in the National Forests and redirected current government logging subsidies, there would be (1) more than $30,000 per year available to hire each timber worker to carry out ecological restoration on deforested public lands and (2) another $500 million per year for other purposes such as funding research on nonwood alternatives, improving recreational facilities in national forests, and providing vocational training and economic development in communities affected by losses of logging jobs.

■ Recreation in national forests provides many more jobs and much more income for local communities than does logging. According to a 2000 study by the economic accounting firm ECONorthwest, recreation, hunting, and fishing in national forests generate about 2.9 million jobs and add about $234 billion to the national economy each year. By contrast, logging, mining, crazing, and other extractive uses in these forests add about 407,000 jobs and $23 billion to the economy per year.

In 1998, Forest Service chief Mike Dombeck announced a policy that shifts the primary emphasis of the agency from maximizing timber harvests (Figure 23-14, left) to putting greater emphasis on (1) recreational use, (2) watershed health and restoration, (3) wildlife conservation (including attempts to control the spread of more than 2,000 nonnative plant, insect, and pathogen species), and (4) sustainable forest management that includes protecting and restoring biodiversity. Since 1988, timber harvesting levels have dropped sharply to about the same level as in 1955 (Figure 23-14, left), and the percentage of timber harvested by clear-cutting dropped from 39% to 12%. However, under pressure from timber companies, Congress may pass legislation to overturn this policy and return to heavy logging in national forests.

In 2000, the Forest Service proposed a permanent ban on road building in up to 223,000 square kilometers (86,000 square miles) of roadless areas in the national forests that are potentially available for protection under the National Wilderness System. Timber, mining, ranching, and recreational vehicle interests (1) have been pushing Congress and the Forest Service for years to build logging roads in these areas to make them ineligible for wilderness protection and **(2)** are pressuring Congress to overturn this executive decision.

The sharp increase in year-round recreational use of national forests (Figure 23-14, right) has led to clashes of **(1)** hikers with mountain bikers, **(2)** horseback riders with all-terrain vehicle users and motorcyclers, **(3)** backcountry skiers with snowmobilers, **(4)** swimmers and fishers with boaters and ski-style watercraft users, and **(5)** hikers and campers seeking quiet with thrillseekers on motorized vehicles that some would like to see banned or greatly restricted in national forests (as they are in many national parks). Heavily used national forests near urban areas are experiencing city problems such as traffic jams, noise, trash, crime, and pollution.

Analysts have proposed a number of reforms in Forest Service policy that would lead to more sustainable use and management of national forests (Solutions, right).

How Can We Reduce the Need to Harvest Trees?

Three ways to reduce pressure to harvest trees on public and private land are to **(1)** improve the efficiency of wood use, **(2)** increase paper recycling (p. 533), and **(3)** use fiber that does not come from trees to make paper.

According to the Worldwatch Institute and forestry analysts, up to 60% of the wood consumed in the United States is used inefficiently or wasted through construction, excess packaging, junk mail, inadequate paper recycling, and failure to reuse wooden shipping containers. Only 3% of the total U.S. production of softwood timber comes from the national forests. Thus, reducing the waste of wood and paper products by only 3% could eliminate the need to remove any timber from the national forests and allow these lands to be used primarily for recreation and biodiversity protection.

Tree-free fibers for making paper come from two sources: **(1)** agricultural residues left over from crops such as wheat, rice, and sugar and **(2)** crops such as kenaf and industrial hemp that can be grown specifically for pulp. Currently, tree-free paper fibers account for about 7% of the world's fiber supply for paper, with 97% of it used in developing countries and less than 1% in the United States. In China, 60% of the paper is made with tree-free pulp such as rice straw and other agricultural wastes left after harvest.

North America produces more than enough rice and wheat straw and other agricultural wastes to meet all its paper needs without having to cut another tree. However, some environmentalists believe that instead of being removed, crop residues such as rice straw should be left to decompose and help fertilize soil and reduce inputs of commercial inorganic fertilizers.

Currently, most of the small amount of tree-free paper produced in the United States is made from the fibers of a rapidly growing woody annual plant called *kenaf* (pronounced "kuh-NAHF"). Compared to pulpwood, kenaf **(1)** needs less herbicide (because it grows

How Can U.S. Federal Forest Management Be Improved?

SOLUTIONS

Foresters and biodiversity researchers have proposed a number of management reforms that would allow national forests in the United States to be used more sustainably without seriously impairing their important ecological functions. These suggestions include the following:

- *Make recreation and protection of watershed and wildlife habitats the primary uses of national forests.*

- *Prohibit timber harvesting from all or most of the remaining old-growth forests on public lands.*

- *Sharply reduce or ban all new road building in national forests and close little-used existing roads.*

- *Allow individuals or groups to buy conservation easements that prevent timber harvesting on designated areas of public old-growth forests.* In such *conservation-for-tax-relief swaps,* purchasers would be given tax breaks for the funds they put up.

- *Require that timber from national forests be sold at a price that includes the costs of road building, site preparation, and site regeneration and that all timber sales in national forests yield a profit for taxpayers.*

- *Do not use money from timber sales in national forests to supplement the Forest Service budget* because it encourages overexploitation of timber resources.

- *Base the provision that returns 25% of gross receipts from national forests to counties containing the forests on recreational user-fee receipts only.*

- *Establish federally funded reforestation and ecological restoration projects on degraded public lands.* In addition to the ecological benefits, this will provide jobs for displaced forest harvesting and wood products workers.

- *Require use of sustainable forestry methods (p. 597).*

Critical Thinking

What might be some disadvantages of implementing these policies?

faster than most weeds) and insecticide (because its outer fibrous covering is nearly insect-proof), **(2)** does not deplete soil nitrogen because it is a nitrogen fixer, **(3)** takes fewer chemicals and less energy to break down its fibers, and **(4)** produces less toxic wastewater.

However, kenaf paper currently costs three to five times more than virgin or recycled paper stock, but prices probably will come down as demand increases and producers cut production costs. Biologists urge caution against the widespread planting of kenaf or industrial hemp because **(1)** they fear that such rapidly growing plants could take over ecosystems, crowd out other plant species, and disrupt food webs, and (2) large plantations of such plants could compete with increasingly scarce land needed to provide food for the world's growing population.

According to the Worldwatch Institute, at least 20% of the world's total supply of fiber for making paper could be supplied from tree-free sources. If such sources are coupled with increased paper recycling, 70–75% of the world's fiber used to make paper would not have to come from wood.

Some analysts have suggested that we can save wood by switching to nonwood home-building and remodeling materials such as (1) steel framing and floor joists, (2) aluminum framing, (3) concrete slabs instead of wooden floor joists, and (4) carpet instead of finished wood floors. However, this involves shifting from a potentially renewable resource (wood) to nonrenewable resources such as limestone, iron and other mineral products, and crude oil used make these products.

Analyses show that switching to these nonrenewable resources usually takes more energy for transportation and manufacture and produces more pollution than use of wood and wood products. For example, (1) steel framing takes 13 times more energy than wood framing, (2) aluminum framing for walls takes 20 times more energy than wood framing, (3) steel floor joists take 50 times more energy than those made of wood, and (4) wall-to-wall carpeting takes four times more energy than wood floors.

23-4 TROPICAL DEFORESTATION AND THE FUELWOOD CRISIS

Why Should We Care About Tropical Forests? Tropical forests cover about 6% of the earth's land area (roughly the area of the lower 48 states of the United States) and grow in equatorial Latin America, Africa, and Asia (Figure 6-16, p. 131). Tropical forests come in several varieties (Figure 6-16, p. 131), including (1) rain forests (which receive rainfall almost daily), (2) deciduous forests with one or two dry seasons each year, (3) dry deciduous forests, and (4) forests on hills and mountains.

Currently, more than 90% of the world's forest loss is occurring in tropical countries (Figure 23-8). Biologists consider the cutting and degradation of most remaining old-growth tropical forests to be a serious global environmental problem, primarily because (1) these forests are home to 50–90% of the earth's terrestrial species (Figure 4-17, p. 82), many with highly specialized niches (Figure 6-28, p. 144) that make them vulnerable to extinction, and (2) they play an important role in removing some of the excess CO_2 we are putting into the atmosphere (Figure 18-5, p. 451). If current rates of tropical deforestation, fragmentation, and degrada-

tion continue, biologist Edward O. Wilson estimates that (1) at least 20% of the world's tropical forest species could be gone by 2022, and by 2042 as many as 50% of these species could become extinct.

Tropical forests have great ecological and economic importance based on various types of values (Section 22-3, p. 559, and Figure 23-15). Some tropical tree species can be used for many purposes (Solutions, p. 607).

How Fast Are Tropical Forests Being Cleared and Degraded? Climatic and biological data suggest that mature tropical forests once covered at least twice as much area as they do today, with most of the destruction occurring since 1950. Satellite scans and ground-level surveys used to estimate forest destruction indicate that large areas of tropical forests are being cut rapidly in parts of South America, Africa, and Asia (see red and orange regions in Figure 23-8). Haiti has lost 99% of its original forest cover, the Philippines 97%, and Madagascar 84%.

Brazil has about half the world's remaining tropical rain forest in the vast Amazon Basin, which is half the size of the continental United States. This basin is a global center of biodiversity that contains about 20% of the world's plant species, 17% of the world's bird species, and 9% of the world's mammal species. Brazil has lost about 40% of its original forests and more than 95% of its Atlantic coastal forest, and its remaining forests are being cut, fragmented, and degraded rapidly (Figures 23-8 and 23-16).

Of all tropical forest types, dry deciduous forests have been the most severely affected by human activities. Most of these dry forests have been cut and fragmented and used to grow crops such as cotton and sorghum and to graze cattle because they are much easier to burn and typically have more fertile soils than wet tropical forests.

There is debate over the current rates of tropical deforestation and degradation because of (1) difficulties in interpreting satellite images, (2) different ways of defining deforestation and forest degradation, and (3) political and economic factors that cause countries to hide or exaggerate deforestation. Using the best available data, it is estimated that between 1972 and 2000, at least 574,000 square kilometers (222,000 square miles) or 14% of the huge Amazon area was deforested.

According to tropical ecologists William Laurance and Lisa Curran, up to 40% of remaining Amazon forest area suffers from the harmful effects of fragmentation (p. 564). Curran's research has show that several factors can interact synergistically to reduce the biodiversity and sustainability of intact forest areas:

■ When areas surrounding protected areas of tropical forest are cleared or destroyed by fire, wildlife in these areas migrate to nearby intact areas and

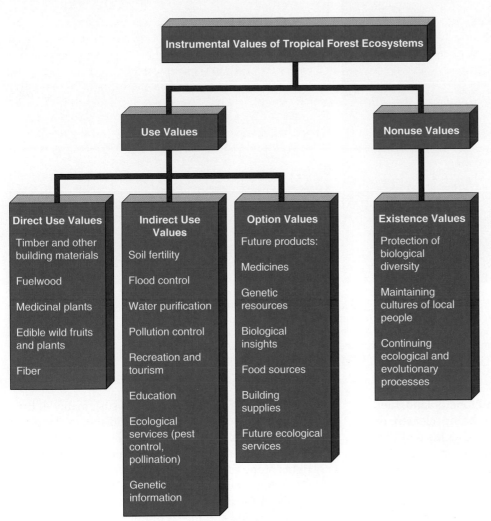

Instrumental Values of Tropical Forest Ecosystems

Use Values

Nonuse Values

Direct Use Values

Timber and other building materials

Fuelwood

Medicinal plants

Edible wild fruits and plants

Fiber

Indirect Use Values

Soil fertility

Flood control

Water purification

Pollution control

Recreation and tourism

Education

Ecological services (pest control, pollination)

Genetic information

Option Values

Future products:

Medicines

Genetic resources

Biological insights

Food sources

Building supplies

Future ecological services

Existence Values

Protection of biological diversity

Maintaining cultures of local people

Continuing ecological and evolutionary processes

Figure 23-15 Some of the many *instrumental values* of tropical forests. These show the importance of preserving the rich diversity of these forests instead of cutting them down or degrading them for short-term economic gain. Some people also believe that tropical forests have *intrinsic value* and thus have an inherent right to exist regardless of their usefulness to people.

devour most of the seed needed to regenerate the intact forests.

- Reduced forest cover over large areas reduces seed production and viability in remaining intact areas.

- Smoke from increased fires intensified by droughts related to El Niño-Southern Oscillations (Figure 6-10, p. 127) kills seedlings in intact forests and can reduce rainfall in areas covered with smoke. This synergistic interaction between drought, fire, and smoke can change the regional climate and lead to more drought, fire, and smoke.

Ecologists warn that current plans by the Brazilian government to pave over the last stretch of highway lining the Amazon River with southern Brazil will hasten this destructive process.

Estimates of total global tropical forest loss per year vary from 50,000 square kilometers (19,300 square miles) to 170,000 square kilometers (65,600 square miles). In 2000, a study by the World Resources Institute estimated that at least 140,000 square kilometers (54,000 square miles) are lost per year. It is estimated that each year an equivalent area of these forests is seriously degraded and fragmented.

However, critics such as biogeographer Philip Stott claim that **(1)** tropical forest losses are exaggerated, **(2)** many deforested areas are regenerating as second-growth forests, **(3)** the loss of tropical forests is no more ecologically important than the clearing of the old-growth forests that once covered Europe and parts of eastern North America, and **(4)** excessive concern over loss of tropical forests diverts attention from more important ecological threats to the world's oceans (Section 24-2, p. 632).

Case Study: Madagascar: A Threatened Jewel of Biodiversity Madagascar, the world's fourth largest island, lies in the Indian Ocean off the East African coast (Figure 23-17). Most of its species have evolved in near isolation for at least 40 million years, after the continental drift of the earth's lithospheric plates (Figure 5-9, p. 114) moved the island far enough from Africa's mainland to prevent migrations to and from Africa.

The result is that about 85% of all plant and animal species found on the island are endemic species unique to this island. Most are found in its vanishing eastern rain forests (Figure 23-17). Because of its astounding biological diversity, this Texas-size island is considered a crown jewel of biodiversity among the earth's ecosystems.

Madagascar's plant and animal species are among the world's most endangered, mostly because of loss of habitat from unsustainable slash-and-burn agriculture (Figure 2-3, p. 25) on poor soils, fueled by rapid population growth. For example, 16 of the island's 31 primate species are threatened with extinction. Officials have not been able to curb illegal smuggling of endangered species—especially frogs, chameleons, and lizards—to Europe and Asia.

Since humans arrived about 1,500 years ago, 84% of the island's tropical seasonal forests and more than

Figure 23-16 Satellite images showing the rapid destruction, degradation, and fragmentation of tropical forest in Brazil's state of Rondonia in the Amazon Basin between 1986 (left) and 1999 (right). This process was promoted by the Brazilian government's program to build the Trans-Amazon Highway and encourage settlement of the remote Amazon Basin by offering farmers cheap land. (Landsat images courtesy of the Tropical Rain Forest Information Center of the Basic Science and Remote Sensing Initiative at Michigan State University, a NASA Earth Science Information Partner. WWW.BSRSI.MSU.EDU)

66% of its rain forests have been cut for cropland, fuel, and lumber, and what is left is being cleared or burned rapidly. The resulting eroded fields and hillsides make Madagascar the world's most eroded country, with huge quantities of eroded sediment flowing in its rivers and emptying into its coastal areas (Figure 7-7, p. 157).

Since 1984, the government, conservation organizations, and scientists worldwide have united to slow the island's ecological degradation. For such efforts to succeed, population growth, projected to double between 1998 and 2025, will have to slow drastically. Even if this internationally funded effort is fully implemented, Madagascar probably will lose half its remaining plant and animal species.

Case Study: Cultural Extinction in Tropical Forests An estimated 250 million people belong to indigenous cultures found in about 70 countries. Many of these peoples, such as the Kuana of Panama, the Kenyah in Indonesia, the Yaneshá in Peru, and the Yanomami in Brazil, have been living in and using tropical and other forests sustainably for centuries. They get most of their food from hunting and gathering, trapping, and sustainable slash-and-burn and shifting cultivation (Figure 2-3, p. 25). Many of the earth's remaining tribal peoples, representing 5,000 cultures, are vanishing as the lands they have lived on for centuries are taken over for economic development. To ecologists and cultural anthropologists, this results in an irreplaceable loss of ecological knowledge and cultural diversity. People in these cultures have impor-

tant knowledge about how to live sustainably in tropical forests (and in other biomes) and about what plants can be used as foods and medicines.

Many analysts urge governments to protect the rights of the earth's remaining indigenous cultures by (1) giving them full ownership of the land they have occupied for centuries and (2) protecting their lands from intrusion and illegal resource extraction.

Other people, mostly resource extractors, strongly disagree. They believe that road building, resource extraction, crop raising, livestock grazing, and other forms of development in tropical forests (and other undeveloped biomes) should not be held back by granting legal rights to land and resources to small numbers of primitive tribal peoples. What do you think?

What Causes Tropical Deforestation? Tropical deforestation results from a number of interconnected causes (Figure 23-18). All of these factors are related to increasing appropriation of the net primary productivity (Figure 4-25, p. 88) and mineral resources of the earth's forests for human use as a result of population growth, poverty, and government policies that encourage deforestation. Population growth and poverty combine to drive subsistence farmers and the landless poor to tropical forests, where they try to grow enough food to survive.

Government subsidies can accelerate deforestation by (1) making timber or other resources cheap relative to their full ecological value and (2) encouraging the poor to colonize tropical forests by giving them title to land they clear, as is done in Indonesia, Mexico, and

The Incredible Neem Tree

Wouldn't it be nice if there were a single plant that could **(1)** quickly reforest bare land, **(2)** provide fuelwood and lumber in dry areas, **(3)** provide alternatives to toxic pesticides, **(4)** be used to treat numerous diseases, and **(5)** help control population growth? There is: the *neem tree*, a broadleaf evergreen member of the mahogany family.

This remarkable tropical species, native to India and Burma, is ideal for reforestation because it can grow to maturity in only 5–7 years. It grows well in poor soil in semiarid lands such as those in Africa, providing an abundance of fuelwood, lumber, and lamp oil.

It also contains various natural pesticides. Chemicals from its leaves and seeds can repel or kill more than 200 insect species, including termites, gypsy moths, locusts, boll weevils, and cockroaches.

Extracts from neem seeds and leaves (Figure 22-12, p. 561) can be used to fight bacterial, viral, and fungal infections. Indeed, the tree's chemicals have allegedly relieved so many different afflictions that the tree has been called a "village pharmacy." Its twigs are used as an antiseptic toothbrush, and oil from its seeds is used to make toothpaste and soap.

That is not all. Neem-seed oil evidently acts as a strong spermicide that might be used in producing a much-needed male birth control pill. According to a study by the U.S. National Academy of Sciences, the neem tree "may eventually benefit every person on the planet."

Despite its numerous advantages, ecologists caution against widespread planting of neem trees outside its native range. If introduced as a nonnative species to other ecosystems, it could take over and displace other species because of its rapid growth and resistance to pests.

Critical Thinking

Assume that you are an elected county official. Someone proposes widespread planting of neem trees on abandoned cropland in the area where you live. Would you favor this proposal? Explain. What tests might you want to run before approving this project?

Brazil (Figure 23-16). International lending agencies encourage developing countries to borrow huge sums of money from developed countries to finance projects such as roads, mines, logging operations, oil drilling, and dams in tropical forests.

The process of degrading a tropical or other type of forest often begins with a road (Figures 23-10 and 23-16),

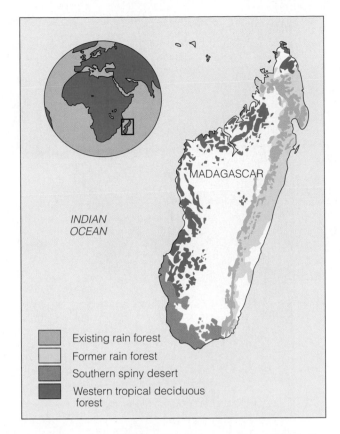

Figure 23-17 Some biomes in Madagascar, a large island off the eastern coast of Africa. The rich biodiversity and numerous endemic species of plants and animals found only on this island are being seriously threatened by a combination of population growth, poverty, and deforestation.

usually cut by logging companies. Once the forest becomes accessible, it can be cleared and degraded by a number of factors, including:

- *Unsustainable forms of small-scale farming.* Instead of practicing various methods of traditional and potentially sustainable shifting cultivation (Figure 2-3, p. 25), many poor people migrating to tropical forests practice unsustainable farming that depletes soils and destroys large tracts of forests (Figure 23-16).

- *Cattle ranching.* Cattle ranches, sometimes supported by government subsidies, often are established on cropland that has been exhausted by small-scale farmers and then abandoned or sold to ranchers. When torrential rains and overgrazing turn the usually thin and nutrient-poor tropical forest soils (Figure 10-15, p. 223) into eroded wastelands, ranchers move to another area and repeat this destructive process.

- *Clearing large areas of tropical forest for raising cash crops.* Large plantations grow crops such as sugarcane, bananas, pineapples, peppers, strawberries, cotton, tea, and coffee, mostly for export to developed countries.

Figure 23-18 Major interconnected causes of the destruction and degradation of tropical forests. These factors ultimately are related to **(1)** population growth, **(2)** poverty, and **(3)** government policies that encourage deforestation.

Labels within figure:

Bromeliad

Primary Causes:
Rapid population growth
Poverty
Exploitive government policies
Exports to developed counties
Failure to include ecological services in evaluating forest resources

Toucan

Scarlet macaw

Golden lion marmoset

Orchid

Secondary Causes:
Roads
Logging
Unsustainable peasant farming
Cash crops
Cattle ranching
Tree plantations
Flooding from dams
Mining
Oil drilling

Blue morpho butterfly

- *Mining (Case Study, p. 328) and oil drilling.*

- *Building dams on rivers that flood large areas of tropical forests.*

- *Commercial logging.* Since 1950, the consumption of tropical timber has risen 15-fold. Japan alone

accounts for 53% of the world's tropical timber imports, followed by Europe (32%) and the United States (15%). After depleting tropical timber in much of Asia, cutting is now shifting to Latin America and Africa (Figure 23-8). According to a 1997 report by the World Resources Institute, logging is the greatest

Sustainable Agriculture and Forestry in Tropical Forests

A combination of the knowledge of indigenous peoples, ecological research, and modern technology can show people how to grow crops and harvest timber in tropical forests more sustainably.

One approach would be to show people migrating to forests how to grow crops using forms of more sustainable agroforestry developed by indigenous peoples throughout the world. The Lacandon Maya Indians of Chiapas, Mexico, for example, have developed a multilayered system of agroforestry that allows them to cultivate up to 75 crop species on 1-hectare (2.5-acre) plots for up to 7 consecutive years. After that, a new plot is planted to allow regeneration of the soil in the original plot.

Another approach being used by Yaneshá Indians in the lush rain forests of Peru's Palcazú Valley is using strip cutting (Figure 23-11e) to harvest tropical trees for lumber. Widely spaced thin trips are harvested in an alternating pattern (for example, strips 1, 3, 5, and then 2, 4, and 6) and any given strip is allowed to regenerate 30–40 years before being harvested again. This type of harvesting reduces environmental degradation and promotes rapid natural regeneration of native tree species from stumps or from seeds of nearby trees.

Oxen haul the cut logs to a gravel road, where a flatbed truck or tractor carries them to the processing plant. There other tribe members convert the logs to charcoal fuel, boards, and fenceposts, most of which are sold locally. Tribe members also act as consultants to help other forest dwellers set up similar systems.

Another way to maintain biodiversity is to grow coffee in shade tree plantations under a diverse canopy of trees instead of in unshaded plantations. However, shade tree coffee plantations in the tropics are being converted rapidly to higher-yield unshaded coffee plantations that need more pesticides and fertilizers.

Some coffee drinkers are fighting this loss of biodiversity by buying only coffee that is certified (by organizations such as the Rainforest Alliance) as being grown under shade trees. In the United States, such coffee is available through more than 24 coffee distributors and mail-order retailers.

Critical Thinking

What applications (if any) might these practices have for growing food and harvesting timber in the United States and other developed counties? How would you encourage use of such approaches for more sustainable use of forests?

---

threat to the world's remaining tropical old-growth forests, threatening 79% of such forests in Africa, 69% in South America, and 54% in Central America. Timber exports to developed countries contribute significantly to tropical forest depletion and degradation, but more than 80% of the trees cut in developing countries are used at home.

- *Increasing forest fires.* Between 1997 and 1999, huge areas of forest were burned in Indonesia, Malaysia, Brazil, Guatemala, Nicaragua, and Mexico, mostly by farmers using fire to prepare fields for planting or cattle-grazing and corporations clearing forests to establish pulp, palm oil, and rubber plantations. The resulting highly polluted air **(1)** sickened tens of millions of people, **(2)** killed hundreds, **(3)** caused billions of dollars in damage, and **(4)** released large amounts of carbon dioxide into the atmosphere.

Clearing tropical forests can also degrade tropical rivers that carry two-thirds of the world's freshwater runoff. When tropical forests near rivers are cleared, soil erosion increases dramatically, often as much as 25-fold. The eroded soil **(1)** converts normally clear water to muddy water, **(2)** silts river bottoms and destroys many forms of aquatic life, **(3)** fills reservoirs, **(4)** overloads estuaries with nutrients and silt (Figure 7-7, p. 157), and **(5)** smothers offshore coral reefs with silt.

Solutions: How Can We Reduce Tropical Deforestation and Degradation? A number of analysts have suggested the following ways to protect tropical forests and use them more sustainably:

- Identify and move rapidly to protect areas of tropical forests that are rich in unique animal and plant species and in imminent danger, called *hot spots* or *critical ecosystems* (Figure 23-25, p. 617).

- Mount global efforts to sharply reduce the poverty (Section 26-6, p. 706) that leads the poor to use forests (and other resources) unsustainably for short-term survival.

- Reduce the flow of the landless poor to tropical forests by slowing population growth.

- Establish programs to help new settlers in tropical forests learn how to practice small-scale sustainable agriculture and forestry (Solutions, above).

- Phase out government subsidies that encourage unsustainable forest use and phase in tariffs, use taxes, user fees, and subsidies that favor more sustainable forestry and biodiversity protection.

- Encourage governments to protect large areas of tropical forests. Currently less than 5% of the world's tropical forests are part of parks and preserves, and

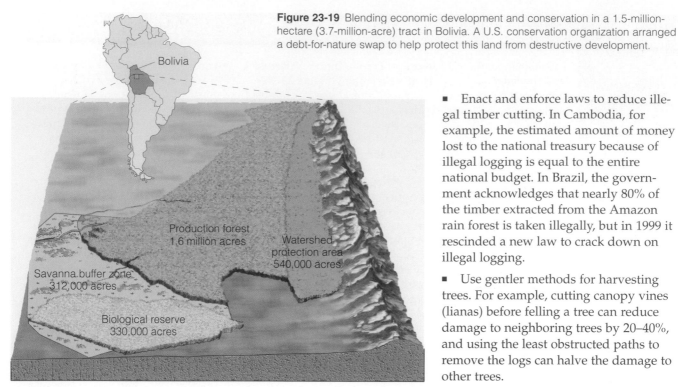

Figure 23-19 Blending economic development and conservation in a 1.5-million-hectare (3.7-million-acre) tract in Bolivia. A U.S. conservation organization arranged a debt-for-nature swap to help protect this land from destructive development.

Bolivia

Production forest
1.6 million acres

Watershed protection area
540,000 acres

Savanna buffer zone
312,000 acres

Biological reserve
330,000 acres

many of these are protected in name only. In 1998, tiny Suriname (a Dutch colony on South America's northern Caribbean coast with a population of 400,000) announced that it was setting aside 12% of its land area as a nature reserve instead of turning it over to Asian loggers. Indonesia, Brazil, Zaire, and Costa Rica (Solutions, p. 615) have established large parks and forest reserves. In 2000, the World Wildlife Fund, the World Bank, and the Brazilian government announced a joint program to preserve 10% of the forests in the Amazon Basin by 2010, with most of the funding provided by the World Bank. However, because of the synergistic effects of forest fragmentation on intact forest (p. 564), tropical ecologist William Laurance estimates that at least 40% of the Amazon forest should be protected to maintain biodiversity and ecosystem function.

■ Use debt-for-nature swaps and conservation easements to encourage countries to protect tropical forests. In a *debt-for-nature swap*, participating countries act as custodians for protected forest reserves in return for foreign aid or debt relief (Figure 23-19); in *conservation easements*, a private organization, country, or group of countries compensates other countries for protecting selected forest areas.

■ Establish an international system for evaluating and labeling timber produced by sustainable methods (Solutions, p. 598). In 1993, the International Forest Stewardship Council (FSC) was established to set standards for sustainable forest production.

■ Enact and enforce laws to reduce illegal timber cutting. In Cambodia, for example, the estimated amount of money lost to the national treasury because of illegal logging is equal to the entire national budget. In Brazil, the government acknowledges that nearly 80% of the timber extracted from the Amazon rain forest is taken illegally, but in 1999 it rescinded a new law to crack down on illegal logging.

■ Use gentler methods for harvesting trees. For example, cutting canopy vines (lianas) before felling a tree can reduce damage to neighboring trees by 20–40%, and using the least obstructed paths to remove the logs can halve the damage to other trees.

■ Concentrate peasant farming, tree and crop plantations, and ranching activities on already-cleared tropical forest areas that are in various stages of secondary ecological succession instead of on old-growth forests.

■ Mount global efforts to reforest and rehabilitate degraded tropical forests and watersheds (Individuals Matter, right). In 1998, China's government instituted a 10-year plan to (1) reduce timber harvests in natural forests, (2) restore natural forests in ecologically sensitive areas, (3) convert marginal farmland to forestland, (4) regenerate natural forests in degraded forest areas, and (5) increase timber production in tree plantations.

■ Transfer control of forests from central governments to communities, as has been done successfully in parts of India.

■ Encourage citizens to work together to protect tropical forests. In Brazil, 500 conservation organizations have formed a coalition to preserve the country's remaining tropical forests.

■ Reduce the waste and overconsumption of industrial timber, paper, and other resources by consumers, especially in developed countries (p. 533).

How Serious Is the Fuelwood Crisis in Developing Countries? According to the UN Food and Agriculture Organization, in 2000 about 2.7 billion people in 77 developing countries (1) did not get enough fuelwood to meet their basic needs or (2) were forced to

Kenya's Green Belt Movement

In Kenya, Wangari Maathai (Figure 23-20) started the Green Belt Movement. The goals of this highly regarded women's self-help group are to establish tree nurseries, raise seedlings, and plant a tree for each of Kenya's almost 30 million people. By 2000, the 100,000 members of this grassroots group had planted and protected more than 10 million trees.

This project's success has sparked the creation of similar programs in more than 30 other African countries. This inspiring leader has said,

I don't really know why I care so much. I just have something inside me that tells me that there is a problem and I have got to do something about it. And I'm sure it's the same voice that is speaking to everyone on this planet, at least everybody who seems to be concerned about the fate of the world, the fate of this planet.

Figure 23-20 Wangari Maathai, the first Kenyan woman to earn a Ph.D. (in anatomy) and to head an academic department (veterinary medicine) at the University of Nairobi, organized the internationally acclaimed Green Belt Movement in 1977. (William Campbell/*TIME* magazine)

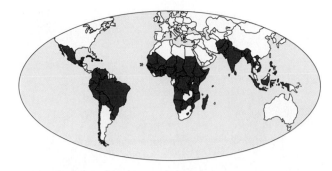

Figure 23-21 Scarcity of fuelwood, 2000. (Data from UN Food and Agriculture Organization)

meet their needs by using wood faster than it is replenished (Figure 23-21).

City dwellers in many developing countries burn charcoal because it is much lighter than fuelwood and is thus much cheaper to transport. However, burning wood in traditional earthen pits to produce charcoal consumes more than half the wood's energy. Thus, each city dweller burning charcoal uses twice as much wood for a given amount of energy as a rural dweller who burns firewood. This helps explain the expanding rings of deforested land that surround many cities in developing countries that use charcoal as the major fuel source.

Besides local deforestation and accelerated soil erosion, fuelwood scarcity has other harmful effects:

- It places a burden on the rural poor, especially women and children, who often must walk long distances searching for firewood.

- Buying fuelwood or charcoal can take 40% of a poor family's income.

- As burning wood (or charcoal derived from wood) to boil water becomes an unaffordable luxury, water-borne infectious diseases and death will spread.

- An estimated 800 million poor people who cannot get enough fuelwood burn dried animal dung and crop residues for cooking and heating (Figure 23-22). Because these natural fertilizers are no longer returned to the soil, cropland productivity is reduced and hunger and malnutrition can increase.

Solutions: What Can We Do About the Fuelwood Crisis? Developing countries can reduce the severity of the fuelwood crisis by **(1)** planting more fast-growing fuelwood trees or shrubs, **(2)** burning wood more efficiently, and **(3)** switching to other fuels.

However, fast-growing tree and shrub species used to establish fuelwood plantations must be selected carefully to prevent harm to local ecosystems. For example, eucalyptus trees are being used to reforest areas threatened by desertification and to establish fuelwood plantations in some parts of the world.

Because these species grow fast even on poor soils, this might seem like a good idea, but ecologists and local villagers see it as an ecological disaster. In their native Australia these trees thrive in areas with good rainfall. However, when planted in arid areas the trees suck up so much of the scarce soil water that most other plants cannot grow. Then farmers do not have fodder to feed their livestock, and groundwater is not replenished.

Eucalyptus trees also deplete the soil of nutrients, produce toxic compounds that accumulate in the soil (because of low rainfall), and inhibit nitrogen uptake.

Figure 23-22 Making fuel briquettes from cow dung in India. As fuelwood becomes scarce, more dung is collected and burned, depriving the soil of an important source of plant nutrients. (Pierre A. Pittet/UN Food and Agriculture Organization)

In Karnata, India, villagers became so enraged over a government-sponsored project to plant these trees that they uprooted the saplings.

Experience shows that planting projects are most successful when local people participate in their planning and implementation. The programs that work best give village farmers incentives, such as ownership of the land or ownership of any trees they grow on village land. Emphasis should be placed on community woodlots, which are easy to tend and harvest, rather than on large fuelwood plantations located far from where the wood is needed.

Another promising method is to encourage villagers to use the sun-dried roots of various gourds and squashes as cooking fuel. These rootfuel plants, which regenerate themselves each year, produce large quantities of burnable biomass per unit area on dry, deforested lands. They also help reduce soil erosion and produce an edible seed with a high protein content.

Using a traditional three-stone fire to burn wood typically wastes about 94% of the wood's energy content. New, cheap, more efficient, and less polluting stoves can provide both heat and light while reducing indoor air pollution, a major health threat. The stoves must be easy and cheap to build, and the materials used to make them must be easily accessible locally (Individuals Matter, right).

Cheap and easily made solar ovens (Figure 15-21d, p. 374) that capture sunlight to provide heat can also be used to reduce wood use and air pollution in sunny and warm areas. However, they often are not accepted by cultures that also use fires for light and heat at night and as centers for social interaction.

Despite encouraging success in some countries (such as China, Nepal, Senegal, and South Korea), most developing countries suffering from fuelwood shortages have inadequate forestry policies and budgets and lack trained foresters. Such countries are cutting trees for fuelwood and forest products 10-20 times faster than new trees are being planted.

23-5 MANAGING AND SUSTAINING NATIONAL PARKS

How Popular Is the Idea of National Parks? Today, more than 1,100 national parks larger than 10 square kilometers (4 square miles) each are located in more than 120 countries. These parks cover a total area equal to that of Alaska, Texas, and California combined.

The U.S. National Park System, established in 1912, is dominated by 55 national parks, most of them in the West (Figure 23-4), sometimes called *America's crown jewels*. Between 1970 and 2000, the area of making up the U.S. National Park System increased almost three-fold and now occupies almost 300,000 square kilometers (116,000 square miles). Its 379 parks, monuments, historic sites, and urban recreation areas are enjoyed by more than 300 million visitors each year and are projected to have 500 million visitors by 2010.

These national parks are supplemented by state, county, and city parks. Most state parks are located near urban areas and have about twice as many visitors per year as the national parks.

How Are Parks Being Threatened? Parks everywhere are under pressure from external and internal threats. A 1999 study by the World Bank and the World Wildlife Fund found that most parks and wildlife reserves in developing countries are poorly managed, and only 1% are protected. The other 99% are mostly *paper parks* that exist in name only and have no protection. Most of these parks are invaded by **(1)** local people who need wood, cropland, game animals, and other natural products for their daily survival and **(2)** loggers, miners, and wildlife poachers who kill animals to obtain and sell items such as rhino horns (Figure 22-20, p. 571), elephant tusks (Figure 22-22, p. 572), and furs. Park services in developing countries typically have too little money and too few personnel to fight these invasions, either by force or by education.

In addition, most of the world's national parks are too small to sustain many large animal species. Many have been invaded by nonnative species that can reduce the populations of native species and cause ecological disruption.

New Stoves Help Save India's Forests and Improve Women's Health

INDIVIDUALS MATTER

Since 1984, Lalita Balakrishnan and the All-India Women's Conference have enlisted 300,000 women to spread the use of efficient smokeless stoves throughout much of India. These women have helped design, build, and install more than 300,000 smokeless wood stoves, called *chulhas*, across the country.

These stoves cost about $5 and are made of cow-dung, mud, and hay slapped together by hand, with a metal pipe leading to the roof. The Indian government pays one-third of the stoves' cost, and all materials (except the pipe) are provided by the stove buyer. Local women are hired to sell, install, and maintain the stoves.

These cheap, fuel-efficient stoves have helped reduce deforestation by reducing fuelwood use. Replacing smoky indoor fires with such stoves has also led to better respiratory health for more than 1 million Indian women.

In recognition of her efforts, the UN Environment Programme added Lalita Balakrishnan to its 1996 Global 500 Roll of Honor.

In the United States, Canada, and most developed countries, many parks have become islands surrounded and threatened by areas of increasing development and population growth. A 2000 study found that all of Canada's 39 national parks were under threat from internal and external stresses.

Popularity is one of the biggest problems of national and state parks in the United States (and other developed countries). Because of increased numbers of roads, cars, and affluent people, annual recreational visits to major national parks increased fourfold (and visits to state parks sevenfold) between 1950 and 1999.

Many parks provide spectacular scenery and enjoyable recreational experiences for visitors. However, because of underfunding, visitors to some of the most heavily used parks find closed campgrounds, uncollected garbage, debris on trails and beaches, dirty toilets, and fewer nature lectures by park rangers.

During the summer, the most popular U.S. national and state parks often are choked with cars and trailers and hour-long entrance backups. Some park units, especially heavily visited ones near urban areas, are also plagued by noise, traffic jams, smog, vandalism, poaching, deteriorating trails and facilities, polluted water, and crime. Many visitors to heavily used national and state parks leave the city to commune with nature, only to find the parks as noisy, congested, and stressful as the places they came from.

In winter, the quiet solitude one might expect at Yellowstone National Park has been replaced with the noise and pollution of snowmobiles. As many as 1,000 snowmobiles per day roar into the park, spewing air pollutants equivalent to the tailpipe emissions of 1.7 million cars.

U.S. Park Service rangers now spend an increasing amount of their time on law enforcement instead of conservation, management, and education. Currently, there is one ranger for every 84,200 visitors to the major national parks. Many overworked and underpaid rangers are leaving for better-paying jobs.

Nonnative species have moved into or been introduced into many parks. Wild boars (imported to North Carolina in 1912 for hunting) are threatening vegetation in part of the Great Smoky Mountains National Park. The Brazilian pepper tree has invaded Florida's Everglades National Park. Mountain goats in Washington's Olympic National Park trample native vegetation and accelerate soil erosion. While some nonnative species have moved into parks, some native species of animals and plants (including many threatened or endangered species) are being killed or removed illegally in almost half of U.S. national parks.

Nearby human activities outside many national parks also threaten the ecological health of such parks. Wildlife and recreational values are threatened by mining, logging, livestock grazing, coal-burning power plants, water diversion, and urban development. Polluted air drifts hundreds of kilometers, killing ancient trees in California's Sequoia National Park and blurring the awesome vistas at Arizona's Grand Canyon. Car smog is damaging scores of plant species in the heavily used Great Smoky Mountains National Park. According to the National Park Service, air pollution affects scenic views in national parks more than 90% of the time.

That is not all. Mountains of trash wash ashore daily at Padre Island National Seashore in Texas. Water use in Las Vegas (Spotlight, p. 314) threatens to shut down geysers in the Death Valley National Monument. Visitors to Sequoia National Park complain of raw sewage flowing through a parking lot; similar contamination has occurred for years at Kentucky's Mammoth Cave. Unless a massive ecological restoration project (Case Study, p. 650) is successful, Florida's Everglades National Park may dry up and make it the country's most endangered national park.

Solutions: How Can Management of U.S. National Parks Be Improved? The 55 major U.S. national parks are managed under the principle of *natural regulation*, as if they were wilderness ecosystems that would adapt and sustain themselves if left alone. Many ecologists consider this a misguided policy. Most parks are far too small to even come close to sustaining themselves. Even the biggest ones, such as Yellowstone, cannot be isolated from the harmful effects caused by activities in nearby areas and from destruction from within by exploding populations of some plant-eating species such as elk and by invading nonnative species.

The U.S. National Park Service has two goals that increasingly conflict: (1) to preserve nature in parks and (2) to make nature more available to the public. It tries to accomplish these goals with a $1.5-billion annual budget and a $3.5-billion backlog of maintenance, repairs, and high-priority construction projects at a time when park usage and external threats to the parks are increasing.

In 1988, the Wilderness Society and the National Parks and Conservation Association made a number of suggestions for sustaining and expanding the national park system, including the following:

- *Require integrated management plans for parks and other nearby federal lands.*

- *Increase the budget for adding new parkland near the most threatened parks and for buying private lands inside parks.*

- *Locate all new and some existing commercial facilities and visitor parking areas outside parks and provide shuttle buses for entering and touring most parks*, as was done in 2000 in Zion National Park, Utah. Shuttle bus systems are available at the Yosemite, Grand Canyon, and Denali National Parks.

- *Require private concessionaires who provide campgrounds, restaurants, hotels, and other services for park visitors to (1) compete for contracts, (2) pay franchise fees equal to 22% of their gross (not net) receipts, and (3) pour some of their receipts back into parks.* Private concessionaires in national parks pay an average of only about 6–7% of their gross receipts in franchise fees to the government, and many large concessionaires with long-term contracts pay little as 0.75% of their gross receipts.

- *Allow concessionaires to lease but not own facilities inside parks.*

- *Provide more funds for park system maintenance and repairs.*

- *Survey the condition and types of wildlife species in parks.*

- *Raise entry fees for park visitors and pour receipts back into parks.*

- *Limit the number of visitors to crowded park areas.*

- *Increase the number and pay of park rangers.*

- *Encourage volunteers to give tours and lectures to visitors.*

- *Encourage individuals and corporations to donate money for park maintenance and repair.*

Use of off-road vehicles in some parks and other public lands is growing rapidly and crushing plants, harming wildlife, polluting the air, and disrupting the calm and quiet of such natural areas. Park officials are considering (1) a ban on the recreational use of snowmobiles, personal watercraft, off-road vehicles, and aircraft flyovers in some or all of the 23 national parks where they are currently allowed and (2) tougher enforcement in the remaining parks where such bans already exist.

23-6 ESTABLISHING, DESIGNING, MANAGING NATURE RESERVES

How Much of the Biosphere Should We Protect from Human Exploitation? Most ecologists and conservation biologists believe that the best way to preserve biodiversity is through a worldwide network of reserves, parks, wildlife sanctuaries, wilderness, and other protected areas. Currently, about 8% of the world's land area is either strictly or partially protected in more than 20,000 nature reserves, parks, wildlife refuges, and other areas.

Tundra makes up about 26% of these global nature reserves (mostly in Alaska, Canada, Scandinavia, and Greenland), followed by tropical dry forests (21%, mostly in Africa), deserts (17%), tropical humid forests (12%), temperate coniferous forests (7%), and temperate deciduous forests (6%). There is little protection of grasslands, islands, lakes, and wetlands.

Switzerland, several countries in Africa, and Costa Rica (Solutions, right) have set aside 10% or more of their land area as protected reserves. Canada has plans to protect 12% of its land area, and Brazil hopes to eventually protect about 18% of its land area in nature reserves.

This is an important beginning, but conservation biologists say that to keep biodiversity from being depleted and degraded, a minimum of 10% of the globe's land area should be strictly protected. Moreover, many existing reserves (1) are too small to provide any real protection for the wild species that live on them and (2) receive so little protection that their resources often are extracted illegally and unsustainably.

Establishing and managing an expanded global system of biodiversity reserves that includes multiple examples of all the earth's biomes will take action and funding by (1) national governments, (2) private groups (Solutions, p. 616), and (3) cooperative ventures involving governments, businesses, and private conservation groups.

Most developers and resource extractors and some economists disagree with protecting even the current 8% of the earth's remaining undisturbed ecosystems. They contend that such areas contain valuable resources that would add to economic growth. Ecologists and conservation biologists disagree and view protected areas as islands of biodiversity that are

Parks in Costa Rica

SOLUTIONS

Costa Rica (Figure 25-3, p. 660), smaller in area than West Virginia, was once almost completely covered with tropical forests. Between 1963 and 1983, however, politically powerful ranching families cleared much of the country's forests to graze cattle, and they exported most of the beef to the United States and western Europe.

Despite such widespread forest loss (which continues today), tiny Costa Rica is a superpower of biodiversity, with an estimated 500,000 plant and animal species. A single park in Costa Rica is home for more bird species than all of North America.

In the mid-1970s, Costa Rica established a system of national parks and conservation reserves that by 2000 included 12% of its land (6% of it in reserves for indigenous peoples). As a result, Costa Rica now has larger proportion of land devoted to biodiversity conservation than any other country.

The country's parks and reserves are consolidated into eight *megareserves* designed to sustain about 80% of the country's biodiversity (Figure 23-23). Each reserve contains a protected inner core surrounded by buffer zones that local and indigenous people use for sustainable logging, food

growing, cattle grazing, hunting, fishing, and ecotourism. The country's largest megareserve, La Amistad Biosphere Reserve (Figure 23-23), includes three national parks, a core biological reserve, two forest reserves, and five Indian reservations.

One reason for this accomplishment in biodiversity protection was the establishment of the Organization of Tropical Studies in 1963. It is a consortium of more than 50 U.S. and Costa Rican universities with the goal of promoting research and education in tropical ecology. The resulting infusion of several thousand scientists has helped Costa Ricans appreciate their country's great biodiversity. It also led to the establishment in 1989 of the National Biodiversity Institute (INBio), a private nonprofit organization set up to survey and catalog

the country's biodiversity. This biodiversity conservation strategy has paid off. Today, the $1 billion a year tourism business (almost two-thirds of it from ecotourists) is the country's largest source of income.

However, legal and illegal deforestation threatens this plan because of the country's population growth, poverty (which affects 10% of its people), and lack of government regulations. Currently, this small country still loses roughly 400 square kilometers (150 square miles) of primary forest per year—four times the rate of loss in Brazil.

In addition, without careful government control, the 1 million tourists visiting Costa Rica each year can degrade some of the protected areas and lead to hotels, resorts, and other potentially harmful forms of development.

Critical Thinking

Why do you think Costa Rica has been able to set aside a much larger percentage of its land for national parks and nature reserves than any other country? Do you agree or disagree that such a plan should be implemented in the country where you live?

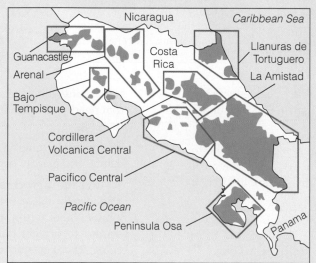

Figure 23-23 Costa Rica has consolidated its parks and reserves into eight *megareserves* designed to sustain about 80% of the country's rich biodiversity.

(1) vital parts of the earth's natural resources that sustains all life and economies (Figure 4-36, p. 99) and (2) centers of evolution.

What Ecological Principles and Goals Should Be Used to Establish and Manage Nature Reserves?

There is general agreement that reserves should be selected, designed, and managed based on three major ecological principles:

■ Ecosystems are rarely at a stable point (the "balance of nature" concept) and thus cannot be locked up and protected from human disturbances. Instead,

they are mostly in an ever-changing *nonequilibrium state* as a result of internal and external disturbances caused by natural processes and human activities (p. 192).

■ Ecosystems and communities that experience fairly frequent but moderate disturbances have the greatest diversity of species, known as the *intermediate disturbance hypothesis* (p. 191 and Figure 8-18, p. 192).

■ Most reserves should be viewed as *habitat islands* surrounded by a sea of developed and fragmented land. As a result, their species diversity and ecological

The Nature Conservancy

Private groups play an important role in establishing wildlife refuges and other protected areas to protect biological diversity. For example, since 1951 the Nature Conservancy has preserved more than 44,500 square kilometers (17,200 square miles) of vital wildlife habitats in the United States and 243,000 square kilometers (94,000 square miles) throughout Canada, Latin America, the Caribbean, Asia, and the Pacific Ocean.

The Nature Conservancy has more than 1 million members throughout the world. It has one of the lowest overhead rates of any nonprofit organization, and 85% of all contributions go directly to its conservation programs.

The Nature Conservancy began in 1951 when an association of professional ecologists wanted to use their scientific knowledge to conserve natural areas. Since then, this science-based organization has used the most sophisticated scientific knowledge available to identify and rank sites that are unique and ecologically significant and whose biodiversity or existence is threatened by development or other human activities (Figure 23-25).

Once sites are identified, a variety of techniques are used to see that they receive legal protection. Then science-based management plans are drawn up to maintain or restore the ecological health of each site and to provide long-term stewardship.

This conservation organization uses private and corporate donations to maintain a fund of more than $160 million, which can be used to buy ecologically important pieces of land or wetlands threatened by development when no other option is available. Sometimes this land is then sold or donated to federal or state agencies.

If it cannot buy land for habitat protection, the Conservancy helps landowners obtain tax benefits in exchange for accepting legal restrictions or conservation easements preventing development. Other techniques include long-term management agreements and debt-for-nature swaps. Landowners have also received sizable tax deductions by donating their land to the Nature Conservancy in exchange for lifetime occupancy rights.

Through such efforts, this organization has created the world's largest system of private natural areas and wildlife sanctuaries, using the guiding principle of land conservation through private action.

Critical Thinking

Do you favor this private-group approach to protecting biodiversity over the government approach to protecting public lands and endangered species? Explain. Would you be in favor of selling or giving some public lands to private groups such as the Nature Conservancy? Explain.

---

functioning can be evaluated to some degree by using the *theory of island biogeography* (Figure 8-6, p. 177, and Connections, p. 618).

The overall goal of nature reserves should be to sustain nature's overall biodiversity and ecological processes. This involves establishing a combination of:

- Large reserves needed to help sustain all of the world's major types of terrestrial ecosystems (Figure 6-16, p. 131).

- Reserves set aside to help protect particular species or groups of species, especially those that are endangered or threatened (Chapter 22).

- Reserves that that have a high degree of biodiversity or that contain many endemic species unique to a particular area.

There is general agreement that *large reserves:*

- Protect a greater number of species, based on the *species-area curve* (Figure 22-9, p. 558).

- Provide a variety of habitats and niches.

- Offer significant protection for species in the core of the reserve by minimizing the area of outside edges exposed to natural disturbances (such as fires and hur-

ricanes), invading species, and human disturbances from nearby communities and developed areas.

- Are the only way to maintain viable populations of large, wide-ranging species such as panthers, elephants, and grizzly bears.

In reality, few countries are physically, politically, or financially able to set aside and protect large reserves. Because there will not be enough money or political support to protect most of the world's biodiversity, ecologists suggest using two approaches:

- Focusing international efforts on establishing a variety of reserves in the world's most biodiverse countries (Figure 23-24) and threatened species-rich areas within such countries. This is a *prevention strategy* designed to reduce the future loss of biodiversity.

- An *emergency action* strategy that identifies and quickly protects *biodiversity hot spots* (Figure 23-25). They are areas that are rich in endemic plant and animal species found nowhere else and that are in great danger of extinction or serious ecological disruption.

Ecologist Norman Myers (Guest Essay, p. 117) proposed the hot spot approach in 1988. In 2000, he and four other ecologists at Conservation International identi-

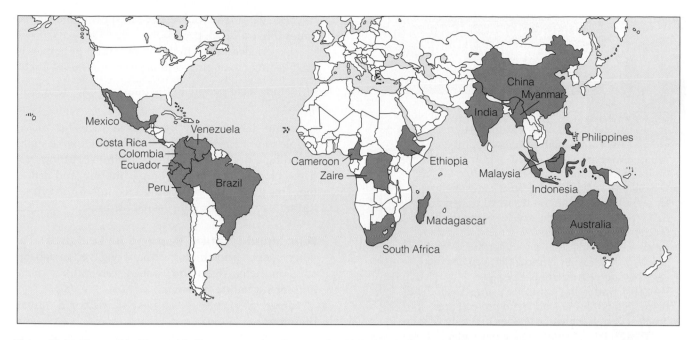

Figure 23-24 The earth's 17 most biodiverse countries. Conservation biologists believe that efforts to preserve repositories of biodiversity as protected wilderness areas and other nature reserves should be concentrated in these 17 countries, which contain at least 60% of the world's known terrestrial species. Protecting the world's biodiversity will take considerable financial and scientific help from developed countries. (Data from Conservation International and World Wildlife Fund)

fied 25 highly vulnerable biodiversity hot spots (15 of them in the tropics) that they believe should be primary focus of international efforts to protect diversity (Figure 23-25).

The researchers defined a *hot spot* as an ecologically distinct bioregion that **(1)** must contain at least 0.5% (or 1,500) of the world's known vascular plants (those with

"interior plumbing") as endemic species found nowhere else and **(2)** has suffered a loss of at least 70% of its original vegetation and thus is under severe pressure. They found that more than a third of the planet's known terrestrial plant and animal species are found only in these 25 hot-spot ecosystems, which cover a mere 1.4% of the earth's land area, made up mostly of tropical forests.

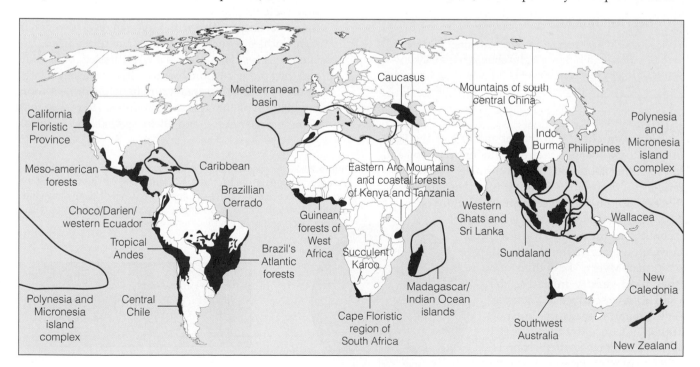

Figure 23-25 Twenty-five hot spots identified by ecologists as being important but endangered centers of biodiversity that contain a large number of endemic plant and animal species found nowhere else. (Data from Conservation International)

CONNECTIONS

Using Island Biogeography Theory to Protect Mainland Communities and Species

Ecologists and conservation biologists are now applying the species equilibrium model or theory of island biogeography (Figure 8-6, p. 177) to communities on the mainland. Most remaining wildlife sanctuaries on land are *habitat islands* surrounded by an inhospitable sea of disturbed or degraded habitat. According to the species equilibrium model, the species diversity in these terrestrial patches or islands should be determined by their size and their distance from other patches that serve as sources of colonist species.

Conservation biologists are using the model to help them **(1)** locate (and try to protect) areas in greatest danger of losing species diversity (Figure 23-25), **(2)** estimate how large a particular nature preserve should be to prevent it from losing species, **(3)** estimate how closely a series of small wildlife reserves should be spaced to allow immigration from one preserve to another if a species in one reserve becomes locally extinct, and **(4)** estimate the size and number of protected corridors needed to connect various reserves and encourage the spread of protected species between them.

Much more research must be done to answer such questions, but progress is being made through this application of ecological theory to wildlife conservation.

Critical Thinking

How would you respond to the statement that the theory of island biogeography should not be taken too seriously because it is only a theory?

---

has also identified biodiversity hot spots in the United States (Figure 22-23, p. 575).

Experience shows that two important principles in establishing, managing, and protecting reserves are to:

- Include local people in the planning and design of a reserve.

- Create "user friendly" reserves that allow local people to use parts of a reserve or a buffer zone surrounding a reserve for sustainable timber cutting, livestock grazing, growing crops, hunting, and fishing. This gives local people a vested interest in protecting a reserve from unsustainable uses.

How Should Nature Reserves Be Designed? For years, conservation biologists have engaged in debates and conducted research to answer several key questions about nature reserves:

Question: What is the best shape for a nature preserve?

Answer: Ideally, a circular reserve is preferable to an elongated reserve because the circular shape allows better protection of the interior area that is further away from the edges of the reserve (Figure 23-26). A long, linear reserve has the most edge, and all of its interior area is close to the edge. In practice, most reserves have irregular shapes because they depend on what land is available.

Question: Is it better to have a *Single Large Or Several Small* reserves of an equal total area (known as the SLOSS debate)?

Answer: If large reserves are available, they are often preferable for the reasons discussed earlier, especially for protecting species with large ranges. However, in some locales several well-placed, medium-sized, isolated reserves **(1)** may have a wider variety of habitats and thus preserve more biodiversity than a single large reserve of the same area, **(2)** may better protect more populations of endemic species with small ranges than a single large reserve, and **(3)** are less likely to be simul-

---

The researchers said that about 40% of the land found in these hot spots is already under protection and that spending about $500 million annually over a 5-year period could go far in safeguarding these hot spots. This expenditure for protecting at least a third of the world's threatened species is equivalent to about what the world's nations spend on weapons every 8 hours. According to Norman Myers, "I can think of no other biodiversity initiative that could achieve so much at a comparatively small cost, as the hotspots strategy." Concentrating efforts to slow population growth (Section 11-3, p. 249) in most of these areas would also help. A coalition of conservation groups

Figure 23-26 A reserve with a generally circular shape offers better protection of species in its core area than a long, linear reserve of the same area. In a linear reserve, all species are fairly close to an edge and are more vulnerable to disturbances such as fire, disease, and predation and invasion by other species.

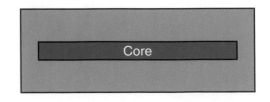

taneously devastated by a single event such as a flood, fire, disease, or invasion by a nonnative species than a single large reserve. If a comprehensive system of reserves is to be established over a large region, a mixture of both large and small reserves (Figure 23-23) may be the best way to protect a variety of species and communities against a variety of threats.

Question: Is it better to have a heterogeneous reserve with a variety of habitats or a homogeneous reserve?

Answer: A heterogeneous reserve generally is favored because nature is rarely at an equilibrium state and is continuously undergoing changes. A heterogeneous reserve **(1)** provides a variety of habitat patches (*patch dynamics*) of different sizes, shapes, and successional stages and fosters continuation of natural processes and change, **(2)** offers different species better protection against intermediate disturbances (Figure 8-18, p. 192), and **(3)** provides opportunities for species in disturbed patches to move to undisturbed patches.

Question: When several small reserves are created, should they be isolated from one another or connected by corridors?

Answer: Establishing protected habitat corridors between reserves can **(1)** help support more species, **(2)** allow migration of vertebrates that need large ranges, **(3)** allow migration of individuals and populations when environmental conditions in a reserve deteriorate, **(4)** reduce loss of genetic diversity from inbreeding by encouraging gene flow between isolated populations, **(5)** help preserve animals that must make seasonal migrations to obtain food, and **(6)** may enable some species to shift their ranges if global warming makes their current ranges uninhabitable. On the other hand, corridors can **(1)** threaten once-isolated populations by allowing the movement of pest species, disease, fire, and exotic species between reserves, **(2)** subject migrating species to greater risks of predation by natural predators and human hunters and to pollution, **(3)** do little to protect endemic species restricted to specific sites, and **(4)** be costly to acquire, protect, and manage.

Question: What should be the role of buffer zones in establishing nature reserves?

Answer: Whenever possible, it is important to use the *buffer zone concept*, in which an inner core of a reserve is protected by two buffer zones in which local people are allowed to extract resources in ways that are sustainable and do not harm the inner core, as is done by the United Nations in establishing a global network of biosphere reserves (Solutions, p. 620).

What Is Gap Analysis? In recent years, a scientific method *called gap analysis* has been developed to determine how adequately native plant and animal species and natural communities are protected by existing networks of nature preserves. Species and communities not adequately represented in existing reserves constitute *conservation gaps*.

Gap analysis involves several steps:

- Combine existing vegetation data and remote-sensing (satellite) data to produce computerized GIS maps (Figure 4-34, p. 98) of the topography, vegetation, hydrology, land ownership, and existing or proposed nature reserves of a region.

- Use bioinformation databases (Spotlight, p. 574) and other sources of data to show the known or estimated geographic distribution of the region's plant and animal species.

- Superimpose the species distribution maps onto the GIS maps of existing vegetation and protected areas to determine unprotected areas or gaps with very high species diversity or pockets of rare species that are not protected.

- Use this information to close the gaps by **(1)** establishing new nature reserves, **(2)** expanding the size and shape of existing reserves, **(3)** adding corridors, or **(4)** persuading private landowners or developers to modify their existing or proposed patterns of land use.

Gap analysis was developed in the mid-1980s by biologist Michael Scott. He used it to compare the geographic distribution of several endangered forest bird species on Hawaii's different islands with the distribution of existing nature reserves (Figure 23-28). After his analysis found that none of Hawaii's national parks were located in areas with significant populations of endangered forest birds, negotiations took place to have most of these sites classified as national park areas.

The Biological Resources Division of the U.S. Geological Survey has initiated a National Gap Analysis Program (GAP) to provide broad geographic information on the status of biodiversity protection throughout the United States.

What Is Wilderness? Protecting biodiversity means protecting *wildness*. One way to do this is to protect undeveloped lands from human exploitation by legally setting them aside as wilderness. According to the U.S. Wilderness Act of 1964, wilderness consists of areas "of undeveloped land affected primarily by the forces of nature, where man is a visitor who does not remain." U.S. president Theodore Roosevelt summarized what we should do with wilderness: "Leave it as it is. You cannot improve it."

The U.S. Wilderness Society estimates that a wilderness area should contain at least 4,000 square kilometers (1,500 square miles); otherwise it can be affected by air, water, and noise pollution from nearby human activities.

Biosphere Reserves

In 1971, the UN Educational, Scientific, and Cultural Organization (UNESCO) created the Man and the Biosphere (MAB) Programme to improve the relationship between people and the environment. The program proposed that at least one (and ideally five or more) *biosphere reserves* be set up in each of the earth's 193 biogeographical zones, such as the one in Costa Rica (Figure 23-23).

Ideally, each reserve should be large enough to prevent gradual species loss and should combine both conservation and sustainable use of natural resources. Today there are more than 350 biosphere reserves in 90 countries.

Each reserve must be nominated by its national government, meet certain size requirements, and contain three zones (Figure 23-27):

■ A *core area* containing an important ecosystem that is legally protected by the government from all human activities except nondestructive research and monitoring.

■ A *buffer zone* that surrounds and protects the core area. In this zone, emphasis is on (1) nondestructive research, education, and recreation and (2) sustainable logging, agriculture, livestock grazing, hunting, and fishing by local people as long as such activities do not harm to the core.

■ A second *buffer* or *transition zone*, which combines conservation and more intensive but sustainable forestry, grazing, hunting, fishing, agriculture, and recreation by local people and ecotourists.

So far, most biosphere reserves fall short of the ideal (Figure 23-27), and too little funding has been provided for their protection and management. An international fund to help countries protect and manage biosphere reserves would cost about $100 million per year—about what the world's nations spend on weapons every 90 minutes.

Figure 23-27 Design of a model biosphere reserve. In traditional parks and wildlife reserves, well-defined boundaries keep people out and wildlife in. By contrast, biosphere reserves recognize people's needs for access to sustainable use of various resources in parts of the reserve.

In the figure:
Biosphere Reserve
Core area
Buffer zone 1
Buffer zone 2

Critical Thinking

Would you be willing to spend $10 a year to help establish and maintain a global network of biosphere reserves? Can you think of any disadvantages of such a system?

Why Preserve Wilderness? According to wilderness supporters, we need wild places where people can experience the beauty of nature and observe natural biological diversity and where they can enhance their mental and physical health by getting away from noise, stress, development, and large numbers of people. Wilderness preservationist John Muir advised us,

> Climb the mountains and get their good tidings. Nature's peace will flow into you as the sunshine into the trees. The winds will blow their freshness into you, and the storms their energy, while cares will drop off like autumn leaves.

Even those who never use the wilderness may want to know it is there, a feeling expressed by novelist Wallace Stegner:

> Save a piece of country...and it does not matter in the slightest that only a few people every year will go into

it. This is precisely its value....We simply need that wild country available to us, even if we never do more than drive to its edge and look in. For it can be a means of reassuring ourselves of our sanity as creatures, a part of the geography of hope.

According to renowned conservationist David Brower, "Without wilderness the world is a cage."

Recently some critics have argued that protecting wilderness for its scenic and recreational value for a small number of people is an outmoded concept that keeps some areas of the planet from being economically useful to humans.

Scientists argue that this idea is not outmoded because it fails to take into account the research that has revealed the ecological importance of wilderness areas for all people and all life. To most biologists, *the most important reason for protecting wilderness and other areas from exploitation and degradation is to preserve the biodiversity they*

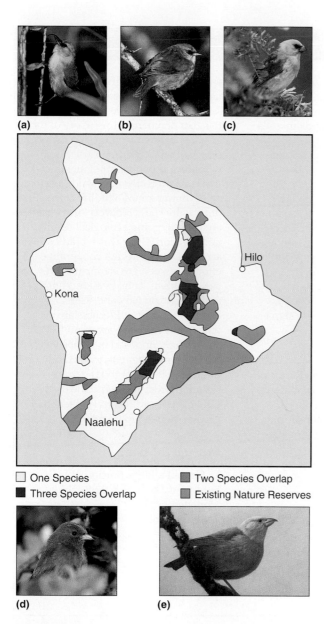

(a)   (b)   (c)

Hilo

Kona

Naalehu

☐ One Species   ■ Two Species Overlap
■ Three Species Overlap   ■ Existing Nature Reserves

(d)   (e)

Figure 23-28 Gap Analysis involves identifying how well species and communities are represented in the existing network of conservation reserves. In 1986, gap analysis was used to determine that less than 10% of the ranges of endangered Hawaiian birds on the island of Kauai were in protected reserves. Since then, the Nature Conservancy and state and federal agencies have protected several areas rich in highly endangered bird species. (Adapted from "Forest Communities of the Hawaiian Islands: their Dynamics, Ecology, and Conservation," by J. M. Scott, S. Mountainspring, F. L. Ramsey, and C. B. Kepler, *Studies in Avian Biology*, vol. 9, 1969. [All images courtesy of Jack Jeffrey except Ou courtesy of Michael Furuay.] (c) Jack Jeffrey [top, all, and bottom, left] (c) Michael Furuay [bottom right])

contribute as a vital part of the earth's natural capital (Figure 4-36, p. 99) *and as centers for evolution in response to mostly unpredictable changes in environmental conditions.*

Wilderness areas **(1)** provide mostly undisturbed habitats for wild plants and animals, especially those that need a large range, **(2)** help protect diverse biomes from damage, and **(3)** provide a natural laboratory in which we can discover more about how nature works in the few remaining areas not seriously disturbed by human activities. In other words, wilderness is a biodiversity and wildness bank and an ecoinsurance policy.

Some analysts also believe that wilderness should be preserved because the wild species it contains have an ethical right to **(1)** exist (or struggle to exist) and **(2)** play their roles in the earth's ongoing saga of biological evolution and ecological processes, without human interference (Section 28-2, p. 742).

Most ecologists and conservation biologists believe that we need to **(1)** protect all the world's remaining wilderness areas, focusing first on the most endangered hot spots in species-rich countries (Figures 23-24 and 23-25), **(2)** emphasize protection of grasslands, lowland forests, and wetlands that are largely absent from the current system, and **(3)** allow some degraded areas to return to a more wild state. They see this as an urgent priority because most of the earth's remaining wild places are being fragmented, invaded, and degraded by human activities. Wilderness advocates also call for more wilderness areas to be protected in the United States (Case Study, p. 622).

How Should Wilderness Be Managed? To protect the most popular areas from damage, wilderness managers increasingly designate sites where camping is allowed and limit the number of people using these sites at any one time. Managers have also increased the number of wilderness rangers to patrol vulnerable areas, and they have enlisted volunteers to pick up trash discarded by thoughtless users who do not follow the Leave No Trace (LNT) wilderness ethic.

Environmental historian and wilderness expert Roderick Nash suggests that wilderness areas be divided into three categories:

- Easily accessible, popular areas that would be intensively managed and have trails, bridges, hiker's huts, outhouses, assigned campsites, and extensive ranger patrols.

- Large, remote wilderness areas that would be used only by people who get a permit by demonstrating their wilderness skills.

- Biologically unique areas that would be left undisturbed as gene pools of plant and animal species, with no human entry allowed.

What Is Adaptive Ecological Management? Ideally, the major management goals of a nature reserve should be to:

- Maintain its variety of native ecosystem types.

- Sustain viable populations of its major native species and prevent or control populations of invading nonnative species.

The U.S. National Wilderness Preservation System

In the United States, preservationists have been trying to save wild areas from development since 1900. On the whole, they have fought a losing battle. Not until 1964 did Congress pass the Wilderness Act, which allowed the government to protect undeveloped tracts of public land from development as part of the National Wilderness Preservation System.

The *good news* is that there was a 10-fold increase in the area of protected wilderness in the United States between 1970 and 2000. The *bad news* is that (1) only about 4.6% of U.S. land is protected as wilderness, with almost three-fourths of it in Alaska, and (2) only 1.8% of the land area of the lower 48 states is protected, most of it in the West. According to a 1999 study by the World Conservation Union (IUCN), the United States ranks 42nd among nations in terms of terrestrial area protected as wilderness, and Canada is in 36th place.

Of the 413 wilderness areas in the lower 48 states, only four are larger than 4,000 square kilometers. Furthermore, the present wilderness preservation system includes only 81 of the country's 233 distinct ecosystems. Like the national parks, most wilderness areas in the lower 48 states are habitat islands in a sea of development.

Almost 400,000 square kilometers (150,000 square miles) in scattered blocks of public lands could qualify for designation as wilderness, about 60% of it in the national forests. In 1999, President Clinton acted to ban road building and logging in 22 million hectares (55 million acres) of roadless but unprotected areas in the national forests. This would keep these areas eligible for possible designation by Congress as part of the National Wilderness System.

Some wilderness advocates also urge that *wilderness recovery areas* be created by (1) closing and obliterating nonessential roads in large areas of public lands, (2) restoring wildlife habitats, (3) allowing natural fires to burn, and (4) reintroducing species that have been driven from such areas.

Some ecologists and conservation biologists have proposed *The Wildlands Project* (TWP) to restore a network of protected wildlands and ecosystems throughout as much of the United States as possible. They believe that current nature reserves are too small, scattered, and threatened to protect biodiversity and natural ecological processes.

This science-based approach has the goal of reconnecting many of the existing fragmented and threatened pieces of wild nature for use by wildlife and people in regional networks of protected core wilderness areas, buffer zones, wildlife corridors, wilderness recovery areas, and areas undergoing ecological restoration (Section 23-7). This would be accomplished through cooperative efforts between government agencies, scientists, conservation groups, and private owners and users.

Expansion of wilderness areas, wilderness recovery areas, and TWP are strongly opposed by timber, mining, ranching, energy, and other interests who want to (1) extract resources from these and most other public lands or (2) convert them to private ownership (p. 589).

Critical Thinking

1. Should a large portion of U.S. Forest Service and Bureau of Land Management land be reclassified as wilderness recovery areas, closed off for many uses, and allowed to undergo natural restoration as wildlife habitat? Explain.

2. Do you support or oppose TWP? Explain.

- Sustain its essential ecological processes such as energy flow, nutrient cycling, and succession.

- Maintain the evolutionary potential of its species and ecosystems.

- Allow sustainable human use of all or part of its area in ways that do not harm its functioning and long-term sustainability.

In the real world, these goals are difficult to achieve because (1) nature reserves are affected by mostly unpredictable biological, cultural, economic, and political changes, and (2) their size, shape, and biological makeup often are determined by political, legal, and economic factors that depend on land ownership and conflicting public demands rather than by ecological principles and considerations.

One way to deal with these uncertainties and conflicts is through *adaptive ecosystem management* that:

- Integrates ecological, economic, and social principles with the goal of restoring and maintaining the sustainability and biological diversity of ecosystems while supporting sustainable economies and communities.

- Seeks ways to get government agencies, private conservation organizations, scientists, business interests, and private landowners to cooperate and reach a consensus on how to achieve common conservation objectives without the use of top-down, chain-of-command decision making.

- Views all decisions and strategies as scientific and social experiments and uses failures as opportunities for learning and improvement.

- Emphasizes continual information gathering, monitoring, reassessment, flexibility, adaptation, and innovation in the face of uncertainty and usually unpredictable change.

Integrating different perspectives and getting people with diverse and sometimes conflicting interests to cooperate in achieving ecological and economic sustainability is not easy and takes time, effort, and patience. However, experience shows that the human and ecological rewards from such a process are worth the effort.

23-7 ECOLOGICAL RESTORATION

How Can We Rehabilitate and Restore Damaged Ecosystems? The *bad news* is that almost every place on the earth has been degraded at least to some degree by human activities. The *good news* is that numerous efforts have shown that much of the environmental damage we have inflicted on nature is at least partially reversible through ecological restoration.

Ecological restoration involves deliberately altering a degraded habitat or ecosystem to restore as much of its ecological structure and function as possible. **Restoration ecology** is the research and scientific study devoted to restoring, repairing, and reconstructing damaged ecosystems. Restoration ecology is a form of applied ecology that makes important contributions to the science of ecology by **(1)** testing how well we understand the structure and function of ecosystems, **(2)** providing an important way of learning more about ecosystems, and **(3)** giving us ways to test various ecological theories and hypotheses.

Farmer and philosopher Wendell Berry says we should try to answer three questions in deciding whether and how to modify or rehabilitate natural ecosystems: **(1)** What is here? **(2)** What will nature permit us to do here? **(3)** What will nature help us do here?

By studying how natural ecosystems recover, scientists are learning how to speed up repair operations by using several approaches:

- *Restoration* involves trying to return a particular degraded habitat or ecosystem to a condition as similar as possible to its predegraded state. Three common difficulties are **(1)** lack of knowledge about the previous composition of a system, **(2)** changes in climate, soil, and species composition that may make it impossible to restore an area to an earlier state, and **(3)** attempting to deal with a moving target because ecosystems are always undergoing change. In many cases, restoration involves creating a different but ideally more sustainable ecosystem by increasing biodiversity and restoring degraded ecological processes.

- *Rehabilitation* involves any attempt to restore at least some of a degraded system's natural species and ecosystem functions. Examples include **(1)** removing pollutants and **(2)** replanting areas such as mining sites, landfills, and clear-cut forests to reduce soil erosion.

- *Replacement* involves replacing a degraded ecosystem with another type of ecosystem. For example, a degraded forest may be replaced with a productive pasture or tree farm.

- *Creating artificial ecosystems* involves using ecological principles to create human-designed ecosystems for specific purposes. Examples are the ecological wastewater treatment systems developed by John Todd (p. 476) and the artificial wetlands developed to treat sewage in Arcata, California (Solutions, p. 496).

The basic steps in all forms of ecological restoration and rehabilitation are as follows:

- Identify what caused the degradation (such as pollution, farming, overgrazing, mining, or invading species), eliminate or sharply reduce these factors, and protect the area from further degradation. For example, **(1)** toxic soil pollutants can be removed, **(2)** nutrients can be added to depleted soil, **(3)** new topsoil can be added, and **(4)** disruptive nonnative species can be eliminated.

- Determine realistic goals and measures of success.

- Develop methods for achieving the goals and implement them.

- Monitor restoration efforts, assess success, and use adaptive ecosystem management to modify strategies as needed.

Methods commonly used in ecological restoration include:

- Eliminating disruptive nonnative species (p. 569) and reintroducing species (especially keystone species, p. 173) to restore natural ecological processes. For example, on its large tall-grass prairie preserve in Oklahoma, the Nature Conservancy (Solutions, p. 616) has reintroduced 1,200 bison as a keystone species and greatly reduced cattle grazing, with the bison herd eventually reaching 3,200. In Hawaii, feral pigs that eat native birds and root out native vegetation are being hunted and killed or captured and removed. Similarly, feral goats and rats on the Galapagos Islands are being killed or trapped to help preserve native plant species.

- Trying to hold an ecosystem at a particular desirable stage of ecological succession. For example, controlled burning can be used to eliminate nonnative grass and tree species, or trees can be cut to restore grassland to an early successional stage.

- Speeding up natural ecological succession (Figure 8-16, p. 189). Examples include planting young trees in clear-cut forest areas and preventing fires that can

Ecological Restoration of a Tropical Dry Forest in Costa Rica

Costa Rica is the site of one of the world's largest *ecological restoration* projects. In the lowlands of the country's Guanacaste National Park (Figure 23-23), a small tropical dry deciduous forest that has been burned, degraded, and fragmented by large-scale conversion to cattle ranches and farms is being restored and relinked to the rain forest on adjacent mountain slopes. The goal is to eliminate damaging nonnative grass and cattle and reestablish a tropical dry forest ecosystem over the next 100–300 years.

The keys to restoring the resulting grasslands to tropical dry forest are to (1) exclude fires and cattle grazing from much of the park and (2) enhance seed dispersal from remaining small patches of native woodland.

Fire prevention methods include (1) frequent cutting and some cattle grazing of nonnative grass species, which fuel fires and prevent the regeneration of native tree and other plant species, (2) closing unnecessary roads, and (3) limiting camping, hunting, and other activities that could lead to accidental fires to park areas where fires can be controlled quickly.

Native tree restoration is being promoted by (1) feeding horses meal containing seeds of important tree species and allowing the horses to wander in surrounding pasture areas and deposit undigested seeds in nutrient-rich piles of manure and (2) hiring local people to transplant seedlings to areas not near natural woodland remnants. Once the seedlings mature, they become attractive foraging and nesting sites for birds that disperse more tree seeds in their droppings.

The strategy seems to be working. Areas of the park that were covered with monoculture expanses of nonnative grasses 10–15 years ago are now healthy secondary growth dry forests of native species. This project illustrates two lessons about restoration: (1) The causes of degradation must be known, and (2) enough remnants of native plant species must be present for seed dispersal and regrowth.

Daniel Janzen, professor of biology at the University of Pennsylvania and a leader in the field of restoration ecology, has helped galvanize international support and has raised more than $10 million for this restoration project.

He recognizes that ecological restoration and protection of the park will fail unless the people in the surrounding area believe that they will benefit from such efforts. Janzen's vision is to make the nearly 40,000 people who live near the park an essential part of the restoration of the degraded forest, a concept he calls *biocultural restoration*.

By actively participating in the project, local residents reap enormous educational, economic, and environmental benefits. Local farmers have been hired to sow large areas with tree seeds and to plant seedlings started in Janzen's lab, and local grade school, high school, and university students and citizens' groups study the ecology of the park and go on field trips to the park. The park's location near the Pan American Highway makes it an ideal area for ecotourism that stimulates the local economy.

The project also serves as a training ground in tropical forest restoration for scientists from all over the world. Research scientists working on the project give guest classroom lectures and lead some of the field trips.

Janzen recognizes that in a few decades today's children will be running the park and the local political system. If they understand the importance of their environment, they are more likely to protect and sustain its biological resources. He believes that education, awareness, and involvement—not guards and fences—are the best ways to restore degraded ecosystems and to protect largely intact ecosystems from unsustainable use.

---

set back succession in areas such as degraded tropical seasonal forests (Individuals Matter, above).

■ Using *natural restoration* by identifying and protecting key plant species and allowing natural ecological succession (Figure 8-16, p. 189) to proceed. Shenandoah National Park in Virginia is an example of fairly rapid recovery from human abuse, mostly through protection and natural restoration. Before the area became a national park in 1935, much of the land had been grazed, mined, logged, and burned for over 200 years. Today, most of this area is a diverse temperate forest dominated by oak and hickory trees (Figure 8-16, p. 189).

Repairing ecosystems requires that a pool of suitable existing plant species and soil organisms be available. For example, recovery of the Shenandoah National Park took place mostly because enough residual seeds (mostly underground) and patches of uncut diverse segments were left for natural regeneration to take place.

Thus, an important step in restoration efforts is to identify the residual native species found above and below the ground of the site or in nearby areas. This is why it is so important to preserve biodiversity throughout the world and to protect sustainable patches of ecosystems that harbor species that can recolonize or be transplanted to disturbed ecosystems.

Private enterprise is getting into ecological restoration. In May 2000, an Australian firm called Earth Sanctuaries, Ltd., was listed on the Australian Stock Exchange. The firm buys degraded land, restores it, and earns income from ecotourism and consulting on ecosystem assessment and ecological restoration.

Case Study: Tall-Grass Prairie Restoration in Illinois In North America, many small parcels of former agricultural land have been restored as prairies. Prairies are ideal subjects for ecological restoration because **(1)** many residual or transplanted native plant species can be established within a few years, **(2)** the technology involved is similar to that in gardening and agriculture, and **(3)** the process is well suited for volunteer labor needed to plant native species and weed out invading species until the natural species can take over.

One of the world's largest tall-grass prairie restoration projects is being carried out at the Fermi National Accelerator Laboratory in Batavia, Illinois, by scientists Ray Schulenberg and Robert Betz. They found remnants of virgin Illinois prairie in old cemeteries, on embankments, and on other patches of land. In 1972, they transplanted these remnants by hand to a 4-hectare (10-acre) patch at the Fermi Laboratory site.

Each year since then, volunteers have carefully prepared more land, sowed it with native prairie plants, weeded it by hand, and used controlled burning to maintain established communities. Today more than 180 hectares (445 acres) have been restored with prairie plants. New native species are introduced each year, with the goal of eventually establishing the 150–200 species that once flourished on the site.

Is Ecological Restoration the Best Approach? Some analysts worry that environmental restoration could encourage continuing environmental destruction and degradation by suggesting that any ecological harm we do can be undone. However, ecologists point out that preventing ecosystem damage in the first place is cheaper and more effective than any form of at best partial ecological restoration.

Another concern is government policies that allow developers to destroy one ecosystem or wetland if they protect, restore, or "create" a similar one of roughly the same size. This trade-off or *mitigation* approach is preferable to wanton destruction of ecosystems. However, according to ecological restoration expert John Berger, "The purpose of ecological restoration is to repair previous damage, not legitimize further destruction."

Because so much of the earth's natural ecosystems have been disturbed and degraded by human activities (Figure 1-4, p. 8), the question is not whether to restore degraded ecosystems. Instead, it is how close we can come to restoring as much of the biodiversity and ecosystem processes in degraded systems as possible through the science and craft of ecological restoration. Some actions you can take to help sustain the earth's biodiversity are listed in Appendix 6.

We abuse land because we regard it as a commodity belonging to us. When we see land as a community to which we belong, we may begin to use it with love and respect.

ALDO LEOPOLD

REVIEW QUESTIONS

1. Define the boldfaced terms in this chapter.

2. List the pros and cons of reintroducing the gray wolf as a keystone predator species in the Yellowstone ecosystem.

3. How is land used in the world and in the United States?

4. What percentage of the land in the United States is owned and managed as public lands by the federal government? What are the five major types of public lands in the United States? What are the major uses allowed on each type?

5. List four principles that most biologists and some economists believe should govern the use of public land in the United States and compare this with how users of mineral and other resources believe these lands should be used, owned, and managed. List the nine goals of the anti-environmental Wise-Use movement in terms of the use of public lands.

6. Describe and list the pros and cons of **(a)** the *takings and property rights* movement in the United States and **(b)** the issue of *regulatory takings*.

7. List the pros and cons of providing government subsidies to ranchers holding permits to graze livestock on public lands in the United States. What are the three major positions on how public rangelands should be used? Summarize ways suggested by some ranchers, environmentalists, and federal range managers to use public rangeland more sustainably.

8. List **(a)** five important ecological services provided by forests and **(b)** five important economic benefits of forests.

9. Distinguish between *old-growth forests, second-growth forests,* and *tree farms*.

10. Under what conditions is a forest a renewable resource?

11. Summarize the state of forest growth and loss in the world's temperate and tropical forests. What are the pros and cons of deforestation?

12. Compare tree farms with monoculture food crops. What are the advantages and disadvantages of tree farms?

13. Describe the *rotation cycle* for harvesting and managing a forest. Distinguish between *even-aged* and *uneven-aged management* of a forest and list the advantages and disadvantages of each type.

14. Describe five major ways trees are harvested and list the advantages and disadvantages of each type.

15. List seven ways suggested by biodiversity researchers, many foresters, and conservation biologists to use forests more sustainably. Describe how industrial forestry based on tree plantations could lead to more sustainable use of forests.

16. List seven ways to help protect forests from diseases and insects.

17. Distinguish between *surface, crown,* and *ground* forest fires. What are the benefits of fire for some plant and animal species? List the pros and cons of **(a)** allowing fires in national parks to burn themselves out instead of fighting them and **(b)** deliberately setting controlled ground fires in some national parks and national forests to reduce flammable ground litter and reduce the chances of harmful crown fires.

18. Summarize the threats to forests from **(a)** air pollution and **(b)** global warming.

19. Describe the general extent and condition of forests in the United States today.

20. How are national forests in the United States used? What are the principles of *sustained yield* and *multiple use* that are supposed to guide the use of national forests in the United States? What are the pros and cons of giving government subsidies to companies cutting timber in national forests? List the pros and cons of making and **(a)** timber cutting and **(b)** recreation the primary use of national forests.

21. Describe and list the advantages of the program for certifying that timber has been grown sustainably.

22. List nine reforms in U.S. Forest Service policy that biodiversity protection advocates and some foresters believe will lead to more sustainable use of national forests in the United States.

23. List three ways to reduce the pressure to harvest trees. List the pros and cons of making more paper from tree-free sources such as rice straw and kenaf fibers.

24. List two reasons why biologists believe that cutting and degrading most of the world's remaining tropical forests is a serious global environmental problem. List major *direct use, indirect use, option, existence,* and *ethical* values of tropical forests. How rapidly are tropical forests being cleared and degraded? Describe the many uses of the tropical *neem tree* and list the pros and cons of widespread planting of neem trees.

25. Describe the importance of Madagascar in terms of biodiversity and what is happening to its biodiversity.

26. Why are some people concerned about the cultural extinction of indigenous peoples in tropical forests and in other biomes?

27. List three underlying causes and seven direct causes of tropical deforestation and degradation.

28. List 15 ways to reduce tropical deforestation and degradation. Describe **(a)** ways to achieve more sustainable farming and logging in tropical forests and **(b)** Kenya's Green Belt movement.

29. What is the *fuelwood crisis*, and how serious is this problem? List three ways to deal with the fuelwood crisis.

30. What are the major threats to national parks in the United States and in other countries? List 12 ways to improve management of national parks in the United States. Describe efforts by Costa Rica to establish reserves to protect its biodiversity.

31. List the pros and cons of protecting 10–12% of the earth's land surface as nature reserves. Describe the role of the Nature Conservancy in establishing nature reserves in the United States and other parts of the world.

32. List three scientific principles and two social principles for guiding the selection, design, and management of nature reserves. Describe the need to establish nature reserves in *megadiversity countries* and *biodiversity hot spots*.

33. What is the best shape for a nature reserve? List the pros and cons of establishing large reserves or several small reserves with the same total area. Why is it usually better to have a heterogeneous reserve with a variety of habitats than a homogeneous reserve? List the pros and cons of establishing corridors between nature reserves. How can the theory of island biogeography be used to help address some of these problems?

34. What is a *biosphere reserve*? What is *gap analysis*, and why is it an important tool for protecting biodiversity?

35. What is *wilderness*, and why is it important? List the pros and cons of protecting more wilderness.

36. List five goals of *adaptive ecosystem management* and four strategies for implementing these goals.

37. Distinguish between *ecological restoration* and *restoration ecology*. Distinguish between ecosystem *restoration, rehabilitation,* and *replacement*. List four basic steps in carrying out ecological restoration or rehabilitation and four techniques for implementing these steps.

38. Describe efforts to restore **(a)** a tall-grass prairie in Illinois and **(b)** degraded tropical dry forest in Costa Rica. What is *biocultural restoration*?

39. What are two concerns some people have about ecological restoration?

CRITICAL THINKING

1. Do you agree or disagree with the program to reintroduce populations of the gray wolf in the Yellowstone ecosystem? Explain. Could the money be better spent on other wildlife programs? If so, what programs would you suggest? Should grizzly bears be reintroduced to the Yellowstone or other ecosystem in the western United States where they were once found? Explain.

2. What specific restrictions, if any, would you impose on how owners can use private land from an environmental or ecological standpoint?

3. Explain why you agree or disagree with (a) the four principles biologists and some economists have suggested for using public land in the United States (pp. 588–89) and (b) the nine suggestions made by developers and resource extractors in the Wise-Use movement for managing and using U.S. public land (p. 589).

4. Should individuals and corporations be compensated financially if they are (a) prevented from using land they own or (b) from extracting resources from land they own because the land is classified by state or federal government as protected wetlands or habitats for endangered or threatened wildlife species? Explain.

5. Explain why you agree or disagree with each of the proposals listed on pp. 590–91 for providing more sustainable use of public rangeland in the United States. What might be some drawbacks of implementing these proposals?

6. Should off-road vehicles and snowmobiles be banned from all public lands? Explain.

7. Should private companies that harvest timber from U.S. national forests continue to be subsidized by federal payments for reforestation and for building and maintaining access roads? Explain.

8. Suppose you are a logger in the Pacific Northwest whose job is being threatened because of a decrease in the cutting of timber from old-growth forests in nearby national forests. Would you support or oppose a ban on logging in remaining old-growth forests in the national forests? Explain.

9. What are the beneficial and harmful environmental effects of decreasing the rotation time for harvesting a forest from 60–100 years to 10 years?

10. In the early 1990s, Miguel Sanchez, a subsistence farmer in Costa Rica, was offered $600,000 by a hotel developer for a piece of land that he and his family had been using sustainably for many years. The land contained an old-growth rain forest and a black sand beach in an area that was being developed rapidly. Sanchez refused the offer. What would you have done if you were a poor subsistence farmer in Miguel Sanchez's position?

11. Should the United States (or the country where you live) meet most of its demand for wood and wood products by growing genetically improved trees on carefully managed tree plantations? Explain. If you agree, what regulations and safeguards would you impose to help ensure that such an approach would be sustainable and have a low environmental impact?

12. Explain why you agree or disagree with each of the proposals listed on p. 609-610 for protecting the world's tropical forests. Should developed countries provide most of the money to preserve remaining tropical forests in developing countries? Explain.

13. What five actions can you take to help preserve some of the world's tropical forests? Which, if any, of these actions do you plan to carry out?

14. Explain why you agree or disagree with each of the proposals listed on p. 614 concerning the U.S. national park system.

15. An ecologist advises that a controlled burn of vegetation in a heavily forested park should be done to help reduce the chance of serious damage from a major forest fire. Many local citizens oppose the burn. If you were an elected official, would you approve or disapprove the proposed controlled burn? If you approved the burn, what restrictions would you impose for ecological and human health reasons?

16. Should more wilderness areas be preserved in the United States, especially in the lower 48 states (or in the country where you live)? Explain. What might be some drawbacks of doing this?

17. List ecological, economic, political, and social reasons that might explain why (a) natural grasslands are rarely protected in nature reserves and (b) many ecological restoration projects are devoted to restoring degraded grasslands.

18. What do you believe is (a) the easiest type of natural ecosystem to restore and (b) the most difficult type to restore? Explain.

19. If ecosystems are undergoing constant change, why should we be concerned about (a) establishing and protecting nature reserves and (b) ecological restoration?

20. Congratulations. You have just been put in charge of the world. List the five most important features of your policies for using and managing of (a) forests, (b) parks, and (c) wilderness and other protected biodiversity reserves.

PROJECTS

1. Obtain a topographic map of the region where you live and use it to identify local, state, and federally owned lands in the form of parks, rangeland, forests, and wilderness areas. Identify the government agency responsible for managing each of these areas and evaluate how well these agencies are preserving the natural resources on this land on your behalf.

2. If possible, try to visit (a) a diverse old-growth forest, (b) an area that has been recently clear-cut, and (c) an area that was clear-cut 5–10 years ago. Compare the biodiversity, soil erosion, and signs of rapid water runoff in each of the three areas.

3. For many decades, New Zealand has had a policy of meeting all its demand for wood and wood products by growing timber on intensively managed tree plantations. Use the library or internet to evaluate the effectiveness of this approach and its major advantages and disadvantages.

4. Evaluate timber harvesting on private and public lands in your local area. What are the most widely used harvesting methods? Try to document any harmful environmental impacts. Have the economic benefits to the community outweighed any harmful environmental effects?

5. Make a survey of the national, state, and local parks within a 97-kilometer (60-mile) radius of your local community. How widely are they used by local residents and outside visitors? What is their condition? Develop a plan to improve their management.

6. Use the library or the internet to find one example of a successful ecological restoration project not discussed in this chapter and one that failed. For each example, describe **(a)** the strategy used, **(b)** the ecological principles that were involved, and **(c)** why the project succeeded or failed.

7. Imagine that you are given the job of restoring a moderate-sized area of degraded prairie. Prepare a list of the three most important **(a)** things you would need to know about the site, **(b)** strategies you would use to restore the site, and **(c)** criteria you would use to evaluate the project's success.

8. Use the library or the internet to find bibliographic information about *Francois-Auguste-René de Chateaubriand* and *Aldo Leopold*, whose quotes appear at the beginning and end of this chapter.

9. Make a concept map of this chapter's major ideas, using the section heads and subheads and the key terms (in boldface). Look at the inside back cover and on the website for this book for information about making concept maps.

INTERNET STUDY RESOURCES AND RESOURCES FOR FURTHER READING AND RESEARCH

The website for this book contains helpful study aids and many ideas for further reading and research. Log on to:

http://www.brookscole.com/product/0534376975s

and click on the Chapter-by-Chapter area. Choose Chapter 23 and select a resource:

- "Flash Cards" allows you to test your mastery of the Terms and Concepts to Remember for this chapter.

- "Tutorial Quizzes" provides a multiple-choice practice quiz.

- "Student Guide to InfoTrac" will lead you to Critical Thinking Projects that use InfoTrac College Edition as a research tool.

- "References" lists the major books and articles consulted in writing this chapter.

- "Hypercontents" takes you to an extensive list of sites with news, research, and images related to individual sections of the chapter.

INFOTRAC COLLEGE EDITION

Improve your skills with InfoTrac College Edition, a searchable online database of articles from more than 700 periodicals. Log on to:

http://www.infotrac-college.com

or access InfoTrac through the website for this book.

Try the following articles:

Goodman, B. 2000. US Forest Service proposes ban on road construction. *BioScience* vol. 50, no. 9, p. 744. (subject guide: Forest Service)

Fordney, C. 2000. Combating the aliens: national parks control of invasive plant species. *National Parks,* Jan 2000, pp. 24–27. (keywords: National Parks, invasive plant species)

24 SUSTAINING AQUATIC BIODIVERSITY

Connections: Can Lake Victoria Be Saved?

Lake Victoria, shared by Kenya, Tanzania, and Uganda in East Africa, is the world's second largest freshwater lake and the source of the Nile River, which drains into the Mediterranean Sea. The lake has been in ecological trouble for more than two decades, and things are getting worse.

Until the early 1980s, Lake Victoria had more than 350 species of fish found nowhere else. About 80% of them were small, algae-eating fishes known as cichlids (pronounced "SIK-lids," Figure 24-3, p. 632), each with a slightly different ecological niche. These fishes were the main source of protein for the more than 30 million people living in the area surrounding the lake and provided a fishing livelihood for many local people.

Currently, only one native minnow species and two introduced fish species dominate the lake. All the remaining fish species are endangered or extinct.

Several factors played a role in this dramatic loss of aquatic biodiversity:

- A large increase in the population of the Nile perch (*Lates nilotica*, Figure 24-1, below), which was deliberately introduced into the lake in 1960 to stimulate local economies and the fishing industry, despite the protests of some biologists. The population of this large, prolific, and voracious fish exploded by preying on the cichlids and by 1985 had wiped out most of these species. With the cichlids gone, the Nile perch now feeds on tiny native shrimp and its own young. There is concern that there may be a collapse of the Nile perch population, which supports a large fishing industry that processes and exports large amounts of the fish to several European countries. The native people who depended on the cichlids for protein cannot afford the perch, and the mechanized fishing industry has put most small-scale fishers and fish vendors out of business. This has increased poverty and protein malnutrition.

- In the 1980s, the lake began experiencing frequent algal blooms and cultural eutrophication (Figure 19-5, p. 482) because of nutrient runoff from surrounding farms, deforested land, untreated sewage, and declines in the populations of the alga-eating cichlids. This greatly decreased oxygen levels in the lower depths of the lake and drove remaining native cichlids and other fish species to shallower waters, where they were more vulnerable to Nile perch and fishing nets.

- Since 1987, the nutrient-rich lake has been invaded by the water hyacinth (Figure 24-3). This rapidly growing plant now carpets large areas of the lake and (1) blocks out sunlight, (2) deprives fish and plankton of oxygen, (3) reduces the diversity of important aquatic plant species, (4) hinders the movement of small fishing boats, and (5) creates stagnant water that is the breeding ground for malaria-spreading mosquitoes and snails that host bilharzia (a human parasite that attacks the liver, lungs, and eyes). Men are vacating villages in search of jobs, often leaving behind women and children who face severe poverty, disease, and protein malnutrition.

- To make matters worse, in 2000 the Tanzanian government constructed a large gold mine next to a river that drains into the lake. The mine uses highly toxic sodium cyanide to recover the gold. If the cyanide ends up in the lake, it will devastate aquatic life, as occurred when cyanide spilled into eastern European rivers in January 2000 (Case Study, p. 328).

Some of the lake's biological diversity and fish productivity could be restored by (1) controlling rapid human population growth (more than 3% a year), nutrient runoff, overfishing, and population of the Nile perch and (2) instituting an aggressive reforestation program along the shores of the lake and the three rivers that feed it. However, restoring a eutrophic lake of this size has never been attempted.

Figure 24-1 Although the Nile perch is a fine food fish, it has played a key role in a major loss of biodiversity in East Africa's Lake Victoria. (Gary Kramer)

The coastal zone may be the single most important portion of our planet. The loss of its biodiversity may have repercussions far beyond our worst fears.

G. CARLETON RAY

This chapter addresses the following questions:

- What is aquatic biodiversity and what is its economic and ecological importance?

- How are human activities affecting aquatic biodiversity?

- How can we protect and sustain marine biodiversity?

- How can we manage and sustain the world's marine fisheries?

- How can we protect, sustain, and restore wetlands?

- How can we protect, sustain, and restore lakes, rivers, and freshwater fisheries?

24-1 THE IMPORTANCE OF AQUATIC BIODIVERSITY

What Do We Know About the Earth's Aquatic Biodiversity? Although we live on a water planet (Figure 7-4, p. 156), we know fairly little about the earth's aquatic systems. Indeed, less than 5% of the earth's global ocean has been explored and mapped with the same level of detail as the surface of the moon and Mars. According to aquatic scientists, the scientific investigation of poorly understood aquatic systems is a greatly underfunded *research frontier* whose study could result in immense ecological and economic benefits.

The world's ocean covers 71% of the planet's surface (p. 1 and Figure 7-4, p. 156), and represents more than 99% of the living space or volume of the earth's biosphere. It consists of a rich mosaic of large and small habitats that support a variety of species at different depths (Figure 7-5, p. 156, and Figure 24-2). A large variety of species is found in the coastal zone, with an array of niches for birds (Figure 8-9, p. 182) and other species at the water's edge (Figure 7-10, p. 160). The three most biologically diverse habitats found in the world's oceans are **(1)** *coral reefs* (Figure 7-1, p. 152, and Figure 7-13, p. 163), **(2)** *estuaries* (Figure 7-8, p. 158, and Figure 7-9, p. 159) and **(3)** *the deep-ocean floor*.

About 63% of roughly 250,000 known fish species are found in marine systems (about 50% in coastal waters, 12% in the deep sea, and 1% in the waters of the open ocean) and 37% in freshwater systems. However, some scientists believe that largely unexplored deep-water marine habitats could harbor at least 10 million species.

Freshwater aquatic systems also contain a variety of visible plant and animal species (Figure 24-3) and microscopic species. Many *lakes* contain a variety of species found in different layers (Figure 7-14, p. 165), although this varies with the amount of nutrients added to a lake's waters (Figure 7-15, p. 166) by natural or human causes (Figure 19-5, p. 482). As they flow from their mountain (headwater) stream to the ocean, *river systems* have a variety of ecological habitats that support different aquatic species (Figure 7-17, p. 168).

Although only 1% of the earth's surface is fresh water, these aquatic systems are home for 12% of all known animal species (including 41% of the known fish species). Most people are unaware how vulnerable freshwater environments are to environmental degradation and depletion.

What Are Some General Patterns of Marine Biodiversity? Despite the lack of knowledge about overall marine biodiversity, scientists have established the following general patterns:

- The greatest marine biodiversity occurs in coral reefs and on the deep-sea floor.

- Biodiversity is higher near coasts than in the open sea because of the greater variety of producers, habitats, and nursery areas in coastal areas.

- Biodiversity is higher in the benthic (bottom) region of the ocean than in the pelagic (surface) region because of the greater variety of habitats and food sources on the ocean bottom.

- Biodiversity increases in the open waters of the ocean in going from the north pole to the tropics, but it is not clear whether a similar pattern exists in the southern hemisphere.

- The lowest marine biodiversity probably is found in the midwaters of the open ocean (Figure 8-4, p. 176).

- Top-level predators, some of which are keystone species (p. 178), often dominate the biodiversity and ecological dynamics of many marine intertidal zones. When key predators are eliminated or their populations drop sharply, trophic cascades from these *top-down food web dynamics* can reduce system biodiversity.

- The biodiversity of coastal upwelling systems (Figure 6-4, p. 124, and Figure 6-9, p. 126) generally is governed by *bottom-up food web dynamics* in which variations at lower trophic levels and nutrient levels have effects that *cascade* to higher trophic levels.

What Is the Ecological and Economic Importance of Aquatic Biodiversity? Aquatic systems and species, like terrestrial ecosystems and species, have instrumental and intrinsic values that provide a variety of important ecological services and goods (Figure 22-11, p. 561, and p. 560). Figure 24-4, (p. 633) lists important ecological and economic services provided by marine systems.

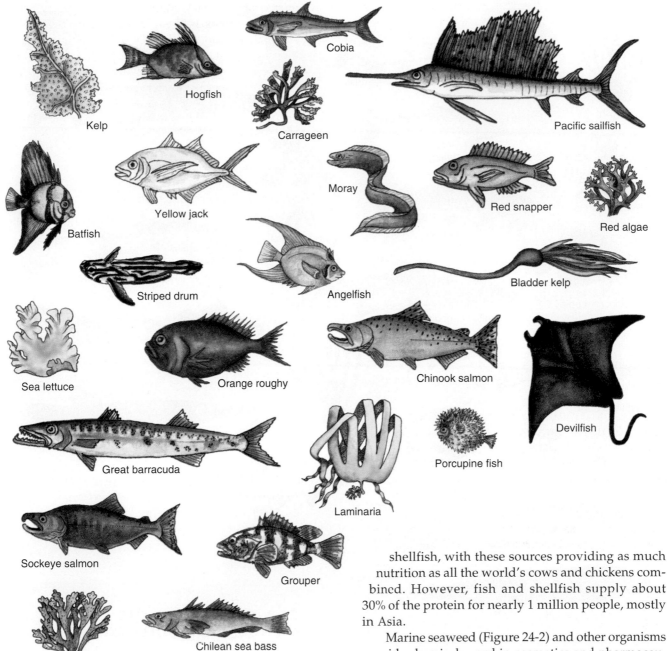

Figure 24-2 Marine biodiversity. Some ocean inhabitants.

In 1997, Robert Costanza and other investigators (Spotlight, p. 562) estimated that the value of the earth's marine resources is at least $22.5 trillion per year, compared with $12 trillion per year for land-based resources. Coastal marine systems were valued at about $12.5 trillion, the open ocean at about $8.5 trillion, and coastal wetlands, which provide many important ecological and economic services (p. 158), at about $1.5 trillion.

On average, people get about 6% of their total protein and 16% of their animal protein from fish and shellfish, with these sources providing as much nutrition as all the world's cows and chickens combined. However, fish and shellfish supply about 30% of the protein for nearly 1 million people, mostly in Asia.

Marine seaweed (Figure 24-2) and other organisms provide chemicals used in cosmetics and pharmaceuticals worth $400 million per year. Examples include:

- Chemicals from several types of algae, sea anemones (Figure 8-13b, p. 187), sponges, and mollusks that have antibiotic and anticancer properties.

- Anticancer chemicals from porcupine fish (Figure 24-2), puffer fish, and shark liver.

- Chemicals to treat hypertension extracted from seaweeds and octopuses.

- Bone reconstruction material from coral.

- A binding adhesive for tooth fillings ("Mother Nature's super glue") from barnacles.

- Chemicals effective against viral encephalitis and herpes (worth $50–100 million per year) from sponges.

Bulrush

Bluegill

White bass

Brook trout

White waterlily

Water lettuce

Rainbow trout

Muskellunge

Rainbow darter

Bowfish

Water hyacinth

Bladderwort

Black crappie

Largemouth black bass

White sturgeon

Walleyed pike

American smelt

Yellow perch

Velvet cichlid

Duckweed

Eelgrass

Longnose gar

Common piranha

Carp

Channel catfish

Egyptian white lotus

African lungfish

Figure 24-3 Freshwater biodiversity. Some inhabitants of freshwater rivers and lakes.

Despite the ecological and economic importance of marine ecosystems, spending on research in the ocean sciences is quite low. In the United States, ocean science research expenditures make up less than 4% of the total federal research budget, down from 7% in 1984.

Important ecological and economic services are provided by freshwater systems (Figure 24-5). Although lakes, rivers, and wetlands occupy only 1% of the earth's surface, they provide ecological and economic services worth trillions of dollars per year (p. 168).

24-2 HUMAN IMPACTS ON AQUATIC BIODIVERSITY

What Are the Major Human Impacts on Aquatic Biodiversity? The indirect and direct causes of the degradation of aquatic ecosystems and premature extinction of aquatic wildlife species are the same as those for terrestrial ecosystems and species (Figure 22-13, p. 564). They include the following:

Species Loss and Endangerment

■ Marine biologists estimate that at least 1,200 marine species have become extinct in the past few hundred years, and many thousands of additional species could disappear during this century as a result of human activities such as (1) overfishing, (2) habitat destruction and degradation, and (3) pollution.

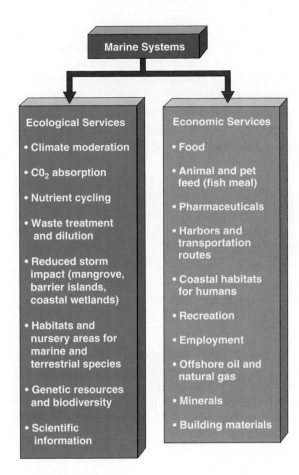

Figure 24-4 Major ecological and economic services provided by marine systems.

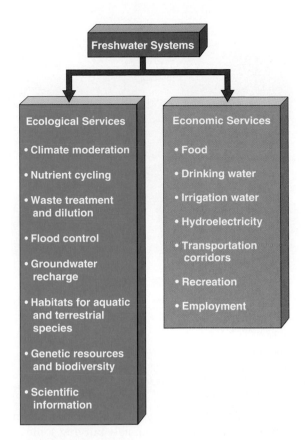

Figure 24-5 Major ecological and economic services provided by freshwater systems.

- As a whole, freshwater species are more at risk of extinction than land-based species. According to the UN Food and Agriculture Organization and the World Wildlife Fund, about one-third of all known fish species and one-half of all known freshwater species are threatened with extinction.

- A 1997 study by the Nature Conservancy and the Natural Heritage Network found that in the United States (1) 67% of freshwater mussel species, (2) 51% of crayfish, (3) 40% of amphibians, and (4) 37% of freshwater fish are at risk of extinction, with the risk varying widely in the lower 48 states (Figure 24-6).

Marine Habitat Loss and Degradation

- Scientists estimate that half of the world's original coastal wetlands and 53% of those in the lower 48 U.S. states have disappeared since 1800, by being filled in for agriculture and coastal development.

- According to a 1999 study by the World Resources Institute, about 58% of the world's coral reefs (Figure 7-2, p. 153 and Figure 7-13, p. 163) are (1) threatened by human activities such as coastal development, overfishing, pollution, and warmer ocean tempera-

tures (p. 460), (2) 27% are gone or seriously threatened, and (3) 70% could be gone within 50 years. Jamaica has lost 90% of its coral reefs, Costa Rica 75%, and Kenya 70%.

- At least half of the world's original mangroves (Figure 7-2, p. 153 and Figure 7-6, p. 157) have disappeared, mostly because of clearing for coastal development and aquaculture shrimp farms, and much of what remains is threatened.

- According to a World Wildlife Fund study, almost 70% of the world's beaches are eroding rapidly because of coastal developments and a rising sea level (caused mostly by global warming).

- Bottom habitats are being degraded and destroyed by dredging operations and trawler boats, which drag huge nets weighted down with chains over ocean bottoms to harvest bottom fish and shellfish (Figure 24-7). Each year, thousands of trawlers scrape and disturb an area of ocean bottom equal to the combined size of Brazil and India and about 150 times larger than the area of forests clear-cut each year. Recovery in heavily trawled areas rarely is possible because of repeated scraping.

- Some biologists warn that deep-sea mining (p. 331) could devastate mostly unknown marine biodiversity

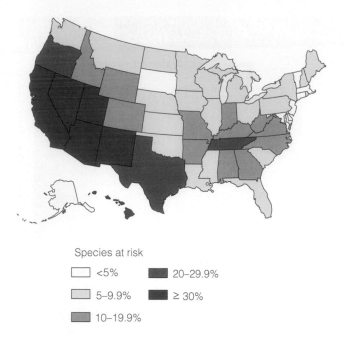

Species at risk

☐ <5%   ■ 20–29.9%

☐ 5–9.9% ■ ≥ 30%

■ 10–19.9%

Figure 24-6 Distribution of freshwater species at risk in the lower 48 states and Hawaii. (Data from the Nature Conservancy and the Natural Heritage Network)

in the deep ocean by destroying or disrupting bottom-dwelling communities and possibly burying some bottom habitats with sediment.

Freshwater Habitat Loss and Degradation

■ According to a 2000 study by the World Wildlife Fund, since 1800 **(1)** the world has lost more than half of its inland wetlands, and **(2)** 53% of inland wetlands in the lower 48 U.S. states have disappeared, mostly by being drained and filled in for agriculture and urban development. This loss disrupts many of the important ecological and economic services provided by inland wetlands (p. 168).

■ According to a 2000 study by the World Resources Institute, almost 60% of the world's 237 large rivers are strongly or moderately fragmented by dams, diversions, or canals. This alters and destroys wildlife habitats along rivers and in coastal deltas and estuaries by reducing water flow (Figure 13-9, p. 301).

■ Flood control levees (Figure 13-22, middle, p. 314) and dikes **(1)** alter and destroy aquatic habitats, **(2)** disconnect rivers from their floodplains, and **(3)** eliminate wetlands and backwaters that are important spawning grounds for fish.

Overfishing

■ According to the UN Food and Agriculture Organization, about 60% of the world's 200 commercially valuable fish species are either overfished or fished to the limit (p. 286), mostly because of too many fishing boats pursuing too few fish—another example of the tragedy of the commons (Connections, right).

■ In most cases, overfishing leads to *commercial extinction*. This is usually only a temporary depletion of fish stocks, as long as depleted areas and fisheries are allowed to recover. However, by depleting prized species and shifting to harvesting species at lower trophic levels, fishers can **(1)** unravel food webs, **(2)** disrupt marine ecosystems, and **(3)** hinder the recovery of fish feeding at higher trophic levels (Figure 24-8, right).

■ Target fish species most threatened with biological extinction are those with long life spans and low reproductive rates such as orange roughy, swordfish, and some shark species (Case Study, p. 185).

■ Overfishing can lead to serious depletion and extinction of species such as sea turtles, dolphins, and other marine mammals that are unintentionally caught as bycatch.

Figure 24-7 Area of ocean bottom before (left) and after (right) a trawler net scraped it (Figure 12-26, p. 286). These ocean floor communities can take decades or centuries to recover. Marine scientist Carl Safina says, "Bottom trawling is akin to harvesting corn with bulldozers." According to marine scientist Elliot Norse, "Bottom trawling is probably the largest human-caused disturbance to the biosphere." Trawler fishers disagree and claim that ocean bottom life recovers after trawling. (Peter J. Auster, National Undersea Research Center/University of Connecticut)

CONNECTIONS

In the 1970s and 1980s, extensive investment in fishing fleets, aided by government and international development agency subsidies, helped to significantly boost the fish catch (Figure 12-27, left, p. 287). However, since 1975 the size of the industrial fishing fleet has expanded twice as fast as the rise in catches.

Thus, there are now too many boats fishing for a declining number of fish. This leads to overfishing, an example of the tragedy of the commons (Connections, p. 12).

A 1998 study by Daniel Pauly warned that the current harvesting of species at increasingly lower trophic levels in ocean food webs (Figure 24-8) can lead to (1) decreased chances for recovery of species at the top of ocean food webs by reducing stocks of the smaller fish they feed on (especially in the North Atlantic), (2) collapse of marine ecosystems, (3) a drop in aquatic biodiversity, and (4) a loss of high-quality protein for humans.

Because of overfishing and the overcapacity of the fishing fleet, it costs the global fishing industry about $125 billion a year to catch $70 billion worth of fish. Most of the $55-billion annual deficit of the industry is made up by government subsidies such as fuel tax exemptions, price controls, low-interest loans, and grants for fishing gear.

Critics contend that such subsidies promote overfishing. They argue that eliminating or significantly lowering these subsidies would reduce the size of the fishing fleet by encouraging free-market competition and would allow some of the economically and biologically depleted stocks to recover.

Eliminating these subsidies will cause of loss of jobs for some fishers and fish processors in coastal communities. However, to fishery biologists the alternative is worse. Continuing to subsidize excess fishing allows fishers to keep their jobs a little longer while making less and less money until the fishery collapses. Then all jobs are gone and fishing communities suffer even more.

Critical Thinking

Do you believe that government subsidies for the fishing industry should be eliminated or sharply reduced? Explain. How would you feel about eliminating such subsidies if your livelihood depended on fishing?

Figure 24-8 Mean trophic levels of the global marine (right) and freshwater (left) fish catch have declined since 1950. (Data from Daniel Pauly)

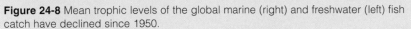

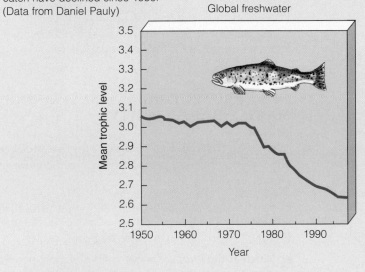

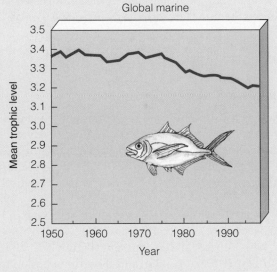

Nonnative Species

■ Hundreds of nonnative species have been deliberately or accidentally introduced into coastal waters, lakes, and wetlands (Spotlight, p. 636) throughout the world. These bioinvaders displace or cause the extinction of native species and disrupt ecosystem functions, as happened to Lake Victoria (p. 629).

■ An example of an accidental invader is the *Asian swamp eel*, a fish that invaded the waterways of southern Florida, probably after escaping from a home aquarium. This prolific eel (1) eats almost anything, including bluegill, bass, crayfish, and many other species, by sucking them in like a vacuum cleaner, (2) can elude cold weather, drought, fires, and predators (including humans with nets) by burrowing into mud banks, (3) is resistant to waterborne poisons because it can breathe air, (4) can wriggle across dry land to invade new waterways, ditches, canals, and marshes, and (5) eventually could take

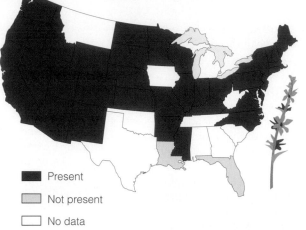
over much of the waterways of the southeastern United States as far north as the Chesapeake Bay.

- Bioinvaders are blamed for about 68% of fish extinctions in the United States between 1900 and 2000.

Pollution and Global Warming

- The major pollution threats to aquatic systems and species are **(1)** oil (p. 491 and Case Study, p. 336), **(2)** acid deposition (Figure 17-9, p. 428, Figure 17-11, p. 429, and Figure 17-13, p. 430), **(3)** excess plant nutrients and oxygen-demanding wastes (Figure 19-3, p. 479, Figure 19-5, p. 482, and Figure 19-12, p. 490), **(4)** toxic chemicals such as cyanide (Case Study, p. 328), PCBs (Figure 19-4, p. 481), and DDT (Figure 16-4, p. 399), **(5)** coastal development (Figure 7-12, p. 162, and Figure 19-11, p. 489), and **(6)** sedimentation from soil erosion (Figure 7-7, p. 157).

- An estimated 44% of marine pollution comes from runoff from developed coastal areas and rivers flowing into estuaries. An additional 33% comes from the atmosphere, and the remaining 23% comes from shipping, offshore oil and gas production, and ocean dumping.

- Global warming (Figure 18-8, p. 453) could **(1)** alter migration and feeding patterns of marine species, **(2)** increase average ocean temperatures, which can destroy many of the world's coral reef sys-

tems (Figure 7-1, p. 152, and Figure 7-2, p. 153) and **(3)** raise average sea levels (Figure 18-11, p. 456) which can flood and destroy coastal wetlands, mangroves (Figure 7-6, p. 157), and coral reefs.

24-3 PROTECTING AND SUSTAINING MARINE BIODIVERSITY

Why Is It Difficult to Protect Marine Biodiversity? Protecting marine biodiversity is difficult for several reasons:

- Coastal development and the accompanying massive inputs of sediment and other wastes from land (Figure 7-7, p. 157) harm shore-hugging species.

- Much of the damage is not visible to most people.

- Many people view the seas as an inexhaustible resource that can absorb an almost infinite amount of waste and pollution.

- Most of the world's ocean area lies outside the legal jurisdiction of any country and is thus an open-access resource, subject to overexploitation because of the tragedy of the commons (Connections, p. 12).

- There are no effective international agreements to protect biodiversity in the open seas, mostly because such agreements are difficult to develop, monitor, and enforce.

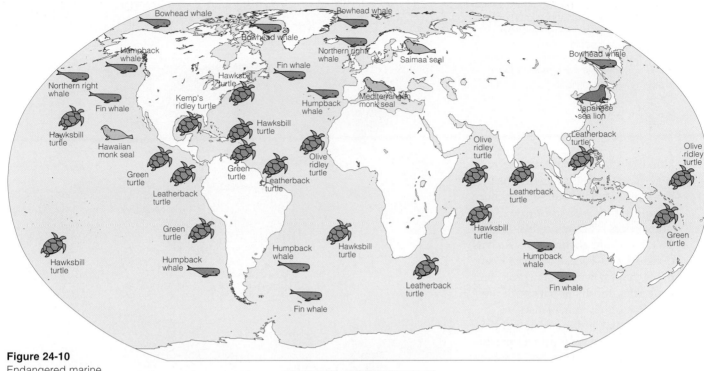

Figure 24-10
Endangered marine mammals (whales, seals, and sea lions) and reptiles (turtles). Many marine fish, seabirds (Figure 8-9, p. 182), and invertebrate species are also threatened.

Whale		Turtle	
Seal		Sea lion	

How Can We Protect and Sustain Marine Biodiversity?

Ways to protect and sustain marine biodiversity include:

- Protecting endangered and threatened species.
- Establishing protected areas.
- Using integrated coastal management.
- Regulating and preventing ocean pollution, as discussed in Section 19-4, p. 488 and p. 492.
- Sustainably managing marine fisheries.

How Can We Protect Endangered and Threatened Marine Species?

One widely used method for protecting biodiversity is identifying and protecting endangered, threatened, and rare species, as has been done to help save a number of endangered terrestrial species (Section 22-5, p. 573).

In addition to fish, this strategy has been used to protected a number of endangered and threatened marine reptiles (turtles) and mammals (especially whales, seals, and sea lions; Figures 24-10 and 24-11). Three of eight major sea turtle species (Figure 24-12) are endangered (Kemp's ridley, leatherbacks, and hawksbills), and the rest are threatened.

The world's sea turtle species are endangered or threatened because of (1) loss or degradation of beach habitat (where they come ashore to lay their eggs), (2) legal and illegal taking of their eggs (Guest Essay, p. 644), (3) their increased use as sources of food, medicinal ingredients, tortoiseshell (for jewelry), and leather from their flippers (with some turtles selling for up to $1,500 in China), and (4) unintentional capture and drowning by commercial fishing boats (especially shrimp trawlers).

Until recently, as many as 55,000 sea turtles (mostly endangered loggerheads and Kemp's ridleys) were

Figure 24-11 Before this discarded piece of plastic was removed, this Hawaiian monk seal was slowly starving to death. Each year plastic items dumped from ships and left as litter on beaches threaten the lives of millions of marine mammals, turtles, and seabirds that ingest, become entangled in, choke on, or are poisoned by such debris. (Doris Alcorn/U. S. National Maritime Fisheries)

Figure 24-12 Major species of sea turtles that have roamed the seas for 150 million years, showing their relative adult sizes. Three of these species (Kemp's ridley, leatherbacks, and hawksbills) are endangered, and the rest are threatened as a result of human activities.

Loggerhead
119 centimeters

Olive ridley
76 centimeters

Leatherback
188 centimeters

killed by U.S. shrimp trawling each year. To reduce this slaughter, the U.S. government has required offshore shrimp trawlers to use turtle exclusion devices (TEDs) since 1989 (Figure 24-13). Certain governments such as Costa Rica and Mexico (Guest Essay, p. 644) have taken a lead in protecting sea turtles, mostly as a result of pressure from environmentalists.

National and international laws and treaties to help protect marine species include **(1)** the 1975 Convention on International Trade in Endangered Species (CITES, p. 573), **(2)** the 1979 Global Treaty on Migratory Species, **(3)** the U.S. Marine Mammal Protection Act of 1972, **(4)** the U.S. Endangered Species Act of 1973 (p. 573), **(5)** the U.S. Whale Conservation and Protection Act of 1976, and **(6)** the 1995 International Convention on Biological Diversity.

Public aquariums that exhibit unusual and attractive fish and some marine animals such as seals and dolphins have been successful in educating the public about the need to protect such species. In the United States, more than 35 million people visit aquariums each year. However, unlike zoos, public aquariums have not yet served as effective gene banks for endangered marine species, especially marine mammals that need large volumes of water.

Two major problems with protecting marine biodiversity by protecting endangered species are **(1)** lack of knowledge about marine species and **(2)** difficulty in monitoring and enforcing treaties to protect marine species, especially in the open ocean.

Case Study: Should Commercial Whaling Be Resumed? *Cetaceans* are an order of mostly marine mammals ranging in size from the 0.9-meter (3-foot) porpoise to the giant 15- to 30-meter (50- to 100-foot)

blue whale. They are divided into two major groups: toothed whales and baleen whales (Figure 24-14, p. 642).

Toothed whales, such as the porpoise, sperm whale, and killer whale (orca), bite and chew their food, and feed mostly on squid, octopus, and other marine animals. *Baleen whales*, such as the blue, gray, humpback, and finback, are filter feeders (Spotlight, p. 154). Instead of teeth, they have several hundred horny plates made of baleen, or whalebone, that hang down from the upper jaw. These plates filter plankton from the seawater, especially tiny shrimplike krill (Figure 4-19, p. 84). Baleen whales are the most abundant group of cetaceans.

Whales are fairly easy to kill because of their large size and their need to come to the surface to

Hawksbill
89 centimeters

Australian
flatback
99 centimeters

Black turtle
99 centimeters

Green turtle
124 centimeters

Kemp's ridley
76 centimeters

breathe. Mass slaughter has become very efficient with the use of fast ships, harpoon guns, and inflation lances (which pump dead whales full of air and make them float).

Whale harvesting, mostly in international waters, has followed the classic pattern of a tragedy of the commons, with whalers killing an estimated 1.5 million whales between 1925 and 1975. This overharvesting **(1)** drove the populations of 8 of the 11 major species to commercial extinction, the point at which it no longer paid to hunt and kill them, and **(2)** some commercially prized species such as the giant blue whale (Figure 24-14) were reduced to the brink of biological extinction (Case Study, p. 641).

In 1946, the International Convention for the Regulation of Whaling established the International Whaling Commission (IWC) to regulate the whaling industry by setting annual quotas to prevent overharvesting and commercial extinction. However, IWC quotas often were based on inadequate data or were ignored by whaling countries. Without any powers of enforcement, the IWC has been unable to stop the decline of most commercially hunted whale species to the point at which they were commercially extinct.

In 1970, the United States stopped all commercial whaling and banned all imports of whale products. Under intense pressure from environmentalists and governments of many nonwhaling countries in the IWC,

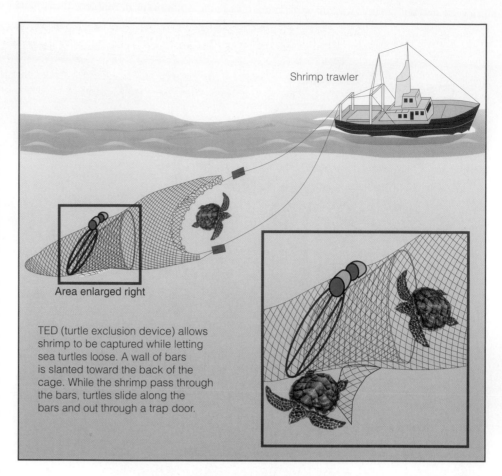

Figure 24-13 Turtle exclusion device (TED) designed to release turtles caught in shrimp trawling nets. Since 1989, the government has required these devices on the nets of all shrimp boats trawling in U.S. waters.

Shrimp trawler

Area enlarged right

TED (turtle exclusion device) allows shrimp to be captured while letting sea turtles loose. A wall of bars is slanted toward the back of the cage. While the shrimp pass through the bars, turtles slide along the bars and out through a trap door.

led by the United States, the IWC has imposed a moratorium on commercial whaling since 1986. As a result, the estimated number of whales killed commercially worldwide dropped from 42,480 in 1970 to about 1,200 in 1999 (taken by Japan and Norway).

The IWC has not banned the harvesting of small cetaceans (porpoises and dolphins, Figure 24-14). However, the United States and several other countries are pressuring the commission to do so because of the significant increase in the Japanese harvest of these species to supplement its large domestic whale meat market.

Japan gets around the IWC ban by calling its annual killing of up to 440 minkes "scientific whaling." Critics contend that the research whaling assertion is a cover for commercial whaling, noting that the whale meat ends up in the Japanese market for human consumption. In 2000, Japan further defied the IWC ban by expanding its annual whale harvest to include sperm whales and Bryde's whales, arguing that their data suggest that the populations of these two species have recovered enough to allow commercial harvesting. Both of these whale species are protected under U.S. law, and the sperm whale is listed as an endangered species.

Norway is allowed to ignore the ban and kills up to 753 minkes per year in the North Atlantic because it filed an objection to the ban when it was first imposed. Iceland has refused to join the IWC and asserts its right to hunt whales when it wishes. Subsistence hunting of a small number of whales for food and survival materials by indigenous populations such as the Inuit in Alaska and some groups in Greenland and Tonga is still allowed.

Japan, Norway, Iceland, Russia, and a growing number of small tropical island countries (which Japan has brought into the IWC to support its position) continue working to **(1)** overthrow the IWC ban on commercial whaling and **(2)** reverse the CITES international ban on buying and selling whale products. Unless the

CITES designation is overturned, any decision by the IWC to resume some commercial whaling would be mostly useless because whaling countries would have no legal market for whale meat (which sells as a gourmet food in Japan for up to $90 per pound) and other whale products.

These and other whaling nations believe that the international ban on commercial whaling of most species and the CITES ban on international trade of whale meat and products should be lifted for several reasons:

- Whaling should be allowed because it has long been a traditional part of the economies and cultures of countries such as Japan, Iceland, and Norway.

- The ban is based on emotion, not updated scientific estimates of whale populations. According to IWC estimates, the population of minke whales (which Japan and Norway continue to hunt) now numbers about 1 million (with an estimated 760,000 in Antarctic waters) and that of pilot whales about 1.4 million. They see no scientific reason for not resuming controlled hunting of these species, along with sperm, Bryde's, and gray whales (in the eastern Pacific).

- DNA testing of whale meat could be used to detect harvesting of species that are still banned.

Near Extinction of the Blue Whale

CASE STUDY

The biologically endangered blue whale (Figure 24-14) is the world's largest animal. Fully grown, it's more than 30 meters (100 feet) long—longer than three train boxcars—and weighs more than 25 elephants. The adult has a heart the size of a Volkswagen Beetle car, and some of its arteries are so big that a child could swim through them.

Blue whales spend about 8 months of the year in Antarctic waters. There they find an abundant supply of krill (Figure 4-19, p. 84), which they filter daily by the trillions from seawater. During the winter, they migrate to warmer waters, where their young are born.

Before commercial whaling began, an estimated 200,000 blue whales roamed the Antarctic Ocean. Today, the species has been hunted to near biological extinction for its oil, meat, and bone.

A combination of prolonged overharvesting and certain natural characteristics of blue whales caused its decline. Their huge size made them easy to spot. They were caught in large numbers because they grouped together in their Antarctic feeding grounds. They also take 25 years to mature sexually and have only one offspring every 2-5 years, a reproductive rate that makes it difficult for the species to recover once its population falls beneath a certain threshold.

Blue whales have not been hunted commercially since 1964

and have been classified as an endangered species since 1975. Despite this protection, some marine biologists believe that too few blue whales—an estimated 1,000–3,000—remain for the species to recover and avoid extinction.

Critical Thinking

Opponents of commercial whaling contend that resuming commercial whaling for some whale species such as minke, pilot, and gray could lead to illegal harvesting of blue whales. Japan contends that excess population on minkes in Antarctic waters is threatening the blue whale population by consuming much of the krill that they eat (Figure 4-19, p. 84). What scientific evidence would you want to have to resolve this issue?

Most environmentalists disagree for several reasons:

- Some argue that whales are peaceful, intelligent, sensitive, and highly social mammals that pose no threat to humans and should be protected for ethical reasons:

- Some question the estimates of minke, Bryde, gray, and pilot whale populations, noting the inaccuracy of past IWC estimates of whale populations.

- Most fear that opening the door to any commercial whaling may eventually lead to widespread harvests of most whale species by weakening current international disapproval and legal sanctions against commercial whaling by the IWC and CITES.

- DNA tests of all whale meat would be difficult to enforce and monitor because **(1)** much of the meat could be consumed before tests are made and illegal meat can be confiscated, and **(2)** fines would be too small to be effective.

- If the ban were lifted for some species, it would be easier to hide illegal taking of protected species. For example, recent DNA tests have uncovered the sale of whale meat from several banned species in the Japanese marketplace.

In the United States, opponents of commercial whaling have urged the U.S. Congress to impose sanctions against countries defying the IWC ban on

commercial whaling. They also urge consumers to **(1)** support economic boycotts of imports of all products from offending nations, **(2)** avoid using air and sea vessels owned by such countries, and **(3)** cease traveling to offending countries. By 1994, a consumer boycott of Norwegian fish products in the United Kingdom and Germany had cost Norway more than $30 million in canceled contracts.

In 2000, the United States banned some Japanese fishing boats from U.S. waters and warned that additional sanctions could be imposed if Japan doesn't curb its illegal whaling program.

What Is the Role of Setting Aside Protected Marine Sanctuaries? Several international treaties, agreements, and actions by individual governments have been used to protect living marine resources in parts of the world. Examples include the following:

- A 1991 treaty in which 26 nations agreed to designate Antarctica as a natural reserve devoted to peace and science. The treaty **(1)** bans mineral and oil exploration and mining in Antarctica for at least 50 years and **(2)** includes regulations for improved wildlife protection, marine pollution, and environmental monitoring. However, much of the ocean surrounding Antarctica, with its rich abundance of biodiversity (Figure 4-19, p. 84) is not covered by the treaty.

Humpback whale

Bowhead whale

Right whale

Minke whale

Blue whale

Feeding on krill

Fin whale

Sei whale

Gray whale

Mysticetes (Baleen Whales)

Figure 24-14 Examples of cetaceans, which can be classified as baleen whales and toothed whales.

■ The United Nations Environment Programme has spearheaded efforts to develop 12 regional agreements to protect large marine areas shared by several countries. The protected areas include parts of the Black Sea, the Persian Gulf, the Red Sea, the South Pacific, the southwestern Atlantic, the Caribbean, and Africa and Asia.

■ About 90 of the world's 350 biosphere reserves (Solutions, p. 620) include coastal or marine habitats.

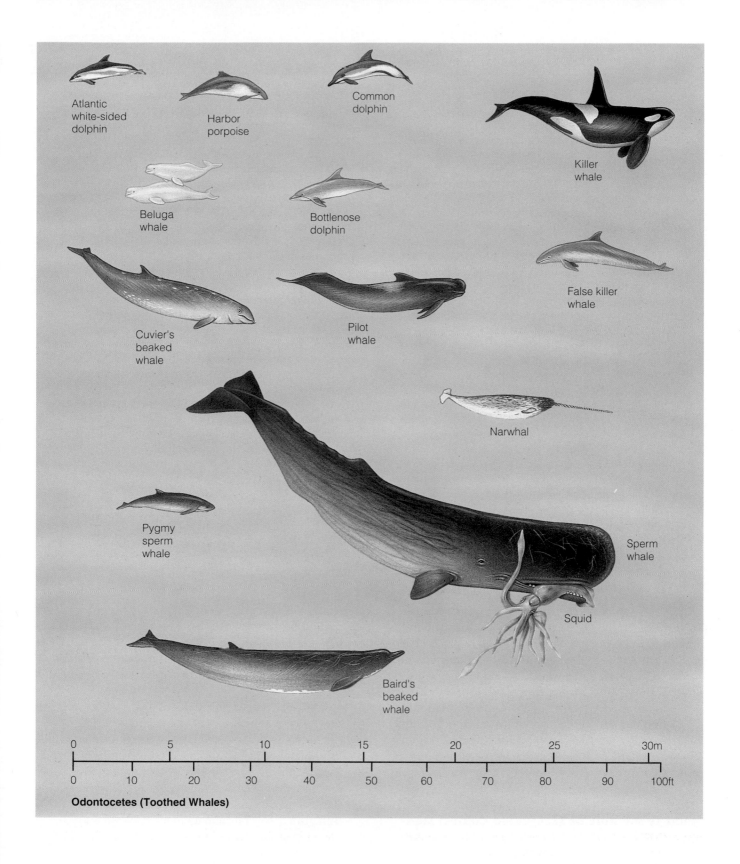

Atlantic
white-sided
dolphin

Harbor
porpoise

Common
dolphin

Killer
whale

Beluga
whale

Bottlenose
dolphin

False killer
whale

Cuvier's
beaked
whale

Pilot
whale

Narwhal

Pygmy
sperm
whale

Sperm
whale

Squid

Baird's
beaked
whale

| 0 | 5 | 10 | 15 | 20 | 25 | 30m |

| 0 | 10 | 20 | 30 | 40 | 50 | 60 | 70 | 80 | 90 | 100ft |

Odontocetes (Toothed Whales)

■ The United States has designated 12 marine sanctuaries, but they make up only 1% of U.S. waters and are much too small to promote effective conservation of marine biodiversity and fisheries.

■ Since 1986, the World Conservation Union (IUCN) has helped establish a global system of *marine protected areas* (MPAs), mostly at the national level. The 1,300 existing MPAs help protect about 0.2% of the earth's total ocean area.

Mazunte: A Farming and Fishing Ecological Reserve in Mexico

Alberto Ruz Buenfil

GUEST ESSAY

Alberto Ruz Buenfil, an international environmental activist, writer, and performer, is the founder of Huehuecoytl, a land-based ecovillage in the mountains of Mexico. His articles on ecology and alternative living have appeared in publications in the United States, Canada, Mexico, Japan, and Europe. His book Rainbow Nation Without Borders: Toward an Ecotopian Millennium *(New Mexico: Bear, 1991) has been published in English, Italian, and Spanish.*

The world's eight known sea turtle species [Figure 24-12] are all officially listed as endangered or threatened. Seven of these species nest on Mexico's Pacific and Atlantic coastlines, making Mexico the world's most important turtle nesting country. The Pacific coasts of southern Mexico, especially the shores of the state of Oaxaca, are the main sites for turtle nesting, reproduction, and conservation; they also contain some of Mexico's last wetland reserves.

During the past few decades, the coasts of Oaxaca have been increasingly exploited, especially after business interests discovered the area's beaches.

The villages of Mazunte and San Agustinilo were founded in the late 1960s to bring in cheap labor from neighboring indigenous villages to work in a slaughterhouse making products from various turtle species nesting in the area. Members of nearly 200 indigenous farming families became fishers and employees of the new turtle meat factory, which operated in the 1970s and 1980s. According to some of these workers, nearly 2,000 turtles were killed and quartered every day during those years. At night, dozens of poachers came to collect turtle eggs from their nests.

In the 1980s, this situation came to the attention of two Mexican environmental organizations. With support from other international organizations, they campaigned for almost 10 years, until a 1990 decree by the president of Mexico made it illegal to harvest the turtles and led to the closing of San Agustinilo's turtle slaughterhouse.

The Mexican government provided some funds, boats, and freezers to compensate for the loss of jobs. However, only about 5% of the indigenous population benefited from this compensation. Since then, most of these indigenous people have been living on the verge of starvation, and some have illegally killed protected turtles to survive. What had seemed to be an important environmental victory turned into a nightmare for a large population of indigenous people. Understandably, these people had no use for ecologists or environmentalists.

In 1990, a group called ECOSOLAR A.C. began efforts to change this situation by implementing a plan for sustainable development of the coast of Oaxaca. They were successful in obtaining funding for this project from different national and international institutions. By 1992, the members of this small but effective group had succeeded in:

- Making a detailed study of the bioregion, which, with the participation of the local people, is being used to define the possible uses of different areas.

A 1995 study by the Great Barrier Reef Marine Park Authority, the IUCN, and the World Bank found that (1) most existing MPAs are to small to protect the species within them, (2) many globally unique marine habits have received no protection, and (3) the boundaries for most MPAs stop at the shoreline and thus do not provide adequate protection from pollution that flows into coastal waters as a result of land use. In 1997, a group of international marine scientists called for governments to increase current MPAs and other protected marine reserves from the current 0.25% of the ocean's surface to 20% by 2020.

What Is the Role of Integrated Coastal Management? *Integrated coastal management* is a community-based attempt to develop and use coastal resources in a sustainable manner that involves cooperative efforts and planning by diverse groups of fishers, scientists, conservationists, citizens, business interests, developers, the general public, and politicians. The overall aim of this tool is for groups competing for the use of coastal resources to (1) identify shared problems and goals and (2) agree to workable and cost-effective solutions that preserve biodiversity and environmental quality while meeting economic and social needs.

Ideally, the overall goal is to zone the land and sea portions of an entire coastal area. Such zoning would include some protected areas where no destructive human activities would be allowed and other zones where different kinds and levels of human activities are permitted. Australia's huge Great Barrier Reef Marine Park is managed in this way. Currently, more than 100 integrated coastal management programs are being developed throughout the world.

In the United States:

- Ninety coastal counties are working to establish coastal management systems, but fewer than 20 of these plans have been implemented.

- Creating a system of credits to help native inhabitants build better houses, establish small family-run restaurants, and manufacture hammocks for rent or sale to visitors.

- Constructing systems for drainage, water collection, and latrines using low-impact technology and local materials and workers, as well as nurseries for local seeds and facilities for reforestation and wildlife preservation projects.

- Working with the community to promote Mazunte as a center for ecotourism where visitors can experience unique ecosystems containing alligators, turtles, and hundreds of bird and fish species.

In only 2 years, the native inhabitants of Mazunte and other neighboring communities completely changed their negative opinion about ecologists and environmentalists. In 1992, Mazunte hosted a general meeting attended by 150 heads of families to discuss ways to get community members to protect turtle nesting areas instead of illegally killing the turtles for food. Out of that meeting came a "Declaration of Mazunte" requesting that higher authorities and the president of Mexico put an immediate end to such destruction, which violates the earlier presidential decree forbidding the annihilation of turtles in Mexico.

Such efforts have paid off. Illegal killing of turtles and removal of their eggs continues but at a much lower rate. According to sea turtle expert Juan Carlos Cantu of Greenpeace Mexico, between 1990 and 1998 the number of sea turtles killed in this area has dropped from about 150,000 to 25,000 per year. The government has assigned a squad of marines around the clock to protect each of the area's sea turtle nesting beaches, and arrests of poachers are widely publicized in the press.

The community went on to declare their village and neighboring environments to be Mexico's first *Farming and Fishing Reserve*. Its goals are to protect the area's forests, water sources, wetlands, wildlife, shores, beaches, and scenic places and to "establish new forms of relationship between humans and nature, for the well-being of today's and tomorrow's generations." This declaration has been presented to the government of Mexico and to many national and international organizations.

Mazunte is taking the lead in demonstrating that cooperation between local people and environmental experts can lead to more ecologically sustainable communities that benefit local people and wildlife alike. This model can show farmers, fishers, and indigenous communities elsewhere how to live more sustainably on the earth and turn things around in a short time. It is a message of hope and empowerment for people seeking a better world for themselves and others.

Critical Thinking

1. What lessons have you learned from this essay that you could apply to your own life?

2. Could the rapid move toward sustainability brought about by environmentalists and local people in Mazunte be accomplished in your own community? If so, how? If not, why not?

- Since the early 1980s, people have worked together with some success to develop an integrated coastal management plan for the Chesapeake Bay (Case Study, p. 490, and Figure 19-13, p. 490).

Case Study: What Can We Do About Beach Erosion? One problem that affects coastal marine biodiversity, (Figure 7-10, p. 160) and the economic health of increasingly populated coastal areas is *beach erosion*. An estimated 70% of the world's beaches are eroding as a result of natural and human-related causes.

Beach erosion is a serious problem along most of the gently sloping beaches of barrier islands and mainland shores (Figure 7-10, bottom, p. 160), with 30% of U.S. shoreline experiencing significant erosion. The main cause of this problem is that the sea level has been rising gradually for the past 12,000 years or so (Figure 18-11, p. 456), mainly because the warmer climate since the most recent ice age has melted much ice and expanded the volume of seawater. Other causes of rising sea levels are (1) extracting groundwater, (2) redirecting rivers, (3) draining wetlands, and (4) other human activities that divert more water to the oceans.

Engineers have tried several methods to halt or reduce beach erosion (Figure 24-15). However, at best these attempts are only temporary solutions because beach erosion in one place and beach buildup in another is a natural process that we can do little to control.

Many coastal zone ecologists call for banning or severely limiting the construction of seawalls, breakwaters, groins, and jetties (essentially long groins used to protect harbors and inlets) and the movement of inlets (which can easily be closed again or moved again by storms) because in the long run they can cause more damage than they prevent. These analysts also favor prohibiting development on most remaining undeveloped beach areas or allowing such development only behind protective dunes (Figure 7-11, p. 161).

According to a 2000 study by the Federal Emergency Management Agency, about 25% of homes and

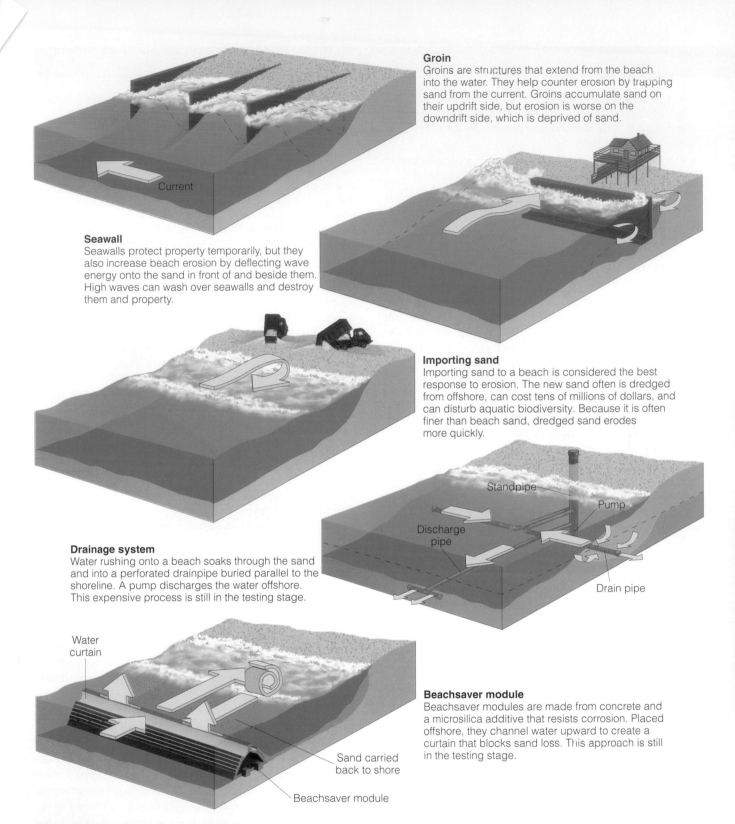

Groin
Groins are structures that extend from the beach into the water. They help counter erosion by trapping sand from the current. Groins accumulate sand on their updrift side, but erosion is worse on the downdrift side, which is deprived of sand.

Seawall
Seawalls protect property temporarily, but they also increase beach erosion by deflecting wave energy onto the sand in front of and beside them. High waves can wash over seawalls and destroy them and property.

Importing sand
Importing sand to a beach is considered the best response to erosion. The new sand often is dredged from offshore, can cost tens of millions of dollars, and can disturb aquatic biodiversity. Because it is often finer than beach sand, dredged sand erodes more quickly.

Drainage system
Water rushing onto a beach soaks through the sand and into a perforated drainpipe buried parallel to the shoreline. A pump discharges the water offshore. This expensive process is still in the testing stage.

Beachsaver module
Beachsaver modules are made from concrete and a microsilica additive that resists corrosion. Placed offshore, they channel water upward to create a curtain that blocks sand loss. This approach is still in the testing stage.

Current

Standpipe

Pump

Discharge pipe

Drain pipe

Water curtain

Sand carried back to shore

Beachsaver module

Figure 24-15 Building groins or seawalls and importing sand to reduce beach erosion can make matters worse or provide only an expensive temporary fix. Two new techniques for reducing beach erosion shown here—a drainage system and beachsaver modules that act as an artificial reef—are being evaluated for effectiveness and cost.

structures within 150 meters (500 feet) of the U.S. coastline will fall victim to the effects of beach erosion between 2000 and 2060. Especially hard hit will be areas along the Atlantic and Gulf of Mexico coastlines, which are expected to account for 60% of nationwide losses.

Since 1965, governments, developers, and communities in the United States have spent almost $4 billion replenishing beaches that have been eroded and will continue to erode. Under present policy the government (taxpayers) usually pays 65% of the cost of beach replenishment. Some critics have called for elimination of this federal subsidy or for reducing the government's share to no more than 35% of the costs.

A growing number of analysts also believe that federal flood insurance subsidies, which greatly reduce the financial risk of building structures at the sea's edge, should be eliminated. They argue that such insurance subsidies encourage **(1)** irresponsible development, **(2)** dune destruction (Figure 7-11, p. 161), and **(3)** beach erosion on barrier beaches and islands (Figure 7-12, p. 162) that are at risk of serious damage from storms, hurricanes, and rising sea levels.

Proponents of federal flood insurance subsidies argue that they are necessary to promote urban development, jobs, and economic growth in coastal areas. Opponents believe that people who choose to live in risky areas should be responsible for their own insurance payments and that the government should take measures such as **(1)** enacting much tougher building codes, **(2)** banning coastal and inland wetland destruction (which reduces flood protection), and **(3)** requiring endangered structures to be elevated, moved inland, or demolished and cleared away to reduce the impact of coastal storms, hurricanes, and rising sea levels on people and marine biodiversity.

24-4 MANAGING AND SUSTAINING THE WORLD'S MARINE FISHERIES

How Can We Project Populations of Marine Fisheries? Overfishing (Figure 12-28, p. 287) is a serious threat to biodiversity in coastal waters and to some marine species in open-ocean waters. Ways to reduce overfishing include **(1)** developing better measurements and models for projecting fish populations and **(2)** controlling fishing methods and access to fisheries.

Methods used to project populations of commercially important fish include **(1)** *maximum sustained yield* (MSY), **(2)** *optimum sustained yield* (OSY), **(3)** multispecies management, **(4)** large marine system management, and **(5)** the precautionary principle.

Until recently, management of commercial fisheries has been based primarily on the MSY, a mathematical model used to project the maximum number of fish that can be harvested annually from a fish stock without driving it into decline.

However, experience has shown that the MSY concept has helped hasten the collapse of most commercially valuable stocks because:

- Populations and growth rates of fish stocks are difficult to measure and predict and usually vary because of changes in ocean temperatures, currents, food supplies, predator-prey relationships, and other mostly unpredictable factors.

- Most fish population estimates are based on fishers reporting their catch, and they may be lying or underreporting their take for financial gain.

- Harvesting a fishery at its estimated maximum sustainable level leaves little margin for error.

- Fishing quotas are difficult to enforce.

- Many groups managing fisheries have ignored projected MSYs for short-term political or economic reasons.

- Maximum sustainable harvesting of one species can affect the populations of other target and nontarget fish species and other marine organisms.

In recent years, fishery biologists and managers have begun placing more emphasis on the *optimum sustained yield* (OSY) concept that attempts to take into account interactions with other species and to provide room for error. This approach can improve the reliability of fish stock estimates but **(1)** it still depends on the poorly understood biology of fish and changing ocean conditions, and **(2)** many bodies governing fisheries ignore OSY estimates for short-term political and economic reasons.

Another approach is *multispecies management* of a number of interacting species that takes into account their competitive and predator-prey interactions. Fishery expert Pauly Daniel of the Fisheries Centre at the University of British Columbia has developed several free computer modeling programs used by scientists in 94 countries (including major fishing nations such as Iceland and Norway). These models allow researchers to keep track of up to 2,500 interactions between as many as 50 different species or groups of species involved in a marine food web. However, such models are still in the development and testing stage and are no better than the data fed into them and the validity of the assumptions built into them.

An even more ambitious approach is to develop models for managing multispecies fisheries at the level of *large marine systems*. Scientists have identified 49 such systems throughout the world, with most being shared by two or more coastal nations.

This approach involves developing even more complex computer models of such systems, and there

Managing Fisheries with Individual Transfer Quotas

SPOTLIGHT

With the ITQ system, the government gives each fishing vessel owner a specified percentage of the total allowable catch (TAC) for a fishery in a given year.

Owners are permitted to buy, sell, or lease their quotas like private property. This approach uses a limited form of private ownership of fish stocks to help prevent overfishing by reducing the number of fishers and fishing boats to more sustainable levels.

Currently, about 50 of the world's fisheries are managed by the ITQ system. Since the ITQ system was introduced in New Zealand in 1986 and in Iceland in 1990, there has been some reduction in overfishing and the overall fishing fleet and an end to government fishing subsidies that encourage overfishing.

However, **(1)** enforcement has been difficult, **(2)** some fishers illegally exceed their quotas, **(3)** the wasteful bycatch has not been reduced, and **(4)** in some cases, TACs have been set at levels that are too high to reduce overfishing.

Some environmentalists generally oppose this free market environmental approach to commercial fishing or have made suggestions for its improvement. They contend that the ITQ system:

- In effect transfers ownership of publicly owned fisheries to the private commercial fishers but still makes the public responsible for the costs of enforcing and managing the system. Reformers suggest that fees (not to exceed 5% of the value of the catch) should be collected from quota holders to pay for the costs of government enforcement and management of the ITQ system.

- Can squeeze out small fishing vessels and companies because they do not have the capital to buy the ITQs from others. For example, 5 years after the ITQ system was implemented in New Zealand, three companies controlled half of all the ITQs. To help resolve this fairness issue, reformers suggest that no fisher or fishing company should be allowed to accumulate more than 20% of the total quota of a fishery.

- Can increase poaching and sales of illegally caught fish on the black market by **(1)** small-scale fishers who receive no quota or too small a quota to make a living or **(2)** larger-scale fishers who deliberately exceed their quotas, as has happened to some degree in New Zealand. Reformers suggest that cheating can be reduced by implementing strict record-keeping and having well-trained observers on every fishing vessel.

- May not reduce overall fishery stress by encouraging fishers to move elsewhere and to select only the highest-quality species.

- Often sets quotas too high. To eliminate overfishing, reformers suggest that 10–50% of the estimated MSY of an ITQ fishery should be left as a buffer to protect the fishery from unexpected decline.

Critical Thinking

Are you for or against using ITQs as the major method for managing fisheries? Explain. What are the alternatives?

---

are political challenges in getting groups of nations to cooperate in their planning and management. Despite the scientific and political difficulties, some limited management of several large marine systems is under way. Examples include **(1)** the Mediterranean Sea by 17 of the 18 nations involved, **(2)** the Great Barrier Reef under the exclusive control of Australia, and **(3)** negotiations between China and South Korea for managing the Yellow Sea.

Because of the inherent uncertainties in all of these approaches, there is growing interest among many fishery scientists and environmentalists in using the *precautionary principle* for managing fisheries and large marine systems to prevent harm to such systems and humans. Recall that this principle states that where there is significant risk of damage to the environment and scientific evidence is inconclusive, we should take precautionary action to limit the potential risk of damage.

Should We Use Public Management or Private Ownership to Control Access to Fisheries? The three major approaches used to control access to fisheries in various parts of the world are **(1)** international and national laws, **(2)** community-based comanagement, and **(3)** *individual transfer quotas* (ITQs) (Spotlight, above).

By international law, a country's offshore fishing zone extends to 370 kilometers (200 nautical miles, or 230 statute miles) from its shores. Foreign fishing vessels can take certain quotas of fish within such zones, called *exclusive economic zones*, but only with a government's permission.

Ocean areas beyond the legal jurisdiction of any country are known as the *high seas*. Any limits on the use of the living and mineral common-property resources in these areas are set by international maritime law and international treaties that are difficult to monitor and enforce.

Traditionally, many coastal fishing communities have developed allotment and enforcement systems that have sustained their fisheries, jobs, and communities for hundreds and sometimes thousands of years. However, with the influx of large modern fishing boats and fleets, the ability of many coastal communities to regulate and sustain local fisheries has been weakened.

Many of these community management systems have been replaced by *comanagement*, in which coastal communities and the government work together to manage fisheries. In this approach, a central government typically (1) sets quotas for various species, (2) divides the quotas between different communities, and (3) may limit fishing seasons and regulate the type of fishing gear that can be used to harvest a particular species.

Each community then allocates and enforces its quota among its members based on its own rules. Often communities focus on managing inshore fisheries, and the central government manages an area's offshore fisheries. Comanagement is being used successfully in Kiribati, Japan, in Norway, and for Alaska's salmon fishery but has had mixed success in Canada, Maine, and the United Kingdom. When it works, community-based comanagement illustrates that the tragedy of the commons is not inevitable.

Solutions: How Can Fisheries Be Used More Sustainably? Analysts suggest the following measures for managing global fisheries more sustainably and protecting marine biodiversity:

Fishery Regulations

■ *Set, monitor, and enforce fishery quotas well below their estimated MSYs.*

■ *Divide up fishing quotas based on fairness and inputs from local communities and fishers.*

■ *Improve monitoring and enforcement of fishing regulations.* Peru, Australia, and New Zealand use satellite-based systems to monitor their waters to prevent overfishing and keep vessels from moving into no-fishing zones or areas off limits to foreign vessels.

Economic Approaches

■ *Sharply reduce or eliminate fishing subsidies* (Connections, p. 635).

■ *Impose fees for harvesting fish and shellfish from publicly owned and managed offshore waters and use the money for government fishery management, as Australia does.*

Bycatch

■ *Reduce bycatch levels.*

■ Ways to do this include (1) fitting trawling nets with devices to exclude seabirds and sea turtles (Figure 24-13), (2) using wider-mesh nets to allow smaller species and smaller individuals of the targeted species to escape, (3) having observers on fishing vessels, (4) licensing boats to catch several species instead of only one target species, (5) finding ways to convert unwanted species to fish meal or fish sticks, (6) using echo-sounding equipment to find schools of a target species, and (7) enacting laws that prohibit throwing edible and marketable fish back to sea (as Nambia and Norway have done, with the law enforced by onboard observers).

Protected Areas

■ *Establish no-fishing marine areas, seasonal fishery closures, and marine protected areas to allow depleted fish species to recover.*

■ *Start by protecting marine habitats that are in good condition or those most likely to benefit from protection instead of using limited funds to protect potentially hopeless cases.*

Nonnative Invasions

■ *Reduce invasions by nonnative aquatic species.*

■ Ways to do this include (1) using heat or disinfectants to kill organisms in ship ballast water, (2) developing filters to trap the organisms when ballast water is taken into or discharged from a ship, and (3) requiring ships to dump their ballast water at least 320 kilometers (200 miles) from shore and replace it with deep-sea water.

Consumer Information

■ *Use labels that allow consumers to identify fish that have been harvested sustainably.** Appendix 6 lists seafood species that consumers should consider not buying and those that are safe for now, as compiled in 2000 by the National Audubon Society and the Monterey Bay Aquarium.

Aquaculture

■ *Restrict location of fish farms to reduce loss of mangrove forests and other threatened coastal environments.*

■ *Enact and enforce stricter pollution regulations for aquaculture operations.*

---

*In 1996, the World Wide Fund for Nature and one of the world's largest seafood product manufacturers (Anglo-Dutch Unilever) formed the Marine Stewardship Council (MSC) to devise criteria for sustainable fish harvesting and identify such fish with an MSC symbol, similar to the Forest Stewardship Council symbol used to identify and label sustainably harvested wood. Unilever, which buys 25% of the world's annual whitefish catch, has pledged not to buy any fish products after 2005 that have not been certified as being sustainably harvested. However, the World Trade Organization (p. 733) may challenge such labeling systems for fish and wood as international trade barriers.

- Raise aquaculture fish that need little or no grain or fish meal in their diets.

24-5 PROTECTING, SUSTAINING, AND RESTORING WETLANDS

How Are Wetlands Protected in the United States?
Coastal wetlands (Figure 7-8, p. 158, and Figure 7-9, p. 159) and inland wetlands (Figure 7-19, p. 170) are important reservoirs of aquatic biodiversity that provide many important ecological and economic services (p. 158).

In the United States, a federal permit is required to fill or to deposit dredged or fill material into wetlands occupying more than 1.2 hectares (3 acres). According to the U.S. Fish and Wildlife Service, this law has helped cut the average annual wetland loss by 75% since 1969.

However, there are continuing attempts to weaken it by using unscientific criteria to classify areas as wetlands. Only about 8% of remaining inland wetlands are under federal protection, and federal, state, and local wetland protection is weak.

The stated goal of current U.S. federal policy is zero net loss in the function and value of coastal and inland wetlands. A policy known as *mitigation banking* allows destruction of existing wetlands as long as an equal area of the same type of wetland is created or restored.

Some wetland restoration projects have been successful (Individuals Matter, right). However, experience has shown that (1) at least half of the attempts to create new wetlands fail to replace lost ones, (2) most of the created wetlands bear little resemblance to natural wetlands, and (3) restoring and creating wetlands is expensive. Wetlands are also being created to serve as sewage treatment plants (Solutions, p. 496) and to treat hog wastes, with the goal of eliminating much of the storage of such wastes in open lagoons that can pollute the air, groundwater, and nearby streams (Case Study, p. 486).

How Can Wetlands Be Sustained and Restored?
Ecologists and environmentalists call for several strategies to protect and sustain existing wetlands, restore degraded ones, and create new ones. They include:

- Using comprehensive land-use planning to steer developers, farmers, and resource extractors away from wetlands.

- Using mitigation banking only as a last resort.

- Requiring that a new wetland be created and evaluated *before* any existing wetland can be destroyed.

- Restoring degraded wetlands.

- Trying to prevent and control invasions of wetlands by nonnative species (Spotlight, p. 636).

Restoring a Wetland

INDIVIDUALS MATTER

Humans have drained, filled in, or covered over swamps, marshes (Figure 7-8, p. 158), and other wetlands for centuries. This has been done to (1) create rice fields and other land to grow crops, (2) create land for urban development and highways, (3) reduce disease such as malaria caused by mosquitoes, and (4) extract minerals, oil, and natural gas.

Some have begun to question such practices as we learn more about the ecological and economic importance of coastal wetlands (p. 158) and inland wetlands (p. 168). Can we turn back the clock to restore or rehabilitate lost marshes?

California rancher Jim Callender decided to try. In 1982, he bought 20 hectares (50 acres) of Sacramento Valley ricefield that had been a marsh until the early 1970s. To grow rice, the previous owner had destroyed the marsh, bulldozing, draining, leveling, uprooting the native plants, and spraying with chemicals to kill the snails and other food of the waterfowl.

Callender and his friends set out to restore the marshland. They hollowed out low areas, built up islands, replanted tules and bulrushes, reintroduced smartweed and other plants needed by birds, and planted fast-growing Peking willows. After 6 years of care, hand planting, and annual seeding with a mixture of watergrass, smartweed, and rice, the marsh is once again a part of the Pacific Flyway used by migratory waterfowl (Figure 22-25, p. 581).

Jim Callender and others have shown that at least part of the continent's degraded or destroyed wetlands can be reclaimed with scientific knowledge and hard work. Such restoration is useful, but to most ecologists the real challenge is to protect remaining wetlands from harm in the first place.

Many developers, farmers, and resource extractors vigorously oppose such wetland protection.

Case Study: Can We Restore the Florida Everglades?
South Florida's Everglades was once a 100-kilometer-wide (60-mile-wide), knee-deep sheet of water flowing slowly south from Lake Okeechobee to Florida Bay (Figure 24-16).

As this shallow body of water trickled south to Florida Bay, it created a vast network of wetlands with a variety of wildlife habitats. Today the Everglades is a haven for 56 endangered or threatened species, including the American alligator (Connections, p. 181) and the highly endangered Florida panther (p. 549).

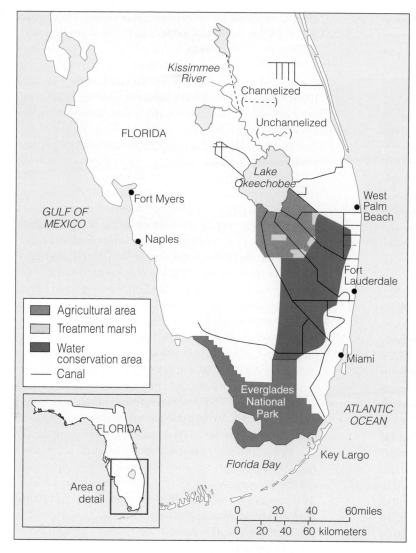

Figure 24-16 The world's largest ecological restoration project is an attempt to undo and redo an engineering project that has been destroying Florida's Everglades and threatening water supplies for south Florida's growing population.

ural Everglades has shrunk to half of its original size and dried out, leaving large areas vulnerable to summer wildfires.

To help preserve the lower end of the system, in 1947 the U.S. government established Everglades National Park, which contains about 20% of the remaining Everglades. However, this did not work because (as environmentalists had predicted) the massive plumbing and land development project to the north cut off much of the water flow needed to sustain the park's wildlife. As a result, (1) 90% of the park's wading birds have vanished, (2) populations of other vertebrates, from deer to turtles, are down 75–95%, and (3) it has become the country's most endangered national park.

Large volumes of fresh water that once flowed through the park into Florida Bay have been diverted for crops and cities, causing the bay to become saltier and warmer. This and increased nutrient input from cropfields and cities have stimulated the growth of large algae blooms in the bay. This large-scale cultural eutrophication (p. 482) threatens the coral reefs and the diving and fishing industries of the Florida Keys.

By the 1970s, state and federal officials recognized that this huge plumbing project threatens (1) wildlife (a major source of tourism income for Florida) and (2) the water supply for the 6 million residents of south Florida and the 6 million more people projected to be living there by 2050.

After more than 20 years of political haggling, in 1990 Florida's state government and the federal government agreed on the world's largest ecological restoration project, to be carried out by the U.S. Army Corps of Engineers between 2000 and 2038. This massive replumbing of south Florida is projected to cost at least $7.8 billion, with the cost to be shared equally by Florida and the federal government. In 2000, Congress approved $1.4 billion for the first phase of the project, and Florida has approved contributing $2 billion between 2000 and 2010.

The project's major goals are to (1) restore the curving flow of more than half of the Kissimmee River, (2) remove 400 kilometers (250 miles) of canals and levees blocking water flow south of Lake Okeechobee, (3) buy 240 square kilometers (93 square miles) of farmland and allow it to flood to create artificial marshes to filter agricultural runoff before it reaches the Everglades National Park, (4) add land adjacent to Everglades National Park's eastern border, (5) create a network of artificial marshes, (6) create 18 large reservoirs (including drilling hundreds of holes in the briny Florida aquifer and pumping fresh water into it during the wet season

Since 1948, much of the Everglades' natural flow south has been diverted and disrupted by 2,250 kilometers (1,400 miles) of canals, levees, spillways, and pumping stations. The most devastating blow came in the 1960s, when the U.S. Army Corps of Engineers transformed the meandering 103-mile-long Kissimmee River (Figure 24-16) into a straight 84-kilometer (56-mile) canal. The canal provided flood control by speeding the flow of water, but it drained water from large wetlands north of Lake Okeechobee, which farmers turned into cow pastures.

Below Lake Okeechobee, vast agricultural fields of sugarcane and vegetables were planted. The runoff of phosphorus and other plant nutrients from these fields has stimulated the growth of cattails, which have taken over and displaced saw grass, choked waterways, and disrupted food webs in a vast area of the Everglades. Mostly as a result of these human alterations, the nat-

and pumping it out during the dry season) to ensure an adequate water supply for south Florida's current and projected population and the lower Everglades, and (7) build new canals, reservoirs, and huge pumping systems to capture 80% of the water currently flowing out to sea and return it to the Everglades (Figure 24-16).

Whether this huge ecological restoration project will work depends not only on the abilities of scientists and engineers but also on prolonged support from citizens and elected state and federal officials. The need to make such expensive efforts to undo some of the damage to the Everglades caused by 120 years of agricultural and urban development is another example of a fundamental lesson from nature: Prevention is the cheapest and best way to go.

24-6 PROTECTING, SUSTAINING, AND RESTORING LAKES AND RIVERS

What Are the Greatest Environmental Threats to Lakes? The major threats to the ecological functioning and biodiversity of lakes are (1) pollution, especially from cultural eutrophication (Figure 19-5, p. 482, Figure 19-6, p. 483, and Figure 19-7, p. 484) and toxic wastes (Figure 19-4, p. 481, and Figure 19-7, p. 484), (2) invasion by nonnative species (Case Study, right), and (3) dropping water levels by diversion of water for irrigation, which has sharply reduced the biodiversity of the Aral Sea (Case Study, p. 305 and Figure 13-12, p. 305). Siberia's Lake Baikal, one of the world's largest and most biologically diverse lakes, is under threat mostly from pollution caused by increased forestry and industrial plants around its shores (Case Study, p. 654).

In many cases, cultural eutrophication of lakes can be reversed by reducing the input of plant nutrients (nitrates and phosphates) through better sewage treatment or by diverting polluted water into less sensitive areas. Lake Washington in the Seattle, Washington, metropolitan area is a success story of recovery from severe cultural eutrophication caused by decades of sewage inputs.

Recovery took place within about 4 years after the sewage was diverted into Puget Sound. This worked for three reasons: (1) A large body of water (Puget Sound) with a rapid rate of exchange with the Pacific Ocean was available to receive and dilute the sewage wastes, (2) the lake had not yet filled with weeds and sediment because of its large size and depth, and (3) preventive corrective action was taken before the lake had become a shallow, highly eutrophic lake (Figure 7-15, bottom, p. 166). Today, the lake's water quality is good, but there is concern about increased urban runoff caused by the area's rapidly growing population.

Since 1972, cultural eutrophication and inputs of toxic chemicals into the Great Lakes have been signif-

icantly reduced by an ongoing joint program carried out by the United States and Canada (Case Study, p. 483, and Figure 19-7, p. 484).

Case Study: Invaders in the Great Lakes Since the 1920s the Great Lakes have been invaded by at least 145 nonnative species including the sea lamprey, zebra mussel, quagga mussel, round goby, Eurasian ruffe, and hydrilla. In 1986, larvae of a nonnative species, the *zebra mussel*, arrived in ballast water discharged from a European ship near Detroit, Michigan. The rapidly reproducing mussel, a native of eastern Europe and western Asia, disperses widely, mostly through canals, in ship ballast, and on the bottoms of recreational boats.

The *bad news* is that with no known natural enemies, these thumbnail-sized mussels have (1) displaced other mussel species, (2) depleted the food supply for other Great Lakes species, (3) clogged irrigation pipes, (4) shut down water intake systems for power plants and city water supplies, (5) fouled beaches, (6) grown in huge masses on boat hulls, piers, pipes, rocks, and almost any exposed aquatic surface, and (7) have cost the Great Lakes basin at least $700 million per year (annual costs could reach $5 billion within a few years). The zebra mussel has spread and is dramatically altering freshwater communities in parts of southern Canada and 18 states in the United States (Figure 24-17).

However, zebra mussels may be *good news* for a number of aquatic plants. By consuming algae and other microorganisms, the mussels increase water clarity. Clearer waters permit deeper penetration of sunlight and more photosynthesis. This allows some native plants to thrive and return the plant composition of Lake Erie (and presumably other lakes) closer to what it was 100 years ago. Because the plants provide food and increase dissolved oxygen, their comeback may benefit certain aquatic animals (including the mussels).

There is more *bad news*, however. In 1991, a larger and potentially more destructive species, the *quagga mussel*, invaded the Great Lakes, probably brought in by a Russian freighter. It can survive at greater depths and tolerate more extreme temperatures than the zebra mussel. There is concern that it may eventually colonize areas such as the Chesapeake Bay and waterways in parts of Florida.

By 1999, a European fish, the *round goby*, had invaded all of the Great Lakes. The *good news* is that these tiny predators have a voracious appetite for zebra mussels. The *bad news* is that these bottom-dwelling fish also devour eggs and the young of any fish sharing their habitat. This includes prized recreational fish such as perch, walleye, and smallmouth bass.

A related problem is that zebra mussels pick up and store toxic pollutants from the water. Mussel-eating gobies can then pass these pollutants on to fish preying on the gobies. This can transfer the poisons further up the food web and contaminate fish that people eat.

Another recent invader from Europe's Black, Caspian, and Azov Seas is the *Eurasian ruffe*. This small bottom-feeding fish preys on bottom insects, perch and trout eggs, and young fish. Because it reproduces rapidly, it overwhelms native populations.

In 2000, biologist John Maden warned that the *hydrilla*, an alien plant species that has invaded waterways in Florida and Connecticut, could soon invade the Great Lakes. This thick weed can form a carpet over water surfaces at the rate of 2.4 centimeters (1 inch) per day. Its spread in the Great Lakes could disrupt biodiversity and cause navigation, flood control, and hydroelectric problems.

Some scientists warn that nonnative microorganisms in ship ballast water pose a greater danger than bigger invading species such as mussels and fish. Each liter of ballast water carries billions of nonnative viruses and bacteria, including some disease-causing bacterial strains of cholera.

What Are the Greatest Environmental Threats to Rivers? Although rivers and streams contain only about 1.3% of the world's nonfrozen fresh water, they provide important ecological services (Figure 13-11, p. 304). Major threats to the ecological services and the biodiversity of rivers are **(1)** pollution (Figure 14-1, p. 320, and Figure 19-3, p. 479), **(2)** disruption of water flows and species composition by dams (Figure 13-9, p. 301, and Figure 13-13, p.306), channelization (Figure 13-22, middle, p. 314), and the diversion of water from rivers for irrigation and urban areas (Figure 13-10, p. 304), and **(3)** overfishing.

Case Study: Managing the Columbia River Basin for People and Salmon The Columbia River, flowing 1,900 kilometers (1,200 miles) through the Pacific Northwest, receives water from a huge basin extending across parts of seven U.S. states and two Canadian provinces (Figure 24-18).

This basin has the world's largest hydroelectric power system, consisting of more than 119 dams, 19 of them large hydroelectric dams. Most of these dams were built fairly cheaply by the U.S. government in the 1930s to **(1)** furnish jobs, **(2)** produce cheap electricity (40% below the national average), **(3)** provide flood control, and **(4)** help stimulate industrial and agricultural development. The river and its reservoirs are also used for

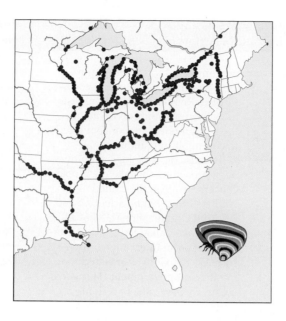

Figure 24-17 Since 1986, the nonnative zebra mussel has established rapidly growing populations in southern Canada and 18 states in the United States. (Data from U.S. Department of Interior)

recreation (such as swimming, boating, and wind surfing) and for sport and commercial fishing for salmon, steelhead trout, and other fish.

Since the dams were built, the Columbia River's migratory wild Pacific salmon population has dropped by 94%. Nine Pacific Northwest salmon species are listed as endangered or threatened under the Endangered Species Act. Major causes of this decline include **(1)** dams and reservoirs that hinder or prevent salmon migration during their life cycles (Figure 12-29, left, p. 288), **(2)** overfishing of salmon in the Pacific Ocean, **(3)** destruction of salmon spawning grounds in streams by sediment from logging and mining, and **(4)** withdrawals of water for irrigation and other human uses.

Commercial fishing operations have modified the salmon's natural cycle by using *salmon ranching*, in which salmon eggs and young are raised in a hatchery and then released (Figure 12-29, p. 288). However, there are problems with hatchery-raised salmon: **(1)** their

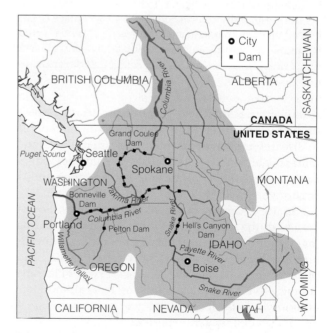

Figure 24-18 The Columbia River basin. (Data from Northwest Power Planning Council)

Protecting Lake Baikal

Lake Baikal in southern Siberia is the world's oldest and deepest lake and contains the world's largest volume of fresh water.

The lake is famous for its breathtaking scenery and its biodiversity. It contains about 1,500 species of plants and animals, 1,200 of them not found anywhere else on the earth (compared to only 4 such endemic species in Lake Superior). The lake's huge watershed consists almost entirely of boreal forest or taiga (Figure 6-30, p. 147).

By global standards most of Lake Baikal's waters are still clean. However, its purity and biodiversity are threatened by (1) extensive logging of some of its surrounding coniferous forests, which has led to soil erosion and landslides, (2) acid deposition from nearby industrial centers, and (3) serious air and water pollution from more than 100 factories on the lake's shores.

The most notorious source of pollution is a large pulp and paper mill at Baykalsk on the lake's southern shore. Since it opened in 1966, the plant has been emitting large quantities of air and water pollutants, especially chlorinated hydrocarbon compounds such as dioxins and PCBs, which can be biologically magnified in food webs (Figure 19-4, p. 481). Since the plant opened in 1966, Russian environmentalists have tried without success to have the plant modernized or converted to a nonpolluting industry.

Because of its biological uniqueness, protecting the lake is a global concern. In 1996, the United Nations Educational, Scientific and Cultural Organization (UNESCO) added Lake Baikal to its World Heritage list as an international treasure. UNESCO recommended that the Russian government (1) enact a law to convert Lake Baikal into a protected ecological zone, (2) convert the Baykalsk pulp and paper plant to a nonpolluting industry, (3) cease logging in the lake's water basin, and (4) improve environmental monitoring of the lake.

By 2001, the Russian government had not put any of these recommendations into effect. Currently, Russian environmentalists are trying protect Lake Baikal by (1) bringing court action to force the Baykalsk plant owners to pay for cleaning up environmental damage they inflict, as required under Russian law, (2) exerting international pressure on the government by pressuring UNESCO to place Lake Baikal on its list of endangered World Heritage Sites, and (3) asking the United States and other developed countries and international environmental organizations to help fund improved environmental monitoring and conversion of the Baykalsk pulp and paper plant to an environmentally safe form of production (at an estimated cost of $100 million).

The plant's 3,000 workers and many people living in nearby Baykalsk, a town of 17,000 near the pulp and paper plant, fear for their jobs, with little opportunity for other work in Russia's troubled economy. Opponents of forcing the plant to modernize contend that (1) environmentalists have exaggerated the pollution of Lake Baikal, (2) the lake is so vast that the pollution from the plant poses no threat to its biodiversity, and (3) a 1998 joint study by Russian and British scientists found no evidence that pollution is harming the lake's biodiversity.

Critical Thinking

1. Should the United States and other developed countries and international environmental organizations provide funds to help protect Lake Baikal? Explain.

2. Use the library and the internet to find information about the latest developments in efforts to protect Lake Baikal.

close quarters permit rapid spread of diseases, (2) because of their genetic uniformity, they are more susceptible to diseases and environmental stress after release, and (3) ranch salmon that interbreed with wild ones reduce the genetic diversity of the wild fish and their ability to survive.

In 1980, Congress passed the Northwest Power Act, with the goal of developing and implementing long-range plans to meet the region's electricity needs and to rebuild wild and hatchery-raised salmon and other fish populations. The Northwest Power Planning Council (NPPC), which combines federal and state authority, administers this act.

The NPPC began experimenting with ways to restore the Columbia River's wild salmon runs by (1) building hatcheries upstream of the dams and releasing juveniles from these hatcheries to underpopulated streams (so they will return to them as adults to reproduce), (2) building fish ladders to enable some of the adult salmon to bypass dams during their upstream migration, (3) using trucks and barges to transport juvenile wild salmon around dams and turning off turbines to allow juveniles to swim safely over dams during periods of heavy downstream migration, (4) releasing extra water from dams to help wash juvenile salmon downstream at a faster rate closer to their natural migration rate, (5) putting more than 64,000 kilometers (40,000 miles) of stream off limits for hydropower development, and (6) obliterating old logging roads and reducing the runoff of silt from existing dirt logging roads above salmon spawning streams.

Environmentalists, Indian tribes, and commercial salmon fishers have urged to government to remove four small hydroelectric dams on the lower Snake River in Washington to restore spawning habitat. This measure is opposed by farmers, barge operators, and aluminum workers who argue that it would devastate local economies by **(1)** reducing the supply of irrigation water, **(2)** eliminating cheap transportation of commodities by ship in the affected areas, and **(3)** reducing the supply of cheap electricity for industries and consumers.

In 2000, the federal government decided against any dam removal for the Columbia River basin. Instead it proposed spending $1 billion to protect salmon by **(1)** restoring streamside habitat and water quality in rivers and streams in which salmon spawn, **(2)** releasing more water from Idaho reservoirs to help young salmon migrate to the ocean, **(3)** limiting salmon harvests for 10 years while maintaining tribal fishing rights, and **(4)** expanding hatcheries designed specifically to rebuild wild salmon populations.

It will take decades to see whether the salmon populations can be rebuilt. Despite problems, this program demonstrates that people with diverse and often conflicting economic, political, and environmental interests can work together to try new ideas and develop potentially sustainable solutions to complex resource management issues.

Critics of this expensive salmon restoration program argue that **(1)** populations of wild salmon are stable in Alaska, so we should not care that wild salmon are declining in the Pacific Northwest, and **(2)** the economic costs to the hydroelectric power, shipping, and timber industries and to farmers and consumers exceed the value of saving the salmon. What do you think?

How Can Freshwater Fisheries Be Managed and Sustained? Managing freshwater fish involves encouraging populations of commercial and sport fish species and reducing or eliminating populations of less desirable species. This is usually done by regulating the time and length of fishing seasons and the number and size of fish that can be taken.

Other techniques include **(1)** building reservoirs and farm ponds and stocking them with fish, **(2)** fertilizing nutrient-poor lakes and ponds, **(3)** protecting and creating spawning sites, **(4)** protecting habitats from sediment buildup and other forms of pollution, **(5)** removing debris, **(6)** preventing excessive growth of aquatic plants from cultural eutrophication (excessive inputs of plant nutrients from human activities, Figure 19-5, p. 482), and **(7)** building small dams to control water flow.

Predators, parasites, and diseases can be controlled by **(1)** improving habitats, **(2)** breeding genetically resistant fish varieties, and **(3)** using antibiotics and disinfectants judiciously. Hatcheries can be used to restock ponds, lakes, and streams with prized species such as trout and salmon, and entire river basins can be managed to protect such valued species as salmon (Figure 12-29, p. 288).

How Can Wild and Scenic Rivers Be Protected and Restored? In 1968, the U.S. Congress passed the National Wild and Scenic Rivers Act. It allows rivers and river segments with outstanding scenic, recreational, geological, wildlife, historical, or cultural values to be protected in the National Wild and Scenic Rivers System.

These waterways are to be kept free of development and cannot be widened, straightened, dredged, filled, or dammed along the designated lengths. The only activities allowed are camping, swimming, nonmotorized boating, sport hunting, and sport and commercial fishing. New mining claims are permitted in some areas, however.

Currently, only 0.2% of the country's 6 million kilometers (3.5 million miles) of waterways along some 150 stretches are protected by the Wild and Scenic Rivers System. In contrast, 17% of the country's wild river length has been tamed and land flooded by dams and reservoirs.

Environmentalists have urged Congress to add 1,500 additional river segments to the system, a goal that is vigorously opposed by some local communities and anti-environmental groups. If this goal were achieved, about 2% of the country's river systems would be protected.

Environmentalists also urge the establishment of a permanent federal administrative body to manage the Wild and Scenic Rivers System and development of wild and scenic river programs by individual states.

There is also a need to focus on improving the condition of the country's most endangered rivers. In 1997, President Clinton launched the American Heritage Rivers program to recognize community-based efforts to restore and protect the environmental, economic, cultural, and historic values of American rivers. Scores of communities in 46 states and the District of Columbia nominated rivers for designation under the program.

In this chapter, we have seen that threats to marine biodiversity are real and growing and are even greater than threats to terrestrial biodiversity (Chapters 22 and 23). Keys to sustaining life in the waters that occupy 71% of the earth's surface include **(1)** increasing research to learn more about aquatic life, **(2)** greatly expanding efforts to protect and restore aquatic biodiversity, and **(3)** promoting integrated ecological management of connected terrestrial and aquatic systems.

To promote conservation fishers and officials need to view fish as a part of a larger ecological system, rather than simply as a commodity to extract.
ANNE PLATT MCGINN

REVIEW QUESTIONS

1. Define the boldfaced terms in this chapter.

2. Describe the ecological and economic problems of Africa's Lake Victoria and list two ways to reduce these threats.

3. What are the three most biologically diverse habitats found in the world's oceans?

4. List seven general patterns of marine biodiversity.

5. List the major ecological and economic services provided by marine systems.

6. List the major ecological and economic services provided by freshwater systems.

7. List the major human impacts on aquatic biodiversity in terms of (a) species loss and endangerment, (b) marine habitat loss and degradation, (c) freshwater habitat loss and degradation, (d) overfishing, (e) nonnative species, and (f) pollution and global warming.

8. Explain how most commercial marine fishing can be viewed as an example of the tragedy of the commons.

9. Describe the threat to wetlands from invasion by purple loosestrife.

10. List five factors that make it difficult to protect marine biodiversity.

11. List five major ways to protect marine biodiversity and describe the two major problems with protecting marine biodiversity.

12. List four threats to marine sea turtles.

13. Describe efforts to protect endangered sea turtles in the Mazunte area of Mexico.

14. Distinguish between *toothed whales* and *baleen whales*. Explain how commercial whaling has been an example of the tragedy of the commons and list the pros and cons of resuming commercial whaling.

15. Describe factors leading to the near extinction of the blue whale.

16. List four examples of marine sanctuaries and four limitations of the current international program to establish *marine protected areas*.

17. List the advantages and difficulties of *integrated coastal management*.

18. Describe methods used to reduce beach erosion and explain why they usually fail.

19. List the pros and cons of federal flood insurance and beach replenishment subsidies in the United States.

20. Describe the strengths and weaknesses of projecting fish populations using (a) maximum sustained yield (MSY), (b) optimum sustained yield (OSY), (c) multi-species management, (d) large marine system management, and (e) the precautionary principle.

21. List the advantages and disadvantages of limiting access to marine fisheries by (a) international and national laws, (b) community-based comanagement, and (c) individual transfer quotients (ITQs).

22. List ways to prevent overfishing and protect marine biodiversity by (a) enacting fishery regulations, (b) using economic approaches, (c) reducing bycatch, (d) setting aside protected areas, (e) dealing with nonnative invasions, (f) providing consumer information, and (g) controlling aquaculture.

23. List the pros and cons of the current wetland protection policy in the United States.

24. Describe five ways to sustain and restore wetlands.

25. Describe the ecological and water supply problems of south Florida's Everglades and describe efforts to deal with these problems.

26. What are the three greatest environmental threats to lakes? Describe the recovery of Washington state's Lake Washington.

27. Describe the biological importance of Siberia's Lake Baikal and the problems that threaten this lake's biodiversity.

28. Describe the threats to the Great Lakes from invasions by nonnative species.

29. What are the three major threats to the world's streams and rivers?

30. Describe efforts to manage the Pacific Northwest's Columbia River for people and wild salmon.

31. Describe the National Wild and Scenic Rivers System and the American Heritage Rivers program in the United States.

CRITICAL THINKING

1. What three things would you do to deal with the ecological and economic problems of Africa's Lake Victoria?

2. Why is marine biodiversity higher (a) near coasts than in the open sea, (b) on the ocean's bottom than at its surface, and (c) in the open ocean waters of the tropics than in the open waters of the north pole?

3. Why is it more difficult to identify and protect endangered marine species than to protect such species on land?

4. Should commercial whaling of selected species with fairly high population levels be resumed? Explain. How would you establish, monitor, and enforce annual commercial whaling quotas?

5. Are you for or against federal subsidies in the United States for (a) beach replenishment and (b) flood insurance? Explain.

6. List five methods for using the precautionary principle as a way to manage marine fisheries and help protect marine biodiversity.

7. Should fishers harvesting fish from a country's publicly owned waters be required to pay the government

(taxpayers) fees for the fish they catch? Explain. If your livelihood depended on commercial fishing, would you be for or against such fees?

8. Are you for or against using *mitigation banking* to help sustain wetlands? Explain.

9. Congratulations. You have just been put in charge of protecting the world's aquatic biodiversity. List the five most important points of your policy to accomplish this goal.

PROJECTS

1. Survey the condition of a nearby wetland, coastal area, river, or stream. Has its condition improved or deteriorated during the last ten years? What local, state, or national efforts are being to protect this aquatic system? Develop a plan for protecting this system.

2. Use the library or the internet to find bibliographic information about *G. Carleton Ray* and *Anne Platt McGinn*, whose quotes appear at the beginning and end of this chapter.

3. Make a concept map of this chapter's major ideas, using the section heads and subheads and the key terms (in boldface). Look at the inside back cover and on the website for this book for information about making concept maps.

INTERNET STUDY RESOURCES AND RESOURCES FOR FURTHER READING AND RESEARCH

The website for this book contains helpful study aids and many ideas for further reading and research. Log on to:

http://www.brookscole.com/product/0534376975s

and click on the Chapter-by-Chapter area. Choose Chapter 24 and select a resource:

- "Flash Cards" allows you to test your mastery of the Terms and Concepts to Remember for this chapter.

- "Tutorial Quizzes" provides a multiple-choice practice quiz.

- "Student Guide to InfoTrac" will lead you to Critical Thinking Projects that use InfoTrac College Edition as a research tool.

- "References" lists the major books and articles consulted in writing this chapter.

- "Hypercontents" takes you to an extensive list of sites with news, research, and images related to individual sections of the chapter.

INFOTRAC COLLEGE EDITION

Improve your skills with InfoTrac College Edition, a searchable online database of articles from more than 700 periodicals. Log on to:

http://www.infotrac-college.com

or access InfoTrac through the website for this book.

Try the following articles:

Wilder, R.J. M.J. Tegner, P.K. Dayton. 2000. Saving marine biodiversity. *Issues in Science and Technology*, Spring 1999 vol. 15, no. 3, pp. 57–64. (keywords: marine biodiversity)

Dudgeon, D. 2000. Large-scale hydrological changes in tropical Asia: prospects for riverine biodiversity. *BioScience* vol. 50, no. 9, p. 793–806. (keywords: rivers, biodiversity)

25 SUSTAINABLE CITIES: URBAN LAND USE AND MANAGEMENT

The Ecocity Concept in Davis, California

Few of today's cities are sustainable. They depend on distant sources for their food, water, energy, and materials, and their massive use of resources damages nearby and distant air, water, soil, and wildlife.

Environmental and urban designers envision the development of more sustainable cities, called *ecocities* or *green cities*. The ecocity is not a futuristic dream. The citizens and elected officials of Davis, California, a city of about 54,000 people about 130 kilometers (80 miles) northeast of San Francisco, committed themselves in the early 1970s to making their city more ecologically sustainable.

Davis's Village Homes, America's first solar neighborhood, was developed in the 1970s by Michael and Judy Corbett (Figure 25-1). It has 240 passive solar and energy-efficient houses facing into a common open space reserved for people and bicycles. Cars are parked around the back on narrow, tree-lined streets that provide shade and natural cooling.

The abundant open space has shared (1) orchards, (2) vineyards, (3) organic gardens (with some of the vegetables sold to help finance upkeep of the community's open space), (4) playgrounds and playing fields, (5) natural surface drainage by vegetation-covered swales (instead of costly underground concrete drains) that allow water to soak in and are used as walking and bike paths, and (6) a solar-heated community center used for day care, meetings, and social gatherings.

Even though it was built in the 1970s, Village Homes is considered one of Davis's most desirable subdivisions.

In Davis, where peak temperatures can reach 45°C (113°F), building codes encourage the use of solar energy for water and space heating, and all new homes must meet high energy efficiency standards. Since 1975, the city has cut its use of energy for heating and cooling in half. It has a solar power plant and has plans to generate all its own electricity using renewable energy.

The city discourages the use of automobiles and encourages the use of bicycles by (1) closing some streets to automobiles, (2) having bike lanes on major streets, and (3) building a network of bicycle paths. As a result, bicycles account for 40% of all in-city transportation, and less land is needed for parking spaces. The city's warm climate and flat terrain aid this heavy dependence on the bicycle.

Davis limits the type and rate of its growth, and it maintains a mix of homes for people with low, medium, and high incomes. Development of the fertile farmland surrounding the city for residential or commercial use is restricted. The city also limits the size of shopping centers to encourage smaller neighborhood shopping centers, each easily reached by foot or bicycle.

Between 2000 and 2100, the percentage of people living in urban areas is expected to increase from 50% to 80–90%. An exciting challenge for the 21st century will be reshape existing cities and design new ones like Davis that are more livable and sustainable and have a lower environmental impact.

Figure 25-1 Passive solar home in Village Homes in Davis, California, a neighborhood of 240 passive solar houses that face into a common open space. This community, developed in the 1970s and completed in 1982, is an experiment in urban sustainability. Direct solar gain provides 50–75% of the heating needs in winter. In the hot summers, most residents rarely need air conditioning because carefully sited trees shade the houses and narrow streets. (Virginia Thigpen)

The test of the quality of life in an advanced economic society is now largely in the quality of urban life. Romance may still belong to the countryside—but the present reality of life abides in the city.

JOHN KENNETH GALBRAITH

This chapter addresses the following questions:

- How is the world's population distributed between rural and urban areas, and what factors determine how urban areas develop?

- What are the major resource and environmental problems of urban areas?

- How do transportation systems shape urban areas and growth, and what are the pros and cons of various forms of transportation?

- What methods are used for planning and controlling urban growth?

- How can cities be made more sustainable and more desirable places to live?

25-1 URBANIZATION AND URBAN GROWTH

What Are Urban and Rural Areas? For more than 6,000 years, cities have been centers of commerce, education, technological developments, culture, social change, vitality, progress, and political power. They have also suffered from crowding, pollution, disease, poverty, and human misery.

An **urban** or **metropolitan area** often is defined as a town or city plus its adjacent suburban fringes with a population of more than 2,500 people (although some countries set the minimum at 10,000–50,000 residents). A **rural area** usually is defined as an area with a population of less than 2,500 people.

A **village** consists of a group of rural households linked together by custom, culture, and family ties, usually surviving by harvesting local natural resources for food, fuel, and other basic needs. By contrast, a **city** is a much larger group of people with a variety of specialized occupations who depend on a flow of resources from other areas to meet most of their needs and wants.

What Causes Urban Growth? A country's **degree of urbanization** is the percentage of its population living in an urban area. **Urban growth** is the rate of increase of urban populations.

Urban areas grow in two ways: **(1)** *natural increase* (more births than deaths) and **(2)** *immigration* (mostly from rural areas) caused by a combination of *push factors* that force people out of rural areas and *pull factors* that draw them into the city.

People can be pushed from rural areas into urban areas by factors such as **(1)** poverty, **(2)** lack of land, **(3)** declining agricultural jobs (because of increased use of mechanized agriculture or government policies that set food prices for urban dwellers so low that rural farmers find it uneconomical to grow crops), **(4)** famine, and **(5)** war.

Rural people are *pulled* to urban areas in search of jobs, food, housing, a better life, entertainment, and freedom. Many are also seeking an escape from social constraints of village cultural life and from religious, racial, and political conflicts. Urban growth in developing countries also is fueled by government policies that **(1)** distribute most income and social services to urban dwellers (especially in capital cities, where a country's leaders live) and **(2)** provide lower-priced food than in rural areas.

What Patterns of Urbanization and Urban Growth Are Occurring Throughout the World? Five trends are important in understanding the problems and challenges of urban growth.

- *The proportion of global population living in urban areas is increasing.* Between 1850 and 2000, the percentage of people living in urban areas increased from 2% to 50% (Figure 25-2). If UN projections are correct, by 2050 about 66% of the world's people will be living in urban areas, with 90% of this urban growth occuring in developing countries. Each day about 160,000 people are added to the world's urban areas.

- *The number of large cities is mushrooming.* In 1900, only 19 cities had a million or more people, and more than 95% of humanity lived in rural communities. Today, more than 400 cities have a million or more people, and there are 19 *megacities* with 10 million or more people (13 of them in developing countries). The world's five largest megacities are **(1)** Tokyo, Japan (28 million), **(2)** Mexico City, Mexico (18.1 million), **(3)** Mumbai (formerly Bombay), India (18 million), **(4)** São Paulo, Brazil (17.8 million), and **(5)** New York in the United States (16.6 million but 20.2 million in the entire metropolitan area).

- *Urbanization is increasing rapidly in developing countries.* Currently, about 38% of the people in developing countries live in urban areas. However, 74% of the people in South America live in cities, mostly along the coasts (Figure 25-2). By 2025, urbanization in developing countries is projected to increase to at least 54%, with the most rapid urban growth expected to take place in Africa and Asia. Many of these cities will be water short, waste filled, and choked with pollution.

Figure 25-2 Major urban areas throughout the world based on satellite images of the earth at night that show city lights. Currently, the 50% of the world's people living in urban areas occupy about 4% of the earth's land area. Note that **(1)** most of the world's urban areas are found along the coasts of continents and **(2)** most of Africa and much of the interior of South America, Asia, and Australia is dark at night. (National Geophysics Data Center, National Oceanic and Atmospheric Administration)

■ *Urban growth is much slower in developed countries* (with 75% urbanization) *than in developing countries.* Still, developed countries are projected to reach 82% urbanization by 2025.

■ *Poverty is becoming increasingly urbanized as more poor people migrate from rural to urban areas.* The United Nations estimates that at least 1 billion people live in the crowded *slums* of inner cities or in vast illegal *squatter settlements* and *shantytowns* in developing countries (Spotlight, p. 665).

Case Study: Mexico City

About 18.1 million people—about one of every five Mexicans—live in Mexico City (Figure 25-3), the world's second most populous city.

Mexico City suffers from **(1)** severe air pollution, **(2)** high unemployment (close to 50%), **(3)** deafening noise, and **(4)** a soaring crime rate. More than one-third of its residents live in crowded slums (called *barrios*) or squatter settlements, without running water or electricity.

At least 8 million people have no sewer facilities. This means that huge amounts of human waste are deposited in gutters and vacant lots every day, attracting armies of rats and swarms of flies. When the winds pick up dried excrement, a *fecal snow* often falls on parts of the city, leading to widespread salmonella and hepatitis infections, especially among children.

Some 4 million motor vehicles and 30,000 factories spew pollutants into the atmosphere. Air pollution is intensified because the city lies in a basin surrounded by mountains, and frequent thermal inversions trap pollutants at ground level (Fig-

Figure 25-3 The locations of Mexico, Brazil, Belize, and Costa Rica. Other countries highlighted here are discussed in other chapters.

ure 17-8, p. 426). Since 1982, the amount of contamination in the city's smog-choked air (Figure 17-5, p. 424) has more than tripled. Indeed, breathing the city's air is said to be roughly equivalent to smoking three packs of cigarettes a day.

The city's air and water pollution cause an estimated 100,000 premature deaths per year. Writer Carlos Fuentes has nicknamed this megacity "Makesicko City."

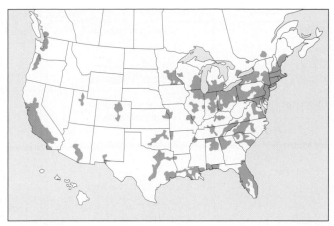

Figure 25-4 Major urban regions in the United States. About 75% of the U.S. population live in urban areas occupying about 3% of the country's land area. Nearly half (48%) of Americans live in *consolidated metropolitan areas* with 1 million or more people. (Data from U.S. Census Bureau)

Excessive groundwater withdrawal has caused parts of Mexico City to sink (subside) more than 9 meters (30 feet) since 1900. Some children mark their height on pipes and buildings to see whether they are growing faster than the ground is sinking.

The Mexican government is industrializing other parts of the country in an attempt to slow migration to Mexico City. Other efforts include **(1)** banning cars from a 50-block central zone, **(2)** taking taxis built before 1985 off the streets, **(3)** having buses and trucks run only on liquefied petroleum gas (LPG), **(4)** planting 25 million trees, **(5)** buying some land for green space, and **(6)** phasing out use of leaded gasoline.

If the city's population continues to grow as projected, these problems, already at crisis levels, will become even worse. If you were in charge of Mexico City, what would you do?

How Urbanized Is the United States? Between 1800 and 2000, the percentage of the U.S. population living in urban areas increased from 5% to 75%. This rural-to-urban population shift has taken place in three phases:

- *Migration to large central cities*. Currently, 75% of Americans live in 350 *metropolitan areas* (cities and towns with at least 50,000 people), and nearly half of the country's population lives in consolidated metropolitan areas containing 1 million or more residents (Figure 25-4).

- *Migration from large central cities to suburbs and smaller cities*. In 1999, about 51% of the U.S. population lived in suburbs.

- *Migration from the North and East to the South and West*. Since 1980, about 80% of the U.S. population increase has occurred in the South and West, particularly near the coasts. This shift is expected to continue. In 1999, the five most populous states were **(1)** California (33 million), **(2)** Texas (20 million), **(3)** New York (18 million), **(4)** Florida (15 million), and **(5)** Illinois (12 million). Between 1990 and 1999,

water-short Las Vegas, Nevada (Spotlight, p. 314) lead the nation in urban growth (62%).

Since the 1970s and especially in the 1990s there has been an *urban-to-rural shift* or *rural rebound* as fewer people have migrated from rural to urban areas and more urban dwellers have moved to rural areas. In the 1990s, 71% of rural counties gained population, mostly from an influx of former urban dwellers.

Reasons for this rural rebound include **(1)** population spillover from metropolitan areas, **(2)** increased jobs in manufacturing and service jobs and lower land costs in rural areas, and **(3)** the desire of retirees and many other people to live in rural areas. However, this shift may be over as the number of people who left rural areas in 1998 and 1999 outnumbered those who moved to metropolitan areas, mostly the suburbs.

What Are the Major Urban Problems in the United States? Here is some *good news*:

- Since 1920, many of the worst urban environmental problems in the United have been reduced significantly.

- Most people have better working and housing conditions, and air and water quality have improved.

- Better sanitation, public water supplies, and medical care have slashed death rates and the prevalence of sickness from malnutrition and transmittable diseases such as measles, diphtheria, typhoid fever, pneumonia, and tuberculosis.

- Concentrating most of the population in urban areas has helped protect the country's biodiversity by reducing the destruction and degradation of wildlife habitat.

Here is some *bad news*. A number of cities in the United States (especially older ones) have **(1)** deteriorating services, **(2)** aging infrastructures (streets, schools, bridges, housing, and sewers), **(3)** budget crunches from rising costs as some businesses and

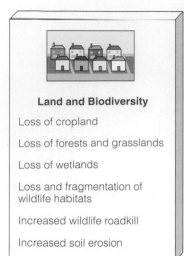

Land and Biodiversity

Loss of cropland

Loss of forests and grasslands

Loss of wetlands

Loss and fragmentation of wildlife habitats

Increased wildlife roadkill

Increased soil erosion

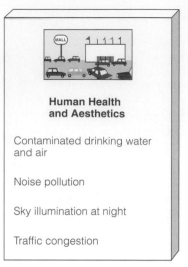

Human Health and Aesthetics

Contaminated drinking water and air

Noise pollution

Sky illumination at night

Traffic congestion

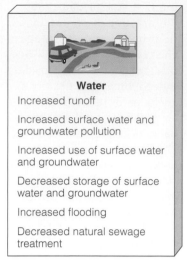

Water

Increased runoff

Increased surface water and groundwater pollution

Increased use of surface water and groundwater

Decreased storage of surface water and groundwater

Increased flooding

Decreased natural sewage treatment

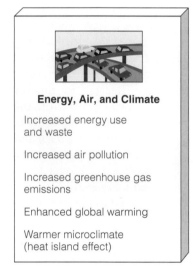

Energy, Air, and Climate

Increased energy use and waste

Increased air pollution

Increased greenhouse gas emissions

Enhanced global warming

Warmer microclimate (heat island effect)

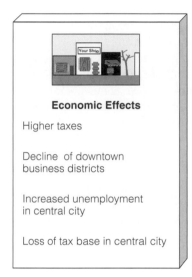

Economic Effects

Higher taxes

Decline of downtown business districts

Increased unemployment in central city

Loss of tax base in central city

Figure 25-5 Some of the undesirable impacts of urban sprawl.

people move to the suburbs or rural areas and reduce revenues from property taxes, and **(4)** rising poverty in many central city areas, where unemployment typically is 50% or higher.

Another major problem in the United States is **urban sprawl**: the growth of low-density development on the edges of cities and towns. Factors promoting urban sprawl in the United States since 1945 include **(1)** ample land for expansion, **(2)** government loan guarantees for new single-family homes for veterans of World War II, **(3)** government and state funding of highways that encourages the development of once-inaccessible outlying tracts of land, **(4)** low-cost gasoline (Figure 15-7, p. 364), which encourages automobile use, **(5)** greater availability of mortgages for homes in new subdivisions than in older cities and suburbs, and **(6)** state and local zoning laws that require large residential lots and separation of resi-

dential and commercial use of land in new communities. Figure 25-5 shows some of the undesirable consequences of urban sprawl.

What Are the Major Spatial Patterns of Urban Development? Three generalized models of urban structure are shown in Figure 25-6.

A *concentric circle city*, such as New York City, develops outward from its central business district (CBD) in a series of rings as the area grows in population and size. Typically industries and businesses in the CBD and poverty-stricken inner-city housing areas are ringed by housing zones that usually become more affluent toward the suburbs. In many developing countries, affluent residents cluster in the central city, and many of the poor live in squatter settlements that spring up on the outskirts.

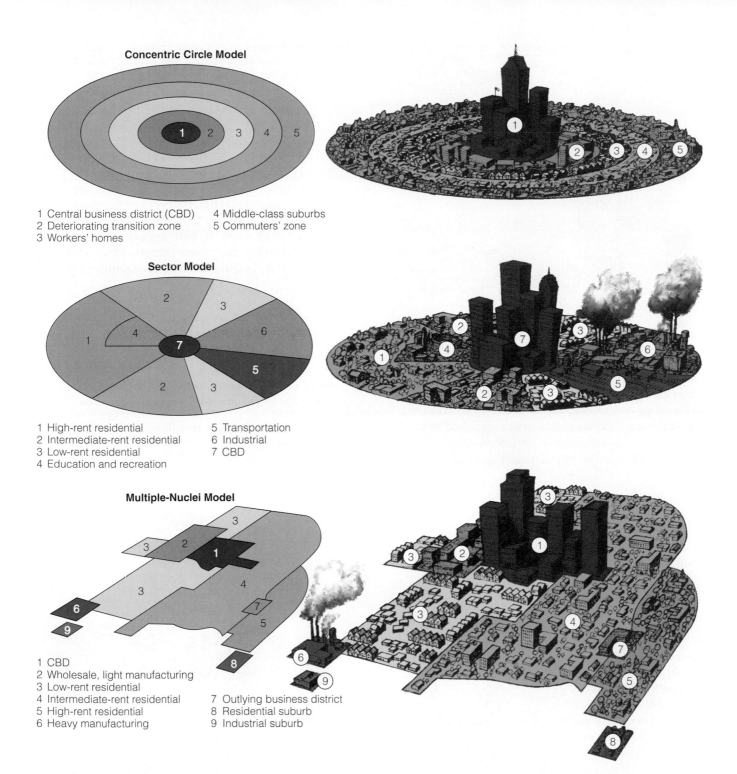

Concentric Circle Model

1 Central business district (CBD) 4 Middle-class suburbs
2 Deteriorating transition zone   5 Commuters' zone
3 Workers' homes

Sector Model

1 High-rent residential   5 Transportation
2 Intermediate-rent residential 6 Industrial
3 Low-rent residential   7 CBD
4 Education and recreation

Multiple-Nuclei Model

1 CBD
2 Wholesale, light manufacturing
3 Low-rent residential
4 Intermediate-rent residential
5 High-rent residential   7 Outlying business district
6 Heavy manufacturing   8 Residential suburb
                             9 Industrial suburb

Figure 25-6 Three models of urban spatial structure. Although no city perfectly matches any of them, these simplified models can be used to identify general patterns of urban development. (Modified with permission from Harm J. de Blij, *Human Geography*, New York: Wiley, 1977)

A *sector city* grows in pie-shaped wedges or strips when commercial, industrial, and housing districts push outward from the CBD along major transportation routes. An example is the large urban area extending from San Francisco to San Jose in California.

A *multiple-nuclei city* develops around a number of independent centers, or satellite cities, rather than a single center. Metropolitan Los Angeles comes fairly close to this pattern. Some cities develop in various combinations of these three patterns.

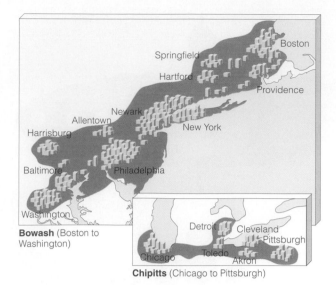

Bowash (Boston to Washington)

Chipitts (Chicago to Pittsburgh)

Figure 25-7 Two megalopolises: Bowash, consisting of urban sprawl and coalescence between Boston and Washington, D.C., and Chipitts, extending from Chicago to Pittsburgh.

As they grow and sprawl outward, separate urban areas may merge to form a *megalopolis*. For example, the remaining open space between Boston, Massachusetts, and Washington, D.C., is rapidly urbanizing and coalescing. This 800-kilometer-long (500-mile-long) urban area, sometimes called *Bowash* (Figure 25-7), contains almost 60 million people, about twice Canada's entire population. Figure 25-8 shows the spread of urbanization and urban sprawl in the Atlanta, Georgia, area between 1980 and 2000.

Megalopolises have developed all over the world. Examples include **(1)** the area between Amsterdam and Paris in Europe, **(2)** the Tokyo-Yokohama-Osaka-Kobe corridor (with nearly 50 million people) in Japan, and **(3)** the Brazilian Industrial Triangle made up of São Paulo, Rio de Janeiro, and Belo Horizonte.

25-2 URBAN RESOURCE AND ENVIRONMENTAL PROBLEMS

What Are the Health and Environmental Pros and Cons of Urbanization? Urbanization has some important health and environmental benefits:

■ In many parts of the world, urban populations live longer and have lower infant mortality rates than do rural populations.

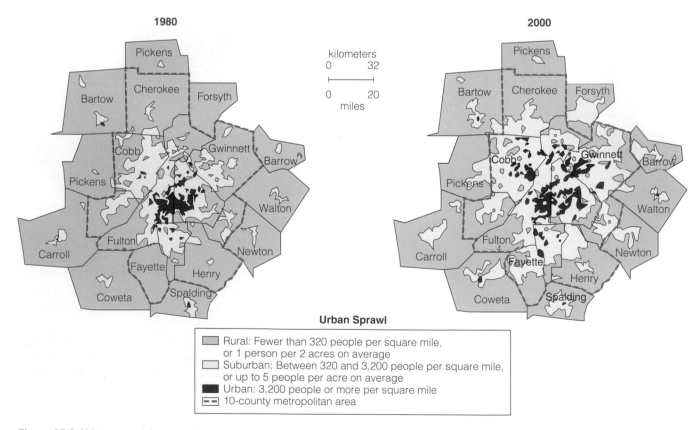

Urban Sprawl

- Rural: Fewer than 320 people per square mile, or 1 person per 2 acres on average
- Suburban: Between 320 and 3,200 people per square mile, or up to 5 people per acre on average
- Urban: 3,200 people or more per square mile
- 10-county metropolitan area

Figure 25-8 Urban sprawl. Increase in the urban and suburban area of Atlanta, Georgia, between 1980 and 2000. (Data from U.S. Bureau of the Census)

The Urban Poor

Squatter settlements and shantytowns in developing countries usually lack clean water supplies, sewers, electricity, and roads. Often the land on which they are built is not suitable for human habitation because of air and water pollution and hazardous wastes from nearby factories (Case Study, p. 509) or because the land is especially subject to landslides, flooding, earthquakes, or volcanic eruptions.

An estimated 100 million people are homeless and sleep on the streets or wherever they can, and the number exceeds 1 billion if squatters and others with temporary or insecure living areas are included. Nearly half of the 18.1 million people in Mexico City, Mexico, and two-thirds of those in Calcutta, India, live in unplanned squatter settlements.

Half of all urban children under age 15 in developing countries live in conditions of extreme poverty, and about one-fifth of them are street children with little or no family support. In Cairo, Egypt, children of kindergarten age can be found digging through clods of ox dung, looking for undigested kernels of corn to eat.

Many cities do not provide squatter settlements and shantytowns with drinking water, sanitation facilities, electricity, food, health care, housing, schools, or jobs. Not only do these cities lack the needed money, but also their officials fear that improving services will attract even more of the rural poor. Many city governments regularly bulldoze squatter shacks and send police to drive the illegal settlers out. The people then move back in or develop another shantytown somewhere else.

Despite joblessness, squalor, overcrowding, environmental hazards, and rampant disease, most squatter and slum residents are better off than the rural poor are. With better access to family-planning programs, they tend to have fewer children, who have better access to schools. Most residents are adaptable and resilient and have hope for a better future. Many squatter settlements provide a sense of community and a vital safety net of neighbors, friends, and relatives for the poor.

A few squatter communities have organized to improve their living conditions. For example, the Orangi district of Karachi, Pakistan, is home for nearly 1 million squatters. In the early 1980s, Akhter Hameed Khan, a dynamic community organizer, formed a self-help citizens' group that since 1981 has organized, collected money, and managed construction of sewers that now serve about 90% of Orangi's residents.

Critical Thinking

1. What three things do you believe should be done to reduce the numbers and improve living conditions for **(a)** the urban poor and **(b)** the rural poor?

2. Should squatters around cities of developing countries be given title to land they do not own? Explain. What are the alternatives?

■ Urban dwellers generally have better access to medical care, family planning, education, social services, and environmental information than do people in rural areas.

■ Environmental pressures from population growth are lower because birth rates in urban areas usually are one-fourth to one-third of those in rural areas.

■ Recycling is more economically feasible because of the large concentration of recyclable materials.

■ Per capita expenditures on environmental protection are higher in urban areas.

■ The 50% of the world's people currently living in urban areas occupy only about 4% of the planet's land area (Figure 25-2).

■ Concentrating people in urban areas helps preserve biodiversity by reducing the stress on wildlife habitats.

On the other hand, urbanization has some harmful health and environmental effects:

■ Although urban dwellers occupy only 4% of the earth's land area, they consume 75% of the earth's resources.

■ Large areas of the earth's land area must be disturbed and degraded (Figure 1-4, p. 8) to provide urban dwellers with food, water, energy, minerals, and other resources. This decreases and degrades the earth's biodiversity.

■ Because of their high resource consumption, urban dwellers produce most of the world's air pollution, water pollution, and solid and hazardous wastes.

■ Pollution levels normally are higher in urban areas because pollutants are produced in a small volume and cannot be as readily dispersed and diluted as those produced in rural areas.

- Most of the world's cities are not self-sustaining systems because of their high resource input and high waste output (Figure 25-9).

- High population densities in urban areas can increase (1) the spread of infectious diseases (especially if adequate drinking water and sewage systems are not available) and (2) physical injuries (mostly from industrial and traffic accidents), and (3) *crime rates* (Connections, right).

Some analysts call for developing a more sustainable relationship between cities and the living world. Doing this would entail converting high-waste, unsustainable cities with a *linear metabolism* (based on an ever-increasing resource throughput and waste output; Figure 3-19, p. 66) to lower-waste, more sustainable cities with a *circular metabolism* (based on more efficient resource use, reuse, recycling, pollution prevention, and waste reduction; Figure 3-20, p. 67).

Why Are Trees and Food Production Important in Cities? *Most cities have few trees, shrubs, or other plants that* (1) absorb air pollutants, (2) give off oxygen, (3) help cool the air as water transpires from their leaves and as they provide shade, (4) reduce soil erosion, (5) muffle noise, (6) provide wildlife habitats, and (7) give aesthetic pleasure.

Population growth, urban sprawl, and poor land-use planning are the major causes of tree loss in urban areas. For example, a combination of satellite photos and geographic information system (GIS) computer mapping show that between 1972 and 1999 the area of heavy tree cover decreased from (1) 37% to 12% in the Washington, D.C., area (Figure 25-7), (2) 42% to 26% in the Puget Sound watershed that surrounds Seattle, Washington, and (3) 47% to 25% in the Atlanta, Georgia, region (Figures 25-8 and 25-10). This loss of tree cover increases (1) air pollution, (2) urban temperatures (by reducing the natural cooling effect of trees), and (3) stormwater runoff.

Most cities produce little of their own food. However, people can grow their own food by (1) planting community gardens in unused lots (as is done on 1,600 plots in Seattle, Washington), (2) using window boxes and balcony planters, (3) creating gardens or greenhouses on the roofs of apartment buildings and on patios, and (4) raising fish in tanks and sewage lagoons.

According to the United Nations, about 800 million urban farmers provide about 15% of the world's food, and this proportion could be increased. Urban farmers in Hong Kong produce two-thirds of the poultry and almost

Figure 25-9 Urban areas are rarely sustainable systems. The typical city is an open system that depends on other areas for large inputs of matter and energy resources and for large outputs of waste matter and heat. Large areas of nonurban land must be used to supply urban areas with resources. For example, according to an analysis by Mathis Wackernagel and William Rees, 58 times the land area of London is needed to supply its residents with resources. Meeting the needs of all the world's people at the same rate of resource use as London would take at least three more earths.

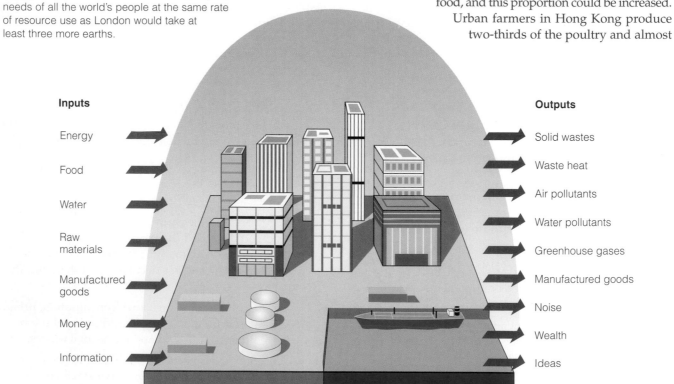

Inputs
- Energy
- Food
- Water
- Raw materials
- Manufactured goods
- Money
- Information

Outputs
- Solid wastes
- Waste heat
- Air pollutants
- Water pollutants
- Greenhouse gases
- Manufactured goods
- Noise
- Wealth
- Ideas

How Can Reducing Crime Help the Environment?

Almost everyone favors reducing crime, but few people realize that doing this can also help improve environmental quality. Crimes such as robbery, assault, and shootings:

■ Drive people out of cities, which are our most energy efficient living arrangements. Every brick in an abandoned urban building represents an energy waste equivalent to burning a 100-watt light bulb for 12 hours. Each new suburb means replacing farmland or reservoirs of natural biodiversity with dispersed, energy- and resource-wasting roads, houses, and shopping centers.

■ Make people less willing to use walking, bicycles, and energy-efficient public transit systems.

■ Force people to use more energy by leaving lights, TVs, and radios on to deter burglars and to clear away trees and bushes near houses that can reduce solar heat gain in the summer and provide windbreaks in the winter.

■ Cause overpackaging of many items to deter shoplifting or poisoning of food or drug items.

Critical Thinking

Can you think of any environmental benefits of certain types of crimes?

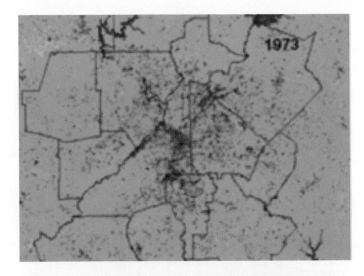

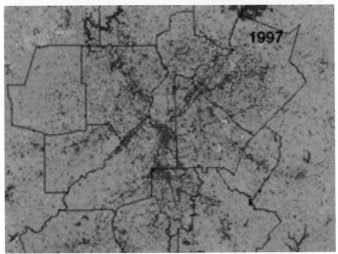

Figure 25-10 A combination of satellite images and computer mapping with geographic information systems (GIS) shows the replacement of heavy tree cover (green) with developed areas (black) in the Atlanta, Georgia, area between 1974 and 1999. During this period, overall tree cover decreased from 47% to 25%. As one observer remarked, "Most cities are places where they cut down the trees and then name the streets after them." (Marc L Imhoff/Goddard Space Flight Center/NASA)

half of the vegetables consumed by the city. Cities can also encourage farmers' markets, which lower food prices by allowing farmers to sell directly to customers. This also helps prevent nearby farmland from being swallowed up by urban sprawl.

What Are the Impacts of Urbanization on the Microclimate of Cities? *Cities generally are warmer, rainier, foggier, and cloudier than suburbs and nearby rural areas.* The enormous amounts of heat generated by cars, factories, furnaces, lights, air conditioners, and heat-absorbing dark roofs and roads in cities create an **urban heat island** (Figure 25-11) surrounded by cooler suburban and rural areas. As urban areas grow and merge (Figure 25-7), individual heat islands also merge. This can affect the climate of a large area and keep polluted air from being diluted and cleansed.

Cities can save money, reduce air-conditioning costs, and partially counteract the heat island effect by:

■ Instituting tree-planting programs. The U.S. Forest Service estimates that planting 95,000 trees in metropolitan Chicago would save $38 million in reduced energy use and other environmental benefits over a 30-year period.

■ Making neighborhood streets narrow enough to be shaded and cooled by trees on each side (p. 658).

■ Using lighter-colored paving, building surfaces, and rooftops to reflect heat away.

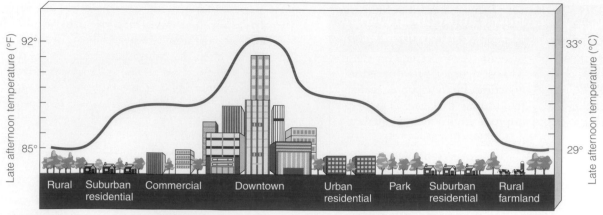

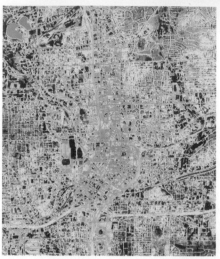

Figure 25-11 Profile of an urban heat island (above) showing how temperature changes as the density of development and trees changes. Most urban areas have few trees and large areas of heat-absorbing paved streets, roofs, and other dark surfaces that absorb and release heat. Thus, urban areas are hotter than surrounding suburban and rural areas with more trees and less heat-absorbing surfaces. At night urban areas are still radiating heat, while rural areas are cooling rapidly. The satellite photo on the left shows that nighttime surface temperatures in downtown Atlanta, Georgia, typically are 8–10°C (14–18°F) warmer (redder) than in surrounding areas. These temperatures have risen as more of the area's trees have been converted to developed areas (Figure 25-10). (Data from U.S. Department of State, above, and Courtesy of the NASA Marshall Space Flight Center, left)

- Reducing inputs of waste heat into the atmosphere by establishing high energy-efficiency standards for vehicles, buildings, and appliances.

- Establishing gardens on the roofs of large buildings (as Chicago, Illinois, plans to do for most of the city's skyscrapers to help cool the city and reduce air pollution). Ecological rooftop gardens needing little or no irrigation or fertilizer are widely used in the Netherlands and Germany.

A 1996 study by researchers at the Lawrence Berkeley National Laboratory estimated that using such measures would **(1)** cool Los Angeles by about 3°C (6°F), **(2)** cut the city's air-conditioning load by about 20% and smog by about 12%, and **(3)** save more than half a billion dollars per year.

What Are the Water Resource Problems of Cities?
Many cities have water supply and flooding problems.

- As cities grow and their water demands increase, expensive reservoirs and canals must be built and deeper wells drilled. This can deprive rural and wild areas of surface water and deplete groundwater faster than it is replenished.

- In urban areas of many developing countries, 50–70% of water is lost or wasted (p. 312). For example, about 58% of the water in Manila is lost because of leaks and poor management, compared to an 8% loss in Singapore, where pipes are better maintained. After Bogor, Indonesia, installed water meters and raised prices to encourage households to conserve water, water use dropped by one-third.

- Flooding tends to be greater in cities because **(1)** many cities are built on floodplain areas subject to natural flooding (Figure 13-22, p. 314, and Figure 13-23, p. 316) and **(2)** covering land with buildings, asphalt, and concrete causes precipitation to run off quickly and overload storm drains.

- Many of the world's largest cities are in coastal areas (Figure 25-2) that could be flooded sometime in this century if sea levels are raised by global warming (p. 460).

What Are the Pollution Problems of Cities?
Urban residents generally are subjected to much higher pollutant concentrations than are rural residents.

- According to the World Health Organization, more than 1.1 billion people (most in developing countries) live in urban areas where air pollution levels exceed healthful levels (p. 438). For example, an estimated 60% of the more than 10 million people living in Calcutta, India, suffer from respiratory diseases linked to air pollution.

- The World Bank estimates that almost two-thirds of urban residents in developing countries do not have adequate sanitation facilities. About 90% of all sewage in developing countries (98% in Latin America) is discharged directly into rivers, lakes, and coastal waters without treatment of any kind.

- According to the World Bank, at least 220 million people in the urban areas of developing countries do not have safe drinking water.

Air pollution is discussed in more detail in Chapter 17, water pollution in Chapter 19, and solid and hazardous wastes in Chapter 21.

What Are the Noise Problems of Cities? *Most urban dwellers also are subjected to excessive noise.* According to the U.S. Environmental Protection Agency, nearly half of all Americans (mostly urban residents) are regularly exposed to **noise pollution**: any unwanted, disturbing, or harmful sound that **(1)** impairs or interferes with hearing, **(2)** causes stress, **(3)** hampers concentration and work efficiency, or **(4)** causes accidents.

In the United States, each year about 9 million workers are exposed to potentially hazardous levels of noise. Millions of people who listen to loud music using home and car stereos ("boom cars"), portable stereos ("boom boxes") held close to the ear, and earphones also are damaging their hearing.

Harmful effects of prolonged exposure to excessive noise include **(1)** permanent hearing loss, **(2)** high blood pressure (hypertension), **(3)** muscle tension, **(4)** migraine headaches, **(5)** gastric ulcers, **(6)** irritability, **(7)** insomnia, and **(8)** psychological disorders, including increased aggression.

Sound pressure is measured in decibel-A (dbA) units (Figure 25-12). Sound pressure becomes damaging at about 75 dbA and painful around 120 dbA. At 180 dbA it can kill. Because the db and dbA scales are logarithmic, sound pressure is multiplied 10-fold with each 10-decibel rise. Thus a rise from 30 dbA (quiet rural area) to 60 dbA (normal restaurant conversation) represents a 1,000-fold increase in sound pressure on the ear.

You are being exposed to a sound level high enough to cause permanent hearing damage if **(1)** you need to raise your voice to be heard above the racket, **(2)** a noise causes your ears to ring, or **(3)** nearby speech seems muffled. Prolonged exposure to lower noise levels and occasional loud sounds may not damage hearing but can greatly increase internal stress.

There are five major ways to control noise: **(1)** Modify noisy activities and devices to produce less noise, **(2)** shield noisy devices or processes, **(3)** shield workers or other receivers from the noise, **(4)** move noisy operations or things away from people, and **(5)** use anti-noise, a new technology that cancels out one noise with another.

How Does Urban Growth Affect Nearby Rural Land and Small Towns? Another problem is the *loss of rural cropland, fertile soil, forests* (Figure 25-10), *wetlands, and wildlife habitats as cities expand.* According to a 2000 survey by the U.S. Department of Agriculture, each year in the United States about 1.3 million hectares (3.2 million acres) of rural land—mostly prime cropland and forestland—is converted to urban development, rights-of-way, highways, and airports. This loss is equivalent in area to building a 3.7-kilometer-wide (2.3-mile-wide) highway across the United States from New York City to Los Angeles, California each year.

A 1998 study by researchers at the U.S. Geological Survey compared the nighttime lighted areas of the United States (Figure 25-4) with digital soil maps. They found that in most places urban expansion (sprawl) is paving or building over the country's best soils, especially in top agricultural states such as California and Illinois. As land values near urban areas rise, taxes on nearby farmland increase so much that many farmers are forced to sell their land. A 1997 study by the American Farmland Trust estimated that the United States might lose 13% of its prime farmland by 2050.

Once prime agricultural land or forestland is paved over or built on, it is lost for

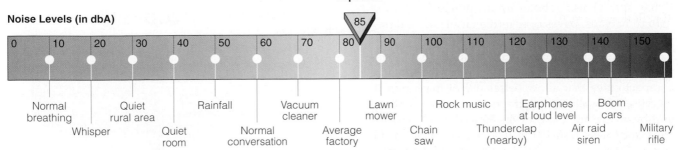

Figure 25-12 Noise levels (in decibel-A [dbA] sound pressure units) of some common sounds.

food production and habitat for most of its former wildlife. In coastal areas, urban growth destroys or pollutes ecologically valuable wetlands.

The outward expansion of cities creates numerous problems for towns in nearby rural areas: **(1)** Streets become congested with traffic, **(2)** air and water pollution, crime (Connections, p. 667), noise, and congestion increase, **(3)** health, school, police, fire, water, sanitation, and other services are overwhelmed, **(4)** taxes must be raised to meet the demand for new public services, and **(5)** some long-time residents are forced out because of rising prices, higher property taxes, decreased environmental quality, and disruption of their way of life. Unless growth is managed carefully (which is rare), old and new residents eventually experience the urban problems they sought to avoid.

25-3 TRANSPORTATION AND URBAN DEVELOPMENT

How Do Transportation Systems Affect Urban Development? Transportation and land-use decisions are linked together and determine **(1)** where people live, **(2)** how far they must go to get to work and buy food and other necessities, **(3)** how much land is paved over, and **(4)** how much air pollution people are exposed to.

If a city cannot spread outward, it must grow vertically—upward and downward (below ground)—so that it occupies a small land area with a high population density. Most people living in such *compact cities* walk, ride bicycles, or use energy-efficient mass transit. Many European cities and urban areas such as Hong Kong and Tokyo are compact and tend to be more energy-efficient than the dispersed cities in the United States (Figures 25-7, 25-8, and 25-10), Canada, and Australia, where ample land often is available for outward expansion.

A combination of cheap gasoline, plentiful land, and a network of highways produces sprawling, automobile-oriented cities with low population density that have a number of undesirable effects (Figure 25-5). Most people in such urban areas live in single-family houses with unshared walls that lose and gain heat rapidly unless they are well insulated and airtight. Urban sprawl also gobbles up unspoiled forests and natural habitats, paves over fertile farmland, and promotes heavy dependence on the automobile.

Dispersed car-centered cities use up to 10 times more energy per person for transportation than more compact cities that allow better use of mass transit, bicycles, and walking. In Europe, for example, walking and bicycling are used for 40–50% of all land-based trips and mass transit for 10%. By contrast, in the United States 95% of all trips are by car, 3% by mass

transit, and 2% by bicycling and walking. In addition, spread-out cities use more building materials, roads, power lines, and water and sewer lines.

Who Has Most of the World's Motor Vehicles? There are two main types of ground transportation: **(1)** *individual* (such as cars, motor scooters, bicycles, and walking) and **(2)** *mass* (mostly buses and rail systems). Only about 10% of the world's people (1% in many developing countries) can afford a car. Thus, about 90% of all travel in the world is by foot, bicycle, or motor scooter.

Between 1999 and 2025, the number of motor vehicles in the world is expected to increase from about 550 million to almost 1 billion, with much of this increase in Asia, Latin America, and eastern Europe.

Despite having only 4.5% of the world's people, the United States has 35% of the world's cars and trucks (212 million vehicles in 1999). In the United States, the car is used for 98% of all urban transportation and 91% of travel to work (with 80% of Americans driving to work alone; Figure 25-13).

Americans drive 3 billion kilometers (2 billion miles) each year—as far as the rest of the world combined. No wonder British author J. B. Priestley remarked, "In America, the cars have become the people." Despite their many advantages, there are a number of drawbacks to relying on motor vehicles as the major form of transportation (Pro/Con, right).

Are Motor Scooters the Answer? A growing number of people in developing countries who cannot afford cars are using motor scooters. Most burn a mixture of oil and kerosene in small, inefficient, and noisy engines that emit clouds of air pollutants.

They could be replaced by quiet electric scooters that produce very little pollution except at power plants supplying the electricity for battery recharging. Recently, Taiwan and Indonesia introduced air-pollution control legislation that may spur the use of electric scooters.

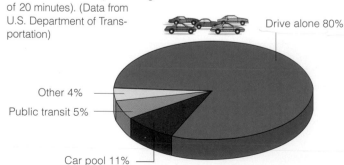

Figure 25-13 How Americans get to and from work. Studies show that Americans who commute using public transportation generally spend about twice as much time traveling (an average of 42 minutes) as those who use their cars (an average of 20 minutes). (Data from U.S. Department of Transportation)

Drive alone 80%
Other 4%
Public transit 5%
Car pool 11%

Good and Bad News About Motor Vehicles

The automobile provides convenience and mobility. To many people, cars are also symbols of power, sex, excitement, social status, and success. Moreover, much of the world's economy is built on producing motor vehicles and supplying roads, services, and repairs for them. In the United States, $1 of every $4 spent and one of every six nonfarm jobs is connected to the automobile, and five of the seven largest U.S. industrial firms produce either cars or their fuel.

Despite their important benefits, motor vehicles have many destructive effects on people and the environment. Since 1885, when Karl Benz built the first automobile, almost 18 million people have been killed in motor vehicle accidents. According to the World Health Organization, car accidents throughout the world annually **(1)** kill an estimated 885,000 people per year (an average of 2,400 deaths per day, equal to eight fatal jumbo jet crashes each day) and **(2)** injure or permanently disable another 15 million people.

In the United States alone, 16 million motor vehicle accidents (up from 7 million in 1970) per year, costing $90 billion, **(1)** kill more than 40,000 people and **(2)** injure another

5 million—at least 300,000 of them severely. *More Americans have been killed by cars than have died in all wars in the country's history.* Each week, motor vehicles in the United States also kill a million wild animals and tens of thousands of pets.

Motor vehicles also are the largest source of air pollution (including 23% of global CO_2 emissions), laying a haze of smog over the world's cities. In the United States, motor vehicles produce at least 50% of the air pollution, even though emission standards are as strict as any in the world. Two-thirds of the oil used in the United States and one-third of the world's total oil consumption are devoted to transportation.

By making long commutes and shopping trips possible, automobiles and highways have helped create urban sprawl and reduced use of more efficient forms of transportation. Worldwide, at least a third of urban land is devoted to roads, parking lots, gasoline stations, and other automobile-related uses.

In the United States, more land is devoted to cars than to housing. Half the land in an average U.S. city is used for cars, prompting urban expert Lewis Mumford to suggest that the U.S. national flower should be the concrete cloverleaf.

In 1907, the average speed of horse-drawn vehicles through the

borough of Manhattan in New York City was 18.5 kilometers (11.5 miles) per hour; today cars and trucks creep along Manhattan streets at an average speed of 5 kilometers (3 miles) per hour. If current trends continue, U.S. motorists will spend an average of 2 years of their lifetimes in traffic jams.

The U.S. economy loses at least $74 billion a year because of wasted fuel and work time lost in traffic delays. Even if the money is available, building more roads is not the answer because, as economist Robert Samuelson put it, "Cars expand to fill available concrete."

According to a recent World Bank study, the high costs of dependence on the automobile hinder economic development. A 1998 survey of U.S. real estate investors and analysts also concluded that more compact cities that promote mass transit, bike paths, and other alternatives to the car offer better investment opportunities than sprawling urban areas.

Critical Thinking

If you own a car (or hope to own one), what conditions (if any) would encourage you to rely less on the automobile and travel to school or work by bicycle, on foot, by mass transit, or by a carpool or vanpool?

Are Riding Bicycles and Walking Alternatives to the Car? Globally, bicycles outsell cars by more than two to one because most people can afford a bicycle. Besides being inexpensive to buy and maintain, bicycles **(1)** produce no pollution, **(2)** are rarely a serious danger to pedestrians or cyclists, **(3)** take few resources to make, and **(4)** are the most energy-efficient form of transportation (including walking).

In 1996, a California firm developed a simple $15 bicycle that eliminates the chain by attaching the pedals to the front wheel. This inexpensive design could greatly increase bicycle use in developing countries.

In urban traffic, cars and bicycles move at about the same average speed. Using separate bike paths or lanes running along roads, cyclists can make most trips shorter than 8 kilometers (5 miles) faster than drivers can.

In China, at least 50% of urban trips are made by bicycle, and the government gives subsidies to those who bicycle to work. In the Netherlands (with more bicycle paths than any other country), bicycle travel makes up 30% of all urban trips, compared to 1% in the United States. In Paris, bikeways now line most of the major boulevards.

In Copenhagen, Denmark, 2,300 bicycles are available for public use at no charge. The system is financed by ads attached to the bicycle frames and wheels. The system is so popular that each bicycle is used on average once every 8 minutes.

For longer trips, secure bike parking spaces can be provided at mass transit stations, and buses and trains can be equipped to carry bicycles. Such *bike-and-ride* systems are widely used in Japan, Germany, the

Netherlands, and Denmark. In Seattle, Washington, all city buses are equipped with bicycle racks.

Only about 2% of Americans bicycle to work, even though half of all U.S. commutes are less than 8 kilometers (5 miles). According to recent polls, 20% of Americans say they would bicycle to work if safe bike lanes were available and if their employers provided secure bike storage and showers at work.

However, as developing countries experience economic growth, some governments are discouraging bicycle use, viewing it as a sign of backwardness. In 1993, for example, Chinese officials banned bicycles from Shanghai's main street and prestigious shopping area so tourists and wealthy Chinese shoppers would not see them. Similarly, officials in Jakarta, Indonesia, have confiscated thousands of cycle rickshaws to project a more modern image for visitors.

In addition, unless efficient and affordable forms of mass transportation are already in place, many urban residents in developing countries abandon bikes and walking as soon as they can afford to buy motorscooters or cars, as is happening in China, India, and Indonesia. Even in the Netherlands, which has a long tradition of bicycle use and ample bicycle paths, more and more people are using cars instead of bicycles.

One development that may increase bike use is the explosive growth in the use of *electric bicycles*. About 1 million electric bikes are on the road (half of them in Japan), and about 400,000 more are added each year. Existing bikes can easily be converted to electric bikes, and new ones can be bought for $600–1,200.

Electric bikes **(1)** weigh about 23 kilograms (50 pounds), **(2)** have speeds up to 29 kilometers per hour (18 miles per hour, limited by a governor so that a motor vehicle license is not required), **(3)** can travel up to 32 kilometers (20 miles) without pedaling at a cost of about 6¢, and **(4)** can be recharged in 2.5–6 hours with a portable charger powered by 110-volt electricity, solar cells, or an automobile engine.

Battery-powered bikes can **(1)** allow cyclists to travel over hilly terrain, **(2)** replace noisy, polluting, and more expensive mopeds and motor scooters, and **(3)** are especially useful for police patrolling downtown areas and for commutes to work or school. To help reduce car use, officials of Santa Cruz County, California give residents discounts, rebates, and interest-free loans for purchase of electric bikes.

Case Study: Mass Transit in the United States In the United States, mass transit accounts for only 3% of all passenger travel, compared with 15% in Germany and 47% in Japan.

In 1917, all major U.S. cities had efficient electrical trolley or streetcar systems. Many people think of Los Angeles, California, as the original car-dominated city, but in the early 20th century Los Angeles had the largest electric rail mass transit system in the United States.

By 1950, a holding company called National City Lines (formed by General Motors, Firestone Tire, Standard Oil of California, Phillips Petroleum, and Mack Truck, which also made buses) had purchased privately owned streetcar systems in 83 major cities. It then dismantled these systems to increase sales of buses and cars. The courts found the companies guilty of conspiracy to eliminate the country's light-rail system, but the damage had already been done. The executives responsible were fined $1 each, and each company paid a fine of $5,000, less than the profit returned by replacing a single streetcar with a bus. Rebuilding these dismantled streetcar (light-rail) systems today would cost at least $300 billion.

During this same period, National City Lines worked to convert electric-powered commuter locomotives to much more expensive and less reliable diesel-powered locomotives. The resulting increased costs contributed significantly to the sharp decline of the nation's railroad system.

In the United States, 80% of federal gasoline tax revenue is used to build and maintain highways, and only 20% is used for mass transit. This encourages states and cities to invest in highways instead of mass transit.

The federal tax code also penalizes mass transit users and those who cycle or walk to work. In the United States, only 10% of commuting employees pay for parking, mainly because employers can deduct from their taxes the expense of providing parking for workers (amounting to a government subsidy of $17 billion per year). This gives such auto commuters a tax-free fringe benefit worth $200–400 a month in major cities. On the other hand, employers can write off only about $15 a month for employees who use public transit and nothing for those who walk or bike.

One remedy for this situation would be for the government to provide a general transportation tax deduction to employers for their employees. Instead of getting free parking, all employees would receive a certain amount of money per month that could be used for any type of transportation to work.

What Are the Pros and Cons of Mass Transit? Rapid-rail, suburban train, and trolley systems can transport large numbers of people at high speed. They also **(1)** are more energy-efficient (Figure 15-8, p. 364), **(2)** produce less air pollution, **(3)** cause fewer injuries and deaths, and **(4)** take up less land than motor vehicles. However, they are efficient and cost-effective only where many people live along a narrow corridor and can easily reach properly spaced stations.

One of the world's most successful rapid-rail systems is in Hong Kong. Several factors contribute to its success: **(1)** The city is densely populated, making it ideal for a rapid-rail system running through its corridor, **(2)** half the population can walk to a subway station in 5 minutes, and **(3)** a car is an economic liability in this crowded city even for those who can afford one.

Over the past two decades, 21 large cities in developing countries, including Cairo, Egypt; Shanghai, China; and Mexico City, Mexico have built rapid-rail systems that have improved transport service in dense city centers, but often at great cost. Critics believe that this money could have been better spent expanding and modernizing bus systems that could carry many more people at a much lower cost.

In the United States, 21 cities—including San Diego, Sacramento, San Jose, Los Angeles, Seattle, Buffalo, and Portland (Oregon)—have built light-rail systems, and seven more cities are planning to build them. Three Canadian cities—Toronto, Edmonton, and Calgary—have built light-rail systems. In Toronto, more than three-fourths of downtown commuters use public transportation to get to work. Zurich, Switzerland, has a highly efficient light-rail system, and new lines are being planned in a number of European cities, including London, England; Dublin, Ireland; Bordeaux, France; and Stockholm, Sweden.

A light-rail line **(1)** costs about one-tenth as much to build per kilometer as a highway or a heavy-rail system, **(2)** has lower operating costs than a comparable bus system, **(3)** can carry up to 400 people for each driver, compared with 40–50 passengers on a typical bus, and **(4)** is quieter and produces less air pollution than a bus system.

What Are the Pros and Cons of High-Speed Regional Trains?

In western Europe and Japan, a new generation of streamlined, comfortable, and low-polluting high-speed rail (HSR) lines is being used for medium-distance travel between cities. These *bullet* or *supertrains* travel on new or upgraded tracks at speeds up to 330 kilometers (200 miles) per hour.

Such trains **(1)** are ideal for trips of approximately 200–1,000 kilometers (120–620 miles) and **(2)** consume only one-third as much energy per rider as a commercial airplane and one-sixth as much as a car carrying only one driver for every kilometer of travel.

However, these systems **(1)** are expensive to run and maintain, **(2)** must operate along heavily used transportation routes to be profitable, **(3)** cause noise and vibration for nearby residents, and **(4)** can have accidents if not adequately maintained and managed.

In Europe, 12 countries plan to spend $76 billion to link major cities with nearly 30,000 kilometers (18,600 miles) of high-speed rail lines. Within the next decade, passengers in Australia, Taiwan, China, Pakistan, and Brazil will also be able to ride new high-speed trains between cities in less time than it takes by plane.

A high-speed train network could replace airplanes, buses, and private cars for most medium-distance travel between major American cities (Figure 25-14). Such a system could be developed at a reasonable cost by upgrading existing intercity tracks and train systems on key routes (Figure 25-14) rather than building new and expensive rail rights-of-way.

Critics say that such a system would cost too much in government subsidies. However, this ignores the fact that motor vehicle transportation receives subsidies of $300–600 billion per year in the United States. Some analysts believe that phasing out some of these automobile subsidies and applying them to developing a national rail system might be a far more efficient use of limited government funds. What do you think?

What Are the Pros and Cons of Buses?

Bus systems are more flexible than rail systems because they can run throughout sprawling cities and be rerouted overnight if transportation patterns change. Bus systems also use less capital and have lower operating costs than heavy-rail systems. Curitiba, Brazil, has developed one of the world's best bus systems and has led the way in becoming a more sustainable city (Solutions, p. 674).

However, bus systems **(1)** often cost more to operate than they bring in because they must offer low fares to attract riders and **(2)** often get caught in traffic unless they operate in separate express lanes. Carpools, vanpools, and jitneys (small vans or minibuses traveling

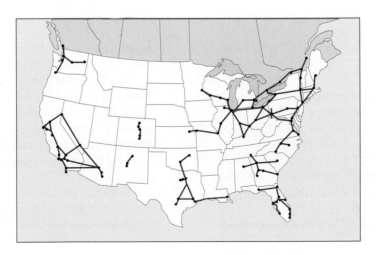

Figure 25-14 Potential routes for high-speed bullet trains in the United States and parts of Canada. Such a system could allow rapid, comfortable, safe, and affordable travel between major cities in a region. It would greatly reduce dependence on cars, buses, and airplanes for trips between these urban areas. (Data from High Speed Rail Association)

Curitiba, Brazil

SOLUTIONS

One of the world's showcase ecocities is Curitiba, Brazil, with a population of 2.2 million. The city is one of Latin America's most livable cities and has a worldwide reputation for its innovative urban planning and environmental protection efforts.

Trees are everywhere in Curitiba because city officials have given neighborhoods more than 1.5 million trees to plant and care for. No tree in the city can be cut down without a permit, and two trees must be planted for each one that is cut down. The city has developed a network of parks where the grass is cut by a municipal shepherd who moves his flock of 30 sheep around as needed.

The air is clean because the city is not built around the car and officials have integrated land-use and transportation planning. There are 160 kilometers (100 miles) of bike paths, and more are being built. With the support of shopkeepers, many streets in the downtown shopping district have been converted to pedestrian zones in which no cars are allowed.

Curitiba probably has the world's best bus system. Each weekday a network of clean and efficient buses carries more than 1.9 million passengers—75% of the city's commuters and shoppers—at a low cost (20–40¢ per ride, with unlimited transfers) on express bus lanes. Since 1974 car traffic has declined by 30% even though the city's population doubled. The city's bus stations are linked to the city's network of bike paths. Special buses, taxis, and other services are provided for the handicapped.

Only high-rise apartment buildings are allowed near major bus routes, and each building must devote the bottom two floors to stores, which reduces the need for residents to travel. Extra-large buses run on popular routes, and the system's speed and convenience are enhanced by tube-shaped bus shelters where riders pay in advance. As a result, Curitiba has one of Brazil's lowest outdoor air pollution rates. The system has also reduced traffic congestion and saved energy.

The city recycles roughly 70% of its paper and 60% of its metal, glass, and plastic, which is sorted by households for collection three times a week. Recovered materials are sold mostly to the city's more than 500 major industries. Litter and graffiti are almost nonexistent because of the civic pride of Curitiba's residents.

Instead of being torn down, existing buildings and sites are recycled for new uses. A glue factory was converted to a creativity center where children make handicrafts, which the city's tourist shops sell to help fund social programs. A garbage dump was converted to a botanical garden that houses 220,000 species, and a quarry was filled and converted to the Free University of the Environment.

The city (1) bought a plot of land downwind of downtown as an industrial park, (2) put in streets, services, housing, and schools, (3) ran a special workers' bus line to the area, and (4) enacted stiff air and water pollution control laws. This development has attracted national and foreign corporations and 500 nonpolluting industries that provide one-fifth of the city's jobs. Most of these workers can walk or bike to work from their nearby homes.

To give the poor basic technical training skills needed for jobs, the city set up old buses as roving technical training schools. Each bus gives courses, costing the equivalent of two bus tokens (less than a dollar), for 3 months in a particular area and then moves to another area. Other retired buses become classrooms, health clinics, food markets, soup kitchens, and some of the city's 200 day-care centers that are open 11 hours a day and are free for low-income parents.

The city provides environmental education for adults and children by using signs along roadways to present environmen-

along regular and stop-on-demand routes) can be used to supplement buses.

Is It Feasible to Reduce Automobile Use? Two recent estimates by economists put the harmful costs of driving in the United States at roughly $300–350 billion per year. These largely hidden costs include (1) deaths and injuries from accidents, (2) higher health insurance costs, (3) air and water pollution, (4) CO_2 emissions that increase the threat of global warming, (5) the value of time wasted in traffic jams, and (6) decreased property values near roads because of noise and congestion. All Americans, whether they drive or not, pay these costs but rarely associate them with driving.

In addition to these costs, American taxpayers also subsidize automobile use. According to a study by the World Resources Institute, federal, state, and local government automobile subsidies in the United States amount to $300–600 billion a year (depending on the costs included), an average subsidy of $1,400–2,800 per vehicle.

Environmentalists and a number of economists suggest that one way to reduce the harmful effects of automobile use is to make drivers pay directly for most

tal information and by having all school children study ecology. Older children are given apprenticeships, job training, and entry-level jobs, often emphasizing environmental skills such as forestry, water pollution control, ecological restoration, and public health.

The city's 700,000 poorest residents can swap sorted trash for locally grown vegetables and fruits or bus tokens and receive free medical, dental, and child care; there are also 40 feeding centers for street children. Poor children get free checkups and regular visits from health workers. Preventive health care is emphasized throughout the city's schools, day care, and teen centers. As a result, the infant mortality rate has fallen by more than 60% since 1977. To help the poor stretch their limited income and to discourage price gouging, the city has a computerized phone system that gives shoppers the current prices of 222 key food items in the 12 largest supermarkets.

In Curitiba, (1) 99.5% of households have electricity and drinking water and 98% have trash collection, (2) the literacy rate is 95%, (3) 83% of the adults have at least a high school education, and (4) the average per capita gross domestic product is about $8,000 (compared to $4,630 for Brazil as a whole).

The city has a *build-it-yourself* system that gives each poor family ownership of a plot of land, building materials, two trees, and an hour's consultation with an architect. A public geographic information system (GIS) gives everyone equal access to information about all the land in the city. The government has an array of telephone- and Web-based systems that respond quickly to the problems and inquiries of its citizens.

All of these things have been accomplished despite Curitiba's enormous population growth, from 300,000 in 1950 to 2.2 million in 1999, as rural poor people have flocked to the city. Another million residents are expected by 2020.

Curitiba has slums, shantytowns, and most of the problems of other cities. However, most of its citizens have a sense of vision, solidarity, pride, and hope, and they are committed to making their city even better. Polls show that 99% of the city's inhabitants would not want to live anywhere else.

The entire program is the brainchild of architect and former college teacher Jaime Lerner, an energetic and charismatic leader who has served as the city's mayor three times since the 1970s. Under his leadership, the municipal government has dedicated itself to finding solutions to problems that are simple, innovative, fast, cheap, and fun and establishing a government that is honest, accountable, and open to public scrutiny. For example, when people pay their property taxes they get to vote for the improvements they would like to see in their neighborhood.

The government innovates and takes risks with the expectation that mistakes will be made and then quickly detected, diagnosed, and corrected (adaptive management). The World Bank cites Curitiba as an example of what can be done to make cities more sustainable and livable through innovative civic leadership and community effort.

Despite its planning success, Curitiba has suffered from a lack of planning in 13 surrounding municipalities in the state of Paraná. However, since 1994 Jamie Lerner has been elected governor of the state of Paraná twice. One sign of his success is the development of a regional bus system.

Critical Thinking

1. Why do you think Curitiba has been so successful in its efforts to become an ecocity, compared to the generally unsustainable cities found in most developed and developing countries?

2. What is the city or area where you live doing to make itself more ecologically and economically sustainable?

of the full costs of automobile use—a *user-pays* approach. This could be done by (1) including the estimated harmful costs of driving as a tax on gasoline, (2) phasing out government subsides for motor vehicle owners, and (3) using the gasoline tax revenues and savings from reduced motor vehicle subsidies to lower taxes on income and wages (Solutions, p. 707) and to help finance mass transit systems, bike paths, and sidewalks.

If drivers had to pay these hidden costs directly in the form of a gasoline tax (as is done in most European countries and Japan), the tax on each gallon would be about $5–7. Such a tax would face intense political opposition unless taxpayers faced an equivalent drop in income or other taxes to compensate for increases in gasoline taxes. Such taxes would also spur the use of more energy-efficient motor vehicles (p. 365).

Other ways to reduce automobile use and congestion are (1) charging tolls on roads, tunnels, and bridges (especially during peak traffic times), (2) raising parking fees, (3) reducing mortgage charges on federally financed loans for homeowners who do not use a car to get to work, and (4) encouraging employers to use staggered hours and have more employees use telecommuting to work at home.

However, increased telecommunication could lead to *telesprawl*, in which people move out of urban areas to the countryside. This deurbanization and population dispersal could threaten biodiversity and ecological services by increasing the loss of forests, grasslands, wetlands, and other natural ecosystems.

Other countries have had success in reducing car use.

- In Hong Kong, motorists are charged by distance for all car travel, with the highest rates during commuter hours. Electronic sensors record highway travel and time of day, and drivers receive a monthly bill.

- Densely populated Singapore is rarely congested because it **(1)** taxes cars heavily, **(2)** auctions the rights to buy a car, **(3)** charges anyone driving downtown a daily user fee of $3–6, and **(4)** uses the revenue from taxes and fees to fund an excellent mass transit system.

- In Rome and Florence, Italy, all vehicles except buses, delivery vehicles, taxis, and cars belonging to local residents are banned between 7:30 A.M. and 7:30 P.M.

- In the central business district of Amsterdam in the Netherlands, existing streets are being made narrower, and the extra space is being used to widen sidewalks and build bike lanes.

- More than 300 cities in Germany, Austria, Italy, Switzerland, and the Netherlands have established *carsharing networks* to disconnect car use from the hassles of renting or owning a car. Each member pays for a card that opens lockers containing keys to cars parked around a city. Members call the network and are directed to the closest locker and car. In Berlin, Germany, carsharing has cut car ownership by 75% and car commuting by nearly 90% without decreasing mobility options.

Making heavy trucks pay for the road damage they cause would shift more freight to energy-efficient rail systems (Figure 15-8, p. 364). It is estimated that heavy trucks cause 95% of all damage to U.S. highways, with one heavy truck causing as much highway wear and tear as 9,600 cars. According to economists, current U.S. government subsidies for trucks give trucking an unfair economic advantage over more efficient and less damaging rail freight.

Including the hidden costs in the market prices of cars, trucks, and gasoline up front may make economic and environmental sense and has worked in a number of other countries. However, most analysts say that it is not feasible in the United States because:

- It faces strong political opposition from the public (mostly because they are unaware of the huge hidden costs they are already paying) and from powerful

transportation-related industries. However, taxpayers might accept sharp increases in gasoline taxes if the extra costs were offset by decreases in taxes on wages and income (Solutions, p. 707).

- Fast, efficient, reliable, and affordable mass transit options and bike paths are not widely available as alternatives to automobile travel.

- The dispersed nature of most urban areas makes people dependent on the car. In 2000, 67% of Americans polled said they could not live at their present dwelling without a car.

- Most people who can afford cars are virtually addicted to them, and most people who cannot afford a car hope to buy one someday.

Analysts point out that developing countries now have a unique economic and environmental opportunity to avoid the "car trap" by not subsidizing motor vehicle transport. Instead, they can invest most of their money in developing modern, efficient, low-polluting public transit systems and providing citywide paths for bicycling and walking, as Curituba (p. 674) has done.

25-4 URBAN LAND-USE PLANNING AND CONTROL

What Is Conventional Land-Use Planning? Most urban areas and some rural areas use some form of **land-use planning** to determine the best present and future use of each parcel of land in the area.

Much land-use planning is based on the assumption that substantial future population growth and economic development should be encouraged, regardless of the environmental and other consequences. Typically this leads to uncontrolled or poorly controlled urban growth and sprawl.

A major reason for this often destructive process is that in the United States 90% of the revenue local governments use to provide schools, police and fire protection, water and sewer systems, and other public services comes from *property taxes* levied on all buildings and property based on their economic value. Thus, local governments often try to raise money by promoting economic growth because they usually cannot raise property taxes enough to meet expanding needs. Typically the long-term result is a destructive positive feedback loop of more poorly managed economic growth leading to more environmental degradation.

Land-use planning can be aided by the use of geographic information system (GIS) technology (Figure 4-34, p. 98). Many cities and counties in the United States have used this technology to convert all their planning maps into digital form (Case Study, right).

GIS: A New Tool for Urban Planning

"The problems we are solving with GIS could not have been solved using non-computerized mapping techniques. Unfortunately many urban planners do not realize the potential," says Santa Cruz County, California, principal planner Gale Conley.

Since 1991, Santa Cruz County has been committed to building a geographic information system (GIS; Figure 4-34, p. 98). It was one of the first counties in the United States to develop GIS capabilities, and its experience illustrates the benefits and problems of investing in this expensive and complex technology.

Feasibility planning for the system started in 1985, and in 1991, custom GIS software was installed on the county's existing mainframe computer. The first maps were produced by a professional firm contracted to scan existing maps and place them in a common coordinate system. Soon the county began using GIS for other routine mapping tasks involving queries about parcel identification, ownership, zoning, land use, and location of associated planning records.

The county's current database has maps of approximately 92,000 land parcels, most of them available to the public. County officials also have approximately 100 layers of other geographic information, including (1) land use (residential, business, agriculture, parks, greenbelt), (2) political and tax jurisdictions, (3) population census data, (4) rare and endangered species locations, (5) watershed boundaries and characteristics, (6) streams, (7) groundwater recharge, (8) biological resources, (9) geology and soil characteristics, and (10) archeological sites.

Applications are continually expanding and include (1) analyzing soils and geology to assess development proposals, (2) mapping election districts, (3) mapping police beats, (4) siting health facilities, (5) compiling census data, (6) locating facilities along bus routes, and (7) protecting greenbelts.

On the negative side, (1) GIS has proven to be expensive to develop and maintain, (2) data quality is only as good as the source maps, and (3) the technology changes so rapidly that current investment is quickly outmoded.

GIS is not the solution to every problem, but it has proven to be a powerful and useful urban planning tool for cities around the world. GIS has been used to spur action by showing citizens and officials local environmental data. For example, the pollution control agency of Rio de Janeiro, Brazil, got public support for improving water quality by using GIS to create maps integrating information on polluting factories, major roadways, and data on air and water quality.

In the 1990s, GIS data from satellite images, historical data, and census data were used to create maps showing snapshots of certain years of urban development in the Baltimore, Maryland, and Washington, D.C., area between 1862 and 1999 (Figure 25-7). These data were presented in an animated video showing how the area had merged into one gigantic urban area by the 1990s. The video helped the governor of Maryland win legislative approval for his antisprawl, smart growth program.

Critical Thinking

Is GIS used for urban planning where you live? If so, how effective has it been?

What Is Ecological Land-Use Planning? Environmentalists urge communities to use comprehensive, regional **ecological land-use** to anticipate a region's present and future needs and problems. It is a complex process that takes into account geological, ecological, economic, health, and social factors (Solutions, p. 678).

Ecological land-use planning sounds good on paper, but it is not widely used for several reasons:

- Local officials seeking reelection every few years usually focus on short-term rather than long-term problems and often can be influenced by economically powerful developers.

- Often, officials and the majority of citizens are unwilling to pay for costly ecological land-use planning and implementation, even though a well-

designed plan can prevent or ease many urban problems and save money in the long run.

- It is difficult to get municipalities within a region to cooperate in planning efforts. As a result, an ecologically sound development plan in one area may be undermined by unsound development in nearby areas.

- In developing countries, most cities do not have the information or funding to carry out ecological land use planning. Urban maps often are 20–30 years old and lack descriptions of large areas of cities, especially those growing rapidly because of squatter settlements.

What Are the Pros and Cons of Using Zoning to Control Land Use? Once a land-use plan is developed, governments control the uses of various parcels

Ecological Land-Use Planning

SOLUTIONS

Six basic steps are involved in ecological land-use planning:

1. *Make an environmental and social inventory.* Experts survey **(a)** geological factors (soil type, floodplains, water availability), **(b)** ecological factors (wildlife habitats, stream quality, pollution), **(c)** economic factors (housing, transportation, and industrial development), and **(d)** health and social factors (disease, crime rates, and poverty). A top priority is to identify and protect areas that are critical for **(a)** preserving water quality, **(b)** supplying drinking water, **(c)** reducing erosion and to identify areas that are most likely to suffer from toxic wastes and natural hazards such as flooding.

2. *Identify and prioritize goals.* For example, goals may be to **(a)** encourage or discourage further economic development (at least some types) and population growth, **(b)** protect prime cropland, forests, and wetlands from development, and **(c)** reduce soil erosion.

3. *Develop individual and composite maps.* Data for each factor surveyed in the environmental and social inventory are plotted on separate transparent plastic maps, sometimes using GIS techniques (Case Study, p. 677). The transparencies are then superimposed or combined by computer into three composite maps, one each for geological, ecological, and socioeconomic factors.

4. *Develop a master composite.* The three composite maps are combined to form a master composite, which shows how the variables interact and indicates the suitability of various areas for different types of land use.

5. *Develop a master plan.* Experts, public officials, and the public evaluate the master composite (or a series of alternative master composites), and a final master plan is drawn up and approved.

6. *Implement the master plan.* The plan is set in motion, monitored, updated, and revised as needed by the appropriate government, legal, environmental, and social agencies.

Critical Thinking

Would you be willing to pay slightly higher local taxes to support ecological land-use planning where you live? Explain.

---

of land by legal and economic methods. The most widely used approach is **zoning**, in which various parcels of land are designated for certain uses.

Zoning can be used to control growth and protect areas from certain types of development. For example, cities such as Portland, Oregon, and Curitiba, Brazil (Solutions, p. 674), have used zoning to encourage high-density development along major mass transit corridors to reduce automobile use and air pollution.

Despite its usefulness, zoning has some drawbacks:

- It can be influenced or modified by developers in ways that cause environmental harm such as destruction of wetlands, prime cropland, forested areas, and open space.

- It often favors high-priced housing and factories, hotels, and other businesses over protecting environmentally sensitive areas because local governments depend on property taxes for revenue.

- Overly strict zoning can discourage innovative approaches to solving urban problems. For example, the pattern in the United States (and in some other countries) has been to prohibit businesses in residential areas, which increases suburban sprawl. There is renewed interest in returning to mixed-use zoning to help reduce urban sprawl, but current zoning laws often prohibits this.

How Is Smart Growth Being Used to Control Growth and Sprawl? There is growing use of the concept of **smart growth**. It recognizes that growth will occur but uses zoning laws and an array of other tools to **(1)** prevent sprawl, **(2)** direct growth to certain areas, **(3)** protect ecologically sensitive and important lands and waterways, and **(4)** develop urban areas that are more environmentally sustainable and more enjoyable places to live. Figure 25-15 list smart growth tools used to prevent and control urban growth and sprawl.

Some developers, real estate firms, and business interests strongly oppose many of the measures listed in Figure 25-15, arguing that they **(1)** hinder economic growth, **(2)** restrict what private landowners can do with their land, and **(3)** involve too much regulation by local, state, and federal agencies.

However, several studies have shown that most forms of smart growth provide more jobs and spur more economic renewal than conventional economic growth. Smart growth forces in the United States are steadily gaining converts. In a 1999 speech, Hugh McColl, CEO of Bank of America, announced his sup-

Figure 25-15 Smart growth tools used to prevent and control urban growth and sprawl.

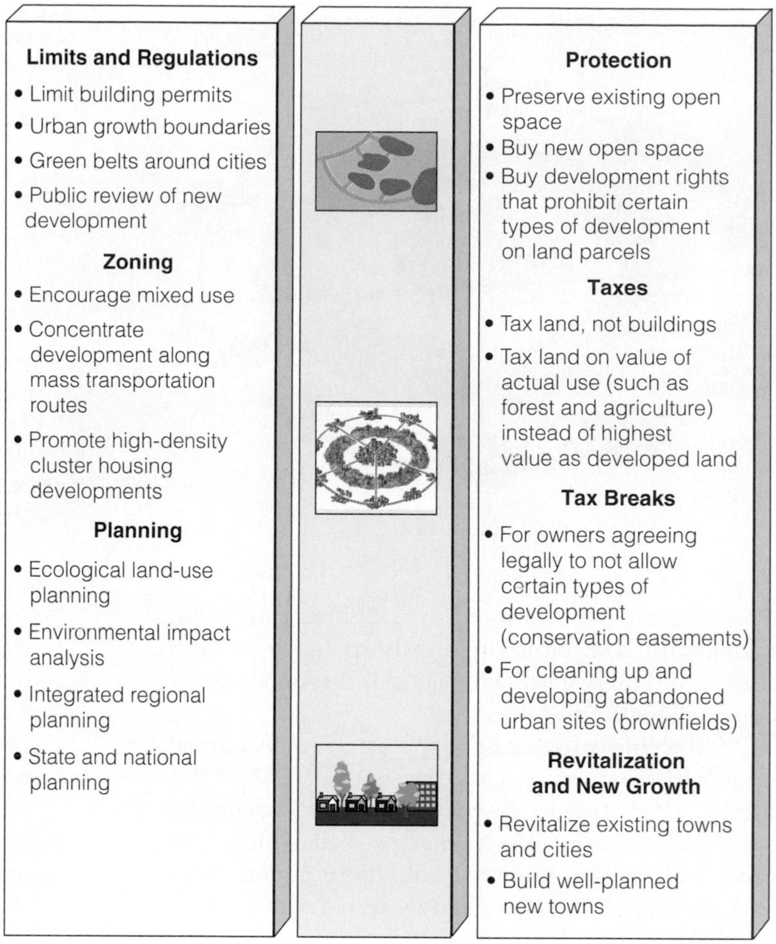

port for smart growth, declaring that the United States can no longer afford sprawl. Also, some leading real estate developers in the United States are jumping on the antisprawl, smart growth bandwagon.

Between 1997 and 2000, 22 states enacted some type of smart growth land-use law. However, by 2000 only 8 states (Arizona, Florida, Georgia, Maryland, Pennsylvania, Tennessee, Utah, and Wisconsin) had passed laws that require local governments to prevent development where roads and sewers don't exist. Tennessee's law, passed in 1998, is the strictest because it requires municipalities to set up urban growth boundaries.

China has taken the strongest stand of any country against sprawl. The government has designated 80% of the country's arable land as *fundamental land.* Building on such land requires approval from local and provincial governments and the State Council—somewhat like having to get congressional approval for a new subdivision in the United States. Developers violating these rules face the death penalty. National land-use planning also is used in Japan and much of western Europe, and state land-use planning is used in Oregon in the United States (Solutions, p. 681).

Most European countries have been successful in discouraging urban sprawl and encouraging compact cities by **(1)** keeping a tight grip on development at the national level, **(2)** imposing high gasoline taxes to encourage people to live closer to work and shops and to discourage car use, **(3)** using high taxes on heating fuel to encourage living in apartments and small houses, and **(4)** using gasoline and heating fuel tax revenues to develop efficient train and other mass transit systems within and between cities. For example, in 1999 France spent $7.7 billion on mass transit and $5 billion on roads, whereas the United States spent $5.5 billion on mass transit and $27.7 billion on roads.

How Can Urban Open Space Be Preserved? Some U.S. cities have had the foresight to preserve significant blocks of open space in the form of municipal parks. Central Park in New York City, Golden Gate Park in San Francisco, Fairmont Park in Philadelphia, and Grant Park in Chicago are examples of large urban parks in the United States.

In 1883, Minneapolis, Minnesota officials vowed to create "the finest and most beautiful system of Public

Limits and Regulations

- Limit building permits
- Urban growth boundaries
- Green belts around cities
- Public review of new development

Zoning

- Encourage mixed use
- Concentrate development along mass transportation routes
- Promote high-density cluster housing developments

Planning

- Ecological land-use planning
- Environmental impact analysis
- Integrated regional planning
- State and national planning

Protection

- Preserve existing open space
- Buy new open space
- Buy development rights that prohibit certain types of development on land parcels

Taxes

- Tax land, not buildings
- Tax land on value of actual use (such as forest and agriculture) instead of highest value as developed land

Tax Breaks

- For owners agreeing legally to not allow certain types of development (conservation easements)
- For cleaning up and developing abandoned urban sites (brownfields)

Revitalization and New Growth

- Revitalize existing towns and cities
- Build well-planned new towns

Parks and Boulevards of any city in America." This goal has been achieved. Today the city has 170 parks spaced so that most homes in Minneapolis are within six blocks of a green space. A 2000 study by the nonprofit Trust for Public Land rated Minneapolis, Minnesota, Cincinnati, Ohio, and Boston, Massachusetts, as having the country's best urban park systems. By contrast, 70% of the surface area of Los Angeles, California is devoted in some way to automobile use and only 5% to parks and open space.

As newer cities expand, they can develop large or medium-size parks. Even older cities can create small or medium-size parks or open spaces by **(1)** removing buildings and streets, **(2)** planting trees and grass, and **(3)** establishing ponds, wetlands, and lakes in areas where buildings have been abandoned.

Another approach is to **(1)** buy land for use as parks and other forms of community open space and to protect environmentally sensitive land or farmland from being developed or **(2)** purchase *development rights* that prohibit certain types of development on environmentally sensitive land. Such purchases can be made by **(1)** private groups such as the Nature

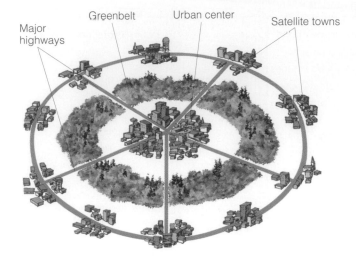

Major highways   Greenbelt   Urban center   Satellite towns

Figure 25-16 Establishing a greenbelt around a large city can control urban growth and provide open space for recreation and other nondestructive uses. Satellite towns sometimes are built outside the belt. Highways or rail systems can be used to transport people around the periphery or into the central city.

Conservancy (Solutions, p. 616), **(2)** local nonprofit, tax-exempt charitable organizations (such as land trusts), and **(3)** government agencies through taxes or bond referendums.

Some cities provide open space and control urban growth by surrounding a large city with a *greenbelt* (Figure 25-16): an open area used for recreation, sustainable forestry, or other nondestructive uses. Satellite towns can be built outside the belt. Ideally, the outlying towns and the central city are linked by an extensive public transport system. Many cities in western Europe and Canadian cities such as Toronto and Vancouver have used this approach.

Another approach to preserving open space outside a city is to set up an *urban growth boundary*: a line surrounding a city beyond which new development is not allowed. This is required for all communities in the states of Washington, Tennessee, and Oregon (Solutions, right).

Since World War II the typical approach to suburban housing development in the United States has been to bulldoze a tract of woods or farmland and build rows of standard houses on standard-size lots (Figure 25-17, middle). Many of these developments and their streets are named after the trees and wildlife they displaced (Oak Lane, Cedar Drive, Pheasant Run, Fox Fields).

In recent years, builders have increasingly used a new pattern, known as *cluster development*, in which housing units are concentrated on one portion of a parcel, with the rest of the land (often 40–50%) used for commonly shared open space (Figure 25-17, right). When done properly, high-density cluster developments are a win-win solution for residents, developers, and the environment. Residents get **(1)** more open and recreational space, **(2)** aesthetically pleasing surroundings, and **(3)** lower heating and cooling costs because some walls are shared. Developers can cut their costs for site preparation, roads, utilities, and other forms of infrastructure by as much as 40% and sell units that have a higher market value.

Some cities have converted abandoned railroad rights-of-way and dry creek beds into bicycle, hiking, and jogging paths, often called *greenways*. More than 500 new greenway projects, developed largely by citizens' groups, are under way in the United States. Many German, Dutch, and Danish cities are connected by extensive networks of footpaths and bike paths.

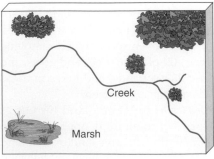

Undeveloped land

Typical housing development

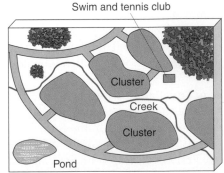

Swim and tennis club

Cluster housing development

Figure 25-17 Conventional and cluster housing developments as they might appear if constructed on the same land area. With cluster development, houses, town houses, condominiums, and two- to six-story apartments are built on part of the tract. The rest, typically 30–50% of the area, is left as open space, parks, and cycling and walking paths.

SOLUTIONS

Since the mid-1970s, Oregon has had a comprehensive statewide land-use planning process based on three principles.

1. *All rural land in Oregon has been permanently zoned as forest, agricultural, or urban land.*

2. *An urban growth line has been drawn around each community in the state, with no urban development allowed outside the boundary.*

3. *Control over the process has been placed in state hands through the Land Conservation and Development Commission.*

Not surprisingly, this last principle has been the most controversial. It is based on the idea that public good takes precedence over private property rights, a well-established principle in most European countries that is generally opposed in the United States.

However, Oregon's plan has worked because it is not designed to "just say no" to development; instead, it encourages certain kinds of development, such as dense urban development that helps prevent destruction of croplands, wetlands, and biodiversity in the surrounding area.

Because of the plan, most of the state's rural areas remain undeveloped, and the state and many of its cities are consistently rated among the best places in the United States to live.

Portland, Oregon, has been very successful in using *smart growth* to control sprawl and reduce auto use by (1) establishing an urban growth boundary around the city, (2) creating large areas of green space, including removing a six-lane expressway to make room for a downtown riverfront park, (3) developing an efficient mass transportation system, (4) using zoning to encourage high-density development along major transit lines (with 85% of all new building to be within a 5-minute walk from a transit stop), (5) allowing mixed development of offices, shops, and residences in the same area, and (6) placing a ceiling on the number of downtown parking spaces.

These policies have helped (1) create 30,000 new inner-city jobs, (2) double the number of people working downtown since 1975, and (3) attract almost $1 billion in private investment.

Critical Thinking

Why don't more states and cities use Oregon's smart growth model for land management and control of urban sprawl? How could you improve on Oregon's model?

25-5 SOLUTIONS: MAKING URBAN AREAS MORE LIVABLE AND SUSTAINABLE

What Are the Pros and Cons of Building New Cities and Towns? Urban problems must be solved in existing cities. However, building new cities and towns could take some of the pressure off overpopulated and economically depressed urban areas and allow development of more environmentally sustainable cities.

Great Britain has built 16 new towns and is building 15 more. New towns have also been built in Singapore, Hong Kong, Japan, Finland, Sweden, France, the Netherlands, Venezuela, Brazil, New Zealand, Philippines, and the United States. They are of three types: (1) *satellite towns*, located fairly close to an existing large city (Figure 25-16), (2) *freestanding new towns*, located far from any major city, and (3) *in-town new towns*, located within existing urban areas.

Typically, new towns are designed for populations of 20,000 to 100,000 people (Case Study, p. 682). Some of the more successful new towns have been funded and developed by a partnership between the public and private sectors.

A new town called Las Colinas has been built 8 kilometers (5 miles) from the Dallas-Fort Worth Airport on land that was once a ranch. The community is built around an urban center and a lake, with water taxis to transport people across it. The town is laced with greenbelts and open spaces to separate high-rise office buildings, warehouses, and residential buildings. People working in high-rise buildings park their cars outside the core and take a computer-controlled personal transit system to their offices. New or rebuilt smaller-scale village communities have been created in (1) Gainesville, Florida, (2) Harbor Town, Tennessee (developed on an island in the Mississippi River, 5 minutes from downtown Memphis), (3) Laguna West, California (a middle-class development on the far fringes of suburban Sacramento), (4) Valencia, California, and (5) Eco-Village, New York (a potentially sustainable village being developed near Ithaca, New York).

New towns and more environmentally sustainable communities offer many benefits for those who can afford to live in such places. However, some critics are concerned that most of these communities do not provide affordable housing for the poor and for many middle-class families.

How Can We Make Cities More Sustainable and More Desirable Places to Live? According to most environmentalists and urban planners, increased

Tapiola, Finland

CASE STUDY

Tapiola, Finland, a satellite new town not far from Helsinki, is internationally acclaimed for its ecological design, beauty, and high quality of life for its residents. Designed in 1951, it is located in a beautiful setting along the shores of the Gulf of Finland.

It is being built gradually in seven sections, with an ultimate projected population of 80,000. Today many of its more than 50,000 residents work in Helsinki, but its long-range goal is industrial and commercial independence.

Tapiola is divided into several villages separated by greenbelts. Each village consists of several neighborhoods clustered around a shopping and cultural center that can be reached on foot. Playgrounds and parks radiate from this center, and walkways lead to the various residential neighborhoods.

Each neighborhood has a social center and contains a mix of about 20% high-rise apartments and 80% single-family houses nestled among lush evergreen forests and rocky hills. Housing is not segregated by income. Because housing is clustered, more land is available for open space, recreation areas, and parks.

Industrial buildings and factories, with strict pollution controls, are located away from residential areas and screened by vegetation to reduce noise and visual pollution, but they are close enough so people can walk or bicycle to work. Finland plans to build six more new satellite towns around Helsinki.

Critical Thinking

What might be some drawbacks to living in a new town such as Tapiola? Would you like to live in such a town? Explain.

---

urbanization and urban density are better than spreading people out over the countryside, which would destroy more of the planet's biodiversity. To these analysts, the primary problem is not urbanization but our failure to make most cities more sustainable and livable and to provide economic support for rural areas.

Over the next few decades, they call for us to make new and existing urban areas more self-reliant, sustainable, and enjoyable places to live through good ecological design (Guest Essay, right). Good ecological urban designers (1) do not plan buildings or cities that they would not like to live in themselves, (2) look for solutions that are desirable enough to attract the necessary funds from both the public and private sector instead of looking primarily for all public or all private

solutions, (3) question part solutions (such as building suburban streets twice as wide as they should be, which attracts more cars traveling faster than they should and prevents shading and cooling of streets and neighborhoods by trees lining both sides of streets), (4) look for *win-win solutions* that provide people with a diverse mixture of livable neighborhoods instead of monocultures of standard development tracts (Figure 25-17, middle), and (5) strive to provide a better life for all people, not just the affluent.

Another goal is to reduce the demands cities make on the earth's resources by having urban systems greatly increase the efficiency of resource use (Solutions, p. 525) and mimic the circular metabolism of nature (Figure 3-20, p. 67). In an environmentally sustainable *ecocity* or *green city* (p. 658), emphasis is placed on (1) preventing pollution and reducing waste, (2) using energy and matter resources efficiently (Section 15-2, p. 362, and Solutions, p. 525), (3) recycling and reusing at least 60% of all municipal solid wastes, (4) using solar and other locally available renewable energy resources, (5) encouraging biodiversity, (6) using composting to help create soil, and (7) using solar-powered living machines (p. 476) to treat sewage.

An ecocity takes advantage of locally available renewable energy sources and requires that all buildings, vehicles, and appliances meet high energy-efficiency standards. Trees and plants adapted to the local climate and soils are planted throughout to (1) provide shade and beauty, (2) reduce pollution, noise, and soil erosion, and (3) supply wildlife habitats. Monoculture grass lawns often are replaced by small organic gardens and a variety of plants adapted to local climate conditions (Figure 13-21, p. 313).

Abandoned inner city and industrial sites (brownfields, p. 544) are recycled to productive uses. Abandoned lots and polluted creeks and rivers are cleaned up and restored. Nearby forests, grasslands, wetlands, and farms are preserved instead of being destroyed or degraded by urban sprawl. Much of an ecocity's food comes from (1) nearby organic farms, (2) solar greenhouses, (3) community gardens, and (4) small gardens on rooftops, in yards, and in window boxes.

An ecocity is *people oriented*, not car oriented. Ideally, the places people live, work, shop, and play and a mass transit station would all be within a 5-minute walk or an even shorter bike ride of one another. Networks of safe walking and bicycle paths connect mixed-zoning residential areas and shopping and workplaces. The goal is to reduce the need for cars by creating better neighborhoods and providing nearby mass transit for travel outside these mostly self-sufficient neighborhoods. In such cities, land-use planning is closely linked to transportation planning and policies.

People designing and living in ecocities take seriously the advice Lewis Mumford gave more than three

The Ecological Design Arts

David W. Orr

GUEST ESSAY

David W. Orr is professor of environmental studies at Oberlin College and one of the nation's most respected environmental educators. He is author of numerous environmental articles and three books, including Ecological Literacy *and* Earth in Mind. *He is education editor for* Conservation Biology *and a member of the editorial advisory board of* Orion Nature Quarterly. *With help from students, faculty, and townspeople he used ecological design to build an innovative environmental studies building at Oberlin College in Ohio (Figure 15-24, p. 376).*

If *Homo sapiens sapiens* entered its industrial civilization in an intergalactic design competition, it would be tossed out in the qualifying round. It does not fit. It will not last. The scale is wrong, and even its defenders admit that it's not very pretty. The most glaring design failures of industrial and technologically driven societies are the loss of diversity of all kinds, impending climate change, pollution, and soil erosion.

Of course, industrial civilization was not designed at all; it was mostly imposed by single-minded individuals, armed with one doctrine of human progress or another, each requiring a homogenization of nature and society. These individuals for the most part had no knowledge of ecological design arts.

Good ecological design incorporates understanding about how nature works into the ways we design, build, and live in just about anything that directly or indirectly uses energy or materials or governs their use.

When human artifacts and systems are well designed, they are in harmony with the ecological patterns in which they are embedded. When poorly designed, they undermine those larger patterns, creating pollution, higher costs, and social stress.

Good ecological design has certain common characteristics, including **(1)** correct scale, **(2)** simplicity, **(3)** efficient use of resources, **(4)** a close fit between means and ends, **(5)** durability, **(6)** redundancy, and **(7)** resilience. These characteristics often are place-specific, or, in John Todd's words, "elegant solutions predicated on the uniqueness of place."

Good design also solves more than one problem at a time and promotes human competence, efficient and frugal resource use, and sound regional economies. Where good design becomes part of the social fabric at all levels, unanticipated positive side effects multiply. When people fail to design with ecological competence, unwanted side effects and disasters multiply.

The pollution, violence, social decay, and waste all around us indicate that we have designed things badly, for, I think, three primary reasons. First, as long as land and energy were cheap and the world was relatively empty, we did not need to master the discipline of good design. The result was **(1)** sprawling cities, **(2)** wasteful economies, **(3)** waste dumped into the environment,

(4) bigger and less efficient automobiles and buildings, and **(5)** conversion of entire forests into junk mail and Kleenex—all in the name of economic growth and convenience.

Second, design intelligence fails when greed, narrow self-interest, and individualism take over. Good design is a cooperative community process requiring people who share common values and goals that bring them together and hold them together, as has been done in Curitiba, Brazil [Solutions, p. 674], and Chattanooga, Tennessee [p. 684]. Most American cities, with their extremes of poverty and opulence, are products of people who believe they have little in common with one another. Greed, suspicion, and fear undermine good community and good design alike.

Third, poor design results from poorly equipped minds. Only people who understand harmony, patterns, and systems can do good design. Industrial cleverness, on the contrary, is evident mostly in the minutiae of things, not in their totality or their overall harmony. Good design requires a breadth of view that causes people to ask how human artifacts and purposes fit within a particular culture and place. It also requires ecological intelligence, by which I mean an intimate familiarity with how nature works.

An example of good ecological design is found in John Todd's living machines, which are carefully orchestrated ensembles of plants, aquatic animals, technology, solar energy, and high-tech materials to purify wastewater, but without the expense, energy use, and chemical hazards of conventional sewage treatment technology. Todd's living machines resemble greenhouses filled with plants and aquatic animals [Figure 19-1, p. 476]. Wastewater enters at one end and purified water leaves at the other. In between, an ensemble of organisms driven by sunlight use and remove nutrients, break down toxins, and incorporate heavy metals in plant tissues.

Ecological design standards also apply to the making of public policy. For example, the Clean Air Act of 1970 required car manufacturers to install catalytic converters to remove air pollutants. Two decades later, emissions per vehicle are down substantially, but because more cars are on the road, air quality is about the same—an example of inadequate ecological design. A sounder design approach to transportation would **(1)** create better access to housing, schools, jobs, stores, and recreation areas, **(2)** build better public transit systems, **(3)** restore and improve railroads, and **(4)** create bike trails and walkways.

An education in the ecological design arts would foster the ability to see things in their ecological context, integrating firsthand experience and practical competence with theoretical knowledge about how nature works. It would equip people to build households, institutions, farms, communities, corporations, and economies that **(1)** do not emit carbon dioxide or other

(continued)

greenhouse gases, (2) operate on renewable energy, (3) preserve biological diversity, (4) recycle material and organic wastes, and (5) promote sustainable local and regional economies.

The outline of a curriculum in ecological design arts can be found in recent work in ecological restoration, ecological engineering, solar design, landscape architecture, sustainable agriculture, sustainable forestry, energy efficiency, ecological economics, and least-cost, end-use analysis. A program in ecological design would weave these and similar elements together around actual design objectives that aim to make students smarter about systems and about how specific things and

processes fit in their ecological context. With such an education we can develop the habits of mind, analytical skills, and practical competence needed to help sustain the earth for us and other species.

Critical Thinking

1. Does your school offer courses or a curriculum in ecological design? If not, suggest some reasons why it does not.

2. Use Orr's three principles of good ecological design to evaluate how well your campus is designed. Suggest ways to improve its design.

decades ago: "Forget the damned motor car and build cities for lovers and friends." Examples of cities that have attempted to become more environmentally sustainable include (1) Curitiba, Brazil (Solutions, p. 674), (2) Davis, California (p. 658), (3) Tapiola, Finland (Case Study, p. 682), (4) Waitakere City, New Zealand, and (5) Chattanooga, Tennessee.

Case Study: Chattanooga, Tennessee In the 1950s, Chattanooga was known as one of the dirtiest cities in the United States. Its air was so polluted that people sometimes had to turn on their headlights in the middle of the day, and the U.S. Environmental Protection Agency warned that when women walked outside, their nylon stockings were likely to disintegrate. The Tennessee River, flowing through the city's industrial wasteland, bubbled with toxic waste.

During the past 15 years Chattanooga has transformed itself from one of the most polluted cities in the United States to one of its most livable cities. Since the mid-1980s, the combined efforts of thousands of Chattanooga citizens have helped clean up the city's air, revitalize its riverfront, and diversify its economy. The overall goal has been to show how environmental improvement and economic development can coexist.

Efforts began in 1984 when civic leaders used town meetings as part of a Vision 2000 process, a 20-week series of community meetings that brought together thousands of citizens from all walks of life to build a consensus about what the city could be like at the turn of the century. Groups of citizens and civic leaders (1) identified the city's main problems, (2) set goals, and (3) brainstormed thousands of ideas for dealing with these problems.

To help achieve the overall goal of more sustainable development, city officials and 1,700 citizens spent more than 5 months shaping the results of the Vision 2000 process into 34 specific goals and 223 projects. The city launched a number of combined public–private partnerships for green development, generating (1) a

total investment of $739 million (two-thirds of it from private investors), (2) 1,300 new permanent jobs, and (3) 7,800 temporary construction jobs.

By 1995, the city had met most of its goals, including (1) encouraging zero-emission industries to locate in Chattanooga, (2) renovating existing low-income housing and building new low-income rental units, (3) building the nation's largest freshwater aquarium, which became the centerpiece for the city's downtown renewal, (4) replacing all its diesel buses with a fleet of quiet, nonpolluting electric buses, made by a new local firm, (5) developing an 8-kilometer-long (5-mile-long) riverfront park, which is being extended 35 kilometers (22 miles) along both sides of the Tennessee River and now draws more than 1 million visitors per year, and (6) launching an innovative recycling program (after citizen activists and environmentalists blocked construction of a new garbage incinerator).

In 1993, the community began the process again in Revision 2000. More than 2,600 participants identified 27 additional goals and more than 120 recommendations for further improvements. One of these goals calls for transforming a blighted, mostly abandoned industrial area in South Chattanooga into an environmentally advanced, mixed community of residences, retail stores, and zero-emission industries so that employees can live near their workplaces.

This new zero-waste ecoindustrial park, modeled after the one in Kalundborg, Denmark (p. 524), will have underground tunnels linking some 30 industrial buildings to share heating, cooling, and water supplies and to use the waste matter and energy of some enterprises as resources for others. For example, excess heat generated by metal foundries will be used to warm water used at a nearby chicken-processing plant. Biomass waste from this plant will be converted to ethanol for use as a fuel in another facility. By-products from the ethanol plant will then be used as nutrients for a tree nursery, greenhouse, and fish farm. The new ecoindustrial area will also have an ecology center that will use a living machine (Figure 19-1, p. 476) to treat sewage, wastewater, and contaminated soils.

These accomplishments show what citizens, environmentalists, and business leaders can do when they work together to develop and achieve common goals. However, the city still faces serious environmental problems. One is the controversy over how to handle at least 11 Superfund toxic waste sites on the once heavily industrialized Chattanooga Creek, where many African-American communities are located.

According to many environmentalists, urban planners, and economists, urban areas that fail to become more ecologically sustainable over the next few decades are inviting economic depression and increased unemployment, pollution, and social tension.

The city is not an ecological monstrosity. It is rather the place where both the problems and the opportunities of modern technological civilization are most potent and visible.

PETER SELF

REVIEW QUESTIONS

1. Define the boldfaced terms in this chapter.

2. Describe how Davis, California, has been designed to be a more ecologically sustainable city.

3. Distinguish between **(a)** an *urban area* and a *rural area*, **(b)** a *village* and a *city*, and **(c)** *degree of urbanization* and *urban growth*.

4. List factors that *push* people and *pull* people to migrate from rural areas to urban areas. Describe the problems faced by the urban poor.

5. What are five major trends in urbanization and urban growth?

6. Summarize the major urban problems of Mexico City.

7. What major shifts in the U.S. population have taken place since 1800? What percentage of the U.S. population lives in urban areas?

8. List four pieces of *good news* and four pieces of *bad news* about urban problems in the United States.

9. What is *urban sprawl*? List six factors that have promoted urban sprawl in the United States. List the major harmful effects of urban sprawl.

10. Describe three major spatial patterns of urban development. What is a *megalopolis*?

11. What are the environmental and health pros and cons of urban areas? Distinguish between unsustainable cities with a linear metabolism and more sustainable cities with a circular metabolism.

12. Describe the major resource and environmental problems of urban areas relating to **(a)** trees and plants, **(b)** food production, **(c)** microclimate (urban heat islands), **(d)** water supply and flooding, **(e)** pollution, **(f)** noise, **(g)** beneficial and harmful effects on human health, and **(h)** loss of rural cropland, fertile soil, forests, wetlands, and wildlife habitats as cities expand. What are four ways cities can grow more of their food? What is an *urban heat island*? What are five ways to save money and counteract the urban heat island effect? List seven ways cities can promote water conservation. What are five ways to reduce noise pollution?

13. How do transportation systems affect urban development? What are two basic types of cities in terms of density and transportation systems?

14. List the pros and cons of **(a)** automobiles, **(b)** motor scooters, **(c)** bicycles, and **(d)** various forms of mass transit such as rail systems and buses?

15. Describe how Curitiba, Brazil, has become a global model of a more environmentally sustainable and livable city.

16. List seven ways to reduce dependence on the automobile. List four factors promoting dependence on the automobile in the United States.

17. Distinguish between *conventional land-use planning* and *ecological land-use planning*. List four factors hindering the use of ecological land-use planning. Describe the cycle that leads to uncontrolled urban growth and sprawl.

18. Describe how geographic information system (GIS) technology can be used in land-use planning.

19. What is *zoning*, and what are its pros and cons?

20. What is *smart growth*? List ten tools cities can use to promote smart growth.

21. Describe how the state of Oregon and the city of Portland, Oregon, have used smart growth to help prevent urban sprawl.

22. Describe five ways to preserve open space. Distinguish between *development rights, greenbelts, urban growth boundaries, cluster development*, and *greenways*.

23. How can new towns and cities be used to help solve urban problems? Describe the new town known as Tapiola, Finland.

24. List six principles of good ecological design.

25. Describe the major characteristics of a more sustainable *ecocity* or *green city*. Briefly describe the progress that Curitiba, Brazil; Davis, California; and Chattanooga, Tennessee, have made in becoming ecocities.

CRITICAL THINKING

1. Do you prefer living in a rural, suburban, small town, or urban environment? Describe the ideal environment in which you would like to live and list the environmental pros and cons of living in such a place.

2. Why is the rate of urbanization **(a)** so high in South America (74%) and **(b)** so low in Africa (33%; Figure 25-2)?

3. Do you believe that the United States (or the country where you live) should develop a comprehensive and integrated mass transit system over the next 20 years, including building an efficient rail network for travel within and between its major cities? How would you pay for such a system?

4. Do you believe that the approach to land-use planning used in Oregon (Solutions, p. 681) should be used in the state or area where you live? Explain your position.

5. In June 1996, representatives from the world's countries met in Istanbul, Turkey, at the Second UN Conference on Human Settlements (nicknamed the City Summit). One of the areas of controversy was whether housing is a universal *right* (a position supported by most developing countries) or just a *need* (supported by the United States and several other developed countries). What is your position on this issue? Defend your choice.

6. Congratulations. You have just been put in charge of the world. List the five most important features of your urban policy.

PROJECTS

1. Consult local officials to determine how land use is decided in your community. What roles do citizens play in this process?

2. For a class or group project, try to borrow one or more decibel meters from your school's physics or engineering department (or from a local electronics repair shop). Make a survey of sound pressure levels at various times of day and at several locations. Plot the results on a map. Also, measure sound levels in a room with a sound system and from earphones at several different volume settings. If possible, measure sound levels at an indoor concert, a club, and inside and outside a boom car at various distances from the speakers. Correlate your findings with those in Figure 25-12.

3. As a class project, **(a)** evaluate land use and land-use planning by your school, **(b)** draw up an improved plan based on ecological principles, and **(c)** submit the plan to school officials.

4. As a class project, use the following criteria to rate your city on a green index from 0 to 100. Are existing trees protected and new ones planted throughout the city? Do you have parks to enjoy? Can you swim in any nearby lakes and rivers? What is the quality of your water and air? Is there an effective noise pollution reduction program? Does your city have a recycling program, a composting program, and a hazardous waste collection program, with the goal of reducing the current solid waste output by at least 60%? Is there an effective mass transit system? Are there bicycle paths? Are all buildings required to meet high energy-efficiency standards? How much of the energy is obtained from locally available renewable resources? Are environmental regulations for existing industry tough enough and enforced well enough to protect citizens? Do local officials look carefully at an industry's environmental record and plans before encouraging it to locate in your city or county? Are land-use decisions made by using ecological planning with active participation by citizens? Are

city officials actively planning to improve the quality of life for all of its citizens? If so, what is their plan?

5. Use the library or the internet to find bibliographic information about *John Kenneth Galbraith* and *Peter Self*, whose quotes appear at the beginning and end of this chapter.

6. Make a concept map of this chapter's major ideas, using the section heads and subheads and the key terms (in boldface). Look at the inside back cover and on the website for this book for information about making concept maps.

INTERNET STUDY RESOURCES AND RESOURCES FOR FURTHER READING AND RESEARCH

The website for this book contains helpful study aids and many ideas for further reading and research. Log on to:

http://www.brookscole.com/product/0534376975s

and click on the Chapter-by-Chapter area. Choose Chapter 25 and select a resource:

- "Flash Cards" allows you to test your mastery of the Terms and Concepts to Remember for this chapter.

- "Tutorial Quizzes" provides a multiple-choice practice quiz.

- "Student Guide to InfoTrac" will lead you to Critical Thinking Projects that use InfoTrac College Edition as a research tool.

- "References" lists the major books and articles consulted in writing this chapter.

- "Hypercontents" takes you to an extensive list of sites with news, research, and images related to individual sections of the chapter.

INFOTRAC COLLEGE EDITION

Improve your skills with InfoTrac College Edition, a searchable online database of articles from more than 700 periodicals. Log on to:

http://www.infotrac-college.com

or access InfoTrac through the website for this book.

Try the following articles:

Grimm, N.B., J.M. Grove, S.T.A. Pickett, C.L. Redman. 2000. Integrated approaches to long-term studies of urban ecological systems. *BioScience* vol. 50, no. 7, pp. 571–584. (subject guide: urban ecology)

McMichael, A.J. 2000. The urban environment and health in a world of increasing globalization: issues for developing countries. *Bulletin of the World Health Organization* vol. 78, no. 9, pp. 1117–1126. (keywords, urban development, technology)

PART VI

ENVIRONMENT AND SOCIETY

When it is asked how much it will cost to protect the environment, one more question should be asked: How much will it cost our civilization if we do not?
GAYLORD NELSON

26 ECONOMICS, ENVIRONMENT, AND SUSTAINABILITY

How Important Are Natural Resources?

Natural resources or *natural capital* (Guest Essay, p. 6) include the earth's air, energy, water, soil, wildlife, biodiversity, and minerals and nature's dilution, waste disposal, pest control, and recycling services that support all life and economies (Figure 4-36, p. 99, and Figure 26-1).

Economic systems transform natural resources into *goods* such as houses, food, household items, highways, and cities, and *services* such as health and education.

Conventional economists acknowledge that our economic systems depend on natural resources. However, they believe that there is an infinite potential for developing new ideas and technologies for providing more of the goods and services we need and want. This can be done by **(1)** using less matter and energy resources (Solutions, p. 525, and Section 15-2, p. 362), **(2)** finding substitutes for scarce matter resources (p. 331 and Solutions, p. 525) and energy resources (Chapter 15), and **(3)** reducing pollution by improving pollution control (Section 17-7, p. 440, Individuals Matter, p. 498, and Section 21-3, p. 522) and mimicking nature (p. 524 and Solutions, p. 209).

These economists see no reason for this process of resource substitution and technological innovation to stop. As a result, they do not believe that depletion and degradation of natural resources will limit future economic growth.

Ecological economists question the validity of unlimited resource substitutability and the ability of some of the earth's natural resources to support a significant increase in certain forms of economic growth. They point out that **(1)** there are no substitutes for many natural resources, such as air, water, fertile soil, and biodiversity, and **(2)** some forms of economic growth deplete and degrade the quantity and quality of these irreplaceable forms of natural capital (Guest Essay, p. 6).

Ecological economists believe that over the next few decades we need to unite ecology and economics by modifying our economic systems to **(1)** encourage earth-sustaining forms of economic development and **(2)** discourage earth-degrading forms of economic growth.

They see this *environmental economic revolution* as a process that can **(1)** greatly increase profits for earth-sustaining business, **(2)** provide more jobs by reversing the trend of substituting machines for labor, **(3)** use much less material and energy by greatly increasing resource productivity (Solutions, p. 525), **(4)** prevent or at least slow depletion and degradation of the earth's natural capital, **(5)** reduce taxes on income and wealth by increasing taxes on pollution and waste, and **(6)** free up financial resources and human creativity for improving environmental quality and human health, reducing poverty and other social ills, and restoring damaged ecosystems.

Figure 26-1 Earth, air, fire, water, and life at a volcanic site in Hawaii. Ecological economists believe that economic systems should be designed to put more emphasis on conserving and sustaining these natural resources, which sustain us and other species and all economies. (Greg Vaughn/Tom Stack & Associates)

Converting the economy of the twenty-first century into one that is environmentally sustainable represents the greatest investment opportunity in history.

LESTER R. BROWN AND CHRISTOPHER FLAVIN

This chapter addresses the following questions:

- What are the different types of economic systems, and what types of resources support them?

- How can we monitor economic and environmental progress?

- How can economics be used to help control pollution and manage resources?

- What are the pros and cons of using regulations and market forces to improve environmental quality?

- How does poverty reduce environmental quality, and how can we reduce poverty?

- How can we shift to more environmentally sustainable economies over the next few decades?

26-1 ECONOMIC RESOURCES AND SYSTEMS AND ENVIRONMENTAL PROBLEMS

What Supports and Drives Economies? An **economy** is a system of production, distribution, and consumption of goods and services that satisfy people's wants or needs. In an economy, individuals, businesses, and governments make **economic decisions** about (1) what goods and services to produce, (2) how to produce them, (3) how much to produce, and (4) how to distribute them.

The kinds of resources (or capital) that produce goods and services in an economy are called **economic resources**. They fall into four groups:

- **Natural resources:** goods and services produced by the earth's natural processes, which support all economies and all life (Figure 4-36, p. 99). There are no substitutes for many of these natural resources, such as air, water, fertile soil, and biodiversity, and the natural income they produce. In 1997, ecological economist Robert Costanza and others estimated conservatively that the global ecosystem services flowing directly into society from the earth's stock of natural capital is worth $17–54 trillion a year, with a best estimate of at least $36 trillion (adjusted for inflation using 1999 dollars, Spotlight, p. 562). This figure is close to the world's total annual economic output of about $40 trillion.

- **Human resources:** people's physical and mental talents that provide labor, innovation, culture, and organization.

- **Financial resources:** cash, investments, and monetary institutions used to support the use of natural resources and human resources to provide goods and services.

- **Manufactured resources:** items made from natural resources with the help of human and financial resources. This type of capital includes tools, machinery, equipment, factories, and transportation and distribution facilities used to provide goods and services.

Somewhere between natural resources and human resources are *cultivated resources* such as tree farms, aquaculture ponds, cropfields, and animal feedlots. However, they provide only some of the services of the natural ecosystems they replace, and they rely on natural capital for their development and maintenance. For example, tree farms provide wood but (1) do not replace the biodiversity and complex ecosystem services of a real forest and (2) usually must be maintained by inputs of pesticides and fertilizers.

What Are the Major Types of Economic Systems? There are two major types of economic systems: command and market. In a **pure command economic system**, the government makes all *economic decisions* about what and how much goods and services are produced, how they are produced, and for whom they are produced.

Market economic systems can be divided into two types: pure free market and capitalist market. In a **pure free-market economic system**, all economic decisions are made in *markets*, in which buyers (demanders) and sellers (suppliers) of economic goods freely interact without any government interference. According to classic economic theory, in a truly free-market system:

- There is perfect competition and no seller or buyer can control or manipulate the market.

- All economic decisions are governed solely by the competitive interactions of (1) *demand* (the amount of a good or service that people want), (2) *supply* (the amount of a good or service that is available), and (3) *price* (the market cost of a good or service). If all factors are constant except price, supply, and demand, the supply and demand curves for a particular good or service intersect at the **market price equilibrium point**: the price at which sellers are willing to sell and buyers are willing to buy a certain quantity of a good or service (Figure 26-2). Changes in supply and demand in the competitive marketplace can shift demand curves (Figure 26-3) and supply curves (Figure 26-4) to new market equilibrium points. In effect, if supply is lower than demand, prices rise, and if the supply is greater than demand, prices fall.

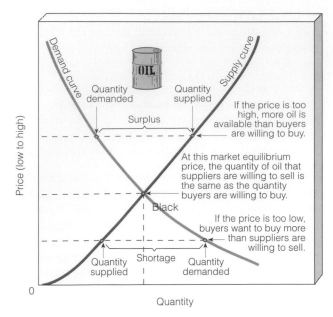

Figure 26-2 Supply, demand, and market equilibrium for oil used to produce gasoline in a pure market system. If all factors except price, supply, and demand are held fixed, market equilibrium occurs at the point at which the demand and supply curves intersect. This represents the price at which sellers are willing to sell and buyers are willing to pay for a particular good or service.

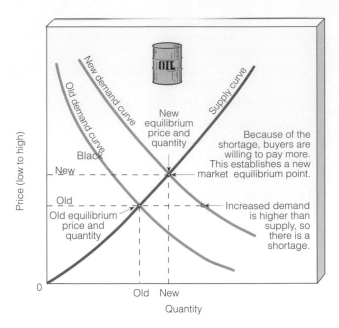

Figure 26-3 An increase in demand for gasoline because of **(1)** more drivers, **(2)** a switch to less fuel-efficient cars, **(3)** more disposable income for travel, or **(4)** decreased use of mass transit shifts the demand curve to the right and establishes a different market equilibrium point. Similarly, if the reverse of one or more of these factors decreased demand, the demand curve would shift to the left and create a different market equilibrium point.

- There are no regulations, taxes, subsidies, or barriers to trade or availability of capital that distort the market or hinder market transactions in any way.

- Economic indicators give full and accurate information about all beneficial and harmful effects of the production and use of each good and service on people and the environment (full-information accounting).

- Prices reflect all harmful costs to society and the environment (full-cost pricing).

It is important to distinguish between the pure free-market economic system just described, which exists only as a theoretical construct, and the capitalist market system we find in the real world.

The **capitalist market economic systems** found in the real world are designed to subvert many of the theoretical conditions of a truly free market. Here are the basic operating rules the world's capitalist market economies:

- Drive out all competition and gain monopolistic control of market prices on a global scale. Any industry in which five firms control 50% or more of a market is considered by economists to be *highly monopolistic*, and 40% control is considered *monopolistic. The Economist* recently reported that five firms control more **(1)** than 50% of the global market in consumer durable goods, automobiles, aerospace, airlines, electronic components, electronic goods, and steel and **(2)** 40% of the global computer, oil, and media markets.

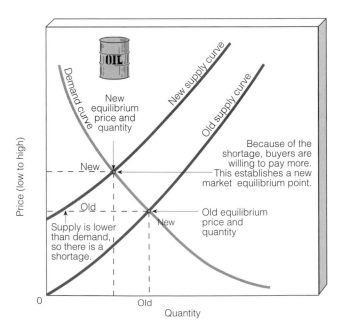

Figure 26-4 A decrease in the supply of gasoline because **(1)** the costs of finding, extracting, and refining crude oil increase or **(2)** existing oil deposits are economically depleted shifts the supply curve to the left and creates a different market equilibrium point. On the other hand, if the gasoline supply increases, the supply curve shifts to the right and creates a different market equilibrium point.

- Lobby for unrestricted *global free trade* that allows anything to be manufactured anywhere in the world and sold anywhere else.

- Lobby for **(1)** government subsidies, tax breaks, or regulations that give a company's products a market advantage over their competitors and **(2)** governments to bail them out if they make bad investments.

- Lobby to reduce corporate taxes. In the 1950s, corporate income taxes in the United States accounted for 39% of all federal income tax revenue, but by 1995 this figure had dropped to only 15%. During the same period, the share of federal income tax revenues from individuals rose from 61% to 81%. Between 1957 and 1987, the percentage of local property tax revenues paid by U.S. corporations dropped from 45% to about 16%.

- Withhold information about dangers posed by products and deny consumers access to information that would allow them to make informed choices.

- Maximize profits by passing harmful costs resulting from production and sale of goods and services on to the public, the environment, and in some cases future generations.

- A business has no legal obligation **(1)** to a particular area or nation, **(2)** to supply any particular good or service, and **(3)** to provide jobs, safe workplaces, or environmental protection.

- A company's primary obligation is to produce the highest profit for the owners or stockholders whose financial capital the company is using to do business.

- Strive to keep democratic and centrally planned governments from interfering in or regulating the marketplace.

Stockholders and owners of corporations do not have to follow these rules. For example, many corporations voluntarily contribute to the well-being of their communities and their nation and feel they have a responsibility to provide jobs and safe workplaces and to improve environmental quality. However, CEOs who spend too much money on such things can be fired and legally sued by their stockholders for failure to maximize profits.

What Kinds of Economic Systems Exist in the Real World? In reality, all countries have **mixed economic systems** that fall somewhere between the command and capitalist market systems and bear little resemblance to the pure free-market and pure command systems. The economic systems of countries such as China and North Korea fall toward the command end of the economic spectrum, and those of countries such as the United States and Canada fall toward the market end of the spectrum. Most other countries have mixed economic systems that fall somewhere between these ends of the economic spectrum.

Both command and market economic systems have an enormous impact on the environment. They also have different views on what causes pollution and environmental degradation and how to reduce these harmful effects (Figure 26-5).

Some economists view the current global economy as consisting of three different, overlapping types of economies:

- The *capitalist market economy* found in developed countries and in middle-income, moderately developed countries as measured by per capita gross national product (GNP; Figure 1-5, p. 9). It consists of about 2.7 billion people, with 1.2 billion of them in the developed countries.

- The *survival economy* found in the rural parts of most developing countries that have a low or very low per capita GNP (Figure 1-5, p. 9). It is made up mostly of about 3.3 billion rural-dwelling Chinese, Indians, and Africans who try to meet their basic survival (subsistence) needs directly from nature.

- *Nature's economy*, which consists of the natural systems and resources that support the market and survival economies.

Why Have Governments Intervened in Market Economic Systems? Governments have intervened in economies to stop or slow down the accomplishment of some of the goals of capitalist market economies by trying to:

- Level the economic playing field by preventing a single seller or buyer (monopoly) or a single group of sellers or buyers (oligopoly) from dominating the market and thus controlling supply or demand and

Type of Economic System	Cause of Environmental Pollution and Degradation	Solution
Command	Economic activity (flow of matter and energy resources)	Regulate economic activity
Market	Too little economic incentive to care for the environment	Put a price on harmful environmental activities so marketplace can respond

Figure 26-5 Command and market economic systems differ on what causes pollution and environmental degradation and how we should deal with these problems.

price. This is becoming increasingly difficult because the current global capitalist market economy is dominated by a small number of financial speculators and large transnational corporations with enough financial and political clout to manipulate prices. Today, for example, 50 of the top 100 economies in the world are transnational corporations, and 75% of all global trade is controlled by 500 transnational corporations.

- Help ensure economic stability by trying to control boom-and-bust cycles that occur in any type of market system. Again, the power of transnational corporations is making it difficult for governments to carry out this role.

- Provide basic services such as national security, education, and health care. However, transnational corporations are rapidly taking control of health care and education services throughout the world.

- Provide an economic safety net for people who because of health, age, and other factors cannot work and meet their basic needs.

- Protect people from fraud, trespass, theft, and bodily harm.

- Protect the health and safety of workers and consumers.

- Help compensate owners for large-scale destruction of their assets by floods, earthquakes, hurricanes, and other natural disasters.

- Protect common property resources such as the atmosphere, open oceans, and ozone layer.

- Prevent or reduce pollution and depletion of natural resources.

- Manage public land resources (p. 588).

What Are the Pros and Cons of a Global Market Economy? Because of rapid globalization since 1950 and especially since 1970 (p. 10), we are moving rapidly toward a global capitalist market economy governed not by nation-states and democratic governments but by institutions such as the World Bank, International Money Fund, and the World Trade Organization (WTO). In 1995, most of the world's national governments granted unprecedented global economic powers to the WTO to establish and enforce rules for the free flow of economic goods and services between countries.

Some see a global market economy run by the WTO as a way to bring economic prosperity to everyone. Others see it as a way to make the rich much richer at the expense of workers, local communities, democratic governments, and the environment.

Proponents say that the WTO will (1) stimulate economic growth in all countries, (2) help bring eco-

nomic prosperity to everyone, (3) allow consumers to buy more things at lower prices, and (4) raise environmental and health standards in developing countries.

Critics contend that the underlying goal of the WTO is to (1) weaken the power of nation-states and democratic governments to interfere in the marketplace, (2) move toward a global capitalist market economy in which large transnational corporations dominate the global marketplace, (3) create a homogenized global economy in which global corporations sell the same goods and services in the same way everywhere, with little or no regard for local customs, tastes, or cultural or religious differences (homogenized mass consumption), and (4) override any economic, environmental, and social policy decisions of the world's nation-states and democratic governments that interfere with the interests the world's transnational corporations.

Instead of increased *economic globalization*, some economists call for increased *economic localization*. Instead of a single global economy that disintegrates communities and homogenizes and destroys local customs and societies, they believe we should create a variety of loosely linked, community-based economies engaged in free regional trade.

These economies would be dominated by a variety of small to medium-sized companies catering mostly (but not exclusively) to local and regional markets. The goal would be to keep as much money as possible circulating in local economies for the benefit of local communities instead of allowing it to flow out of such communities to large national and transnational corporations that are largely indifferent to local cultures and concerns.

Many people feel that they have lost control over their economic, social, and ecological future. However, opponents contend that economic globalization is not inevitable if enough people band together and elect people to national offices that will reign in the powers of the WTO and make it more responsive to governments, communities, workers, and the environment (p. 733).

How Do Conventional and Ecological Economists Differ in Their View of Market-Based Economic Systems? Conventional economists often depict a market-based economic system as a circular flow of economic goods and money between households and businesses operating essentially independently of the earth's life-support systems or natural resources (Figure 26-6). They (1) assume that the economy is the total system and nature's economy is a subsystem and (2) consider the earth's natural resources important but not vital because of our ability to find substitutes for scarce resources and ecosystem services.

Ecological economists (Guest Essay, p. 698) believe that this conventional economic view is backwards. They consider all economies as human subsystems that are supported by the earth's natural resources (materials and ecological services), many of which have no substitutes (Figure 26-7).

Ecological economists distinguish between **(1)** unsustainable economic growth and **(2)** environmentally sustainable economic development (Figure 26-8). **Economic growth** consists of an increase in the *quantity* of the goods and services produced and consumed (quantitative growth), as measured by an increase in the flow (throughput) of high-quality energy and matter resources through a human economy (Figure 3-19, p. 66). Such growth or throughput is increased by population growth (more consumers), more consumption per person, or both. It is usually measured in terms of an increase in a country's GNP and per capita GNP (p. 5).

Conventional economists believe that because of human ingenuity there are no limits to such economic growth. Ecological economists believe that such growth is eventually unsustainable because it depletes or degrades the natural resources on which the economy depends.

Ecological economists believe that we should make the economy better, not bigger. They call for **environmentally sustainable economic development**, which involves increasing the *quality* of goods and services without depleting or degrading the quality of natural resources to unsustainable levels for current and future generations (qualitative growth). With such development,

■ The flow (throughput) of materials and energy through

economies would not exceed the ability of the environment to **(1)** provide us with the high-quality energy (p. 58) and matter resources (p. 57) we need to make things and **(2)** break down our wastes and turn them back into raw materials (Figure 3-20, p. 67).

■ The natural capital stored in nonrenewable resources (stock resources) would not be used faster than substitutes can be developed and phased in.

■ The natural capital stored in renewable resources (flow resources) would not be used faster than natural processes can replenish it. In other words, we would live on the annual natural income provided by the world's renewable resources (p. 4).

Ecological economists urge developing countries to **(1)** leapfrog to environmentally sustainable economic development based on low pollution and low waste instead of following the pollute-first-and-clean-up-later economic path of developed countries and **(2)** emphasize resource-saving, labor-intensive *appropriate technology* (Solutions, p. 696).

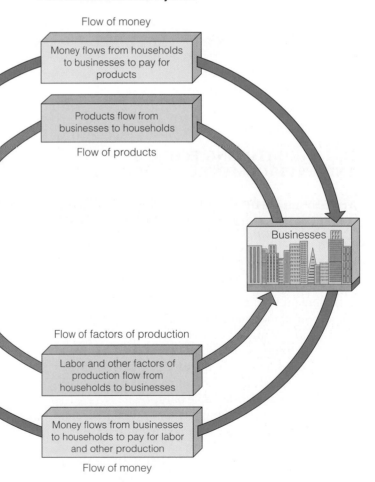

Pure Market Economic System

Flow of money

Money flows from households to businesses to pay for products

Products flow from businesses to households

Flow of products

Households

Businesses

Flow of factors of production

Labor and other factors of production flow from households to businesses

Money flows from businesses to households to pay for labor and other production

Flow of money

Figure 26-6 Conventional view of economic activity. In a market economic system, economic goods and money flow between households and businesses in a closed loop. People in households spend money to buy goods that firms produce, and firms spend money to buy factors of production (natural, human, financial, and manufactured resources). In many economics textbooks, such market economic systems are shown, as here, as if they were self-contained and thus independent of the natural resources that support all economies and all life. This model reinforces the idea that unlimited economic growth of any kind is sustainable.

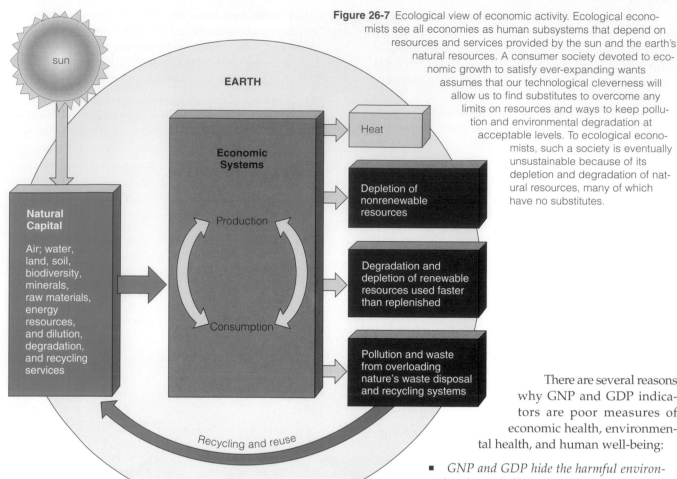

Figure 26-7 Ecological view of economic activity. Ecological economists see all economies as human subsystems that depend on resources and services provided by the sun and the earth's natural resources. A consumer society devoted to economic growth to satisfy ever-expanding wants assumes that our technological cleverness will allow us to find substitutes to overcome any limits on resources and ways to keep pollution and environmental degradation at acceptable levels. To ecological economists, such a society is eventually unsustainable because of its depletion and degradation of natural resources, many of which have no substitutes.

26-2 MONITORING ECONOMIC AND ENVIRONMENTAL PROGRESS

Are GNP and GDP Useful Measures of Economic and Environmental Health and Human Well-Being? GNP, *gross domestic product* (GDP), and *per capita GNP and GDP* indicators (p. 5) provide a standardized method for comparing the economic outputs of nations regardless of their size, stage of economic development (Figure 1-5, p. 9), and national currency. These indicators are measures of the gross market value of all a nation's money transactions for goods and services, and they assume that everything we spend money for is good. They do not distinguish between beneficial and harmful goods and activities or between sustainable and unsustainable ones.

The economists who developed these indicators many decades ago never intended them to be used as measures of economic health, environmental health, or human well-being. However, most governments and business leaders use them for such purposes.

There are several reasons why GNP and GDP indicators are poor measures of economic health, environmental health, and human well-being:

- *GNP and GDP hide the harmful environmental and social effects of producing goods and services.* Pollution, crime, sickness, and death are counted as positive gains in the GDP or GNP. Thus, the GNP increases for a country with rising cancer, crime, and divorce rates and increasingly polluted air and water because money is spent to deal with these problems. In other words, these harmful effects are counted as benefits that are added to these indicators instead of as costs that should be subtracted.

- *GNP and GDP do not include the depletion and degradation of natural resources or assets on which all economies depend.* When an irreplaceable old-growth forest is cut down or a wetland is filled, the GDP and GNP go up, and the harmful economic, environmental, and social costs are not subtracted from these indicators. The faster we junk and replace our cars, computers, TVs, and appliances and cut down forests, the faster the GDP and GNP grow. A country can be exhausting its mineral resources, eroding its soils, cutting down its forests, destroying its wetlands and estuaries, and depleting its wildlife and fisheries. At the same time, it can have a rapidly rising GNP and GDP, at least until the country no longer has its environmental assets and the income

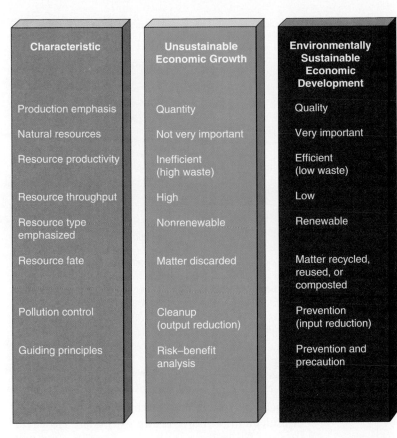

Characteristic	Unsustainable Economic Growth	Environmentally Sustainable Economic Development
Production emphasis	Quantity	Quality
Natural resources	Not very important	Very important
Resource productivity	Inefficient (high waste)	Efficient (low waste)
Resource throughput	High	Low
Resource type emphasized	Nonrenewable	Renewable
Resource fate	Matter discarded	Matter recycled, reused, or composted
Pollution control	Cleanup (output reduction)	Prevention (input reduction)
Guiding principles	Risk–benefit analysis	Prevention and precaution

Figure 26-8 Comparison of unsustainable economic growth and environmentally sustainable economic development.

from these assets. In other words, GNP and GDP count depletion and degradation of natural capital as a benefit instead of as a cost. This is like having car dashboard gauges that tell us how fast we are going but not how much gas is in the tank.

■ *GNP and GDP fail to include many nonmonetized benefits that meet basic needs.* GDP and GNP indicators exclude beneficial transactions in which no money changes hands such as the **(1)** labor we put into volunteer work, **(2)** health care and child care we give loved ones, **(3)** food we grow for ourselves, and **(4)** cooking, cleaning, and repairs we do for ourselves.

■ *GNP and GDP hide the enormous waste of natural and human resources because they measure only money spent, not value received.* According to Paul Hawken (Guest Essay, p. 6) and Amory Lovins (Guest Essay, p. 361), at least $2 trillion of the roughly $9 trillion spent every year in the United States provides consumers with no value. For example, money, time, and resources spent sitting in traffic jams produce zero value for drivers and cost the United States about $100 billion a year in lost productivity.

Another $200 billion a year spent on energy is wasted because the United States does not use energy as efficiently as other developed countries such as Japan.

■ *GNP and GDP tell us nothing about income distribution and economic justice.* They do not reveal how resources, income, or the harmful effects of economic growth (pollution, waste dumps, and land degradation) are distributed among the people in a country. The United Nations Children's Fund (UNICEF) suggests that countries should be ranked not by average per capita GNP or GDP but by average or median income of the poorest 40% of their people.

Solutions: Can Environmental Indicators Help? Ecological economists believe that GNP and GDP indicators should be supplemented with widely used and publicized *environmental indicators* that give a more realistic picture of environmental health, human welfare, and economic health.

Basically, such indicators would be developed by **(1)** subtracting from the GDP and GNP things that lead to a lower quality of life and depletion of natural resources and **(2)** adding to the GNP and GDP things that enhance environmental quality and human well-being but are currently being left out.

Economists Herman E. Daly (Guest Essay, p. 698-699), John B. Cobb, Jr., and Clifford W. Cobb have developed an *index of sustainable economic welfare (ISEW)* and applied it to the United States. This comprehensive indicator of human welfare measures **(1)** per capita GNP adjusted for inequalities in income distribution, **(2)** depletion of nonrenewable resources, **(3)** loss of wetlands, **(4)** loss of farmland from soil erosion and urbanization, **(5)** the cost of air and water pollution, and **(6)** estimates of long-term environmental damage from ozone depletion and possible global warming. Figure 26-9 shows the relationship between per capita GNP and ISEW for the United States between 1950 and 1990 (the latest year for which this information is available).

A newer indicator is called the *genuine progress indicator (GPI)*. This indicator, developed by Redefining Progress (a nonprofit organization), evaluates economic output by **(1)** subtracting expenses that do not improve environmental quality and human well-being from the GDP and **(2)** adding services that improve environmental quality and human well-being not currently included in the GDP. Figure 26-10

Appropriate Technology

SOLUTIONS

In his 1973 classic book *Small Is Beautiful: Economics as if People Mattered*, the late E. F. Schumacher described principles for developing environmentally sustainable economies and societies. One of his proposals was for developing countries to make greater use of what he called *appropriate technology* that is small scale, efficient, and labor intensive and uses locally available resources to produce goods that benefit local communities.

Some specific characteristics of appropriate technology are as follows:

- Human labor is favored over automation.

- Machines are small to medium-sized and easy to understand and repair.

- Production is decentralized, and whenever possible factories use local resources and renewable resources to produce products mostly for local consumption.

- Factories use energy and materials efficiently, produce little pollution and wastes, and do not disrupt the local culture.

- Management stresses meaningful work by allowing workers to perform a variety of tasks instead of a single production-line task.

Examples of appropriate technology include **(1)** simple stoves for use in rural villages (Individuals Matter, p. 613), **(2)** passive solar heating, **(3)** small wind turbines, **(4)** solar cells, **(5)** conventional and electric bicycles, **(6)** sewage gardens (Figure 19-17, p. 497), **(7)** drip irrigation systems (Solutions, p. 312), **(8)** pumps for tubewells, **(9)** small-scale water-purification systems (Individuals Matter, p. 498), and **(10)** compost bins.

Appropriate technology is not a panacea, but it is an important tool that can be used to provide jobs, produce more environmentally sustainable products, and benefit local economies, especially in developing countries.

Critical Thinking

1. What are some disadvantages of appropriate technology?

2. What types of goods might be made by appropriate technology for use in your local community?

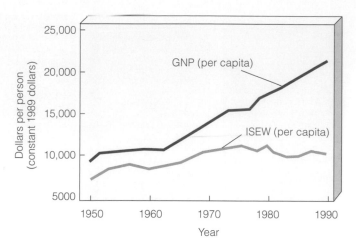

Figure 26-9 Comparison of per capita gross national product (GNP) and per capita index of sustainable economic welfare (ISEW) for the United States between 1950 and 1990. After rising by 42% between 1950 and 1976, the per capita ISEW fell 14% between 1977 and 1990. (Adapted from *Environmental Science*, 5/E by Chiras, p. 513, fig 26-6. Copyright ©1998 by Wadsworth)

Such environmental indicators are far from perfect and include many crude estimates. However, without such indicators, we do not **(1)** know much about what is happening to people, the environment, and the planet's natural resource base, and **(2)** have an effective way to measure what policies work. In effect, we are trying to guide national and global economies through treacherous economic and environmental waters at ever-increasing speeds using faulty radar.

26-3 HARMFUL EXTERNAL COSTS AND FULL-COST PRICING

What Are Internal and External Costs? All economic goods and services have both internal and external costs. For example, the price a consumer pays for a car reflects the costs of the factory, raw materials, labor, marketing, and shipping, as well as a markup to allow the car company and its dealers some profits. After a car is purchased, the buyer must pay for gasoline, maintenance, and repair. All these direct and indirect costs, which are paid for by the seller and the buyer of an economic good, are called **internal costs**.

Making, distributing, and using any economic good or service also involve external costs or benefits not included in the market price. For example, if a car dealer builds an aesthetically pleasing showroom, it is an **external benefit** to people who enjoy the sight at no cost.

compares the total and per capita values of the GNP and GPI for the United States between 1950 and 1998 and shows a steady decline in both GPI indicators since 1980.

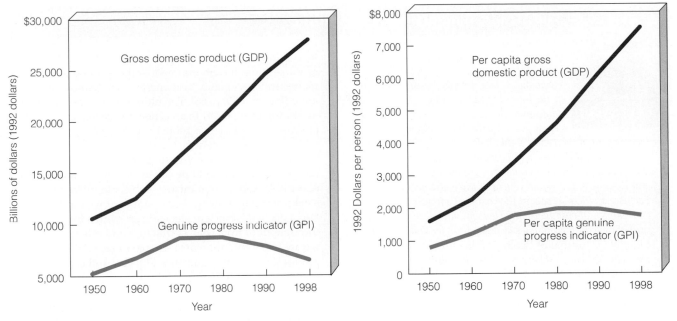

Figure 26-10 Comparison of the gross domestic product (GDP) and genuine progress indicator (GPI, left) and the per capita values for these indicators (right) in the United States between 1950 and 1998. (Data from Clifford Cobb, Gary Sue Goodman, and Mathis Wackernagel)

On the other hand, extracting and processing raw materials to make and propel cars **(1)** depletes non-renewable energy and mineral resources, **(2)** produces solid and hazardous wastes, **(3)** disturbs land, **(4)** pollutes the air and water, **(5)** contributes to global climate change, and **(6)** reduces biodiversity. These harmful effects are **external costs** passed on to the public, the environment, and in some cases future generations.

Because these harmful costs are not included in the market price, people do not connect them with car ownership. Still, everyone pays these hidden costs sooner or later, in the form of **(1)** poor health, **(2)** higher costs for health care and health insurance, and **(3)** higher taxes for pollution control. For example, in the United States gasoline costs about $1.60 per gallon (42¢ per liter). However, when you add all the external costs of pollution, health, insurance, and other social costs, American consumers are paying $5–7 per gallon ($1.30–1.90 per liter). Most vehicle owners are unaware that they are paying these hidden costs and do not associate them with driving.

Should We Shift to Full-Cost Pricing? For most economists, the solution to the harmful costs of goods and services is to include such costs in the market prices of goods and services. This process, called *internalizing the external costs*, is one of the goals of a pure free-market economy (p. 689) but is resisted in a capitalist market economy (p. 691).

According to economist Harold M. Hubbard, "The progress of civilization can be charted in terms of the internalization of costs formerly viewed as external." For example, the once essentially free cost of using slaves for labor has been internalized by including the cost of labor in the price of goods and services.

As long as businesses receive subsidies and tax breaks for extracting and using virgin resources and are not taxed for the pollutants they produce, few will volunteer to reduce short-term profits by becoming more environmentally responsible. For example, assume you own a company and believe it's wrong to subject your workers to hazardous conditions and pollute the environment beyond what natural processes can handle. Suppose you voluntarily improve safety conditions for your workers and install pollution controls, but your competitors do not. Then your product will cost more to produce than similar products produced by your competitors, and you will be at a competitive disadvantage. Your profits will decline; you may eventually go bankrupt and have to lay off your employees.

One way to deal with the problem of harmful external costs is for the government to levy taxes, pass laws and develop regulations, provide subsidies, or use other strategies that encourage or force producers to include all or most of these costs in the market prices of economic goods and services. Then the market price would be the **full cost** of these goods and services: internal costs plus short- and long-term external costs.

The Steady-State Economy in Outline

Herman E. Daly

Herman E. Daly is senior research scholar at the School of Public Affairs, University of Maryland. Between 1989 and 1993 he was senior economist at the Environmental Department of the World Bank. Before joining the World Bank he was Alumni Professor of Economics at Louisiana State University. He has been a member of the Committee on Mineral Resources and Environment of the National Academy of Sciences and has served on the boards of advisers of numerous environmental organizations. He is founder of the emerging field of ecological economics and is cofounder and associate editor of the Journal of Ecological Economics. *He has written many articles and several books, including* Steady-State Economics *(2d ed., 1991),* Beyond Growth: The Economics of Sustainable Development *(1996),* Ecological Economics and the Ecology of Economics: Essays in Criticism *(2000) and, with coauthor John Cobb Jr.,* For the Common Good: Redirecting the Economy Towards Community, the Environment, and a Sustainable Future *(1994). He is considered one of the foremost thinkers in ecological economics.*

A steady-state economic system is characterized by balanced, opposing forces that maintain a constant stock of physical wealth and people through a system of dynamic interactions and feedback loops. A low rate of flow or throughput of matter and energy resources maintains this wealth and population size at some desirable and sustainable level.

In such systems, emphasis is on increasing the quality of goods and services without depleting or degrading natural resources to unsustainable levels for current and future generations (sustainable or steady-state economic development). In other words, *sustainable economic development* is development without growth in resource throughput and depletion and degradation of natural resources [Figure 26-8].

The worldview underlying conventional economics is that an economy is a system that is essentially isolated from the natural world and involves a circular exchange of goods and services between businesses and households [Figure 26-6]. This model ignores the origin of natural resources flowing into the system and the fate of wastes flowing out of the system. It is as if a biologist had a model of an animal that contained a circulatory system but had no digestive system that tied it firmly to the environment at both ends.

The steady-state economic view recognizes that economic systems are not isolated from the natural world but are fully dependent on ecosystems for the natural goods and services they provide [Figure 26-7].

Throughput is roughly equivalent to GNP, the annual flow of new production of goods and services. It is the cost of maintaining the stocks of final goods and services by continually importing high-quality matter and energy resources from the environment and exporting waste matter and low-quality heat energy back to the environment [Figure 3-19, p. 66]. The ultimate cost of continually increasing throughput in an economic system is the loss of ecosystem services as more irreplaceable natural capital is converted to manufactured capital.

The reasoning just given suggests that the gross national product should be called the gross national cost (GNC). We should minimize it, subject to maintaining stocks of essential items. For example, if we can maintain a desired, sufficient stock of cars with a lower throughput of iron, coal, petroleum, and other resources, we are better off, not worse. To maximize GNP throughput for its own sake is absurd. Physical and ecological limits to the volume of throughput suggest that sooner or later a steady-state economy will be necessary.

Another problem with conventional economic theory is that it assumes that manufactured capital based on human ingenuity can be substituted for natural capital. In fact, however, manufactured capital loses its economic value without an input of natural capital. What good are more fishing boats without fish, more chain saws and sawmills without forests, more water pumps without aquifers, and more crop-harvesting machines without fertile soil? Thus, humankind is facing a historic juncture. For the first time, the limits to increased prosperity will increasingly be limited by lack of natural capital as a growing number of consumers use more of the world's natural resources to produce more manufactured goods.

The greatest challenges facing us today are:

- For physical and biological scientists to define more clearly the limits and interactions within ecosystems and the biosphere (which determine the feasible levels of the steady state) and to develop technologies more in conformity with such limits.

- For social scientists to design institutions that will bring about the transition to a steady-state or environmentally sustainable economy and permit its continuance.

- For philosophers, theologians, and educators to stress the neglected traditions of stewardship and distributive justice that exist in our cultural and religious heritage.

This last item is of paramount importance because the ethical and political problem of sharing a fixed amount of resources and goods is much greater than that of sharing a growing amount. Indeed, this has been the primary reason for giving top priority to economic growth. If the pie is always growing, it is said there will always be crumbs—and the hope of a slice—for the poor. This avoids the moral question of a more equitable distribution of the world's resources and wealth.

(continued)

The kinds of economic institutions needed to make this transition follow directly from the definition of a steady-state economy:

- An institution for maintaining a constant population size within the limits of available resources. For example, economic incentives can be used to encourage each woman or couple to have no more than a certain number of children, or each woman or couple could be given a marketable license to have a certain number of children, as the late economist Kenneth Boulding suggested.

- An institution for maintaining a constant stock of physical wealth and for limiting resource throughput. For example, the government could set and auction off transferable annual depletion quotas for key resources.

- An institution for limiting inequalities in the distribution of the constant physical wealth among the constant population in a steady-state economy. For example, there might be minimum and maximum limits on personal income and maximum limits on personal wealth. In 1997, for example, the average annual compensation (including stock options) for CEOs of U.S. corporations was $7.8 million—326 times the earnings of the average factory worker. One formula for helping

achieve a fairer distribution of wealth might be that high-income people should make only about 10 times as much as low-income people. This would give high-income people an incentive to raise the incomes of low-income people because it would be the only way they could increase their incomes. Another way to distribute income is to tax the haves and transfer the money to the have-nots.

Many such institutions could be imagined. The problem is to achieve the necessary global and societal (macro) control with the least sacrifice of freedom at the individual (micro) level.

Critical Thinking

1. What are the pros and cons of establishing institutions for maintaining population size and resource throughput to avoid depleting and degrading natural resources?

2. What are the pros and cons of establishing minimum and maximum limits on personal income? Do you favor doing this? Explain.

This approach requires government action because few companies will intentionally increase their cost of doing business unless their competitors must do so as well. Internalizing the external costs of pollution and environmental degradation would make (1) preventing pollution more profitable than cleaning it up and (2) waste reduction, recycling, and reuse more profitable than burying or burning most of the waste we produce.

The *bad news* is that when external costs are internalized, the market prices for most goods and services would rise. However, the *good news* is that total price people pay would be about the same because the hidden external costs related to each product would be included in its market price. In addition, full-cost pricing provides consumers with information needed to make informed economic decisions about the effects of their lifestyles on the planet's life-support systems and on human health.

However, as external costs are internalized, economists and environmentalists warn that governments need to (1) reduce income, payroll, and other taxes and (2) withdraw subsidies formerly used to hide and pay for these external costs. Otherwise, consumers will face higher market prices without tax relief—a politically unacceptable policy guaranteed to fail.

Some more *good news* is that some goods and services would cost less because internalizing external costs encourages producers to (1) find ways to cut costs by inventing more resource-efficient and less-polluting methods of production and (2) offer more earth-sustaining (or *green*) products. Jobs would be lost in earth-degrading businesses, but at least as many and probably more jobs could be created in earth-sustaining businesses. If a shift to full-cost pricing took place over several decades, most current earth-degrading businesses would have time to transform themselves into profitable earth-sustaining businesses.

Full-cost pricing seems to make a lot of sense. Why is it not used more widely? There are several reasons:

- Many producers of harmful and wasteful goods would have to charge so much that they could not stay in business.

- Huge government subsidies distort the marketplace and hide many of the harmful environmental and social costs caused by producing and using some goods and services.

- Producers would have to give up government subsidies and tax breaks that have helped hide the harmful external costs of their goods and services and distort the marketplace. Studies by Norman Myers

(Guest Essay, p. 117) and other analysts estimate that governments around the globe spend about $1.5 trillion per year—about twice as much as all military spending—to help subsidize (1) deforestation, (2) overfishing, (3) overgrazing, (4) unsustainable agriculture, (5) use of nonrenewable fossil fuels and nuclear energy, (6) groundwater depletion, and (7) other environmentally harmful activities.

- The prices of harmful but desirable goods and services would rise.

- It is hard to put a price tag on many harmful environmental and health costs.

- Many business interests, political leaders, and consumers are unaware that they are paying these costs in ways not connected to the market prices of goods and services.

Despite the difficulties, proponents believe that full-cost pricing for harmful environmental and health effects should be phased in over the next 20 years. They argue that doing the best we can to estimate and internalize current external costs is far better than continuing the current pricing system, which gives too little or misleading information about the environmental and health effects of goods and services. The key question is whether a *tell-the-truth full-cost* pricing system would be worse than the current *hide-the-true-cost* pricing system. What do you think?

26-4 THE ECONOMICS OF POLLUTION CONTROL AND RESOURCE MANAGEMENT

How Much Are We Prepared to Pay to Control Pollution The cleanup cost for removing specific pollutant gases or wastewater being discharged into the environment rises with each additional unit of pollutant that is removed. As Figure 26-11 shows, a large percentage of the pollutants emitted by a smokestack or wastewater discharge can be removed fairly cheaply.

However, as more and more pollutants are removed, the cost to remove each additional unit of pollution rises sharply. Beyond a certain point, the cleanup costs exceed the harmful costs of pollution. The *breakeven point* is the level of pollution control at which the harmful costs to society and the costs of cleanup are equal. Pollution control beyond this point costs more than it is worth, and not controlling pollution to this level causes pollution that we can afford to avoid.

To find the breakeven point, economists plot two curves: (1) a curve of the estimated economic costs of

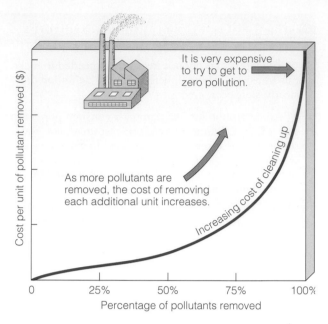

Figure 26-11 The cost of removing each additional unit of pollution rises. Cleaning up a certain amount of pollution is affordable, but at some point the cost of pollution control is greater than the harmful costs of the pollution to society.

cleaning up pollution and (2) a curve of the estimated harmful external costs of pollution to society. They then add the two curves together to get a third curve showing the total costs. The lowest point on this third curve is the breakeven point, or the *optimum level of pollution* (Figure 26-12).

On a graph, determining the optimum value of pollution looks neat and simple, but there are problems with this approach:

- Ecological economists, health scientists, and business leaders often disagree in their estimates of the harmful costs of pollution, and such costs are difficult to determine.

- Some critics raise environmental justice questions about who benefits and who suffers from allowing optimum levels of pollution. The levels may be optimum for the entire country or an area, but not for the people who live near or downwind or downriver from a polluting power plant or factory and are exposed to high levels of pollution.

- Assigning monetary values to things such as lost lives, forests, wetlands, and ecological services is difficult and controversial. For example, how much is a person's life worth, and what is the economic value of the ecological services provided by a patch of forest (Spotlight, p. 562)? On the other hand, assigning little or no value to such things means

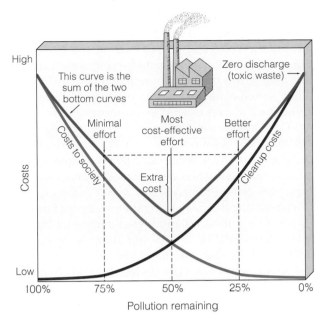

Figure 26-12 Finding the optimum level of pollution. This graph shows the optimum level at 50%, but the actual level varies depending on the pollutant.

that they will not be counted in determining optimum pollution levels.

Three ways used to estimate the monetary value of resources that cannot be priced by conventional means are to determine:

- *Mitigation costs:* the costs of offsetting damages. For example, how much would it cost to protect a forest from cutting, move an endangered species to a new habitat, or restore a statue damaged by air pollution?

- *Willingness to pay:* using a survey to determine how much people would be willing to pay to keep a particular species from becoming extinct, a particular forest from being cut down, or a specific river or beach from being polluted.

- *Maintenance and protection costs:* how much it would cost to maintain and protect the quality of various natural resources.

There are also problems in assessing the costs of pollution. Economists identify three types of costs associated with pollution:

- *Direct costs* that a polluting company incurs when it has to add pollution control equipment or clean up an oil spill or toxic waste dump.

- *Indirect costs* incurred by local, federal, and state governments in regulating pollution and in damages to local economies from reductions in tourism, jobs, and revenues from fishing and other activities curtailed because of pollution.

- *Repercussion costs,* in which a polluting industry can have an unfavorable image that causes consumers to boycott or reduce their purchases the company's products.

Some analysts point out that we should not be unduly influenced by curves showing the sharply increasing costs of improving pollution control (Figure 26-11) because such curves are based on using a particular form of pollution control. Stricter regulations and higher pollution control costs can stimulate business to find cheaper ways to control pollution or to redesign manufacturing processes to sharply reduce or eliminate pollution (p. 522). This use of innovation to solve pollution problems can lead to significant savings and increase profits and competitiveness by *tunneling through* the cost barrier in curves such as the one in Figure 26-11.

For example, in the 1970s the U.S. chemical industry predicted that controlling benzene emissions would cost $350,000 per plant. Shortly after these predictions were made, however, the companies developed a process that substituted other chemicals for benzene and almost eliminated control costs.

Who Should Pay for Pollution Control? There are basically two strategies for paying for pollution control:

- *The consumer or user pays.* Government regulations require businesses to pay for devices and processes used to clean up pollution, and businesses pass these costs on to consumers of their products.

- *The taxpayer pays.* The government through tax revenue pays for the costs of pollution control. Examples are government payments **(1)** to coal miners suffering from black lung and **(2)** for cleanup of contaminated waste sites on public land (p. 320), military bases, and government-run facilities for producing nuclear weapons.

Those favoring the consumer-pays approach believe that people who use products that cause pollution and environmental degradation should pay for the costs as part of the market price of such goods. This is a way of internalizing the harmful external costs of pollution and environmental degradation and coming closer to using full-cost pricing for goods and services.

Business interests favoring the taxpayer-pays approach believe that when governments change regulations and require more stringent pollution control, then society (taxpayers) should be responsible for the costs involved. In this case, taxpayers pay indirectly for the costs of pollution and environmental degradation through higher taxes but do not connect these costs with the products and services they use.

What Economic Factors Affect How a Natural Resource Is Used or Managed? Other economic factors influencing how a resource is used or managed are **(1)** *discount rates,* **(2)** *time preferences,* **(3)** *opportunity costs,* **(4)** *government subsidies and tax breaks,* **(5)** *taxes,* and **(6)** *ethics.* The **discount rate** is an estimate of a resource's future economic value compared to its present value. The size of the discount rate (usually given as a percentage) is a primary factor affecting how a resource such as a forest or fishery is used or managed.

At a zero discount rate, for example, a stand of redwood trees worth $1 million today will still be worth $1 million 50 years from now. However, most businesses, the U.S. Office of Management and Budget, and the World Bank typically use a 10% annual discount rate to evaluate how resources should be used. At this rate, the stand of redwood trees will be worth only $10,000 in 50 years. With this discount rate, it makes sense from an economic standpoint to cut these trees down as quickly as possible and invest the money in something else.

Proponents of high (5–10%) discount rates argue that **(1)** inflation will reduce the value of their future earnings, **(2)** innovation or changes in consumer preferences could make a product or service obsolete, and **(3)** without a high discount rate, they can make more money by investing their capital in some other venture.

Critics point out that high discount rates encourage such rapid exploitation of resources for immediate payoffs that sustainable use of most renewable natural resources is virtually impossible. These critics believe that a 0% or even a negative discount rate should be used to protect unique and scarce resources and that discount rates of 1–3% would make it profitable to use other nonrenewable and renewable resources sustainably or slowly.

Time preference is a measure of how willing one is to postpone some current income for the possibility of greater income in the future. For example, farmers have two options in deciding how to manage the soil on their land:

- A *depletion strategy,* in which artificial fertilizers, irrigation, and pesticides are used to achieve high yields and profits for a short period until soil fertility is depleted or degraded (p. 224) and aquifers are depleted or polluted (Case Study, p. 309).

- A *conservation strategy,* in which investments are made in techniques to conserve topsoil and maintain fertility (Section 10-6, p. 228) and reduce waste of irrigation water (Section 13-7, p. 310) and thus sustain profits in the future.

Opportunity cost is the cost from not putting money in something that could give a higher yield. For example, forest owners and farmers who spend money on using a conservation strategy to harvest crops and trees on a long-term sustainable basis might have the opportunity to make more money by investing their money in the stock market or another business venture.

Government subsidies and tax breaks can strongly influence how resources are used. Resource depletion and degradation are accelerated by **(1)** giving mining, oil, and coal companies depletion allowances and tax breaks for getting minerals and oil out of the ground, **(2)** subsidizing the cost of irrigation water for farmers, **(3)** providing subsidies and low-cost loans for buying fishing boats, and **(4)** subsidizing livestock grazing on public land. On the other hand, resource conservation can be encouraged by **(1)** withdrawing such resource-depletion subsidies and tax breaks or **(2)** giving resource users subsidies and tax breaks for practicing resource conservation and not selling farmland, ranchland, and forestland to developers.

High taxes on raw materials, pollution, and manufactured goods can be used to discourage resource depletion and degradation, and *low taxes* on resource conservation can encourage use of this strategy. Of course, the reverse is also true.

Economic return is not always the determining factor in how resources are used or managed. In some cases, farmers and owners of forests, wetlands, and other resources use *ethical concerns* in determining how they use and manage such resources. Their respect for the land and nature or what they believe to be their responsibility to future generations can override their desire for short-term profit at the expense of long-term resource sustainability.

What Is Benefit-Cost Analysis, and How Can It Be Improved? A related useful and widely used tool for making economic decisions about how to control pollution and manage resources is **benefit-cost analysis**. It involves comparing the estimated short-term and long-term benefits (gains) and costs (losses) for various courses of action such as implementing a particular pollution control regulation or deciding whether a dam should be built, a wetland filled, or a forest cleared.

However, such analyses **(1)** have some serious limitations, **(2)** usually are very rough approximations, and **(3)** can be manipulated easily. There are controversies over **(1)** assigning discount rates (left), **(2)** determining who benefits and who is harmed, and **(3)** putting estimated price tags on human life, good health, clean air and water, wilderness, and various forms of natural resources.

Because estimates of benefits and costs are so variable, *figures can be weighted easily to achieve the out-*

come desired by proponents or opponents of a proposed project or environmental regulation. For example, one U.S. industry-sponsored cost–benefit study estimated that compliance with a standard to protect U.S. workers from vinyl chloride would cost $65–90 billion; in fact, less than $1 billion was actually needed to comply with the standard.

Critics also point out that many benefit–cost analyses fail to evaluate alternative ways to control or prevent a particular form of pollution, manage a certain resource, or find alternative uses or a substitute for a resource.

Benefit–cost analysis causes us to think about a proposed regulation or type of resource use carefully, and it can provide very useful information in helping us make such decisions. However, critics contend that analysis of various benefit–cost analyses indicates that they are too imprecise for use as the *primary way* to make decisions about environmental regulations and resource.

To these critics, using this tool is somewhat like trying to detect a car speeding at 160 kilometers per hour (100 miles per hour) with a radar device so unreliable that at best it can tell us only that the car's speed is somewhere between 80 kph (50 mph) and 8,000 mph (5,000 mph).

Their advice is to use such analyses but **(1)** use critical thinking skills (Guest Essay, p. 48) to determine what assumptions are being used, evaluate the quality of the data, and check the math and conclusions, **(2)** realize that at best they provide only rough estimates about projected benefits and costs, and **(3)** push for use of standardized guidelines in making such analyses.

To minimize possible abuses, environmentalists and economists advocate the following guidelines for all benefit–cost analyses: **(1)** Use uniform standards, **(2)** state all assumptions, **(3)** evaluate the reliability of all data inputs as high, medium, or low, **(4)** make projections using low, medium, and high discount rates, **(5)** show the estimated range of costs and benefits based on various sets of assumptions, **(6)** estimate the short- and long-term benefits and costs to all affected population groups, and **(7)** evaluate and compare all alternative courses of action.

26-5 USING REGULATIONS AND MARKET FORCES TO IMPROVE ENVIRONMENTAL QUALITY

What Are the Pros and Cons of Using Regulations to Improve Environmental Quality? Most economists agree that controlling or preventing pollution and reducing resource waste require government intervention in the marketplace. Such government action can take the form of regulation, the use of market forces, or some combination of these approaches.

Regulation is a *command-and-control* approach. It involves enacting and enforcing laws that **(1)** set pollution standards, **(2)** regulate harmful activities, **(3)** ban the release of toxic chemicals into the environment, and **(4)** require that certain irreplaceable or slowly replenished resources be protected from unsustainable use.

Depending on how they are crafted, environmental regulations can raise or lower the costs for businesses and discourage or encourage looking for innovative solutions to environmental problems. Both environmentalists and business leaders agree that *innovation-friendly regulations* can encourage companies to **(1)** find innovative ways to prevent pollution and improve resource productivity (which reduce costs) and **(2)** develop innovative products and industrial processes (which can increase profits and competitiveness).

On the other hand, business leaders and many environmentalists agree that some pollution control regulations are unnecessarily costly and discourage innovation by **(1)** concentrating on cleanup instead of prevention, **(2)** being too prescriptive (for example, by mandating specific technologies), **(3)** setting compliance deadlines that are too short to allow companies to find innovative solutions, and **(4)** discouraging risk taking and experimentation.

Consider the difference between pollution regulation of the pulp and paper industries in the United States and Sweden. In the 1970s, strict regulations in the United States with short compliance deadlines forced U.S. companies to adopt the best available but costly end-of-pipe pollution treatment systems.

By contrast, in Sweden the government started out with slightly less strict standards and longer compliance deadlines but clearly indicated that tougher standards would follow. This more flexible and innovation-friendly approach **(1)** gave companies time to focus on redesigning the production process itself instead of relying mostly on waste treatment and **(2)** spurred them to look for innovative ways to prevent pollution and improve resource productivity to meet stricter future standards. The Swedish companies were able to develop innovative pulping processes and chlorine-free bleaching processes that **(1)** met the emission standards, **(2)** lowered operating costs, and **(3)** gave them a competitive advantage in international markets.

An innovation-friendly regulatory process

- *Emphasizes pollution control and waste reduction*
- *Requires industry and environmental interests to participate in developing realistic standards and timetables*

- *Sets goals but frees industries to meet such goals in any way that works*

- *Sets standards strict enough to promote real innovation*

- *Establishes well-defined deadlines that allow companies enough time to innovate*

- *Uses market incentives such as emission and resource use charges and tradable pollution and resource use permits to encourage compliance and innovation*

How Can Economic Incentives Be Used to Improve Environmental Quality and Reduce Resource Waste? *Market forces* can help improve environmental quality and reduce resource waste, mostly by encouraging the internalization of external costs by using **(1)** *economic incentives* (rewards) or **(2)** *economic disincentives* (punishments). This is based on a fundamental principle of the marketplace in today's mixed economic systems: *What we reward (mostly by government subsidies and tax breaks) we tend to get more of, and what we discourage (mostly by regulations and taxes) we tend to get less of.*

Incentive methods include:

- *Phasing in government subsidies and tax breaks that encourage earth-sustaining behavior.* The difficulty is that adding such subsidies and tax breaks (or phasing out harmful subsidies that encourage or pay people to pollute and deplete or degrade resources) involves political decisions that often are opposed successfully by powerful economic interests.

- *Removing market barriers* such as regulations, quotas, and preferential rates for transporting some types of materials that create an uneven economic playing field and hinder free-market competition. For example, in the United States, freight rates established by federal regulations hinder recycling by making it cheaper to haul virgin materials than recovered scrap. At the international level, controversial agreements have been enacted to help lower barriers to international trade, as discussed on page 733.

- Use *product or ecolabeling programs* to encourage companies to develop green products and services and to help consumers select the most environmentally sustainable products and services. In the early 1980s, the government of the former West Germany instituted an ecolabeling program known as the *Blue Angel* program to show that a product has been produced in an environmentally acceptable manner and meets certain environmental standards. In some cases, companies with products receiving poor ratings altered their manufacturing processes or product designs to receive the coveted green seal of approval.

Ecolabeling programs have also been developed in Europe, Japan, Canada, and the United States (for example, the *Green Seal* labeling program that has certified more than 300 products).

How Can Economic Disincentives Be Used to Improve Environmental Quality and Reduce Resource Waste? Economic *disincentive* approaches include:

- *Using green taxes or effluent fees* to help internalize many of the harmful environmental costs of production and consumption. Taxes can be levied on each unit of **(1)** pollution discharged into the air or water, **(2)** hazardous or nuclear waste produced, **(3)** virgin resources used, and **(4)** fossil fuel used. Economists point out that such *green taxes* work if **(1)** they reduce or replace income, payroll, or other taxes and **(2)** the poor and middle class are given a safety net to reduce the regressive nature of consumption taxes on essentials such as food, fuel, and housing. In other words, to be politically acceptable environmental taxes must be seen as a *tax-shifting* instead of a *tax-burden* approach (Solutions, p. 707).

- *Charging user fees* that cover all or most costs for activities such as **(1)** extracting lumber and minerals from public lands, **(2)** using water provided by government-financed projects, and **(3)** using public lands for recreation or livestock grazing.

- *Requiring businesses to post a pollution prevention or assurance bond* when they plan to develop a new mine, plant, incinerator, landfill, or development and before they introduce a new chemical or new technology. After a set length of time, the deposit (with interest) would be returned, minus actual or estimated environmental costs. This approach is similar to the performance bonds contractors are often required to post for major construction projects.

What Are the Pros and Cons of Using Tradable Pollution and Resource-Use Rights? Another market approach is for the government to *grant tradable pollution and resource-use rights*. For example, a total limit on emissions of a pollutant or use of a resource such as a fishery could be set, and permits would be used to allocate or auction the total among manufacturers or users.

Permit holders not using their entire allocation could **(1)** use it as a credit against future expansion, **(2)** use it in another part of their operation, or **(3)** sell it to other companies. Tradable rights could also be established among countries to help preserve biodiversity and reduce emissions of greenhouse gases and other pollutants with harmful regional (or global) effects.

Some environmentalists support tradable pollution rights as an improvement over the current regulatory approach. Ecological economist Herman Daly (Guest Essay, p. 698) likes the idea of issuing tradable pollution permits because it requires society to confront the political and ethical problems of (1) how much total pollution is acceptable and (2) how the distribution of rights to pollute should be distributed (fair distribution).

Other environmentalists oppose allowing companies to buy and trade rights to pollute is because it (1) allows the wealthiest companies to continue polluting (thereby excluding smaller companies from the market), (2) tends to concentrate pollutants at the dirtiest plants (thereby jeopardizing the health of people downwind of the plant), (3) creates an incentive for fraud because most pollution control regulations are based on self-reporting of pollution outputs, and government monitoring and enforcement of such outputs is inadequate, and (4) has no built-in incentives to reduce overall pollution unless the system requires an annual decrease in the total pollution allowed (which is rarely the case).

Should We Rely Mostly on Regulations or Market Approaches? Most analysts would answer that both regulations and market approaches are appropriate, depending on the situation. Table 26-1 shows that each of the major approaches discussed in this section has advantages and disadvantages. Currently, in the United States private industry and local, state, and federal government spend about $130 billion a year to comply with federal environmental regulations (compared to about $55 billion in 1972). Studies estimate that greater reliance on a variety of market-based policies could cut these expenditures by one-third to one-half.

Evolving Eras of Environmental Management

Figure 26-13 shows the evolution of several phases of environmental management, with the long-term goal of achieving environmentally sustainable economies and societies.

The period between 1970 and 1985 can be viewed as the *resistance-to-change management era.* In the 1970s, many companies and government environmental regulators developed an adversarial approach in which companies resented and actively resisted environmental regulations, and government regulators thought they had to prescribe ways for reluctant companies to clean up their pollution emissions. At this resistant adaptation stage, most companies dealt with environmental regulations by (1) hiring outside environmental consultants, who usually favored end-of-pipe pollution control solutions, (2) using lawyers to oppose or find legal loopholes in the regulations, and (3) lobbying elected officials to have environmental laws and regulations overthrown or weakened.

By 1985 most company managers accepted environmental regulations and continued to rely mostly on pollution control. However, they placed little emphasis on trying to find innovative solutions to pollution and resource waste problems because they were believed to be too costly.

In 1990s, a growing number of company managers began to realize that environmental improvement is an economic and competitive opportunity instead of a cost to be resisted. This was the beginning of the *innovation management era* that environmental and business visionaries project will go through several phases over the next 40–50 years (Figure 26-13).

An increasing number of companies are instituting *total quality management*, with an emphasis on preventing pollution and more recently on greatly improving resource productivity (Solutions, p. 525). In the initial stages, this involves making minor changes that save money and are viewed as win–win solutions. However,

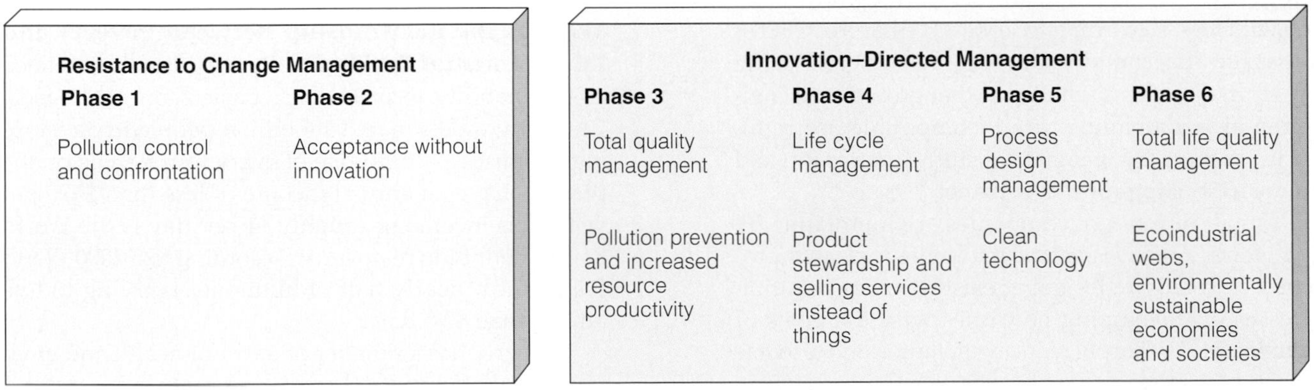

Resistance to Change Management		Innovation–Directed Management			
Phase 1	**Phase 2**	**Phase 3**	**Phase 4**	**Phase 5**	**Phase 6**
Pollution control and confrontation	Acceptance without innovation	Total quality management	Life cycle management	Process design management	Total life quality management
		Pollution prevention and increased resource productivity	Product stewardship and selling services instead of things	Clean technology	Ecoindustrial webs, environmentally sustainable economies and societies

Figure 26-13 Evolving eras of environmental management.

Table 26-1 Economic Solutions to Pollution and Resource Waste

Solution	Internalizes External Costs	Innovation	International Competitiveness	Administrative Costs	Increases Government Revenue
Regulation	Partially	Can encourage	Decreased*	High	No
Subsidies	No	Can encourage	Increased	Low	No
Withdrawing harmful subsidies	Yes	Can encourage	Decreased*	Low	Yes
Tradable rights	Yes	Encourages	Decreased*	Low	Yes
Green taxes	Yes	Encourages	Decreased*	Low	Yes
User fees	Yes	Can encourage	Decreased*	Low	Yes
Pollution-prevention bonds	Yes	Encourages	Decreased*	Low	No

*Unless more cost-effective and productive technologies are developed.

once the easy changes are made, saving money will take more innovative approaches.

A small number of companies are beginning to implement *life cycle management*, assuming environmental stewardship of a product throughout its entire life cycle. After products are sold or leased to customers, companies are responsible for maintaining them and taking them back for repair, upgrading, recycling, or remanufacturing (p. 527 and Individuals Matter, p. 528). A few companies are switching their emphasis from selling products (such as heating and air conditioning equipment) to selling the services (such as heating and cooling) their products provide (p. 527).

This involves a shift from growing by producing more stuff to using information and knowledge to increase resource productivity by (1) avoiding waste, (2) miniaturizing things (bigger is not better), (3) replacing chemical plants with green plants capable of producing chemicals we need, and (4) learning how to do more with less. According to Robert B. Shapiro, CEO of Monsanto, "If economic development means using more stuff, then those who argue that growth and environmental sustainability are incompatible are right. And if we grow by using more stuff, I'm afraid we'd better start looking for a new planet."

A small number of companies are innovating by using *process design management* to achieve cleaner production by totally redesigning existing manufacturing processes or developing new ones with the goals of (1) eliminating or sharply reducing pollution and resource waste and (2) decreasing production, waste management, product liability, and pollution compliance costs.

In the last phase of *total life quality management*, many companies are projected to form or become involved in *ecoindustrial networks* (p. 524), where they exchange resources and wastes in industrial webs not unlike the food webs found in nature. In this phase, companies will also focus on (1) learning how to develop integrated and adaptable management plans and (2) using such plans to integrate their goals with those of governments and individuals to find innovative ways to develop environmentally sustainable and just local, national, and global economies and societies. At this stage, people are driven by a vision of environmental sustainability that improves life for everyone without degrading nature's economy that sustains us all.

26-6 REDUCING POVERTY TO IMPROVE ENVIRONMENTAL QUALITY AND HUMAN WELL-BEING

What Is the Relationship Between Poverty and Environmental Problems? Poverty usually is defined as the inability to meet one's basic economic needs. Currently, an estimated 1.4 billion people in developing countries—almost one of every four people on the planet—have an annual income of less than $370 per year. This income of roughly $1 per day is the World Bank's definition of poverty. According to a 2000 World Bank study, nearly half of humanity is trying to live on less than $2 a day.

Poverty has a number of harmful health and environmental effects. It (1) causes premature deaths and preventable health problems, (2) tends to increase birth rates because the poor have more children to help them grow food or work and to take care of them in their

Shifting the Tax Burden from Wages and Profits to Pollution and Waste

An important goal of a more sustainable economic system would be to tax labor, income, and wealth less and tax throughput of matter and energy resources more. The current tax structure in most countries encourages businesses to increase productivity by substituting manufactured capital (machines) and throughput (energy and materials) for workers (human capital).

People of almost all political and economic persuasions agree that such a tax system is backwards. It discourages what we want more of (jobs and income), encourages what we want less of (pollution and depletion), and leads to tremendous waste of natural and human capital. Such tax systems remain primarily because of political resistance to change and the political power of businesses who profit from them.

Fully taxing pollution would raise about $1 trillion per year worldwide, enough to allow a 15% cut in taxes on wages and profits. Shifting more of the tax burden from wages and profits to pollution and waste (1) decreases depletion and degradation of natural capital, (2) improves environmental quality by making polluters and resource depleters and degraders pay for the harm they cause (full-cost pricing), (3) stimulates resource productivity (Solutions, p. 525) by encouraging pollution prevention, waste reduction, and creativity in solving environmental problems (to avoid paying pollution taxes), (4) rewards recycling and reuse instead of use of nonrenewable virgin resources, (5) uses the marketplace, rather than regulations, to help protect the environment, (6) reduces unemployment, and (7) allows cuts in highly resented and often regressive income, payroll, and sales taxes.

Such a tax shift would have to be phased in over a 15- to 20-year period. This would allow businesses to plan for the future and depreciate existing capital investments over their useful lives.

Small trial versions of such tax shifts are taking place in Germany, Sweden, Great Britain, Norway, Denmark, Italy, France, and the Netherlands. If Europe and other countries continue moving toward this type of tax shift over the next two decades, their labor costs will be lowered and their resource productivity will increase. To remain competitive in the global marketplace, the United States and other countries may be forced to follow.

Critical Thinking

What are disadvantages of making such a tax shift for you and for the country where you live? Do you favor making such a shift in what is taxed? Explain.

---

old age, and (3) often pushes poor people to use potentially renewable resources unsustainably to survive. For example, a major factor causing deforestation in developing countries is hundreds of millions of poor farmers who are driven to feed their families by cutting and burning patches of forest to grow crops.

In addition to its important health and environmental benefits, reducing poverty reduces human suffering. Economists also see it as a way to increase economic growth by creating more consumers able to buy more things in an increasingly global economy.

Has Economic Growth Reduced Poverty? Most economists believe that a growing economy is the best way to help the poor because such growth (1) creates more jobs, (2) enables more of the increased wealth to reach workers, and (3) provides greater tax revenues that can be used to help the poor help themselves.

However, since 1960 most of the benefits of global economic growth as measured by income have flowed up to the rich rather than down to the poor. This has made the top one-fifth of the world's people much richer and the bottom one-fifth poorer in terms of total real income and real per capita GNP (adjusted for inflation; Figure 26-14). Most people in between have

lost, or gained only slightly, in real per capita GNP. Since 1980, this wealth gap has accelerated (Figure 1-5, p. 9).

Government studies also show that current forms of economic growth have not led to large increases in jobs for the unemployed and underemployed poor. According to the United Nations Development Program, worldwide there are more than 1 billion willing workers who are involuntarily unemployed (150 million) or underemployed (900 million).

A major reason for this is that since 1980 most economic systems have been increasing productivity by replacing human capital (which is an abundant resource, especially in developing countries) with manufactured capital (machines). If emphasis shifts to increasing resource productivity (Solutions, p. 525), jobs will increase because this takes more labor and less natural capital, and environmental quality will improve.

How Can Poverty Be Reduced? Analysts point out that the governments of most developing countries must make policy changes to reduce poverty. Two examples are (1) shifting more of the national budget to help the rural and urban poor work their way out of poverty and (2) giving villages, villagers, and the

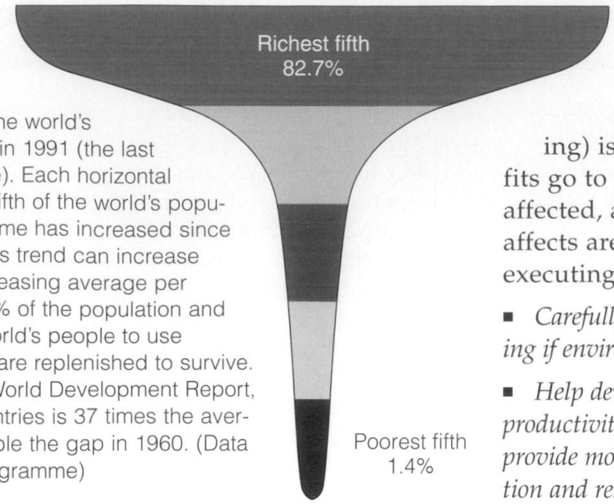

Figure 26-14 Data on the global distribution of income show that most of the world's income has flowed up, with the richest 20% of the world's population receiving more of the world's income than all of the remaining 80% in 1991 (the last year for which such data are available). Each horizontal band in this diagram represents one-fifth of the world's population. This upward flow of global income has increased since 1960 and accelerated since 1980. This trend can increase environmental degradation by (1) increasing average per capita consumption by the richest 20% of the population and (2) causing the poorest 20% of the world's people to use renewable resources faster than they are replenished to survive. According to the World Bank's 2000 World Development Report, average income in the 20 richest countries is 37 times the average in the 20 poorest countries—double the gap in 1960. (Data from United Nations Development Programme)

Richest fifth
82.7%

Poorest fifth
1.4%

urban poor title to common lands and to crops and trees they plant on them.

Analysts also urge developed countries and the wealthy in developing countries to help reduce poverty. Several ways to do this have been suggested:

- *Forgive at least 60% of the $2.5 trillion debt that developing countries owe to developed countries and international lending agencies and all of the $213 billion debt of the most heavily indebted poor countries.* Currently, developing countries pay almost $300 billion per year in interest to developed countries to service this debt. Some of this debt can be forgiven in exchange for carefully monitored agreements by the governments of developing countries to increase expenditures for family planning, health care, education, land redistribution, biodiversity protection and restoration, and more sustainable use of renewable resources. In 2000, the U.S. government announced that it would cancel all the debt owed by the world's poorest countries if the money is spent on basic human needs.

- *Increase nonmilitary government and private aid to developing countries from developed countries, with the aid going directly to the poor to help them become more self-reliant.*

- *Shift most international aid from large-scale to small-scale projects intended to benefit local communities of the poor.*

- *Encourage banks and other organizations to make small loans to poor people wanting to increase their income* (Solutions, right).

- *Require international lending agencies to use standard environmental and social impact analysis to evaluate any proposed development project.* Some analysts

believe that no project should be supported unless (1) its net environmental impact (using full-cost accounting) is favorable, (2) most of its benefits go to the poorest 40% of the people affected, and (3) the local people it affects are involved in planning and executing the project.

- *Carefully monitor all projects and halt funding if environmental safeguards are not followed.*

- *Help developing countries increase resource productivity* (Solutions, p. 525) *as a way to provide more jobs and income and reduce pollution and resource waste.* According to Robert B. Shapiro, CEO of Monsanto, "If emerging economies have to relive the entire industrial revolution with all its waste, its energy use, and its pollution, I think it's all over."

- *Establish policies that encourage both developed countries and developing countries to slow population growth and stabilize their populations.*

How Much Will It Cost? According to the United Nations Development Program (UNDP), it will cost about $40 billion a year to provide universal access to basic services such as education, health, nutrition, family planning, reproductive health, safe water, and sanitation. The UNDP notes that this is less than 0.1% of the world income, amounting to "barely more than a rounding error."

This expenditure for providing everyone with basic services is only a fraction of the almost $800 billion per year devoted to military spending and less than the $50 billion spent on cigarettes each year in Europe. For example, providing:

- Universal basic health and nutrition would cost about $10 billion per year—much less than the $17 billion spent each year on pet foods in the United States and Europe.

- Universal access to sanitation and safe drinking water would cost about $12 billion per year—about what Americans and Europeans spend each year on perfumes.

26-7 MAKING THE TRANSITION TO ENVIRONMENTALLY SUSTAINABLE ECONOMIES

How Can We Make Working with the Earth Profitable? Greening Business Paul Hawken (Guest Essay, p. 6) and several other business leaders and

Microloans to the Poor

Most of the world's poor desperately want to earn more, become more self-reliant, and have a better life. However, they have no credit record and few if any assets to use for collateral to secure a loan. Thus, they cannot go to traditional banks or other money-lending institutions to buy seeds and fertilizer or to buy tools and materials for a small business.

During the last 25 years an innovative tool called *microlending* has increasingly helped deal with this problem. For example, since economist Muhammad Yunus started it in 1976, the Grammeen (Village) Bank in Bangladesh has provided more than $2 billion in microloans to mostly poor, rural, and landless women in 35,000 villages. The loans vary from $50 to $500 but average around $100. About 94% of the loans are to women who start their own small businesses as sewers,

weavers, bookbinders, peanut fryers, or vendors.

To stimulate repayment and provide support, the Grammeen Bank organizes microborrowers into five-member "solidarity" groups. If one member of the group misses a weekly payment or defaults on the loan, the other members of the group must make the payments.

The Grammeen Bank's experience has shown that microlending is both successful and profitable. For example, less than 3% of microloan repayments to the Grammeen Bank are late, and the repayment rate on its loans is an astounding 98%, compared to a repayment rate of conventional loans by commercial banks of only 30%.

About half of Grameen's borrowers move above the poverty line within 5 years, and domestic violence, divorce, and birth rates are lower among borrowers. Microloans to the poor by the Grammeen Bank are being used to develop day-care

centers, health clinics, reforestation projects, drinking water supply projects, literacy programs, and group insurance programs and to bring solar and wind micropower systems to rural villages.

Grameen's model has inspired the development of microcredit projects in more than 52 countries that have reached 36 million people (including dependents). In 1997, some 2,500 representatives of microlending organizations from 113 countries met at a Microcredit Summit in Washington, D.C., and adopted a goal of reaching 100 million of the world's poorest people by 2005.

Critical Thinking

Why do you think there has been little use of microloans by international development and lending agencies such as the World Bank and the International Monetary Fund? How might this situation be changed?

---

economists have laid out the following principles for making the transition from environmentally unsustainable to environmentally sustainable (Figure 26-8) economies over the next several decades:

- *Reward (subsidize) earth-sustaining behavior.*

- *Discourage (tax and do not subsidize) earth-degrading behavior.*

- *Use environmental and social indicators to measure progress toward environmental and economic sustainability and human well being* (Figures 26-9, and 26-10).

- *Use full-cost pricing to include the external costs of goods and services in their market prices.*

- *Replace taxes on wages and profits with taxes on matter and energy throughput* (Solutions, p. 707).

- *Use low discount rates* (p. 702) *for evaluating future worth of irreplaceable or vulnerable natural resources.*

- *Drastically increase resource productivity* (Solutions, p. 525).

- *Sell services instead of things* (p. 527).

- *Do not use renewable resources (soil, water, forests, grasslands, and wildlife) faster than they can be replenished.*

- *Do not use nonrenewable resources faster than substitutes can be developed and phased in.*

- *Do not release pollutants into the environment faster than the earth's natural processes can dilute or assimilate them.*

- *Improve environmental management in business* (Figure 26-13 and Spotlight, p. 711).

- *Reduce poverty.*

- *Slow population growth.*

Paul Hawken's simple golden rule for such an economy is, "*Leave the world better than you found it, take no more than you need, try not to harm life or the environment, and make amends if you do.*"

Case Study: How Are Germany and the Netherlands Working to Achieve an Environmentally Sustainable Economy? Germany (Case Study, p. 710) and the Netherlands are working to make their economies more environmentally sustainable. The Netherlands is a

German political and business leaders see sales of environmental protection goods and services— already more than $600 billion per year—as a major source of new markets and future income because environmental standards and concerns are expected to rise everywhere.

Mostly because of stricter air pollution regulations, German companies have developed some of the world's cleanest and most efficient gas turbines, and they have invented the world's first steel mill that uses no coal to make steel. Germany sells these and other improved environmental technologies globally.

On the other hand, Germany continues to heavily subsidize its coal-mining industry, which increases environmental degradation and pollution and adds large amounts of carbon dioxide to the atmosphere. According to environ-mental expert Norman Myers (Guest Essay, p. 117), these subsidies are so high that the German government would save money by closing all of its coal mines and sending the miners home on full pay for the rest of their lives.

In 1977, the German government started the Blue Angel product-labeling program to inform consumers about products that cause the least environmental harm. Most international companies use the German market to test and evaluate green products.

Germany has also revolutionized the recycling business. German car companies are required to pick up and recycle all domestic cars they make. Bar-coded parts enable disassembly plants to dismantle an auto for recycling in 20 minutes. Such *take-back* requirements are being extended to almost all products to reduce use of energy and virgin raw materials. Germany plans to sell its newly developed recycling technologies to other countries.

The German government has also supported research and development aimed at making Germany the world's leader in solar-cell technology (p. 373) and hydrogen fuel (Section 15-7, p. 384), which it expects to provide a rapidly increasing share of the world's energy.

Finally, Germany provides about $1 billion per year in green foreign aid to developing countries. Much of the aid is designed to stimulate demand for German technologies such as solar-powered lights, solar cells, and wind-powered water pumps.

Critical Thinking

What major steps (if any) are the government and businesses in the country where you live taking to make the transition to a more environmentally sustainable economy? What three major things could you do to help promote such a shift in your country and in your local community?

tiny country about one-fourth the size of the state of Illinois, with a population of about 16 million.

In 1989, the Netherlands began implementing a National Environmental Policy Plan (or Green Plan) as a result of widespread public alarm over declining environmental quality. The goal of this plan is to slash production of many types of pollution by 70–90% and achieve the world's first environmentally sustainable economy by 2010.

The government identified eight major areas for improvement: **(1)** climate change, **(2)** acid deposition, **(3)** eutrophication, **(4)** toxic chemicals, **(5)** waste disposal, **(6)** groundwater depletion, **(7)** unsustainable use of renewable and nonrenewable resources, and **(8)** local nuisances (mostly noise and odor pollution).

Target groups, consisting of major industrial, government, and citizens' groups, were formed for each of the eight areas, and each group was asked to develop a voluntary agreement on establishing targets and timetables for drastically reducing pollution. Each group was free to pursue whatever policies or technologies it wanted. However, if a group could not agree the government would impose its own targets and timetables and stiff penalties for industries not meeting certain pollution reduction goals.

The government identified four general themes for each group to focus on: **(1)** implementing life cycle management, which makes companies responsible for the remains of their products after users are through with them (a goal that increased the design of products that can be reused or recycled), **(2)** improving energy efficiency (Section 15-2, p. 362), with the government committing $385 million per year to energy conservation programs, **(3)** inventing new or improved more sustainable technologies, supported by a government program to help develop such technologies, and **(4)** improving public awareness through a massive government-sponsored public education program.

Many of the country's leading industrialists like the Green Plan because:

■ They can make investments in pollution prevention and pollution control with less financial risk because they have a high degree of certainty about long-term environmental policy.

- They are free to deal with the problems in ways that make the most sense for their businesses. This has encouraged cooperation between environmentalists and industrial leaders.

- Developing and implementing the plan have helped industrial leaders in the Netherlands (like those in Germany) learn that creating more efficient and environmentally sound products and processes can often reduce costs and increase profits as such innovations are sold at home and abroad.

Is the plan working? The news is mixed but encouraging. Many of the target groups are meeting their goals on schedule, and some have even exceeded them. A huge amount of environmental research by the government and private sector has taken place. This has led to (1) an increase in organic agriculture, (2) greater reliance on bicycles in some cities, and (3) more ecologically sound new housing developments.

However, there have been some setbacks:

- Some of the more ambitious goals such as decreasing CO_2 levels may have to be revised downward or even abandoned.

- Some environmentalists who strongly support the plan are not happy with (1) the compromises they have had to make and (2) the mild backlash in some industry circles against an energy tax to reduce CO_2 emissions.

Despite its shortcomings, the Netherlands plan is the first attempt by any country to (1) foster a national debate on the issue of environmental sustainability and (2) encourage innovative solutions to environmental problems.

Can We Make the Transition to an Environmentally Sustainable Economy? Even if people believe that an environmentally sustainable economy is desirable, is it possible to make such a drastic change in the way we think and act? Some environmentalists, economists, and business leaders (Individuals Matter, p. 528) say that it's not only possible but imperative and that it can be done over the next 40–50 years.

According to Paul Hawken, this new approach to economic thinking and actions recognizes that most business leaders are not evil, earth-degrading ogres. Instead, they are trapped in a system that by design rewards them (with the highest profits and salaries and best chances for promotion) for maximizing short-term profits for owners and investors, regardless of the harmful short- and long-term environmental and social impacts.

Hawken argues that environmentally sustainable economies throughout the world would free business leaders, workers, and investors from this ethical dilemma. In such economies, they would be financially compensated and respected for doing socially and eco-

What Is Good Environmental Management?

SPOTLIGHT

According to business leaders and business management analysts, key practices in good corporate environmental management are:

- Providing leadership, beginning with the board of directors.

- Giving a clear statement of the company's environmental principles and objectives, which have full backing of the board.

- Involving employees, environmental groups, customers, and members of local communities in developing and evaluating the business's environmental policies, strategies for improvement, and progress.

- Making the improvement of environmental quality and worker safety and health a major priority for every employee.

- Conducting an annual cradle-to-grave environmental audit of all operations and products, including a detailed strategy for making improvements. Disseminate the results to employees, stockholders, and the public. Currently, some U.S. companies shun environmental audits because they fear lawsuits or prosecution by state and federal environmental agencies. However, several states have passed laws protecting the results of a company's voluntary environmental audit as privileged information.

- Helping customers safely distribute, store, use, and dispose of or recycle company products.

- Recognizing that carrying out these policies is the best way to (1) encourage innovation, (2) expand markets, (3) improve profit margins, (4) develop happy and loyal customers, (5) attract and keep the best-qualified employees, (6) and help sustain companies and the economy.

Critical Thinking

Why have many companies not instituted the environmental management principles listed here? What could be done to change this situation?

logically responsible work, improving environmental quality, and still making hefty profits for owners and stockholders. Making this shift should also create jobs (Connections, p. 712).

The problem in making this shift is not economics but politics. It involves the difficult task of convincing more business leaders, elected officials, and voters to begin

Jobs and the Environment

CONNECTIONS

Environmental protection is a major growth industry that creates new jobs. According to Worldwatch Institute estimates, annual sales of global ecotechnology industries are $600 billion—on a par with the global car industry—and these industries employ about 11 million people.

In 2000, the environmental industry in the United States employed nearly 1.4 million people and generated annual revenues of more than $185 billion. It is projected that at least another 500,000 environmental jobs will be added between 2000 and 2005.

If countries begin placing more emphasis on increasing resource productivity (Solutions, p. 525) and shifting taxes from wages and income to pollution and waste (Solutions, p. 707), there will be a sharp increase in environmental-related jobs. Germany (Case Study, p. 710), the Netherlands (pp. 709-710), and several other European countries have begun making such shifts.

Studies by the EPA show that environmental laws create far more jobs than have been lost. Indeed, the U.S. Clean Air and Clean Water Acts have created more than 300,000 jobs in pollution control.

Improving resource productivity by reducing the waste of matter and energy resources and using solar energy, wind energy, recycling, reuse, remanufacturing, and cleaner production (Section 21-3, p. 522 and Guest Essay, p. 523) are labor-intensive activities that provide many more jobs than traditional resource-intensive industries. Reforestation, ecological restoration, sustainable agriculture, and integrated pest management are also labor-intensive activities.

In the United States,

- Nearly twice as many people are employed in aluminum recycling as in aluminum production.

- The Worldwatch Institute estimates that weatherizing all U.S. homes to improve their energy efficiency would create 300,000 jobs, each lasting 20 years.

- Remanufacturing is a $53-billion-a-year business directly employing some 480,000 people.

- A congressional study concluded that investing $115 billion per year in solar energy and improving energy efficiency in the United States would **(1)** eliminate about 1 million jobs in oil, gas, coal, and electricity production but **(2)** create 2 million other new jobs. Investing the money saved by reducing energy waste could create another 2 million jobs.

In Europe,

- Producing, installing, and maintaining solar cell systems could create 70,000 to 294,000 new jobs, depending on global sales of such cells.

- Projected global increases in wind power between 2000 and 2010 should create 190,000 to 320,000 jobs.

Globally, about 10% of all existing jobs vanish each year and are replaced by different jobs and new occupations. Such economic restructuring is a crisis for displaced workers who do not have the skills to find new jobs. However, it is an opportunity for workers who have the needed skills or who can be retrained.

Shifting to environmentally sustainable economies over the next several decades would create a host of new jobs. However, jobs would be lost in industries such as mining, logging, fossil fuels, and types of manufacturing that do not convert to cleaner production and in regions and communities dependent on such industries.

Ways suggested by various analysts to ease this transition include **(1)** providing tax breaks to make it more profitable for companies to keep or hire more workers instead of replacing them with machines, **(2)** using incentives to encourage location of new, emerging industries in hard-hit communities, **(3)** helping such areas diversify their economic bases, and **(4)** providing income and retraining assistance for workers displaced from environmentally destructive businesses (a *Superfund for Workers*).

Critical Thinking

1. What major things (if any) have the national and local governments of the country where you live done to stimulate the growth of environmental jobs? What major things (if any) have these governments done to discourage the growth of environmental jobs?

2. Do you believe that government should have a significant role in stimulating environmental jobs, or should this be left up mostly to private enterprise? Explain.

changing current government systems of economic rewards and penalties.

Here are three pieces of *great news*:

- Making the shift to environmentally sustainable economies could be an extremely profitable enterprise that will create many jobs, greatly improve environmental quality, and sharply reduce poverty that helps retard environmental progress.

- We already have most of the technologies needed to implement this economic shift.

- Governments would not have to spend more money. Such a shift would be revenue neutral if governments **(1)** shifted environmentally harmful subsidies to environmentally beneficial enterprises and **(2)** taxed pollution and resource waste instead of wages and income (Solutions, p. 707).

Forward looking investors, corporate executives (Individuals Matter, p. 528) and political leaders recognize that earth-sustaining businesses with good environmental management will prosper as the environmental revolution proceeds. *They recognize that the environmental revolution is also an economic revolution.*

I'm fascinated with the concept of distinctions that transform people. Once you learn certain things—once you learn to ride a bike, say—your life has changed forever. You cannot unlearn it. For me sustainability is one of those distinctions. Once you get it, it changes how you think. It becomes automatic and is a part of who you are.

ROBERT B. SHAPIRO

REVIEW QUESTIONS

1. Describe how conventional economists and ecological economists disagree on the importance of natural resources in sustaining economies.

2. Distinguish between *natural, human, financial,* and *manufactured resources* used in an economic system.

3. Distinguish between *pure command, pure free-market, capitalist market,* and *mixed* economic systems.

4. Distinguish between *the capitalist market economy, the survival economy,* and *nature's economy.*

5. List ten reasons why governments have to intervene in economic systems.

6. List the pros and cons of a global market economy. Distinguish between *economic globalization* and *economic localization.*

7. Distinguish between *economic growth* and *environmentally sustainable economic development.* What are three characteristics of environmentally sustainable economic development?

8. Explain how conventional economists and ecological economists differ in their view of market-based economic systems. Describe Herman Daly's concept of the *steady-state economy.*

9. List the major characteristics of and give three examples of *appropriate technology.*

10. List five reasons why GNP and GDP are not useful measures of economic health, environmental health, or human well-being. Describe two environmental indicators that could be used to provide such information.

11. Distinguish between *internal costs* and *external costs* and give an example of each. What is *full-cost pricing,* and what are the pros and cons of using this approach to internalize external environmental costs? List five reasons why full-cost pricing has not been widely used.

12. How do economists determine the *optimum level of pollution* for a particular chemical, and what are the pros and cons of using this approach?

13. Distinguish between using *mitigation, willingness to pay,* and *maintenance and protection cost methods* to estimate the monetary value of natural resources.

14. Distinguish between *direct, indirect,* and *repercussion costs* associated with pollution.

15. Distinguish between the *consumer- or user-pays* and the *taxpayer-pays* approaches for paying for pollution control.

16. Distinguish between the use of *discount rates, time preferences, opportunity costs, government subsidies and tax breaks,* and *taxes,* in managing natural resources.

17. What is *benefit–cost analysis,* and what are the pros and cons of using this tool to evaluate alternative courses of action in controlling pollution and managing natural resources? List seven ways to improve benefit–cost analysis.

18. What are the pros and cons of using regulations to improve environmental quality? List six characteristics of *innovation-friendly regulations.*

19. Describe three types of *economic incentives* (rewards) that can be used to improve environmental quality and reduce resource waste. List the pros and cons of each type.

20. Describe three types of *economic disincentives* (punishments) that can be used to improve environmental quality and reduce resource waste and list the pros and cons of each type.

21. What are the pros and cons of using tradable pollution and resource-use rights to reduce pollution and resource waste?

22. List six phases or eras in the evolution of environmental management.

23. What is *poverty,* and what are its harmful health and environmental effects?

24. List two ways in which the governments of developing countries can reduce poverty. List eight ways in which governments of developed countries can help reduce poverty. What are *microloans,* and how are they being used to reduce poverty?

25. List fourteen principles for shifting to environmentally sustainable economies over the next several decades.

26. List seven principles of good environmental management.

27. Describe attempts to develop more environmentally sustainable economies in **(a)** Germany and **(b)** the Netherlands.

28. List three great pieces of news about making the shift to environmentally sustainable economies over the next several decades and explain why the environmental revolution is also an economic revolution.

CRITICAL THINKING

1. Conventional and ecological economists disagree on the importance of natural resources in supporting and sustaining economies (p. 688). What are the underlying assumptions or environmental worldviews of the people with these two opposing beliefs (See p. 742 and 744)? Which position do you support? Why?

2. According to one definition, *sustainable development* involves meeting the needs of the present human generation without compromising the ability of future generations to meet their needs. What do you believe are the needs used in this definition? Compare this definition with the definition of environmentally sustainable economic development given on p. 693.

3. Suppose that over the next 20 years the current harmful environmental and health costs of goods and services are internalized so that their market prices reflect their total costs. What harmful and beneficial effects might this have on **(a)** your lifestyle and **(b)** any child you might have?

4. Do you believe that we should establish optimum levels or zero-discharge levels for toxic chemicals we release into the environment? Explain.

5. Do you agree or disagree with the proposals various analysts have made for sharply reducing poverty as discussed on pp. 707–708? Explain.

6. Do you agree or disagree with the guidelines for an environmentally sustainable economy listed on p. 708-709? Explain.

7. What are the major pros and cons of shifting from our current economy to a more environmentally sustainable economy over the next 40 years?

PROJECTS

1. List all the goods you use, then identify those that meet your basic needs and those that satisfy your wants. Identify any economic wants you **(a)** would be willing to give up, **(b)** you believe you should give up but are unwilling to give up, and **(c)** you hope to give up in the future. Relate the results of this analysis to your personal impact on the environment. Compare your results with those of your classmates.

2. Use the library or the internet to find bibliographic information about *Gaylord Nelson, Lester R. Brown, Christopher Flavin,* and *Robert B. Shapiro,* whose quotes appear at the beginning and end of this chapter.

3. Make a concept map of this chapter's major ideas, using the section heads and subheads and the key terms (in boldface). Look at the inside back cover and on the website for this book for information about making concept maps.

INTERNET STUDY RESOURCES AND RESOURCES FOR FURTHER READING AND RESEARCH

The website for this book contains helpful study aids and many ideas for further reading and research. Log on to:

http://www.brookscole.com/product/0534376975s

and click on the Chapter-by-Chapter area. Choose Chapter 26 and select a resource:

- "Flash Cards" allows you to test your mastery of the Terms and Concepts to Remember for this chapter.

- "Tutorial Quizzes" provides a multiple-choice practice quiz.

- "Student Guide to InfoTrac" will lead you to Critical Thinking Projects that use InfoTrac College Edition as a research tool.

- "References" lists the major books and articles consulted in writing this chapter.

- "Hypercontents" takes you to an extensive list of sites with news, research, and images related to individual sections of the chapter.

INFOTRAC COLLEGE EDITION

Improve your skills with InfoTrac College Edition, a searchable online database of articles from more than 700 periodicals. Log on to:

http://www.infotrac-college.com

or access InfoTrac through the website for this book.

Try the following articles:

Barrett, G.W., and A. Farina. 2000. Integrating ecology and economics. *BioScience* vol. 50, no. 4, pp. 311–312. (keywords: economics, ecology)

Lenhardt, W.C. 2000. Green trade on the web. *Environment* vol. 42, no. 9, pp. 3–4. (keywords: green trade)

27 POLITICS, ENVIRONMENT, AND SUSTAINABILITY

Rescuing a River

In the 1960s, Marion Stoddart (Figure 27-1) moved to Groton, Massachusetts, on the Nashua River, then considered one of the nation's filthiest rivers. For decades, industries and towns along the river had used it as a dump. Dead fish bobbed on its waves, and at times the water was red, green, or blue from pigments discharged by paper mills.

Instead of thinking nothing could be done, she committed herself to restoring the Nashua and establishing public parklands along its banks.

She did not start by filing lawsuits or organizing demonstrations. Instead she created a careful cleanup plan and approached state officials with it in 1962. They laughed, but she was not deterred and began practicing the most time-honored skill of politics: one-on-one persuasion. She identified the power brokers in the riverside communities and began to educate them, win them over, and get them to cooperate in cleaning up the river.

She also got the state to ban open dumping in the river. When promised federal matching funds for building the treatment plant failed to materialize, Stoddart gathered 13,000 signatures on a petition sent to president Richard Nixon. The funds arrived in a hurry.

Stoddart's next success was getting a federal grant to beautify the river. She hired high school dropouts to clear away mounds of debris. When the river cleanup was completed, she persuaded communities along the river to create some 2,400 hectares (6,000 acres) of riverside park and woodlands along both banks.

Now, almost four decades later, the Nashua is still clean. Several new water treatment plants have been built, and a citizens' group founded by Stoddart keeps watch on water quality. The river supports many kinds of fish and other wildlife, and its waters are used for canoeing and other kinds of recreation. The project is considered a model for other states and is testimony to what a committed individual can do to bring about change from the bottom up by getting people to work together.

For her efforts, the UN Environment Programme has named Stoddart as an outstanding worldwide worker for the environment. However, she might say that the blue and canoeable Nashua itself is her best reward.

Politics is the process by which individuals and groups try to influence or control the policies and actions of governments at the local, state, national, or international levels. Politics is concerned with who has power over the distribution of resources and who gets what, when, and how. Thus, it plays a significant role in **(1)** regulating and influencing economic decisions (Chapter 26) and **(2)** persuading people to work together toward a common goal, as Marion Stoddart did.

Figure 27-1 Marion Stoddart canoeing on the Nashua River near Groton, Massachusetts. She spent more than two decades spearheading successful efforts to have this river cleaned up. (Seth Resnick)

Politics is the art of making good decisions on insufficient evidence.

LORD KENNET

This chapter addresses the following questions:

- How do democracies work, and what factors hinder the ability of democracies to deal with environmental problems?
- What are guidelines for making environmental policy, and how can people affect such decisions?
- How is environmental policy made in the United States, and how can such policy decisions be improved?
- What are the major types and roles of environmental groups and anti-environmental groups, and how can we evaluate the claims of these opposing forces?
- What types of global environmental policies exist, and how might they be improved?

27-1 POLITICS AND ENVIRONMENTAL POLICY

What Is a Democracy, and How Do Democratic Governments Work? **Democracy** is government by the people through elected officials and representatives. In a *constitutional democracy*, a constitution (1) provides the basis of government authority, (2) limits government power by mandating free elections, and (3) guarantees freely expressed public opinion. In a functioning constitutional democratic system, the people elect representatives to legislatures who make laws implemented by government.

Political institutions in constitutional democracies are designed to allow gradual change to ensure economic and political stability. In the United States, for example, rapid and destabilizing change is curbed by the system of checks and balances that distributes power between the three branches of government—*legislative, executive,* and *judicial*—and between federal, state, and local governments.

In passing laws, developing budgets, and formulating regulations, elected and appointed government officials must deal with pressure from many competing *special-interest groups*. Each group advocates passing laws or establishing regulations favorable to its cause and weakening or repealing laws and regulations unfavorable to its position.

Some special-interest groups (such as corporations) are *profit-making organizations*, and others are *nonprofit, nongovernment organizations (NGOs)*. Examples of NGOs are educational institutions, labor unions, and mainstream and grassroots environmental organizations.

Most political decisions made by democratic governments result from bargaining, accommodation, and compromise between leaders of competing *elites*, or power brokers. The primary goal of government by competing elites is to maintain the overall economic and political stability of the system (status quo) by making only gradual or *incremental* change.

What Factors Hinder the Ability of Democracies to Deal with Environmental Problems? The deliberately stable design of democracies is desirable but has several related disadvantages for dealing with environmental problems:

- In democracies, the emphasis is on reacting to short-term environmental problems in isolation from one another instead of acting to prevent them from occurring in the future. However, many important environmental problems such as climate change, biodiversity loss, and long-lived hazardous waste (1) have long-range effects, (2) are related to one another, and (3) require integrated solutions emphasizing prevention.
- Because elections are held every few years, most politicians focus on short-term individual problems rather than on the more complex and time-consuming job of finding integrated solutions to long-term problems.
- Children, unborn generations, and wild species do not vote, and most politicians will no longer be in office when any harmful long-term effects from environmental problems appear. Thus, there is no powerful political constituency for the future or for long-term environmental sustainability.
- Whether they like it or not, most elected officials must spend much of their time raising money to get reelected.

Some politicians with vision and a highly developed sense of justice and ethical concern overcome this focus on the short term, but they are rare. Moreover, those who want to do something about these problems often do not get enough support from financial backers or the public to remain in office.

27-2 DEVELOPING AND INFLUENCING ENVIRONMENTAL POLICY

What Principles Can Be Used as Guidelines in Making Environmental Policy Decisions? Analysts have suggested that legislators and individuals evaluating existing or proposed environmental policy should be guided by the following principles:

- *The humility principle:* Recognize and accept that we have a limited capacity to manage nature because our understanding of nature and of the consequences of our actions will always be limited.

- *The reversibility principle:* Try not to do something that cannot be reversed later if the decision turns out to be wrong. For example, ecologists believe that the current large-scale destruction and degradation of forests, wetlands, wild species, and other components of the earth's biodiversity is unwise because much of it could be irreversible on a human time scale.

- *The precautionary principle:* Be cautious and think deeply when making decisions about problems that we do not understand and that could have potentially serious effects on current and future generations. In such cases, it is better to be safe than sorry.

- *The prevention principle:* Whenever possible, make decisions that help prevent a problem from occurring or becoming worse.

- *The integrative principle:* Make decisions that involve integrated solutions to environmental and other problems.

- *The ecological design principle:* Pass laws and develop regulations that are to be implemented by using good ecological design (Guest Essay, p. 683).

- *The environmental justice principle:* Establish environmental policy so that no group of people bears a disproportionate share of the harmful environmental risks from industrial, municipal, and commercial operations or from the execution of laws, regulations, and policies. The EPA defines **environmental justice** as "the fair treatment and meaningful involvement of all people regardless of race, color, national origin, or income with respect to the development, implementation, and enforcement of environmental laws, regulations, and policies."

How Can Individuals Affect Environmental Policy? A major theme of this book is that *individuals matter.* History shows that significant change usually comes from the *bottom up* when individuals join with others to bring about change (p. 715). Without grassroots political action by millions of individual citizens and organized groups, the air you breathe and the water you drink today would be much more polluted, and much more of the earth's biodiversity would have disappeared.

Individuals can influence and change government policies in constitutional democracies by:

- Running for office (especially local offices).

- Trying to get appointed to local planning and zoning boards and environmental commissions that study and make recommendations to elected officials.

- Appearing at hearings before local planning and zoning boards, environmental commissions, and public meetings of elected officials to make their views known.

- Voting for candidates and ballot measures.

Environmental Careers

In the United States (and in other developed countries), economists claim that the *green job market* is one of the fastest-growing segments of the economy.

Many employers are actively seeking environmentally educated graduates. They are especially interested in people with scientific and engineering backgrounds and those with double majors (business and ecology, for example) or double minors.

Environmental career opportunities exist in a large number of fields: environmental engineering (currently the fastest-growing job market), sustainable forestry and range management, parks and recreation, air and water quality control, solid-waste and hazardous-waste management, recycling, urban and rural land-use planning, computer modeling, ecological restoration, and soil, water, fishery, and wildlife conservation and management.

Environmental careers can also be found in education, environmental planning, environmental management, environmental health, toxicology, geology, ecology, conservation biology, chemistry, climatology, population dynamics and regulation (demography), law, risk analysis, risk management, accounting, environmental journalism, design and architecture, energy conservation and analysis, renewable-energy technologies, hydrology, consulting, public relations, activism and lobbying, economics, diplomacy, development and marketing, publishing (environmental magazines and books), teaching, and law enforcement (pollution detection and enforcement teams).

Critical Thinking

Have you considered an environmental career? Why or why not?

- Contributing money and time to candidates seeking office.

- Writing, faxing, e-mailing, calling, or meeting with elected representatives, asking them to pass or oppose certain laws, establish certain policies, and fund various programs (Appendix 6).

- Forming or joining NGOs of individuals that lobby elected and regulatory officials to support a particular position (photo on p. 687).

- Using education and persuasion to convert elected officials and other citizens to a particular position (p. 715).

- Exposing fraud, waste, and illegal activities in government (whistleblowing).

What Is Environmental Leadership? It is important to distinguish between leaders, rulers, and managers:

- A *leader* is a person other people follow voluntarily because of his or her vision, credibility, or charisma.

- A *ruler* is a person who has enough power to make people follow against their will.

- A *manager* is a person who knows how to organize things, get things done, pay attention to detail, and delegate responsibility. Some leaders and rulers can also be good managers and vice versa.

Individuals can provide leadership on environmental (or other) issues by:

- *Leading by example*, using one's lifestyle to show others that change is possible and beneficial.

- *Working within existing economic and political systems to bring about environmental improvement.* People can influence political elites by campaigning and voting for candidates and by communicating with elected officials (Appendix 6). They can also work within the system by choosing environmental careers (Individuals Matter, p. 717).

- *Challenging the system and basic societal values.*

- *Proposing and working for better solutions to environmental problems.* Leadership is more than being against something; it also involves coming up with better ways to accomplish various goals and getting people to work together to achieve such goals.

27-3 CASE STUDY: ENVIRONMENTAL POLICY IN THE UNITED STATES

What Are the Three Branches of Government in the United States? The federal government of the United States consists of three separate but interconnected branches: legislative, executive, and judicial.

- The *legislative branch* is the Congress, composed of the House of Representatives and the Senate. Its goal is to approve and oversee government policy by **(1)** passing laws that establish a government agency or instruct an existing agency to take on new tasks or programs and **(2)** overseeing the functioning and funding of various agencies of the executive branch concerned with carrying out government policies.

- The *executive branch* consists of a chief executive, the president, who with his or her staff oversees the various agencies authorized by Congress to carry out

government policies such as environmental policy (Figure 27-2). The president also **(1)** proposes annual budgets, legislation, and appointees for executive positions that must be approved by Congress and **(2)** tries to persuade Congress and the public to support his or her policy proposals.

- The *judicial branch* consists of a complex and layered series of courts at the local, state, and federal levels (Supreme Court). These courts enforce and interpret different categories of laws such as constitutional law, administrative law, and laws passed by legislative bodies (statutory laws).

The major function of the federal government in the United States is to develop and implement *policy* for dealing with various issues. This policy typically is composed of various **(1)** *laws* passed by the legislative branch, **(2)** *regulations* instituted by the executive branch to put laws into effect (Figure 27-2), and **(3)** *funding* to implement and enforce the laws and regulations. Figure 27-3 is a greatly simplified overview of how individuals and lobbyists for and against a particular environmental law interact with the three branches of government in the United States.

How Is Environmental Policy Made in the United States? There are several steps in establishing federal environmental policy (or any other policy) in the United States:

- Persuade lawmakers that an environmental problem exists and that the government has a responsibility to address it.

- Try to influence how laws are written and to pass laws to deal with the problem. Converting a bill into a law is a complex process (Figure 27-4). Most environmental bills are evaluated by as many as 10 committees in the House of Representatives and the Senate. Effective proposals often are weakened by this fragmentation and by lobbying from groups opposing the law. Nonetheless, since the 1970s a number of environmental laws have been passed in the United States (Appendix 5 and Solutions, p. 722).

- Appropriate enough funds to implement and enforce each law. Indeed, developing and adopting a budget is the most important and controversial thing the executive and legislative branches do. Developing a budget involves answering two key questions: **(1)** What programs will be funded, and **(2)** how much money will be used to address each problem?

- Have the appropriate government department or agency (Figure 27-2) draw up regulations for implementing each law. Groups try to influence how the regulations are written and enforced and sometimes challenge the final regulations in court.

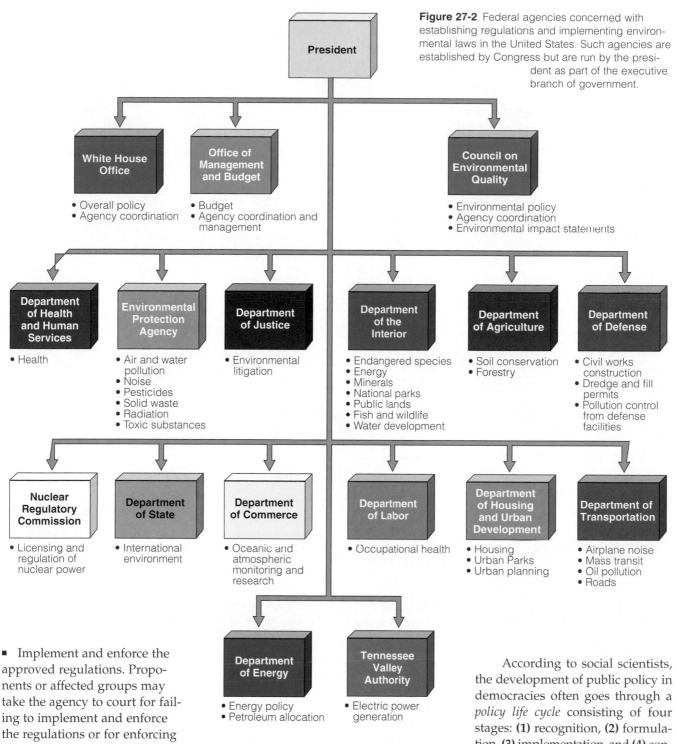

Figure 27-2 Federal agencies concerned with establishing regulations and implementing environmental laws in the United States. Such agencies are established by Congress but are run by the president as part of the executive branch of government.

President

White House Office
- Overall policy
- Agency coordination

Office of Management and Budget
- Budget
- Agency coordination and management

Council on Environmental Quality
- Environmental policy
- Agency coordination
- Environmental impact statements

Department of Health and Human Services
- Health

Environmental Protection Agency
- Air and water pollution
- Noise
- Pesticides
- Solid waste
- Radiation
- Toxic substances

Department of Justice
- Environmental litigation

Department of the Interior
- Endangered species
- Energy
- Minerals
- National parks
- Public lands
- Fish and wildlife
- Water development

Department of Agriculture
- Soil conservation
- Forestry

Department of Defense
- Civil works construction
- Dredge and fill permits
- Pollution control from defense facilities

Nuclear Regulatory Commission
- Licensing and regulation of nuclear power

Department of State
- International environment

Department of Commerce
- Oceanic and atmospheric monitoring and research

Department of Labor
- Occupational health

Department of Housing and Urban Development
- Housing
- Urban Parks
- Urban planning

Department of Transportation
- Airplane noise
- Mass transit
- Oil pollution
- Roads

Department of Energy
- Energy policy
- Petroleum allocation

Tennessee Valley Authority
- Electric power generation

■ Implement and enforce the approved regulations. Proponents or affected groups may take the agency to court for failing to implement and enforce the regulations or for enforcing them too rigidly.

Businesses affected by regulations try to influence regulatory agencies by setting up a *revolving-door* relationship between government and business officials. They try to have people sympathetic to their cause appointed to administrative positions in regulatory agencies. Some also offer regulatory officials lucrative jobs and use their inside knowledge to find ways to weaken or get around the regulations.

According to social scientists, the development of public policy in democracies often goes through a *policy life cycle* consisting of four stages: **(1)** recognition, **(2)** formulation, **(3)** implementation, and **(4)** control. Figure 27-5 shows the general position of several major environmental problems in the policy life cycle in most developed countries.

During the recognition and formulation stages there is often a great deal of controversy over what should be done, but such dissension usually increases in the implementation and control phases. Sometimes existing environmental laws or regulations can be **(1)** too strict, **(2)** imposed unfairly, or **(3)** considered an

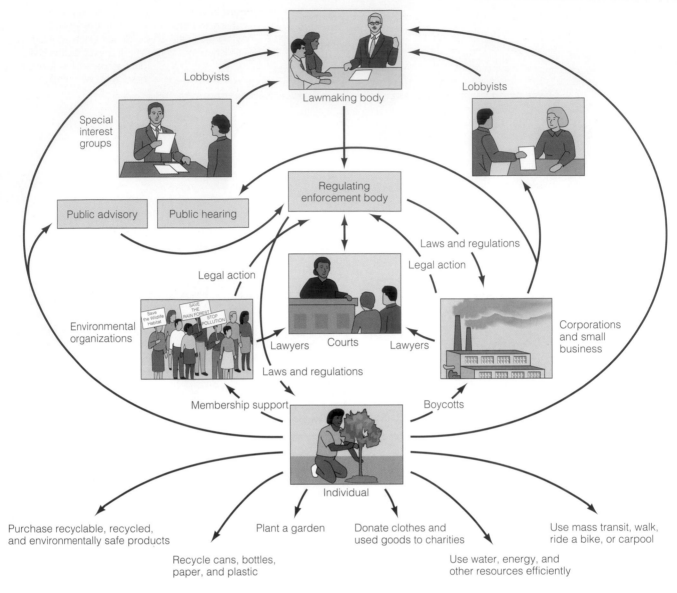

Figure 27-3 Greatly simplified overview of how individuals and lobbyists for and against a particular environmental law interact with the legislative, executive, and judicial branches of government in the United States. The bottom of this diagram also shows some ways in which individuals can bring about environmental change through their own lifestyles. More details on specific things you can do to reduce your environmental impact and details for contacting elected representatives are given in Appendix 6. (Adapted from *Environmental Science*, 5/E by Chiras, p. 542, fig 28.1. Copyright (c) 1998 by Wadsworth)

economic threat to some businesses. This can lead to an anti-environmental backlash as people opposed to the laws and regulations organize to have them weakened or overturned.

How Can We Make Government More Responsive to Citizens' Environmental Concerns? The nature and effectiveness of environmental laws and regulations at the federal, state, and local levels in the United States are determined mostly by elected officials. According to most political analysts, *campaign financing* is the biggest problem that keeps elected offi-

cials from being more responsive to the environmental and other needs and problems of ordinary citizens.

The candidates raising the most money win about 90% of federal, state, and local elections. Unless they are wealthy and willing to spend their own money, candidates can get this kind of money only from wealthy individuals and corporations.

Most analysts and about 80% of citizens polled agree that the U.S. political system is based more on money than on citizens' votes—one reason so many Americans do not bother to vote. For example, voter turnout for federal elections in the United States

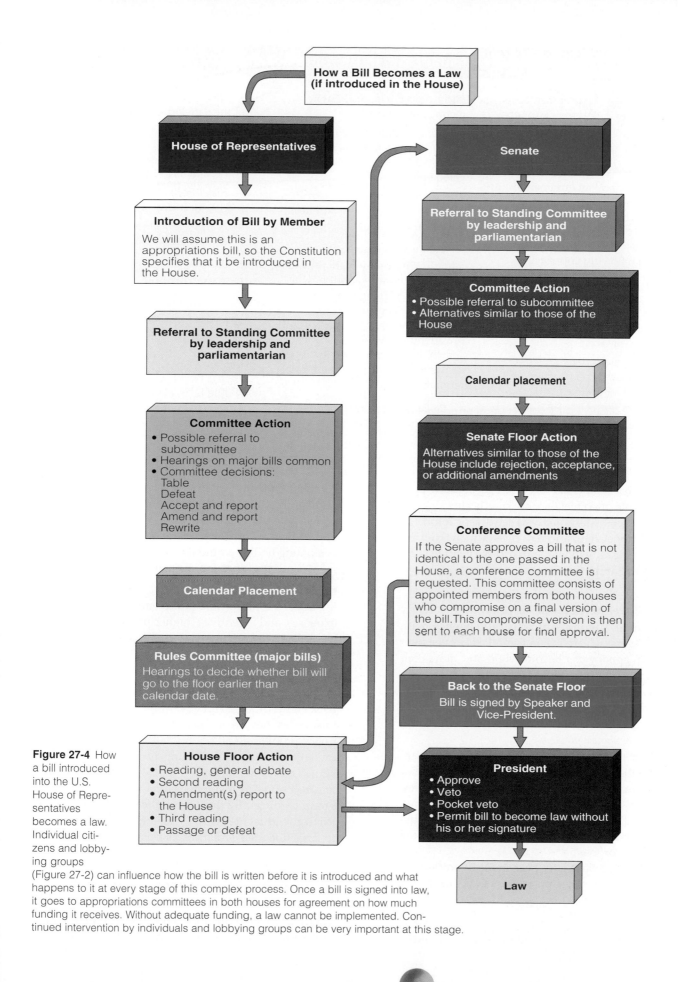

**How a Bill Becomes a Law
(if introduced in the House)**

House of Representatives

Senate

Introduction of Bill by Member

We will assume this is an appropriations bill, so the Constitution specifies that it be introduced in the House.

Referral to Standing Committee by leadership and parliamentarian

Committee Action
• Possible referral to subcommittee
• Alternatives similar to those of the House

Referral to Standing Committee by leadership and parliamentarian

Calendar placement

Committee Action
• Possible referral to subcommittee
• Hearings on major bills common
• Committee decisions:
Table
Defeat
Accept and report
Amend and report
Rewrite

Senate Floor Action

Alternatives similar to those of the House include rejection, acceptance, or additional amendments

Calendar Placement

Conference Committee

If the Senate approves a bill that is not identical to the one passed in the House, a conference committee is requested. This committee consists of appointed members from both houses who compromise on a final version of the bill. This compromise version is then sent to each house for final approval.

Rules Committee (major bills)
Hearings to decide whether bill will go to the floor earlier than calendar date.

Back to the Senate Floor
Bill is signed by Speaker and Vice-President.

Figure 27-4 How a bill introduced into the U.S. House of Representatives becomes a law. Individual citizens and lobbying groups

House Floor Action
• Reading, general debate
• Second reading
• Amendment(s) report to the House
• Third reading
• Passage or defeat

President
• Approve
• Veto
• Pocket veto
• Permit bill to become law without his or her signature

(Figure 27-2) can influence how the bill is written before it is introduced and what happens to it at every stage of this complex process. Once a bill is signed into law, it goes to appropriations committees in both houses for agreement on how much funding it receives. Without adequate funding, a law cannot be implemented. Continued intervention by individuals and lobbying groups can be very important at this stage.

Law

Types of Environmental Laws in the United States

Environmentalists and their supporters have persuaded the U.S. Congress to enact a number of important federal environmental and resource protection laws, as discussed throughout this text and listed in Appendix 5. These laws seek to protect the environment using the following approaches:

- *Setting standards for pollution levels or limiting emissions or effluents for various classes of pollutants* (Federal Water Pollution Control Act and Clean Air Acts).

- *Screening new substances for safety before they are widely used* (Toxic Substances Control Act).

- *Requiring comprehensive evaluation of the environmental impact of an activity before it is undertaken by a*

federal agency (National Environmental Policy Act).

- *Setting aside or protecting various ecosystems, resources, and species from harm* (Wilderness Act and Endangered Species Act).

- *Encouraging resource conservation* (Resource Conservation and Recovery Act and National Energy Act).

Some environmental laws contain glowing rhetoric about goals but little guidance about how to meet them, leaving this task to regulatory agencies and the courts. In other cases, the laws or presidential executive orders specify one or more of the following general principles for setting regulations:

- *No unreasonable risk:* food regulations in the Food, Drug, and Cosmetic Act.

- *No risk:* the zero-discharge goals of the Safe Drinking Water and Clean Water Acts.

- *Standards based on best available technology:* the Clean Air, Clean Water, and Safe Drinking Water Acts.

- *Risk-benefit analysis* (Section 16-5, p. 410): pesticide regulations.

- *Benefit-cost analysis* (p. 702): the Toxic Substances Control Act.

Critical Thinking

Pick one of the U.S. environmental laws listed in Appendix 5 (or a law in the country where you live). Use the library or the internet to evaluate the law's major strengths and weaknesses. Decide whether the law should be weakened, strengthened, or abolished and explain why. List the three most important ways you believe the law should be changed.

---

dropped from 74% in 1900 to 49% in 2000. Many people who do vote often feel they are simply choosing the lesser of two evils. Many analysts see drastic reform in the way elections are financed as the key to making the government more responsive to ordinary citizens on environmental and other issues (Solutions, p. 724).

How Can Bureaucracies Be Improved? In its early stages, a small, vigorous bureaucratic agency with dynamic leadership can do some good. However, over the years an agency can become rigid and more concerned with its own survival and getting an increasing share of the federal or state budget than with its original mission. It may also come under the influence of the businesses it is supposed to regulate—the *foxes-guarding-the-henhouse problem*.

Another problem is that responsibility for managing the nation's environmental and resource policy is divided among many federal and state agencies. This situation often leads to (1) contradictory policies, (2) duplicated efforts, (3) wasted funds, and (4) inability to develop an integrated approach to interrelated environmental problems.

Several suggestions have been made for improving government agencies' responsiveness to the people:

- *Pass a sunset law that automatically terminates any government agency or program after, say, 6 years.* After 5 years, the General Accounting Office and an outside

commission would evaluate each agency and program and recommend to Congress whether it should be reinstated, reorganized, assigned a new mission, or terminated. If renewal were recommended, the evaluating bodies would list suggested improvements and deadlines for implementation.

- *Reward whistle-blowers.* This would involve providing better job protection, government-paid legal expenses, and incentives (higher pay, promotion, cash rewards up to $100,000, well-publicized award ceremonies) for workers who expose fraud, waste, or illegal activities in government.

- *Slow the revolving door* through which people move back and forth between government agencies and the businesses they are supposed to regulate. Establish a policy prohibiting all political appointees and senior executive managers (including lawyers) of any regulatory or oversight agency from accepting any form of direct or indirect compensation from any person or group regulated by their agency for 5 years after they leave government service. Violators would be subject to large fines, jail sentences, or both.

27-4 ENVIRONMENTAL LAW

What Is the Difference Between Statutory and Common Laws? Almost every major environmental

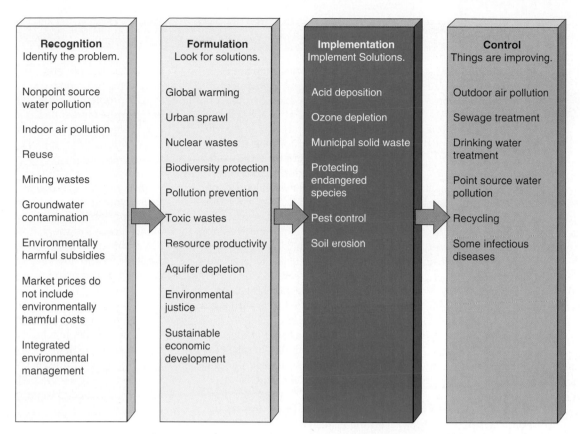

Recognition	Formulation	Implementation	Control
Identify the problem.	Look for solutions.	Implement Solutions.	Things are improving.
Nonpoint source water pollution	Global warming	Acid deposition	Outdoor air pollution
Indoor air pollution	Urban sprawl	Ozone depletion	Sewage treatment
Reuse	Nuclear wastes	Municipal solid waste	Drinking water treatment
Mining wastes	Biodiversity protection	Protecting endangered species	Point source water pollution
Groundwater contamination	Pollution prevention	Pest control	Recycling
Environmentally harmful subsidies	Toxic wastes	Soil erosion	Some infectious diseases
Market prices do not include environmentally harmful costs	Resource productivity		
	Aquifer depletion		
Integrated environmental management	Environmental justice		
	Sustainable economic development		

Figure 27-5 General position of several major environmental problems in the policy life cycle in most developed countries.

regulation is challenged in court by industries, environmental organizations, individuals, or groups of individuals. In any court case **(1)** the **plaintiff** is the party bringing the charge and **(2)** the **defendant** is the party being charged.

Environmental lawsuits involve statutory laws and common laws. **Statutory laws** are those developed and passed by legislative bodies such as federal and state governments. **Common law** is a body of unwritten rules and principles derived from thousands of past legal decisions. It is based on evaluation of what is reasonable behavior in attempting to balance competing social interests.

What Is the Difference Between Nuisance and Negligence? Many common law cases are settled using the legal principles of *nuisance* and *negligence*. A *nuisance* occurs when someone uses his or her property in a way that causes annoyance or injury to others. For example, a homeowner may bring a nuisance suit against a nearby factory because of the noise it generates.

In such a *civil suit*, the plaintiff seeks to collect damages for injuries to health or for economic loss, to have the court issue a permanent injunction against any further wrongful action, or both. An individual or a clearly identified group may bring such suits. A *class action suit* is a civil suit filed by a group, often a public interest or environmental group, on behalf of a larger number of citizens who allege similar damages but who need not be listed and represented individually.

Using the principles of common law, the court may side with the plaintiff if it finds that the loss of sleep, health problems, or other damage from the noise is greater than the cost of preventing the risk by eliminating or reducing the noise or having to close the factory. Often the court tries to find a reasonable or balanced solution to the problem. For example, it may order the factory to reduce the noise to certain levels or eliminate the noise during certain periods such as at night.

Negligence occurs if a person causes damage by knowingly acting in an unlawful or unreasonable manner. For example, a company may be found negligent if it fails to handle hazardous waste in a way required by a statutory law. A court may also find a company negligent if it fails to do something that a reasonable person would do, such as testing wastes for certain harmful chemicals before dumping them into a sewer, landfill, or river.

Generally, negligence is harder to prove than nuisance. For example, a company may be found not guilty because it argues that it did not know that its wastes were harmful and therefore did not act in a negligent or unreasonable manner.

A growing number of analysts of all political persuasions see *election finance reform* as the single most important way to reduce the influence of special-interest money in local, state, and federal elections in the United States. They urge American citizens to focus their efforts on this crucial issue as the key to making government more responsive to ordinary people on environmental and other matters.

One suggestion for reducing undue influence by powerful special interests would be to let the people (taxpayers) *alone* finance all federal, state, and local election campaigns, with low spending limits. Candidates could use their own money because it would be unconstitutional to forbid such use. However, they could not accept direct or indirect donations from any other individuals, groups, or parties, *with absolutely no exceptions*.

Once elected, officials could not (1) use free mailing, staff employ-

ees, or other privileges to aid their election campaigns and (2) accept donations or any kind of direct or indirect financial aid from any individual, corporation, political party, or interest group for their future election campaigns or for any other reason that could even remotely influence their votes on legislation.

Violators of this new *Public Funding Elections Act* would be barred from the campaign or removed from office. Anyone making illegal donations would be subject to large fines and possible jail sentences.

Having all elections financed entirely by public funds would cost each U.S. taxpayer only about $5–10 per year (as part of their income taxes) for all federal elections and a much smaller amount for state and local elections.

With such a reform, elected officials could spend their time governing instead of raising money and catering to powerful special interests. Office seekers would not need to be wealthy. Special-interest groups would be heard because of the validity of their ideas, not the size of their pocketbooks.

Proponents of this reform contend that this is an issue that ordinary people of all political persuasions could work together on. From this fundamental political reform, other

political, economic, and environmental reforms could flow.

The problem is getting members of Congress (and state legislatures) to pass a virtually foolproof and constitutionally acceptable plan that would put them on equal financial footing with their challengers and thus could decrease their chances of getting reelected.

Supporters of public financing of all elections argue that the way out of this dilemma is to band together to find and elect candidates who pledge to bring about this fundamental political reform, and then vote them out of office if they do not. They believe that many of the people who have given up exercising their right to vote because of the undue financial influence of special interest groups could be persuaded to participate in such efforts.

Critical Thinking

1. Do you agree or disagree with this approach to election reform? Explain. What role would you take in bringing about or opposing such a reform?

2. Why might the mainstream environmental groups in the United States not support such a reform? Should they?

*For an in-depth analysis of this issue, see Moti Nissani, "Brass-Tacks Ecology," *The Trumpeter* 14, no. 3 (Summer 1997), pp. 143–48, and Chapter 10 in Moti Nissani, *Lives in the Balance: The Cold War and American Politics, 1945–1991* (Carson City, Nev.: Dowser Publishing Group, 1992).

What Factors Hinder the Effectiveness of Environmental Lawsuits? Several factors limit the effectiveness of environmental lawsuits:

- Permission to file a damage suit is granted only if the harm to an individual plaintiff is clearly unique or different enough to be distinguished from that to the general public. For example, you could not sue the Department of the Interior for actions leading to the commercialization of a wilderness area because the harm to you could not be separated from that to the general public. However, if the government damaged property you own, you would have grounds to sue.

- Bringing any suit is expensive.

- Public interest law firms cannot recover attorneys' fees unless Congress has specifically autho-

rized such recovery in the laws the firms seek to have enforced. By contrast, corporations can reduce their taxes by deducting their legal expenses and in effect have the public pay for part of their legal fees.

- To stop a nuisance or to collect damages from a nuisance or act of negligence, plaintiffs must prove that they have been harmed in some significant way and that the defendant caused the harm. This is often very difficult and costly to do. Suppose that a company (the defendant) is charged with harming individuals by polluting a river. If hundreds of other industries and cities dump waste into that river, establishing that the defendant is the culprit is very difficult and entails costly investigation, scientific research, and expert testimony. In addition, it is difficult to establish that a particular chemical is what caused the plaintiffs to come down with a disease such as cancer.

- The *statute of limitations* limits the length of time within which a plaintiff can sue after a particular event in most states. In such states, this makes it essentially impossible for victims of cancer, which may take 10–20 years to develop, to file or win a negligence suit.

- The court (or series of courts if the case is appealed) may take years to reach a decision. During that time a defendant may continue the allegedly damaging action unless the court issues a temporary injunction against the action until the case is decided.

- Plaintiffs sometimes abuse the system by bringing frivolous suits that delay and run up the costs of projects. In recent years, some corporations and developers have begun filing lawsuits for damages against citizen activists (Spotlight, right).

Because of such difficulties, an increasing number of environmental lawsuits are being settled out of court, either privately or by *mediation*, in which a neutral party tries to resolve the dispute in a way that is acceptable to both parties. Mediation is much less costly and time-consuming and may bring about a more satisfactory resolution of a dispute than going to court. On the other hand, a settlement drawn up by mediation is not legally binding. Thus, months of expensive mediation can result in an agreement that polluters may ignore.

Despite many obstacles, proponents of environmental law have accomplished a great deal since the 1960s. In the United States, more than 20,000 attorneys in 100 public interest law firms and groups now specialize partly or entirely in environmental and consumer law. In addition, many other lawyers and scientific experts participate in environmental and consumer lawsuits as needed.

How Can We Level the Legal Playing Field for Ordinary Citizens? Reforms that have been suggested for making the legal playing field more level for citizens suffering environmental damage include the following:

- Allowing citizens to sue violators of environmental laws for triple damages.

- Awarding citizens attorney fees in successful lawsuits.

- Establishing rules and procedures for identifying frivolous SLAPP suits (Spotlight, right) so that cases without factual or legal merit could be dismissed within a few weeks rather than years.

- Raising fines for violators of environmental laws and punishing more violators with jail sentences. Polls indicate that 84% of Americans consider damaging the environment a serious crime.

SPOTLIGHT

SLAPPs: Intimidating Environmental Activists

In recent years, some corporations and developers have begun filing *strategic lawsuits against public participation (SLAPPs)* against individuals and activist environmental groups.* Such suits have two goals: **(1)** to use up the time and financial resources of private citizens and environmental groups and **(2)** to deter citizens from becoming involved.

SLAPPs range from $100,000 to $100 million but average $9 million per suit. Judges who recognize them for what they are throw out about 90% of the SLAPPs that go to court. However, individuals and groups hit with SLAPPs must spend much money on lawyers and typically spend 1–3 years defending themselves.

Most SLAPPs are not meant to be won. Instead, they are intended to intimidate individual citizens and activist groups so that they will not exercise their democratic rights. Once fear and rising defense costs shake the victim of a SLAPP, the plaintiff often offers a voluntary dismissal, provided that the citizen or group agrees never to discuss the case or oppose the plaintiff again.

Some citizen activists have fought back with countersuits and have been awarded damages. For example, a Missouri woman who was sued for criticizing a medical waste incinerator won an $86.5 million judgment against the incinerator's owner.

Even after paying such awards, corporations and developers generally save money by filing such suits. Unlike the people they are suing, these businesses can count such legal and insurance liability costs as a business expense and write them off on their taxes. In other words, they get all taxpayers to pay much of the cost of such lawsuits against a few taxpayers who are merely exercising their rights as citizens.

Critical Thinking

Are you in favor of SLAPPs? Explain. If not, what do you think should be done to discourage such tactics?

*Free information on SLAPPs can be obtained from the SLAPP Resource Center, University of Denver College of Law, 1900 Olive St., Denver, CO 80220; 303-871-6266.

27 5 ENVIRONMENTAL AND ANTI-ENVIRONMENTAL GROUPS AND CLAIMS

What Are the Roles of Mainstream Environmental Groups? In the United States, more than 8 million citizens belong to at least 30,000 NGOs dealing with environmental issues at the international, national, state, and local levels.

Some of these environmental organizations are multimillion-dollar *mainstream* groups, led by chief executive officers and staffed by expert lawyers, scientists, and economists.

Mainstream environmental groups are active primarily at the national level and to a lesser extent at the state level; sometimes they form coalitions to work together on issues. Some mainstream organizations such as the Sierra Club funnel substantial funds to local activists and projects.

Some groups focus much of their efforts on specific issues, such as population (Zero Population Growth), protecting habitats (local land trusts, Wilderness Society, and Nature Conservancy; Solutions, p. 616), and wildlife conservation (National Audubon Society, National Wildlife Federation, and the World Wildlife Fund). Other organizations concentrate on education and research (Worldwatch Institute, Rocky Mountain Institute, Population Reference Bureau, and World Resources Institute). Still other groups provide information, training, and assistance to localities and grassroots organizations (Center for Health, Environment, and Justice and the Institute for Local Self-Reliance).

Mainstream groups work within the political system. Many of these NGOs have been major forces in persuading Congress to pass and strengthen environmental laws (Appendix 5) and fighting off attempts to weaken or repeal such laws.

All of the 10 largest U.S. mainstream environmental organizations—the "Group of 10"—rely heavily on corporate donations, and many of them have corporate executives as board members, trustees, or council members. Proponents of this corporate involvement argue that it is a way to raise much-needed funds and to influence industry, whereas critics believe that it is a way for corporations to influence environmental organizations.

Instead of acting as adversaries, some industries and environmental groups have worked together to find solutions to environmental problems. For example, the Environmental Defense Fund has worked with **(1)** McDonald's to redesign its packaging system to eliminate its polyethylene-foam clamshell hamburger containers, **(2)** General Motors to remove high-pollution cars from the road, and **(3)** a number of businesses to promote the use of recycled paper.

Some environmental groups have shifted some of their resources away from demonstrating and litigating to publicizing research on innovative solutions to environmental problems. For example, Greenpeace Germany ran a successful advertising campaign endorsing German-made refrigerators that use a mixture of propane and butane as a safer cooling agent than ozone-depleting hydrofluorocarbons and hydrochlorofluorocarbons (Table 18-3, p. 471). To promote the use of chlorine-free paper, Greenpeace Germany printed a magazine using such paper and encouraged readers to demand that magazine publishers switch to chlorine-free paper. Shortly thereafter, several major magazines made such a shift.

There are also more than 3,000 international NGOs working on environmental issues. They include mainstream groups such as the Worldwide Fund for Nature (nearly 5 million members), Greenpeace (2.5 million members), and Friends of the Earth (FOE). Using e-mail and the internet, environmental NGOs have organized themselves into an array of powerful international networks. Examples include the Pesticide Action, Climate Action, International Rivers, Women's Environment and Development, and Biodiversity Action Networks.

These international networks **(1)** monitor the environmental activities of governments, corporations, and international agencies such as the United Nations, World Bank, International Money Fund, and the World Trade Organization (WTO, p. 733) and **(2)** push for improved environmental performance by such organizations and by governments.

What Are the Roles of Grassroots Environmental Groups? The base of the environmental movement in the United States and throughout the world consists of thousands of grassroots citizens' groups organized to improve environmental quality, often at the local level. According to political analyst Konrad von Moltke, "There isn't a government in the world that would have done anything for the environment if it weren't for the citizen groups."

These groups carry out a number of environmental roles such as:

- Preventing environmental harm to their members and their local communities by opposing projects such as landfills, waste incinerators, nuclear waste dumps, clear-cutting of forests, and harmful development projects.

- Getting government officials to take action because they have been victims of environmental harm (p. 546) or environmental injustice because of the unequal distribution of environmental risks (Guest Essay, right, and photo, p. 687).

Environmental Justice for All

Robert D. Bullard

GUEST ESSAY

Robert D. Bullard is professor of sociology and director of the Environmental Justice Resource Center at Clark Atlanta University. For more than a decade, he has conducted research in the areas of urban land use, housing, community development, industrial facility siting, and environmental justice. He is the author of four books and more than three dozen articles, monographs, and scholarly papers that address concerns about environmental justice. His book Dumping in Dixie: Race, Class, and Environmental Quality, *2d ed. (Westview Press, 1994), has become a standard text in the field. Other books are* Confronting Environmental Racism *(South End Press, 1993) and* Unequal Protection: Environmental Justice and Communities of Color *(Sierra Club Books, 1994).*

Despite widespread media coverage and volumes written on the U.S. environmental movement, environmentalism and social justice have seldom been linked. Nevertheless, an environmental revolution has been taking shape in the United States that combines the environmental and social justice movements into one framework.

People of color (African-Americans, Latinos, Asians, Pacific Islanders, and Native Americans), working-class people, and poor people in the United States suffer disproportionately from industrial toxins, dirty air and drinking water, unsafe work conditions, and the location of noxious facilities such as municipal landfills, incinerators, and toxic-waste dumps.

The *environmental justice* movement attempts to dismantle **(1)** exclusionary zoning ordinances, **(2)** discriminatory land-use practices, **(3)** differential enforcement of environmental regulations, **(4)** disparate siting of risky technologies, and **(5)** the dumping of toxic waste on the poor and people of color in the United States and in developing countries.

Despite the government's attempts to level the playing field, all communities are not created equal when it comes to resolving environmental and public health concerns. More than 300,000 farm workers (more than 90% of whom are people of color) and their children are poisoned by pesticides sprayed on crops in the United States. Some 3–4 million children (many of them African-Americans or Latinos living in the inner city) are poisoned by lead-based paint in old buildings, lead-soldered pipes and water mains, lead-tainted soil contaminated by industry, and air pollutants from smelters.

All communities do not bear the same burden or reap the same benefits from industrial expansion. This is true in the case of the mostly African-American Emelle, Alabama (home of the nation's largest hazardous-waste landfill), Navajo lands in Arizona where uranium is mined, and the 2,000 factories, known as *maquiladores*, located just across the U.S. border in Mexico.

Nationally, 60% of African-Americans and 50% of Latinos live in communities with at least one uncontrolled toxic-waste site. Three of the five largest hazardous-waste landfills are located in communities that are predominantly African-American or Latino.

Environmental justice does not stop at the U.S. border. Environmental injustices exist from the *favelas* of Rio de Janeiro, Brazil, to the shantytowns of Johannesburg, South Africa. Members of the environmental justice movement are also questioning the wasteful and non-sustainable development models being exported to the developing world.

Grassroots leaders are demanding justice. Residents of communities such as West Dallas and Texarkana (Texas), West Harlem (New York), Rosebud (South Dakota), Kettleman City (California), and Sunrise, Lions, and Wallace (Louisiana) see their struggle for environmental justice as a life-and-death matter. Unfortunately, their stories of environmental racism are not broadcast into the nation's living rooms during the nightly news, nor are they splashed across the front pages of national newspapers and magazines. To a large extent, the communities that are the victims of environmental injustice remain invisible to the larger society.

The environmental justice movement is led, planned, and to a large extent funded by people who are not part of the established environmental community or the "Big 10" environmental organizations. Most environmental justice groups are small and operate with resources generated from the local community.

For too long these groups and their leaders have been invisible and their stories muted. This is changing as these grassroots groups are forcing their issues onto the nation's environmental agenda.

The United States has a long way to go in achieving environmental justice for all its citizens. The membership of decision-making boards and commissions still does not reflect the racial, ethnic, and cultural diversity of the country, and token inclusion of people of color on boards and commissions does not necessarily mean that their voices will be heard or their cultures respected. The ultimate goal of any inclusion strategy should be to democratize the decision-making process and empower disenfranchised people to speak and do for themselves.

Critical Thinking

1. How would you define environmental injustice? Can you identify any examples of environmental injustice in your community?

2. Have you been a victim of environmental injustice? Compare your answers with those of other members of your class.

- Voicing the concerns of people who are often not heard in state and national policy debates on environmental (Guest Essay, p. 727)

- Forming land trusts and other local organizations to save wetlands, forests, farmland, and ranchland from development, restore degraded rivers (p. 715) and wetlands, and convert abandoned urban lots into community gardens and parks.

- Forming coalitions of workers and environmentalists to improve worker safety and health.

- Setting up internet service providers and networks to (1) improve environmental education and health, (2) provide individuals and NGOs with information on toxic releases and other environmentally harmful activities by local industries, and (3) publicize successful environmental projects. Emphasis is often on providing such information for the poor in rural areas in developing countries. Groups have also set up community telecenters in rural areas that can include public telephones, fax machines, computers, and internet access.

- Working to make colleges and public schools more environmentally sustainable (Individuals Matter, right).

- Playing key roles in experimenting with new approaches to environmental problems. For example, in Walthill, Nebraska, the Center for Rural Affairs has developed an innovative way to preserve farmland and promote environmentally sustainable agriculture. The group uses funds raised by a bond issue to help young farmers buy land from retiring farmers and also gives them technical assistance for implementing environmentally sustainable farming methods.

John W. Gardner, former cabinet official and founder of Common Cause, summarized the basic rules for effective political action by grassroots organizations:

- *Have a full-time continuing organization.*

- *Limit the number of targets and hit them hard.* Groups dilute their effectiveness by taking on too many issues.

- *Organize for action, not just for study, discussion, or education.*

- *Form alliances with other organizations on a particular issue.*

- *Communicate your positions in an accurate, concise, and moving way.*

- *Persuade and use positive reinforcement.*

- *Concentrate efforts mostly at the state and local levels.*

How Successful Have Environmental Groups Been? Opponents sometimes portray the environmental movement as a well-organized and well-funded effort to bring about fundamental changes in how we perceive and act in the world. In truth, what is called the environmental movement has not really been a movement at all.

Instead, it has consisted of a hodgepodge of efforts on many fronts without overall direction and coherence. During the past 30 years, this variety of groups has (1) raised understanding of environmental issues by the general public and some business leaders (Individuals Matter, p. 528), (2) gained public support for passage of an array of environmental and resource use laws in the United States (Appendix 5) and other developed countries, and (3) helped individuals deal with a number of local environmental problems.

Polls show that three-fourths of the U.S. public are strong supporters of environmental laws and regulations and do not want them weakened. However, polls also show that less than 5% of the U.S. public views the environment as one of the nation's most pressing problems. As a result, environmental concerns increasingly do not get transferred to the ballot box. As one political scientist put it, "Environ- mental concerns are like the Everglades, a mile wide but only a few inches deep."

In one sense, the diversity of environmental groups and concerns is a strength because they analyze and provide a range of possible solutions to the complex environmental problems we face in a largely unpredictable world. On the other hand, it is a weakness because it has not led to deep convictions and environmental concerns on the part of most of the public.

This diversity has also allowed a well-organized and well-funded anti-environmental movement to develop an array of tactics* for undermining much of the improvement in environmental understanding and support for environmental concerns. Since 1980, mainline environmental groups in the United States have spent most of their time and money trying to prevent existing environmental laws and regulations from being weakened or repealed.

One problem is that many environmentalists have concentrated mostly on making the public aware of environmental bad news. However, history shows that bearers of bad news are not received well, and anti-environmentalists have used this fact to undermine environmental concerns.

History also shows that people are moved to bring about change mostly by an inspiring, positive vision of what the world could be like that provides them with a sense of hope for the future. So far, environmental-

*See the website for this book for a summary of the tactics of the anti-environmental movement.

Environmental Action on Campuses

Since 1988, there has been a boom in environmental awareness on college campuses and some public schools across the United States. Much of this momentum began in 1989, when the Student Environmental Action Coalition (SEAC), then at the University of North Carolina at Chapel Hill (UNC), held the first national student environmental conference on the UNC campus.

SEAC groups are active on 700 campuses, and the National Wildlife Federation's Campus Ecology Program (launched in 1989) has groups on about 600 campuses.* Most student environmental groups work with members of the faculty and administration to bring about environmental improvements on their own campuses and in their local communities.

Many of these groups focus on making an environmental audit of their own campuses or schools; they then use the data gathered to propose changes that will make their campuses or schools more ecologically sustainable, usually saving them money in the process.**

Such audits have resulted in numerous improvements. For example, Morris A. Pierce, a graduate student at the University of Rochester in New York, developed an energy management plan adopted by that school's board of trustees. Under this plan, a capital investment of $33 million is projected to save the university $60 million over 20 years. Students have also induced almost 80% of universities and colleges in the United States to develop recycling programs.

At Bowdin College in Maine, chemistry professor Dana Mayo and student Caroline Foote developed the concept of *microscale experiments*, in which smaller amounts of chemicals are used. This has reduced toxic wastes and saved the chemistry department more than $34,000. Today more than 50% of all undergraduates in chemistry in the United States use such microscale techniques, as do universities in a growing number of other countries.

Students at Oberlin College in Ohio helped design a new sustainable environmental studies building (Figure 15-24, p. 376). At Northland College in Wisconsin students helped design a "green" dorm which features a large wind generator, panels of solar cells, recycled furniture, and waterless (composting) toilets.

A 1997 report by the National Wildlife Foundation's Campus Ecology Program found that 23 student-researched and student-motivated projects had saved the participating universities and colleges $16.3 million. According to this study, implementing similar programs in the nation's 3,700 universities and colleges could **(1)** help improve environmental quality and environmental education and **(2)** lead to a savings of more than $27.6 billion. Such student-spurred environmental activities and research studies are spreading to universities in at least 42 other countries.

*See *Ecodemia: Campus Environmental Stewardship at the Turn of the 21st Century* (Washington, D.C.: National Wildlife Federation, 1995) and the *Campus Environmental Yearbook*, published annually by the National Wildlife Federation.

**Details for conducting such audits are found in April Smith and the Student Environmental Action Coalition, *Campus Ecology: A Guide to Assessing Environmental Quality and Creating Strategies for Change* (Los Angeles, Calif.: Living Planet Press, 1993), and Jane Heinze-Fry, *Green Lives, Green Campuses*, available free on the internet site for this textbook.

---

ists with a variety of beliefs and goals have not worked together to develop one or a series of broad, compelling, coherent, and positive visions that can be used to as a roadmap for a more sustainable future for humans and other species. Such a vision or series of such visions should **(1)** recognize our immense and growing power over the other forms of life on the earth that support our lifestyles and economies and **(2)** inspire us to grapple with the fundamental questions of the purpose of our species and how we should live. This will require us to think deeply about our environmental worldviews, as discussed in Chapter 28.

What Are the Goals of the Anti-Environmental Movement in the United States? Despite general public approval, there is strong opposition to many environmental proposals by:

- Leaders of some corporations and people in positions of economic and political power who see environmental laws and regulations as threats to their wealth and power.

- Citizens who see environmental laws and regulations as threats to their private property rights and jobs.

- People who disagree with the basic beliefs behind some environmental worldviews (Section 28-2, p. 742).

- Some state and local government officials who are fed up with having to implement federal environmental laws and regulations without federal funding (unfunded mandates) or who disagree with certain federal environmental regulations.

Since 1980, businesses, individuals, and grassroots groups in the United States have mounted a strong campaign to weaken or repeal existing environmental

Covering the Environment

Andrew C. Revkin

GUEST ESSAY

Andrew (Andy) C. Revkin, an environmental reporter for The New York Times, *has written about the environment and science since 1982. He is author of two widely acclaimed environmental books:* The Burning Season: The Murder of Chico Mendes and the Fight for the Amazon Basin Forest *(Plume Paperback, 1994), which was the basis of the Emmy-winning HBO film of the same name, and* Global Warming: Understanding the Forecast *(Abbeville Press, 1992). He is formerly a senior editor of* Discover *magazine and senior writer at* Science Digest. *He has won more than half a dozen national writing awards, including the Robert F. Kennedy award for* The Burning Season *and the American Association for the Advancement of Science/Westinghouse Science Journalism Award. He has taught a course on environment and energy reporting at the Columbia University Graduate School of Journalism.*

When I traveled to Brazil in 1989 to write a biography of Chico Mendes, the slain leader of the movement to save the Amazon rain forest, Chico's friends were at first very suspicious. "Why should we talk to you?" they asked me. "How do we know you are going to tell the truth?"

They said they were skeptical because, shortly after the murder, dozens of journalists had flown in for a day or two and then left. As a result, many of the articles or television reports were simplistic or inaccurate. Only after I had stayed for months did people open up. And only then did the complexities of the story of the invasion of the rain forest and the struggle to defend it start to make a little sense.

I was very lucky. For the first time in my career, I had the luxury of being able to spend substantial time on the story. Most often, a reporter has a few hours, or maybe a few days, or—rarely—a week or two to figure out how to explain a complicated subject in an interesting and accurate way.

The lack of time affects journalists covering everything from war to sports. Something that is news today does not stay news for very long. But for a writer covering the environment, other factors combine with deadline pressure to make the challenge of effective communication particularly difficult.

First, there is the complexity and unfamiliarity of science. A baseball writer or a political writer can assume at least some basic knowledge on the reader's part of the rules of the game. But when writing about the deterioration of a diffuse layer of ozone high in the atmosphere that helps block harmful ultraviolet radiation, the writer must explain almost every step. It is no easy task to fit in all the essential ideas while still telling a story that does not cause readers to flip to the comics or movie reviews.

Making the task more difficult is the widespread lack of understanding of some of the most basic scientific concepts—not just among readers, but also among the editors who decide what stories make front pages or evening broadcasts. Recently, a biology professor at Oberlin College, Michael Zimmerman, designed a survey to gauge the basic scientific literacy of Americans. He decided to send a little scientific quiz to the managing editors of the nation's 1,563 daily newspapers. He was not trying to pick on the editors. He just figured that they represented a decent cross-section of educated America.

The results of the survey were disquieting, to say the least. When asked whether it were true or false that "dinosaurs and humans lived contemporaneously," only 51% of the editors disagreed strongly with the statement. Thirty-seven percent either agreed or had no opinion on the matter. That means a big chunk of our newspapers are being edited by people who subscribe, at least tacitly, to what amounts to the "Flintstones Theory" of evolution.

Another impediment to good environmental reporting is something one of my journalism professors used to call "the MEGO factor," with the acronym standing

laws and regulations, change the way in which public lands are used (p. 589), and destroy the reputation and effectiveness of the environmental movement. A list of some of the tactics used by this anti-environmental movement is found on the website for this book.

Whom Should We Believe? As you have seen throughout this book, there is much controversy over most environmental issues. A key reason for such controversy is that these are important political, economic, social, and ethical issues that will not go away and that must be dealt with.

One problem is that the focus of environmental issues has shifted to more complex and controversial

environmental problems that **(1)** are harder to understand and solve and **(2)** have long-range harmful effects instead of easily visible short-term effects. Examples are global climate change, ozone depletion, biodiversity protection, nonpoint water pollution (such as runoff from farms and lawns), and protection of unseen groundwater. Explaining such complex issues to the public and mobilizing support for often controversial, long-range solutions to such problems are difficult (Guest Essay, above).

Polarization and distortion of opposing views on environmental issues often leads to deadlock. Some people on both sides of such issues are working to mediate such disputes by getting each side to **(1)** listen to one another's concerns, **(2)** try to find areas of

for "my eyes glaze over." Editors are bombarded with so much information every day that they become numb to its significance. If a reporter asks to write an update on toxic chemicals in a river or the latest findings on climate change, at many papers, magazines, and television networks, the response is likely to be along the lines of, "Haven't we done that story?" Nothing causes an editor's eyes to glaze faster than a complicated, subtle topic.

Another impediment to effective environmental journalism is the endless appetite for the sound bite or snappy quote. Journalists often try to create balance in a story by quoting a yea-sayer and a nay-sayer. As one journalist put it, echoing a famous law of physics, "For every Ph.D., there is an equal and opposite Ph.D." This is a quick and easy way of establishing that the reporter has no bias. The problem is that the loudest voices on an issue often are the most suspect. Scientists who can deliver well-honed sound bites may have been spending more time in front of microphones than microscopes. Too often, such stories confuse instead of inform.

Finally, there is the need for articles to be timely, to have a "news peg" to hang on, in journalistic parlance. In stark contrast to traditional news events, most environmental problems develop in a creeping, almost imperceptible fashion.

In the 1980s, networks and many national publications gave top billing to a report on risks to children from Alar, a pesticide used on apple crops. Apple growers in the Northwest were nearly bankrupted as apple juice sales plummeted. Soon it became clear that the threat was drastically overstated, and most apples were not even treated with the compound. This kind of reactive reporting can end up confusing readers and creating an air of cynical skepticism about environmental threats.

One result, according to many scientists and health experts, is that the public has become overly fearful of such things as toxic dumps and does not take seriously such threats to humanity's future as the continuing growth in concentrations of greenhouse gases in the atmosphere. We fret about parts per billion of certain synthetic chemicals in food while happily heading to the beach for a new dose of ultraviolet radiation—and increasingly getting there in fleets of gas-guzzling Blazers and Broncos and thus contributing to global warming.

Many journalists have worked long and hard to overcome some of these impediments to coverage of environmental issues. Editors are becoming more informed. Reporters have created a Society of Environmental Journalists, which is fostering a daily debate on the internet and its newsletter on ways to do a better job. The result is a steady improvement in the national debate on environmental and resource use issues.

Ultimately, though, part of the responsibility rests with readers as well as writers. Consumers of news on television or in the print media become better informed when they treat reports with a skeptical eye and seek a variety of sources. Just as good journalism results when a reporter seeks a multiplicity of sources for a story, good citizenship results when people seek to understand an issue by relying on more than one medium for their information—magazines, newspapers, television reports, books, the internet, and more.

Critical Thinking

1. Do your eyes generally glaze over when you read a newspaper or magazine story covering some complex environmental issue such as ozone depletion, global warming, biodiversity, or environmental economics?

2. How well do you meet your responsibility as a consumer of information to seek a variety of sources on key environmental issues? Have you become skeptical of environmental stories? Why? Have you become too skeptical of such stories by assuming that almost all of them are suspect?

---

agreement, and **(3)** work together to find solutions, as is being done in the Netherlands (pp. 709–710). For example, both environmentalists and anti-environmentalists agree that some environmental regulations in the United States go too far or are unjustly enforced and could work together to improve environmental laws and regulations (Solutions, p. 732).

Because of the importance of these issues, analysts urge us to try to understand both sides of these issues and decide what should be done. In doing this, we need to:

- Gather and carefully evaluate the evidence for each position by using the techniques of critical thinking (Guest Essay, p. 48).

- Distinguish between frontier and consensus science (p. 48) when claims are made.

- Identify the consensus of most scientists in the particular field involved and not give equal weight to a small minority of scientists in such fields or those working outside these fields who disagree with the current consensus view.

- Become better informed about the relative degree of potential harm posed by various risks and try to rank risks as best we can (Figure 16-13, p. 411, and Figure 16-15, p. 414).

- Support the use of the pollution prevention and precautionary principles to deal with potentially

Environmentalists agree that some government laws and regulations go too far and that bureaucrats sometimes develop and impose ridiculous and excessively costly regulations. They also agree that one of the functions of government is to steer a course toward a desired goal, such as improving environmental quality, without rowing.

This is often accomplished by **(1)** establishing laws and regulations with general goals and guidelines, **(2)** allowing the marketplace and local governments leeway in finding the best ways to meet these goals, and **(3)** carefully monitoring progress.

Environmentalists argue that the solution is to stop regulatory abuse, not throw out or seriously weaken the body of laws and regulations that help protect the public good.

According to environmental economist William Ashworth,

If we wish to make progress, it will do us no good to replace one failed system with another

that failed just as badly. Government regulation, after all, did not fall out of the sky; it was erected, piece by piece, as an attempt to deal with the damage caused by unrestrained property rights and the unregulated free-market system....We do not need to deconstruct regulation, but to reconstruct it.

To accomplish this, a growing number of analysts urge environmentalists to take a hard look at existing environmental laws and regulations. Which laws (or parts of laws) have worked, and why? Which ones have failed, and why? Which government bureaucracies concerned with developing and enforcing environmental and resource regulations have abused their power or have not been responsive enough to the needs of ordinary people? How can such abuses be corrected? What existing environmental laws (or parts of laws) and regulations should be repealed or modified?

How can a balanced program of regulation and market-based

approaches (Table 26-1, p. 706) be used to achieve environmental goals? What environmental problems lend themselves to market-based approaches (free-market environmentalism), and which ones do not? How can pollution prevention and waste reduction and the precautionary principle become guidelines for environmental legislation and regulation? What is the minimum amount of environmental legislation and regulation that is needed?

These are important issues that environmentalists, business leaders, elected officials, and government regulators need to address with a cooperative, problem-solving spirit.

Critical Thinking

Analyze a particular environmental law in the United States (Appendix 5) or in the country where you live to come up with ways in which it could be improved. Develop a strategy for bringing about such changes.

---

serious problems for which there is inadequate scientific information.

27-6 GLOBAL ENVIRONMENTAL POLICY

Should We Expand the Concept of National and Global Security? Countries are legitimately concerned with *military security* and *economic security*. However, ecologists point out that without adequate soil, water, clean air, and biodiversity, no nation can be physically or economically secure. All economies are supported by natural resources (Figure 26-7, p. 694, and Guest Essay, p. 6), and many environmental problems do not recognize political boundaries. Thus, military and economic security also depends on national and global environmental security.

According to Sam Nunn, former senator and chair of the U.S. Senate Armed Services Committee, "I believe that one of our key national security objectives must be to reverse the accelerating pace of environmental

destruction around the globe." According to environmental expert Norman Myers (Guest Essay, p. 117),

If a nation's environmental foundations are degraded or depleted, its economy may well decline, its social fabric deteriorate, and its political structure become destabilized as growing numbers of people seek to sustain themselves from declining resource stocks. Thus, national security is no longer about fighting forces and weaponry alone. It relates increasingly to watersheds, croplands, forests, genetic resources, climate, and other factors that, taken together, are as crucial to a nation's security as are military factors.

Examples of conflicts arising from environmental shortages include:

- Clashes between nations over declining fishery stocks.

- Disputes over access to shared water supplies (p. 294). Of 200 major river systems, almost 150 are shared by 2 nations, and more than 50 by 3 to 10 nations. Some 80 countries, with 40% of the world's population, already experience water shortages, which are likely to intensify in the future (Section 13-3, p. 299).

- Clashes over deforestation when water-absorbing tree cover removed by one nation leads to extensive flooding of downstream communities in other nations (Connections, p. 315).

Proponents call for all countries to make environmental security a major focus of diplomacy and government policy at all levels. They propose that national governments have a council of advisers made up of highly qualified experts in environmental, economic, and military security. Any major decision would entail integrating all three security concerns.

What Progress Has Been Made in Developing International Environmental Cooperation and Policy? Since the 1972 UN Conference on the Human Environment in Stockholm, Sweden, some progress has been made in addressing environmental issues at the global level. Today, 115 nations have environmental protection agencies, and nearly 240 international environmental treaties and agreements between various countries have been signed. They address issues such as endangered species, ozone depletion, ocean pollution, climate change, ozone layer depletion, biodiversity, and hazardous waste export. The 1972 conference also created the UN Environment Programme (UNEP) to negotiate environmental treaties and to help implement them.

Developing international treaties is important, but little is accomplished if they are not put into effect. This involves setting up and adequately funding agencies such as the UNEP to monitor the progress of participating countries and publicize the results to create international pressure to live up to treaty commitments. Currently, such organizations do not have enough funding to carry out this job effectively.

In June 1992, the second UN Conference on the Human Environment, known as the *Rio Earth Summit*, was held in Rio de Janeiro, Brazil. More than 100 heads of state, thousands of officials, and more than 1,400 accredited NGOs from 178 nations met to develop plans to address environmental issues.

The major official results included **(1)** an *Earth Charter*, a nonbinding statement of broad principles for guiding environmental policy that commits countries that sign it to pursue sustainable development and work toward eradicating poverty, **(2)** *Agenda 21*, a nonbinding detailed action plan to guide countries toward sustainable development and protection of the global environment during the 21st century, **(3)** a *forestry agreement* that is a broad, nonbinding statement of principles of forest management and protection, **(4)** a *convention on climate change* that requires countries to use their best efforts to reduce their emissions of greenhouse gases, **(5)** a *convention on protecting biodiversity* that calls for countries to develop

strategies for the conservation and sustainable use of biological diversity, and **(6)** the *UN Commission on Sustainable Development*, composed of high-level government representatives charged with carrying out and overseeing the implementation of these agreements.

Most environmentalists were disappointed because these accomplishments consisted of nonbinding agreements without sufficient incentives or funding for their implementation. According to analyses by the United Nations, by 2000 there was little improvement in the major environmental problems discussed at the Rio summit.

However, there is hope for greater progress in the slowly moving arena of international cooperation because:

- The conference gave the world a forum for discussing and seeking solutions to environmental problems. This led to general agreement on some key principles, which with enough political pressure from citizens and NGOs could be implemented or improved.

- Paralleling the official meeting was a Global Forum that brought together 20,000 concerned citizens and activists from more than 1,400 NGOs in 178 countries. These individuals outnumbered the conference's official representatives by at least two to one. These NGOs **(1)** worked behind the scenes to influence official policy, **(2)** formulated their own agendas and treaties for environmental sustainability, **(3)** learned from one another, and **(4)** developed a series of new global networks, alliances, and projects and developed key goals (Solutions, p. 734). In the long run, these newly formed networks and alliances may play the greatest role in helping **(1)** monitor, support, and implement the commitments and plans developed by the 1992 conference and **(2)** set the agenda for the second UN Conference on the Human Environment in 2002.

Is Encouraging Global Free Trade Environmentally Helpful or Harmful? Like it or not, we are in an age of rapid economic, political, and social globalization (p. 10). Countries (or, more accurately, transnational corporations) that are involved in international trade want to eliminate trade barriers that prevent the free flow of goods and services from one place to another.

On April 15, 1994, representatives of 120 nations signed the Uruguay Round of the General Agreement on Tariffs and Trade (GATT). This is a revised version of the 1948 GATT convention, which attempted to lower tariff barriers to world trade between member nations. The new GATT established a World Trade Organization (WTO) and gave it the status of a major international organization (similar to the United Nations and the World Bank). The WTO, which came into existence in 1995, is charged with **(1)** enforcing the

Goals of Global and National Environmental NGOs

SOLUTIONS

Some key goals of global and national environmental NGOs are to:

■ Pressure powerful international organizations such as the World Bank, International Monetary Fund, and World Trade Organization to **(1)** give the public more information about their often secret meetings, policies, and deliberations and

(2) allow more input from NGOs concerned with environment, public health, and worker health and safety issues into their policies and decisions and be more responsive to these issues.

■ Pressure the United Nations to create a new assembly within its overall body in which the views of the people of the world could be more directly represented than under the current system. Such a

body could similar to the directly elected European Forum.

■ Exert pressure for upgrading the UNEP into a World Environment Organization (WEO) on a par with the WTO that would oversee and integrate global environmental policies.

Critical Thinking

Do you agree or disagree with each of these goals? Explain.

---

new GATT rules of world trade and **(2)** settling any disputes about these rules between nations.

Currently, the WTO has 138 member countries (nations that are not members can be frozen out of international trade). Most WTO policy is determined by WTO representatives of the Quad Countries: the United States, Canada, Japan, and the European Union. WTO officials consist mostly of trade experts and corporate lawyers, representing primarily the interests of transnational corporations.

The member countries have granted the WTO unprecedented power to **(1)** govern world trade and **(2)** make binding decisions about whether environmental, health, and worker safety laws and regulations by its member countries are illegal because they restrict trade as defined by WTO rules.

There are 700 pages of GATT rules for international trade that WTO member countries must follow. Any member country can charge another member country with violating one of the trade rules. When this occurs,

■ The case is decided by a tribunal of three anonymous WTO judges, usually corporate lawyers with no particular expertise in the issues being decided. There are no conflict-of-interest restraints on tribunal members, and information about their possible conflicts of interest is not available to the public.

■ All cases are decided in secret at unannounced times and places and are fully insulated from ordinary citizens, the press, representatives of state and local governments, and groups and experts concerned with environmental protection, health, safety, and worker issues. Only official government representatives of the countries involved can submit documents or appear before the tribunal. This weakens the ability of the governments of poor developing countries without the necessary government experts to present their case.

■ All documents, transcripts, and details of the proceedings and are kept secret, and only the results are announced. Governments involved in a case can release information about what they submitted to the tribunal, but all other details of the deliberations of the tribunal members are kept secret.

■ Decisions are binding worldwide and can be appealed only to another tribunal of judges within the WTO. A final panel ruling can be appealed to the entire WTO but can be overturned only by consensus of all WTO members. This is virtually impossible because the winning country is unlikely to vote to overturn a ruling in its favor.

■ Any country (or part of a country) that violates a ruling of WTO panels has four choices: **(1)** Amend its laws to comply with WTO rules, **(2)** pay annual compensation to the winning country, **(3)** pay high tariffs imposed on the disputed goods by the WTO, or **(4)** find itself shunned and locked out of global commerce. According to critics, in effect, countries (and states and local communities) are coerced into giving up their sovereignty and amending any of their laws or regulations that restrict global free trade according to WTO rules.

According to critics, the WTO is the closest thing we have to world government that is **(1)** radically undemocratic and **(2)** designed primarily to serve the interests and needs of transnational corporations, not the citizens of countries their decisions affect.

Proponents argue that this significant transfer of power from nations to the WTO is necessary and beneficial because:

■ Globalization of trade is inevitable and we have to guide it.

■ Reducing global trade barriers will benefit developing countries, whose products often are at a competitive disadvantage in the global marketplace because of trade barriers erected by developed countries.

- Reducing global trade barriers can stimulate economic growth in all countries by allowing consumers to buy more things at lower prices.

- Globalization of trade will raise the environmental and health standards of developing countries.

- Globalization of trade may involve some sacrifices, but in the long run the economic benefits will outweigh the costs.

Many environmental groups, and those concerned with consumer protection and worker health and safety, agree that global trade barriers are needed to help level the economic playing field for nations involved in international trade. However, they are strongly opposed to WTO rules that weaken **(1)** environmental protection and protection for consumers and workers and **(2)** the power of governments to interfere in the marketplace in the interests of their citizens. They contend that current WTO rules:

- *Will increase the economic and political power of transnational corporations and decrease the power of small businesses, citizens, and democratically elected governments.*

- *Will eliminate many jobs and lower wages in developed countries and eventually in developing countries.* Transnational companies are increasingly **(1)** moving their operations throughout the world in search of cheap labor and natural resources and **(2)** substituting machines for human labor in manufacturing and service industries. This will **(1)** eliminate many manufacturing and service jobs in developing countries, **(2)** drag the wages of those with jobs down to the lowest global common denominator, and **(3)** require most people lucky enough to find a job to work without job security or benefits. This competition between communities throughout the world for jobs for their citizens is the true meaning of *global competitiveness.* Eventually, jobs in developing countries will also be lost as businesses cut costs by substituting machines for human labor. The result is *jobless economic growth* and increased fear and insecurity among the world's workers. These critics say that it does consumers little good to have access to cheaper products if they lose their jobs, get paid less for their work, and face higher taxes to cover the social cost of increased unemployment.

- *Will weaken environmental and health and safety standards in developed countries.* Countries are free to adopt environmental, health, and other laws that affect products or manufacturing processes within their borders but cannot impose such rules on products imported from other countries.

WTO rulings have found that the following government laws or regulations are illegal barriers to free international trade:

- U.S. laws banning imports of tuna caught by methods not designed to prevent dolphins from drowning in tuna nets. Congress weakened the law.

- A U.S. ban on imports of shrimp from countries not requiring their shrimpers to use devices on their nets that let sea turtles escape (Figure 24-13, p. 640). The government altered the way the law was implemented by banning specific shipments of shrimp not using turtle excluder devices while not banning all shipments from a country that does not have such requirements. Malaysia has asked the WTO to overturn this solution.

- European Union (EU) laws banning U.S. imports of beef treated with growth hormones suspected of causing cancer and hormone disruption (Connections, p. 404) because it was based on inadequate risk assessment. When the EU disregarded the ruling, the WTO ordered the EU to pay the United States $117 million in compensatory tariffs. To critics this is a particularly worrisome decision because almost everyone agrees that the EU enacted this ban as a precautionary measure to protect consumers, not as a trade barrier to protect its domestic cattle industry.

- In a pending case, an EU law requiring that imports of genetically modified food (Pro/Con, p. 275) from the United States or other countries carry identifying labels to protect consumers' right to know about the possible health and environmental impact of products they purchase. The EU contends that such labeling requirements are a prudent and precautionary response to a new technology that has potentially serious health and ecological effects that are still clouded by many scientific uncertainties.

- EPA regulations requiring that gasoline imported into the United States must be formulated in a way that reduces air pollution. The EPA weakened its standards.

- Japan's use of stricter limits on pesticide residues in imported agricultural products than those of other countries such as the United States. Japan reluctantly agreed to force its consumers to ingest more pesticides than their government considered safe.

Here are a few WTO rules or omissions of principles that critics say affect the ability of national, state, and local governments to protect the environment, the health of its citizens, and worker health and safety:

- Governments cannot set standards for how imported products are produced or harvested. This means, for example, that government purchasing policies cannot **(1)** discriminate against materials produced by child labor or slave labor, (2) require that items be manufactured from recycled materials or use cleaner production methods, or (3) require fish-harvesting methods that help protect other species such as dolphins or turtles.

- National, state, or local governments cannot ban products from countries that are recognized as violating universally recognized human rights such as using torture and murder to suppress political opposition or using slave labor.

- Current WTO rules do not acknowledge the rights of countries to take action to protect the atmosphere, the oceans, and other parts of the global commons. There is concern that some provisions of international treaties to protect biodiversity and the ozone layer and reduce the threats of global warming might be ruled as illegal under WTO rules.

- All national, state, or local environmental, health, and safety laws and regulations must be based on globally accepted scientific evidence and risk analysis showing that there is a worldwide scientific consensus on the danger. Otherwise, they are considered to be trade barriers that exceed the international standards set by corporations through the WTO. Because of the inherent scientific and other uncertainties in determining health risks (Section 16-5, p. 410) and carrying out risk analysis (Section 16-2, p. 398), this is an almost impossible standard to meet. In other words, any chemical, product, or technology that is traded internationally is assumed to be harmless unless it can proven to be harmful by extensive scientific research and by risk analysis, which most corporations prefer because of their many uncertainties and opportunities for manipulation and legal challenges (p. 412). In other words, the burden of proof falls on those trying to prevent pollution instead of the polluters (Figure 27-6). The effect of this rule is that a chemical, product, or technology cannot be banned on the basis of the pollution prevention or precautionary principles, which are two of the foundations of modern environmental protection.

- National laws covering packaging, recycling, and eco-labeling of items involved in international markets are ille-

gal barriers to trade. This rule effectively cancels the third mainstay of modern environmental protection: *consumers' right to know* about the safety and content of products through labeling.

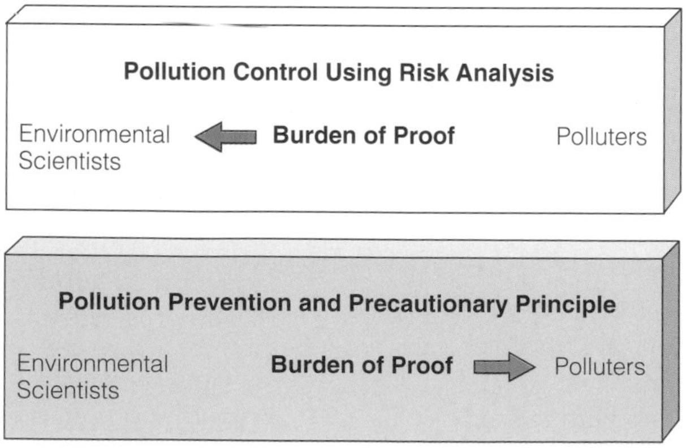

If allowed to stand, environmentalists contend that the last three WTO rules just described will force us back to an earlier and less effective era of using end-of-pipe pollution cleanup based on uncertain and easily manipulated risk assessment (p. 410) as the primary way for dealing with pollution (Guest Essay, p. 526).

On a more positive side, nations may be able to use the WTO to reduce environmentally harmful subsidies that distort the economic playing field. However, this would also prevent using subsidies to reward companies producing environmentally beneficial goods and services.

Critics contend that main goal of the people who created the WTO is not to promote free trade. Instead, it is to create a true free-market economy (p. 689) worldwide that is run by business interests through the WTO without any interference from governments, citizens, or other parties. In other words, the underlying purpose is to weaken governments by getting rid of any government rules that restrict the freedom of transnational corporations to operate in the global marketplace as they see fit.

Critics of the latest version of GATT call for it to be improved (Solutions, right).

Can We Develop More Environmentally Sustainable Political and Economic Systems in the Next Few Decades? Environmentalists call for people from all political persuasions and walks of life to work together to develop a positive vision for making the transition to environmentally sustainable societies throughout the world.

Two major goals would be **(1)** promoting the development of creative experiments at local levels, such as the one in Curitiba, Brazil (Solutions, p. 674), that could be spread to other areas over the next few decades and **(2)** getting citizens, business leaders, and elected officials to cooperate in trying to find and implement innovative solutions to local, national, and global environmental, economic, and social problems.

Proponents recognize that making such a cultural shift over the next 40–50 years will be controversial, and like all significant change it will not be predictable, orderly, or painless.

Figure 27-6 Burden of proof under pollution control and pollution prevention and precautionary systems.

Pollution Control Using Risk Analysis

Environmental Scientists ← **Burden of Proof** Polluters

Pollution Prevention and Precautionary Principle

Environmental Scientists **Burden of Proof** → Polluters

Improving Trade Agreements

Critics of current trade agreements would rewrite and correct what they believe are serious weaknesses in GATT and turn it into GAST: the *General Agreement for Sustainable Trade*. They offer the following suggestions for doing this:

- Judging GATT or any trade agreement primarily on how it benefits the environment, workers, and the poorest 40% of humanity and changing WTO rules as needed to meet these goals.

- Setting minimum environmental, consumer protection, and worker health and safety standards for all participating countries.

- Requiring all panels or bodies setting and enforcing WTO rules to have environmental, labor, consumer, and health representatives from developed countries and developing countries alike.

- Opening all discussions and findings of any GATT panel or other WTO body to (1) global public scrutiny and (2) inputs from experts on the issues involved. Critics argue that by closing its doors to the public, the WTO (1) denies itself important information that could help it make better decisions about public health, worker safety, and the environment and (2) loses public support.

- Incorporating the precautionary and pollution prevention principles into WTO rules.

- Protecting the rights of consumers to know about the health and environmental impact of imported products they purchase by allowing ecolabeling programs.

- Recognizing the right of countries to use trade measures to protect the global commons.

- Allowing countries to require that items be manufactured totally or partially from recycled materials.

- Allowing national, state, or local governments to restrict imports of products from countries shown by international investigation to (1) use child or slave labor or (2) violate universally recognized human rights.

- Allowing international environmental agreements and treaties to prevail when they conflict with WTO rules or the rules of any other trade agreement.

Unless citizens and NGOs exert intense pressure on legislators, critics warn that such safeguards will not be incorporated into GATT and international trade agreements.

Critical Thinking

Explain why you agree or disagree with each of the suggestions in this box. If you agree with all or most of these proposals, how could they be implemented?

According to business leader Paul Hawken (Guest Essay, p. 6), making this change means

Thinking big and long into the future. It also means doing something now. It means electing people who really want to make things work [Solutions, p. 724], and who can imagine a better world. It means writing to companies and telling them what you think. It means never forgetting that the cash register is the daily voting booth in democratic capitalism.

In working with the earth we should be guided by historian Arnold Toynbee's observation, "If you make the world ever so little better, you will have done splendidly, and your life will have been worthwhile," and by George Bernard Shaw's reminder that "indifference is the essence of inhumanity."

As the wagon driver said when they came to a long, hard hill, "Them that's going on with us, get out and push. Them that ain't, get out of the way."

ROBERT FULGHUM

REVIEW QUESTIONS

1. Explain how Marion Stoddart's actions in cleaning up a river illustrate the importance of individual political action.

2. What is *politics*? What is a *democracy*?

3. List three factors that hinder the ability of democracies to deal with environmental problems.

4. List seven principles that can be used as guidelines in making environmental policy decisions.

5. List nine ways in which individuals can influence the environmental policies of local, state, and federal governments.

6. Distinguish between *leaders*, *rulers*, and *managers*. Describe three types of environmental leadership.

7. Describe the role of each of the three branches of the federal government in the United States.

8. What are the three main functions of the federal government in the United States?

9. Describe the five steps used to develop federal environmental policy in the United States.

10. Describe the four phases of a *policy life cycle*.

11. List five types of federal environmental laws in the United States and four principles used to establish regulations for implementing such laws.

12. Describe and evaluate the usefulness of using only public funds to finance all election campaigns in the United States.

13. What are two problems government bureaucracies face in developing and implementing environmental regulations? List three ways that could be used to improve the responsiveness and effectiveness of government bureaucracies.

14. Distinguish between **(a)** *plaintiffs* and *defendants*, **(b)** *statutory law* and *common law*, and **(c)** *nuisance* and *negligence*. List six factors that hinder the effectiveness of environmental law in the United States.

15. What are SLAPPs, and how are they used?

16. List three ways to level the legal playing field for ordinary citizens suffering from environmental damage.

17. Describe the key roles of mainstream and grassroots environmental groups. List five rules for effective political action by grassroots organizations.

18. What is *environmental justice*? What types of environmental injustice did Robert Bullard describe in his Guest Essay on p. 727?

19. Describe some of the environmentally beneficial activities that have been carried out by high school and college students.

20. Describe three major accomplishments of environmental groups over the past 30 years. What are the major strengths and weaknesses of the diverse array of groups involved in bringing about environmental improvement?

21. List four reasons why some people oppose environmental reform.

22. Describe the problems environmental journalists face in explaining environmental issues to the public as discussed by Andrew Revkin in his Guest Essay on p. 730.

23. List five ways to evaluate the conflicting claims of environmentalists and anti-environmentalists.

24. Describe ways in which environmentalists could work to improve environmental laws and regulations.

25. Distinguish between environmental, economic, and military security and explain the importance of making environmental security a key priority of governments.

26. What is the World Trade Organization (WTO)? Describe the procedure it uses to determine whether a member nation of this group has violated one of its international trade rules.

27. List the major pros and cons of the international GATT treaty as enforced by the WTO.

28. List five WTO trade rules that critics say will weaken the ability of governments to protect the environment, the health of their citizens, and the health and safety of their workers.

29. List ten measures critics have suggested for improving the GATT treaty as administered by the World Trade Organization.

CRITICAL THINKING

1. What are the greatest strengths and weaknesses of the system of government in your country with respect to protecting the environment and ensuring environmental justice for all? What three major changes, if any, would you make in this system?

2. Explain why you agree or disagree with the seven principles used in making environmental policy decisions listed on p. 716-717.

3. Rate the last four presidents of the United States (or leaders of the country where you live) on a scale of 1 to 10 in terms of their ability to act as a **(a)** leader and **(b)** manager.

4. Suppose that a presidential candidate ran on a platform calling for the federal government to phase in a tax on gasoline so that, over 5–10 years, the price of gasoline would rise to $5–7 a gallon (as is the case in Japan and most western European nations). The candidate argues that this tax increase is necessary to encourage oil and gasoline conservation, reduce air pollution, and enhance future economic, environmental, and military security. The candidate also says that the tax revenue should be used to **(1)** reduce income taxes on the poor and middle class by an amount roughly equal to the increase in gasoline taxes and **(2)** reduce taxes on wages and profits (Solutions, p. 707). Would you vote for this candidate who wants to triple the price of gasoline? Explain.

5. What are the pros and cons of using only public funds to finance all election campaigns? Explain why you support or oppose such an idea.

6. Explain why you agree or disagree with each of the four solutions given on p. 725 for leveling the legal playing field for citizens who have suffered environmental harm. Try to have an environmental lawyer and a corporate lawyer discuss their views of such ideas with your class.

7. Do you agree or disagree with the position that we need to place much more emphasis on environmental security? Should we treat it with the same degree of seriousness, analysis, and funding as we do economic security and military security, or should environmental security have higher or lower priority? Defend your answers.

8. Explain why you agree or disagree with the charges **(a)** that the World Trade Organization was set up primarily to create a true free market economy worldwide that is run by business interests through the WTO without interference from governments, citizens, or any other parties and **(b)** that its current rules weaken the ability of governments to protect the environment, the health of their citizens, and the health and safety of their workers.

9. Congratulations. You have been put in charge of the World Trade Organization. What five changes would you make in the way this organization operates and in the rules it uses to regulate international trade?

PROJECTS

1. A 1990 national survey by the Roper Organization found that even though 78% of Americans believe that a major national effort is needed for environmental improvement (ranking it fourth among national priorities), only 22% were making significant efforts to improve the environment. The poll identified five categories of citizens: **(1)** *true-blue greens* (11%) involved in a wide range of environmental activities, **(2)** *greenback greens* (11%) who do not have time to be involved but are willing pay more for a cleaner environment, **(3)** *grousers* (24%) who are not involved in environmental action, mainly because they do not see why they should be if everybody else is not involved, **(4)** *sprouts* (26%) who are concerned but do not believe individual action will make much difference, and **(5)** *basic browns* (28%) who strongly oppose the environmental movement. To which category do you belong? As a class, conduct a similar poll on your campus.

2. Have each member of your class select a particular environmental law (Appendix 5) and evaluate the law in terms of **(a)** its use of or failure to use the principles listed on p. 717 and **(b)** the role environmental organizations and citizen actions played in its development. Compare the results of these analyses.

3. Have each member of your class select a particular environmental legal case and evaluate its outcome in terms of the limiting factors discussed on pp. 724–725. Compare the results of these analyses.

4. What student environmental groups (if any) are active at your school? How many people actively participate in these groups? What environmentally beneficial things have they done? What actions (if any) taken by such groups do you disagree with? Why?

5. Try to interview **(a)** a lobbyist for an industry seeking to keep a specific environmental law from being strengthened, **(b)** a lobbyist for an environmental group seeking to strengthen the law, **(c)** an EPA official supporting strengthening the law, and **(d)** an elected representative who must make a decision about the law. Compare their views and perspectives and come to a conclusion about what should be done.

6. Have each member of your class use the library or the internet to learn about and evaluate the effectiveness of a different major environmental group.

7. Have each member of your class use the library or the internet to learn about and evaluate a different decision made by the World Trade Organization in determining whether a member nation violate one of its international trade rules.

8. Use the library or the internet to find bibliographic information about *Lord Kennet* and *Robert Fulghum*, whose quotes appear at the beginning and end of this chapter.

9. Make a concept map of this chapter's major ideas, using the section heads and subheads and the key terms (in boldface). Look at the inside back cover and on the website for this book for information about making concept maps.

INTERNET STUDY RESOURCES AND RESOURCES FOR FURTHER READING AND RESEARCH

The website for this book contains helpful study aids and many ideas for further reading and research. Log on to:

http://www.brookscole.com/product/0534376975s

and click on the Chapter-by-Chapter area. Choose Chapter 27 and select a resource:

- "Flash Cards" allows you to test your mastery of the Terms and Concepts to Remember for this chapter.

- "Tutorial Quizzes" provides a multiple-choice practice quiz.

- "Student Guide to InfoTrac" will lead you to Critical Thinking Projects that use InfoTrac College Edition as a research tool.

- "References" lists the major books and articles consulted in writing this chapter.

- "Hypercontents" takes you to an extensive list of sites with news, research, and images related to individual sections of the chapter.

INFOTRAC COLLEGE EDITION

Improve your skills with InfoTrac College Edition, a searchable online database of articles from more than 700 periodicals. Log on to:

http://www.infotrac-college.com

or access InfoTrac through the website for this book.

Try the following articles:

Pouyat, R.V. 1999. Science and environmental policy—making them compatible. *BioScience* vol. 49, no. 4, pp. 281–286. (keywords: science, environmental policy)

Solomon, B.D., and R. Lee. 2000. Emissions trade systems and environmental justice. *Environment* vol. 42, no. 8, pp. 32–45. (keywords: environmental justice)

28 ENVIRONMENTAL WORLDVIEWS, ETHICS, AND SUSTAINABILITY

Biosphere 2: A Lesson in Humility

In 1991, eight scientists (four men and four women) were sealed into Biosphere 2, a $200-million facility designed to be a self-sustaining life-support system designed to mimic the earth's nutrient-cycling systems (Figure 28-1).

The project, financed with private venture capital, was designed to (1) provide information and experience in designing self-sustaining stations in space or on the Moon or other planets and (2) increase our understanding of the earth's biosphere: Biosphere 1.

The 1.3-hectare (3.2-acre) closed and sealed system was built in the desert near Tucson, Arizona. It had a variety of ecosystems, each built from scratch. They included a tropical rain forest, lakes, a desert, streams, freshwater and saltwater wetlands, and a mini-ocean with a coral reef.

The facility was stocked with more than 4,000 species of plants, animals, and microorganisms selected to maintain ecosystem functions. It also had living quarters for its crew, who would get their food from intensive organic farming, raising a few goats and chickens, and fish farming in ponds and tanks. Energy was provided by sunlight and external natural gas-powered generators.

The Biospherians were supposed to be isolated for 2 years and to (1) raise their own food, (2) breathe air recirculated by plants, and (3) drink water cleansed by natural nutrient-cycling processes. From the beginning they encountered numerous unexpected problems.

The life-support system began unraveling. When some oxygen disappeared mysteriously, more had to be pumped in from the outside to keep the Biospherians from suffocating.

The nitrogen and carbon cycling systems also failed to function properly. Levels of nitrous oxide rose high enough to threaten the occupants with brain damage and had to be controlled by outside intervention. Carbon dioxide skyrocketed to levels that threatened to poison the humans and spurred the growth of weedy vines that choked out food crops. Nutrients leached from the soil and polluted the water systems.

Tropical birds disappeared after the first freeze. An Arizona ant species got into the enclosure, proliferated, and killed off all other soft-bodied insects. After the majority of the introduced insect species became extinct, the facility was overrun with cockroaches and katydids.

All together, 19 of the Biosphere's 25 small animal species became extinct. Before the 2-year period was up all plant-pollinating insects became extinct, thereby dooming to extinction most of the plant species. Scientists Joel Cohen and David Tilman, who evaluated the project, concluded, "No one yet knows how to engineer systems that provide humans with life-supporting services that natural ecosystems provide for free." In other words, an expenditure of $200 million failed to maintain a life-support system for eight people. The earth—Biosphere 1—does this every day for 6.1 billion people and millions of other species at no cost. If we had to pay for these services at the same annual cost of $12.5 million per person in Biosphere 2, the total bill for the earth's 6.1 billion people would be 1,900 times the annual world national product.

This world's largest ecological laboratory is now used by Columbia University's Lamont-Doherty Earth Observatory to carry out climate and ecological research.

Figure 28-1 Biosphere 2, constructed near Tucson, Arizona, was designed to be a self-sustaining life-support system for eight people sealed into the facility in 1991. The experiment failed because of a breakdown in its nutrient cycling systems. (Stone/Russell Kaye)

The main ingredients of an environmental ethic are caring about the planet and all of its inhabitants, allowing unselfishness to control the immediate self-interest that harms others, and living each day so as to leave the lightest possible footprints on the planet.

ROBERT CAHN

This chapter addresses the following questions:

- What human-centered environmental worldviews guide most industrial societies?
- What are some life-centered and earth-centered environmental worldviews?
- What ethical guidelines might we use to help us work with the earth?
- How can we live more sustainably?

28-1 ENVIRONMENTAL WORLDVIEWS IN INDUSTRIAL SOCIETIES

How Shall We Live? A Clash of Cultures and Values There are conflicting views about how serious our environmental problems are and what we should do about them. These conflicts arise mostly out of differing **environmental worldviews: (1)** how people think the world works, **(2)** what they think their role in the world should be, and **(3)** what they believe is right and wrong environmental behavior (**environmental ethics**).

People with widely differing environmental worldviews can take the same data, be logically consistent, and arrive at quite different conclusions (Appendix 2) because they start with different assumptions and values.

There are many different types of environmental worldviews, as summarized in Figure 28-2. Most can be divided into two groups according to whether they are *individual-centered* (atomistic) or *earth-centered* (holistic). Atomistic environmental worldviews tend to be *human-centered* (anthropocentric) or *life-centered* (biocentric, with the primary focus on either individual species or individual organisms). Holistic or ecocentric environmental worldviews are either *ecosystem-centered* or *biosphere* (life-support system)-*centered*.

What Is the Difference Between Instrumental and Intrinsic Values? How we act is determined largely by what we value. Environmental philosophers normally divide values into two types: instrumental and intrinsic. An **instrumental** or **utilitarian value** is a value something has because of its usefulness to us or to the biosphere, for example. An **intrinsic** or **inherent value** is the value something has just because it exists, regardless of whether it has any instrumental value to us.

Most of us believe that human beings have intrinsic value. However, there is much controversy over whether nonhuman forms of life and nature as a whole have intrinsic value. Many people view other species, forests, rocks and soil, rivers, biodiversity, and the biosphere as having instrumental value based on how useful they are to us. For example, the concept of preserving natural capital and biodiversity because they sustain life and support economies is an instrumental value based mostly on the usefulness of these natural goods and services to us. However, many people do not believe that the earth's nonhuman species, biotic communities, ecosystems, biodiversity, and the biosphere have intrinsic value and thus should be protected merely because they exist.

The view that a wild species, a biotic community ecosystem, biodiversity, or the biosphere has value only because of its usefulness to us is called an **anthropocentric** (human-centered) instrumental value. According to an *anthropocentric worldview*, **(1)** humans have intrinsic value, **(2)** the rest of nature has instrumental value, and **(3)** we are in charge of the earth and can act as its masters or caretakers.

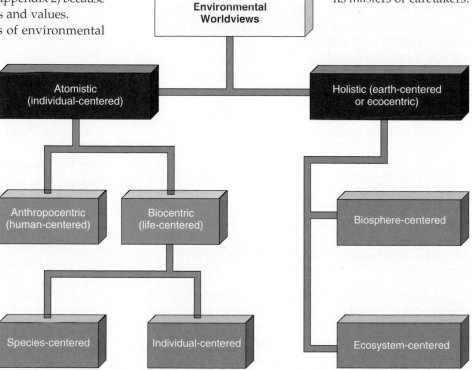

Figure 28-2 General types of environmental worldviews. (Diagram developed by Jane Heinze-Fry)

On the other hand, the view that these forms of life are valuable simply because they exist, independently of their use to human beings, is called a **biocentric** (life-centered) intrinsic value. According to a *biocentric worldview*, **(1)** all species and ecosystems and the biosphere have both intrinsic and instrumental value, **(2)** we are just one of many species, and **(3)** we have an ethical responsibility not to impair the long-term sustainability and adaptability of the earth's natural systems for all life.

What Are the Major Human-Centered Environmental Worldviews? Most people in today's industrial consumer societies have a **planetary management worldview**, which has become increasingly common during the past 50 years. According to this human-centered environmental worldview, human beings, as the planet's most important and dominant species, can and should manage the planet mostly for their own benefit. Other species and parts of nature are seen as having only *instrumental value* based on how useful they are to us.

The basic environmental beliefs of this worldview include the following:

- *We are the planet's most important species, and we are apart from and in charge of the rest of nature.* This idea crops up when people talk about "our" planet, "our" earth, or "saving the earth."

- *There is always more.* The earth has an essentially unlimited supply of resources for use by us through science and technology. If we deplete a resource, we will find substitutes. To deal with pollutants, we can invent technology to clean them up, dump them into space, or move into space ourselves. If we extinguish other species, we can use genetic engineering to create new and better ones.

- *All economic growth is good, and the potential for global economic growth is essentially limitless.*

- *Our success depends on how well we can understand, control, and manage the earth's life-support systems for our benefit.*

All or most aspects of worldview are widely supported because it is said to be the primary driving force behind the major improvements in the human condition since the beginning of the industrial revolution (Appendix 2, left).

There are several variations of this environmental worldview:

- The *no-problem school.* There are no environmental, population, or resource problems that cannot be solved by more economic growth and development, better management, and better technology.

- The *free-market school.* The best way to manage the planet for human benefit is through a free-market global economy with minimal government interference and regulations. Free-market advocates would convert all public property resources to private property resources and let the global marketplace, governed by pure free-market competition (p. 689), decide essentially everything.

- The *responsible planetary management school.* We have serious environmental problems, but we can sustain our species with a mixture of market-based competition, better technology, and some government intervention that **(1)** promotes environmentally sustainable forms of economic development, **(2)** protects environmental quality and private property rights, and **(3)** protects and manages public and common property resources. People holding this view follow the pragmatic principle of *enlightened self-interest:* Better earth care is better self-care.

- The *spaceship-earth school.* Earth is seen as a spaceship: a complex machine that we can understand, dominate, change, and manage to prevent environmental overload and provide a good life for everyone. This view developed as a result of photographs taken from space showing the earth as a finite planet or island "floating" in space (Figure 5-1, p. 102). This powerful image led many people to see that the earth is our only home and that we had better treat it right.

- The *stewardship school.* We have an ethical responsibility to be caring and responsible managers or stewards, of the earth. According to this view, we can and should make the world a better place for our species and other species through love, care, knowledge, and technology.

28-2 LIFE-CENTERED AND EARTH-CENTERED ENVIRONMENTAL WORLDVIEWS

Can We Manage the Planet? Some people believe that any human-centered worldview will eventually fail because it wrongly assumes that we now have (or can gain) enough knowledge to become effective managers or stewards of the earth.

These people argue that the unregulated global free-market approach will not work because it **(1)** is based on increased losses of natural capital (Figure 4-36, p. 99, and Figure 26-7, p. 694) that support all life and economies and **(2)** focuses on short-term economic benefits regardless of the harmful long-term environmental and social consequences (Guest Essay, right).

Nature and the Market Revolution

Donald Worster

Donald Worster is Hall Distinguished Professor of American History at the University of Kansas and one of the founders of the new field of environmental history. His books
include Nature's Economy: A History of Ecological Ideas; Dust Bowl: The Southern Plains in the 1930s; Rivers of Empire: Water, Aridity, and the Growth of the American West; The Wealth of Nature: Environmental History, the Ecological Imagination, *and a biography of scientist and explorer John Wesley Powell. He has held numerous fellowships, lectured widely in the United States and abroad, and served on the board of the Land Institute and other environmental organizations. He believes that a fuller understanding of the history of the human relationship to nature is essential to effective policy making and social change.*

Rapid economic growth over the past decade has produced a rush of good feeling. The rich industrial nations feel liberated from the gloomy mood of the 1960s and 1970s, when ominous signs of environmental collapse appeared at home and abroad. Political and economic leaders now regularly insist that there is no deep or real conflict between the ecology of the planet and the economy of industrial capitalism. But the conflict is real, and it will not go away soon.

Hope, uncritical and unbounded, has been a persistent emotion throughout the modern period. The market, or capitalist, revolution, which began in earnest in the 18th century, came on a floodtide of extravagant promises. To be sure, it spoke the language of hard-headed realism, practical problem solving, and careful bookkeeping, but underneath that language lay a powerful utopian impulse that would not die.

The new economics promised unlimited wealth and happiness. It maintained that humans are capable of mass-producing their way to earthly paradise. Even greed, which the Judeo-Christian religion and others had singled out as the root of all evil, went through a miraculous transformation into virtue. This new virtue would lead to new wealth—a wealth that would flow first to the most virtuous citizens and then to every other member of society.

It is not a big leap from that historical revolution in values and institutions to the current panaceas of "free-market environmentalism." Now we are told that the market economy offers the perfect solution to all resource depletion and pollution. If left unfettered, it will bring us greater efficiency and more benign substitutes: wood fibers that get recycled a hundred times or more, automobiles without tailpipes, and crops that need no pesticides.

Private property, we hear repeatedly, is the best remedy for the tragedy of the commons, in which community property relations supposedly leads to land abuse. Privatizing all land, including national parks and forests, as well as the more elusive water or the ubiquitous air will lead to environmental responsibility. More wealth, the promise goes, will turn everybody into environmentalists. As money accumulates, societies will save their endangered species or clean up their streams. Poor people do not care about the earth; wealthy people do.

Those beguiling claims are based on a too casual reading of modern history, which shows a more troubled, complicated relationship between environmentalism and economic growth. Environmental reform generally has not come from those who have been most successful in the market: the *Fortune* 500 top executives in the present and their counterparts in the past.

Like many reform movements, environmentalism has spread through increasing education and economic security. However, all along it has encountered unbending resistance within the business community, which even today bankrolls heavily the anti-environmental lobby. Environmentalists have tried to reform, restrain, or even overthrow the market economy because they have realized that its dominant ethos of acquisitive self-interest is a threat to the natural world.

A careful reading of history shows that capitalism was not created to save the earth; it was created to turn nature into wealth, as fast and thoroughly as possible. Only a utopian would think that it can now be turned easily from exploitation to preservation or that eventually, by its own dynamics, without change or reform, it will lead us into the Garden of Eden.

We did not reach our current condition of widespread ecological crisis and degradation by a single, universal cause. Yet one fact is striking: A vast acceleration in environmental change, much of it very damaging, began in the 18th century, exactly during the period when the market revolution was gathering momentum.

That revolution did not reach the United States in full force until 1830. From that point on, we can easily trace its bad effects. Air and water pollution began to show up, first in the larger cities, then across the country. Vast populations of plant and animal life diminished or even disappeared. Resource consumption shot up dramatically. Soils washed away at an unprecedented rate. Behind that acceleration of destructive change lay the rise of capitalism, its free-market ideology, and its cornucopian vision of endless wealth.

Since the work of economist Adam Smith, nature has not been included in the accounting system of capitalism [Figure 26-6, p. 693]. Capital has referred only to the money an entrepreneur could amass and invest. It has not included the natural capital of the planet, the diverse ecosystems on which all life depends [Figure 4-36, p. 99, and Guest Essay, p. 6].

Consequently, most people accumulating wealth do not believe that they owe anything to nature, to the vast and intricate web of life, and to future generations. That narrowing of value and obligation lies at the very core of modern economic culture. This culture has been astonishingly powerful and creative but also profoundly disruptive to both society and the natural world.

We can argue over whether the market revolution has, on balance, been good or bad for people [Appendix 2]. Some of its promises have been realized; others have not. We can see around us an extraordinary increase in consumer goods, from foodstuffs to telephones, spreading even into the poorest households. We can also see an increase in social ills, caused largely by huge disparities of income and power that have resulted.

But whether the revolution has been a curse or blessing for humanity, for nature it has been a disaster. The costs have far outweighed the benefits. After two centuries of growth, the world has lost an immense portion of its natural capital—soils, groundwater, forests, grasslands, and species—and is sure to lose more.

Largely because of the market revolution, the human population has increased more than sixfold; the impact of those numbers has turned large stretches of the planet into a wasteland. For all the material gains it has brought to our species, the market revolution is unsustainable ecologically.

The giddy mood we are experiencing today will not last, nor will the notion of the market, privatization, or business profit as a panacea. When the current mood passes, we will get back to hard reality and tough choices. It will become obvious once more that only a profound change in the nature of capitalism will make it more environmentally friendly—a change so profound as to constitute another revolution.

Critical Thinking

Explain why you agree or disagree with the viewpoint presented by the author of this essay.

The image of the earth as an island or spaceship in space has played an important role in raising global environmental awareness. However, critics argue that thinking of the earth as a spaceship that we should and can manage is an oversimplified and misleading way to view an incredibly complex and ever-changing planet, as the failure of Biosphere 2 demonstrated (p. 740). For example, these critics point out that we do not even know how many species live on the earth, much less what their roles are and how they interact with one another and their nonliving environment. We have only an inkling of what goes on in a handful of soil, a meadow, a patch of forest, a pond, or any other part of the earth.

As biologist David Ehrenfeld puts it, "In no important instance have we been able to demonstrate comprehensive successful management of the world, nor do we understand it well enough to manage it even in theory." Environmental educator David Orr (Guest Essay, p. 683) says we are losing rather than gaining the knowledge and wisdom needed to adapt creatively to continually changing environmental conditions: "On balance, I think, we are becoming more ignorant because we are losing cultural knowledge about how to inhabit our places on the planet sustainably, while impoverishing the genetic knowledge accumulated through millions of years of evolution."

Even if we had enough knowledge and wisdom to manage spaceship earth, some critics see this approach as requiring us to give up individual freedom to survive. Life on spaceship earth under a comprehensive system of planetary management or world government might be very much like the regimented life of astronauts in their capsule. The astronauts have almost no individual freedom; essentially all of their actions are dictated by a central command (ground control).

What Are Some Major Biocentric and Ecocentric Worldviews? People disagree over how far we should extend our ethical concerns for various forms or levels of life (Figure 28-3). Critics of human-centered environmental worldviews believe that such worldviews should be expanded to recognize the *inherent value* of all forms of life. According to this view, all species are part of a community of living things and have just as much right to exist as humans do. In other words, each species has *intrinsic value* unrelated to its potential or actual use to us.

Most people with a life-centered (biocentric) worldview believe that we have an ethical responsibility to (1) not cause the premature extinction of a species and (2) actively protect a species from going extinct because of our activities.

Some people believe that each individual organism, not just each species, has an inherent right to survive. Some apply this only to animal species and others only to animal species that are believed to be capable of having feelings.

However, most people give protection of a species priority over protection of an individual member of a species. The reason is that when an individual of a species dies, another one replaces it, but when a species becomes biologically extinct, its genetic lineage ends. Each species is a unique storehouse of genetic information that **(1)** should be respected and protected because it exists (intrinsic value), **(2)** is a potential economic good for human use (instrumental value), and **(3)** is capable through evolution and speciation of adapting to changing environmental conditions. In this sense, **(1)** individuals are temporary representatives of a species, and **(2)** the premature extinction of a species can be regarded as killing future generations of a species and eliminating its possibility for future evolutionary adaptation or speciation.

Trying to decide whether all or only some species should be protected from premature extinction resulting from human activities is a difficult and controversial ethical problem. It is hard to know where to draw the line and be ethically consistent. For example:

- Should all species be protected from premature extinction because of their intrinsic value, or should only certain ones be preserved because of their known or potential instrumental value to us or to their ecosystems?

- Should all insect and bacterial species be protected, or should we attempt to exterminate those that eat our crops, harm us, or transmit disease organisms?

- Should we emphasize protecting keystone species (p. 178) in ecosystems over other species that play lesser ecological roles?

Others believe that we must go beyond this biocentric worldview, which focuses on species and individual organisms. They contend that it is not just a species that we should preserve (for example, in a zoo or arboretum) but a species in a functioning ecosystem (Figure 28-3). They believe that we have an ethical responsibility not to degrade the earth's ecosystems, biodiversity, and life-support systems (biosphere) for this and future generations of humans (Connections, p. 746) and other forms of life based on their intrinsic and instrumental values. In other words, they have an *earth-centered*, or *ecocentric*, environmental worldview, devoted to preserving the earth's biodiversity and the functioning of its life-support systems (Figure 28-3) for all forms of life.

Why should we care about the earth's biodiversity? According to environmentalist and systems expert Donella Meadows.

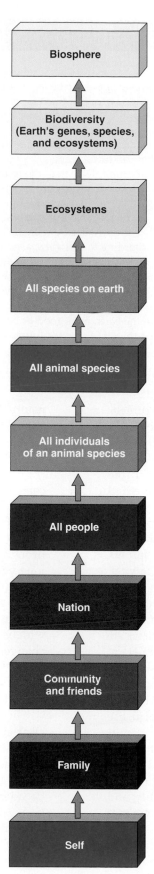

Figure 28-3 Levels of ethical concern. People disagree over how far we should extend out ethical concerns on this scale.

Biodiversity contains the accumulated wisdom of nature and the key to its future. If you wanted to destroy a society, you would burn its libraries and kill its intellectuals. You would destroy its knowledge. Nature's knowledge is contained in the DNA within living cells. The variety of genetic information is the driving engine of evolution and the source of adaptability.

According to the ecocentric worldview, we are part of, not apart from, the community of life and the ecological processes that sustain all life. Aldo Leopold (p. 39) summed up this idea in 1948: "All ethics rest upon a single premise: that the individual is a member of a community of interdependent parts."

There are many life-centered and earth-centered environmental worldviews, and several of them overlap in some of their beliefs. One ecocentric environmental worldview is the **environmental-wisdom worldview**. It is based on the following major beliefs, which are the opposite of those making up the planetary management worldview:

- *We are part of nature, and nature does not exist just for us.* We need the earth, but the earth does not need us (Guest Essay, p. 750). Each species has an inherent right to exist.

- *There is not always more.* The earth's resources are limited, should not be wasted, and should be used efficiently and sustainably for us and all other species.

- *Some forms of technology and economic growth are environmentally beneficial and should be encouraged, but some are environmentally harmful and should be discouraged.*

- *Our success depends on **(1)** learning how the earth sustains itself and adapts to ever-changing environmental conditions and **(2)** integrating such scientific lessons from nature (environmental wisdom) into the ways we think and act* (Solutions, p. 209).

A related ecocentric environmental worldview is the *deep ecology worldview,*

Some people believe that our only ethical obligation is to the present human generation. They ask, "What has the future done for me?" or believe that we cannot know enough about the condition of the earth for future generations to be concerned about it.

According to biologist David W. Ehrenfeld, caring about future generations enough not to degrade the earth's life-support systems is important because it gives future generations options for dealing with the problems they will face. He points out that if our ancestors had left for us the ecological degradation we appear to be leaving our descendants, our options for enjoyment—perhaps even for survival—would be quite limited.

And in response to the question, "What can future generations do for us?" Ehrenfeld gives the following answer: "They give us a reason for treating our ecological home respectfully, so that our lives as well as theirs will be enriched."

According to this view, as we use the earth's natural resources we **(1)** are borrowing from the earth and from future generations and **(2)** have an ethical responsibility to leave the earth in as good or better shape than it is now for future generations. In thinking about our responsibility toward future generations, some analysts believe that we should consider the wisdom given to us in the 18th century by the Iroquois Confederation of Native Americans: *In our every deliberation, we must consider the impact of our decisions on the next seven generations.*

Critical Thinking

What obligations, if any, concerning the environment do you have to future generations? To how many future generations do you have responsibilities? Be honest about your feelings.

---

which recognizes both the intrinsic and instrumental values of species, ecosystems, and the biosphere (Spotlight, p. 748).

These and other life-centered and earth-centered environmental worldviews have their roots in the ways of life of many primal peoples that have had respect for other forms of life. Earth-wisdom principles have also been articulated by Saint Francis of Assisi, Benedict de Spinoza, Henry David Thoreau, Ralph Waldo Emerson, John Muir, Aldo Leopold (p. 39), Rachel Carson (p. 36), Alan Watts, Gary Snyder, Charles Reich, Theodore Roszak, Arne Naess (developer of what is called the *deep ecology worldview*, Spotlight, p. 748), Bill Devall, George Sessions, and many others. Many of the basic ideas of these worldviews are also found in teachings of Hinduism, Taoism, Islam, Confucianism, Buddhism, and the Judeo-Christian tradition. Various biocentric and ecocentric environmental worldviews also emphasize beliefs found in the ecofeminism, social ecology, and environmental justice movements (Guest Essay, p. 727).

Much of modern ecocentric ethics rests on the principles developed in the 1940s by Aldo Leopold (p. 39). One his major ethical principles is, "A thing is right when it tends to preserve the integrity, stability, and beauty of the biotic community. It is wrong when it tends otherwise."

Some analysts point out that shifts in the principles of contemporary ecology may invalidate Leopold's land ethic or require that it be updated. According to modern ecological theory, instead of a static "balance of nature" that should be conserved and preserved, nature is in dynamic, ever-changing state of flux. If such a model is correct, it is not possible to preserve the integrity and stability of a biotic community.

Some ecologists say that the way out of this is to consider the time period, or *temporal scale*, and the size, or *spatial scale*, being affected by human ecological impacts. Violent natural disturbances such as volcanic eruptions, tornadoes, hurricanes, fires, floods, and droughts occur every now and then and took place long before the human species appeared on the evolutionary scene.

In general, these natural disturbances occur infrequently and affect small areas of the earth's surface in widely spaced areas. Natural global and regional changes such as climate change and the spread and retreat of glaciers that can affect large areas of the earth's surface take place slowly on a human time scale, over thousands to millions of years.

In contrast, human ecological disturbances such as clear-cutting, agriculture, urban development, and overharvesting of fish and other natural resources generally are far more frequent and take place over a much larger area (Figure 1-4, p. 8) and on a time scale of decades to hundreds of years. In other words, **(1)** the human *temporal scale* of environmental change is much faster than nature's temporal scale, and **(2)** the *spatial scale* of our activities is much larger than nature's spatial scale. Thus, an updating of Leopold's basic land ethic based on the modern ecological concepts might be, "*A thing is right when it tends to disturb a biotic com-*

munity at nature's normal temporal and spatial scales. It is wrong when it tends otherwise."

Others say we do not need to be biocentrists or ecocentrists to value life or the earth because human-centered stewardship and planetary management environmental worldviews also call for us to value individuals, species, and the earth's life-support systems as part of our responsibility as earth's caretakers. What do you think?

What Is the Ecofeminist Worldview? The term *ecofeminism*, coined in 1974 by French writer Francoise d'Eaubonne, includes a spectrum of views on the relationships of women to the earth and to male-dominated societies (patriarchies). Although most ecofeminists agree that we need a life-centered or earth-centered environmental worldview, they believe that a main cause of our environmental problems is not just human-centeredness, but specifically male-centeredness (*androcentrism*).

Many ecofeminists argue that the rise of male-dominated societies and environmental worldviews since the advent of agriculture is primarily responsible for our violence against nature (and for the oppression of women and minorities as well). To such ecofeminists, this led to a shift from an image of nature as a nurturing mother to a foe to be conquered.

As evidence of male domination, ecofeminists note that women earn less than 10% of all wages, own less than 1% of all property, and in most societies have far fewer rights than men. These analysts argue that to become primary players in the male power-and-domination game, most women are forced to emphasize the characteristics deemed masculine and become "honorary men."

Some ecofeminists suggest that oppression by men has driven women closer to nature and made them more compassionate and nurturing. As oppressed members of society, they argue, women have more experience in **(1)** dealing with interpersonal conflicts, **(2)** bringing people together, **(3)** acting as caregivers, and **(4)** identifying emotionally with injustice, pain, and suffering.

Ecofeminists argue that women should be **(1)** given the same rights as men, **(2)** allowed to have their views heard and respected, and **(3)** treated as equal partners. They do not want just a fair share of the patriarchal pie; they want to work with men to bake an entirely new pie that helps heal the rift between humans and nature and ends oppression based on sex, race, class, and cultural and religious beliefs. In doing this, they do not want to be given token roles or co-opted into the male power game.

Ecofeminists are not alone in calling for us to encourage the rise of *life-centered people* who emphasize the best human characteristics: gentleness, caring, compassion, nonviolence, cooperation, and love.

What Is the Social Ecology Worldview? According to anarchist philosopher Murray Bookchin, the ecological crisis we currently face results from the power of our hierarchical and authoritarian social, economic, and political structures and from the various technologies used to dominate people and nature. In other words, our current environmental situation has been created by industrialized societies driven by the conventional planetary management environmental worldview (p. 742).

To alleviate the ecological crisis, Bookchin believes we must adopt a *social ecology environmental worldview*, which would involve decentralizing political and economic systems and corporations and changing the types of technology we use. Bookchin urges us to create **(1)** better versions of democratic communities, **(2)** new forms of earth-sustaining production, and **(3)** types of appropriate ecotechnology that are smaller in scale, consume fewer resources and less earth capital, and do not cause environmental degradation of local ecological regions (Solutions, p. 696).

Are There Really Physical and Biological Limits to Human Economic Growth? The planetary management and earth-wisdom worldviews (and related versions of these two types of worldviews) differ over whether there are physical and biological limits to economic growth, beyond which both ecological and economic collapse is likely to occur.

This argument over limits has been going on since Thomas Malthus published his book *The Principles of Political Economy* in 1836, with each side insisting that it is right at least in the long term. Some economists believe that that there are no limits to economic growth and that environmental problems either are not serious or can be dealt with by technological innovation. They appeal to people's natural desire to believe that everything will be OK—be happy, don't worry. Some environmentalists believe that there are ecological limits to economic growth, and they appeal to people's rational desire to avoid a crash as a way to motivate political change to stop environmental destruction.

In 2000, conservation biologist Carlos Davidson (with a background in economics) proposed a way to bridge the gap between these two opposing viewpoints and help motivate the political changes needed to halt the spread of environmental degradation.

He strongly disagrees with the view of some economists that technology will allow continuing economic growth without causing serious environmental damage. However, he also disagrees with the concept that continuing economic growth based on consuming

Deep Ecology

Deep ecology consists of the following ecocentric beliefs developed in 1972 by Norwegian philosopher Arne Naess, in conjunction with philosopher George Sessions and sociologist Bill Devall:

- The well-being and flourishing of human and nonhuman life on earth have inherent value. These values are independent of the usefulness of the nonhuman world for human purposes.

- The fundamental interdependence, richness, and diversity of life forms contribute to the flourishing of human and nonhuman life on earth.

- Humans have no right to reduce this interdependence, richness, and diversity except to satisfy vital needs.

- Present human interference with the nonhuman world is excessive, and the situation is worsening rapidly.

- Because of the damage caused by human interference in the nonhuman world, it would be better for humans, and much better for nonhumans, if there were a substantial decrease in the human population.

- Basic economic, technological, and ideological policies must therefore be changed.

- The ideological change is mainly that of appreciating *life quality* (involving situations of inherent value) rather than adhering to an ever-higher material standard of living.

- Those who subscribe to these points have an obligation directly or indirectly to try to implement the necessary changes.

Naess has also described some lifestyle guidelines compatible with the basic beliefs of deep ecology. They include (1) appreciating all forms of life, (2) protecting or restoring local ecosystems, (3) consuming less, (4) emphasizing satisfying vital needs rather than

desires, (5) attempting to live *in* nature and promote community, (6) appreciating ethnic and cultural differences, (7) working to improve the standard of living for the world's poor, (8) working to eliminate injustice toward fellow humans or other species, and (9) acting nonviolently.

Deep ecology is not an ecoreligion, nor is it antireligious or antihuman, as some of its critics have claimed. Instead, it is a set of beliefs designed to have us think more deeply about the inherent value of all life on the earth and about our obligations toward both human and nonhuman life.

Critical Thinking

1. Which, if any, of the eight basic beliefs of the deep ecology environmental worldview do you agree with? Explain.

2. List five major changes that would occur in your life if you and most people lived by the beliefs of deep ecology.

---

and degrading natural capital will lead to ecological and economic crashes on both ecological and political grounds.

Instead of crashes, he suggests that we use the metaphor of the gradual unraveling of some of the threads in a woven *tapestry* to describe the effects of environmental degradation. Nature is viewed as an incredibly diverse and interwoven tapestry of threads consisting of a variety of patterns (biomes, aquatic systems, and ecosystems). As environmental degradation removes threads from different parts of the biosphere, the overall integrity of the tapestry in these areas is reduced, and it becomes worn and tattered.

Clearly, if too many of the threads are removed the tapestry can be destroyed. However, he contends that as threads are pulled from various parts of the tapestry, there are multiple local losses of ecological function and occasional local and regional tears rather than overall collapse. Ecological crashes such as the collapse of a fishery are (1) rare, (2) local or regional rather than global, and (3) can recover eventually if the stresses are removed. Instead of clear disaster thresholds, there

is a continuum of environmental degradation. Global problems such as climate change, ozone depletion, and biodiversity loss may be exceptions to this analogy. However, even with these problems some areas of the tapestry are damaged more than other areas.

Davidson strongly agrees with the views of ecologists and environmentalists that degradation in parts of the earth's ecological tapestry (1) is occurring, (2) is spreading, and (3) must be prevented. He supports doing this by using the pollution prevention and precautionary principles and protecting areas of the biosphere where damage is not severe from further damage.

However, he believes that using catastrophe metaphors such as "ecological collapse" and "going over a cliff" can hinder achievement of these important goals. He argues that repeated predictions of catastrophe, like the boy who cries wolf, initially motivate people's concern. However, when the threats turn out to be less severe than predicted, people tend to ignore future warnings. As a result, they are not politically motivated to (1) prevent more tears in nature's tapestry and (2) look more deeply at the economic, political, and social forces that are responsible for environmental degradation.

Biologist Kevin S. McCann at McGill University in Montreal, Canada, contends that the tapestry metaphor assumes that nature is in a static equilibrium state when in fact it undergoes constant dynamic change. He points out that tearing nature's fabric sends "waves of dynamic change through an ecosystem, and the waves get bigger as biological diversity declines."

28-3 SOLUTIONS: LIVING SUSTAINABLY

How Should We Evaluate Sustainability Proposals? *Sustainability* has become a buzzword that means different things to different people. Lester Brown (Guest Essay, p. 18) has this simple test for any sustainability proposal: "Does this policy or action lower carbon emissions? Does it reduce the generation of toxic wastes? Does it slow population growth? Does it increase the earth's tree cover? Does it cut emissions of ozone-depleting chemicals? Does it reduce pollution? Does it reduce radioactive waste generation? Does it lead to less soil erosion? Does it protect the planet's biodiversity?"

Additional questions also help us evaluate sustainability proposals. Does it deplete the earth's natural capital? Does it entail getting most of our energy from current sunlight instead of ancient sunlight stored as fossil fuels? Does it promote full-cost pricing of goods and services (p. 697)? Does it diminish cultural diversity? Does it reduce poverty, hunger, and disease? Does it promote individual and community self-reliance? Does it prevent pollution? Does it reduce resource waste? Does it save energy? Does it transfer the most resource-efficient and environmentally benign technologies to developing countries? Does it keep wealth in the local community? What are its true costs, and who pays? Does it enhance environmental and economic justice for all?

Solutions: What Are Some Ethical Guidelines for Working with the Earth? Ethicists and philosophers have developed a variety of ethical guidelines for living more sustainably or lightly on the earth. Such guidelines can be used by anyone, regardless of their environmental worldview.

Biosphere and Ecosystems

- We should try to understand and work with the rest of nature to help sustain the natural capital, biodiversity, and adaptability of the earth's life-support systems.

- When we alter nature, we should do it no faster and not over a larger area than the typical time and spatial scales of the earth's natural changes.

- When we alter nature to meet our needs or wants, we should carefully evaluate our proposed actions and choose methods that do the least possible short- and long-term environmental harm. This ethical concept of *ahimsa*, or avoiding unnecessary harm, is a key element of Hinduism, Buddhism, and Janism.

Species and Cultures

- We should work to preserve as much of the earth's genetic variety as possible because it is the raw material for all future evolution, speciation, and genetic engineering.

- We have the right to defend ourselves against individuals of species that do us harm and to use individuals of species to meet our vital needs, but we should strive not to cause the premature extinction of any wild species.

- The best ways to protect species and individuals of species are to protect the places where they live and to help restore places we have degraded.

- No human culture should become extinct because of our actions.

Individual Responsibility

- We should not inflict unnecessary suffering or pain on any animal we raise or hunt for food or use for scientific or other purposes.

- We should leave the earth as good as or better than we found it.

- We should use no more of the earth's resources than we need.

- We should work with the earth to help heal ecological wounds we have inflicted.

In March 2000, the Earth Charter was finalized. More than 100,000 people in 51 countries and 25 global leaders in environment, business, politics, religion, and education took part in creating this charter. It is a document creating an ethical and moral framework to guide the conduct of people and nations to each other and to the earth. Its four guiding principles are to:

- Respect earth and life in all its diversity.

- Care for life with understanding, love, and compassion.

- Build societies that are free, just, participatory, sustainable, and peaceful.

Envisioning a Sustainable Society

Lester W. Milbrath

Lester W. Milbrath is director of the Research Program in Environment and Society and professor emeritus of political science and sociology at the State University of New York at Buffalo. During his distinguished career he has served as director of the Environmental Studies Center at SUNY/Buffalo (1976–87) and taught at Northwestern University, Duke University, the University of Tennessee, National Taiwan University in Taipei, and Aarhus University in Denmark. He has also been a visiting research scholar at the Australian National University and at Mannheim University in Germany. His research has focused on the relationships between science, society, and citizen participation in environmental policy decisions, with emphasis on environmental perceptions, beliefs, attitudes, and values. He has written numerous articles and books. His book Envisioning a Sustainable Society: Learning Our Way Out *(1989) summarizes a lifetime of studying our environmental predicament. It is considered one of the best analyses of what we can do to learn how to work with the earth. His most recent book is* Learning to Think Environmentally While There Is Still Time *(1995).*

GUEST ESSAY

The 1992 Earth Summit at Rio popularized the goal of sustainable development. Most of the heads of state meeting there believed that goal could be achieved by developing better technology and by writing better laws, agreements, and treaties, and enforcing them. Unfortunately, their approach was flawed and will not achieve sustainability because they do not understand the nature of the crisis in our earthly home.

Try this thought experiment: Imagine that, suddenly, all the humans disappeared, but all the buildings, roads, shopping malls, factories, automobiles, and other artifacts of modern civilization were left behind. What then? After three or four centuries, buildings would have crumbled, vehicles would have rusted and fallen apart, and plants would have recolonized fields, roads, parking lots, even buildings. Water, air, and soil would gradually clear up; some endangered species would flourish. Nature would thrive splendidly without us.

That mental experiment makes it clear that we do not have an environmental crisis; we have a crisis of civilization. Heads of state meeting at the Earth Summit neither understood nor dealt with civilization's most crucial

problems: **(1)** Humans are reproducing at such epidemic rates that world population could rise to 9 billion by 2054, **(2)** resource depletion and waste generation could easily triple or even quadruple over that period, **(3)** waste discharges are already beginning to change the way the biosphere works, and **(4)** climate change and ozone loss will reduce the productivity of ecosystems just when hordes of new humans will be looking for sustenance and will destroy the confidence people need to invest in the future.

Without intending to, we have created a civilization that is headed for destruction. Either we learn to control our growth in population and in economic activity, or nature will use death to control it for us.

Present-day society is not capable of producing a solution because it is disabled by the values our leaders constantly trumpet: economic growth, jobs, consumption, competitiveness, power, and domination. Societies pursuing these goals cannot avoid depleting their resources, degrading nature, poisoning life with wastes, and upsetting biospheric systems. *We have no choice but to change, and resisting change will make us victims of change.*

But how do we transform to a sustainable society? My answer, which I believe is the only answer, is that *we must learn our way.* Nature, and the imperatives of its laws, will be our most powerful teacher as we learn our way to a new society. Most crucially, we must learn how to think about values.

Life in a viable ecosystem must become the core value of a sustainable society; that means all life, not just human life. Ecosystems function splendidly without humans, but human society would die without viable ecosystems. People seeking life quality need a well-functioning society living in well-functioning ecosystems. We must give top priority to the ecosystems that support us and second priority to human societies.

A sustainable society would affirm love as a primary value and extend it not only to those near and dear but to people in other lands, to future generations, and to other species. A sustainable society emphasizes partnership rather than domination, cooperation over competition, love over power. A sustainable society affirms justice and security as primary values.

A sustainable society would encourage self-realization—helping people to become all they are capable of being, rather than spending and consuming—as the key to a fulfilling life. A sustainable society would make

- Secure earth's bounty and beauty for present and future generations.

What Is Earth Education? Most environmentalists believe that learning how to live sustainably takes a foundation of environmental or earth education that relies heavily on an interdisciplinary and holistic

approach to learning (Guest Essay, above). Achieving such ecological literacy means having the ability to synthesize and connect knowledge from a variety of disciplines to see the big picture (holistic thinking).

According to its proponents, the most important goals of such an education are to:

- *Develop respect or reverence for all life.*

long-lasting products to be cherished and conserved. People would learn a love of beauty and simplicity.

A sustainable society would use both planning and markets as basic and supplementary information systems. Markets fail us because they can neither anticipate the future nor make moral choices between objects and between policies. Markets also cannot provide public goods such as schools, parks, and environmental protection, which are just as important for life quality as private goods.

A sustainable society would continue further development of science and technology because we need practical creative solutions that are both environmentally sound and economically feasible. However, we should recognize that those who control science and technology could use them to dominate all other creatures; we must learn to develop social controls of science and technology to make our society more sustainable. We should not allow the deployment of powerful new technologies that can induce sweeping changes in economic patterns, lifestyles, governance, and social values without careful forethought regarding their long-term impacts.

Conscious social learning would become the dynamic of social change in a sustainable society, not only to deal with pressing problems but also to realize a vision of a good society. Meaningful and lasting social change occurs when nearly everyone learns the necessity of change and the value of working toward it.

Ecological thinking is different from most thinking that guides modern society. For example, the following key maxims derived from the law of conservation and matter [p. 60], the laws of energy or thermodynamics [p. 65], and the workings of ecosystems [Chapters 4-10] are routinely violated in contemporary thinking and discourse: **(1)** *Everything must go somewhere (there is no away)*, **(2)** *energy should not be wasted because all use of energy produces disorder in the environment*, **(3)** *we can never do just one thing (everything is connected)*, and **(4)** *we must constantly ask, "And then what?"*

Every schoolchild and every adult should learn these simple truths; we need to reaffirm the tradition that knowledge of nature's workings and a respect for all life are basic to a true education. We should require such environmental education of all students, just as we now require every student to study history.

Ecological thinking recognizes that a proper understanding of the world requires people to learn how to think holistically, systematically, and futuristically. Because everything is connected to everything else, we must learn to anticipate second-, third-, and higher-order consequences for any contemplated major societal action.

A society learning to be sustainable would redesign government to maximize its ability to learn. It would require that people who govern listen to citizens, not only to keep the process open for public participation but also to cultivate mutual learning between officials and citizens.

A sustainable society would strive for an effective system of planetary politics because our health and welfare are vitally affected by how people, businesses, and governments in other lands behave. It would nurture planetwide social learning.

Learning our way to a new society cannot occur until enough people become aware of the need for major societal change. As long as contemporary society is working reasonably well and leaders keep telling us that society is on the right track, most people will not listen to a message urging significant change. For that reason, urgently needed change probably will be delayed, and conditions on our planet are likely to get worse before they can get better. Nature will be our most powerful teacher, especially when biospheric systems no longer work the way they used to. In times of great system turbulence, social learning can be extraordinarily swift.

Our species has a special gift: the ability to recall the past and foresee the future. Once we have a vision of the future, every decision becomes a moral decision. Even the decision not to act becomes a moral judgment. Those who understand what is happening to the only home for us and other species are not free to shrink from the responsibility to help make the transition to a more sustainable society.

Critical Thinking

1. Do you agree or disagree that we can only learn our way to a sustainable society? Explain.

2. Do you think we will learn our way to a sustainable society? Explain. What role, if any, do you intend to play in this process?

- *Understand as much as we can about how the earth works and sustains itself, and use such knowledge* (Solutions, p. 209) *to guide our lives, communities, and societies.*

- *Understand as much as we can about connections and interactions.* This includes those **(1)** within nature, **(2)** between people and the rest of nature, **(3)** between people with different cultures and beliefs, **(4)** between generations, **(5)** between the problems we face, and **(6)** between the solutions to these problems.

- *Use critical thinking skills* (Guest Essay, p. 48) *to become wisdom seekers instead of vessels of information*

- *Understand and evaluate one's worldview and see this as a lifelong process* (Individuals Matter, p. 752).

Mindquake: Evaluating One's Environmental Worldview

INDIVIDUALS MATTER

Questioning and perhaps changing one's environmental worldview can be difficult and threatening. It can set off a cultural *mindquake* that involves examining many of one's most basic beliefs. However, once people change their worldviews, it no longer makes sense for them to do things in the old ways. If enough people do this, then tremendous cultural change, once considered impossible, can take place rapidly.

Most environmentalists urge us to think about what our basic environmental beliefs are and why we have them. They believe that evaluating our beliefs, and being open to the possibility of changing them, should be one of our most important lifelong activities.

As this book emphasizes, most environmental issues are filled with controversy and uncertainty. A clearly right or wrong path is not easy to discover and usually is strongly influenced by one's environmental worldview. As philosopher Georg Hegel pointed out nearly two centuries ago, tragedy is not the conflict between right and wrong, but the conflict between right and right.

Critical Thinking

What are the basic beliefs of your current environmental worldview?

- *Learn how to evaluate the beneficial and harmful consequences of one's lifestyle and profession on the earth, today and in the future.*

- *Use critical thinking skills to evaluate advertising that encourages us to buy more and more things.* As humorist Will Rogers put it, "Too many people spend money they haven't earned to buy things they don't want, to impress people they don't like." David Orr points out, "Our children, consumers-in-training, can identify more than a thousand corporate logos but only a dozen or so plants and animals native to their region." The average U.S. TV viewer watches 22,000 commercials per year. Each year U.S. corporations spend more than $150 billion on advertising, far more than is spent on all secondary education in the country.

- *Foster a desire to make the world a better place and act on this desire.* As David Orr puts it, education should help students "make the leap from 'I know' to 'I care' to 'I'll do something.'"

According to environmental educator Mitchell Thomashow, four basic questions should be at the heart of environmental education:

- Where do the things I consume come from?

- What do I know about the place where I live?

- How am I connected to the earth and other living things?

- What are my purpose and responsibility as a human being?

How we answer these questions determines our *ecological identity*.

In addition to formal education, some analysts believe that we need to experience nature directly to help us learn to walk more lightly on the earth (Connections, right).

How Can We Live More Simply? Many analysts urge us to *learn how to live more simply*. Although seeking happiness through the pursuit of material things is considered folly by almost every major religion and philosophy, it is preached incessantly by modern advertising.

Some affluent people in developed countries are adopting a lifestyle of *voluntary simplicity*, doing and enjoying more with less by learning to live more simply. Voluntary simplicity is based on Mahatma Gandhi's *principle of enoughness:* "The earth provides enough to satisfy every person's need but not every person's greed….When we take more than we need, we are simply taking from each other, borrowing from the future, or destroying the environment and other species."

Implementing this principle means asking oneself, "How much is enough?" This is not an easy thing to do because people in affluent societies are conditioned to want more and more, and they often think of such wants as vital needs (Spotlight, p. 754).

Voluntary simplicity begins by asking a series of questions before buying anything: **(1)** Do I really need this? **(2)** Can I buy it secondhand (reuse)? **(3)** Can I borrow, rent, lease, or share it? **(4)** Can I build it myself?

The decision to buy something then triggers another set of questions: **(1)** Is the product produced in an environmentally sustainable manner? **(2)** Did the workers producing it get fair wages for their work, and did they have safe and healthful working conditions? **(3)** Is it designed to last as long as possible? **(4)** Is it easy to repair, upgrade, reuse, and recycle?

Choosing voluntary simplicity means **(1)** spending less time working for money, **(2)** leading lives less driven to accumulate stuff, and **(3)** spending more time living. Instead of seeing this as a sacrifice, those practicing voluntary simplicity view it as a more satisfying, meaningful, and sustainable way to live.

Learning from the Earth

Formal earth education is important, but many earth thinkers believe that it is not enough. They urge us to take the time to escape the cultural and technological body armor we use to insulate ourselves from nature and to experience nature directly.

They suggest that we reenchant our senses and kindle a sense of awe, wonder, and humility by standing under the stars, sitting in a forest, taking in the majesty and power of an ocean, or experiencing a stream, lake, or other part of nature.

We might pick up a handful of soil and try to sense the teeming microscopic life in it that keeps us alive. We might look at a tree, mountain, rock, or bee and try to sense how they are a part of us, and we a part of them as interdependent participants in the earth's life-sustaining recycling processes.

Earth thinker Michael J. Cohen suggests that we recognize who we really are by saying,

I am a desire for water, air, food, love, warmth, beauty, freedom, sensations, life, community, place, and spirit in the natural world….I have two

mothers: my human mother and my planet mother, Earth. The planet is my womb of life.

According to theologian Thomas Berry,

We need to be outdoors, to see the clouds, to feel the rain, to run across meadows. The wild expands the human soul….No being is nourished by itself. Everything is nourished by something outside of itself.

Many psychologists believe that consciously or unconsciously we spend much of our lives in a search for roots: something to anchor us in a bewildering and frightening sea of change. As philosopher Simone Weil observed, "To be rooted is perhaps the most important and least recognized need of the human soul."

Earth philosophers say that to be rooted, each of us needs to find a *sense of place:* a stream, a mountain, a yard, a neighborhood lot, or any piece of the earth we feel at one with as a place we know, experience emotionally, and love. It can be a place where we live or a place we occasionally visit and experience in our inner being. When we become part of a place, it becomes a part of us. Then we are driven to

defend it from harm and to help heal its wounds.

According to David Orr (Guest Essay, p. 683), "It is possible for us to live well without consuming the world's loveliness along with our children's legacy. But we must be inspired to act by examples that we can see, touch, and experience."

To many earth thinkers, emotionally experiencing our connectedness with the earth leads us to recognize that the healing of the earth and the healing of the human spirit are one and the same. They call for us to discover and tap into what Aldo Leopold calls "the green fire that burns in our hearts" and use this as a force for respecting and working with the earth and with one another.

Critical Thinking

Some analysts believe that learning earth wisdom by experiencing the earth and forming an emotional bond with its life forms and processes is unscientific, mystical poppycock based on a romanticized view of nature. They believe that better scientific understanding of how the earth works and improved technology are the only ways to achieve sustainability. Do you agree or disagree? Explain.

Surveys by the Trends Research Institute found that the shift toward voluntary simplicity was one of the top trends in America during 1997 and estimated that 12–15% of adult Americans are participating in this social trend. Another survey by the Harwood Group found that between 1992 and 1997, 28% of the respondents had voluntarily reduced their incomes as part of a change in their personal priorities. Later surveys indicate that this trend is increasing.

Voluntary simplicity by those who have more than they need should not be confused with the *forced simplicity* of the poor, who do not have enough to meet their basic needs for food, clothing, shelter, clean water and air, and good health.

After a lifetime of studying the growth and decline of the world's human civilizations, historian Arnold Toynbee summarized the true measure of a civilization's

growth in what he called the *law of progressive simplification:* "True growth occurs as civilizations transfer an increasing proportion of energy and attention from the material side of life to the nonmaterial side and thereby develop their culture, capacity for compassion, sense of community, and strength of democracy."

Is a Cultural Shift Beginning to Take Place in the United States? Researchers Paul Ray and Sherry Anderson have been gathering and analyzing survey data indicating cultural patterns and beliefs in the United States. Using surveys, they have identified three major cultural groupings in the United States:

- *Moderns* (about 48% of the adult U.S. population), who **(1)** actively seek materialism, **(2)** take a cynical view of idealism and caring, and **(3)** accept some form of the planetary management worldview (p. 742).

What Are Our Basic Needs

Obviously, each of us has a basic need for enough food, clean air, clean water, shelter, and clothing to keep us alive and in good health. According to various psychologists and other social scientists, each of us also has other basic needs:

- A secure and meaningful livelihood to provide our basic material needs

- Good physical and mental health

- The opportunity to learn and give expression to our intellectual, mechanical, and artistic talents

- A nurturing family and friends and a peaceful and secure community that help us develop our capacity for caring and loving relationships while giving us the freedom to make personal choices

- A clean and healthy environment that is vibrant with biological and cultural diversity

- A sense of belonging to and caring for a particular place and community (Connections, p. 753)

- An assurance that our children and grandchildren will have access to these same basic needs

A difficult but fundamental question is asking how much of the stuff we are all urged to buy helps us meet any of these basic needs. Indeed, psychologists point out that many people buy things in the hope or belief that they will make up for not having some of the basic needs listed here.

Critical Thinking

1. What basic needs, if any, would you add to or remove from the list given here?

2. Which of the basic needs listed here (or additional ones you would add) do you feel are being met? What are your plans for trying to fulfill any of your unfulfilled needs? Relate these plans to your environmental worldview.

- *Traditionals* (about 25% of the adult U.S. population), who tend toward religious conservatism, fundamentalism, and strong nationalism and want to return to traditional ways of life. They believe in **(1)** family, church and community, **(2)** helping others, **(3)** having caring relationships, and **(4)** working to create a better society. Many are pro-environment and

anti-big business. They tend to be older, poorer, and less educated than others in the U.S.

- *Cultural Creatives* (about 26% of the adult U.S. population), who have a strong commitment to family, community, the environment, education, internationalism, equality, personal growth and spiritual development, concern for human relationships, helping other people, and making a contribution to society. They have an optimistic outlook and are open to change and different beliefs and points of view. However, they reject the hedonism, materialism, and cynicism of the modernists and the intolerance of the religious right.

National public polls by the Harwood Group indicate that most Americans believe that materialism, greed, and selfishness increasingly dominate American life and are crowding out more meaningful values centered on family, responsibility, and community. This and other surveys indicate that the Cultural Creatives are the leading edge of cultural change in the United States. They are crafting a new environmental and spiritual worldview and a new agenda to help solve the environmental and social problems the United States faces.

These surveys indicate that the environmental beliefs of Moderns are beginning to weaken and to shift more to some of the environmental views of the Cultural Creatives. Evidence of this shift comes from surveys indicating that **(1)** 87% of adult Americans believe that we need to treat the planet as a living system, **(2)** 83% say that we need to rebuild our neighborhoods and small communities, **(3)** 82% say we should be stewards over nature and protect it, **(4)** 75% believe that humans are part of nature, and not its ruler, **(5)** 68% want to return to a simpler life with less emphasis on consumption and wealth, and **(6)** 63% believe that the government should shut down industries that keep polluting the air.

These surveys suggest a shift in values away from some of the beliefs of the planetary management environmental worldview (p. 742) toward some of the beliefs of the environmental-wisdom environmental worldview (p. 745). The real test of such a shift in values is **(1)** whether it is accompanied by changes in human behavior and **(2)** whether an existing or new political party can mobilize these individuals into an effective force for political change.

How Can We Move Beyond Blame, Guilt, and Denial to Responsibility? According to many psychologists, when we first encounter an environmental problem, our initial response often is to find someone or something to blame: greedy industrialists, uncaring politicians, environmentalists, and misguided worldviews. It is the fault of such villains, and we are the victims.

This can lead to despair, denial, and inaction because we feel powerless to stop or influence these forces. There are also many complex and interconnected environmental problems and conflicting views about their seriousness and possible solutions. As a result, we feel overwhelmed and wonder whether there is any way out—another emotion leading to denial and inaction.

Upon closer examination we may realize that we all make some direct or indirect contributions to the environmental problems we face. As Pogo said, "We have met the enemy and it is us." We do not want to feel guilty or bad about all of the things we are not doing, so we avoid thinking about them—another path to denial and inaction.

How do we move beyond immobilizing blame, fear, and guilt to engaging in more responsible environmental actions in our daily lives? Analysts have suggested several ways to do this:

- Recognize and avoid common mental traps that lead to denial, indifference, and inaction. These traps include **(1)** *gloom-and-doom pessimism* (it's hopeless), **(2)** *blind technological optimism* (science and technofixes will save us), **(3)** *fatalism* (we have no control over our actions and the future), **(4)** *extrapolation to infinity* (if I cannot change the entire world quickly, I will not try to change any of it), **(5)** *paralysis by analysis* (searching for the perfect worldview, philosophy, solutions, and scientific information before doing anything), and **(6)** *faith in simple, easy answers*.

- Understand that no one can even come close to doing all of the things people suggest (or that we know we should be doing) to work with the earth. Try to determine what things you do have the greatest environmental impact and concentrate on them. Focus on things you feel strongly about and that you can do something about. Rejoice in the good things you have done and continually expanded your efforts make the earth a better place (Appendix 6).

- Keep your empowering feelings of hope slightly ahead of your immobilizing feelings of despair.

- Do not use guilt and fear to motivate other people to work with the earth and other people. We need to nurture, reassure, understand, and care for one another.

- Recognize that there is no single correct or best solution to the environmental problems we face. Indeed, one of nature's most important lessons (from evolution) is that preserving diversity or a rainbow of flexible and adaptable solutions to the problems we face is the best way to adapt to earth's largely unpredictable, ever-changing conditions.

- Expect the unexpected and hedge your bets. The future will always have some unpredictable surprises for us. This is where the precautionary principle comes in.

- Have fun and take time to enjoy life. Every day we should laugh and enjoy nature, beauty, friendship, and love.

What Are the Major Components of the Environmental Revolution? The *environmental revolution* (Guest Essay, p. 18) that many environmentalists call for us to bring about would have several components:

- An *efficiency revolution* that involves not wasting matter and energy resources and relying on the earth's natural income without depleting its natural capital. Analysts point out that we already have the knowledge and technology to increase *resource productivity* by getting 75–90% more work or service from each unit of material resource we use (Solutions, p. 525).

- A *pollution prevention revolution* that reduces pollution and environmental degradation by **(1)** reducing the waste of matter and energy resources, **(2)** using cleaner production (Section 21-3, p. 522, and Guest Essay, p. 526) to keep highly toxic substances from being released into the environment by recycling or reusing them within industrial processes, and **(3)** trying to find less harmful or easily biodegradable substitutes for such substances or not producing them at all.

- A *sufficiency revolution*. This involves trying to meet the basic needs of all people on the planet and asking how many material things we really need to have a decent and meaningful life (Spotlight, left).

- A *demographic revolution* based on bringing the size and growth rate of the human population into balance with the earth's ability to support humans and other species without serious environmental degradation.

- An *economic revolution* in which we use economic systems to reward environmentally beneficial behavior and discourage environmentally harmful behavior (Section 26-7, p. 708).

Opponents of such a cultural change like to paint environmentalists as messengers of gloom, doom, and hopelessness. However, *the real message of environmentalism is not gloom and doom, fear, and catastrophe but hope and a positive vision of the future.* This is an exciting message of challenge and adventure as we struggle to find better and more responsible ways to live on this planet.

Envision the earth's life-sustaining processes as a beautiful and diverse web of interrelationships—a kaleidoscope of patterns, rhythms, and connections whose very complexity and multitude of possibilities remind us that cooperation, sharing, honesty, humility, and compassionate love should be the guidelines for our behavior toward one another and the earth.

When there is no dream, the people perish.

PROVERBS 29:18

REVIEW QUESTIONS

1. Describe the Biosphere 2 experiment and the lessons it taught us.

2. Distinguish between **(a)** *environmental worldviews* and *environmental ethics*, **(b)** *instrumental values* and *intrinsic values*, and **(c)** *anthropocentric instrumental values* and *biocentric intrinsic values*.

3. What are the five major beliefs of the *planetary management environmental worldview*? Describe the **(a)** *no-problem*, **(b)** *free-market*, **(c)** *responsible planetary management*, **(d)** *spaceship earth*, and **(e)** *stewardship* variations of the planetary management worldview.

4. Explain why many environmentalists believe that planetary management worldviews will not work.

5. Summarize the major viewpoint presented by Donald Worster in his Guest Essay on p. 743.

6. Give two reasons for caring about future generations.

7. What are **(a)** the five major beliefs of the *earth-wisdom environmental worldview* and **(b)** the eight major beliefs of the *deep ecology environmental worldview*?

8. Explain how Aldo Leopold's land ethic could be modified to account for modern ecological theory that views nature as being in a state of constant flux instead of in a static balance.

9. Summarize the major beliefs of the **(a)** *ecofeminist* and **(b)** *social ecology* environmental worldviews.

10. Describe Carlos Davidson's tapestry metaphor as it relates to environmental degradation.

11. Describe ways to evaluate sustainability proposals.

12. List eleven ethical guidelines for living more sustainably.

13. List the four guiding principles of the Earth Charter.

14. Describe the viewpoint given by Lester Milbrath in his Guest Essay on p. 750.

15. List eight guidelines of earth education and four questions that should be answered to determine one's *ecological identity*.

16. Explain why it is important to learn more about the earth from direct experience.

17. What is *voluntary simplicity*?

18. Distinguish between *Moderns, Traditionals,* and *Cultural Creatives* as major cultural groups in the United States.

19. Describe evidence that a shift in cultural values may be taking place in the United States.

20. What are seven ways to move beyond blame, guilt, and denial and become more environmentally responsible? Describe six mental traps that can lead to denial, indifference, and inaction about environmental (and other) problems.

21. What are five major components of an environmental revolution that most environmentalists believe we should bring about over the next 50 years?

CRITICAL THINKING

1. One of the primary lessons from the Biosphere 2 experiment (p. 740) is that some of the world's best scientific minds had trouble constructing an ecosystem for keeping eight people (equal to the number of people we are adding to the planet every 3 seconds) alive for 2 years. Were the lessons learned from this experiment worth its $200-million cost? Explain.

2. This chapter has summarized a number of different environmental worldviews. Go through these worldviews and find the beliefs you agree with to describe your environmental worldview. Which of your beliefs were added or modified as a result of taking this course?

3. List the five most important environmental benefits and the five most important harmful conditions passed on to you by the previous two generations.

4. Theologian Thomas Berry calls the industrial consumer society built on the human-centered, planetary management environmental worldview the "supreme pathology of all history." He says, "We can break the mountains apart; we can drain the rivers and flood the valleys. We can turn the most luxuriant forests into throwaway paper products. We can tear apart the great grass cover of the western plains, and pour toxic chemicals into the soil and pesticides onto the fields, until the soil is dead and blows away in the wind. We can pollute the air with acids, the rivers with sewage, the seas with oil. . . . We can invent computers capable of processing ten million calculations per second. And why? To increase the volume and speed with which we move natural resources through the consumer economy to the junk pile or the waste heap....If, in these activities, the topography of the planet is damaged, if the environment is made inhospitable for a multitude of living species, then so be it. We are, supposedly, creating a technological wonderworld....But our supposed progress...is bringing us to a wasteworld instead of a wonderworld." Explain why you agree or disagree with this assessment.

5. List the pros and cons of viewing nature as a tapestry for helping prevent environmental degradation. Are you for or against using this approach? Explain.

6. Do you agree with the cartoon character Pogo that "we have met the enemy and he is us"? Explain. Criticize this statement from the viewpoint of the poor. Criticize it from the viewpoint that many large corporations are big polluters and resource depleters and degraders.

7. Try to answer the following fundamental ecological questions about the corner of the world you inhabit: Where does your water come from? Where does the energy you use come from? What kinds of soils are under your feet? What types of wildlife are your neighbors? Where does your food come from? Where does your waste go?

8. Would you classify yourself as a *Modern*, *Traditional*, or *Cultural Creative* (p. 753)? Compare your answer with those of your classmates.

9. (a) Would you accept employment on a project that you knew would degrade or destroy a wild habitat? Explain. **(b)** If you were granted three wishes, what would they be? **(c)** If you did not have to work for a living, what would you do with your time? **(d)** Could you live without an automobile? Explain. **(e)** Could you live without TV? Explain. **(f)** If you won $100 million in a lottery, how would you spend the money? **(g)** Do you believe that wolves have as much right to eat sheep as people do? Explain.

10. How do you feel about **(a)** carving huge faces of people in mountains, **(b)** carving your initials into a tree, **(c)** driving an off-road motorized vehicle in a desert, grassland, or forest, **(d)** using throwaway paper towels, tissues, napkins, and plates, **(e)** wearing furs, **(f)** hunting for recreation, **(g)** fishing for recreation, and **(h)** having tropical fish, birds, snakes, or other wild animals as pets?

11. Would it be better if government tax dollars currently being spent on space travel and exploration were used instead to deal with the earth's environmental and social problems? Explain.

12. Explain why you agree or disagree with the following ideas: **(a)** Everyone has the right to have as many children as they want, **(b)** each member of the human species has a right to pollute and to use as many resources as they want, and **(c)** individuals should have the right to do anything they want with land they own. Relate your answers to the beliefs of your environmental worldview that you listed in question 2.

13. Would you or do you use the principle of voluntary simplicity in your life? How?

14. Which (if any) of the ethical guidelines on p. 749 do you disagree with? Explain. Can you suggest any additional ethical guidelines for working with the earth?

15. Review your experience with the mental traps described on p. 755. Which of these traps have you fallen into? Were you aware that you had been ensnared by any of these mental traps? Do you plan to free yourself from these traps? How?

PROJECTS

1. Use a combination of the major existing societal, economic, and environmental trends, possible new trends, and your imagination to construct three different scenarios of what the world might be like in 2060. Identify the scenario you favor and outline a program for achieving this alternative future. Compare your scenarios with those of your classmates.

2. Which of the goods and services you use meet primary needs, and which fulfill wants? Which ones could you do without? Which ones are you willing to do without? Which goods that you now throw away could be reused or recycled? What percentage of those do you actually reuse or recycle?

3. Make an environmental audit of your school. What proportion of each of the major types of matter resources it uses are recycled, reused, or composted? What priority does your school give to buying recycled materials? How much emphasis does it put on energy efficiency, use of renewable forms of solar energy, and environmental design in developing new buildings and renovating existing ones? Does it use ecologically sound planning in deciding how its grounds and buildings are managed and used? Does your school limit the use of toxic chemicals in its buildings and on its grounds? What proportion of its food purchases comes from nearby farmers? What proportion of the food it purchases is grown by sustainable agriculture?

4. Does your school's curriculum provide *all* graduates with the basics elements of ecological literacy? To what extent are the funds in its financial endowments invested in enterprises that are working to develop or encourage environmental sustainability? Over the past 20 years, what important roles have its graduates played in making the world a better and more sustainable place to live? Using such information, rate your school on a 1–10 scale in terms of its contributions to environmental awareness and sustainability. Develop a detailed plan illustrating how your school could become better at achieving such goals and present this information to school officials, alumni, parents, and financial backers.

5. If you knew you were going to die and had an opportunity to address everyone in the world for 5 minutes, what would you say? Write out your 5-minute speech and compare it with those of other members of your class.

6. Write an essay in which you try to identify key environmental experiences that have influenced your life and thus helped form your current ecological identity. Examples may include **(a)** fond childhood memories of special places where you connected with the earth through emotional experiences, **(b)** places that you knew and cherished that have been polluted, developed, or destroyed, **(c)** key events that forced you to think about environmental values or worldviews, **(d)** people or educational experiences that influenced your understanding and concern about environmental problems and challenges, and **(e)** direct experience and contemplation of wild places. Share your experiences with other members of your class.

7. Use the library or the internet to find bibliographic information about *Robert Cahn*, whose quote is found at the beginning of this chapter.

8. Make a concept map of this chapter's major ideas, using the section heads and subheads and the key terms (in boldface). Look at the inside back cover and on the website for this book for information about making concept maps.

INTERNET STUDY RESOURCES AND RESOURCES FOR FURTHER READING AND RESEARCH

The website for this book contains helpful study aids and many ideas for further reading and research. Log on to:

http://www.brookscole.com/product/0534376975s

and click on the Chapter-by-Chapter area. Choose Chapter 28 and select a resource:

- "Flash Cards" allows you to test your mastery of the Terms and Concepts to Remember for this chapter.

- "Tutorial Quizzes" provides a multiple-choice practice quiz.

- "Student Guide to InfoTrac" will lead you to Critical Thinking Projects that use InfoTrac College Edition as a research tool.

- "References" lists the major books and articles consulted in writing this chapter.

- "Hypercontents" takes you to an extensive list of sites with news, research, and images related to individual sections of the chapter.

INFOTRAC COLLEGE EDITION

Improve your skills with InfoTrac College Edition, a searchable online database of articles from more than 700 periodicals. Log on to:

http://www.infotrac-college.com

or access InfoTrac through the website for this book.

Try the following articles:

Dasgupta, P., S. Levin, J. Lubchenco. 2000. Economic pathways to ecological sustainability. *BioScience* vol. 50, no. 4, pp. 339–346. (keywords: ecological sustainability)

Annan, K.A. 2000. Sustaining the earth in the new millennium. *Environment* vol. 42, no. 8, pp. 20–30. (keywords: Kofi Annan)

UNITS OF MEASUREMENT

LENGTH

Metric

1 kilometer (km) = 1,000 meters (m)
1 meter (m) = 100 centimeters (cm)
1 meter (m) = 1,000 millimeters (mm)
1 centimeter (cm) = 0.01 meter (m)
1 millimeter (mm) = 0.001 meter (m)

English

1 foot (ft) = 12 inches (in)
1 yard (yd) = 3 feet (ft)
1 mile (mi) = 5,280 feet (ft)
1 nautical mile = 1.15 miles

Metric–English

1 kilometer (km) = 0.621 mile (mi)
1 meter (m) = 39.4 inches (in)
1 inch (in) = 2.54 centimeters (cm)
1 foot (ft) = 0.305 meter (m)
1 yard (yd) = 0.914 meter (m)
1 nautical mile = 1.85 kilometers (km)

AREA

Metric

1 square kilometer (km^2) = 1,000,000 square meters (m^2)
1 square meter (m^2) = 1,000,000 square millimeters (mm^2)
1 hectare (ha) = 10,000 square meters (m^2)
1 hectare (ha) = 0.01 square kilometer (km^2)

English

1 square foot (ft^2) = 144 square inches (in^2)
1 square yard (yd^2) = 9 square feet (ft^2)
1 square mile (mi^2) = 27,880,000 square feet (ft^2)
1 acre (ac) = 43,560 square feet (ft^2)

Metric–English

1 hectare (ha) = 2.471 acres (ac)
1 square kilometer (km^2) = 0.386 square mile (mi^2)
1 square meter (m^2) = 1.196 square yards (yd^2)
1 square meter (m^2) = 10.76 square feet (ft^2)
1 square centimeter (cm^2) = 0.155 square inch (in^2)

VOLUME

Metric

1 cubic kilometer (km^3) = 1,000,000,000 cubic meters (m^3)
1 cubic meter (m^3) = 1,000,000 cubic centimeters (cm^3)
1 liter (L) = 1,000 milliliters (mL) = 1,000 cubic centimeters (cm^3)
1 milliliter (mL) = 0.001 liter (L)
1 milliliter (mL) = 1 cubic centimeter (cm^3)

English

1 gallon (gal) = 4 quarts (qt)
1 quart (qt) = 2 pints (pt)

Metric–English

1 liter (L) = 0.265 gallon (gal)
1 liter (L) = 1.06 quarts (qt)
1 liter (L) = 0.0353 cubic foot (ft^3)
1 cubic meter (m^3) = 35.3 cubic feet (ft^3)
1 cubic meter (m^3) = 1.30 cubic yards (yd^3)
1 cubic kilometer (km^3) = 0.24 cubic mile (mi^3)
1 barrel (bbl) = 159 liters (L)
1 barrel (bbl) = 42 U.S. gallons (gal)

MASS

Metric

1 kilogram (kg) = 1,000 grams (g)
1 gram (g) = 1,000 milligrams (mg)
1 gram (g) = 1,000,000 micrograms (µg)
1 milligram (mg) = 0.001 gram (g)
1 microgram (µg) = 0.000001 gram (g)
1 metric ton (mt) = 1,000 kilograms (kg)

English

1 ton (t) = 2,000 pounds (lb)
1 pound (lb) = 16 ounces (oz)

Metric–English

1 metric ton (mt) = 2,200 pounds (lb) = 1.1 tons (t)
1 kilogram (kg) = 2.20 pounds (lb)
1 pound (lb) = 454 grams (g)
1 gram (g) = 0.035 ounce (oz)

ENERGY AND POWER

Metric

1 kilojoule (kJ) = 1,000 joules (J)
1 kilocalorie (kcal) = 1,000 calories (cal)
1 calorie (cal) = 4,184 joules (J)

Metric–English

1 kilojoule (kJ) = 0.949 British thermal unit (Btu)
1 kilojoule (kJ) = 0.000278 kilowatt-hour (kW-h)
1 kilocalorie (kcal) = 3.97 British thermal units (Btu)
1 kilocalorie (kcal) = 0.00116 kilowatt-hour (kW-h)
1 kilowatt-hour (kW-h) = 860 kilocalories (kcal)
1 kilowatt-hour (kW-h) = 3,400 British thermal units (Btu)
1 quad (Q) = 1,050,000,000,000,000 kilojoules (kJ)
1 quad (Q) = 2,930,000,000,000 kilowatt-hours (kW-h)

TEMPERATURE CONVERSIONS

Fahrenheit (°F) to Celsius (°C): $°C = (°F - 32.0) \div 1.80$
Celsius (°C) to Fahrenheit (°F): $°F = (°C \times 1.80) + 32.0$

Appendix 2

GOOD AND BAD NEWS ABOUT ENVIRONMENTAL PROBLEMS

SOME GOOD NEWS

Annual world population growth slowed from 2.2% to 1.35% between 1963 and 2000.

Between 1950 and 2000, the average number of children per woman dropped from 6 to 2.9.

Global life expectancy rose from 33 to 66 years between 1900 and 2000.

Global infant mortality dropped by 65% between 1900 and 2000.

Global grain production has outpaced population growth since 1978.

The percentage of the world's population that is hungry has decreased since 1960.

Biodiversity loss has been reduced by higher crop yields on less land.

Biodiversity loss has been reduced by increased urbanization, with more people living on less land.

About 8% of the world's land area has been set aside to protect wildlife and wildlife habitats.

Total forest area in temperate industrial countries increased during the 1980s.

Some analysts believe that tropical deforestation has been exaggerated.

Estimates of premature extinction of wildlife are mostly guesses and probably overstate the seriousness of biodiversity loss.

Conservation tillage and no-till cultivation can cut soil erosion by about 75%.

A handful of scientists contend that global warming is not a serious problem.

A handful of scientists say that depletion of ozone in the stratosphere (the atmosphere's second layer) is not a serious problem.

The earth has plenty of water.

Since 1950, proven supplies (reserves) of almost all nonrenewable fossil fuel and key mineral resources have increased significantly.

Adjusted for inflation, most nonrenewable fossil fuels and key mineral resources cost less today than in 1950.

Some scientists contend that the health risks from exposure to toxic and hazardous chemicals are exaggerated.

Since 1970, air and water pollution levels in most industrialized countries have dropped significantly for most pollutants.

SOME BAD BUT CHALLENGING NEWS

The world's population is still projected to increase from 6.1 billion to 8 billion between 2000 and 2028.

This is still far above the average of 2.1 children per woman needed for population growth to level off.

Life expectancy in developing countries is 11 years less than in developed countries.

Infant mortality in developing countries is eight times higher than in developed countries and could be improved in many developed countries.

Since 1980 population growth has exceeded grain production in 69 developing countries. Future food production may be limited by its harmful environmental impacts.

Because of population growth, there are still 826 million malnourished people.

The challenge is to increase high yields while sharply reducing the growing environmental impacts of such intensive crop production.

This figure is misleading because biodiversity is often decreased and environmental degradation is increased in the croplands, forests, grasslands, and fisheries elsewhere that supply resources for urban dwellers.

This is less than the 10% estimated minimum needed. Many wildlife sanctuaries exist on paper only and receive little protection, and many important types of biological communities are not protected.

Much of this increase came from replacing biologically diverse old-growth forests with simplified and vulnerable tree farms and less diverse second-growth forests.

If current deforestation rates continue, within 30-50 years little of these forests will remain, and their loss will lead to a widespread loss of biodiversity.

Most biologists believe that the evidence indicates that human actions are bringing about an extinction of species greater than anything the earth has seen since the fall of the dinosaurs and that this biodiversity loss is one of the most serious environmental problems we face.

These techniques are widely used in the United States but are rarely used in other countries. Topsoil is eroding faster than it forms on about one-third of the world's cropland.

The vast majority of scientists in this field contend that global warming is one of the world's most serious environmental and economic problems.

The vast majority of scientists in this field contend that ozone depletion in the stratosphere is a very serious problem.

The number of people facing a chronic water shortage is expected to grow from 1.2 billion to at least 3 billion between 2000 and 2025.

The exponential increase in the use of fossil fuel and mineral resources is causing widespread land degradation, water pollution, and air pollution that may limit future use of such resources.

These resource prices are low because most of the harmful environmental and health costs associated with their production and use are not included in their market prices.

The harmful health effects of most chemicals are unknown, and a growing body of evidence suggests that their harmful effects (especially to the nervous, immune, and endocrine systems) have been underestimated.

More people using more resources and a lack of emphasis on pollution prevention are threatening such gains in pollution control in industrialized countries. Little progress has been made in reducing air and water pollution in most developing countries.

SOME BASIC CHEMISTRY

How Can Elements Be Arranged In the Periodic Table According to Their Chemical Properties? Chemists have developed a way to classify the elements according to their chemical behavior, in what is called the *periodic table of elements* (Figure 1). Each of the horizontal rows in the table is called a *period*. Each vertical column lists elements with similar chemical properties and is called a *group*.

The partial periodic table in Figure 1 shows how the elements can be classified as *metals*, *nonmetals*, and *metalloids*. Most of the elements found to the left and at the bottom of the table are *metals*, which usually conduct electricity and heat, and are shiny. Examples are sodium (Na), calcium (Ca), aluminum (Al), iron (Fe), lead (Pb), and mercury (Hg). Atoms of such metals achieve a more stable state by losing one or more of their electrons to form posi-

tively charged ions such as Na^+, Ca^{2+}, and Al^{3+}.

Nonmetals, found in the upper right of the table, do not conduct electricity very well and are usually not shiny. Examples are hydrogen (H)*, carbon (C), nitrogen (N), oxygen (O), phosphorus (P), sulfur (S), chlorine (Cl), and fluorine (F). Atoms of some nonmetals such as chlorine, oxygen, and sulfur tend to gain one or more electrons lost by metallic atoms to form negatively charged ions such as O^{2-}, S^{2-}, and Cl^-. Atoms of nonmetals can also combine with one another to form molecules in which they share one or more pairs of their electrons.

The elements arranged in a diagonal staircase pattern between the metals and

---

* Hydrogen, a nonmetal, is placed by itself above the center of the table because it does not fit very well into any of the groups.

nonmetals have a mixture of metallic and nonmetallic properties and are called *metalloids*. Figure 1 also identifies the elements required as *nutrients* for all or some forms of life and elements that are moderately or highly toxic to all or most forms of life. Six nonmetallic elements—carbon (C) oxygen (O), hydrogen (H), nitrogen (N), sulfur (S), and phosphorus (P)—make up about 99% of the atoms of all living things.

What Are Ionic and Covalent Bonds? Sodium chloride (NaCl) consists of a three-dimensional network of oppositely charged *ions* (Na^+ and Cl^-) held together by the forces of attraction between opposite charges (Figure 2). The strong forces of attraction between such oppositely charged ions are called *ionic bonds*. Because ionic compounds consist of ions formed from atoms of metallic (positive

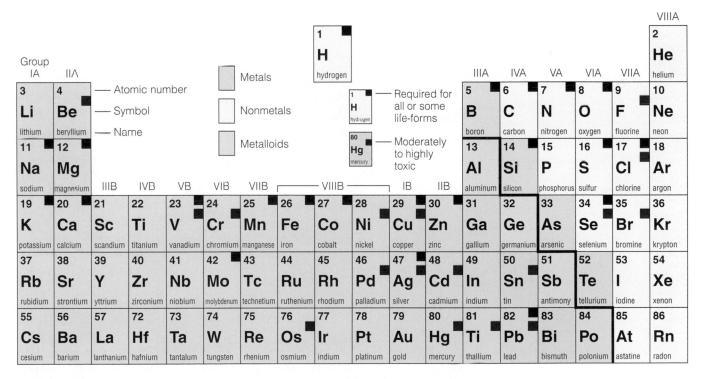

Figure 1 Abbreviated periodic table of elements. Elements in the same vertical column, called a group, have similar chemical properties. To simplify matters at this introductory level, only 72 of the 118 known elements are shown.

Figure 2 A solid crystal of an ionic compound such as sodium chloride consists of a three-dimensional array of opposite charged ions held together by ionic bonds resulting from the strong forces of attraction between opposite electrical charges. They are formed when an electron is transferred from a metallic atom such as sodium (Na) to a nonmetallic element such as chlorine (Cl). Such compounds tend to exist as solids at normal room temperature and atmospheric pressure.

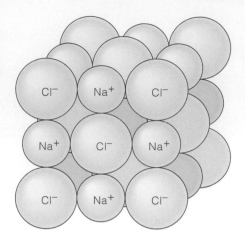

What Are Hydrogen Bonds? Ionic and covalent bonds form between the ions or atoms *within* a compound. There are also weaker forces of attraction *between* the molecules of covalent compounds (such as water) when the electrons shared by two atoms are shared unequally.

For example, an oxygen atom has a much greater attraction for electrons than does a hydrogen atom. Thus, in a water molecule the electrons shared between the oxygen atom and its two hydrogen atoms are pulled closer to the oxygen atom, but not actually transferred to the oxygen atom. As a result, the oxygen atom in a water molecule has a slightly negative partial charge and its two hydrogen atoms have a slightly positive partial charge (Figure 4, top). The slightly positive hydrogen atoms in one water molecule are then attracted to the slightly negative oxygen atoms in another water molecules. These forces of attraction *between* water molecules are called *hydrogen bonds* (Figure 4, bottom). Hydrogen bonds also form between other covalent molecules or portions of such molecules containing hydrogen and

ions) and nonmetallic (negative ions) elements, they can be described as *metal-nonmetal compounds*.

Figure 3 shows the chemical formulas and shapes of the molecules for several common covalent compounds, formed when atoms of one or more nonmetallic elements (Figure 1) combine with one another.

The bonds between the atoms in such molecules are called *covalent bonds* and are formed when the atoms in the molecule share one or more pairs of their electrons. Because they are formed from atoms of nonmetallic elements, molecular or covalent compounds can be described as *nonmetal-nonmetal compounds*.

Figure 3 Chemical formulas and shapes for some molecular compounds formed when atoms of one or more nonmetallic elements combine with one another. The bonds between the atoms in such molecules are called *covalent bonds*. Molecular compounds tend to exist as gases or liquids at normal room temperature and atmospheric pressure.

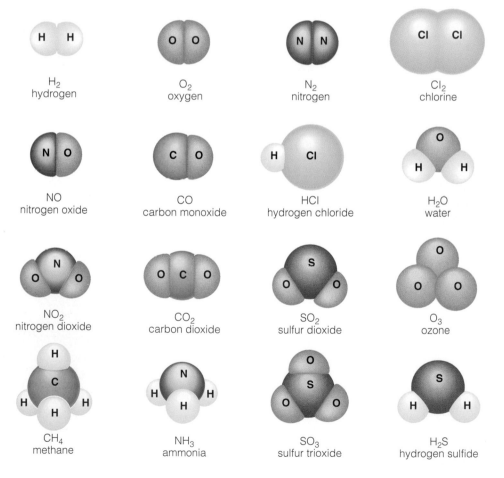

Figure 4 *Hydrogen bonds.* Slightly unequal sharing of electrons in the water molecule creates a molecule with a slightly negatively charged end and a slightly positively charged end. Because of this polarity, hydrogen atoms of one water molecule are attracted to oxygen atoms of another water molecule. These forces of attraction *between* water molecules are called *hydrogen bonds*.

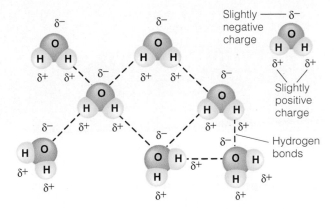

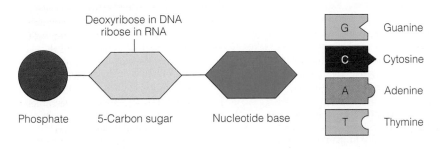

Phosphate   5-Carbon sugar   Nucleotide base

Deoxyribose in DNA
ribose in RNA

G — Guanine
C — Cytosine
A — Adenine
T — Thymine

Figure 5 Generalized structure of nucleotide molecules linked in various numbers and sequences to form large nucleic acid molecules such as various types of DNA (deoxyribose nucleic acid) and RNA (ribose nucleic acid). In DNA the 5-carbon sugar in each nucleotide is deoxyribose; in RNA it is ribose. The four basic nucleotides used to make various forms of DNA molecules differ in the types of nucleotide bases they contain–guanine (G), cytosine (C), adenine (A), and thymine (T).

nonmetallic atoms with a strong ability to attract electrons.

What Are Nucleic Acids? *Nucleic acids,* such as DNA and RNA, are made by linking hundreds to thousands of five different types of monomers, called *nucleotide.* Examples of nucleic acids are DNA and RNA. Each nucleotide consists of **(1)** a phosphate group, **(2)** a sugar molecule containing five carbon atoms (deoxyribose in DNA molecules and ribose in RNA molecules), and **(3)** one of four different nucleotide bases (represented by A, G, C, and T, the first letter in each of their names) (Figure 5).

In the cells of living organisms, these nucleotide units combine in different numbers and sequences to form *nucleic acids* such as various types of DNA and RNA. Hydrogen bonds formed between parts of the four nucleotides in DNA hold two DNA strands together like a spiral staircase, forming a double helix (Figure 6). DNA molecules can unwind and replicate themselves.

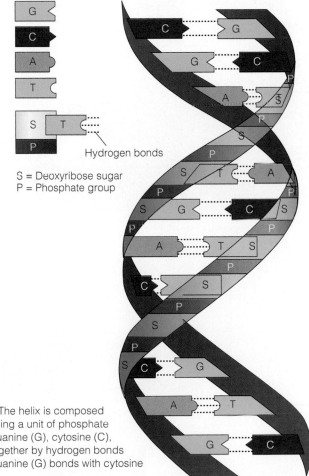

Hydrogen bonds

S = Deoxyribose sugar
P = Phosphate group

Figure 6 Portion of the double helix of a DNA molecule. The helix is composed of two spiral (helical) strands of nucleotides, each containing a unit of phosphate (P), deoxyribose (S), and one of four nucleotide bases: guanine (G), cytosine (C), adenine (A), and thymine (T). The two strands are held together by hydrogen bonds formed between various pairs of the nucleotide bases. Guanine (G) bonds with cytosine (C), and adenine (A) with thymine (T).

How Can Species Be Classified? Biologists classify species into different *kingdoms*, on the basis of similarities and differences in characteristics such as their modes of nutrition, cell structure, appearance, and developmental features.

In this book, the earth's organisms are classified into six kingdoms: *eubacteria, archaebacteria, protists, fungi, plants,* and *animals* Most bacteria, fungi, and protists are *microorganisms*: organisms so small that they cannot be seen with the naked eye.

Eubacteria consist of all single-celled prokaryotic bacteria except archaebacteria. Examples are various cyanobacteria and bacteria such as *Staphylococcus and Streptococcus.*

Archaebacterai are single-celled bacteria that are evolutionarily closer to eukaryotic cells than to eubacteria, Examples are (1) methanogens that live in anaerobic sediments of lakes and swamps and in animal guts, (2) halophiles that live in extremely salty water, and (3) thermophiles that live in hot springs, hydrothermal vents, and acidic soil.

Protists (Protista) are mostly single-celled eukaryotic organisms such as diatoms, dinoflagellates, amoebas, golden brown and yellow-green algae, and protozoans. Some protists cause human diseases such as malaria and sleeping sickness.

Fungi are mostly many-celled, sometimes microscopic, eukaryotic organisms such as mushrooms, molds, mildews, and yeasts. Many fungi are decomposers. Other fungi kill various plants and cause huge losses of crops and valuable trees.

Plants (Plantae) are mostly many-celled eukaryotic organisms such as red, brown, and green algae and mosses, ferns, and flowering plants (whose flowers produce seeds that perpetuate the species). Some plants such as corn and marigolds are *annuals*, which complete their life cycles in one growing season; others are *perennials*, which can live for more than 2 years, such as roses, grapes, elms, and magnolias.

Animals (Animalia) are also many-celled, eukaryotic organisms. Most, called *invertebrates*, have no backbones. They include sponges, jellyfish, worms, arthropods (insects, shrimp, and spiders), mollusks (snails, clams, and octopuses), and echinoderms (sea urchins and sea stars). Insects play roles that are vital to our existence. *Vertebrates* (animals with backbones and a brain protected by skull bones) include fishes (sharks and tuna), amphibians (frogs and salamanders), reptiles (crocodiles and snakes), birds (eagles and robins), and mammals (bats, elephants, whales, and humans).

How Are Species Named? Within each kingdom, biologists have created subcategories based on anatomical, physiological, and behavioral characteristics. Kingdoms are divided into *phyla*, which are divided into subgroups called *classes*. Classes are subdivided into *orders*, which are further divided into *families*. Families consist of *genera* (singular, *genus*), and each genus contains one or more *species*. Note that the word species is both singular and plural. Figure 1 shows this detailed taxonomic classification for the current human species.

Most people call a species by its common name, such as robin or grizzly bear Biologists use scientific names (derived from Latin) consisting of two parts (printed in italics, or underlined) to describe a species. The first word is the capitalized name (or abbreviation) for the genus to which the organism belongs. This is followed by a lowercase name that distinguishes the species from other members of the same genus. For example, the scientific name of the robin is *Turdus migratorius* (Latin for "migratory thrush") and the grizzly bear goes by the scientific name *Ursus horribilis* (Latin for "horrible bear").

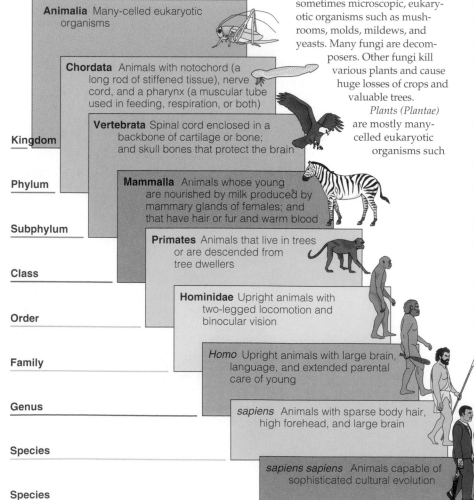

Kingdom

Animalia Many-celled eukaryotic organisms

Phylum

Chordata Animals with notochord (a long rod of stiffened tissue), nerve cord, and a pharynx (a muscular tube used in feeding, respiration, or both)

Subphylum

Vertebrata Spinal cord enclosed in a backbone of cartilage or bone; and skull bones that protect the brain

Class

Mammalla Animals whose young are nourished by milk produced by mammary glands of females; and that have hair or fur and warm blood

Order

Primates Animals that live in trees or are descended from tree dwellers

Family

Hominidae Upright animals with two-legged locomotion and binocular vision

Genus

Homo Upright animals with large brain, language, and extended parental care of young

Species

sapiens Animals with sparse body hair, high forehead, and large brain

Species

sapiens sapiens Animals capable of sophisticated cultural evolution

Figure 1 Taxonomic classification of the latest human species, *Homo sapiens sapiens.*

MAJOR U.S. RESOURCE CONSERVATION AND ENVIRONMENTAL LEGISLATION

GENERAL

National Environmental Policy Act of 1969 (NEPA)
International Environmental Protection Act of 1983
Environmental Education Act of 1990

ENERGY

Energy Policy and Conservation Act of 1975
National Energy Act of 1978, 1980
National Appliance Energy Conservation Act of 1987
Energy Policy Act of 1992

WATER QUALITY

Water Quality Act of 1965
Water Resources Planning Act of 1965
Federal Water Pollution Control Acts of 1965, 1972
Ocean Dumping Act of 1972
Safe Drinking Water Act of 1974, 1984, 1996
Water Resources Development Act of 1986
Clean Water Act of 1977, 1981, 1987
Ocean Dumping Ban Act of 1988
Great Lakes Critical Programs Act of 1990
Oil Spill Prevention and Liability Act of 1990

AIR QUALITY

Clean Air Act of 1963, 1965, 1970, 1977, 1990
Asbestos Hazard and Emergency Response Act of 1986
Pollution Prevention Act of 1990

NOISE CONTROL

Noise Control Act of 1965
Quiet Communities Act of 1978

RESOURCES AND SOLID WASTE MANAGEMENT

Mining Act of 1872
Solid Waste Disposal Act of 1965
Resource Recovery Act of 1970
Resource Conservation and Recovery Acts (RCRA) of 1976, 1989
Marine Plastic Pollution Research and Control Act of 1987
Waste Reduction Act of 1990

TOXIC SUBSTANCES

Hazardous Materials Transportation Act of 1975
Toxic Substances Control Act (TOSCA) of 1976
Resource Conservation and Recovery Acts (RCRA) of 1976, 1984
Comprehensive Environmental Response, Compensation, and Liability (Superfund) Act of 1980, 1986, 1990
Low-Level Radioactive Policy Act of 1980
Nuclear Waste Policy Act of 1982

PESTICIDES

Food, Drug, and Cosmetics Acts of 1938, 1954, 1958
Federal Insecticide, Fungicide, and Rodenticide Control Act of 1972, 1988
Food Quality Protection Act of 1996

WILDLIFE CONSERVATION

Lacey Act of 1900
Migratory Bird Treaty Act of 1918
Migratory Bird Conservation Act of 1929
Migratory Bird Hunting Stamp Act of 1934
Pittman–Robertson Act of 1937
Anadromous Fish Conservation Act of 1965
Fur Seal Act of 1966
Species Conservation Act of 1966, 1969
National Wildlife Refuge System Act of 1966, 1976, 1978
Marine Mammal Protection Act of 1972
Marine Protection, Research, and Sanctuaries Act of 1972
Endangered Species Act of 1973, 1982, 1985, 1988
Whale Conservation and Protection Study Act of 1976
Fishery Conservation and Management Act of 1976, 1978, 1982
Fish and Wildlife Improvement Act of 1978
Fish and Wildlife Conservation Act of 1980 (Nongame Act)

LAND USE AND CONSERVATION

Homestead Act of 1862
Taylor Grazing Act of 1934
Soil Conservation Act of 1935
Wilderness Act of 1964
Land and Water Conservation Act of 1965
Multiple Use Sustained Yield Act of 1968
Wild and Scenic Rivers Act of 1968
National Trails System Act of 1968
National Coastal Zone Management Act of 1972, 1980
Forest Reserves Management Act of 1974, 1976
Forest and Rangeland Renewable Resources Act of 1974, 1978
Federal Land Policy and Management Act of 1976
National Forest Management Act of 1976
Soil and Water Conservation Act of 1977
Surface Mining Control and Reclamation Act of 1977
Antarctic Conservation Act of 1978
Public Rangelands Improvement Act of 1978
Endangered American Wilderness Act of 1978
Alaskan National Interests Lands Conservation Act of 1980
Coastal Barrier Resources Act of 1982
Food Security Act of 1985
Emergency Wetlands Resources Act of 1986
Coastal Development Act of 1990
California Desert Protection Act of 1994

APPENDIX 6

INDIVIDUALS MATTER: WHAT CAN YOU DO?

This is a list of things individuals can do based on suggestions from a wide variety of environmentalists. It is not meant to be a list of things you must do but a list of actions you might consider. Start by picking the ones that **(1)** you are willing to do, **(2)** you feel will have the most impact, and **(3)** fit in with your beliefs. Each year, look over the list and try to add several new items.

PRESERVING BIODIVERSITY AND PROTECTING THE SOIL

- *Plant trees regularly and take care of them.*
- *Reduce your use of wood and paper products, recycle paper products, and buy recycled paper products.*
- *Don't buy furniture, doors, flooring, window frames, paneling, or other products made from tropical hardwoods such as teak or mahogany.* Look for the Good-Wood seal given by Friends of the Earth.
- *Don't buy wood and paper products produced by cutting remaining old-growth forests in the tropics and elsewhere.* Information on such products can be obtained from the Rainforest Action Network, Rainforest Alliance, and Friends of the Earth.
- *Help rehabilitate or restore a degraded area of forest near your home.*
- *Don't buy furs, ivory products, items made of reptile skin, tortoiseshell jewelry, and materials made from endangered or threatened animal species.*
- *When building a home, save all the trees possible.* Require that the contractor **(1)** disturb as little soil as possible, **(2)** set up barriers that catch any soil eroded during construction, and **(3)** save and replace any topsoil removed instead of hauling it off and selling it.
- *Landscape areas not used for gardening with a mix of wildflowers, herbs (for cooking and for repelling insects), low-growing ground cover, small bushes, and other forms of vegetation natural to the area.*
- *Set up a compost bin and use it to produce soil conditioner for yard and garden plants.*

PROMOTING SUSTAINABLE AGRICULTURE AND REDUCING PESTICIDE USE

- *Waste less food.*
- *Eat lower on the food chain* by reducing or eliminating meat consumption to reduce its environmental impact.
- *If you have a dog or a cat, don't feed it canned meat products.* Balanced-grain pet foods are available.
- *Reduce the use of pesticides on agricultural products by asking grocery stores to stock fresh produce and meat produced by organic methods.*
- *Grow some of your own food using organic farming techniques and drip irrigation to water your crops.*
- *Compost your food wastes.*
- *Reduce the threat of overfishing by eating only certain types of fish and shellfish (Table 1).*
- *Think globally, eat locally.* Whenever possible, eat food that is locally grown and in season.
- *Give up the idea that the only good bug is a dead bug.* Recognize that insect species keep most of the populations of pest insects in check and that full-scale chemical warfare on insect pests wipes out many beneficial insects.
- *Don't insist on perfect-looking fruits and vegetables.* These are more likely to contain high levels of pesticide residues.
- *Use pesticides in or around your home only when absolutely necessary, and use them in the smallest amount possible.*
- *Don't become obsessed with having the perfect lawn.* About 40% of U.S. lawns are treated with pesticides. These chemicals can cause **(1)** headaches, **(2)** dizziness, **(3)** nausea, **(4)** eye trouble, and **(5)** even more acute effects in sensitive people (including children who play on treated lawns and in parks).
- *If you hire a lawn care company, use one that relies only on organic methods, and get its claims in writing.*

USING NATURAL CHEMICALS OR METHODS TO CONTROL COMMON INSECT PEST AND WEEDS

- You can usually persuade *ants* to leave within about 4 days by **(1)** sprinkling repellents such as red or cayenne pepper, crushed mint leaves, or boric acid (with an anticaking agent) along their trails inside a house and **(2)** wiping off countertops with vinegar. (However, boric acid is poisonous and should not be placed in areas accessible to small children and pets.)
- Repel *mosquitoes* by **(1)** planting basil outside windows and doors and **(2)** rubbing a bit of vinegar, basil oil, lime juice, or mugwort oil on exposed skin. You can also reduce mosquito attacks by not using scented soaps or wearing perfumes, colognes, and other scented products outdoors during mosquito season. Researchers have found that the $30 million U.S. consumers spend each year on electric bug zappers to kill mosquitoes is mostly wasted. Only about 3% of the insects they kill on an average night are female mosquitoes—the kind that bite.
- Kill *cockroaches* (Spotlight, p. 112) by **(1)** sprinkling boric acid under sinks and ranges, behind refrigerators, and in cabinets, closets, and other dark, warm places (but not in areas accessible to children and pets) or **(2)** establishing populations of banana spiders.
- Trap *cockroaches* by **(1)** filling a bottle or large jar with raw potato, stale beer, banana skins, or other food scraps (especially fruits), **(2)** greasing the inner neck of the bottle with petroleum jelly, and **(3)** placing a small ramp leading to it.
- Repel *flies* by planting sweet basil and tansy (a common herb) near doorways and patios.
- Trap *flies* by making nontoxic flypaper by applying honey to strips of yellow paper (their favorite color) and hanging it from the ceiling in the center of rooms.
- Keep *fleas* off pets by **(1)** using green dye or flea-repellent soaps, **(2)** feeding them

brewer's yeast or vitamin B, **(3)** using flea powders made from eucalyptus, sage, tobacco, wormwood, or vetiver, or **(4)** dipping or shampooing pets in a mixture of water and essential oils such as citronella, cedarwood, eucalyptus, fleabane, sassafras, geranium, clove, or mint.

- Trap *fleas* by **(1)** using green-yellow light to attract them to an adhesive-coated surface or **(2)** putting a light over a shallow pan of water before going to bed at night and turning out all other lights, emptying the water every morning, and continuing this for a month. (Fleas are attracted to heat and light but they cannot swim.)

- Rid a house of fleas by sprinkling carpets with **(1)** desiccant powders, such as Dri-Die, Perma-Guard, or SG-67 to dry them or **(2)** diatomaceous earth (or diatom powder) to kill them. Diatom powder can be purchased in bulk at stores that sell it for use in swimming pool filters. It is also found in some gardening stores under the name of Permatex. Because it contains fine particles of silicate, you should wear a dust mask when applying this powder to avoid inhaling tiny particles of silicate.

- Control *lawn weeds* by raising the cutting level of your lawn mower so the grass can grow 8-10 centimeters (3-4 inches) high. This gives it a strong root system that can hinder weed growth; the higher grass also provides habitats for spiders and other insects that eat insect pests. Pull weeds and douse the hole with soap solution or human urine (which is high enough in nitrogen to kill the weed).

SAVING ENERGY AND REDUCING OUTDOOR AIR POLLUTION

- *Reduce use of fossil fuels.* Drive a car that gets at least 15 kilometers per liter (35 miles per gallon), join a carpool, and use mass transit, walking, and bicycling as much as possible. This reduces emissions of CO_2 and other air pollutants, saves energy and money, and can improve your health.

- *Plant and care for trees to help absorb CO_2.*

- *Insulate new or existing houses heavily, caulk and weatherstrip to reduce air infiltration and heat loss, and use energy-efficient windows.* Add an air-to-air heat exchanger to minimize indoor air pollution.

- *Obtain as much heat and cooling as possible from natural sources,* especially sun, wind, geothermal energy, and trees.

- *Buy the most energy-efficient homes, lights, cars, and appliances available. Evaluate them only in terms of lifetime cost.*

- *Turn down the thermostat on water heaters to 43-49°C (110-120[°F]) and insulate hot water pipes.*

- *Lower the cooling load on an air conditioner by* **(1)** *increasing the thermostat setting,* **(2)** *installing energy-efficient lighting,* **(3)** *using floor and ceiling fans, and* **(4)** *using whole-house window or attic fans to bring in outside air (especially at night, when temperatures are cooler).*

REDUCING EXPOSURE TO INDOOR AIR POLLUTANTS

- *Test for radon and take corrective measures as needed.*

- *Install air-to-air heat exchangers or regularly ventilate your house by opening windows.*

- *At the beginning of the winter heating season, test your indoor air for formaldehyde when the house is closed up.* To locate a testing laboratory in your area, write to the Consumer Product Safety Commission, Washington, DC 20207, or call 301-492-6800.

- *Don't buy furniture and other products containing formaldehyde. Use low-emitting formaldehyde or nonformaldehyde building materials.*

- *Reduce indoor levels of formaldehyde and other toxic gases by growing certain house plants.* Examples are the spider or airplane plant (removes 96% of carbon monoxide), aloe vera (90% of formaldehyde), banana (89% of formaldehyde), elephant ear philodendron (86% of formaldehyde), ficus (weeping fig, 47% of formaldehyde), golden porthos (67% of formaldehyde and benzene and 75% of carbon monoxide), Chinese evergreen (92% of toluene and 81% of benzene), English ivy (90% of benzene), peace lily (80% of benzene and 50% of trichloroethylene), and Janet Craig (corn plant, 79% of benzene). (Toxic removal figures indicate percentage of toxin removed by one plant in a 24-hour period in a 3.4-cubic-meter [12-cubic-foot] space.) Plants should be potted with a mixture of soil and granular charcoal (which absorbs organic air pollutants).

- *Consider not using carpeting and using wood or linoleum floors instead.* New synthetic carpeting releases vapors from more than 100 volatile organic compounds. New and old carpeting is a haven for microbes (many of them highly allergenic), dust, and traces of lead and pesticides brought in by shoes.

- *Remove your shoes before entering your house.* This reduces inputs of dust, lead, and pesticides.

- *Test your house or workplace for asbestos fiber levels and for any crumbling asbestos materials if it was built before 1980. Don't buy a pre-1980 house without having its indoor air tested for asbestos and lead.* To get a free list of certified asbestos laboratories that charge $25-50 to test a sample, call the EPA's Toxic Substances Control Hotline at 202-554-1404.

- *Don't store gasoline, solvents, or other volatile hazardous chemicals inside a home or attached garage.*

- *Don't use aerosol spray products and commercial room deodorizers or air fresheners.*

- *If you smoke, do it outside or in a closed room vented to the outside.*

- *Make sure that wood-burning stoves, fireplaces, and kerosene- and gas-burning heaters are properly installed, vented, and maintained.* Install carbon monoxide detectors in all sleeping areas.

SAVING WATER

- *For existing toilets, reduce the amount of water used per flush by installing a toilet dam.*

- *Install water-saving toilets that use no more than 6 liters (1.6 gallons) per flush.*

- *Flush toilets only when necessary.* Consider using the advice found on a bathroom wall in a drought-stricken area: "If it's yellow, let it mellow; if it's brown, flush it down."

- *Install water-saving showerheads and flow restrictors on all faucets.* If a 3.8-liter (1-gallon) jug can be filled by your showerhead in less than 15 seconds, you need a more efficient fixture.

- *Check frequently for water leaks in toilets and pipes, and repair them promptly.* A toilet must be leaking more than 940 liters (250 gallons) *per day* before you can hear the leak. To test for toilet leaks, add a water-soluble vegetable dye to the water in the tank but don't flush. If you have a leak, some color will show up in the bowl's water within a few minutes.

- *Turn off sink faucets while brushing teeth, shaving, or washing.*

- *Wash only full loads of clothes;* if smaller loads must be used, use the lowest possible water-level setting.

- *When buying a new washer, choose one that uses the least amount of water and that fills up to different levels for loads of different sizes.* Front-loading models use less water and energy than comparable top-loading models.

- *Use automatic dishwashers for full loads only.* Also, use the short cycle and

Table 1 Good and Bad Fish Choices*

Good Choices	Bad Choices
Albacore/Tombo tuna (Pacific)	Bluefin tuna
Calamari/Squid (Pacific)	Chilean seabass/Patagonian toothfish
Catfish (farmed)	Cod (Atlantic)
Clams (farmed; U.S., Canada, New Zealand)	Lingcod
Dungeness crab	Monkfish
Halibut (Alaska)	Orange roughy
Mahi-mahi/Dolphinfish/Dorado	Rockfish/Pacific red snapper/Rock cod
Mussels (farmed: U.S., Canada, New Zealand)	Sablefish/Butterfish/Rock cod
New Zealand cod/Hoki	Salmon
Oysters (farmed: U.S., Canada, New Zealand)	Sea scallops (Atlantic)
Rainbow trout (farmed)	Shark (all)
Salmon (wild-caught from California/Alaska)	Shrimp/Prawns (wild-caught, farmed)
Striped bass (farmed)	Spot prawns (trawl-caught)
Sturgeon (farmed)	Swordfish
Tilapia (farmed)	

*Data from the Monterey Bay Aquarium. For more information and updated lists, see www.montereybayaquarium.org)

let dishes air dry to save energy and money.

- *When washing many dishes by hand, don't let the faucet run.* Instead, use one filled dishpan or sink for washing and another for rinsing.

- *Keep one or more large bottles of water in the refrigerator rather than running water from the tap until it gets cold enough for drinking.*

- *Don't use a garbage disposal system, a large user of water.* Instead, consider composting your food wastes.

- *Wash a car from a bucket of soapy water, and use the hose for rinsing only. If you use a commercial car wash, find one that recycles its water.*

- *Sweep walks and driveways with brooms (not energy-using, noisy, and polluting leaf blowers) instead of hosing them off.*

- *Reduce evaporation losses by watering lawns and gardens in the early morning or evening rather than in the heat of midday or when it's windy.*

- *Use drip irrigation and mulch for gardens and flower beds.* Better yet, landscape with native plants adapted to local aver-age annual precipitation so that watering is unnecessary.

REDUCING WATER POLLUTION

- *Use manure or compost instead of commercial inorganic fertilizers to fertilize garden and yard plants.*

- *Use biological methods or integrated pest management to control garden, yard, and household pests.*

- *Use low-phosphate, phosphate-free, or biodegradable dishwashing liquid, laundry detergent, and shampoo.*

- *Don't use water fresheners in toilets.*

- *Don't pour pesticides, paints, solvents, oil, antifreeze, or other products containing harmful chemicals down the drain or onto the ground.* Contact your local health department about disposal.

- *If you get water from a private well or suspect that municipal water is contaminated, have it tested by an EPA-certified laboratory for lead, nitrates, trihalomethanes, radon, volatile organic compounds, and pesticides.*

- *If you have a septic tank, have it cleaned out every 3-5 years by a reputable contrac-tor so that it won't contribute to ground-water pollution.*

REDUCING SOLID WASTE AND HAZARDOUS WASTE

- *Buy less by asking yourself whether you really need a particular item.*

- *Buy things that are reusable, recyclable, or compostable, and be sure to reuse, recycle, and compost them.*

- *Buy beverages in refillable glass containers instead of cans or throwaway bottles.*

- *Use reusable plastic or metal lunchboxes and metal or plastic garbage containers without throwaway plastic liners (unless such liners are required for garbage collection).*

- *Carry sandwiches and store food in the refrigerator in reusable containers instead of wrapping them in aluminum foil or plastic wrap.*

- *Use rechargeable batteries and recycle them when their useful life is over.*

- *Carry groceries and other items in a reusable basket, a canvas or string bag, or a small cart.*

- *Use sponges and washable cloth napkins, dish towels, and handkerchiefs instead of paper ones.*

- *Do not use throwaway paper and plastic plates, cups, and eating utensils, and other disposable items when reusable or refillable versions are available.*

- *Buy recycled goods, especially those made by primary recycling, and then make an effort to recycle them.*

- *Reduce the amount of junk mail you get.* Do this (as several million Americans have done) at no charge by contacting the Mail Preference Service, Direct Marketing Association, 11 West 42nd St., P.O. Box 3681, New York, NY 10163-3861 (212-768-7277) and asking that your name not be sold to large mailing-list companies. Of the junk mail you do receive, recycle as much of the paper as possible.

- *Buy products in concentrated form whenever possible.*

- *Choose items that have the least packaging-or better yet, no packaging ("nude products").*

- *Do not buy helium-filled balloons that end up as litter. Urge elected officials and school administrators to ban balloon releases except for atmospheric research and monitoring.*

- *Lobby local officials to set up a community composting program.*

- *Use pesticides and other hazardous chemicals only when absolutely necessary and in the smallest amount possible.*

- *Don't dispose of hazardous chemicals by* (1) *flushing them down the toilet,* (2) *pouring them down the drain,* (3) *burying them,* (4) *throwing them into the garbage, or* (5) *dumping them down storm drains.* Consult your local health department or environmental agency for safe disposal methods.

EVALUATING ECOTOURS

Before embarking on an ecotour, seek answers *in writing* to the following questions:

- What precautions are taken to reduce the tour's impact on local ecosystems?

- How much time is spent in the field, versus in the city or traveling in a vehicle?

- What happens to the tour's garbage?

- What percentage of the people involved in planning, organizing, and guiding tours are local?

- Are the guides trained naturalists?

- Will you stay in locally owned hotels or other facilities, or will you be staying in accommodations owned by national or international companies?

- Does the tour operation respect local customs and cultures? If so, how?

- What percentage of the tour's *gross income* goes into the salaries and businesses of local residents?

- What percentage of tour's *gross income* does the tour company donate to local conservation and social projects?

COMMUNICATING WITH ELECTED OFFICIALS

- Find out their names and addresses. Then write, call, fax, or e-mail them. Contact a senator by writing to The Honorable ___, U.S. Senate, Washington, DC 20510; Tel: 202-224-3121, Web: http://thomas.loc.gov. Contact a representative by writing to The Honorable ___, U.S. House of Representatives, Washington, DC 20510; Tel: 202-225-3121, Web: http://thomas.loc.gov. Contact the president by writing to President ___, The White House, 1600 Pennsylvania Ave. NW, Washington, DC 20500; Tel: 202-456-1414, Comment line: 202-456-1111, Fax: 202-456-2461, e-mail: president@whitehouse.gov, Web: www.whitehouse.gov.

- When you write a letter or e-mail, (1) use your own words, (2) be brief and courteous, (3) address only one issue, and (4) ask the elected official to do something specific (such as cosponsoring, supporting, or opposing certain bills). Give reasons for your position, explain its effects on you and your district, try to offer alternatives, share any expert knowledge you have, and ask for a response. Be sure to include your name and return address.

- After your representatives have cast votes* supporting your position, send them a short note of thanks.

- Call and ask to speak to a staff member who works on the issue you are concerned about: for the White House, call 202-456-1414 (Web: www.whitehouse.gov); for the U.S. Senate, call 202-224-3121; for the House of Representatives, call 202-225-3121. The website address for the U.S. Congress is http://thomas.loc.gov.

- Once a desirable bill is passed, call or write to urge the president not to veto it. Urge the members of the appropriations committee to appropriate enough money to implement the law—a crucial decision.

- Monitor and influence action at the state and local levels, where all federal and state laws are either ignored or enforced. As Thomas Jefferson said, "The execution of laws is more important than the making of them."

- Get others who agree with your position to contact their elected officials.

---

*Each year the League of Conservation Voters (see address, phone, and website in Appendix 1) publishes an *Environmental Scorecard* that rates all members of Congress on how they voted on environmental issues.

GLOSSARY

abiotic Nonliving. Compare *biotic*.

absolute humidity Amount of water vapor found in a certain mass of air (usually expressed as grams of water per kilogram of air). Compare *relative humidity*.

acclimation Adjustment to slowly changing new conditions. Compare *threshold effect*.

accuracy Extent to which a measurement agrees with the accepted or correct value for that quantity, based on careful measurements by many people over a long time. Compare *precision*.

acid See *acid solution*.

acid deposition The falling of acids and acid-forming compounds from the atmosphere to the earth's surface. Acid deposition is commonly known as *acid rain*, a term that refers only to wet deposition of droplets of acids and acid-forming compounds.

acid rain See *acid deposition*.

acid solution Any water solution that has more hydrogen ions (H^+) than hydroxide ions (OH^-); any water solution with a pH less than 7. Compare *basic solution, neutral solution*.

active solar heating system System that uses solar collectors to capture energy from the sun and store it as heat for space heating and water heating. Liquid or air pumped through the collectors transfers the captured heat to a storage system such as an insulated water tank or rock bed. Pumps or fans then distribute the stored heat or hot water throughout a dwelling as needed. Compare *passive solar heating system*.

adaptation Any genetically controlled structural, physiological, or behavioral characteristic that helps an organism survive and reproduce under a given set of environmental conditions. It usually results from a beneficial mutation. See *biological evolution, differential reproduction, mutation, natural selection*.

adaptive development Methods used to improve our ability to adapt to inevitable but largely unexpected and unpredictable changes in environmental conditions. See *adaptive management*.

adaptive management Flexible management that views attempts to solve problems as experiments, analyzes failures to see what went wrong, and tries to modify and improve an approach before abandoning it. Because of the inherent unpredictability of complex systems, it often uses the precautionary principle as a management tool. See *precautionary principle*.

adaptive radiation Period of time (usually millions of years) during which numerous new species evolve to fill vacant and new ecological niches in changed environments, usually after a mass extinction.

adaptive trait See *adaptation*.

advanced sewage treatment Specialized chemical and physical processes that reduce the amount of specific pollutants left in wastewater after primary and secondary sewage treatment. This type of treatment usually is expensive. See also *primary sewage treatment, secondary sewage treatment*.

aerobic respiration Complex process that occurs in the cells of most living organisms, in which nutrient organic molecules such as glucose ($C_6H_{12}O_6$) combine with oxygen (O_2) and produce carbon dioxide (CO_2), water (H_2O), and energy. Compare *photosynthesis*.

age structure Percentage of the population (or number of people of each sex) at each age level in a population.

agricultural revolution Gradual shift from small, mobile hunting and gathering bands to settled agricultural communities in which people survived by learning how to breed and raise wild animals and to cultivate wild plants near where they lived. It began 10,000-12,000 years ago. Compare *environmental revolution, hunter-gatherers, industrial revolution, information and globalization revolution*.

agroforestry Planting trees and crops together.

air pollution One or more chemicals in high enough concentrations in the air to harm humans, other animals, vegetation, or materials. Excess heat and noise can also be considered forms of air pollution. Such chemicals or physical conditions are called air pollutants. See *primary pollutant, secondary pollutant*.

albedo Ability of a surface to reflect light.

alien species See *nonnative species*.

allele Slightly different molecular form found in a particular gene.

alley cropping Planting of crops in strips with rows of trees or shrubs on each side.

alpha particle Positively charged matter, consisting of two neutrons and two protons, that is emitted as a form of radioactivity from the nuclei of some radioisotopes. See also *beta particle, gamma rays*.

altitude Height above sea level. Compare *latitude*.

anaerobic respiration Form of cellular respiration in which some decomposers get the energy they need through the breakdown of glucose (or other nutrients) in the absence of oxygen. Compare *aerobic respiration*.

ancient forest See *old-growth forest*.

animal manure Dung and urine of animals that can be used as a form of organic fertilizer. Compare *green manure*.

animals Eukaryotic, multicelled organisms such as sponges, jellyfishes, arthropods (insects, shrimp, lobsters), mollusks (snails, clams, oysters, octopuses), fish, amphibians (frogs, toads, salamanders), reptiles (turtles, lizards, alligators, crocodiles, snakes), birds, and mammals (kangaroos, bats, cats, rabbits, elephants, whales, porpoises, monkeys, apes, humans). See *carnivores, herbivores, omnivores*.

annual Plant that grows, sets seed, and dies in one growing season. Compare *perennial*.

anthropocentric Human-centered. Compare *biocentric*.

appropriate technology Forms of technology that are small scale, efficient, and labor intensive, and use locally available resources to produce goods that benefit local communities.

aquaculture Growing and harvesting of fish and shellfish for human use in freshwater ponds, irrigation ditches, and lakes, or in cages or fenced-in areas of coastal lagoons and estuaries. See *fish farming, fish ranching*.

aquatic Pertaining to water. Compare *terrestrial*.

aquatic life zone Marine and freshwater portions of the ecosphere. Examples include freshwater life zones (such as lakes and streams) and ocean or marine life zones (such as estuaries, coastlines, coral reefs, and the deep ocean).

aquifer Porous, water-saturated layers of sand, gravel, or bedrock that can yield an economically significant amount of water.

arable land Land that can be cultivated to grow crops.

area strip mining Type of surface mining used where the terrain is fairly flat. An earthmover strips away the overburden, and a power shovel digs a cut to remove the mineral deposit. After the mineral is removed, the trench is filled with overburden, and a new cut is made parallel to the previous one. The process is repeated over the entire site. Compare *dredging, open-pit mining, subsurface mining*.

arid Dry. A desert or other area with an arid climate has little precipitation.

artificial selection Process by which humans select one or more desirable genetic traits in the population of a plant or animal and then use *selective breeding* to end up with populations of the species containing large numbers of individuals with the desired traits. Compare *genetic engineering, natural selection*.

asexual reproduction Reproduction in which a mother cell divides to produce two identical daughter cells that are clones of the mother cell. This type of reproduction is common in single-celled organisms. Compare *sexual reproduction*.

atmosphere The whole mass of air surrounding the earth. See *stratosphere, troposphere*.

atom Minute unit made of subatomic particles that is the basic building block of all chemical elements and thus all matter; the smallest unit of an element that can exist and still have the unique characteristics of that element. Compare *ion, molecule*.

atomic number Number of protons in the nucleus of an atom. Compare *mass number*.

autotroph See *producer*.

background extinction Normal extinction of various species as a result of changes in local environmental conditions. Compare *mass extinction*.

bacteria Prokaryotic, one-celled organisms. Some transmit diseases. Most act as decomposers and get the nutrients they need by breaking down complex organic compounds in the tissues of living or dead organisms into simpler inorganic nutrient compounds.

barrier islands Long, thin, low offshore islands of sediment that generally run parallel to the shore along some coasts.

basic solution Water solution with more hydroxide ions (OH^-) than hydrogen ions (H^+); water solution with a pH greater than 7. Compare *acid solution, neutral solution*.

beneficiation Separation of an ore mineral from the waste mineral material (gangue). See *tailings*.

benefit-cost analysis Estimates and comparison of short-term and long-term costs (losses) and benefits (gains) from an economic decision. If the estimated benefits exceed the estimated costs, the decision to buy an economic good or provide a public good is considered worthwhile.

benthos Bottom-dwelling organisms. Compare *decomposer, nekton, plankton*.

beta particle Swiftly moving electron emitted by the nucleus of a radioactive isotope. See also *alpha particle, gamma rays*.

bioaccumulation An increase in the concentration of a chemical in specific organs or tissues at a level higher than would normally be expected. Compare *biomagnification*.

biocentric Worldview with primary focus on individual species or organisms. Compare *anthropocentric*.

biodegradable Capable of being broken down by decomposers.

biodegradable pollutant Material that can be broken down into simpler substances (elements and compounds) by bacteria or other decomposers. Paper and most organic wastes such as animal manure are biodegradable but can take decades to biodegrade in modern landfills. Compare *degradable pollutant, nondegradable pollutant, slowly degradable pollutant*.

biodiversity Variety of different species (*species diversity*), genetic variability among individuals within each species (*genetic diversity*), variety of ecosystems (*ecological diversity*), and functions such as energy flow and matter cycling needed for the survival of species and biological communities (*functional diversity*).

biofuel Gas or liquid fuel (such as ethyl alcohol) made from plant material (biomass).

biogeochemical cycle Natural processes that recycle nutrients in various chemical forms from the nonliving environment to living organisms and then back to the nonliving environment. Examples are the carbon, oxygen, nitrogen, phosphorus, sulfur, and hydrologic cycles.

bioinformatics Applied science of managing, analyzing, and communicating biological information.

biological amplification See *biomagnification*.

biological community See *community*.

biological diversity See *biodiversity*.

biological evolution Change in the genetic makeup of a population of a species in successive generations. If continued long enough, it can lead to the formation of a new species. Note that populations-not individuals-evolve. See also *adaptation, differential reproduction, natural selection, theory of evolution*.

biological oxygen demand (BOD) Amount of dissolved oxygen needed by aerobic decomposers to break down the organic materials in a given volume of water at a certain temperature over a specified time period.

biological pest control Control of pest populations by natural predators, parasites, or disease-causing bacteria and viruses (pathogens).

biomagnification Increase in concentration of DDT, PCBs, and other slowly degradable, fat-soluble chemicals in organisms at successively higher trophic levels of a food chain or web. Compare *bioaccumulation*.

biomass Organic matter produced by plants and other photosynthetic producers; total dry weight of all living organisms that can be supported at each trophic level in a food chain or web; dry weight of all organic matter in plants and animals in an ecosystem; plant materials and animal wastes used as fuel.

biome Terrestrial regions inhabited by certain types of life, especially vegetation. Examples are various types of deserts, grasslands, and forests.

biosphere Zone of earth where life is found. It consists of parts of the atmosphere (the troposphere), hydrosphere (mostly surface water and groundwater), and lithosphere (mostly soil and surface rocks and sediments on the bottoms of oceans and other bodies of water) where life is found. Also called the *ecosphere*.

biotic Living organisms make up the biotic parts of ecosystems. Compare *abiotic*.

biotic potential Maximum rate at which the population of a given species can increase when there are no limits on its rate of growth. See *environmental resistance*.

birth rate See *crude birth rate*.

bitumen Gooey, black, high-sulfur, heavy oil extracted from tar sand and then upgraded to synthetic fuel oil. See *tar sand*.

breeder nuclear fission reactor Nuclear fission reactor that produces more nuclear fuel than it consumes by converting nonfissionable uranium-238 into fissionable plutonium-239.

broadleaf deciduous plants Plants such as oak and maple trees that survive drought and cold by shedding their leaves and becoming dormant. Compare *broadleaf evergreen plants, coniferous evergreen plants*.

broadleaf evergreen plants Plants that keep most of their broad leaves year-round. Examples are the trees found in the canopies of tropical rain forests. Compare *broadleaf deciduous plants, coniferous evergreen plants*.

buffer Substance that can react with hydrogen ions in a solution and thus hold the acidity or pH of a solution fairly constant. See *pH*.

calorie Unit of energy; amount of energy needed to raise the temperature of 1 gram of water $1°C$ (unit on Celsius temperature scale). See also *kilocalorie*.

cancer Group of more than 120 different diseases, one for each type of cell in the human body. Each type of cancer produces a tumor in which cells multiply uncontrollably and invade surrounding tissue.

capitalism See *capitalist market economic system*. Compare *pure command economic system, pure free-market economic system*.

capitalist market economic system Economic system built around controlling market prices of goods and services, global free trade, and maximizing profits for the owners or stockholders whose financial capital the company is using to do business. Compare *pure command economic system, pure free-market economic system*.

carbon cycle Cyclic movement of carbon in different chemical forms from the environment to organisms and then back to the environment.

carcinogen Chemicals, ionizing radiation, and viruses that cause or promote the development of cancer. See *cancer, mutagen, teratogen*.

carnivore Animal that feeds on other animals. Compare *herbivore, omnivore*.

carrying capacity (K) Maximum population of a particular species that a given habitat can support over a given period of time.

cell Smallest living unit of an organism. Each cell is encased in an outer membrane or wall and contains genetic material (DNA) and other parts to perform its life function. Organisms such as bacteria consist of only one cell, but most of the organisms we are familiar with contain many cells. See *eukaryotic cell, prokaryotic cell*.

centrally planned economy See *pure command economic system*.

CFCs See *chlorofluorocarbons*.

chain reaction Multiple nuclear fissions, taking place within a certain mass of a fissionable isotope, that release an enormous amount of energy in a short time.

chaos Behavior that never repeats itself exactly. Examples of chaotic behavior are the waves of an ocean, the movement of leaves in the wind, and day-to-day variations in weather.

chemical One of the millions of different elements and compounds found naturally and synthesized by humans. See *compound, element*.

chemical change Interaction between chemicals in which there is a change in the chemical composition of the elements or compounds involved. Compare *nuclear change, physical change*.

chemical evolution Formation of the earth and its early crust and atmosphere, evolution of the biological molecules necessary for life, and evolution of systems of chemical reactions needed to produce the first living cells. These processes are believed to have occurred about 1 billion years before biological evolution. Compare *biological evolution*.

chemical formula Shorthand way to show the number of atoms (or ions) in the basic structural unit of a compound. Examples are H_2O, $NaCl$, and $C_6H_{12}O_6$.

chemical reaction See *chemical change*.

chemosynthesis Process in which certain organisms (mostly specialized bacteria) extract inorganic compounds from their environment and convert them into organic nutrient compounds without the presence of sunlight. Compare *photosynthesis*.

chlorinated hydrocarbon Organic compound made up of atoms of carbon, hydrogen, and chlorine. Examples are DDT and PCBs.

chlorofluorocarbons (CFCs) Organic compounds made up of atoms of carbon, chlorine, and fluorine. An example is Freon-12 (CCl_2F_2), used as a refrigerant in refrigerators and air conditioners and in making plastics such as Styrofoam. Gaseous CFCs can deplete the ozone layer when they slowly rise into the stratosphere and their chlorine atoms react with ozone molecules. Use of these molecules is being phased out.

chromosome A grouping of various genes and associated proteins in plant and animal cells that carry certain types of genetic information. See *genes*.

city Large group of people with a variety of specialized occupations who live in a specific area and depend on a flow of resources from other areas to meet most of their needs and wants. Compare *village*. See *rural area, urban area*.

civil suit Lawsuit in which a plaintiff seeks to collect damages for injuries or for economic loss or to have the court issue a permanent injunction against further wrongful action. Compare *class action suit*.

class action suit Civil lawsuit in which a group files a suit on behalf of a larger number of citizens who allege similar damages but who need not be listed and represented individually. Compare *civil suit*.

clear-cutting Method of timber harvesting in which all trees in a forested area are removed in a single cutting. Compare *selective cutting, seed-tree cutting, shelterwood cutting, strip cutting*.

climate Physical properties of the troposphere of an area based on analysis of its weather records over a long period (at least 30 years). The two main factors determining an area's climate are *temperature*, with its seasonal variations, and the amount and distribution of *precipitation*. Compare *weather*.

climax community See *mature community*.

closed system System in which energy but not matter is exchanged between the system and its environment. Compare *open system*.

coal Solid, combustible mixture of organic compounds with 30-98% carbon by weight, mixed with various amounts of water and small amounts of sulfur and nitrogen compounds. It is formed in several stages as the remains of plants are subjected to heat and pressure over millions of years.

coal gasification   Conversion of solid coal to synthetic natural gas (SNG).

coal liquefaction   Conversion of solid coal to a liquid hydrocarbon fuel such as synthetic gasoline or methanol.

coastal wetland   Land along a coastline, extending inland from an estuary that is covered with salt water all or part of the year. Examples are marshes, bays, lagoons, tidal flats, and mangrove swamps. Compare *inland wetland*.

coastal zone   Warm, nutrient-rich, shallow part of the ocean that extends from the high-tide mark on land to the edge of a shelflike extension of continental land masses known as the continental shelf. Compare *open sea*.

coevolution   Evolution in which two or more species interact and exert selective pressures on each other that can lead each species to undergo various adaptations. See *evolution, natural selection*.

cogeneration   Production of two useful forms of energy, such as high-temperature heat or steam and electricity, from the same fuel source.

cold front   Leading edge of an advancing mass of cold air. Compare *warm front*.

commensalism   An interaction between organisms of different species in which one type of organism benefits and the other type is neither helped nor harmed to any great degree. Compare *mutualism*.

commercial extinction   Depletion of the population of a wild species used as a resource to a level at which it is no longer profitable to harvest the species.

commercial inorganic fertilizer   Commercially prepared mixture of plant nutrients such as nitrates, phosphates, and potassium applied to the soil to restore fertility and increase crop yields. Compare *organic fertilizer*.

common law   Body of unwritten rules and principles derived from thousands of past legal decisions. It is based on evaluation of what is reasonable behavior in attempting to balance competing social interests. Compare *statutory law*.

common-property resource   Resource that people are normally free to use; each user can deplete or degrade the available supply. Most are potentially renewable and are owned by no one. Examples are clean air, fish in parts of the ocean not under the control of a coastal country, migratory birds, gases of the lower atmosphere, and the ozone content of the upper atmosphere (stratosphere). See *tragedy of the commons*.

community   Populations of all species living and interacting in an area at a particular time.

community development   See *ecological succession*.

competition   Two or more individual organisms of a single species (*intraspecific competition*) or two or more individuals of different species (*interspecific competition*) attempting to use the same scarce resources in the same ecosystem.

competitive exclusion   No two species can occupy exactly the same fundamental niche indefinitely in a habitat where there is not enough of a particular resource to meet the needs of both species. See *ecological niche, fundamental niche, realized niche*.

compost   Partially decomposed organic plant and animal matter that can be used as a soil conditioner or fertilizer.

compound   Combination of atoms, or oppositely charged ions, of two or more different elements held together by attractive forces called chemical bonds. Compare *element*.

concentration   Amount of a chemical in a particular volume or weight of air, water, soil, or other medium.

condensation nuclei   Tiny particles on which droplets of water vapor can collect.

conditions   Physical or chemical attributes of the environment that, without being consumed, influence biological processes and population growth. Examples are temperature, salinity, and acidity. Compare *resources*.

coniferous evergreen plants   Cone-bearing plants (such as spruces, pines, and firs) that keep some of their narrow, pointed leaves (needles) all year. Compare *broadleaf deciduous plants, broadleaf evergreen plants*.

coniferous trees   Cone-bearing trees, mostly evergreens, that have needle-shaped or scalelike leaves. They produce wood known commercially as softwood. Compare *deciduous plants*.

consensus science   Scientific data, models, theories, and laws that are widely accepted by scientists considered experts in the area of study. These results of science are very reliable. Compare *frontier science*.

conservation   Sensible and careful use of natural resources by humans. People with this view are called *conservationists*.

conservation biologist   Biologist who investigates human impacts on the diversity of life found on the earth (biodiversity) and develops practical plans for preserving such biodiversity. Compare *conservationist, ecologist, environmentalist, environmental scientist, preservationist, restorationist*.

conservation biology   Multidisciplinary science created to deal with the crisis of maintaining the genes, species, communities, and ecosystems that make up earth's biological diversity. Its goals are to investigate human impacts on biodiversity and to develop practical approaches to preserving biodiversity and ecological integrity.

conservationist   Person concerned with using natural areas and wildlife in ways that sustain them for current and future generations of humans and other forms of life. Compare *conservation biologist, ecologist, environmentalist, environmental scientist, preservationist, restorationist*.

conservation-tillage farming   Crop cultivation in which the soil is disturbed little (minimum-tillage farming) or not at all (no-till farming) to reduce soil erosion, lower labor costs, and save energy. Compare *conventional-tillage farming*.

constancy   Ability of a living system, such as a population, to maintain a certain size. Compare *inertia, resilience*. See *homeostasis*.

consumer   Organism that cannot synthesize the organic nutrients it needs and gets its organic nutrients by feeding on the tissues of producers or of other consumers; generally divided into *primary consumers* (herbivores), *secondary consumers* (carnivores), *tertiary (higher-level) consumers, omnivores*, and *detritivores* (decomposers and detritus feeders). In economics, one who uses economic goods.

contour farming   Plowing and planting across the changing slope of land, rather than in straight lines, to help retain water and reduce soil erosion.

contour strip mining   Form of mining used on hilly or mountainous terrain. A power shovel cuts a series of terraces into the side of a hill. An earth-mover removes the overburden, and a power shovel extracts the coal, with the overburden from each new terrace dumped onto the one below. Compare *area strip mining, dredging, subsurface mining*.

controlled burning   Deliberately set, carefully controlled surface fires to reduce flammable litter and decrease the chances of damaging crown fires.

conventional-tillage farming   Crop cultivation method in which a planting surface is made by plowing land, breaking up the exposed soil, and then smoothing the surface. Compare *conservation-tillage farming*.

convergence   Resemblance between species belonging to different taxonomic groups as the result of adaptation to similar environments.

convergent plate boundary   Area where earth's lithospheric plates are pushed together. See *sub-*

duction zone. Compare *divergent plate boundary, transform fault*.

coral reef   Formation produced by massive colonies containing billions of tiny coral animals, called polyps, that secrete a stony substance (calcium carbonate) around themselves for protection. When the corals die, their empty outer skeletons form layers that cause the reef to grow. They are found in the coastal zones of warm tropical and subtropical oceans.

core   Inner zone of the earth. It consists of a solid inner core and a liquid outer core. Compare *crust, mantle*.

corridors   Long areas of land that connect habitat that would otherwise become fragmented.

cost-benefit analysis   See *benefit-cost analysis*.

critical mass   Amount of fissionable nuclei needed to sustain a nuclear fission chain reaction.

crop rotation   Planting a field, or an area of a field, with different crops from year to year to reduce soil nutrient depletion. A plant such as corn, tobacco, or cotton, which removes large amounts of nitrogen from the soil, is planted one year. The next year a legume such as soybeans, which adds nitrogen to the soil, is planted.

crown fire   Extremely hot forest fire that burns ground vegetation and treetops. Compare *controlled burning, ground fire, surface fire*.

crude birth rate   Annual number of live births per 1,000 people in the population of a geographic area at the midpoint of a given year. Compare *crude death rate*.

crude death rate   Annual number of deaths per 1,000 people in the population of a geographic area at the midpoint of a given year. Compare *crude birth rate*.

crude oil   Gooey liquid consisting mostly of hydrocarbon compounds and small amounts of compounds containing oxygen, sulfur, and nitrogen. Extracted from underground accumulations, it is sent to oil refineries, where it is converted to heating oil, diesel fuel, gasoline, tar, and other materials.

crust   Solid outer zone of the earth. It consists of oceanic crust and continental crust. Compare *core, mantle*.

cultural eutrophication   Overnourishment of aquatic ecosystems with plant nutrients (mostly nitrates and phosphates) because of human activities such as agriculture, urbanization, and discharges from industrial plants and sewage treatment plants. See *eutrophication*.

cyanobacteria   Single-celled, prokaryotic, microscopic organisms. Before being reclassified as monera, they were called blue-green algae.

DDT   Dichlorodiphenyltrichloroethane, a chlorinated hydrocarbon that has been widely used as a pesticide but is now banned in some countries.

death rate   See *crude death rate*.

debt-for-nature swap   Agreement in which a certain amount of foreign debt is canceled in exchange for local currency investments that will improve natural resource management or protect certain areas in the debtor country from harmful development.

deciduous plants   Trees, such as oaks and maples, and other plants that survive during dry seasons or cold seasons by shedding their leaves. Compare *coniferous trees, succulent plants*.

decomposer   Organism that digests parts of dead organisms and cast-off fragments and wastes of living organisms by breaking down the complex organic molecules in those materials into simpler inorganic compounds and then absorbing the soluble nutrients. Producers return most of these chemicals to the soil and water for reuse. Decomposers consist of various bacteria and fungi. Compare *consumer, detritivore, producer*.

deductive reasoning   Using logic to arrive at a specific conclusion based on a generalization or premise. It goes from the general to the specific. Compare *inductive reasoning*.

defendant   The individual, group of individuals, corporation, or government agency being charged in a lawsuit. Compare *plaintiff*.

deforestation   Removal of trees from a forested area without adequate replanting.

degradable pollutant   Potentially polluting chemical that is broken down completely or reduced to acceptable levels by natural physical, chemical, and biological processes. Compare *biodegradable pollutant, nondegradable pollutant, slowly degradable pollutant*.

degree of urbanization   Percentage of the population in the world, or a country, living in areas with a population of more than 2,500 people (higher in some countries). Compare *urban growth*.

democracy   Government by the people through their elected officials and appointed representatives. In a *constitutional democracy*, a constitution provides the basis of government authority and puts restraints on government power through free elections and freely expressed public opinion.

demographic transition   Hypothesis that countries, as they become industrialized, have declines in death rates followed by declines in birth rates.

depletion time   How long it takes to use a certain fraction-usually 80%-of the known or estimated supply of a nonrenewable resource at an assumed rate of use. Finding and extracting the remaining 20% usually costs more than it is worth.

desalination   Purification of salt water or brackish (slightly salty) water by removal of dissolved salts.

desert   Biome in which evaporation exceeds precipitation and the average amount of precipitation is less than 25 centimeters (10 inches) a year. Such areas have little vegetation or have widely spaced, mostly low vegetation. Compare *forest, grassland*.

desertification   Conversion of rangeland, rainfed cropland, or irrigated cropland to desertlike land, with a drop in agricultural productivity of 10% or more. It usually is caused by a combination of overgrazing, soil erosion, prolonged drought, and climate change.

detritivore   Consumer organism that feeds on detritus, parts of dead organisms, and cast-off fragments and wastes of living organisms. The two principal types are *detritus feeders* and *decomposers*.

detritus   Parts of dead organisms and cast-off fragments and wastes of living organisms.

detritus feeder   Organism that extracts nutrients from fragments of dead organisms and their cast-off parts and organic wastes. Examples are earthworms, termites, and crabs. Compare *decomposer*.

deuterium (D; hydrogen-2)   Isotope of the element hydrogen, with a nucleus containing one proton and one neutron and a mass number of 2.

developed country   Country that is highly industrialized and has a high per capita GNP. Compare *developing country*.

developing country   Country that has low to moderate industrialization and low to moderate per capita GNP. Most are located in Africa, Asia, and Latin America. Compare *developed country*.

dew point   Temperature at which condensation occurs for a given amount of water vapor.

dieback   Sharp reduction in the population of a species when its numbers exceed the carrying capacity of its habitat. See *carrying capacity*.

differential reproduction   Phenomenon in which individuals with adaptive genetic traits produce more living offspring than do individuals without such traits. See *natural selection*.

dioxins   Family of 75 different chlorinated hydrocarbon compounds formed as unwanted by-products in chemical reactions involving chlorine and hydrocarbons, usually at high temperatures.

discount rate   The economic value a resource will have in the future compared with its present value.

dissolved oxygen (DO) content   Amount of oxygen gas (O_2) dissolved in a given volume of water at a particular temperature and pressure, often expressed as a concentration in parts of oxygen per million parts of water.

disturbance   A discrete event that disrupts an ecosystem or community. Examples of *natural disturbances* include fires, hurricanes, tornadoes, droughts, and floods. Examples of *human-caused disturbances* include deforestation, overgrazing, and plowing.

divergent plate boundary   Area where earth's lithospheric plates move apart in opposite directions. Compare *convergent plate boundary, transform fault*.

DNA (deoxyribonucleic acid)   Large molecules in the cells of organisms that carry genetic information in living organisms.

domesticated species   Wild species tamed or genetically altered by crossbreeding for use by humans for food (cattle, sheep, and food crops), pets (dogs and cats), or enjoyment (animals in zoos and plants in gardens).

dose   The amount of a potentially harmful substance an individual ingests, inhales, or absorbs through the skin. Compare *response*. See *dose-response curve, median lethal dose*.

dose-response curve   Plot of data showing effects of various doses of a toxic agent on a group of test organisms. See *dose, median lethal dose, response*.

doubling time   The time it takes (usually in years) for the quantity of something growing exponentially to double. It can be calculated by dividing the annual percentage growth rate into 70.

drainage basin   See *watershed*.

dredge spoils   Materials scraped from the bottoms of harbors and streams to maintain shipping channels. They are often contaminated with high levels of toxic substances that have settled out of the water. See *dredging*.

dredging   Type of surface mining in which chain buckets and draglines scrape up sand, gravel, and other surface deposits covered with water. It is also used to remove sediment from streams and harbors to maintain shipping channels. See *dredge spoils*.

drift-net fishing   Catching fish in huge nets that drift in the water.

drought   Condition in which an area does not get enough water because of lower-than-normal precipitation or higher-than-normal temperatures that increase evaporation.

dust dome   Dome of heated air that surrounds an urban area and traps pollutants, especially suspended particulate matter. See also *urban heat island*.

dust plume   Elongation of a dust dome by winds that can spread a city's pollutants hundreds of kilometers downwind.

early successional plant species   Plant species found in the early stages of succession that (1) grow close to the ground, (2) can establish large populations quickly under harsh conditions, and (3) have short lives. Compare *late successional plant species, midsuccessional plant species*.

earth capital   See *natural resources*.

earthquake   Shaking of the ground resulting from the fracturing and displacement of rock, which produces a fault, or from subsequent movement along the fault.

earth resources   See *natural resources*.

earth-sustaining economy   See *low-throughput economy*. Compare *high-throughput economy*.

earth-wisdom revolution   See *environmental revolution*.

earth-wisdom society   See *environmentally sustainable society*.

earth-wisdom worldview   See *environmental wisdom worldview*.

ecological diversity   The variety of forests, deserts, grasslands, oceans, streams, lakes, and other biological communities interacting with one another and with their nonliving environment. See *biological diversity*. Compare *genetic diversity, species diversity*.

ecological efficiency   Percentage of energy transferred from one trophic level to another in a food chain or web.

ecological footprint   A measure of the ecological impact of the (1) consumption of food, wood products, and other resources, (2) use of buildings, roads, garbage dumps, and other things that consume land space, and (3) destruction of the forests needed to absorb the CO_2 produced by burning fossil fuels.

ecological land-use planning   Method for deciding how land should be used; development of an integrated model that considers geological, ecological, health, and social variables.

ecologically sustainable development   Development in which the total human population size and resource use in the world (or in a region) are limited to a level that does not exceed the carrying capacity of the existing natural capital and is therefore sustainable. Compare *economic development, economic growth, sustainable economic development*.

ecological niche   Total way of life or role of a species in an ecosystem. It includes all physical, chemical, and biological conditions a species needs to live and reproduce in an ecosystem. See *fundamental niche, realized niche*.

ecological population density   Number of individuals of a population per unit area of habitat. Compare *population density*.

ecological resource   Anything needed by an organism for normal maintenance, growth, and reproduction. Examples include habitat, food, water, and shelter. Compare *economic resource*.

ecological restoration   Deliberate alteration of a degraded habitat or ecosystem to restore as much of its ecological structure and function as possible.

ecological succession   Process in which communities of plant and animal species in a particular area are replaced over time by a series of different and often more complex communities. See *primary succession, secondary succession*.

ecologist   Biological scientist who studies relationships between living organisms and their environment. Compare *conservation biologist, conservationist, environmentalist, environmental scientist, preservationist, restorationist*.

ecology   Study of the interactions of living organisms with one another and with their nonliving environment of matter and energy; study of the structure and functions of nature.

economic decision   Deciding (1) what goods and services to produce, (2) how to produce them, (3) how much to produce, and (4) how to distribute them to people.

economic depletion   Exhaustion of 80% of the estimated supply of a nonrenewable resource. Finding, extracting, and processing the remaining 20% usually costs more than it is worth; may also apply to the depletion of a potentially renewable resource, such as a species of fish or tree.

economic development   Improvement of living standards by economic growth. Compare *ecologically sustainable development, economic growth, sustainable development*.

economic growth   Increase in the real value of all final goods and services produced by an economy; an increase in real GNP. Compare *economic development, environmentally sustainable economic development, sustainable economic development*.

economic resource   Anything obtained from the environment (the earth's life-support systems) to meet human needs and wants. Examples include food, water, shelter, manufactured goods, transportation, communication, and recreation. They include natural resources, human resources, financial resources, and manufactured resources. Compare *ecological resource*.

economic resources Natural resources, capital goods, and labor used in an economy to produce material goods and services. See *natural resources*.

economic system Method that a group of people uses to choose (1) what goods and services to produce, (2) how to produce them, (3) how much to produce, and (4) how to distribute them to people. See *mixed economic system, pure command economic system, pure free-market economic system*.

economic threshold Point at which the economic loss caused by pest damage outweighs the cost of applying a pesticide.

economy System of production, distribution, and consumption of economic goods.

ecosphere See *biosphere*.

ecosystem Community of different species interacting with one another and with the chemical and physical factors making up its nonliving environment.

ecosystem services Natural services or natural capital that support life on the earth and are essential to the quality of human life and the functioning of the world's economies. See *natural resources*.

ecotone Transitional zone in which one type of ecosystem tends to merge with another ecosystem. See *edge effect*.

edge effect The existence of a greater number of species and a higher population density in an ecotone than in either adjacent ecosystem. See *ecotone*.

electromagnetic radiation Forms of kinetic energy traveling as electromagnetic waves. Examples are radio waves, TV waves, microwaves, infrared radiation, visible light, ultraviolet radiation, X rays, and gamma rays. Compare *ionizing radiation, nonionizing radiation*.

electron (e) Tiny particle moving around outside the nucleus of an atom. Each electron has one unit of negative charge (−) and almost no mass. Compare *neutron, proton*.

element Chemical, such as hydrogen (H), iron (Fe), sodium (Na), carbon (C), nitrogen (N), or oxygen (O), whose distinctly different atoms serve as the basic building blocks of all matter. There are 92 naturally occurring elements. Another 26 have been made in laboratories. Two or more elements combine to form compounds that make up most of the world's matter. Compare *compound*.

endangered species Wild species with so few individual survivors that the species could soon become extinct in all or most of its natural range. Compare *threatened species*.

endemic species Species that is found in only one area. Such species are especially vulnerable to extinction.

energy Capacity to do work by performing mechanical, physical, chemical, or electrical tasks or to cause a heat transfer between two objects at different temperatures.

energy efficiency Percentage of the total energy input that does useful work and is not converted into low-quality, usually useless heat in an energy conversion system or process. See *energy quality, net energy*. Compare *material efficiency*.

energy productivity See *energy efficiency*.

energy quality Ability of a form of energy to do useful work. High-temperature heat and the chemical energy in fossil fuels and nuclear fuels are concentrated high-quality energy. Low-quality energy such as low-temperature heat is dispersed or diluted and cannot do much useful work. See *high-quality energy, low-quality energy*.

enhanced oil recovery Removal of some of the heavy oil left in an oil well after primary and secondary recovery. Compare *primary oil recovery, secondary oil recovery*.

entropy A measure of the disorder or randomness of a system. The greater the disorder of a system, the higher its entropy; the greater its order, the lower its entropy.

environment All external conditions and factors, living and nonliving (chemicals and energy), that affect an organism or other specified system during its lifetime.

environmental degradation Depletion or destruction of a potentially renewable resource such as soil, grassland, forest, or wildlife that is used faster than it is naturally replenished. If such use continues, the resource can become nonrenewable (on a human time scale) or nonexistent (extinct). See also *sustainable yield*.

environmental ethics Our beliefs about what is right or wrong environmental behavior.

environmentalist Person who is concerned about the impact of people on environmental quality and believe that some human actions are degrading parts of the earth's life-support systems for humans and many other forms of life. Some of their beliefs and proposals for dealing with environmental problems are based on scientific information and concepts and some are based on their social and ethical environmental beliefs (environmental worldviews). Compare *conservation biologist, conservationist, ecologist, environmental scientist, preservationist, restorationist*.

environmental justice Fair treatment and meaningful involvement of all people regardless of race, color, national origin, or income with respect to the development, implementation, and enforcement of environmental laws, regulations, and policies.

environmentally sustainable economic development Development that (1) *encourages* environmentally sustainable forms of economic growth that meet the basic needs of the current generations of humans and other species without preventing future generations of humans and other species from meeting their basic needs and (2) *discourages* environmentally harmful and unsustainable forms of economic growth. It is the economic component of an *environmentally sustainable society*.

environmentally sustainable society Society that satisfies the basic needs of its people without depleting or degrading its natural resources and thereby preventing current and future generations of humans and other species from meeting their basic needs.

environmental movement Efforts by citizens at the grassroots level to demand that political leaders enact laws and develop policies to (1) curtail pollution, (2) clean up polluted environments, and (3) protect pristine areas from environmental degradation.

environmental resistance All the limiting factors that act together to limit the growth of a population. See *biotic potential, limiting factor*.

environmental revolution Cultural change involving halting population growth and altering lifestyles, political and economic systems, and the way we treat the environment so that we can help sustain the earth for ourselves and other species. This involves working with the rest of nature by learning more about how nature sustains itself. See *environmental wisdom worldview*. Compare *agricultural revolution, hunter-gatherers, industrial revolution, information and globalization revolution*.

environmental science Study of how we and other species interact with one another and with the nonliving environment (matter and energy). It is a physical and social science that integrates knowledge from a wide range of disciplines, including physics, chemistry, biology (especially ecology), geology, geography, resource technology and engineering, resource conservation and management, demography (the study of population dynamics), economics, politics, sociology, psychology, and ethics. In other words, it is a study of how the parts of nature and human societies operate and interact—a study of connections and interactions.

environmental scientist Scientist who uses information from the physical sciences and social sciences to (1) understand how the earth works, (2) learn how humans interact with the earth, and (3) develop solutions to environmental problems. Compare *conservation biologist, conservationist, ecologist, preservationist, restorationist*.

environmental wisdom worldview Beliefs that (1) nature exists for all the earth's species, not just for us, and we are not in charge of the rest of nature; (2) there is not always more, and it's not all for us; (3) some forms of economic growth are beneficial and some are harmful, and our goals should be to design economic and political systems that encourage earth-sustaining forms of growth and discourage or prohibit earth-degrading forms; and (4) our success depends on learning to cooperate with one another and with the rest of nature instead of trying to dominate and manage earth's life-support systems primarily for our own use. Compare *frontier worldview, planetary management worldview, spaceship-earth worldview*.

environmental worldview How individuals think the world works, what they think their role in the world should be, and what they believe is right and wrong environmental behavior (environmental ethics).

EPA Environmental Protection Agency; responsible for managing federal efforts in the United States to control air and water pollution, radiation and pesticide hazards, environmental research, hazardous waste, and solid-waste disposal.

epidemiology Study of the patterns of disease or other harmful effects from toxic exposure within defined groups of people to find out why some people get sick and some do not.

epiphyte Plant that uses its roots to attach itself to branches high in trees, especially in tropical forests.

erosion Process or group of processes by which loose or consolidated earth materials are dissolved, loosened, or worn away and removed from one place and deposited in another. See *weathering*.

estuary Partially enclosed coastal area at the mouth of a river where its fresh water, carrying fertile silt and runoff from the land, mixes with salty seawater.

eukaryotic cell Cell containing a *nucleus*, a region of genetic material surrounded by a membrane. Membranes also enclose several of the other internal parts found in a eukaryotic cell. Compare *prokaryotic cell*.

euphotic zone Upper layer of a body of water through which sunlight can penetrate and support photosynthesis.

eutrophication Physical, chemical, and biological changes that take place after a lake, estuary, or slow-flowing stream receives inputs of plant nutrients-mostly nitrates and phosphates-from natural erosion and runoff from the surrounding land basin. See *cultural eutrophication*.

eutrophic lake Lake with a large or excessive supply of plant nutrients, mostly nitrates and phosphates. Compare *mesotrophic lake, oligotrophic lake*.

evaporation Conversion of a liquid into a gas.

even-aged management Method of forest management in which trees, sometimes of a single species in a given stand, are maintained at about the same age and size and are harvested all at once. Compare *uneven-aged management*.

evergreen plants Plants that keep some of their leaves or needles throughout the year. Examples are ferns and cone-bearing trees (conifers) such as firs, spruces, pines, redwoods, and sequoias. Compare *deciduous plants, succulent plants*.

evolution See *biological evolution*.

exhaustible resource See *nonrenewable resource*.

exotic species See *nonnative species*.

experiment Procedure a scientist uses to study some phenomenon under known conditions. Some experiments are conducted in the laboratory, but others are conducted in nature. The resulting scientific data or facts must be verified or confirmed by repeated observations and measurements, ideally by several different investigators.

exploitation competition Situation in which two competing species have equal access to a specific resource but differ in how quickly or efficiently they exploit it. See *interference competition, interspecific competition*.

exponential growth Growth in which some quantity, such as population size or economic output, increases by a fixed percentage of the whole in a given time period; when the increase in quantity over time is plotted, this type of growth yields a curve shaped like the letter *J*. Compare *linear growth*.

external benefit Beneficial social effect of producing and using an economic good that is not included in the market price of the good. Compare *external cost, full cost*.

external cost Harmful social effect of producing and using an economic good that is not included in the market price of the good. Compare *external benefit, full cost*.

externalities Social benefits ("goods") and social costs ("bads") not included in the market price of an economic good. See *external benefit, external cost*. Compare *full cost, internal cost*.

extinction Complete disappearance of a species from the earth. This happens when a species cannot adapt and successfully reproduce under new environmental conditions or when it evolves into one or more new species. Compare *speciation*. See also *endangered species, threatened species*.

family planning Providing information, clinical services, and contraceptives to help people choose the number and spacing of children they want to have.

famine Widespread malnutrition and starvation in a particular area because of a shortage of food, usually caused by drought, war, flood, earthquake, or other catastrophic events that disrupt food production and distribution.

feedback loop Circuit of sensing, evaluating, and reacting to changes in environmental conditions as a result of information fed back into a system; it occurs when one change leads to some other change, which eventually reinforces or slows the original change. See *negative feedback loop, positive feedback loop*.

feedlot Confined outdoor or indoor space used to raise hundreds to thousands of domesticated livestock. Compare *rangeland*.

fermentation See *anaerobic respiration*.

fertilizer Substance that adds inorganic or organic plant nutrients to soil and improves its ability to grow crops, trees, or other vegetation. See *commercial inorganic fertilizer, organic fertilizer*.

financial resources Cash, investments, and monetary institutions used to support the use of natural resources and human resources to provide economic goods and services. Compare *human resources, manufactured resources, natural resources*.

first law of energy See *first law of thermodynamics*.

first law of thermodynamics In any physical or chemical change, no detectable amount of energy is created or destroyed, but in these processes energy can be changed from one form to another; you can't get more energy out of something than you put in; in terms of energy quantity, you can't get something for nothing (there is no free lunch). This law does not apply to nuclear changes, in which energy can be produced from small amounts of matter. See also *second law of thermodynamics*.

fishery Concentrations of particular aquatic species suitable for commercial harvesting in a given ocean area or inland body of water.

fish farming Form of aquaculture in which fish are cultivated in a controlled pond or other environment and harvested when they reach the desired size. See also *fish ranching*.

fish ranching Form of aquaculture in which members of a fish species such as salmon are held in captivity for the first few years of their lives, released, and then harvested as adults when they return from the ocean to their freshwater birthplace to spawn. See also *fish farming*.

fissionable isotope Isotope that can split apart when hit by a neutron at the right speed and thus undergo nuclear fission. Examples are uranium-235 and plutonium-239.

floodplain Flat valley floor next to a stream channel. For legal purposes the term is often applied to any low area that has the potential for flooding, including certain coastal areas.

flows See *throughputs*.

flyway Generally fixed route along which waterfowl migrate from one area to another at certain seasons of the year.

food chain Series of organisms in which each eats or decomposes the preceding one. Compare *food web*.

food web Complex network of many interconnected food chains and feeding relationships. Compare *food chain*.

forest Biome with enough average annual precipitation (at least 76 centimeters, or 30 inches) to support growth of various species of trees and smaller forms of vegetation. Compare *desert, grassland*.

fossil fuel Products of partial or complete decomposition of plants and animals that occur as crude oil, coal, natural gas, or heavy oils as a result of exposure to heat and pressure in earth's crust over millions of years. See *coal, crude oil, natural gas*.

fossils Skeletons, bones, shells, body parts, leaves, seeds, or impressions of such items that provide recognizable evidence of organisms that lived long ago.

free-access resources See *common-property resource*.

Freons See *chlorofluorocarbons*.

freshwater life zones Aquatic systems where water with a dissolved salt concentration of less than 1% by volume accumulates on or flows through the surfaces of terrestrial biomes. Examples are **(1)** *standing* (lentic) bodies of fresh water such as lakes, ponds, and inland wetlands and **(2)** *flowing* (lotic) systems such as streams and rivers. Compare *biomes*.

front The boundary between two air masses with different temperatures and densities. See *cold front, warm front*.

frontier science Preliminary scientific data, hypotheses, and models that have not been widely tested and accepted. Compare *consensus science*.

frontier worldview Viewing undeveloped land as a hostile wilderness to be conquered (cleared, planted) and exploited for its resources as quickly as possible. Compare *environmental wisdom worldview, planetary management worldview, spaceship-earth worldview*.

full cost Cost of a good when its internal costs and its estimated short- and long-term external costs are included in its market price. Compare *external cost, internal cost*.

functional diversity Biological and chemical processes or functions such as energy flow and matter cycling needed for the survival of species and biological communities. See *biodiversity, ecosystem diversity, genetic diversity, species diversity*.

fundamental niche The full potential range of the physical, chemical, and biological factors a species can use if there is no competition from other species. See *ecological niche*. Compare *realized niche*.

fungi Eukaryotic, mostly multicelled organisms such as mushrooms, molds, and yeasts. As decomposers, they get the nutrients they need by secreting enzymes that speed up the breakdown of organic matter in the tissue of other living or dead organisms. Then they absorb the resulting nutrients.

fungicide Chemical that kills fungi.

Gaia hypothesis Hypothesis that earth is alive and can be considered a system that operates and changes by feedback of information between its living and nonliving components.

game species Type of wild animal that people hunt or fish for, for sport and recreation and sometimes for food.

gamma rays A form of ionizing electromagnetic radiation with a high energy content emitted by some radioisotopes. They readily penetrate body tissues. See also *alpha particles, beta particles*.

gangue Waste or undesired material in an ore. See *ore*.

gap analysis Scientific method used to determine how adequately native plant and animal species and the existing network of conservation lands protects natural communities. Species and communities not adequately represented in existing conservation lands constitute conservation gaps. The idea is to identify these gaps and then eliminate them by establishing new reserves or changing land management practices.

GDP See *gross domestic product*.

gene flow Movement of genes between populations, which can lead to changes in the genetic composition of local populations.

gene mutation See *mutation*.

gene pool The sum total of all genes found in the individuals of the population of a particular species.

generalist species Species with a broad ecological niche. They can live in many different places, eat a variety of foods, and tolerate a wide range of environmental conditions. Examples are flies, cockroaches, mice, rats, and human beings. Compare *specialist species*.

genes Coded units of information about specific traits that are passed on from parents to offspring during reproduction. They consist of segments of DNA molecules found in chromosomes.

gene splicing See *genetic engineering*.

genetic adaptation Changes in the genetic makeup of organisms of a species that allow the species to reproduce and gain a competitive advantage under changed environmental conditions. See *differential reproduction, evolution, mutation, natural selection*.

genetically modified organism (GMO) Organism whose genetic makeup has been modified by genetic engineering.

genetic diversity Variability in the genetic makeup among individuals within a single species. See *biodiversity*. Compare *ecological diversity, species diversity*.

genetic drift Change in the genetic composition of a population by chance. It is especially important for small populations.

genetic engineering Insertion of an alien gene into an organism to give it a beneficial genetic trait. Compare *artificial selection, natural selection*.

genome Complete set of genetic information for an organism.

geographic isolation Separation of populations of a species for fairly long times into areas with different environmental conditions.

geology Study of the earth's dynamic history. Geologists study and analyze rocks and the features and processes of the earth's interior and surface.

geothermal energy Heat transferred from the earth's underground concentrations of **(1)** dry steam (steam with no water droplets), **(2)** wet steam (a mixture of steam and water droplets), or **(3)** hot water trapped in fractured or porous rock.

globalization Broad process of global social, economic, and environmental change that leads to an increasingly integrated world.

global warming Warming of the earth's atmosphere as a result of increases in the concentrations of one or more greenhouse gases. See *greenhouse effect, greenhouse gases.*

GNP See *gross national product.*

grassland Biome found in regions where moderate annual average precipitation (25 to 76 centimeters, or 10 to 30 inches) is enough to support the growth of grass and small plants but not enough to support large stands of trees. Compare *desert, forest.*

greenhouse effect A natural effect that releases heat in the atmosphere (troposphere) near the earth's surface. Water vapor, carbon dioxide, ozone, and several other gases in the lower atmosphere (troposphere) absorb some of the infrared radiation (heat) radiated by the earth's surface and then radiate it back toward the earth's surface. This causes their molecules to vibrate and transform the absorbed energy into longer-wavelength infrared radiation (heat) in the troposphere. If the atmospheric concentrations of these greenhouse gases rise and they are not removed by other natural processes, the average temperature of the lower atmosphere will increase gradually. Because the energy release into the atmosphere by greenhouse gases is not the same wavelength as the energy they absorbed, it is scientifically incorrect to say that greenhouse gases trap and reradiate energy released from the earth's surface.

greenhouse gases Gases in the earth's lower atmosphere (troposphere) that cause the greenhouse effect. Examples are carbon dioxide, chlorofluorocarbons, ozone, methane, water vapor, and nitrous oxide.

green manure Freshly cut or still-growing green vegetation that is plowed into the soil to increase the organic matter and humus available to support crop growth. Compare *animal manure.*

green revolution Popular term for introduction of scientifically bred or selected varieties of grain (rice, wheat, maize) that, with high enough inputs of fertilizer and water, can greatly increase crop yields.

gross domestic product (GDP) Total market value in current dollars of all goods and services produced *within a country* for final use, usually during a year. Compare *gross national product, gross world product.*

gross national product (GNP) Total market value in current dollars of all goods and services produced by an economy for final use usually during a year. Compare *gross domestic product, gross world product.*

gross primary productivity (GPP) The rate at which an ecosystem's producers capture and store a given amount of chemical energy as biomass in a given length of time. Compare *net primary productivity.*

gross world product (GWP) Market value in current dollars of all goods and services produced in the world each year. Compare *gross domestic product, gross national product.*

ground fire Fire that burns decayed leaves or peat deep below the ground surface. Compare *crown fire, surface fire.*

groundwater Water that sinks into the soil and is stored in slowly flowing and slowly renewed underground reservoirs called aquifers; underground water in the zone of saturation, below the water table. Compare *runoff, surface water.*

gully reclamation Restoring land suffering from gully erosion by seeding gullies with quick-growing plants, building small dams to collect silt and gradually fill in the channels, and building channels to divert water away from the gully.

habitat Place or type of place where an organism or population of organisms lives. Compare *ecological niche.*

habitat fragmentation Breakup of a habitat into smaller pieces, usually as a result of human activities.

half-life Time needed for one-half of the nuclei in a radioisotope to emit its radiation. Each radioisotope has a characteristic half-life, which may range from a few millionths of a second to several billion years.

hazard Something that can cause injury, disease, economic loss, or environmental damage. See also *risk.*

hazardous chemical Chemical that can cause harm because it (1) is flammable or explosive, (2) can irritate or damage the skin or lungs (such as strong acidic or alkaline substances), or (3) can cause allergic reactions of the immune system (allergens). See *toxic chemical.*

hazardous waste Any solid, liquid, or containerized gas that (1) can catch fire easily, (2) is corrosive to skin tissue or metals, (3) is unstable and can explode or release toxic fumes, or (4) has harmful concentrations of one or more toxic materials that can leach out. See also *toxic waste.*

heat Total kinetic energy of all the randomly moving atoms, ions, or molecules within a given substance, excluding the overall motion of the whole object. This form of kinetic energy flows from one body to another when there is a temperature difference between the two bodies. Heat always flows spontaneously from a hot sample of matter to a colder sample of matter. This is one way to state the second law of thermodynamics. Compare *temperature.*

herbicide Chemical that kills a plant or inhibits its growth.

herbivore Plant-eating organism. Examples are deer, sheep, grasshoppers, and zooplankton. Compare *carnivore, omnivore.*

heterotroph See *consumer.*

high-input agriculture See *industrialized agriculture.*

high-quality energy Energy that is organized or concentrated and has great ability to perform useful work. Examples are high-temperature heat and the energy in electricity, coal, oil, gasoline, sunlight, and nuclei of uranium-235. Compare *low-quality energy.*

high-quality matter Matter that is organized and concentrated and contains a high concentration of a useful resource. Compare *low-quality matter.*

high-throughput economy The situation in most advanced industrialized countries, in which ever-increasing economic growth is sustained by maximizing the rate at which matter and energy resources are used, with little emphasis on pollution prevention, recycling, reuse, reduction of unnecessary waste, and other forms of resource conservation. Compare *low-throughput economy, matter-recycling economy.*

high-throughput society See *high-throughput economy.*

high-waste society See *high-throughput economy.*

homeostasis Maintenance of favorable internal conditions in a system despite fluctuations in external conditions. See *constancy, inertia, resilience.*

host Plant or animal on which a parasite feeds.

human capital See *human resources.*

human resources Physical and mental talents of people used to produce, distribute, and sell an economic good. Compare *financial resources, manufactured resources, natural resources.*

humus Slightly soluble residue of undigested or partially decomposed organic material in topsoil. This material helps retain water and water-soluble nutrients, which can be taken up by plant roots.

hunter-gatherers People who get their food by gathering edible wild plants and other materials and by hunting wild animals and fish. Compare *agricultural revolution, environmental revolution,*

industrial revolution, information and globalization revolution.

hydrocarbon Organic compound of hydrogen and carbon atoms. The simplest hydrocarbon is methane (CH_4), the major component of natural gas.

hydroelectric power plant Structure in which the energy of falling or flowing water spins a turbine generator to produce electricity.

hydrologic cycle Biogeochemical cycle that collects, purifies, and distributes the earth's fixed supply of water from the environment to living organisms and then back to the environment.

hydropower Electrical energy produced by falling or flowing water. See *hydroelectric power plant.*

hydrosphere The earth's (1) liquid water (oceans, lakes, other bodies of surface water, and underground water), (2) frozen water (polar ice caps, floating ice caps, and ice in soil, known as permafrost), and (3) small amounts of water vapor in the atmosphere. See also *hydrologic cycle.*

identified resources Deposits of a particular mineral-bearing material of which the location, quantity, and quality are known or have been estimated from direct geological evidence and measurements. Compare *undiscovered resources.*

igneous rock Rock formed when molten rock material (magma) wells up from the earth's interior, cools, and solidifies into rock masses. Compare *metamorphic rock, sedimentary rock.* See *rock cycle.*

immature community Community at an early stage of ecological succession. It usually has a low number of species and ecological niches and cannot capture and use energy and cycle critical nutrients as efficiently as more complex, mature ecosystems. Compare *mature community.*

immigrant species Species that migrate into an ecosystem or are deliberately or accidentally introduced into an ecosystem by humans. Some of these species are beneficial, whereas others can take over and eliminate many native species. Compare *indicator species, keystone species, native species.*

immigration Migration of people into a country or area to take up permanent residence.

indicator species Species that serve as early warnings that a community or ecosystem is being degraded. Compare *immigrant species, keystone species, native species.*

inductive reasoning Using observations and facts to arrive at generalizations or hypotheses. It goes from the specific to the general and is widely used in science. Compare *deductive reasoning.*

industrialized agriculture Using large inputs of energy from fossil fuels (especially oil and natural gas), water, fertilizer, and pesticides to produce large quantities of crops and livestock for domestic and foreign sale. Compare *subsistence farming.*

industrial revolution Use of new sources of energy from fossil fuels and later from nuclear fuels, and use of new technologies, to grow food and manufacture products. Compare *agricultural revolution, environmental revolution, hunter-gatherers, information and globalization revolution.*

industrial smog Type of air pollution consisting mostly of a mixture of sulfur dioxide, suspended droplets of sulfuric acid formed from some of the sulfur dioxide, and a variety of suspended solid particles. Compare *photochemical smog.*

inertia Ability of a living system to resist being disturbed or altered. Compare *constancy, resilience.*

infant mortality rate Number of babies out of every 1,000 born each year that die before their first birthday.

infiltration Downward movement of water through soil.

information and globalization revolution Use of new technologies such as the telephone, radio, television, computers, the Internet, automated databases, and remote sensing satellites to enable people to have increasingly rapid access to much more information on a global scale. Compare *agri-*

cultural revolution, environmental revolution, hunter-gatherers, industrial revolution.

inherent value See *intrinsic value.*

inland wetland Land away from the coast, such as a swamp, marsh, or bog, that is covered all or part of the time with fresh water. Compare *coastal wetland.*

inorganic compounds All compounds not classified as organic compounds. See *organic compounds.*

inorganic fertilizer See *commercial inorganic fertilizer.*

input Matter, energy, or information entering a system. Compare *output, throughput.*

input pollution control See *pollution prevention.*

insecticide Chemical that kills insects.

instrumental value Value of an organism, species, ecosystem, or the earth's biodiversity based on its usefulness to us. Compare *intrinsic value.*

integrated pest management (IPM) Combined use of biological, chemical, and cultivation methods in proper sequence and timing to keep the size of a pest population below the size that causes economically unacceptable loss of a crop or livestock animal.

intercropping Growing two or more different crops at the same time on a plot. For example, a carbohydrate-rich grain that depletes soil nitrogen and a protein-rich legume that adds nitrogen to the soil may be intercropped. Compare *monoculture, polyculture, polyvarietal cultivation.*

interference competition Situation in which one species limits access of another species to a resource, regardless of whether the resource is abundant or scarce. See *exploitation competition, interspecific competition.*

intermediate goods See *manufactured resources.*

internal cost Direct cost paid by the producer and the buyer of an economic good. Compare *external cost.*

interplanting Simultaneously growing a variety of crops on the same plot. See *agroforestry, intercropping, polyculture, polyvarietal cultivation.*

interspecific competition Members of two or more species trying to use the same limited resources in an ecosystem. See *competition, competitive exclusion, exploitation competition, interference competition.*

intertidal zone The area of shoreline between low and high tides.

intraspecific competition Two or more organisms of a single species trying to use the same limited resources in an ecosystem. See *competition, interspecific competition.*

intrinsic rate of increase (r) Rate at which a population could grow if it had unlimited resources. Compare *environmental resistance.*

intrinsic value Value of an organism, species, ecosystem, or the earth's biodiversity based on its existence, regardless of whether it has any usefulness to us. Compare *instrumental value.*

inversion See *temperature inversion.*

invertebrates Animals that have no backbones. Compare *vertebrates.*

ion Atom or group of atoms with one or more positive (+) or negative (−) electrical charges. Compare *atom, molecule.*

ionizing radiation Fast-moving alpha or beta particles or high-energy radiation (gamma rays) emitted by radioisotopes. They have enough energy to dislodge one or more electrons from atoms they hit, forming charged ions in tissue that can react with and damage living tissue. Compare *nonionizing radiation.*

isotopes Two or more forms of a chemical element that have the same number of protons but different mass numbers because they have different numbers of neutrons in their nuclei.

J-shaped curve Curve with a shape similar to that of the letter J; can represent prolonged exponential growth.

kerogen Solid, waxy mixture of hydrocarbons found in oil shale rock. When the rock is heated to high temperatures, the kerogen is vaporized. The vapor is condensed, purified, and then sent to a refinery to produce gasoline, heating oil, and other products. See also *oil shale, shale oil.*

keystone species Species that play roles affecting many other organisms in an ecosystem. Compare *immigrant species, indicator species, native species.*

kilocalorie (kcal) Unit of energy equal to 1,000 calories. See *calorie.*

kilowatt (kW) Unit of electrical power equal to 1,000 watts. See *watt.*

kinetic energy Energy that matter has because of its mass and speed or velocity. Compare *potential energy.*

K-selected species Species that produce a few, often fairly large offspring but invest a great deal of time and energy to ensure that most of those offspring reach reproductive age. Compare *r-selected species.*

K-strategists See *K-selected species.*

kwashiorkor Type of malnutrition that occurs in infants and very young children when they are weaned from mother's milk to a starchy diet low in protein. See *marasmus.*

lake Large natural body of standing fresh water formed when water from precipitation, land runoff, or groundwater flow fills a depression in the earth created by (1) glaciation, (2) earth movement, (3) volcanic activity, or (4) a giant meteorite. See *eutrophic lake, mesotrophic lake, oligotrophic lake.*

land classification Method for reducing soil erosion that identifies easily erodible land that should be neither planted in crops nor cleared of vegetation.

landfill See *sanitary landfill.*

land-use planning Process for deciding the best present and future use of each parcel of land in an area. See *ecological land-use planning.*

late successional plant species Mostly trees that can tolerate shade and form a fairly stable complex forest community. Compare *midsuccessional plant species.*

latitude Distance from the equator. Compare *altitude.*

law of conservation of energy See *first law of thermodynamics.*

law of conservation of matter In any physical or chemical change, matter is neither created nor destroyed but merely changed from one form to another; in physical and chemical changes, existing atoms are rearranged into different spatial patterns (physical changes) or different combinations (chemical changes).

law of conservation of matter and energy In any nuclear change, the total amount of matter and energy involved remains the same.

law of tolerance The existence, abundance, and distribution of a species in an ecosystem are determined by whether the levels of one or more physical or chemical factors fall within the range tolerated by the species. See *threshold effect.*

LD$_{50}$ See *median lethal dose.*

LDC See *developing country.*

leaching Process in which various chemicals in upper layers of soil are dissolved and carried to lower layers and, in some cases, to groundwater.

less developed country (LDC) See *developing country.*

life cycle cost Initial cost plus lifetime operating costs of an economic good. Compare *full cost.*

life expectancy Average number of years a newborn infant can be expected to live.

limiting factor Single factor that limits the growth, abundance, or distribution of the population of a species in an ecosystem. See *limiting factor principle.*

limiting factor principle Too much or too little of any abiotic factor can limit or prevent growth of a population of a species in an ecosystem, even if all other factors are at or near the optimum range of tolerance for the species.

linear growth Growth in which a quantity increases by some fixed amount during each unit of time. Compare *exponential growth.*

liquefied natural gas (LNG) Natural gas converted to liquid form by cooling to a very low temperature.

liquefied petroleum gas (LPG) Mixture of liquefied propane (C_3H_8) and butane (C_4H_{10}) gas removed from natural gas and used as a fuel.

lithosphere Outer shell of the earth, composed of the crust and the rigid, outermost part of the mantle outside of the asthenosphere; material found in earth's plates. See *crust, mantle.*

loams Soils containing a mixture of clay, sand, silt, and humus. Good for growing most crops.

logistic growth Exponential population growth when the population is small and results in a steady decrease in population growth with time as the population approaches the carrying capacity.

low-input agriculture See *sustainable agriculture.*

low-quality energy Energy that is disorganized or dispersed and has little ability to do useful work. An example is low-temperature heat. Compare *high-quality energy.*

low-quality matter Matter that is disorganized, dilute, or dispersed or contains a low concentration of a useful resource. Compare *high-quality matter.*

low-throughput economy Economy based on working with nature by (1) recycling and reusing discarded matter; (2) preventing pollution; (3) conserving matter and energy resources by reducing unnecessary waste and use; (4) not degrading renewable resources; (5) building things that are easy to recycle, reuse, and repair; (6) not allowing population size to exceed the carrying capacity of the environment; and (7) preserving biodiversity and ecological integrity. See *environmental worldview.* Compare *high-throughput economy, matter-recycling economy.*

low-waste society See *low-throughput economy.*

LPG See *liquefied petroleum gas.*

macroevolution Long-term, large-scale evolutionary changes among groups of species. Compare *microevolution.*

macronutrients Chemical elements organisms need in fairly large amounts to live, grow, or reproduce. Examples are carbon, oxygen, hydrogen, nitrogen, phosphorus, sulfur, potassium, calcium, magnesium, and iron. Compare *micronutrients.*

magma Molten rock below the earth's surface.

malnutrition Faulty nutrition, caused by a diet that does not supply an individual with enough protein, essential fats, vitamins, minerals, and other nutrients needed for good health. Compare *overnutrition, undernutrition.*

mangrove swamps Swamps found on the coastlines in warm tropical climates. They are dominated by mangrove trees, any of about 55 species of trees and shrubs that can live partly submerged in the salty environment of coastal swamps.

mantle Zone of the earth's interior between its core and its crust. Compare *core, crust.* See *lithosphere.*

manufactured capital See *manufactured resources.*

manufactured resources Manufactured items made from natural resources and used to produce and distribute economic goods and services bought by consumers. These include tools, machinery, equipment, factory buildings, and transportation and distribution facilities. Compare *human resources, financial resources, natural resources.*

manure See *animal manure, green manure.*

marasmus Nutritional deficiency disease caused by a diet that does not have enough calories and protein to maintain good health. See *kwashiorkor, malnutrition.*

market equilibrium See *market price equilibrium point.*

market price equilibrium point State in which sellers and buyers of an economic good agree on the quantity to be produced and the price to be paid.

mass The amount of material in an object.

mass extinction A catastrophic, widespread, often global event in which major groups of species are wiped out over a short time compared to normal (background) extinctions. Compare *background extinction.*

mass number Sum of the number of neutrons (n) and the number of protons (p) in the nucleus of an atom. It gives the approximate mass of that atom. Compare *atomic number.*

mass transit Buses, trains, trolleys, and other forms of transportation that carry large numbers of people.

material efficiency Total amount of material needed to produce each unit of goods or services. Also called *resource productivity.* Compare *energy efficiency.*

matter Anything that has mass (the amount of material in an object) and takes up space. On the earth, where gravity is present, we weigh an object to determine its mass.

matter quality Measure of how useful a matter resource is, based on its availability and concentration. See *high-quality matter, low-quality matter.*

matter-recycling economy Economy that emphasizes recycling the maximum amount of all resources that can be recycled. The goal is to allow economic growth to continue without depleting matter resources and without producing excessive pollution and environmental degradation. Compare *high-throughput economy, low-throughput economy.*

mature community Fairly stable, self-sustaining community in an advanced stage of ecological succession; usually has a diverse array of species and ecological niches; captures and uses energy and cycles critical chemicals more efficiently than simpler, immature communities. Compare *immature community.*

maximum sustainable yield See *sustainable yield.*

MDC See *developed country.*

median lethal dose (LD₅₀) Amount of a toxic material per unit of body weight of test animals that kills half the test population in a certain time.

megacity City with 10 million or more people.

meltdown The melting of the core of a nuclear reactor.

mesosphere Third layer of the atmosphere; found above the stratosphere. Compare *stratosphere, troposphere.*

mesotrophic lake Lake with a moderate supply of plant nutrients. Compare *eutrophic lake, oligotrophic lake.*

metabolism Ability of a living cell or organism to capture and transform matter and energy from its environment to supply its needs for survival, growth, and reproduction.

metamorphic rock Rock produced when a preexisting rock is subjected to high temperatures (which may cause it to melt partially), high pressures, chemically active fluids, or a combination of these agents. Compare *igneous rock, sedimentary rock.* See *rock cycle.*

metastasis Spread of malignant (cancerous) cells from a cancer to other parts of the body.

metropolitan area See *urban area.*

microclimates Local climatic conditions that differ from the general climate of a region. Various topographic features of the earth's surface such as mountains and cities typically cause them.

microevolution The small genetic changes a population undergoes. Compare *macroevolution.*

micronutrients Chemical elements organisms need in small or even trace amounts to live, grow, or reproduce. Examples are sodium, zinc, copper, chlorine, and iodine. Compare *macronutrients.*

microorganisms Organisms such as bacteria that are so small that they can be seen only by using a microscope.

micropower systems Systems of small-scale decentralized units that generate 1–10,000 kilowatts of electricity. Examples include (1) microturbines, (2) fuel cells, and (3) household solar panels and solar roofs.

midsuccessional plant species Grasses and low shrubs that are less hardy than early successional plant species. Compare *early successional plant species, late successional plant species.*

mineral Any naturally occurring inorganic substance found in the earth's crust as a crystalline solid. See *mineral resource.*

mineral resource Concentration of naturally occurring solid, liquid, or gaseous material in or on the earth's crust in a form and amount such that extracting and converting it into useful materials or items is currently or potentially profitable. Mineral resources are classified as *metallic* (such as iron and tin ores) or *nonmetallic* (such as fossil fuels, sand, and salt).

minimum dynamic area (MDA) Minimum area of suitable habitat needed to maintain the minimum viable population. See *minimum viable population.*

minimum-tillage farming See *conservation-tillage farming.*

minimum viable population (MVP) Estimate of the smallest number of individuals necessary to ensure the survival of a population in a region for a specified time period, typically ranging from decades to 100 years. See *population viability analysis.*

mixed economic system Economic system that falls somewhere between pure market and pure command economic systems. Almost all the world's economic systems fall into this category, with some closer to a pure market system and some closer to a pure command system. Compare *capitalist market economic system, pure command economic system, pure free-market economic system.*

mixture Combination of one or more elements and compounds.

model An approximate representation or simulation of a system being studied.

molecule Combination of two or more atoms of the same chemical element (such as O₂) or different chemical elements (such as H₂O) held together by chemical bonds. Compare *atom, ion.*

monera See *bacteria, cyanobacteria.*

monoculture Cultivation of a single crop, usually on a large area of land. Compare *polyculture, polyvarietal cultivation.*

more developed country (MDC) See *developed country.*

multiple use Use of an ecosystem such as a forest for a variety of purposes such as timber harvesting, wildlife habitat, watershed protection, and recreation. Compare *sustainable yield.*

municipal solid waste Solid materials discarded by homes and businesses in or near urban areas. See *solid waste.*

mutagen Chemical or form of ionizing radiation that causes inheritable changes (mutations) in the DNA molecules in the genes found in chromosomes. See *carcinogen, mutation, teratogen.*

mutation A random change in DNA molecules making up genes that can yield changes in anatomy, physiology, or behavior in offspring. See *mutagen.*

mutualism Type of species interaction in which both participating species generally benefit. Compare *commensalism.*

native species Species that normally live and thrive in a particular ecosystem. Compare *immigrant species, indicator species, keystone species.*

natural capital See *natural resources.*

natural gas Underground deposits of gases consisting of 50–90% by weight methane gas (CH₄) and small amounts of heavier gaseous hydrocarbon compounds such as propane (C₃H₈) and butane (C₄H₁₀).

natural greenhouse effect Heat buildup in the troposphere because of the presence of certain gases, called greenhouse gases. Without this effect, the earth would be nearly as cold as Mars, and life as we know it could not exist. There is much evidence that we are enhancing this natural effect by excess additions of greenhouse gases from human activities.

natural ionizing radiation Ionizing radiation in the environment from natural sources.

natural law See *scientific law.*

natural radioactive decay Nuclear change in which unstable nuclei of atoms spontaneously shoot out particles (usually alpha or beta particles) or energy (gamma rays) at a fixed rate.

natural rate of extinction See *background extinction.*

natural recharge Natural replenishment of an aquifer by precipitation, which percolates downward through soil and rock. See *recharge area.*

natural resources The earth's natural materials and processes that sustain other species and us. Compare *financial resources, human resources, manufactured resources.*

natural selection Process by which a particular beneficial gene (or set of genes) is reproduced in succeeding generations more than other genes. The result of natural selection is a population that contains a greater proportion of organisms better adapted to certain environmental conditions. See *adaptation, biological evolution, differential reproduction, mutation.*

negative feedback loop Situation in which a change in a certain direction provides information that causes a system to change less in that direction. Compare *positive feedback loop.*

negawatt A watt of electrical power saved by improving energy efficiency. See *watt.*

nekton Strongly swimming organisms found in aquatic systems. Compare *benthos, plankton.*

nematocide Chemical that kills nematodes (roundworms).

net economic welfare (NEW) Measure of annual change in quality of life in a country. It is obtained by subtracting the value of all final products and services that decrease the quality of life from a country's GNP.

net energy Total amount of useful energy available from an energy resource or energy system over its lifetime, minus the amount of energy (1) used (the first energy law), (2) automatically wasted (the second energy law), and (3) unnecessarily wasted in finding, processing, concentrating, and transporting it to users.

net primary productivity (NPP) Rate at which all the plants in an ecosystem produce net useful chemical energy; equal to the difference between the rate at which the plants in an ecosystem produce useful chemical energy (primary productivity) and the rate at which they use some of that energy through cellular respiration. Compare *gross primary productivity.*

neutral solution Water solution containing an equal number of hydrogen ions (H⁺) and hydroxide ions (OH⁻); water solution with a pH of 7. Compare *acid solution, basic solution.*

neutron (n) Elementary particle in the nuclei of all atoms (except hydrogen-1). It has a relative mass of 1 and no electric charge. Compare *electron, proton.*

niche See *ecological niche.*

nitrogen cycle Cyclic movement of nitrogen in different chemical forms from the environment to organisms and then back to the environment.

nitrogen fixation Conversion of atmospheric nitrogen gas into forms useful to plants by lightning, bacteria, and cyanobacteria; it is part of the nitrogen cycle.

noise pollution Any unwanted, disturbing, or harmful sound that **(1)** impairs or interferes with hearing, **(2)** causes stress, **(3)** hampers concentration and work efficiency, or **(4)** causes accidents.

nondegradable pollutant Material that is not broken down by natural processes. Examples are the toxic elements lead and mercury. Compare *biodegradable pollutant, degradable pollutant, slowly degradable pollutant.*

nonionizing radiation Forms of radiant energy such as radio waves, microwaves, infrared light, and ordinary light that do not have enough energy to cause ionization of atoms in living tissue. Compare *ionizing radiation.*

nonnative species Species that migrate into an ecosystem or are deliberately or accidentally introduced into an ecosystem by humans. Compare *native species.*

nonpersistent pollutant See *degradable pollutant.*

nonpoint source Large or dispersed land areas such as cropfields, streets, and lawns that discharge pollutants into the environment over a large area. Compare *point source.*

nonrenewable resource Resource that exists in a fixed amount (stock) in various places in the earth's crust and has the potential for renewal only by geological, physical, and chemical processes taking place over hundreds of millions to billions of years. Examples are copper, aluminum, coal, and oil. We classify these resources as exhaustible because we are extracting and using them at a much faster rate than they were formed. Compare *potentially renewable resource.*

nontransmissible disease A disease that is not caused by living organisms and does not spread from one person to another. Examples are most cancers, diabetes, cardiovascular disease, and malnutrition. Compare *transmissible disease.*

no-till farming See *conservation-tillage farming.*

nuclear change Process in which nuclei of certain isotopes spontaneously change, or are forced to change, into one or more different isotopes. The three principal types of nuclear change are natural radioactivity, nuclear fission, and nuclear fusion. Compare *chemical change, physical change.*

nuclear energy Energy released when atomic nuclei undergo a nuclear reaction such as the spontaneous emission of radioactivity, nuclear fission, or nuclear fusion.

nuclear fission Nuclear change in which the nuclei of certain isotopes with large mass numbers (such as uranium-235 and plutonium-239) are split apart into lighter nuclei when struck by a neutron. This process releases more neutrons and a large amount of energy. Compare *nuclear fusion.*

nuclear fusion Nuclear change in which two nuclei of isotopes of elements with a low mass number (such as hydrogen-2 and hydrogen-3) are forced together at extremely high temperatures until they fuse to form a heavier nucleus (such as helium-4). This process releases a large amount of energy. Compare *nuclear fission.*

nucleus Extremely tiny center of an atom, making up most of the atom's mass. It contains one or more positively charged protons and one or more neutrons with no electrical charge (except for a hydrogen-1 atom, which has one proton and no neutrons in its nucleus).

nutrient Any food or element an organism must take in to live, grow, or reproduce.

nutrient cycle See *biogeochemical cycle.*

oil See *crude oil.*

oil shale Fine-grained rock containing various amounts of kerogen, a solid, waxy mixture of hydrocarbon compounds. Heating the rock to high temperatures converts the kerogen into a vapor that can be condensed to form a slow-flowing heavy oil called shale oil. See *kerogen, shale oil.*

old-growth forest Virgin and old, second-growth forests containing trees that are often hundreds, sometimes thousands of years old. Examples include forests of Douglas fir, western hemlock, giant sequoia, and coastal redwoods in the western United States. Compare *second-growth forest, tree farm.*

oligotrophic lake Lake with a low supply of plant nutrients. Compare *eutrophic lake, mesotrophic lake.*

omnivore Animal that can use both plants and other animals as food sources. Examples are pigs, rats, cockroaches, and people. Compare *carnivore, herbivore.*

open-pit mining Removing minerals such as gravel, sand, and metal ores by digging them out of the earth's surface and leaving an open pit.

open sea The part of an ocean that is beyond the continental shelf. Compare *coastal zone.*

open system A system, such as a living organism, in which both matter and energy are exchanged between the system and the environment. Compare *closed system.*

ore Part of a metal-yielding material that can be economically and legally extracted at a given time. An ore typically contains two parts: the ore mineral, which contains the desired metal, and waste mineral material (gangue).

organic compounds Compounds containing carbon atoms combined with each other and with atoms of one or more other elements such as hydrogen, oxygen, nitrogen, sulfur, phosphorus, chlorine, and fluorine. All other compounds are called *inorganic compounds.*

organic farming Producing crops and livestock naturally by using organic fertilizer (manure, legumes, compost) and natural pest control (bugs that eat harmful bugs, plants that repel bugs, and environmental controls such as crop rotation) instead of using commercial inorganic fertilizers and synthetic pesticides and herbicides. See *sustainable agriculture.*

organic fertilizer Organic material such as animal manure, green manure, and compost, applied to cropland as a source of plant nutrients. Compare *commercial inorganic fertilizer.*

organism Any form of life.

other resources Identified and undiscovered resources not classified as reserves.

output Matter, energy, or information leaving a system. Compare *input, throughput.*

output pollution control See *pollution cleanup.*

overburden Layer of soil and rock overlying a mineral deposit. It is removed during surface mining.

overfishing Harvesting so many fish of a species, especially immature fish, that there is not enough breeding stock left to replenish the species, such that it is not profitable to harvest them.

overgrazing Destruction of vegetation when too many grazing animals feed too long and exceed the carrying capacity of a rangeland area.

overnutrition Diet so high in calories, saturated (animal) fats, salt, sugar, and processed foods and so low in vegetables and fruits that the consumer runs high risks of diabetes, hypertension, heart disease, and other health hazards. Compare *malnutrition, undernutrition.*

oxygen-demanding wastes Organic materials that are usually biodegraded by aerobic (oxygen-consuming) bacteria if there is enough dissolved oxygen in the water. See also *biological oxygen demand.*

ozone depletion Decrease in concentration of ozone (O_3) in the stratosphere. See *ozone layer.*

ozone layer Layer of gaseous ozone (O_3) in the stratosphere that protects life on earth by filtering out harmful ultraviolet radiation from the sun.

PANs Peroxyacyl nitrates. Group of chemicals found in photochemical smog.

parasite Consumer organism that lives on or in and feeds on a living plant or animal, known as the host, over an extended period of time. The parasite draws nourishment from and gradually weakens its host; it may or may not kill the host. See *parasitism.*

parasitism Interaction between species in which one organism, called the parasite, preys on another organism, called the host, by living on or in the host. See *host, parasite.*

parts per billion (ppb) Number of parts of a chemical found in one billion parts of a particular gas, liquid, or solid.

parts per million (ppm) Number of parts of a chemical found in one million parts of a particular gas, liquid, or solid.

parts per trillion (ppt) Number of parts of a chemical found in one trillion parts of a particular gas, liquid, or solid.

passive solar heating system System that captures sunlight directly within a structure and converts it into low-temperature heat for space heating or for heating water for domestic use without the use of mechanical devices. Compare *active solar heating system.*

pasture Managed grassland or enclosed meadow that usually is planted with domesticated grasses or other forage to be grazed by livestock. Compare *feedlot, rangeland.*

pathogen Organism that produces disease.

PCBs See *polychlorinated biphenyls.*

per capita GNP Annual gross national product (GNP) of a country divided by its total population. See *gross national product, real per capita GNP.*

percolation Passage of a liquid through the spaces of a porous material such as soil.

perennial Plant that can live for more than 2 years. Compare *annual.*

permafrost Perennially frozen layer of the soil that forms when the water there freezes. It is found in arctic tundra.

permeability The degree to which underground rock and soil pores are interconnected with each other, and thus a measure of the degree to which water can flow freely from one pore to another. Compare *porosity.*

perpetual resource See *renewable resource.*

persistence How long a pollutant stays in the air, water, soil, or body. See also *inertia.*

persistent pollutant See *slowly degradable pollutant.*

pest Unwanted organism that directly or indirectly interferes with human activities.

pesticide Any chemical designed to kill or inhibit the growth of an organism that people consider to be undesirable. See *fungicide, herbicide, insecticide.*

pesticide treadmill Situation in which the cost of using pesticides increases while their effectiveness decreases, mostly because the pest species develop genetic resistance to the pesticides.

petrochemicals Chemicals obtained by refining (distilling) crude oil. They are used as raw materials in the manufacture of most industrial chemicals, fertilizers, pesticides, plastics, synthetic fibers, paints, medicines, and many other products.

petroleum See *crude oil.*

pH Numeric value that indicates the relative acidity or alkalinity of a substance on a scale of 0 to 14, with the neutral point at 7. Acid solutions have pH values lower than 7, and basic or alkaline solutions have pH values greater than 7.

phosphorus cycle Cyclic movement of phosphorus in different chemical forms from the environment to organisms and then back to the environment.

photochemical smog Complex mixture of air pollutants produced in the lower atmosphere by the reaction of hydrocarbons and nitrogen oxides under the influence of sunlight. Especially harmful components include ozone, peroxyacyl nitrates (PANs), and various aldehydes. Compare *industrial smog.*

photosynthesis Complex process that takes place in cells of green plants. Radiant energy from the sun is used to combine carbon dioxide (CO_2) and water (H_2O) to produce oxygen (O_2) and carbohydrates (such as glucose, $C_6H_{12}O_6$) and other nutrient molecules. Compare *aerobic respiration, chemosynthesis.*

photovoltaic cell (solar cell) Device in which radiant (solar) energy is converted directly into electrical energy.

physical change Process that alters one or more physical properties of an element or a compound without altering its chemical composition. Examples are changing the size and shape of a sample of matter (crushing ice and cutting aluminum foil) and changing a sample of matter from one physical state to another (boiling and freezing water). Compare *chemical change, nuclear change.*

phytoplankton Small, drifting plants, mostly algae and bacteria, found in aquatic ecosystems. Compare *plankton, zooplankton.*

pioneer community First integrated set of plants, animals, and decomposers found in an area undergoing primary ecological succession. See *immature community, mature community.*

pioneer species First hardy species, often microbes, mosses, and lichens, which begin colonizing a site as the first stage of ecological succession. See *ecological succession, pioneer community.*

plaintiff The individual, group of individuals, corporation, or government agency bringing the charges in a lawsuit. Compare *defendant.*

planetary management worldview Beliefs that (1) we are the planet's most important species; (2) there is always more, and it's all for us; (3) all economic growth is good, more economic growth is better, and the potential for economic growth is limitless; and (4) our success depends on how well we can understand, control, and manage the earth's life-support systems for our own benefit. See *spaceship-earth worldview.* Compare *environmental wisdom worldview.*

plankton Small plant organisms (phytoplankton) and animal organisms (zooplankton) that float in aquatic ecosystems.

plantation agriculture Growing specialized crops such as bananas, coffee, and cacao in tropical developing countries, primarily for sale to developed countries.

plants (plantae) Eukaryotic, mostly multicelled organisms such as algae (red, blue, and green), mosses, ferns, flowers, cacti, grasses, beans, wheat, rice, and trees. These organisms use photosynthesis to produce organic nutrients for themselves and for other organisms feeding on them. Water and other inorganic nutrients are obtained from the soil for terrestrial plants and from the water for aquatic plants.

plates Various-sized areas of the earth's lithosphere that move slowly around with the mantle's flowing asthenosphere. Most earthquakes and volcanoes occur around the boundaries of these plates. See *lithosphere, plate tectonics.*

plate tectonics Theory of geophysical processes that explains the movements of lithospheric plates and the processes that occur at their boundaries. See *lithosphere, plates.*

point source Single identifiable source that discharges pollutants into the environment. Examples are the **(1)** smokestack of a power plant or an industrial plant, **(2)** drainpipe of a meatpacking plant, **(3)** chimney of a house, or **(4)** exhaust pipe of an automobile. Compare *nonpoint source.*

poison A chemical that in one dose kills exactly 50% of the animals (usually rats and mice) in a test population (usually 60 to 200 animals) within a 14-day period. See *median lethal dose.*

politics Process through which individuals and groups try to influence or control government policies and actions that affect the local, state, national, and international communities.

pollutant A particular chemical or form of energy that can adversely affect the health, survival, or activities of humans or other living organisms. See *pollution.*

pollution An undesirable change in the physical, chemical, or biological characteristics of air, water, soil, or food that can adversely affect the health, survival, or activities of humans or other living organisms.

pollution cleanup Device or process that removes or reduces the level of a pollutant after it has been produced or has entered the environment. Examples are automobile emission control devices and sewage treatment plants. Compare *pollution prevention.*

pollution prevention Device or process that prevents a potential pollutant from forming or entering the environment or that sharply reduces the amount entering the environment. Compare *pollution cleanup.*

polychlorinated biphenyls (PCBs) Group of 209 different toxic, oily, synthetic chlorinated hydrocarbon compounds that can be biologically amplified in food chains and webs.

polyculture Complex form of intercropping in which a large number of different plants maturing at different times are planted together. See also *intercropping.* Compare *monoculture, polyvarietal cultivation.*

polyvarietal cultivation Planting a plot of land with several varieties of the same crop. Compare *intercropping, monoculture, polyculture.*

population Group of individual organisms of the same species living within a particular area.

population change An increase or decrease in the size of a population. It is equal to (Births + Immigration) − (Deaths + Emigration).

population density Number of organisms in a particular population found in a specified area.

population dispersion General pattern in which the members of a population are arranged throughout its habitat.

population distribution Variation of population density over a particular geographic area. For example, a country has a high population density in its urban areas and a much lower population density in rural areas.

population dynamics Major abiotic and biotic factors that tend to increase or decrease the population size and age and sex composition of a species.

population size Number of individuals making up a population's gene pool.

population viability analysis (PVA) Use of mathematical models to estimate a population's risk of extinction. See *minimum viable population.*

porosity Percentage of space in rock or soil occupied by voids, whether the voids are isolated or connected. Compare *permeability.*

positive feedback loop Situation in which a change in a certain direction provides information that causes a system to change further in the same direction. This can lead to a runaway or vicious cycle. Compare *negative feedback loop.*

potential energy Energy stored in an object because of its position or the position of its parts. Compare *kinetic energy.*

potentially renewable resource Resource that can be replenished fairly rapidly (hours to several decades) through natural processes. Examples are trees in forests, grasses in grasslands, wild animals, fresh surface water in lakes and streams, most groundwater, fresh air, and fertile soil. If such a resource is used faster than it is replenished, it can be depleted and converted into a nonrenewable resource. Compare *nonrenewable resource* and *renewable resource.* See also *environmental degradation.*

poverty Inability to meet basic needs for food, clothing, and shelter.

ppb See *parts per billion.*

ppm See *parts per million.*

ppt See *parts per trillion.*

precautionary principle When there is scientific uncertainty about potentially serious harm from chemicals or technologies, decision-makers should act to prevent harm to humans and the environment. See *pollution prevention.*

precipitation Water in the form of rain, sleet, hail, and snow that falls from the atmosphere onto the land and bodies of water.

precision A measure of reproducibility, or how closely a series of measurements of the same quantity agree with one another. Compare *accuracy.*

predation Situation in which an organism of one species (the predator) captures and feeds on parts or all of an organism of another species (the prey).

predator Organism that captures and feeds on parts or all of an organism of another species (the prey).

predator-prey relationship Interaction between two organisms of different species in which one organism, called the *predator*, captures and feeds on parts or all of another organism, called the *prey*.

preservationist Person concerned primarily with setting aside or protecting undisturbed natural areas from harmful human activities. Compare *conservation biologist, conservationist, ecologist, environmentalist, environmental scientist, restorationist.*

prey Organism that is captured and serves as a source of food for an organism of another species (the predator).

primary consumer Organism that feeds on all or part of plants (herbivore) or on other producers. Compare *detritivore, omnivore, secondary consumer.*

primary oil recovery Pumping out the crude oil that flows by gravity or under gas pressure into the bottom of an oil well. Compare *enhanced oil recovery, secondary oil recovery.*

primary pollutant Chemical that has been added directly to the air by natural events or human activities and occurs in a harmful concentration. Compare *secondary pollutant.*

primary productivity See *gross primary productivity, net primary productivity.*

primary sewage treatment Mechanical treatment of sewage in which large solids are filtered out by screens and suspended solids settle out as sludge in a sedimentation tank. Compare *advanced sewage treatment, secondary sewage treatment.*

primary succession Sequential development of communities in a bare area that has never been occupied by a community of organisms. Compare *secondary succession.*

prior appropriation Legal principle by which the first user of water from a stream establishes a legal right to continued use of the amount originally withdrawn. Compare *riparian rights.*

probability A mathematical statement about how likely it is that something will happen.

producer Organism that uses solar energy (green plant) or chemical energy (some bacteria) to manufacture the organic compounds it needs as nutrients from simple inorganic compounds obtained from its environment. Compare *consumer, decomposer.*

prokaryotic cell Cell that doesn't have a distinct nucleus. Other internal parts are also not enclosed by membranes. Compare *eukaryotic cell.*

protists (protista) Eukaryotic, mostly single-celled organisms such as diatoms, amoebas, some algae (golden brown and yellow-green), protozoans, and slime molds. Some protists produce their own organic nutrients through photosynthesis. Others are decomposers and some feed on bacteria, other protists, or cells of multicellular organisms.

proton (p) Positively charged particle in the nuclei of all atoms. Each proton has a relative mass of 1 and a single positive charge. Compare *electron, neutron.*

pure capitalism See *pure free-market economic system.*

pure command economic system System in which all economic decisions are made by the government or some other central authority. Compare *capitalist market economic system, mixed economic system, pure free-market economic system.*

pure free-market economic system System in which all economic decisions are made in the market, where buyers and sellers of economic goods interact freely, with no government or other interference. Compare *capitalist market economic system, mixed economic system, pure command economic system.*

pyramid of biomass Diagram representing the biomass, or total dry weight of all living organisms, that can be supported at each trophic level in a food chain or food web. See *pyramid of energy flow, pyramid of numbers.*

pyramid of energy flow Diagram representing the flow of energy through each trophic level in a food chain or food web. With each energy transfer, only a small part (typically 10%) of the usable energy entering one trophic level is transferred to the organisms at the next trophic level. Compare *pyramid of biomass, pyramid of numbers.*

pyramid of numbers Diagram representing the number of organisms of a particular type that can be supported at each trophic level from a given input of solar energy at the producer trophic level in a food chain or food web. Compare *pyramid of biomass, pyramid of energy flow.*

radiation Fast-moving particles (particulate radiation) or waves of energy (electromagnetic radiation). See *alpha particle, beta particle, gamma rays.*

radiation temperature inversion Temperature inversion that typically occurs at night in which a layer of warm air lies atop a layer of cooler air nearer the ground as the air near the ground cools faster than the air above it. As the sun rises and warms the earth's surface, the inversion normally disappears by noon and disperses the pollutants built up during the night. See *temperature inversion.* Compare *subsidence temperature inversion.*

radioactive decay Change of a radioisotope to a different isotope by the emission of radioactivity.

radioactive isotope See *radioisotope.*

radioactive waste Waste products of nuclear power plants, research, medicine, weapon production, or other processes involving nuclear reactions. See *radioactivity.*

radioactivity Nuclear change in which unstable nuclei of atoms spontaneously shoot out "chunks" of mass, energy, or both, at a fixed rate. The three principal types of radioactivity are gamma rays and fast-moving alpha particles and beta particles.

radioisotope Isotope of an atom that spontaneously emits one or more types of radioactivity (alpha particles, beta particles, gamma rays).

rain shadow effect Low precipitation on the far side (leeward side) of a mountain when prevailing winds flow up and over a high mountain or range of high mountains. This creates semiarid

and arid conditions on the leeward side of a high mountain range.

rangeland Land that supplies forage or vegetation (grasses, grasslike plants, and shrubs) for grazing and browsing animals and is not intensively managed. Compare *feedlot, pasture.*

range of tolerance Range of chemical and physical conditions that must be maintained for populations of a particular species to stay alive and grow, develop, and function normally. See *law of tolerance.*

rare species A species that (1) has naturally small numbers of individuals, often because of limited geographic ranges or low population densities, or (2) has been locally depleted by human activities.

real GDP Gross domestic product adjusted for inflation.

real GNP Gross national product adjusted for inflation. Compare *per capita GNP, gross national product, real per capita GNP.*

realized niche Parts of the fundamental niche of a species that are actually used by that species. See *ecological niche, fundamental niche.*

real per capita GDP Per capita GDP adjusted for inflation.

real per capita GNP Per capita GNP adjusted for inflation.

recharge area Any area of land allowing water to pass through it and into an aquifer. See *aquifer, natural recharge.*

recycling Collecting and reprocessing a resource so that it can be made into new products. An example is collecting aluminum cans, melting them down, and using the aluminum to make new cans or other aluminum products. Compare *reuse.*

reforestation Renewal of trees and other types of vegetation on land where trees have been removed; can be done naturally by seeds from nearby trees or artificially by planting seeds or seedlings.

relative humidity The amount of water vapor in a certain mass of air, expressed as a percentage of the maximum amount it could hold at that temperature. Compare *absolute humidity.*

reliable runoff Surface runoff of water that generally can be counted on as a stable source of water from year to year.

renewable resource An essentially inexhaustible resource on a human time scale. Solar energy is an example. See *potentially renewable resource.* Compare *nonrenewable resource.*

replacement-level fertility Number of children a couple must have to replace them. The average for a country or the world usually is slightly higher than 2 children per couple (2.1 in the United States and 2.5 in some developing countries) because some children die before reaching their reproductive years. See also *total fertility rate.*

reproduction Production of offspring by one or more parents.

reproductive isolation Long-term geographic separation of members of a particular sexually reproducing species.

reproductive potential See *biotic potential.*

reserves Resources that have been identified and from which a usable mineral can be extracted profitably at present prices with current mining technology. See *identified resources, undiscovered resources.*

reserve-to-production ratio Number of years reserves of a particular nonrenewable mineral will last at current annual production rates. See *reserves.*

resilience Ability of a living system to restore itself to original condition after being exposed to an outside disturbance that is not too drastic. See *constancy, inertia.*

resource Anything obtained from the living and nonliving environment to meet human needs and wants. It can also be applied to other species. Compare *ecological resource, economic resource.*

resource partitioning Process of dividing up resources in an ecosystem so that species with

similar needs (overlapping ecological niches) use the same scarce resources at different times, in different ways, or in different places. See *ecological niche, fundamental niche, realized niche.*

resource productivity See *material efficiency.*

respiration See *aerobic respiration.*

response The amount of health damage caused by exposure to a certain dose of a harmful substance or form of radiation. See *dose, dose-response curve, median lethal dose.*

restoration ecology Research and scientific study devoted to restoring, repairing, and reconstructing damaged ecosystems.

restorationist Scientist or other person devoted to the partial or complete restoration of natural areas that have been degraded by human activities. Compare *conservation biologist, conservationist, ecologist, environmental scientist, preservationist.*

reuse Using a product over and over again in the same form. An example is collecting, washing, and refilling glass beverage bottles. Compare *recycling.*

riparian rights System of water law that gives anyone whose land adjoins a flowing stream the right to use water from the stream, as long as some is left for downstream users. Compare *prior appropriation.*

riparian zones Thin strips and patches of vegetation that surround streams. They are very important habitats and resources for wildlife.

risk The probability that something undesirable will happen from deliberate or accidental exposure to a hazard. See *risk analysis, risk assessment, risk-benefit analysis, risk management.*

risk analysis Identifying hazards, evaluating the nature and severity of risks (*risk assessment*), using this and other information to determine options and make decisions about reducing or eliminating risks (*risk management*), and communicating information about risks to decision makers and the public (*risk communication*).

risk assessment Process of gathering data and making assumptions to estimate short- and long-term harmful effects on human health or the environment from exposure to hazards associated with the use of a particular product or technology. See *risk, risk-benefit analysis.*

risk-benefit analysis Estimate of the short- and long-term risks and benefits of using a particular product or technology. See *risk.*

risk communication Communicating information about risks to decision makers and the public. See *risk, risk analysis, risk-benefit analysis.*

risk management Using risk assessment and other information to determine options and make decisions about reducing or eliminating risks. See *risk, risk analysis, risk-benefit analysis, risk communication.*

rock Any material that makes up a large, natural, continuous part of earth's crust. See *mineral.*

rock cycle Largest and slowest of the earth's cycles, consisting of geologic, physical, and chemical processes that form and modify rocks and soil in the earth's crust over millions of years.

rodenticide Chemical that kills rodents.

r-selected species Species that reproduce early in their life span and produce large numbers of usually small and short-lived offspring in a short period of time. Compare *K-selected species.*

r-strategists See *r-selected species.*

rule of 70 Doubling time (in years) = 70/percentage growth rate. See *doubling time, exponential growth.*

runoff Fresh water from precipitation and melting ice that flows on the earth's surface into nearby streams, lakes, wetlands, and reservoirs. See *surface runoff, surface water.* Compare *groundwater.*

rural area Geographic area in the United States with a population of less than 2,500. The number of people used in this definition may vary in different countries. Compare *urban area.*

salinity Amount of various salts dissolved in a given volume of water.

salinization Accumulation of salts in soil that can eventually make the soil unable to support plant growth.

saltwater intrusion Movement of salt water into freshwater aquifers in coastal and inland areas as groundwater is withdrawn faster than it is recharged by precipitation.

sanitary landfill Waste disposal site on land in which waste is spread in thin layers, compacted, and covered with a fresh layer of clay or plastic foam each day.

scavenger Organism that feeds on dead organisms that were killed by other organisms or died naturally. Examples are vultures, flies, and crows. Compare *detritivore*.

science Attempts to discover order in nature and use that knowledge to make predictions about what should happen in nature. See *consensus science, frontier science, scientific data, scientific hypothesis, scientific law, scientific methods, scientific model, scientific theory*.

scientific data Facts obtained by making observations and measurements. Compare *model, scientific hypothesis, scientific law, scientific methods, scientific theory*.

scientific hypothesis An educated guess that attempts to explain a scientific law or certain scientific observations. Compare *model, scientific data, scientific law, scientific methods, scientific theory*.

scientific law Description of what scientists find happening in nature over and over in the same way, without known exception. See *first law of thermodynamics, second law of thermodynamics, law of conservation of matter*. Compare *scientific data, scientific hypothesis, scientific methods, scientific model, scientific theory*.

scientific methods The ways scientists gather data and formulate and test scientific hypotheses, models, theories, and laws. See *model, scientific data, scientific hypothesis, scientific law, scientific theory*.

scientific model A simulation of complex processes and systems. Many are mathematical models that are run and tested using computers.

scientific theory A well-tested and widely accepted scientific hypothesis. Compare *scientific data, scientific hypothesis, scientific law, scientific methods, scientific model*.

secondary consumer Organism that feeds only on primary consumers. Most secondary consumers are animals, but some are plants. Compare *detritivore, omnivore, primary consumer*.

secondary oil recovery Injection of water into an oil well after primary oil recovery to force out some of the remaining, usually thicker, crude oil. Compare *enhanced oil recovery, primary oil recovery*.

secondary pollutant Harmful chemical formed in the atmosphere when a primary air pollutant reacts with normal air components or other air pollutants. Compare *primary pollutant*.

secondary sewage treatment Second step in most waste treatment systems, in which aerobic bacteria break down up to 90% of degradable, oxygen-demanding organic wastes in wastewater. This usually is done by bringing sewage and bacteria together in trickling filters or in the activated sludge process. Compare *advanced sewage treatment, primary sewage treatment*.

secondary succession Sequential development of communities in an area in which natural vegetation has been removed or destroyed but the soil is not destroyed. Compare *primary succession*.

second-growth forest Stands of trees resulting from secondary ecological succession. Compare *old-growth forest, tree farm*.

second law of energy See *second law of thermodynamics*.

second law of thermodynamics In any conversion of heat energy to useful work, some of the initial energy input is always degraded to a lower-quality, more dispersed, less useful energy, usually low-temperature heat that flows into the environment; you can't break even in terms of energy quality. See *first law of thermodynamics*.

sedimentary rock Rock that forms from the accumulated products of erosion and in some cases from the compacted shells, skeletons, and other remains of dead organisms. Compare *igneous rock, metamorphic rock*. See *rock cycle*.

seed-tree cutting Removal of nearly all trees on a site in one cutting, with a few seed-producing trees left uniformly distributed to regenerate the forest. Compare *clear-cutting, selective cutting, shelterwood cutting, strip cutting*.

selective cutting Cutting of intermediate-aged, mature, or diseased trees in an uneven-aged forest stand, either singly or in small groups. This encourages the growth of younger trees and maintains an uneven-aged stand. Compare *clear-cutting, seed-tree cutting, shelterwood cutting, strip cutting*.

septic tank Underground tank for treatment of wastewater from a home in rural and suburban areas. Bacteria in the tank decompose organic wastes and the sludge settles to the bottom of the tank. The effluent flows out of the tank into the ground through a field of drainpipes.

sexual reproduction Reproduction in organisms that produce offspring by combining sex cells or *gametes* (such as ovum and sperm) from both parents. This produces offspring that have combinations of traits from their parents. Compare *asexual reproduction*.

shale oil Slow-flowing, dark brown, heavy oil obtained when kerogen in oil shale is vaporized at high temperatures and then condensed. Shale oil can be refined to yield gasoline, heating oil, and other petroleum products. See *kerogen, oil shale*.

shelterbelt See *windbreak*.

shelterwood cutting Removal of mature, marketable trees in an area in a series of partial cuttings to allow regeneration of a new stand under the partial shade of older trees, which are later removed. Typically, this is done by making two or three cuts over a decade. Compare *clear-cutting, seed-tree cutting, selective cutting, strip cutting*.

shifting cultivation Clearing a plot of ground in a forest, especially in tropical areas, and planting crops on it for a few years (typically 2-5 years) until the soil is depleted of nutrients or the plot has been invaded by a dense growth of vegetation from the surrounding forest. Then a new plot is cleared and the process is repeated. The abandoned plot cannot successfully grow crops for 10-30 years. See also *slash-and-burn cultivation*.

slash-and-burn cultivation Cutting down trees and other vegetation in a patch of forest, leaving the cut vegetation on the ground to dry, and then burning it. The ashes that are left add nutrients to the nutrient-poor soils found in most tropical forest areas. Crops are planted between tree stumps. Plots must be abandoned after a few years (typically 2-5 years) because of loss of soil fertility or invasion of vegetation from the surrounding forest. See also *shifting cultivation*.

slowly degradable pollutant Material that is slowly broken down into simpler chemicals or reduced to acceptable levels by natural physical, chemical, and biological processes. Compare *biodegradable pollutant, degradable pollutant, nondegradable pollutant*.

sludge Gooey mixture of toxic chemicals, infectious agents, and settled solids, removed from wastewater at a sewage treatment plant.

smart growth Form of urban planning that recognizes that urban growth will occur but uses zoning laws and an array of other tools to (1) prevent sprawl, (2) direct growth to certain areas, (3) protect ecologically sensitive and important lands and waterways, and (4) develop urban areas that are more environmentally sustainable and more enjoyable places to live.

smelting Process in which a desired metal is separated from the other elements in an ore mineral.

smog Originally a combination of smoke and fog but now used to describe other mixtures of pollutants in the atmosphere. See *industrial smog, photochemical smog*.

soil Complex mixture of inorganic minerals (clay, silt, pebbles, and sand), decaying organic matter, water, air, and living organisms.

soil conservation Methods used to reduce soil erosion, prevent depletion of soil nutrients, and restore nutrients already lost by erosion, leaching, and excessive crop harvesting.

soil erosion Movement of soil components, especially topsoil, from one place to another, usually by wind, flowing water, or both. This natural process can be greatly accelerated by human activities that remove vegetation from soil.

soil horizons Horizontal zones that make up a particular mature soil. Each horizon has a distinct texture and composition that vary with different types of soils.

soil permeability Rate at which water and air move from upper to lower soil layers. Compare *porosity*.

soil porosity See *porosity*.

soil profile Cross-sectional view of the horizons in a soil.

soil structure How the particles that make up a soil are organized and clumped together. See also *soil permeability, soil texture*.

soil texture Relative amounts of the different types and sizes of mineral particles in a sample of soil.

solar capital Solar energy from the sun reaching earth. Compare *natural resources*.

solar cell See *photovoltaic cell*.

solar collector Device for collecting radiant energy from the sun and converting it into heat. See *active solar heating system, passive solar heating system*.

solar energy Direct radiant energy from the sun and a number of indirect forms of energy produced by the direct input. Principal indirect forms of solar energy include wind, falling and flowing water (hydropower), and biomass (solar energy converted into chemical energy stored in the chemical bonds of organic compounds in trees and other plants).

solid waste Any unwanted or discarded material that is not a liquid or a gas. See *municipal solid waste*.

spaceship-earth worldview View of the earth as a spaceship: a machine that we can understand, control, and change at will by using advanced technology. See *planetary management worldview*. Compare *environmental wisdom worldview*.

specialist species Species with a narrow ecological niche. They may be able to (1) live in only one type of habitat, (2) tolerate only a narrow range of climatic and other environmental conditions, or (3) use only one or a few types of food. Compare *generalist species*.

speciation Formation of two species from one species as a result of divergent natural selection in response to changes in environmental conditions; usually takes thousands of years. Compare *extinction*.

species Group of organisms that resemble one another in appearance, behavior, chemical makeup and processes, and genetic structure. Organisms that reproduce sexually are classified as members of the same species only if they can actually or potentially interbreed with one another and produce fertile offspring.

species diversity Number of different species and their relative abundances in a given area. See *biodiversity*. Compare *ecological diversity, genetic diversity*.

species equilibrium model See *theory of island biogeography.*

spoils Unwanted rock and other waste materials produced when a material is removed from the earth's surface or subsurface by mining, dredging, quarrying, and excavation.

S-shaped curve Leveling off of an exponential, *J*-shaped curve when a rapidly growing population exceeds the carrying capacity of its environment and ceases to grow.

stability Ability of a living system to withstand or recover from externally imposed changes or stresses. See *constancy, inertia, resilience.*

statutory law Law developed and passed by legislative bodies such as federal and state governments. Compare *common law.*

stewardship View that because of our superior intellect and power or because of our religious beliefs, we have an ethical responsibility to manage and care for domesticated plants and animals and the rest of nature. Compare *environmental wisdom worldview, planetary management worldview.*

storage area Place within a system where energy, matter, or information can accumulate for various lengths of time before being released. Compare, *input, output, throughput.*

store See *storage area.*

stratosphere Second layer of the atmosphere, extending about 17-48 kilometers (11-30 miles) above the earth's surface. It contains small amounts of gaseous ozone (O_3), which filters out about 99% of the incoming harmful ultraviolet (UV) radiation emitted by the sun. Compare *troposphere.*

stream Flowing body of surface water. Examples are creeks and rivers.

strip cropping Planting regular crops and close-growing plants, such as hay or nitrogen-fixing legumes, in alternating rows or bands to help reduce depletion of soil nutrients.

strip cutting A variation of clear-cutting in which a strip of trees is clear-cut along the contour of the land, with the corridor narrow enough to allow natural regeneration within a few years. After regeneration, another strip is cut above the first, and so on. Compare *clear-cutting, seed-tree cutting, selective cutting, shelterwood cutting.*

strip mining Form of surface mining in which bulldozers, power shovels, or stripping wheels remove large chunks of the earth's surface in strips. See *surface mining.* Compare *subsurface mining.*

subatomic particles Extremely small particles-electrons, protons, and neutrons-that make up the internal structure of atoms.

subduction zone Area in which oceanic lithosphere is carried downward (subducted) under the island arc or continent at a convergent plate boundary. A trench ordinarily forms at the boundary between the two converging plates. See *convergent plate boundary.*

subsidence Slow or rapid sinking down of part of the earth's crust that is not slope-related.

subsidence temperature inversion Inversion of normal air temperature layers when a large mass of warm air moves into a region at a high altitude and floats over a mass of colder air near the ground. This keeps the air over a city stagnant and prevents vertical mixing and dispersion of air pollutants. See *temperature inversion.* Compare *radiation temperature inversion.*

subsistence farming Supplementing solar energy with energy from human labor and draft animals to produce enough food to feed oneself and family members; in good years there may be enough food left over to sell or put aside for hard times. Compare *industrialized agriculture.*

subsurface mining Extraction of a metal ore or fuel resource such as coal from a deep underground deposit. Compare *surface mining.*

succession See *ecological succession.*

succulent plants Plants, such as desert cacti, that survive in dry climates by having no leaves, thus reducing the loss of scarce water. They store water and use sunlight to produce the food they need in the thick, fleshy tissue of their green stems and branches. Compare *deciduous plants, evergreen plants.*

sulfur cycle Cyclic movement of sulfur in different chemical forms from the environment to organisms and then back to the environment.

superinsulated house House that is heavily insulated and extremely airtight. Typically, active or passive solar collectors are used to heat water, and an air-to-air heat exchanger is used to prevent buildup of excessive moisture and indoor air pollutants.

surface fire Forest fire that burns only undergrowth and leaf litter on the forest floor. Compare *crown fire, ground fire.*

surface mining Removing soil, subsoil, and other strata and then extracting a mineral deposit found fairly close to the earth's surface. Compare *subsurface mining.*

surface runoff Water flowing off the land into bodies of surface water.

surface water Precipitation that does not infiltrate the ground or return to the atmosphere by evaporation or transpiration. See *runoff.* Compare *groundwater.*

survivorship curve Graph showing the number of survivors in different age groups for a particular species.

sustainability Ability of a system to survive for some specified (finite) time. See *sustainable system.*

sustainable agriculture Method of growing crops and raising livestock based on organic fertilizers, soil conservation, water conservation, biological control of pests, and minimal use of nonrenewable fossil-fuel energy.

sustainable development See *sustainable economic development, ecologically sustainable development.*

sustainable economic development Increasing the *quality* of goods and services without depleting or degrading the quality of natural resources to unsustainable levels for current and future generations (qualitative growth). Compare *economic growth, economic development, ecologically sustainable development.*

sustainable living Taking no more potentially renewable resources from the natural world than can be replenished naturally and not overloading the capacity of the environment to cleanse and renew itself by natural processes.

sustainable society A society that manages its economy and population size without doing irreparable environmental harm by overloading the planet's ability to absorb environmental insults, replenish its resources, and sustain human and other forms of life over a specified period, usually hundreds to thousands of years. During this period, it satisfies the needs of its people without depleting natural capital and thereby jeopardizing the prospects of current and future generations of humans and other species.

sustainable system A system that survives and functions over some specified (finite) time; a system that attains its full expected lifetime.

sustainable yield (sustained yield) Highest rate at which a potentially renewable resource can be used without reducing its available supply throughout the world or in a particular area. See also *environmental degradation.*

symbiosis Any intimate relationship or association between members of two or more species. See *symbiotic relationship.*

symbiotic relationship Species interaction in which two kinds of organisms live together in an intimate association. Members of the participating species may be harmed by, benefit from, or be unaffected by the interaction. See *commensalism, interspecific competition, mutualism, parasitism, predation.*

synergistic interaction Interaction of two or more factors or processes so that the combined effect is greater than the sum of their separate effects.

synergy See *synergistic interaction.*

synfuels Synthetic gaseous and liquid fuels produced from solid coal or sources other than natural gas or crude oil.

synthetic natural gas (SNG) Gaseous fuel containing mostly methane produced from solid coal.

system A set of components that function and interact in some regular and theoretically predictable manner.

tailings Rock and other waste materials removed as impurities when waste mineral material is separated from the metal in an ore.

tar sand Deposit of a mixture of clay, sand, water, and varying amounts of a tarlike heavy oil known as bitumen. Bitumen can be extracted from tar sand by heating. It is then purified and upgraded to synthetic crude oil. See *bitumen.*

technology Creation of new products and processes intended to improve our efficiency, chances for survival, comfort level, and quality of life. Compare *science.*

temperature Measure of the average speed of motion of the atoms, ions, or molecules in a substance or combination of substances at a given moment. Compare *heat.*

temperature inversion Layer of dense, cool air trapped under a layer of less dense, warm air. This prevents upward-flowing air currents from developing. In a prolonged inversion, air pollution in the trapped layer may build up to harmful levels. See *radiation temperature inversion, subsidence temperature inversion.*

teratogen Chemical, ionizing agent, or virus that causes birth defects. See *carcinogen, mutagen.*

terracing Planting crops on a long, steep slope that has been converted into a series of broad, nearly level terraces with short vertical drops from one to another that run along the contour of the land to retain water and reduce soil erosion.

terrestrial Pertaining to land. Compare *aquatic.*

territoriality Process in which organisms patrol or mark an area around their home, nesting, or major feeding site and defend it against members of their own species.

tertiary (higher-level) consumers Animals that feed on animal-eating animals. They feed at high trophic levels in food chains and webs. Examples are hawks, lions, bass, and sharks. Compare *detritivore, primary consumer, secondary consumer.*

tertiary oil recovery See *enhanced oil recovery.*

tertiary sewage treatment See *advanced sewage treatment.*

theory of evolution Widely accepted scientific idea that all life forms developed from earlier life forms. Although this theory conflicts with the creation stories of many religions, it is the way biologists explain how life has changed over the past 3.6-3.8 billion years and why it is so diverse today.

theory of island biogeography The number of species found on an island is determined by a balance between two factors: the (1) *immigration rate* (of species new to the island) from other inhabited areas and (2) *extinction rate* (of species established on the island). The model predicts that at some point the rates of immigration and extinction will reach an equilibrium point that determines the island's average number of different species (species diversity).

thermal inversion See *temperature inversion.*

thermocline Zone of gradual temperature decrease between warm surface water and colder deep water in a lake, reservoir, or ocean.

threatened species Wild species that is still abundant in its natural range but is likely to become endangered because of a decline in numbers. Compare *endangered species.*

threshold effect The harmful or fatal effect of a small change in environmental conditions that exceeds the limit of tolerance of an organism or population of a species. See *law of tolerance*.

throughput Rate of flow of matter, energy, or information through a system. Compare *input, output, storage area*.

throwaway society See *high-throughput economy*.

time delay Time lag between the input of a stimulus into a system and the response to the stimulus.

tolerance limits Minimum and maximum limits for physical conditions (such as temperature) and concentrations of chemical substances beyond which no members of a particular species can survive. See *law of tolerance*.

total fertility rate (TFR) Estimate of the average number of children that will be born alive to a woman during her lifetime if she passes through all her childbearing years (ages 15-44) conforming to age-specific fertility rates of a given year. In simpler terms, it is an estimate of the average number of children a woman will have during her childbearing years.

totally planned economy See *pure command economic system*.

toxic chemical Chemical that is fatal to humans in low doses, or fatal to more than 50% of test animals at stated concentrations. Most are neurotoxins, which attack nerve cells. See *carcinogen, hazardous chemical, mutagen, teratogen*.

toxicity Measure of how harmful a substance is.

toxicology Study of the adverse effects of chemicals on health.

toxic waste Form of hazardous waste that causes death or serious injury (such as burns, respiratory diseases, cancers, or genetic mutations). See *hazardous waste*.

traditional intensive agriculture Producing enough food for a farm family's survival and perhaps a surplus that can be sold. This type of agriculture uses higher inputs of labor, fertilizer, and water than traditional subsistence agriculture. See *traditional subsistence agriculture*.

traditional subsistence agriculture Production of enough crops or livestock for a farm family's survival and, in good years, a surplus to sell or put aside for hard times. Compare *traditional intensive agriculture*.

tragedy of the commons Depletion or degradation of a potentially renewable resource to which people have free and unmanaged access. An example is the depletion of commercially desirable species of fish in the open ocean beyond areas controlled by coastal countries. See *common-property resource*.

transform fault Area where the earth's lithospheric plates move in opposite but parallel directions along a fracture (fault) in the lithosphere. Compare *convergent plate boundary, divergent plate boundary*.

transmissible disease A disease that is caused by living organisms (such as bacteria, viruses, and parasitic worms) and can spread from one person to another by air, water, food, or body fluids (or in some cases by insects or other organisms). Compare *nontransmissible disease*.

transpiration Process in which water (1) is absorbed by the root systems of plants, (2) moves up through the plants, (3) passes through pores (stomata) in their leaves or other parts, and (4) evaporates into the atmosphere as water vapor.

tree farm Site planted with one or only a few tree species in an even-aged stand. When the stand matures it is usually harvested by clear-cutting and then replanted. These farms normally are used to grow rapidly growing tree species for fuelwood, timber, or pulpwood. See *even-aged management*. Compare *old-growth forest, second-growth forest, uneven-aged management*.

tree plantation See *tree farm*.

trophic level All organisms that are the same number of energy transfers away from the original source of energy (for example, sunlight) that enters an ecosystem. For example, all producers belong to the first trophic level, and all herbivores belong to the second trophic level in a food chain or a food web.

troposphere Innermost layer of the atmosphere. It contains about 75% of the mass of earth's air and extends about 17 kilometers (11 miles) above sea level. Compare *stratosphere*.

true cost See *full cost*.

undergrazing Reduction of the net primary productivity of grassland vegetation and grass cover from absence of grazing for long periods (at least 5 years). Compare *overgrazing*.

undernutrition Consuming insufficient food to meet one's minimum daily energy needs for a long enough time to cause harmful effects. Compare *malnutrition, overnutrition*.

undiscovered resources Potential supplies of a particular mineral resource, believed to exist because of geologic knowledge and theory, although specific locations, quality, and amounts are unknown. Compare *identified resources, reserves*.

uneven-aged management Method of forest management in which trees of different species in a given stand are maintained at many ages and sizes to permit continuous natural regeneration. Compare *even-aged management*.

upwelling Movement of nutrient-rich bottom water to the ocean's surface. This can occur far from shore but usually occurs along certain steep coastal areas where the surface layer of ocean water is pushed away from shore and replaced by cold, nutrient-rich bottom water.

urban area Geographic area with a population of 2,500 or more. The number of people used in this definition may vary, with some countries setting the minimum number of people at 10,000-50,000.

urban growth Rate of growth of an urban population. Compare *degree of urbanization*.

urban heat island Buildup of heat in the atmosphere above an urban area. The large concentration of cars, buildings, factories, and other heat-producing activities produces this heat. See also *dust dome*.

urbanization See *degree of urbanization*.

urban sprawl Growth of low-density development on the edges of cities and towns. See *smart growth*.

utilitarian value See *instrumental value*.

vertebrates Animals with backbones. Compare *invertebrates*.

village Group of rural households linked together by custom, culture, and family ties, usually surviving by harvesting local natural resources for food, fuel, and other basic needs. Compare *city*. See *rural area, urban area*.

volcano Vent or fissure in the earth's surface through which magma, liquid lava, and gases are released into the environment.

vulnerable species See *threatened species*.

warm front The boundary between an advancing warm air mass and the cooler one it is replacing. Because warm air is less dense than cool air, an advancing warm front rises up over a mass of cool air. Compare *cold front*.

water cycle See *hydrologic cycle*.

waterlogging Saturation of soil with irrigation water or excessive precipitation so that the water table rises close to the surface.

water pollution Any physical or chemical change in surface water or groundwater that can harm living organisms or make water unfit for certain uses.

watershed Land area that delivers the water, sediment, and dissolved substances via small streams to a major stream (river).

water table Upper surface of the zone of saturation, in which all available pores in the soil and rock in the earth's crust are filled with water.

watt Unit of power, or rate at which electrical work is done. See *kilowatt*.

weather Short-term changes in the temperature, barometric pressure, humidity, precipitation, sunshine, cloud cover, wind direction and speed, and other conditions in the troposphere at a given place and time. Compare *climate*.

weathering Physical and chemical processes in which solid rock exposed at earth's surface is changed to separate solid particles and dissolved material, which can then be moved to another place as sediment. See *erosion*.

wetland Land that is covered all or part of the time with salt water or fresh water, excluding streams, lakes, and the open ocean. See *coastal wetland, inland wetland*.

wilderness Area where the earth and its community of life have not been seriously disturbed by humans and where humans are only temporary visitors.

wildlife All free, undomesticated species. Sometimes the term is used to describe only free, undomesticated animal species.

wildlife management Manipulation of populations of wild species (especially game species) and their habitats for (1) human benefit, (2) the welfare of other species, and (3) the preservation of threatened and endangered wildlife species.

wildlife resources Wildlife species that have actual or potential economic value to people.

wildness Existence of wild gene pools, species, and ecosystems that are completely or mostly undisturbed by human activities. Another term for biodiversity.

wild species Species found in the natural environment.

windbreak Row of trees or hedges planted to partially block wind flow and reduce soil erosion on cultivated land.

wind farm Cluster of small to medium-sized wind turbines in a windy area to capture wind energy and convert it into electrical energy.

worldview How individuals think the world works and what they think their role in the world should be. See *environmental wisdom worldview, planetary management worldview, spaceship-earth worldview*.

zero population growth (ZPG) State in which the birth rate (plus immigration) equals the death rate (plus emigration) so that the population of a geographic area is no longer increasing.

zone of aeration Zone in soil that is not saturated with water and that lies above the water table. See *water table, zone of saturation*.

zone of saturation Area where all available pores in soil and rock in the earth's crust are filled by water. See *water table, zone of aeration*.

zoning Regulating how various parcels of land can be used.

zooplankton Animal plankton. Small floating herbivores that feed on plant plankton (phytoplankton). Compare *phytoplankton*.

INDEX

Note: Page references followed by *f*, *t*, or *n* indicate figures, tables, or notes, respectively.